D1798595

Autodesk Inventor 5/5.3

 # Autodesk Inventor 5/5.3

Basics Through Advanced

David P. Madsen

Upper Saddle River, New Jersey
Columbus, Ohio

Editor in Chief: Stephen Helba
Executive Editor: Debbie Yarnell
Editorial Assistant: Sam Goffinet
Media Development Editor: Michelle Churma
Production Editor: Louise N. Sette
Production Supervision: Karen Fortgang, bookworks
Design Coordinator: Diane Ernsberger
Cover Designer: Jeff Vanik
Cover art: Jeff Vanik
Production Manager: Brian Fox
Marketing Manager: Jimmy Stephens

This book was set in Times Roman by STELLARViSIONs. It was printed and bound by Courier Kendallville, Inc. The cover was printed by Phoenix Color Corp.

Pearson Education Ltd.
Pearson Education Australia Pty. Limited
Pearson Education Singapore Pte. Ltd.
Pearson Education North Asia Ltd.
Pearson Education Canada, Ltd.
Pearson Educación de Mexico, S. A. de C.V.
Pearson Education—Japan
Pearson Education Malaysia Pte. Ltd.
Pearson Education, *Upper Saddle River, New Jersey*

Copyright © 2003 by Pearson Education, Inc., Upper Saddle River, New Jersey 07458. All rights reserved. Printed in the United States of America. This publication is protected by Copyright and permission should be obtained from the publisher prior to any prohibited reproduction, storage in a retrieval system, or transmission in any form or by any means, electronic, mechanical, photocopying, recording, or likewise. For information regarding permission(s), write to: Rights and Permissions Department.

10 9 8 7 6 5 4 3 2 1

ISBN: 0-13-098514-7

|||||||||||||||||||||||||||||||||||||||||||||||||||||||||||||||||||||||||||||||||||||||

Preface

Autodesk Inventor

Autodesk Inventor mechanical design software is an assembly-centric solid modeling (3D) and drawing production (2D) system built with two goals in mind:

- It eliminates the limitations imposed by other Mechanical CAD (MCAD) software, especially pure parametric, and to capitalize on the process developments now emerging in the manufacturing industry.

- It adapts to the way you work, so work continues the way you think, and provides:

  Single-Day Productivity—Autodesk Inventor was built to help you get more done from Day One and to ensure your everyday ease of use.

  Large-Model Performance—With Autodesk Inventor, you work efficiently in the context of complete assemblies. In fact, you load, view, edit, and save large models two to ten times faster than with other top-selling mechanical design solutions.

  Adaptive Design—Autodesk Inventor introduces the concept of adaptive design, a design methodology that enables you to work the way you think, rather than forcing you to adapt to arbitrary processes.

Autodesk Inventor brochure, by Autodesk, Inc.

Autodesk Inventor 5/5.3: Basics Through Advanced is a text and workbook that provides complete coverage of Autodesk Inventor™. This text is designed to introduce you to solid modeling objects and the tools and commands used in Autodesk Inventor to complete fully parametric three-dimensional parts, assemblies, and presentations, and two-dimensional drawings. The chapters are arranged in an easy-to-understand format and begin with basic topics, working toward advanced subjects. To use this text, you should have a basic understanding of computer-aided and mechanical drafting. This text provides a detailed explanation of the Autodesk Inventor tools that aid in producing professional and accurate solid models and drawings. This text is intended for anyone who wants to learn how to use Autodesk Inventor effectively to create real-world parametric models and two-dimensional drawings. The text can be used in secondary, postsecondary, and technical schools. It provides the beginning student and the drafting or engineering professional with a complete understanding of every Autodesk Inventor command using professional methods and techniques. All software commands and tools are presented in a manner that shows each prompt with examples of the exact input you should use. This text also contains a variety of valuable features that help make learning Autodesk Inventor easy for you:

- It is easy to read, use, and understand.

- Learning Goals are provided at the beginning of each chapter to identify what you will learn as a result of completing the chapter.

- Commands and tools are presented in a manner that shows every access technique, application, and dialog box, with the exact input you should use.

- Field Notes explain special applications, tricks, and professional tips for using Autodesk Inventor.

- Exercises are provided throughout for you to practice as you learn Autodesk Inventor.

- Projects allow you to learn how to use Autodesk Inventor in your drafting or engineering field.

- Chapter tests provide you with the opportunity to do a comprehensive review of each chapter by answering questions related to the chapter content.

- The Instructor's Resource Manual contains chapter test answers, exercise solutions, and project solutions.

Format of This Text

The format of this text helps you learn how to use Autodesk Inventor by example and complete explanations of each feature. Your learning is also supported by exercises that allow you to practice while you learn specific content, chapter tests that provide you with an excellent way to review chapter content, and projects that pull together everything you have learned in each chapter.

Accessing Autodesk Inventor Commands

In addition to the quick reference mapping of available command selection methods, all options are further described within the text. For example, to access the **Line** command, use one of the following techniques:

✓ Pick the **Line** button on the panel bar.

✓ Pick the **Line** button on the **Sketch** toolbar.

✓ Type the **+** key and **L** key on your keyboard.

One goal of this book is for you to see all the methods of using Autodesk Inventor commands and to decide which methods work best for you.

Understanding the Command Format Presented in This Text

Command instructions provide you with the techniques used to access and complete the command. For example: "To draw a line, pick the point where you want to begin the line, and then pick the next point. See Figure 3.15. If you want to continue the line, pick the next point. If you do not want to continue the line, press the **Esc** key on your keyboard, or right-click and pick the **Done** option. If you do not want to continue the current line, but would like to draw another line at another location, either close the current **Line** command and access it again, or right-click and select **Restart** from the shortcut menu."

After you have learned the different ways to respond to Autodesk Inventor prompts, you can use the method that works best for you.

Identifying Key Words

Key words are ***bold italic*** for your ease in identification. These words relate to Autodesk Inventor terminology and professional applications. These words are defined in the text. Key words are an important part of your Autodesk Inventor learning process, because they are part of the world of computer-aided drafting, drafting standards, and Autodesk Inventor communication.

Special Features

Autodesk Inventor 5/5.3: Basics Through Advanced contains several special features that help you learn and use Autodesk Inventor, as explained in the following paragraphs. Most features are shown here as they appear in the book.

LEARNING GOALS

◎   Use Autodesk Inventor to prepare parametric solid models and drawings for the manufacturing industry.

◎   Answer questions related to Autodesk Inventor.

◎   Use Autodesk Inventor and drafting related terminology.

◎   Do exercises as you learn Autodesk Inventor.

◎   Use projects to learn Autodesk Inventor and create solid models and two-dimensional drawings.

Exercises

1. An exercise is provided after each Autodesk Inventor topic or command lesson.
2. The exercises allow you to practice what you have just learned. Practicing Autodesk Inventor applications is one of the most important keys to learning the program effectively.
3. Exercises should be completed at a computer while using Autodesk Inventor to reinforce what you have just studied.
4. Exercises build on each other, allowing you to develop understanding as you work with Autodesk Inventor.
5. Exercises are saved to disk and can be used only as practice or as classroom assignments.

 Field Notes

Field Notes are another special feature of *Autodesk Inventor 5/5.3: Basics Through Advanced.* Field Notes are placed throughout the text to provide you with any one or more of the following advantages:

- Professional tips and applications.
- Special features of Autodesk Inventor.
- Advanced applications.
- Additional instruction about how certain features work.

CHAPTER TESTS

There is a comprehensive test at the end of each chapter. The test can be used to check your understanding of the chapter content or it can be used as review. Answering the test questions is an excellent way to go back through the material and reinforce what you have learned.

PROJECTS

Projects are one of the most important ways to complete and solidify your learning and understanding of Autodesk Inventor. It has been said many times, in terms of learning AutoCAD and Autodesk Inventor: "If you do not practice you will forget." The projects allow you to put to practice what you have just learned. The projects are different from the exercises, because they combine a variety of commands that are used in the current chapter and in past chapters. Exercises focus only on using the currently discussed command. The projects at the end of every chapter are designed to provide you with practice, using Autodesk Inventor tools and applications discussed in the chapter and earlier chapters. You or your instructor can select the projects that relate directly to your specific course objectives. The projects provided in this text are real-world solid model and drawing tasks.

Instructor's Resource Manual

The Instructor's Resource Manual for this text provides the following items:

- Introduction: How to Use This Text
- Chapter Test Answers
- Chapter Exercise Solutions
- Chapter Project Solutions

Acknowledgments

I would like to give special thanks to my father, David A. Madsen, for his contributions, support, and professional advice during the development of this textbook.

I would also like to acknowledge the reviewers of this text: Alex Devereux, ITT Technical Institute (AZ); Louis A. Moegenburg, Wisconsin Indianhead Technical College; and Marsha Walton, Finger Lakes Community College (NY).

Warning and Disclaimer

This book is designed to provide tutorial information about the Autodesk Inventor computer program. Every effort has been made to make this book complete and as accurate as possible, but no warranty or fitness is implied.

The information is provided on an "as-is" basis. The author and Pearson Education, Inc. shall have neither liability nor responsibility to any person or entity with respect to any loss or damage in connection with or arising from information contained in this book.

|||||||||||||||||||||||||||||||||||||||||||||||||||||||||||||||||||||||||||||||||||||||||||||

Contents

Chapter 4

Developing Basic Part Model Features   141

☐ C h a p t e r   5

Creating Placed Features   195

☐ C h a p t e r   6

Patterning Features   229

☐ C h a p t e r 1 2

Creating Part Drawings **413**

☐ C h a p t e r 1 3

Dimensioning Drawings **493**

Chapter 14

Working with Assemblies   563

☐ **C h a p t e r 1 5**

Adapting Parts and Assemblies   641

☐ **C h a p t e r 1 6**

Working with Presentations   667

CHAPTER 1

Starting Autodesk Inventor

LEARNING GOALS

After completing this chapter, you will be able to:

◎ Use the **Autodesk Inventor** dialog box.

◎ Access and use the **Getting Started** option of the **What To Do** column.

◎ Use the **Learn how to work with Autodesk Inventor** link.

◎ Open and use the **Learn how to build models quickly** link.

◎ Use the **See "What's New" in this release** link.

◎ Access the **Autodesk Point A** and **Autodesk Streamline** Websites.

◎ Work with the **New** option of the **What To Do** column.

◎ Open a **Part, Assembly, Presentation,** and **Drawing** file.

◎ Use and create custom tabs and templates in the **New** option.

◎ Operate the **Open** option of the **What To Do** column.

◎ Use the **Projects** option **(Project Editor)** of the **What To Do** column.

◎ Create a new **Project.**

Using the Autodesk Inventor Dialog Box

Initial launching of Autodesk Inventor™ opens the **Autodesk Inventor** dialog box shown in Figure 1.1.

Four main options are available under the **What To Do** column, on the left side of the **Autodesk Inventor** dialog box. The options are: **Getting Started, New, Open,** and **Projects.**

 Field Notes

> Although the **Getting Started** window is the first to open upon initial launch of Autodesk Inventor, whichever of the four **What To Do** options is selected last, opens the next time Autodesk Inventor is launched.

The **Autodesk Inventor** dialog box can also be accessed by selecting one of the following options from the **File** pull-down menu, by typing the designated accelerator key, or by selecting the appropriate toolbar button:

✓ **New...** or **Ctrl + N** or the **New** button on the **Standard** toolbar

✓ **Open...** or **Ctrl + O** or the **Open** button on the **Standard** toolbar

✓ **Projects...**

✓ **Getting Started...**

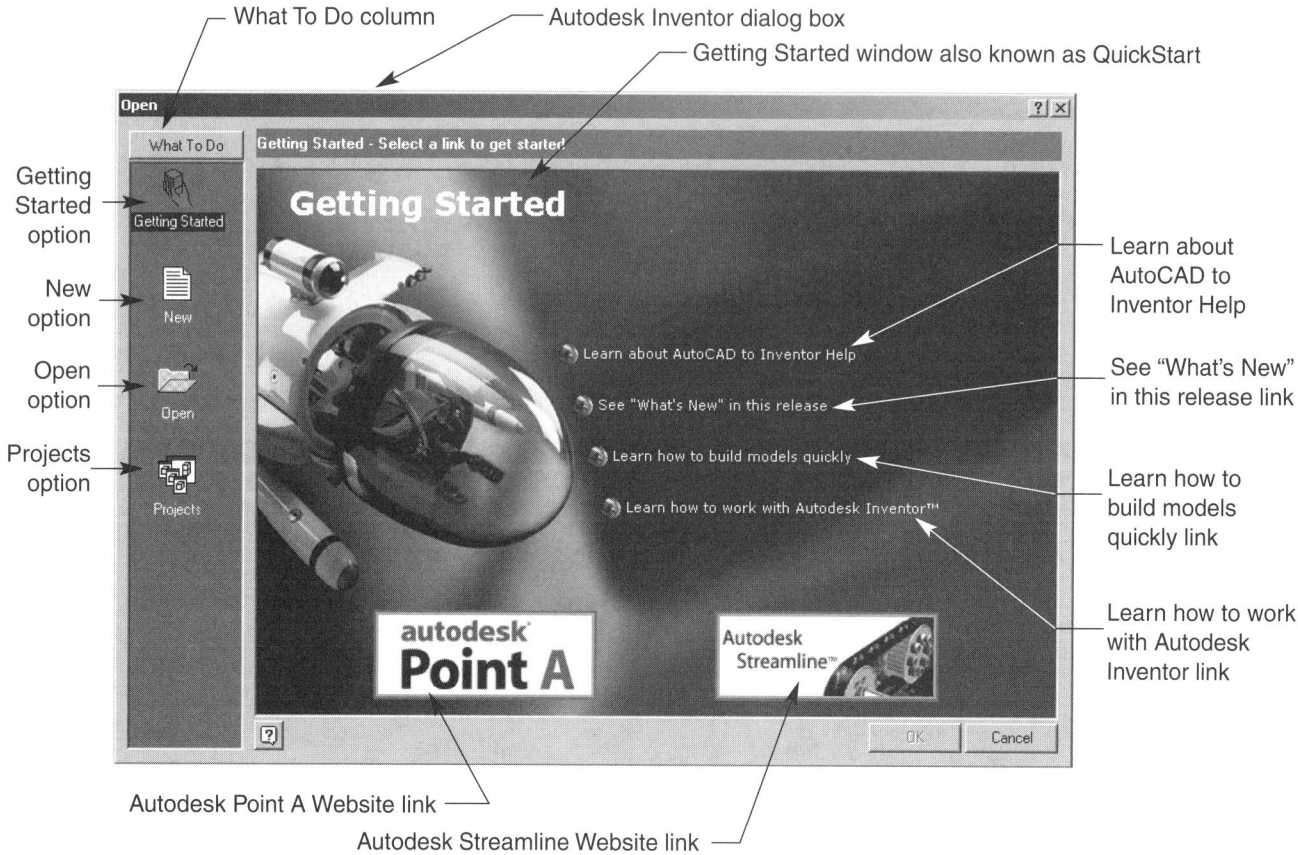

What To Do column — Autodesk Inventor dialog box

Getting Started window also known as QuickStart

Getting Started option

New option

Open option

Projects option

Learn about AutoCAD to Inventor Help

See "What's New" in this release link

Learn how to build models quickly link

Learn how to work with Autodesk Inventor link

Autodesk Point A Website link

Autodesk Streamline Website link

FIGURE 1.1   The **Autodesk Inventor** dialog box, with the **Getting Started** window.

Getting Started

The **Getting Started** choice, also known as **QuickStart,** of the **What To Do** menu allows you to pick from three learning tools. These tools include: **Learn about AutoCAD to Inventor Help, See "What's New" in this release, Learn how to build models quickly,** and **Learn how to work with Autodesk Inventor.** In addition, the **Autodesk Point A** and **Autodesk Streamline** Website links are available. See Figure 1.1. Hovering over each of these areas with the mouse pointer provides a simple description of the services these tools provide in the upper left corner of the **Quick Start** window. **Getting Started** is also accessed by:

✓ Selecting **Getting Started...** from the **File** pull down-menu.

Field Notes

The purpose of this chapter is to fully explain the **Autodesk Inventor** dialog box, which is accessed when you launch Autodesk Inventor. The **Getting Started** window guides you through tutorials on learning how to work with Autodesk Inventor, learning how to build models quickly, and understanding the new features of the program. The tutorials are also accessible from the **Help** pull-down menu. However, you must first close the **Autodesk Inventor** dialog box to access the **Help** pull-down menu, and that is not the objective of this chapter.

Learn How to Work with Autodesk Inventor

Clicking on the **Learn how to work with Autodesk Inventor** option opens the **Autodesk Inventor** page of the **Autodesk Inventor – Tutorials** window. Inside this area, you can click on the **Help, Main Menu,** or **Next** buttons. See Figure 1.2.

The **Autodesk Inventor – Tutorials** window is also accessed by:

✓ Selecting **Tutorials** from the **Help** pull-down menu.

Picking the **Help** button opens the **Autodesk Inventor Help** window, as shown in Figure 1.3. This section of the program is similar to other AutoCAD Help files, allowing you to further explore areas of Autodesk Inventor and receive answers to any questions you may encounter. The **Autodesk Inventor Help** window is also accessed by:

✓ Selecting **Help Topics** from the **Help** pull-down menu.

✓ Selecting the **Help Topics** button from the **Standard** toolbar.

FIGURE 1.2 The **Autodesk Inventor – Tutorials** window with the **Help** button, the **Main Menu** button, and the **Next** button.

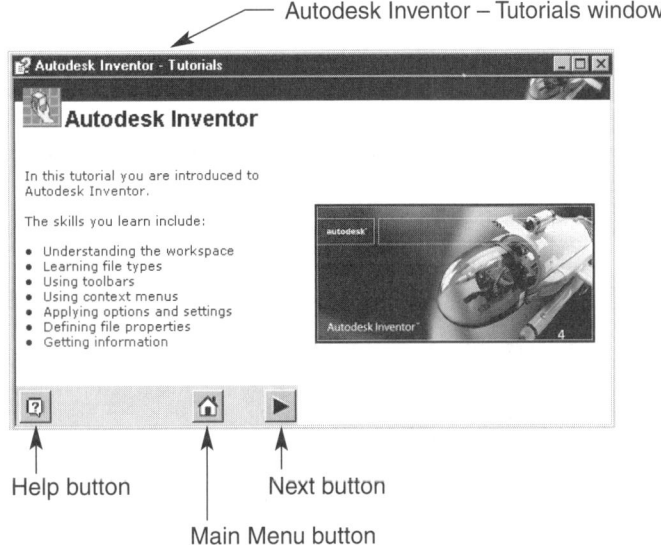

Autodesk Inventor – Tutorials window

Help button   Next button

Main Menu button

FIGURE 1.3 The **Autodesk Inventor** help file window.

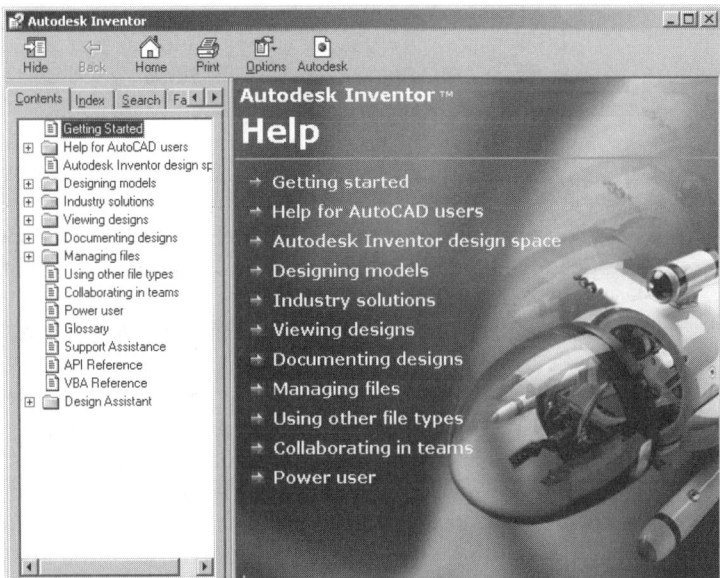

FIGURE 1.4   The
QuickStart page of the
**Autodesk Inventor –
Tutorials** window.

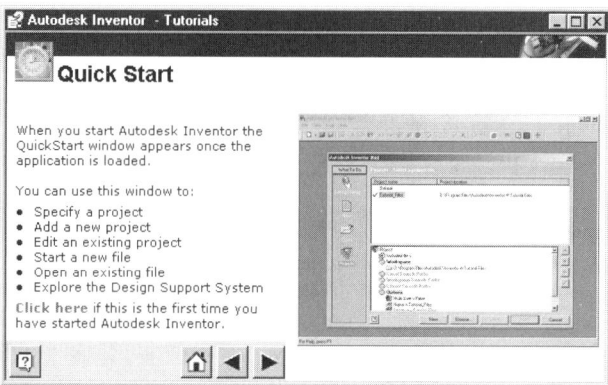

✓ Picking the **Help Topics** button from a **Browser.**

✓ Typing the **F1** key on your keyboard.

✓ Right clicking and selecting the **How To...** option.

Clicking on the **Main Menu** button takes you to the **Single Day Productivity** menu of the **Autodesk Inventor – Tutorials** window. This section is discussed later in this chapter.

Clicking the **Next** button continues the "introduction tour" of Autodesk Inventor, which contains many pages. The **Quick Start** page is an example of one of these **Autodesk Inventor – Tutorials** window pages. See Figure 1.4.

Additional clicks of the **Next** button takes you to more introductory tutorials.

Exercise 1.1

1. Use one of the options previously described to launch Autodesk Inventor.
2. Pick the **Getting Started** option of the **What To Do** menu if the **QuickStart** page is not currently displayed.
3. Select the **Learn how to work with Autodesk Inventor** link.
4. Use the techniques previously discussed to navigate through the introduction to Autodesk Inventor tutorial at your own pace.
5. When finished navigating through the tutorial, choose the **Help** button.
6. Select and read each of the options available in the Autodesk Inventor Home **Help** menu.
7. Close the **Autodesk Inventor – Tutorials** window and continue exploring Autodesk Inventor, or exit Autodesk Inventor if necessary.

Learn How to Build Models Quickly

Choosing the **Learn how to build models quickly** option opens the **Autodesk Inventor – Tutorials** window. See Figure 1.5. Inside this window you can select options including **Using the Tutorials, Single-Day Productivity, Collaboration,** and **Share and Reuse Data.** When you pick the **Single-Day Productivity** button, the **Single-Day Productivity** menu opens. See Figure 1.6. Inside this area, you can select the **Using Autodesk Inventor** button, previously described, or choose one of the guided tutorial buttons including **Creating a Part, Building Sheet Metal Parts, Creating Assemblies,** and **Preparing Final Drawings.**

The **Creating a Part** button takes you to a tutorial that shows you how to develop a simple part. Selecting the **Next** button advances the tutorial. The **Click for more information** and **Main Menu** buttons are also available. Choosing the **Main Menu** button in this section takes you back to the **Single-Day Productivity** menu. Creating parts is explored in great depth in Chapters 2 through 11.

The **Building Sheet Metal Parts** button takes you to a tutorial that guides you through the development of a simple sheet metal part. Selecting the **Next** button advances the tutorial. The **Click for more**

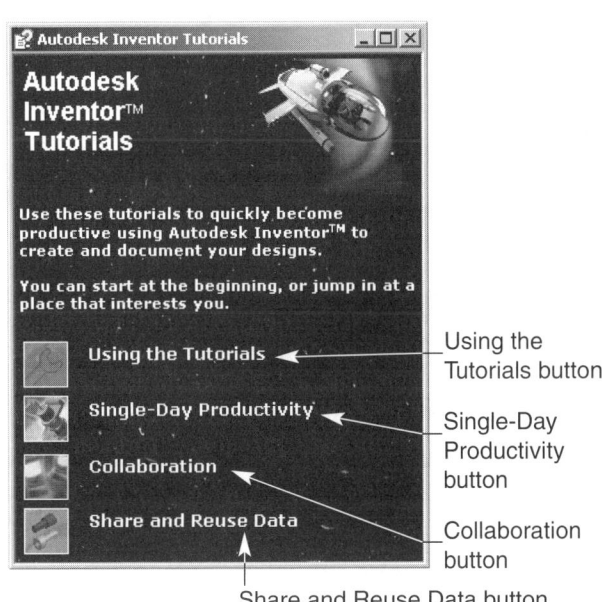

FIGURE 1.5   The **Autodesk Inventor – Tutorials** window.

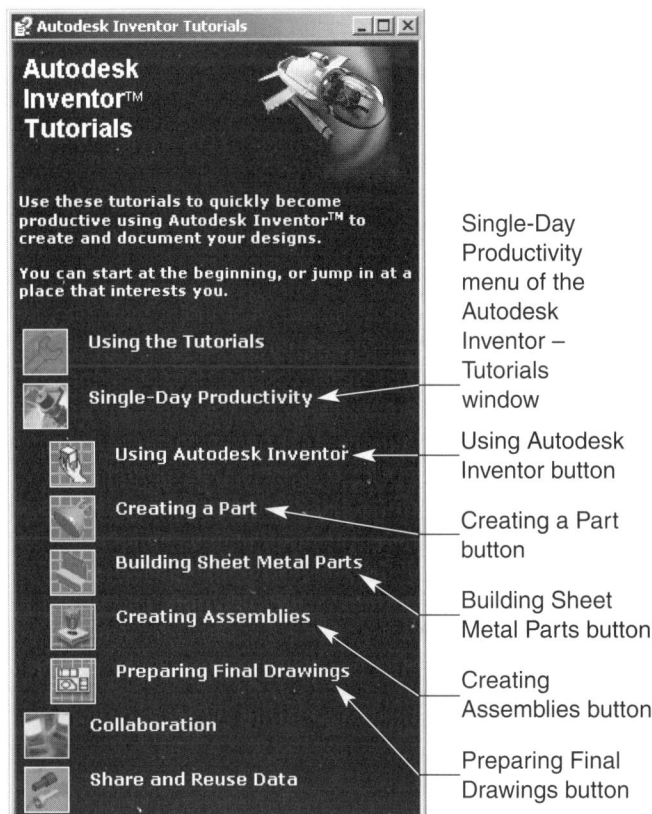

FIGURE 1.6   The **Single-Day Productivity** menu of the **Autodesk Inventor – Tutorials** window.

information and **Main Menu** buttons are also available. Choosing the **Main Menu** button in this section takes you back to the **Single-Day Productivity** menu. Creating sheet metal parts is explored in great depth in Chapter 10.

The **Creating Assemblies** button takes you to a tutorial showing you how to develop a simple assembly. Selecting the **Next** button advances the tutorial. The **Click for more information** and **Main Menu** buttons are also available. Creating assemblies is explored in depth in Chapters 14 and 15.

The **Preparing Final Drawings** button takes you to a tutorial that guides you through the development of a simple final drawing. Selecting the **Next** button advances the tutorial. The **Click for more information** and **Main Menu** buttons are also available. Preparing final drawings is explored in depth in Chapters 12, 13, and 17.

Using the Collaboration Menu

The **Collaboration** menu, shown in Figure 1.7, provides you with three options: **Working with Projects, Creating Adaptive Parts,** and **Creating Presentations.** The **Main Menu, Back,** and **Next** buttons are also available.

The **Working with Projects** button takes you to a tutorial that guides you through using projects. Working with projects is explored in depth later in this chapter.

The **Creating Adaptive Parts** button takes you to a tutorial that guides you through developing adaptive parts. Selecting the **Next** button advances the tutorial. The **Click for more information** and **Main Menu** buttons are also available. Creating adaptive parts is explored in depth in Chapter 15.

The **Creating Presentations** button takes you to a tutorial that guides you through developing presentations. Selecting the **Next** button advances the tutorial. The **Click for more information** and **Main Menu** buttons are also available. Creating presentations is explored in depth in Chapter 16.

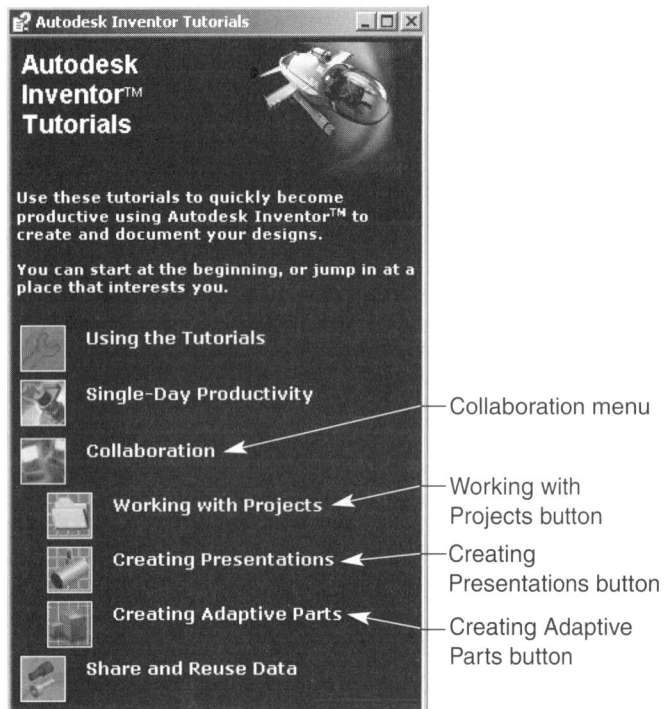

FIGURE 1.7   The **Collaboration** menu of the
Autodesk Inventor – Tutorials window.

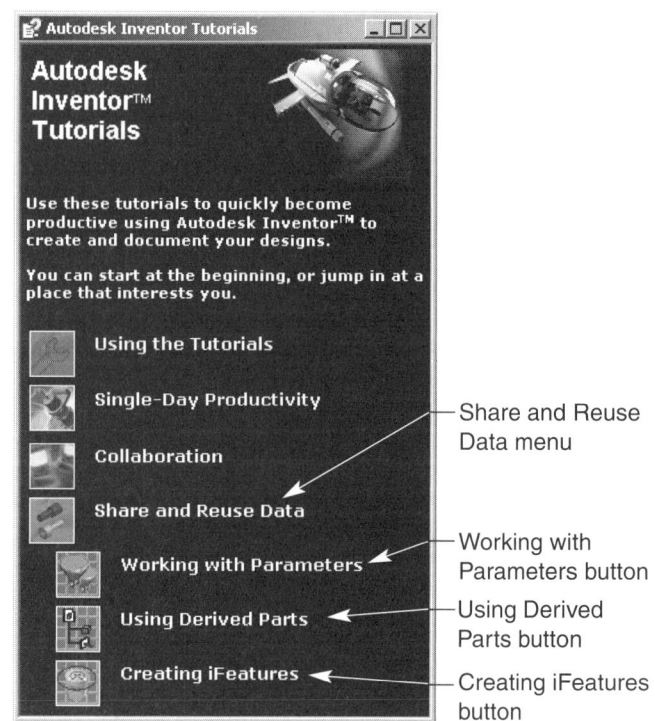

FIGURE 1.8   The **Share and Reuse Data** page
of the **Autodesk Inventor – Tutorials** window.

Using the Share and Reuse Data Menu

The **Share and Reuse Data** menu, shown in Figure 1.8, provides three options: **Working with Parameters, Using Derived Parts,** and **Creating iFeatures.**

The **Working with Parameters** button takes you to a tutorial that guides you through using parameters. Selecting the **Next** button advances the tutorial. The **Click for more information** and **Main Menu** buttons are also available. Parameters are explored throughout this text.

The **Using Derived Parts** button takes you to a tutorial that guides you through working with derived parts. Selecting the **Next** button advances the tutorial. The **Click for more information** and **Main Menu** buttons are also available. Using derived parts is discussed in depth in Chapter 18.

The **Creating iFeatures** button takes you to a tutorial that guides you through developing design elements. Selecting the **Next** button advances the tutorial. The **Click for more information** and **Main Menu** buttons are also available. Creating design elements and working with iFeatures are explored in depth in Chapter 9.

Using the Tutorials Page

Selecting the **Using the Tutorials** button opens the **Using the Tutorials** page of the **Autodesk Inventor – Tutorials** window. See Figure 1.9.

Here you can read how Autodesk Inventor tutorials work and view some of the different options that may be available in the tutorials.

FIGURE 1.9 The **Using the Tutorials** page of the **Autodesk Inventor – Tutorials** window.

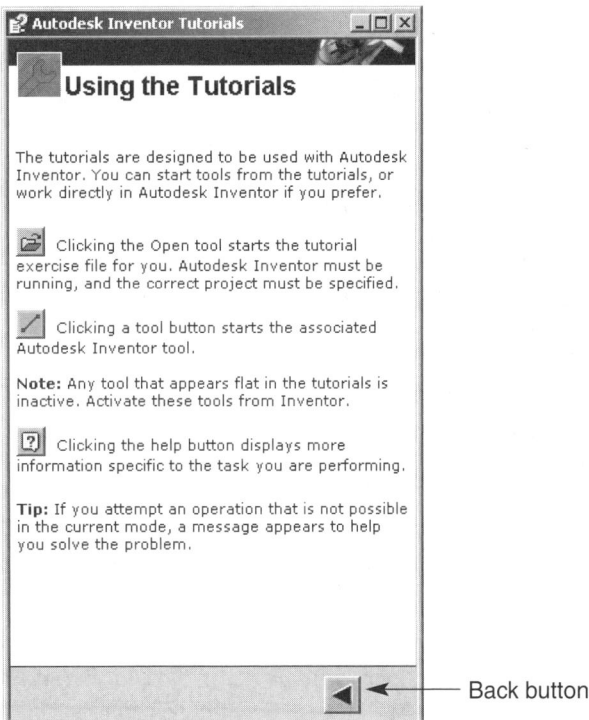

Back button

Exercise 1.2

1. Continue from Exercise 1.1, or launch Autodesk Inventor.
2. Select the **Learn how to build models quickly** link, or pick the **Getting Started** option of the **What To Do** menu if the **QuickStart** page is not currently displayed, and select the link.
3. Navigate through the options of the **Table of Contents** menu of the **Autodesk Inventor – Tutorials** window to explore the **Single-Day Productivity, Collaboration, Share and Reuse Data,** and **Using the Tutorials** menu options and tutorials.
4. Close the **Autodesk Inventor – Tutorials** window and continue exploring Autodesk Inventor, or exit Autodesk Inventor if necessary.

See "What's New" in This Release

Selecting the **See "What's New" in this release** button opens the **What's new in Autodesk Inventor** window, shown in Figure 1.10.

Inside this window, you find the **What's New** menu and a list of Autodesk Inventor tools, commands, and options that have been enhanced or have new features from previous releases. From this menu, you can choose **Assembly Design, Sketch, Part Design, Sheet Metal Design, Drawing Production, Import and Export, Visualization, Design space, Design Assistant, Development API,** and **Release 5.3 Enhance-**

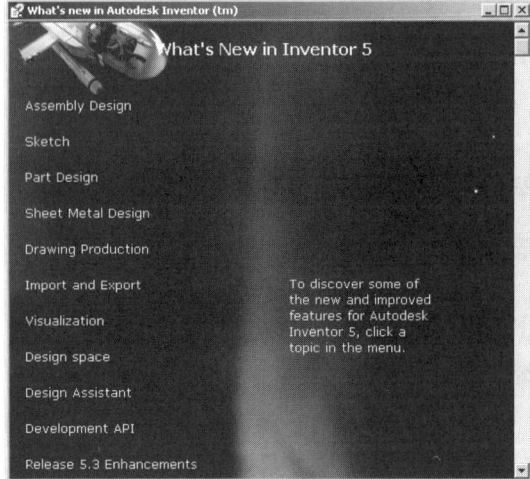

FIGURE 1.10   The **What's new in Autodesk Inventor** window.

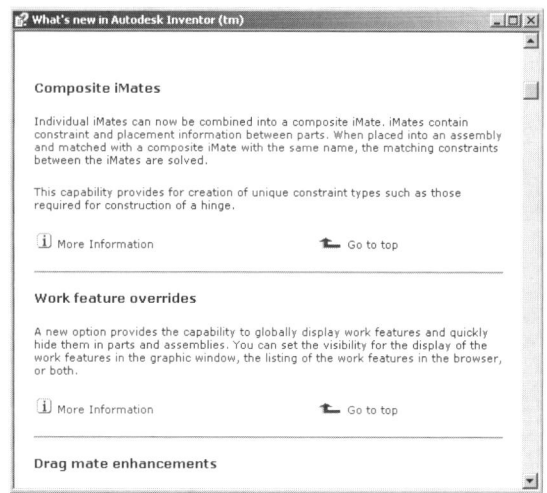

FIGURE 1.11   The **What's new in Autodesk Inventor** information.

ments. Selecting one of these options takes you to the information associated with the options, or opens another menu listing specific topics. See Figure 1.11.

You can also go through a small tutorial describing changes for this release by selecting the corresponding **More Information** button. While in the tutorial, the buttons previously described are available, including **Help, Main Menu, Back,** and **Next.**

Field Notes

Another way of reviewing the **What's New** information is to scroll through the topics.

Exercise 1.3

1. Continue from Exercise 1.2, or launch Autodesk Inventor.
2. Pick the **Getting Started** option of the **What To Do** menu if the **QuickStart** page is not currently displayed.
3. Select the **See "What's New" in this release** link.
4. Navigate through the **What's New** menu options at your own pace.
5. Close the **What's new in Autodesk Inventor** window and continue exploring Autodesk Inventor, or exit Autodesk Inventor if necessary.

Accessing the Autodesk Point A Website

The **Autodesk Point A** Website is accessed by selecting the direct **Autodesk Point A** link at the bottom of the **Getting Started** window. This Website provides news, assets, and multiple drafting industry–related services. **Autodesk Point A** contains links to resource materials and is a source of software updates. You can tailor your access to the Website by addressing only engineering drafting and design.

Accessing the Autodesk Streamline Website

The **Autodesk Streamline site** is accessed by selecting the direct **Autodesk Streamline** link at the bottom of the **Getting Started** window. If you do not have a project hosting account on **Autodesk Streamline,** picking the **Autodesk Streamline** link launches your default Web browser and displays your **Autodesk Streamline** project hosting page. This page allows you to set up an account. If you have an account, picking the **Autodesk Streamline** link displays all your projects in the **Files** list. **Autodesk Streamline** provides access to projects hosted in a manufacturing industry business-to-business marketplace.

Working With the New Option

Selecting **New** in the **What To Do** menu opens the **New File** section of the **Autodesk Inventor** dialog box, shown in Figure 1.12. Initially three tabs are available, including **Default, English,** and **Metric.** You can also access the **New File** window by using one of the following options:

✓ Select **New** from the **File** pull-down menu.

✓ Type the **Ctrl** key and the **N** key on your keyboard.

✓ Pick the **New** button on the **Standard** toolbar.

Inside each of the tabs are new *template* files for each of the four Autodesk Inventor file types, including: parts, assemblies, presentations, and drawings. As shown in Figure 1.12, the four Autodesk Inventor file types are represented by a symbol. A template is a file with preestablished settings, often also referred to as a prototype. You can specify the **Default** tab as English or Metric when you install the program, based on the particular units you use, and which you would like to designate as default. In addition to the **Default** tab, you can select other tabs with different unit templates including ANSI from the **English** tab, and BSI, DIN, GB, ISO, and JIS from the **Metric** tab.

FIGURE 1.12   The **New File** section of the **Autodesk Inventor** window.

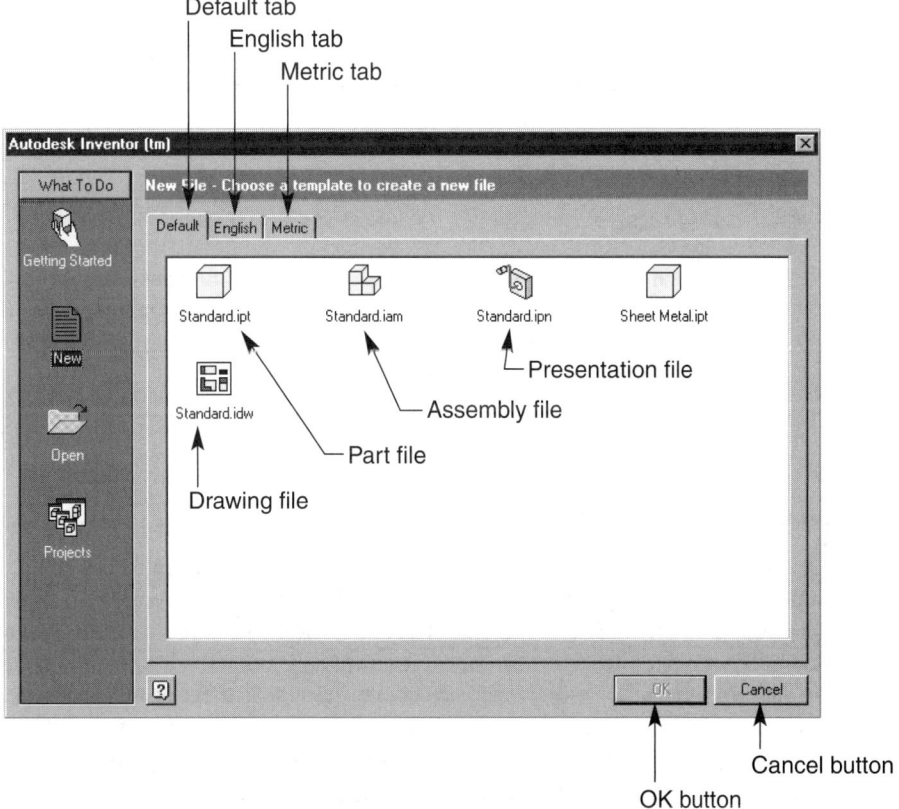

FIGURE 1.13 An example of a simple part.

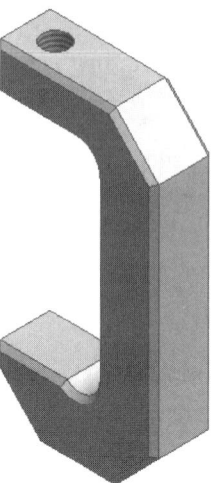

 Field Notes

> You may want to manipulate the Standard.ipt, Standard.iam, Standard.ipn, and Standard.idw template files to create user specific template files in the **Default** tab. These files should include any settings and options you consistently use, because when you pick the **New** button on the **Standard** toolbar, the Standard.ipt, .iam, ipn, or .idw template file opens.

What Are Part Files?

Part files carry the extension .ipt, which stands for Inventor part. For example, a part named Part1 is Part1.ipt. Part files contain only one component. Part file templates are used to create single parts, sheet metal parts, component parts for an assembly, and geometry to be used in two-dimensional drawings. Parts can be created separately in individual part file templates or can be created in an assembly file template with other parts. An example of a part is shown in Figure 1.13. Parts are discussed in detail in Chapters 2 through 11.

 Field Notes

> Sheet metal part templates are similar to other part templates and carry the same file extension, .ipt. Opening a new Sheet Metal.ipt file from the **New File** menu creates a part as a sheet metal subfile and opens the sheet metal tools necessary to create sheet metal parts when required. Sheet metal parts are discussed in detail in Chapter 10.

What Are Assembly Files?

Assembly files carry the extension .iam, which stands for Inventor assembly. For example, an assembly named Assembly1 is Assembly1.iam. Assembly template files are used to create assemblies and subassemblies. Unlike part files, assembly files contain one or multiple parts. You can develop assemblies using assembly template files two ways. One, insert parts from part files into an assembly template file or assembly file, and create an assembly. Two, create individual parts in the assembly template file, to create an assembly. Though either option is effective, it is often faster, easier, and more productive to create parts within an assembly template file or assembly file. Building parts inside an assembly file creates both an assembly and a separate part file. Therefore, you can build and modify individual parts directly in the assembly, and open separate part files for further use, such as developing single part drawings. In addition, you do not have to constantly open part files or search for an individual part. An example of an assembly is shown in Figure 1.14. Assemblies are discussed in detail in Chapters 14 and 15.

FIGURE 1.14 An example of a simple assembly.

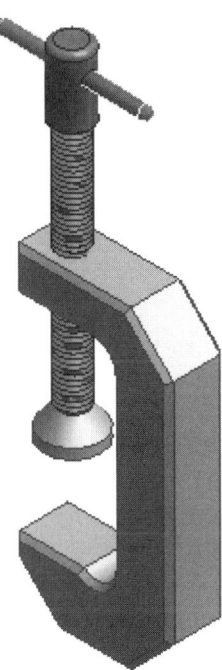

Field Notes

In Chapter 15, you will see another reason why the concept of creating parts in an assembly file is effective, when you create adaptive parts and assemblies.

What Are Presentation Files?

Presentation files have an extension of .ipn, which stands for Inventor presentation. For example, a presentation named Presentation1 is Presentation1.ipn. As the name implies, presentations allow you to present assemblies in a way that shows how separate parts interact together within the full assembly. Presentation template files are used to create presentations of assemblies including exploded, animated, and stylized views of assemblies. The presentations may include connection graphics between parts (trails) and part modification (tweaking) of components. An example of a presentation is shown in Figure 1.15. Presentations are discussed in detail in Chapter 16.

What Are Drawing Files?

Drawing files carry the extension .idw, which stands for Inventor drawing. For example, a drawing named Drawing1 is Drawing1.idw. Drawing file templates are used to create working drawings from part, assembly, and presentation files. An example of a drawing file is shown in Figure 1.16. Drawings are discussed in detail in Chapters 12, 13, and 17.

Exercise 1.4

1. Continue from Exercise 1.3, or launch Autodesk Inventor.
2. Pick the **New** option of the **What To Do** menu if the **New File** page is not currently displayed.
3. Select the **Default, English,** and **Metric** tabs, and observe the contents of each.
4. Continue exploring the **Autodesk Inventor** dialog box, or exit Autodesk Inventor if necessary.

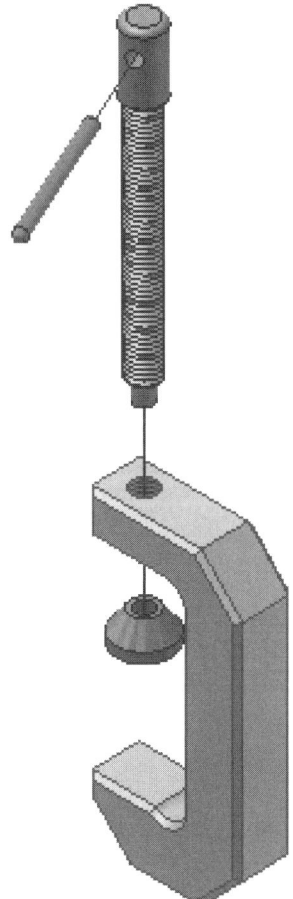

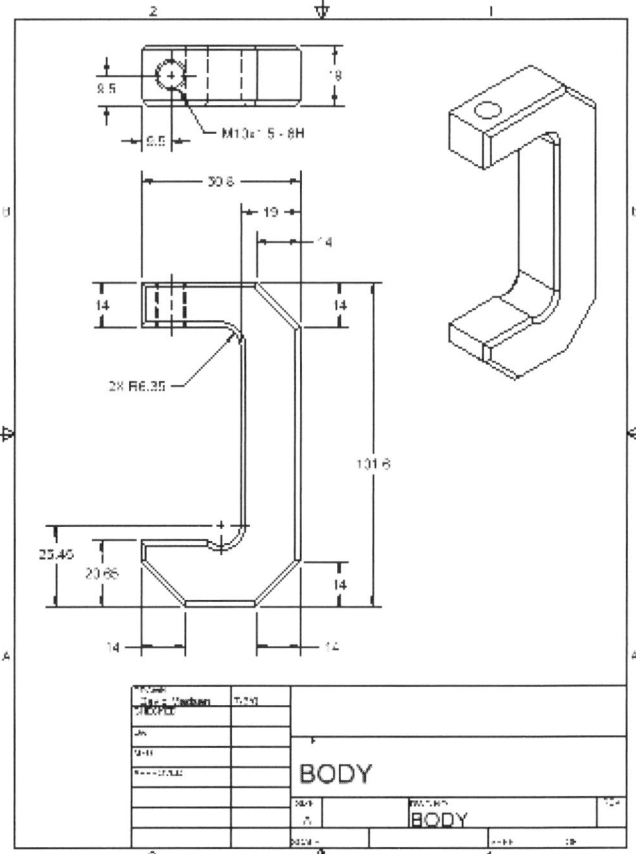

FIGURE 1.15   An example of a presentation.

FIGURE 1.16   An example of a simple drawing.

Using Custom Tabs and Templates

In addition to the **Default, English,** and **Metric** tabs available in the **New File** window of the **Autodesk Inventor** dialog box, it is possible to create your own tabs and templates. Though the various templates provided by Autodesk Inventor are adequate for many applications, you may want to customize a template for your specific use. Developing your own template can save time and effort, because the original work environment settings you have established are already set, every time you open the template. You can develop a template for each of the four Autodesk Inventor file types. For example, a custom drawing file template may include unit settings, a title block, and a border. A custom part file template may include unit settings and various work environment settings such as grid spacing, and colors. Each of these components will be discussed in later chapters.

Creating a Custom Tab and Template: An Introduction

The process of creating a custom tab and template involves first opening an existing template file. The next step is to adjust the work environment and template settings. These settings will be discussed in detail in future chapters. Still, you can begin to understand user-defined tabs and templates, and initially create your own, before you manipulate the work environment, settings and options.

All the Autodesk Inventor tabs and templates found in the **New File** section of the **Autodesk Inventor** dialog box are stored in folders in the path Autodesk/Inventor 5.3/Templates. See Figure 1.17. You can access these folders using your Windows Explorer. The folders correspond to the tabs shown in the **New File** section of the **Autodesk Inventor** dialog box. The **Default** tab does not have a separate folder. Consequently, the part, assembly, presentation, and drawing file templates, found under the **Default** tab, are located in the **Templates** folder, in the path Autodesk/Inventor 5.3/Templates. In contrast to the **Default** tab, the **English** and **Metric** tabs each correspond to separate folders. The part, assembly, presentation, and drawing file templates are found in the **English** and **Metric** folders, located in the paths Autodesk/Inventor 5.3/Templates/English, and Autodesk/Inventor 5.3/Templates/Metric.

FIGURE 1.17 Using
Windows Explorer, you
can see the location
and contents of the
Templates folder.

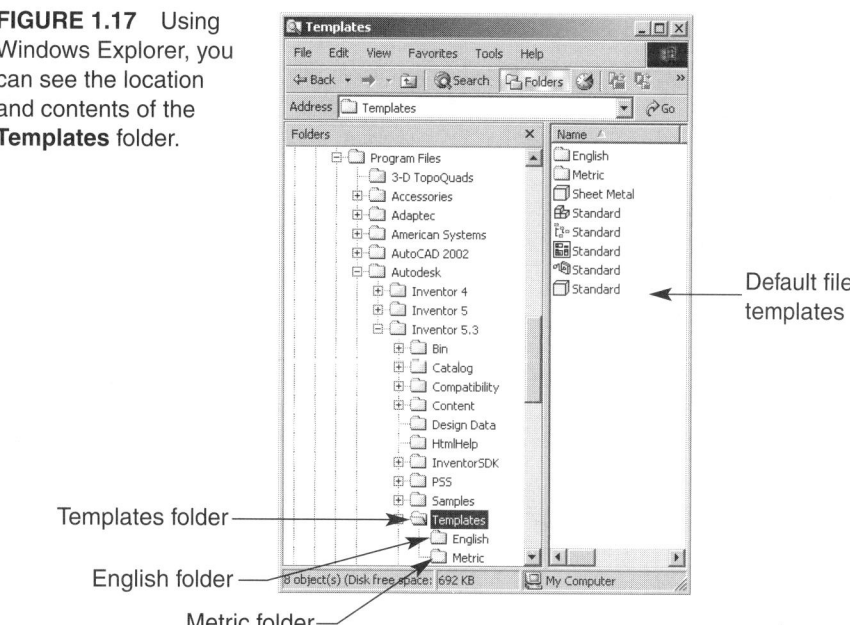

Templates folder

English folder

Metric folder

Default file
templates

Creating a Custom Tab and Template

Probably the easiest and most effective way to create a template is to start with a preexisting Autodesk
Inventor part, assembly, presentation, or drawing file template. For example, you may want to begin with
Standard.ipt for a part or Standard.iam for an assembly.

First, open one of the available template files from one of the tabs located in the **New File** window. This
is done by selecting the file and choosing the **OK** button, or double-clicking on the template file. After opening one of the available template files, select **Save** under the **File** pull-down menu, as shown in Figure 1.18.

FIGURE 1.18 The
Save option of the **File**
pull-down menu.

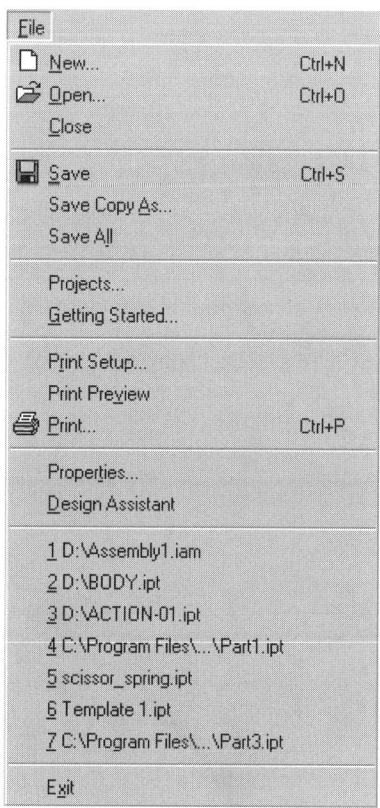

FIGURE 1.19 The **Save As** dialog box.

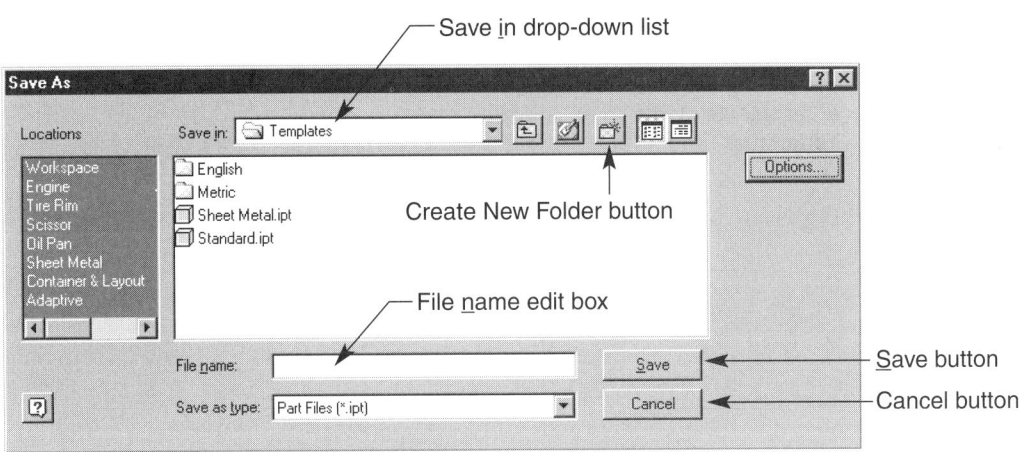

Selecting **Save** opens the **Save As** dialog box. See Figure 1.19. Now, locate Templates in the designated file path, most often, Autodesk/Inventor 5.3/Templates. Navigating to the Templates location is accomplished by opening drives and folders in the **Save in:** drop-down list. Selecting the down arrow next to the drop-down list box allows you to navigate through drives and folders to find a certain file. Selecting the **Return** button to the right of the down arrow allows you to return to the previously opened drive or folder. Once you have found the location of the Templates folder, select the **Create New Folder** button, which is the third button to the right of the **Save in:** pull-down menu box. Next, type a name, such as Template, over the existing **New Folder** default name. Now, double-click on the folder you have just created, or type the **Enter** key on your keyboard twice. Finally, specify a template file name, such as Template 1, in the **File name:** edit box, and select the **Save** button, depress the **alt** and **s** keys on your keyboard, or type the **Enter** key on your keyboard.

Now when you launch Autodesk Inventor and select the **New** option of the **What To Do** menu, in the **Autodesk Inventor** dialog box, you see the template tab and file you have created. See Figure 1.20.

You will also see the tab as a folder under Autodesk/Inventor 5.3/Templates, with the template file you have created, as shown in Figure 1.21.

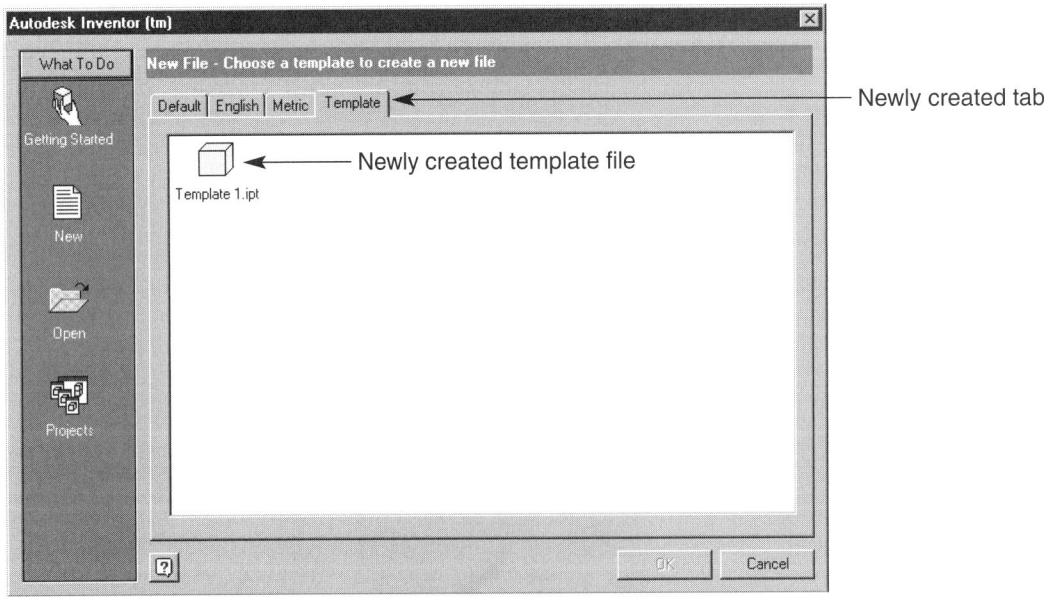

FIGURE 1.20 A custom tab and template file in the **New** option of the **What To Do** menu.

FIGURE 1.21 The custom tab as a folder and the custom template file located in the folder.

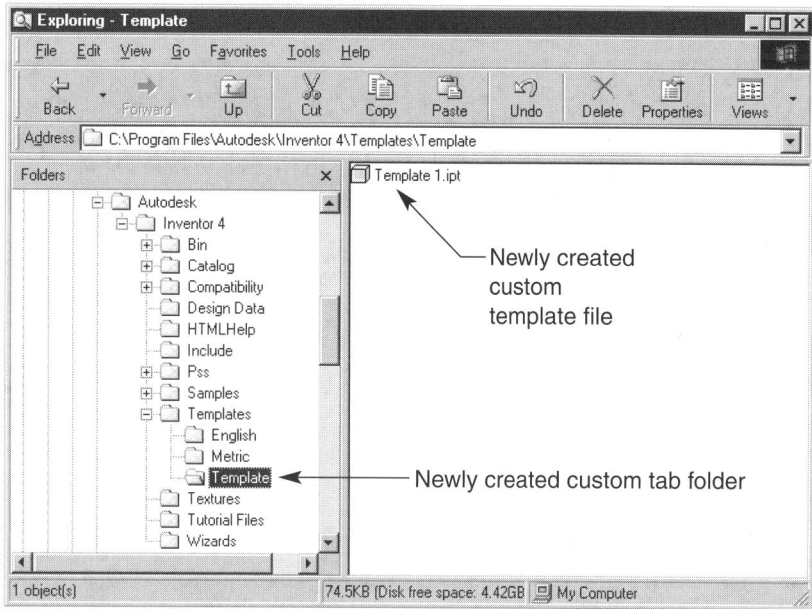

Exercise 1.5

1. Continue from Exercise 1.4, or launch Autodesk Inventor.
2. Pick the **New** option of the **What To Do** menu if the **New File** page is not currently displayed.
3. Create a custom tab and template file using the method previously described. Name the tab, Template, and the template file Template 1.
4. Open your Explorer and locate the tab folder and template file you just created.
5. If you are in a classroom setting, remove the Template folder and the contents (Template 1) from the Autodesk Inventor Templates folder by dragging and dropping it to an area on your network directory, diskette, or other location specified by your instructor.
6. Continue exploring the **Autodesk Inventor** dialog box, or exit Autodesk Inventor if necessary.

Using the Open Option

Use the **Open** option in the **What To Do** menu to open the **Open File** section, shown in Figure 1.22, of the **Autodesk Inventor** dialog box. This area is used to access existing files that you have worked with previously. The **Open File** window can also be accessed by using one of the following techniques:

✓ Select **Open** from the **File** pull-down menu.

✓ Type the **Ctrl** key and the **O** key on your keyboard.

✓ Pick the **Open** button on the **Standard** toolbar.

The following options are located inside the **Open File** section:

■ **Locations** This list box shows a list of the file paths in reference to the active project. You can select a path to display the files and subfolders. Projects are discussed later in this chapter.

■ **Look in:** This drop-down list allows you to navigate through drives and folders to find a certain file by selecting the down arrow next to the drop-down list box. The list box is located below the **Look in:** drop-down list and shows the names of the files located in the current directory.

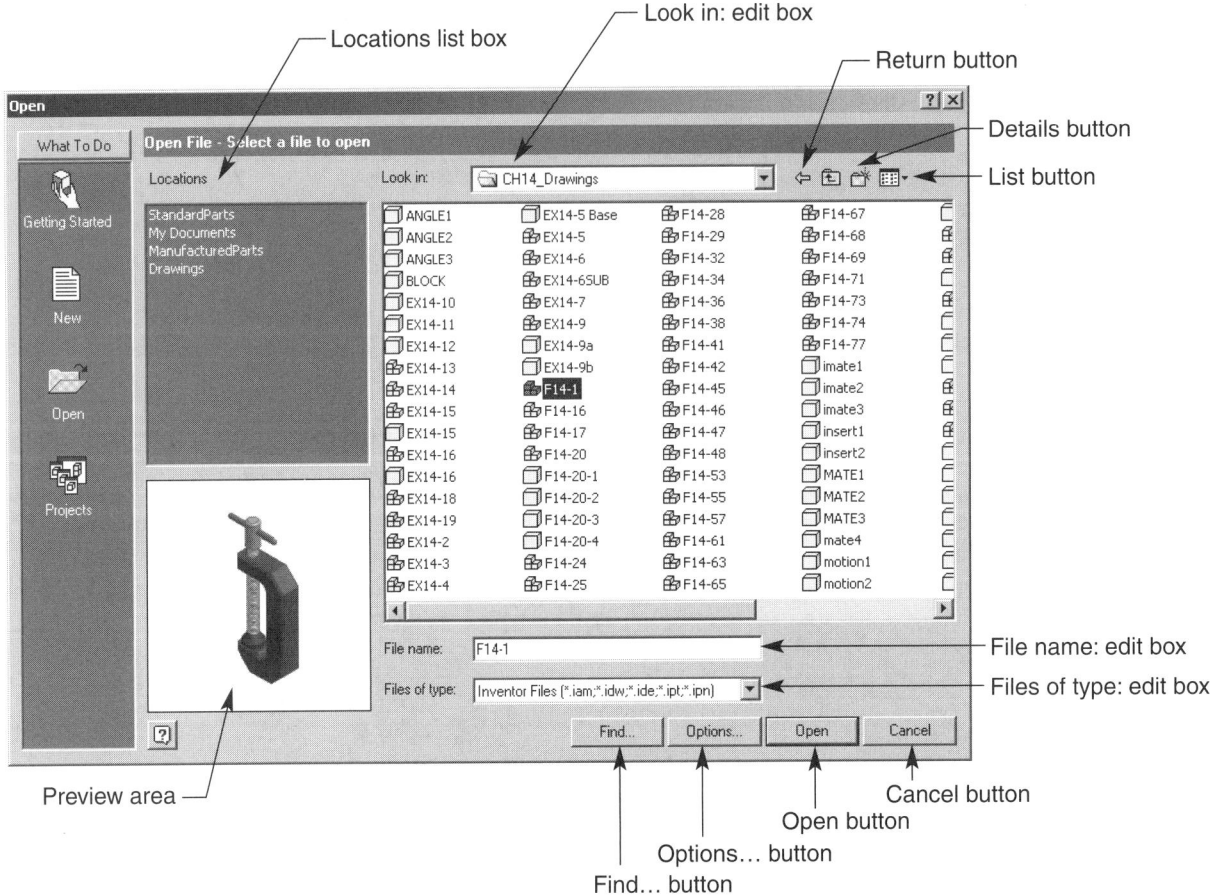

FIGURE 1.22   The **Open** option in the **What To Do** menu opens the **Open File** section of the **Autodesk Inventor** dialog box.

- **Return**   This button, to the right of the down arrow, allows you to return to the previously opened drive or folder.

- **List**   This button, to the right of the **Return** button, allows you to view the drives and folders in a list format.

- **Details**   This button, to the right of the **List** button, provides you with drive, folder, and file details. The information displayed includes: **Name,** which shows the folder or file name, **Size,** which displays the size of the file, **Modified,** which shows when the folder or file was last modified, and **Attributes,** which displays any attributes the file has.

- **File name:**   This edit box shows the currently selected and highlighted Inventor file. To display the file name, pick the file from the list box, or type the name of the file in the **File name:** edit box.

- Preview area   To the left of the **File name:** edit box is a preview area that displays an image of the currently selected Inventor file. Previewing a file can be helpful if you do not know the file name, or have files with similar names, and would like to see a preview of the file before opening it.

- **Files of type:**   This drop-down list allows you to choose which files you want to have visible in the list box and provides you with a variety of options. You can choose to have all files listed (*.*), only Inventor files listed (*.ipt,*.ide,*.iam,*.idw,*.ipn), or only certain files listed such as Part files (*.ipt). These options can be especially helpful if you are looking for a certain type of file and do not what to see every other type of file in the list box.

FIGURE 1.23 The
File Open Options
dialog box.

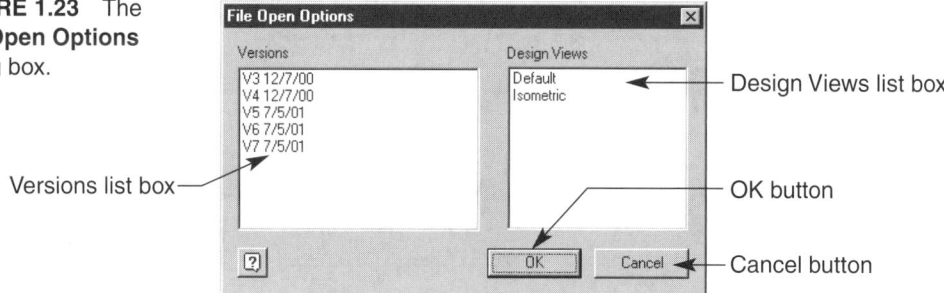

Versions list box

Design Views list box

OK button

Cancel button

■ **Options...**   Picking this button opens the **File Open Options** dialog box. See Figure 1.23. To use this box, select a file located in the **Open** dialog box and pick the **Options** button. The **Versions** list box shows the specified file versions specific to the selection. You can adjust the number of versions listed in this box by manipulating the number shown in the **Number of versions to keep** edit box. This box is found in the **General** tab of the **Options** dialog box. To access the **Options** dialog box, select **Application Options...** from the **Tools** pull-down menu. The **Design Views** list box displays the available design views if the file is an assembly. You can select a design view to display when the assembly file is opened.

■ **Browse...**   Selecting this button opens the standard **Open** dialog box, shown in Figure 1.24. The **Open** dialog box provides you with more options than the **Open File** section of the **Autodesk Inventor** dialog box. The additional features are described next.

  ■ **View Desktop**   This button allows you to see your desktop options and select a drive or initial folder location. You may find this helpful if you need to navigate quickly back from a path to your desktop.

  ■ **Create New Folder**   This button allows you to create a new folder if desired. To use the **Create New Folder** button, locate the specific area where you would like a new file, and then select the **Create New Folder** button. Now, type a name—for example, New1—over the existing **New Folder** default name. Pick the folder you have just created, or type the **Enter** key on your keyboard.

  ■ **Find...**   This button opens the **Find: Autodesk Inventor Files** dialog box, shown in Figure 1.25, which contains the following options:

    ■ **Find files that match these criteria**   This area describes the search that is currently specified. The default search displays in the window all Inventor files (*.ipt, *.ide, *.iam, *.idw, *.ipn).

FIGURE 1.24 The
Open dialog box.

View Desktop button

Create New Folder button

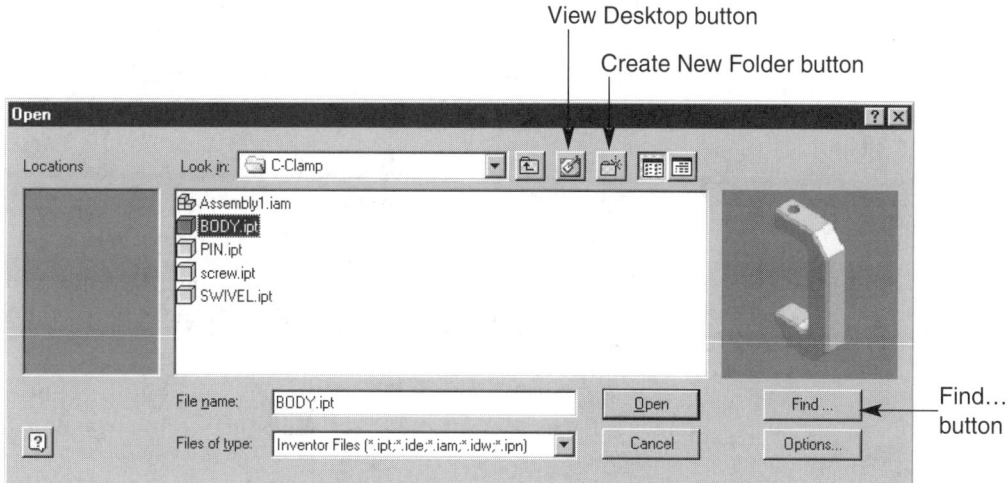

Find...
button

Find files that match these criteria area

Match Case check box

Define more criteria area

And radio button

Or radio button

Property: edit box

Condition: edit box

Current Folders radio button

Look in: drop-down list

Open Search... button

Save Search... button

Clear Search button

Delete Line button

Add to list button

Value: edit box

Search Location area

Sub Folders radio button

Related Files radio button

Browse button

Find Now button

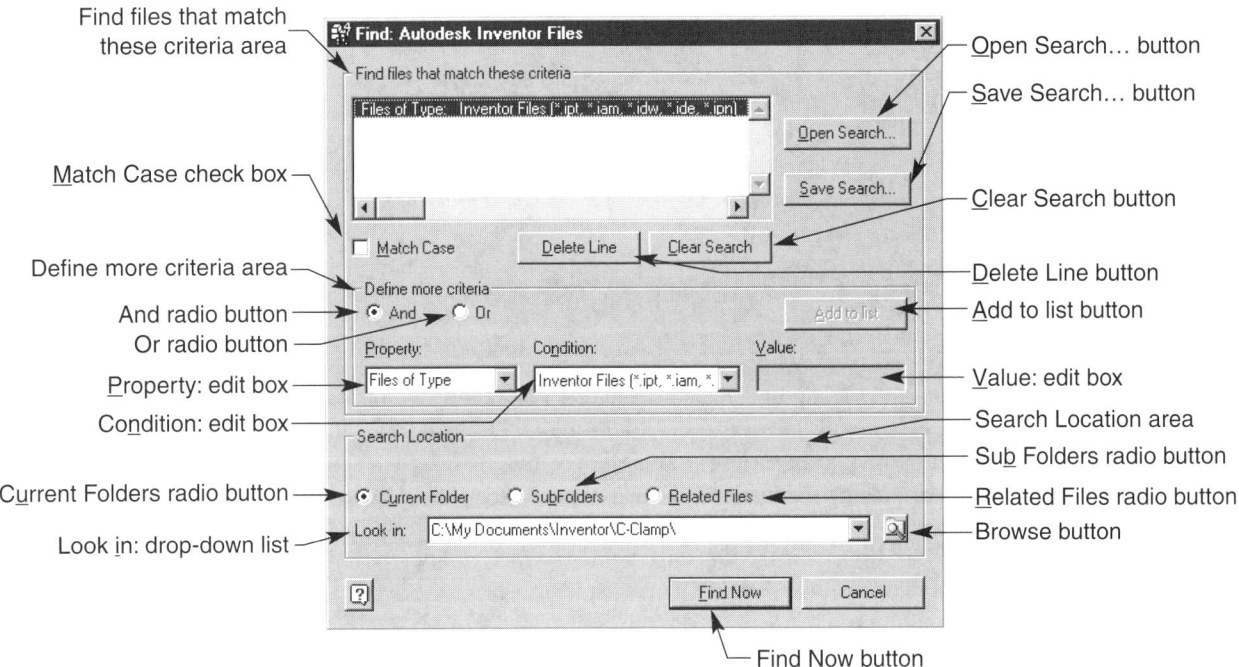

FIGURE 1.25   The **Find: Autodesk Inventor Files** dialog box.

- **Delete Line**   This button deletes the file type currently displayed.

- **Clear Search**   This button returns all the currently selected and displayed options to the default selections and options.

- **Match Case**   This check box is used to find files with the same upper- or lowercase text you define in your search parameters.

Field Notes

Selecting the **Clear Search** button does not remove the check from the **Match Case** check box.

- **Define more criteria**   This area enables you to develop search criteria, by constraining your search for a file, and has the following features:

  - **And**   Use this radio button to search files that include only the specified criteria that are found in other searches and earlier defined search criteria.

  - **Or**   This radio button allows you to find files that include any of the specified criteria.

  - **Property:**   Use this drop-down list to select from a variety of search options—for example, Files of type, Author, or Keywords. You can also specify a property by selecting the edit box and typing a property not listed.

  - **Condition:**   This drop-down list allows you to select from a number of conditions that apply to the selected property value. Some of the conditions require user definitions to be applied in the **Value:** edit box.

  - **Value:**   This edit box enables you to type in values for conditions that require values. For example, if you select Author in the **Property:** drop-down list, and "begins with" in the **Con_d_ition:** drop-down list, the **_V_alue:** edit box is now available for you to enter text.

■ **_A_dd to list**   Once you have selected from the various options, pick the **_A_dd to list** button, to add the list to the **Find files that match these criteria** display box.

■ **Save Search...**   This button opens the **Autodesk Inventor Find: Save Search** dialog box, shown in Figure 1.26. This dialog box enables you to save a search for further use. This can be very helpful if you have previous search criteria you would like to use for another files search. Simply enter a name for the search in the **Name for this search:** edit box, and select the **_O_K** button, or type the **Enter** key on your keyboard.

■ **_O_pen Search...**   Selecting this button opens the **Autodesk Inventor Find: Open Search** dialog box, shown in Figure 1.27. Here you can open a search you previously saved by selecting the search you would like to open and picking the **_O_pen** button, or double-clicking the search name. You can also delete a search if it is no longer needed by picking the search name and selecting the **_D_elete** button.

■ **Search Location**   This area contains three radio buttons, a **Look _in_:** drop-down list, and a **Browse for folder** button.

   ■ **_C_urrent Folders**   Selecting this radio button confines the search to the currently selected folder.

   ■ **Su_b_ Folders**   Choosing this radio button enables the search to look through the designated folder and its subfolders.

   ■ **_R_elated Folders**   This radio button allows the search to reference files from a related folder you select.

   ■ **Look _in_:**   The **Look _in_:** drop-down list allows you to navigate through drives and folders to find a certain file.

   ■ **Browse**   The **Browse** button opens the **Browse for Folder** dialog box, which provides you with further access to drive, folder, file, and network folder locations.

Once you have specified all your search options, choose the **Find Now** button, located at the bottom of the **Find: Autodesk Inventor Files** dialog box, to begin the file search.

FIGURE 1.26 The **Autodesk Inventor Find: Save Search** dialog box.

FIGURE 1.27 The **Autodesk Inventor Find: Open Search** dialog box.

Exercise 1.6

1. Continue from Exercise 1.5, or launch Autodesk Inventor.
2. Pick the **Open** option of the **What To Do** menu if the **Open File** window is not currently displayed.
3. Use the **Look in:** drop-down list to navigate to the directory where the Autodesk Inventor program files are located, usually Program Files/Autodesk/Inventor 5. From there, find the blade_main.ipt part file from the following directory: Samples/Models/Scissor/ Components. Do not open the file.
4. Use the **List** and **Details** buttons to view the file in a list and details format.
5. Pick the **Return** button to return to the Scissor folder. Select the assembly file, scissor.iam, and pick the **Options...** button to view its options. Now select Part Files (*.ipt) from the **Files of type:** drop-down list. Notice the Assembly file is no longer visible.
6. Select the **Browse...** button and locate the file: Oil Pan.ipt, from the Oil pan folder of the Models folder.
7. Next, pick the View Desktop button to view your desktop. Then create a new folder, named Folder 1, by selecting the **Create New Folder** button. Delete Folder 1 by right clicking and selecting **Delete.**
8. Select the **Find...** button and use the methods previously described to find only the Inventor part files located in the following directory: Program Files/Autodesk/Inventor 5/Models.
9. Close the **Open** dialog box.
10. Continue exploring the **Autodesk Inventor** dialog box, or exit Autodesk Inventor if necessary.

Using Projects

Projects are used to manage and organize Autodesk Inventor files, and allow Autodesk Inventor to find and provide connections between files. For example, an assembly references all the parts used to create the assembly through the active project.

 Field Notes

An error message tells you if the assembly components are not located in the active project.

You can create and use as many projects as necessary to manage your work in an efficient manor. The **Projects** section of the **Autodesk Inventor** dialog box allows you to manipulate projects including editing existing projects, creating new projects, and activating the project you want to be current.

Using the Project Editor

Select the **Projects** option in the **What To Do** menu to open the **Projects** section (also known as the **Project Editor**) of the **Autodesk Inventor** dialog box. See Figure 1.28. The **Project Editor** can also be accessed by using one of the following options:

✓ Select the **Windows Start** menu button, followed by the **Programs** option, then the **Inventor 5.3** option, followed by **Tools,** and finally **Project Editor.**

✓ Right-click on a project file (project files carry the extension .ipj) and select **Edit.**

Opening the **Project Editor** using one of these two options does not open the **Autodesk Inventor** dialog box. Instead, it opens the **Inventor Project Editor** dialog box. See Figure 1.29. Notice that the only

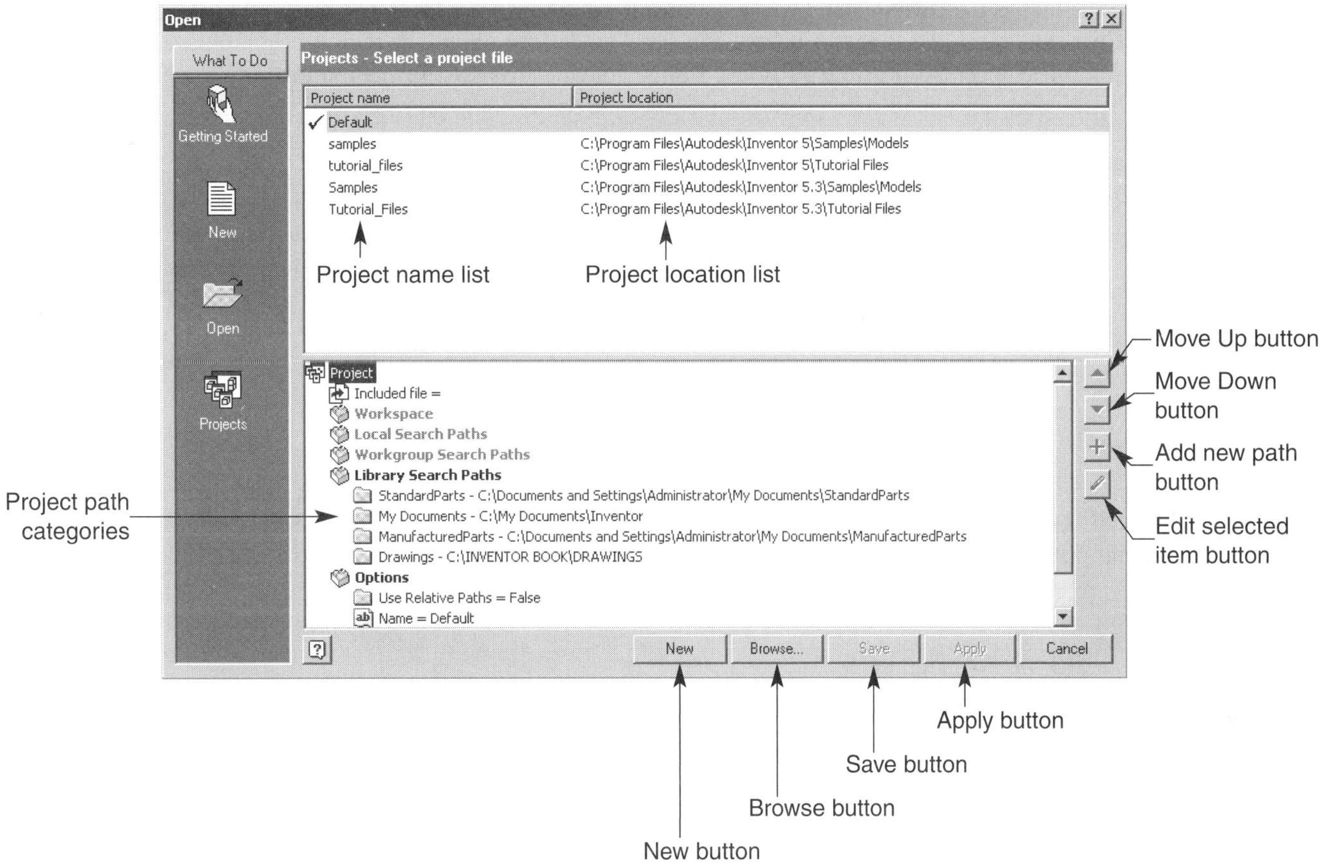

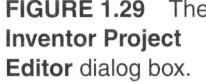

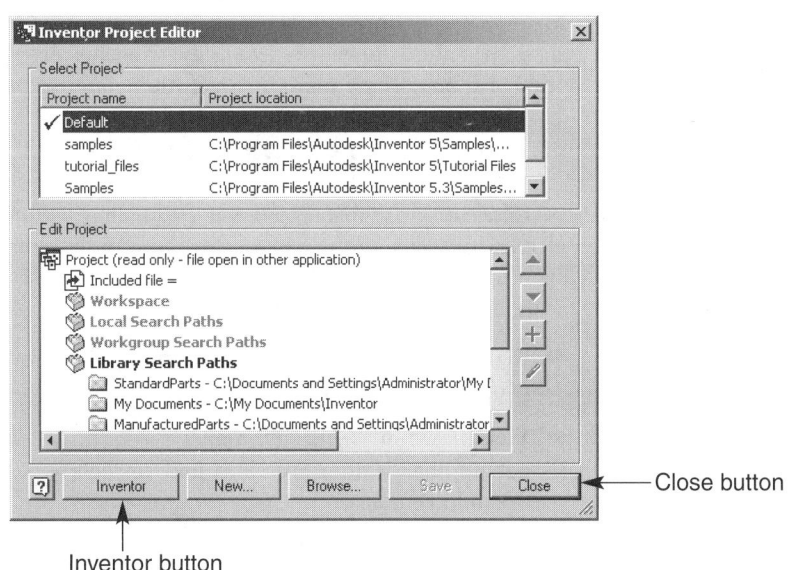

FIGURE 1.28   The **Projects** section (also known as the **Project Editor**) of the **Autodesk Inventor** dialog box.

FIGURE 1.29   The **Inventor Project Editor** dialog box.

difference is the **Inventor** and **Close** buttons, located below the lower list box, and the lack of **Apply** and **Cancel** buttons. Choosing this button launches Autodesk Inventor with the selected project active.

The upper list box in the **Project Editor** contains two lists. **Project name** provides the names of the existing projects, and **Project location** provides the path, which describes the location of the project file.

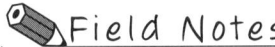**Field Notes**

Initially, only the three Autodesk Inventor project files are available: **Default, Samples,** and **Tutorial_Files.** The **Default** project is initially active when you open the **Project Editor** for the first time. It contains no search paths, and will require you to resolve links to component folders if used for an assembly.

Clicking once on the name of the project allows you to view the contents of the path in the lower list box. Double-clicking the project name, or selecting the project and picking the **Apply** button, activates the project. When a project is active, a check mark is visible to the left of the project name.

**Field Notes**

The project that is active when the **Project Editor** is closed will be the active project when the **Project Editor** is opened again. You must close all Autodesk Inventor files before you can activate a different project.

When you right-click on a project name or on the project path location, the **Rename, Browse, New,** and **Delete** options are available. The function of these options is explained as follows:

■ **Rename**   This option allows you to change the name of the project. This is accomplished by typing a new name over the existing name, and picking the new name or selecting **Enter** on your keyboard. Renaming the project is also done by clicking the highlighted project name once, and typing a new name for the project as previously mentioned.

■ **Browse**   This option opens the **Choose project file** dialog box. See Figure 1.30. You can also open the **Choose project file** dialog box by selecting the **Browse** button below the lower list box. The **Choose project file** dialog box allows you to select a project that is not currently shown under **Project name,** in the upper list box. The following options are available inside the **Choose project file** dialog box:

   ■ **Look in:**   Use this drop-down list to navigate through drives and folders to find a certain file, by selecting the down arrow next to the drop-down list box. The list box is located below the **Look in:** drop-down list and shows the names of the files located in the current directory.

   ■ **Return**   This button, to the right of the down arrow, allows you to return to the previously opened drive or folder.

   ■ **View Desktop**   This button allows you to see your desktop options and select a drive or initial folder location. You may find this helpful if you need to navigate quickly back from a path to your desktop.

   ■ **Create New Folder**   This button allows you to create a new folder if desired. To use the **Create New Folder** button, locate the specific area where you would like a new file, and then select the **Create New Folder** button. Now, type a name—for example, New1—over the existing **New Folder** default name. Pick the folder you have just created, or press the **Enter** key on your keyboard.

FIGURE 1.30 The **Choose project file** dialog box.

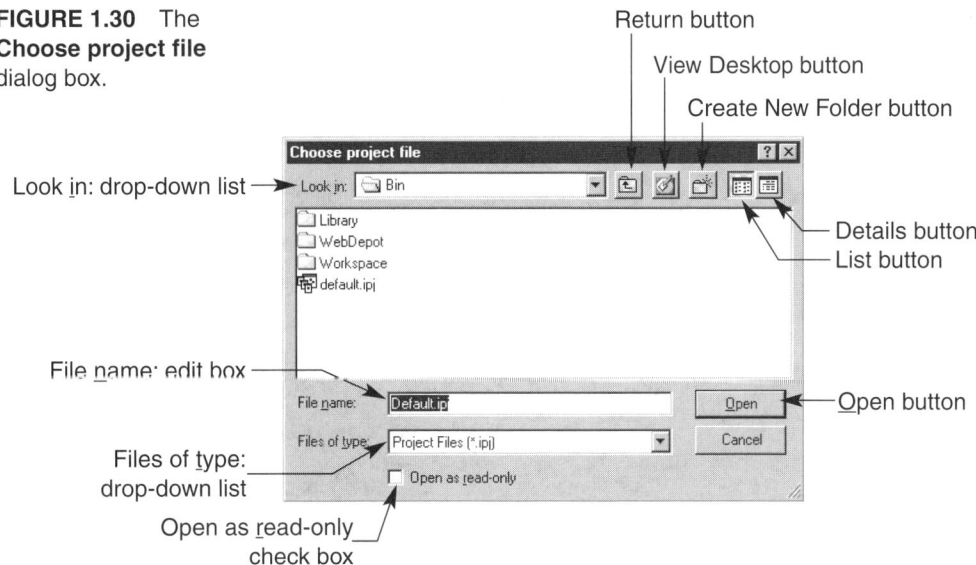

- **List**   Use this button, located to the right of the **Return** button, to view the drives and folders in a list format.

- **Details**   This button, to the right of the **List** button, provides you with drive, folder, and file details. The information displayed includes: **Name,** which shows the folder or file name; **Size,** which displays the size of the file; **Modified,** which shows when the folder or file was last modified; and **Attributes,** which displays any attributes the file has.

- **File name:**   This edit box shows the currently selected and highlighted Inventor file. To display the file name, pick the file from the list box, or type the name of the file in the **File name:** edit box.

- **Files of type:**   This drop-down list allows you to choose which files you want to have visible in the list box, and provides you with a variety of options. You can choose to have all files listed (*.*), only Inventor files listed (*.ipt,*.ide,*.iam,*.idw,*.ipn), or only certain files listed such as Part files (*.ipt). These options can be especially helpful if you are looking for a certain type of file and do not what to see every other type of file in the list box.

- **Open as read-only**   If selected, this check box allows you to open a file as read-only. Read-only allows you to view a file, but you cannot change or add any new information to the file.

After finding the project, double-click on the project name, or select the file and pick **Open.** Accomplishing this task allows Autodesk Inventor to create a shortcut to the file in the **Projects** folder, and lists the project in the upper list box.

- **New**   Selecting this button opens the **Inventor project wizard** dialog box, shown in Figure 1.31. This dialog box allows you to create a new project file. The **Inventor project wizard** and creating a new project file are discussed later in this chapter.

- **Delete**   Selecting this button deletes the specified Project file.

✎ Field Notes

You cannot rename or delete the active Project. Only editing tools that apply to the currently selected Project are available (other tools are shaded).

FIGURE 1.31 The **Inventor project wizard** dialog box.

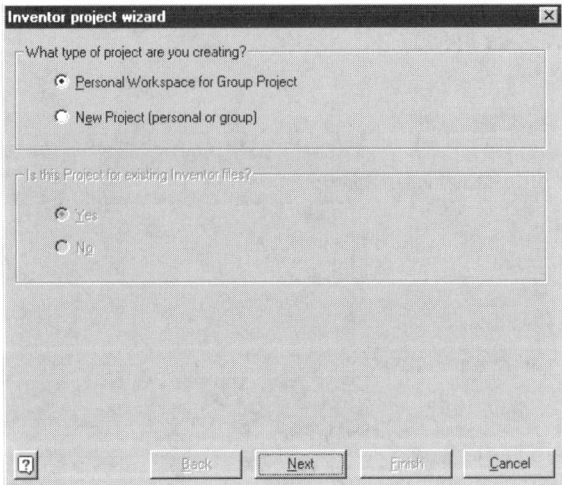

The lower list box of the **Project Editor** is composed of a **Project** folder that contains shortcuts to an **Included file** path, **Library Search Paths, Workspace, Local Search Paths, Workgroup Search Paths,** and **Options.** These shortcuts are described as follows.

Field Notes

Double-click any of these options to show or hide the paths.

- **Included file**   An **Included file** allows one additional project file and its paths to be included in a given project by combining the information in the **Included file** and the project file together. Often the **Included file** is a workgroup or library file that allows many users, such as a design team, to access files that are commonly used for a certain project. Right-clicking on an **Included file** reveals the **Open, Edit,** and **Delete** options described as follows:

  - **Open**   Use this option to open the specified **Included file** path.
  - **Edit**   This option allows you to edit the **Included file** path location.

Field Notes

The **Edit selected item** button, located to the right of the lower list box, also allows you to edit the **Included file** path location.

  - **Delete**   Pick this option to delete the **Included file** path.

- **Library Search Paths**   These paths allow Autodesk Inventor to locate files contained in a library that are used in a project or several different projects. The elements of a library are generally shared and may include standard, often-used parts, components from other programs, such as Mechanical Desktop, or any other libraries that you may need to reference. Right-clicking on **Library Search Paths** reveals the following options:

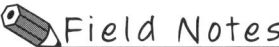Field Notes

When you open a file, the specified **Library Search Paths** are searched first.

■ **<u>A</u>dd Path** This option allows you to add a path to the specified selection. After choosing the **<u>A</u>dd Path** option, you can change the default name and find a folder using the **Browse** button located to the right of the path. The **Browse for Folder** dialog box, shown in Figure 1.32, allows you to add a path to the specified selection from an existing file.

Field Notes

The **Add new path** button, located to the right of the lower list box, also allows you to add a new path to the specified selection.

■ **Add Paths from <u>F</u>ile...** Selecting this option opens the **Choose project file** dialog box. As previously described, and shown in Figure 1.30, the **Choose project file** dialog box allows you to select a project that is not currently shown under **Project name,** in the upper list box. After finding the project, double-click on the project name, or select the file and pick **<u>O</u>pen.**

■ **<u>P</u>aste Paths** This option allows you to paste a path from one location to another. To use this tool, right-click on the path that you want to paste to the new location. Next, select either **<u>C</u>ut** or **<u>C</u>opy** from the menu items. Finally, locate the area where you want to paste the path, right-click and select **<u>P</u>aste Paths.**

■ **Delete Section Paths** Selecting this choice deletes every path located under the specified section.

Right-clicking on one of the folders located under **Library Search Paths** reveals these menu options.

■ **<u>E</u>dit** This option allows you to edit the path name and location. Selecting **<u>E</u>dit** also displays the **Browse for Folder** button, located to the right of the edit box, previously discussed.

FIGURE 1.32 The **Browse for Folder** dialog box.

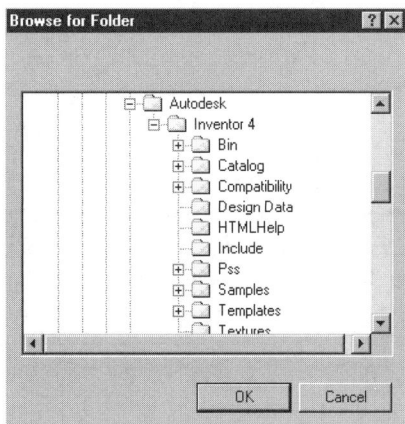

Field Notes

> The **Edit selected item** button, located to the right of the lower list box, also allows you to edit the path name and location.

- **Add Path**   This option allows you to add a path to the specified selection. After choosing the **Add Path** option, you can change the default name and find a folder using the **Browse** button located to the right of the path. The **Browse** button opens the **Browse for Folder** dialog box.

Field Notes

> The **Add new path** button, located to the right of the lower list box, also allows you to add a new path to the specified selection.

- **Cut**   Use this option to remove the specified path without deleting. The path can then be pasted in a new location.

- **Copy**   Select this option to copy the specified path for use in a new location, without removing the path from the initial directory.

- **Paste**   This option allows you to paste a file from one location to another. To use this tool, right-click on the path that you want to paste to the new location. Next, select either **Cut** or **Copy** from the menu items. Finally, locate the area you want to paste the path, right-click, and select **Paste**.

- **Delete**   Pick this choice to delete the specified path from the directory.

- **Workspace**   The **Workspace** is the default directory where Autodesk Inventor files are located. If all the files are located in one place, a **Workspace** is the only working directory in a given project. However, the **Workspace** may be one of many directories if files are located in several areas. In addition, the **Workspace** is your exclusive area to operate if you are working on a design team.

Field Notes

> A project can only contain one **Workspace**. When you open a file and it is not found in the specified **Library Search Paths,** the specified **Workspace** is searched second.

Right-clicking on **Workspace** reveals the same options available as when you right-click on **Library Search Paths,** as previously described. If a workspace section path already exists, **Delete Section Paths** is the only option available, because a project can contain only one **Workspace**. Right-clicking on a path located under **Workspace** reveals the following four options:

- **Edit**   This option allows you to edit the name and location of the specified path and find a folder using the **Browse** button located to the right of the path. The **Browse** button opens the **Browse for Folder** dialog box.

Field Notes

The **Edit selected** item button, located to the right of the lower list box, also allows you to edit the path name and location.

- **Cut**   This option allows you to remove the specified path without deleting. The path can then be pasted in a new location.

- **Copy**   This option allows you to copy the specified path for use in a new location, without removing the path from the initial directory.

- **Delete**   Select this option to delete the specified path from the directory.

- **Local Search Paths**   These paths locate additional files you may need for your personal use. The files are located either on your computer or the network.

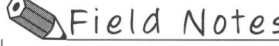Field Notes

When you open a file, and it is not found in the specified **Workspace,** the specified **Local Search Paths** are searched third.

Right-clicking on **Local Search Paths** displays the same options available as when right-clicking on **Workspace**. However, in addition the **Add Paths from Directory** option is also available.

- **Add Paths from Directory**   Selecting this option opens the **Browse for Folder** dialog box, as shown in Figure 1.32. This box allows you to locate a folder from anywhere on your computer or the network.

Right-clicking on a path located under **Local Search Paths** reveals eight menu options. The **Edit, Add Path, Cut, Copy, Paste,** and **Delete** options were previously discussed. Only one path can be specified in **Workspace,** while many can be specified in **Local Search Paths**. Therefore, you have the option of moving a local search path up or down in the list of search paths using the **Move Up** and **Move Down** selections.

- **Move Up**   This option allows you to move the path above other paths in the path list.

Field Notes

The path can also be moved up by selecting the **Move selected path one step up** button, located to the right of the lower list box.

- **Move Down**   Pick this option to move the path below other paths in the path list.

Field Notes

The path can also be moved down by selecting the **Move selected path one step down** button, located to the right of the lower list box.

■ **Workgroup Search Paths** These paths allow projects to locate files on a server or other computers on a network. The files are found in the shared environment of the network and are used primarily in a design team setting.

Field Notes

When you open a file, and it is not found in the specified **Local Search Paths,** the specified Workgroup Search Paths are searched fourth. If the file is still not found, a final search is done in the folder that contains the project file.

Right-clicking on **Workgroup Search Paths** provides you with the same options available as when right-clicking on **Local Search Paths**. Right-clicking on a folder located in **Workgroup Search Paths** reveals the same options available as when right-clicking on a **Local Search Paths** folder.

■ **Options** This selection establishes the default settings for the specified project. The options may include **MultiUser, Name, Shortcut, Location,** and **Default File**. Each of these options is described below.

Field Notes

Only options specific to the currently selected project are available.

■ **MultiUser** Selecting this choice establishes the **MultiUser** option of Autodesk Inventor. Use this option when multiple design team members are working on a project.

Field Notes

Setting **MultiUser** to **True** initiates a file reservation. Autodesk Inventor displays an error message if you open a file that another design team member is currently working with.

■ **Name** This option simply displays the name of the project.
■ **Shortcut** This option displays the name given to the project shortcut. To edit the name of the shortcut, right-click on the shortcut name and select **Edit,** or select the shortcut name a second time. To delete the specified shortcut, right-click on the shortcut name and select **Delete**.

Field Notes

Right-click the **Options** selection of the lower list box and select <u>C</u>**reate shortcut** to create a
new shortcut.

■ **Location**   This option simply displays the location and path of the current project.

Creating a New Project

As previously discussed, you can create a new project from within the **Projects** section **(Project Editor)** of
the **Autodesk Inventor** dialog box. To create a project using this technique, right-click on a project name or
a project location and select the <u>N</u>**ew** option. An alternative method is to select the **New** button, located
under the lower list box of the **Project Editor**. Both techniques open the **Inventor project wizard,** shown
in Figure 1.33, which allows you to create a new project file.

    The first prompt on the initial page of the Inventor project wizard asks, "What type of project are you
creating?" Two possible radio buttons are available, <u>P</u>**ersonal Workspace for Group Project,** and **N**<u>e</u>**w
Project (personal and group).** These options are as follows:

■ <u>P</u>**ersonal Workspace for Group Project**   Selecting this radio button creates a project that
contains only a **Workspace**.

■ **N**<u>e</u>**w Project (personal or group)**   Selecting this radio button creates a project that does
not contain a **Workspace**. This option creates a project that accesses workgroup and library
search paths.

    If you select <u>P</u>**ersonal Workspace for Group Project,** the only option is to select the <u>N</u>**ext** button,
which opens the second page of the **Inventor project wizard**. See Figure 1.34. This page contains the following options:

Field Notes

You have the option of canceling the **Inventor project wizard** at any time by picking the **Cancel** button.

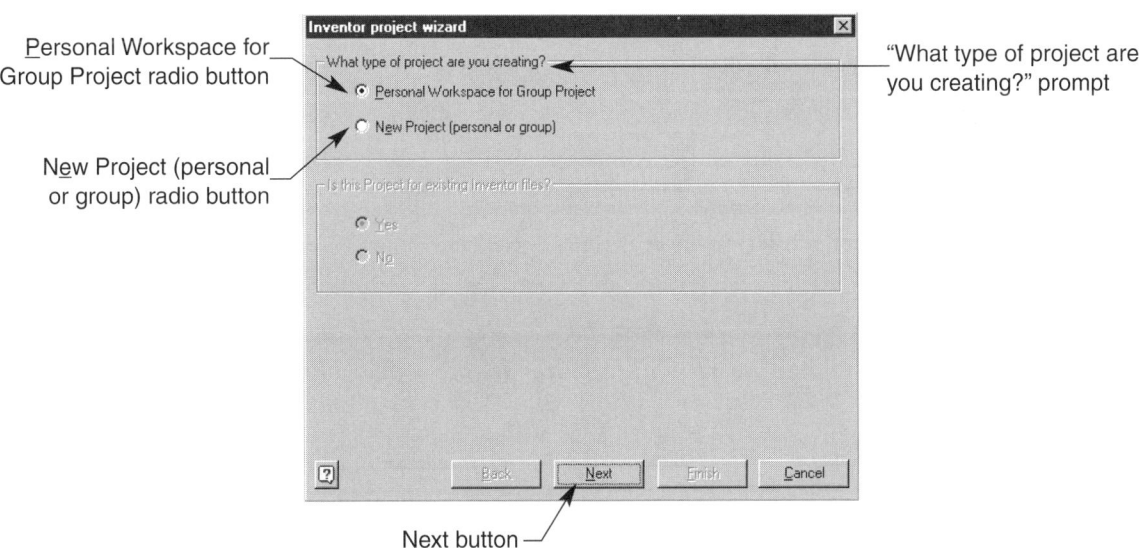

FIGURE 1.33   The Autodesk Inventor, **Inventor project wizard** (first page).

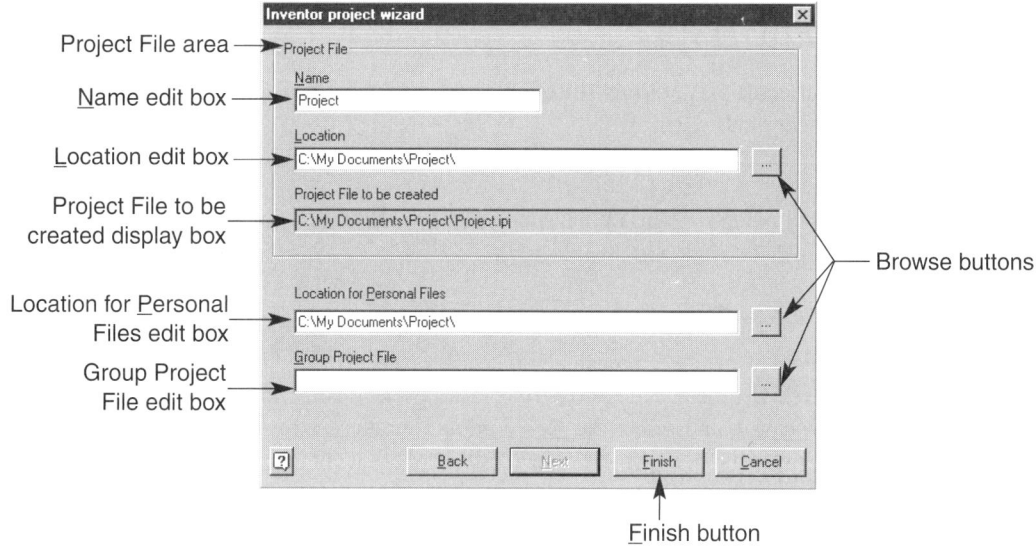

FIGURE 1.34 The second page of the **Inventor project wizard** if you select the **Personal Workspace for Group Project** radio button on the first page.

- **Name**   Use this edit box to specify a name for your new project.

- **Location**   Use this edit box to specify where the project file is located.

- **Browse**   This button, located to the right of the **Location** edit box, opens the **Browse for folder** dialog box. Here you can locate a folder from anywhere on your computer or the network.

Field Notes

The default home folder carries the same name as the project and is a subfolder of the folder first listed in the **Location** edit box.

- **Project File to be created**   This box simply displays the path and file name of the newly created project file.

- **Location for Personal Files**   This edit box allows you to specify your workspace folder.

Field Notes

The default location for **Personal Files** is the project home folder.

- **Group Project File**   Use this edit box to specify your **Included Project File.**

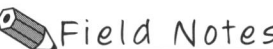Field Notes

> You cannot place an **Included Project File** or **Workspace** folder in the **Group Project File**. As in the **Location** edit box previously described, you can also select the **Browse** button located to the right of the **Location for Personal Files,** and **Group Project File** edit boxes. This option opens the **Browse for folder** dialog box, which enables you to locate a folder from anywhere on your computer or the network.

After you have specified all your options, pick the **Finish** button to complete the creation of a project file.

If you select the **New Project (personal and group)** radio button, the prompt, "Is this Project for existing Inventor files?" becomes available. See Figure 1.35. If you would like to add a folder to **Workgroup Search Paths** that includes existing Inventor files, select **Yes**. If you do not want to add a folder to your **Workgroup Search Paths** that includes existing Inventor files, select **No**. Now you can pick the **Next** button on the lower section of the initial page of the **Inventor project wizard.**

If you choose the **Yes** radio button for the prompt, "Is this Project for existing Inventor files?" and pick the **Next** button, you see the **Project File** area, including the **Name, Location,** and **Project File to be created** edit boxes, and the **Browse** button. Each of these options has been previously described in the **Personal Workspace for Group Project** selection. In addition, the **Location of Existing Files (Workgroup)** edit box is available. See Figure 1.36. This edit box allows you to add a workgroup search path to the specified existing file.

Field Notes

> The default home folder is the folder first listed in the **Location** edit box, but a subfolder is not added when the project name is edited.

Once you have specified all the options for this page of the **Inventor project wizard,** pick the **Next** button, which brings you to the third page of the **Inventor project wizard,** as shown in Figure 1.37. This

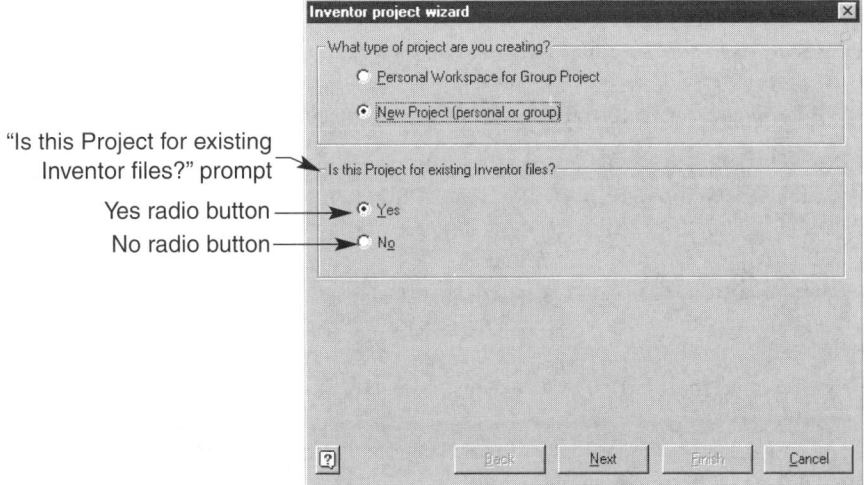

FIGURE 1.35 The prompt, "Is this Project for existing Inventor files?" becomes available when the **New Project (personal or group)** radio button is selected.

FIGURE 1.36 The **Location of Existing Files (Workgroup)** edit box is available when you choose the **Yes** radio button for the prompt, "Is this Project for existing Inventor files?"

Location of Existing Files (Workgroup) edit box

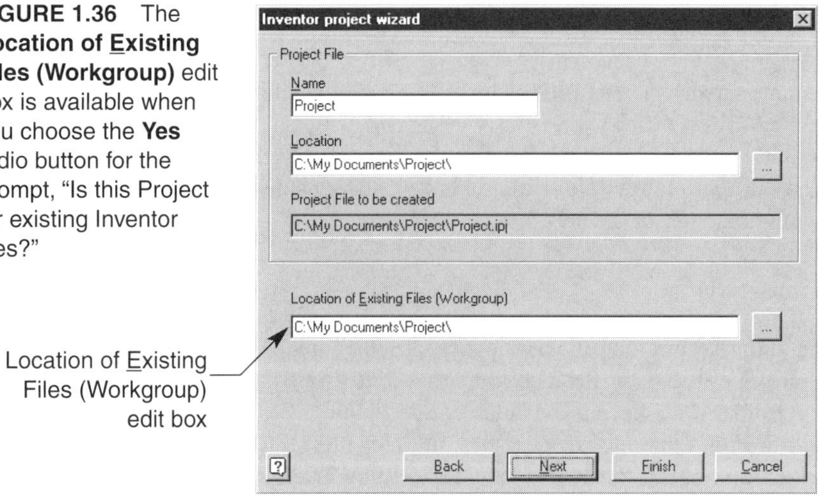

FIGURE 1.37 The **Select Libraries** page of the **Inventor project wizard.**

All Projects: list box

New Project: list box

Add selected libraries button

Delete selected libraries button

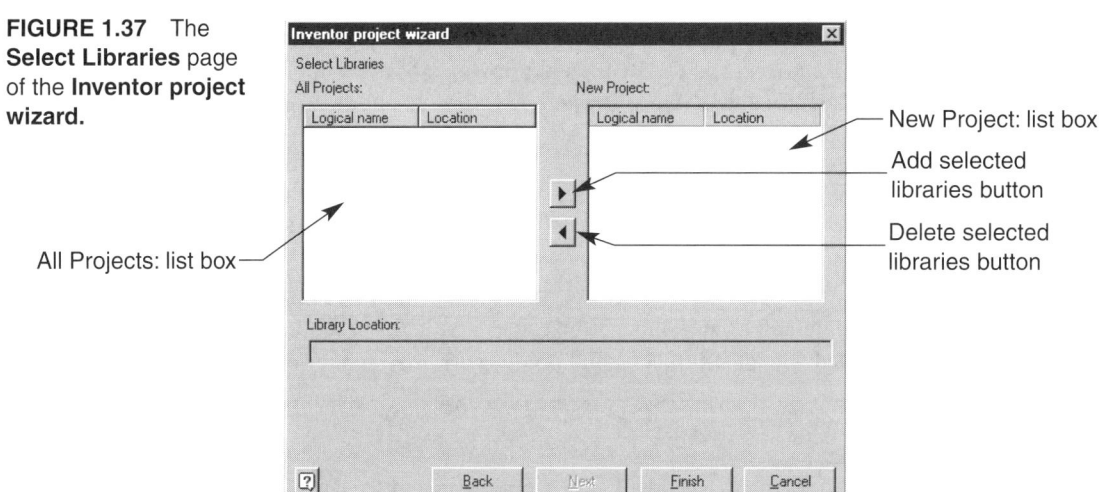

page allows you to add existing project files to your new project file with **Library Search Paths**. The **Select Libraries** page contains the following two list boxes:

- **All Projects:** This list box displays all the projects in your projects folder.

- **New Project:** This list box displays all the projects to be added to the new project. To add a library search path, select the libraries you would like to add from the **All Projects:** list box. Then pick the **Add selected libraries** button, located between the **All Projects:** and **New Project:** list boxes. To remove a library search path, select the libraries you would like to remove from the **New Project:** list box. Then pick the **Delete selected libraries** button, located under the **Add selected libraries** button.

After you have specified all your options, pick the **Finish** button to complete the creation of the new project file.

If you choose the **No** radio button for the prompt, "Is this Project for existing Inventor files?" you see the **Project File** area as previously described when you select **Yes**. However, the **Storage Location for New Files (Workgroup)** edit box is now available. See Figure 1.38. This edit box allows you to add a workgroup search path to your new project file.

FIGURE 1.38 The **Storage Location for New Files (Workgroup)** edit box is available when you choose the **No** radio button for the prompt, "Is this Project for existing Inventor files?"

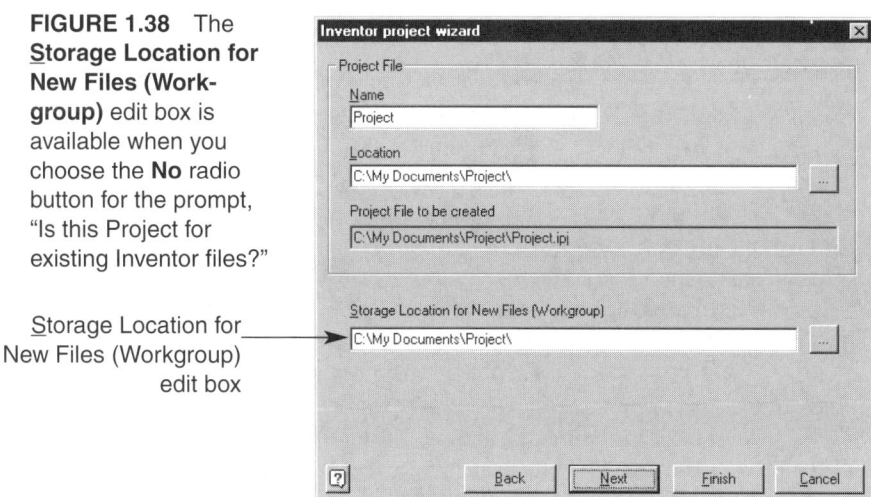

Storage Location for New Files (Workgroup) edit box

The project home folder is the default path for the **Workgroup Search Path**.

Once you have specified all the options for this page of the Inventor project wizard, pick the **Next** button on the lower section of the page, which brings you to the third page of the **Inventor project wizard**. This page is the same as the third page discussed in the **Yes** option.

Picking the **Back** button allows you to return to the last **Inventor project wizard** page.

After you have all your options for this page set, pick the **Finish** button to complete the creation of the new project file.

Exercise 1.7

1. Continue from Exercise 1.6, or launch Autodesk Inventor.
2. Pick the **Projects** option of the **What To Do** menu if the **Projects** window is not currently displayed.
3. Create a new project using the methods described above, selecting **Personal Workspace for Group Project** radio button.
4. Name the project EX1.7.
5. Pick the OK button when prompted "The project path you entered does not exist. Should it be created?"
6. Activate project EX1.7 using one of the previously discussed techniques.
7. Test project EX1.7 by selecting the **Open** option of the **What To Do** menu. Notice the EX7.1 folder opens automatically, and the **Locations** area contains your **Workspace**.
8. Add a library path named Samples using the following directory: Program Files/Autodesk/Inventor 5/Samples.
9. Delete project EX1.7, and open it using the **Choose project file** dialog box. EX1.7 should again be available in the **Projects** window.
10. Rename EX1.7 to Inventor Projects.
11. Continue exploring the **Autodesk Inventor** dialog box, or exit Autodesk Inventor if necessary.

||||||||||| CHAPTER TEST

Answer the following questions on a separate sheet of paper.

1. What do you get when you initially launch Autodesk Inventor?

2. Give the function of the Autodesk Inventor Help window.

3. Discuss the function of the Creating a Part button in the Single-Day Productivity menu.

4. What is the purpose of the Next button when using a tutorial?

5. Explain the function of the Building Sheet Metal Parts button in the Single-Day Productivity menu.

6. Give the function of the Creating Assemblies button in the Single-Day Productivity menu.

7. Explain the function of the Preparing Final Drawings button in the Single-Day Productivity menu.

8. Discuss the purpose of the Autodesk Point A Website.

9. What is the purpose of the Autodesk Streamline Website?

10. Define template.

11. What are Part Files, and what file extension is used?

12. How many components are there in a part file?

13. Identify at least three uses for Part file templates.

14. What are Assembly files, and what file extension is used?

15. Give the primary use of Assembly template files.

16. Identify the two ways that you can use to develop assemblies using assembly template files.

17. What are Presentation files, and what file extension is used?

18. Identify the common use of Presentation template files.

19. What are Drawing files, and what file extension is used?

20. How does developing your own template save time and effort?

21. Give the steps used to create your own template.

22. How do you find the template that you created when you launch Autodesk Inventor?

23. Give the function of the Open File area of the Open option in the What To Do menu.

24. Discuss the function of the Locations list box in the Open File area of the Autodesk Inventor dialog box.

25. Explain the function of the Look in: drop-down list in the Open File area of the Autodesk Inventor dialog box.

26. Give the function of the Return button in the Open File area of the Autodesk Inventor dialog box.

27. Explain the function of the List button in the Open File area of the Autodesk Inventor dialog box.

28. Discuss the function of the Details button in the Open File area of the Autodesk Inventor dialog box.

29. Give the function of the File name: edit box in the Open File area of the Autodesk Inventor dialog box.

30. Discuss the function of the Preview area in the Open File area of the Autodesk Inventor dialog box.

31. Explain the function of the Files of type: drop-down list in the Open File area of the Autodesk Inventor dialog box.

32. Discuss the function of the Show only files in active project: check box in the Open File area of the Autodesk Inventor dialog box.

33. Briefly explain how to use the File Open Options dialog box.

34. Explain the function of the Browse… button in the Open File area of the Autodesk Inventor dialog box.

35. Give the function of the View Desktop: button in the Open dialog box.

36. Discuss the function of the Create New Folder: button in the Open dialog box.

37. How do you use the Create New Folder button?

38. Give the function of the Find files that match these criteria area in the Find: Autodesk Inventor Files dialog box.

39. Explain the function of the Delete Line button in the Find: Autodesk Inventor Files dialog box.

40. Discuss the function of the Clear Search button in the Find: Autodesk Inventor Files dialog box.

41. Give the function of the Match Case check box in the Find: Autodesk Inventor Files dialog box.

42. Explain the function of the Define more criteria area in the Find: Autodesk Inventor Files dialog box.

43. Discuss the function of the And radio button in the Define more criteria area in the Find: Autodesk Inventor Files dialog box.

44. Give the function of the Or radio button in the Define more criteria area in the Find: Autodesk Inventor Files dialog box.

45. Explain the function of the Property drop-down list in the Define more criteria area in the Find: Autodesk Inventor Files dialog box.

46. Explain the function of the Condition drop-down list in the Define more criteria area in the Find: Autodesk Inventor Files dialog box.

47. Give the function of the Value edit box in the Define more criteria area in the Find: Autodesk Inventor Files dialog box.

48. Discuss the function of the Add to list button in the Define more criteria area in the Find: Autodesk Inventor Files dialog box.

49. Give the function of the Autodesk Inventor Find: Save Search dialog box.

50. Discuss the function of the Autodesk Inventor Find: Open Search dialog box.

51. Discuss the function of the Current Folders radio button in the Search Location area of the Find: Autodesk Inventor Files dialog box.

52. Explain the function of the Sub Folders radio button in the Search Location area of the Find: Autodesk Inventor Files dialog box.

53. Give the function of the Related Folders radio button in the Search Location area of the Find: Autodesk Inventor Files dialog box.

54. Discuss the function of the Look in drop-down list in the Search Location area of the Find: Autodesk Inventor Files dialog box.

55. Give the function of the Browse button in the Search Location area of the Find: Autodesk Inventor Files dialog box.

56. What do you do to begin the file search once you have specified all your search options?

57. Define projects.

58. Give the function of the Project name in the Project Editor.

59. Give the function of the Project location in the Project Editor.

60. What do you get when you click once on the name of a project?

61. What do you get when you double-click on the name of a project?

62. How do you know when a project is active?

63. Give the function of the Look in: drop-down list in the Choose project file dialog box.

64. Discuss the function of the Return button in the Choose project file dialog box.

65. Explain the function of the View Desktop button in the Choose project file dialog box.

66. Give the function of the Create New Folder button in the Choose project file dialog box.

67. Give the function of the List button in the Choose project file dialog box.

68. Discuss the function and features of the Details button in the Choose project file dialog box.

69. Explain the function of the File name edit box in the Choose project file dialog box.

70. Give the function of the Files of type drop-down list in the Choose project file dialog box.

71. Give the function of the Open as read-only check box in the Choose project file dialog box.

72. What happens after you find a desired project, double-click on the project name, or select the file and pick Open?

73. Give the function of Library Search Paths.

74. What are the elements of a library?

75. Define Workplace and give its function.

76. Identify two ways to open the Inventor project wizard.

77. Give the function of the Personal Workspace for Group Project: radio button.

78. Give the function of the New Project (personal and group): radio button.

79. If you select the New Project (personal and group) radio button, the prompt, "Is this Project for existing Inventor files?" becomes available. Explain what happens if you select Yes and if you select No.

80. Discuss the function of the Select Libraries page of the Inventor project wizard.

|||||||||||||||||||||||||||||||||||||||||||||||||||||||||||||||||||||||||||||||||||||||

CHAPTER 2

Introduction to Modeling and the Autodesk Inventor Interface

LEARNING GOALS

After completing this chapter, you will be able to:

◎ Explain how to work with models and define the elements of a part model.

◎ Describe parametric modeling and explain how to develop a parametric model.

◎ Define constraints.

◎ Use the pull-down menu system.

◎ Work with the panel bar, and the browser bar.

◎ Use toolbars and toolbar buttons.

◎ Describe the command bar.

◎ Use shortcut keys.

Working with Models: Basics

Unlike traditional CAD programs, Autodesk Inventor combines three-dimensional solid modeling with two-dimensional drawing capabilities. The concepts and applications of the working Autodesk Inventor environment may be different from other drafting you have done and other drafting programs you have used. As a result, it is important to begin to understand some basic information about how to develop models using Autodesk Inventor, and working in parametric situations. In addition, you should start to recognize and control the Autodesk Inventor interface.

What Are the Elements of a Part Model?

Every part model consists of at least one *sketch,* and at least one *sketched feature.* A more complex model may include *placed, work, pattern,* and *catalog* features. Each of these elements is described as follows:

- **Sketches** Sketches represent the first step in the creation of a model. Sketches provide the foundation, pattern, and profile for developing models that contain sketched features. See Figure 2.1.

- **Sketched Features** Sketched features are the product of a sketch. They are created as an extrusion, revolution, sweep, loft, or coil, depending on the particular part and sketch geometry. Most of the time, the initial feature, known as the base feature, is created as a sketched feature. Holes and ribs are also sketched features. See Figure 2.2.

- **Placed Features** Placed features are created without using a sketch. Shells, fillets, chamfers, and face drafts are considered placed features. As the name implies, placed features are placed onto a part model. Generating one of these features requires dimensions and a location, such as a point or an edge. Often placed features are created on existing features towards the end of the design process. See Figure 2.3.

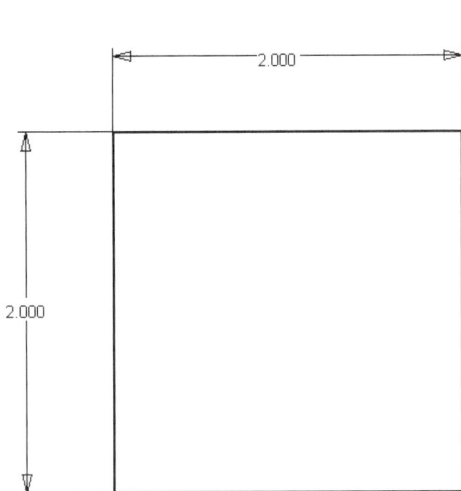

FIGURE 2.1   An example of a simple sketch.

FIGURE 2.2   An example of a simple sketched feature, an extrusion.

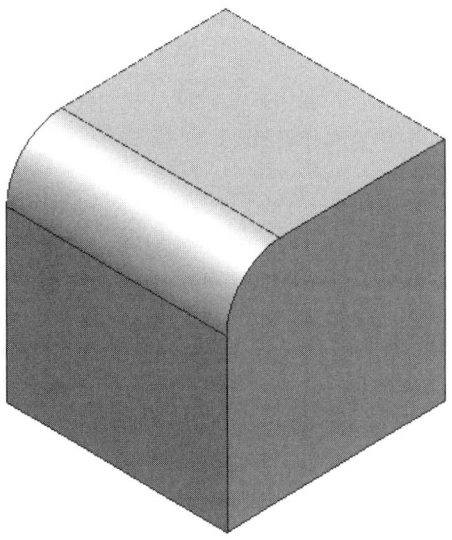

FIGURE 2.3   An example of a simple placed feature, a fillet.

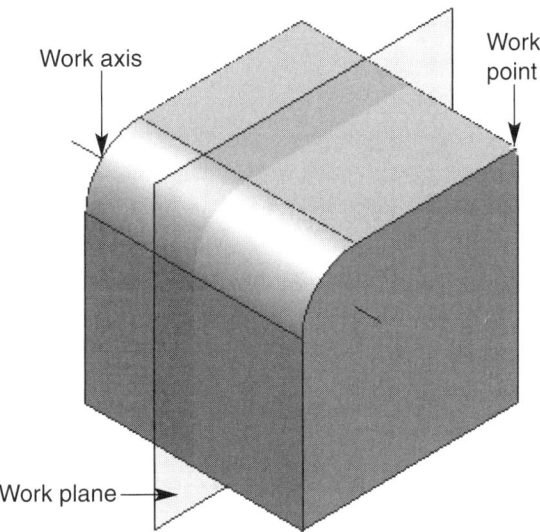

FIGURE 2.4   Autodesk Inventor work features.

- **Work Features**   Work features are used to direct the location and arrangement of other features. They are construction references used to develop additional sketches and features in areas where no geometry is available or can be created. As shown in Figure 2.4, the three Autodesk Inventor work features are work planes, work axes, and work points. ***Work planes*** are used to create a construction plane, ***work axes*** are used to create construction lines, or axes, and ***work points*** are used to create construction points, anywhere on a feature or in three-dimensional space. Each of these features is attached to a model. If attached, they change when the part feature or sketch is modified. In the same respect, if the work plane is modified, the feature attached to the plane will also reflect the changes.

- **Catalog Features**   Catalog features are similar to placed features, because you can place them onto an existing feature. However, these features may be far more complex than a standard shell, fillet, chamfer, or face draft. See Figure 2.5. Catalog features include design elements and derived parts. ***Design Elements,*** also known as ***iFeatures,*** are existing features

FIGURE 2.5 An example of a simple catalog feature.

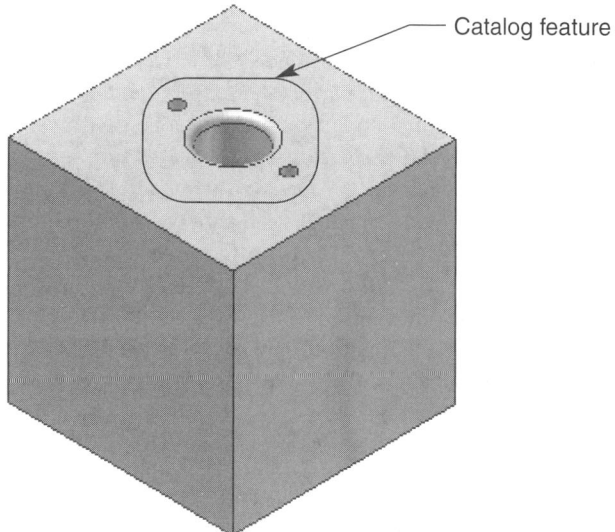

Catalog feature

you have created, then save and store in a catalog to be used in other models. ***Derived components*** are more complicated than design elements and may actually contain a complete model consisting of several features, or even multiple parts. Derived components are often used as base features, but can also be inserted into existing models.

■ **Pattern Features**   Pattern features are developed using existing features such as extrusions or holes to make a pattern of that feature. Unlike sketched or placed features, pattern features do not create a completely new feature, but simply copy an existing feature. Autodesk Inventor allows you to build three types of pattern features including rectangular and circular patterns, and mirrored features. A ***rectangular pattern,*** shown in Figure 2.6A, copies and organizes a feature into a designated number of rows and columns and places them a specified distance apart from each other. A ***circular pattern,*** shown in Figure 2.6B, copies and organizes a feature around an imaginary circle a designated number of times and places each one a specified distance from the other. A ***mirrored feature,*** shown in Figure 2.6C, simply mirrors an existing feature about a work axis or part edge.

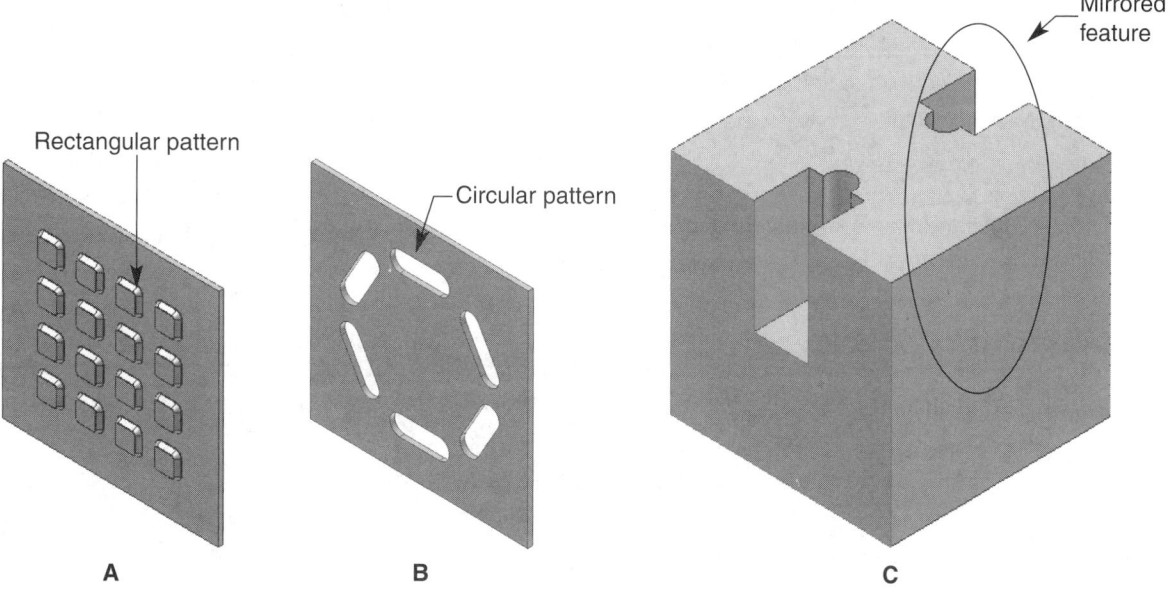

Rectangular pattern

Circular pattern

Mirrored feature

A   B   C

FIGURE 2.6   Examples of the three Autodesk Inventor pattern features.

What Is Parametric Modeling?

Autodesk Inventor works on the concept of *parametric solid modeling*. Parametric solid modeling involves the basic idea of developing part and assembly models that contain controls, limits, and checks, and allow you to make changes and updates. This means that when you define the size, shape, and position of model geometry using specific parameters, you have to modify the previously defined size, shape, and position specifications for changes to be made to the model. In addition, if the modifications or new parameters conflict with any other geometry, the model cannot be built without errors attached.

The parametric concept is also referred to as *intelligence,* because parametric modeling occurs as a result of Autodesk Inventor's ability to store model information in the database and manage the model information. All the information on how a model was created is stored in this area. The data includes knowledge of every aspect of the part, such as sketches, features, dimensions, geometric constraints, and when each piece of the model was created.

Parametric modeling allows you to control every aspect of a model during and after the design process. Unlike other CAD programs, in the parametric modeling world of Autodesk Inventor, you begin the design process with a simple two-dimensional sketched shape and develop it into a complex three-dimensional model. It is important to think about the parametric relationships between sketches and features, parts and assemblies, and the entire design process. As a result, planning how a model should be made—and thinking about how it relates to future models and assemblies—is an important step in beginning a design.

Developing a Parametric Model

Certain basic steps are required to develop a three-dimensional parametric model in Autodesk Inventor. Only a portion of these steps may be required to construct a very simple model, while additional steps may be required to build a complex model. You may form your own specific way of creating a model that may vary from the typical process. However, the general process of developing a model occurs as follows:

1. Open or create a project to be used for the model. Refer to Chapter 1 for project information.

2. Open a part or assembly file template, and create a new part or assembly file. See Chapter 1 for opening a new or existing file.

3. Define work and sketch environment settings.

4. If applicable, project the origin (0,0,0) onto the sketch. This allows the sketch to become fully constrained about the origin when completed.

5. Draft sketch geometry to create a general outline of the initial base feature. You should plan and develop your sketch in terms of the three-dimensional model it will create.

6. Add geometric constraints and dimensions to further define the sketch. *Constraints* and *dimensions* are parametric design parameters that are held constant throughout the model development. Still, these parameters can be altered if necessary. Constraints are discussed later in this chapter.

7. Use sketch feature tools, such as **Extrude** and **Revolve,** to create an initial feature from the sketch profile. The initial base feature is the basic foundation of the model.

Field Notes

In some situations, derived parts, discussed in Chapter 18, may be used in place of a sketched feature.

8. Option 1: Draft sketch geometry on the base feature and use feature tools to create additional features that join, cut, or intersect the base feature.

   Option 2: Use placed feature tools such as **Hole, Shell, Fillet,** and **Chamfer,** to create placed features on the sketched features.

   Option 3: Apply any catalog features to your model.

9. If necessary, create additional objects using pattern features.

10. Edit any sketches or features as required to change the model or assess different design ideas.

Each of these steps is discussed in detail in the chapter where they apply. At this point, you should begin to understand the fundamental steps required in developing a model.

Working with Constraints and Parametric Models

Once you understand how to begin the design process, you first develop an initial sketch. Usually the initial sketches and shapes are easy to develop and are meant as the beginning part model concept. You then further specify the initial shape using constraints. ***Constraints,*** including dimensions and geometric constraints, are an important principle in Autodesk Inventor and in parametric modeling. Constraints are parameters required for design development. They are used throughout the design process, from the initial sketch to defining features. You must constrain the part as accurately as possible to preserve your design intensions. Constraints and parametric modeling allow you to:

- Identify and define your part by using dimensions and geometric constraining tools.

- Enable design elements to remain the same, protecting your model and design intention. To accomplish this, Autodesk Inventor will tell you if a sketch is overconstrained or if a modeling failure will result. An ***overconstrained*** sketch has too many dimensions or geometric constraint controls and does not allow Autodesk Inventor to recognize the sketch geometry. A modeling failure is the result of dimensions or geometric constraints that are impossible to apply to the model. These conditions must be resolved for a feature to be created, and the model to be parametric.

   A sketch or model can also not contain enough constraints. An ***underconstrained*** sketch or model has elements that have ambiguity, can be changed or moved, and remain undefined. As you progress through the design process, you must define your sketch or feature enough so that it is fully constrained. This will ensure your model is parametric and your design intensions are not violated.

- Make any necessary changes to the design of a model that allow you almost immediately to assess design alternatives by changing, adding, or deleting sketches, features, dimensions, and geometric constraints.

- Develop assemblies, drawings, and presentations. Each of these Autodesk Inventor components are created almost instantaneously because all the dimensions and geometric constraints have already been made in the part or the assembly. These elements were introduced in Chapter 1 and are discussed in detail in future chapters.

- Develop ***parameter-driven assemblies*** that allow changes made to individual parts to be reproduced automatically as changes in the assembly and assembly drawings.

- Use ***adaptive parts*** in assemblies. Adaptive parts are extremely effective when you may not know the exact dimensions of a part, or fully understand the relationship between assembly components. Adaptive parts modify automatically if another part is changed.

- Develop equations. Equations can be set in place that direct your parts, making it easier for you to allow a few dimensions of a part to drive the entire model, or even to create a family of related parts.

Using the Autodesk Inventor Interface

Interface is a term used to describe the tools and techniques used to provide and receive information to and from a computer program. The Autodesk Inventor interface includes a menu system, panel bar, browser bar, toolbars, command bar, and status bar. See Figure 2.7. These interface tools are discussed later in this chapter.

The interface is always changing as you progress through a part, assembly, presentation, or drawing file, because only the tools and options that apply to the current work environment are available. For example, the interface for developing a sketch is different than the interface for creating a feature, and the interface for working on a part is different than the interface for working on a drawing. Tools such as the panel bar and menu options automatically change when you progress to different aspects of the design process, such as going from sketch development to feature creation, or parts to assemblies.

Though some of the interface tools and options are the same for each of the four Autodesk Inventor file types, many of the options are specific to a certain file type. For example, a part contains the tools and options available for creating parts, because the interface is in part mode. Consequently, only menu items, browser and panel bar options, toolbars, and toolbar buttons specific to parts are available. As a result, you can have different file types open at the same time. Moving from one file to the next will change the interface to reflect the currently opened file.

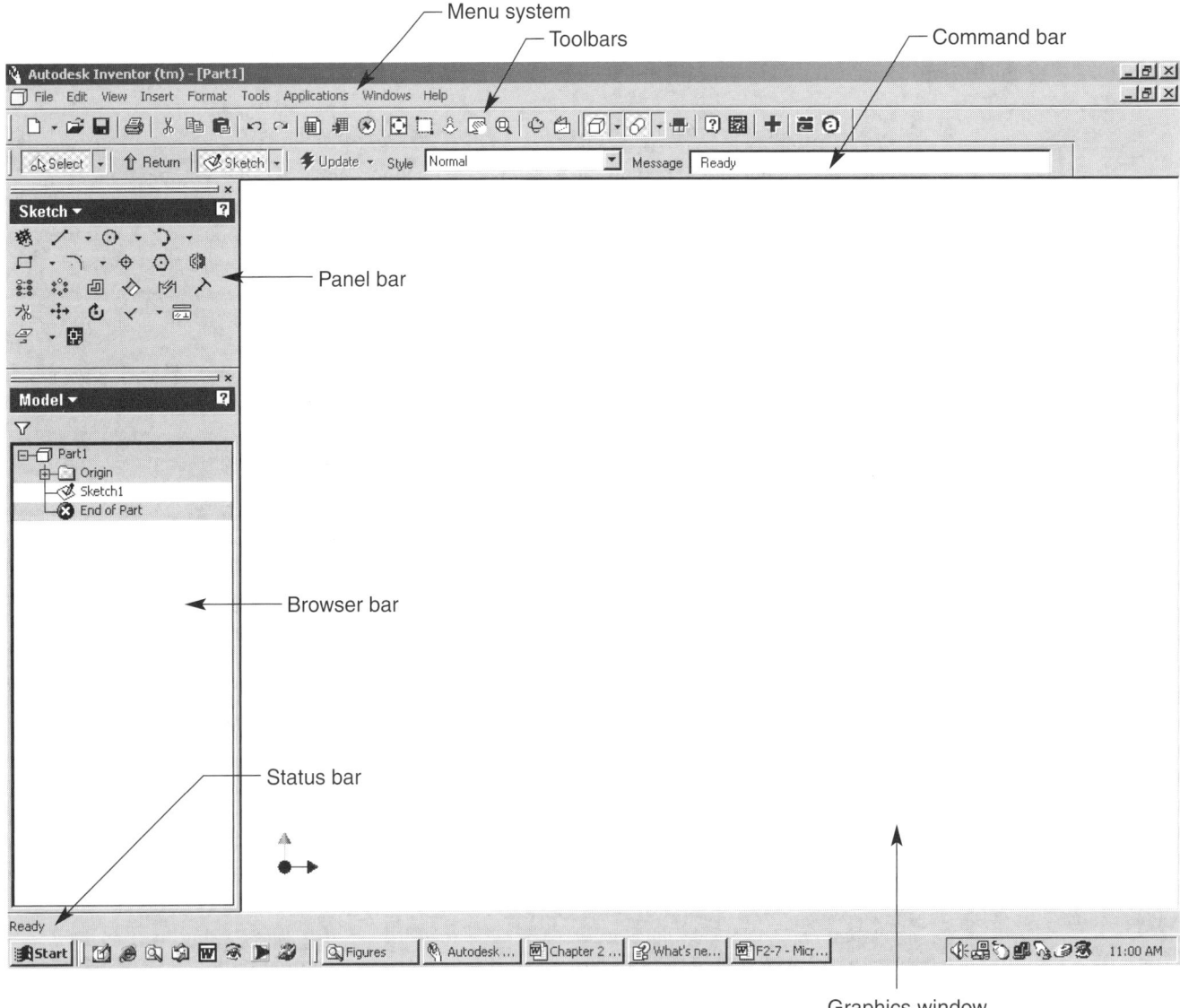

FIGURE 2.7   The Autodesk Inventor Interface.

Autodesk Inventor is Microsoft Windows logo compliant. This means that many of the pull-down menus, toolbars, and accelerator keys are similar to those available in other Microsoft Windows programs, such as Word or Excel. If you are currently using, or have used, any of these programs, you may already be familiar with some of the Autodesk Inventor interface options.

Exercise 2.1

1. Launch Autodesk Inventor.
2. Open a new Autodesk Inventor part file, and explore the default part interface.
3. Close the part file without saving.
4. Open a new Autodesk Inventor assembly file, and explore the default assembly interface.
5. Close the assembly file without saving.
6. Open a new Autodesk Inventor drawing file, and explore the default drawing interface.
7. Close the drawing file without saving.
8. Open a new Autodesk Inventor presentation file, and explore the default presentation interface.
9. Close the presentation file without saving.
10. Exit Autodesk Inventor or continue working with the program.

Using the Autodesk Inventor Pull-Down Menu System

The pull-down menu system is one of the interface options in Autodesk Inventor and provides you with a text-based, menu configuration. The pull-down menu bar, shown in Figure 2.8, houses a number of pull-down menus that are currently available for the particular work environment. You can pick any of these pull-down menus to reveal the options for that menu, shown in Figure 2.9, and move your cursor down and up to highlight any of the menu options.

File   Edit   View   Insert   Format   Tools   Applications   Windows   Help

FIGURE 2.8   The Autodesk Inventor pull-down menu bar.

FIGURE 2.9 Menu options are displayed when you pick one of the menus.

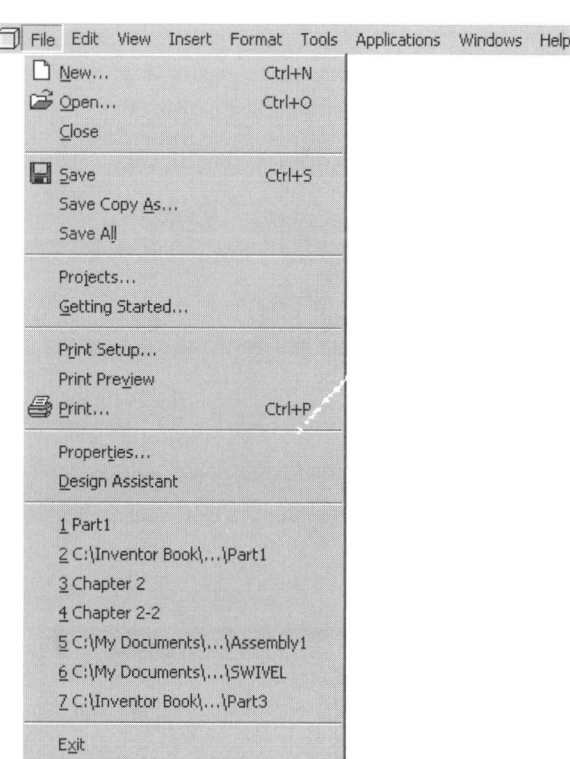

As you select different options, the status bar, located towards the bottom of the screen, provides you with information about the option by displaying a help string. A *help string* displays a short description of what happens if you select the particular option over which your cursor is hovering. As shown in Figure 2.10, if you highlight the **New...** option of the **File** pull-down menu the help string in the status bar displays: Create a new document. After you know what particular menu option you would like to access, pick that option. Picking the highlighted option starts the command or applies the setting.

If you press the **Alt** key on your keyboard, an underlined letter appears for each of the pull-down menu names. An underlined character is also available for each of the menu options. The underlined character in the menu names and options is called a menu access key. *Menu access keys* allow you to open a menu using your keyboard by pressing the **Alt** key and the key of the underlined character. For example, you can access the **File** pull-down menu by pressing the **Alt** and **F** keystroke. You can then access the **Open...** option of the **File** pull-down menu, for example, by pressing the **O** key.

Some of the options in the **File** and **Edit** pull-down menus have an accelerator key displayed to the right of the option name. See Figure 2.11. An *accelerator key* is similar to an access key and allows you to access menu options using your keyboard, without opening the pull-down menu. To use an accelerator key, press the keystroke displayed for the particular option you would like to access. The keystroke includes pressing the **Ctrl** key and the particular letter of the option you would like to open. For example, to access

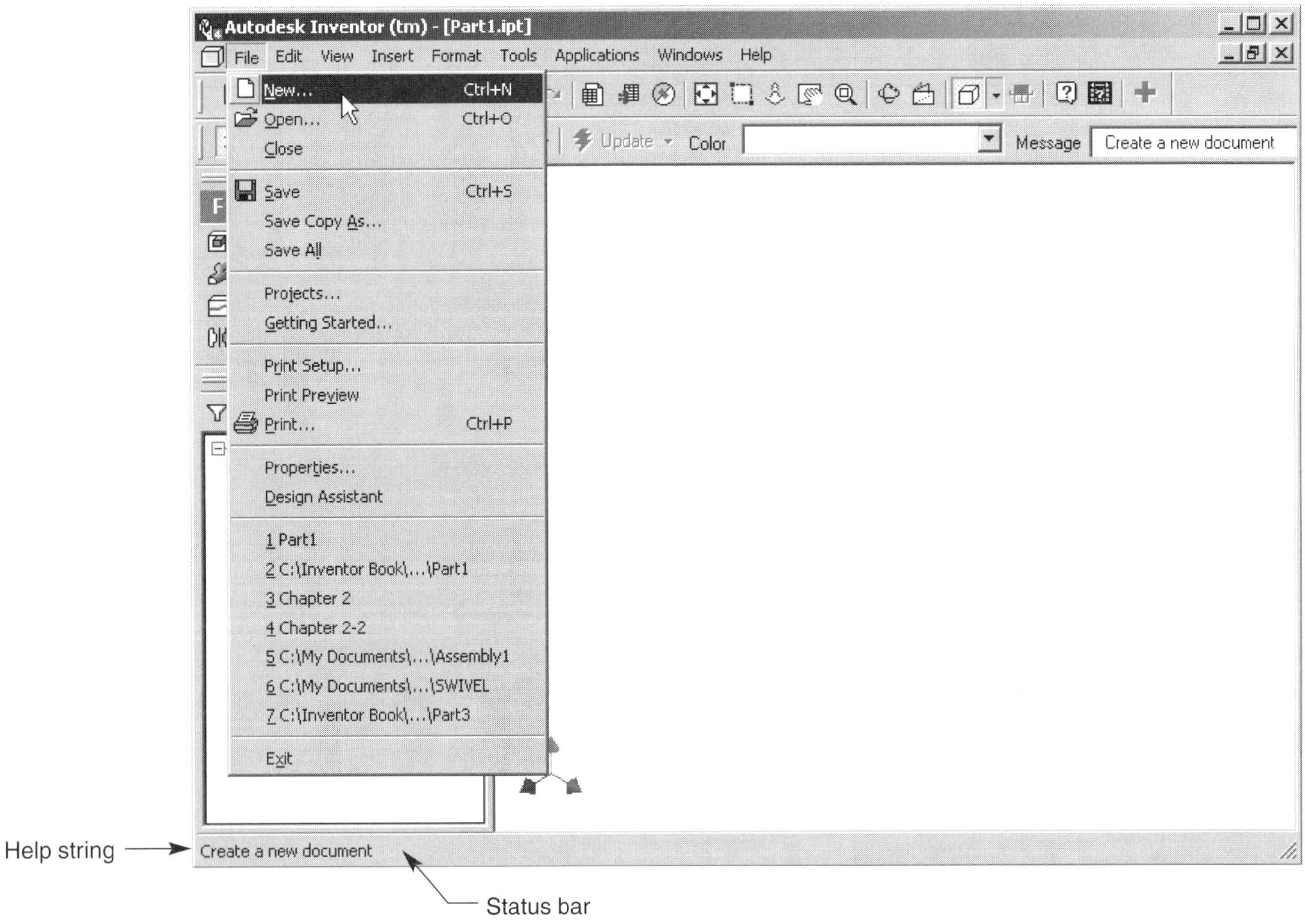

Help string → Create a new document

Status bar

FIGURE 2.10   An example of a menu option help string.

FIGURE 2.11
Example of accelerator keys from the **File** pull-down menu.

FIGURE 2.12 An example of a cascading submenu, the toolbar cascading submenu.

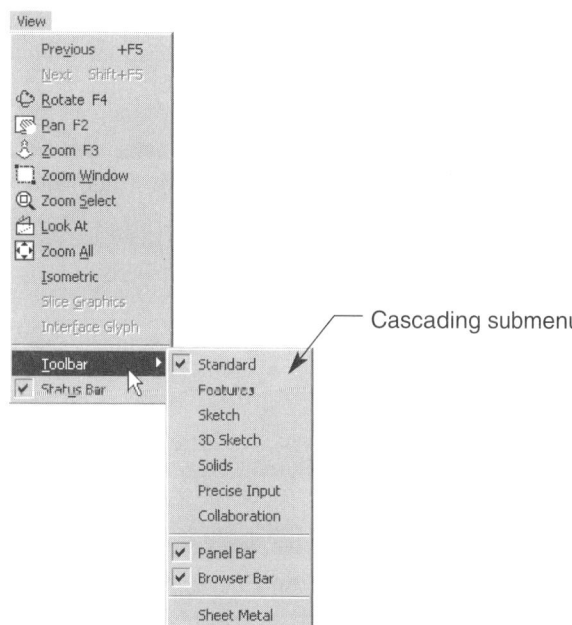

Cascading submenu

the **New File** window of the **Autodesk Inventor** dialog box, use the **Ctrl** and **O** keystroke, by holding the **Ctrl** key down and pressing the **O** key.

Some pull-down menu options that display an arrow to the right of the submenu name contain *cascading submenus*. These options are menus inside of menus, and display additional selections. See Figure 2.12. To use a cascading submenu, also known as a hierarchical menu, select or right-click the submenu name, or move your cursor over the arrow to the right of the submenu name. Next, pick the option you want.

If you need to close the currently selected menu without opening another menu, simply press the **Esc** key, or pick outside of the menu list or on the Menu bar. If you would like to change to a different menu, move your cursor over the specified menu name or pick the menu name.

Exercise 2.2

1. Continue from Exercise 2.1, or launch Autodesk Inventor.
2. Open a new Autodesk Inventor file of your choice.
3. Explore the Autodesk Inventor pull-down menu system by picking any of the menu titles, and moving your cursor down and up through the menu options.
4. Pick the **File** pull-down, and highlight the **New...** option. Notice the help string displayed in the status bar.
5. Utilize a menu access key to access the **Open...** option of the **File** pull-down menu, and open the **Open File** window of the **Autodesk Inventor** dialog box.
6. Pick the **Cancel** button to close the **Autodesk Inventor** dialog box.
7. Explore cascading submenus by highlighting the **Toolbars** option of the **View** pull-down menu.
8. Press the Esc key on you keyboard, or pick outside of the menu to close the View pull-down menu.
9. Exit Autodesk Inventor or continue working with the program.

Using Specific Autodesk Inventor Pull-Down Menus

As mentioned previously, the Autodesk interface changes. An example of this is the different menus and menu options available throughout the design process. The only menus and menu options that do not change are the **Windows** and **Help** pull-down menus, discussed later in this chapter. These two pull-down menus remain the same whether you are working on a part, assembly, presentation, or drawing. The **Applications** pull-down menu is available only in a part environment. This menu allows you to change from a **Modeling** design to a **Sheet metal** design while in a part work environment and is further discussed in Chapter 10. The other pull-down menus have different options for each of the Autodesk Inventor file types. The file specific options are discussed in the chapter where they apply. The Autodesk Inventor pull-down menus and the standard pull-down menu options available in each of the four Autodesk Inventor file types are further described in this chapter.

Working with the File Pull-Down Menu

The **File** pull-down menu, shown in Figure 2.9, contains the following options in each of the four Autodesk Inventor file types:

- **New...**   This option opens the **New File** window of the **Autodesk Inventor** dialog box, discussed in Chapter 1. Here you can create a new part, assembly, drawing, or presentation file.

- **Open...**   This option opens the **Open File** window of the Autodesk Inventor dialog box, discussed in Chapter 1. Here you can open an existing Autodesk Inventor file.

- **Close**   This selection closes the current file.

- **Save**   This selection resaves the file if it has already been saved, or opens the **Save As** dialog box if the file has not been previously saved. See Figure 2.13. The following options are located inside the **Save As** dialog box:

  - **Locations**   This list box shows a list of the file paths in reference to the active project. You can select a path to display the files and subfolders. Refer to Chapter 1 for Project information.

  - **Save in:**   Use this drop-down list to navigate through drives and folders to find a certain file by selecting the down arrow next to the drop-down list box. The list box is located below the **Save in:** drop-down list and shows the names of the files located in the current directory.

  - **Last**   This button, to the right of the down arrow, allows you to go back to the last folder visited.

  - **Return**   This button is similar to the **Last** button and allows you to return to the previously opened drive or folder.

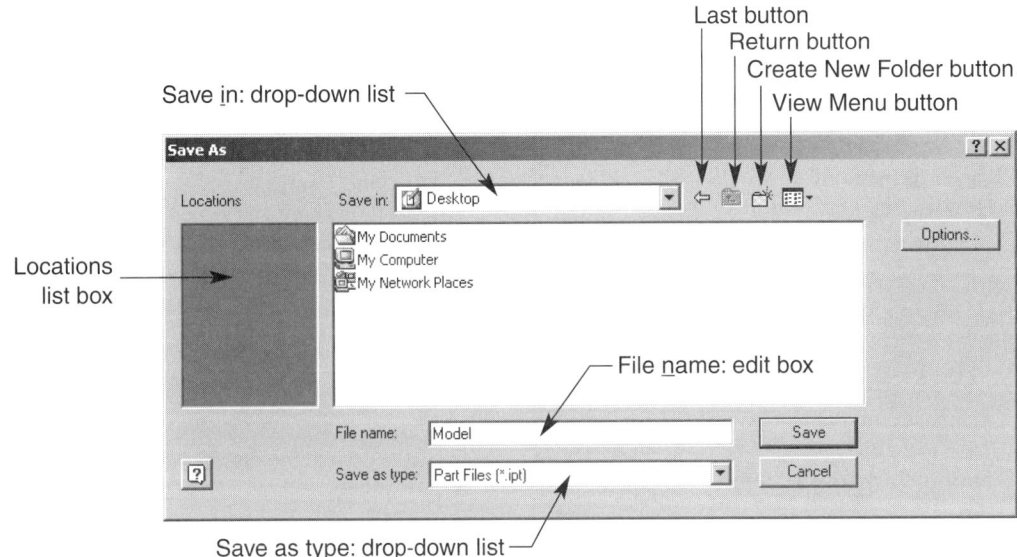

FIGURE 2.13   The **Save As** dialog box.

■ **Create New Folder**   This button allows you to create a new folder if desired. To use the **Create New Folder** button, locate the specific area where you would like a new file and then select the **Create New Folder** button. Now, type a name—for example, New1—over the existing **New Folder** default name. Pick the folder you have just created, or press the **Enter** key on your keyboard.

■ **View Menu**   Select this button to view the drives and folders in a large icon, small icon, list, detail, or thumbnail format.

■ **File name:**   This edit box shows the currently selected and highlighted Inventor file. To display the file name, pick the file from the list box, or type the name of the file in the **File name:** edit box.

■ **Save as type:**   This drop-down list allows you to choose how you would like to save the file. You can save a file only as whatever Autodesk Inventor file it is: *.ipt, *.ide, *.iam, *.idw, or *.ipn. Still you have the option of selecting the save as all files (*.*) option, or the particular file type the document happens to be (*.ipt, *.ide, *.iam, *.idw, or *.ipn).

■ **Options...**   Picking this button opens the **File Save Options** dialog box. See Figure 2.14. The **File Save Options** dialog box allows you to specify how you would like to save the current models image displayed in the preview area of the **File Open** dialog box, discussed in Chapter 1. The following options are available:

  ■ **Save Preview Picture** check box   This check box enables you to save a thumbnail image of the file to be shown in the preview area of the **File Open** dialog box.

  ■ **Active Window on Save** radio button   This radio button allows you to save the file at its active window state, by establishing a thumbnail image to be displayed in the graphics window, when you select the save button.

  ■ **Active Window** radio button   This radio button allows you to save the file at its active window state, by establishing a thumbnail image to be displayed in the graphics window. To accomplish this, select the **Active Window on Save** radio button, and pick the **Capture** button.

  ■ **Import From File** radio button   This allows you to import a thumbnail bitmap file. Select the **Import From File** radio button and pick the **Import** button. This displays the **Open** dialog box discussed in Chapter 1. Once you have specified your entire save options, choose the **Save** button, located at the bottom of the **Save As** dialog box, to save the file. You can also pick the **Cancel** button at any time to terminate the save.

■ **Save Copy As...**   This option enables you to save a copy of the current file, to be used in other applications. Selecting this option opens the **Save Copy As** dialog box, shown in Figure 2.15, which is very similar to the **Save As** dialog box previously discussed. To use the **Save Copy As** dialog box, select the type of file you would like to save the file as from the **Save as type:** drop-down list, and/or type a new name for the file in the **File name:** edit box.

■ **Save All**   Selecting this option saves every file that is currently opened in Autodesk Inventor. If the files have been saved in the past, **Save All** resaves the file. If they have not, the **Save As** dialog box appears, as discussed previously for the **Save** option.

FIGURE 2.14 The **File Save Options** dialog box.

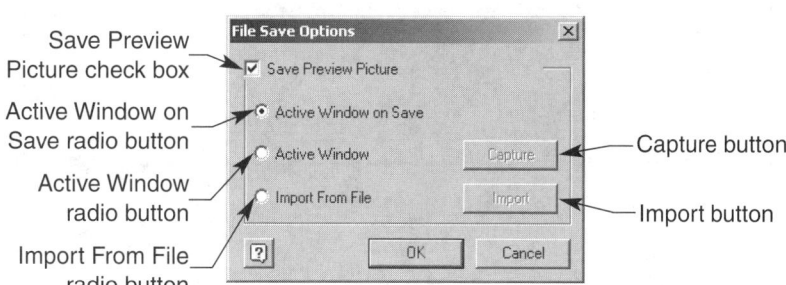

Save Preview Picture check box

Active Window on Save radio button

Active Window radio button

Import From File radio button

Capture button

Import button

FIGURE 2.15   The **Save Copy As** dialog box.

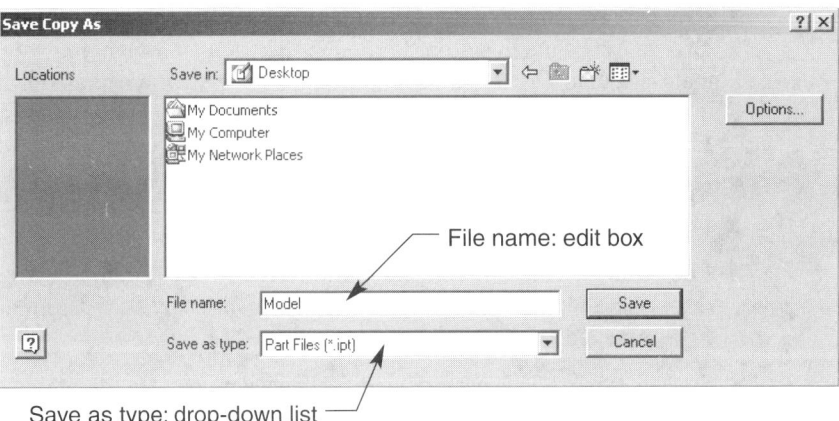

Save as type: drop-down list

Exercise 2.3

1. Continue from Exercise 2.2, or launch Autodesk Inventor.
2. Use the **New...** option of the **File** pull-down menu to open a new part, file.
3. Finish the sketch by picking the **Return** button on the **Command** bar. Typically, you cannot save a file while in sketch mode. If you are in sketch mode and try to save a file, you may be prompted to first finish the sketch.
4. Save the file as EX 2.3, using the **Save** option.
5. Save a copy of the file as EX2.3A, using the **Save Copy As...** option.
6. Close both of the files using the **Close** option of the **File** pull-down menu.
7. Exit Autodesk Inventor or continue working with the program.

■ **Projects...**   Pick this option to open the **Projects** window of the **Autodesk Inventor** dialog box. Refer to Chapter 1 for information about projects.

■ **Getting Started...**   Selecting this choice opens the **Getting Started,** or **QuickStart** window of the **Autodesk Inventor** dialog box, as discussed in Chapter 1.

■ **Print Setup...**   Selecting this option opens the **Print Setup** dialog box. See Figure 2.16. The following options are available inside the **Print Setup** dialog box:

FIGURE 2.16   The **Print Setup** dialog box.

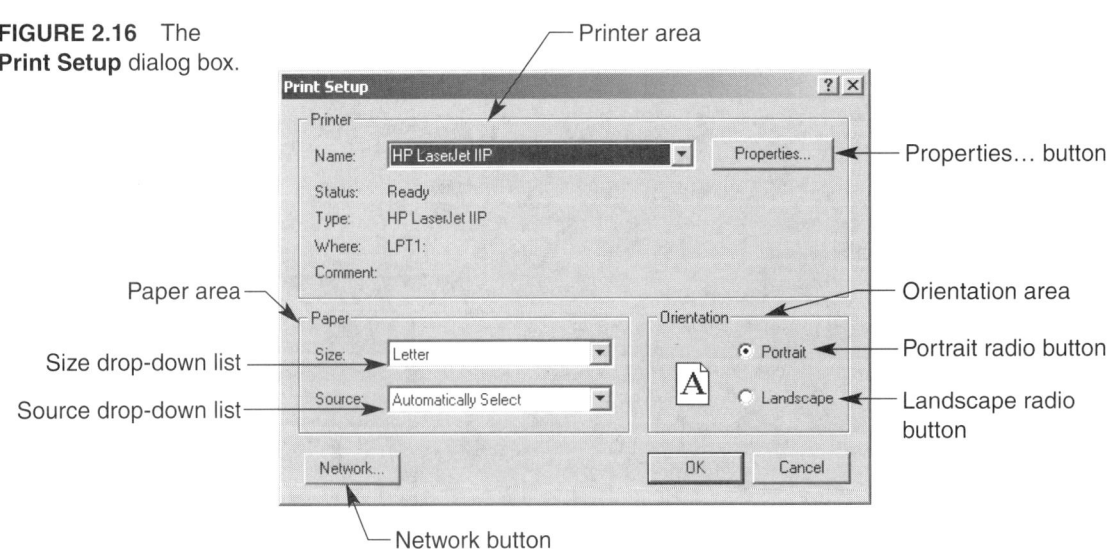

- **Printer** area   This section of the **Print Setup** dialog box contains several options as follows:

  - **Name** drop-down list   This allows you to select the printer you would like to use to plot your document.

  - **Status**   This displays the status of the print job—for example, printing, canceled, complete, etc.

  - **Type**   This displays the type of printer currently being used.

  - **Where**   This displays the printer port currently being used by the specified printer.

  - **Properties** button   Selecting the **Properties** button opens the **Document Properties** dialog box for the specified printer. See Figure 2.17. Here you can adjust the page layout, paper source, print quality, and other options that may apply to the specified print device.

- **Paper** area   The following options are available in the **Paper** area of the **Print Setup** dialog box:

  - **Size** drop-down list   This lets you select the size of paper to use for the document.

  - **Source** drop-down list   This lets you select where the paper comes from.

- **Orientation** area   Inside this area of the **Print Setup** dialog box, you can select the **Portrait** radio button or the **Landscape** radio button. The **Portrait** radio button allows you to rotate and print the document in a portrait orientation. The **Landscape** radio button allows you to rotate and print the document in a landscape orientation.

- **Network** button   Selecting this button opens the **Connect to Printer** dialog box, shown in Figure 2.18. If you are working in a networked setting and are not connected to a local printer, use this dialog box to connect to a shared network printer, and plot to that device.

- **Print Preview**   This option provides you with a preview of future print job. See Figure 2.19. From the print preview area, you can initiate the print by selecting the **Print...** button, review other pages that will print with the current print job by selecting the **Next Page, Prev Page,** or **Two Page,** and zoom in or zoom out to further review the print preview, by choosing the **Zoom In** or **Zoom Out** buttons. In addition, you can select the **Close** button to terminate the print preview.

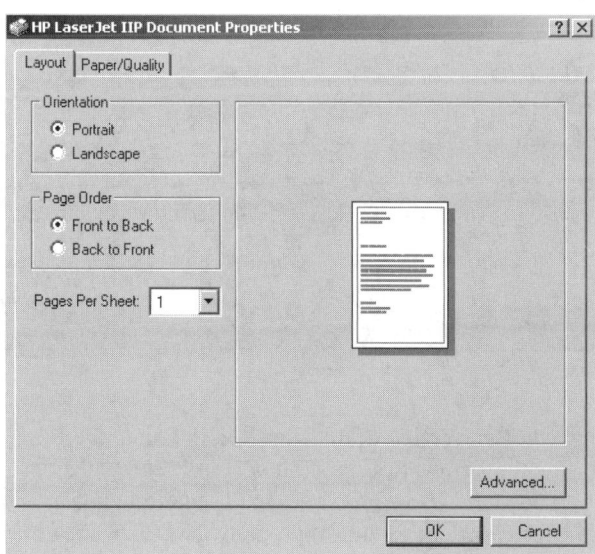

FIGURE 2.17   The **Document Properties** dialog box.

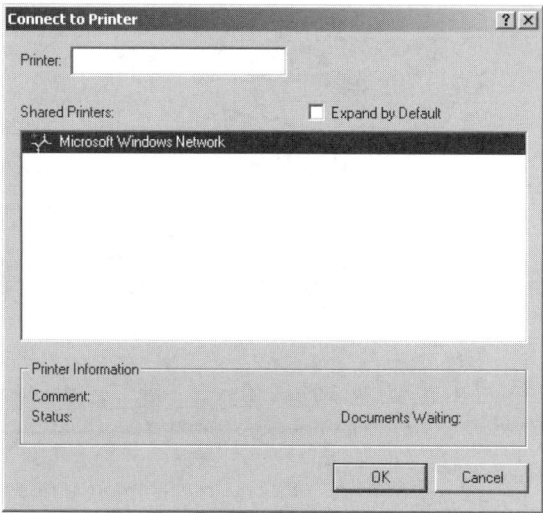

FIGURE 2.18   The **Connect to Printer** dialog box.

Pre<u>v</u>ious Page button   Zoom <u>I</u>n button   <u>C</u>lose button

<u>N</u>ext Page button   <u>T</u>wo Page button   Zoom <u>O</u>ut button

Print button ——→

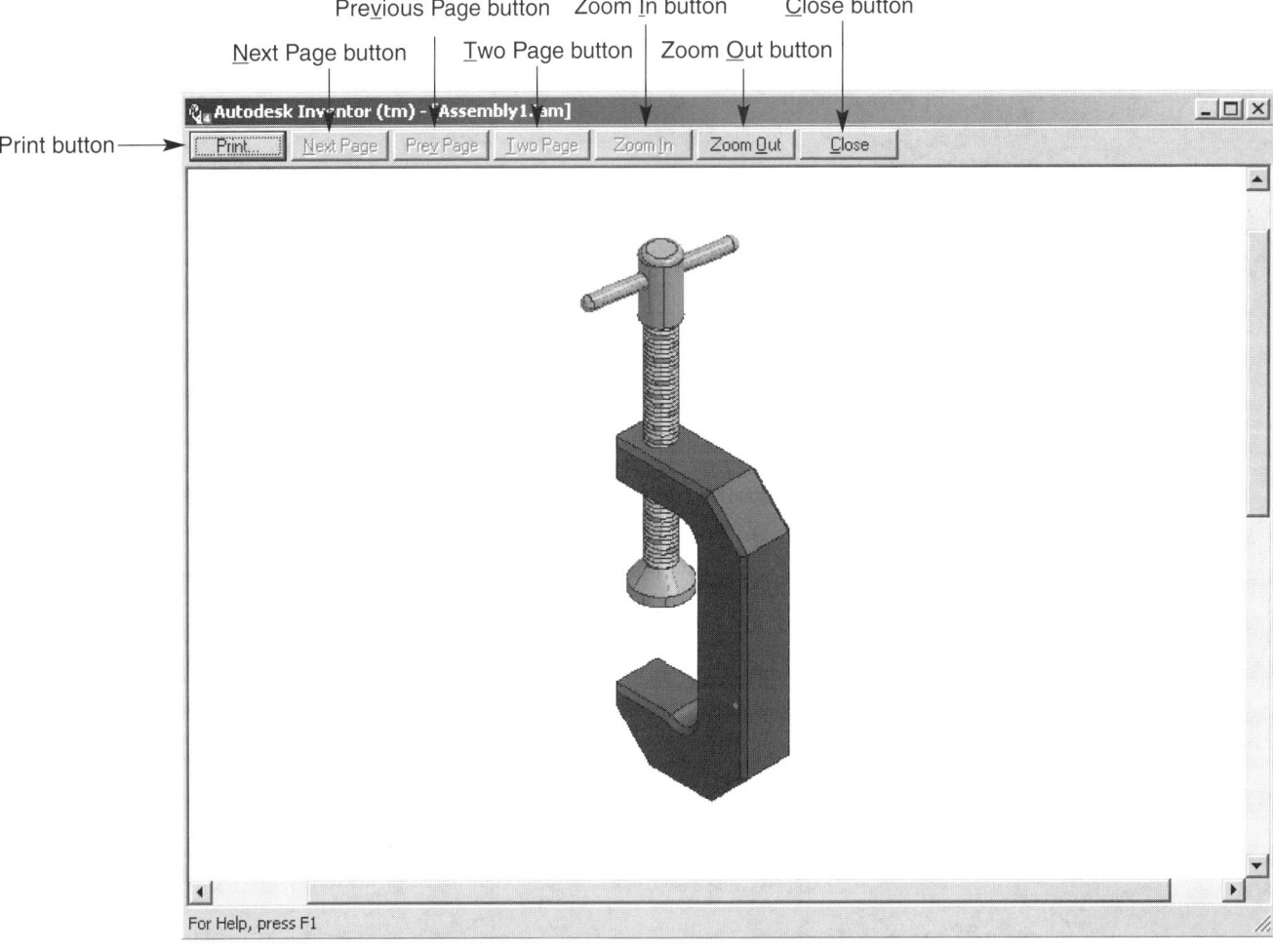

FIGURE 2.19   A preview of your print job.

- **Print...**   Selecting this option opens the **Print** dialog box. The **Print** dialog box is specific to each Autodesk Inventor file type and will be discussed in a later chapter.

Exercise 2.4

1. Continue from Exercise 2.3, or launch Autodesk Inventor.
2. Open Knob.ipt from the following directory: Autodesk/Inventor 5/Samples/Models/Adjustable Bracket/Components/Knob.ipt.
3. Print the part model using the techniques discussed.
4. Exit Autodesk Inventor or continue working with the program.

- **Properties...**   Picking this option opens the **Properties** dialog box for the specified Autodesk Inventor file. See Figure 2.20. Inside the **Properties** dialog box are different tabs that represent sets of Autodesk Inventor file properties. These properties are used throughout the design process to organize, manage, and document specific model property data. They are also used for developing title blocks, part lists, and bills of materials. The **Properties** dialog box and the particular properties and tabs available are further discussed in later chapters.

FIGURE 2.20 The **Properties** dialog box (part and assembly files).

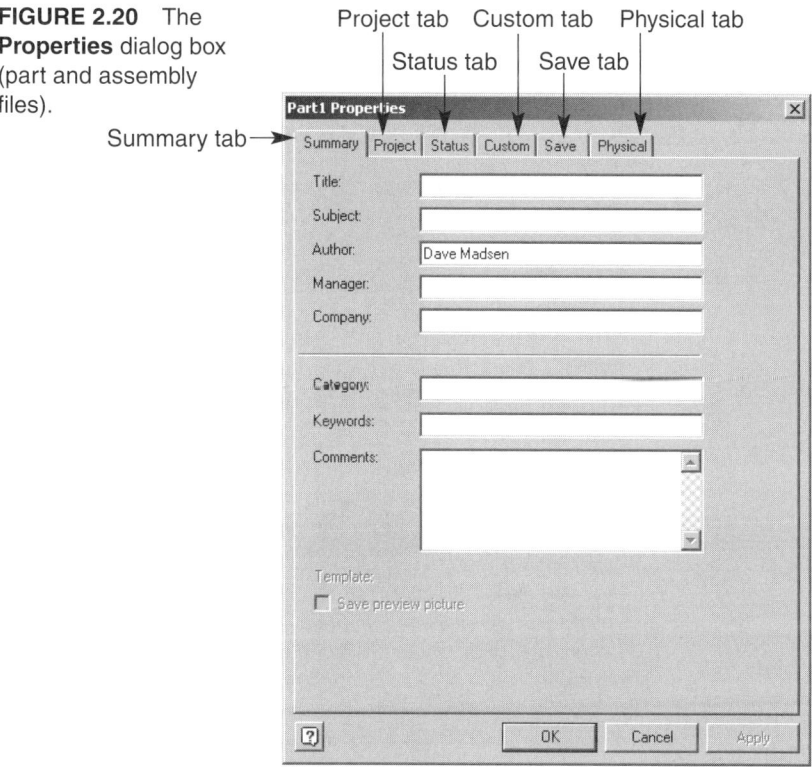

Summary tab →
Project tab   Custom tab   Physical tab
Status tab   Save tab

■ **Design Assistant**   Selecting this option opens the Autodesk Inventor **Design Assistant,** shown in Figure 2.21. The **Design Assistant** is used to locate, manage, and help modify Autodesk Inventor files. You can search for files, create file reports, work with the links between Autodesk Inventor files, copy properties, and manipulate design properties. Any additional files from other programs that may relate to your Autodesk Inventor files are also available. The **Design Assistant,** and the particular functions available inside **Design Assistant,** are further discussed in later chapters.

FIGURE 2.21 The Autodesk Inventor **Design Assistant** (properties mode).

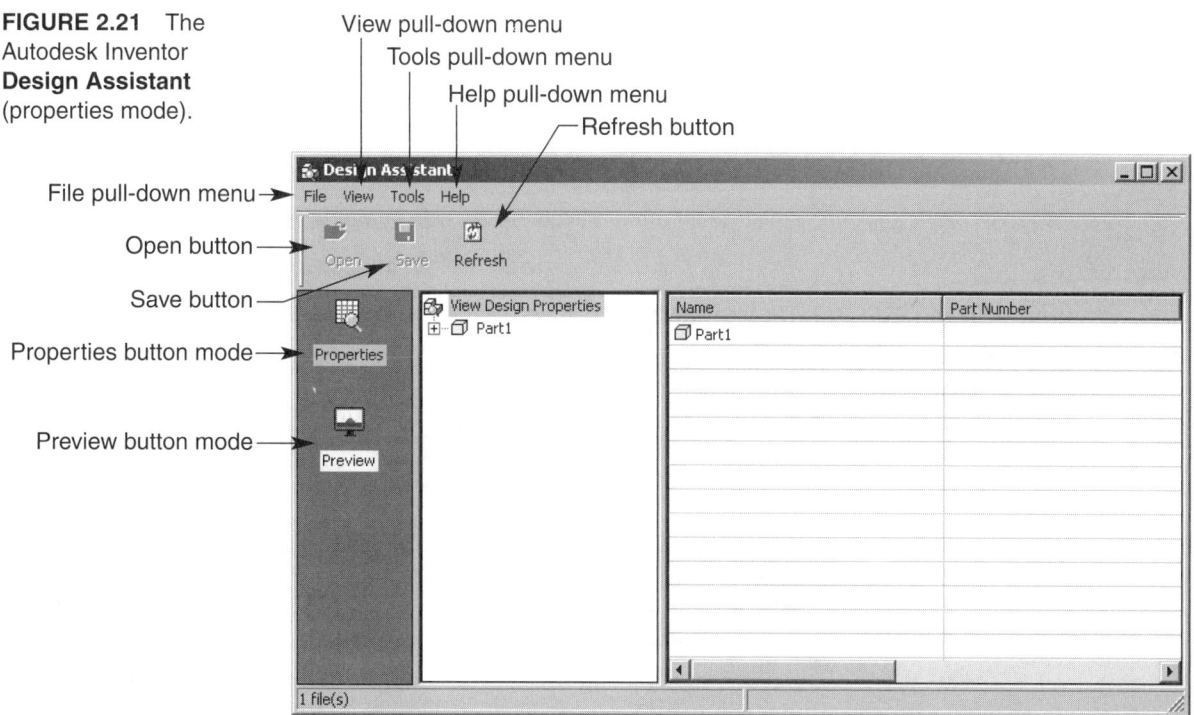

View pull-down menu
Tools pull-down menu
Help pull-down menu
Refresh button

File pull-down menu →
Open button →
Save button →
Properties button mode →
Preview button mode →

FIGURE 2.22 The **quick document access** area of the **File** pull-down menu.

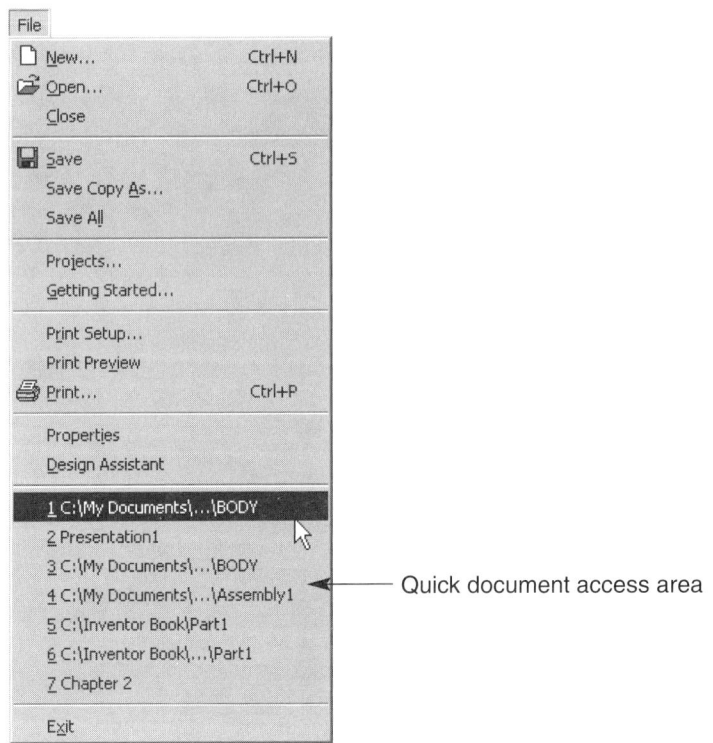

Quick document access area

■ **Quick document access**   The path and file name of several previously used documents is presented towards the bottom of the **File** pull-down menu. See Figure 2.22. This option is helpful if you are constantly opening the same recently used files, because these are the files displayed. Each of the paths is associated with a number. Number 1 is the most recently used, while the highest number is the file opened longest ago. To use this option, locate the path you would like to open, and then pick the path.

Field Notes

Using one of the quick document access options to open a file does not close any other files currently opened.

■ **E**x**it**   Pick this option to close all files and completely exit Autodesk Inventor.

Working with the **E**dit Pull-Down Menu

The **Edit** pull-down menu, shown in Figure 2.23, contains the following menu options for each of the four Autodesk Inventor file types:

FIGURE 2.23 The Autodesk Inventor **Edit** pull-down menu.

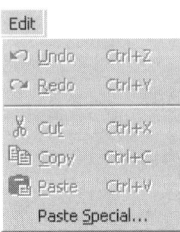

- **Undo**   Selecting this option cancels out, or undoes, the command entered last. All information and changes made by the command will revert back to those present before the command.

- **Redo**   Selecting this option cancels out the **Undo** option. All the information lost and changes made by the **Undo** option are returned.

- **Cut**   This option allows you to remove certain objects without deleting. The object can then be pasted in a new location.

- **Copy**   This option allows you to copy certain objects for use in a new location, without removing the object from the initial area.

- **Paste**   This option allows you to paste an object from one location to another. To use this tool, select the object or object name that you want to paste to the new location, and select either **Cut** or **Copy** from the menu items. Finally, locate the area where you want to paste the object, pick **Paste** from the **Edit** pull-down menu, or right-click and select **Paste**.

- **Paste Special...**   Selecting this option opens the **Paste Special** dialog box. See Figure 2.24. Here you can paste the contents of the clipboard into your Autodesk Inventor file. To use this option, place an item on the clipboard, using the **Cut** or **Copy** options. Then select the **Paste** radio button to paste the information, using one of the options provided in the **As:** display box. You can also pick the **Paste Link** radio button and pick one of the options provided in the **As:** list box. This option pastes a picture into your Autodesk Inventor file, which is linked to the original document. As a result, if the original document is modified, changes are also seen in the Autodesk Inventor file.

Exercise 2.5

1. Continue from Exercise 2.4, or launch Autodesk Inventor.
2. Open Knob.ipt from the following directory: Autodesk/Inventor 5/Samples/Models/ Adjustable Bracket/Components/Knob.ipt.
3. Use this part model to explore the **Edit** pull-down menu options discussed.
4. Exit Autodesk Inventor or continue working with the program.

Working with the Ⅴiew Pull-Down Menu

The **Ⅴiew** pull-down menu, shown in Figure 2.25, contains the following options for each of the four Autodesk Inventor file types.

- **Preⅴious**   Select this option to return to the previously displayed view. You can also access the **Previous View** command by right-clicking in the graphics window and selecting **Previous View,** or by pressing the **F5** key on your keyboard.

FIGURE 2.24   The **Paste Special** dialog box.

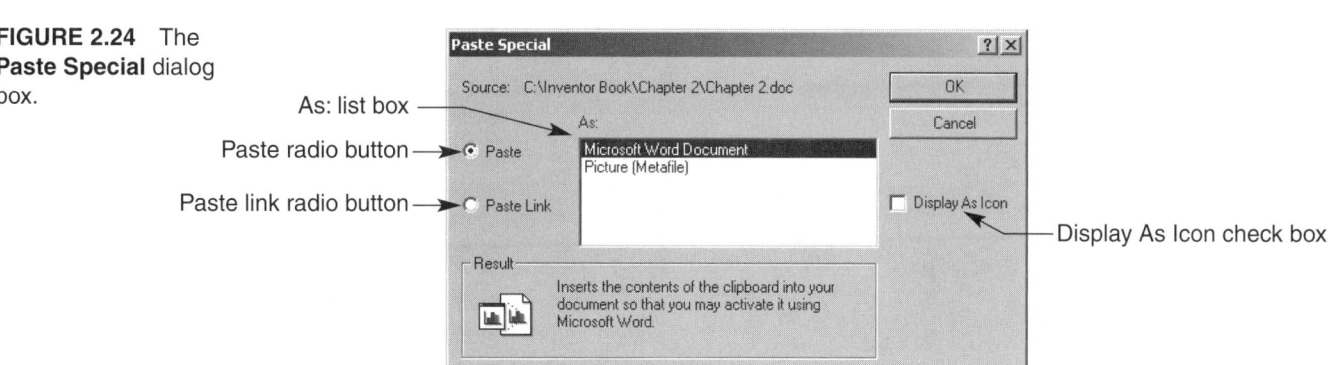

As: list box

Paste radio button

Paste link radio button

Display As Icon check box

FIGURE 2.25 The
Autodesk Inventor **View**
pull-down menu.

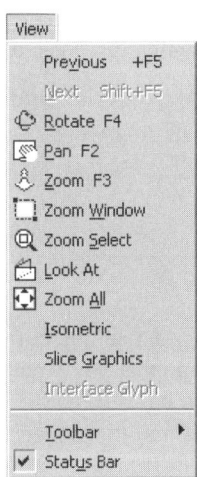

- **Pan**   This option lets you reposition the objects in the graphics window in order to see areas that might be hidden. To pan, click and hold while moving the pan icon around the graphics window. You can also access the **Pan** command by pressing and holding the wheel of a wheel mouse, by pressing and holding the **F2** key on your keyboard, or by selecting the **Pan** button on the **Standard** toolbar, discussed later in this chapter.

- **Zoom**   This selection allows you to zoom in to and out from objects displayed in the graphics window. You can also access this command by rolling the wheel of a wheel mouse forward to zoom in, and backwards to zoom out, picking the **Zoom In-Out** button on the **Standard** toolbar, or by pressing the **F3** key on your keyboard.

- **Zoom Window**   Select this option to zoom into a particular area by creating a window. You can also access this command by picking the **Zoom Window** button on the **Standard** toolbar. To use the **Zoom Window** option, access the command, and create a window around the area you would like to zoom in on, by picking a start corner and then picking the opposite, diagonal corner. See Figure 2.26.

- **Zoom All**   This selection lets you see all the objects and the entire design in the graphics window. The **Zoom All** tool is also accessed by picking the **Zoom All** button on the **Standard** toolbar.

- **Toolbar**   This option contains a cascading submenu that lists all the toolbars available for the particular Autodesk Inventor file. You can also access this list by right-clicking any of the currently displayed toolbars or toolbar buttons. To display one of the toolbars, pick the name presented in the cascading submenu. Toolbars are discussed later in this chapter.

- **Status Bar**   Use this option to turn the **Status Bar** on and off.

FIGURE 2.26 Using
the **Zoom Window**
option.

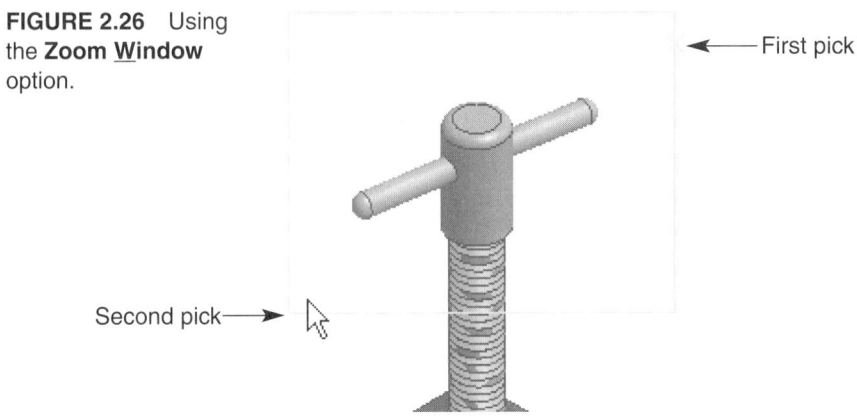

First pick

Second pick

Exercise 2.6

1. Continue from Exercise 2.5, or launch Autodesk Inventor.
2. Open Knob.ipt from the following directory: Autodesk/Inventor 5/Samples/Models/ Adjustable Bracket/Components/Knob.ipt.
3. Use this part model to explore the **View** pull-down menu options discussed.
4. Exit Autodesk Inventor or continue working with the program.

Working with the Insert Pull-Down Menu

The **Insert** pull-down menu, shown in Figure 2.27, contains the **Object** option for each of the four Autodesk Inventor file types. Selecting the **Object** option opens the **Insert Object** dialog box, shown in Figure 2.28. Inside this dialog box, you can insert a new image or document into your Autodesk Inventor file by selecting the **Create New** radio button and picking an object type from the **Object Type:** display box. You can also insert an existing image or document into your Autodesk Inventor file by selecting the **Create from File** radio button. Here you can search for a file to insert using the **Browse** button.

Working with the Format Pull-Down Menu

The **Format** pull-down menu, shown in Figure 2.29, contains the **Organizer...** option for each of the four Autodesk Inventor file types. Choosing the **Organizer...** option opens the **Organizer** dialog box. See Figure 2.30. This dialog box allows you to copy the types of materials, color styles, and lighting styles from one Autodesk Inventor file to another. To use the **Organizer** dialog box, first locate the source file by selecting one of the files from the **Source File:** drop-down list, or by browsing for a file using the **Browse** button. Next, copy any materials, color styles, and lighting styles from the source document to the active document by selecting the particular item or group of items and picking the **Copy** button. You can also delete any of the items by picking the item or items and selecting the **Delete** button.

Working with the Tools Pull-Down Menu

The **Tools** pull-down menu, shown in Figure 2.31, contains the following options for each of the four Autodesk Inventor file types:

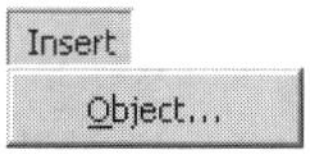

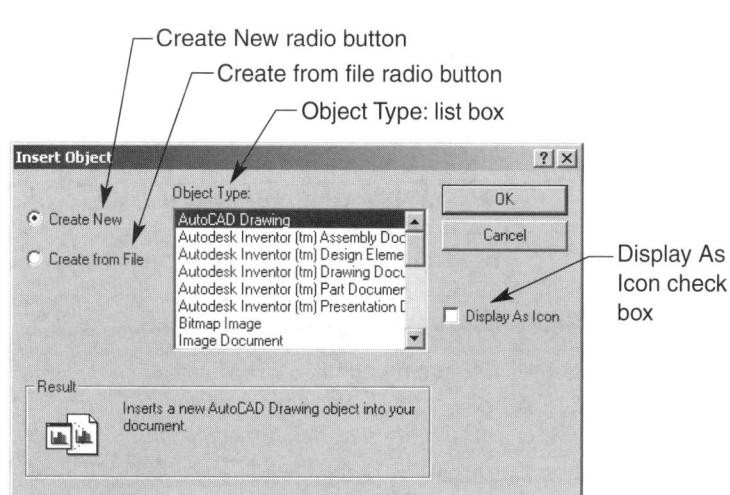

FIGURE 2.27 The Autodesk Inventor **Insert** pull-down menu.

FIGURE 2.28 The **Insert Object** dialog box.

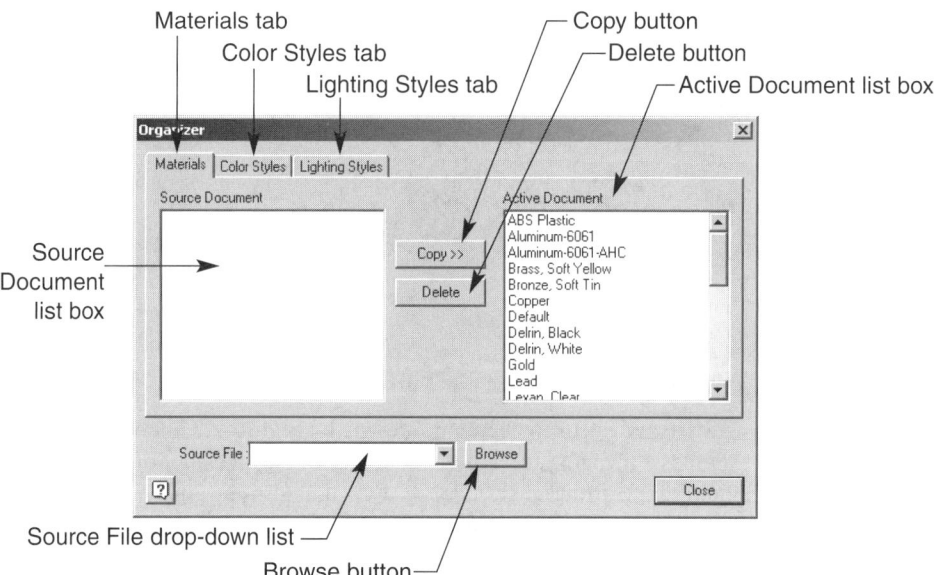

FIGURE 2.30   The **Organizer** dialog box.

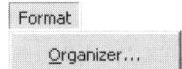

FIGURE 2.29   The
Autodesk Inventor
Format pull-down menu.

FIGURE 2.31   The
Autodesk Inventor
Tools pull-down menu.

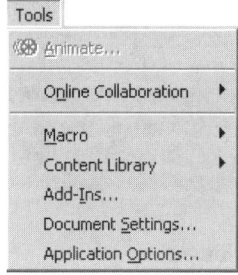

- **Online Collaboration**   This selection contains the **Meet Now...** cascading submenu option, which allows you to connect to another Autodesk Inventor user via the Internet. (The collaborative capabilities of Autodesk Inventor are discussed in Chapter 19.)

- **Content Library**   This cascading submenu controls options for inserting content into assemblies as discussed in Chapter 14.

- **Add-Ins...**   Selecting this option opens the **Add-In Manager** dialog box, which is used to list, load, and unload available add-ins. See Figure 2.32. Add-Ins available on your system are displayed in the **Available Add-Ins** list box, and the activities of each are specified under the **Load Behavior** list box. Below these lists is a **Description** display box that provides information regarding the currently selected add-in. Next to the **Description** display box is the **Load Behavior** area where you can manipulate the **Loaded/Unloaded** and **Load On Startup** check boxes.

- **Documents Settings...**   Pick this option to display the **Document Settings** dialog box for the specified Autodesk Inventor file. See Figure 2.33. This dialog box controls the default units, unit precision, dimension precision, and sketch and model work environment settings for the active document. The following three tabs are available inside the **Document Settings** dialog box:

  - **Units**   This tab, shown in Figure 2.33, contains **Units** and **Precision** areas. Inside the **Units** area are drop-down lists that allow you to manipulate the unit system for dimen-

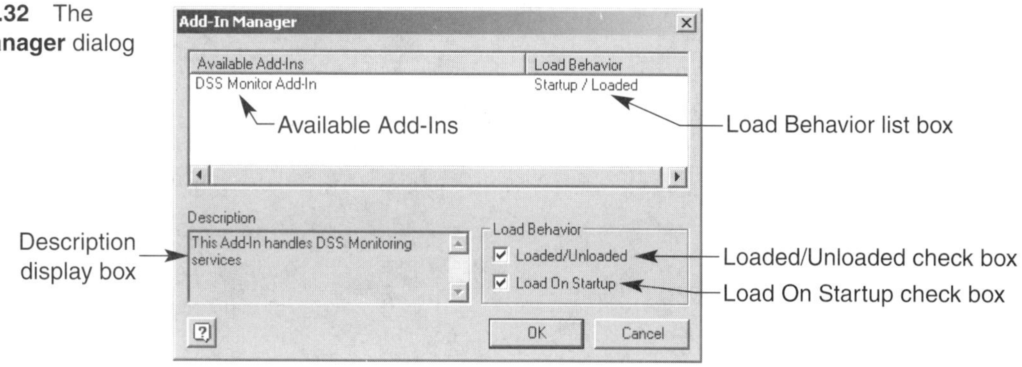

FIGURE 2.32 The **Add-In Manager** dialog box.

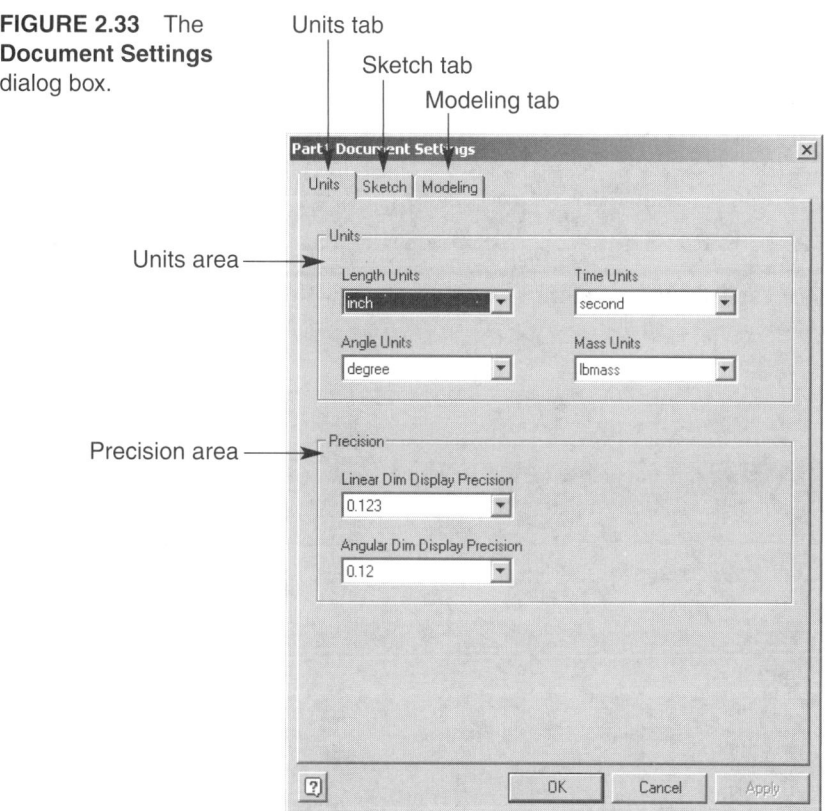

FIGURE 2.33 The **Document Settings** dialog box.

sions, mass, and time. Inside the **Precision** area are drop-down lists that allow you to manipulate the precision of dimension units.

- **Sketch**   This tab, shown in Figure 2.34, contains snap and grip spacing options for two-dimensional sketches and an **Auto-Bend Radius** edit box for three-dimensional sketches. These options are discussed in further detail in a later chapter.

- **Modeling**   This tab, shown in Figure 2.35, allows you to control the three-dimensional snap spacing in the modeling environment. In addition, the **Compact Model History** check box is used to purge or compact the model information.

- **Application Options...**   Selecting this choice opens the **Options** dialog box. See Figure 2.36. This dialog box controls the performance, settings, and options for various Autodesk Inventor functions. The following tabs are available in the **Options** dialog box:

  - **General**   This tab, shown in Figure 2.36, sets various general Autodesk Inventor options. These choices are described as follows:

2D Sketch area →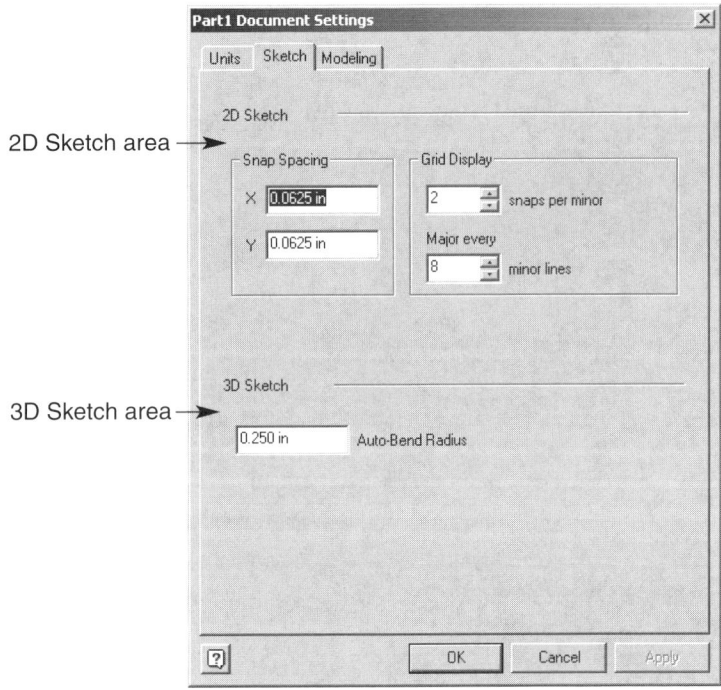

3D Sketch area →

FIGURE 2.34   The **Document Settings** dialog box, **Sketch** tab.

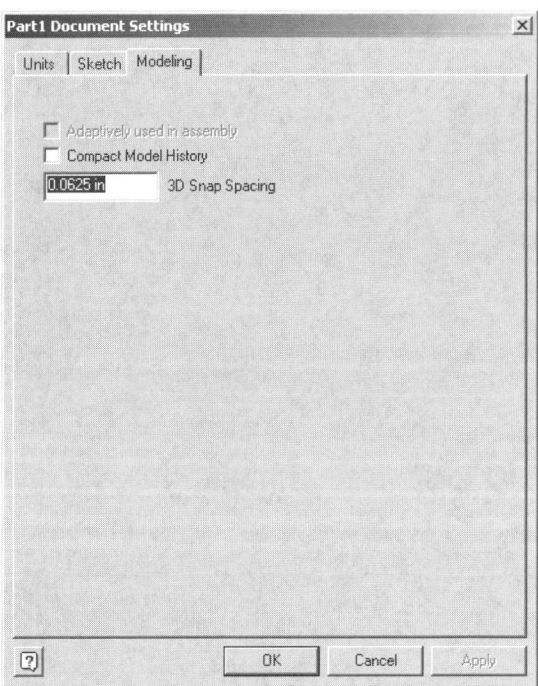

FIGURE 2.35   The **Document Settings** dialog box, **Modeling** tab.

Maximum size of undo file (Mb) edit box

Number of versions to keep edit box

Locate tolerance edit box

Select Other time delay (sec) edit box

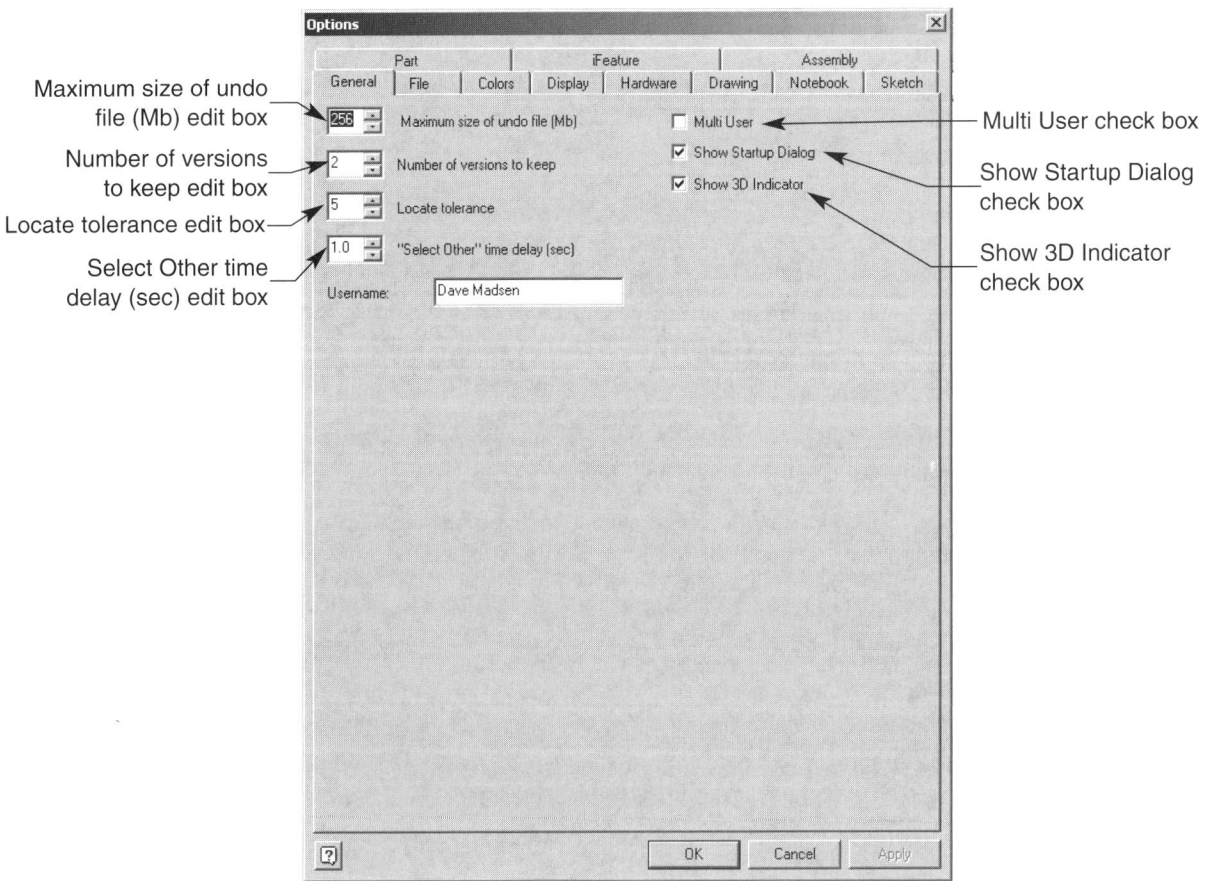

Multi User check box

Show Startup Dialog check box

Show 3D Indicator check box

FIGURE 2.36   The **Options** dialog box (**General** tab shown).

- **Maximum size of undo file (Mb)**   Use this edit box to set the number of times you can use the **Undo** option. Pick or right-click inside the edit box to type in a number, or use the up and down arrows to increase or decrease the number of megabytes.

- **Number of versions to keep**   Every time you save a model, a version of the model is stored in the file. The **Number of versions to keep** edit box controls the number of model versions to store. You can store as many as 10, and as few as 1. When Autodesk Inventor reaches the specified storage number, the oldest model version is removed. You can select or double-click inside the edit box and type a number, or pick the up or down arrows to increase or decrease the number of versions.

- **Locate tolerance**   This edit box controls how far away from an object you can pick and still select the object. The distance is defined as the number of pixels from 1 to 10. You can select or double-click inside the edit box and type a distance, or pick the up or down arrows to increase or decrease the distance.

- **"Select Other" time delay (sec)**   Use this edit box to specify the number of seconds you can hover over model geometry before the **Select Other** tool is shown.

- **Username**   Use this edit box to enter the username for use in a number of tools, commands, and options.

- **Multiuser**   This check box controls Multiuser protection options, discussed in Chapter 1. If the check box is clear, no warnings or file reservation effects are used. If the check box is selected, multiuser warnings and file reservation effects are seen.

- **Show Startup Dialog**   This check box specifies whether the Autodesk Inventor dialog box, discussed in Chapter 1, opens automatically. If the check box is clear, the Autodesk Inventor dialog box will not automatically open when Autodesk Inventor is launched.

- **Show 3D Indicator**   An XYZ axis, shown in Figure 2.37, is displayed in the graphics window of every Autodesk Inventor three-dimensional work environment. The X axis is portrayed as a red arrow, the Y axis as a green arrow, and the Z axis as a blue arrow. The Show 3D Indicator check box is used to turn the 3D indicator on and off. If the check box is clear, the 3D indicator will not be displayed on the graphics window.

- **File**   This tab, shown in Figure 2.38, controls the location of files that control various Autodesk Inventor functions. The **Undo** area controls the location of temporary files created by undoing commands, the **Templates** area allows you to define the location of templates, the **Projects Folder** area identifies the location of the active project folder, the **Workgroup Design Data** area controls the location of external work group files such as spreadsheets and thread tables, and finally, the **Default VBA Project** area allows you to define the default location of visual basic files.

    Each of the file areas contains an edit box and a **Browse** button. You can enter a path in the edit box or locate a path using the **Browser** button.

- **Colors**   This tab, shown in Figure 2.39, controls the background color displayed in the graphics window. Inside this tab you can specify the background color for the modeling environment using the **Design** button, and the drawing environment using the **Drawing** button. You can select a color from the list of colors found in the **Color Scheme** display list. In addition, you can pick the **Background Gradient** check box to display the background color as a gradient, and pick the **Show Reflections** check box to display background color reflections.

FIGURE 2.37 The Autodesk Inventor **3D Indicator.**

FIGURE 2.38   The
Options dialog box
(**File** tab shown).

Undo area

Templates area

Project Folder area

Workgroup Design Data area

Default VBA Project area

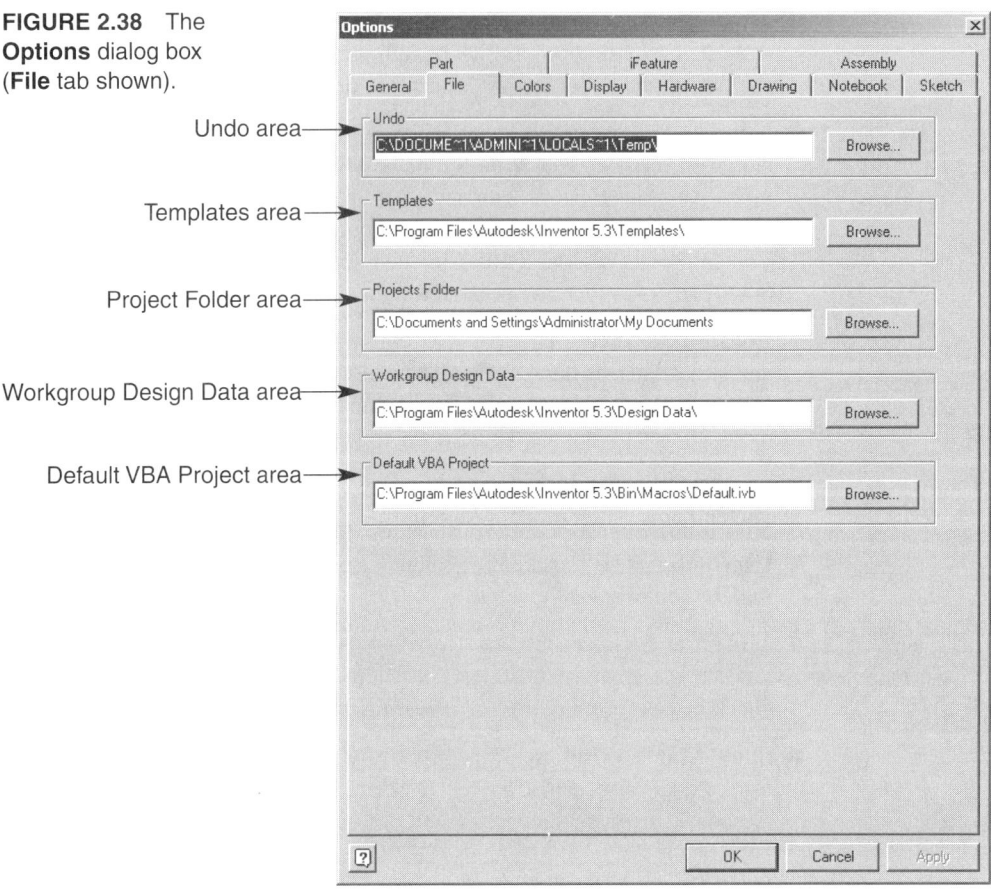

FIGURE 2.39   The
Colors tab of the
Options dialog box.

Design button

Drafting button

Color Scheme list box

Background
Gradient check box

Show Reflections
check box

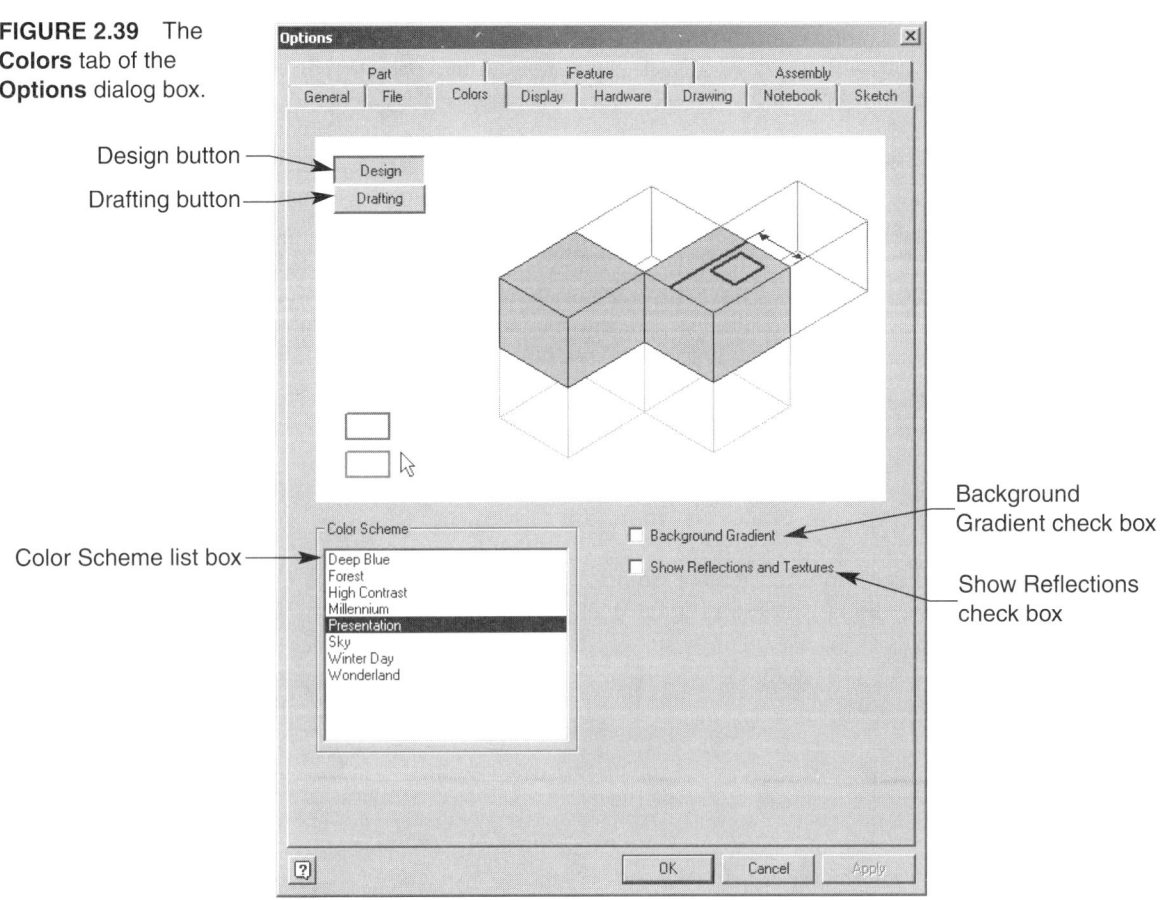

■ **Display**   This tab, shown in Figure 2.40, controls the wireframe and shaded display model display. Inside this tab you have access to the following areas and options:

- ■ **Wireframe Display Mode**   This area specifies the wireframe display options for a model.

- ■ **Display Quality**   This area controls the model display resolution. Consequently, the regeneration time is also adjusted here.

- ■ **Shaded Display Modes**   This area specifies the shaded display options for a model.

- ■ **View Transition Time (seconds)**   Move the slide bar left or right to control the time it takes to change between various views, such as zooming in and out, panning, and looking at a specific object. Seconds can be any number from 0 to 3. Setting transition time to 0 seconds allows changing between views to happen almost instantaneously, while setting transition time to 3 seconds allows for the longest time between view transitions.

- ■ **Minimum Frame Rate (Hz)**   Move the slide bar left or right to control the rate at which frames regenerate between various views, such as zooming in and out, panning, and rotating. Frame generation can be specified as any number from 0 to 5. Setting minimum frame rate to 0 allows all the frames with no time limit, while setting minimum frame rate to 5 allows at least 5 frames do be developed per second. Though you can adjust these settings, generally the frame rate is adjusted much faster than what you specify.

- ■ **% Hidden Line Dimming**   This edit box allows you to control the dimming percentage for hidden edges. The **% Hidden Line Dimming** edit box is only available when

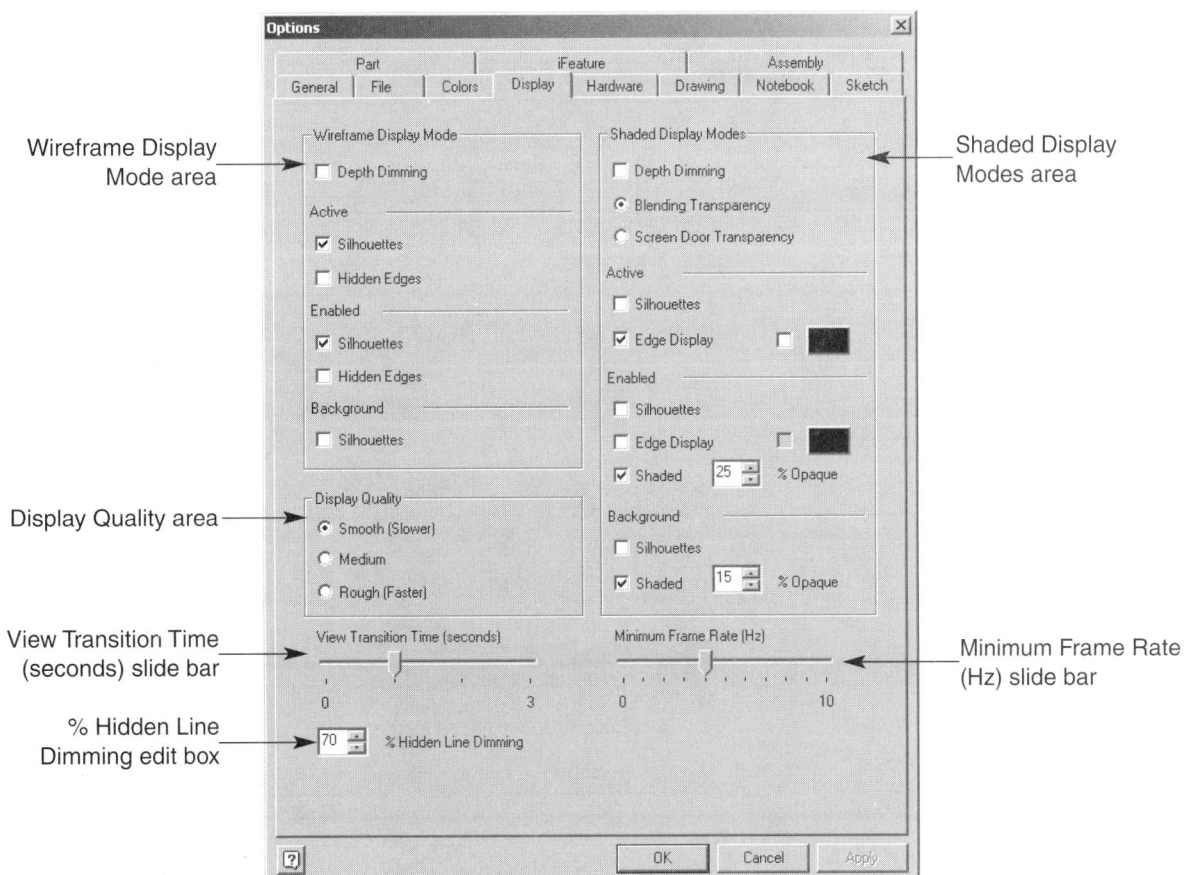

FIGURE 2.40   The **Display** tab of the **Options** dialog box.

at least one of the **Hidden Edges** check boxes is picked. You can enter a new percentage or scroll through available percentages using the up and down arrows.

■ **Hardware**   This tab, shown in Figure 2.41, contains options concerning your computers graphics hardware and its relation to Autodesk Inventor. For more information on graphics and the graphics hardware specifications, refer to the Autodesk Inventor help files.

■ **Drawing**   Use the options in this tab, shown in Figure 2.42, to control the options for drawings. This tab and drawings are discussed in Chapters 12, 13, and 17.

■ **Notebook**   Use the options in this tab, shown in Figure 2.43, to control the display options of design notes in the **Engineer's Notebook**. This tab, and the Engineer's Notebook, are discussed in Chapter 19.

■ **Sketch**   This tab enables you to specify two-dimensional and three-dimensional sketch options. See Figure 2.44. This tab, and using sketches, are discussed in sketching chapters.

■ **Part**   This tab, shown in Figure 2.45, allows you to specify the default options for creating new parts. This tab, and working with parts, are discussed in future chapters.

■ **iFeature**   This tab, shown in Figure 2.46, allows you to control design elements (also known as iFeatures) and file locations, and specify a viewer to manage and look at the files. iFeatures are catalog features and objects used numerous times for developing different models. iFeatures and catalog files are discussed in Chapter 9.

■ **Assembly**   The options in this tab allow you to specify preferences for working with assemblies. See Figure 2.47. This tab, and assemblies, are discussed in Chapters 14 and 15.

FIGURE 2.41   The **Hardware** tab of the **Options** dialog box.

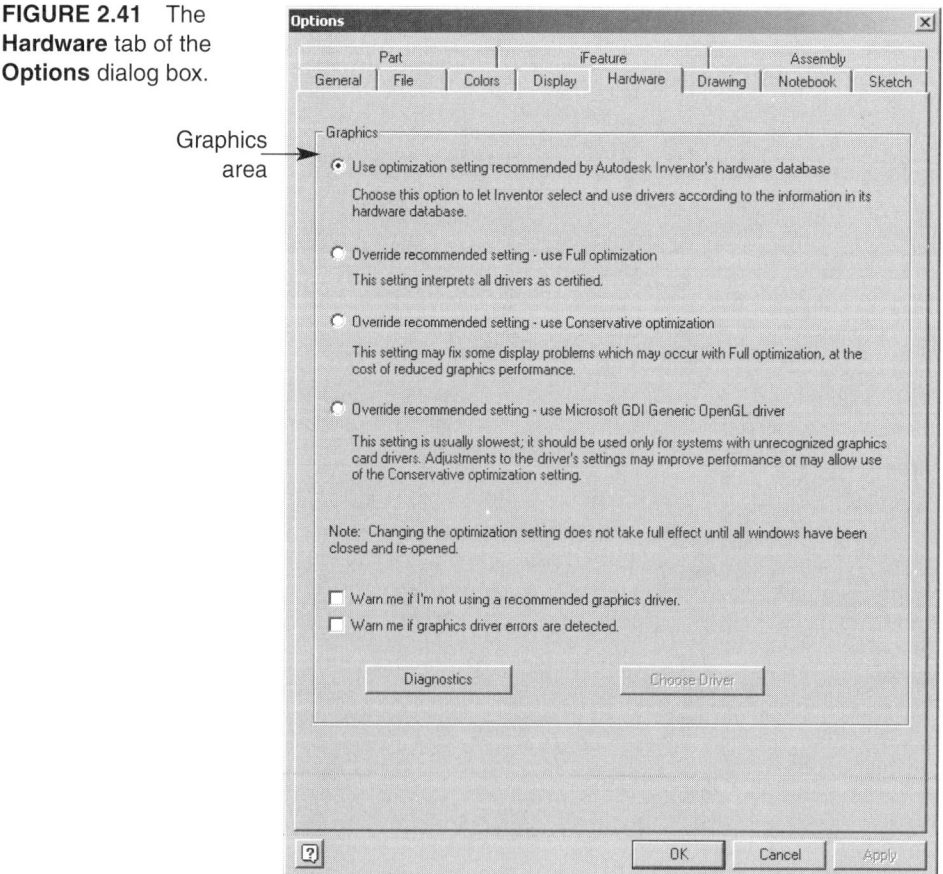

FIGURE 2.42   The
Drawing tab of the
Options dialog box.

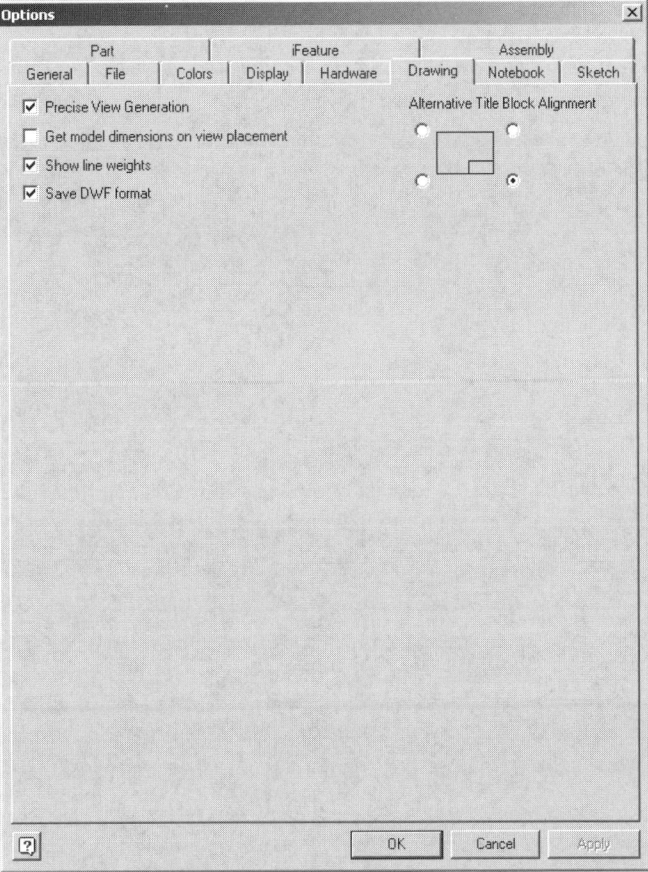

FIGURE 2.43   The
Notebook tab of the
Options dialog box.

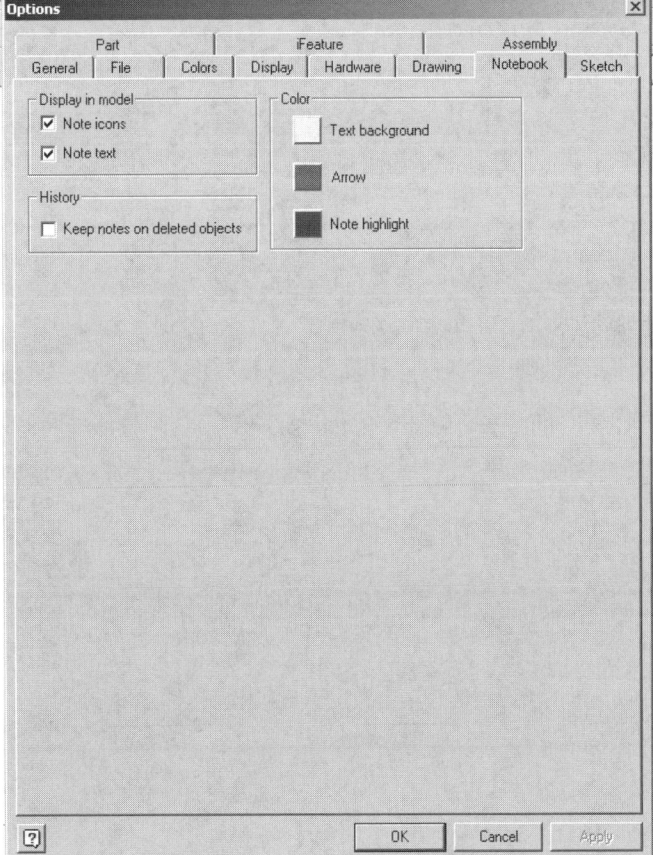

FIGURE 2.44 The **Sketch** tab of the **Options** dialog box.

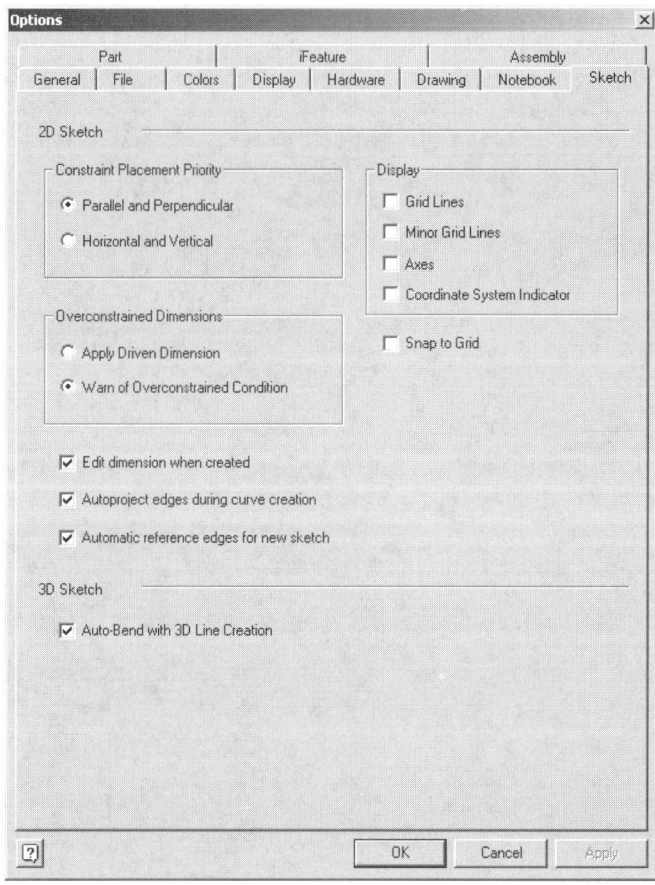

FIGURE 2.45 The **Part** tab of the **Options** dialog box.

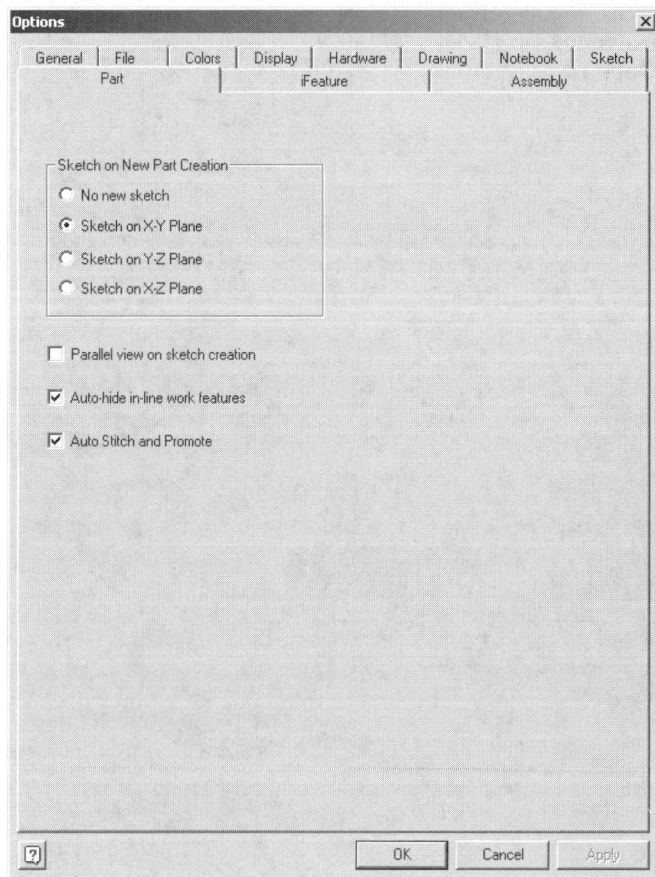

FIGURE 2.46 The **iFeature** tab of the **Options** dialog box.

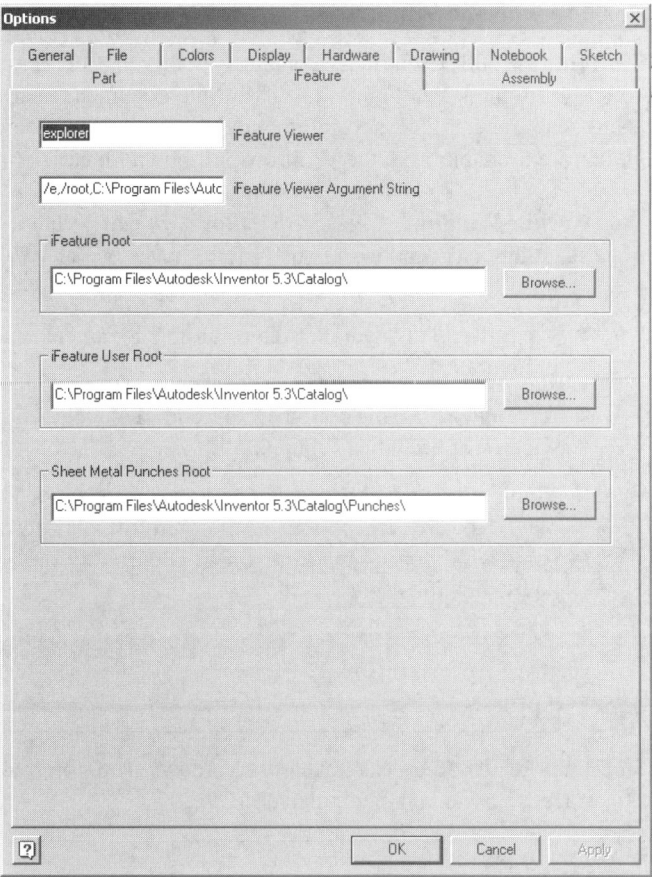

FIGURE 2.47 The **Assembly** tab of the **Options** dialog box.

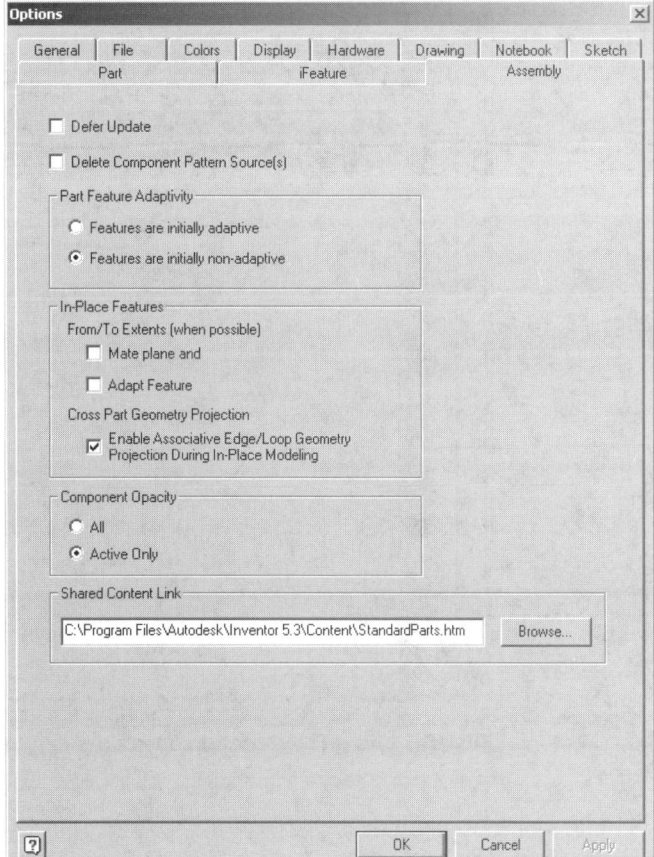

Working with the **W**indow Pull-Down Menu

The **Window** pull-down menu contains options that help you control the multiple document interface of Autodesk Inventor. See Figure 2.48. Multiple document interface, also known as multiple design interface, means that you can have several documents or views of documents open at the same time. The following options are available from the **Window** pull-down menu:

- **New Window**   This option allows you to create a new window for the active document. Additional windows enable you to have several views of a design opened and accessible at once.

- **Cascade**   This option arranges all the document windows in a cascading fashion, as shown in Figure 2.49.

- **Arrange All**   This option arranges all the document windows in a way that shows the entire window of each. See Figure 2.50.

- Multiple document access   The names of each of the currently opened documents are displayed towards the bottom of the **Window** pull-down menu. See Figure 2.48. The window name with a check mark next to it is currently the active window. To open one of the windows, pick the name.

Field Notes

Each of the Autodesk Inventor Windows available in a document can be minimized, maximized, closed, and manipulated in size.

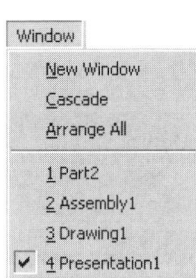

FIGURE 2.48   The **Window** pull-down menu.

FIGURE 2.49   The **Cascade** option of the **Window** pull-down menu.

FIGURE 2.50 The **Arrange All** option of the **Window** pull-down menu.

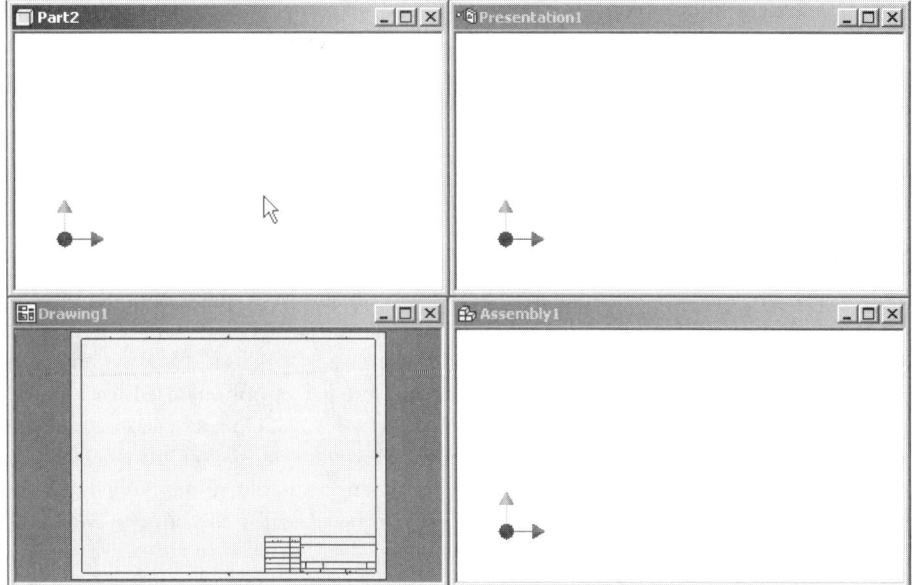

Exercise 2.7

1. Continue from Exercise 2.6, or launch Autodesk Inventor.
2. Open a new part, assembly, drawing, and presentation file.
3. Make the part file current by selecting Part1 from the **View** pull-down menu, multiple document access area.
4. Create a new window for the part file.
5. View the documents in a cascade and arrange all format.
6. Exit Autodesk Inventor or continue working with the program.

Working with the Help Pull-Down Menu

The **Help** pull-down menu provides you with options that help you better understand Autodesk Inventor. See Figure 2.51.

Pick the **Help Topics** option to open the Autodesk Inventor help files. Selecting the **What's New** option opens the **What's new in Autodesk Inventor, What's New** menu. When you choose **Tutorials,** the **Autodesk Inventor – Tutorial** window opens. In addition, menu options are available that direct you to specific help for working with Autodesk Inventor for AutoCAD users, programming assistance, and graph-

FIGURE 2.51 The **Help** pull-down menu.

ics drivers information. Pick **Autodesk <u>O</u>nline** to open Autodesk Inventor Help files, with the **Autodesk Online** window active. This area allows you to locate Autodesk information via the Internet. Each of these tools was discussed in Chapter 1.

    About Autodesk Inventor™ is the last option in the **<u>H</u>elp** pull-down menu. Here you can view information about Autodesk Inventor including release information, dates, serial number, third part credits, and license information.

Using the Panel Bar

The *panel bar,* shown in Figure 2.52, provides you with a list of options. Inside the list of options are several design tools available for creating models in the specific work environment where you are currently working. For example, only the list of options and tools used for parts are available in the panel bar while you are working on a part. In addition, only the list of options used for a particular design stage are available. Autodesk Inventor accomplishes this in two ways. One, as you move from one design stage to another, the menu and tools automatically change. For example, if you are working on a part and are currently in sketch mode, only the sketch list of options are available. When you finish the sketch, the features menu becomes available. Two, only tools that can be used for the current work environment are available for selection. All others are shaded.

    As previously mentioned, exiting one design stage and entering another automatically opens a new list of options. If you would like to change from one list of options to another yourself, apply one of the following techniques:

- Right-click directly on, or beside, any of the tools provided in the current list of options. Next, select the desired menu from the list.

- Click on the current list of options name, and select the desired menu from the list.

FIGURE 2.52 The Autodesk Inventor **Panel Bar** (sketch mode).

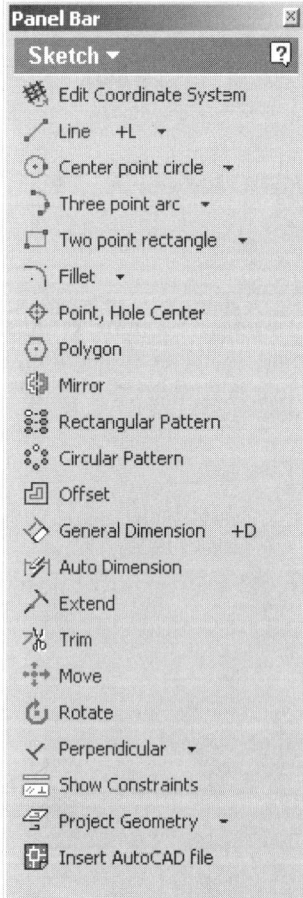

FIGURE 2.53 Panel Bar menu options (sketch mode).

Applying one of the previously mentioned techniques opens a list of menus that includes the **Expert** option. See Figure 2.53. By default, the panel bar is set on *Learning mode.* Learning mode provides you with tool icons and names. Learning mode is helpful when you first begin to work with Autodesk Inventor and do not want to wait for a tooltip, discussed later in this chapter. Selecting the **Expert** option sets the panel bar to *Expert mode.* A panel bar in Expert mode looks similar to a toolbar, discussed later in this chapter, and contains only tool buttons. This can be helpful if you want a smaller panel bar displayed.

You can resize the panel bar using the resizing arrows accessible from the panel bar edges. By default the panel bar is docked in the upper left corner of the work environment. To undock or redock the panel bar, simply double-click above the list of options name, or pick the area above the list of options name and drag the panel bar to a new location.

Using the Browser Bar

The *browser bar,* or *browser,* displays all the objects in your model. See Figure 2.54A. The number and type of objects vary depending on the current design stage, model type, and work environment. Typically, the name of the file and an icon is displayed on top. Below the name is the **Origin** folder and all the origin options, followed by the sketches, features, or components of the file. These components are listed in the

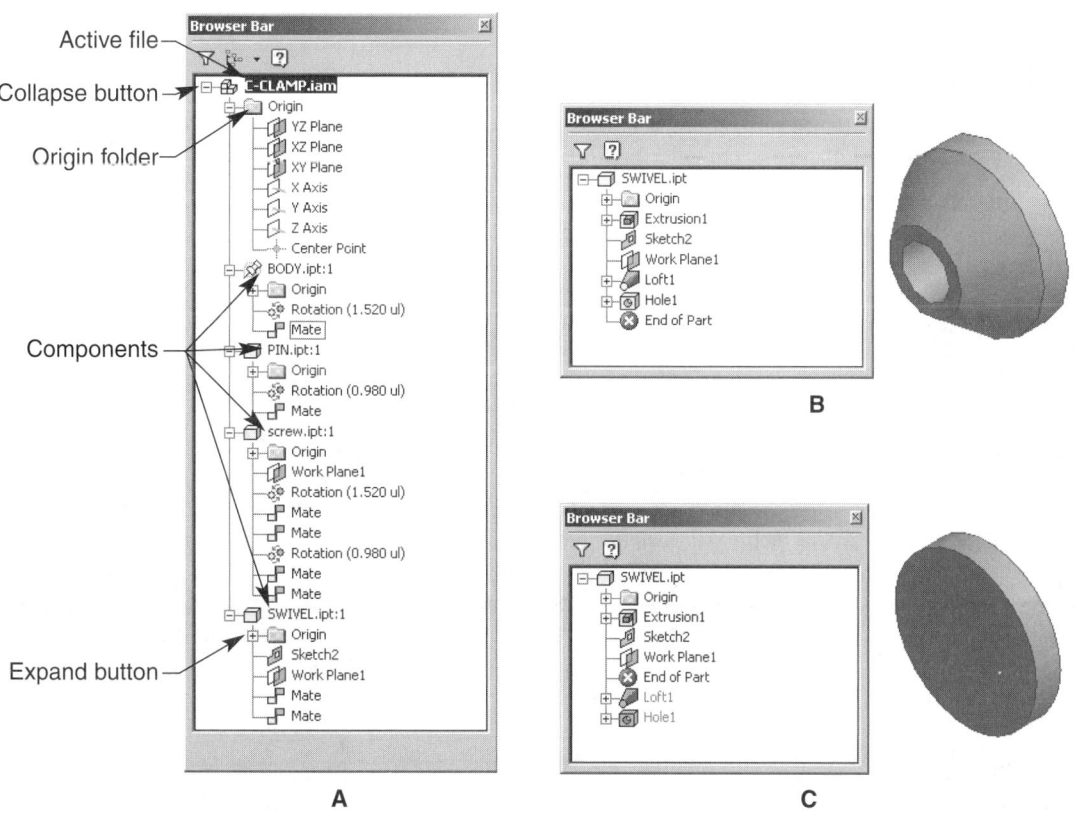

FIGURE 2.54　(A) An example of the Autodesk Inventor **Browser Bar.** (B) Manipulating **Browser Bar** components. (C) Dragging and dropping **End of Part.**

order they were created or inserted. Finally, the **End of Part** displays last, unless moved. However, it may be possible in some applications to drag and drop the object up or down in the list. As an example, Figure 2.54B shows a completed part. Figure 2.54C shows how dragging and dropping **End of Part** up above two features changes the display and work environment of a part.

Many of the objects displayed in the browser may contain *child nodes*. See Figure 2.54A. These are sketches, features, or components that create the main object. To display the child nodes, pick the **Expand** button to the left of the object name or right-click on the object name and select **Expand All Children**. To hide the child nodes, pick the **Collapse** button to the left of the object name or right-click on the object name and select **Collapse All Children**.

The browser bar interface is specific to each type of Autodesk Inventor document. Using the browser and the browser bar variations in parts, assemblies, drawings, and presentations is discussed later.

Exercise 2.8

1. Continue from Exercise 2.7, or launch Autodesk Inventor.
2. Open blade_main.ipt from the following directory: Autodesk/Inventor 5/Samples/Models/ Scissor/Components/blade_main.ipt.
3. Locate and explore the panel bar.
4. Pick the panel bar title **(Features)** and select the **Expert** option from the shortcut menu. Observe the changes made to the panel bar.
5. Locate and explore the default browser bar.
6. Drag the **End of Part** indicator up and above **Sketch4,** and notice the changes.
7. Pick the **Expand** button to the left of the **Origin** folder to expand all children.
8. Right-click the **Origin** folder, and select **Collapse All Children**.
9. Close blade_main.ipt without saving.
10. Exit Autodesk Inventor or continue working with the program.

Using Autodesk Inventor Toolbars and Toolbar Buttons

The toolbars and toolbar buttons are important Autodesk Inventor interface elements. There are a number toolbars that apply to specific tasks, but by default, the first time you start Autodesk Inventor, only one toolbar is visible: the **Standard** toolbar. See Figure 2.55. Toolbars display several buttons, each with a specific icon, that allow access to Autodesk Inventor commands. Toolbars have features similar to pull-down menus and the panel bar. However, unlike pull-down menus, which may require you to select a number of options to begin a command or apply a setting, toolbars usually require you to select only one toolbar button. Unlike the panel bar, toolbars usually take up less space on your screen.

Using Toolbar Button Tooltips and Help Strings

Autodesk Inventor toolbars have tooltips and help strings that help you understand and recognize different toolbar buttons. A *tooltip* is a small text box that pops up under the toolbar button when you point to the button with your cursor. Inside the text box is the button name that helps you remember what the button icon represents. A *help string,* which describes the button function, appears in the status bar when the tooltip pops up. The status bar is usually located towards the bottom of your screen. It displays the tooltips, help strings, and other information associated with the button, menu, option, or selection, your cursor is pointing to. The function of the toolbar button help strings is similar to that of the pull-down menu previously discussed. Figure 2.56 shows the tooltip for the **Save** button on the **Standard** toolbar. The tooltip tells

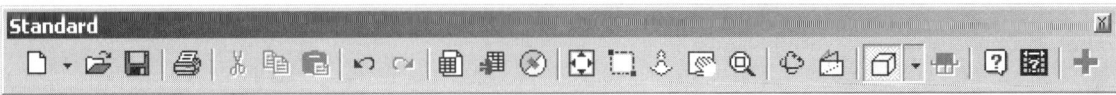

FIGURE 2.55   The Autodesk Inventor **Standard** toolbar.

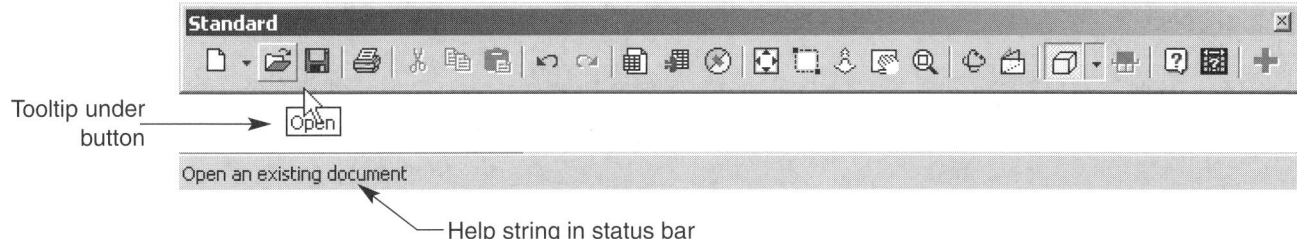

Tooltip under button

Help string in status bar

FIGURE 2.56   An example of a tooltip and help string.

you that the icon you see is the **Save** command. The help string is displayed on the status bar and tells you the function of the **Save** button: Save the active document.

Once you have reviewed the toolbar button icon, seen the tooltip, and read the help string, pick the toolbar button to activate the command.

Using Flyout Toolbar Buttons

Some toolbar buttons have an arrow to the right of the button, known as a flyout. See Figure 2.57. A *flyout* is a button that contains additional, related command buttons; it is similar to a cascading submenu, discussed previously. To use a flyout toolbar button, select the arrow next to the button. This will open the other toolbar button options associated with the specified button. These buttons work exactly the same as other toolbar buttons and have tooltips and help strings.

Field Notes

Most of the time, when you pick one of the flyout options, that button becomes active and is displayed on the toolbar on the top level of the flyout. In this case, the rest of the buttons then become invisible.

Working with Commonly Used Toolbars and Toolbar Buttons

Only toolbars and toolbar buttons applicable to the current work environment, Autodesk Inventor file type, and current stage of design, are available for you to use. These specific toolbars are discussed in the chapters where they apply.

Each of the four Autodesk Inventor files can display the **Standard** and **Collaboration** toolbars. The following describes these two toolbars and their options:

- **Standard** toolbar   This toolbar, shown in Figure 2.55, is composed of buttons that are commonly used throughout the design process and in each of the Autodesk Inventor file types. Toolbars have thin lines that separate groups of similar tools. The **Standard** toolbar has the following tool groups (some of the tools are discussed in later chapters):

  - File-handling tools   These tools include **New, Open,** and **Save,** and are the same as the **New, Open,** and **Save** menu options previously described. However, the **New** button con-

FIGURE 2.57   An example of a flyout button, the **New** button of the **Standard** toolbar.

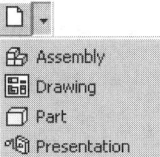

tains a flyout that enables you to quickly access a new part, assembly, drawing, or presentation file.

- **Print**   Select this button to open the **Print** dialog box and quickly print a document.

- Editing tools   These tools include the standard **Cut, Copy,** and **Paste** options previously described.

- **Undo/Redo**   The **Undo** and **Redo** buttons are exactly the same as the menu options previously discussed.

- **Parameters** tool   This tool opens the **Parameters** dialog box, which enables you to control and add model parameters.

- **iPart Author** tool   This tool creates an **iPart Factory** from the specified part. An **iPart Factory** is a part preserved on a spreadsheet with several configurations (discussed in Chapter 18).

- **Create iMate**   This tool allows you to create constraints, called **iMates**. **iMates** can define how parts go together in an assembly (discussed in Chapter 14).

- **View** tools   These tools include **Zoom All, Zoom Window, Zoom,** and **Pan** buttons. These buttons are the same as the menu options previously discussed. In addition, the **Zoom selected** button is available. This button allows you to pick a specific object and zoom in on it.

- Additional viewing tools   The **Rotate** and **Look At** commands are selected from the **Standard** toolbar. The **Rotate** button allows you to rotate an object in the graphics window. The **Look At** button zooms in on and rotates the specified object. These tools are discussed later.

- **Display** tools   Use these flyout buttons to specify the display characteristics of the model. These options are discussed in later chapters.

- Support tools   The **Help** button opens the Autodesk Inventor Help Files, discussed in Chapter 1. The **Visual Syllabus** button opens the **Visual Syllabus** dialog box. See Figure 2.58. Here you can select a specific topic, such as creating an extruded feature, and the Autodesk Inventor Help Files will display information on how to create an extruded feature. In addition, the **Standard** toolbar has direct link buttons to the **Autodesk Streamline** and **Autodesk Point A** Websites.

- **Collaboration** toolbar   The **Collaboration** toolbar, shown in Figure 2.59, is also available in each of the Autodesk Inventor files. This toolbar is discussed in later chapters.

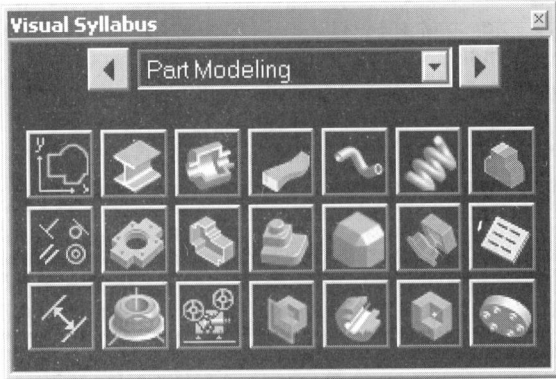

FIGURE 2.58   The Autodesk Inventor **Visual Syllabus.**

FIGURE 2.59   The **Collaboration** toolbar.

Exercise 2.9

1. Continue from Exercise 2.8, or launch Autodesk Inventor.
2. Open a new Autodesk Inventor file of you choice.
3. Locate and explore the **Standard** toolbar.
4. Using one of the techniques described, open the collaboration toolbar, and dock it by double-clicking the toolbar name.
5. Place your cursor over several toolbar buttons, and observe the tooltip and help strings provided.
6. Pick one of the flyout buttons located in the **Standard** toolbar, and pick one of the flyout options to initiate a command, and/or modify the flyout button displayed.
7. Exit Autodesk Inventor or continue working with the program.

Working with the Command Bar

The *Command* bar, shown in Figure 2.60, is similar to a toolbar but is not actually considered one. It is a very important piece of the Autodesk Inventor interface. The **Command** bar contains **Select, Sketch, Update,** and **Color** or **Style** drop-down lists. It is also home to a **Return** button and a message display box. You will become very familiar with the **Command Bar** as you work with Autodesk Inventor. The **Command Bar** and its components are discussed in future chapters.

Exercise 2.10

1. Continue from Exercise 2.9, or launch Autodesk Inventor.
2. Open a new Autodesk Inventor file of your choice.
3. Locate the **Command Bar** and explore some of the options.
4. Close the Auotdesk Inventor file without saving.
5. Exit Autodesk Inventor or continue working with the program.

Using Shortcut Keys

Autodesk Inventor *shortcut keys* allow you to initiate certain predefined commands by simply pressing a single button on your keyboard. Like many of the Autodesk Inventor interface components, particular shortcut keys are available only while in a certain work environment. The following shortcut keys are available while in the each of the four Autodesk Inventor file types (part, assembly, drawing, presentation):

- F1   **Help**
  Opens the **Show Me** dialog box. This area is similar to the tutorials discussed in Chapter 1. Here you can view a small animation that shows you how to complete a certain task.

- F2   **Pan**
  Starts the **Pan** command, which allows you to view additional areas of an object in the graphics window. (You must hold the F2 key down.)

- F3   **Zoom**
  Initiates the **Zoom** command and allows you to zoom in and out of an area. (You must hold the F3 key down.)

FIGURE 2.60   The Autodesk Inventor **Command Bar.**

- F5   **Previous view**
  Starts the **Previous view** command and allows you to return to the last display.

Only while in a part, assembly, or presentation file, the following shortcut keys are available:

- F4   **Rotate**
  Initiates the **Rotate** command and allows you to view all model areas.

- D   **Sketch/Drawing Dimensions**
  Begins the Dimensions command and allows you to place drawing dimensions on a drawing, or sketch dimensions on a sketch.

The following shortcut keys are available only while in a part or assembly file:

- E   **Extrude**
  Opens the **Extrude** dialog box and allows you to extrude a sketch.

- R   **Revolve**
  Opens the **Revolve** dialog box and allows you to revolve a sketch.

- H   **Hole**
  Displays the **Hole** dialog box and allows you to create a hole in a feature.

- S   **Sketch**
  Begins the Sketch command and allows you to create a sketch.

- L   **Line**
  Initiates the line command and allows you to sketch a line.

The following shortcut keys are available only while in an assembly file:

- P   **Place Part**
  Opens the **Open** dialog box and allows you to insert a part into an assembly.

- C   **Constraints**
  Opens the **Place Constraints** dialog box and allows you to place constraints between parts.

The following shortcut keys are available only while in a drawing file:

- O   **Ordinate Dimensions**
  Initiates the **Ordinate Dimensioning** command and allows you to place ordinate dimensions.

- B   **Balloon**
  Begins the **Balloon** command and allows you to place a balloon for an assembly drawing

- F   **Feature Control Frame**
  Initiates the **Feature Control Frame** command and allows you to place a feature control frame.

||||||||||| CHAPTER TEST

Answer the following questions on a separate sheet of paper.

1. What are sketches and what do they provide?

2. Define and describe sketched features.

3. Describe placed features.

4. Describe work features.

5. What are work planes?

6. What are work axes?

7. What are work points?

8. Describe catalog features.

9. Define design elements.

10. Explain the function of derived parts.

11. Describe pattern features.

12. Name the three types of pattern features that you can create with Autodesk Inventor.

13. Explain how parametric solid modeling works and the concept of intelligence.

14. What are constraints and dimensions?

15. Give a general definition of constraints.

16. What does it mean when a sketch is overconstrained?

17. When is a sketch underconstrained?

18. What do parameter-driven assemblies allow you to do?

19. What are adaptive parts?

20. Define interface.

21. List at least five features of the Autodesk Inventor interface.

22. When do you get a help string and where is it displayed?

23. What do menu access keys allow you to do and how do they work?

24. What do accelerator keys allow you to do and how do they work?

25. How do you know when a pull-down menu option has cascading submenus?

26. How do you use cascading submenus?

27. What do you do if you need to close the currently selected menu without opening another menu?

28. Give the function of the Locations list box in the Save As dialog box.

29. Discuss the function of the Save in: drop-down list in the Save As dialog box.

30. Explain the function of the Last button in the Save As dialog box.

31. Give the function of the Return button in the Save As dialog box.

32. Discuss the function of the Create New Folder button in the Save As dialog box.

33. Give the function of the View Menu button in the Save As dialog box.

34. Discuss the function of the File name edit box in the Save As dialog box.

35. Give the function of the Save as type drop-down list in the Save As dialog box.

36. Give the function of the Options… button in the Save As dialog box.

37. Give the function of the Files Save Options dialog box.

38. Explain the function of the Save Copy As… option in the Save As dialog box.

39. Explain the function of the Save All option in the Save As dialog box.

40. Discuss the function of the Projects option in the Save As dialog box.

41. Give the function of the Getting Started… option in the Save As dialog box.

42. Explain the function of the Print Setup option in the Save As dialog box.

43. Explain the function of the Design Assistant.

44. Give the function of the Undo option in the Edit pull-down menu.

45. Discuss the function of the Redo option in the Edit pull-down menu.

46. Give the function of the Cut option in the Edit pull-down menu.

47. Explain the function of the Copy option in the Edit pull-down menu.

48. Explain how the Paste option in the Edit pull-down menu works.

49. Give the function of the Paste Special… option in the Edit pull-down menu.

50. How do you use the Paste Special option?

51. Give the function of the Previous option in the View pull-down menu.

52. In addition to the View pull-down menu, give two ways to access the Previous View command.

53. Give the function and how to use the Pan option in the View pull-down menu.

54. Discuss the function and how to use the Zoom option in the View pull-down menu.

55. Explain the function and how to use the Zoom Window option in the View pull-down menu.

56. Give the function of the Zoom All option in the View pull-down menu.

57. Discuss the function of the Toolbar option in the View pull-down menu.

58. How do you turn the Status Bar on and off?

59. Explain how to insert a new image into your Autodesk Inventor file using the Insert Object dialog box.

60. Discuss how to insert an existing image into your Autodesk Inventor file using the Insert Object dialog box.

61. Explain the function and how to use the Organizer dialog box.

62. Give the function of the Add-In Manager dialog box.

63. Explain the function of the Document Settings dialog box.

64. Discuss the function of the Options dialog box.

65. Explain the function of the Locate Tolerance edit box in the General tab of the Options dialog box.

66. Give the function of the Number of versions to keep edit box in the General tab of the Options dialog box.

67. Give the function of the Multiuser check box in the General tab of the Options dialog box.

68. How do you get the Autodesk Inventor dialog box to not show up automatically when you launch Autodesk Inventor?

69. Give the function of the Show 3D Indicator check box in the General tab of the Options dialog box.

70. Explain the function of the Templates area in the General tab of the Options dialog box.

71. What is transcripting?

72. Give the function of the Transcripting area in the General tab of the Options dialog box.

73. Explain the function of the Workgroup Design Data area in the General tab of the Options dialog box.

74. Discuss the function of the Projects Folder area in the General tab of the Options dialog box.

75. Give the function of the Colors tab of the Options dialog box.

76. Explain the function of the Display tab of the Options dialog box.

77. Give the function of the Drawing tab of the Options dialog box.

78. Discuss the function of the Notebook tab of the Options dialog box.

79. Give the function of the Sketch tab of the Options dialog box.

80. Discuss the function of the Part tab of the Options dialog box.

81. Give the function of the iFeature tab of the Options dialog box.

82. Explain the function of the Assembly tab of the Options dialog box.

83. What does multiple document interface mean?

84. The panel bar provides you with a list of options. Inside the list of options are several design tools available for creating models in the specific work environment where you are currently working. Give the two ways that Autodesk Inventor lists the options and tools used for parts that are available in the panel bar while you are working on a part.

85. If you exit one design stage and enter another, Autodesk Inventor automatically opens a new list of options. Identify two ways to change from one list of options to another.

86. Describe the panel bar Learning mode.

87. Describe the panel bar Expert mode.

88. Many of the objects displayed in the browser may contain child nodes. Explain what child nodes are and how to display and hide them.

89. What is a tooltip and how is it activated?

90. What is a flyout toolbar button and how does it work?

91. Give a general description of the Standard toolbar and its use.

92. Briefly describe the Command bar and list its contents.

93. Explain the function of Autodesk Inventor shortcut keys.

94. Identify the shortcut key for Help.

95. Identify the shortcut key for Zoom.

96. Identify the shortcut key for Sketch/Drawing Dimensions.

97. Identify the shortcut key for the Extrude dialog box.

98. Identify the shortcut key that begins the Sketch command.

99. Identify the shortcut key that opens the Place Constraints dialog box.

|||||||||||||||||||||||||||||||||||||||||||||||||||||||||||||||||||||||||||||||

CHAPTER 3

Initial Part Modeling Drafting Techniques

LEARNING GOALS

After completing this chapter, you will be able to:

◎ Work in the sketch environment and explain specific sketch interface options.

◎ Reference and project geometry for sketching.

◎ Create sketched lines, splines, circles, ellipses, and arcs.

◎ Sketch rectangles and polygons.

◎ Place fillets and chamfers on sketch geometry.

◎ Place sketch points and hole centers.

◎ Mirror, offset, and pattern sketch geometry.

◎ Constrain sketches using geometric constraints and dimensions.

◎ Use the **Precise Input** toolbar to place sketch geometry.

◎ Use the **Parameters** tool for sketches.

◎ Edit sketches.

◎ Insert AutoCAD and Mechanical Desktop files into Autodesk Inventor for use in sketches.

◎ Use the Sketch Doctor.

Developing Part Sketches

Part sketches are very important Autodesk Inventor components and usually represent the first step in the creation of a model. Sketches provide the foundation, pattern, and profile for developing models that contain sketched features; they are used in part and assembly files.

Sketches are generally easy to make and should not require very much time to complete. You may want to use the following concepts when developing sketches:

■ You may find it helpful to project the **Center Point,** located in the **Origin** folder, before you sketch any geometry. Begin your sketch from the projected center point to allow the sketch to become fully defined, and ensure the sketch is referenced at a particular point.

■ Initial sketch geometry is usually approximate and contains very few constraints. The idea is to draw the basic shape or outline of the sketch, quickly and easily, and then add dimensions and geometric constraints.

■ Fully close a sketch loop when developing a profile. If a gap or opened loop exists, you will not be able to create some features from the sketch.

■ Develop simple, often incomplete sketch geometry, and create as many objects as possible in the feature environment. For example, place a chamfer on a feature using feature tools,

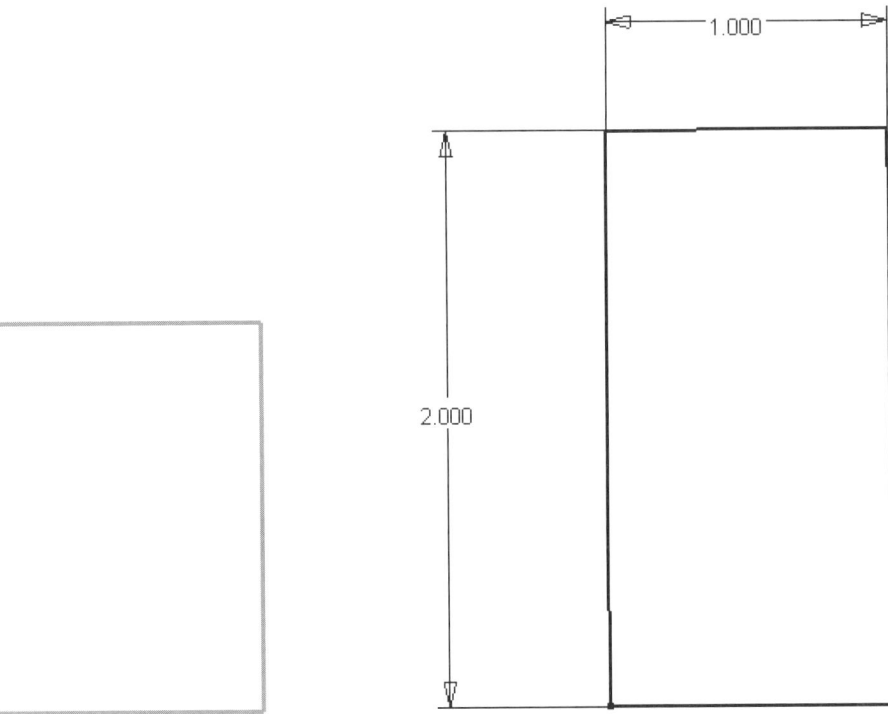

FIGURE 3.1   Initial sketch using the sketching tools.

FIGURE 3.2   Constrained sketch.

instead of drawing and dimensioning a chamfer in the sketch environment, using sketch tools.

- Fully constrain your sketch if applicable, and do not allow the sketch to become over constrained.

Typically, the first step for creating a sketch involves drawing a simple sketch profile using sketch tools such as **Line, Circle, Arc,** and **Rectangle**. See Figure 3.1. Next, you constrain the sketch using dimensions or geometric constraints. See Figure 3.2. These are the only two steps involved in sketch development. However, for some sketches you may want to use additional sketch tools to create more geometry, or edit existing sketch data.

Working in the Sketch Environment and Interface

The work environment and interface are different for each stage of the design process. Some of the tools and interface components remain constant or are similar throughout the entire design process. However, the sketch environment and many of the interface components are specific to working with sketches.

Pull-Down Menu System Options

Most of the pull-down menus discussed in Chapter 2 apply to the sketch environment. However, there are a few specific options that relate only to sketches:

- **Document Settings...**   Located in the **Tools** pull-down menu, the **Document Settings...** option opens the **Document Settings** dialog box. This dialog box contains a **Units** tab, a **Sketch** tab, and a **Modeling** tab. The **Units** tab, shown in Figure 3.3, is used throughout the design process, but should be set initially for sketch development. This tab allows you to specify unit type and precision. In addition, you can define how to display sketch dimensions in the graphics window by selecting either the **Display as value, Display as name,** or **Display as expression** radio button.

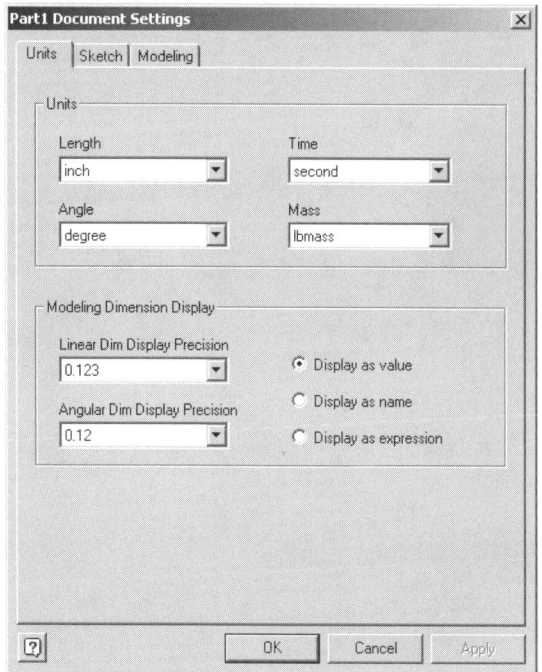

FIGURE 3.3 The **Units** tab of the **Document Settings** dialog box.

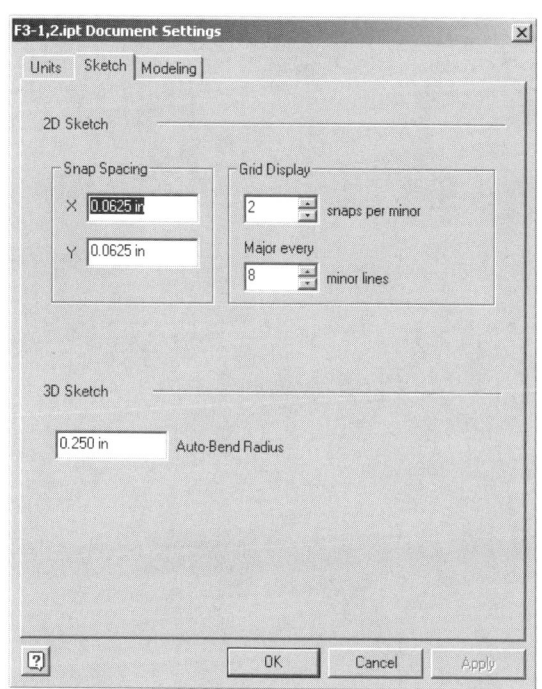

FIGURE 3.4 The **Sketch** tab of the **Document Settings** dialog box.

The **Sketch** tab, shown in Figure 3.4, is specific to the sketch environment. This tab allows you to adjust your snap spacing and grid display for a 2D sketch, and the Auto-Bend Radius for a 3D sketch.

■ **Application Options…** Located in the **Tools** pull-down menu, the **Application Options…** selection opens the **Options** dialog box. This dialog box contains a **Sketch** tab and a **Part** tab, specific to the sketch environment. The **Sketch** tab of the **Options** dialog box, shown in Figure 3.5, contains a **2D Sketch** and a **3D Sketch** section with the following areas:

■ **Constraint Placement Priority** Autodesk Inventor has the ability to automatically create some designated constraints if applied. This area allows you to give constraint priority to parallel and perpendicular constraints instead of the grid system, or the horizontal and vertical grid system instead of existing sketch geometry.

■ **Display** This area allows you to turn on or off grid lines, axes, and the coordinate system indicator.

■ **Overconstrained Dimensions** This section houses two radio buttons that allow you to apply overconstrained dimensions to a sketch or warn you if a sketch is overconstrained. If you select the **Apply overconstrained dimensions to sketch** radio button, a nonparametric, driven dimension is created. If you select the **Warn of Overconstrained Condition** radio button, a warning is displayed if a sketch will become overconstrained.

Field Notes

Usually it is most helpful to pick the **Warn of Overconstrained Condition** radio button to ensure a sketch does not become overconstrained and nonparametric.

■ **Snap to Grid** Picking this check box allows you to snap directly to a grid point when sketching geometry or locating a point.

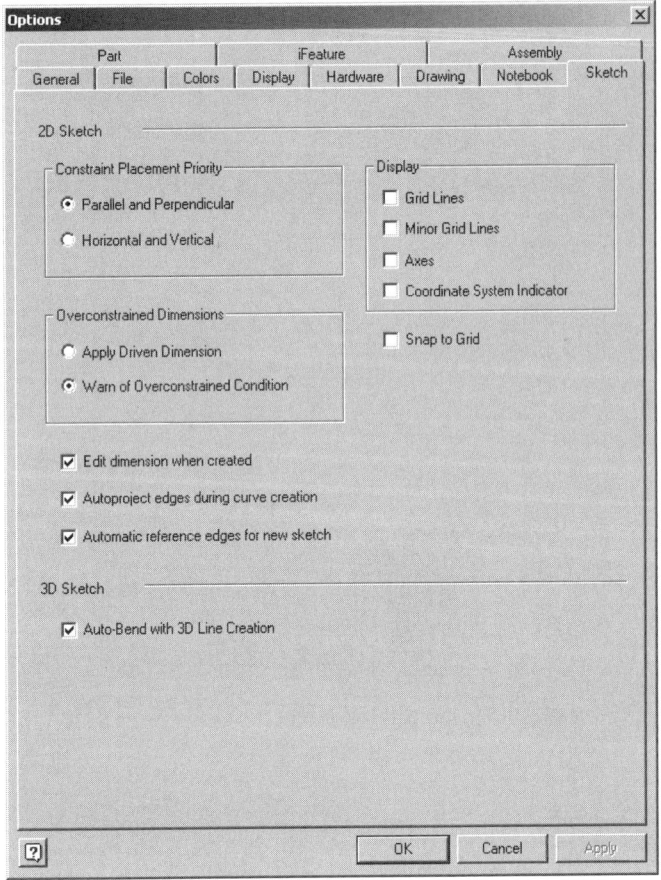

FIGURE 3.5   The **Sketch** tab of the **Options** dialog box.

FIGURE 3.6   The **Edit Dimension** dialog box.

- **Edit dimension when created check box**   If you pick this check box, an **Edit Dimension** dialog box appears. See Figure 3.6. This is a valuable tool that allows you to enter a specific dimension for the object when you place a dimension.

- **Autoproject edges during curve creation**   Selecting this check box automatically projects existing geometry from a sketch or feature onto a new sketch.

- **Automatic reference edges for new sketch**   Selecting this check box automatically projects the referenced feature face geometry edges to the new sketch.

- **Auto-Bend with 3D Line Creation**   When creating lines in a 3D sketch environment, selecting this check box creates an arc or corner bend at the corners of lines.

The **Part** tab of the **Options** dialog box, shown in Figure 3.7, contains the following options:

- **Sketch on New Part Creation**   When you develop a new part file, a sketch is automatically created on a designated plane. Inside the **Sketch on New Part Creation** area you can specify which plane the sketch is created. You can select the X-Y, Y-Z, or X-Z planes. An alternative is to pick the **No new sketch** radio button and not have a sketch automatically appear.

- **Parallel view on sketch creation**   Pick this check box if you want to automatically rotate the current view to a two-dimensional profile view when you begin a sketch. For example, when this check box is selected and you are currently in an isometric view, the display reorients to show the two-dimensional profile view of the sketch or the plane you are going to sketch on.

FIGURE 3.7 The **Part** tab of the **Options** dialog box.

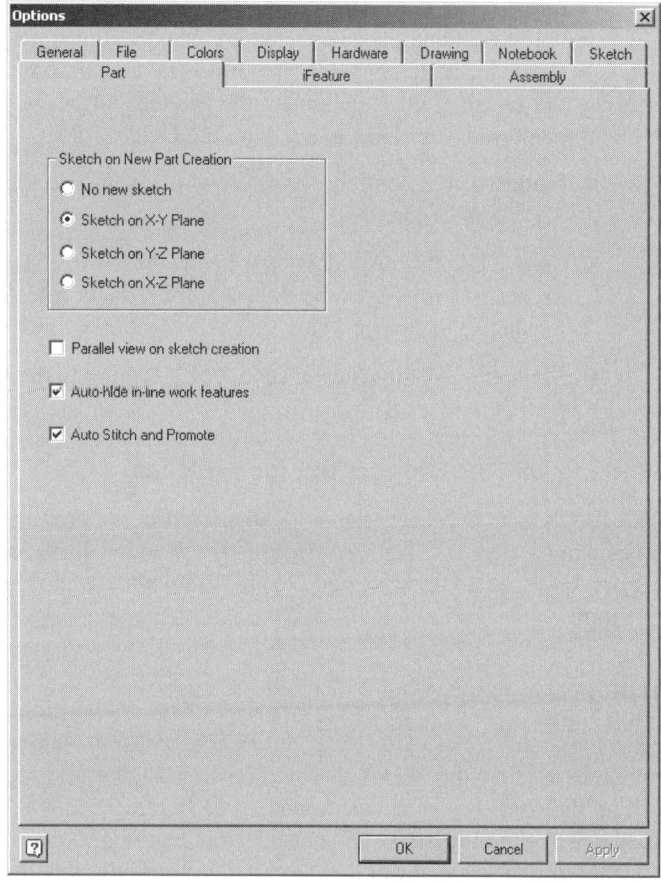

■ **Auto-hide in-line work features**   Selecting this check box initiates the **Auto-hide in-line work features** command and automatically hides a consumed work feature.

■ **Auto Stitch and Promote**   Select this check box to automatically stitch and promote solids, which are typically inserted as STEP files. Solids and STEP files are discussed in Chapter 11.

Using the Command Bar for Sketches

The command bar, shown in Figure 3.8, contains some important sketch environment tools. Each of these tools is discussed as follows:

■ **Select**   By default the **Select** button is set to **<u>Sketch Priority.</u>** This allows you to pick objects while in sketch mode.

■ **Return**   Picking this button finishes the current sketch and takes you from sketch mode to feature mode.

■ **Sketch**   This button is selected when you create a new part, and you can instantly begin developing sketch geometry. This button also activates the sketch environment, if not cur-

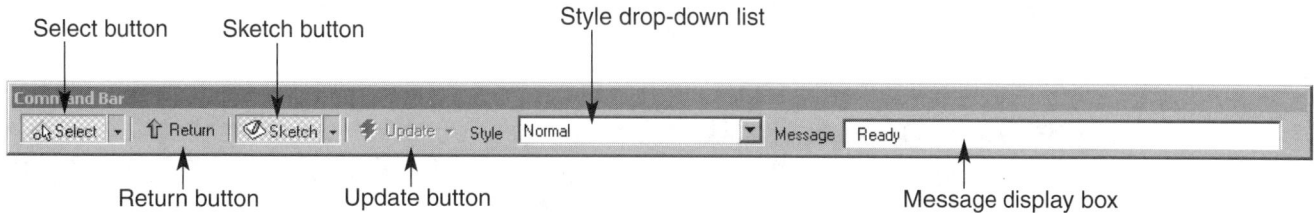

FIGURE 3.8   The command bar (sketch mode).

rently active, and allows you to choose from a 2D or 3D sketch. To activate a sketch using the **Sketch** button, select a feature plane and pick the **Sketch** button, or pick the **Sketch** button and pick the feature plane. If no feature exists, pick the plane you would like to use from the **Origin** folder and select the **Sketch** button, or pick the **Sketch** button and select the plane you would like to use from the **Origin** folder.

- ■ **Update**   This button allows you to regenerate the current part or subassembly and its dependent children.

- ■ **Style**   Use this drop-down list to change the line style of sketch geometry. You may choose to specify a line style and then draw objects, or select existing objects and pick the style you would like to change.

- ■ **Message**   The message display box is similar to the status bar discussed in Chapter 2.

Working with the Browser Bar for Sketches

The sketch icon and sketch name are displayed in the browser bar. See Figure 3.9. A sketch used to create an existing feature is called a ***consumed sketch*** and is displayed in the browser bar as shown in Figure 3.10. Figure 3.9 shows an unconsumed sketch. An ***unconsumed sketch*** is one that has not yet been used to create a feature.

Using Sketching Tools

Sketching tools are used to create, dimension, constrain, and manipulate geometry for sketches. These tools are accessed from the **Sketch** toolbar, shown in Figure 3.11, or from the **Sketch** panel bar. When a sketch is active, the panel bar is in sketch mode. See Figure 3.12. By default, the panel bar is displayed in learning mode, as seen in Figure 3.8. To change from learning to expert mode, seen in Figure 3.13 and discussed in Chapter 2, right-click any of the panel bar tools or left-click the panel bar title and select the **Expert** option.

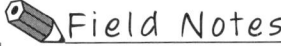

> When creating sketch geometry, you may want to use the coordinates and measurements displayed in the right of the status bar. These can help you draw your initial sketch and understand geometry placement. However, they do not constrain or dimension a sketch.

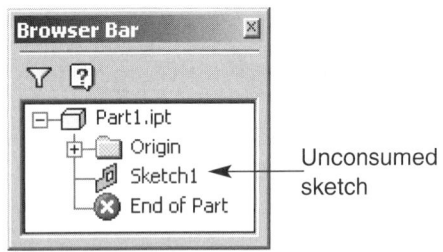

FIGURE 3.9 An unconsumed sketch in the browser bar.

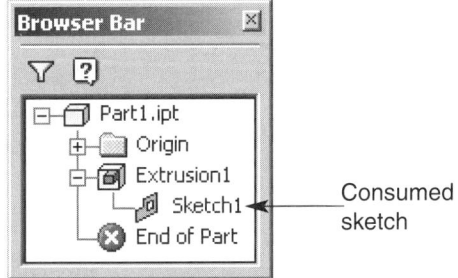

FIGURE 3.10 A consumed sketch in the browser bar.

FIGURE 3.11 The **Sketch** toolbar.

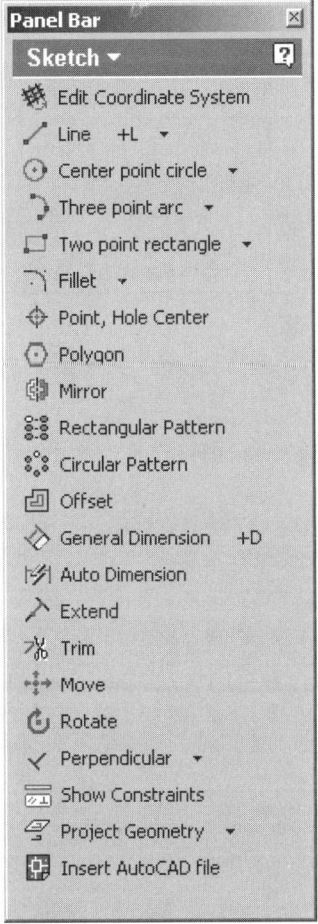

FIGURE 3.12 The panel bar in sketch mode (learning mode).

FIGURE 3.13 The panel bar in sketch mode (expert mode).

Exercise 3.1

1. Launch Autodesk Inventor.
2. Open a new Autodesk Inventor part file, and explore the sketch interface options previously discussed.
3. Open the **Document Settings** dialog box and manipulate some of the values.
4. Close the part file without saving.
5. Exit Autodesk Inventor or continue working with the program.

Referencing and Projecting Geometry

Autodesk Inventor provides you with a few different options for using existing geometry for sketch development. Many of the reference and projection options involve using feature edges and are discussed in future chapters. However, you can reference the work features available in the **Origin** folder of the **Browser** bar. As discussed in Chapter 2, projecting the center point, or 0,0,0 origin, onto the current sketch plane is an effective way to begin a sketch and ensure the sketch becomes fully constrained. Figure 3.14A shows a rectangle that appears to be fully constrained, but is still actually floating in space. Figure 3.14B shows the same rectangle referenced to the projected center point. The rectangle is now fully constrained.

The process for projecting the work features available in the **Origin** folder is exactly the same as projecting feature edges. First, select the **Project Geometry** button on the panel bar or **Sketch** toolbar. Then

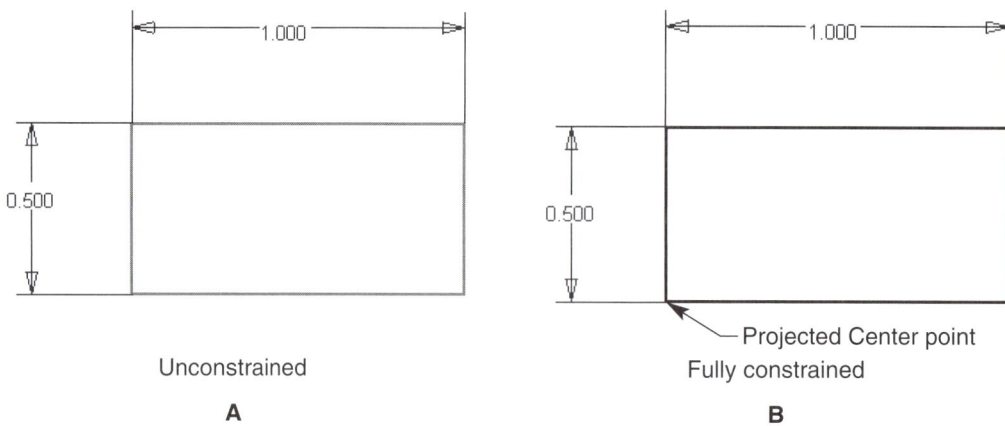

FIGURE 3.14   Projecting the center point onto the current sketch plane to fully constrain a sketch.

simply pick the geometry or work feature you would like to reference. For example, to project the center point, access the **Project Geometry** command and pick **Center Point,** located in the **Origin** folder of the **Browser** bar.

 Field Notes

> Using the **Fix** constraint tool, discussed later in this chapter, also secures a point to a position in space. However, this technique does not necessarily fix any point to a work feature such as the center point.

Exercise 3.2

1. Continue from Exercise 3.2, or launch Autodesk Inventor.
2. Open a new Autodesk Inventor part file.
3. Project the center point, located in the **Origin** folder.
4. Save the part file as EX3.2.
5. Exit Autodesk Inventor or continue working with the program.

Creating Lines and Splines

Lines and splines are used to draw simple geometric sketches. To access the **Line** command, use one of the following techniques:

✓ Pick the **Line** button on the panel bar.

✓ Pick the **Line** button on the **Sketch** toolbar.

✓ Type the **+** and **L** keys on your keyboard.

To draw a line, simply pick the point where you want to begin the line, and then pick the next point. See Figure 3.15. If you want to continue the line, pick the next point. If you do not want to continue the line, press the **Esc** key on your keyboard, or right-click and pick the **Done** option. If you do not want to continue the current line, but would like to draw another line at another location, either close the current **Line** command and access it again, or right-click and select **Restart** from the shortcut menu.

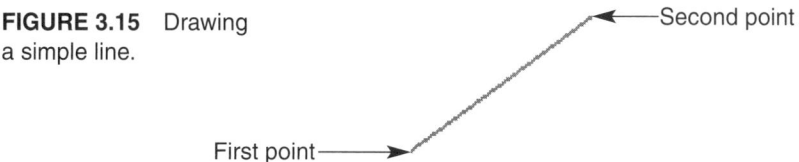

FIGURE 3.15 Drawing
a simple line.

If you would like to draw a line that references existing sketch geometry, you have the following options:

■ **Endpoint**   Use this option to draw a line to or from an existing endpoint. Access the line command and move the cross and yellow dot over the apparent endpoint of an existing object. When the yellow dot turns green, gets larger, and the endpoint, or coincident, icon is visible, pick the point. See Figure 3.16.

■ **Midpoint**   Use this option to draw a line to or from an existing midpoint. First, access the line command and move the cross and yellow dot over the apparent midpoint of an existing object. When the yellow dot turns green, gets larger, and the midpoint icon is visible, pick the point. See Figure 3.17. An alternative method of locating the midpoint is to right-click and select the **Midpoint** option from the shortcut menu, and pick the object, not the midpoint, you would like to draw the line to or from.

■ **Center point**   Use this option to draw a line to or from an existing circle, arc, fillet, or other radius. First, access the line command and move the cross and yellow dot over the apparent center of an existing round object. When the yellow dot turns green, gets larger, and the coincident icon is visible, pick the point. An alternative method of locating the center point is to right-click and select the **Center** option from the shortcut menu, and pick the object, not the center point, you would like to draw the line to or from. See Figure 3.18.

■ **Intersection**   Use this option to draw a line to or from an existing intersection of two objects. First, access the line command and move the cross and yellow dot over the apparent intersection of the existing objects. When the yellow dot turns into a yellow cross and the intersection icon appears, pick the point. See Figure 3.19. An alternative method of locating the intersection is to right-click and select the **Intersection** option from the shortcut menu, and then pick the two intersecting objects you would like to draw the line to or from.

■ **AutoProject**   Use this option to draw a line from a point in space to a referenced point on an existing object. First, access the line command and move your cursor and yellow dot over, or near, the existing object you would like to draw a line to or from. Then move the cursor away from the object (a dotted line should trail your cursor). Once you are far enough away from the existing geometry, and the dotted line is still visible, pick the point. Finally, pick your second point. See Figure 3.20. An alternative method for locating the AutoProject point is to right-click and select the **AutoProject** option from the shortcut menu and then follow the steps specified earlier.

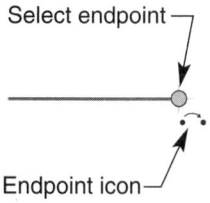

FIGURE 3.16   Using the **Endpoint** selection tool.

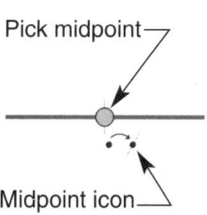

FIGURE 3.17   Using the **Midpoint** selection tool.

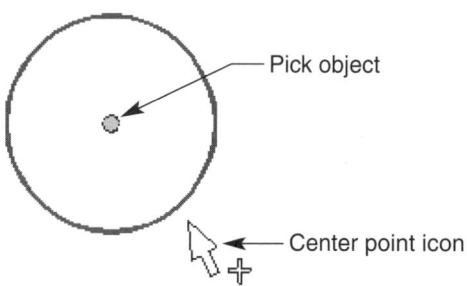

FIGURE 3.18   Using the **Center point** selection tool.

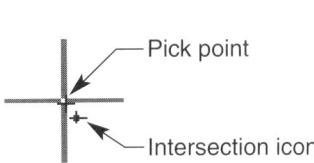

FIGURE 3.19 Using the **Intersection** selection tool.

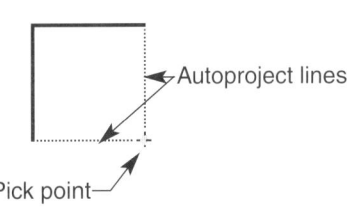

FIGURE 3.20 Using the **AutoProject** selection tool.

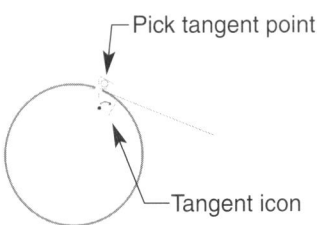

FIGURE 3.21 Using the tangent option to draw a line tangent to an existing radius.

- **Tangent**   Use this option to draw an object from the tangent edge of an existing circle, arc, fillet, or other radius. First, access the line command and move your cursor and yellow dot over, or near, the existing radius edge you would like to draw a line to or from. When you are close to tangent, the tangent icon, shown in Figure 3.21, appears. When you see the tangent icon, pick the point and continue drawing your line or cancel the command.

- **Coincident**   Often when you move your cursor over existing geometry while in the line command, the coincident icon appears. See Figure 3.22. When this icon appears, the point you are about to specify as the start or endpoint of the line is coincident, or exactly corresponds to some point on the existing geometry.

Field Notes

If you do not want to use any of the automatic referencing tools previously discussed, hold down the **Ctrl** key on your keyboard while in the **Line** command.

You may have already noticed the **Line** option of the **Sketch** panel bar is a flyout button that contains the **Spline** command. This command is also accessed from the **Line** flyout button of the **Sketch** toolbar. A *spline* is similar to a line, but creates a single line with rounded corners instead of angled corners. To draw a spline, pick the point where you want to begin the spline, then pick the next point and right-click and pick the **Continue** menu option, or double-click the last point. See Figure 3.23.

To create a more complicated spline, pick more points followed by right-clicking and choosing the **Continue** menu option, or double clicking the last point. See Figure 3.24.

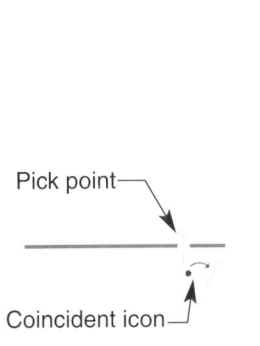

FIGURE 3.22 Using the coincident, precise position option to draw a line coincident to an existing line.

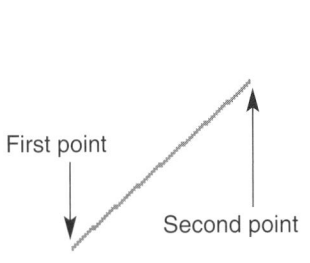

FIGURE 3.23 A simple two-point spline.

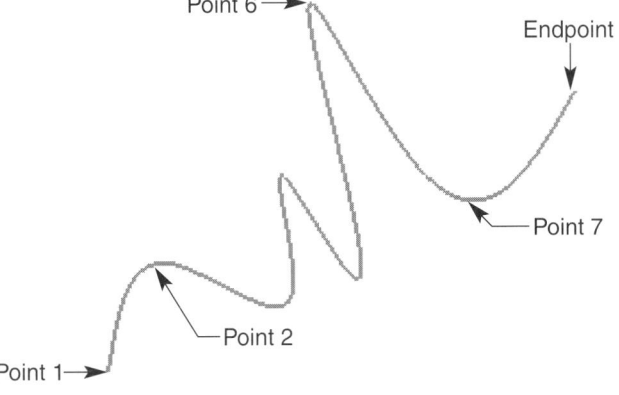

FIGURE 3.24 A multiple-point spline.

If you would like to sketch another spline, either close the current **Spline** command and access it again, or right-click and select **Restart** from the shortcut menu. Then repeat the process as previously described. If you do not want to continue the spline command, press the **Esc** key on your keyboard, right-click and pick the **Done** option, or pick another tool. Once you have sketched a spline, the following menu options are available when you right-click on the spline:

- **Fit Method**   Pick this option to access a cascading submenu with three menu options. Select the **Smooth** option to display a spline with smooth curves, the **Sweat** option to display a spleen with the smoothest curves, or the **AutoCAD** option to define AutoCAD's spline fit technique.

- **Insert Point**   Pick this option to add another point to the spline. Once you access this command, pick a point on an existing spline to insert the additional spline joint.

- **Close Spline**   Select this option to close an open spline, in which the start point and the endpoint do not connect.

- **Display Curvature**   When you choose this menu option, curvature lines are added to the spline to illustrate the curvature of the spline. If you select a different fit method from the **Fit Method** cascading submenu, you may also want to pick the **Display Curvature** option to help visualize the fit.

Field Notes

Each of the techniques used to reference existing geometry with the **Line** tool apply to the **Spline** tool. Though you may pick several times to create a spline, only one object is created.

Exercise 3.3

1. Continue from Exercise 3.2, or launch Autodesk Inventor.
2. Open a new Autodesk Inventor part file.
3. Create the lines in the order shown, with the specified autoconstrain tools.
4. Save the part file as EX3.3.
5. Exit Autodesk Inventor or continue working with the program.

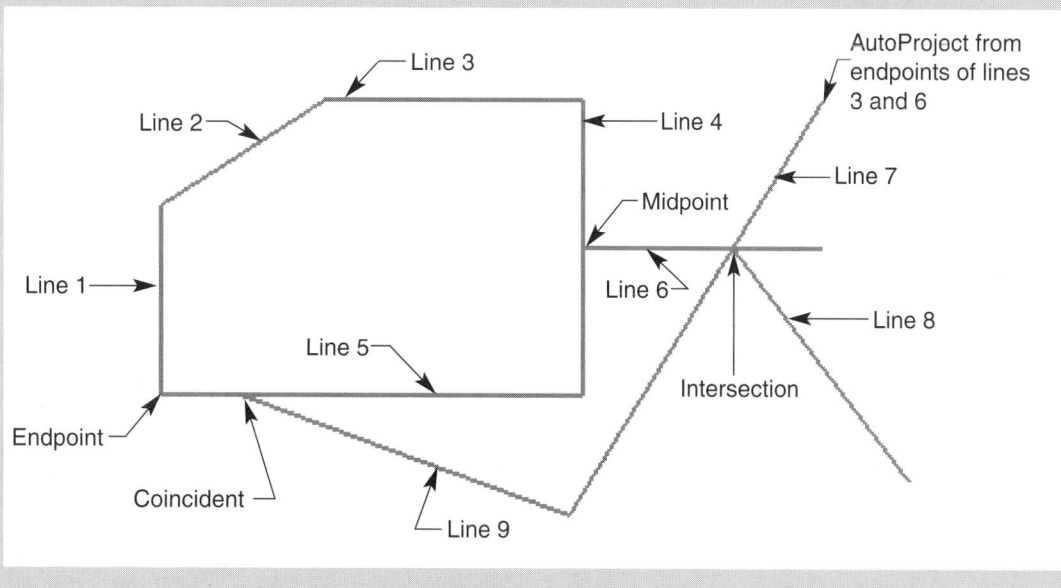

Exercise 3.4

1. Continue from Exercise 3.3, or launch Autodesk Inventor.
2. Open a new Autodesk Inventor part file.
3. Sketch a spline with 2 points, and a spline with 7 points.
4. Right-click on the splines and explore the menu options discussed.
5. Add spline points to the existing splines.
6. Save the part file as EX3.4.
7. Exit Autodesk Inventor or continue working with the program.

Exercise 3.5

1. Continue from Exercise 3.4, or launch Autodesk Inventor.
2. Open a new Autodesk Inventor inch part file.
3. Open a new sketch on the **XY Plane.**
4. Project the **Center Point.**
5. Sketch the lines shown, with the **Line** tool.
6. Save the part file as EX3.5.
7. Exit Autodesk Inventor or continue working with the program.

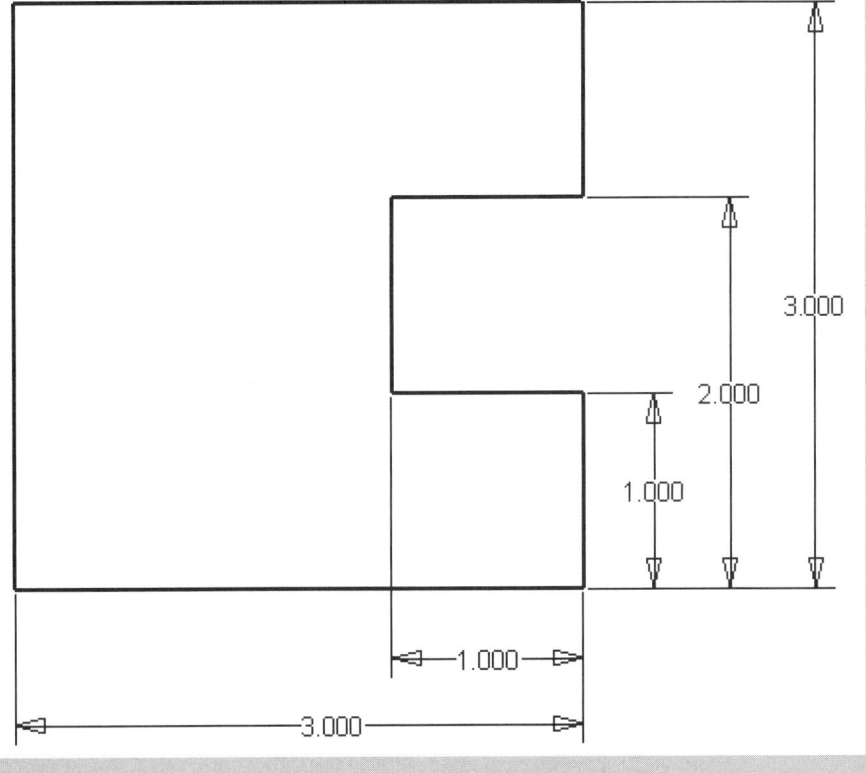

Creating Circles and Ellipses

Circles and ellipses are used to draw circular geometric sketches. The default circle option is the center point circle. To access the **Center Point Circle** command, use one of the following techniques:

✓ Pick the **Center Point Circle** button on the panel bar.

✓ Pick the **Center Point Circle** button on the **Sketch** toolbar.

To draw a center point circle, pick the point you would like to designate as the center of the circle, then drag the circle away from the center and pick the endpoint. See Figure 3.25. If you would like to draw another circle, pick a new point. If you do not want to draw another circle, press the **Esc** key on your keyboard, right-click and pick the **Done** option, or pick another tool.

The default **Center Point Circle** is a flyout button that contains options to draw a tangent circle or an ellipse. To draw a tangent circle, you must have three reference objects available, such as lines, to define the circle. If these objects are present, access the **Tangent Circle** command from the circle flyout button on the panel bar or **Sketch** toolbar. Then pick the three existing reference objects. A circle is automatically placed between the objects. See Figure 3.26. If you would like to draw another tangent circle, pick a new group of objects. If you do not want to draw another circle, press the **Esc** key on your keyboard, right-click and pick the **Done** option, or pick another tool.

An *ellipse* is similar to a circle, but contains a major and a minor axis. While a circle is perfectly round, an ellipse is more of an oval shape. To draw an ellipse, access the **Ellipse** command from the circle flyout button on the panel bar or **Sketch** toolbar. First, specify the center point of the ellipse. Then, choose the first axis endpoint, followed by the second axis endpoint. See Figure 3.27. If you would like to draw another ellipse, pick a new center point. If you do not want to draw another ellipse, press the **Esc** key on your keyboard, right-click and pick the **Done** option, or pick another tool.

Field Notes

Each of the previously discussed techniques used to reference existing geometry with the **Line** tool apply to the **Center Point Circle, Tangent Circle,** and **Ellipse** tools. Use these referencing options to select the center point and outside radius/endpoint of your radius and axes. If you do not want to utilize any of these automatic referencing tools, hold down the **Ctrl** key on your keyboard while in the **Circle** command.

If you begin to draw a circle or ellipse at an incorrect location, right-click and select the **Restart** option from the shortcut menu while the command is still active. Then pick where you would like to replace the first point.

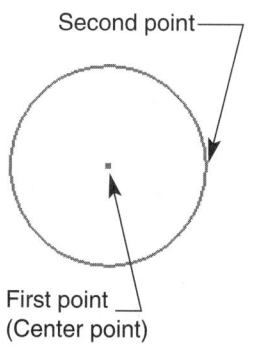

FIGURE 3.25 Drawing a circle using the **Center Point Circle** button.

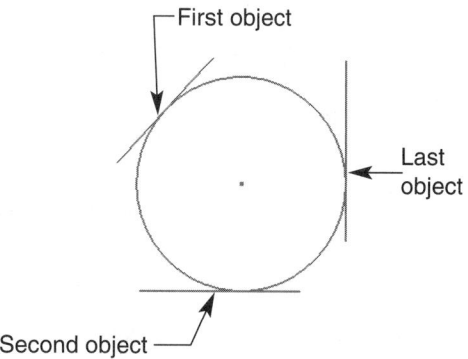

FIGURE 3.26 Drawing a circle using the **Tangent Circle** button.

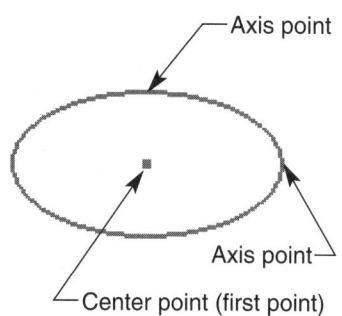

FIGURE 3.27 Drawing an ellipse.

Exercise 3.6

1. Continue from Exercise 3.5, or launch Autodesk Inventor.
2. Open a new Autodesk Inventor part file.
3. Sketch Circle 1 using the **Center Point Circle** tool.
4. Sketch the following lines using the specified tools.
5. Create Circle 2 using the **Tangent Circle** tool.
6. Save the part file as EX3.6.
7. Exit Autodesk Inventor or continue working with the program.

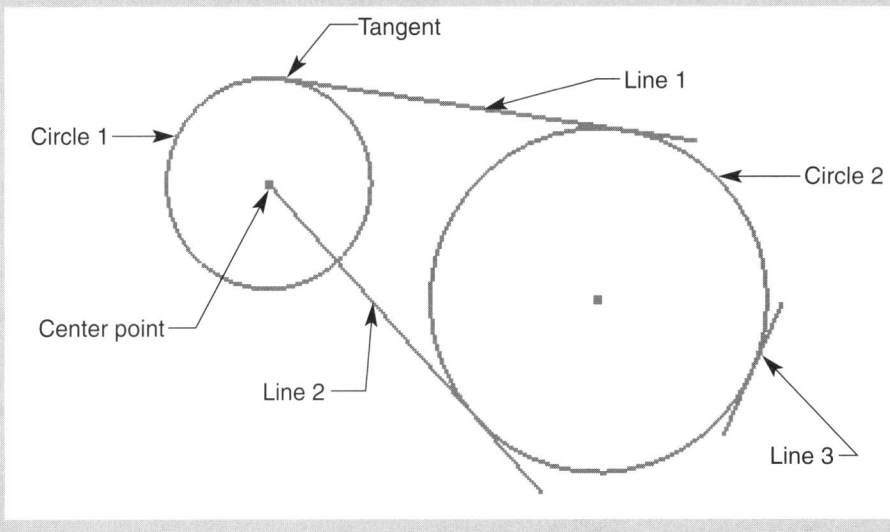

Creating Arcs

Autodesk Inventor provides you with three different ways to draw arcs. The default arc option is the three point arc. To access the **Three point arc** command, use one of the following techniques:

✓ Pick the **Three point arc** button on the panel bar.

✓ Pick the **Three point arc** button on the **Sketch** toolbar.

To draw a three point arc, pick the first point of the arc, followed by the last point (endpoint), and finally the midpoint of the arc. See Figure 3.28. If you want to draw another three point arc, pick a new point. If you do not want to draw another three point arc, press the **Esc** key on your keyboard, right-click and pick the **Done** option, or pick another tool.

Picking the arc flyout button allows you to access the other two arc commands. One of the options is the center point arc. Access the **Center point arc** command from the arc flyout button on the panel bar or **Sketch** toolbar. To create an arc using this technique, first pick the center point of the arc radius. Then drag the dashed line out and specify the first arc endpoint. Finally, drag the center line out and specify an endpoint for the arc. See Figure 3.29. If you would like to draw another three point arc, pick a new group of objects. If you do not want to draw another three point arc, press the **Esc** key on your keyboard, right-click and pick the **Done** option, or pick another tool.

The third method for drawing an arc is to use the tangent arc option. To draw a tangent arc, you must have two reference objects available, such as lines, to define the arc. If these objects are present, access the **Tangent arc** command from the arc flyout button on the panel bar or **Sketch** toolbar. Then pick the existing reference objects. An arc is automatically placed between the objects. See Figure 3.30. If you would like to draw another tangent arc, pick a new group of objects. If you do not want to draw another arc, press the **Esc** key on your keyboard, right-click and pick the **Done** option, or pick another tool.

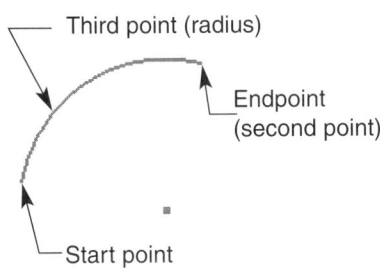

FIGURE 3.28 Drawing a three point arc.

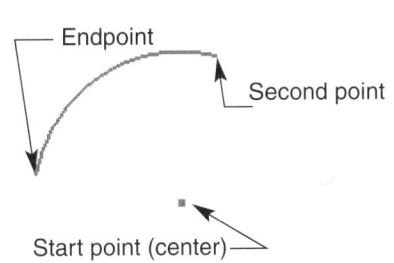

FIGURE 3.29 Drawing an arc using the **Center point arc** button.

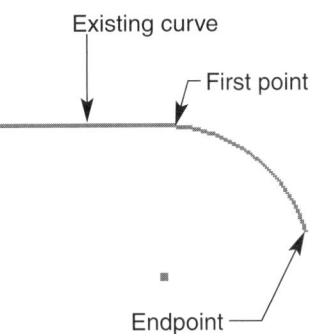

FIGURE 3.30 Drawing an arc using the **Tangent arc** button.

Field Notes

Each of the previously discussed techniques used to reference existing geometry with the **Line** tool apply to the **Three point arc, Center point arc,** and **Tangent arc** tools. Use these referencing options to select the center point and outside radius/endpoint of your radius and axes. If you do not want to use any of these automatic referencing tools, hold down the **Ctrl** key on your keyboard while in the **Arc** command.

If you begin to draw an arc at an incorrect location, right-click and select the **Restart** option from the shortcut menu while the command is still active. Then pick where you would like to replace the first point.

Exercise 3.7

1. Continue from Exercise 3.6, or launch Autodesk Inventor.
2. Open a new Autodesk Inventor part file.
3. Create Arc 1 using the **Three point arc** tool.
4. Sketch Line1 from the endpoint of Arc 1.
5. Sketch Arc 2 using the **Center point arc** tool. Locate the center point at the start point of Line 1, and the second point at the start point of Arc 1.
6. Create Arc 3 using the **Tangent arc** tool. Located the first point at the endpoint of Line 1, and the second point at the endpoint of Arc 2.
7. Save the part file as EX3.7.
8. Exit Autodesk Inventor or continue working with the program.

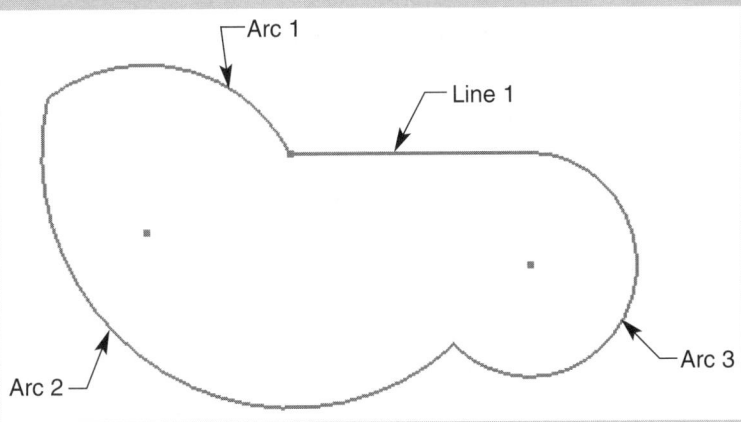

FIGURE 3.31 Using
the line command to
create an arc.

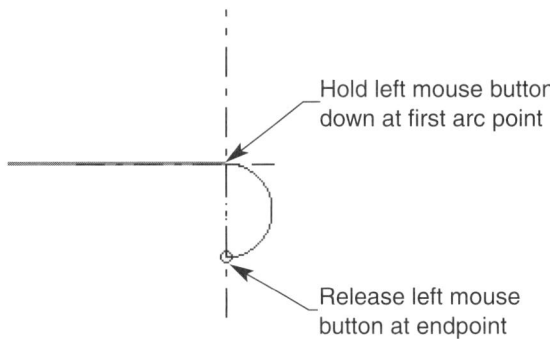

Hold left mouse button
down at first arc point

Release left mouse
button at endpoint

Using the Line Command to Create Arcs

Autodesk Inventor contains a powerful tool that allows you to create arcs while in the **Line** command. To use this tool, access the **Line** command by picking the **Line** button from the panel bar or **Sketch** toolbar. First, create a line or number of lines, but do not exit the **Line** command. While still in the **Line** command, move your cursor over the last specified point (the dot should turn gray). When the dot is gray, hold down the left mouse button, move the arc endpoint to its location and release the left mouse button. See Figure 3.31. If you would like to draw another arc, use the same method by moving your cursor over the arc endpoint. You can also continue to draw a straight line from the arc. If you do not want to draw another arc or line, press the **Esc** key on your keyboard, right-click and pick the **Done** option, or pick another tool.

Exercise 3.8

1. Continue from Exercise 3.7, or launch Autodesk Inventor.
2. Open a new Autodesk Inventor part file.
3. Create the following sketch using the techniques previously discussed.
4. Save the part file as EX3.8.
5. Exit Autodesk Inventor or continue working with the program.

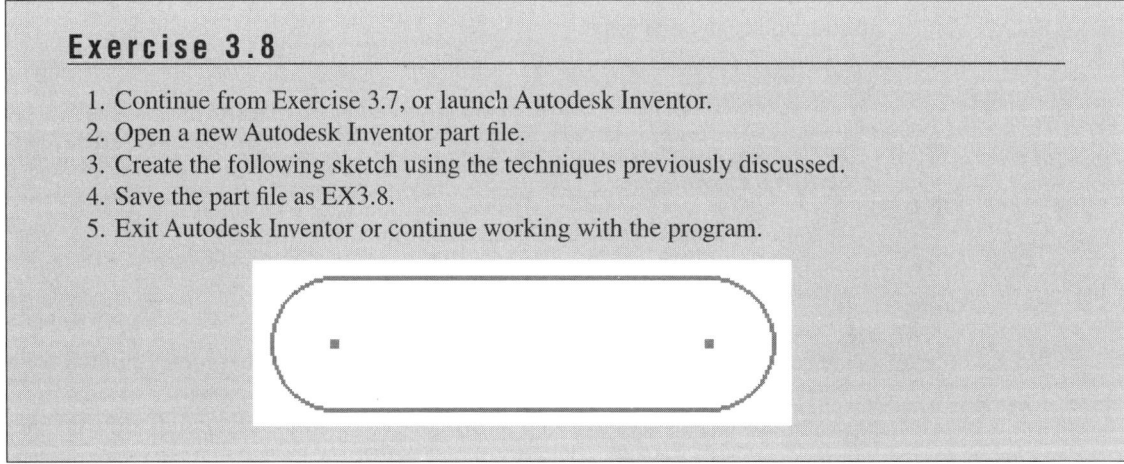

Sketching Rectangles

Autodesk Inventor provides you with two different ways to draw rectangles. The default rectangle option is the two point rectangle. To access the **Two point rectangle** command, use one of the following techniques:

✓ Pick the **Two point rectangle** button on the panel bar.

✓ Pick the **Two point rectangle** button on the **Sketch** toolbar.

To draw a two point rectangle, pick the first point of the rectangle, followed by the opposite, diagonal corner. See Figure 3.32. If you would like to draw another two point rectangle, pick a new point. If you do not want to draw another two point rectangle, press the **Esc** key on your keyboard, right-click and pick the **Done** option, or select another tool.

Picking the rectangle flyout button allows you to access the second rectangle drawing tool: three point rectangle. Access the **Three point rectangle** command from the rectangle flyout button on the panel bar or **Sketch** toolbar. To create a rectangle using this technique, pick a starting corner for the rectangle. Then drag the line and cursor out, and specify the second rectangle point. Finally, drag the rectangle out and

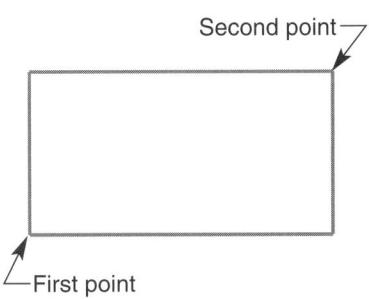

FIGURE 3.32 Drawing a two point rectangle.

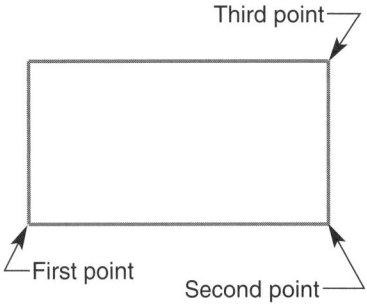

FIGURE 3.33 Drawing a three point rectangle.

specify an endpoint for the width or length. See Figure 3.33. If you would like to draw another tangent circle, pick a new group of objects. If you do not want to draw another circle, press the **Esc** key on your keyboard, right-click and pick the **Done** option, or select another tool.

Field Notes

> Though you can create a rectangle using the **Line** command, it is often easier to use one of the rectangle options, because it only requires two or three picks, and the parallel, and perpendicular constraints are already added.
>
> If you begin to draw a rectangle at an incorrect location, right-click and select the **Restart** option from the shortcut menu while the command is still active. Then pick where you would like to replace the first point.

Exercise 3.9

1. Continue from Exercise 3.8, or launch Autodesk Inventor.
2. Open a new Autodesk Inventor part file.
3. Create a rectangle using the **Two point rectangle** tool.
4. Create a rectangle using the **Three point rectangle** tool.
5. Save the part file as EX3.9.
6. Exit Autodesk Inventor or continue working with the program.

Placing Sketch Fillets and Chamfers

As previously discussed, rarely do you want to apply geometry such as fillets or chamfers to a sketch. Often it is easier to place a fillet or chamfer on a feature rather than a sketch. In addition, complicated sketch geometry that includes objects such as fillets and chamfers may become difficult to handle as you progress through different design stages. Still, you do have the ability to fillet or chamfer a sketched corner, if necessary.

To access the **Fillet** command, pick the **Fillet** button on the panel bar or **Sketch** toolbar. Selecting the **Fillet** button opens the **2D Fillet** dialog box, shown in Figure 3.34. To create a fillet, first specify a radius in the drop-down list by typing in a radius or picking a radius from the list. Then select the **Equal** button if you would like to create fillets with equal radii without adding additional dimensions. Next, you can see a preview of the fillet by moving your cursor to the corner of the lines, or by selecting one line and moving your cursor over the next line. If the fillet looks acceptable in the preview, select the intersecting corners of the lines or the two lines that you would like to fillet.

FIGURE 3.34
Creating a two-dimensional sketch fillet.

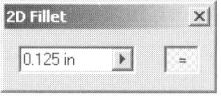

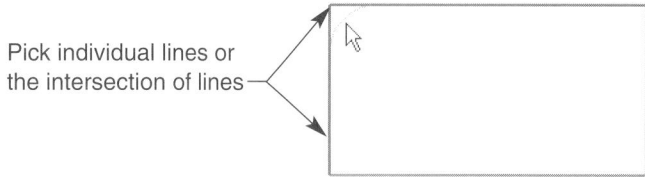

Pick individual lines or the intersection of lines

Field Notes

You can also create a fillet by first selecting the intersecting corner of two lines, or by selecting one of the lines and then accessing the **Fillet** command.

To access the **Chamfer** command, pick the **Chamfer** button from the **Fillet/Chamfer** flyout button on the panel bar or **Sketch** toolbar. Selecting the **Chamfer** button opens the **2D Chamfer** dialog box, shown in Figure 3.35.

The following options are located inside the **Chamfer** dialog box:

■ **Create dimensions**   If you pick this button, Autodesk Inventor will place dimensions on the chamfer when you create it, to indicate the chamfer size and help constrain the sketch.

■ **Equal**   Use this button when you have already developed a chamfer in the current chamfer command instance, and would like subsequent chamfers to contain dimensions equal to the first.

■ **Equal distance**   This button allows you to set the two chamfer dimensions to the same value and creates a 45° chamfer with the specified distance. To set the distance, select or type a distance in the **Distance** drop-down list.

■ **Unequal distance**   Pick this button to create a chamfer with two different chamfer distances. This chamfer option creates a chamfer with a less than or greater than 45° angle. To set the distances, select or type a distance in the **Distance1** drop-down list and the **Distance2** drop-down list.

■ **Distance and angle**   This button allows you to specify a distance for the chamfer to begin from the intersection of the first selected line, and an angle that intersects the second selected line. To set the distance and angle, select or type a distance in the **Distance** drop-down list, and select or type an angle in the **Angle** drop-down list.

To create a chamfer, access the **Chamfer** tool, and specify your options in the **2D Chamfer** dialog box. Next, you can see a preview of the chamfer by moving your cursor to the corner of the lines, or by selecting one line and moving your cursor over the next line. If the chamfer looks acceptable in the preview,

FIGURE 3.35   The **2D Chamfer** dialog box.

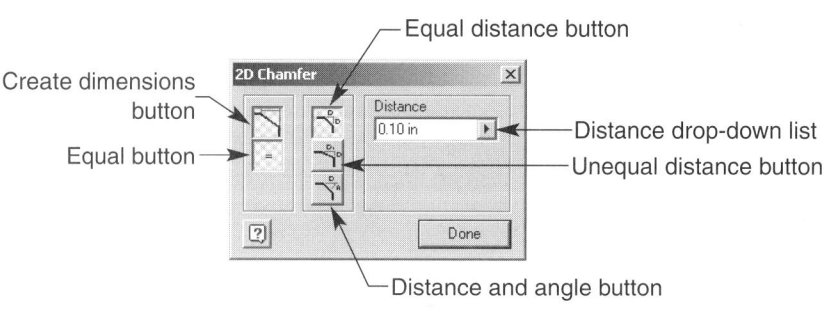

Create dimensions button

Equal button

Equal distance button

Distance drop-down list

Unequal distance button

Distance and angle button

FIGURE 3.36
Creating a two-
dimensional sketch
chamfer.

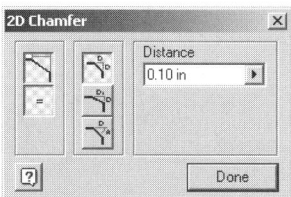

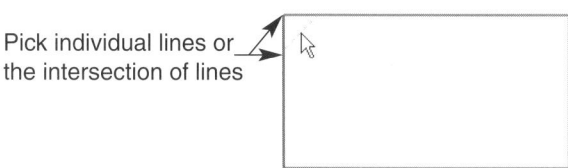

Pick individual lines or
the intersection of lines

select the intersecting corners of the lines or the two lines that you would like to chamfer. See Figure 3.36. For unequal distance chamfers, the first line selected corresponds to **Distance2,** while the second line selected corresponds to **Distance1.** Similarly, for distance and angle chamfers, the first line selected corresponds to **Angle,** while the second line selected corresponds to **Distance.**

Exercise 3.10

1. Continue from Exercise 3.10, or launch Autodesk Inventor.
2. Open EX3.9.
3. Save a copy of EX3.9 as EX3.10.
4. Close EX3.9 without saving, and open EX3.10.
5. Fillet all four corners of one of the rectangles, using your own desired radii. Pick the **Equal** button, and fillet two of the rectangle corners using the same radius. Unpick the **Equal** button, and fillet the third corner to the same radius. Fillet the last corner of the rectangle using a different radius.
6. Chamfer all four corners of the other rectangles using your own desired chamfer distances and angles. Pick the **Equal** and **Equal distance** buttons, and chamfer two of the rectangle corners using the same distance. Unpick the **Equal** button, and pick the **Unequal distance** button. Chamfer the third corner of the rectangle using two different distances. Use the **Distance and angle** button to chamfer the last rectangle corner to a desired distance and angle.
7. Resave EX3.10.
8. Exit Autodesk Inventor or continue working with the program.

Placing Sketch Points and Hole Centers

Use the **Point, Hole Center** tool to create sketch points or to place a hole center point. *Sketch points* are used for construction purposes and help you develop sketch geometry. *Hole centers* are used to define the center of a hole, and allow you to create hole feature when you are working in a feature creation environment.

The default **Point, Hole Center** tool style is a hole center. A hole center must have some type of reference feature. As a result, hole centers are placed on a feature face sketch. To place a hole center, pick the **Point, Hole Center** button on the panel bar or **Sketch** toolbar. Then pick the spot you would like to place the hole center point. See Figure 3.37. If you would like to draw another hole center, pick a new point. As seen in Figure 3.38, a dotted line indicates the new hole center is horizontally or vertically aligned with another hole center or sketch point. If you do not want to draw another hole center, press the **Esc** key on your keyboard, right-click and pick the **Done** option, or choose another tool.

In order to change the default **Point, Hole Center** tool style from a hole center to a sketch point, you must specify **Sketch Point** from the **Style** drop-down list of the **Command Bar.** See Figure 3.39. To specify **Sketch Point** from the **Style** drop-down list, either access the **Point, Hole Center** tool and then select **Sketch Point,** or pick an existing hole center and select **Sketch Point.** Sketch points are placed in a sketch exactly the same as a hole center.

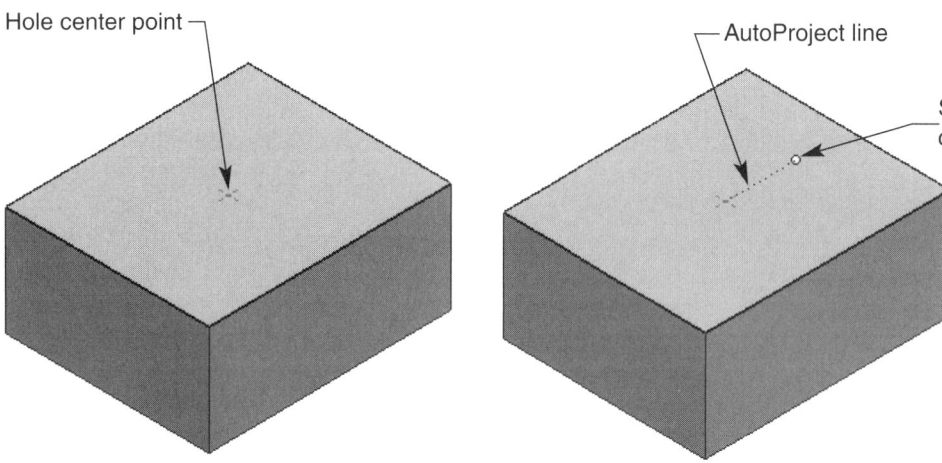

Hole center point

AutoProject line

Second hole center point

Style Hole Center
Hole Center
Sketch Point

FIGURE 3.37 An example of a hole center point on a feature face sketch.

FIGURE 3.38 Aligning an additional hole center or sketch point with an existing hole center or sketch point.

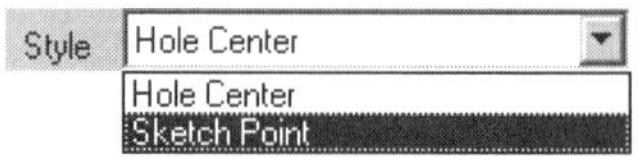

FIGURE 3.39 Specify **Sketch Point** or **Hole Center** from the **Style** drop-down list of the **Command Bar.**

 Field Notes

Each of the previously discussed techniques used to reference existing geometry with the **Line** tool applies to the **Hole Center** and **Sketch Point** tools. Use these referencing options to place a hole center or sketch point precisely in reference to existing geometry. If you do not want to use any of these automatic referencing tools, hold down the **Ctrl** key on your keyboard while in the **Point, Hole Center** command.

Exercise 3.11

1. Continue from Exercise 3.10, or launch Autodesk Inventor.
2. Use the **Point, Hole Center** tool to place several points anywhere in sketch space.
3. Close the part file without saving.
4. Exit Autodesk Inventor or continue working with the program.

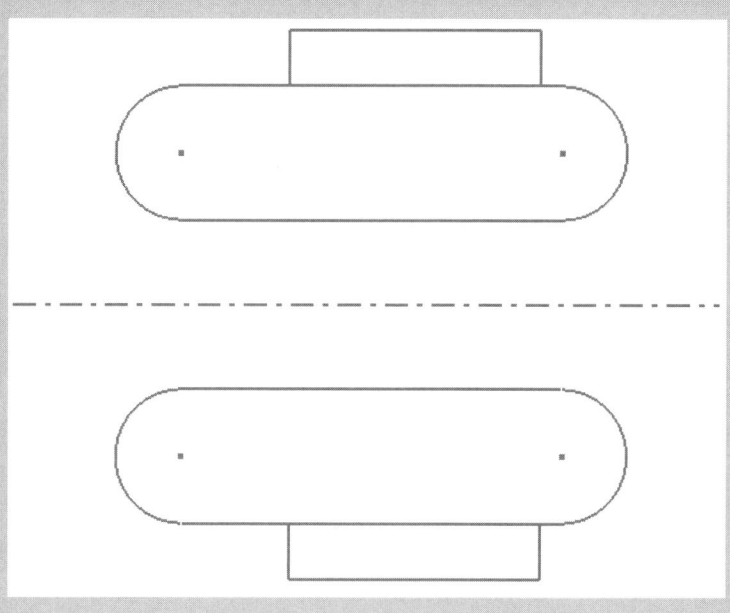

Sketching Polygons

A *polygon* is a geometric shape with three or more sides, such as a triangle, square, or hexagon. The **Polygon** tool is used to sketch polygons. Although you can sketch a four-sided polygon, such as a square or rectangle, using the **Polygon** sketch tool, the **Polygon** sketch tool is typically used to place a shape with three, or five or more sides, because Autodesk Inventor provides you with two different ways to draw rectangles. To access the **Polygon** command, use one of the following techniques:

✓ Pick the **Polygon** button on the panel bar.

✓ Pick the **Polygon** button on the **Sketch** toolbar.

Selecting the **Polygon** tool opens the **Polygon** dialog box as shown in Figure 3.40.

Use the **Polygon** dialog box to specify how you want to create the polygon by selecting either the **Inscribed** or **Circumscribed** button. An inscribed polygon is measured from the polygon corners, while a circumscribed polygon is measured from the polygon flats. Then define the type of polygon you want to sketch, by entering or selecting the number of sides from the **Number of sides** drop-down list.

Once you have established the creation method and number of polygon sides in the **Polygon** drop-down list, pick the center point of the polygon. Finally, select the edge of the shape. If you chose the **Inscribed** button, your cursor and the second point you pick is attached to the corner of the polygon. If you chose the **Circumscribed** button, your cursor and the second point you pick is attached to the flats of the polygon. See Figure 3.41. After you sketch a polygon, the **Polygon** command remains open allowing you to continue creating polygons. If you do not want to draw another polygon, press the **Esc** key on your keyboard, right-click and pick the **Done** option, choose the **Done** button on the **Polygon** dialog box, or select another tool.

 Field Notes

> Each of the previously discussed techniques used to reference existing geometry with the **Line** tool apply to the **Polygon** tool. Use these referencing options to select the center point and outside radius/endpoint of your radius and axes. If you do not want to utilize any of these automatic referencing tools, hold down the **Ctrl** key on your keyboard while in the **Circle** command.

Exercise 3.12

1. Continue from Exercise 3.11, or launch Autodesk Inventor.
2. Open a new Autodesk Inventor part file.
3. Create a 3-, 6-, and 12-sided polygon, using the **Polygon** tool. Experiment with the **Inscribed** and **Circumscribed** options.
4. Save the part file as EX3.12.
5. Exit Autodesk Inventor or continue working with the program.

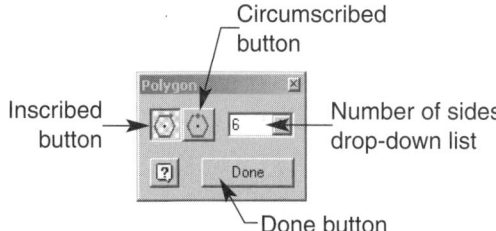

FIGURE 3.40   The **Polygon** dialog box.

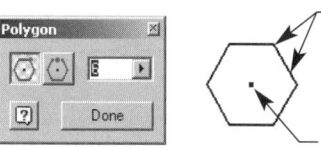 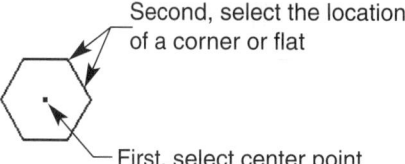

FIGURE 3.41   Creating a two-dimensional sketched polygon.

Mirroring Sketch Geometry

The **Mirror** command allows you to create a mirrored copy of sketch geometry over a centerline. In order to mirror an object, you must first draw a centerline in addition to the geometry you would like to mirror. See Figure 3.42.

The centerline acts as the line of symmetry between the existing objects and mirrored copies of the objects. Centerlines are drawn exactly the same as the lines discussed earlier in this chapter. In order to change a normal or construction line to a centerline, you must choose **Centerline** from the **Style** drop-down list of the **Command Bar.** See Figure 3.43. To specify **Centerline** from the **Style** drop-down list, either access the **line** tool and then select **Centerline,** or pick an existing geometry and select **Centerline.**

To mirror an object or group of objects, access the **Mirror** command by picking the **Mirror** button on the panel bar or **Sketch** toolbar. Then select all the objects you would like to mirror. Finally, pick the centerline to mirror the selected objects. See Figure 3.44.

If you would like to mirror additional sketch geometry, repeat the steps discussed. If you are finished mirroring, press the **Esc** key on your keyboard, right-click and pick the **Done** option, or pick another tool.

 Field Notes

You can also mirror geometry by picking the centerline first and then selecting the objects. However, if you use this technique, you will only be able to pick one object at a time.

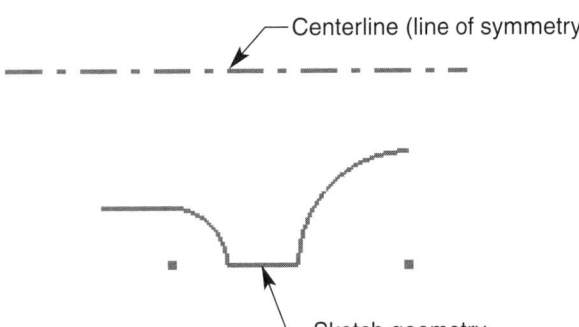

FIGURE 3.42 An object or objects and a centerline are required for mirroring sketch geometry.

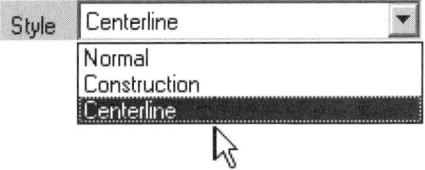

FIGURE 3.43 Changing line styles with the **Style** drop-down list of the **Command Bar.**

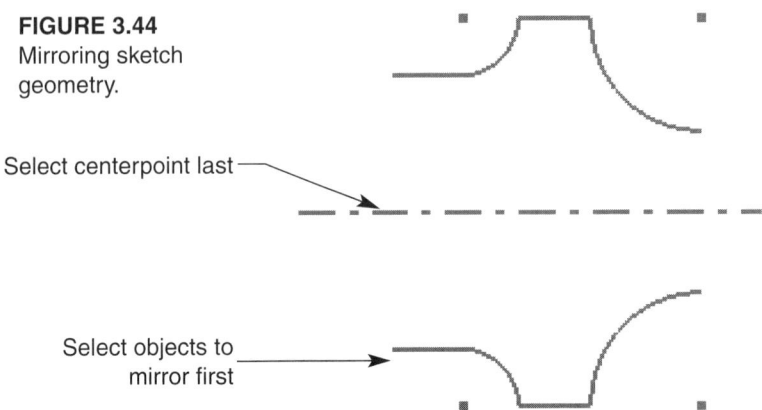

FIGURE 3.44 Mirroring sketch geometry.

Exercise 3.13

1. Continue from Exercise 3.12, or launch Autodesk Inventor.
2. Open EX3.8.
3. Save a copy of EX3.8 as EX3.13.
4. Close EX3.8 without saving, and open EX3.13.
5. Right-click the name of the sketch (Sketch 1 by default), and select the **Edit Sketch** menu option.
6. Sketch a rectangle on the existing object as shown.
7. Create a centerline above the sketched objects as shown.
8. Mirror the objects as shown.
9. Resave EX3.13.
10. Exit Autodesk Inventor or continue working with the program.

Patterning Sketch Geometry

Autodesk Inventor provides you with three tools that allow you to pattern sketch geometry. A *sketched pattern* is created any time you place multiple arranged copies, or patterns, of sketches. Although typically features, not sketches, are patterned, you may find it helpful to pattern sketched shapes for some applications. The **Mirror** tool previously discussed is one type of patterning command. In addition, you can array sketches in a rectangular fashion using the **Rectangular Pattern** tool, and in a circular fashion using the **Circular Pattern** tool. See Figure 3.45.

Using the Rectangular Pattern Tool

Rectangular sketch patterns are created using the **Rectangular Pattern** tool. Access the **Rectangular Pattern** command using one of the following techniques:

✓ Pick the **Rectangular Pattern** button on the **Sketch** panel bar.

✓ Pick the **Rectangular Pattern** button on the **Sketch** toolbar.

The **Rectangular Pattern** dialog box is displayed when you access the **Rectangular Pattern** command. See Figure 3.46.

To create a rectangular pattern of a feature or series of features, first pick the **Geometry** button if it is not already active. Then select the sketch geometry you want to array. Once you have selected all the shapes you want to pattern, pick the **Direction 1** button, located in the **Direction 1** area, and select the direction of

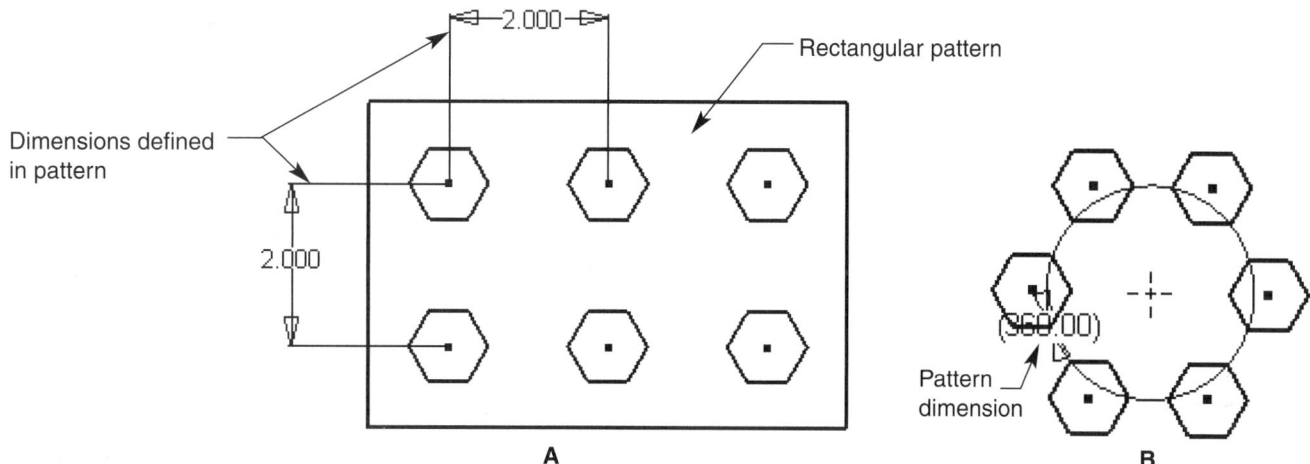

FIGURE 3.45   An example of (A) a rectangular sketch pattern and (B) a circular sketch pattern.

FIGURE 3.46   The
Rectangular Pattern
dialog box.

Geometry button —

Direction 1 area —

Direction 1 button —

Flip direction button —

Count drop-down list —

Spacing drop-down list —

Suppress button —

OK button —

— Direction 2 area

— Flip direction button

— Direction 2 button

— Count drop-down list

— Spacing drop-down list

— More button

— Associative check box

— Fitted check box

the row or column of patterned geometry. You can select any edge, axis, or work axis available in the model. When you choose a direction, an arrow shows you the specified pattern path, and a preview displays the pattern operation. See Figure 3.47A. If the direction is not correct, pick the **Flip** button to reverse the direction.

When you have defined the direction of the first pattern, specify how many copies of the sketch or series of sketch shapes you want to create by entering or selecting a count value from the **Count** drop-down list. Then define the distance between the copies by entering or selecting a spacing value from the **Spacing** drop-down list. Spacing is measured by the width of the selected features and the distance between the copies. It is not just the space between patterns. For example, if you want to pattern a rectangle that is 1″ wide, and you want a ½″ space between pattern copies, you must specify a 1½″ spacing. Another way to understand spacing is the distance from a point on the parent geometry to the corresponding point on each pattern occurrence. For example, if you create a rectangular pattern of circles, and you want the distance between the circle centers to be 10mm, specify a spacing of 10mm.

Once you have fully defined **Direction 1,** repeat the steps previously described for **Direction 2.** Pick the **Direction 2** button, located in the **Direction 2** area, and select the desired direction. If the first direction defined the rows, **Direction 2** will define the columns. If the first direction defined the columns, **Direction**

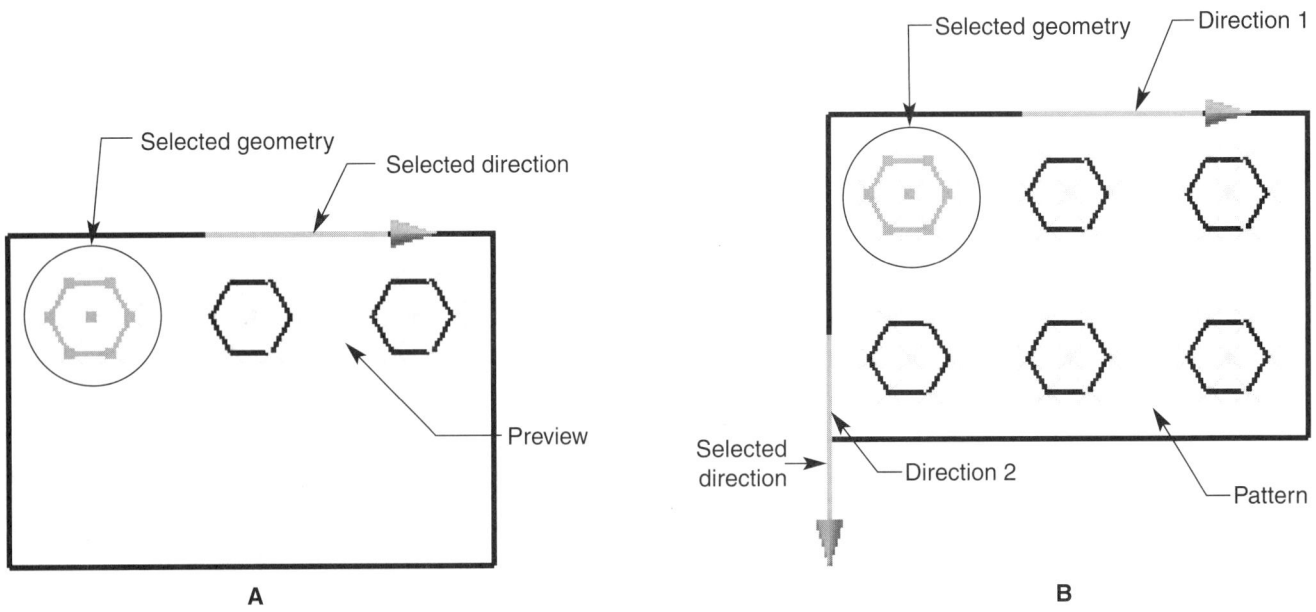

FIGURE 3.47   (A) The pattern direction arrow and pattern preview for a rectangular pattern. (B) The second pattern direction arrow and pattern preview.

2 will define the rows. You can select any edge, axis, or work axis available in the model. Theoretically you can choose the same edge or axis you chose for **Direction 1** or a parallel edge or axis. However, for most applications, you should select an edge or axis that is perpendicular, or at least nonparallel to the first direction. When you choose the second direction, an arrow shows you the specified pattern path, and a preview displays the pattern operation. See Figure 3.47B. If the direction is not correct, pick the **Flip** button to reverse the direction.

When you have defined the direction of the second pattern, specify how many copies of the shapes or series of shapes you want to create by entering or selecting a count value from the **Count** drop-down list. Then define the distance between the copies by entering or selecting a spacing value from the **Spacing** drop-down list.

After you have selected the sketch geometry and specified the direction, count, and spacing information, you may want to change the rectangular pattern creation method by picking the **More** button and accessing the following options:

- **Suppress**   Pick this button and then select pattern occurrences that you do not want to be part of the pattern.

- **Associative**   Select this check box if you want the pattern to be associative. This means that when changes are made to the part, the pattern automatically updates to reflect the new geometry.

- **Fitted**   When you select this check box, the spacing characteristics previously discussed are overridden. The pattern occurrences are equally fitted within the specified spacing. The space will no longer define the distance between points on each of the pattern occurrences.

Finally, pick the **OK** button to generate the rectangular pattern.

Using the Circular Pattern Tool

Circular sketch patterns are created using the **Circular Pattern** tool. Access the **Circular Pattern** command using one of the following techniques:

✓ Pick the **Circular Pattern** button on the **Sketch** panel bar.

✓ Pick the **Circular Pattern** button on the **Sketch** toolbar.

The **Circular Pattern** dialog box is displayed when you access the **Circular Pattern** command. See Figure 3.48.

To create a circular pattern of a sketch or series of sketch shapes, pick the **Geometry** button if it is not already active. Then select the geometry you want to array. When all the sketched shapes you want to pattern are selected, pick the **Rotation Axis** button, and select the axis of rotation. The axis of rotation is the center, or pivot point, which selected geometry orbits around. You can select any edge, axis, or work axis available in the model. When you choose a rotation axis, an arrow shows you the specified pattern path, and a preview displays the pattern operation. See Figure 3.49. If the direction is not correct, pick the **Flip** button to reverse the direction.

FIGURE 3.48   The **Circular Pattern** dialog box.

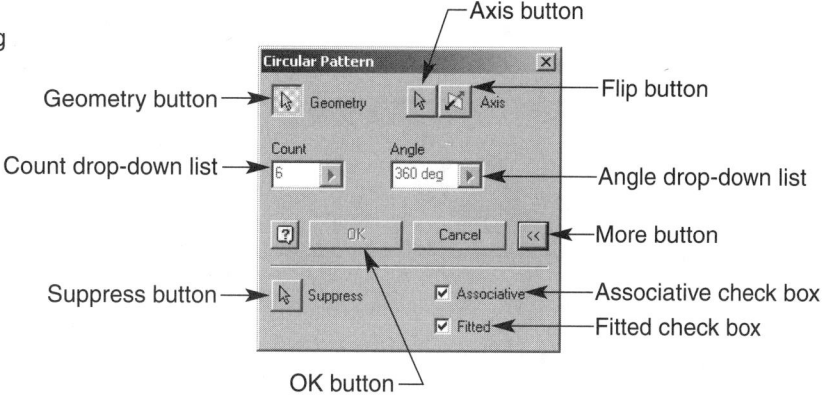

FIGURE 3.49 The pattern direction arrow and pattern preview for a circular pattern.

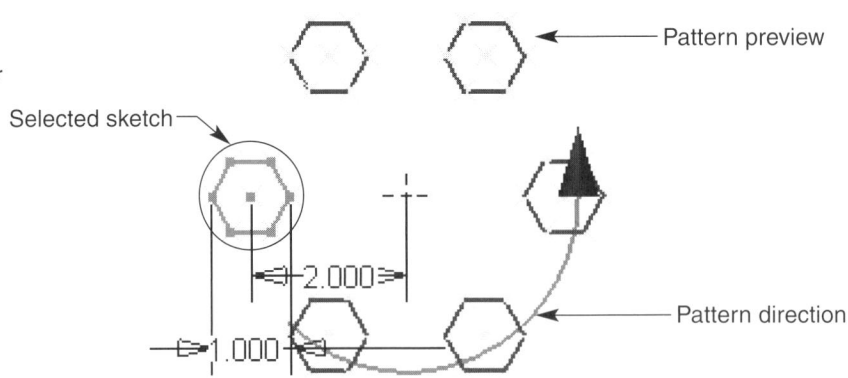

Once you have selected the geometry and defined the axis of rotation, specify how may copies of the sketch you want to create by entering or selecting a count value from the **Count** drop-down list. Then you may want to change the circular pattern creation method by picking the **More** button and accessing the following options:

- **Suppress**　Pick this button and then select pattern occurrences that you do not want to be part of the pattern.

- **Associative**　Select this check box if you want the pattern to be associative. This means that when changes are made to the part, the pattern automatically updates to reflect the new geometry.

- **Fitted**　When you select this check box, the pattern occurrences are equally fitted within the specified angle. This check box allows you to specify the distance between features by defining the total amount of rotation. The angle between each occurrence is defined by the number of occurrences and the entire revolution angle of all the pattern features. A 360° angle is a full rotation.

Finally, pick the **OK** button to generate the circular pattern.

Field Notes

> You will notice that dimensions that correspond to the pattern specifications are automatically placed when you pattern sketch geometry.

Exercise 3.14

1. Continue from Exercise 3.13, or launch Autodesk Inventor.
2. Open a new inch part file.
3. Open a sketch on the XY plane, and sketch the polygon shown.
4. Add a **Point, Hole Center.**
5. Access the **Circular Pattern** tool, and pattern the polygon. Select the **Point, Hole Center** as the axis, enter a count of 6, and an angle of 360°.
6. Suppress the pattern occurrence shown, and pick the **OK** button to generate the pattern.
7. Access the **Rectangular Pattern** tool, and pattern the specified polygon. Select the directions shown, enter a count of 3 and a spacing of 1″ for the first direction, and a count of 4 and a spacing of .5″ for the second direction.
8. Suppress the pattern occurrences shown, and pick the **OK** button to generate the pattern.
9. Save the part as EX3.14.
10. Exit Autodesk Inventor or continue working with the program.

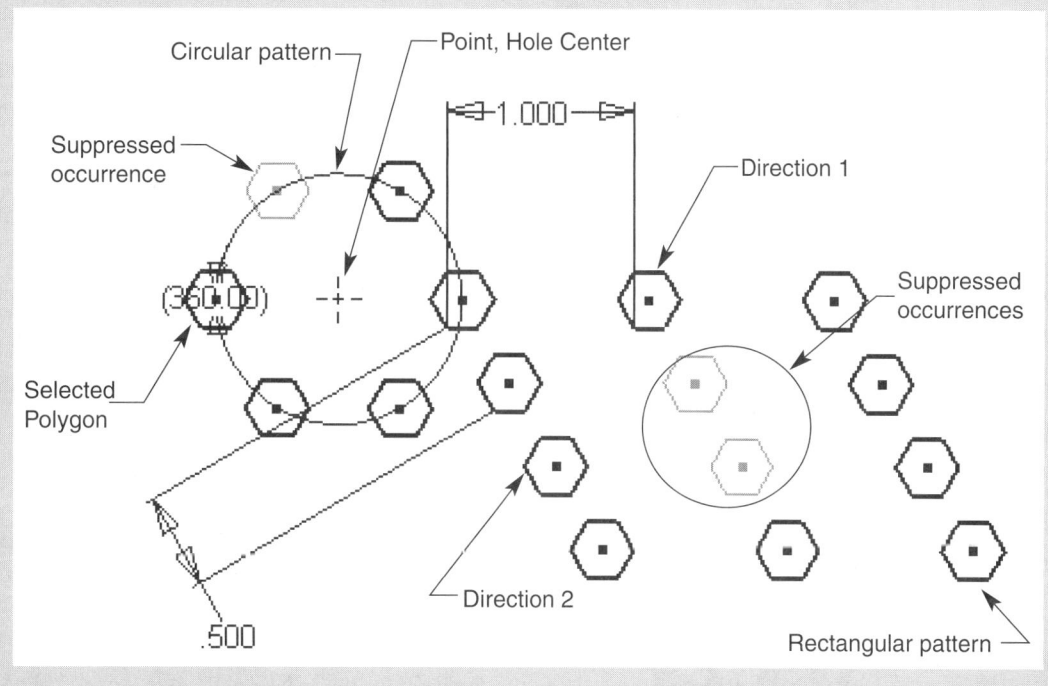

Offsetting Sketch Geometry

Use the **Offset** command to *offset* most existing sketch geometry, including lines, splines, circles, arcs, rectangles, fillets, and chamfers. The **Offset** tool is very effective for creating parallel copies of objects at a specified distance from the existing object. See Figure 3.50. Each of the geometric constraints associated with the parent object is duplicated in the offset object, if desired.

To offset sketch geometry, access the **Offset** command by picking the **Offset** button on the panel bar or **Sketch** toolbar. Then select the objects you would like to offset, and drag the offset object to the desired location. Notice the offset distance is displayed in the right corner of the **Status** bar. When you are satisfied with the location of the offset, pick that point. See Figure 3.51. If you would like to offset additional sketch

FIGURE 3.50 Various **Offset** command applications.

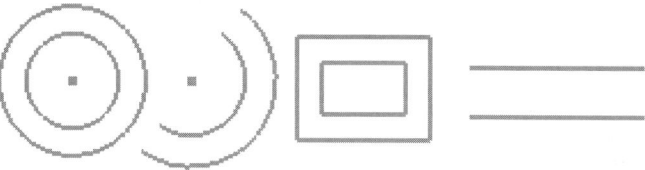

FIGURE 3.51   Using
the **Offset** tool.

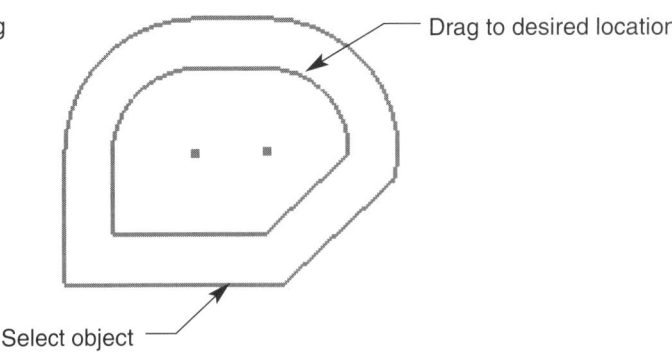

Drag to desired location

Select object

geometry, repeat the steps discussed. If you are finished offsetting, press the **Esc** key on your keyboard, right-click and pick the **Done** option, or choose another tool.

Field Notes

By default, when you offset objects, the entire loop is selected, as shown by the rectangle in Figure 3.50, and the offset object is constrained the same as the parent object. If you want to select individual pieces of an object instead of the entire loop, right-click and unselect the **Loop Select** option from the shortcut menu. For example, this option will allow you to offset only one line of a rectangle instead of the entire rectangle. If you do not want to apply the parent object constraints to the offset object, right-click and unselect the **Constrain Offset** option from the shortcut menu.

Exercise 3.15

1. Continue from Exercise 3.14, or launch Autodesk Inventor.
2. Open EX3.7.
3. Save a copy of EX3.7 as EX3.12.
4. Close EX3.7 without saving, and open EX3.12.
5. Right-click the name of the sketch (Sketch 1 by default), and select the **Edit Sketch** menu option.
6. Offset the entire object loop as shown.
7. While still in the **Offset** command, right-click and unselect the **Loop Select** menu option.
8. Offset the left arc as shown.
9. Resave EX3.15.
10. Exit Autodesk Inventor or continue working with the program.

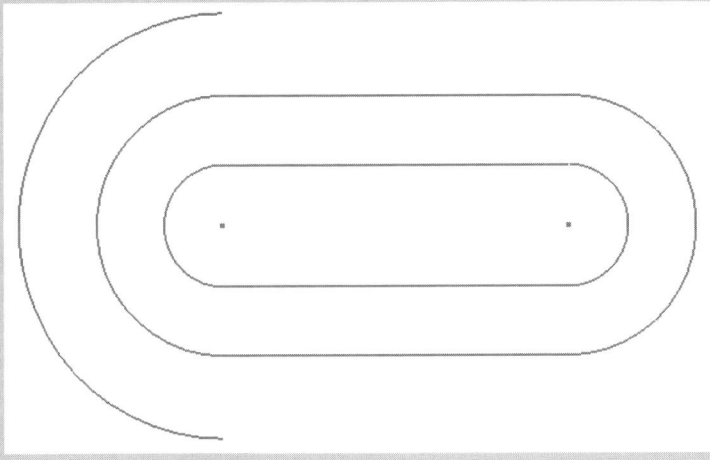

Constraining Sketches Using Geometric Constraints

As discussed in Chapter 2, sketches can be constrained by placing geometric constraints or dimensions on an object. Geometric constraints, often just referred to as constraints, are used to define sketch geometry in reference to other sketch geometry. Constraints are used for common geometric relationships, such as two perpendicular lines, equal-sized objects, or a line tangent to a circle. Usually, at least some geometric constraints are used to fully constrain a sketch. Autodesk Inventor provides you with a couple different options for constraining sketch objects. These options are discussed as follows:

- *Automatic constraints*   Automatic constraints are constraints applied to geometry as you sketch. They are attached to sketch geometry when you select a point. You can choose which types of constraints are automatically available by selecting either the **Parallel and Perpendicular** radio button or the **Horizontal and Vertical** radio button in the **Constraint Placement Priority** area of the **Options** dialog box, **Sketch** tab. Selecting one of these radio buttons enables you to select a parallel, perpendicular, horizontal, or vertical point before any other point, such as a grid or snap coordinate.

  Automatic constraints are similar to the automatic referencing tools discussed in the section "Creating Lines and Splines" of this chapter. In fact, the endpoint, midpoint, center point, intersection, AutoProject, tangent, and coincident options discussed earlier are all types of automatic constraints that create relationships between two or more objects. In addition, horizontal, vertical, parallel, and perpendicular automatic constraints are also available. See Figure 3.52. To use one of these options, drag new sketch geometry until you see the appropriate automatic geometric constraint symbol, and pick the point. To confirm that a constraint is present, attempt to drag the object. You should observe less freedom of movement.

Field Notes

> If you do not want automatic constraints to be placed on sketch geometry, hold down the **Ctrl** key on your keyboard while sketching objects.
>
> Some sketching tools automatically apply certain constraints to your sketch. For example, when you use the rectangle tool to create a rectangle, parallel, perpendicular, coincident, and horizontal constraints are placed on the sketch to make the a true rectangle.

- Constraint buttons   The automatic geometric constraint tools previously discussed are very effective for quickly constraining geometry while you sketch. However, often you may need to apply additional constraints after the general sketch is developed. To accomplish this task,

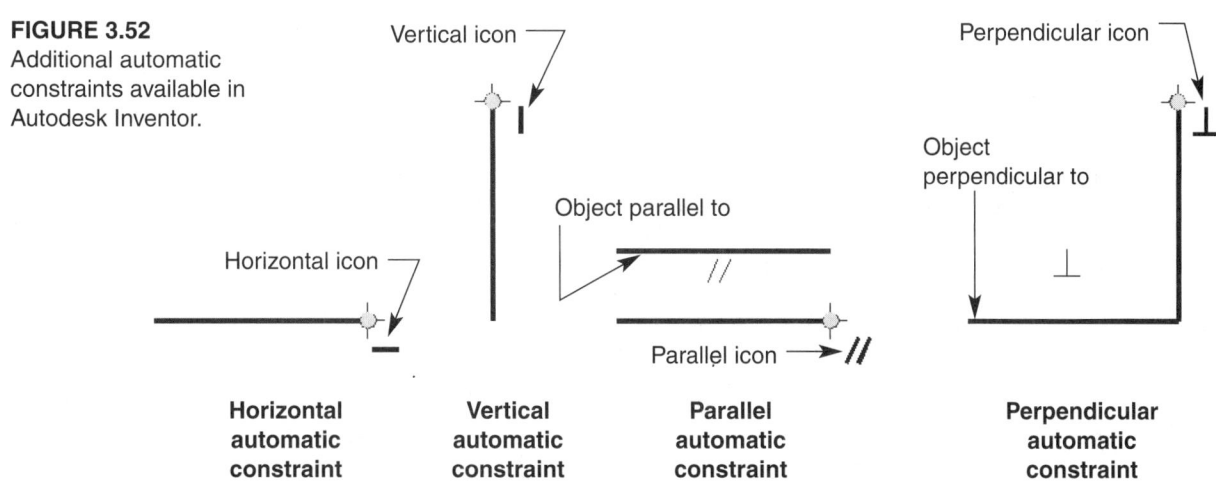

FIGURE 3.52
Additional automatic constraints available in Autodesk Inventor.

use one of the Autodesk Inventor constraint buttons. Use one of the following options to locate the desired constraint button:

✓ Pick the desired constraint button on the sketch panel bar.

✓ Pick the desired constraint button on **Sketch** toolbar.

✓ Right-click while in the sketch environment, and select the desired constraint from the **Create Constraint** cascading submenu.

The active constraint button is displayed in the panel bar and **Sketch** toolbar. To access additional options, pick the flyout button arrow and select the desired tool. See Figure 3.53. The following options are available inside the constraints flyout button:

■ **Perpendicular**   This constraint allows you to create lines or ellipse axes that are perpendicular, or at a 90° angle to each other. Unless the constraint is removed, you are not able to increase or decrease the angle between objects.

■ **Parallel**   This constraint allows you to create objects that are parallel to each other. Parallel objects, such as lines and ellipse axes, will never intersect, no matter how long they become.

■ **Tangent**   This constraint is used to define the tangent edge of one curve such as a line, spline, circle, arc, or fillet in reference to the tangent edge of an existing curve.

■ **Coincident**   Use this tool to constrain two points, or a curve and a point, to the same location.

■ **Concentric**   Use this tool to constrain the center point of an ellipse, circle, or arc to the center point of another ellipse, circle, or arc.

■ **Collinear**   This constraint is used to align two lines or ellipse axes along the same line.

■ **Horizontal**   Use this tool to horizontally align lines, points, or ellipse axes. A horizontal constraint positions geometry along the X axis.

■ **Vertical**   Use this tool to vertically align lines, points, or ellipse axes. A vertical constraint positions geometry along the Y axis.

■ **Equal**   This constraint allows you to size an object in reference to another object.

■ **Fix**   This constraint allows you to secure a point or object to its current location in space.

■ **Symmetric**   This constraint allows you to establish symmetry by selecting one object, followed by another object, and finally, a line of symmetry.

To use the constraining tools previously discussed, access the desired tool using one of the access options. For constraint tools that require you to pick two objects, the first object you select always remains the same. The object selected second changes in relation to the initially selected reference object. For example, say you want to make two lines the same

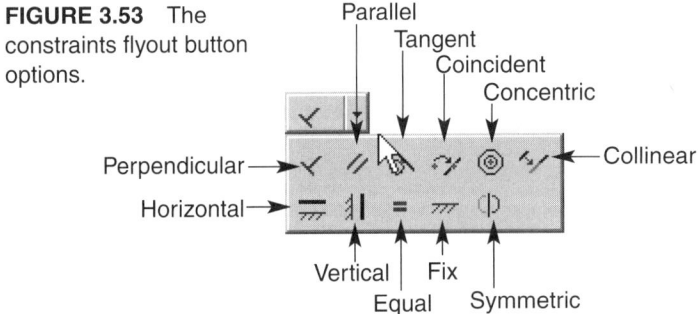

FIGURE 3.53 The constraints flyout button options.

length using the equal constraint tool. If you select a 2″ line first, the second line you choose also becomes 2″.

To constrain more geometry using the same tool, continue selecting objects. If you want to use a different constraining tool, access that tool and begin selecting objects. If you are finished applying geometric constraints, press the **Esc** key on your keyboard, right-click and pick the **Done** option, or choose another tool.

Exercise 3.16

1. Continue from Exercise 3.15, or launch Autodesk Inventor.
2. Open a new Autodesk Inventor part file.
3. Create a sketch similar to sketch A. Do not allow any automatic constraints to be placed on the sketch geometry.
4. Using the constraint tool buttons, constrain sketch A to look like sketch B. Use each type of constraint at least once.
5. Create a sketch similar to sketch B, using automatic constraints.
6. Save the part file as EX3.16.
7. Exit Autodesk Inventor or continue working with the program.

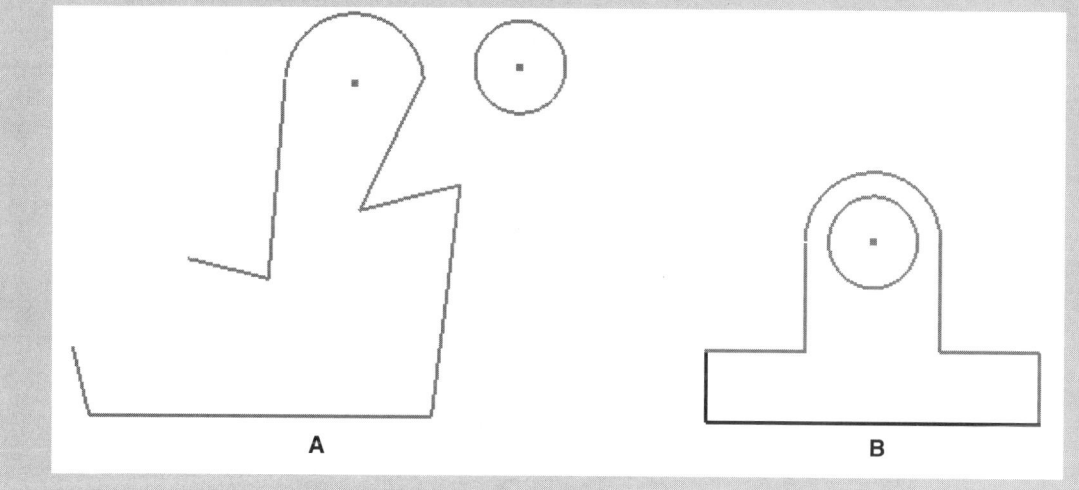

Using the Show Constraints Tool

Often you may want to see and delete constraints associated with a particular piece or group of geometry, so you can remove relationships or apply different constraints. The **Show Constraints** tool allows you to display all the geometric constraints placed on an object or group of objects. To show all the constraints for only certain objects, pick the **Show Constraints** button located in the panel bar or **Sketch** toolbar, and move your cursor over the desired object. After a short time, all the constraints are displayed, represented by their symbols. See Figure 3.54. If you want to keep the constraint symbols on the screen, pick the object instead of just moving your cursor over the object. You can select as many objects as desired using this technique. Notice when you move your cursor over, or select, a constraint symbol, the constraint and constrained objects become highlighted.

To show all the sketch constraints without selecting individual objects, right-click on the **Browser** bar or anywhere on the sketch environment graphics window, and choose **Show All Constraints.**

When the constraint display symbols are visible, you have the option of deleting certain constraints in order to apply a different relationship. To delete constraints, right-click the constraint you would like to remove and select **Delete** from the shortcut menu. If the constraint display symbols are blocking your view and need to be moved pick the double vertical lines to the right of the symbols and drag the box to the new location. If you want to close a constraint display symbols box, pick the hide button located to the right of

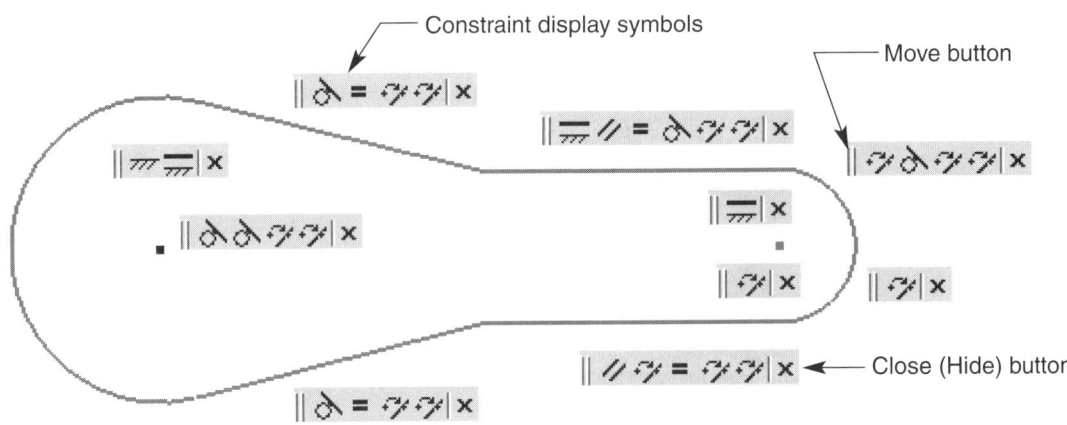

FIGURE 3.54 The constraint display symbols.

the symbols. To close all the sketch constraints without selecting individual hide buttons, right-click on the **Browser** bar or anywhere on the sketch environment graphics window, and choose **Hide All Constraints.**

Field Notes

Generally, you should apply geometric constraints before dimensions. Placing too many constraints may cause problems as you progress through the design process. Apply only the constraints necessary for the particular sketch.

You can constrain sketch objects as previously discussed, or constrain a sketch or sketch object to an existing part feature edge.

Exercise 3.17

1. Continue from Exercise 3.16, or launch Autodesk Inventor.
2. Open EX 3.16.
3. Use the **Show Constraints** button to show individual constraints of some of the sketch objects.
4. Use the **Show All Constraints** tool to view all the sketch constraints.
5. Move, and close some of the constraint symbol displays.
6. Delete some of the constraints.
7. Close EX3.16 without saving.
8. Exit Autodesk Inventor or continue working with the program.

Applying Sketch Dimensions

In addition to placing geometric constraints, another way to constrain a sketch is by applying dimensions. Unless you want to make a part adaptive, you should try to fully constrain a sketch using these two constraining tools. You know when your sketch is fully constrained because the object color changes, and you are no longer able to drag any of the sketched objects.

Dimensions are used to numerically define sketch geometry, such as the length of a line, diameter of a circle, and radius of an arc. Usually, at least some dimensions must be used to fully constrain, or define, an object. Autodesk Inventor provides you with the **General Dimension** and **Auto Dimension** tools to define sketch geometry.

■ **General Dimension**   This tool allows you to place aligned, linear, diameter, radius, or angular dimensions on a sketch by selecting individual objects. To apply a general dimension, access the **General Dimension** command by using one of the following options:

✓ Pick the **General Dimension** button on the panel bar.

✓ Pick the **General Dimension** button on the **Sketch** toolbar.

✓ Right-click inside the graphics window or **Browser** bar, and select the **Create Dimension** option.

✓ Press the **+** and **D** keys on your keyboard.

Then pick the geometry you want to define with a dimension, and drag the dimension text to a designated area.

If you do not select the **Edit dimensions when created** check box from the **Sketch** tab of the **Options** dialog box, the object is dimensioned to its current measurement. You must then pick the displayed dimension to open the **Edit Dimension** dialog box, shown in Figure 3.55, and enter a value. If the **Edit dimensions when created** check box is selected, the **Edit Dimension** dialog box automatically appears when you create a dimension. You can also access the edit dimensions when created command by right-clicking inside the graphics window while using the **General Dimension** tool, and select **Edit Dimension,** as shown in Figure 3.56.

Field Notes

To edit sketch **dimensions** any time, double-click a dimension to access the **Edit Dimension** dialog box.

There are a few different ways to specify dimension values in the **Edit Dimension** dialog box. You can type a specified numerical value in the edit box, or you can type in an equation. Usually this is done when you do not know the exact numerical value of the dimension. Use the following keys to create an expression:

■ The + key adds. For example, 2+2 or 3in+5mm.

■ The - key subtracts. For example, 5-2 or 7in-5mm

■ The * key multiplies. For example, 5*2 or 7in*5mm

■ The / key divides. For example, 6/2 or 7in/5mm

■ The ^ key allows for power operations. For example 3^5.

■ The parentheses keys ( ) separate expressions. For example, (5*3)+47.25/26.

In addition to these standard operations, you can also use functions such as sin, cos, log, and tan.

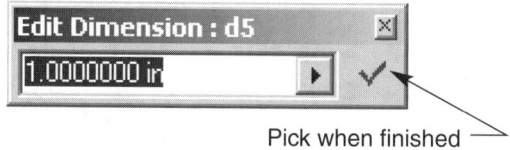

Pick when finished

FIGURE 3.55 The **Edit** dimension dialog box allows you to specify a numerical dimension value, another dimension, or an equation.

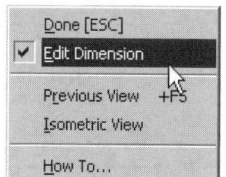

FIGURE 3.56 The **Edit Dimension** option of the **General Dimension** shortcut menu.

Dimension units reflect the current work environment and unit settings, such as inch or metric. Therefore, the value entered is represented in these units. For example, typing 1 means 1″ when working with inch units. You can also enter units after a numerical value to convert the value to the specified units. For example, you can enter 20mm even while in an inch-unit system.

Field Notes

You can enter equations for most Autodesk Inventor edit boxes. You may want to refer to the **Equations** reference section of the Autodesk Inventor Help files for further information regarding expressions, equation prefixes, units, and procedures for entering more complicated expressions.

Another way to specify a dimension value is to reference existing sketch dimensions. To apply this technique, access the **Edit Dimension** dialog box for the new dimension. Then select the existing dimension you want to reference. Notice the title of the selected dimension is displayed in the edit box. See Figure 3.57. If you pick the check mark, the object becomes the same size as the reference object. It is easier and more effective to use an **Equal** geometric constraint for this process. Consequently, usually when you reference an existing dimension to define an object, you include an equation, as shown in Figure 3.58.

The **General Dimension** tool has a number of options that allow you to dimension and define all sketch geometry. The following describes the general dimension options available:

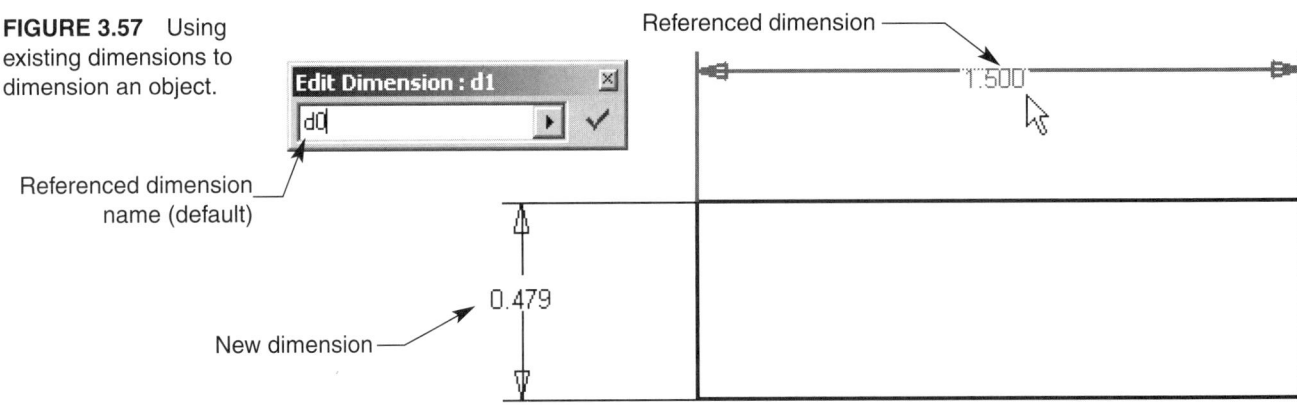

FIGURE 3.57 Using existing dimensions to dimension an object.

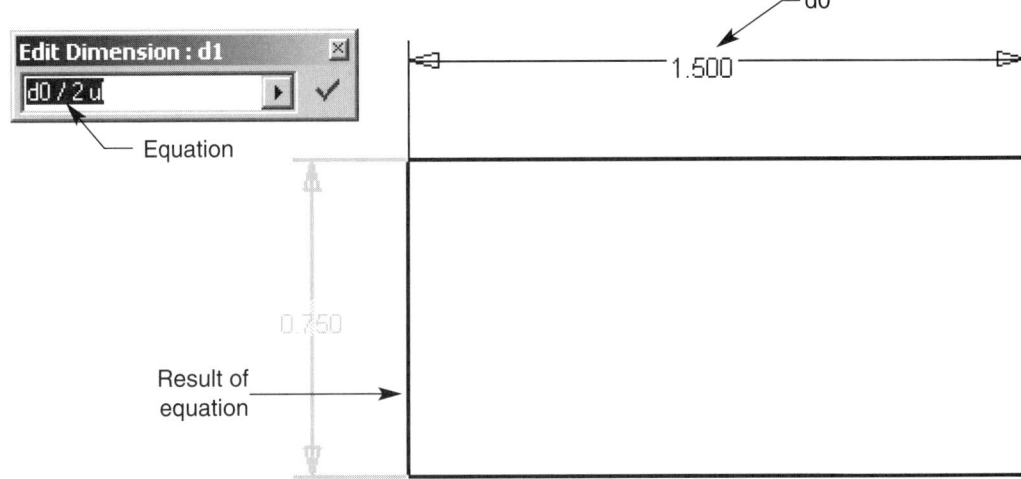

FIGURE 3.58 Using existing dimensions and equations to dimension an object (ul means unitless and is applied to values that do not contain specified units).

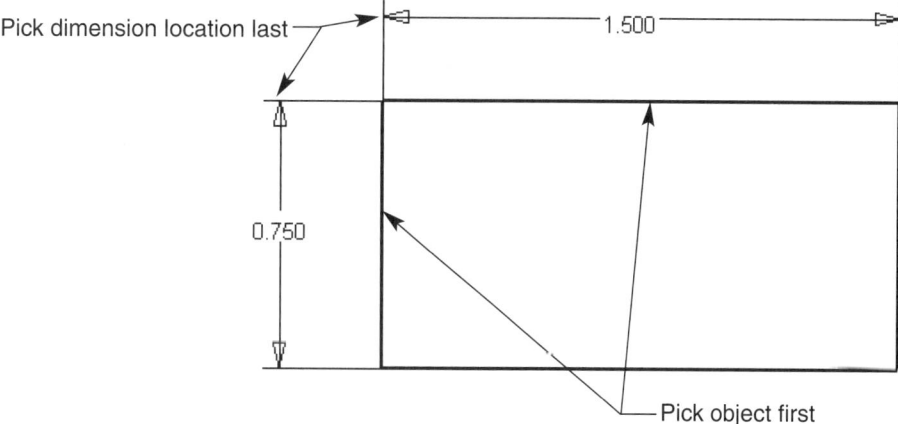

FIGURE 3.59
Applying linear
dimensions to a sketch.

Pick dimension location last
1.500
0.750
Pick object first

- Linear dimensions   These dimensions are used to define the vertical or horizontal measurement of an object. To place a linear dimension, access the **General Dimension** tool and pick the object you would like to dimension. If the object is vertical, drag the dimension to the left or right and pick the desired point. If the object is horizontal, drag the dimension up or down, and pick the desired point. See Figure 3.59.

  If you want to create another linear dimension, repeat the steps discussed. If you are finished, press the **Esc** key on your keyboard, right-click and pick the **Done** option, or select another tool.

- Aligned dimensions   These dimensions are used to define an object that is not vertical or horizontal. To place an aligned dimension, access the **General Dimension** tool and pick the object you would like to dimension. The default setting for placing dimensions is linear. As a result, to create an aligned dimension, you must select the object a second time. Then drag the dimension to the desired location and pick the point. See Figure 3.60. An alternative way to create an aligned dimension is to pick the object and right-click. Then select **Aligned, Vertical,** or **Horizontal** from the shortcut menu.

  If you would like to create another aligned dimension, repeat the steps discussed. If you are finished, press the **Esc** key on your keyboard, right-click and pick the **Done** option, or pick another tool.

- Angular dimensions   These dimensions define the angle between two lines. To place an angular dimension, access the **General Dimension** tool, and select two lines. Then drag the dimension to the desired location, and pick that point. See Figure 3.61.

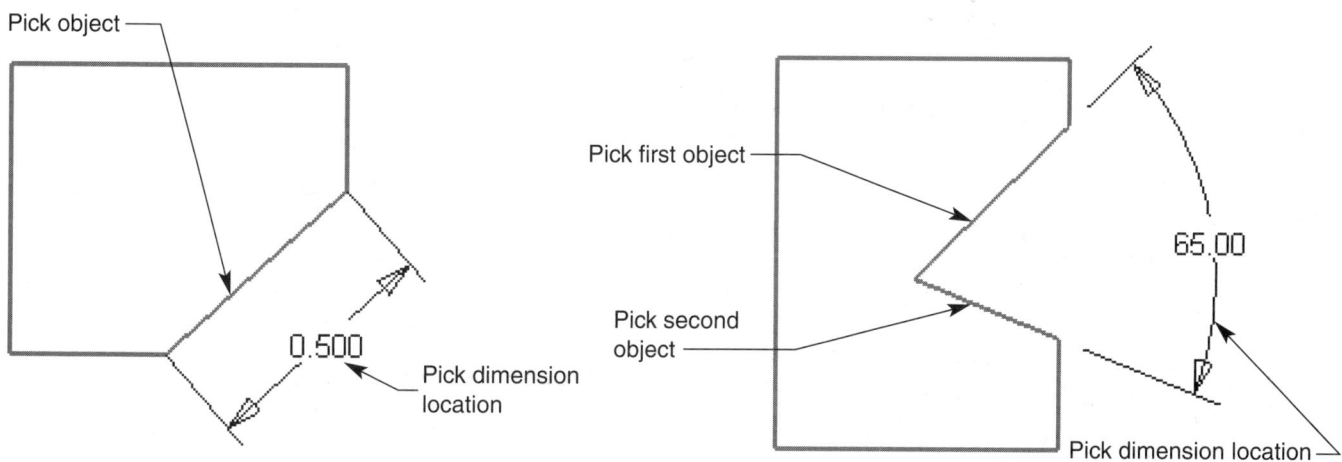

Pick object

0.500

Pick dimension location

FIGURE 3.60   Applying aligned dimensions to a sketch.

Pick first object

Pick second object

65.00

Pick dimension location

FIGURE 3.61   Applying angular dimensions to a sketch.

If you would like to create another angular dimension, repeat the steps discussed. If you are finished, press the **Esc** key on your keyboard, right-click and pick the **Done** option, or select another tool.

■ Diameter dimensions   These dimensions are used to define the diameter of a circle. To place a diameter dimension, access the **General Dimension** tool and select a circle. Then drag the dimension to the desired location and pick the point. See Figure 3.62. If for some reason you would like to dimension a circle as a radius instead of a diameter, select the circle, right-click, and choose **Radius** from the shortcut menu.

   If you would like to create another diameter dimension, repeat the steps discussed. If you are finished, press the **Esc** key on your keyboard, right-click and pick the **Done** option, or select another tool.

■ Radius   These dimensions are used to define the radius of an arc. To place a radius dimension, access the **General Dimension** tool and select an arc. Then drag the dimension to the desired location and pick the point. See Figure 3.63. If for some reason you would like to dimension an arc as a diameter instead of a radius, select the circle, right-click, and choose **Diameter** from the shortcut menu.

   If you would like to create another radius dimension, repeat the steps discussed. If you are finished, press the **Esc** key on your keyboard, right-click and pick the **Done** option, or choose another tool.

Field Notes

Sketch dimensions are used to define and constrain sketch geometry. Do not be overly concerned about the placement or display of these dimensions. However, you may want to move and manipulate sketch dimensions so the sketch environment is as uncluttered as possible. To move dimensions, select the dimension you want to move and drag it to a new location.

Sketch dimensions are created using Autodesk Inventor and may not comply with ASME standards.

Right-click on dimensions and select **Show Value, Show Name,** or **Show Expression,** as previously discussed to set the display characteristics of the dimension text.

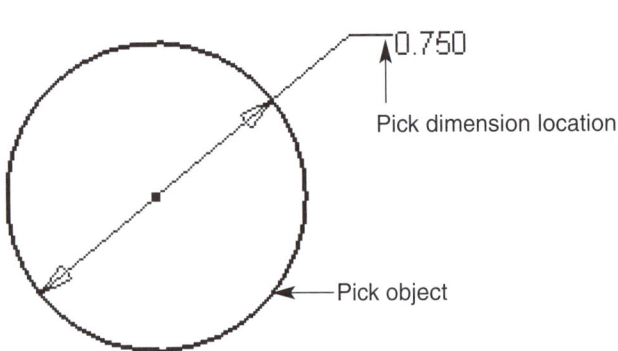

FIGURE 3.62   Applying a diameter dimension to a sketched circle.

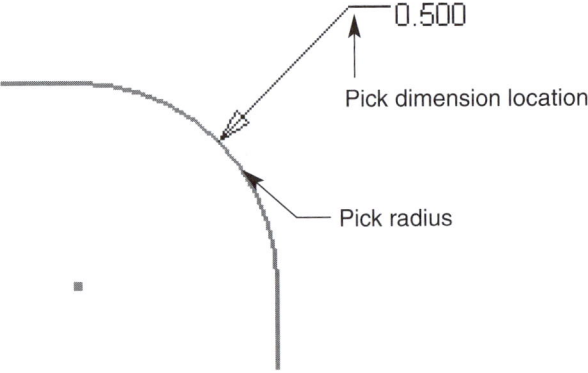

FIGURE 3.63   Applying a radius dimension to a sketched arc.

Exercise 3.18

1. Continue from Exercise 3.18 or launch Autodesk Inventor.
2. Open EX3.16
3. Save a copy of EX3.16 as EX3.18.
4. Close EX3.16 without saving, and open EX3.18.
5. Use the **General Dimension** tool to apply the dimensions shown for sketch A. The sketch should become fully constrained if the center point was projected or a point was fixed.
6. Create a sketch similar to sketch B, and apply the dimensions shown, using the **General Dimension** tool.
7. Resave EX3.18.
8. Exit Autodesk Inventor or continue working with the program.

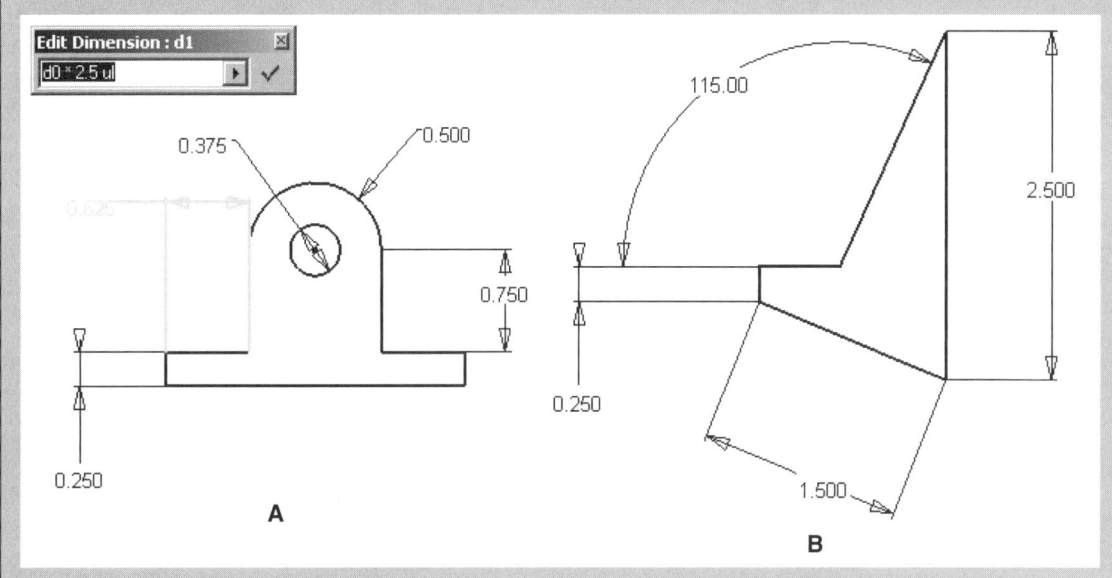

- **Auto Dimension** This tool allows you to dimension and constrain several objects at once. To access the **Auto Dimension** tool and open the **Auto Dimension** dialog box shown in Figure 3.64, pick the **Auto Dimension** button on the panel bar or **Sketch** toolbar. The following options are available inside the **Auto Dimension** dialog box:

  - **Dimensions Required** This box shows the number of dimensions required to constrain the sketch fully.

  - **Curves** By default, if you pick the **Apply** button, all the sketch geometry is automatically dimensioned and constrained. However, if you would like to select individual objects, pick the **Curves** button and pick only the geometry you want to define with the **Auto Dimension** tool.

FIGURE 3.64 The **Auto Dimension** dialog box.

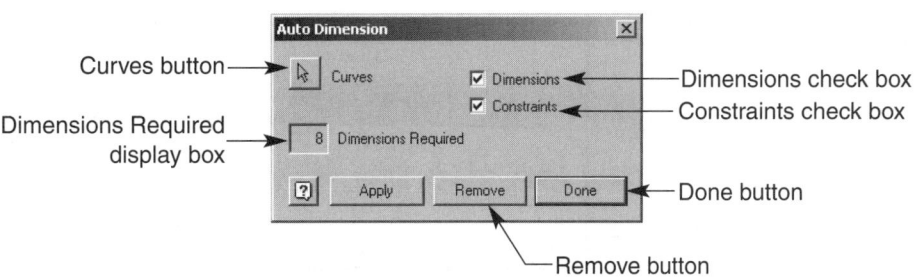

■ **Dimensions**   By default, the **Dimensions** check box is selected. Unselect this check box if you do not want to place any dimensions on the selected geometry.

■ **Constraints**   By default, the **Constraints** check box is selected. Pick this check box if you do not want to place any geometric constraints on the selected geometry.

■ **Remove**   This button allows you to remove dimensions and constraints you have already placed on the sketch.

■ **Done**   Select this button when you are finished using the **Auto Dimension** tool, and want to close the dialog box.

Field Notes

The dimensions you create using the **General Dimensions** tool are not replaced when you use the **Auto Dimension** tool. The **Edit dimensions when created** check box, previously discussed, has no function in the **Auto Dimension** tool. Therefore, dimensions placed using the **Auto Dimension** command represent the current object measurements. To modify a dimension value, you must double-click a dimension and enter a value in the **Edit Dimension** dialog box.

You can dimension sketch objects as previously discussed, or dimension a sketch or sketch object in reference to an existing part feature edge.

Exercise 3.19

1. Continue from Exercise 3.18, or launch Autodesk Inventor.
2. Open EX3.16.
3. Save a copy of EX3.16 as EX3.19.
4. Close EX3.16 without saving, and open EX3.19.
5. Use the **Auto Dimension** tool to apply dimensions to the sketch.
6. Resave EX3.19.
7. Exit Autodesk Inventor or continue working with the program.

What Are Driven Dimensions?

When you place a dimension on a sketch that defines an object that already contains a dimension, the sketch should become overconstrained. However, it is not really possible to overconstrain a sketch, because Autodesk Inventor does not allow overconstraining to occur. As a result, you either do not accept the dimension and cancel the command, or accept the additional dimension and allow the dimension to become *driven.* Driven dimensions are identified by being enclosed in parentheses. See Figure 3.65. Driven dimen-

FIGURE 3.65   An example of a driven dimension.

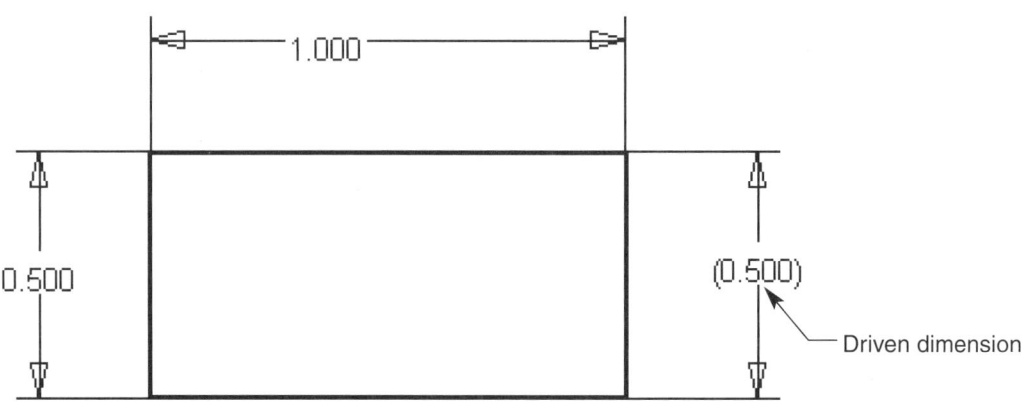

sions are used for reference only and are not parametric. You cannot edit a driven dimension to change the size of an object, but a driven dimension changes if the same parametric dimension is modified.

You can also create a driven dimension without attempting to overconstrain a sketch. Usually these types of driven dimensions are used only for reference when creating an adaptive part, discussed in later chapters. A nondriven, parametric dimension is created in the **Normal** style. If you want to change a non-driven dimension to driven, select the dimension and choose the **(Driven)** option from the **Style** drop-down list of the **Command** bar.

Field Notes

If you do not want Autodesk Inventor to warn you of an overconstrained condition, pick the **Apply Driven Dimension** radio button. This button is found in the **Overconstrained Dimensions** area of the **Options** dialog box, **Sketch** tab.

Exercise 3.20

1. Continue from Exercise 3.19, or launch Autodesk Inventor.
2. Open a new Autodesk Inventor part file.
3. Create the rectangle shown below.
4. Fully dimension the rectangle, and apply the driven dimension using the two techniques discussed.
5. Save the part file as EX3.20.
6. Exit Autodesk Inventor or continue working with the program.

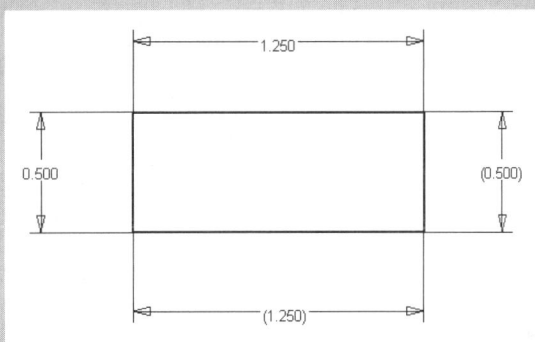

Using the Precise Input Toolbar

The **Precise Input** toolbar, shown in Figure 3.66, has a number of options that allow you to identify the size of sketch geometry when it is initially created, instead of after it is drawn as discussed previously using the geometric constraint and dimensioning options. For example, you can define the start or endpoint of a line, center point of a circle, the points of a three point arc, or the corners of a rectangle using points.

Using precise input techniques allows you to define the initial location of objects. However, using precise input tools does not constrain or dimension any of the sketch geometry. In order to constrain the

FIGURE 3.66 The **Precise Input** toolbar.

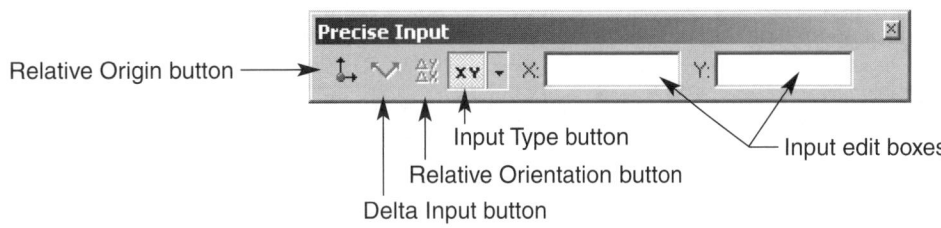

sketch you must still place dimensions and geometric constraints. Consequently, most of the time, using precise inputs to create objects takes longer and is more complicated than simply drawing a general sketch and placing dimensions and geometric constraints. Still, the **Precise Input** toolbar may be effective for some applications.

To access the **Precise Input** toolbar, select the **Precise Input** menu option from the **Toolbar** cascading submenu, located in the **View** pull-down menu. You must activate a sketching tool command, such as line, arc, or circle, to use the **Precise Input** toolbar. The following tools and options are available inside the **Precise Input** toolbar:

- **Input Type**   Use this button to select how you would like to input coordinate entries from the origin. By default the input type is XY. To activate a different input type, pick the **Input Type** flyout button, and choose the desired **Input Type** button. See Figure 3.67. You can choose to enter an X and Y point, an X point and an angle from the specified X point, a Y point and an angle from the specified Y point, or a distance and an angle from the X axis.

- **Input**   These edit boxes change depending on the current input type selection, previously discussed. To use these edit boxes, type the value of the first point in the left box. This box is initially active when using the precise input tool. Then press the **Tab** key on your keyboard, or select inside the right box to activate the right box, and enter the second value. Finally, pick anywhere inside the graphics window, or press the **Enter** key on your keyboard, to accept the specified point.

- **Relative Origin**   This button is used to specify the location of a temporary origin and is displayed by the coordinate icon. See Figure 3.68. To create a relative origin, access the **Relative Origin** button from the **Precise Input** toolbar. Then select an available sketch point. Once the relative origin is positioned, each new point you enter in the coordinate input edit boxes is referenced from this origin.

- **Delta Input**   This button is used to specify automatically the location of a temporary origin at the last selected point. This means an origin indicator is placed on the sketch geometry previously created. As a result, when the **Delta Input** button is selected, the first point of any new object is located at the temporary origin. When you pick the **Delta Input** button, the **Relative Origin** button is automatically selected.

- **Relative Orientation**   This button rotates the coordinate system axes and is only available while in a drawing environment, discussed in future chapters.

✎**Field Notes**

> In addition to standard numerical values, you can enter equations in most Autodesk Inventor edit boxes, including the **Input** edit boxes.
>
> Precise input units reflect the current work environment and unit settings. However, you can enter units after a numerical value to convert the value to the specified units.

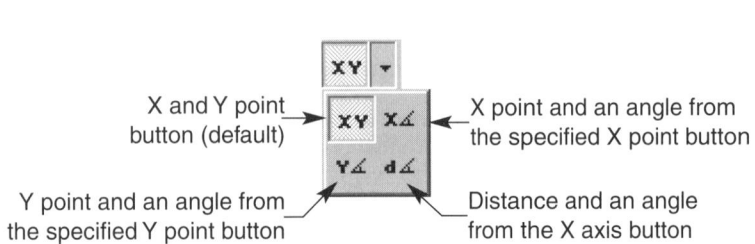

X and Y point button (default)

X point and an angle from the specified X point button

Y point and an angle from the specified Y point button

Distance and an angle from the X axis button

FIGURE 3.67   The **Input Type** flyout button options.

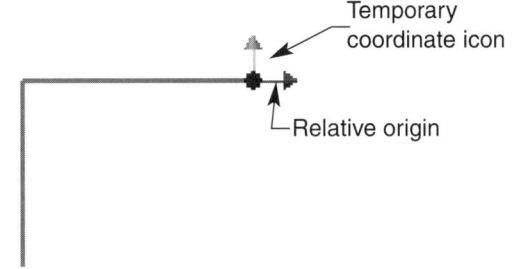

Temporary coordinate icon

Relative origin

FIGURE 3.68   A relative origin and the temporary coordinate icon.

Exercise 3.21

1. Continue from Exercise 3.20, or launch Autodesk Inventor.
2. Open a new Autodesk Inventor part file.
3. Create the object shown using the **Precise Input** tools and entry techniques described. Select the **Delta Input** button to create lines 1 through 6. Unselect the **Delta Input** button to create line 7.
4. Save the part file as EX3.21.
5. Exit Autodesk Inventor or continue working with the program.

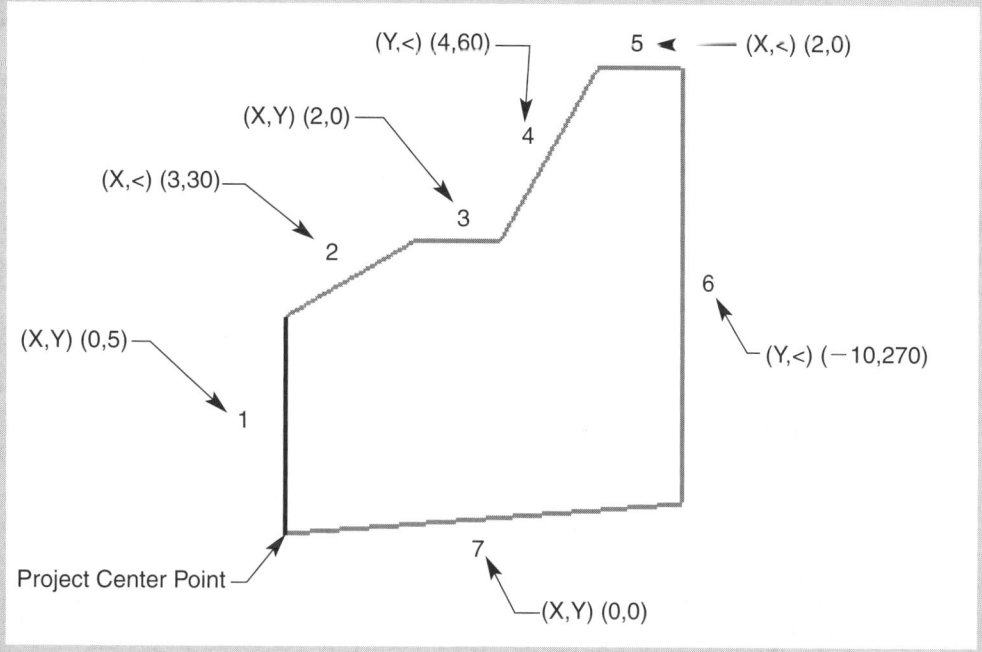

Using Parameters in Sketches

Parameters are size and structural limits placed on sketches and features; they are available inside an Autodesk Inventor part or assembly file. Parameters in the sketch environment refer to the dimensions placed on sketch geometry. To access the **Parameters** dialog box/spreadsheet, select the **Parameters** button from the **Standard** toolbar or choose the **Parameters** option from the **Tools** pull-down menu. A model parameter is automatically created every time you dimension a sketched object. Figure 3.69 shows the model parameters for a sketch with three dimensions. The following information is displayed under the **Model Parameters** section of the **Parameters** dialog box:

- **Parameter Name** This edit box displays the default name of the parameters. The first dimension applied to a sketch receives the name d0, the next dimension d1, and so on. To change a default parameter name to something meaningful to the sketch, such as length, width, or diameter, pick inside the edit box. Then type the desired name and press the **Enter** key, or select the **Done** button if you are finished using the **Parameters** dialog box.

- **Unit** This display box shows the measurement units used to dimension the object.

- **Equation** This edit box displays the value or equation specified when dimensioning a sketch. To change an equation, pick inside the edit box and type the desired value or equation. Then press the **Enter** key, or select the **Done** button if you are finished using the **Parameters** dialog box.

- **Value** This display box shows the current value of a parameter. If the dimension was created using an equation, the calculated value is displayed.

FIGURE 3.69   The
Parameters dialog box.

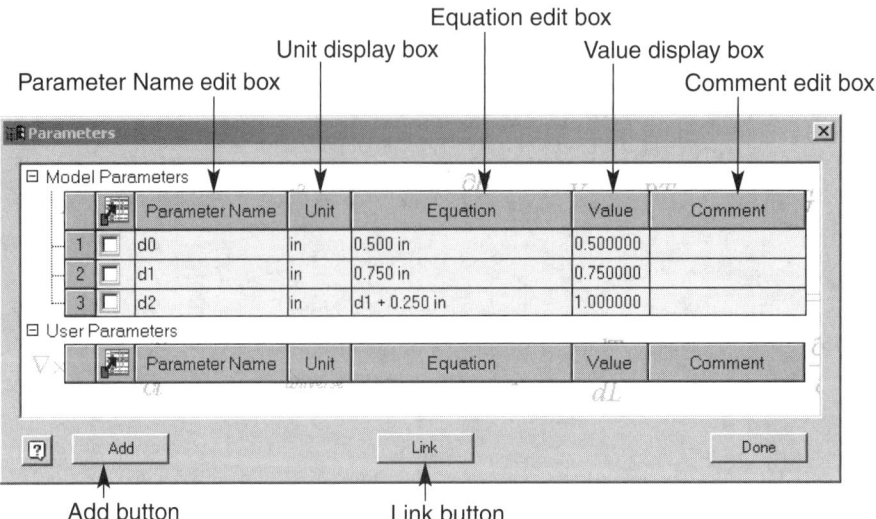

Add button   Link button

- **Comment**   This edit box allows you to enter a parameter comment. To create a comment, pick inside the edit box and type the desired information. Then press the **Enter** key, or select the **Done** button if you are finished using the **Parameters** dialog box.

- **Add**   This button is used to add user-defined parameters, discussed in future chapters.

- **Link**   Pick the **Link** button to link or embed a specified spreadsheet to use for parameters.

In addition to the automatically specified model parameters, you can also specify your own parameters in the **User Parameters** section of the **Parameters** dialog box.

Field Notes

> Parameters and the Parameters dialog box can be used throughout the design process, from the most simple sketch to the most complex assembly. For more information of parameters you may want to refer to Autodesk Inventor help resources.

Exercise 3.22

1. Continue from Exercise 3.21, or launch Autodesk Inventor.
2. Open EX3.18.
3. View the object parameters and explore the parameter options discussed in this chapter.
4. Close EX3.18 without saving.
5. Exit Autodesk Inventor or continue working with the program.

Editing Sketches

Autodesk Inventor provides you with a number of sketch editing tools. Use these commands during and after sketch creation to modify geometry. Some of these sketch editing commands, such as modifying line type in the **Style** drop-down list and changing dimensions or geometric constraints, have already been discussed. The following information describes additional sketch editing options.

- **Dragging geometry**   Sketch objects that are not constrained can be moved or dragged to a new location or size. To drag an object, pick and hold down the left mouse button. The drag

FIGURE 3.70
Dragging sketch
geometry, and the drag
icon.

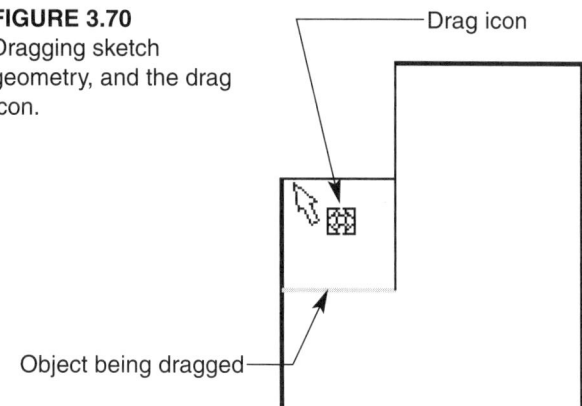

icon, shown in Figure 3.70, becomes visible. Then move the object to the desired location and release the left mouse button. You may find a number of uses for this dragging technique. Often, dragging or attempting to drag an object allows you to see where a constraint is still required, and assess design options.

 Field Notes

You can utilize all the automatic referencing tools previously discussed in the Line command when dragging objects.

■ **Extend**   Use this button to increase the size of a curve, such as a line, spline, or arc, to the nearest intersection. In order to create a feature a sketch must have a closed loop, which means it does not contain any gaps. Often you can close a gap by using the **Extend** tool to extend a curve to the nearest intersection. To extend a curve, shown in Figure 3.71, access the **Extend** command by picking the **Extend** button on the panel bar or **Sketch** toolbar. Then, move your cursor over the object you would like to extend to display a preview of the operation. If the preview looks acceptable, pick the curve to complete the command. If desired, pick a new curve to extend. If you do not want to extend another object, press the **Esc** key on your keyboard, right-click and pick the **Done** option, or pick another tool.

 Field Notes

If you just extended a curve, and now need to trim a segment from your sketch, press the **Shift** key on your keyboard to access the **Trim** command, or right-click and choose the **Trim** option.

FIGURE 3.71
Extending a sketched
curve.

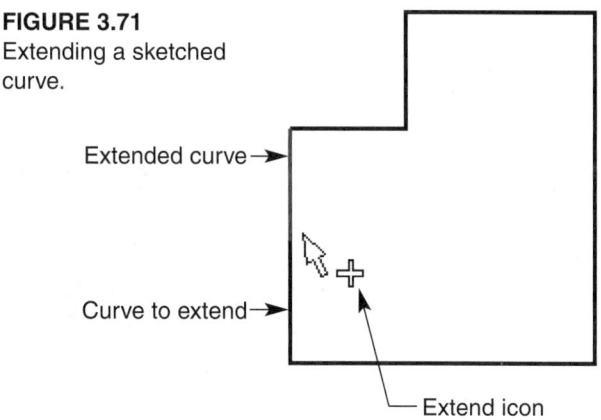

■ **Trim** Use this button to remove unwanted sketch objects that extend past an intersection. See Figure 3.72. To trim an object, access the **Trim** command by picking the **Trim** button on the panel bar or **Sketch** toolbar. Then move your cursor over the object you would like to remove to display a preview of the operation. If the preview looks acceptable, pick the sketch segment to complete the command. If desired, pick a new object to trim. If you do not want to extend another object, press the **Esc** key on your keyboard, right-click and pick the **Done** option, or pick another tool.

Field Notes

If you just trimmed an object and now need to extend a curve in your sketch, press the **Shift** key on your keyboard to access the **Extend** command, or right-click and choose the **Extend** option.

■ **Move** Use this tool to move an entire sketch or individual sketch objects form one location to another. To move an object, access the **Move** command by picking the **Move** button on the panel bar or **Sketch** toolbar. Selecting the **Move** button opens the **Move** dialog box, shown in Figure 3.73. Now, select the objects you would like to move, by picking them individually or utilizing a crossing window. Once all the geometry you would like to move is selected, right-click and pick the **Continue** option, or pick the **From Point** button of the **Move** dialog box, and select the base point you would like to move the objects from. Once the from point is selected, the **To Point** button is activated. This allows you to select the endpoint, where you would like to move the objects to. If you would like to copy the objects in addition to moving them to a new location, pick the **Copy** check box. Once the to point is selected and all other options are specified, press the **Enter** key or pick the **Apply** button to move the designated objects.

Field Notes

When you move individual sketch objects, those pieces of the sketch connected to the object being moved change to accept the new location of the object, but only if constraints are placed on the sketch geometry. See Figure 3.74. This process is similar to a stretch command.

You can also copy sketch geometry by selecting the object you would like to duplicate, right-clicking, and choosing the **Copy** option. Then right-clock again, and pick the **Paste** option.

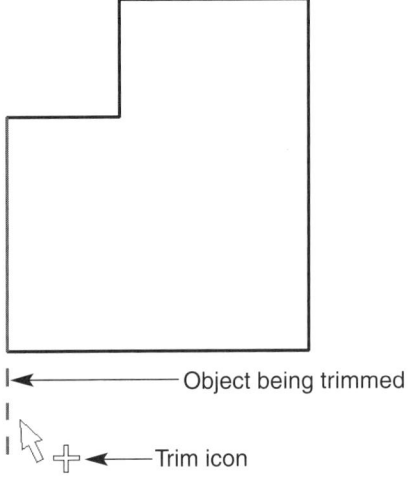

FIGURE 3.72 Trimming a sketched object piece.

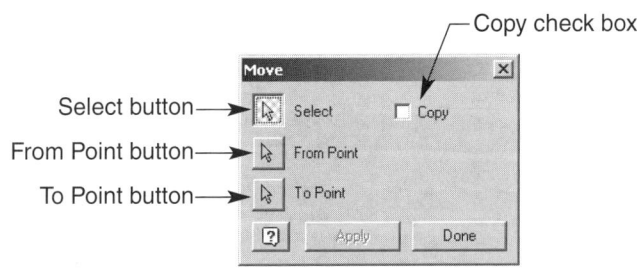

FIGURE 3.73 The Autodesk Inventor **Move** dialog box.

FIGURE 3.74 Moving individual sketch components.

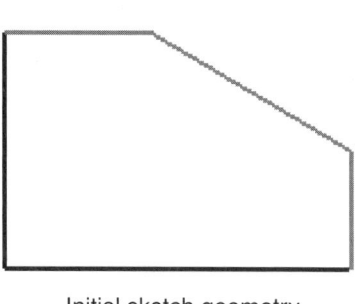
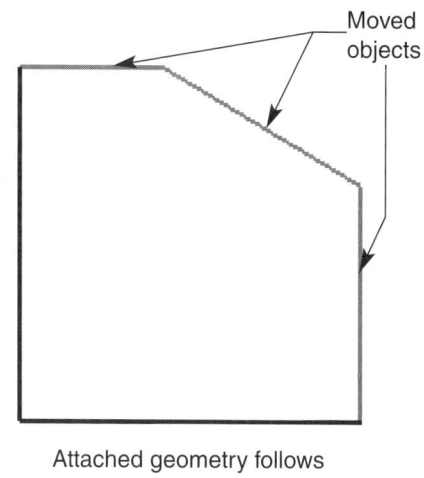

Moved objects

Initial sketch geometry

Attached geometry follows moved objects

■ **Rotate**   Use this tool to rotate an entire sketch or individual sketch objects from one location to another. To rotate an object, access the **Rotate** command by picking the **Rotate** button on the panel bar or **Sketch** toolbar. This opens the **Rotate** dialog box, shown in Figure 3.75. Now select the objects you would like to rotate by picking them individually or utilizing a crossing window. Once all the geometry you would like to move is selected, right-click and pick the **Continue** option, or pick the **Center Point** button of the **Rotate** dialog box, and select the center, or base, point you would like to rotate the objects around. Once the center point is selected, enter a desired rotation angle in the **Angle** edit box, or select one from the drop-down list. If you would like to copy the objects in addition to rotating them, pick the **Copy** check box. Once all your rotation options are specified, press the **Enter** key or pick the **Apply** button to rotate the designated objects.

Field Notes

When you rotate individual sketch objects, those pieces of the sketch connected to the object being rotated change to accept the new location of the object, but only if constraints are placed on the sketch geometry.

FIGURE 3.75 The Autodesk Inventor **Rotate** dialog box.

Select button
Center Point button

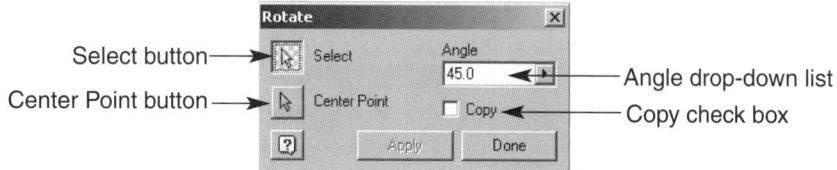

Angle drop-down list
Copy check box

Exercise 3.23

1. Continue from Exercise 3.22, or launch Autodesk Inventor.
2. Open a new Autodesk Inventor part file.
3. Create a sketch similar to the objects labeled A.
4. Use the **Copy** option of the **Move** tool to create a copy of A. Use point 1 as the **From Point,** and point 2 as the **To Point.**
5. Drag the points and lines of sketch B to create a sketch similar to sketch C.
6. Extend and trim sketch geometry as shown in D.
7. Rotate the sketch 45°, using point 3 as the center point.
8. Save the part file as EX3.23.
9. Exit Autodesk Inventor or continue working with the program.

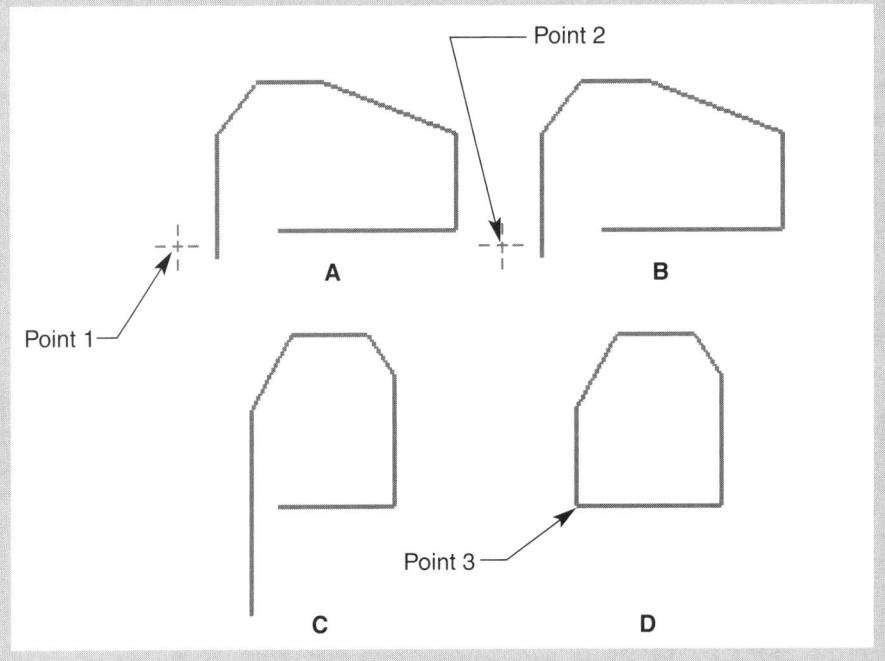

Inserting AutoCAD Drawings for Sketches

In many situations, you may want to use existing AutoCAD drawings to create Autodesk Inventor sketches, or Mechanical Desktop parts or assemblies to create Autodesk Inventor parts or assemblies. To accomplish this task, access the **Insert AutoCAD File** button from the panel bar or **Sketch** toolbar. Selecting the **Insert AutoCAD File** button opens the **Open** dialog box, discussed in Chapter 1. See Figure 3.76. The **Open** dia-

FIGURE 3.76 The Autodesk Inventor **Open** dialog box.

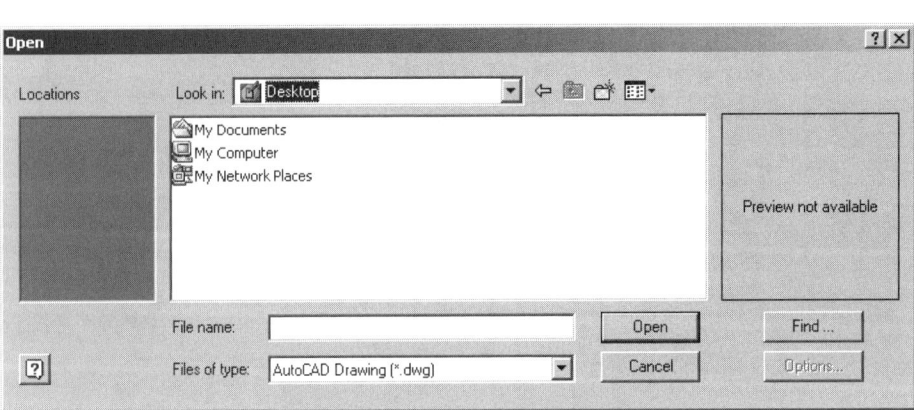

FIGURE 3.77 The
Insert AutoCAD File
wizard (AutoCAD files,
first page).

log box allows you to locate and open a particular AutoCAD or Mechanical Desktop file. Once you find the file you would like to insert, pick the **Open** button to access the **Insert AutoCAD File** wizard. See Figure 3.77. The **Insert AutoCAD File** wizard contains pages specific to the type of file you are importing and allows you to define the inserted file characteristics. Once you define the options in one page, access additional pages in the wizard by selecting the **Next** button. To return to a previous page, pick the **Back** button, and when you have fully defined the import options, pick the **Finish** button. The following pages are available in the **Insert AutoCAD File** wizard:

- **DWG File Import Options** This page, shown in Figure 3.77, is used to specify AutoCAD, Mechanical Desktop, or DXF file import options. The selected **Read Content From** radio button identifies the type of file you selected in the **Open** dialog box for importing. Select the file units, inches, cm, or, mm from the **Units** drop-down list. Then, select a configuration option from the **Configuration** drop-down list, or pick the **Browse** button to locate an alternative configuration. The specified configuration defines the file open options for the selected file.

- **Layers and Objects Import Options** This page, shown in Figure 3.78, is used to define what information is transferred from the drawing file when imported into Autodesk Inventor. When inserting AutoCAD files, this page allows you to choose whether to import from **Model Space** or **Paper Space** and to define which layers are translated. The selected layers are displayed in the **Selected layers to read** list box and can be selected or unselected depending on the application. In addition, you can allow 3D solids to be imported by selecting the **3DSOLIDS** check box; you can pick the **Constrain endpoints** check box if you want constraints to be applied to sketch geometry as it is imported.

- **Import Destinations Options** This page, shown in Figure 3.79, allows you to specify imported file destination options and contains the following areas:

  - **Destination for 2D data** This area contains several radio buttons. Pick the radio button that corresponds to the location where you want to place the imported 2D data.

  - **3D Solids options** This area allows you to specify where solids are placed when converted to Autodesk Inventor.

FIGURE 3.78 The **Insert AutoCAD File** wizard (AutoCAD files, second page).

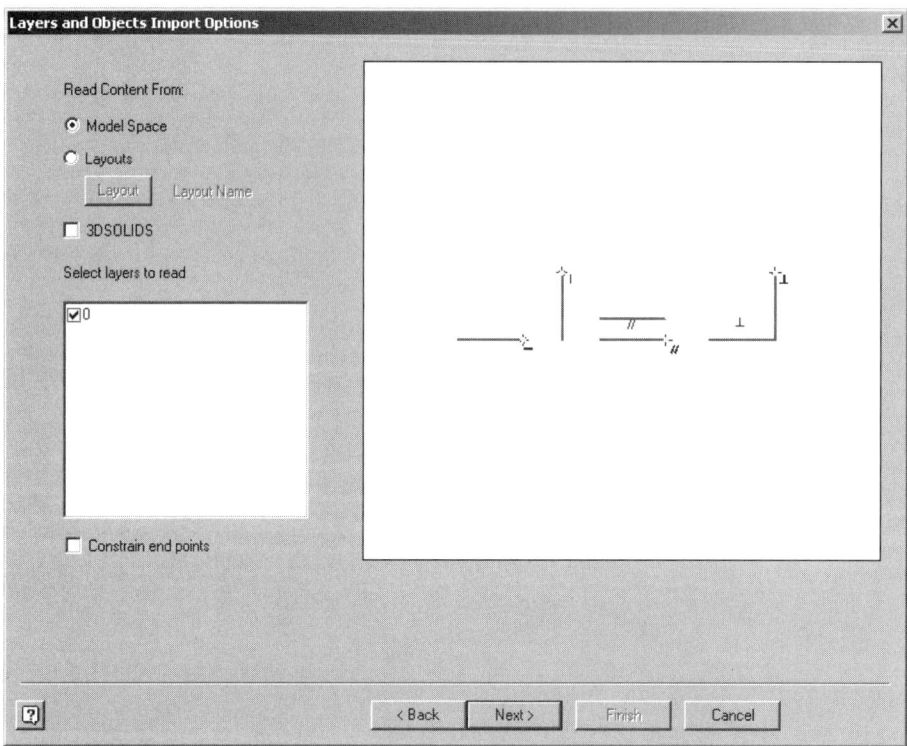

FIGURE 3.79 The **Insert AutoCAD File** wizard (third page).

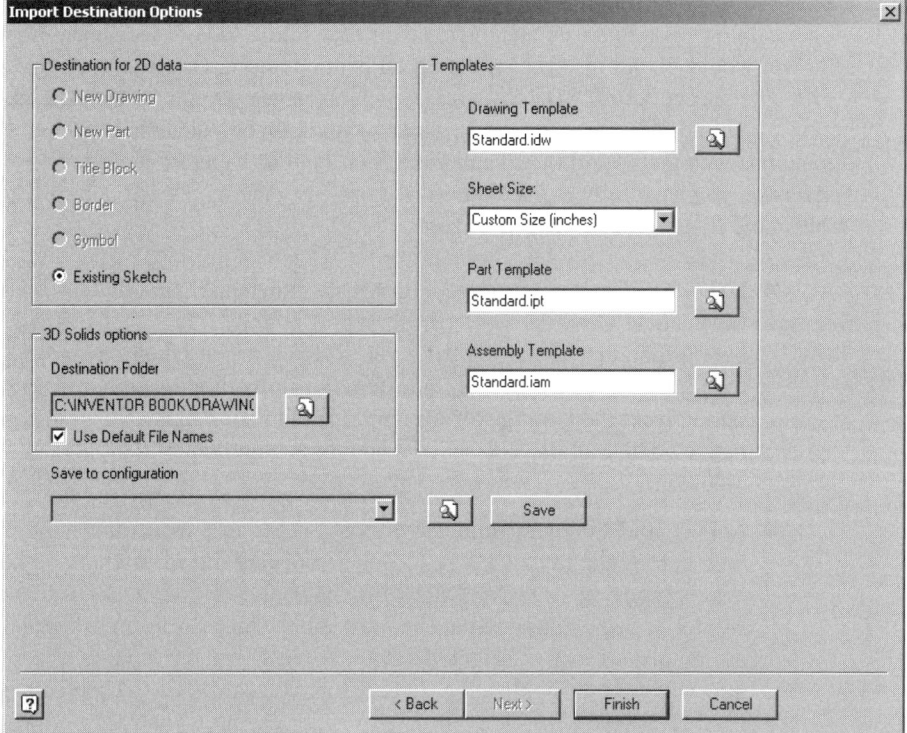

- **Templates** This area contains edit boxes and Browse button for defining templates to use for specific operations including drawings, parts, and assemblies.

- **Save to configuration** Use this Browse button to locate and specify an import wizard configuration location and name. Then pick the **Save** button to save the configuration for later DWG importing.

Field Notes

You may want to create a custom template for inserting AutoCAD files.

Using the Sketch Doctor

As previously discussed, the next step in developing a model is to create a feature from the sketch. In order to create a feature, most of the time your sketch should result in a closed loop and be free of defects. Sketch problems or defects are typically the result of unneeded points, a missing coincident constraint, overlapping curves, open loops, or self-intersecting loops. The Sketch Doctor is a tool that helps you resolve these types of sketch problems. To access the Sketch Doctor use one of the following techniques:

✓ Right-click anywhere in the graphics window or **Browser** bar, and select the **Sketch Doctor** option while in a sketch environment.

✓ Pick the **Recover** button on the **Standard** toolbar.

✓ When creating a sketched feature, such as an extrusion, pick the **Examine Profile Problems** button on the sketch feature dialog box.

If you access the Sketch Doctor from a sketched feature dialog box, you automatically see the **Sketch Doctor – Examine** section of the **Sketch Doctor** dialog box, discussed later. Otherwise, the **Select** section of the Sketch Doctor is active. See Figure 3.80. Here you can choose the **Diagnose Sketch** button to open the **Diagnose Sketch** dialog box, shown in Figure 3.81. The **Diagnose Sketch** dialog box allows you to choose which of the previously mentioned diagnostic tests to use. By default, all the diagnostic tests are selected. Once you determine which diagnostic tests you would like to use, pick the **OK** button. As seen in Figure 3.82, the sketch icons and sketch problems are displayed in the Select a problem to recover list box. Also notice your sketch and the sketch problems are highlighted.

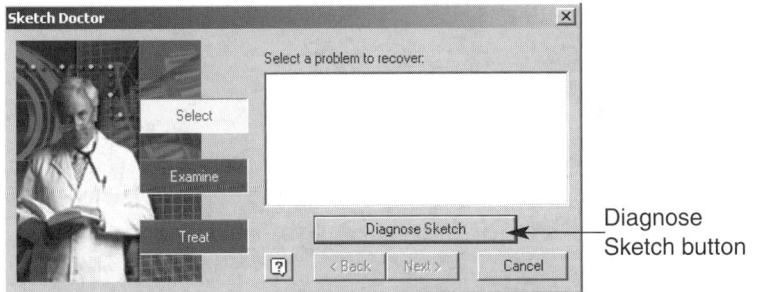

FIGURE 3.80 The **Select** section of the Sketch Doctor.

FIGURE 3.81 The **Diagnose Sketch** dialog box.

FIGURE 3.82 The **Select** section of the Sketch Doctor, and highlighted sketch.

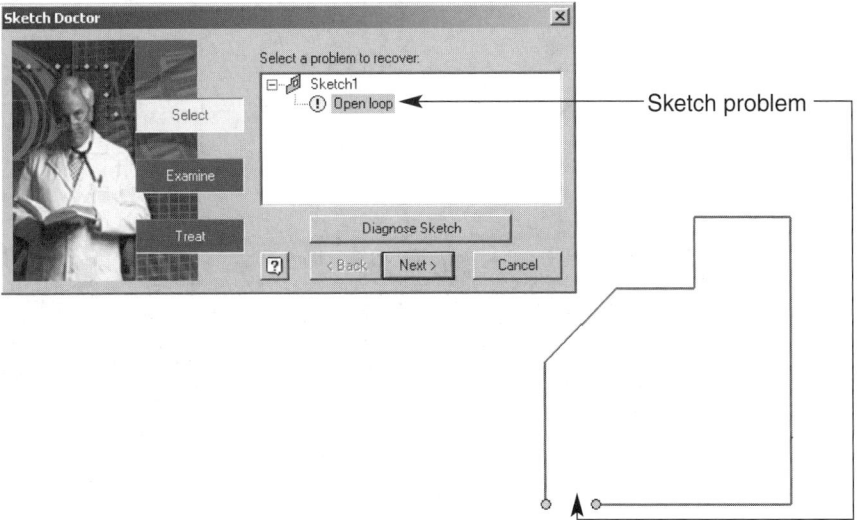

FIGURE 3.83 The **Sketch Doctor – Examine** section of the **Sketch Doctor** dialog box.

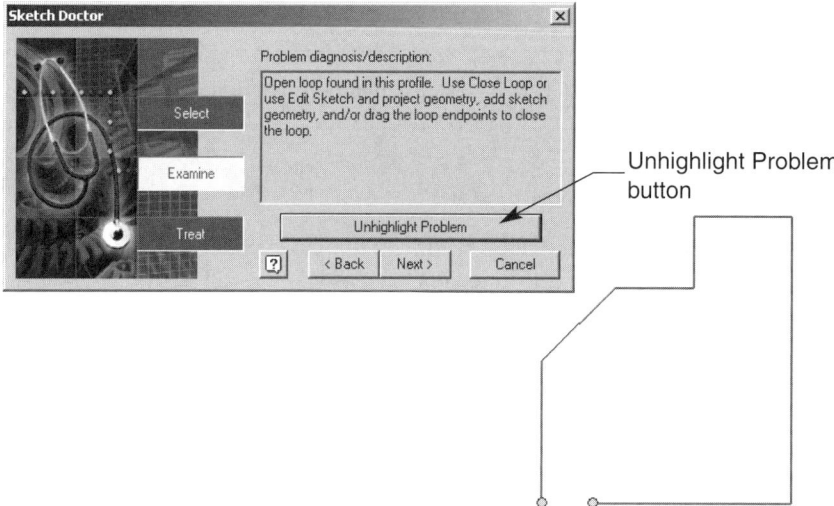

Pick the **Next** button to continue the Sketch Doctor application and display the **Sketch Doctor – Examine** section of the **Sketch Doctor** dialog box. See Figure 3.83. The **Sketch Doctor – Examine** section displays a description of the problem, and the diagnosis. You can also highlight or unhighlight the sketch if desired. Pick the **Next** button to access the **Treat** section of the **Sketch Doctor** dialog box. See Figure 3.84. The treatment, or problem-solving, options are displayed in the **Select a treatment** list box. Choose the option you would like to use to solve the problem and pick the **Finish** button. As shown in Figure 3.85, a solution is displayed in the graphics window. Follow the solution directions to solve the sketch problem.

FIGURE 3.84 The **Treat** section of the **Sketch Doctor** dialog box.

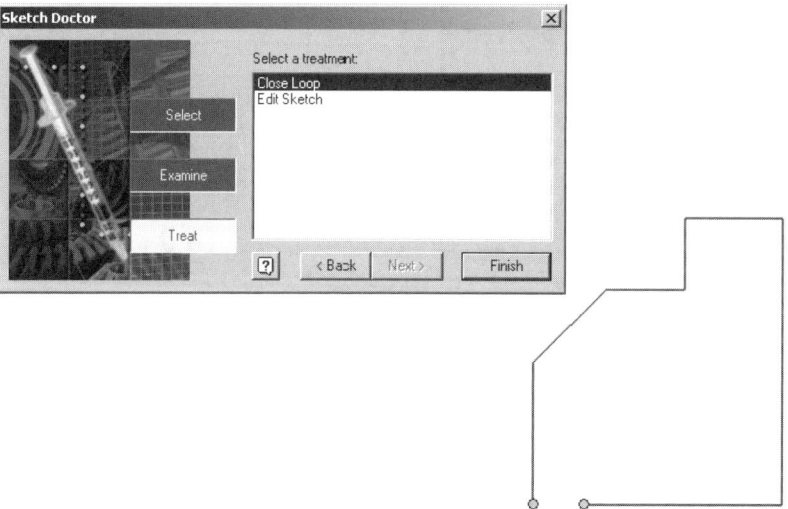

FIGURE 3.85 An example of a Sketch Doctor solution.

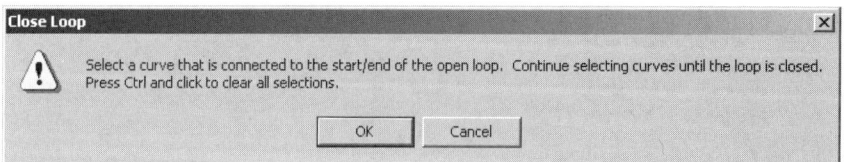

Exercise 3.24

1. Continue from Exercise 3.23, or launch Autodesk Inventor.
2. Open a new Autodesk Inventor part file.
3. Create a sketch similar to the sketch shown in Figure 3.82.
4. Use the Autodesk Inventor Sketch Doctor to diagnose, select, examine, and treat the open loop sketch problem.
5. Save the part file as EX3.24.
6. Exit Autodesk Inventor or continue working with the program.

Using the Properties Dialog Box

Picking the **Properties...** option from the **File** pull-down menu opens the **Properties** dialog box for the specified Autodesk Inventor file. See Figure 3.86. You can also open the **Properties** dialog box using one of the following techniques:

✓ In your Windows Explorer, right-click any Autodesk Inventor file and pick **Properties** from the shortcut menu.

✓ In the Autodesk Inventor **Design Assistant,** right-click any Autodesk Inventor file and pick **Properties** from the shortcut menu. The Autodesk Inventor **Design Assistant** is discussed in Chapter 19.

Inside the **Properties** dialog box are different tabs that represent sets of Autodesk Inventor file properties. These properties are used throughout the design process to organize, manage, and document specific model property data. They are also used for developing title blocks, part lists, and bills of materials.

Field Notes

> You can specify properties at any time during the design process. However, it may be most effective to define properties when you begin to develop a model, either in a part or assembly file.
>   Specify many of the common properties used throughout a project in a template file. Every time you create a new part, the properties will be the same.

The following describes each of the tabs located in the **Properties** dialog box:

■ **Summary**   This tab, shown in Figure 3.86, contains a number of edit boxes that allow you to define summary properties for the specified file. Summary properties are used to organize and help control specific file information such as the title, author, and company name. In addition, these properties help you develop and manipulate title blocks, part lists, reports, and bills of materials. To use the edit boxes in the **Summary** tab, pick inside the edit box and type your information.

Field Notes

> The author is automatically specified as the name provided when Autodesk Inventor was loaded, or as the name of the individual who initially created the file if you are working in a networked environment. Still, the default name can be modified.

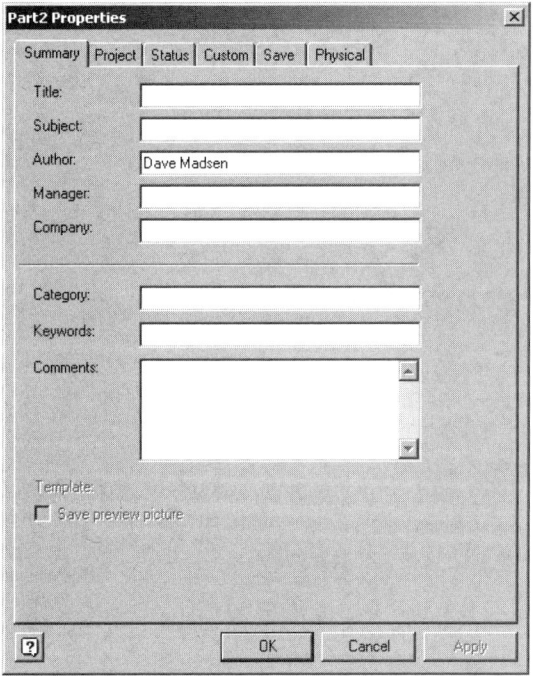

FIGURE 3.86 The **Properties** dialog box (part and assembly files).

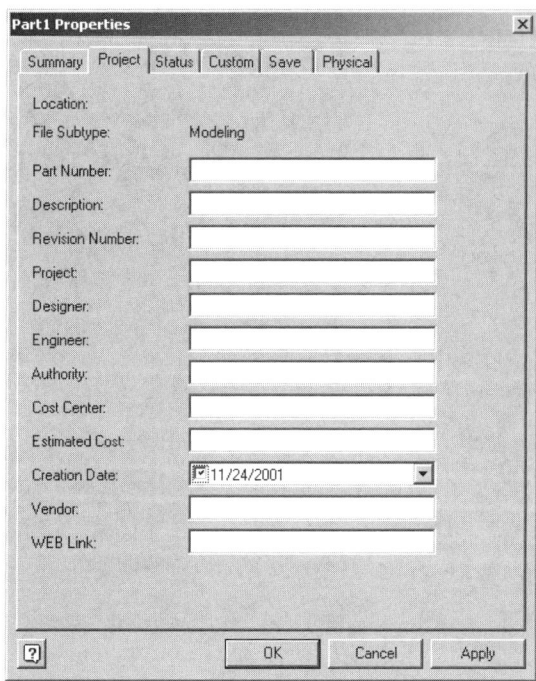

FIGURE 3.87 The **Project** tab of the **Properties** dialog box.

■ **Project**   This tab contains a number of edit boxes and a creation date drop-down list that allow you to define project properties for the specified file. See Figure 3.87. Project properties are used to organize and help control additional file information such as the part number and engineer's name. These properties also help you develop and manipulate title blocks, part lists, reports, and bills of materials. To use the edit boxes, pick inside the edit box and type in your information. To use the **Creation Date:** drop-down list, pick the arrow and select a date.

Field Notes

The file name is automatically assigned as the part number if you do not define another number for the part.
A real number must be used for the estimated cost.

■ **Status**   This tab, shown in Figure 3.88, contains a number of edit boxes and drop-down lists that allow you to define the status properties of the specified file. Status properties are used to organize and help control additional file information regarding design standing. Information such as the design state, checks, approvals, and file status are available. These properties also help you develop and manipulate title blocks, part lists, reports, and bills of materials. To use the edit boxes, pick inside the edit box and type in your information. To use the date drop-down lists, pick the arrow and select a date.

The **File Status** area of the **Status** tab provides information about the file condition for use in a collaborative work environment. To use this area you must initiate the **Multiuser** option, introduced in Chapter 1. The **Multiuser** check box is available from the **Options** dialog box described later in this chapter. The reserve information regarding the file is shown in the edit boxes. To reserve the file, pick the **Reserve** button.

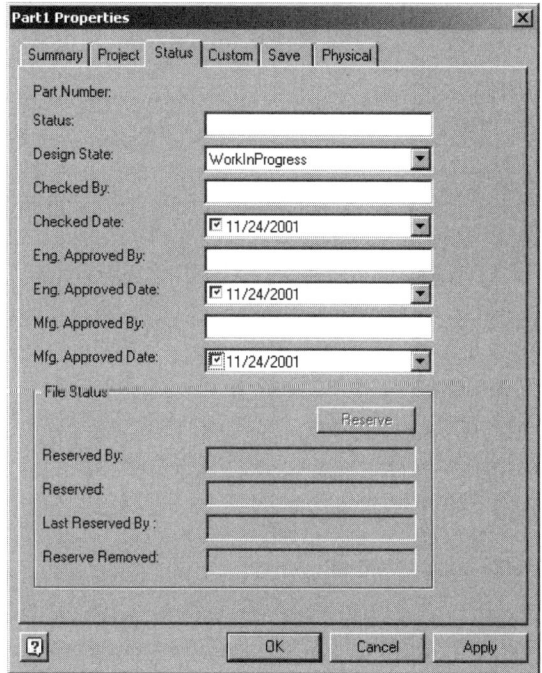

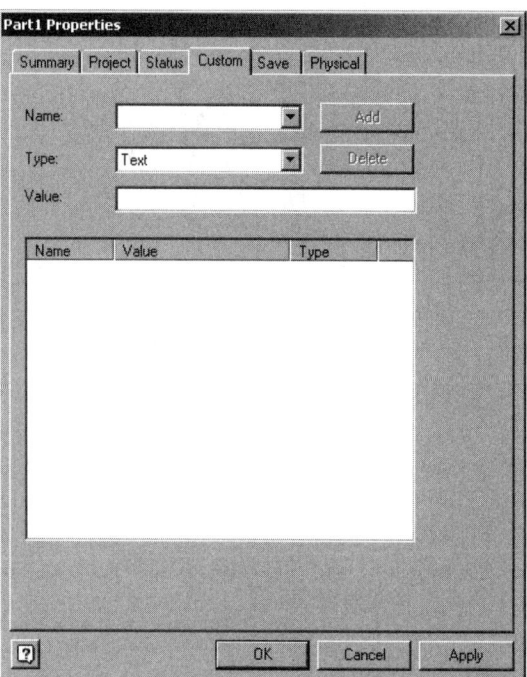

FIGURE 3.88 The **Status** tab of the **Properties** dialog box.

FIGURE 3.89 The **Custom** tab of the **Properties** dialog box.

- **Custom** This tab contains a **Name** drop-down list/edit box, a **Type** and **Value** drop-down list, **Add** and **Delete** buttons, and a display window. See Figure 3.89. This tab allows you to create and delete custom properties for the specified file. Custom properties are used to develop any additional user defined properties that may be required to further organize and help control file information. These properties also help you add information to title blocks, part lists, reports, and bills of materials.

  To create a custom property, pick the **Name** edit box and type in a property name. Then, select the type of property from the **Type** drop-down list. Next, enter a value, or use the drop-down list, if available, to specify the property value. The particular value corresponds to the property type. Once you have completed the custom property information, pick the **Add** button and notice the property shown in the display window. If you would like to delete a custom property, pick the property in the display window and select the **Delete** button.

- **Save** This tab, shown in Figure 3.90, is exactly the same as the **File Save Options** dialog box discussed earlier in this chapter. The **Save** tab allows you to adjust the save settings of the current models image, displayed in the preview area of the **File Open** dialog box, discussed in Chapter 1.

- **Physical** This tab contains a number of options that allow you to set the type of material and adjust and analyze the physical properties of a part or assembly file. See Figure 3.91. You can use this tab to show how different materials, dimensions, and tolerances influence the physical and inertial properties of the specified model.

 Field Notes

When you select a material in the **Physical** tab of the **Properties** dialog box, the part becomes this material, and the color displayed in the **Color** drop-down list of the **Command** bar is: As Material.

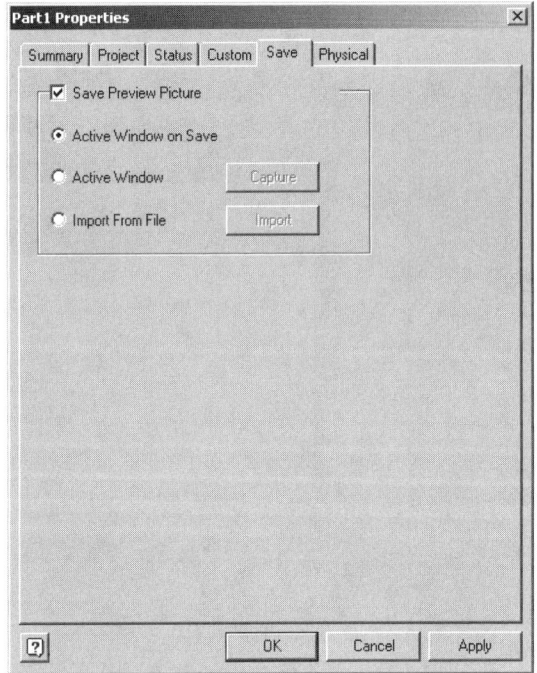

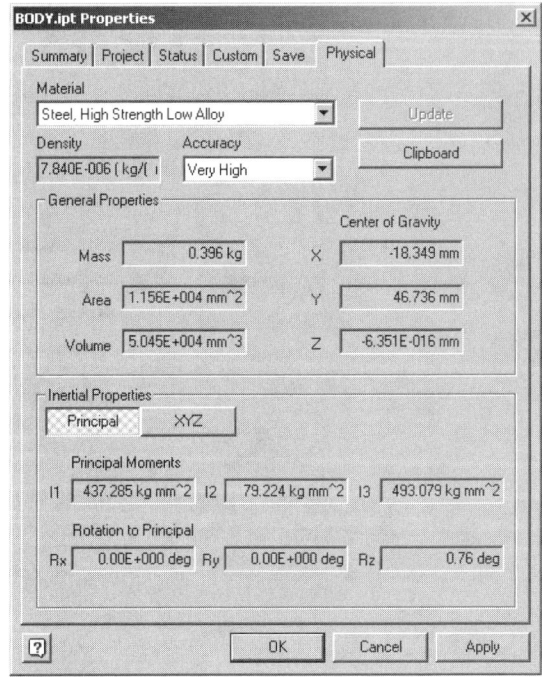

FIGURE 3.90   The **Save** tab of the **Properties** dialog box.

FIGURE 3.91   The **Physical** tab of the **Properties** dialog box.

CHAPTER TEST

Answer the following questions on a separate sheet of paper.

Part 1: Developing Sketches

1. What is the typical first step in creating a sketch?

2. What is typically the second step in creating a sketch?

3. Give the function of the Units tab in the Document Settings dialog box.

4. Discuss the function of the Sketch tab in the Document Settings dialog box.

5. Explain the function of the Constraint Placement Priority area of the Sketch tab in the Options dialog box.

6. Give the function of the Display area of the Sketch tab in the Options dialog box.

7. Discuss the function of the Overconstrained Dimensions section and its two radio buttons of the Sketch tab in the Options dialog box.

8. Explain the function of the Snap to Grid check box of the Sketch tab in the Options dialog box.

9. Give the function of the Edit dimensions when created check box of the Sketch tab in the Options dialog box.

10. Discuss the function of the Autoproject edges during curve creation check box of the Sketch tab in the Options dialog box.

11. Explain the function of the Automatic reference edges for new sketch check box of the Sketch tab in the Options dialog box.

12. Give the function of the Auto-Bend with 3D Line Creation check box of the Sketch tab in the Options dialog box.

13. Explain the function of the Sketch on New Part Creation area of the Part tab in the Options dialog box.

14. Discuss the function of the Parallel View of Sketch Criteria check box of the Part tab in the Options dialog box.

15. Give the function of the Auto-hide in-line work features check box of the Part tab in the Options dialog box.

16. Give the function of the Sketch Priority setting of the Select button in the command bar.

17. Explain the function of the Return button in the command bar.

18. Discuss the function of the Sketch button in the command bar.

19. Give the function of the Update option in the command bar.

20. Explain the function of the Style drop-down list in the command bar.

21. What is a consumed sketch?

22. What is an unconsumed sketch?

23. Explain the process for projecting the work features available in the Origin folder.

24. By default, the panel bar is displayed in learning mode. How do you change from learning to expert mode?

25. To draw a line, pick the point where you want to begin the line, and then pick the next point. If you want to continue the line, pick the next point. What do you do if you do not want to continue the line?

26. What do you do if you do not want to continue the current line, but would like to draw another line at another location?

27. Explain the function of the AutoProject option.

28. Discuss the function of the Coincident icon.

29. What is a spline?

30. How do you deactivate the automatic referencing tools while in the Line command?

31. What must exist before you can use the tangent arc option?

32. Autodesk Inventor contains a powerful tool that allows you to create arcs while in the Line command. Explain how this works?

33. Name the two options for sketching a rectangle.

34. Why do you rarely want to apply geometry such as fillets or chamfers to a sketch?

35. What are sketch points?

36. What are hole centers?

37. What does the dotted line indicate that you get after drawing one or more hole centers and you want to draw another hole center?

38. Outline how to sketch a polygon.

39. What do you need to do before you can mirror an object?

40. Name two patterning tools in addition to the Mirror tool.

41. What is an associative pattern?

42. What is the axis of rotation of a pattern?

43. How do you change from a normal line to a centerline?

44. Name the command that allows you to create parallel copies of objects at a specified distance from the existing object.

45. Give the function of the Perpendicular constraint tool.

46. Discuss the function of the Parallel constraint tool.

47. Explain the function of the Tangent constraint tool.

48. Give the function of the Coincident constraint tool.

49. Discuss the function of the Concentric constraint tool.

50. Give the function of the Collinear constraint tool.

51. Explain the function of the Horizontal constraint tool.

52. Discuss the function of the Vertical constraint tool.

53. Give the function of the Equal constraint tool.

54. Explain the function of the Fix constraint tool.

55. Give the function of the Show Constraints tool.

56. How do you show all the constraints for only certain objects?

57. What do you do if you want to keep the constraint symbols on the screen?

58. What happens when you move your cursor over, or select, a constraint symbol?

59. How do you show all the sketch constraints without selecting individual objects?

60. When the constraint display symbols are visible, you have the option of deleting certain constraints in order to apply a different relationship. How do you delete constraints?

61. How do you move the constraint display symbols if they are blocking your view?

62. How do you close the constraint display symbols box?

63. How do you close all the sketch constraints without selecting individual hide buttons?

64. How do you know when your sketch is fully constrained?

65. Give the basic function of the General Dimension tool.

66. What do you get when placing dimensions if the Edit dimensions when created check box is selected?

67. If you do not select the Edit dimensions when created check box from the Sketch tab of the Options dialog box, the object is dimensioned to its current measurement. How do you open the Edit Dimension dialog box in this case?

68. Where do you type a specific numerical value when using the Edit Dimension dialog box?

69. Give the function of the following keys when using the Edit Dimension dialog box:

    a. The + key.

    b. The - key.

    c. The * key.

    d. The / key.

    e. The ^ key.

70. What do dimension units reflect?

71. What are linear dimensions and how are they placed?

72. What are aligned dimensions and how are they placed?

73. What are angular dimensions and how are they placed?

74. Where are diameter dimensions used and how are they placed?

75. Where are radius dimensions used and how are they placed?

76. How do you move sketch dimensions after they have been placed?

77. Explain the function of the Auto Dimension tool and how it is accessed.

78. Give the function of the Dimensions Required display box of the Auto Dimension dialog box.

79. Discuss the function of the Curves button of the Auto Dimension dialog box.

80. Give the function of the Dimensions check box of the Auto Dimension dialog box.

81. Explain the function of the Constraints check box of the Auto Dimension dialog box.

82. Give the function of the Remove button of the Auto Dimension dialog box.

83. Give the function of the Done button of the Auto Dimension dialog box.

84. How are driven dimension identified?

85. What do you do if you do not want Autodesk Inventor to warn you of an overconstrained condition?

86. What must you do before you can use the Precise Input toolbar?

87. Give the function of the Input Type button in the Precise Input toolbar.

88. Explain the function of the Input edit boxes in the Precise Input toolbar.

89. Give the function of the Relative Origin button in the Precise Input toolbar.

90. Discuss the function of the Delta Input button in the Precise Input toolbar.

91. Give the function of the Relative Orientation button in the Precise Input toolbar.

92. What are parameters?

93. Give the function of the Parameter Name edit box in the Model Parameters section of the Parameters dialog box.

94. Give the function of the Unit display box in the Model Parameters section of the Parameters dialog box.

95. Explain the function of the Equation edit box in the Model Parameters section of the Parameters dialog box.

96. Discuss the function of the Value display box in the Model Parameters section of the Parameters dialog box.

97. Give the function of the Comment edit box in the Model Parameters section of the Parameters dialog box.

98. Give the function of the Add button in the Model Parameters section of the Parameters dialog box.

99. Discuss the function of the Link button in the Model Parameters section of the Parameters dialog box.

100. Explain how dragging geometry works when editing a sketch.

101. Discuss the function of the Extend button when editing a sketch.

102. Give the function of the Trim button when editing a sketch.

103. Explain the function of the Move button when editing a sketch.

104. Discuss the function of the Rotate button when editing a sketch.

Part 2: Inserting AutoCAD Drawings for Sketches
    1. How do you access the Open DWG File Options dialog box?

2. Briefly give the function of the Insert AutoCAD File wizard.

3. Once you define the options in one page of the Insert AutoCAD File wizard, how do you access additional pages?

4. How do you return to a previous page in the Insert AutoCAD File wizard?

5. What do you do when you have fully defined the import options when using the Insert Auto-CAD File wizard?

6. Give the function of the DWG File Import Options page in the Insert AutoCAD File wizard.

7. Give the function of the Layers and Objects Import Options page in the Insert AutoCAD File wizard.

8. Discuss the function of the Destination for 2D data area of the Import Destinations Options page in the Insert AutoCAD File wizard.

9. Explain the function of the Templates area of the Import Destinations Options page in the Insert AutoCAD File wizard.

10. Give the function of the Save to configuration area of the Import Destinations Options page in the Insert AutoCAD File wizard.

Part 3: The Sketch Doctor

1. The Sketch Doctor is a tool that helps you resolve what types of sketch problems?

2. Name the button on the Standard toolbar where you can access the Sketch Doctor.

3. Give the function of the Diagnose Sketch dialog box.

4. Which diagnostic tests are selected by default in the Diagnose Sketch dialog box?

5. Explain the function of the Sketch Doctor – Examine section.

6. How do you select a treatment list box and get solution directions to solve the sketch problem?

Part 4: The Properties Dialog Box

1. Give the function of the Summary tab in the Properties dialog box.

2. Who is automatically specified as the author?

3. Explain the function of the Project tab in the Properties dialog box.

4. Discuss the function of the Status tab in the Properties dialog box.

5. Give the function of the File Status area in the Summary tab in the Properties dialog box.

6. What must you do first before you can use the File Status area in the Summary tab?

7. Discuss the function of the Custom tab in the Properties dialog box.

8. Give the steps used to create a custom property.

9. Give the function of the Save tab in the Properties dialog box.

10. Discuss the function of the Physical tab in the Properties dialog box and what happens when you select a material in the Physical tab.

|||||||||||| PROJECTS

Instructions:

- Create sketches of the following objects using the techniques discussed in this chapter.

- Develop sketch geometry from the projected center point.

- Fully constrain each sketch.

- Use an Equal constraint for like objects not dimensioned in the project figure.

- *Note:* Sketch dimensions shown for reference are created using Autodesk Inventor and may not comply with ASME standards.

1. Name: C-Clamp Pin

   Units: Metric

   Save As: P3.1

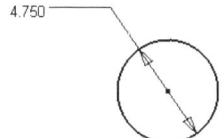

2. Name: C-Clamp Swivel

   Units: Metric

   Save As: P3.2

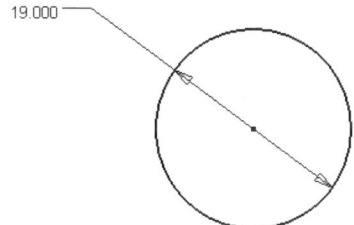

3. Name: C-Clamp Screw

   Units: Metric

   Save As: P3.3

4. Name: Support Bracket

Save As: P3.4

Units: Inch

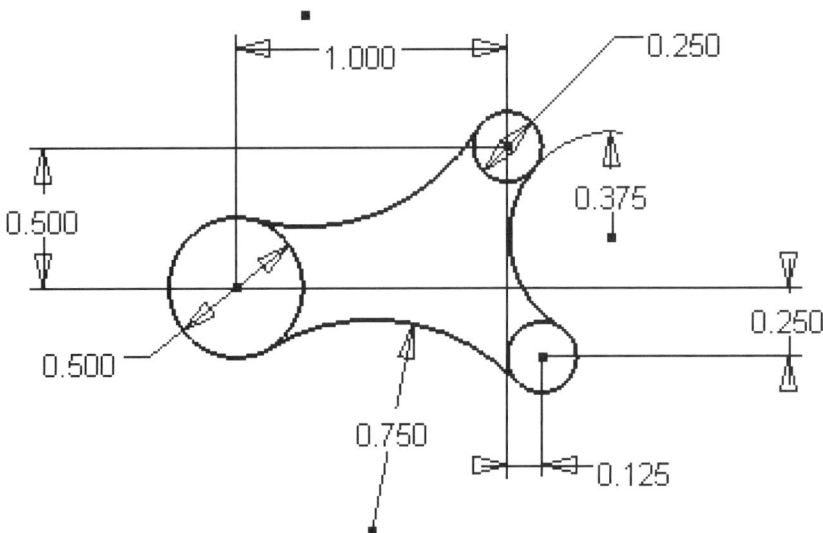

5. Name: Handle

Save As: P3.5

Units: Inch

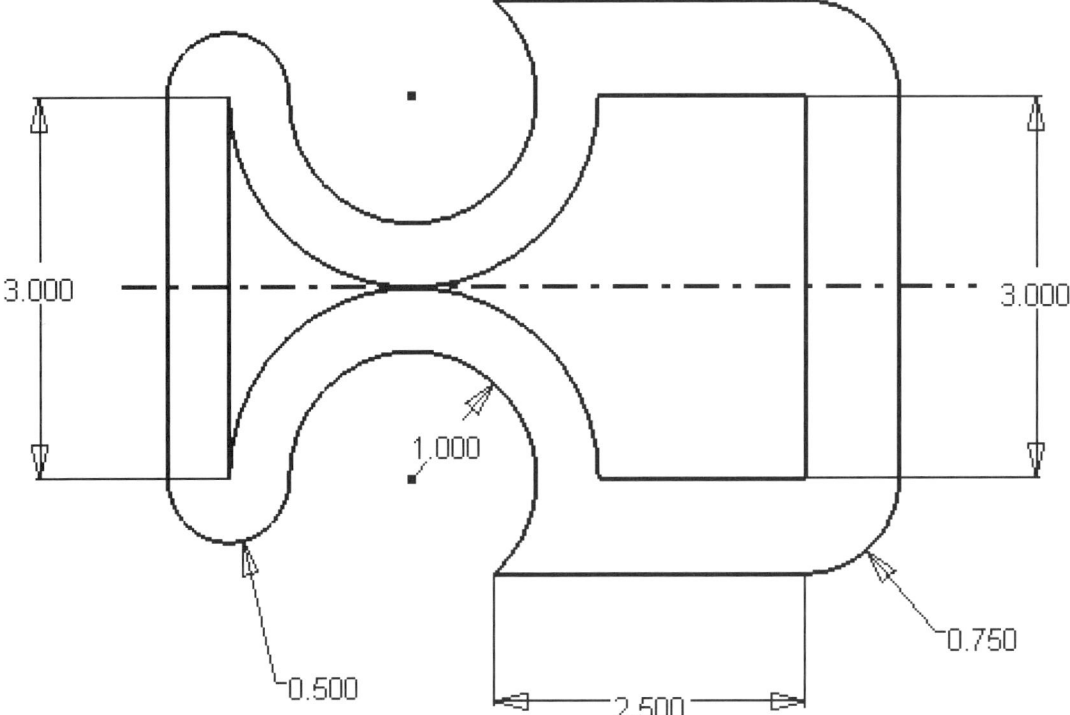

6. Name: C-Clamp Body

Save As: P3.6

Units: Metric

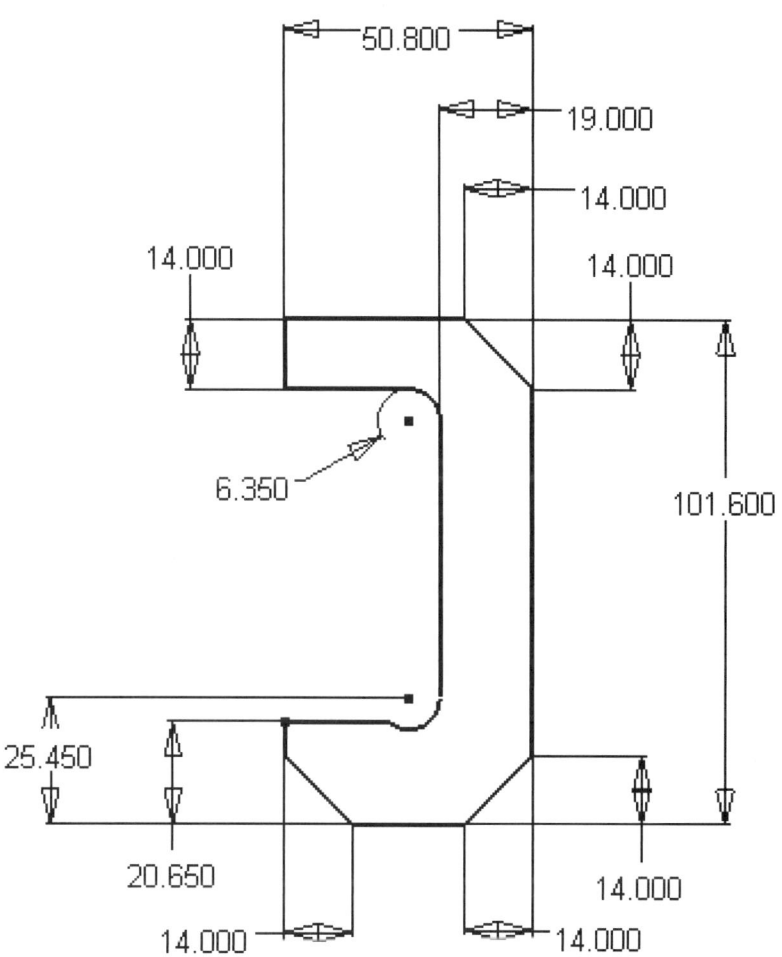

|||||||||||||||||||||||||||||||||||||||||||||||||||||||||||||||||||||||||||||||||||||||||||||||

CHAPTER 4

Developing Basic Part Model Features

LEARNING GOALS

After completing this chapter, you will be able to:

◎ Work in the part and feature environment and interface.

◎ Create extruded base features with the **Extrude** command.

◎ Create revolved features with the **Revolve** tool.

◎ Sketch on features.

◎ Use automatic model edge referencing techniques.

◎ Generate additional extruded features.

◎ Add holes to a part feature using the **Hole** tool.

◎ Create ribs and webs using the **Rib** tool.

◎ Develop coils using the **Coil** command.

Developing Sketched Features

Once you have created a sketch, you are ready to develop a three-dimensional part using Autodesk Inventor feature tools. Usually the first type of feature you create is a sketched feature. As the name implies, sketched features are created from sketch geometry. They are typically simple and do not require a great deal of time to create. Often the initial feature, known as the base feature, is created as a sketched feature. Still, sketches and sketched features can be created any time during part model development. Sketch features include extrusions, revolutions, sweeps, lofts, coils, holes, and ribs.

Working in the Part and Feature Environment

The work environment and interface is different for each stage of the design process. Some of the tools and interface components remain constant or are similar throughout the entire design process. However, the sketched feature and part environment and many of the interface components are specific to working with parts and features.

To enter the feature interface and work environment after you have created a sketch, use one of the following options:

✓ Right-click and select the **Finish Sketch** option from the shortcut menu.

✓ Use the **+S** accelerator key option by pressing the **+** and **S** keys on your keyboard.

✓ Pick the available feature tool from the **Feature** panel bar or **Feature** toolbar.

✓ Select the **Return** button on the **Command** bar.

Pull-Down Menu System Options

Most of the pull-down menus discussed in Chapter 2 apply to the feature environment. However, there are a few specific options that relate to parts and features:

- **View**   As discussed in Chapter 2, the **View** pull-down menu contains a number of tools that allow you to manipulate the display of objects in the graphics window. Viewing tools available in a part are also available in the sketch environment, discussed in Chapter 3. However, it is not until you have created a feature that the following tools become most useful:

- **Rotate**   Select the **Rotate** option to rotate an object and view it from different angles. You can also access the **Rotate** command by picking the **Rotate** button on the **Standard** toolbar or by holding down the **F4** key on your keyboard. There are two rotation modes available for viewing an object. When you access the **Rotate** command, by default, you are in **Free Rotate** mode. See Figure 4.1. To rotate an object while in **Free Rotate** mode, hold down the left mouse button and drag the rotation icon and object to the desired display. If you choose inside the circle shown in Figure 4.1, you can rotate the object any direction. You can also rotate an object around one of the following axes:

  - **Y axis**   To rotate the object around the Y axis, move your cursor over one of the horizontal lines, hold down the left mouse button, and drag the rotation icon left or right.

  - **X axis**   To rotate the object around the X axis, move your cursor over one of the vertical lines, hold down the left mouse button, and drag the rotation icon up or down.

  - **Z axis**   To rotate the object around the Z axis, move your cursor outside of the circle, near one of the quadrants. Then hold down the left mouse button, and drag the rotation icon in a circular pattern.

  When you are done using the **Rotate** tool, press the Esc key, pick outside the rotation area, or right-click and choose the **Done** menu option.

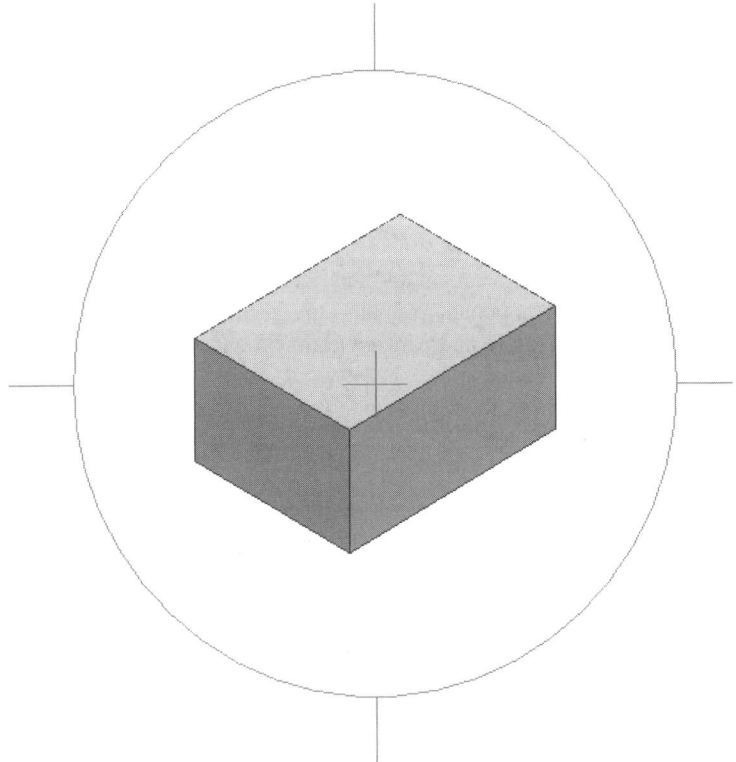

FIGURE 4.1 The **Free Rotate** mode of the **Rotate** command.

FIGURE 4.2 The **Common View** mode of the **Rotate** command.

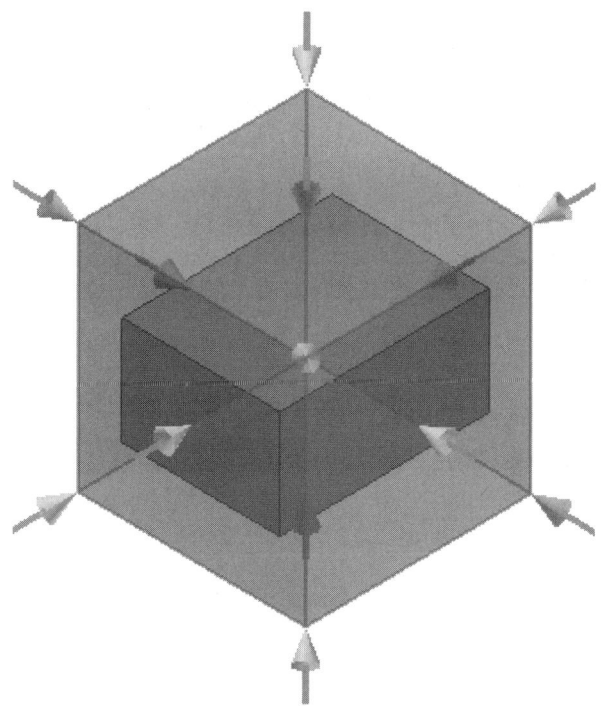

The second mode available for rotating an object is the **Common View** mode. See Figure 4.2. Access the **Common View** tool by first accessing the **Rotate** command. Then right click and select the **C̲ommon View [Space]** menu option. To rotate an object while in **Common View** mode, simply pick the desired green arrow. These arrows point to the view that they display.

Field Notes

The rotate options previously discussed rotate the entire graphics space, including the display of the objects. They do not actually change the location of any object.

- **Zoom S̲elect**   Use the **Zoom S̲elect** tool to zoom in on a selected piece of geometry. The object you can select is determined by the selection option you choose from the **Select** flyout button of the **Command** bar. See Figure 4.3. You can choose to select a face, feature, or sketch. The sketch option and **Sketch Priority** was discussed in Chapter 2. Setting the **Select** button to **F̲ace Priority** allows you to pick a feature face; setting the **Select** button to **F̲eature Priority** allows you to pick an entire feature.

    To use the **Zoom S̲elect** tool, simply pick the object you would like to zoom in on. You can also access the **Zoom Select** command by picking the **Zoom Select** button on the **Standard** toolbar.

FIGURE 4.3 The **Select** flyout button of the **Command** bar.

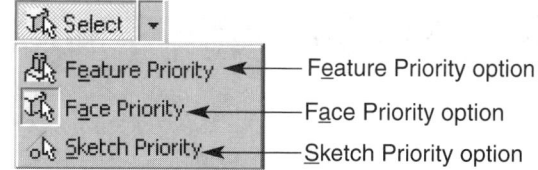

Field Notes

Sometimes, you may not be able to select the piece of geometry you desire. To locate the geometry you may first need to rotate or change the display options. An alternative way to access the object you want to select is to use the wheel on a wheel mouse. To use this tool, move your cursor over a face, point, edge, or feature, and wait for the wheel mouse icon to appear. Then roll the wheel on your wheel mouse until the geometry you want to select is highlighted, and pick the object.

- **Look At**   Use this tool to reorient the display to the selected object. You can also access the **Look At** command by picking the **Look At** button on the **Standard** toolbar, or by right-clicking on an object and choosing the **Look At** menu option.

- **Isometric**   Use this tool to display the isometric view of an object. You can also access the **Isometric** command by right-clicking and selecting the **Isometric View** menu option.

Field Notes

If you want to redefine the default isometric view, first rotate the object to the desired display using the **Common View** tool. Then, while still in the **Common View** command, right-click and choose the **Redefine Isometric** menu option.

- **Format**   This pull-down menu allows you to modify certain display and material information. The following options are available inside the **Format** pull-down menu while in a part:

- **Lighting...**   Picking this option opens the **Lighting** dialog box, shown in Figure 4.4. Here you can adjust and create model display lighting preferences. The following options are available inside the **Lighting** dialog box:

- **Style Name**   All the current lighting styles are displayed in the **Style Name** list box. You can select different style names to view lighting style settings, but you cannot activate a style from this location. By default, two lighting styles are available: Default and Two Lights. To create a new style, pick the **New** button. A new style is created from the active style. To change the name of the style, highlight the style name inside

FIGURE 4.4   The **Lighting** dialog box.

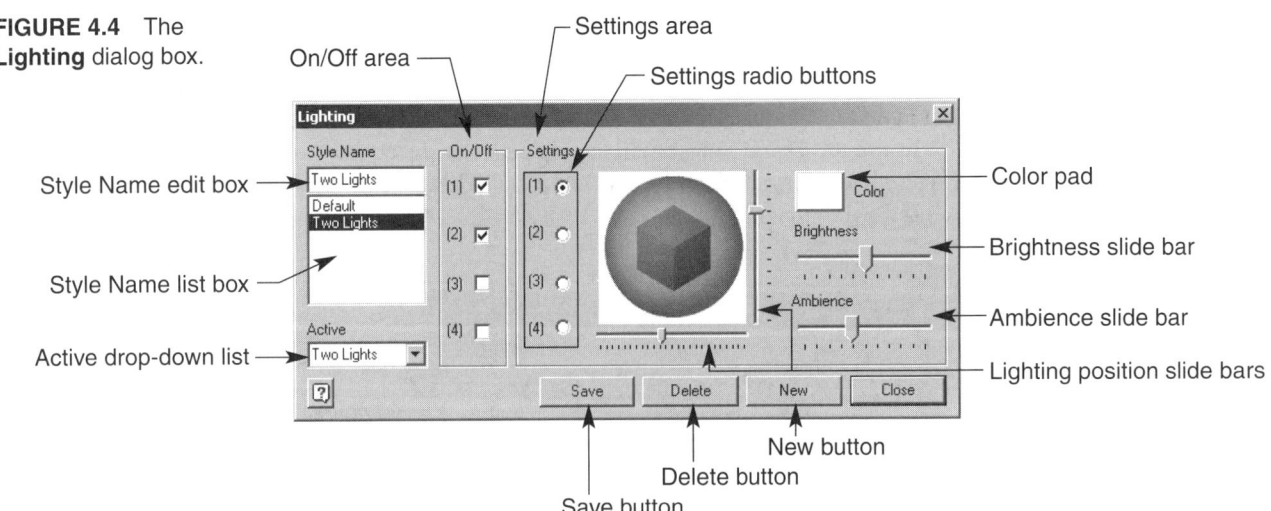

the **Style Name** edit box and enter the desired name. Finish the new style creation by pressing **Enter** on your keyboard, or picking the **Save** button. To delete a style, highlight the style you would like to remove and pick the **Delete** button.

- **Active**   This drop-down list allows you to activate an available style by simply picking a style name.

- **On/Off**   The **On/Off** area contains four check boxes that allow you to activate or suppress lighting settings. When you activate more than one of the check boxes, the lighting effects of each of the settings are combined together.

- **Settings**   The **Settings** area controls lighting preferences. The four radio buttons correspond to the **On/Off** check boxes previously discussed and allow you to create different lighting options for the same style. Pick the desired radio button and apply the following settings:

  - Lighting position slide bars   The vertical lighting position slide bar controls the amount of light provided above and below the model. The horizontal lighting position slide bar controls the amount of light provided left and right of the model.

  - **Color**   Pick this button to open the **Color** dialog box. See Figure 4.5. Here you can specify a light color.

  - **Brightness**   This slide bar allows you to change the brightness of the light source for all settings. If you would like to change the brightness for just one setting, you need to change the color of the light source.

  - **Ambience**   This slide bar controls the ambience and contrast of the available lighting.

Field Notes

> If you change the color of the light source, all model and material colors will reflect the new color.
>
> When changing lighting options, have a model in the graphics window to instantly see the changes made by your selections.

- **Materials**   Picking this option opens the **Materials** dialog box, shown in Figure 4.6. Here you can create user-defined materials and edit the properties of existing materials. These are the materials specified in the **Physical** tab of the **Properties** dialog box,

FIGURE 4.5   The
Color dialog box.

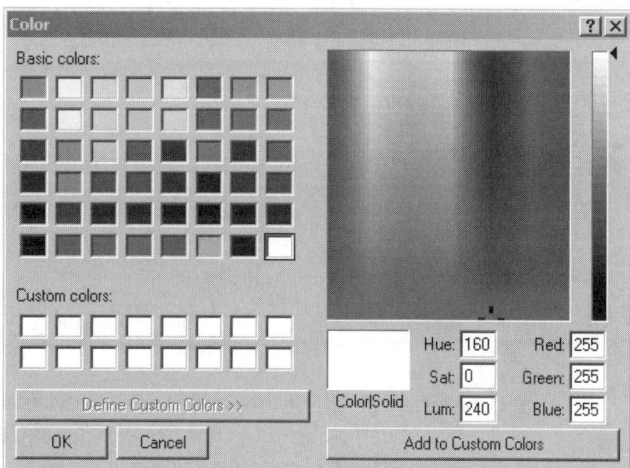

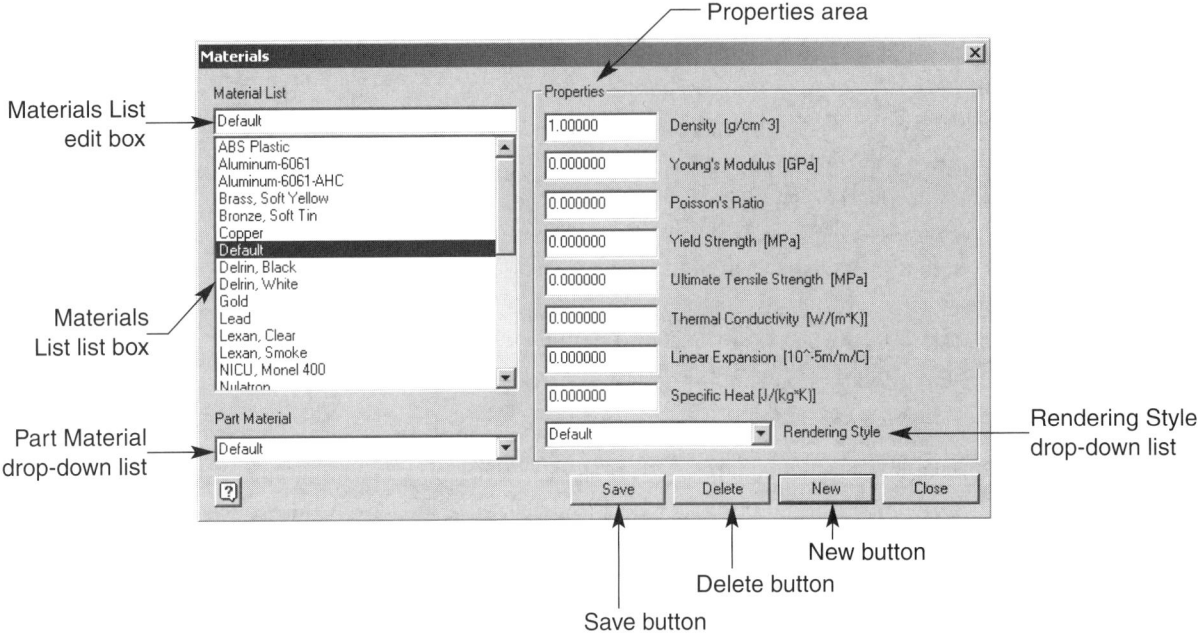

FIGURE 4.6 The **Materials** dialog box.

discussed in Chapter 3. The following options are available inside the **Materials** dialog box:

- **Materials List**   All currently available materials are displayed in the **Materials List** list box. Select a material to view its properties in the **Properties** area. To create a new material, pick the **New** button. A new material is created from the active material. To change the name of the material, highlight the material name and enter the desired name. Finish the new material creation by pressing **Enter** on your keyboard or picking the **Save** button. To delete a material, highlight the material you would like to remove and pick the **Delete** button.

- **Part Material**   This drop-down list allows you to specify the material for the current part by simply picking a material name.

- **Properties**   This area contains a number of edit boxes that allow you to adjust the physical properties of a material. These properties are used to calculate the physical properties of a part, as discussed in Chapter 3. To modify an existing property, highlight the property value and enter a new value in the edit box. You can also adjust the rendering style by selecting a color or material from the **Rendering Style** drop-down list. This can also be accomplished by picking a part and selecting a new color from the **Color** drop-down list of the **Command** bar.

Field Notes

For more information regarding the specific physical properties available inside the **Materials** dialog box, you may want to refer to the **Material Formats** section of the Autodesk Inventor Help Files.

- **Colors...**   Picking this option opens the **Colors** dialog box, shown in Figure 4.7. Here you can adjust and create model display colors. The following options are available inside the **Colors** dialog box:

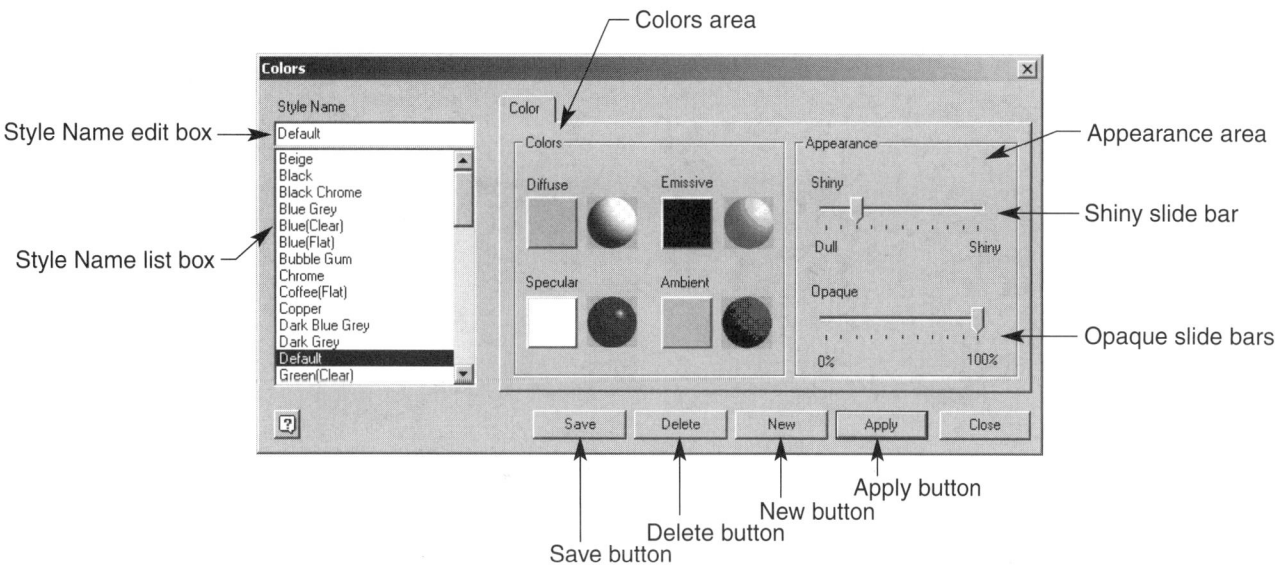

FIGURE 4.7   The **Colors** dialog box.

■ **Style Name**   All current color style options are displayed in the **Style Name** list box. You can select different color style names to view their settings. To apply a color style to the active part, pick the **Apply** button. This can also be accomplished by picking a part and selecting a new color from the **Color** drop-down list of the **Command** bar. To create a new color style from the active color style, pick the **New** button. To change the name of the color style, highlight the color style name inside the **Style Name** edit box and enter the desired name. Finish the new color style creation by pressing **Enter** on your keyboard or picking the **Save** button. To delete a color style, highlight the style you would like to remove and pick the **Delete** button.

■ **Colors**   This section of the **Colors** dialog box contains four color pads that open the **Color** dialog box and allow you to modify the diffuse, emissive, specular, and ambient colors.

■ **Appearance**   This section allows you to modify the shiny and opaque characteristics of a part surface in reference to the specified color. To change the appearance of a color, move the **Shiny** and **Opaque** slide bars to the desired setting.

Using the Display Flyout Buttons

The display flyout buttons are located on the **Standard** toolbar and are available in part, assembly, and presentation files. See Figure 4.8. By default, the **Shaded Display** button is active, and features are displayed in a shaded format. See Figure 4.9. The **Shaded Display** option looks realistic and allows you to observe a model as if it is an actual part.

When you select the **Hidden Edge Display** button, features are displayed in a shaded format. In addition, the feature edges are visible. See Figure 4.10. The **Hidden Edge Display** option has all the features of the **Shaded Display** option.

FIGURE 4.8   The display flyout buttons of the **Standard** toolbar.

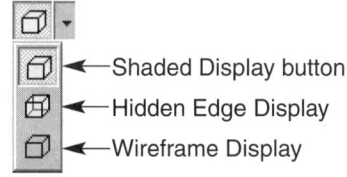

FIGURE 4.9 The shaded display of a part.   **FIGURE 4.10**   The **Hidden Edge Display** option.

FIGURE 4.11   The
Wireframe Display
option.

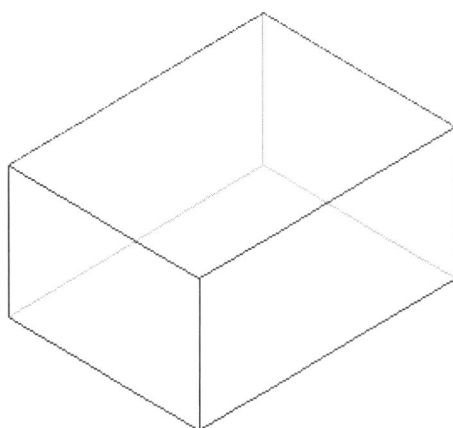

When you select the **Wireframe Display** button, features are displayed in a wireframe format. See Figure 4.11. A wireframe display removes all colors and lighting and allows you to view only the edges of a feature. This may be helpful when working with more complex features.

Using the Camera View Buttons

The camera view flyout buttons allow you to define the model view characteristics. These buttons are located on the **Standard** toolbar and are available in part, assembly, and presentation files. See Figure 4.12. By default, the **Orthographic Camera** button is active, and features are displayed in an orthographic appearance. See Figure 4.13. The **Orthographic Camera** option displays features in their true size and shape and allows you to observe the actual dimensions and position of geometry.

When you select the **Perspective Camera** button, features are displayed in a perspective appearance. See Figure 4.14. Features shown in a perspective view look more realistic because depth and distortion are added in an effort to illustrate what the human eye would actually see if the feature was a real object. For example, a pipe edge that is closer to you will look larger than the more distant edge.

While observing an object using the **Perspective Camera** option, you can use view tools such as **Zoom, Pan,** and **Rotate** to apply what is known as a *fly-through.* A fly-through is what you do when you

FIGURE 4.12   The
camera view buttons of
the **Standard** toolbar.

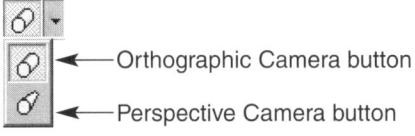

◄————Orthographic Camera button

◄————Perspective Camera button

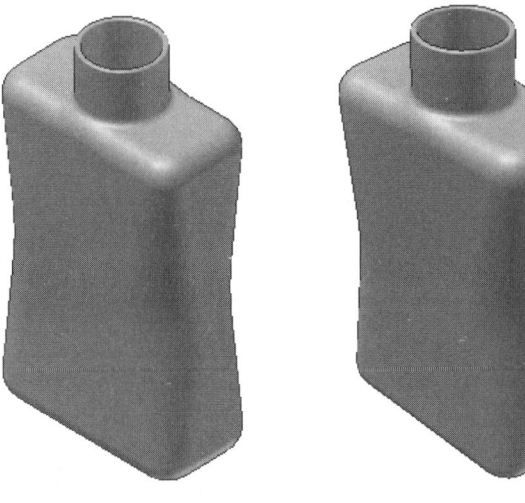

FIGURE 4.13 The orthographic display of a part.

FIGURE 4.14 The **Perspective Camera** option.

zoom and move in, out, and around an object in perspective camera display. When you change the position of the object by zooming, panning, and rotating, the view depth and distortion also changes to reflect the new position. Although, you can fly through features while in orthographic camera mode, typically a fly-through looks more realistic while in perspective camera mode.

Using the Browser in a Part

The browser for a typical part displays the part icon, and part name, the **Origin** folder, each of the features, and **End of Part.** Figure 4.15 shows the browser for a c-clamp part named BODY.ipt. The first feature under the **Origin** folder is the base feature, as previously discussed in this chapter. The base feature in this example is an extrusion named Extrusion1.

When you right-click on Extrusion1 or any sketched feature in a part, the following options are available:

- **Copy**   This option allows you to copy a feature to paste into the current part or another document.

- **Paste**   Using this option, you can insert a feature that has been previously copied using the **Copy** tool.

Field Notes

> You can also copy and paste a feature by dragging the feature icon from the **Browser** bar to the graphics window.

FIGURE 4.15 The browser display for a typical part.

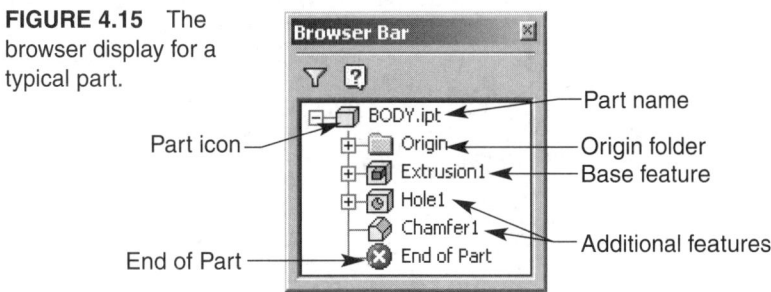

- **Delete**   This menu option completely removes the feature from the part.

- **Show Dimensions**   Selecting this menu option displays the dimensions of the sketch used to create the feature.

- **Edit Sketch**   This selection allows you to reopen the sketch and modify any sketch geometry, dimensions, constraints, or any other sketch information.

- **Edit Feature**   This menu option reopens the specified feature dialog box and allows you to modify the dimensions and geometry of the feature.

- **Create Note**   This selection allows you to add a note using the **Engineer's Notebook,** discussed in Chapter 19.

- **Suppress Features**   The **Suppress Features** selection enables you to suppress, or hide, the specified feature. This menu option does not actually remove or delete the feature. You will notice the feature and any related geometry turns gray and is crossed out.

- **Adaptive**   Selecting this option defines the feature and the relative sketch as adaptive. Adapting parts is discussed in Chapter 15.

- **Expand all Children**   As mentioned in Chapter 2, the **Expand all Children** selection expands, or makes visible, all the child nodes.

- **Collapse all Children**   As mentioned in Chapter 2, the **Collapse all Children** selection collapses, or hides, all the child nodes.

- **Find in Window**   The **Find in Window** option finds the specified feature on the part. Autodesk Inventor zooms in and outlines the feature in the graphics window.

- **Properties**   Selecting the **Properties** option opens the **Feature Properties** dialog box for the specified feature. See Figure 4.16. Different types of features will display different feature properties in the **Feature Properties** dialog box. For example, a hole has different properties than an extrusion. The following choices may be available inside the **Feature Properties** dialog box:

  - **Name**   This edit box allows you to change the name of the feature. If the current name in the edit box is not highlighted, double-click inside the edit box to highlight the name. Then type in the new name.

  - **Suppress**   This check box accomplishes the same task as the **Suppress Features** option. It enables you to suppress or hide the specified feature without removing or deleting the feature.

  - **Adaptive**   This area may or may not be available for some features. If it is displayed, a specific number of check boxes are offered. These check boxes pertain to the specified feature and allow you to set aspects of the feature as adaptive.

  - **Feature Color**   This drop-down list allows you to change the color of the specified feature.

- **How To...**   This option opens the Autodesk Inventor Help window, discussed in Chapter 1.

FIGURE 4.16 The **Feature Properties** dialog box.

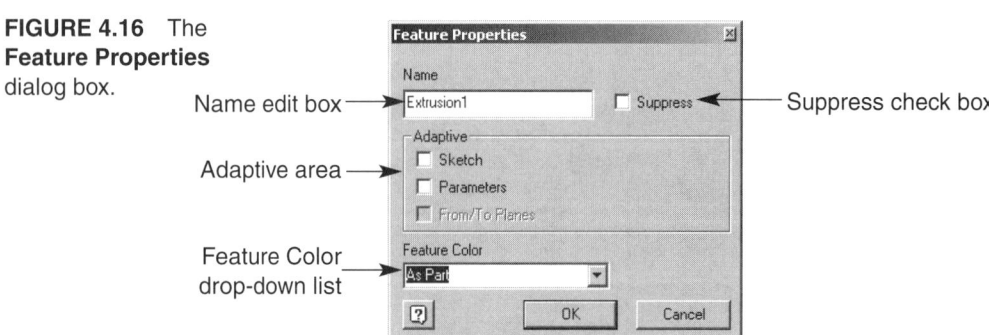

FIGURE 4.17 The consumed Sketch1.

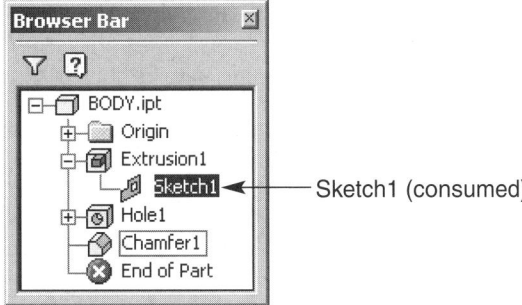

Sketch1 (consumed)

Inside Extrusion1 is a sketch, named Sketch1. See Figure 4.17. If you right-click on Sketch1, many of the previously discussed options are again available. In addition, the following options are offered:

- **Edit Coordinate System**   This option allows you to move or rotate the coordinate system to a different location on the sketch plane, by dragging or rotating the coordinates.

- **Reattach Sketch**   Use this option to move the specified sketch to a different work plane or planar model face. Once the sketch is reattached to the new area, use geometric constraints and dimensions to modify and further locate the sketch.

- **Share Sketch**   Select this option to reuse a sketch that has already been used to create a feature. This option opens the specified sketch as a new sketch and allows you to create another feature.

- **Visibility**   This option controls the visibility of the sketch. It allows you to see the sketch geometry and dimensions, even if you are not in the sketching environment.

If you right-click on a placed feature, such as Chamfer1 in Figure 4.14, the previously described: **Delete, Edit Feature, Create Note, Suppress Feature, Find in Window, Properties,** and **How To...** options are displayed. Placed features are discussed in Chapter 5.

If you right-click on a work feature, such as Work Axis1 in Figure 4.18, many of the previously described options are again available. In addition, the following two options are displayed.

- **Redefine Feature**   This option allows you to edit and remake the specified work feature. This allows you to change an existing work feature without deleting and creating a new work feature.

- **Flip Normal**   Selecting this option turns over, or flips, the Z axis direction.

- **Ground**   This option grounds the feature. A ground feature is not able to be moved.

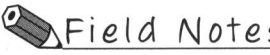

Field Notes

Only when right-clicking on a work plane are all the options previously described available.

FIGURE 4.18 Work Axis1.

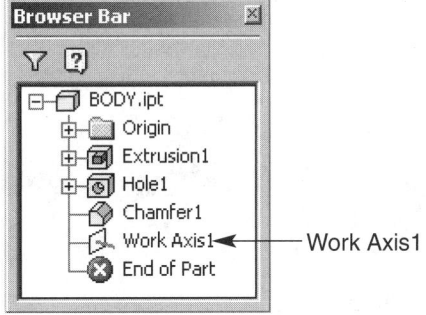

Work Axis1

In addition to the **Origin** folder and the features available in the part Browser, the **Browser Filter** button is also available, which allows you to specify the Browser display characteristics. The Following menu options are offered when you pick the **Browser Filter** button:

- **Hide Work Features**   Pick this option to hide all work features inside the Browser. Selecting this option does not hide or remove the work features in the actual model.

- **Hide Notes**   Pick this option to hide all notes inside the Browser. Selecting this option does not hide or remove the note icon in the model.

- **Hide Documents**   Choose this option if you do not want to display inserted documents inside the Browser.

- **Hide Warnings**   Select this option if you do not want to display warnings about model constraints.

Field Notes

> Additional part browser options are discussed throughout this chapter.

Using Sketched Feature Tools

Sketched feature tools are used to create sketched features, including extrusions, revolutions, sweeps, lofts, coils, holes, and ribs. These tools are accessed from the **Features** toolbar, shown in Figure 4.19, or from the **Sketch** panel bar. When a sketch is complete and you are in the feature work environment, the panel bar is in feature mode. See Figure 4.20. By default, the panel bar is displayed in learning mode, as seen in Figure 4.20. To change from learning to expert mode, seen in Figure 4.21 and discussed in Chapter 2, right-click any of the panel bar tools or left-click the panel bar title and select the **Expert** option.

Exercise 4.1
_____

1. Launch Autodesk Inventor.
2. Open Autodesk/Inventor 5.3/Samples/Models/TireRim/Tire Rim.
3. Explore the feature interface options previously discussed.
4. Close the part file without saving.
5. Exit Autodesk Inventor or continue working with the program.

Creating Extruded Base Features

An *extruded* base feature is one of the most basic and usually least complicated Autodesk Inventor features. An example of an extrusion is shown in Figure 4.22. Access the **Extrude** command using one of the following techniques:

✓ Pick the **Extrude** button on the panel bar.

✓ Pick the **Extrude** button on the **Features** toolbar.

✓ Type the **+** and **E** keys on your keyboard.

FIGURE 4.19   The **Features** toolbar.

FIGURE 4.20 The panel bar in features mode (learning mode).

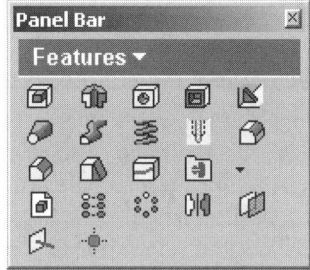

FIGURE 4.21 The panel bar in features mode (expert mode).

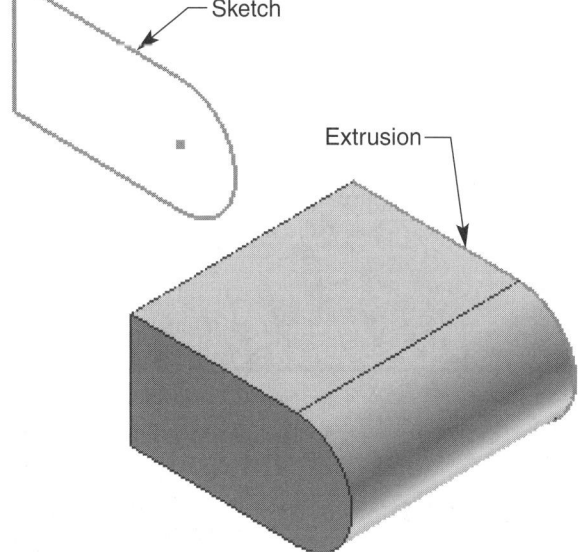

FIGURE 4.22 An example of an extrusion and the sketch used to create the extrusion.

The **Extrude** dialog box is displayed when you access the **Extrude** command. See Figure 4.23.

If only one sketch loop exists, as is often the case when creating base features, the sketch profile is already selected. To extrude a sketch with more than one profile available, pick the **Profile** button, from the **Shape** area of the **Extrude** dialog box, if it is not already active. Then pick the profile of the sketch you want to extrude. When the sketch profile is selected, a preview of the extrusion is displayed. See Figure 4.24.

Field Notes

The extrusion shown in Figure 4.21 contains only one sketch profile. However, sketches can include many profiles. To create different features from a sketch with several profiles, use the **Share Sketch** option, discussed earlier. This tool is used later in this chapter.

FIGURE 4.23 The **Extrude** dialog box (distance termination option).

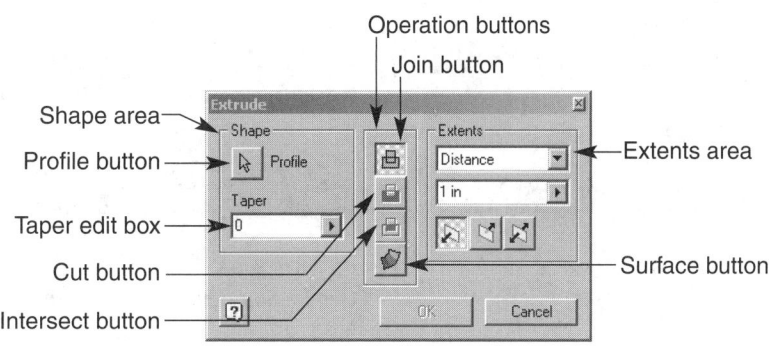

FIGURE 4.24 An extrusion preview.

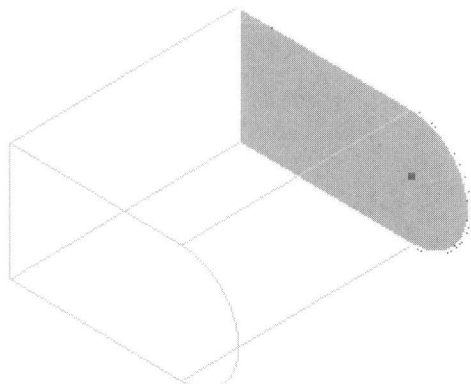

When you create a more complicated part, and multiple features exist, more options are available for developing extrusions. For a simple base feature, as shown in Figure 4.24, the following options are usable inside the **Extrude** dialog box:

- Operation type buttons   These buttons allow you to choose whether the extrusion will join, cut, or intersect another feature, or create a volumeless surface extrusion, as shown in Figure 4.25. Only the **Join** and **Surface** buttons are available for a base feature.

- **Extents**   This area contains options that allow you to specify the extrusion size and boundaries. When creating a base feature, the **Distance** ending option is the only usable extent parameter. See Figure 4.26. Using **Distance** to end an extrusion allows the extrusion to be created along the sketch plane at a desired length. Specify a distance termination from the distance drop-down list, or enter the desired value. You also have the option of picking **Measure,** as shown in Figure 4.27. This allows you to reference an existing edge to create an extrusion of the same length.

    Once you have specified an extrusion distance, pick the desired direction button. The direction buttons control the side of the sketch the extrusion depth occurs. You can choose to extrude in a positive or negative direction, or extrude equal amounts on both sides of the sketch profile.

    If you want to taper the extrusion, as shown in Figure 4.28, specify a taper angle in the **Taper** edit box, or pick one from the drop-down list. You can also use the measure option previously discussed. When you have specified all the extrusion options, pick the **OK** button to complete the extruded feature.

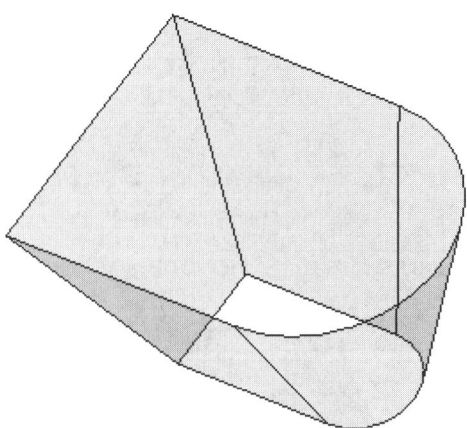

FIGURE 4.25 An example of a surface extrusion.

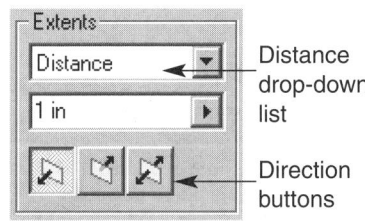

FIGURE 4.26 The Distance termination option of the **Extents** area.

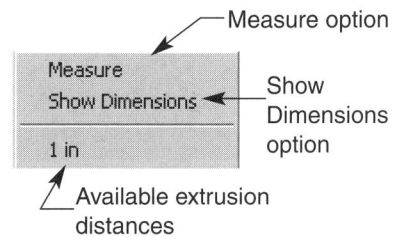

FIGURE 4.27 Using the **Measure** tool to reference an edge for the distance.

FIGURE 4.28
Tapering an extrusion.

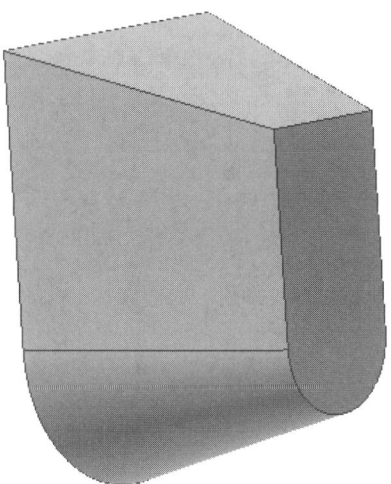

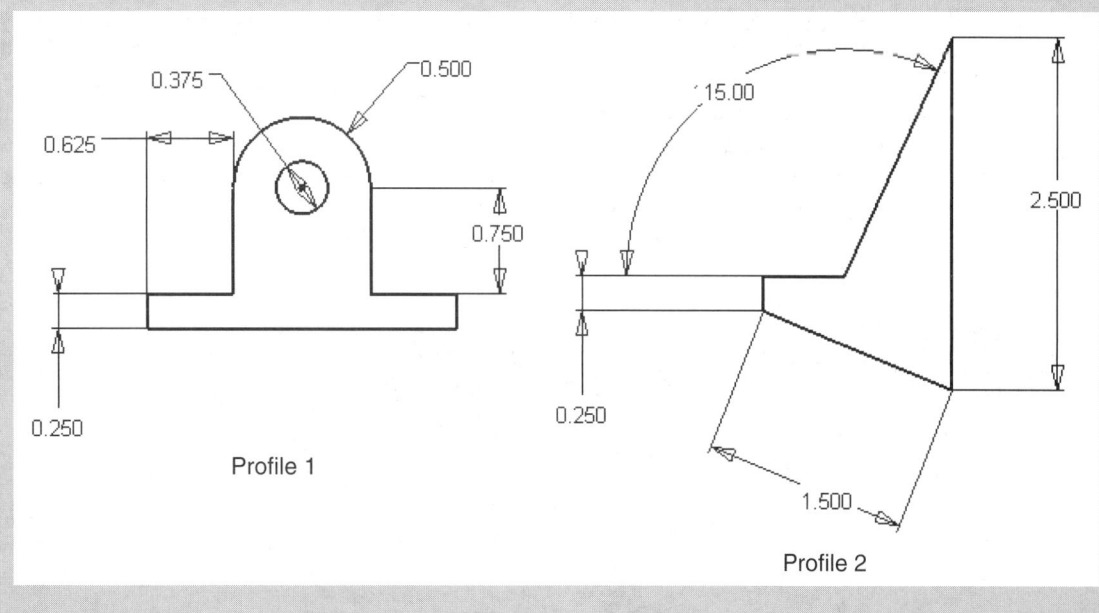

Exercise 4.2

1. Continue from Exercise 4.1, or launch Autodesk Inventor.
2. Open EX 3.18.
3. Save a copy of EX3.18 as EX4.2.
4. Close EX3.18 without saving, and open EX4.2.
5. Extrude Profile1 2″ in a positive direction.
6. Edit **Extrusion1** to extrude Profile1, 1″ in a negative direction.
7. Create a shared sketch of **Sketch1** in order to apply different extrusions to the two different sketch shapes contained in one sketch.
8. Extrude Profile2, 1.5″ in a positive direction.
9. Edit **Extrusion2** to extrude Profile2, 1.5″, equal amounts on each side of the sketch, with a taper angle of 20°.
10. Resave EX4.2.
11. Exit Autodesk Inventor or continue working with the program.

0.375   0.500
0.625
0.750
0.250
Profile 1

15.00
2.500
0.250
1.500
Profile 2

FIGURE 4.29 An example of a revolved feature and the sketch used to create the revolved feature.

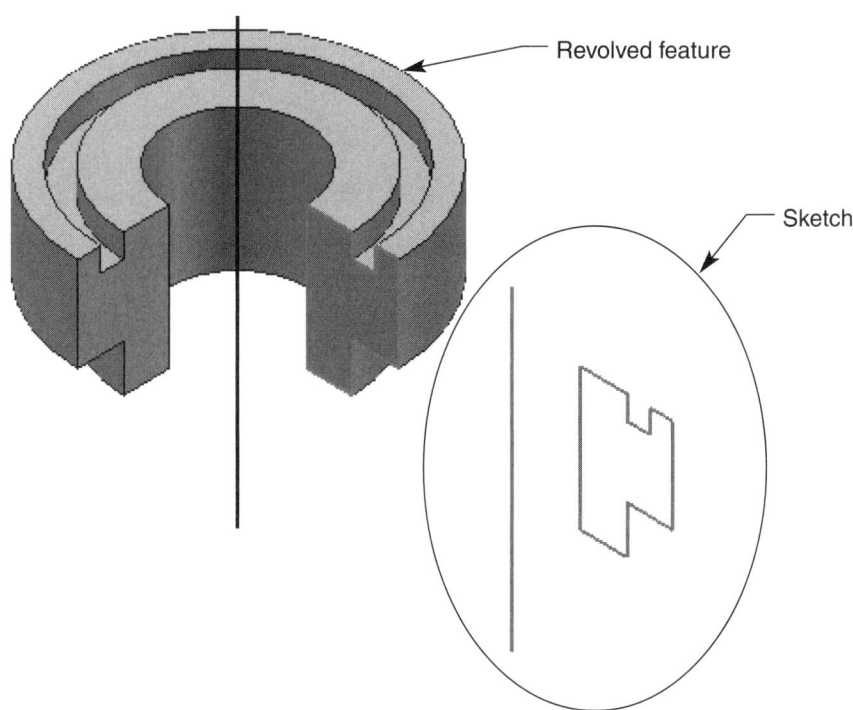

Revolved feature

Sketch

Creating Revolved Features

A *revolved* feature is very similar to an extrusion. Revolutions create a feature around an axis, in a circular path, instead of creating a feature in a linear direction, along the sketch plane, like an extrusion. An example of a revolution is shown in Figure 4.29.

Access the **Revolve** command using one of the following techniques:

✓ Pick the **Revolve** button on the panel bar.

✓ Pick the **Revolve** button on the **Features** toolbar.

✓ Type the **+** and **R** keys on your keyboard.

The **Revolve** dialog box is displayed when you access the **Revolve** command. See Figure 4.30.

If only one sketch loop exists, as is often the case when creating base features, the sketch profile is already selected. To revolve a sketch with more than one profile available, pick the **Profile** button from the **Shape** area of the **Revolve** dialog box if it is not already active. Then pick the profile of the sketch you want to revolve. Once the profile is selected, you must pick an axis. You may need to pick the **Axis** button in order to select an axis. In contrast to an extrusion, an axis is required to create a revolved feature, because the axis defines where the revolution occurs, the line of symmetry, and the diameter of the revolved feature. Axes are created in the sketch environment.

FIGURE 4.30 The **Revolve** dialog box (full rotation option).

Shape area

Profile button

Axis button

Operation buttons

Extents area

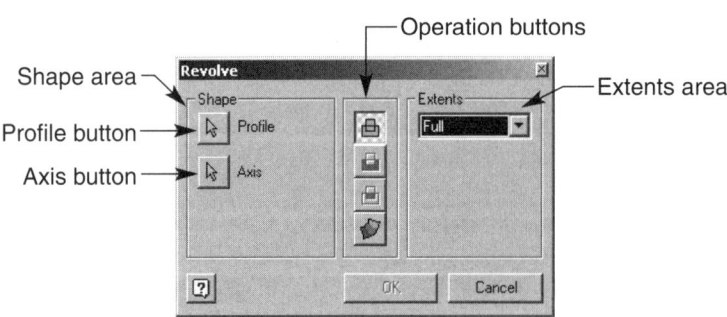

Field Notes

Project the **Center Point** and create a constraint between the origin and sketch geometry. This allows the sketch profile to become fully constrained to the origin, not just a point in space. It also allows you to define the diameter of the revolved feature.

Field Notes

Although you can use a sketched line or a projected axis from the **Origin** folder as an axis, you may want to pick one of the axes from the **Origin** folder as the line of revolution.

When the sketch profile and the axis are selected, a preview of the extrusion is displayed. See Figure 4.31.

When you create a more complicated part, and multiple features exist, more options are available for developing revolved features. For a simple base feature, the following options are usable inside the **Revolve** dialog box:

- **Operation type buttons** Only the **Join** button is available for a base feature, as previously discussed. When you create additional features all operation buttons are offered for use. These options are discussed later in this chapter.

- **Extents** The **Extents** area contains options that allow you to specify the angle or amount of the revolution that occurs. By default, as shown in Figure 4.30, the **Full** option is active. The **Full** option allows you to revolve a feature 360°, or completely around the axis. If you do not want to make a full revolution, pick the **Angle** option from the drop-down list. See Figure 4.32. Using **Angle** termination option allows you to specify a revolution angle of greater than 0° and less than 360°. Specify an angle from the drop-down list or enter the desired value. You also have the option of picking **Measure,** as previously discussed. This allows you to reference an existing edge and revolve around the measurement angle.

  Once you have specified the angle you want to revolve the feature, pick the desired direction button. The direction buttons are the same as those used in the **Extrude** command and control the side of the sketch the revolution occurs. When you have specified all the revolve options, pick the **OK** button to complete the revolved feature.

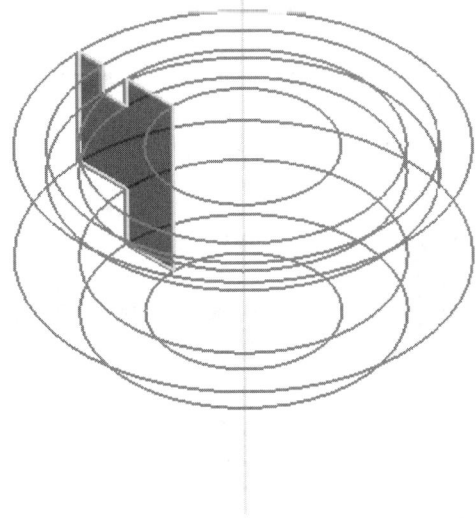

FIGURE 4.31 A revolution preview.

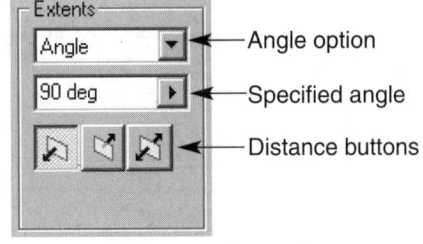

FIGURE 4.32 The **Angle** option of the **Extents** area.

Exercise 4.3

1. Continue from Exercise 4.2, or launch Autodesk Inventor.
2. Open a new part file, and create the sketch shown. Use equal constraints to fully constrain the sketch in addition to the dimensions shown.
3. Revolve the sketch profile fully around the **Y Axis.**
4. Edit **Revolution1** to revolve around the **Y Axis,** in a positive direction at an angle of 180°.
5. Save the part file as EX4.3.
6. Exit Autodesk Inventor or continue working with the program.

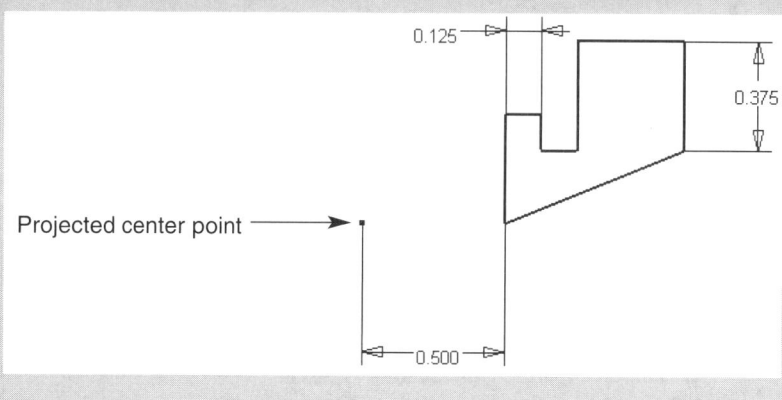

Sketching on a Base Feature

Once you have created a base feature using a sketched feature tool, such as **Extrude** or **Revolve,** you can do one of the following to make a more complex part:

- Position placed features such as shells, threads, fillets, chamfers, and face drafts.

- Insert catalog features.

- Pattern features.

Each of these choices is discussed in future chapters. Another alternative is to create a sketch on the base feature, to create more sketched features. The sketched features may be developed using one of the following techniques:

- Extruded or revolve a sketch profile to develop a feature that joins, cuts, or intersects the base feature.

- Create a hole in the base feature using a center point sketch and the **Hole** tool.

- Make a rib or web from one feature to another.

- Use another sketched feature tool such as loft, sweep, or coil.

To create a sketch on a base feature or any existing feature, use one of the following techniques:

✓ Pick the face you want to open a sketch on, and pick the **Sketch** button on the **Command** bar.

✓ Right-click on the face you want to open a sketch on, and pick the **New Sketch** menu option.

✓ Use the **+S** accelerator key, and pick the face you want to sketch on, or pick the face and press the **+** and **S** keys on your keyboard.

Field Notes

> You can also create a new sketch on any of the reference planes located in the **Origin** folder, including the YZ Plane, XZ Plane, and XY Plane.

Figure 4.33 shows a new sketch plane opened on an extruded base feature. When you open a new sketch on an existing feature, you are in the sketching work environment discussed in Chapter 3. All the tools and applications described in the sketching chapter apply.

 Field Notes

> The techniques and applications used to open a sketch on a base feature apply to any feature, face, and plane.

Automatic Model Edge Referencing

Autodesk Inventor contains tools that allow you automatically to use existing part edges for the creation of new sketch geometry. Two of these tools are accessed by selecting the following check boxes from the **Sketch** tab of the **Options** dialog, located in the **Tools** pull-down menu:

- **Autoproject edges during curve creation**   Picking this check box allows you automatically to project the edge of an existing feature onto the sketch plane. See Figure 4.34. To use the **Autoproject edges during curve creation** tool, you must first access a sketching tool such as **Line, Spline, Circle,** or **Arc.** Then move your cursor over the edge you want to project. The edge should turn black to signify it is projected and constrained. If you want to project another edge, repeat the process.

- **Automatic reference edges for new sketch**   Picking this check box enables Autodesk Inventor automatically to project all the edges of an existing part to the new sketch plane when you open a new sketch. See Figure 4.35. You know when the edges have been projected when the edges are black and constrained.

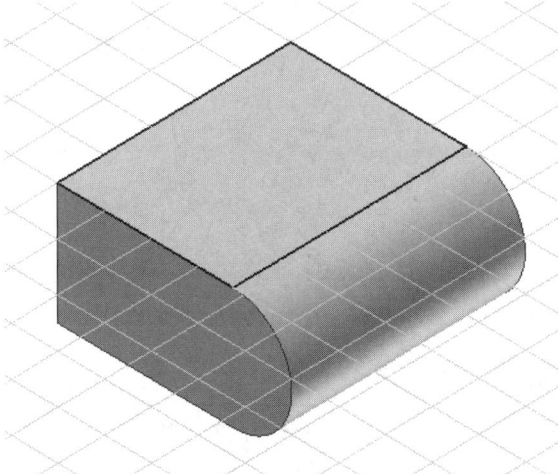

FIGURE 4.33   An example of a sketch plane opened on a feature face.

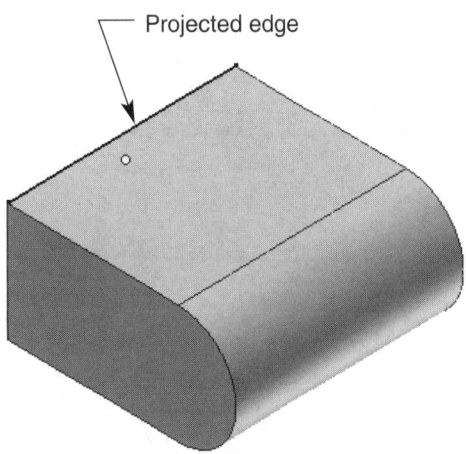

FIGURE 4.34   An example of an autoprojected feature edge.

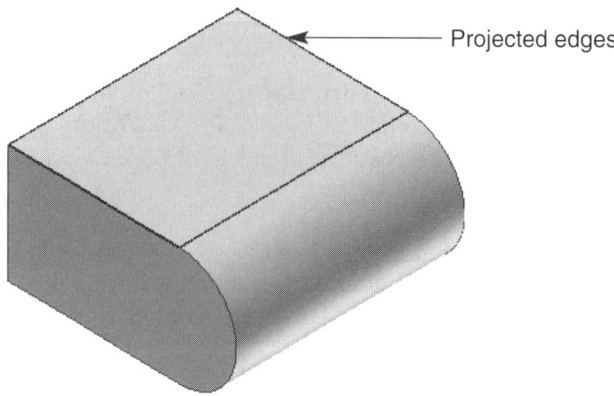

FIGURE 4.35 An autoprojection of edges of a feature.

Unless one of the Autoproject check boxes is selected, you will not be able to use existing feature edges for sketch development. However, you can still constrain and dimension your sketch to the edges of a feature because constraining and adding dimensions also automatically projects the referenced edges.

Exercise 4.4

1. Continue from Exercise 4.3, or launch Autodesk Inventor.
2. Open a new part file.
3. Create a $1'' \times 2'' \times 2''$ extrusion.
4. Activate the **Autoproject edges during curve creation** tool, but not the **Automatic reference edges for new sketch** tool.
5. Open a new sketch on the top face.
6. Create the sketch shown by automatically projecting two lines and sketching the other two lines.
7. Select **Properties...** from the **File** pull-down menu to open the **Properties** dialog box for the model.
8. Enter EXERCISE 4.4 in the **Title** edit box, AUTODESK INVENTOR in the **Company** edit box, and EX4.4 in the **Part Number** edit box.
9. Save the part file as EX4.4.
10. Exit Autodesk Inventor or continue working with the program.

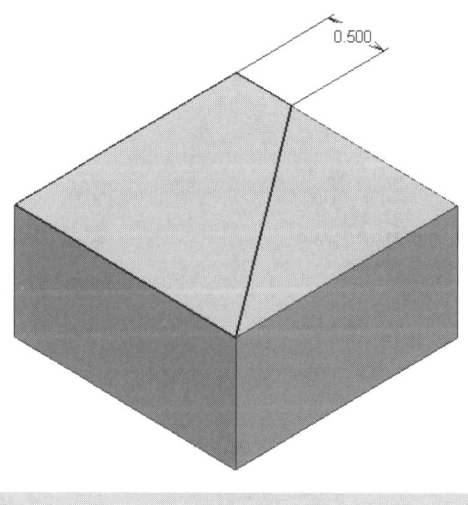

Exercise 4.5

1. Continue from Exercise 4.3, or launch Autodesk Inventor.
2. Open EX4.3.
3. Save a copy of EX4.3 as EX4.5
4. Close EX4.3 without saving, and open EX4.5.
5. Activate the **Automatic reference edges for new sketch** tool.
6. Open a new sketch on one of the feature faces. The entire sketch should be automatically created. Add an axis line .5″ from the furthest edge of the sketch geometry.
7. Revolve the sketch face 180° in a positive direction, as shown.
8. Resave EX4.5.
9. Exit Autodesk Inventor or continue working with the program.

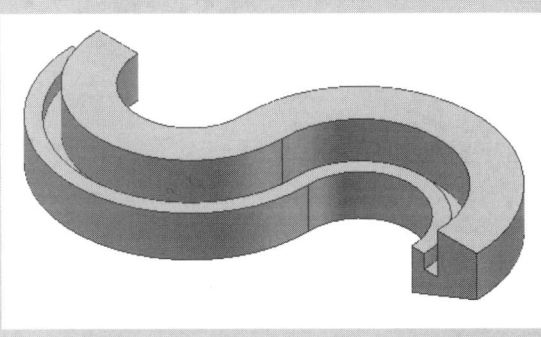

Creating Additional Extruded Features

Now that you have created a base feature for your part and opened a sketch on an existing feature, you are ready to create additional extrusions, if needed. See Figure 4.36. When you access the **Extrude** command when one or more features exist, the **Extrude** dialog box contains additional usable options because reference faces are present. See Figure 4.37.

The following preferences are available inside the **Extrude** dialog box:

■ Operation type buttons   In addition to the **Join** and **Surface** buttons, the **Cut** and **Intersect** buttons are also available when creating more features. Picking the **Cut** button allows you take away material from an existing feature by cutting through the existing feature with your sketch. See Figure 4.38. Picking the **Intersect** button allows you to create a feature by using existing intersecting features. When you apply an intersect operation to two or more fea-

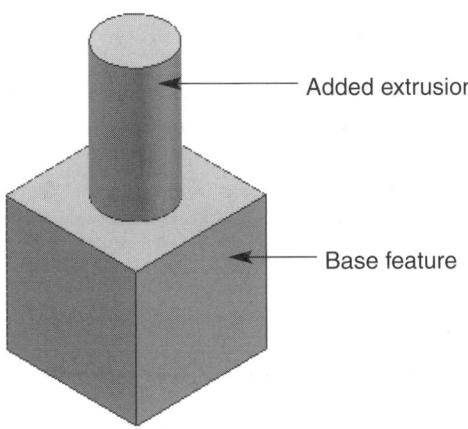

Added extrusion

Base feature

FIGURE 4.36 An example of adding extruded features to a base feature.

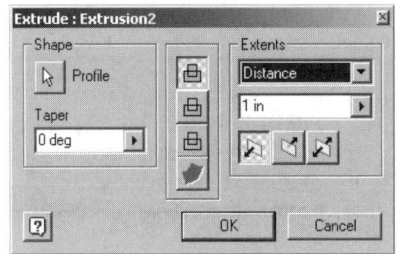

FIGURE 4.37 The **Extrude** dialog box (all options available).

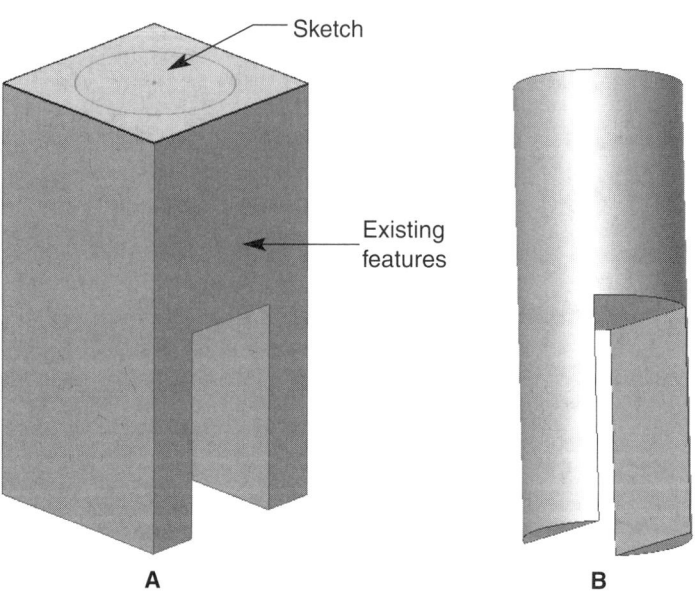

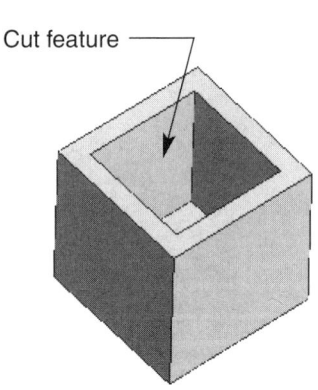

FIGURE 4.38   An example of a cut extrusion.

FIGURE 4.39   An example of an intersection extrusion.

tures, the intersecting portion of the features is retained, and all other material is removed. Figure 4.39A shows the existing features, and Figure 4.39B shows the intersection extrusion.

■ **Extents**   As shown in Figure 4.40, the following extrusion termination options are now usable inside the **Extents** area:

■ **To Next**   Selecting this option enables you to extrude a feature to the next possible face or plane. See Figure 4.41. To extrude a feature to the next face or plane, pick the sketch profile, and select the direction you want the extrusion to occur. Then pick the **OK** button to create the extruded feature.

Field Notes

There is no equal side, or midplane, direction option available when extruding **To Next** face or plane.
    If a face or plane does not exist that can end the extrusion, the feature cannot be created.

■ **To**   This extent option is similar to the **To Next** option. However, selecting **To** allows you to choose a face or plane to end the extrusion, instead of the extrusion ending at the next possible face or plane. See Figure 4.42. To terminate an extrusion at a specified face or plane using the **To** option, first pick the sketch profile. Then pick the face or plane you

FIGURE 4.40   The extrusion extent options of the **Extents** area, for parts with multiple features.

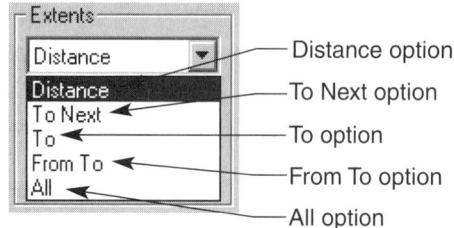

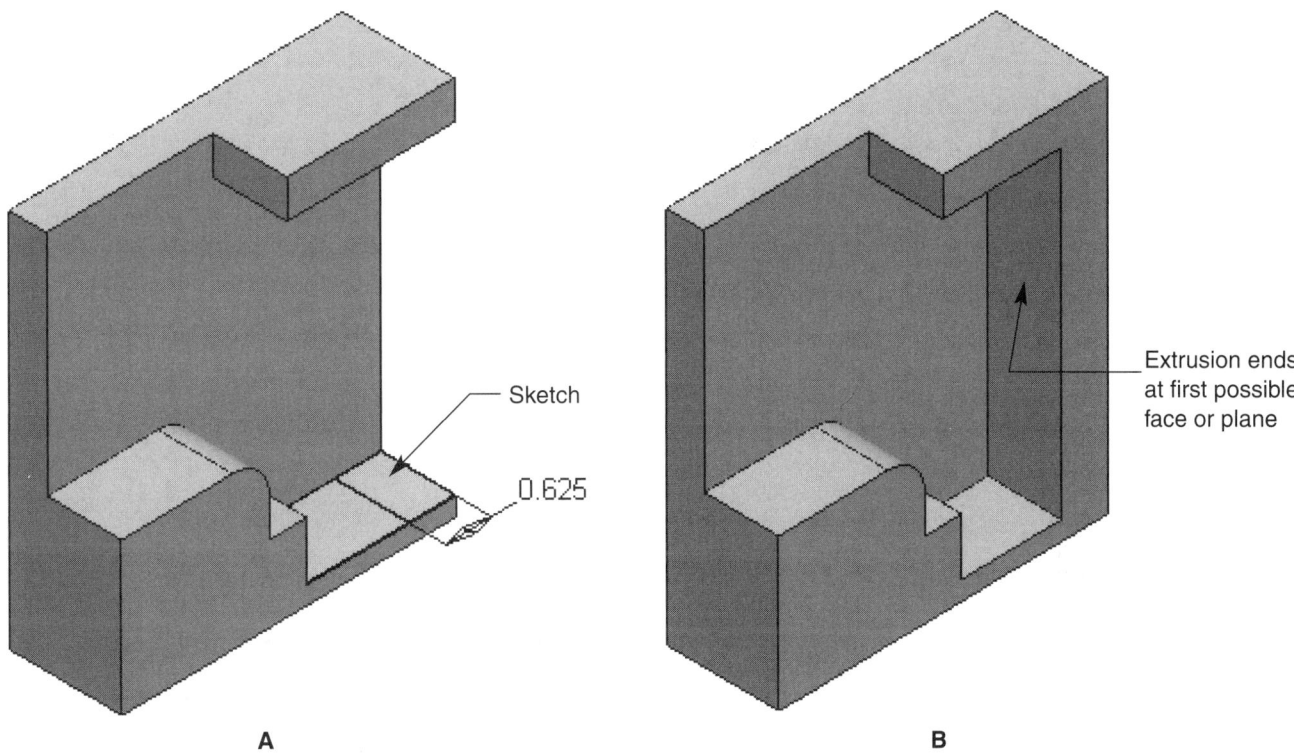

FIGURE 4.41   Creating an extrusion using the **To Next** extents option.

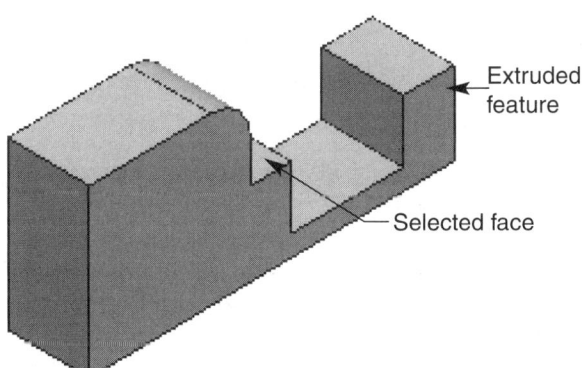

FIGURE 4.42   Creating an extrusion using the **To** extents option.

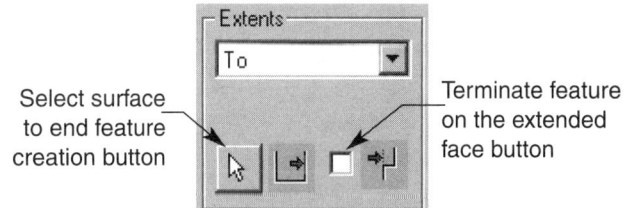

FIGURE 4.43   The **Extents** area of the **Extrusion** dialog box (**To** option).

want to extrude the feature to. You may need to choose the **Select surface to end the feature creation** button if it is not already activated. See Figure 4.43. If the face or plane you choose does not intersect the extrusion path, as shown in Figure 4.42, you need to check the **Terminate feature on extended face** check box.

■ **From To**   This extent option is similar to the **To** option previously described. However, selecting **From To** allows you to choose a face or plane to begin and end the extrusion, instead of just the ending surface. See Figure 4.44. To begin and terminate an extrusion at a specified surface using the **From To** option, first pick the sketch profile. Then pick the face or plane you want to begin the extrusion from. You may need to choose the **Select surface to start the feature creation** button, if it is not already activated. See Figure 4.45. Next, pick the face or plane you want to end the extrusion. You may need to choose the **Select surface to end the feature creation** button if it is not already activated. If the

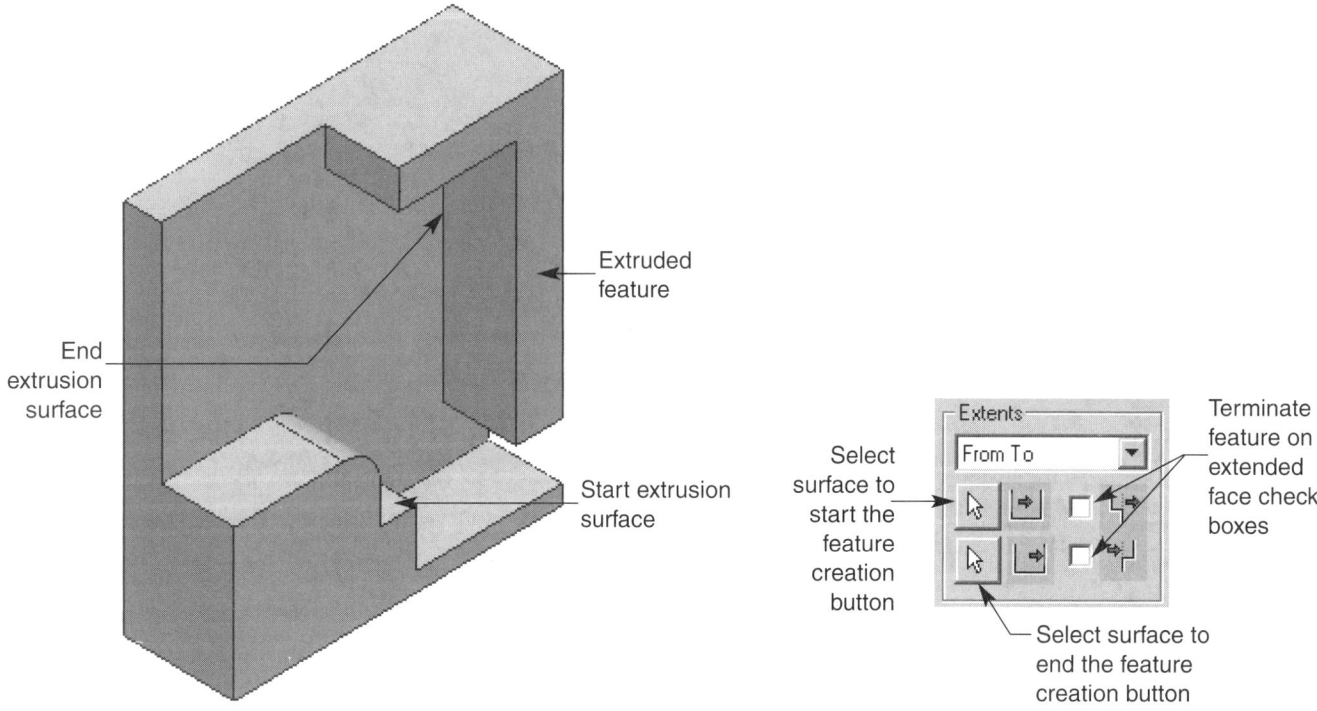

FIGURE 4.44　Creating an extrusion using the **From To** extents option.

FIGURE 4.45　The **Extents** area of the **Extrusion** dialog box (**From To** option).

surfaces you choose to start and end the extrusion do not intersect the extrusion path, you need to check the **Terminate feature on extended face** check boxes.

- **All**　Selecting this option enables you to extrude a feature directly through all other features. See Figure 4.46. If you know for sure you want a feature to extrude the entire way through existing features, use the All option. It does not require any distances or surface selections, and may save some time. To extrude a feature through all existing features, pick the sketch profile and select the direction you want the extrusion to occur. Then pick the **OK** button to create the extruded feature.

FIGURE 4.46
Creating an extrusion using the **All** extents option.

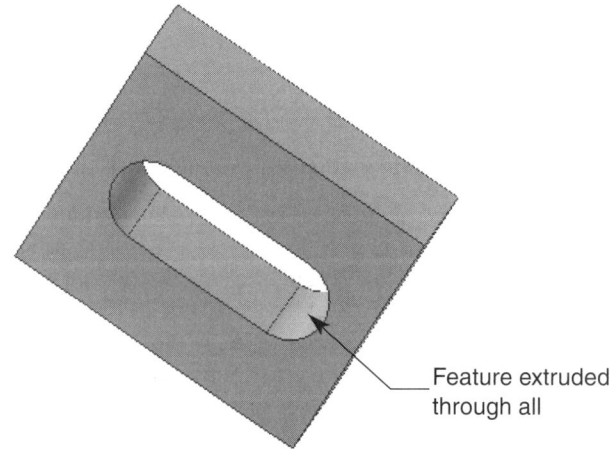

Feature extruded through all

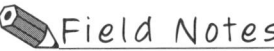Field Notes

Use any of the termination options previously discussed to end a feature at any feature, face, or plane.

The operation type and extent buttons discussed for the extrude command function the same for additional revolved features.

Exercise 4.6

1. Continue from Exercise 4.5, or launch Autodesk Inventor.
2. Open EX4.4.
3. Save a copy of EX4.4 as EX4.6.
4. Close EX4.4 without saving, and open EX4.6.
5. Extrude the sketch profile .75″ in a positive direction.
6. Open a new sketch on the top face, and develop the sketch shown.
7. Cut extrude Profile1 in a negative direction, .5″.
8. Undo, to reverse the effects of the cut extrusion.
9. Cut extrude Profile2, in a negative direction, using the **To Next** termination option.
10. Edit Extrusion3 to cut extrude Profile2 in a negative direction, using the **From To** termination option. Specify Face2 as the start face, and Face3 as the end face.
11. Edit Extrusion3 to cut extrude Profile2 in a negative direction, using the **To** termination option. Specify Face2 as the end face.
12. Resave EX4.6.
13. Exit Autodesk Inventor or continue working with the program.

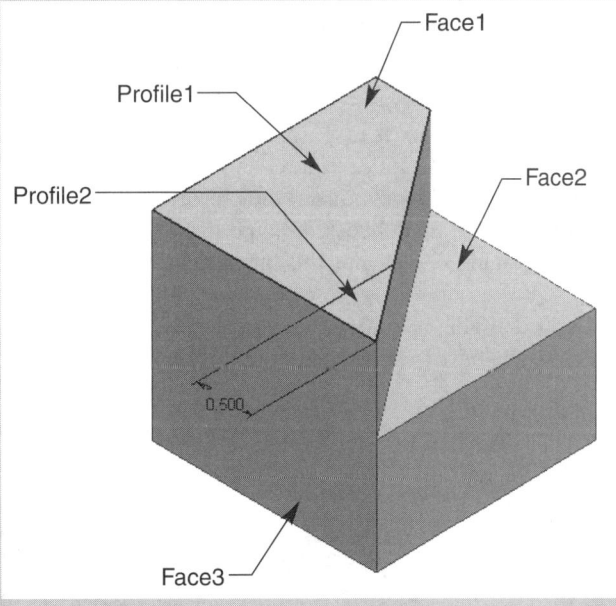

Creating Holes

Holes are used to create specific cylindrical extrusions through a feature. Applying holes to a feature is similar to extruding a circle. However, the **Hole** command provides you with many more options and allows you to create complex feature geometry in a single step. You can produce the following types of holes using the **Hole** tool:

- *Drilled,* as shown in Figure 4.47A. Drilled holes are the simplest types of holes. They require only one diameter and depth to create.

FIGURE 4.47   Using the **Hole** tool to (A) create drilled holes, (B) create counterbored holes, and (C) create countersunk holes.

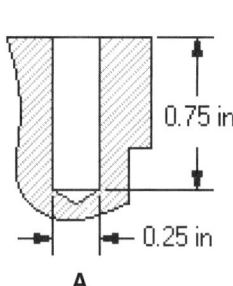
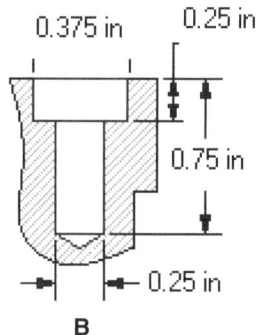
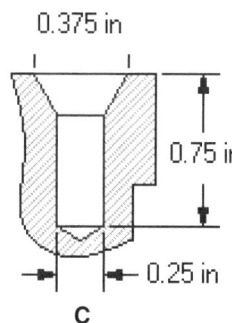

A   **B**   **C**

- *Counterbored,* as shown in Figure 4.47B. Counterbored holes require a hole diameter and depth and a counter-bore diameter and depth.

- *Countersunk,* as shown in Figure 4.47C. Countersunk holes require a hole diameter and depth, and countersink diameter and angle.

You can also apply threads to any of the three types of holes previously discussed. See Figure 4.48.

The first step in creating a hole is to make a sketch. Most of the time, points, or hole centers, are used to define the center of a hole, as mentioned in Chapter 3. These points should be constrained and dimensioned, just like any other sketch. See Figure 4.49. You can also use any other point on a sketch, such as the endpoint of a curve, or the center point of a fillet, arc, circle, or other radius. See Figure 4.49.

Once you have developed a sketch, access the **Hole** command using one of the following techniques:

✓ Pick the **Hole** button on the panel bar.

✓ Pick the **Hole** button on the **Features** toolbar.

✓ Type the **+** key and **H** key on your keyboard.

The **Holes** dialog box is displayed when you access the **Hole** command and is used to specify your hole options and parameters. See Figure 4.50. You can view any changes or selections you make to the hole before the hole is created by looking at the preview area.

To create a hole, first choose the type of hole you want to produce using the **Type** tab of the **Holes** dialog box. See Figure 4.50. Inside this tab, you have the following options:

- **Centers**   This button allows you to select the points where you want holes to be created. When you have sketched center points on a feature and access the **Hole** command, the center

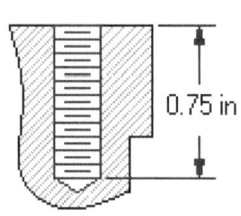

FIGURE 4.48   Applying threads to holes.

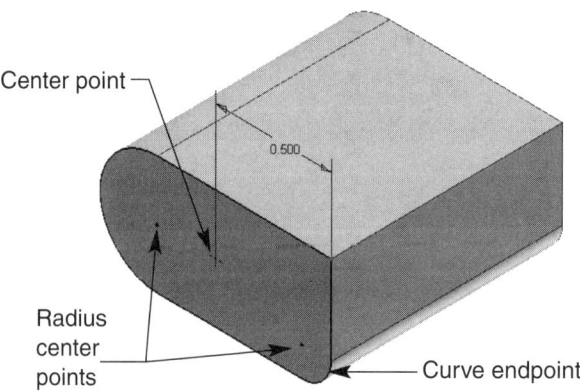

FIGURE 4.49   Sketching on a feature face to create holes.

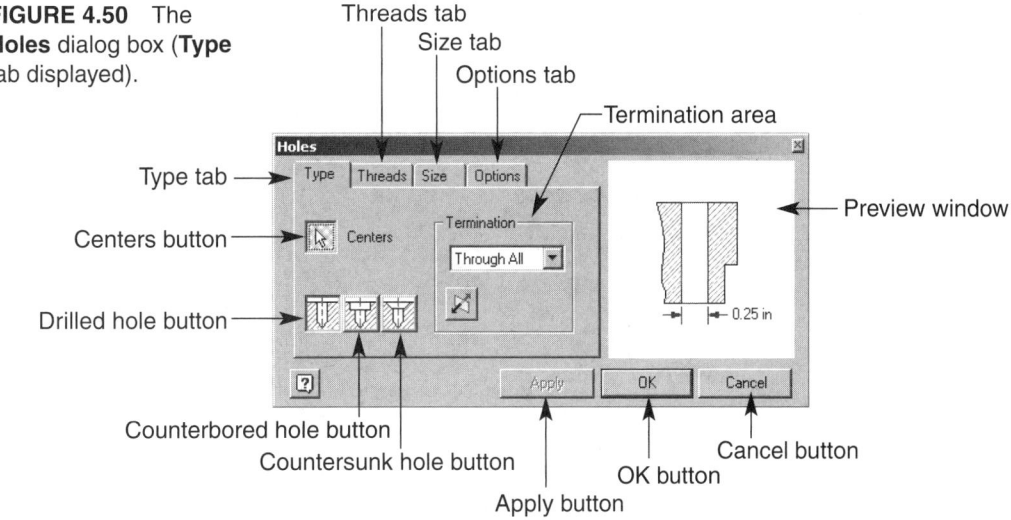

FIGURE 4.50 The **Holes** dialog box (**Type** tab displayed).

points are automatically selected. However, if you would like to choose other center points or endpoints, pick the **Centers** button and pick the points.

Field Notes

If you want to deselect points, pick the **Centers** button, hold down the **Ctrl** key on your keyboard, and choose the selected points.

- **Drilled**   This button specifies the hole as drilled. To define the hole diameter, highlight the default diameter in the preview area and enter the desired value. To define the hole depth, highlight the default depth in the preview area and enter the desired value.

- **Counterbore**   This button specifies the hole as counterbored. Define the hole diameter and depth, and the counterbore diameter and depth inside the preview area, by highlighting the default numbers and entering the desired values.

- **Countersink**   This button specifies the hole as countersunk. Define the hole diameter and depth, and the countersink diameter inside the preview area, by highlighting the default numbers and entering the desired values. The countersink angle is defined in the **Options** tab of the **Holes** dialog box, discussed later in this chapter.

- **Termination**   The termination area is very similar to the **Extents** area of the **Extrude** dialog box. As shown in Figure 4.51, the following termination options are available:

  - **Distance**   Using the **Distance** choice allows you to end a hole at a desired depth. Specify a distance by highlighting the default hole depth in the preview area and entering the desired value. If necessary, use the **Flip** button to flip the cutting direction of the hole.

  - **Through All**   Selecting this option enables you to drill a hole directly through all other features. If necessary, use the **Flip** button to flip the cutting direction of the hole.

FIGURE 4.51 Hole termination options.

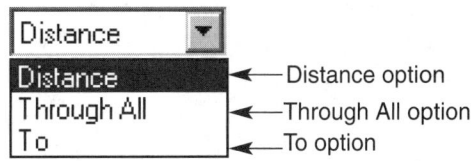

FIGURE 4.52
Creating a hole using the **To** termination option (part is split for clarity).

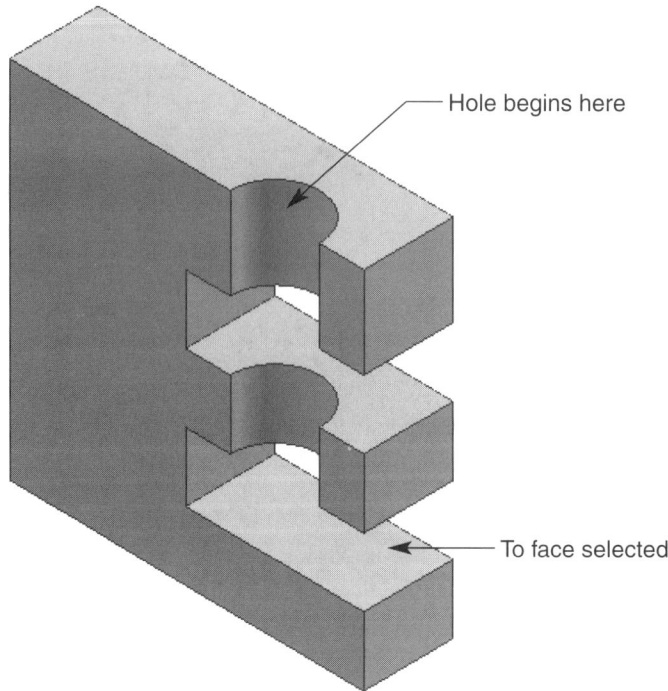

Hole begins here

To face selected

■ **To** This option allows you to choose a face or plane to end the extrusion. See Figure 4.52. To terminate a hole at a specified face or plane, first pick the sketch point. Then pick the face or plane you want to extrude the feature to. You may need to choose the **Select surface to end the feature creation** button if it is not already activated. If the face or plane you choose does not intersect the hole path, you need to check the **Terminate feature on extended face** check box.

Once you have specified all the hole type options, pick the **Threads** tab to created tapped holes, if required. See Figure 4.53. Inside the **Threads** tab, you have the following options:

■ **Tapped** Pick this check box to tap the hole. Then highlight the default thread depth in the preview window, and specify the thread depth in the hole.

■ **Full Depth** Pick this check box to apply threads to the entire depth of the hole.

■ **Thread Type** This drop-down list allows you to specify the kind of threads applied to the hole. You can choose **ANSI Unified Screw Threads** for inch parts or **ANSI Metric M Profile** for metric parts.

■ **Right Hand** and **Left Hand** radio buttons You can specify whether the threads are right or left hand by choosing the appropriate radio button. Right-hand threads move a right-hand

FIGURE 4.53 The **Threads** tab of the **Holes** dialog box.

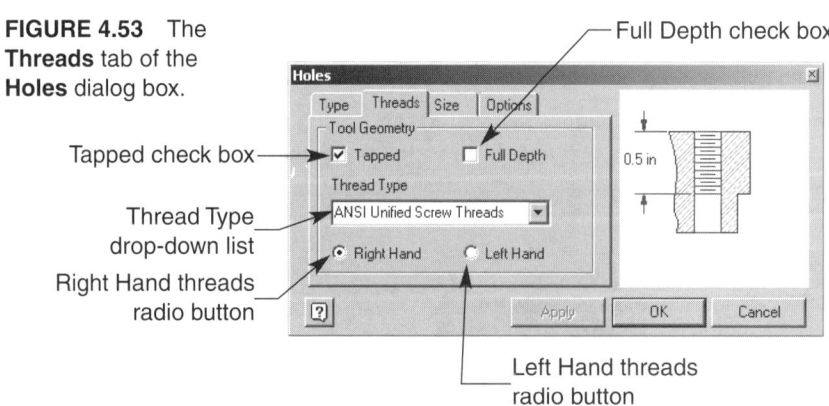

Full Depth check box

Tapped check box

Thread Type drop-down list

Right Hand threads radio button

Left Hand threads radio button

threaded bolt forward in a clockwise fashion, while left-hand threads move a left-hand threaded bolt forward in a counterclockwise fashion.

Field Notes

Use **ANSI Unified Screw Threads** for an inch hole, and **ANSI Metric M Profile** threads for a metric hole.

Once you have specified the thread preferences, select the **Size** tab to set the thread size and additional thread information. See Figure 4.54. The following drop-down lists are available inside the **Size** tab:

■ **Nominal Size**   This drop-down list allows you to specify the nominal size of the threads. The nominal size is the designated size of a commercial product and is usually the same as the major diameter.

■ **Pitch**   This drop-down list allows you to specify the pitch of the threads. Pitch is distance parallel to the axis between a point on one thread to the corresponding point on the next thread. For ANSI Unified Screw Threads, the pitch determines and is specified by the number of threads per inch. For ANSI Metric M Profile threads, the pitch is specifically specified as the pitch.

Field Notes

Only pitches that correspond to the designated nominal size are available.

■ **Class**   This drop-down list allows you to specify the thread class. Class is the designated amount of tolerance specified for the thread. For ANSI Unified Screw Threads, 3, 4, 5, 6, 7, 8, or 9 may be available depending on the specified nominal size and pitch. These numbers identify the grade of tolerance from fine to coarse threads. For ANSI Metric M Profile threads, 1, 2, or 3 may be available depending on the specified nominal size and pitch. 1 is a coarse tolerance, 2 is a moderate tolerance, and 3 is a fine thread tolerance. For holes, these numbers will be followed by B, because B identifies internal threads.

■ **Diameter**   This drop-down list allows you to identify the diameter of the nominal size as **Minor, Pitch, Major,** or **Tap Drill.**

FIGURE 4.54 The **Size** tab of the **Hole** dialog box.

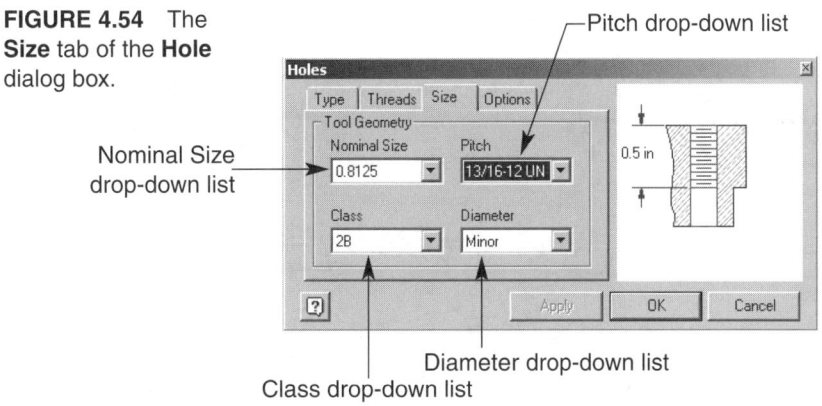

Field Notes

For most thread applications, the **Major** option of the **Diameter** drop-down list should be used.

When you apply threads to a hole, a bitmap thread image is placed in the hole. These thread representations are adequate for most applications, but if you need to create actual detailed threads, you must use the coil tool, discussed later in this chapter.

The **Options** tab, shown in Figure 4.55, allows you to specify the drill point and countersink angles. Pick the **Flat** radio button to specify a flat, 0°, drill point angle. See Figure 4.56A. To specify an angled drill point, pick the **Angle** radio button and choose an angle from the drop-down list, or specify a desired angle. See Figure 4.56B. If you are creating a countersunk hole, the **Countersink Angle** drop-down list is available. To specify the countersink angle, choose an angle from the drop-down list or specify a desired angle.

FIGURE 4.55 The **Options** tab of the **Holes** dialog box.

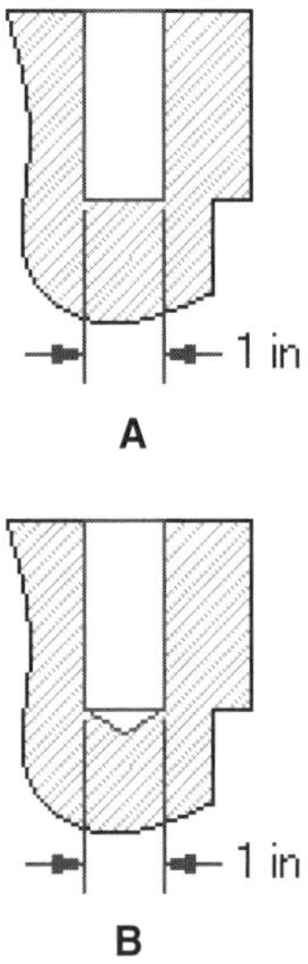

Flat drill point radio button

Angle drill point radio button

Countersink Angle drop-down list

Drill point Angle drop-down list

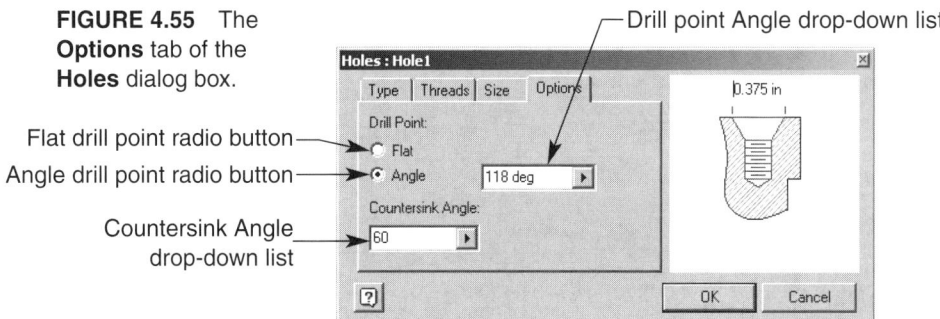

FIGURE 4.56 Drill point options.

Exercise 4.7

1. Continue from Exercise 4.6, or launch Autodesk Inventor.
2. Open a new part file.
3. Create the extrusion and holes shown in the engineer's sketch. Place the holes at a location of your choice.
4. Save the part file as EX4.7.
5. Exit Autodesk Inventor or continue working with the program.

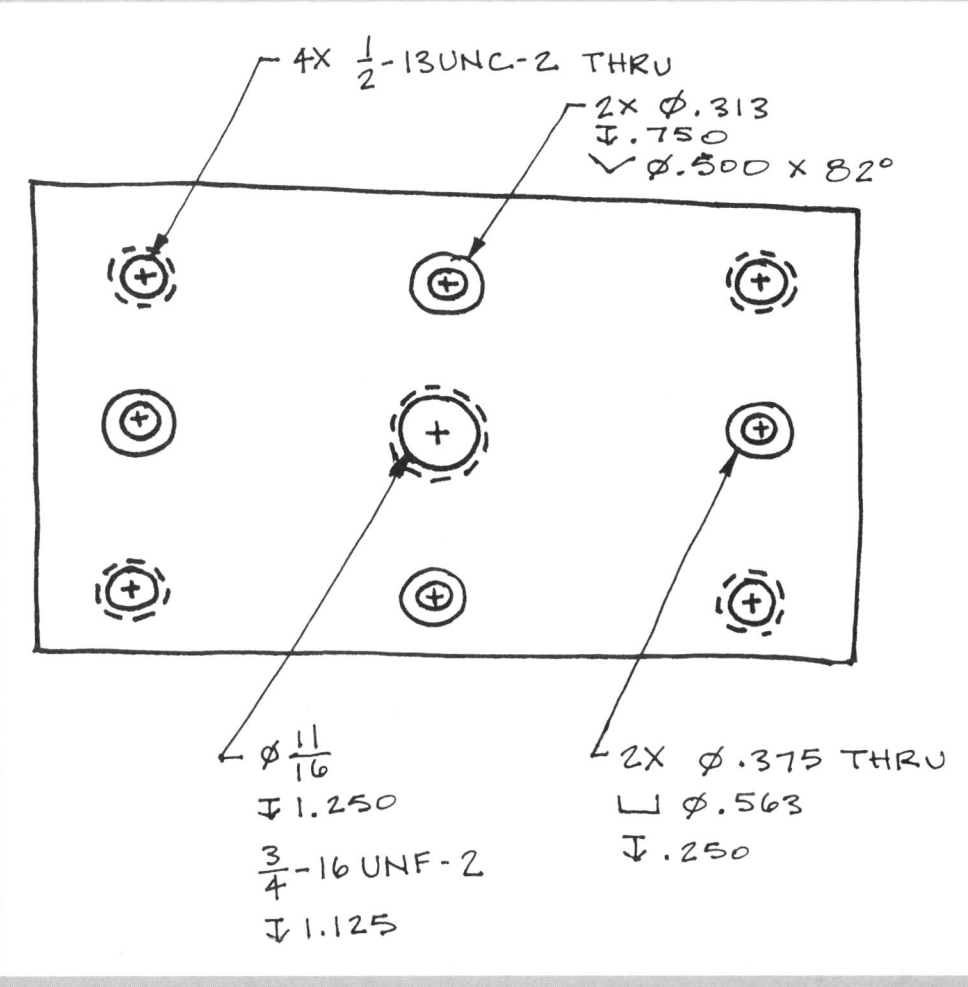

Creating Ribs and Webs

Ribs and *webs* are thin sections of material placed between part features to reinforce the part without adding too much weight. Ribs and webs are similar and can be created from the same sketch geometry. Ribs are closed features; webs are open. Figure 4.57 shows an example of ribs and webs. The ribs and webs displayed in Figure 4.57 constitute a ***rib*** and ***web network***. A rib or web network consists of several ribs and webs created using the same direction and thickness.

Field Notes

A network can be created using a single sketch, as shown in Figure 4.54, or patterned sketch. Patterning features is discussed in future chapters.

FIGURE 4.57 An
example of (A) ribs and
(B) webs used in the
same part.

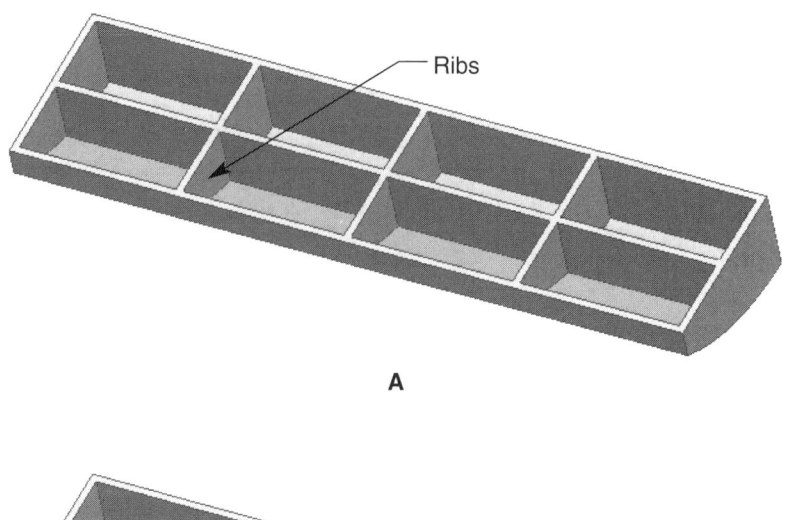

Ribs

A

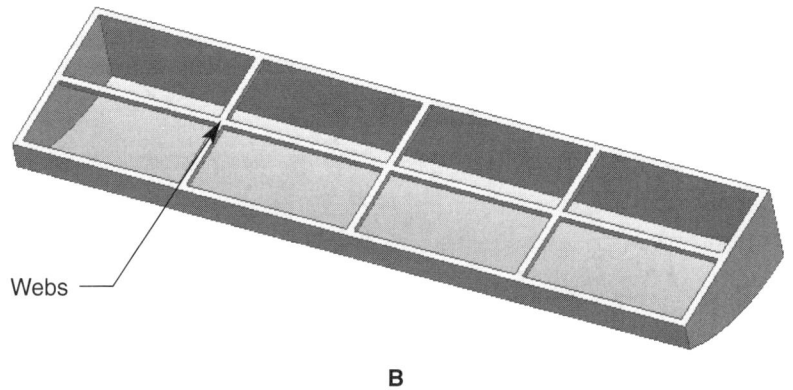

Webs

B

The first step in creating a rib or web is developing a sketch. Unlike most sketches, you do not need to fully constrain and dimension rib and web sketch geometry, because many ribs and webs are created from open loop sketches. You can even use sketches that do not touch or intersect other features. See Figure 4.58. However, you may still want to situate your sketch in reference to existing features so your sketch is defined and stable. See Figure 4.59. If you are creating a rib or web network, sketch all profiles on the same sketch plane. See Figure 4.60.

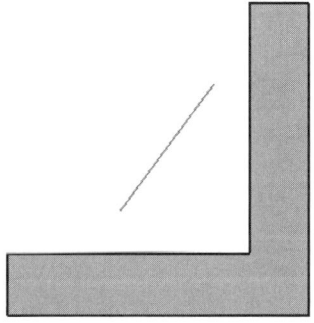

FIGURE 4.58   An example of
an open loop, unconstrained
sketch used to create ribs and
webs.

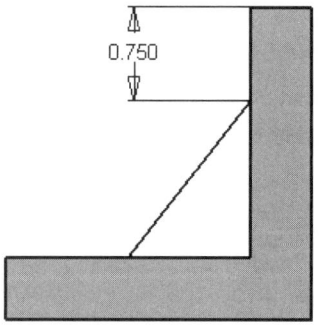

0.750

FIGURE 4.59   An example of a
constrained sketch, referencing
existing geometry, used to create
ribs and webs.

FIGURE 4.60 An example of a rib or web sketch.

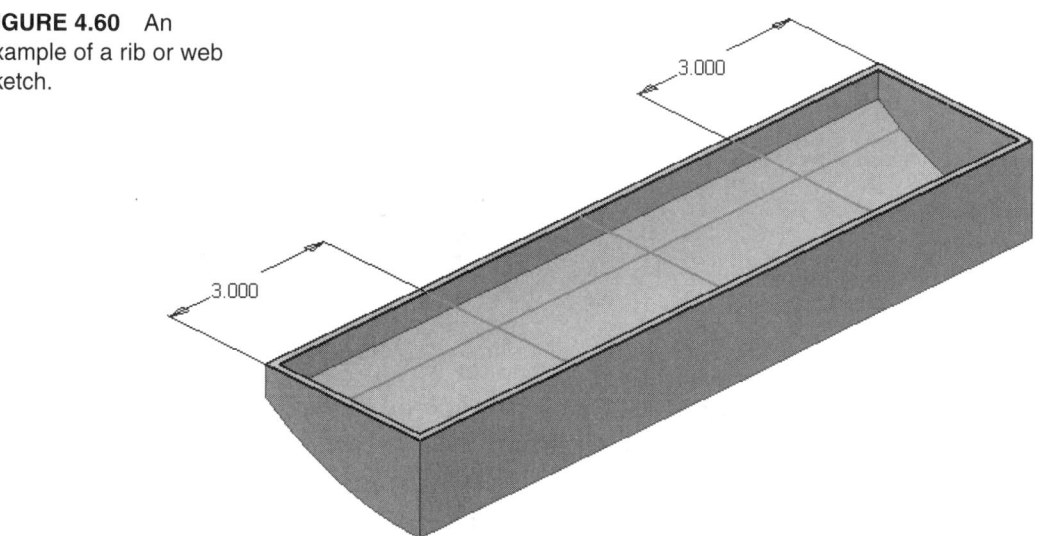

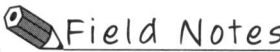

Use any of the sketching tools and applications discussed in Chapter 3 to create rib and web sketches.

To create ribs or webs at locations where no face is present, use a work plane. Work planes and other work features are discussed in subsequent chapters.

Once you have developed a sketch, access the **Rib** command using one of the following techniques:

✓ Pick the **Rib** button on the panel bar.

✓ Pick the **Rib** button on the **Features** toolbar.

The **Rib** dialog box is displayed when you access the **Rib** command and is used to specify your rib or web options and parameters. See Figure 4.61.

To create a rib or web, pick the **Profile** button, if it is not active, and select the sketch profile. Then pick the **Direction** button and specify where you want the rib or web to be created. Direction arrows and a preview of the rib or web are shown when you move your cursor. See Figure 4.62. If the arrows are pointing to the correct location, and the preview looks okay, pick to enter the direction. Next, specify a rib or web width from the **Thickness** drop-down list, or enter the desired value. Then pick the side of the profile you want the rib to be created by selecting one of the flip buttons. By default, the **To Next** button is active when you open the **Rib** dialog box. As a result, when all the previously discussed preferences are set, pick the **OK** button to create a rib.

FIGURE 4.61 The **Rib** dialog box (**To Next** extents button selected).

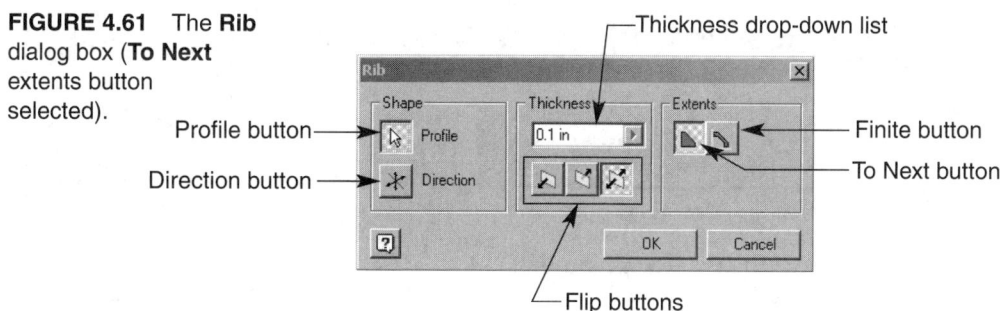

FIGURE 4.62 Rib direction arrow and preview.

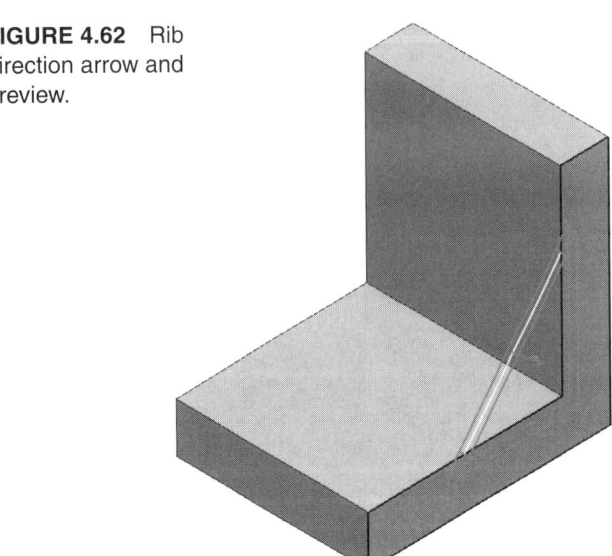

Occasionally, it may be difficult to locate the direction you want the rib or web to be created. Do not attempt to generate a rib or web until a full preview of the feature is displayed, and you are sure the rib will be created successfully.

If you want to create a web, use the same process as previously described for creating a rib. However, you must also pick the **Finite** button. When the **Finite** button is active, as shown in Figure 4.63, an **Extents Depth** drop-down list and an **Extend Profile** check box are available. The **Extent Thickness** drop-down list allows you to specify the extents, or depth. Define the depth by choosing a distance from the drop-down list or by entering a value. If selected, the **Extend Profile** check box allows you to extend the ends of the web to the intersecting features. See Figure 4.64. If unselected, the web will terminate at the ends of the sketch profile. When you have specified all the web options, pick the **OK** button to create the web.

Finite button

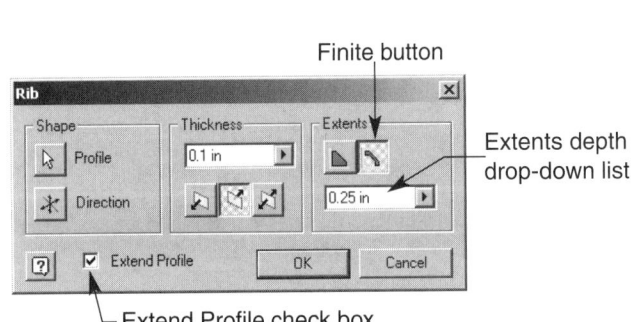

Extents depth drop-down list

Extend Profile check box

FIGURE 4.63 The **Rib** dialog box (**Finite** extents button selected).

FIGURE 4.64 An intersecting web created by extending the sketch profile.

Exercise 4.8

1. Continue from Exercise 4.7, or launch Autodesk Inventor.
2. Open a new part file.
3. Create a part similar to the one shown as A, using two extrusions.
4. Open two different sketches, and sketch the lines shown in B.
5. Generate a rib on one of the faces as shown in C.
6. Create a web on the opposite face as shown in C. The depth of the web should be greater than the thickness.
7. Save the part file as EX4.8.
8. Exit Autodesk Inventor or continue working with the program.

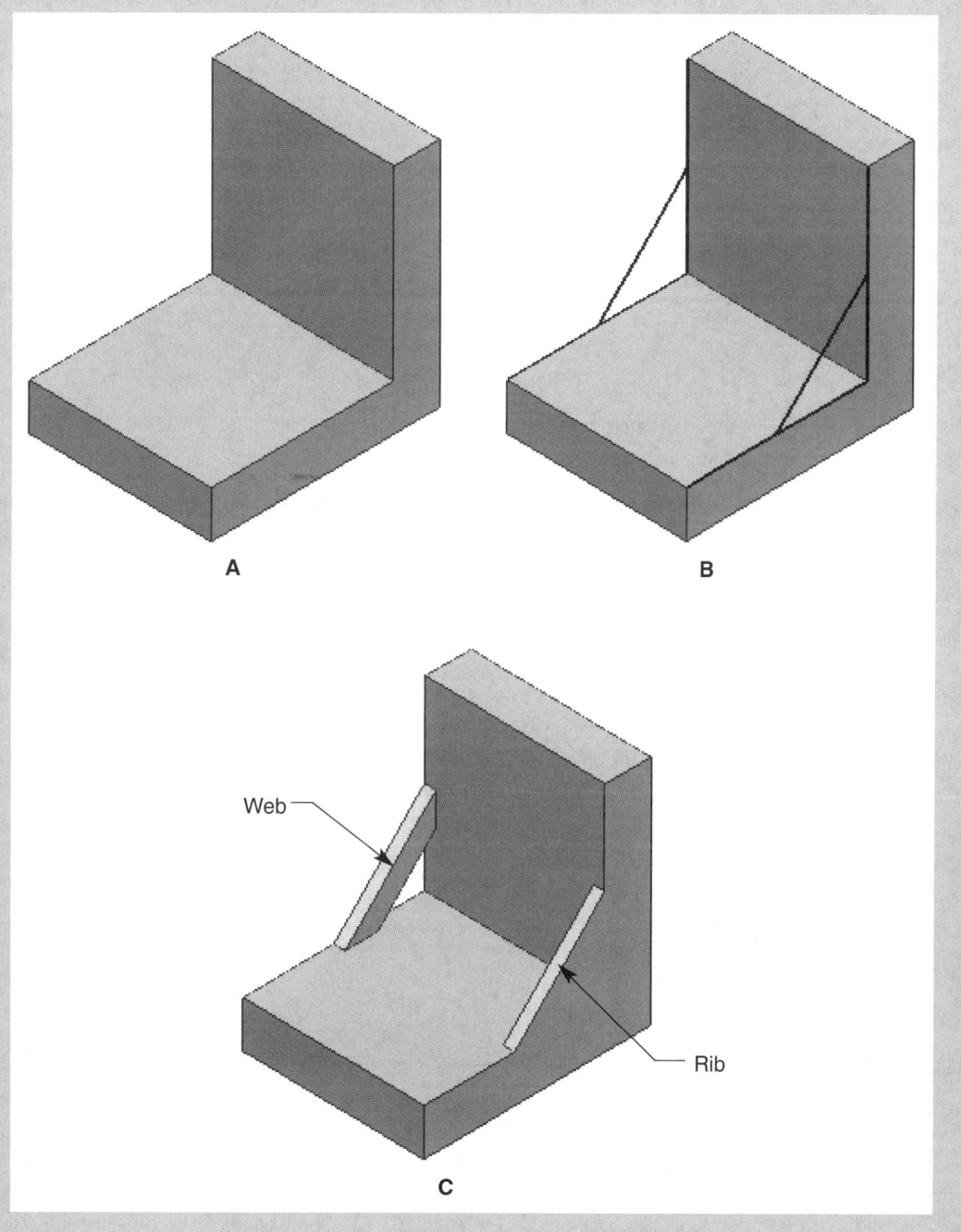

A

B

Web

Rib

C

Developing Coils

Coils are similar to revolved features, but the sketch profile rotates both around and up or down an axis. *Coils* are spiral, or helix, features used to create springs or threads. See Figure 4.65. A *spring* is a mechanism that expands or contracts when force is applied, and returns to its normal form when force is released. Springs and threads are created approximately the same way using the **Coil** tool.

The first step in creating a coil feature is to generate a sketch profile. Typically, for threads, sketches are developed on a work plane. The profile is then cut into an extruded cylinder. Work planes and work features are discussed in Chapter 6. However, threaded fasteners can be created exactly the same way as a spring.

Once you have developed a sketch profile, as shown in Figure 4.66, you are ready to create a coil. Access the **Coil** command using one of the following techniques:

✓ Pick the **Coil** button on the panel bar.

✓ Pick the **Coil** button on the **Features** toolbar.

The **Coil** dialog box is displayed when you access the **Coil** command. See Figure 4.67. The **Coil Shape** tab of the **Coil** dialog box is initially displayed and contains a **Shape** area and a **Rotation** area. The following buttons are available inside the **Coil Shape** tab:

■ **Profile**   This button allows you to select a sketch profile. If only one sketch loop exists, as is often the case when creating a coil, the sketch profile is already selected. To choose a sketch with more than one profile available, pick the **Profile** button if it is not already active. Then pick the sketch profile you want to spiral.

Field Notes

> Project the **Center Point** and create a constraint between the origin and sketch geometry. This allows the sketch profile to become fully constrained to the origin, not just a point in space. It also allows you to define the diameter of the coil.

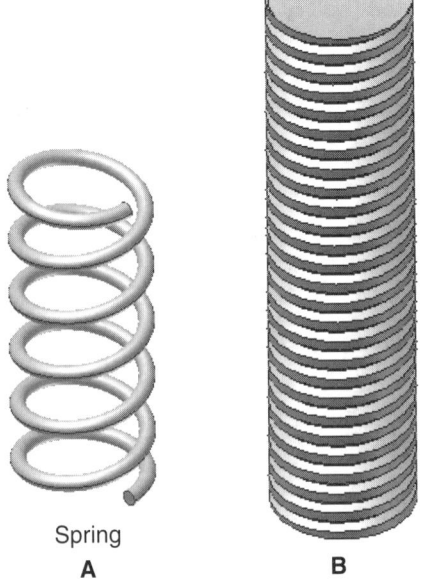

Spring
A

B

FIGURE 4.65   Examples of (A) a spring and (B) threads created with the **Coil** tool.

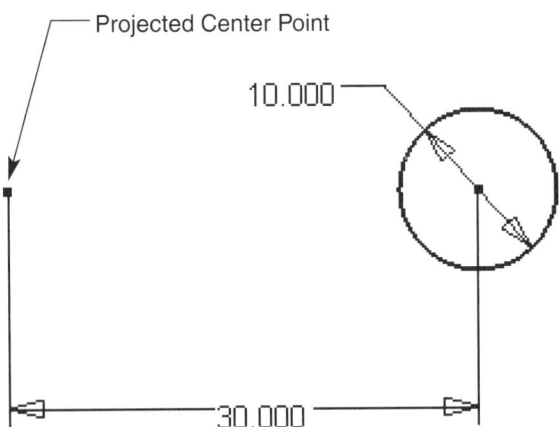

Projected Center Point

10.000

30.000

FIGURE 4.66   An example of a simple spring sketch.

FIGURE 4.67 The
Coil dialog box (**Coil
Shape** tab).

Coil Size tab

Coil Ends tab

Coil Shape tab

Profile button

Axis button

Flip button

Rotation
buttons

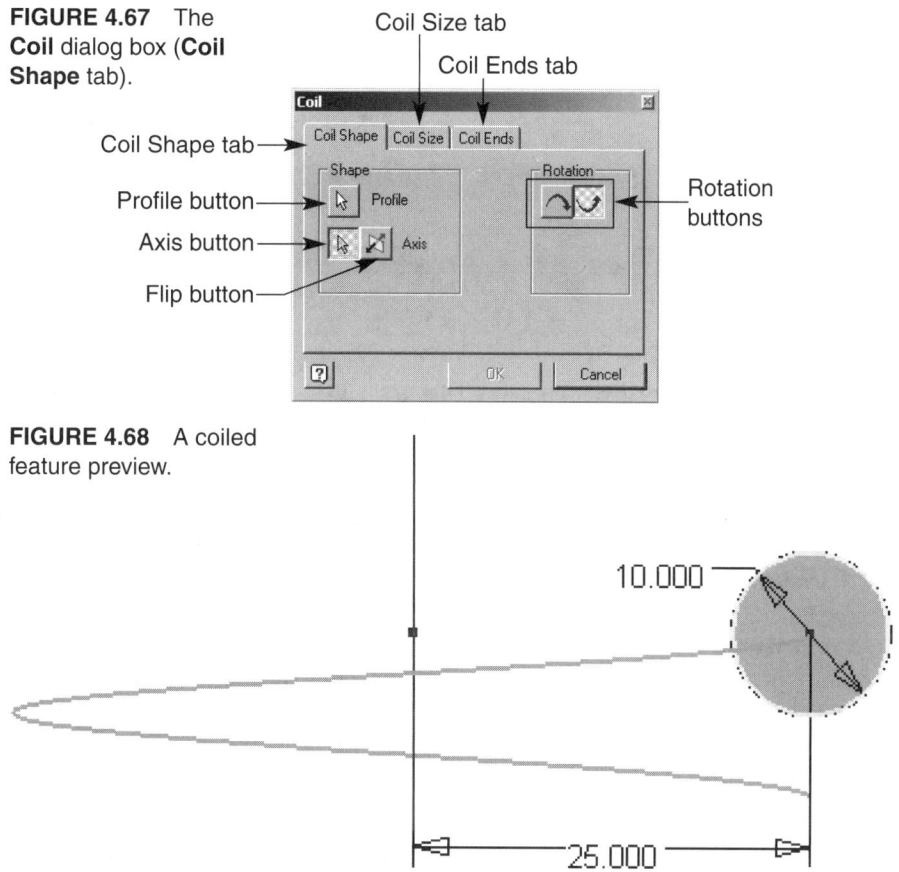

FIGURE 4.68 A coiled
feature preview.

10.000

25.000

- **Axis**   This button allows you to select an axis. Typically, the **Axis** button is automatically
  available when you select the profile. If not, you must pick the **Axis** button in order to select
  an axis. An axis is required to create a coil, because the axis defines where the spiral occurs
  and the diameter of the coil. Axes are created in the sketch environment. When the sketch
  profile and the axis are selected, a preview of the coil is displayed. See Figure 4.68.

Field Notes

Although you can use a sketched line or a projected axis from the **Origin** folder as an axis, you
may want to pick one of the axes from the **Origin** folder as the axis.

- **Flip**   This button allows you to flip the coil spiral direction.
- **Rotation**   These buttons allow you to specify the coil's direction of rotation—clockwise or
  counterclockwise.

Field Notes

A counterclockwise rotation corresponds to right-hand threads; a clockwise rotation corresponds to left-hand threads.
    When you create a coil on a part with existing features, the **Join, Cut,** and **Intersect**
operation buttons are available.

FIGURE 4.69 The **Coil** dialog box (**Coil Size** tab).

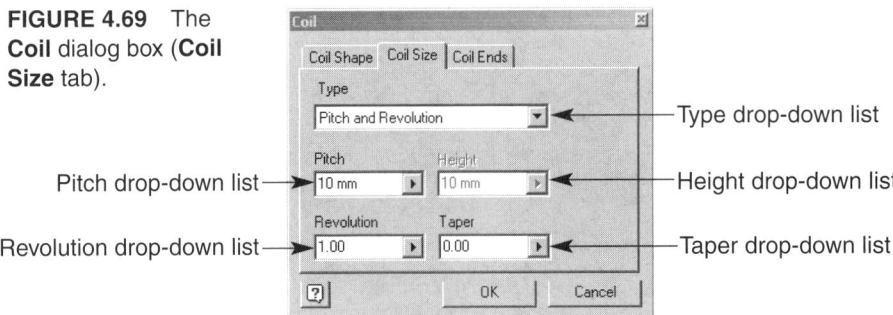

Type drop-down list
Pitch drop-down list
Revolution drop-down list
Height drop-down list
Taper drop-down list

When you have fully defined the coil shape, pick the **Coil Size** tab of the **Coil** dialog box to specify the size of the coil. See Figure 4.69. Then pick one of the following options from the **Type** drop-down list to choose how you want to define the coil size:

■ **Pitch and Revolution**   This option allows you to define the coil size using pitch and revolution input values. The height of the coils is calculated form the pitch and revolution values. Pitch is the distance parallel to the axis between a point on one coil spiral to the corresponding point on the next coil spiral. Revolution is the number of spirals, or 360° loops, the coil creates. Selecting the **Pitch and Revolution** option activates the **Pitch** and **Revolution** drop-down lists. Pick the arrow and select a pitch and revolution, or enter the desired values. In addition to the **Pitch** and **Revolution** drop-down lists, the **Taper** drop-down list is available. This list allows you to specify a taper angle for the coil. See Figure 4.70. Pick the arrow and select an angle, or enter the desired value.

■ **Revolution and Height**   This choice allows you to define the coil size using revolution and height input values. The pitch is calculated from the revolution and height values. Height is the total depth of the coil from the center of the starting profile to the center of the ending profile. Selecting the **Revolution and Height** option activates the **Revolution** and **Height** drop-down lists. Pick the arrow and select a revolution and height, or enter the desired values. The **Taper** drop-down list is also available for the **Revolution and Height** option.

■ **Pitch and Height**   This choice allows you to define the coil size using pitch and height input values. The revolution is calculated from the pitch and height values. Selecting the **Pitch and Height** option activates the **Pitch** and **Height** drop-down lists. Pick the arrow and select a revolution and height, or enter the desired values. The **Taper** drop-down list is also available for the **Pitch and Height** option.

■ **Spiral**   This choice allows you to create a coil without any height. See Figure 4.71. The coil size is defined using pitch and revolution input values, and no height is calculated. Selecting the **Spiral** option activates the **Pitch** and **Revolution** drop-down lists. Pick the arrow and select a pitch and revolution, or enter the desired values. The **Taper** drop-down list is not available.

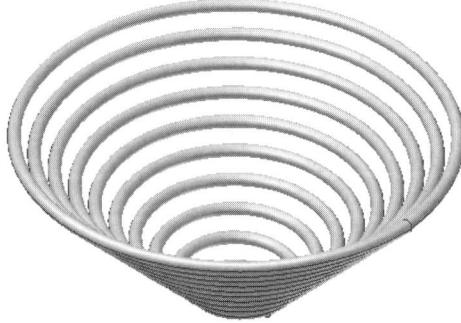

FIGURE 4.70   An example of a tapered coil (45° taper).

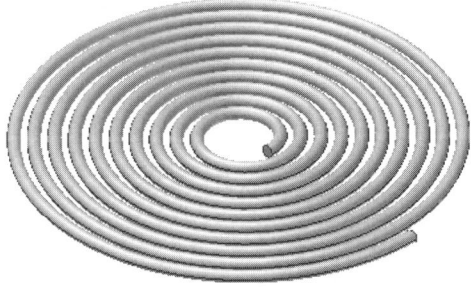

FIGURE 4.71   An example of a spiral coil.

FIGURE 4.72 The
Coil dialog box (**Coil
Ends** tab).

Start end area

Natural/Flat
drop-down list

Transition Angle
drop-down list

Flat Angle
drop-down list

End end area

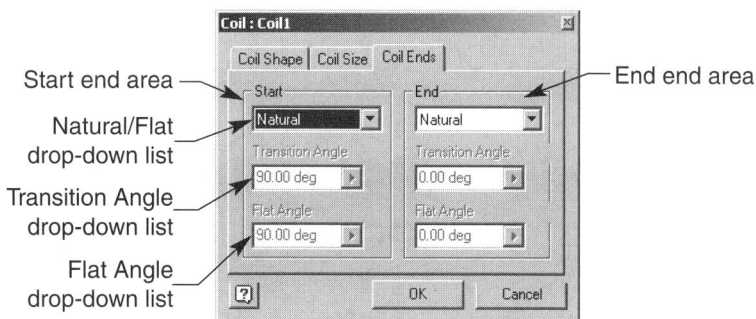

Once the coil size is identified, pick the **Coil Ends** tab of the **Coil** dialog box to specify the ends of the coil. See Figure 4.72. You can define the form of the coil ends for the start and end of the coil, as natural or flat, by selecting the desired option from the drop-down list. A natural end is the result of the pitch, revolution, height, and profile of the coil. This type of end does not allow you to define a transition or flat angle, because these angles are 0. As shown in Figure 4.73, a natural end is considered unaltered or untreated. Selecting the **Flat** end option activates the **Transition Angle** and **Flat Angle** drop-down lists. The transition angle is the number of degrees a coil end travels, or transitions, with pitch. Coil A in Figure 4.74 shows the difference between a natural end and an end with a transition angle of 90°, and 0° flat angle. The flat angle is the number of degrees a coil end travels without pitch. Coil B in Figure 4.74 shows a flat end of 180°, and a transition angle of 0°. Also shown is an end with a transition and flat end.

Once you have completely specified your coil, pick the **OK** button to create the coiled feature.

Field Notes

Applying transition and flat angles to a coil does not change the profile geometry.

As previously discussed, to create threads using the **Coil** tool, typically you sketch on a work plane and then cut the sketch profile around an extruded cylinder. This process is discussed in Chapter 6. An alternative method is to use the techniques previously discussed for creating a spring, but use a thread profile sketch. See Figure 4.75. Be sure you specify the distance shown in Figure 4.75 as the pitch.

Transition angle
of 90°

Flat and
Transition
angles
applied

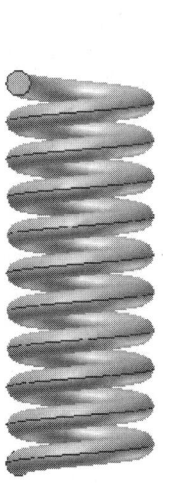

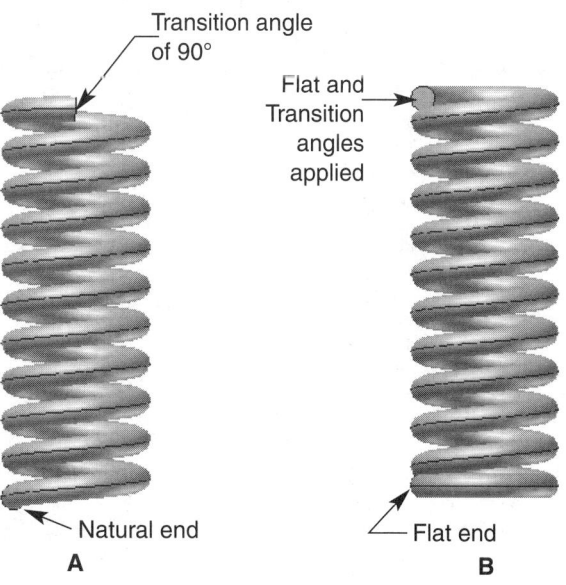

Natural end

A

Flat end

B

FIGURE 4.73 An example
of natural coil ends.

FIGURE 4.74 Examples of various transition
and flat angles applied to coil ends.

FIGURE 4.75 A thread profile sketch used to create threads with the coil command.

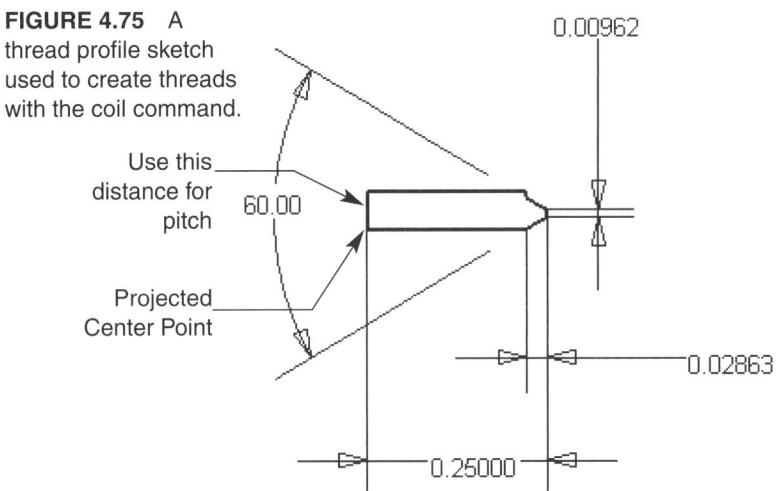

 Field Notes

To create realistic, detailed threads, you may want to use the **Coil** tool. However, for most applications, the **Thread** tool, discussed in Chapter 5, is adequate for generating thread representations.

Exercise 4.9

1. Continue from Exercise 4.8, or launch Autodesk Inventor.
2. Open a new part file.
3. Develop the sketch shown on the XY Plane.
4. Create a coil around the Y axis using the following specifications:

   Left-hand rotation.
   .5″ pitch.
   2″ height.
   3 Natural Start end.
   4 Flat End end, with 0° Transition angle and 180° Flat angle.

5. Save the part file as EX4.9.
6. Exit Autodesk Inventor or continue working with the program.

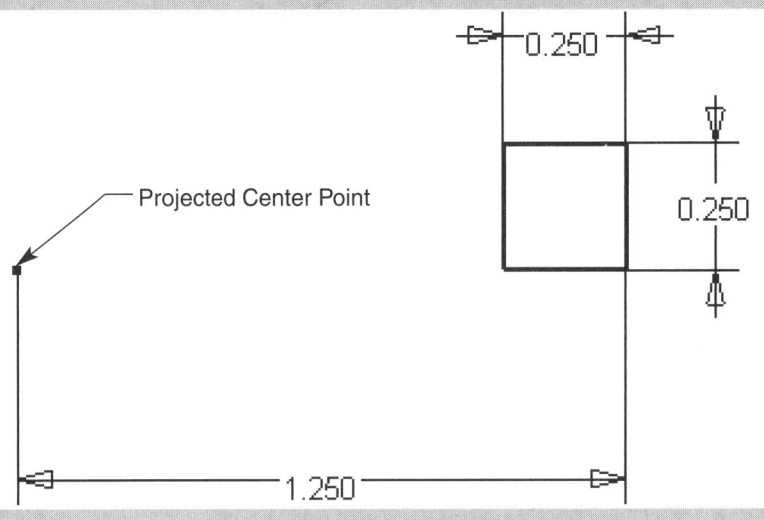

||||||||||| CHAPTER TEST

Answer the following questions on a separate sheet of paper.

1. Define and describe sketched features.

2. Give the general function of the View pull-down menu options.

3. Explain how to rotate an object while in Free Rotate mode.

4. Sometimes, you may not be able to select the piece of geometry you desire. To locate the geometry you may first need to rotate or change the display options. Identify an alternative way to access the object you want to select using the wheel on a wheel mouse.

5. How do you redefine the default isometric view?

6. Explain the function of the Style Name list box in the Lighting dialog box.

7. Give the function of the Active drop-down list box in the Lighting dialog box.

8. Discuss the function of the On/Off area in the Lighting dialog box.

9. Give the function of the Settings area in the Lighting dialog box.

10. Explain the function of the Materials dialog box.

11. Give the function of the Colors dialog box.

12. What does the Shaded Display option provide?

13. What does the Hidden Edge Display option provide?

14. What does the Wireframe Display option provide?

15. Name the Camera option that displays features in their true size and shape and allows you to observe the actual dimensions and position of geometry.

16. What is a fly-through?

17. Give the function of the Copy option when you right-click on Extrusion1 or any sketched feature in a part.

18. Give the function of the Paste option when you right-click on Extrusion1 or any sketched feature in a part

19. Give the function of the Delete option when you right-click on Extrusion1 or any sketched feature in a part.

20. Give the function of the Show Dimensions option when you right-click on Extrusion1 or any sketched feature in a part.

21. Give the function of the Edit Sketch option when you right-click on Extrusion1 or any sketched feature in a part.

22. Give the function of the Edit Feature option when you right-click on Extrusion1 or any sketched feature in a part.

23. Give the function of the Create Note option when you right-click on Extrusion1 or any sketched feature in a part.

24. Give the function of the Suppress Features option when you right-click on Extrusion1 or any sketched feature in a part.

25. Give the function of the Find in Window option when you right-click on Extrusion1 or any sketched feature in a part

26. Explain the function of the Name edit box in the Feature Properties dialog box.

27. Explain the function of the Suppress box in the Feature Properties dialog box.

28. Give the function of the Adaptive area in the Feature Properties dialog box.

29. Give the function of the Edit Coordinate System option when you right-click on Sketch1.

30. Give the function of the Reattach Sketch option when you right-click on Sketch1.

31. Give the function of the Share Sketch option when you right-click on Sketch1.

32. Give the function of the Visibility option when you right-click on Sketch1.

33. Give the function of the Redefine Feature option when you right-click on a work feature, such as Work Axis1.

34. Give the function of the Flip Normal option when you right-click on a work feature, such as Work Axis1.

35. Give the function of the Ground option when you right-click on a work feature, such as Work Axis1.

36. Name at least five features that sketched feature tools are used to create.

37. How do you change the panel bar display from learning to expert mode?

38. Give the function of the Extents area in the Extrude dialog box.

39. Give the function of the Distance option in the Extrude dialog box.

40. Give the function of the Measure option in the Extrude dialog box.

41. How do you create an extrusion with a taper?

42. What is a revolved feature?

43. Why is an axis required to create a revolved feature?

44. Give the function of the Full option in the Revolve dialog box.

45. Give the function of the Angle option in the Revolve dialog box.

46. What must you do first in order to use the Autoproject edges during curve creation tool?

47. Give the function of the Automatic reference edges for new sketch option.

48. How do you know when edges have been projected?

49. How do you extrude a feature to the next possible face or plane?

50. How is the To option different from the To Next option?

51. Explain the function of the From To option.

52. How do you extrude a feature through all existing features?

53. Name at least three types of hole features.

54. List the three termination options for holes.

55. Identify the Thread Type drop-down list options that allow you to specify threads for inch parts and metric parts.

56. What are ribs and webs?

57. Give the steps used to create a rib or web

58. Why is an axis required to create a coil?

59. Explain the function of the Pitch and Revolution option in the Coil dialog box.

60. Discuss the function of the Revolution and Height option in the Coil dialog box.

61. Explain the function of the Pitch and Height option in the Coil dialog box.

62. What is the difference between creating threads with the Coil tool and the Thread tool?

|||||||||||| PROJECTS

Instructions:

- Open the following Chapter 3 part files, and follow the specific instructions provided for each project.

1. Name: C-Clamp Pin

   Units: Metric

   Specific Instructions:

   - Open P3.1, and Save Copy As P4.1.

   - Close P3.1 without saving, and Open P4.1.

   - Extrude the sketch profile 52mm in a positive direction as shown.

   - Save As: P4.1.

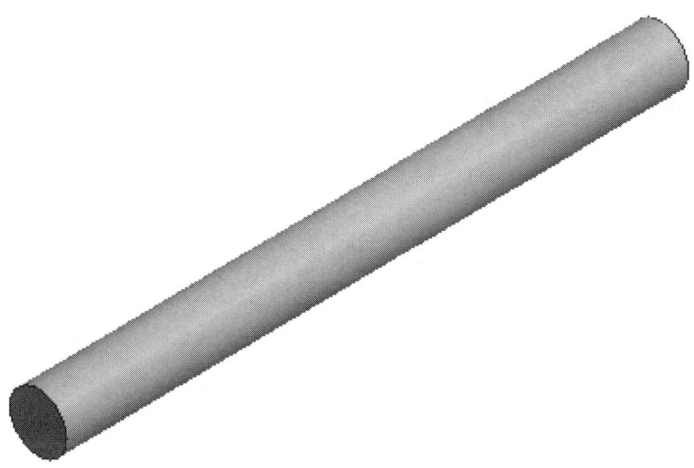

2. Name: C-Clamp Swivel

   Units: Metric

   Specific Instructions:

   ■ Open P3.2, and Save Copy As P4.2.

   ■ Close P3.2 without saving, and Open P4.2.

   ■ Extrude the sketch profile 9.5mm in a negative direction as shown.

   ■ Apply the counterbored hole using the dimensions provided, to the center of the base extrusion as shown.

   ■ Save As: P4.2.

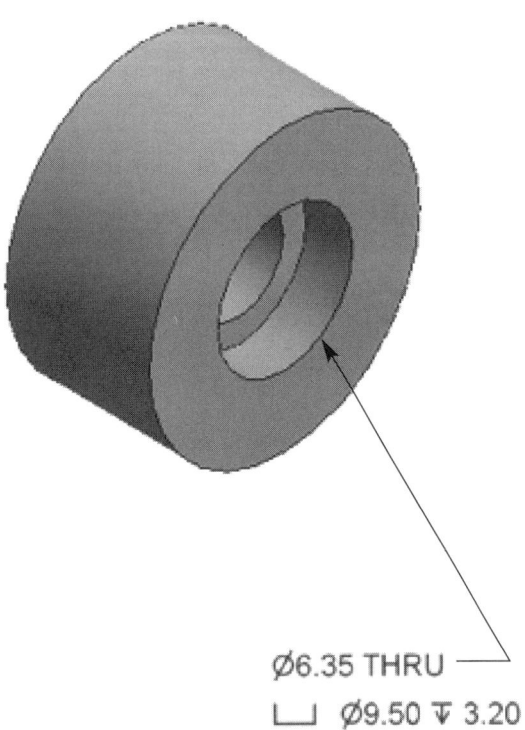

Ø6.35 THRU
⌴ Ø9.50 ▼ 3.20

3. Name: C-Clamp Screw

Units: Metric

Specific Instructions:

- Open P3.3, and Save Copy As P4.3.

- Close P3.3 without saving, and Open P4.3.

- Extrude the sketch profile 80mm in a positive direction as shown.

- Open a sketch on one end and sketch a ∅6.35 circle at the center point.

- Extrude the sketch 7.2mm as shown.

- Open a sketch on the opposite end of Extrusion1, and sketch a ∅12.7 circle at the center point.

- Extrude the sketch 22mm as shown.

- Save As: P4.3.

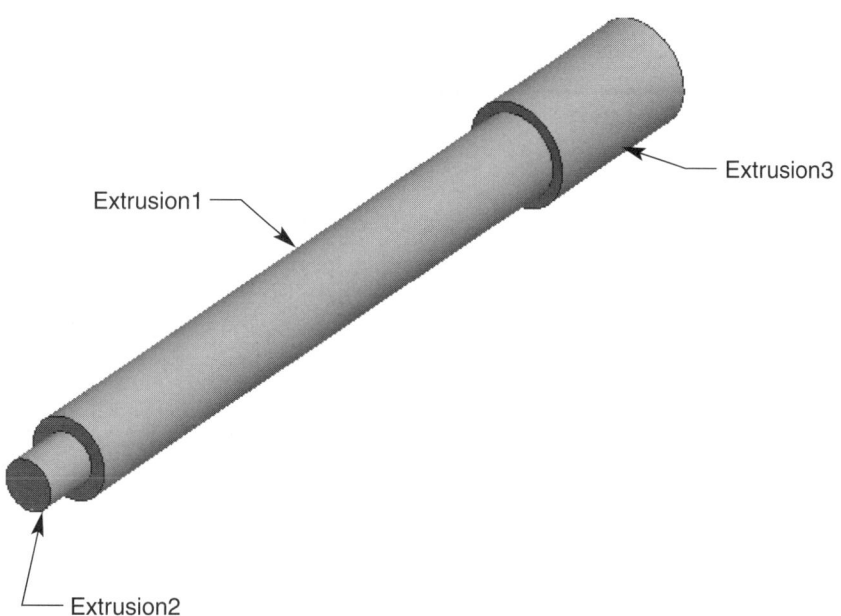

4. Name: Support Bracket

Units: Inch

Specific Instructions:

- Open P3.4, and Save Copy As P4.4.

- Close P3.4 without saving, and Open P4.4.

- Extrude Profile1, .125″ midplane.

- Share Sketch1.

- Extrude Profile2, .375″ midplane.

- Open a sketch on Face1, and offset the automatically projected geometry .625″. You may need to edit and constrain sketch geometry.

- Cut extrude the sketch through the part as shown.

- Apply the holes shown using the dimensions provided.

- Save As: P4.4.

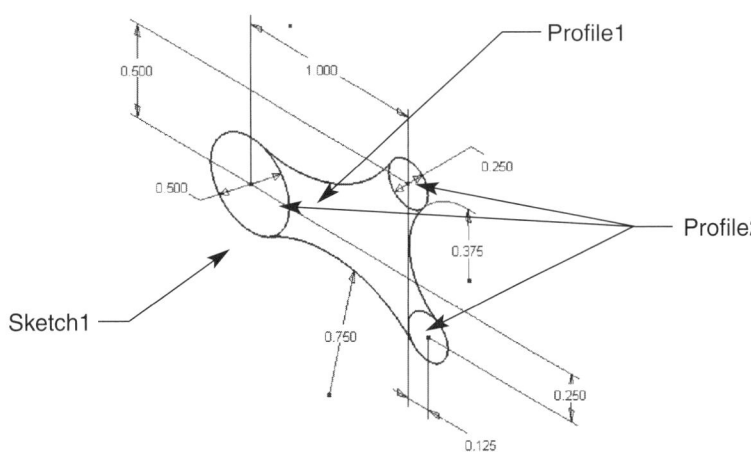

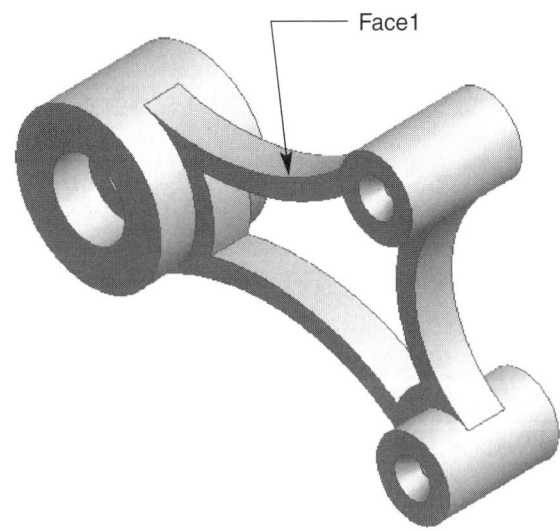

5. Name: Funnel

   Units: Inch

   Specific Instructions:

   ■ Open a new part file.

   ■ Fully revolve the sketch shown around the Y Axis.

   ■ Sketch a rectangle tangent to the radius of the circle, on the base of the revolved feature.

   ■ Extrude the sketch .125″ in a negative direction as shown.

   ■ Save As: P4.5.

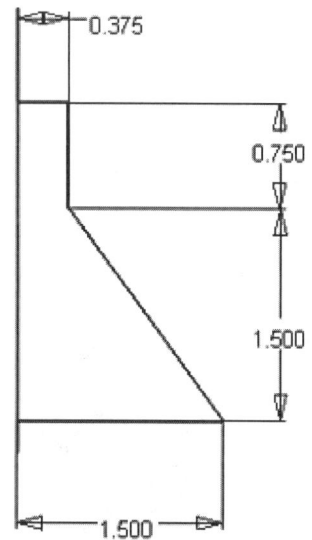

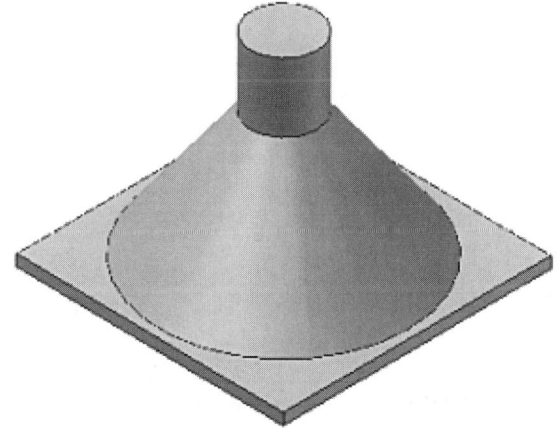

6. Name: C-Clamp Body

   Units: Metric

   Specific Instructions:

   - Open P3.6, and Save Copy As P4.6.

   - Close P3.6 without saving, and Open P4.6.

   - Extrude the sketch 19mm, midplane.

   - Create the center point sketch shown.

   - Apply the hole shown using the dimensions provided.

   - Save As: P4.6.

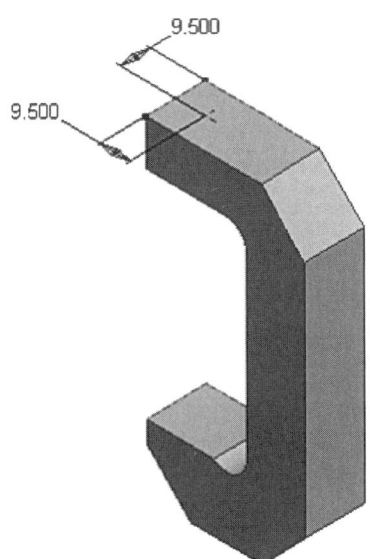
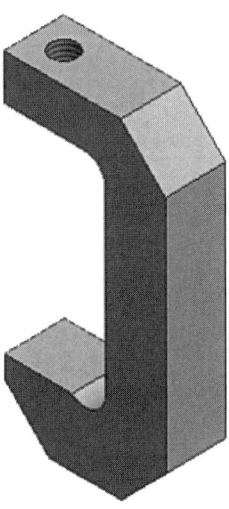

7. Name: Torsion Spring

Units: Inch

Specific Instructions:

- Open a new part file.

- Create the sketch shown on the XY Plane.

- Coil the sketch in a positive direction, using the following parameters:

  Axis: Y Axis

  Rotation: Right-hand

  Pitch: .125″

  Revolutions: 5

  Ends: Natural

- Use the **Automatic reference edges for new sketch** tool, to create a sketch of the bottom coil face.

- Coil the sketch in a negative direction, using the following parameters:

  Axis: Y Axis

  Rotation: Right-hand

  Pitch: .125″

  Revolutions: 5

  Taper: −30°

  Ends: Start=Natural, End=Flat with 90° transition angle.

- Use the **Automatic reference edges for new sketch** tool, to create a sketch of the top coil face of the Coil1.

- Coil the sketch in a positive direction, using the following parameters:

  Axis: Y Axis

  Rotation: Right-hand

  Pitch: .25″

  Revolutions: 5

  Taper: 30°

  Ends: Natural

- Use the **Automatic reference edges for new sketch** tool, to create a sketch of the top coil face of the Coil3.

- Coil the sketch in a positive direction, using the following parameters:

  Axis: Y Axis

  Rotation: Right-hand

  Pitch: .25″

  Revolutions: 5

  Taper: −30°

  Ends: Natural

- Use the **Automatic reference edges for new sketch** tool to create a sketch of the top coil face of the Coil4.

- Coil the sketch in a positive direction, using the following parameters:

  Axis: Y Axis

  Rotation: Right-hand

  Pitch: .125″

  Taper: 0°

  Revolutions: 5

  Ends: Natural

- Use the **Automatic reference edges for new sketch** tool, to create a sketch of the top coil face of the Coil5.

- Coil the sketch in a positive direction, using the following parameters:

  Axis: Y Axis

  Rotation: Right-hand

  Pitch: .125″

  Revolutions: 3

  Taper: −30°

  Ends: Start=Natural, End=Flat with 45° flat angle.

- The final project should look like the part shown.

- Save As: P4.7

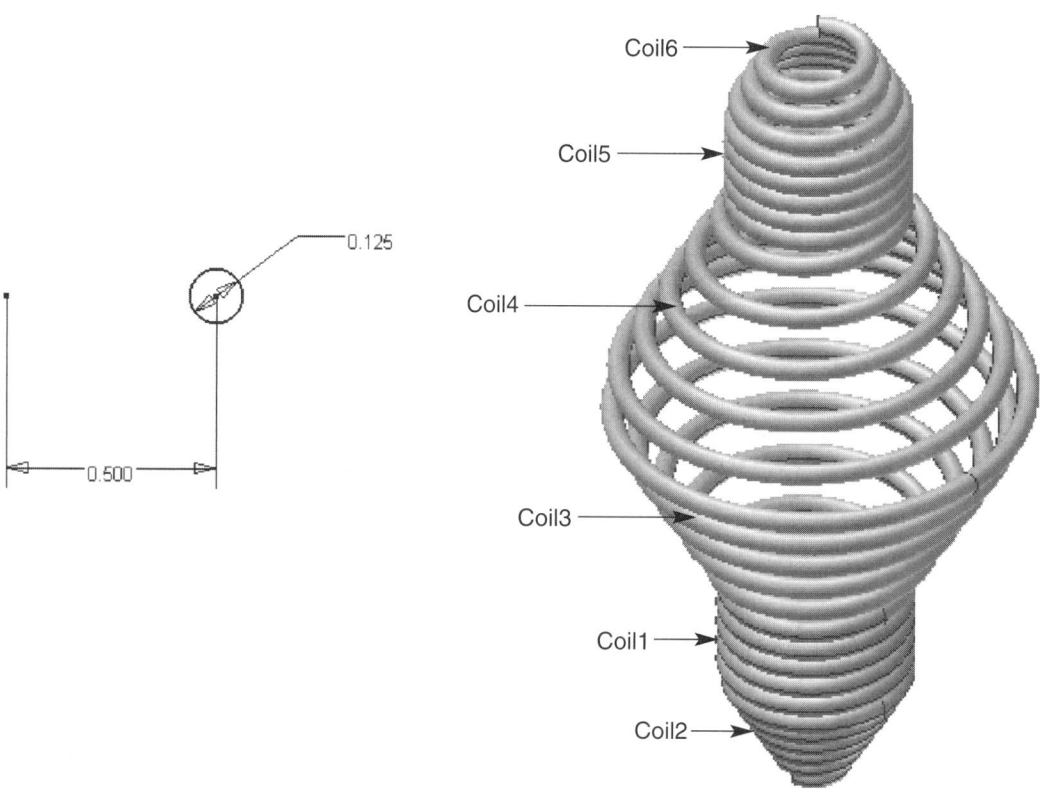

8. Name: Torsion Spring Swivel Eye

Units: Inch

Specific Instructions:

- Open a new part file.

- Create the sketch shown on the XY Axis.

- Revolve the sketch, 305° around the Y Axis.

- Use the **Automatic reference edges for new sketch** tool, to create a sketch of Face1.

- Extrude the sketch .75″ in a positive direction.

- Use the **Automatic reference edges for new sketch** tool to create a sketch of Face2. Add an axis line .5″ from the center of the sketched circle as shown.

- Revolve the sketch, 320° around the sketched axis line.

- The final project should look like the part shown.

- Save As: P4.8.

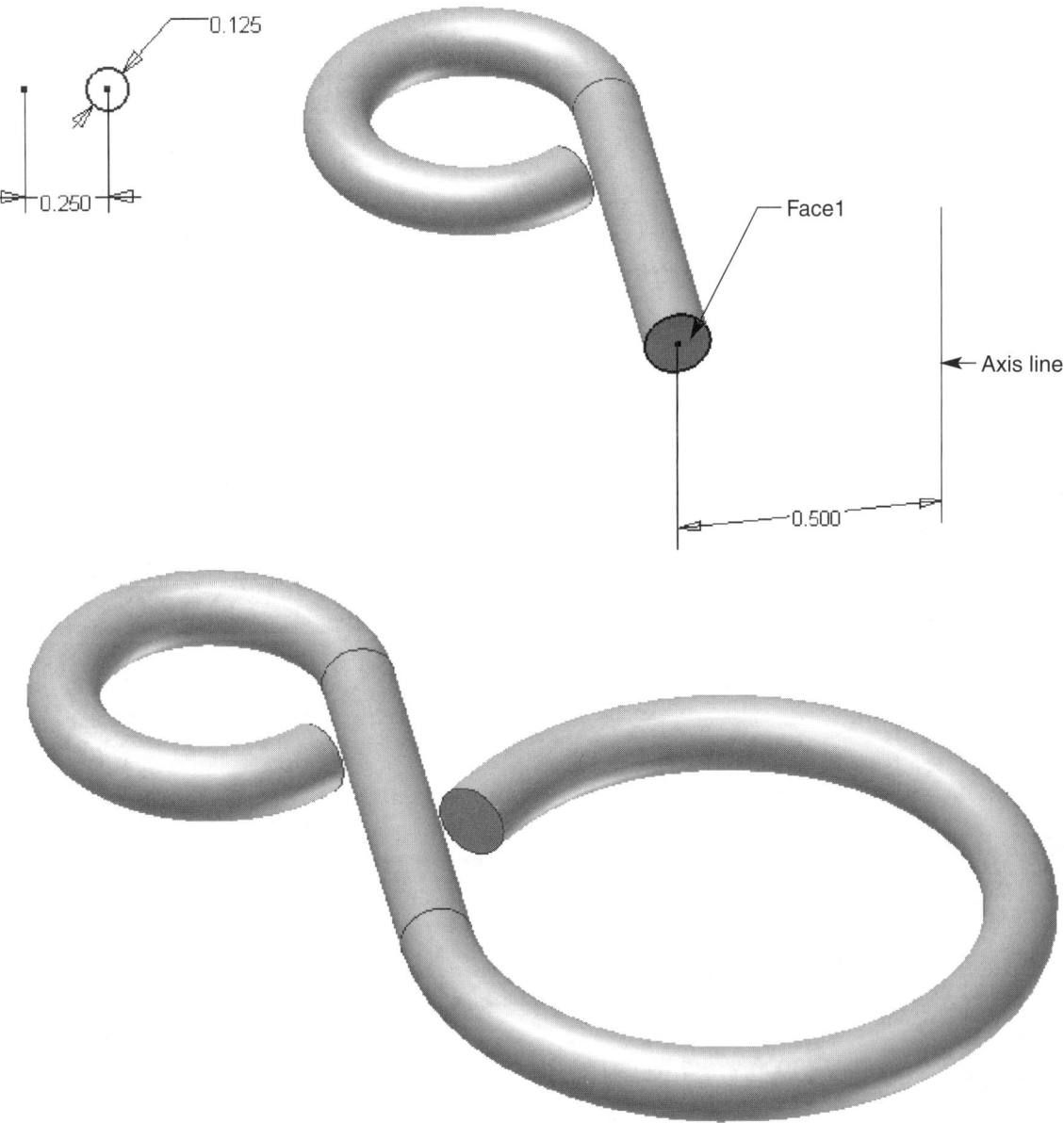

9. Name: Torsion Spring Swivel Bolt

   Units: Inch

   Specific Instructions:

   ■ Open a new part file.

   ■ Create the sketch shown on the XY Axis.

   ■ Fully revolve the sketch around the Y Axis.

   ■ The final project should look like the part shown.

   ■ Save As: P4.9.

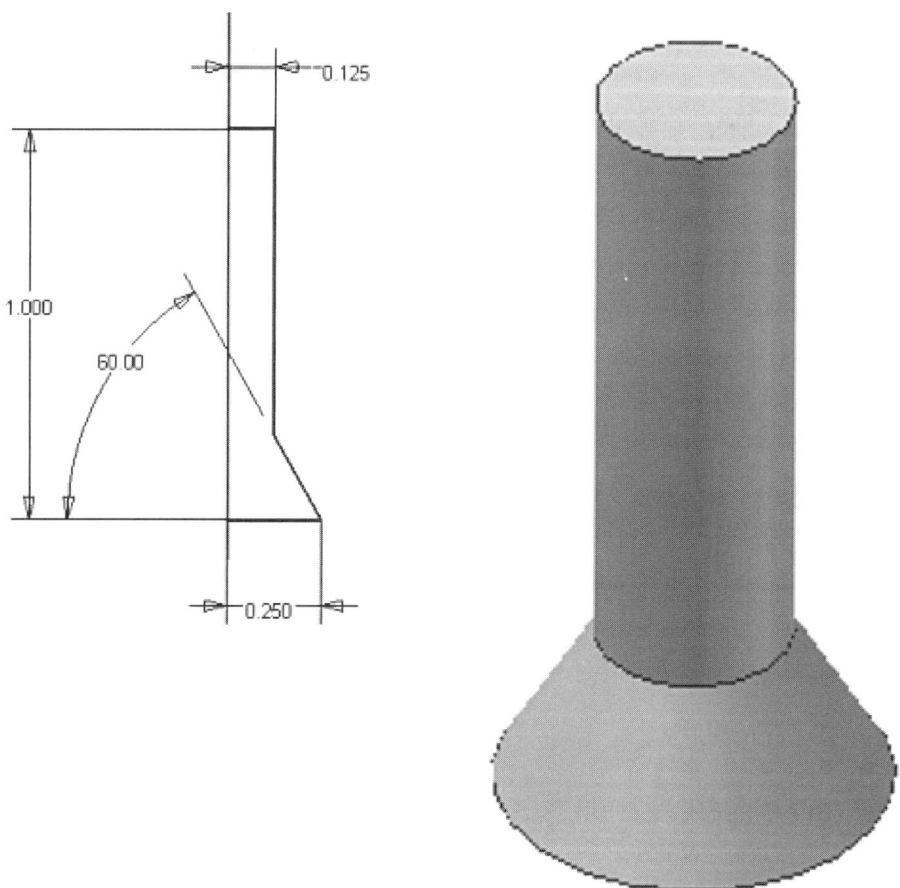

10. Name: Bracket

Units: Metric

Specific Instructions:

- Open a new part file.

- Create the part shown using the features specified.

- When complete, your part should contain at most: 3 extrusions from 3 sketches, 3 ribs from 3 sketches, 2 webs from 1 sketch, and 2 holes from 1 sketch.

- Save As: P4.10.

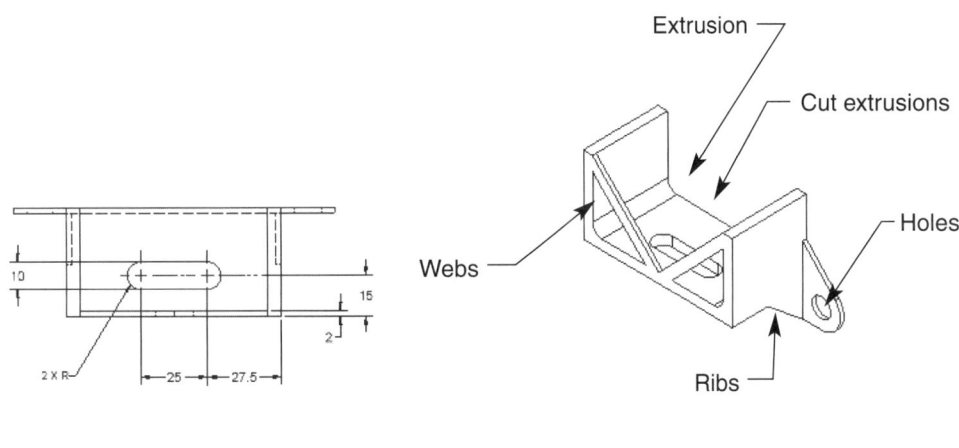

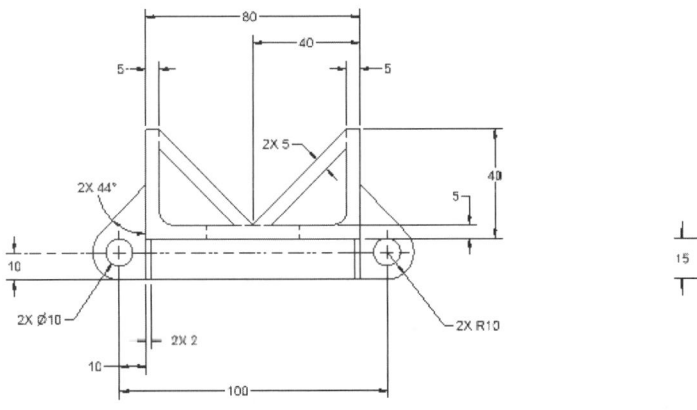

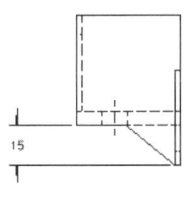

CHAPTER 5

Creating Placed Features

LEARNING GOALS

After completing this chapter, you will be able to:

◎ Add placed features to a part, and use placed feature tools.

◎ Add fillets and rounds to feature edges.

◎ Create chamfered feature edges.

◎ Place shell features.

◎ Apply face drafts.

◎ Place threads on cylindrical, conical, and hole features.

Developing Placed Features

Once you have created a sketch and have generated one or multiple sketched features, you are ready to create placed features. **Placed features** can be added to the most basic feature, such as a simple extruded rectangle, or to very complex, multifeature parts. As the name implies, placed features are placed on existing feature geometry. For example, a fillet is placed on the corner of an extruded cylinder. See Figure 5.1. In contrast to sketched features, placed features do not require a sketch to be created. However, they do require an acceptable sketched, placed, catalog, or pattern feature to be present. Placed features can be created any time during part model development, as long as one of these suitable features exists. Placed features include: fillets, chamfers, shells, face drafts, and threads.

FIGURE 5.1 An example of a fillet placed feature on a cylinder sketched feature.

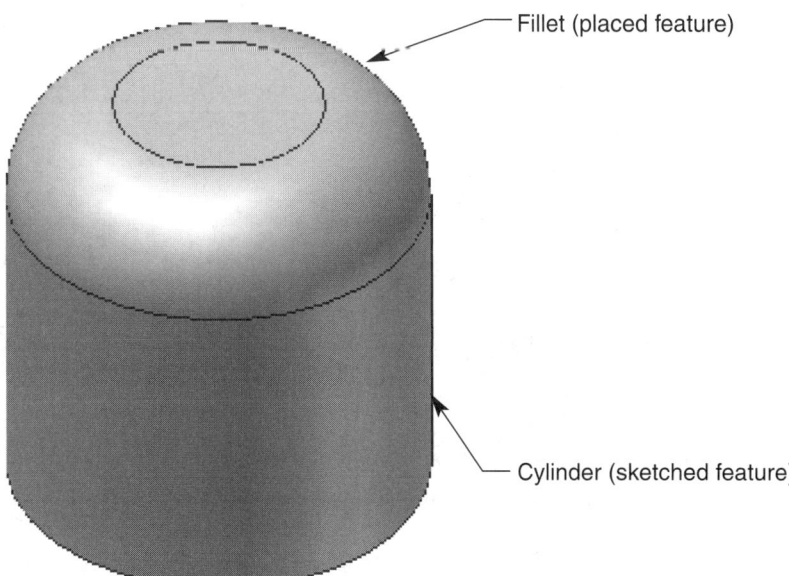

Fillet (placed feature)

Cylinder (sketched feature)

Field Notes

> All the interface and work environment options and applications discussed in Chapter 4 apply to working with placed features.

Using Placed Feature Tools

Placed feature tools are used to create placed features, including fillets, chamfers, shells, face drafts, and threads. These tools are accessed from the **Features** toolbar, shown in Figure 5.2, or from the **Features** panel bar. When a feature is present and you are in the feature work environment, the panel bar is in features mode. See Figure 5.3. By default, the panel bar is displayed in learning mode, as seen in Figure 5.3. To change from learning to expert mode, seen in Figure 5.4 and discussed in Chapter 2, right-click any of the panel bar tools or left-click the panel bar title and select the **Expert** option.

FIGURE 5.2 The **Features** toolbar.

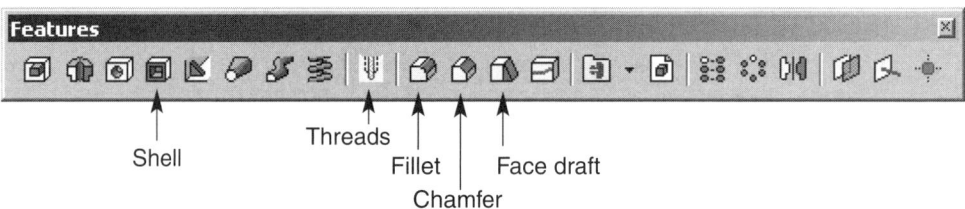

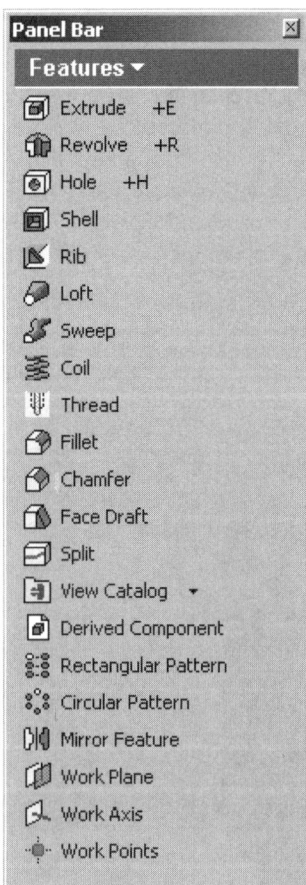

FIGURE 5.3 The panel bar, in features (learning) mode.

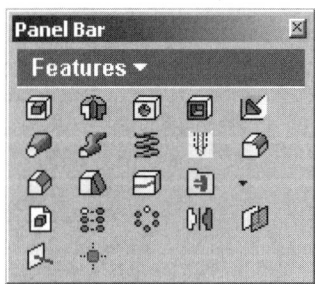

FIGURE 5.4 The panel bar, in features (expert) mode.

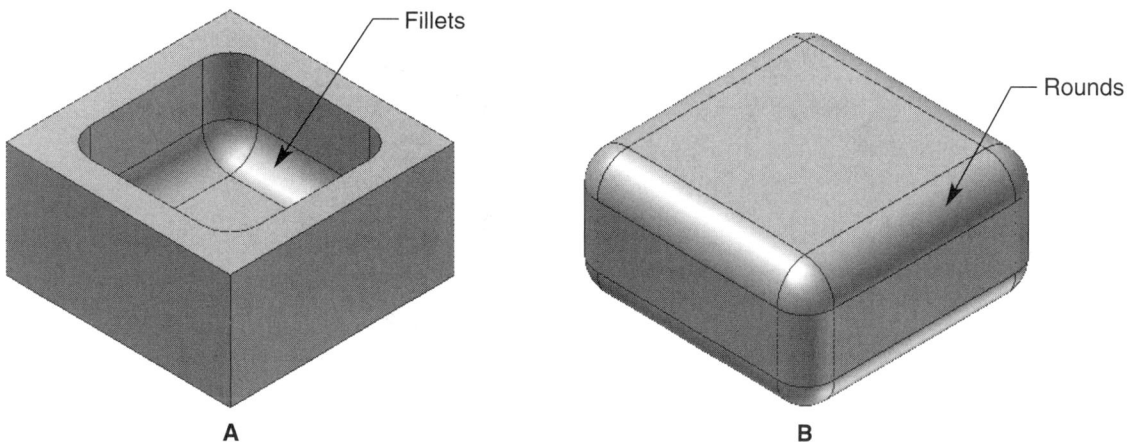

FIGURE 5.5   An example of (A) a filleted feature and (B) a rounded feature.

Adding Fillets and Rounds to a Feature

A *fillet* is a curve placed at the inside intersection of two or more surfaces, adding material to a feature. See Figure 5.5A. Typically, fillets are used to help to release patterns from forgings and castings. They are also created to relieve stress and ease the machining of inside corners. A *round* is a curve placed on the exterior intersection of two or more surfaces, removing material from a feature. See Figure 5.5B. Rounds remove sharp machined edges and help to release patterns from forgings and castings.

Both fillets and rounds are created using the **Fillet** tool. Access the **Fillet** command using one of the following techniques:

✓ Pick the **Fillet** button on the panel bar.

✓ Pick the **Fillet** button on the **Features** toolbar.

The **Fillet** dialog box is displayed when you access the **Fillet** command. See Figure 5.6.

Often, you can fully fillet or round a part using only a single **Fillet** command because the **Fillet** dialog box allows you to fillet or round a part completely with different fillet styles and radii. You do not need to continually access the **Fillet** command to create a different type of fillet or round. Figure 5.7 shows an example of a part with different fillet and round styles and radii. All the fillets and rounds were created using a single fillet command, and each of the fillets placed during one operation create a single feature.

FIGURE 5.6   The **Fillet** dialog box (**Constant** tab opened).

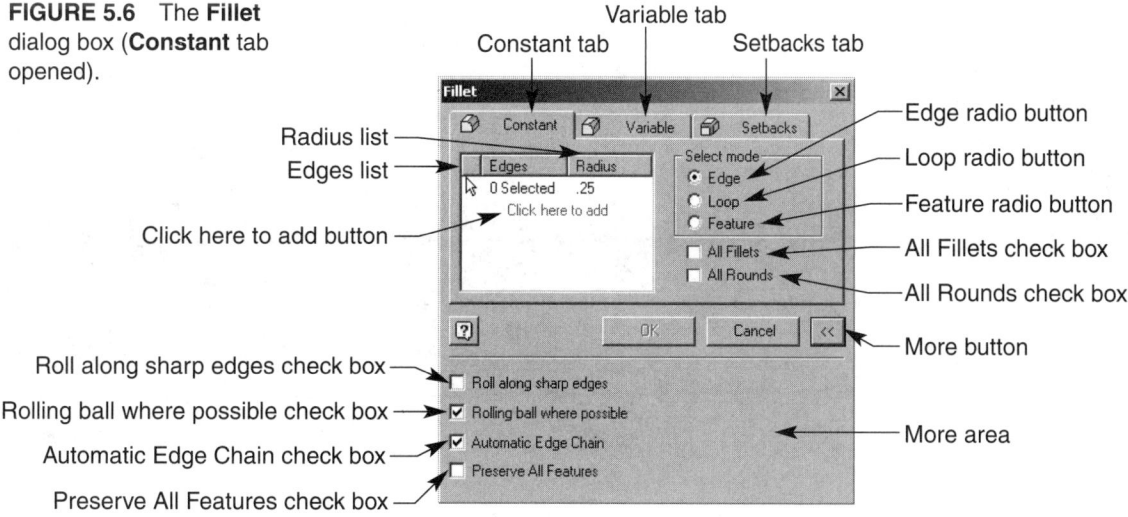

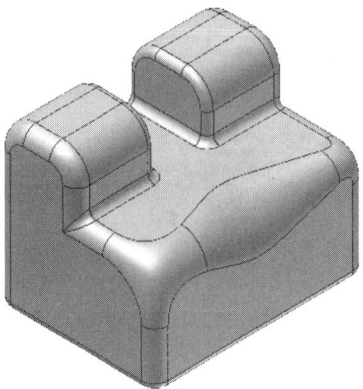

FIGURE 5.7 By accessing the **Fillet** tool once, you can apply several fillet and round styles and radii.

Field Notes

If you plan to suppress only certain fillets or want to have additional filleted features, you may want to access the **Fillet** command more than once.

Before you begin to create a fillet or round, you may want to adjust some of the options located in the **More** area of the **Fillet** dialog box. The following check boxes are available when you pick the **More** button to access the **More** area:

■ **Roll along sharp edges**   When this check box is not selected, a lip is formed when filleting an object around a sharp edge, and the fillet radius changes. See Figure 5.8A. If you want the fillet radius to remain constant and the feature faces to stay the same, select this check box. See Figure 5.8B.

■ **Rolling ball where possible**   This check box allows you to specify the corner style for your fillets and rounds. When the **Rolling ball where possible** check box is selected, fillets and rounds are placed on feature edges in a way that looks like a ball rolled around them. See Figure 5.9A. If unselected, fillets and rounds are blended together, assuming at least one X, Y, and Z axis edge is selected. See Figure 5.9B.

■ **Automatic Edge Chain**   This option allows you to select the entire edge chain of a tangent loop. See Figure 5.10A. When the **Automatic Edge Chain** check box is not selected, you can pick the individual pieces of a tangent loop edge. See Figure 5.10B.

FIGURE 5.8 (A) An example of a fillet when the **Roll along sharp edges** check box is not activated and (B) when it is.

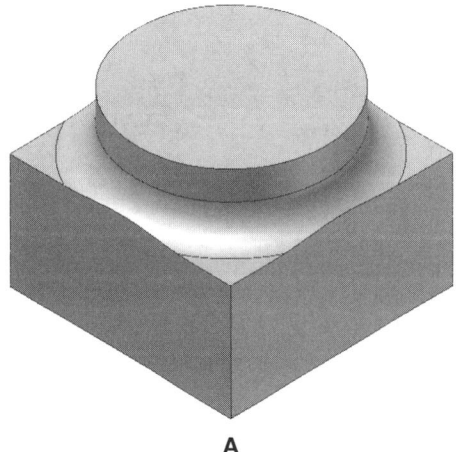

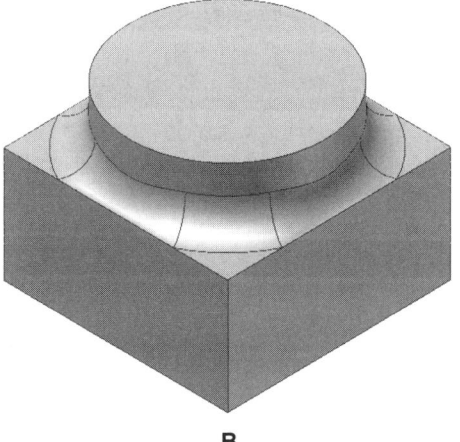

A   **B**

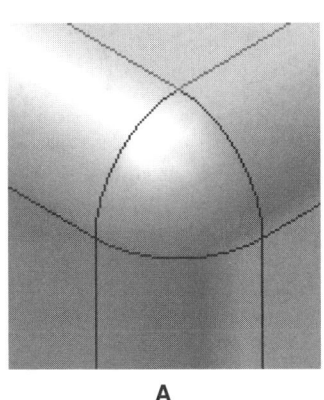

A

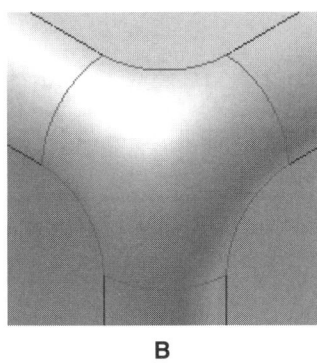

B

FIGURE 5.9   (A) An example of a round corner when the **Rolling ball where possible** check box is activated and (B) when it is not.

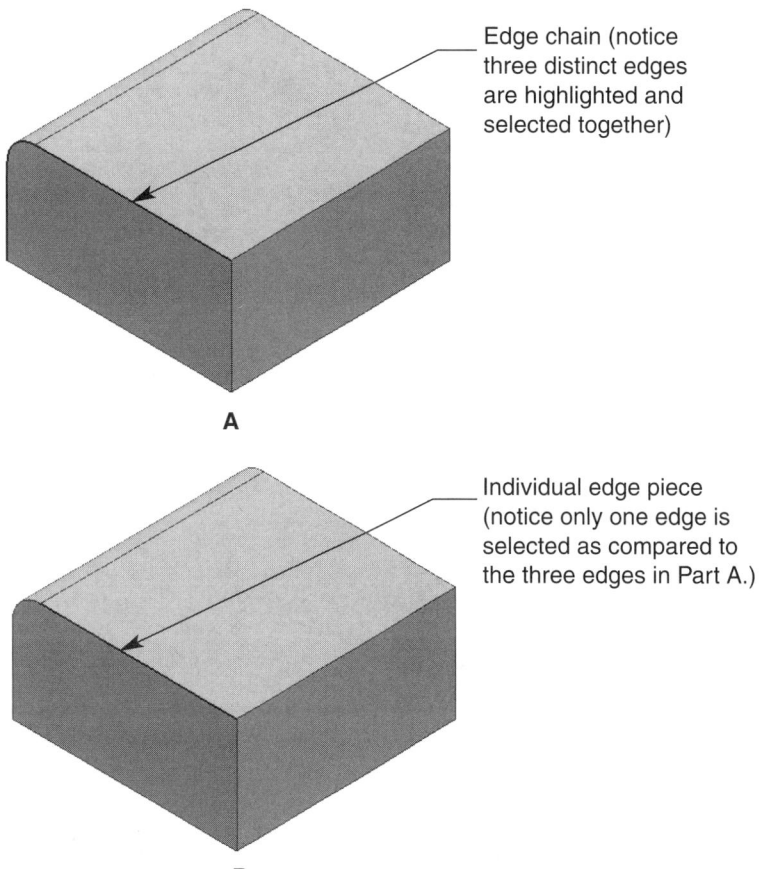

Edge chain (notice three distinct edges are highlighted and selected together)

A

Individual edge piece (notice only one edge is selected as compared to the three edges in Part A.)

B

FIGURE 5.10   (A) Selecting an entire edge chain. (B) Selecting an individual edge piece.

- **Preserve all features**   When this check box is selected, Autodesk Inventor ensures that the features that intersect the fillet are accounted for, and the intersection information is determined when the fillet is created.

Field Notes

Typical fillets and rounds are created using the default selected check boxes located in the **More** area of **Fillet** dialog box. Still, these options may be required for some applications.

Autodesk Inventor allows you to create two types of fillets, constant and variable. A ***constant fillet*** or ***round*** has a curve radius that does not change. They are the simplest fillets and rounds to place on a feature and do not allow you to apply setbacks, as discussed later in this chapter. See Figure 5.11. A ***variable fillet*** or round has different sized curved radii placed at precise points between the start and end of a fillet or round. See Figure 5.12.

To apply a constant fillet or round, use the options available inside the **Constant** tab of the **Fillet** dialog box (Figure 5.6). First, specify a fillet or round radius by highlighting the default radius and entering the desired value, or by selecting a value from the drop-down list. Next, identify which objects you want to fillet or round. If you want to fillet all the interior part edges with the same style and radius of fillet, pick the **All Fillets** check box. Similarly, if you want to round all the exterior part edges with the same style and

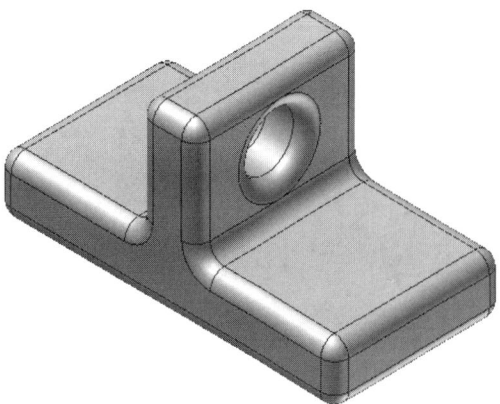

FIGURE 5.11   An example of constant fillets and rounds.

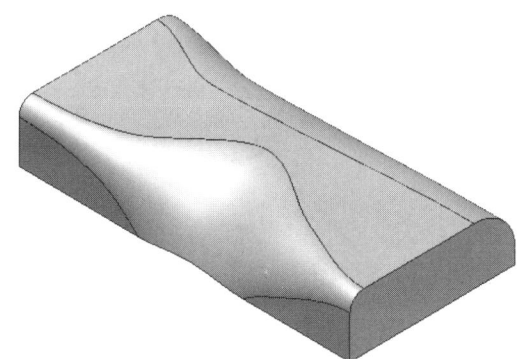

FIGURE 5.12   An example of variable fillets and rounds.

radius of round, pick the **All Rounds** check box. When you pick one of these check boxes, all feature edges are automatically selected.

To apply fillets or rounds to specific edges and to use different styles and radii, use one of the following **Select Mode** radio buttons:

- **Edge**   Pick this radio button to select individual feature edges.
- **Loop**   Pick this radio button to select a feature edge loop.
- **Feature**   Pick this radio button to select an entire feature.

Once you have chosen a select mode, pick the object you want to fillet or round. A preview of the feature is displayed.

Field Notes

You can deselect any of the selected objects in any of the **Fillet** dialog box drop-down lists by holding down the **Ctrl** key on your keyboard and picking the object.

If the preview looks correct and you do not want to place any additional fillets or rounds, pick the **OK** button to accept the curves. If you would like to apply more constant fillets or rounds, pick the **Click here to add** button. Then, as previously discussed, enter the desired fillet or round radius and select the edge. Continue adding fillets and rounds as required. When finished, pick the **OK** button.

Field Notes

To delete a fillet or round edge listed in the **Edges** list box, pick the edge you would like to remove and press the **Delete** key on your keyboard.

To add variable fillets or rounds, and setbacks, in addition to constant fillets and rounds, pick the **Variable** tab.

Exercise 5.1

1. Launch Autodesk Inventor.
2. Open a new part file.
3. Create a 1″×2″×4″ extruded box as shown.
4. Extrude a ⌀1.75″ cylinder 1″ from one of the box edges as shown.
5. Apply constant fillets and rounds to the corners using the radii and selection modes specified. Be sure the **Roll along sharp edges** check box is unselected.
6. You should need to access the Fillet command only once for this exercise.
7. Right click **Fillet1** in the panel bar, and select the **Edit Feature** menu option.
8. Access the **More** area of the **Fillet** dialog box, and select the **Roll along sharp edges** check box.
9. The final part should look like the part shown.
10. Save the part as EX5.1.
11. Exit Autodesk Inventor or continue working with the program.

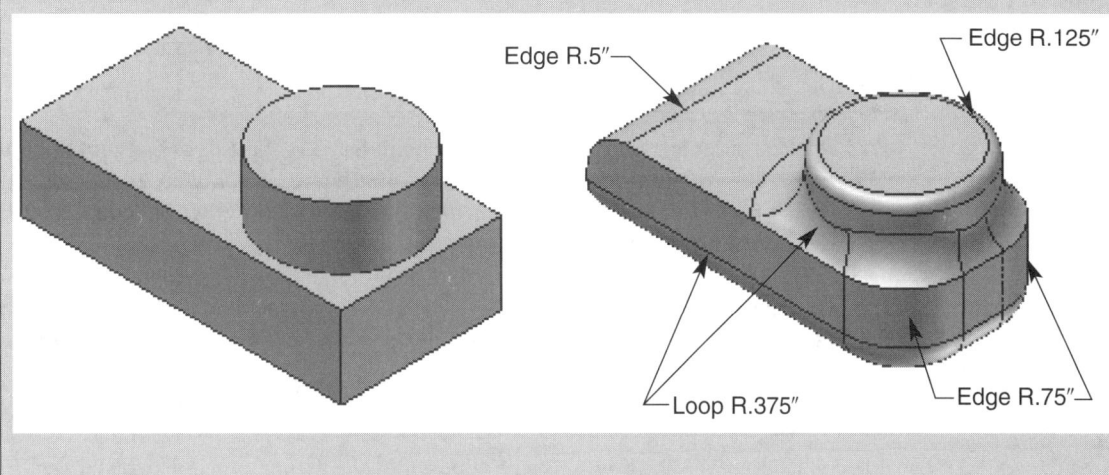

To apply a variable fillet or round, use the options available inside the **Variable** tab of the **Fillet** dialog box. See Figure 5.13.

Variable fillets and rounds are created by specifying an edge, different radii, and the location of the radii along the edge. To place a variable fillet or round on a feature edge, you must first select the desired feature edge. The start point is initially active and is identified by a blue dot, as shown in Figure 5.14. As you progress through the variable fillet or round process, other points become active and display the blue dot. Specify the radius of the start point by entering or selecting a value from the **Radius** drop-down list. Next, pick **End** from the **Point** list box, and specify the radius of the endpoint by entering or selecting a value from the **Radius** drop-down list. Now you can choose additional points along the specified edge by moving your cursor to the desired

FIGURE 5.13 The **Variable** tab of the **Fillet** dialog box.

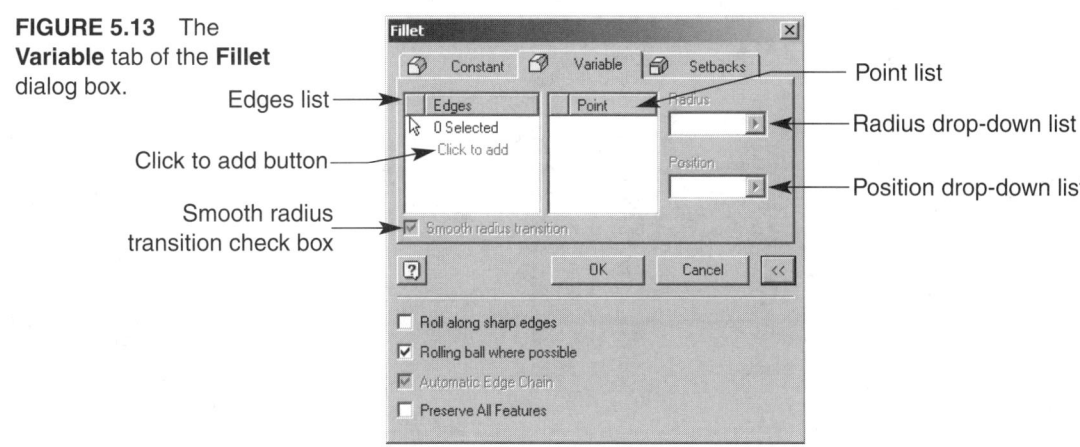

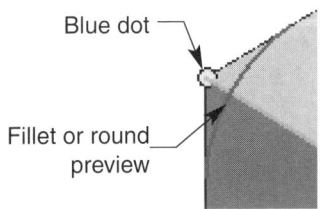

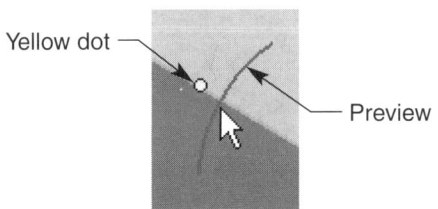

FIGURE 5.14   The active point and preview shows the location and radius of the current fillet or round point.

FIGURE 5.15   Use your cursor to drag the yellow dot and fillet or round preview to the desired location.

location. As shown in Figure 5.15, a yellow dot and preview identifies the point. When you are satisfied with the location of the radius, pick the point. Then specify the radius of the endpoint by entering or selecting a value from the **Radius** drop-down list. If you would like to further define the location of the selected point, enter or select a value from the **Position** drop-down list. If required, continue adding points along the feature edge using the techniques discussed. When finished, pick the **OK** button to create the fillets or rounds.

Field Notes

To delete a variable fillet or round edge or point listed in the **Edges** or **Point** list boxes, pick the edge or point you want to remove and press the **Delete** key on your keyboard. The position of additional points is relative to the start point of the variable fillet or round, and the individual edge pieces, not a tangent loop.

By default, the **Smooth radius transition** check box is selected. This option creates a steadily increasing or decreasing fillet or round blend between variable points. If you want a direct, linear arc between variable points, unselect the **Smooth radius transition** check box.

If you unselect the **Rolling ball where possible** check box, previously discussed, fillet and round corners are blended together. The blend of a non-rolling ball corner is defined by setbacks. *Setbacks* are points where a fillet or round on one edge begins to combine with a fillet or round of at least two other edges. Setbacks can be any value, but only a setback that is greater than the fillet or round radius will blend corners. A setback that is less than or equal to the fillet or round radius creates a rolling ball effect.

To modify the setbacks of a fillet or round, you must first unselect the **Rolling ball where possible** check box, located in the **More** area of the **Fillet** dialog box. Then fillet the feature using the techniques previously discussed. Now pick the **Setbacks** tab of the **Fillet** dialog box. See Figure 5.16. Once you access the **Setbacks** tab of the **Fillet** dialog box, pick the corner, or vertex, of the feature edges. The first vertex is identified as **Vertex1** in the **Vertex** list box, and once selected, displays a blue dot. See Figure 5.17. Once the vertex is specified,

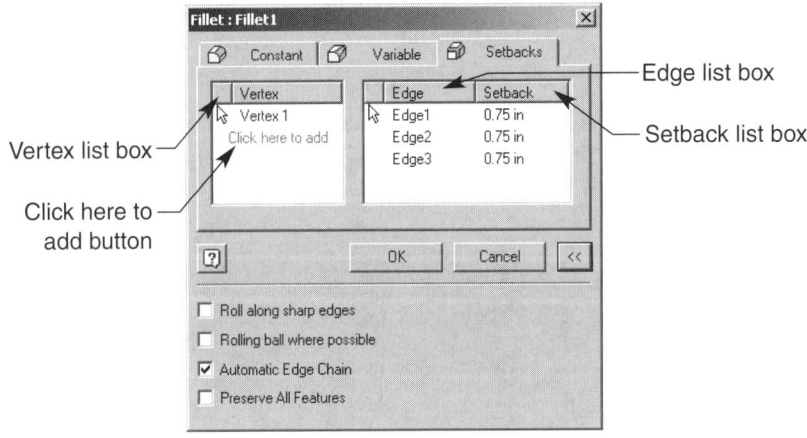

FIGURE 5.16   The **Setbacks** tab of the **Fillet** dialog box.

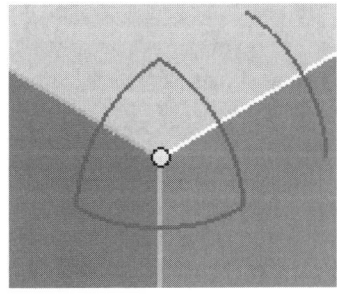

FIGURE 5.17   An example of a selected vertex.

define the setback sizes. Each of the edges corresponding to the selected vertex is listed in the **Edge** list box. If you pick one of the edge names, the matching edge on the feature is highlighted. If you pick one of the edges on the feature, the corresponding edge name in the **Edge** list box is identified by a cursor to the left of the name. Specify the setback for **Edge1** by highlighting the default setback and entering or selecting a value from the **Setback** drop-down list. Then pick **Edge2** and specify the desired setback. Finally, pick **Edge3** and specify the desired setback. If you would like to modify additional setbacks, choose another vertex by picking the **Click here to add** button and using the techniques previously discussed. When complete, pick the **OK** button.

Field Notes

> To delete a setback vertex or edge listed in the **Vertex** or **Edge** list boxes, pick the vertex or edge you would like to remove and press the **Delete** key on your keyboard.

Exercise 5.2

1. Continue from Exercise 5.1, or launch Autodesk Inventor.
2. Open EX 4.4.
3. Save a copy of EX4.4 as EX5.2.
4. Close EX4.4 without saving, and open EX5.2.
5. Access the **Fillet** dialog box, and open the **Variable** tab.
6. Unselect the **Rolling ball where possible** check box.
7. Apply the variable fillets and rounds to the edges shown, using the following specifications:

   Variable Edge1: Start (R.75″), End (R.125″), Point1 (position=.5ul, R.5″)
   Variable Edge2: Start (R.75″), End (R.125″), Point 1 (position=.5ul, R.5″)

8. Apply a constant R.25″ round to the specified constant edges.
9. Access the **Setbacks** tab, and modify the setbacks of the constant fillets using the following specifications:

   Constant Edge1: .5″ setback
   Constant Edge2: .75″ setback
   Constant Edge3: .25″ setback

10. The final part should look like the part shown.
11. Resave EX5.2.
12. Exit Autodesk Inventor or continue working with the program.

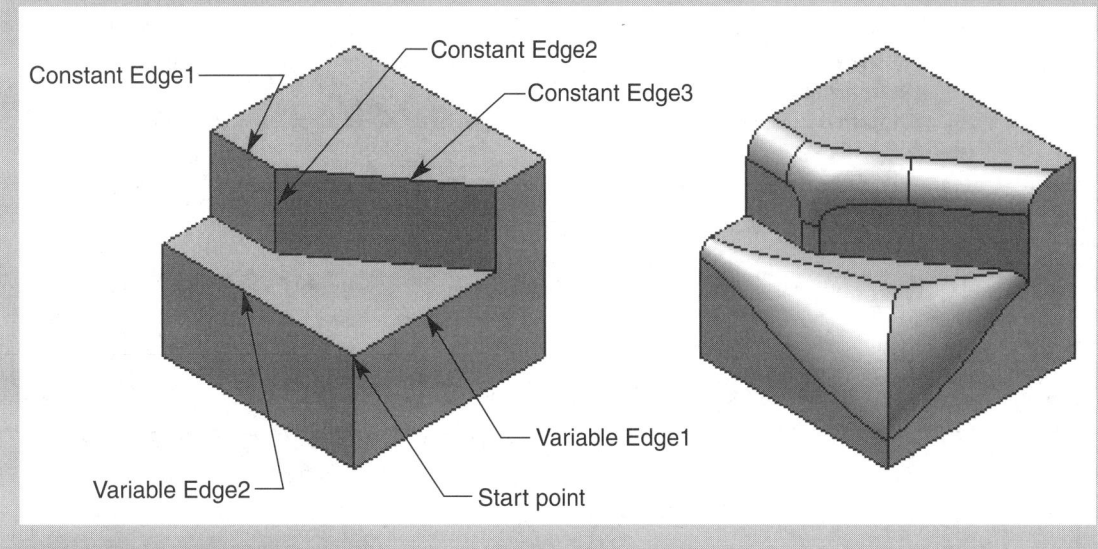

Adding Chamfers to a Feature

A *chamfer* is an angled planar face, placed on a feature edge. Chamfers cut through, or remove, the corner of two existing, intersecting, nonparallel planar faces. See Figure 5.18. Chamfers are used to relieve sharp edges, aid the entry of a pin or thread into a hole, or ease inside corners. Typically chamfers remove material from the external or internal corner of an angle. However, chamfers can also add material to an internal corner.

Chamfers are created using the **Chamfer** tool. Access the **Chamfer** command by using one of the following techniques:

✓ Pick the **Chamfer** button on the panel bar.

✓ Pick the **Chamfer** button on the **Features** toolbar.

The **Chamfer** dialog box is displayed when you access the **Chamfer** command. See Figure 5.19.

FIGURE 5.18
Examples of several chamfers placed on two parts.

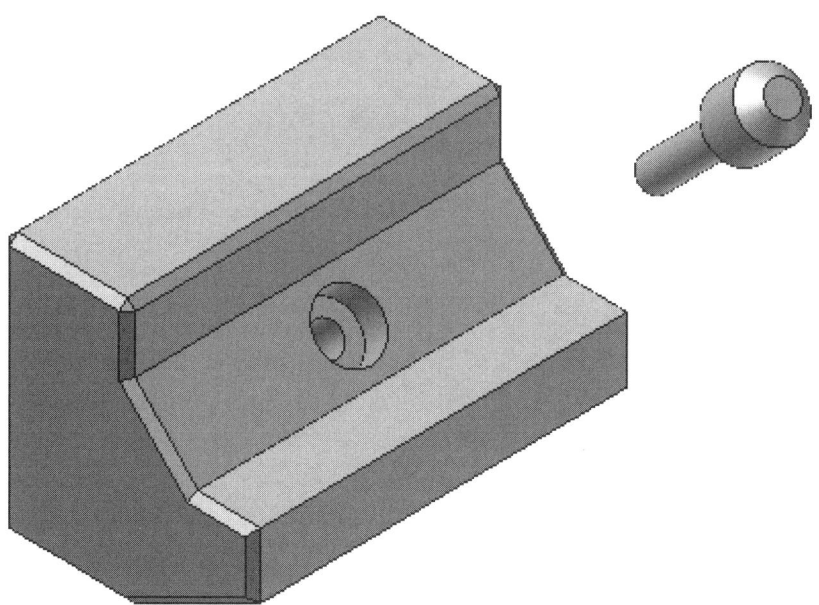

FIGURE 5.19 The **Chamfer** dialog box (**Distance** option).

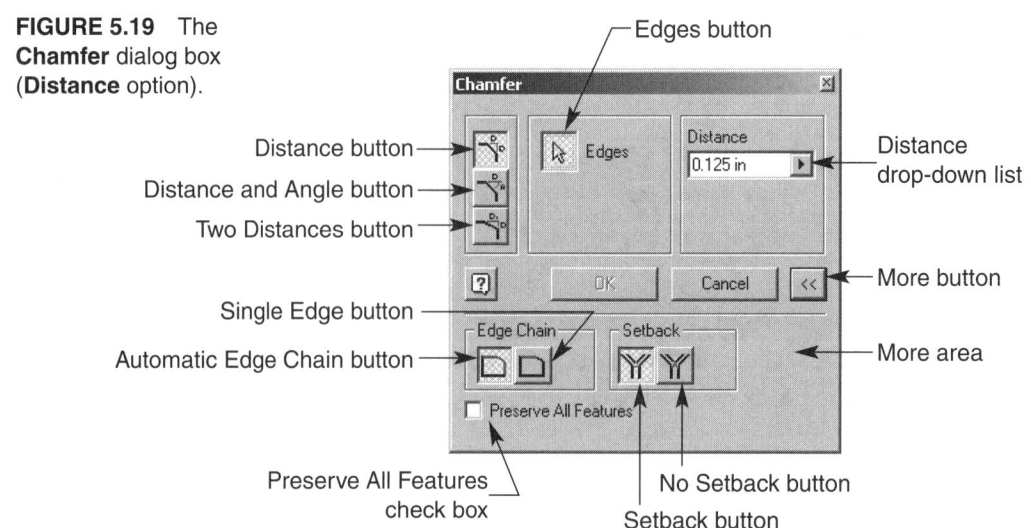

Edges button

Distance button
Distance and Angle button
Two Distances button

Distance drop-down list

More button

More area

Single Edge button
Automatic Edge Chain button

Preserve All Features check box

No Setback button
Setback button

✎ Field Notes

> Unlike the **Fillet** tool, you need to access the **Chamfer** command every time you want to create a different size chamfer. However, you can fully chamfer a part using only one **Chamfer** command if the distances are constant for each edge. Still, each of the chamfers placed during one operation create a single feature.

Before you begin to create a chamfer, you may want to adjust the options located in the **More** area of the **Chamfer** dialog box. The following buttons are available when you pick the **More** button to access the **More** area:

■ **Edge Chain** This area allows you to set the type of edge selection. Pick the **Automatic Edge Chain** button to select the entire edge chain of a tangent loop. See Figure 5.20A. Pick the **Single Edge** button to select an individual piece of a tangent loop edge. See Figure 5.20B.

■ **Setback** This area allows you to define the chamfer corner style when chamfering parts with three or more intersecting edges. Pick the **Setback** button to apply a setback to the intersecting chamfer edges. See Figure 5.21A. Pick the **No Setback** button if you do not

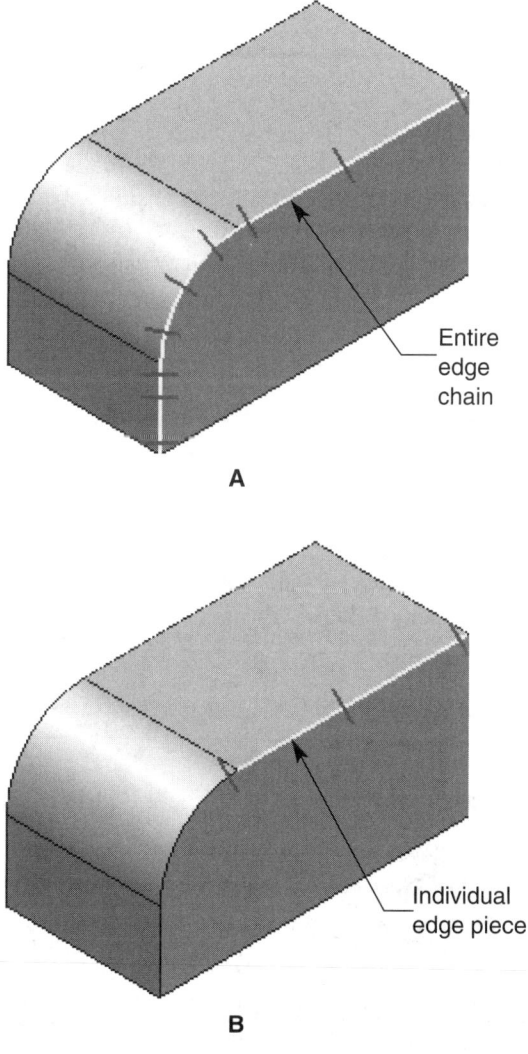

FIGURE 5.20 (A) Selecting an entire edge chain. (B) Selecting an individual edge piece.

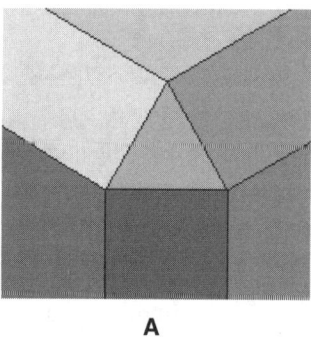

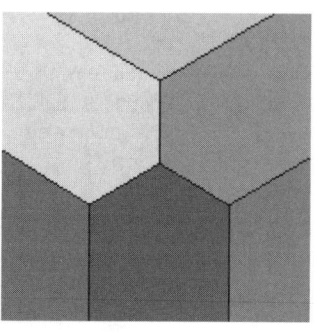

FIGURE 5.21 (A) An example of a chamfered corner with the **Setback** button selected and (B) with the **No Setback** button selected.

FIGURE 5.22
Examples of the effects
of each type of chamfer
creation method.

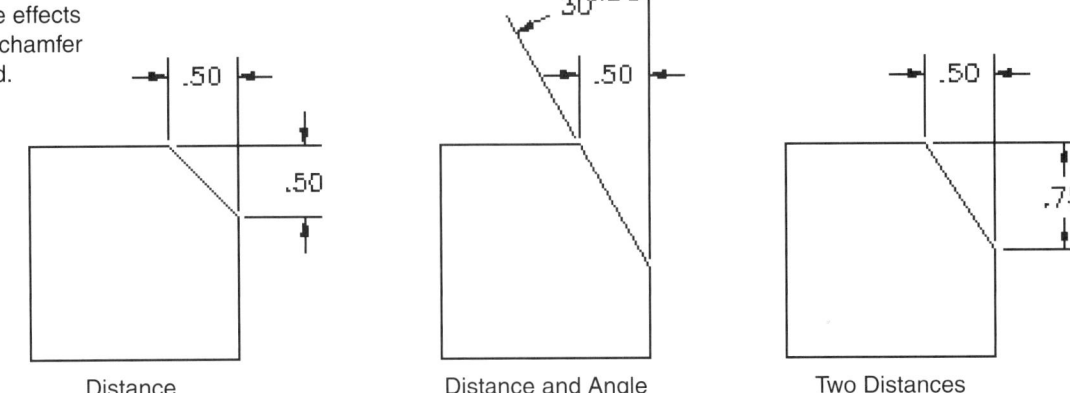

Distance   Distance and Angle   Two Distances

want to create a setback. This option establishes a chamfer point at the intersection of the
edges instead of a flat surface. See Figure 5.21B.

■ **Preserve All Features**   Pick this check box when features intersect a corner that you want
to chamfer. When this check box is selected, existing feature geometry is not modified by a
new chafer that intersects the feature.

To create a chamfer, first select one of the following chamfer method buttons: **Distance, Distance
and Angle,** and **Two Distances.** The techniques used by each of these buttons to create chamfers are shown
in Figure 5.22. The method button selected by default is **Distance,** which is the simplest option to use. The
Distance option allows you to create a chamfer with equal-distance sides from the selected edge, or a 45°
angle. To place a chamfer using this option, you must first be sure the **Distance** button is selected. Then
pick the edges you want to chamfer and enter or select a chamfer distance from the **Distance** drop-down
list. If the chamfer preview looks correct, pick the **OK** button to create the chamfer.

Field Notes

You can deselect any of the selected objects in any of the **Chamfer** dialog box drop-down lists
by holding down the **Ctrl** key on your keyboard and picking the object.

If you want to specify one chamfer side as a distance from the edge and the other side as an angle
from the edge, pick the **Distance and Angle** method button. Figure 5.23 shows the **Chamfer** dialog box

FIGURE 5.23   The
Chamfer dialog box
(**Distance and Angle**
option).

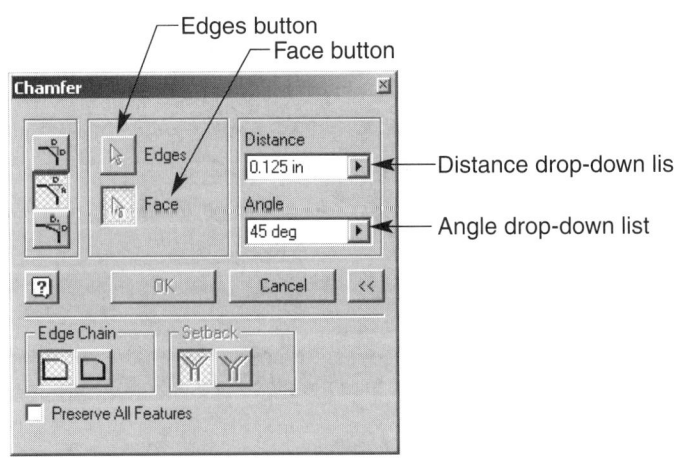

FIGURE 5.24
Determining which
Face and **Edge** to
select.

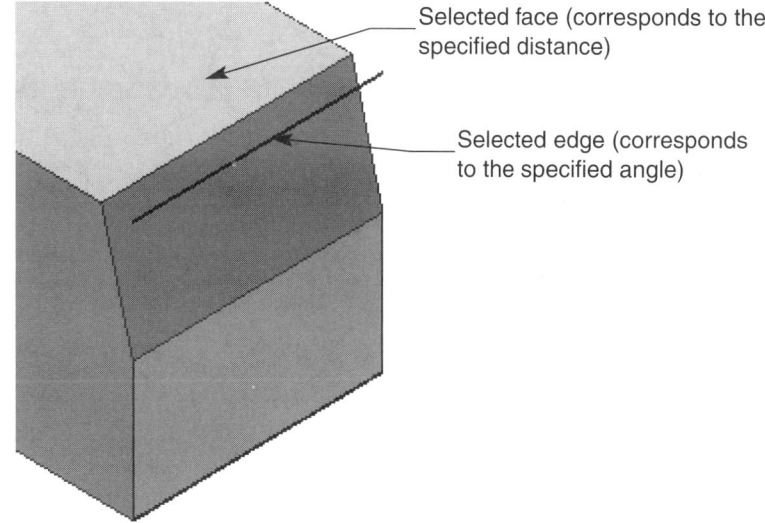

Selected face (corresponds to the
specified distance)

Selected edge (corresponds
to the specified angle)

with the **Distance and Angle** button selected. Initially, the **Face** button should be active. The **Face** button allows you to select the face from which you want the chamfer distance from the edge to be measured. See Figure 5.24. Once you have defined the face, the **Edge** button should become active. The **Edge** button allows you to select the edges from which you want the angle to be measured. See Figure 5.24. To change the face, pick the **Face** button again and then reselect the edges. Once you have established the face and edges you want to chamfer, enter or select a chamfer distance from the **Distance** drop-down list and enter or select a chamfer angle from the **Angle** drop-down list. If the chamfer preview looks correct, pick the **OK** button to create the chamfer.

If you want to specify one chamfer side with a distance from the edge, and the other side with a different distance from the edge, pick the **Two Distances** method button. Figure 5.25 shows the **Chamfer** dialog box with the **Two Distances** button selected. To place a chamfer using the **Two Distances** method, pick the edge you want to chamfer. Then enter or select a chamfer distance from the **Distance1** drop-down list, and enter or select a chamfer distance from the **Distance2** drop-down list. If the chamfer preview looks correct, pick the **OK** button to create the chamfer. However, if the distances you enter need to be reversed, pick the **Flip** button to redefine the direction of the chamfer distances from the edge. See Figure 5.26. Then pick the **OK** button to create the chamfer.

FIGURE 5.25   The
Chamfer dialog box
(**Two Distances**
option).

Edge button

Flip button

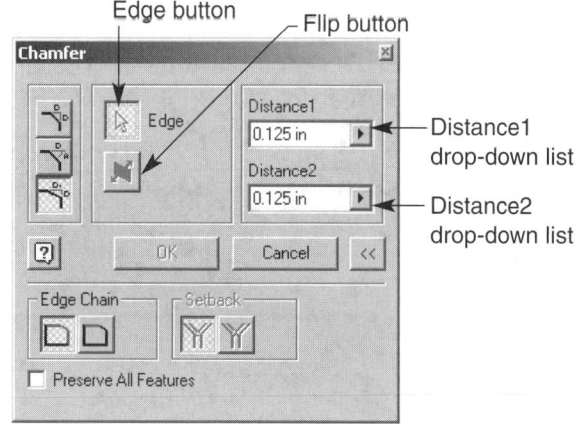

Distance1
drop-down list

Distance2
drop-down list

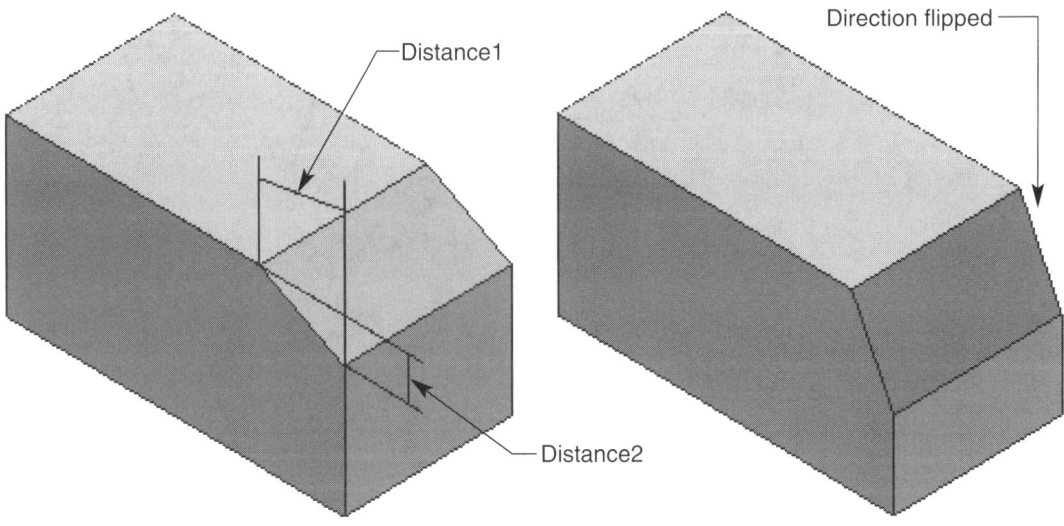

FIGURE 5.26   Using the **Flip** button to reverse the direction of the specified distances.

Exercise 5.3

1. Continue from Exercise 5.2, or launch Autodesk Inventor.
2. Open EX 4.4.
3. Save a copy of EX4.4 as EX5.3.
4. Close EX4.4 without saving, and open EX5.3.
5. Access the **Chamfer Dialog** box.
6. Chamfer the edges shown using the specified techniques.
7. The part should look like the part shown when complete.
8. Resave EX5.3.
9. Exit Autodesk Inventor or continue working with the program.

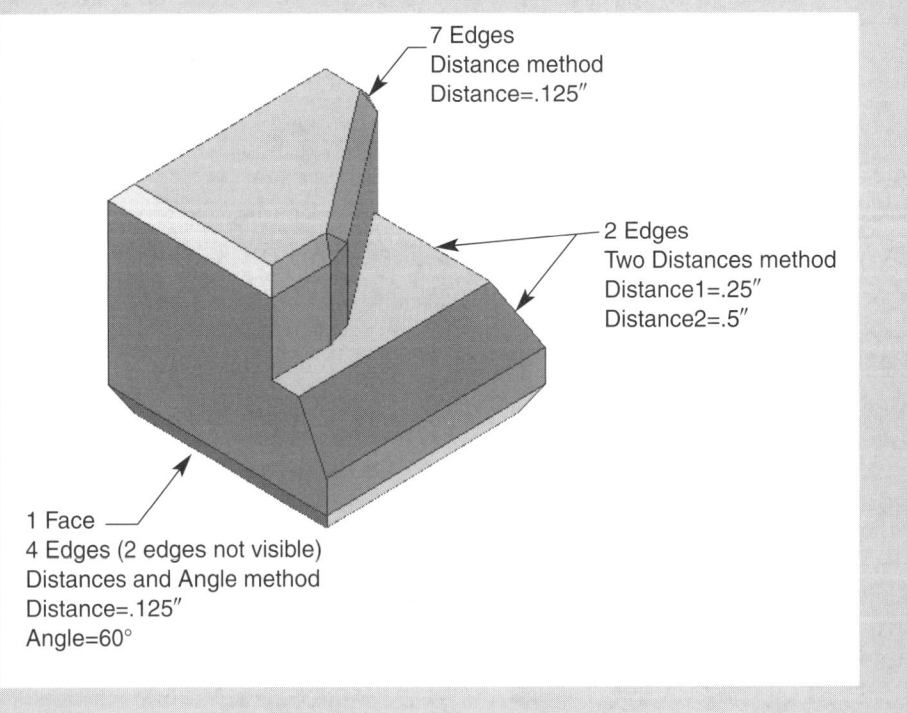

Placing Shell Features

A *shell* removes material from a feature and creates a hollow space or opening. See Figure 5.27. Shells are typically used to design parts for casting or forging. However, shells can be used for any application where it is necessary to remove material from a feature, leaving behind a specified wall thickness. The material removed by a shell is specified by one or more selected feature faces. Shells modify features by removing material and creating new faces on the inside, outside, or inside and outside of existing feature geometry.

Shell features are created using the **Shell** tool. Access the **Shell** command by using one of the following techniques:

✓ Pick the **Shell** button on the panel bar.

✓ Pick the **Shell** button on the **Features** toolbar.

The **Shell** dialog box is displayed when you access the **Shell** command. See Figure 5.28.

To place a shell, first pick the **Remove Faces** button if it is not already selected. The **Remove Faces** button allows you to specify which of feature faces you want to remove during the shelling process. You can remove one or more feature faces. See Figure 5.29. Once you have specified the faces you want to remove, determine which direction to shell the feature using one of the following buttons:

■ **Inside**   Selecting this button allows you to create shell walls on the inside of the existing feature walls. See Figure 5.30A.

■ **Outside**   Picking this button allows you to create shell walls on the outside of the existing feature walls. See Figure 5.30B.

■ **Both**   Selecting this button allows you to create shell walls equally thick on the inside and the outside of the existing feature walls. See Figure 5.30C.

FIGURE 5.27
Examples of a shelled feature.

FIGURE 5.28 The **Shell** dialog box (**Distance** option).

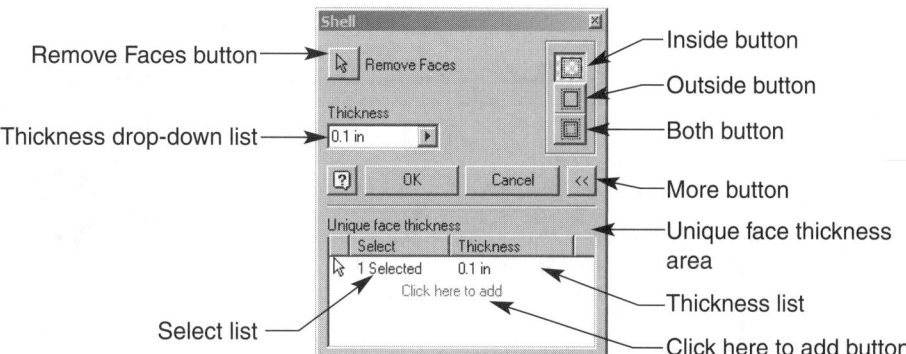

Remove Faces button

Thickness drop-down list

Select list

Inside button

Outside button

Both button

More button

Unique face thickness area

Thickness list

Click here to add button

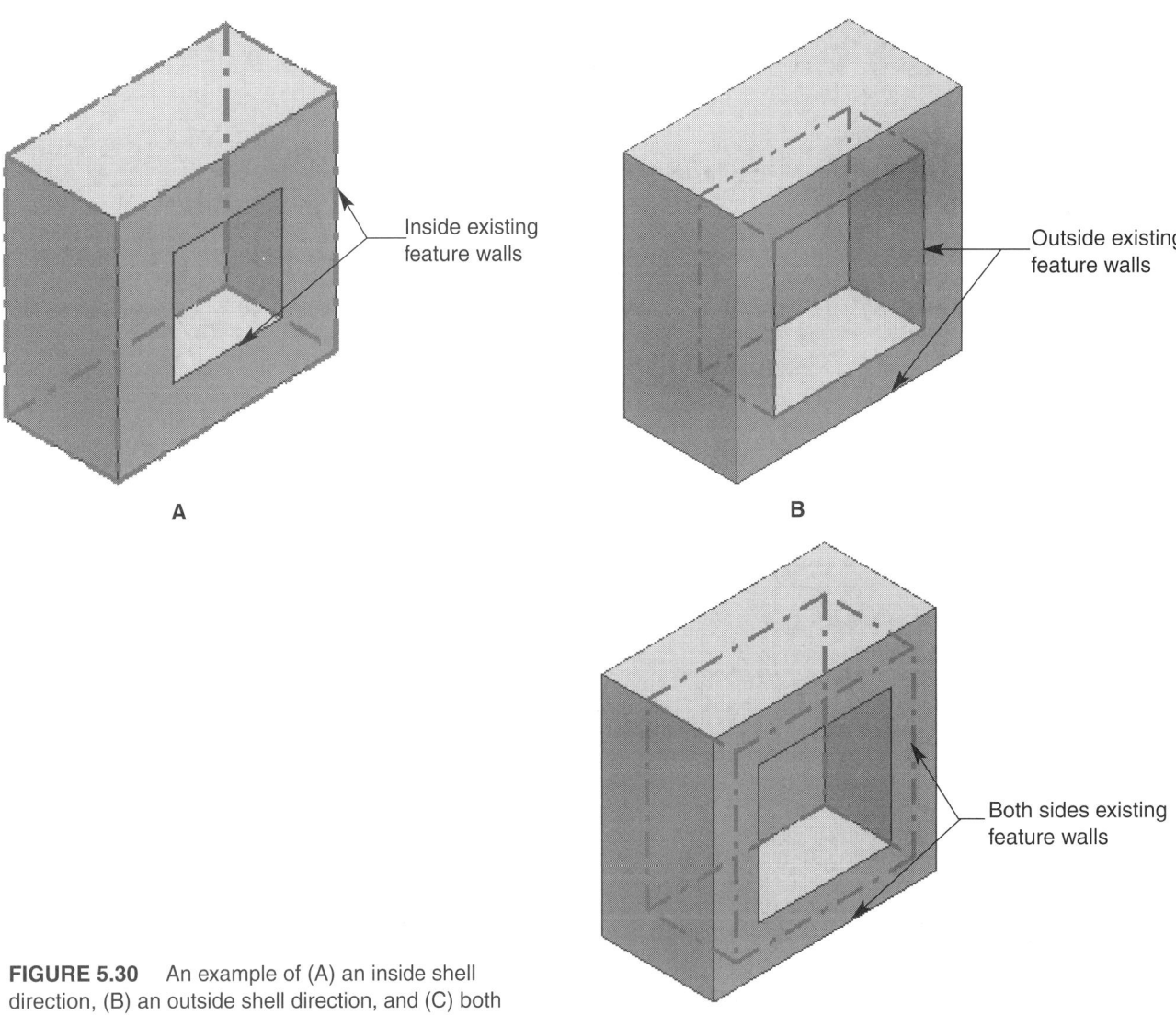

FIGURE 5.29 (A) Removing one feature face or (B) multiple feature faces.

Inside existing feature walls

A

Outside existing feature walls

B

Both sides existing feature walls

C

FIGURE 5.30 An example of (A) an inside shell direction, (B) an outside shell direction, and (C) both inside and outside shell direction.

210

FIGURE 5.31 An
example of a shell
feature with a single,
uniform wall thickness.

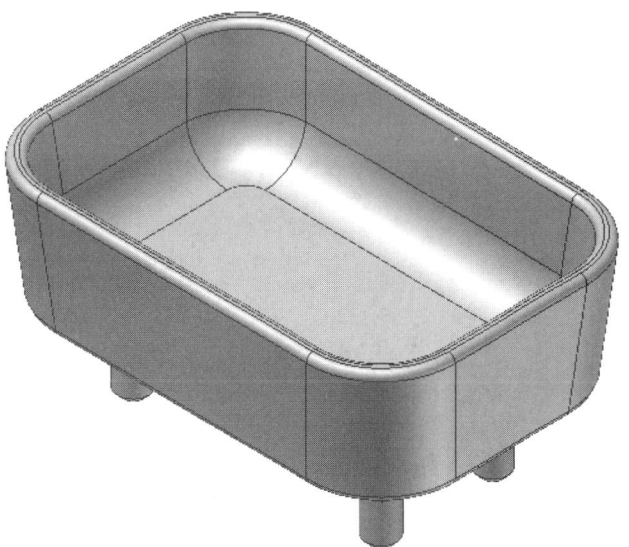

The next step in creating a shell feature is defining the wall thickness, which determines the amount of material evenly removed by the operation. To define and apply a uniform thickness throughout the feature, enter or select a value from the **Thickness** drop-down list. Then pick the **OK** button to create the shell. See Figure 5.31.

If you want to apply different thicknesses to certain walls, as shown in Figure 5.32, pick the **More** button to access the **Unique face thickness** area. Then, pick the **Click here to add** button and select the face where you want to place a unique wall thickness. Finally, enter or select the desired thickness from the **Thickness** drop-down list, and pick the **OK** button to create the shell.

FIGURE 5.32 An
example of a shell
feature with different
wall thickness.

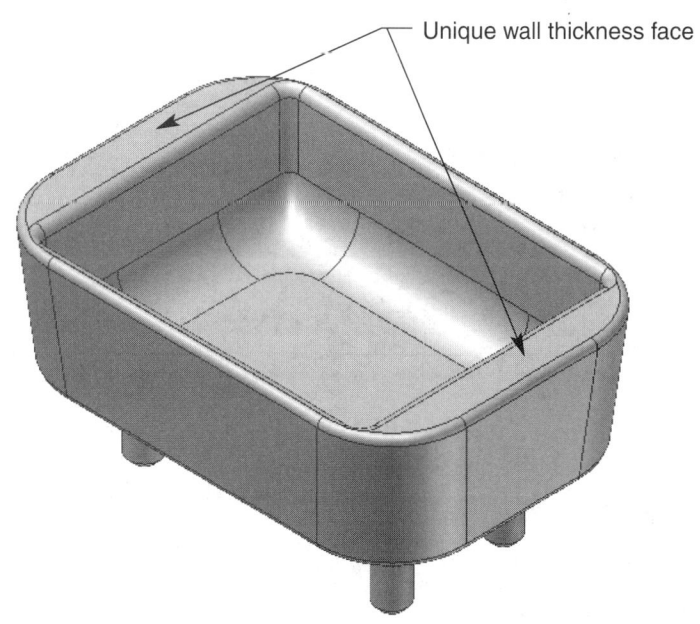

Unique wall thickness face

Exercise 5.4

1. Continue from Exercise 5.3, or launch Autodesk Inventor.
2. Open a new part file.
3. Open a sketch on the XY plane, and sketch a rectangle 24mm×38mm
4. Extruded the sketch 38mm in a positive direction.
5. Apply a shell feature using the following specifications:

   Remove: Face1
   Direction: Inside
   Thickness: 3mm

6. Right-click **Shell1** and select **Edit Feature** from the shortcut menu.
7. Modify the existing shell using the following specifications:

   Remove: Face1 and Face2
   Direction: Outside
   Thickness: 5mm

8. Right-click **Shell1** and select **Edit Feature** from the shortcut menu.
9. Modify the existing shell using the following specifications:

   Remove: Face1 and Face2
   Direction: Both
   Thickness: 2
   Unique Thicknesses: Face3=6mm, Face4=12mm

10. When complete, the part should look like the part shown.
11. Save the file as EX5.4.
12. Exit Autodesk Inventor or continue working with the program.

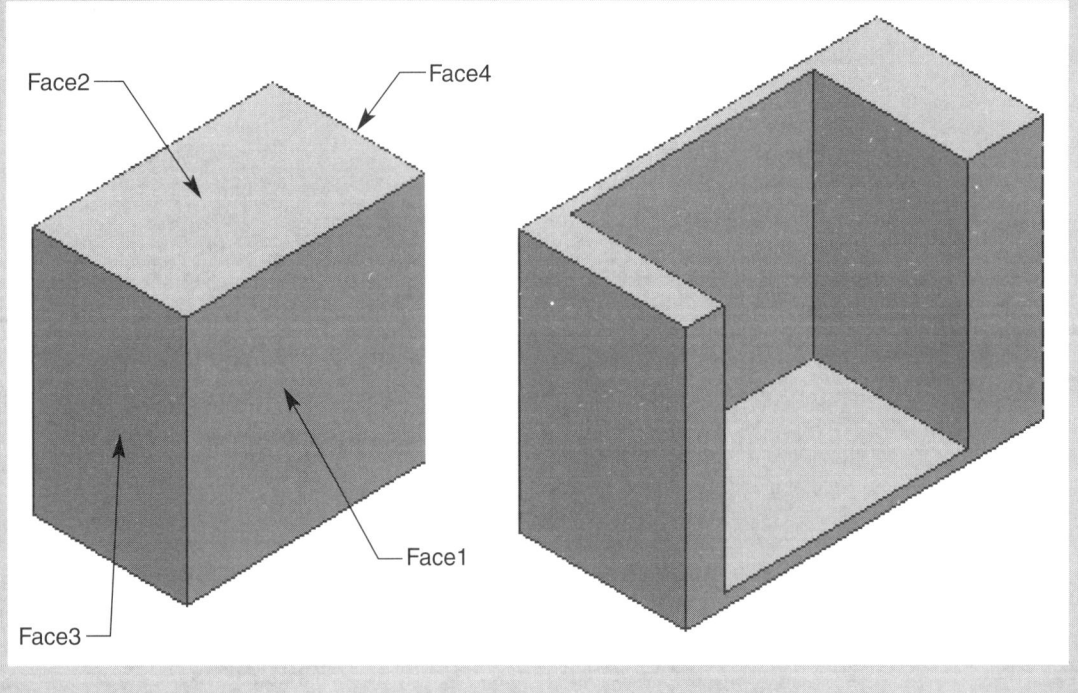

Applying Face Drafts

A *face draft* is a taper allowance placed on part surfaces to aid in the removal of a pattern from a mold. See Figure 5.33. You can apply draft to any acceptable feature face, including horizontal part surfaces, but typically for part casting, draft is added to vertical part surfaces. Draft and the Autodesk Inventor **Face Draft** tool are usually used to create mold and pattern designs for casting applications. However, you may want to use this tool anytime a face requires a specified slant angle.

Draft features are created using the **Face Draft** tool. Access the **Face Draft** command by using one of the following techniques:

✓ Pick the **Face Draft** button on the panel bar.

✓ Pick the **Face Draft** button on the **Features** toolbar.

The **Face Draft** dialog box is displayed when you access the **Face Draft** command. See Figure 5.34.

To draft a face, first pick the **Pull Direction** button, if it is not already selected, and pick the face draft pull direction. The *pull direction* is the direction from which the part is pulled or removed from the casting mold. See Figure 5.33. When you select the pull direction, an arrow identifies the which way the part is pulled. See Figure 5.35. If you want to reverse the specified pull direction, pick the **Flip** button.

Once you have designated the pull direction, the **Faces** button should automatically become active. The **Faces** button allows you to select the faces you want to draft. When you move your cursor over feature faces, a vector preview of the face draft operation is displayed. See Figure 5.35. The vector indicates the fixed edge and angle of the face draft. Once the vector displays the correct orientation, pick the face.

When you have identified the pull direction and the faces you want to draft, enter or select the desired draft angle from the **Draft Angle** drop-down list and pick the **OK** button. Figure 5.36 shows the drafted face.

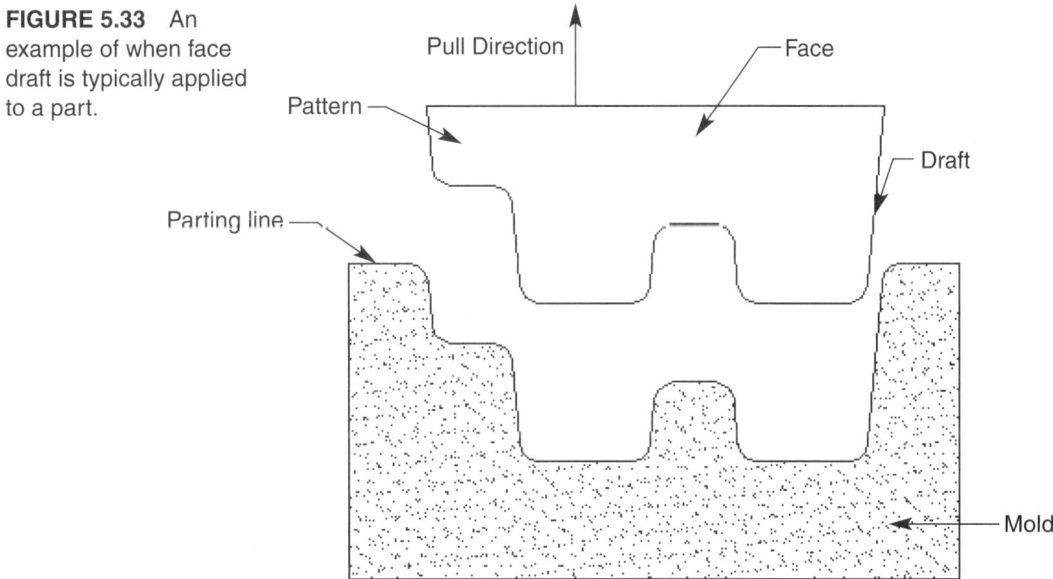

FIGURE 5.33 An example of when face draft is typically applied to a part.

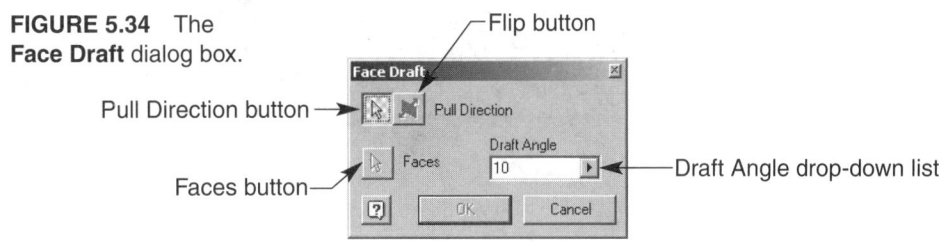

FIGURE 5.34 The **Face Draft** dialog box.

FIGURE 5.35   The pull direction arrow.

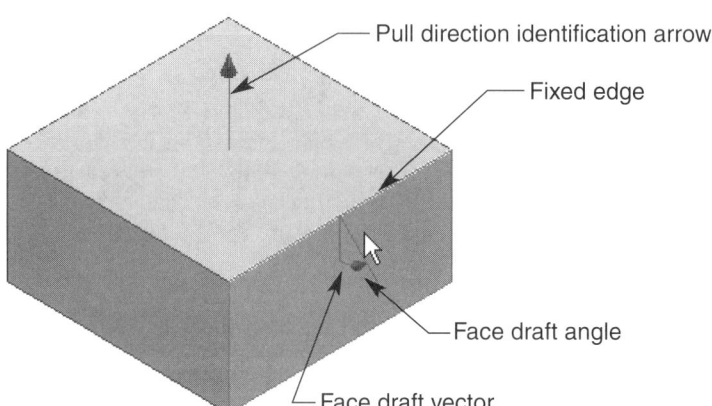

Pull direction identification arrow

Fixed edge

Face draft angle

Face draft vector

FIGURE 5.36   An example of a face draft.

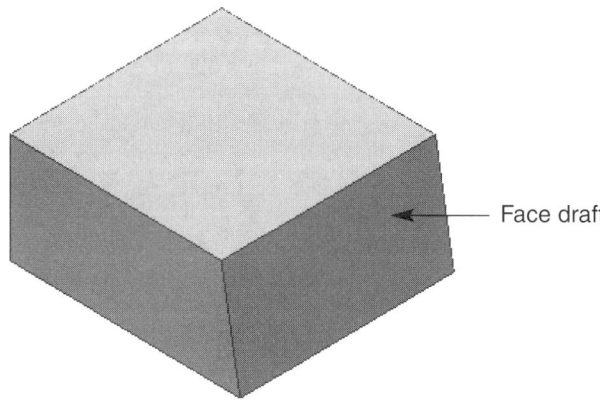

Face draft

✏️ Field Notes

When you apply a face draft to a single face, which is not connected to any tangent features, as shown in Figures 5.35–5.36, the draft is labeled **TaperEdge** in the **Browser.** When you apply a face draft to a face that is connected to a tangent feature such as a fillet, as shown in Figure 5.37, the face draft is labeled **TaperShadow** in the **Browser.** If a part contains tangent features such as an arc or fillet, you are required to select multiple faces, depending on how many tangencies exist.

FIGURE 5.37   An example of a **TaperShadow** face draft.

Exercise 5.5

1. Continue from Exercise 5.4, or launch Autodesk Inventor.
2. Open a new part file.
3. Open a sketch on the XY plane, and sketch a rectangle 1″×2″
4. Extruded the sketch 3″ in a positive direction.
5. Fillet the specified corner using the dimensions shown.
6. Add the face drafts shown.
7. Save the file as EX5.5.
8. Exit Autodesk Inventor or continue working with the program.

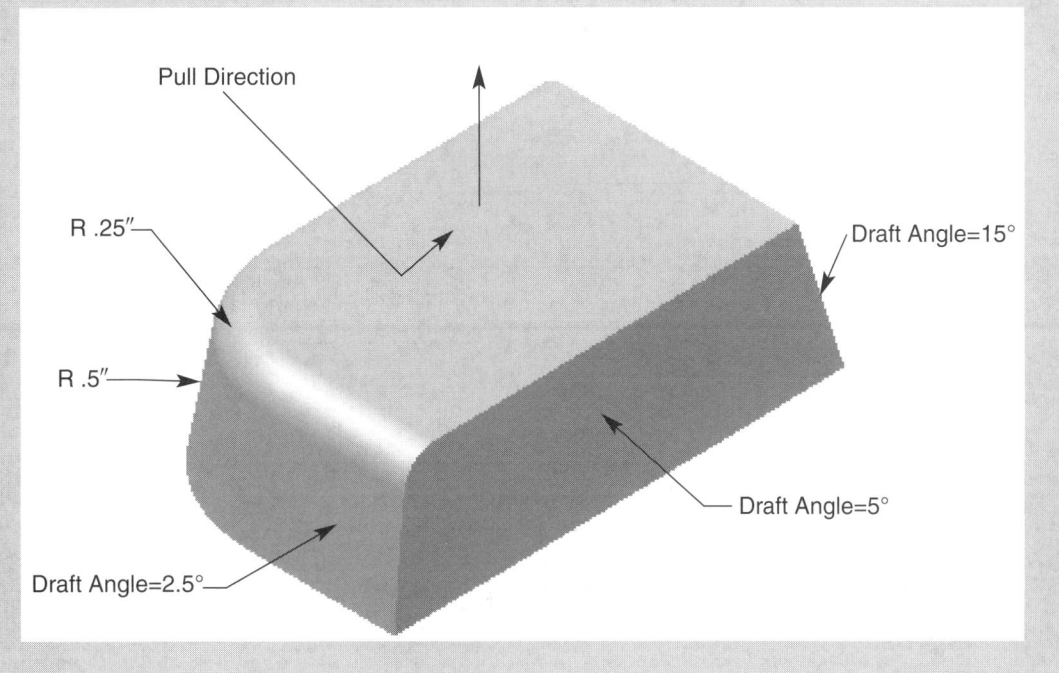

Pull Direction

R .25″

R .5″

Draft Angle=15°

Draft Angle=5°

Draft Angle=2.5°

Placing Threads

Threads are grooves cut in a spiral fashion in or around the face of a cylindrical or conical feature. Threads can be placed both externally and internally. *External threads* are thread forms on an external feature such as a pin, shaft, bolt, or screw. See Figure 5.38A. *Internal threads* are thread forms on an internal hole feature. See Figure 5.38B.

FIGURE 5.38 An example of (A) an externally threaded feature and (B) a threaded hole.

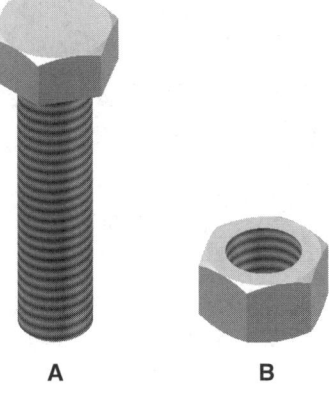

A   B

⬛ Field Notes

> Although the **Thread** tool allows you to add threads to a hole, you may want to use the thread options available in the **Holes** dialog box, discussed in Chapter 4, to create threaded holes.

Threads are created using the **Thread** tool. This tool does not actually cut threads into a feature. It displays a bitmap image of threads on the cylinder, cone, or hole. These thread representations are adequate for most applications, but if you need to create actual detailed threads, you must use the coil tool, discussed in Chapter 4. Access the **Thread** command by using one of the following techniques:

✓ Pick the **Thread** button on the panel bar.

✓ Pick the **Thread** button on the **Features** toolbar.

The **Thread Feature** dialog box is displayed when you access the **Thread** command. See Figure 5.39.

⬛ Field Notes

> Thread representations created using the **Thread** command are fully parametric, which means the information you use to create the threads is used to annotate drawings and is available for other applications.

To apply threads to an interior or exterior cylindrical or conical face, pick the **Face** button if it is not automatically selected. Then specify the thread length, using the **Thread Length** area of the **Thread Feature** dialog box. If you want the threads to encompass the entire length of the feature face, pick the **Full Length** check box. See Figure 5.40.

If you want to specify a certain length of threads and create threads that do not run the entire length of the face, unselect the **Full Length** check box. Then enter or select the desired thread length from the **Length** drop-down list. Once you specify the thread length, enter or select an offset distance from the **Offset** drop-down list. *Offset* is the distance from the edge of the face to the beginning of threads. See Figure 5.41. Once you determine the thread length and offset, decide if you want to hide the bitmap thread representation on the part. By default the thread representation is displayed. However, if you want to hide the image, unselect the **Display in Model** check box.

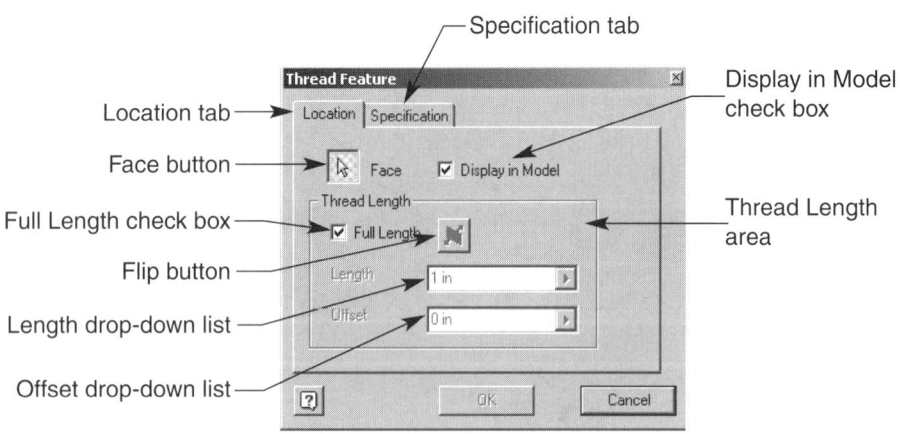

FIGURE 5.39   The **Thread Feature** dialog box (**Location** tab selected).

FIGURE 5.40   An example of full-length threads.

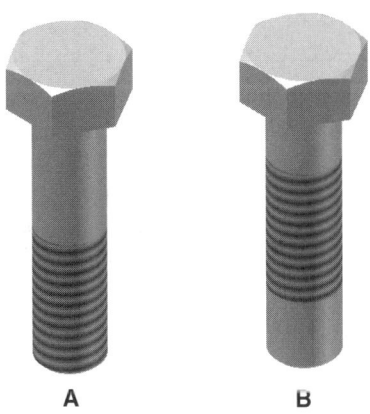

FIGURE 5.41   A threaded feature with (A) a thread length of 1″ and an offset of 0 and (B) a thread length of 1″, and an offset of .5″.

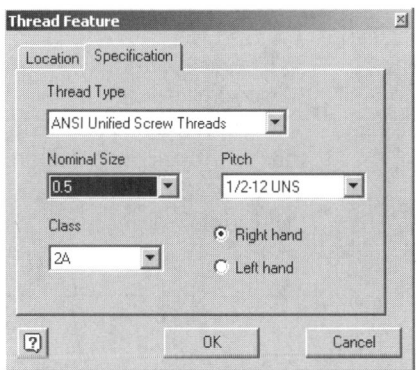

FIGURE 5.42   The **Specification** tab of the **Thread** dialog box.

When you have defined all the thread location options in the **Location** tab, pick the **Specification** tab of the **Thread Feature** dialog box to specify the thread properties. See Figure 5.42. The following options are available:

- **Thread Type**   This drop-down list allows you to specify the kind of threads applied to the feature. You can choose **ANSI Unified Screw Threads, ANSI Metric M profile, NPT,** or **JIS Taper** from the drop-down list.

- **Nominal Size**   This drop-down list allows you to specify the nominal size of the threads. The nominal size is the designated size of a commercial product; it is usually the same as the major diameter.

- **Pitch**   This drop-down list allows you to specify the pitch of ANSI Unified Screw Threads and ANSI Metric M Profile threads. Pitch is the distance parallel to the axis between a point on one thread to the corresponding point on the next thread. For ANSI Unified Screw Threads, the pitch determines and is specified by the number of threads per inch. For ANSI Metric M Profile threads, the pitch is specifically specified as the pitch.

- **Class**   This drop-down list allows you to specify the thread class for ANSI Unified Screw Threads and ANSI Metric M Profile threads. Class is the designated amount of tolerance specified for the thread. For ANSI Unified Screw Threads, 3, 4, 5, 6, 7, 8, or 9 may be available depending on the specified nominal size and pitch. These numbers identify the grade of tolerance from fine to coarse threads. For ANSI Metric M Profile threads, 1, 2, or 3 may be available depending on the specified nominal size and pitch. 1 is a coarse tolerance, 2 is a moderate tolerance, and 3 is a fine thread tolerance. For holes, these numbers will be followed by B, because B identifies internal threads.

- **Right hand** and **Left hand** radio buttons   You can specify whether the threads are right or left hand by choosing the appropriate radio button. Right-hand threads move a right-hand threaded bolt forward in a clockwise fashion, while left-hand threads move a left-hand threaded bolt forward in a counterclockwise fashion.

✎ Field Notes

The thread information available inside the **Specification** tab is outlined by Autodesk Inventor on a spreadsheet. The location of the spreadsheet is: Program Files\Autodesk\Inventor 5\Design Data folder\Thread.xls. Use this spreadsheet to modify and add thread types and thread information for your specific applications. Changes made to the **Thread** spreadsheet do not alter existing thread specifications. For more information regarding adding threads to the **Thread** spreadsheet, refer to the "To add thread data to a spreadsheet" section of the Autodesk Inventor help files.

Once you have specified the thread preferences in the **Specification** tab, pick the **OK** button to apply the threads to the feature.

Exercise 5.6

1. Continue from Exercise 5.5, or launch Autodesk Inventor.
2. Open a new part file.
3. Extrude a 1.25″ circle, 3″ in a positive direction.
4. Cut extrude a .75″ circle through all of the first cylinder at the center point.
5. Place the threads on the features, using the specifications shown.
6. The final part should look like the one shown.
7. Save the file as EX5.6.
8. Exit Autodesk Inventor or continue working with the program.

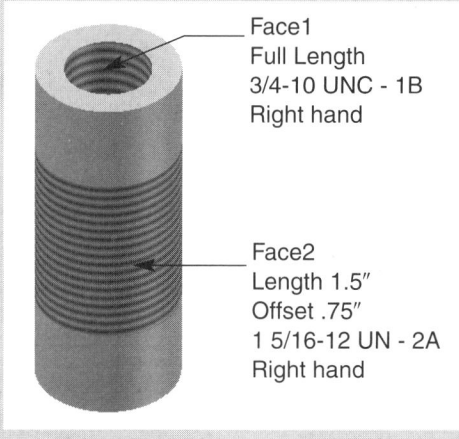

Face1
Full Length
3/4-10 UNC - 1B
Right hand

Face2
Length 1.5″
Offset .75″
1 5/16-12 UN - 2A
Right hand

||||||||||| CHAPTER TEST

1. Explain what placed features are and give an example of one.

2. Do placed features require a sketch to be created?

3. What is required before you can create placed features?

4. Explain what fillets are and give an example of their use.

5. Explain what rounds are and give an example of their use.

6. Give the function of the Roll along sharp edges check box in the More area of the Fillet dialog box.

7. Explain the function of the Roll ball where possible check box in the More area of the Fillet dialog box.

8. Give the function of the Automatic Edge Change option in the More area of the Fillet dialog box.

9. Give the function of the Preserve all features check box in the More area of the Fillet dialog box.

10. What is a constant fillet or round?

11. Give the steps used to apply a constant fillet and a constant round.

12. Give the function of the Select Mode Edge radio button.

13. Give the function of the Select Mode Loop radio button.

14. Give the function of the Select Mode Feature radio button.

15. How are variable fillet active points displayed on the selected feature edge?

16. Give the steps used to create a variable fillet or round.

17. How do you delete a variable fillet or round?

18. What are setbacks?

19. Identify the specifications required for a setback that will blend corners.

20. What happens if a setback is less than or equal to the fillet or round radius?

21. What is a chamfer and what are chamfers used for?

22. Give the function of the Edge Chain area of the More area of the Chamfer dialog box.

23. Give the function of the Setback area of the More area of the Chamfer dialog box.

24. Give the steps used to create a chamfer with the Distance option.

25. Explain the use of the Distance1 and Distance2 drop-down lists when using the Two Distance method to place a chamfer.

26. What is a shell and how are shells used?

27. Give the function of the Remove Faces button in the Shell dialog box.

28. Give the function of the Inside button in the Shell dialog box.

29. Give the function of the Outside button in the Shell dialog box.

30. Give the function of the Both button in the Shell dialog box.

31. How do you define a uniform wall thickness for a shell?

32. How do you apply different wall thickness for a shell?

33. What is a face draft?

34. What is the pull direction?

35. What do you get when you select the pull direction?

36. Give the function of the Faces button in the Face Draft dialog box.

37. What happens when you move your cursor over feature faces while using the Faces button?

38. When and how do you enter a desired draft angle?

39. What do you get when you use the Thread tool?

40. The Thread tool thread representations are adequate for most applications, but what do you use if you need to create actual detailed threads?

41. What is the offset as related to threads?

42. Explain the function of the Thread Type drop-down list in the Specification tab of the Thread dialog box.

43. Give the function of the Nominal Size drop-down list in the Specification tab of the Thread dialog box.

44. Discuss the function of the Pitch drop-down list in the Specification tab of the Thread Feature dialog box.

45. Give the function of the Class drop-down list in the Specification tab of the Thread Feature dialog box.

|||||||||||| PROJECTS

Open the following Chapter 4 part files and follow the specific instructions provided for each project.

1. Name: C-Clamp Pin

   Units: Metric

   Specific Instructions:

   ■ Open P4.1, and Save Copy As P5.1.

   ■ Close P4.1 without saving, and Open P5.1.

   ■ Place 2.375mm fillets on both ends of the pin as shown.

   ■ Save As: P5.1.

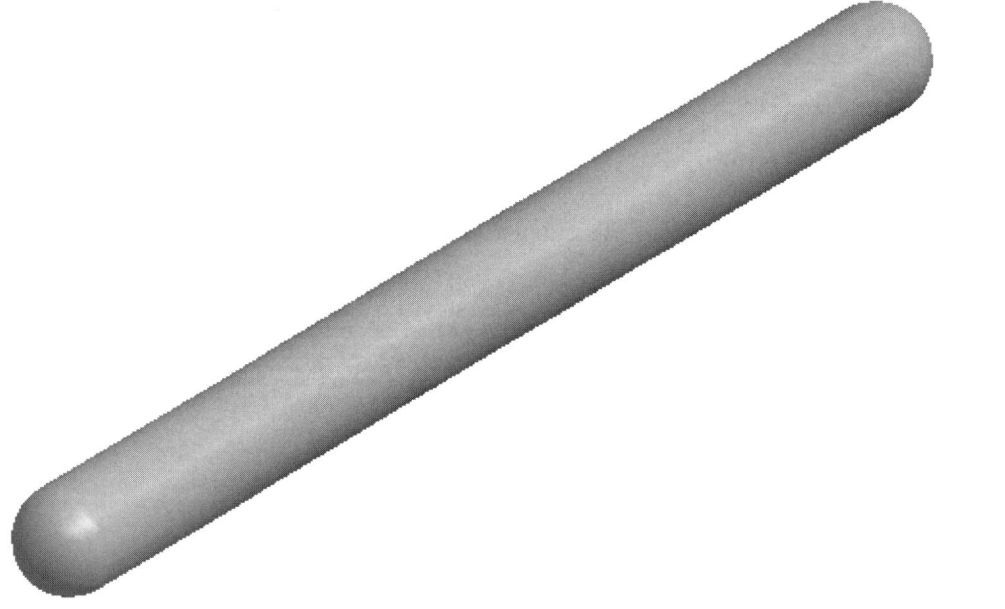

2. Name: C-Clamp Swivel

Units: Metric

Specific Instructions:

- Open P4.2, and Save Copy As P5.2.

- Close P4.2 without saving, and Open P5.2.

- Place a chamfer on the part as shown. *Note:* Dimensions are for reference only.

- Save As: P5.2.

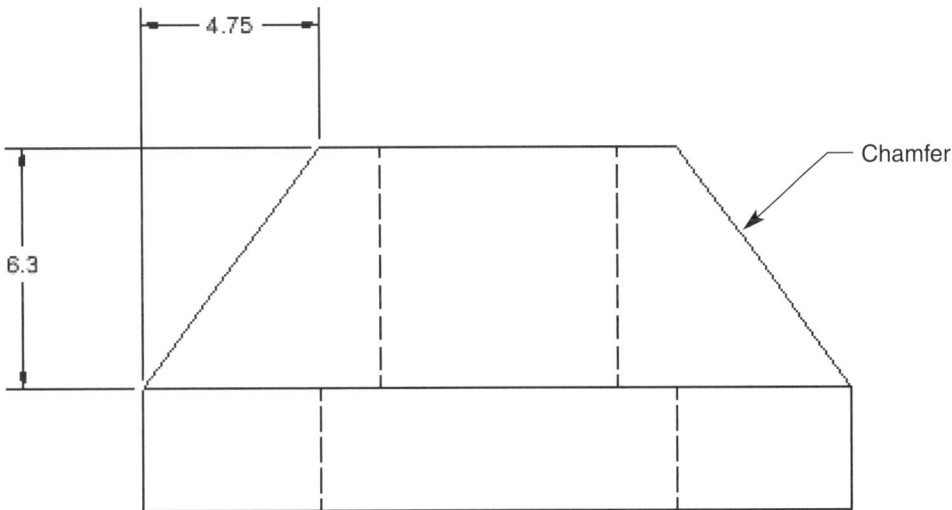

3. Name: C-Clamp Screw

Units: Metric

Specific Instructions:

- Open P4.3, and Save Copy As P5.3.

- Close P4.3 without saving, and Open P5.3.

- Place a 2mm fillet on the edge shown.

- Apply full-length, right-hand threads using the specifications shown.

- Save As: P5.3.

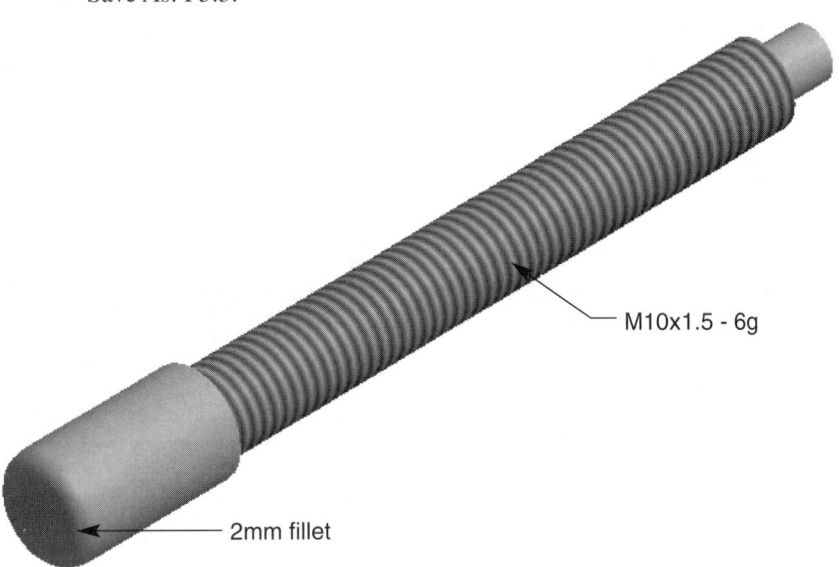

4. Name: Support Bracket

   Units: Inch

   Specific Instructions:

   ■ Open P4.4, and Save Copy As P5.4.

   ■ Close P4.4 without saving, and Open P5.4.

   ■ Apply the fillets as shown.

   ■ Apply the chamfers as shown.

   ■ Save As: P5.4.

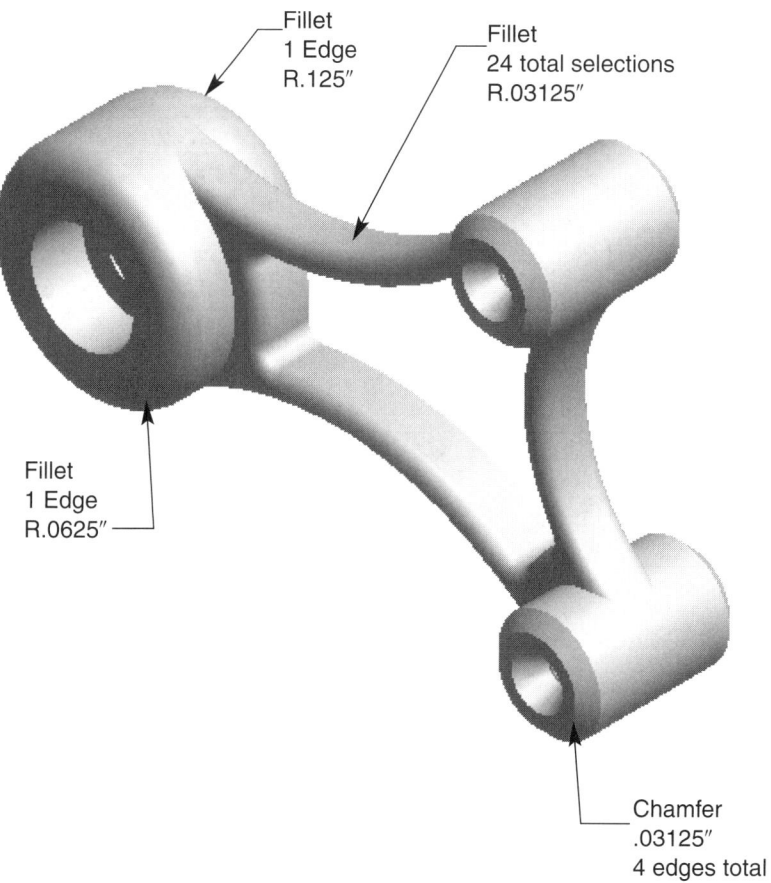

Fillet
1 Edge
R.125″

Fillet
24 total selections
R.03125″

Fillet
1 Edge
R.0625″

Chamfer
.03125″
4 edges total

5. Name: Funnel

Units: Inch

Specific Instructions:

- Open P4.5, and Save Copy As P5.5.

- Close P4.5 without saving, and Open P5.5.

- Place four .75″ chamfers as shown.

- Shell the part using the following techniques:

  Remove Faces1 and 2.

  Use an Inside shell method.

  Apply a Thickness of .0625″

  Apply a Unique Thickness of .125″ to Extrusion1, and Face3.

- Place .03125″ fillets and rounds using the **All Fillets** and **All Rounds** check boxes.

- The final part should look like the part shown.

- Save As: P5.5.

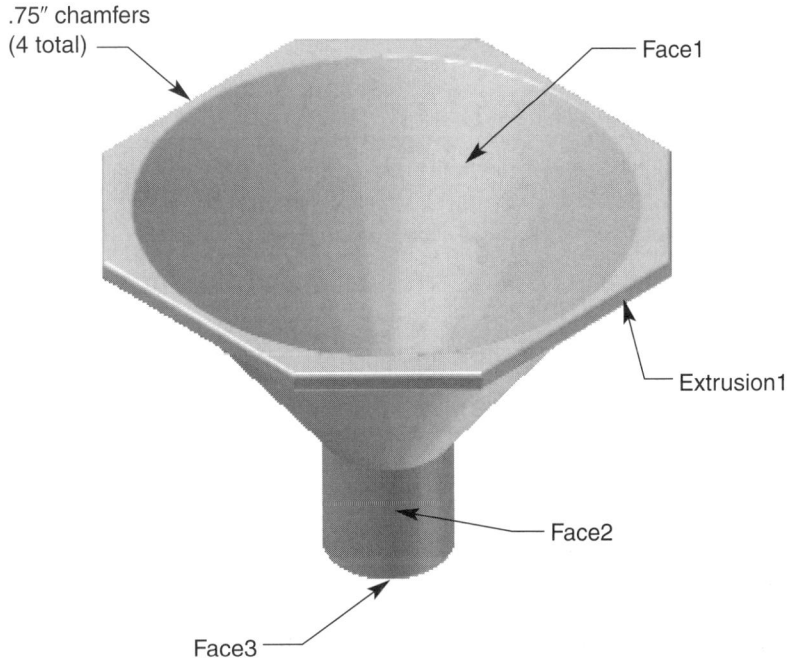

.75″ chamfers
(4 total)

Face1

Extrusion1

Face2

Face3

6. Name: C-Clamp Body

   Units: Metric

   Specific Instructions:

   - Open P4.6, and Save Copy As P5.6.

   - Close P4.6 without saving, and Open P5.6.

   - Place 1.6mm chamfers on the edges shown, 26 total.

   - Save As: P5.6.

Chamfers

7. Name: Torsion Spring Swivel Bolt

   Units: Inch

   Specific Instructions:

   - Open P4.9, and Save Copy As P5.7.

   - Close P4.9 without saving, and Open P5.7.

   - Thread the bolt stud using the specifications shown. The threads should be right hand, .5″ in length, and have 0 offset.

   - Place a .025″ chamfer on the edge shown.

   - The final project should look like the part shown.

   - Save As: P5.7.

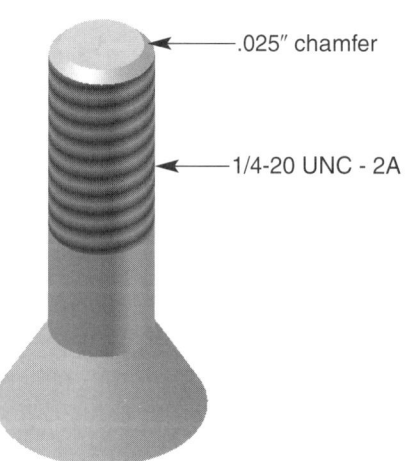

.025″ chamfer

1/4-20 UNC - 2A

8. Name: Bottle

   Units: Metric or Inch

   Specific Instructions:

   - Use the following tools to create a bottle similar to the one shown:

     2 Extrusions

     Constant fillets

     Variable fillets

     1 Shell

     1 Threaded feature

   - Save As: P5.8.

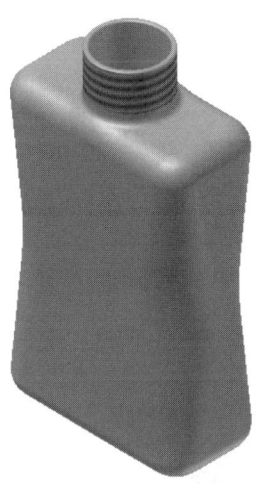

9. Name: Housing

   Units: Metric

   Specific Instructions:

   - Open a new part file.

   - Extrude a 150mm×145mm rectangle 50mm in a positive direction.

   - Sketch a ⌀70mm circle in the center of the top face of the first extrusion.

   - Extrude the circle 35mm in a positive direction.

   - Fillet the specified edges using the provided information.

   - Shell the features using the following information:

     Remove Faces1 and 2

     Thickness: 10mm

     Direction: Inside

   - Sketch a 20mm radius arc in the center of the face shown.

   - Cut extrude the arc using the **To Next** termination option.

   - The final project should look like the part shown.

   - Save As: P5.9.

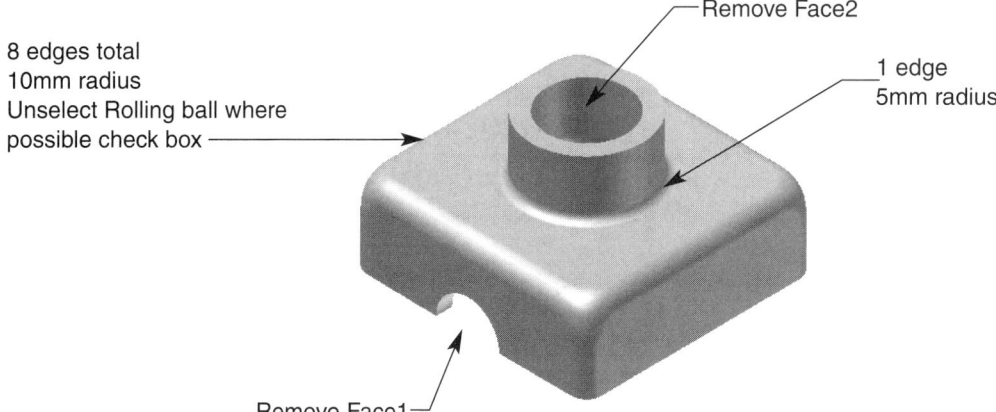

8 edges total
10mm radius
Unselect Rolling ball where
possible check box

Remove Face2

1 edge
5mm radius

Remove Face1

Open the specified Chapter 3 part file and follow the specific instructions provided for the project.

10. Name: Support Block

    Units: Inch

    Specific Instructions:

    - Open EX3-4, and Save Copy As P5.10.
    - Close EX3-4 without saving, and Open P5.10.
    - Extrude the sketch 3″ in a positive direction as shown.
    - Place a 2″ chamfer on the edges shown.
    - Open a sketch on Face1, and sketch the geometry shown.
    - Add a ⌀.75″ hole through the entire part.
    - The final part should look like the part shown.
    - Save As: P5.10.

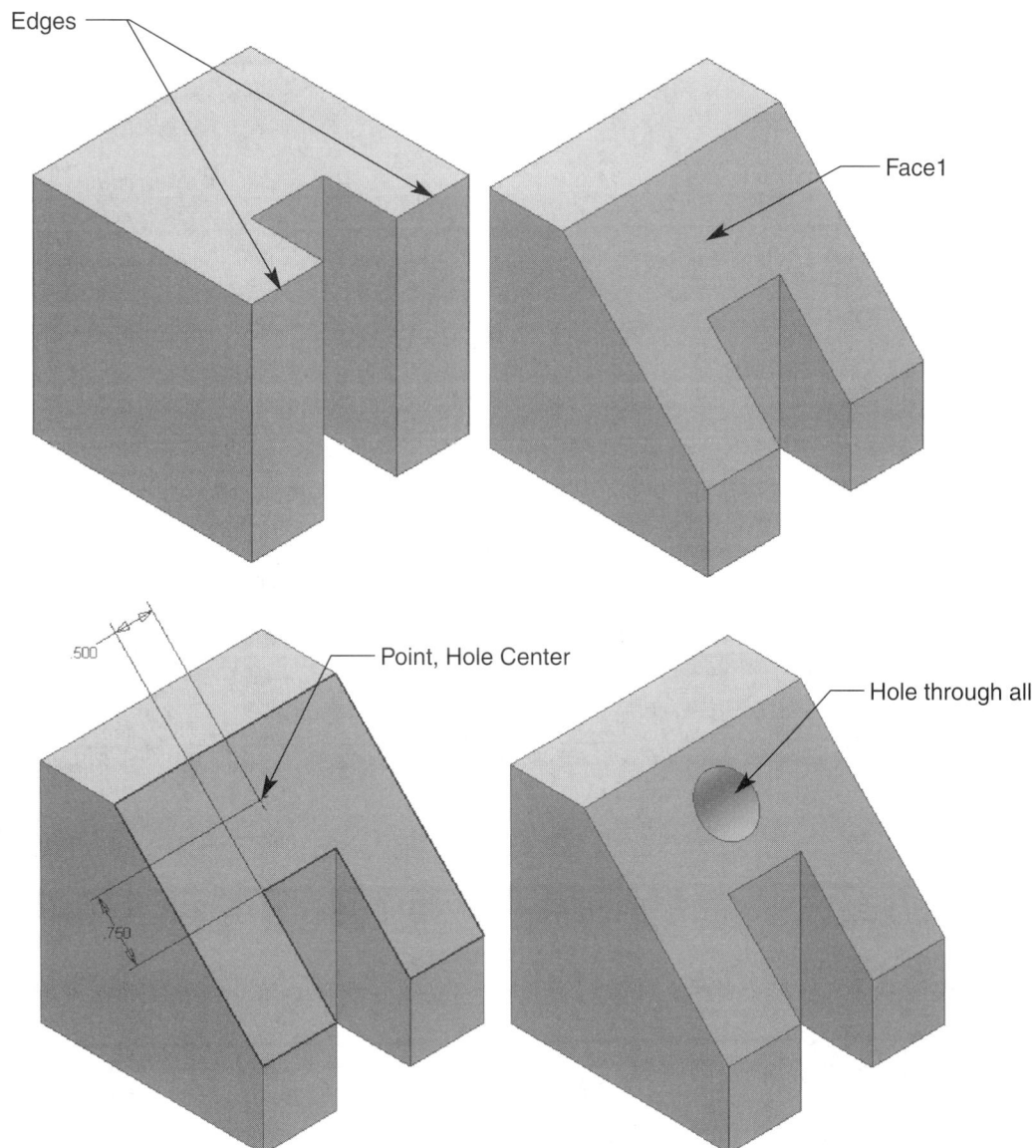

||||||||||||||||||||||||||||||||||||||||||||||||||||||||||||||||||||||||||||||||||||||||||||||||

CHAPTER 6

Patterning Features

LEARNING GOALS

After completing this chapter, you will be able to:

◎   Pattern features, and use pattern feature tools.

◎   Mirror features.

◎   Create rectangular feature patterns.

◎   Create circular feature patterns.

Patterning Features

As you continue developing your part models, you may need to create multiple arranged copies, or patterns, of features. A reoccurrence of features, in a designated configuration, is known as a *pattern feature.* Autodesk Inventor provides you with tools that allow you to pattern a feature or series of features by mirroring or arraying in a circular or rectangular fashion. See Figure 6.1.

Pattern features can be created anytime during part model development, as long as suitable features exist. Typically, you pattern features when they are complete, such as patterning an extrusion that contains a chamfer and a hole. You cannot pattern placed features such as fillets, chamfers, shells, face drafts, and

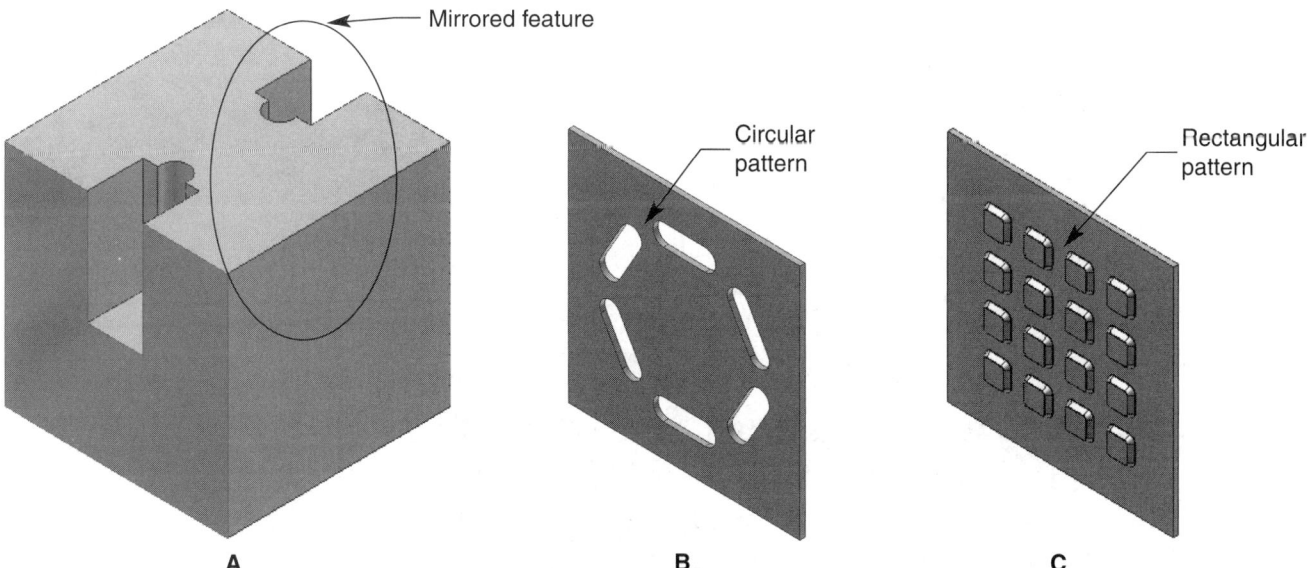

FIGURE 6.1   An example of (A) a mirrored feature, (B) a circular pattern, and (C) a rectangular pattern.

threads by themselves. However, when you pattern a feature that contains placed features, the placed features are reproduced in the patterns.

Field Notes

All the interface and work environment options and applications discussed in Chapter 4 apply to working with placed features.

Using Pattern Feature Tools

Pattern feature tools are used to create mirrored features, circular patterns, and rectangular patterns. These tools are accessed from the **Features** toolbar, shown in Figure 6.2, or from the **Features** panel bar. When a feature is present, and you are in the feature work environment, the panel bar is in feature mode. See Figure 6.3. By default, the panel bar is displayed in learning mode, as seen in Figure 6.3. To change from learning to expert mode, seen in Figure 6.4 and discussed in Chapter 2, right-click any of the panel bar tools or left-click the panel bar title and select the **Expert** option.

FIGURE 6.2 The
Features toolbar.

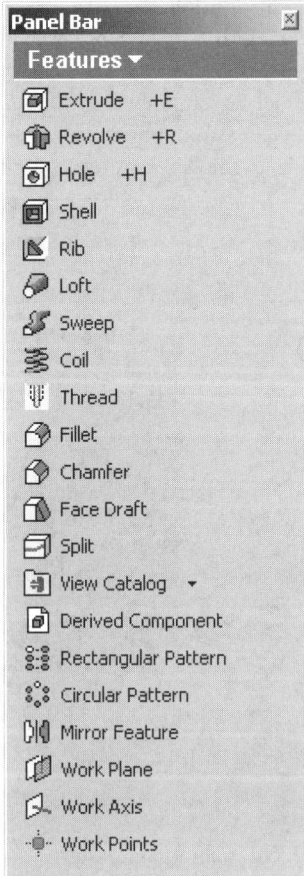

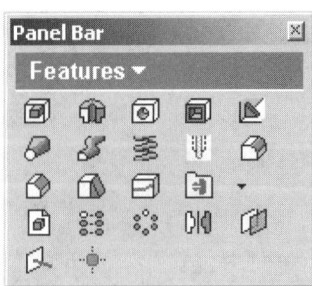

FIGURE 6.3 The panel bar,
in features (learning) mode.

FIGURE 6.4 The panel bar,
in features (expert) mode.

Mirroring Feature

A *mirrored feature* consists of one or more features that are mirrored symmetrically over a specified plane. Mirroring is used to create a backwards copy of features for symmetrical parts, without having to develop an additional sketch, placed, or catalog feature geometry. Figure 6.5 shows an example of a part created by mirroring an extrusion and a hole.

Mirrored features are created using the **Mirror Feature** tool. Access the **Mirror Feature** command using one of the following options:

✓ Pick the **Mirror Feature** button on the panel bar.

✓ Pick the **Mirror Feature** button on the **Features** toolbar.

The **Mirror Pattern** dialog box is displayed when you access the **Mirror Feature** command. See Figure 6.6.

To mirror a feature or series of features, first pick the **Features** button if it is not already active. Then select the geometry you want to mirror. You may need to pick certain features from the **Browser** bar if you cannot access them from the graphics window.

FIGURE 6.5 An example of mirroring features.

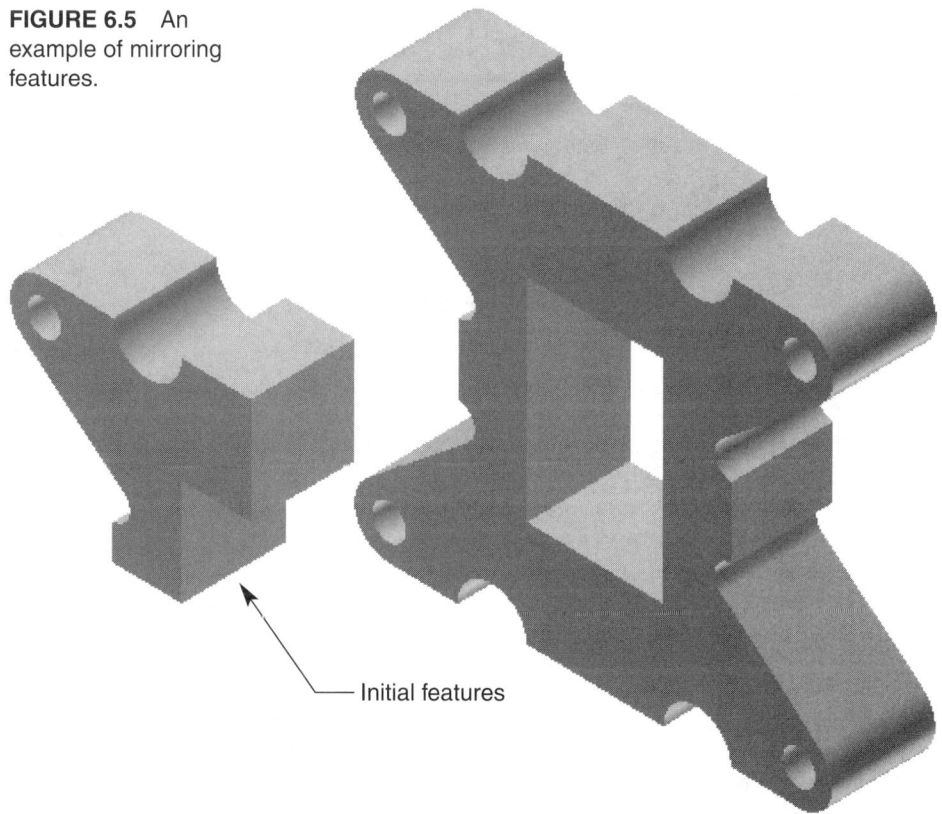

Initial features

FIGURE 6.6 The **Mirror Pattern** dialog box.

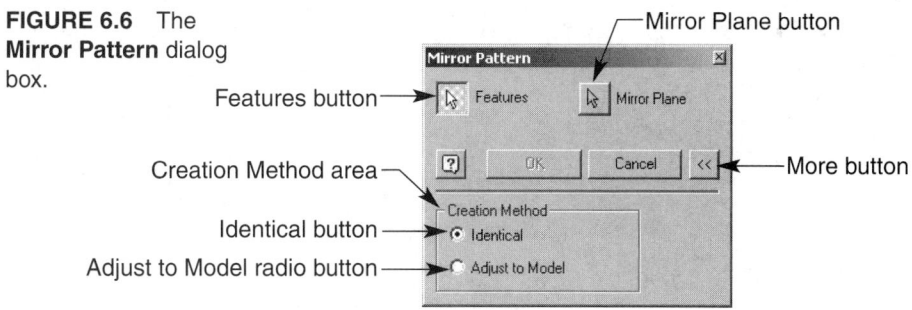

Once you have selected all the features you want to mirror, pick the **Mirror Plane** button and select the mirror plane. The *mirror plane* acts as the line of symmetry, defining where the specified features are mirrored across. You can choose any available plane, including planar feature faces, work planes, and the default reference planes located in the **Browser** bar **Origin** folder.

After you specify a mirror plane, you may want to change the mirror creation method by picking the **More** button and accessing the **Creation Method** area. The following radio buttons are available inside:

- **Identical**   This radio button allows you to create mirrored features that are exactly the same as the parent features. When this radio button is selected all aspects of the parent feature, including the specified termination option, are reproduced in the mirrored feature. See Figure 6.7. Select the **Identical** radio button when the features you are mirroring terminate on a work plane or the same planar face.

Field Notes

> The **Identical** creation method generates mirrored features the quickest. As a result, you should select this radio button for mirroring complex geometry, unless the features must change termination as they are reproduced.

- **Adjust to Model**   This radio button allows you to create mirrored features that are based on the parent features, but change size to adjust for variations in the model. See Figure 6.8. For most typical mirroring operations you do not need to select the **Adjust to Model** radio button.

Field Notes

> The **Adjust to Model** creation method takes longer than the **Identical** creation method to generate mirrored features. Select this radio button only when mirrored features must modify to fit changes in model geometry.

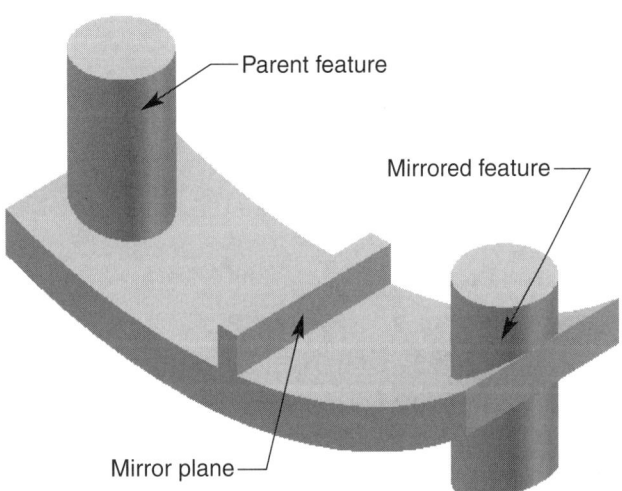

FIGURE 6.7   Mirroring a feature with the **Identical** radio button selected.

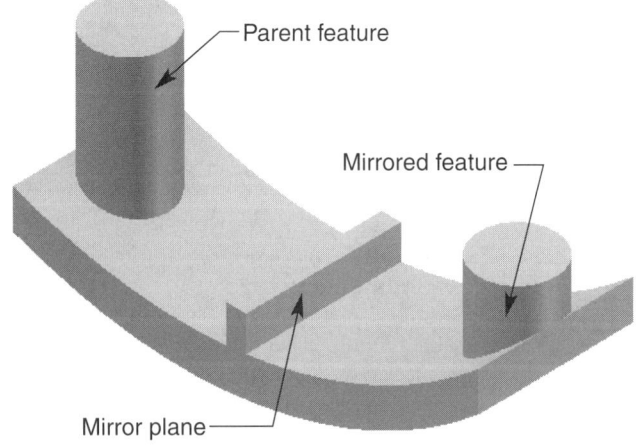

FIGURE 6.8   Mirroring a feature with the **Adjust to Model** radio button selected.

FIGURE 6.9 An example of the information displayed in the **Browser** when a mirror feature is created.

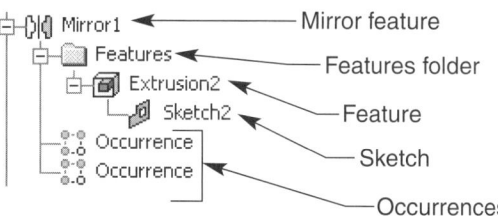

Once you have selected the features you want to mirror, the mirror plane, and the desired creation method, pick the **OK** button to mirror the features.

Figure 6.9 shows an example of the information displayed in the **Browser** bar when you mirror a feature. The mirror feature is identified by the mirror icon and is named **Mirror1,** by default, for the first mirror feature. As you create additional mirror features, the numbers next to Mirror increase. The mirror feature contains the following information used to generate the mirrored geometry:

- **Features** This folder contains the parent features and sketches that have been mirrored. The feature mirrored in Figure 6.9 contains an extrusion, and the extrusion contains the sketch used to create the feature.

- **Occurrences** Every time you pattern a feature, including mirror features, the occurrences are listed under the mirror feature. The *occurrences* show how many features are present as a result of the pattern operation. For mirror features, only two occurrences are present, the parent feature and the mirrored feature. When you move your cursor over the occurrence, the feature is highlighted in the graphics window. If you want to suppress, or "turn off," the mirror feature, right-click the correct occurrence and pick the **Suppress** menu option.

Field Notes

You cannot suppress the parent feature.

Many of the additional **Browser** bar options and applications discussed in Chapter 4 apply to mirror features.

Exercise 6.1

1. Launch Autodesk Inventor.
2. Open a new part file.
3. Open a sketch on the XY plane, and sketch the geometry shown. Use equal constraints to fully constrain the sketch, in addition to the dimensions shown.
4. Extrude the sketch .75″ in both directions.
5. Add a .25″⌀ hole as shown.
6. Mirror the extrusion and hole using Plane1.
7. Mirror **Mirror1**, using Plane2.
8. The final part should look like the part shown.
9. Save the part as EX6.1.
10. Exit Autodesk Inventor or continue working with the program.

Exercise 6.1 continued on next page

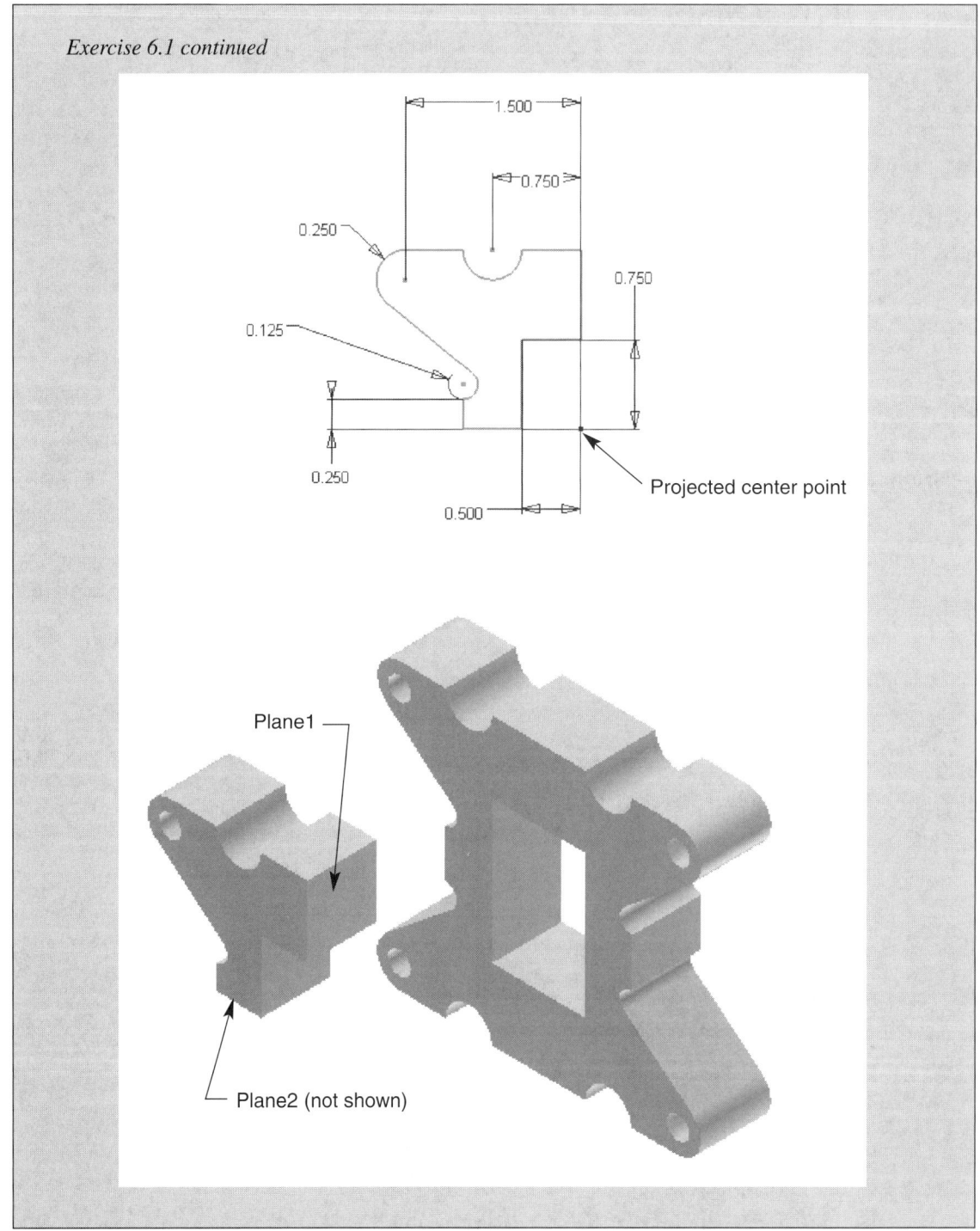

Exercise 6.1 continued

Creating Rectangular Feature Patterns

A *rectangular pattern* is one or more objects arranged in a rectangular configuration. A rectangular pattern consists of a number of features, or series of features, positioned a specified distance apart, in rows and columns. Rectangular patterning is used to arrange multiple copies of feature geometry, without having to develop an additional sketch, placed, or catalog features. Figure 6.10 shows an example of a part created by rectangular patterning a cut extrusion and fillets.

Rectangular patterns are created using the **Rectangular Pattern** tool. Access the **Rectangular Pattern** command using one of the following techniques:

FIGURE 6.10 An example of a rectangular pattern.

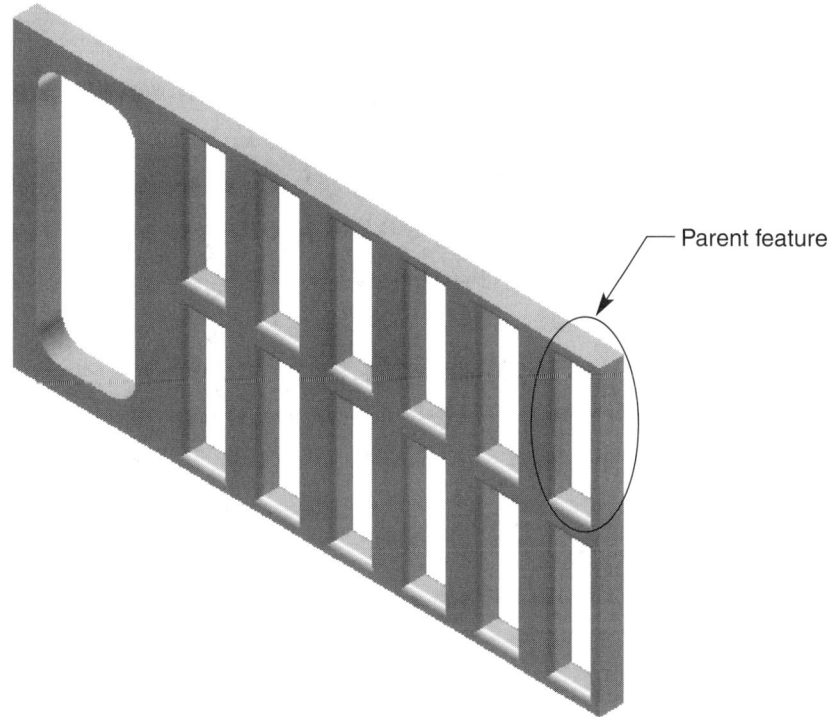

Parent feature

✓ Pick the **Rectangular Pattern** button on the panel bar.

✓ Pick the **Rectangular Pattern** button on the **Features** toolbar.

The **Rectangular Pattern** dialog box is displayed when you access the **Rectangular Pattern** command. See Figure 6.11.

To create a rectangular pattern of a feature or series of features, first pick the **Features** button if it is not already active. Then select the geometry you want to pattern. You may need to pick certain features from the **Browser** bar if you cannot access them from the graphics window.

Once you have selected all the features you want to pattern, pick the **Direction** button, located in the **Direction 1** area, and select the direction of the row or column of patterned features. You can select any feature edge, axis, or work axis available in the model, including the default reference axes located in the **Browser** bar **Origin** folder. When you choose a direction, an arrow shows you the specified pattern path, and a preview displays the pattern operation. See Figure 6.12.

When you have defined the direction of the first pattern, specify how many copies of the feature or series of feature you want to create by entering or selecting a count value from the **Count** drop-down list. Then define the distance between the copies by entering or selecting a spacing value from the **Spacing**

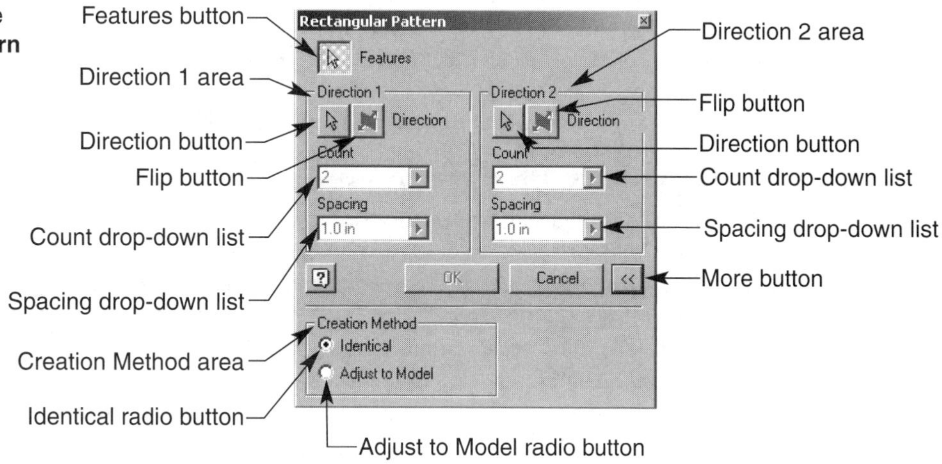

FIGURE 6.11 The **Rectangular Pattern** dialog box.

Features button

Direction 1 area

Direction button

Flip button

Count drop-down list

Spacing drop-down list

Creation Method area

Identical radio button

Direction 2 area

Flip button

Direction button

Count drop-down list

Spacing drop-down list

More button

Adjust to Model radio button

FIGURE 6.12 The
pattern direction arrow
and pattern preview.

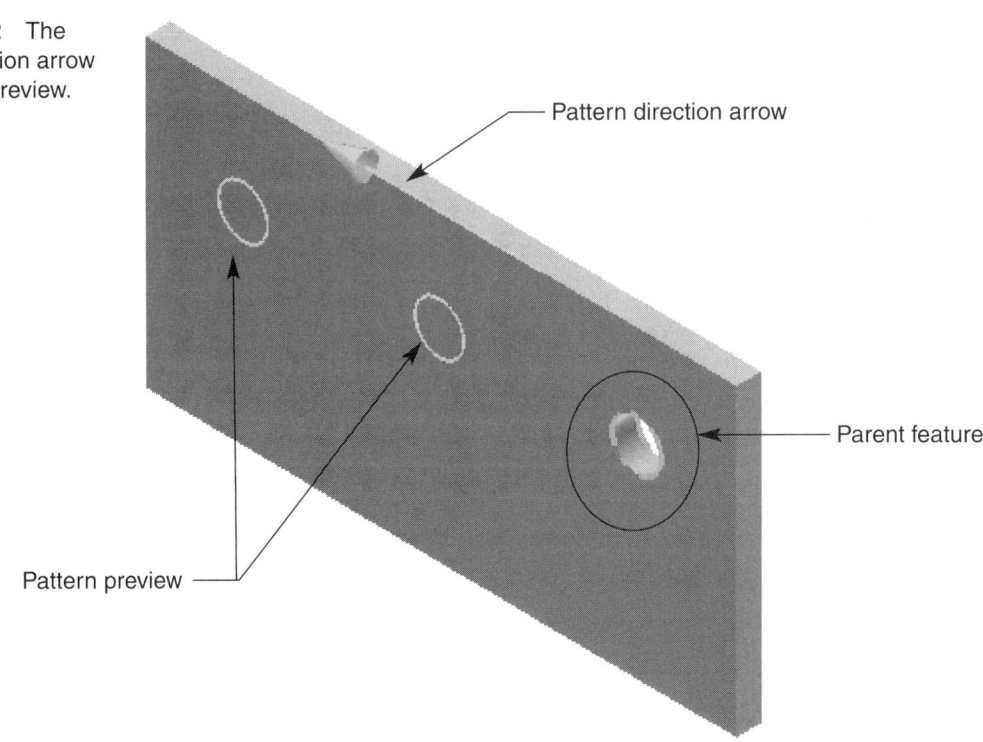

drop-down list. Spacing is measured by the width of the selected features and the distance between the copies. It is not just the space between features. For example, if you want to pattern an extrusion that is 1″ wide, and you want a ½″ space between pattern copies, you must specify a 1½″ spacing. Another way to understand spacing is this: it is the distance from a point on the parent feature to the corresponding point on each pattern occurrence. For example, if you want a rectangular pattern of cylinders, and you want the distance between the cylinder centers to be 10mm, specify a spacing of 10mm.

Once you have fully defined **Direction 1,** repeat the steps previously described for **Direction 2.** Pick the **Direction** button, located in the **Direction 2** area, and select the desired direction. If the first direction defined the rows, **Direction 2** will define the columns. If the first direction defined the columns, **Direction 2** will define the rows. You can select any feature edge, axis, or work axis available in the model, including the default reference axes located in the **Browser** bar **Origin** folder. Theoretically you can choose the same edge or axis you chose for **Direction 1,** or a parallel edge or axis. However, for most applications you should select an edge or axis that is perpendicular, or at least nonparallel to the first direction. When you choose the second direction, an arrow shows you the specified pattern path, and a preview displays the pattern operation. See Figure 6.13.

When you have defined the direction of the second pattern, specify how many copies of the feature or series of feature you want to create by entering or selecting a count value from the **Count** drop-down list. Then define the distance between the copies by entering or selecting a spacing value from the **Spacing** drop-down list.

After you have selected the features and specified the direction, count, and spacing information, you may want to change the rectangular pattern creation method by picking the **More** button and accessing the **Creation Method** area. The **Creation Method** area contains the **Identical** and **Adjust to Model** radio buttons previously discussed. Finally, pick the **OK** button to generate the rectangular pattern.

The **Browser** bar options and applications previously discussed for a mirror feature also apply to rectangular patterns. Rectangular patterns are identified by the rectangular pattern icon and the default **Rectangular Pattern1** name. In addition, the **Features** folder and each of the **Occurrences** are listed. See Figure 6.14. Like a mirror feature, the occurrences show how many features are present as a result of the pattern operation, including the parent feature. Often you may want to "turn off" certain rectangular pattern occurrences. To suppress a feature, right-click the desired occurrence, and pick the **Suppress** menu option. See Figure 6.15.

FIGURE 6.13 The pattern direction arrow and pattern preview.

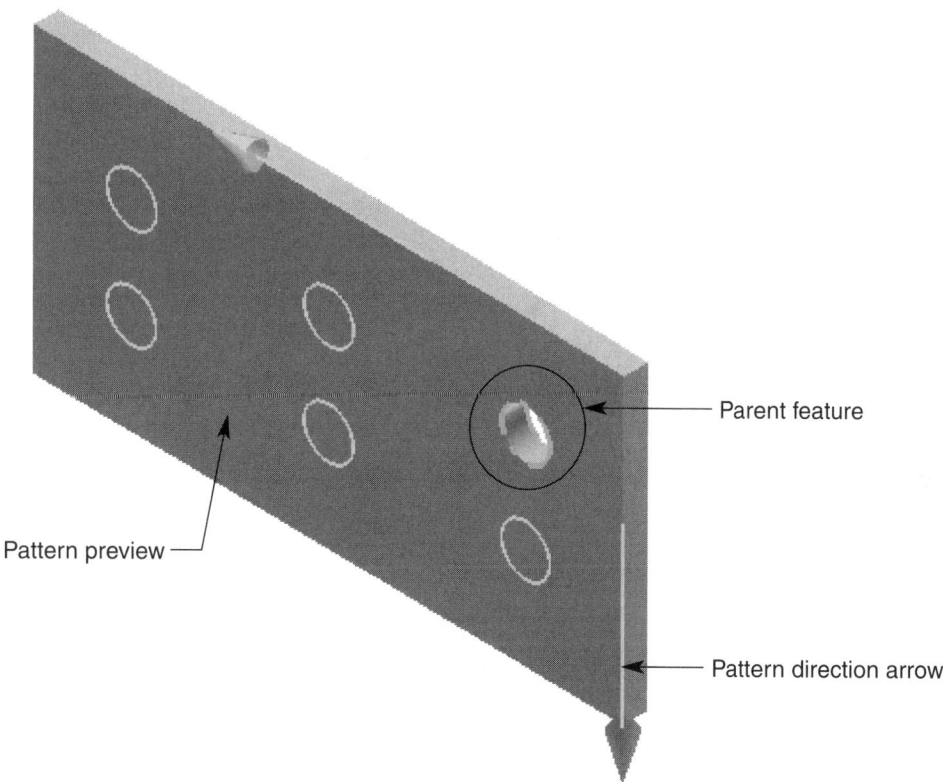

Parent feature

Pattern preview

Pattern direction arrow

FIGURE 6.14 An example of the information displayed in the **Browser** when a rectangular pattern is created.

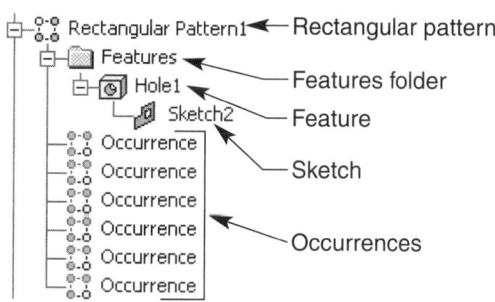

Rectangular pattern

Features folder

Feature

Sketch

Occurrences

Field Notes

You cannot suppress the parent feature.

Many of the additional **Browser** bar options and applications discussed in Chapter 4 also apply to rectangular patterns.

FIGURE 6.15 The effects of suppressing a feature.

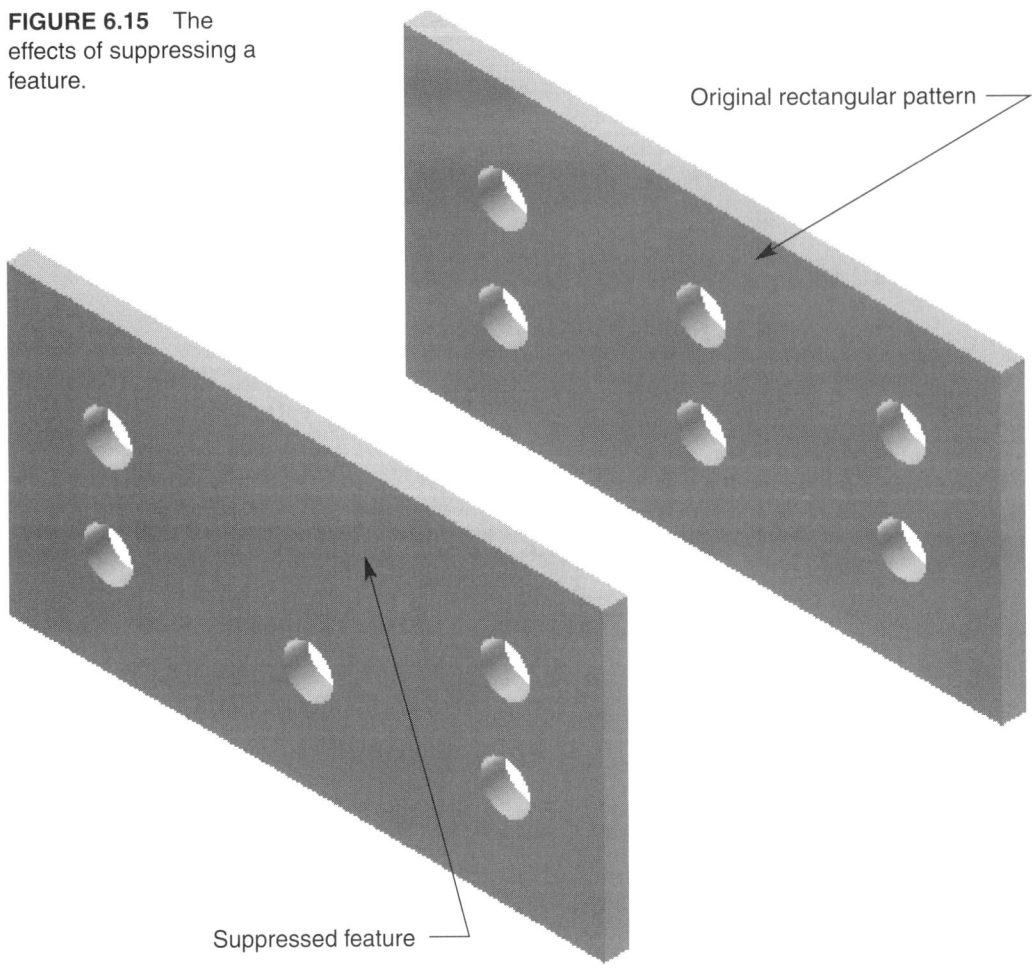

Original rectangular pattern

Suppressed feature

Exercise 6.2

1. Continue from Exercise 6.1, or launch Autodesk Inventor.
2. Open a new part file.
3. Open a sketch on the XY plane, and sketch the geometry shown. Use an equal constraint to fully constrain the sketch, in addition to the dimensions shown.
4. Extrude the sketch 10mm in a positive direction.
5. Add a ∅10mm hole as shown.
6. Use the **Rectangular Pattern** tool to pattern the hole as shown. Direction 1 has a count of 5 and a spacing of 20mm. Direction 2 has a count of 5 and a spacing of 13.75mm.
7. Save the part as EX6.2.
8. Exit Autodesk Inventor or continue working with the program.

Exercise 6.2 continued on next page

Exercise 6.2 continued

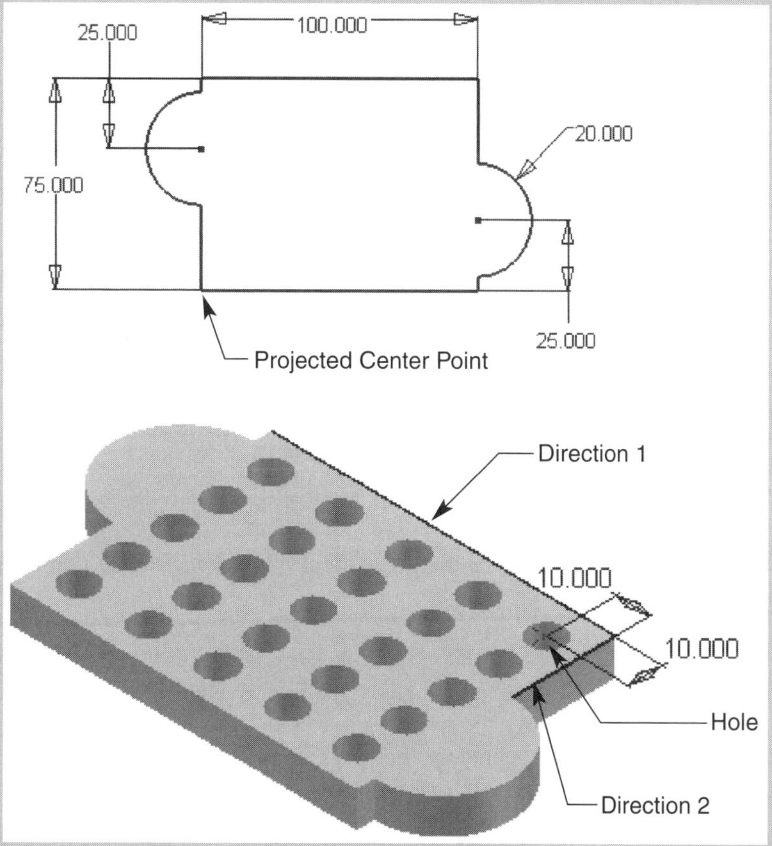

25.000

100.000

20.000

75.000

25.000

Projected Center Point

Direction 1

10.000

10.000

Hole

Direction 2

Creating Circular Feature Patterns

A *circular pattern* is one or more objects arranged in a circular configuration. A circular pattern consists of a number of features, or series of features, positioned a specified distance apart around a specified axis. Circular patterning is used to arrange multiple copies of feature geometry without having to develop additional sketch, placed, or catalog features. Figure 6.16 shows an example of a part created by circular patterning a cut extrusion and fillets.

FIGURE 6.16 An example of a circular pattern.

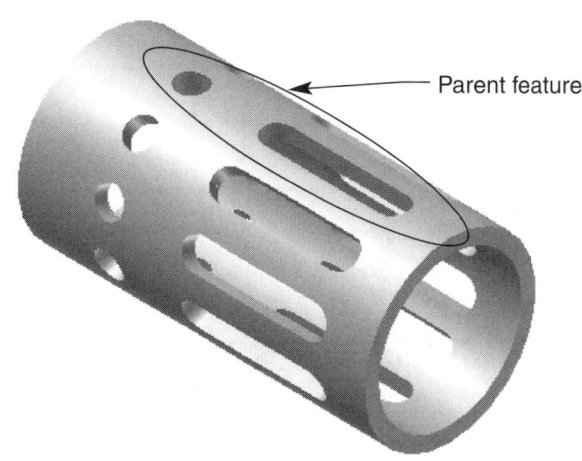

Parent feature

FIGURE 6.17 The
Circular Pattern dialog
box.

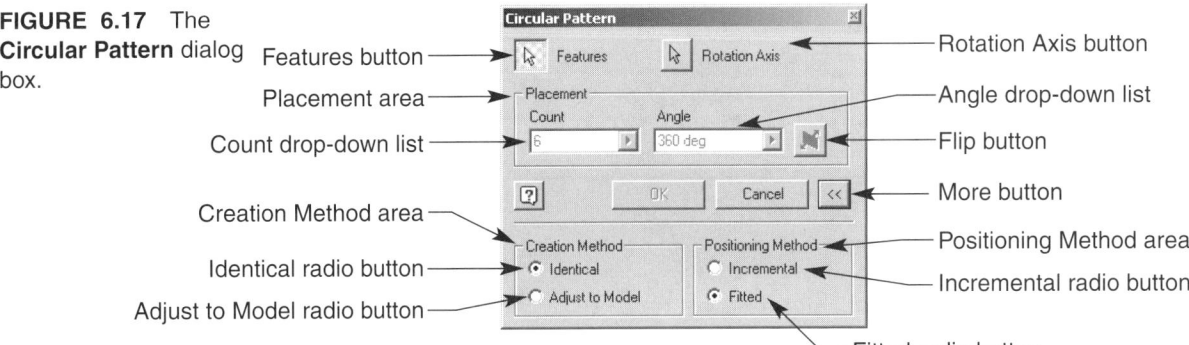

Circular patterns are created using the **Circular Pattern** tool. Access the **Circular Pattern** command using one of the following techniques:

✓ Pick the **Circular Pattern** button on the panel bar.

✓ Pick the **Circular Pattern** button on the **Features** toolbar.

The **Circular Pattern** dialog box is displayed when you access the **Circular Pattern** command. See Figure 6.17.

To create a circular pattern of a feature or series of features you may first want to change the circular pattern creation and positioning methods by picking the **More** button and accessing the **Creation Method** and **Positioning Method** areas. The **Creation Method** area contains the **Identical** and **Adjust to Model** radio buttons previously discussed. The **Positioning Method** area allows you to specify how the pattern features are placed and contains the following radio buttons:

■ **Incremental** This radio button allows you to specify the distance between features by defining the angle each pattern feature rotates around the axis. The total amount of rotation is defined by the number of occurrences and the angle between occurrences. Figure 6.18 shows an example of patterning a feature five times, using a 60° incremental angle.

■ **Fitted** This radio button allows you to specify the distance between features, by defining the total amount of rotation. The angle between each occurrence is defined by the number of occurrences and the entire revolution angle of all the pattern features. A 360° angle is a full rotation. Figure 6.19 shows an example of patterning a feature five times, using a 360° fitted angle.

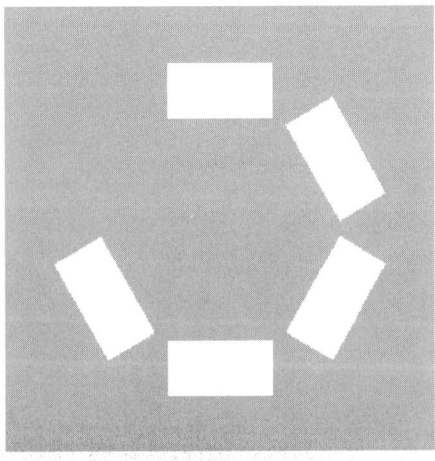

FIGURE 6.18 Patterning a feature using the incremental option.

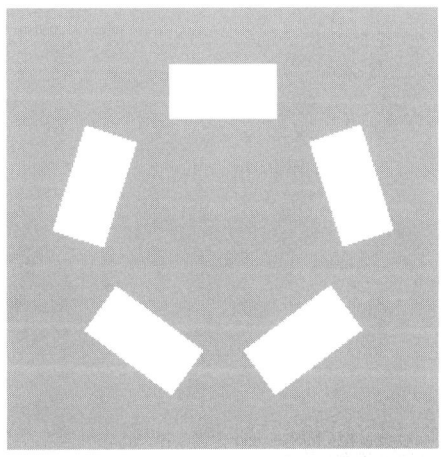

FIGURE 6.19 Patterning a feature using the fitted option.

Field Notes

If the **Fitted** option is used to create the pattern, the spacing of pattern feature occurrences change if the design is modified.

Once you have selected the desired creation and positioning methods, pick the **Features** button if it is not already active. Then select the geometry you want to pattern. You may need to pick certain features from the **Browser** bar if you cannot access them from the graphics window.

When all the features you want to pattern are selected, pick the **Rotation Axis** button and select the axis of rotation. The axis of rotation is the center or pivot point where selected features orbit around. You can select any feature edge, axis, or work axis available in the model, including the default reference axes located in the **Browser** bar **Origin** folder. When you choose a rotation axis, an arrow shows you the specified pattern path, and a preview displays the pattern operation. See Figure 6.20.

Once you have selected the features and defined the axis of rotation, specify how many copies of the feature or series of feature you want to create by entering or selecting a count value from the **Count** drop-down list. Then define either the incremental or fitted angle depending on the radio button you choose by entering or selecting an angle value from the **Angle** drop-down list. If the total rotation of all occurrences is not 360, and you want the circular pattern to revolve around the opposite direction, pick the **Flip** button. Finally, pick the **OK** button to generate the circular pattern.

The **Browser** bar options and applications previously discussed for a rectangular feature also apply to circular patterns. Circular patterns are identified by the circular pattern icon and the default **Circular Pattern1** name. The **Features** folder and each of the **Occurrences** are also listed. See Figure 6.21. Like a mirror feature and rectangular patterns, the occurrences define how many features are present, as a result of the pattern operation, including the parent feature. Again, to suppress a circular pattern feature, right click the desired occurrence, and pick the **Suppress** menu option. See Figure 6.22.

Field Notes

You cannot suppress the parent feature.

Many of the additional **Browser** bar options and applications discussed in Chapter 4 also apply to rectangular patterns.

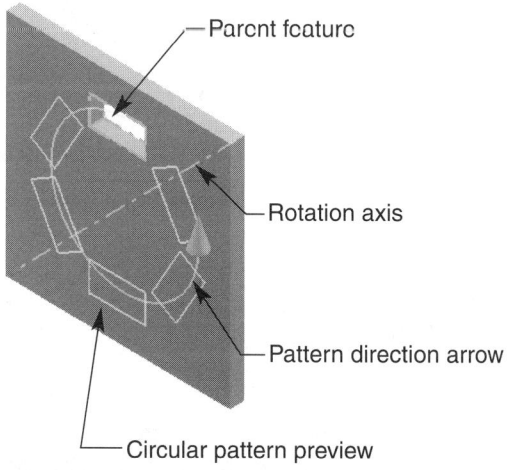

FIGURE 6.20 The pattern direction arrow and pattern preview.

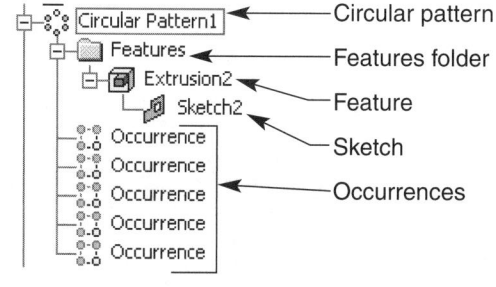

FIGURE 6.21 An example of the information displayed in the **Browser** when a circular pattern is created.

FIGURE 6.22 The
effects of suppressing
a feature.

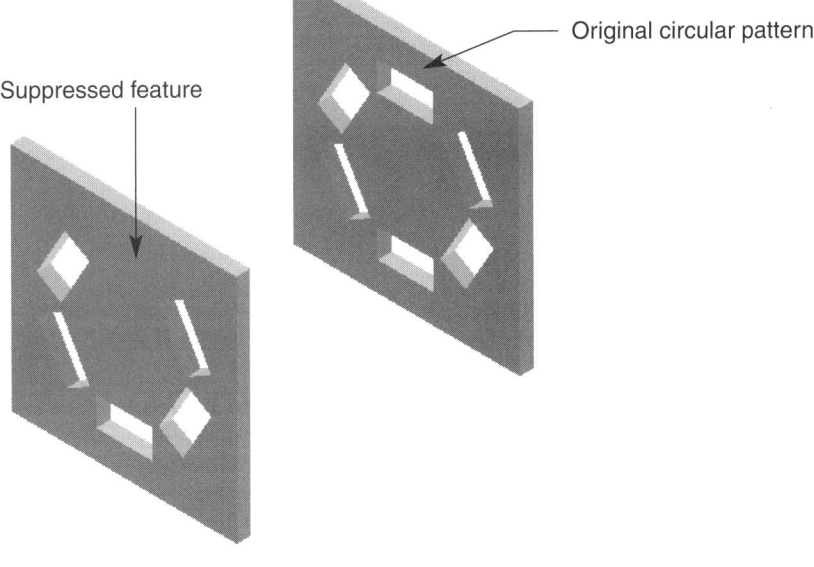

Suppressed feature

Original circular pattern

Exercise 6.3

1. Continue from Exercise 6.2, or launch Autodesk Inventor.
2. Open a new part file.
3. Open a sketch on the XZ plane.
4. Extrude a ∅5″ circle .5″ both sides.
5. Place the specified hole as shown on the top face of the extruded cylinder. *Note:* Place the hole center vertical to the **Center Point.**
6. Use the **Circular Pattern** tool to pattern the hole, using the following information:

   Rotation axis: Y Axis
   Count: 6
   Angle: 360
   Positioning Method: Fitted

7. Save the part as EX6.3
8. Exit Autodesk Inventor or continue working with the program.

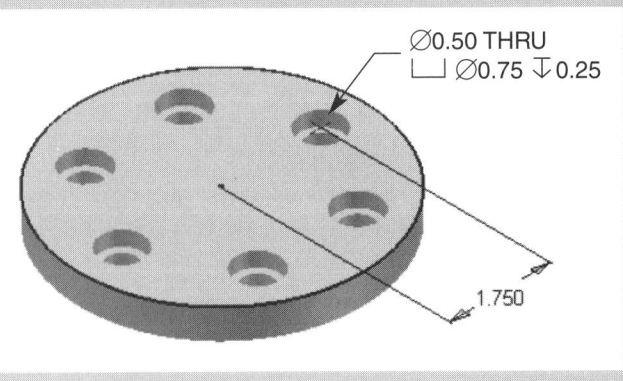

∅0.50 THRU
⌴ ∅0.75 ⤓0.25

1.750

|||||||||||| CHAPTER TEST

1. What is a pattern feature?

2. What kind of features cannot be patterned by themselves?

3. What is a mirrored feature?

4. What is the mirror plane?

5. Discuss the function of the Identical radio button in the Creation Method area of the Mirror Pattern dialog box.

6. When should you select the Identical radio button?

7. Give the function of the Adjust to Model radio button in the Creation Method area of the Mirror Pattern dialog box.

8. What do the occurrences show?

9. What is a rectangular pattern?

10. Describe how spacing is measured when making a rectangular pattern, and give an example.

11. What is a circular pattern?

12. Give the function of the Incremental radio button in the Positioning Method area of the Circular Pattern dialog box.

13. Explain the function of the Fitted radio button in the Positioning Method area of the Circular Pattern dialog box.

14. How do you suppress a circular pattern feature?

15. How are circular patterns identified?

|||||||||||| PROJECTS

Follow the specific instructions provided for each project.

1. Name: Support

   Units: Inch

   Specific Instructions:

   ■ Open a new part file, and create the sketch shown on the **XY Plane.**

   ■ Extrude the sketch 2″ in both directions.

   ■ Create the sketch shown on the top face.

   ■ Cut extrude the sketch 1″ in a negative direction, as shown.

   ■ Mirror the cut extrusion using the **XY Plane** located under the **Origin** folder, as the **Mirror Plane.**

   ■ Add the specified hole .5″ from the bottom edge and 1″ from the vertical edge, as shown.

   ■ Place a .028625″ chamfer at the hole edges.

   ■ Place a .0625″ fillet on **Extrusion1.**

   ■ The final part should look like the part shown.

   ■ Save As: P6.1.

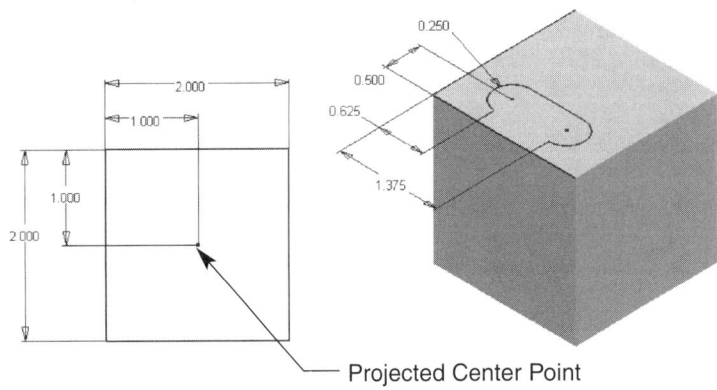

Projected Center Point

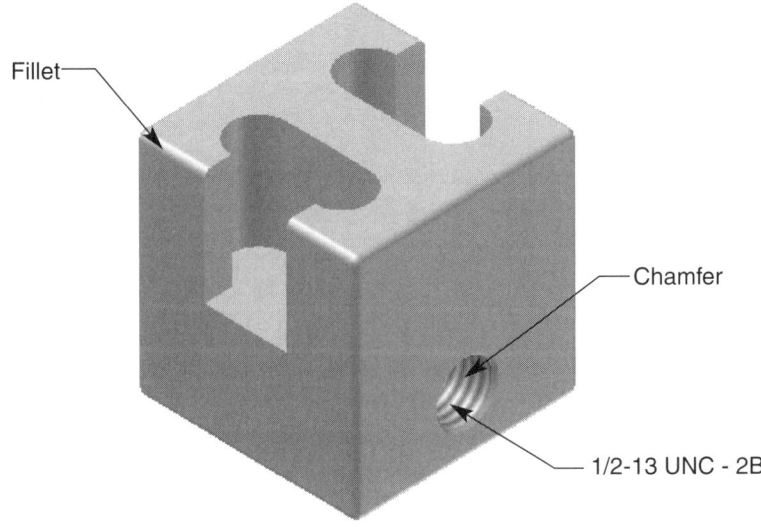

Fillet

Chamfer

1/2-13 UNC - 2B

2. Name: Cover Plate

Units: Inch or Metric

Specific Instructions:

- Create features similar to the ones shown, using your own specifications.
- Mirror the features twice to create the part shown.
- Place a .125″ fillet on the 2 loops shown.
- If .125″ fillets cannot be created, use a smaller radius.
- Save As: P6.2.

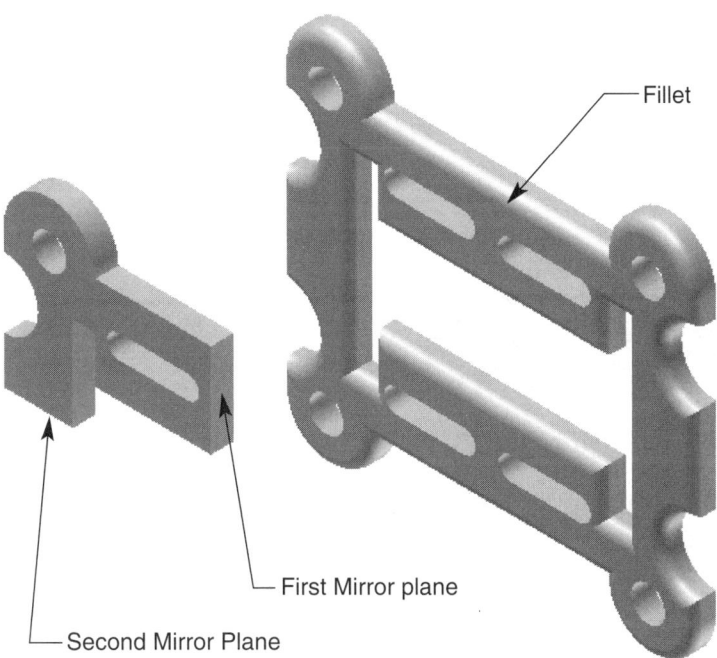

3. Name: Selector Bracket

   Units: Metric

   Specific Instructions:

   - Open EX6.2, and Save Copy As P6.3.

   - Close EX6.2 without saving, and Open P6.3.

   - Sketch the ∅27.5 mm circles shown on the top face, and extrude them 10mm.

   - Suppress 6 of the rectangular pattern holes, as shown.

   - Use the **Fillet** tool to fillet and round **All Fillets** and **All Rounds**, 2mm.

   - Add the specified holes.

   - Save As: P6.3.

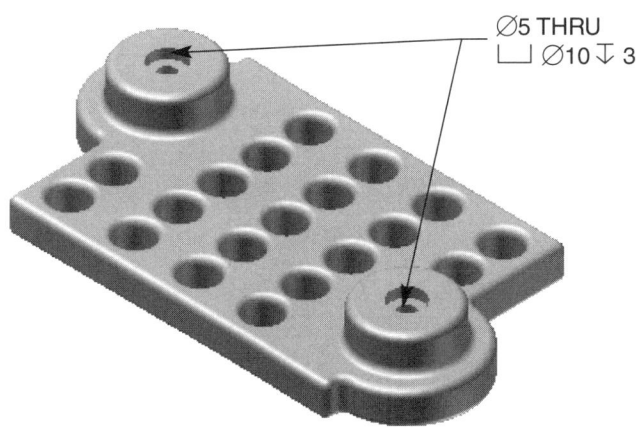

4. Name: Slide Bar Hinge Connector

   Units: Metric

   Specific Instructions:

   - Open a new part file, and create the sketch shown on the **YZ Plane.**

   - Extrude the sketch 100mm in a positive direction.

   - Place a 5mm fillet as shown.

   - Create the sketch shown on the specified face.

   - Cut extrude the sketch 10mm in a negative direction.

   - Use the **Rectangular Pattern** tool to pattern the extrusion as shown, with the following information:

     Count: 5

     Spacing: 22.5mm

   - Add a10mm hole through all the feature geometry, as shown.

   - Place a 10mm fillet on the edges shown.

   - Mirror all the features, and add the specified hole as shown aligned with the edge shown using a vertical constraint.

- Use the **Rectangular Pattern** tool to pattern the hole as shown, with the following information:

  Count: 5

  Spacing: 13.75mm

- The final part should look like the part shown.

- Save As: P6.4.

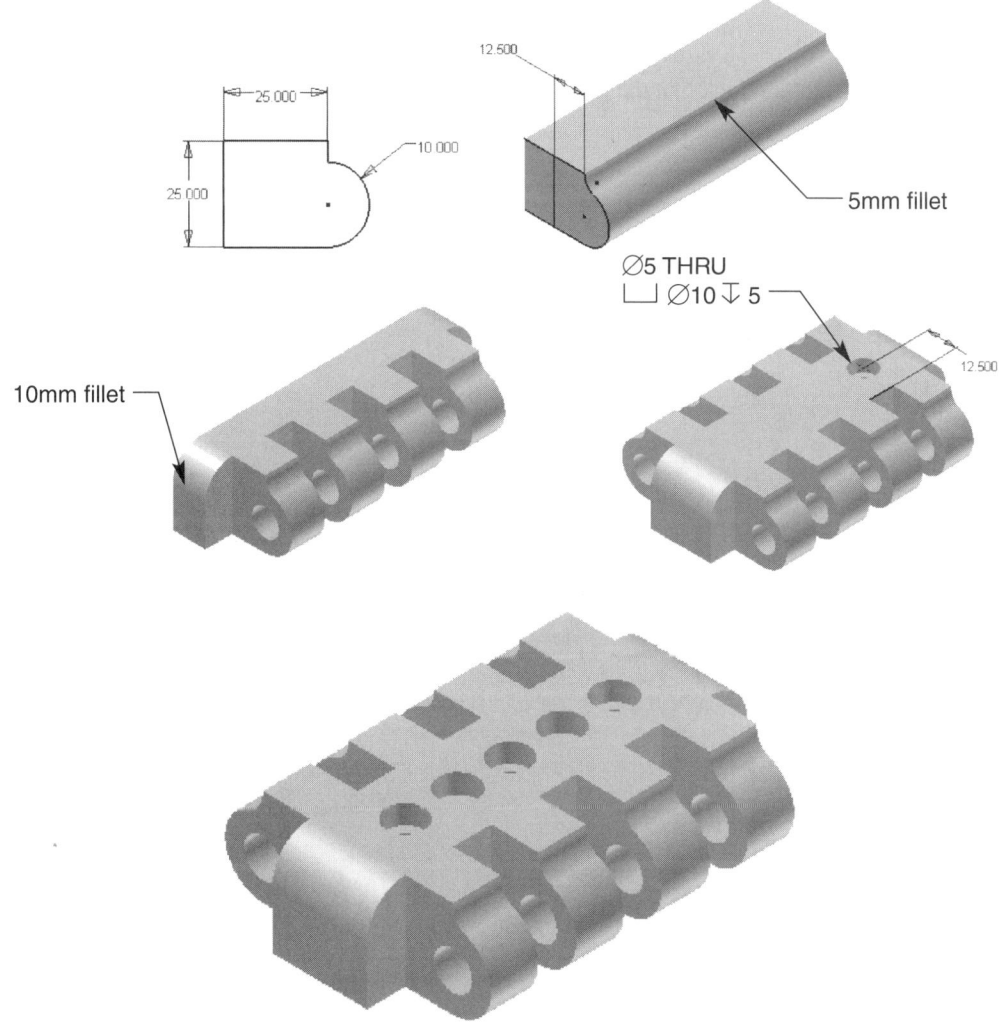

5. Name: Support

Units: Inch

Specific Instructions:

- Open P5.4, and Save Copy As P6.5.

- Close P5.4 without saving, and Open P6.5.

- Use the **Circular Pattern** tool to pattern all the features as shown, with the following information:

  Rotation Axis: Z Axis

  Count: 2

  Position Method: Incremental

  Angle: 135°

- The final part should look like the part shown.

- Save As: P6.5.

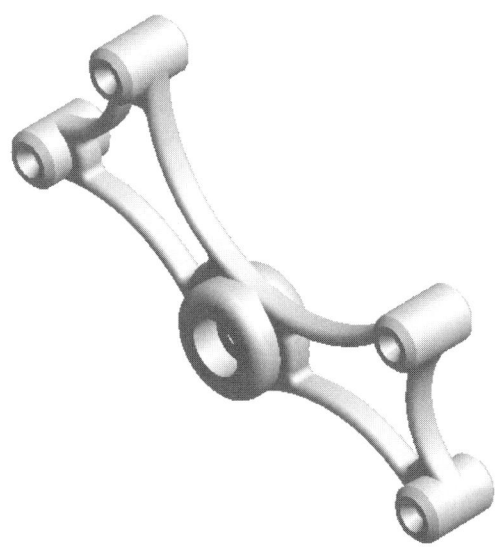

6. Name: Wheel

Units: Inch

Specific Instructions:

- Open a new part file.

- Sketch a ⌀2″ circle around the projected **Center Point,** on the **XY Plane.**

- Extrude the sketch 2″ in both directions.

- Sketch the geometry shown on the **XZ Plane.** *Note:* You may need to use the **Slice Graphics** option to clarify the sketch environment.

- Extrude the sketch 6″ in a positive direction.

- Sketch the circle shown on the **YZ Plane.**

- Fully revolve the sketch around the **Z Axis.**

- Use the **Circular Pattern** tool to pattern Extrusion2 as shown, with the following information:

  Rotation Axis: Z Axis

  Count: 6

  Position Method: Fitted

  Angle: 360°

- Add the specified hole to **Extrusion1.**

- The final project should look like the part shown.

- Save As: P6.6.

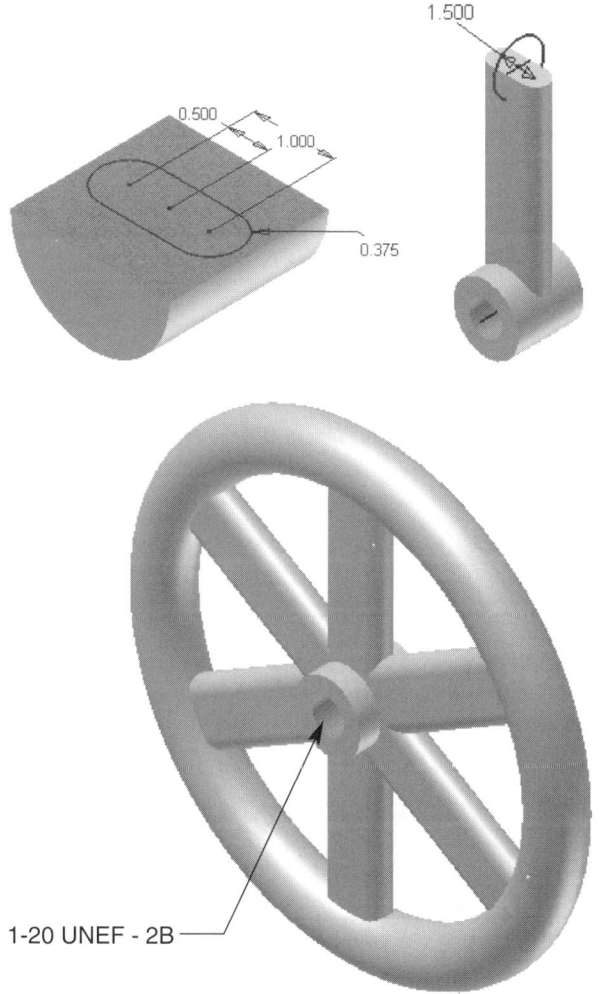

1-20 UNEF - 2B

7. Name: Hub

   Units: Inch

   Specific Instructions:

   ■ Open a new part file.

   ■ Sketch a circle ∅1.5″ circle on the **XZ Plane.**

   ■ Extrude the sketch 5″ in a positive direction.

   ■ Add the specified hole as shown.

   ■ Save As: P6.7.

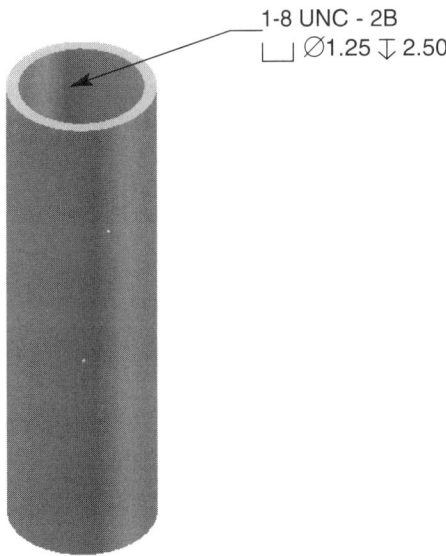

1-8 UNC - 2B
⊔ ∅1.25 ⊤ 2.50

8. Name: Bottle Cap

   Units: Metric or Inch

   Specific Instructions:

   ■ Develop a cap for the bottle you created in Chapter 5 (P5.8).

   ■ Use the Circular Pattern tool to apply straight knurls to the cap as shown.

   ■ *Note:* Sketch the parent knurl feature from one of reference planes located in the **Origin** folder, and extrude through the cap. Then apply the circular pattern, followed by the hole.

   ■ Save As: P5.8.

9. Name: 45° Elbow

Units: Metric

Specific Instructions:

- Open a new part file.

- Create the sketch shown on the **XY Plane.**

- Revolve both circles, at a 45° angle, in a negative direction.

- Sketch a ⌀108mm circle on the bottom face of the revolved feature.

- Extrude the sketch 11.5mm in a negative direction.

- Add a 16.66mm hole as shown. *Note:* The hole is vertical to, and 44mm from, the center point.

- Use the **Circular Pattern** tool to pattern the hole as shown, with the following information:

  Rotation Axis: Z Axis

  Count: 8

  Position Method: Fitted

  Angle: 360°

- Copy **Extrusion1** to the top face of the revolved feature by dragging **Extrusion1** from the Browser into the graphics window. To paste, select the top face of the revolved feature and specify a rotation angle of 0°. When ready, pick the finish button.

- Pattern a hole on **Extrusion2** as shown, using the same techniques as before. *Note:* This time, select **Extrusion2** as the rotation axis.

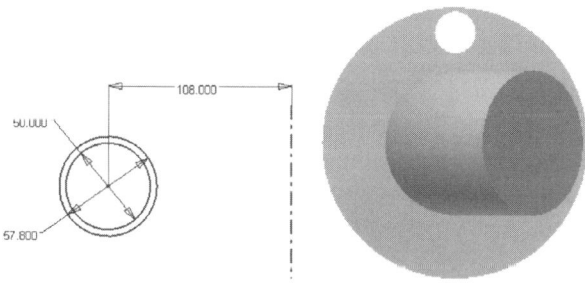

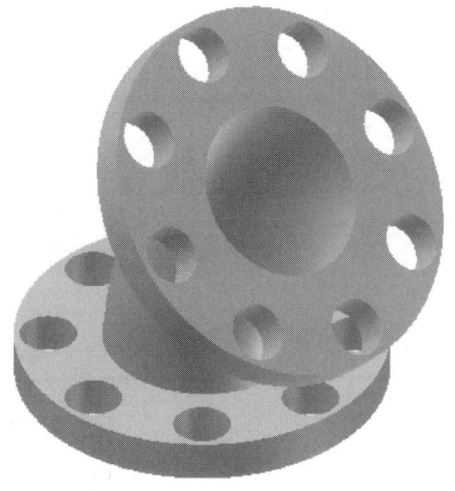

- Share **Sketch1.**
- Cut revolve the ⌀50mm circle.
- The final project should look like the part shown.
- Save As: P6.9.

||||||||||||||||||||||||||||||||||||||||||||||||||||||||||||||||||||||||||||||||||||||||||||||

CHAPTER 7

Applying Work Features

LEARNING GOALS

After completing this chapter, you will be able to:

◎   Apply and use work features for part modeling.

◎   Access and make use of work feature tools.

◎   Create work planes.

◎   Establish work axes.

◎   Use work points.

Work Features

Often while developing a part model, you may not have sufficient pieces of geometry to sketch on, dimension and constrain to, and build features on. As a result, Autodesk Inventor provides you with work features. *Work features* are similar to construction lines and allow you to create reference elements anywhere in space to help you position and generate features. This ensures that you can create complex parametric geometry that is connected to other features of the part.

     You are already familiar with the default reference work features located in the **Origin** folder of the **Browser** bar. Work features function the same as the default work planes, axes, and **Center Point.** However, work features allow you to define the location of planes, axes, and points. See Figure 7.1.

FIGURE 7.1
Examples of work planes, work axes, and work points.

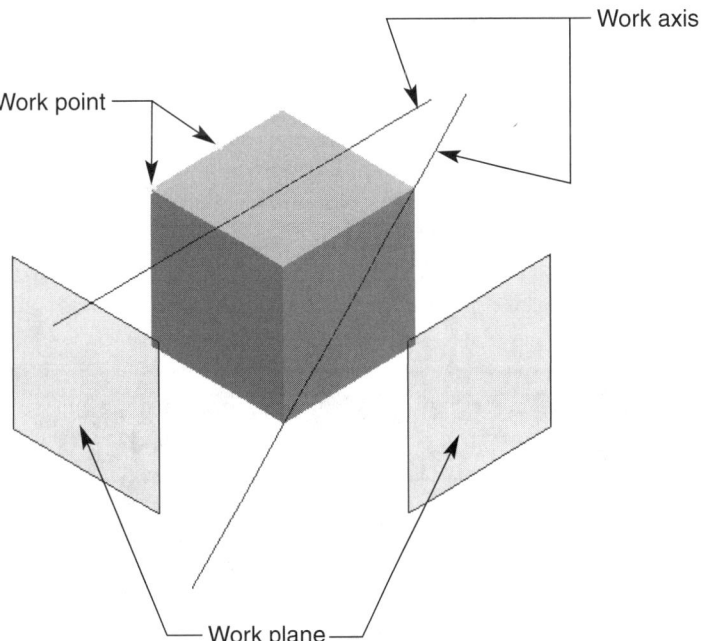

Work axis

Work point

Work plane

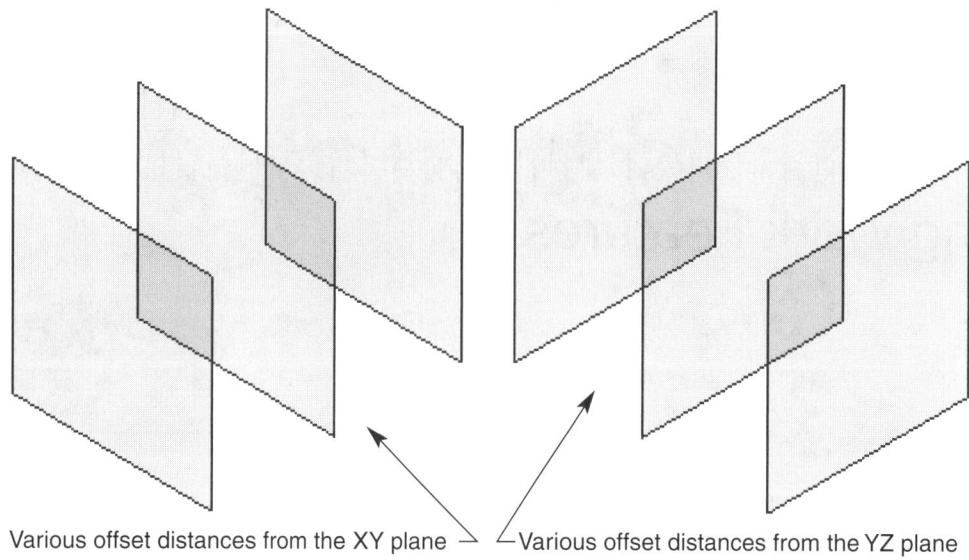

Various offset distances from the XY plane   Various offset distances from the YZ plane

FIGURE 7.2   Creating work features by referencing existing feature geometry.

Work features can be created anytime during part model development. However, most work features, especially axes and points, are added when at least one feature currently exists. The only way to create work features without existing geometry is to reference the default work features from the **Origin** folder of the **Browser** bar. Figure 7.2 shows how several work planes can be generated by referencing the default work planes.

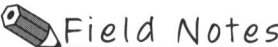 Field Notes

All the interface and work environment options and applications discussed in Chapter 4 apply to working with work features.

Using Work Feature Tools

Work feature tools are used to create work planes, work axes, and work points. These tools are accessed from the **Features** toolbar, shown in Figure 7.3, or from the **Features** panel bar. When a feature is present and you are in the feature work environment, the panel bar is in features mode. See Figure 7.4. By default, the panel bar is displayed in learning mode, as seen in Figure 7.4. To change from learning to expert mode, seen in Figure 7.5 and discussed in Chapter 2, right-click any of the panel bar tools or left-click the panel bar title and select the **Expert** option.

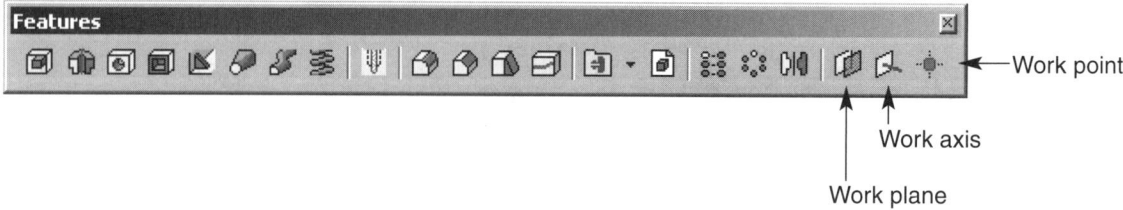

Work point

Work axis

Work plane

FIGURE 7.3   The **Features** toolbar

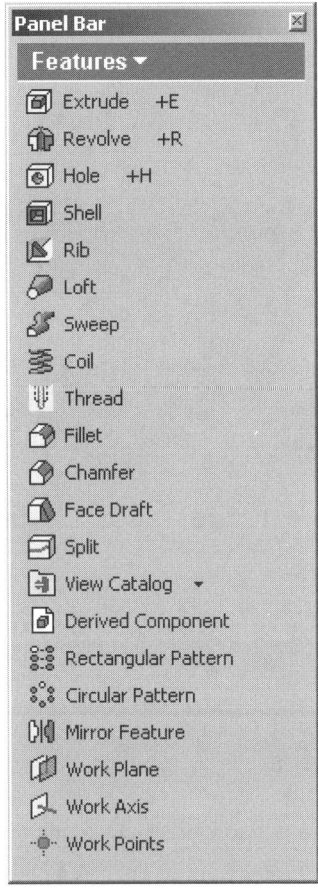

FIGURE 7.4 The panel bar in features (learning) mode.

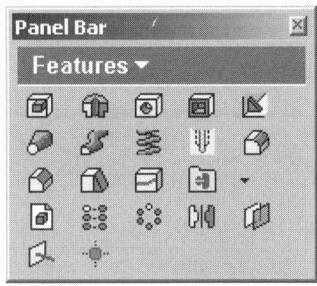

FIGURE 7.5 The panel bar in features (expert) mode.

Using Work Planes

A *work plane* is a flat reference surface that can be located anywhere in space. Though you may only see part of a work plane in the graphics window, work planes are infinite. This means that they extend, or cover, the entire plane of the location where they are placed. As you work with Autodesk Inventor and continue to develop parts and assemblies, you will find many uses for work planes. Generally, you may want to use work planes for the following applications:

- To create sketches, when no default reference planes or feature faces are available.

- To define another work feature.

- To add a termination plane for creating features such as extrusions.

- To create constraint planes for assemblies.

Field Notes

Many work plane uses are explored in the projects at the end of the chapter.

Work planes appear semitransparent in the graphics window. See Figure 7.6. If at any time you want to turn off the visibility, or hide, a work plane, right-click the work plane in the **Browser** and unselect the

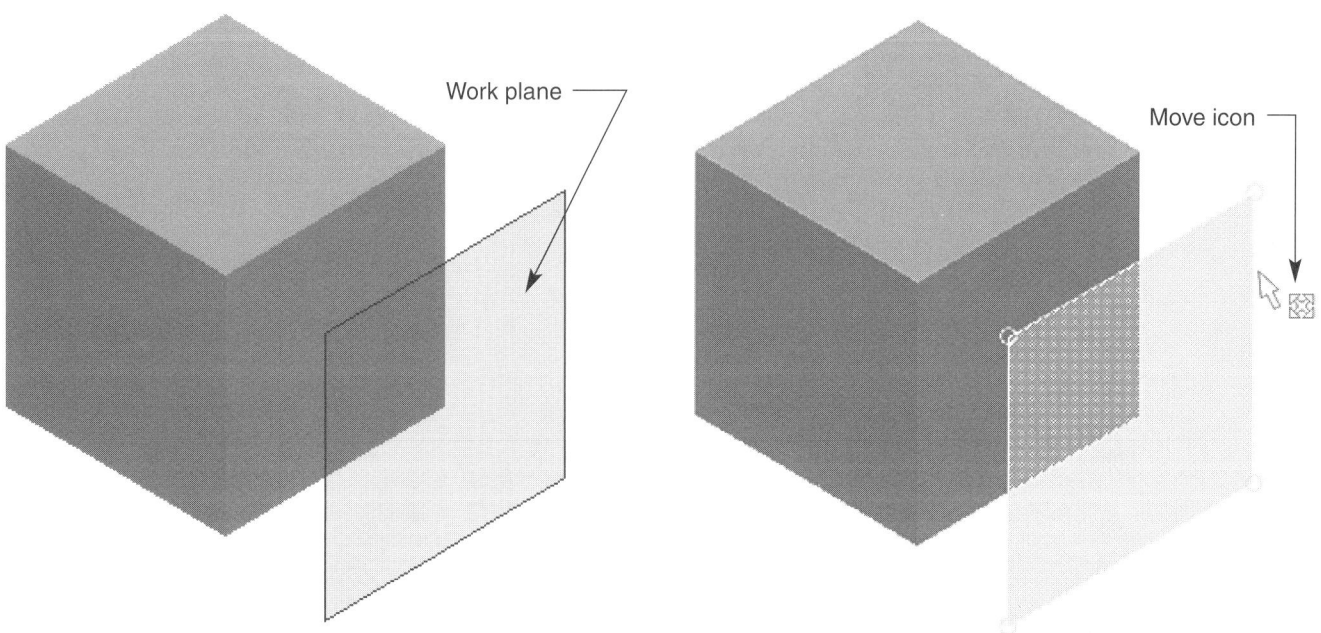

FIGURE 7.6 A visible work plane.   **FIGURE 7.7** Moving a work plane.

Visibility menu option, or pick the work plane in the graphics window, right-click, and unselect the **Visibility** menu option. You can also move and manipulate the displayed size of a work plane. To move a work plane, move your cursor over the edge until you see the **Move** icon. See Figure 7.7. Then hold down the left mouse button and drag the work plane to the desired location. When you finish moving the work plane, release the mouse button. To change the displayed size of the work plane, move your cursor over one of the work plane corners until you see the **Resize** icon. See Figure 7.8. Then hold down the left mouse button and drag the corner of the work plane to the desired location and size. When you finish resizing the work plane, release the mouse button.

FIGURE 7.8 Resizing
a work plane.

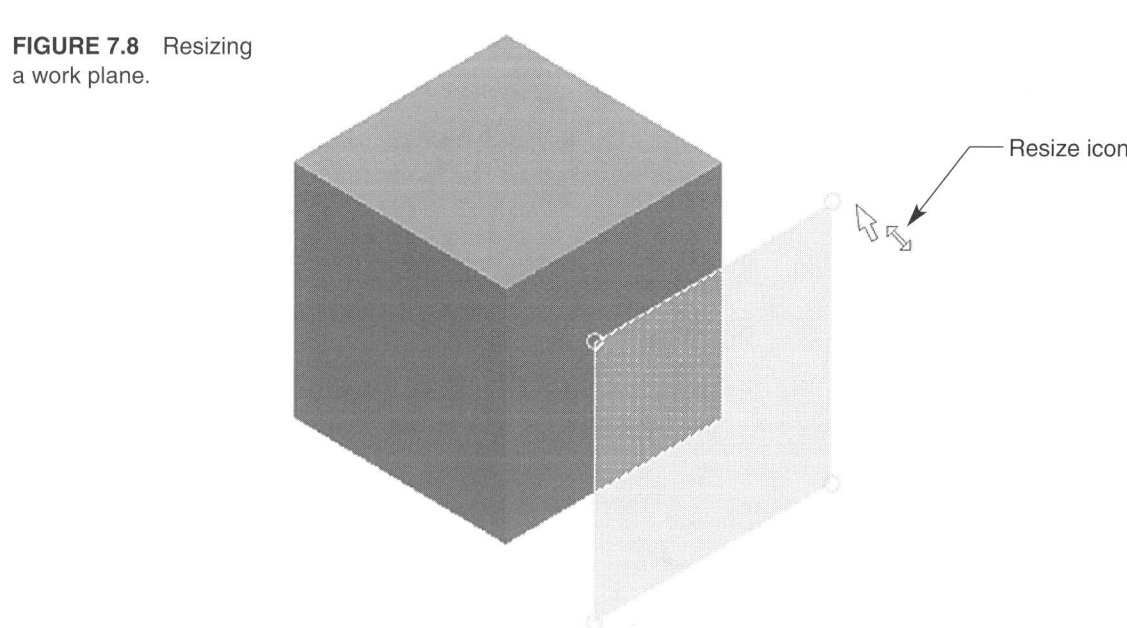

Field Notes

Moving or resizing the work plane display does not change the specified location, constraints, or area of the work feature. To physically relocate the work plane in space, you must right-click the work plane in the **Browser** and select the **Redefine Feature** menu option, or pick the work plane in the graphics window, right-click, and select the **Redefine Feature** menu option. You can then change the work plane extents. Once you are finished using a work feature, right-click the feature inside the **Browser** and unselect the **Visibility** menu option. This process hides the work feature from view.

Work planes are created using the **Work Plane** tool. Access the **Work Plane** command using one of the following techniques:

✓ Pick the **Work Plane** button on the panel bar.

✓ Pick the **Work Plane** button on the **Features** toolbar.

Unlike other features, when you access the **Work Plane** command or any other work feature, no dialog box opens. Instead, you define where you want to place the work plane by selecting existing or reference geometry. Once you have accessed the **Work Plane** tool, use one of the following techniques to create a work plane:

■ Using default reference work features   Use one of the following options to create a work plane using the default reference work features located in the **Origin** folder of the **Browser:**

    ■ Select the **YZ Plane, XZ plane,** or **XY Plane.** Selecting one of the default planes displays a preview of the plane and the drag icon. See Figure 7.9. To locate the plane, hold down the left mouse button and drag the work plane to the desired location. While you are dragging the work plane, the **Offset** dialog box appears. See Figure 7.10. Use the **Offset** dialog box to help you place the work plane. Once you release the mouse button, you can enter a value in the edit box. Use a value of 0 to create the work plane at its current location.

    ■ Select two available axes. Selecting any pair of default reference work axis creates a work plane. For example, if you select the **X Axis** and the **Y Axis,** an XY work plane is created.

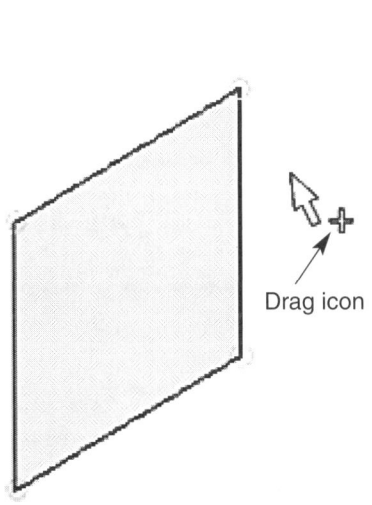

Drag icon

FIGURE 7.9 Offsetting a work plane.

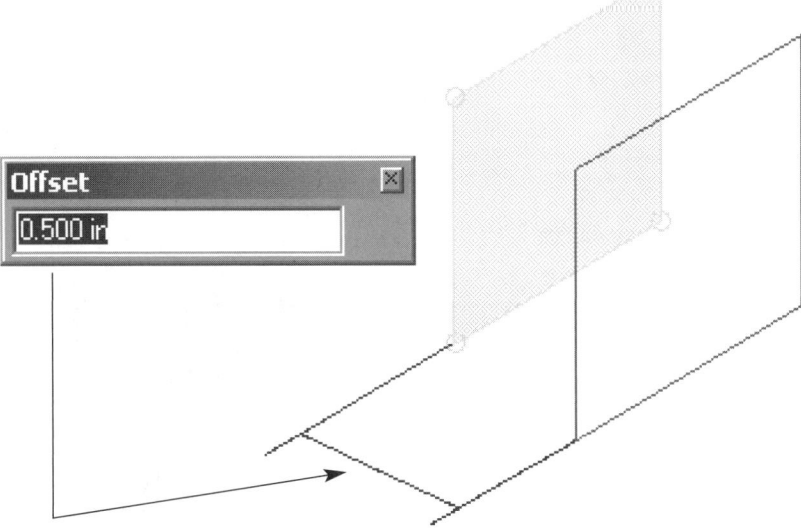

FIGURE 7.10 The **Offset** dialog box.

You can accomplish the same task by simply referencing the one of the default work planes, such as the **XY Plane.**

■ Select the **Center Point** and one of the available axes or planes. Selecting the **Center Point** and a default work axis or plane creates a work plane. For example, if you select **Center Point** and the **X Axis,** an YZ plane is created. Again, you can accomplish the same task by simply referencing one of the default work planes, such as the **YZ Plane.**

Field Notes

Use the offsetting techniques discussed to offset a new plane from an existing plane.

■ From an existing feature face   Use this option to define a work plane parallel to, and a specified distance from, a face. To create a work plane using an existing feature face, pick the face you want to offset the work plane from. Then offset the work plane using the offsetting technique previously discussed. See Figure 7.11.

■ From two feature edges or axes   This option generates a work plane coplanar to two edges or axes. To create a work plane by referencing two edges, pick the edges or axes where you want the work plane to be located. See Figure 7.12.

■ Using 3 existing points   This option, often referred to as the 3-point work plane, generates a work plane by referencing any three points, including feature edge endpoints, intersections, midpoints, and work points. To create a 3-point work plane, pick three different points. See Figure 7.13.

■ From an edge or axis and a point   Use this option to define a work plane perpendicular to the selected edge, through a specified point. To create a work plane by referencing an edge or axis and a point, pick the edge or axis, and then pick any point including feature edge endpoints, intersections, midpoints, center points, and work points. Or select the point, followed by the edge or axis. See Figure 7.14.

FIGURE 7.11
Creating a work plane
at a specified location
by referencing an
existing feature face.

Selected face

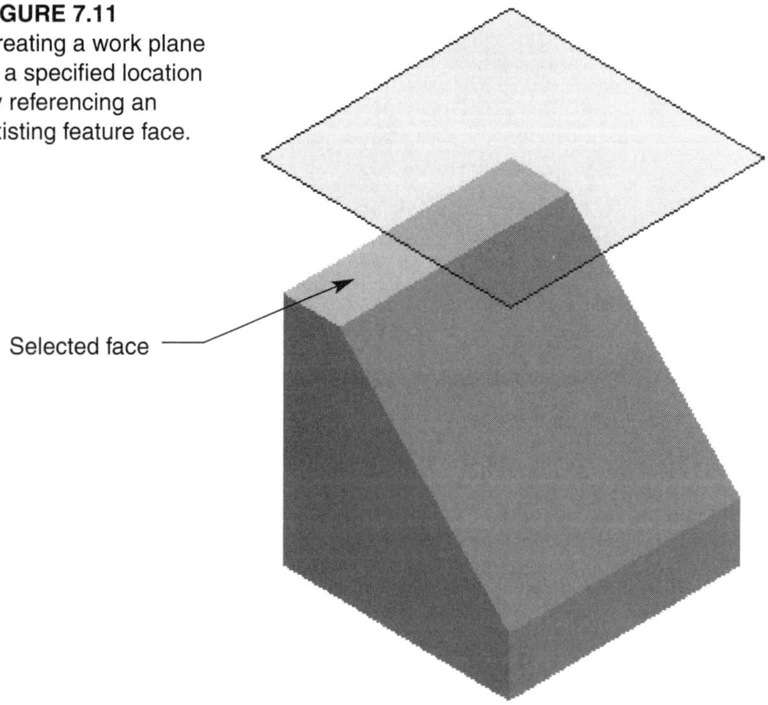

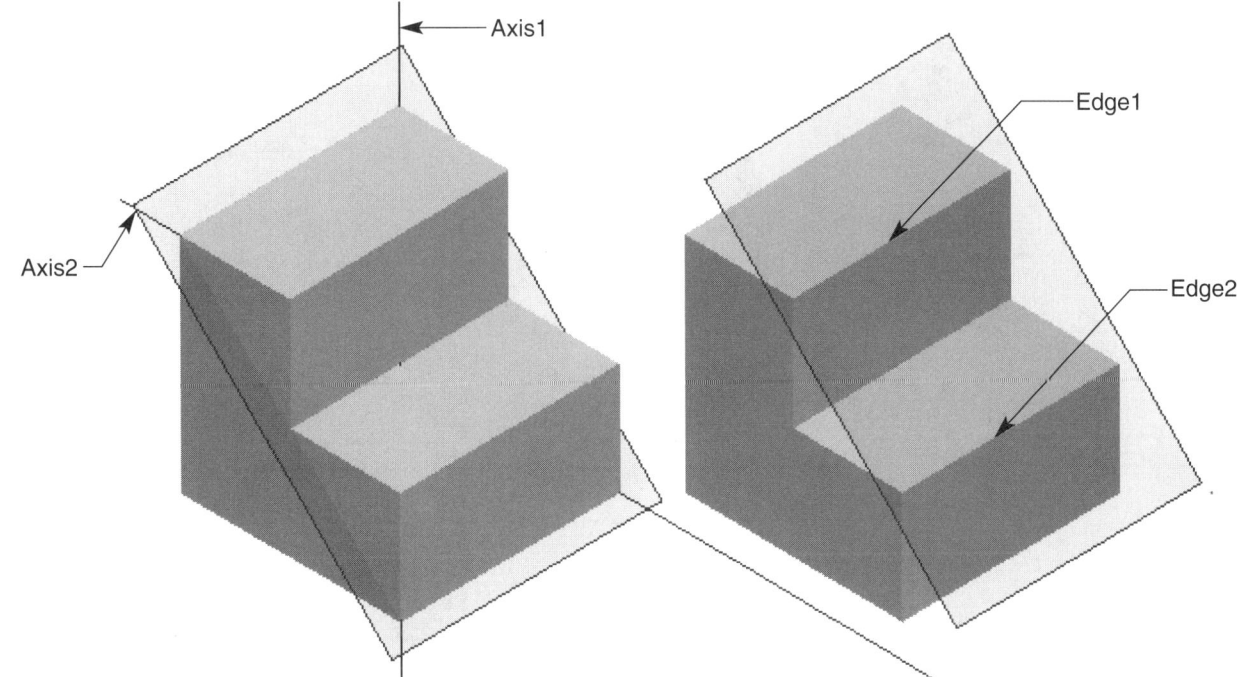

FIGURE 7.12 Creating a work plane coplanar to two edges or axes.

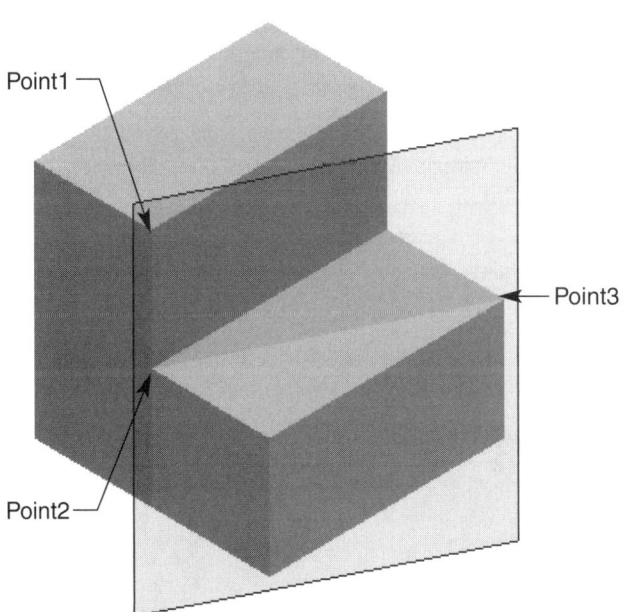

FIGURE 7.13 Creating a work plane by selecting three points.

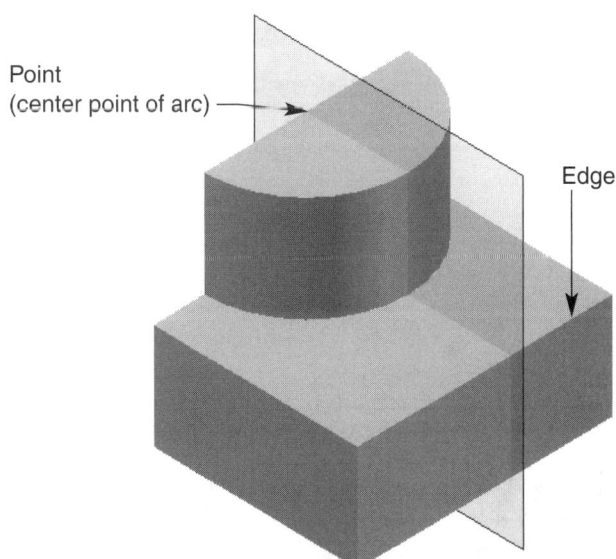

FIGURE 7.14 Creating a work plane by selecting an edge or axis and a point.

FIGURE 7.15
Creating a work plane
by selecting a face and
a point.

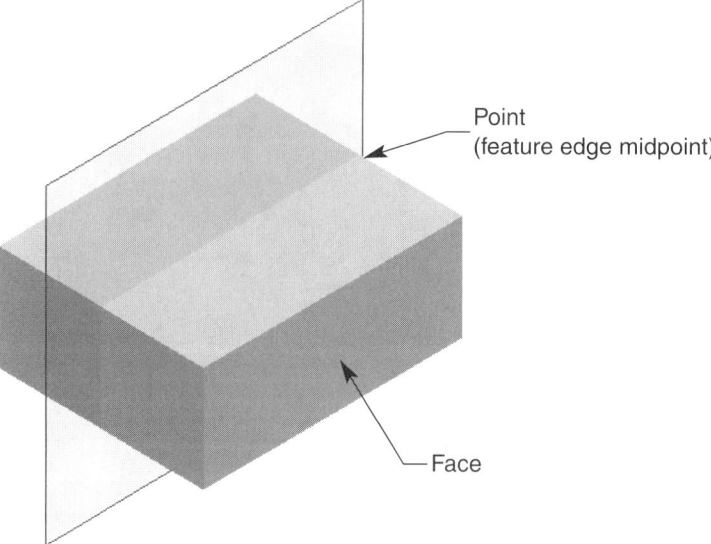

Point
(feature edge midpoint)

Face

- From a face and a point   This option allows you to define a work plane parallel to the selected face through a specified point. To create a work plane by referencing a face and a point, pick the face and then pick any point including feature edge endpoints, intersections, midpoints, center points, and work points. Or select the point, followed by the face. See Figure 7.15.

- Using a tangent face and an edge or axis   This option defines a work plane tangent to the selected curved feature, such as a cylinder or fillet, and coplanar to the selected edge or axis. To create a work plane by referencing a tangent face and an edge or axis, pick the face and then pick the edge or axis. Or select the edge or axis, followed by the face. See Figure 7.16.

- Using a tangent face and a planar face   This option defines a work plane tangent to the selected curved feature, such as a cylinder or fillet, and parallel to a different face. To create a work plane by referencing a tangent face and a planar face, pick the tangent face and then pick the other face. Or select the planar face, followed by the tangent face. See Figure 7.17.

- Angled from a planar edge or axis and a face   Use this option to define a work plane at an angle to the selected face, through a specified edge or axis. To create a work plane by referencing a face and an edge or axis, pick the face and then pick an edge or axis. Or select the

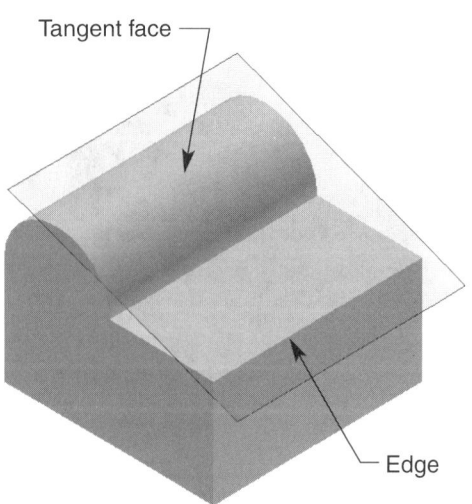

Tangent face

Edge

FIGURE 7.16   Creating a work plane by
selecting a tangent face and an edge or axis.

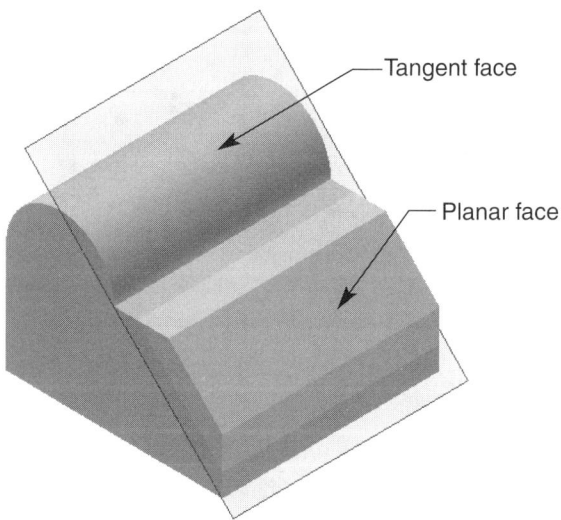

Tangent face

Planar face

FIGURE 7.17   Creating a work plane by selecting a
tangent face and a planar face.

FIGURE 7.18
Creating a work plane
by selecting a face and
an edge or axis.

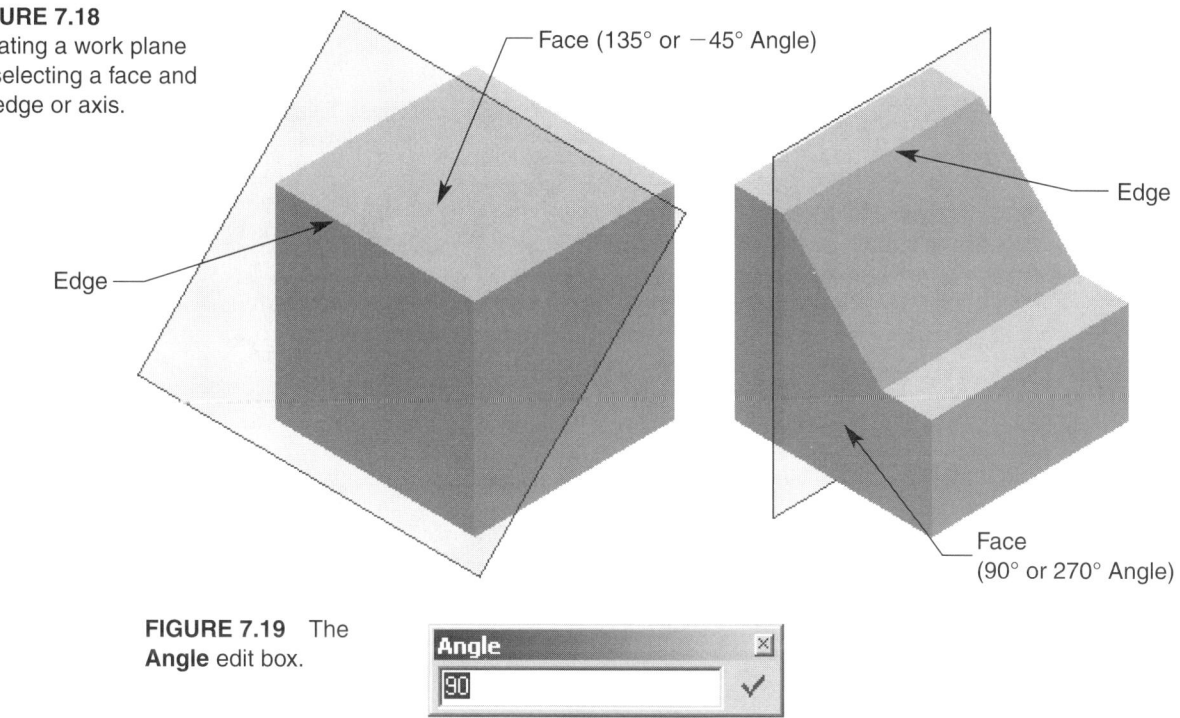

FIGURE 7.19   The
Angle edit box.

edge or axis, followed by the face. See Figure 7.18. Once you have picked an edge or axis and a face, the **Angle** edit box will appear. See Figure 7.19. The **Angle** edit box allows you to specify the angle of work plane. The selected edge acts as the angle pivot point.

 Field Notes

The **Angle** edit box defines the angle of the work plane. This means that you can select any face parallel to the selected edge and create the same work plane by entering different angles.

■ Tangent to a cylinder   This option defines a work plane tangent to the selected curved feature, such as a cylinder or fillet. To create a work plane by referencing a tangent face, another feature or sketch must be available to select. If the cylinder was created by referencing the **Center Point,** and you want the work plane to be parallel to a default work plane, you can select the tangent feature, and one of the default reference work planes. See Figure 7.20. If you want to create a work plane tangent to a cylinder, but cannot reference one of the default work planes, or you do not want the work plane to be parallel to one of the default work planes, you must access alternative geometry. One option is to create a sketch, using construction lines, on one of the cylinder faces, as shown in Figure 7.21. Once you have sketched the line or lines, pick the tangent point on the cylinder edge and then select the sketched line, or pick the sketched line followed by the tangent point of the cylinder edge. See Figure 7.22.

 Field Notes

The construction line endpoints must be coincident to the tangent edge and axis of the cylinder. To fully constrain the sketch, generate lines from the center point of the cylinder, and apply any needed dimensions or geometric constraints.

FIGURE 7.20
Creating a work plane
tangent to a cylinder.

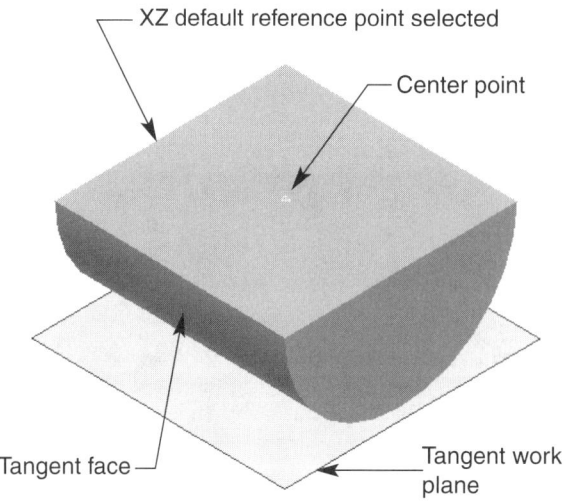

XZ default reference point selected

Center point

Tangent face

Tangent work
plane

(Graphics sliced for clarity)

FIGURE 7.21
Sketching constructions
for reference geometry.

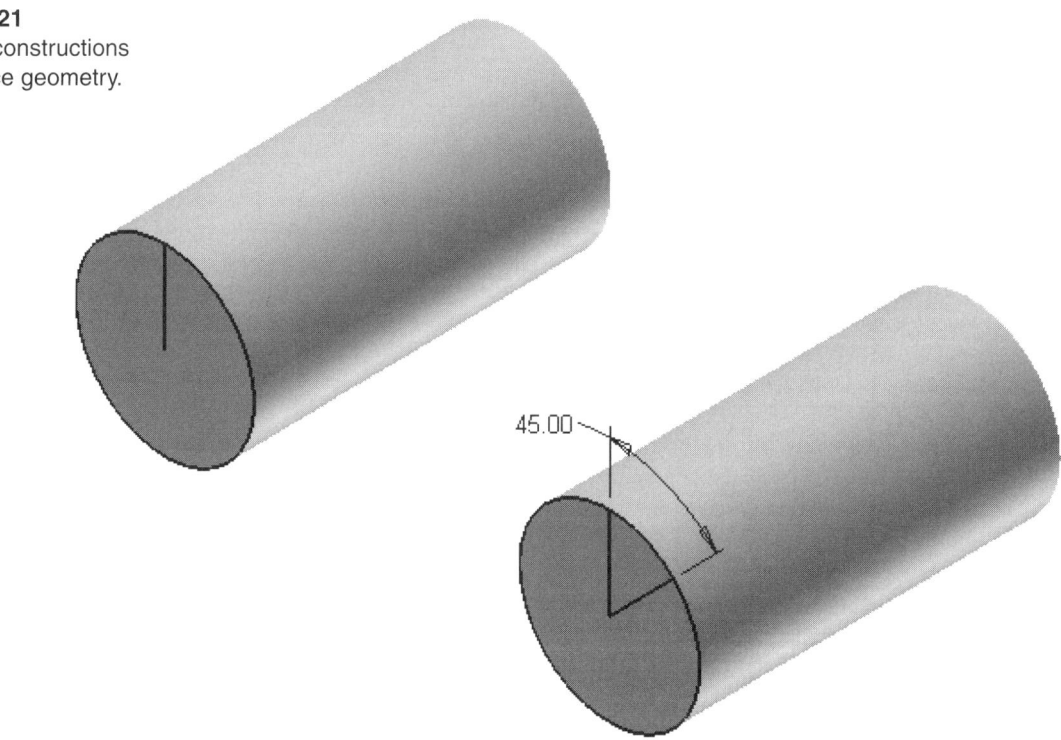

45.00

 Field Notes

When finished, you may want to hide the sketch geometry by right-clicking the sketch in the
Browser, and unselect the **Visibility** menu option.

Once you have added any of the work planes previously discussed, you can offset another
work plane. To accomplish this task, access the **Work Plane** tool, pick the existing work plane,
and apply the offsetting technique discussed.

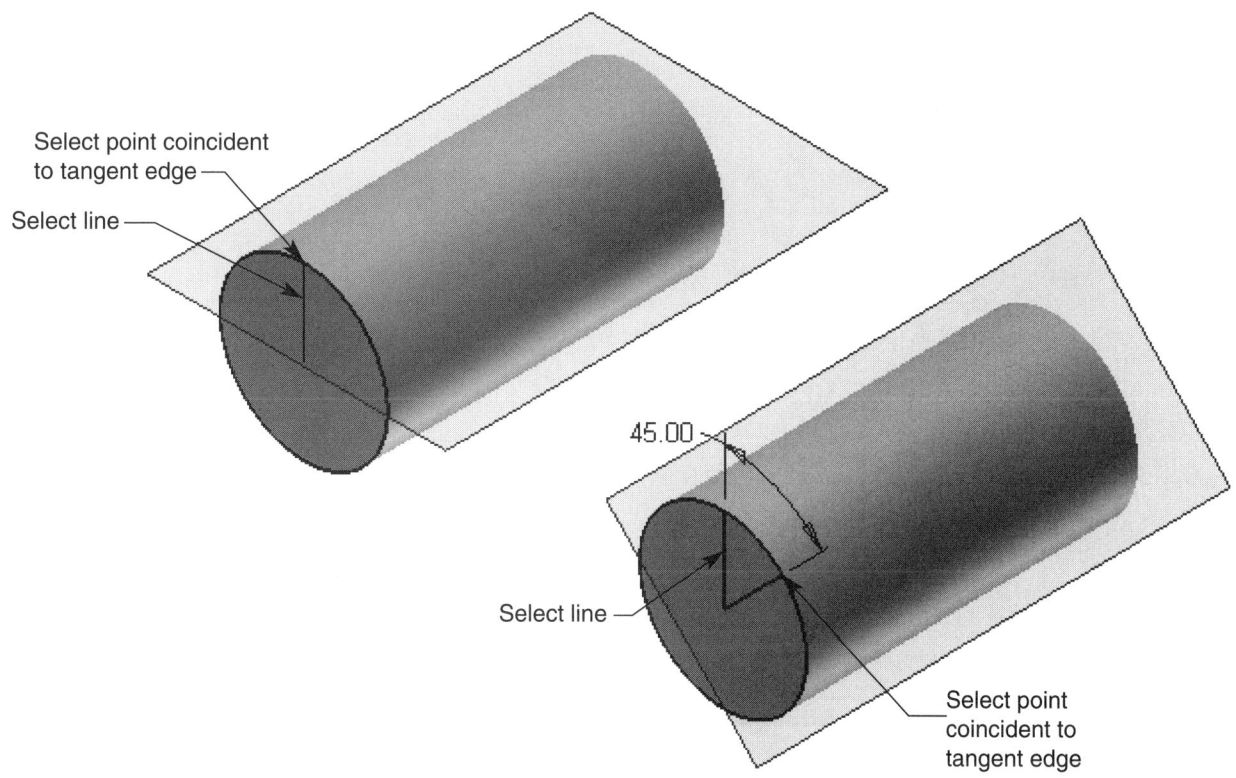

FIGURE 7.22 Creating a work plane tangent to a cylinder.

Exercise 7.1

1. Launch Autodesk Inventor.
2. Open a new part file.
3. Open a sketch on the **XY Plane,** and sketch the geometry shown.
4. Extrude the sketch, 2″ in both directions.
5. Add the work planes as shown, using the techniques discussed. *Note:* You may need to hide some work features for clarity.
6. Save the part as EX7.1.
7. Exit Autodesk Inventor or continue working with the program.

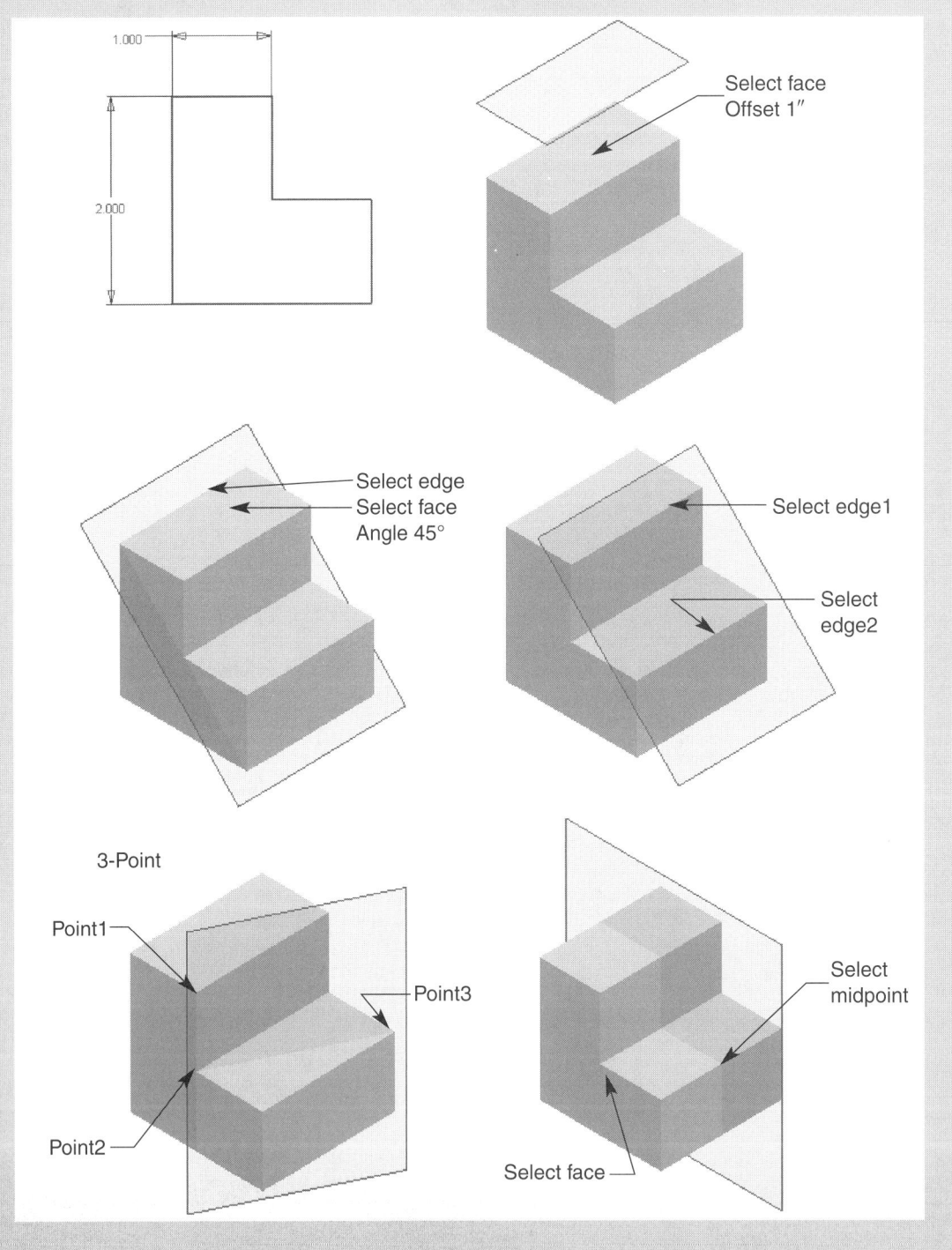

Exercise 7.2

1. Continue from Exercise 7.2, or launch Autodesk Inventor.
2. Open a new part file.
3. Open a sketch on the **XZ Plane,** and sketch the geometry shown.
4. Extrude the sketch 3″ both sides.
5. Fillet and chamfer the figures as shown, using the specified dimensions.
6. Add the work planes as shown, using the techniques discussed. *Note:* You may need to hide some work features for clarity.
7. Save the part as EX7.2.
8. Exit Autodesk Inventor or continue working with the program.

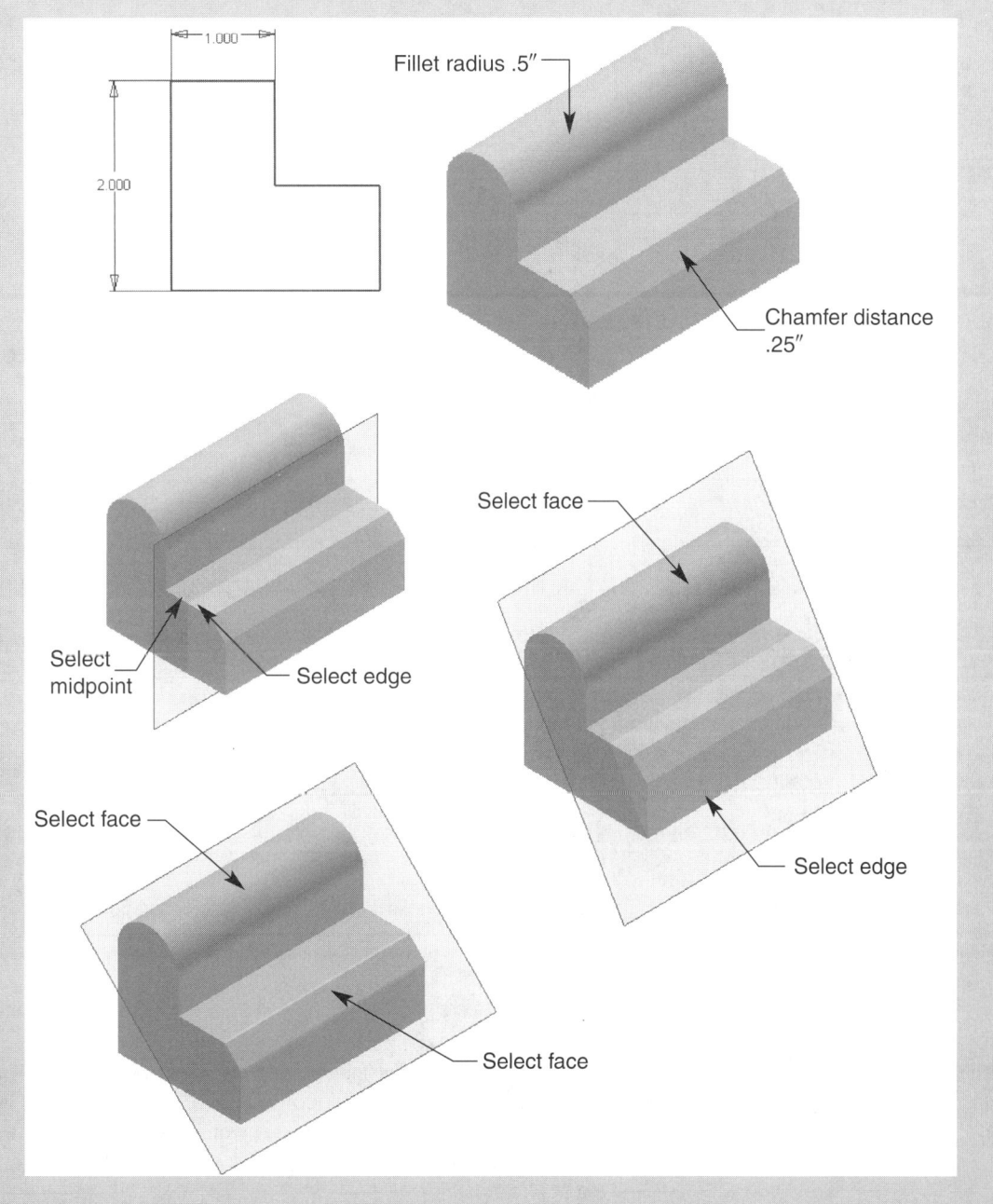

Exercise 7.3

1. Continue from Exercise 7.2, or launch Autodesk Inventor.
2. Open a new part file.
3. Open a sketch on the **XZ Plane,** and sketch a 2″ circle around the projected **Center Point.**
4. Extrude the sketch 4″ both sides.
5. Create work plane1, as shown, by picking the cylinder face, followed by **XY Plane,** from the **Origin** folder of the **Browser.**
6. Sketch the line as shown.
7. Create work plane2, as shown, by picking the sketched line, followed by the endpoint coincident to the tangent edge.
8. Save the part as EX7.3.
9. Exit Autodesk Inventor or continue working with the program.

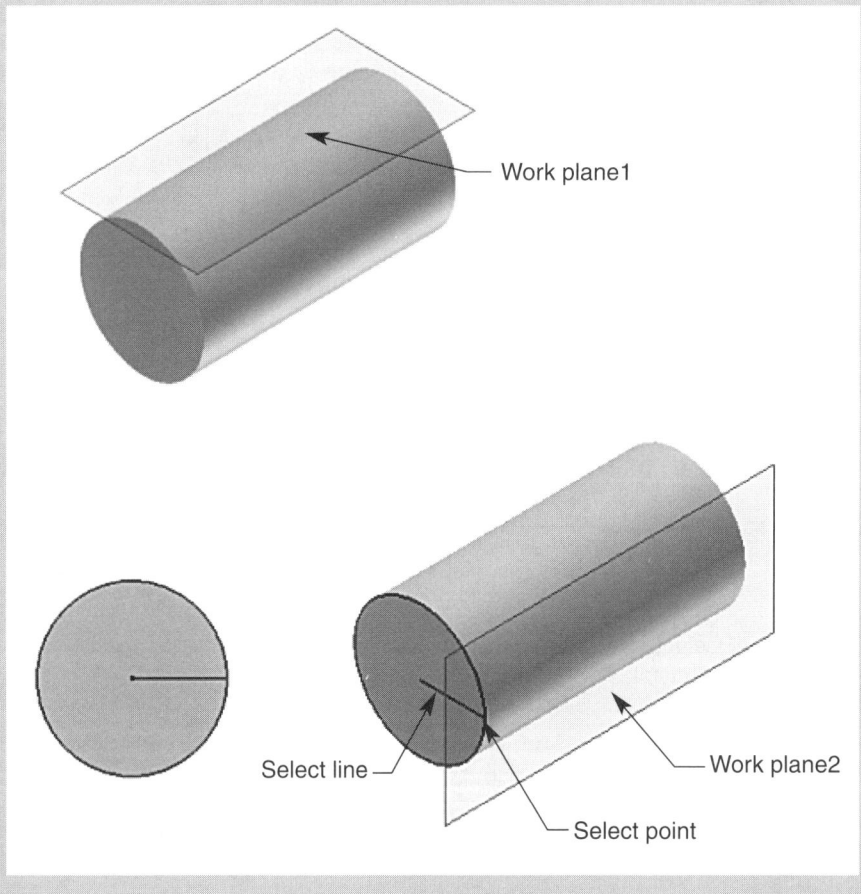

Work plane1

Select line

Work plane2

Select point

Using Work Axes

A *work axis* is a parametric reference line that can be located anywhere in space. Like work planes, you may see only part of a work axis in the graphics window. However, work axes are infinite and extend the entire length of the location where they are placed. As you work with Autodesk Inventor and continue to develop parts and assemblies, you will find numerous uses for work axes. Generally, you may want to use work axes for the following applications:

■ To create work planes, when no default reference planes, feature edges, or points are available.

■ To create an axis for revolved and coiled features.

■ To create a direction reference line for rectangular patterns, and a rotation axis for circular patterns.

■ To create constraint axes for assemblies.

Field Notes

Many uses of the work axes are explored in the projects at the end of the chapter.

Work axes are displayed as thin lines in the graphics window. They extend past referenced features, and may be partially hidden. See Figure 7.23. If at any time you want to turn off the visibility of, or hide, a work axis, right-click the work axis in the **Browser** and unselect the **Visibility** menu option, or pick the work axis in the graphics window, right-click, and unselect the **Visibility** menu option.

Work axes are created using the **Work Axis** tool. Access the **Work Axis** command using one of the following techniques:

✓ Pick the **Work Axis** button on the panel bar.

✓ Pick the **Work Axis** button on the **Features** toolbar.

Unlike other features when you access the **Work Axis** command or any other work feature, no dialog box opens. Instead, you define where you want to place the work axis by selecting existing or reference geometry. Once you have accessed the **Work Axis** tool, use one of the following techniques to create a work axis:

■ Using default reference work features   Though you can select the default work features located in the **Origin** folder of the **Browser** to create a work axis, you can accomplish the same task by simply referencing one of the default work axes. For example, if you pick the **Center Point** and the **YZ Plane,** an axis identical to the default **X Axis** is created.

■ From two existing points   Use this option to generate a work plane by referencing any two points, including feature edge endpoints, intersections, midpoints, and work points. To create a work axis by referencing points, pick two different points. See Figure 7.24.

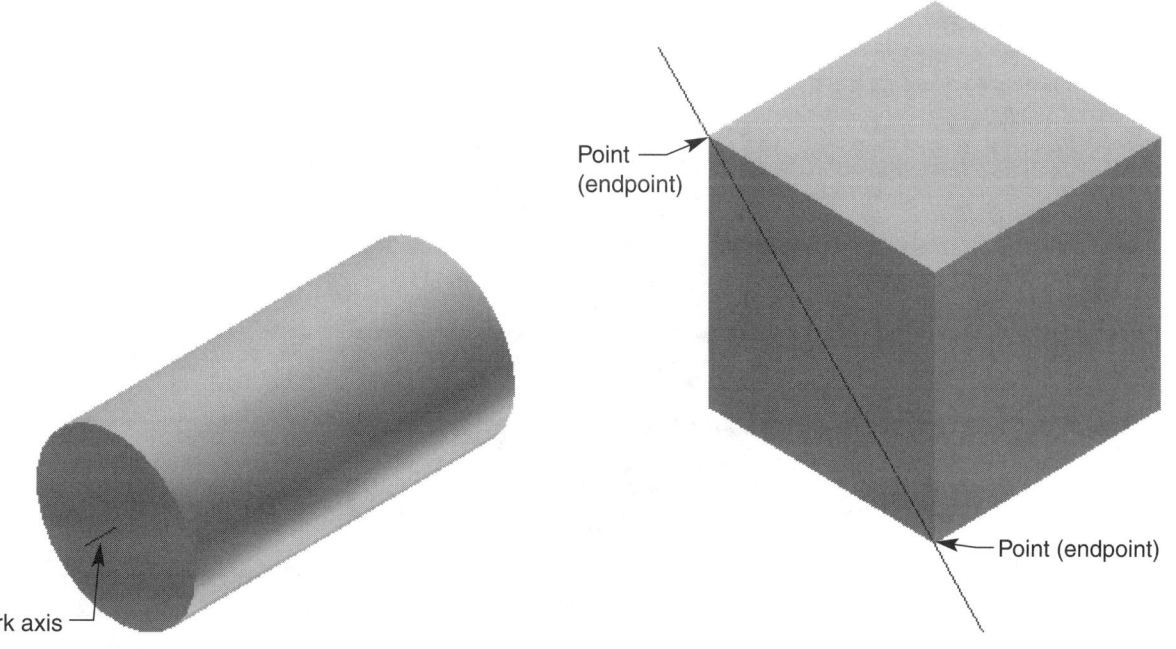

FIGURE 7.23 A visible work axis.   **FIGURE 7.24** Creating a work axis by selecting two points.

- Using two intersecting work planes or faces   This option generates a work axis coplanar to, and at the intersection of, two work planes or faces. To create a work axis by referencing two work planes or faces, pick the work planes or faces where you want the work axis to be located. See Figure 7.25.

- Using a revolved or curved feature   This option places a work axis through the middle of a revolved or curved feature, such as an extruded circle or a fillet, and is accomplished by picking the feature. See Figure 7.26.

- From a feature edge   Use this option to place a work axis collinear to the selected edge. To create a work axis by referencing an edge, pick the edge where you want the work axis to be located. See Figure 7.27.

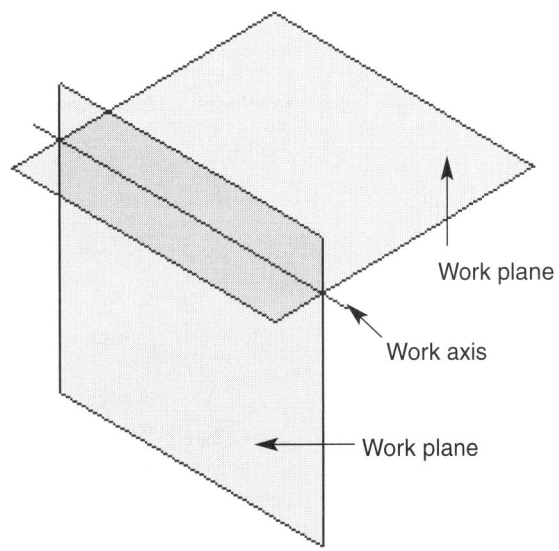

FIGURE 7.25   Creating a work axis at the intersection of two work planes or feature faces.

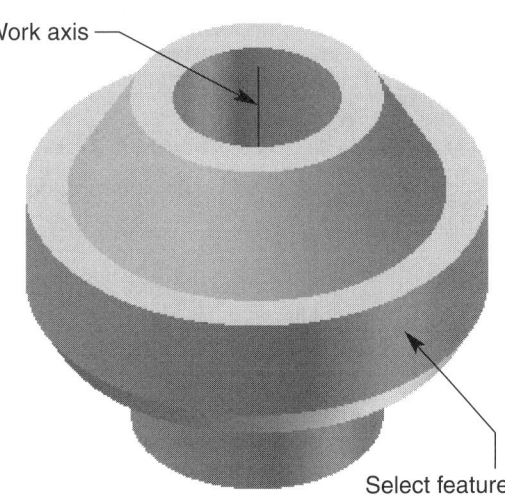

FIGURE 7.26   Creating a work axis by picking a revolved or cylindrical feature.

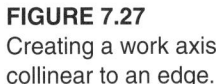

FIGURE 7.27
Creating a work axis collinear to an edge.

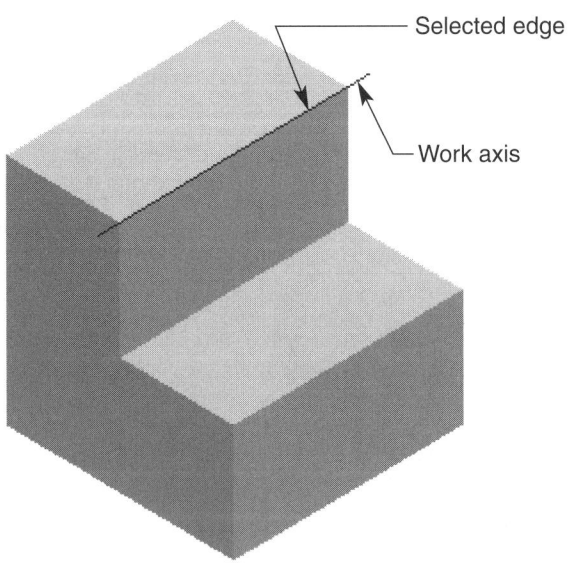

■ From a face or plane and a point   This option allows you to define a work axis perpendicular to the selected face or work plane, through a specified point. To create a work plane by referencing a face or plane and a point, pick the face or plane, and then pick any point including feature edge endpoints, intersections, midpoints, center points, and work points. Or select the point, followed by the face. See Figure 7.28.

■ Using a planar edge and a face   This option allows you to define a work axis at an edge of the selected face, parallel to a specified edge. To create a work axis by referencing a face and an edge, pick the face and then pick an edge. Or select the edge, followed by the face. See Figure 7.29.

FIGURE 7.28
Creating a work axis by selecting a face or plane, and a point.

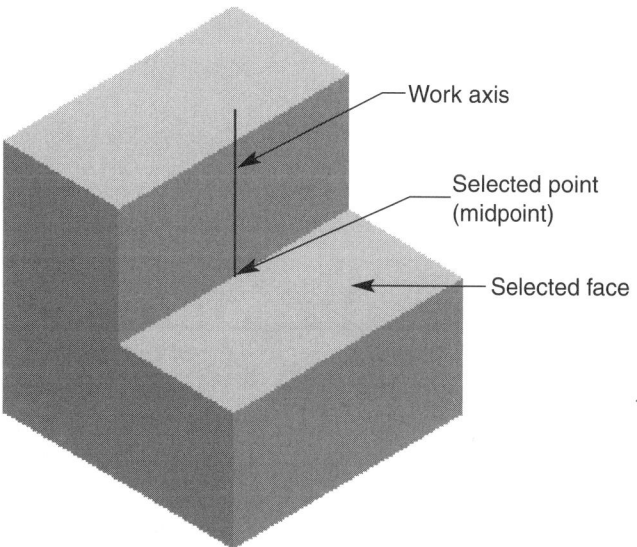

Work axis

Selected point (midpoint)

Selected face

FIGURE 7.29
Creating a work axis by selecting a face and an edge.

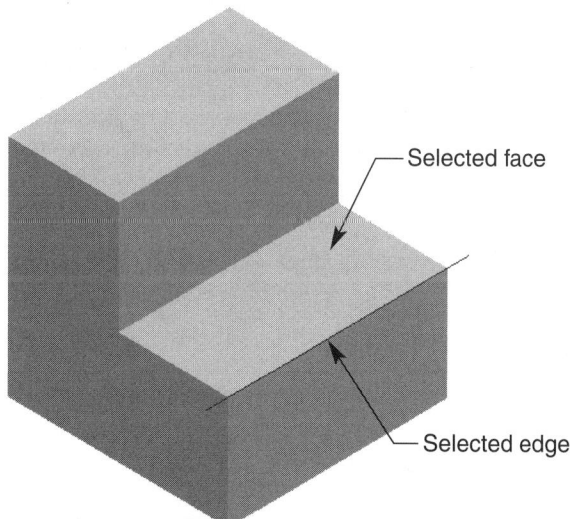

Selected face

Selected edge

Exercise 7.4

1. Continue from Exercise 7.3, or launch Autodesk Inventor.
2. Open a new part file.
3. Open a sketch on the **XZ Plane,** and sketch the geometry shown.
4. Fully revolve the sketch around the **Z Axis.**
5. Add the sketch shown, and extrude 15mm.
6. Add the work axes as shown, using the techniques discussed. *Note:* You may need to hide some work features for clarity.
7. Save the part as EX7.4.
8. Exit Autodesk Inventor or continue working with the program.

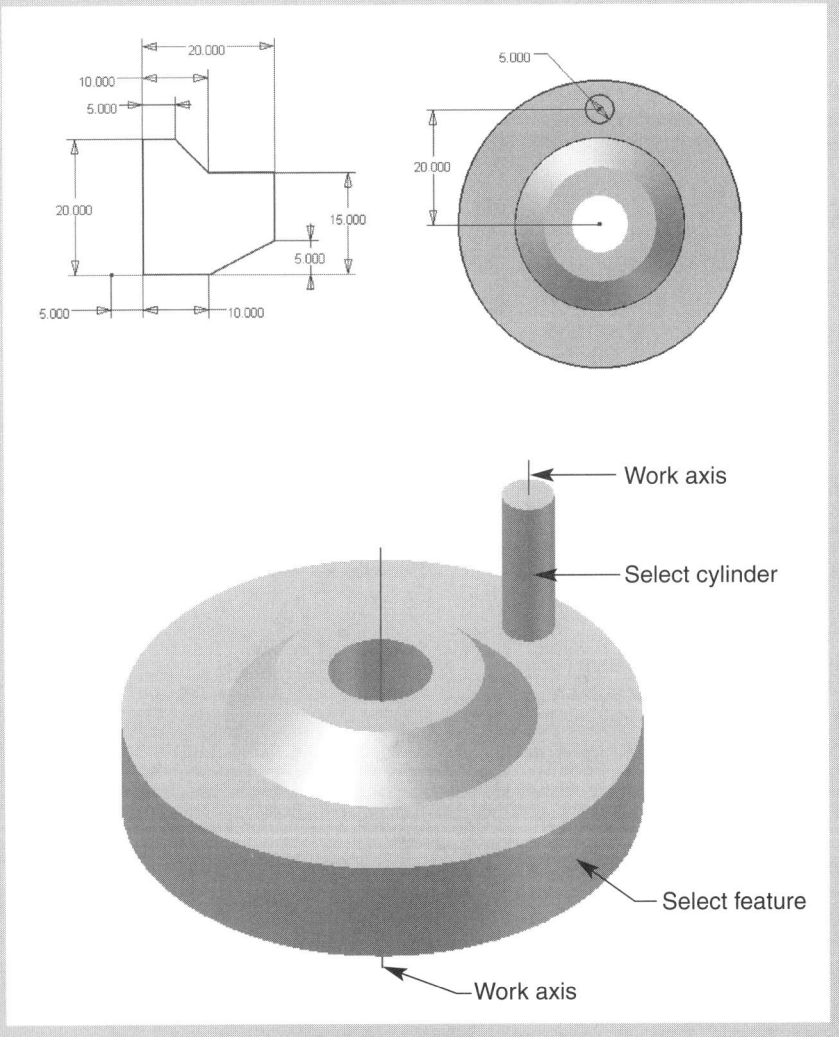

Exercise 7.5

1. Continue from Exercise 7.4, or launch Autodesk Inventor.
2. Open a new part file.
3. Open a sketch on the **XZ Plane,** and sketch the geometry shown.
4. Extrude the sketch 20mm in both directions.
5. Offset a work plane 15mm from the back face, and one 10mm from the bottom face as shown.
6. Add the work axes as shown, using the techniques discussed. *Note:* You may need to hide some work features for clarity.
7. Save the part as EX7.5.
8. Exit Autodesk Inventor or continue working with the program.

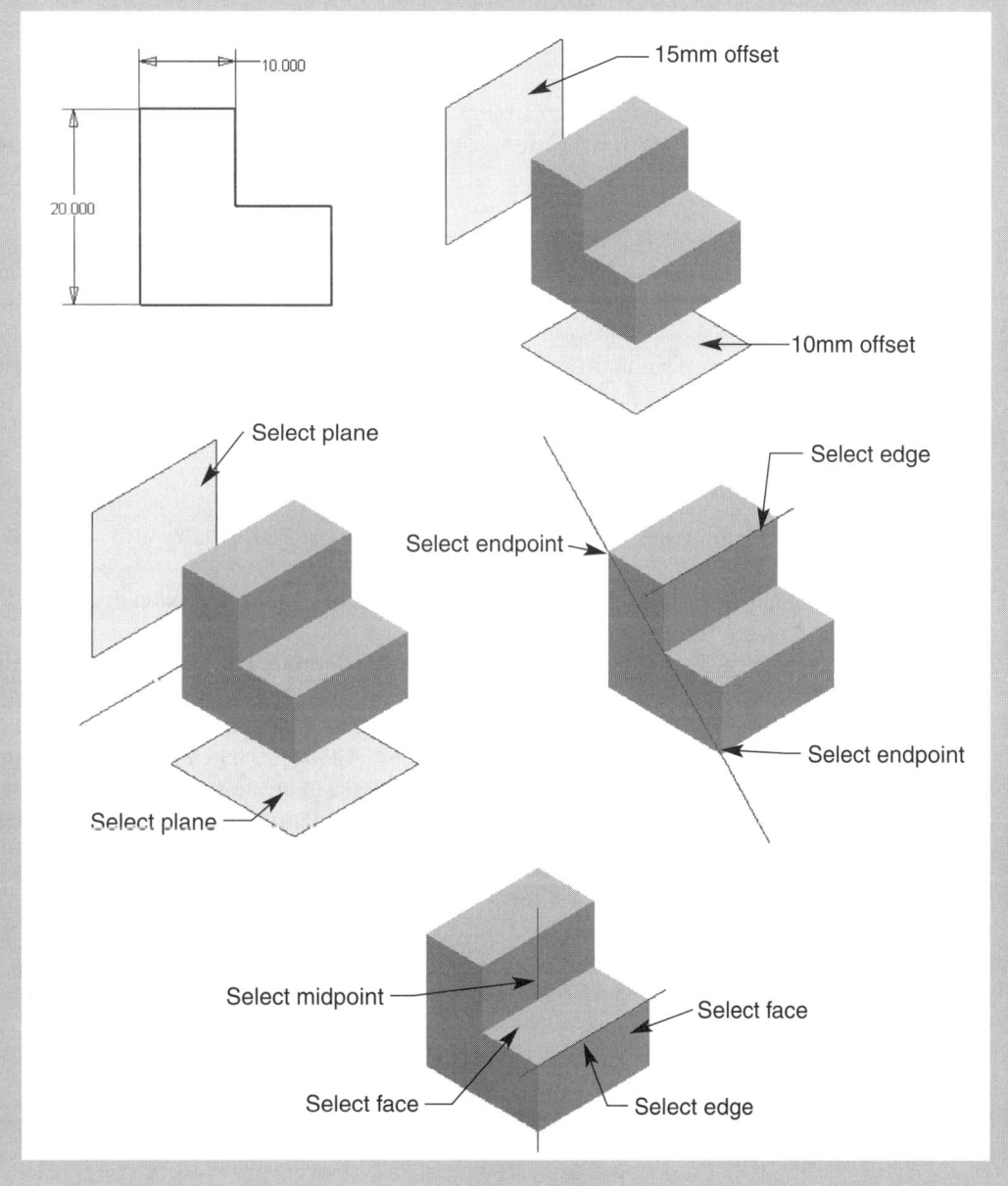

Using Work Points

A *work point* is a parametric reference point that can be located on any part feature or in three-dimensional space. As you work with Autodesk Inventor and continue to develop parts and assemblies, you will find numerous uses for work points. Generally, you may want to use work points for the following applications:

- To establish points at work plane and work axis intersections.

- To place points at the intersection of three feature faces or planes.

- To locate part vertices, such as edge endpoints, midpoints, or curve center points.

- To help place a work axis or work plane.

- To create constraint points for assemblies.

 Field Notes

Many of the uses of work axes are explored in the projects at the end of the chapter.

Work points are displayed as highlighted work point icons in the graphics window and may be partially hidden by feature geometry. See Figure 7.30. If at any time you want to turn off the visibility, or hide, a work point, right-click the work point in the **Browser** and unselect the **Visibility** menu option, or right-click on the work point in the graphics window and unselect the **Visibility** menu option.

Work points are created using the **Work Point** tool. Access the **Work Point** command using one of the following techniques:

✓ Pick the **Work Point** button on the panel bar.

✓ Pick the **Work Point** button on the **Features** toolbar.

Unlike other features when you access the **Work Point** command or any other work feature, no dialog box opens. Instead, you define where you want to place the work point by selecting existing or reference geometry. Once you have accessed the **Work Point** tool, use one of the following techniques to create a work point:

- Using default reference work features   Although you can select the default work features located in the **Origin** folder of the **Browser** to create a work point, you can accomplish the same task by simply referencing the **Center Point.** For example, if you pick the **YZ Plane, XZ Plane,** and **XY Plane,** a work point identical to the default **Center Point** is created.

- At existing feature vertices   Use this option to generate a work point by referencing any available work point location, including feature edge and curve endpoints, intersections, midpoints, and center points. To create a work point by referencing vertices, pick the point location. See Figure 7.31.

FIGURE 7.30 A visible work point.

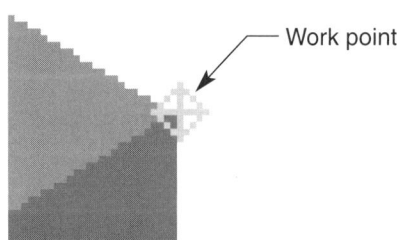

— Work point

FIGURE 7.31
Creating a work point
by selecting a single
point.

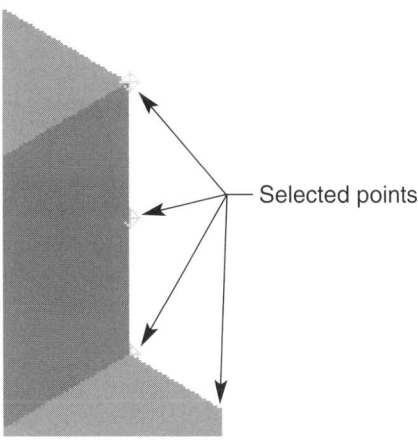

Selected points

Field Notes

You can also place work points at the vertices of sketch geometry.

- Using three intersecting work planes or faces   This option generates a work point at the intersection of three work planes or faces. To create a work point by referencing three work planes or faces, pick the work planes or faces where you want the work point to be located. See Figure 7.32.

- From an edge or axis and a face or plane   This option allows you to define a work point at the intersection of the selected face or plane, and edge or axis. To create a work point by ref-

FIGURE 7.32
Creating a work point
at the intersection of
three work planes or
feature faces.

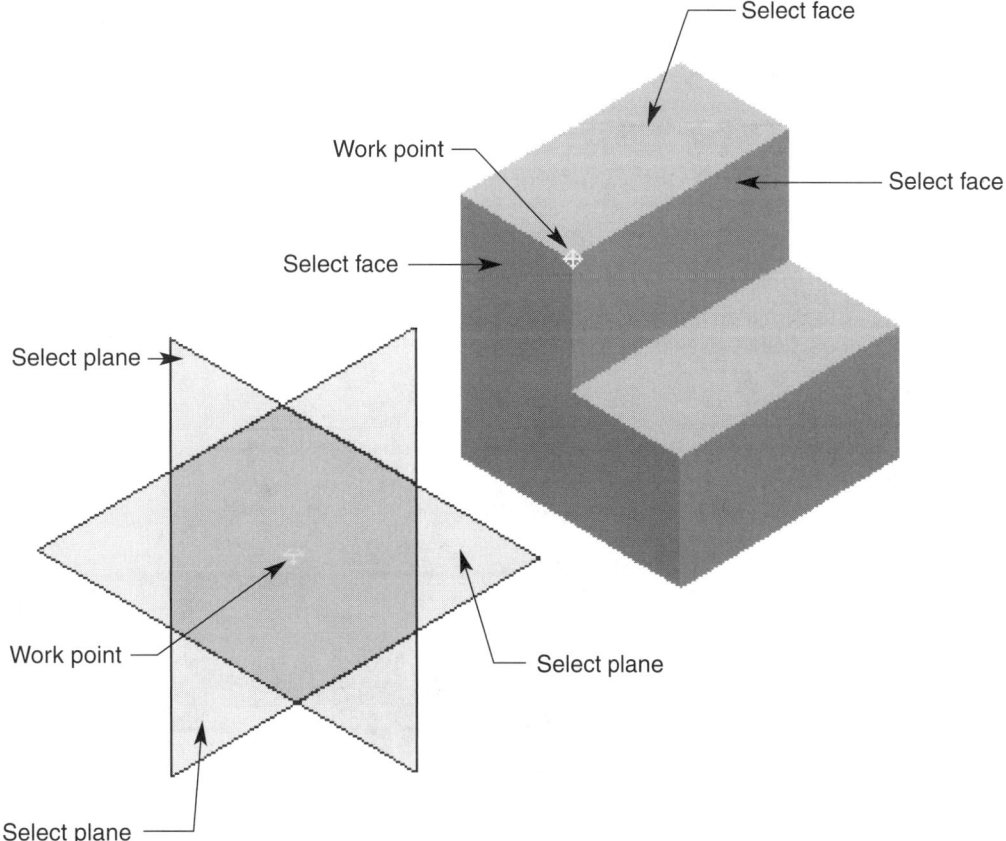

Select face

Work point

Select face

Select face

Select plane

Work point

Select plane

Select plane

erencing a face or plane and an edge or axis, pick the face or plane, and then pick an edge or axis. Or select the edge or axis, followed by the face or plane. See Figure 7.33.

- **From two feature edges or axes** Use this option to define a work point at the intersection of two selected edges or axes. To create a work point by referencing edge or axes, pick the two edges or axes where you want to place the point. See Figure 7.34.

- **At the intersection of two projected coplanar edges** Use this option to define a work point at the projected intersection of two selected coplanar feature edges. To create a projected work point by referencing edges, pick the two edges where you want to place the point. See Figure 7.35.

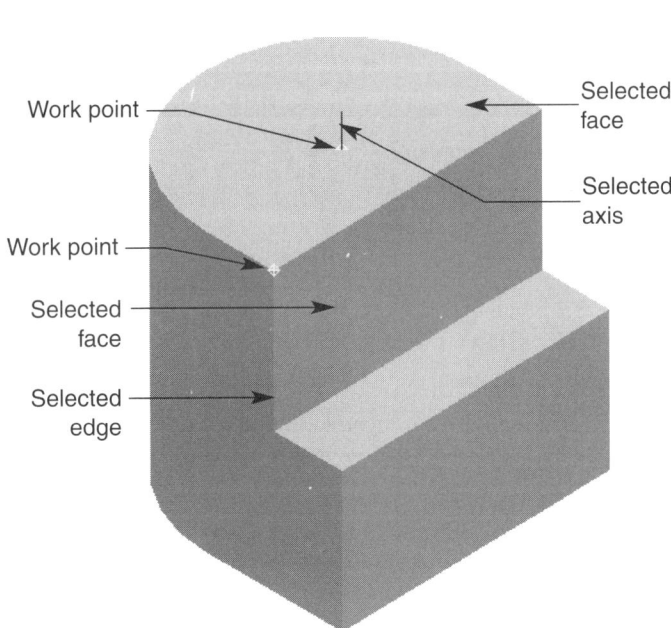

FIGURE 7.33 Creating a work point by selecting a face or plane and an edge or axis.

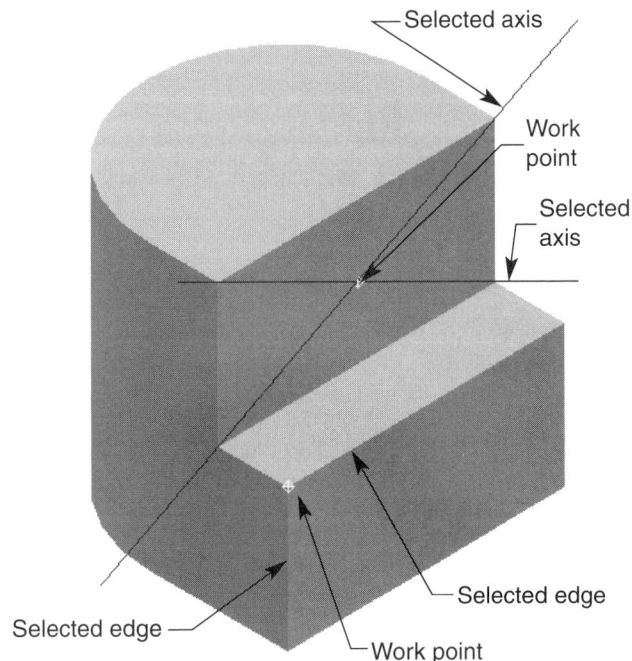

FIGURE 7.34 Creating a work point by selecting two edge or axes.

FIGURE 7.35
Creating a projected work point by selecting two edges.

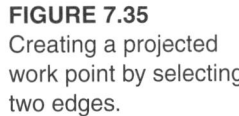

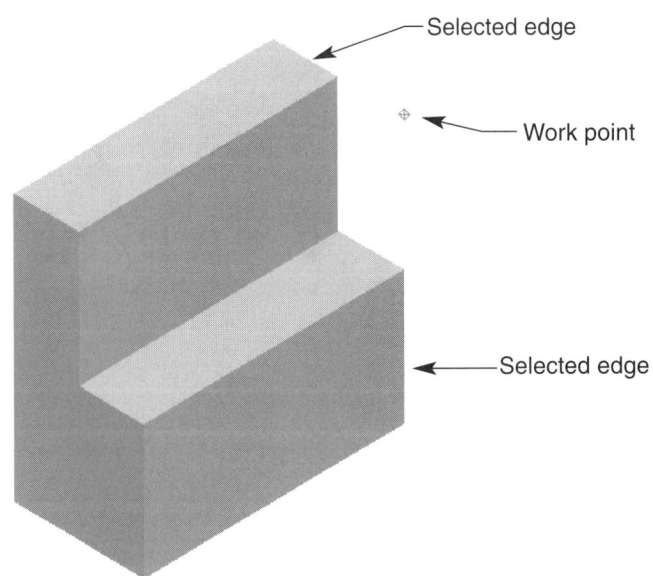

Exercise 7.6

1. Continue from Exercise 7.5, or launch Autodesk Inventor.
2. Open a new part file.
3. Open a sketch on the **XZ Plane,** and sketch the geometry shown.
4. Extrude the sketch 2″ in both directions.
5. Add the work points as shown, using the techniques discussed. *Note:* You may need to hide some work features for clarity.
6. Save the part as EX7.6.
7. Exit Autodesk Inventor or continue working with the program.

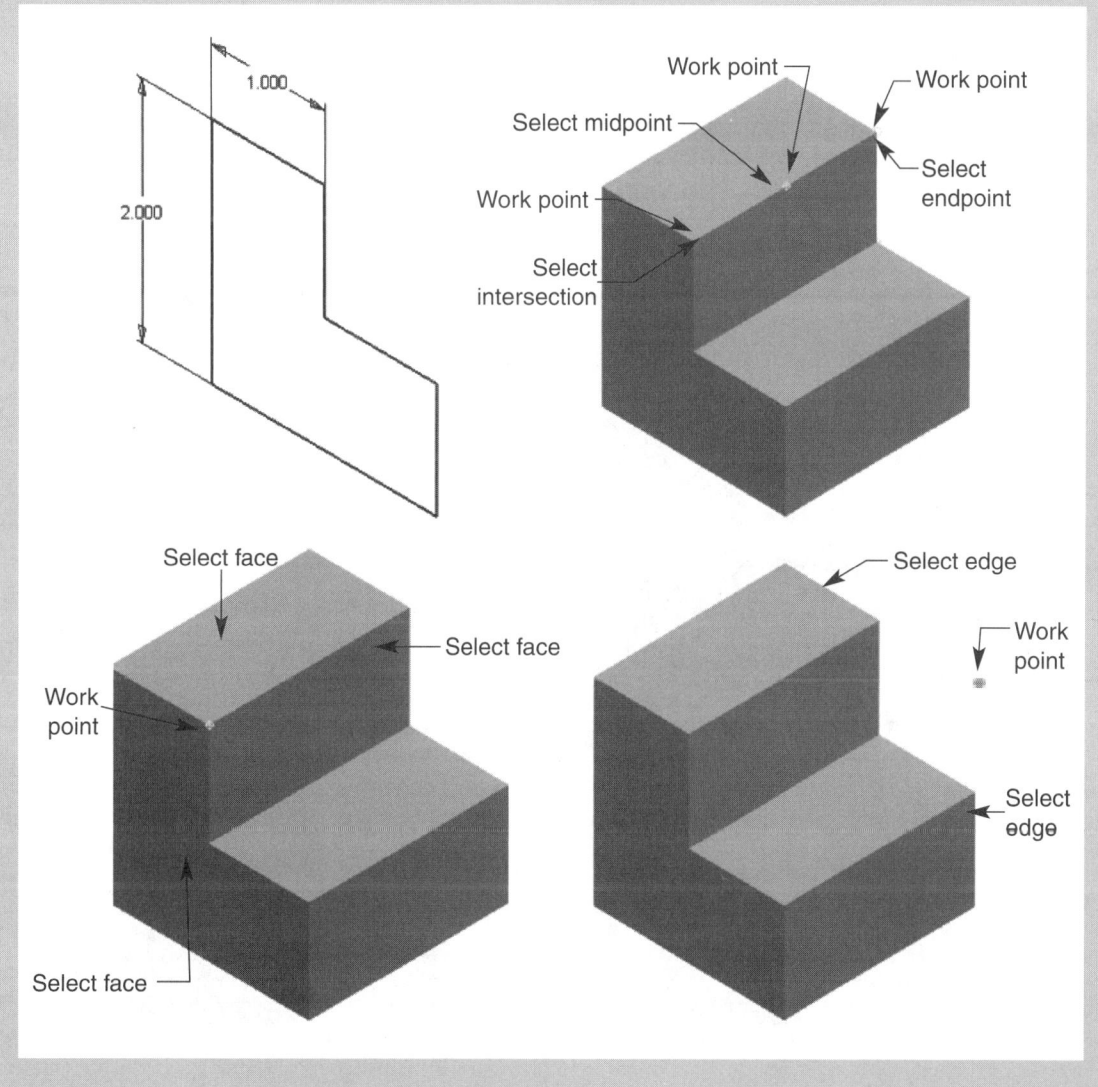

Exercise 7.7

1. Continue from Exercise 7.6, or launch Autodesk Inventor.
2. Open a new part file.
3. Open a sketch on the **XY Plane,** and sketch the geometry shown.
4. Extrude the sketch 1″ in a positive direction.
5. Add the work axes as shown, using the techniques previously discussed.
6. Add the work points as shown, using the techniques discussed. *Note:* You may need to hide some work features for clarity.
7. Save the part as EX7.7.
8. Exit Autodesk Inventor or continue working with the program.

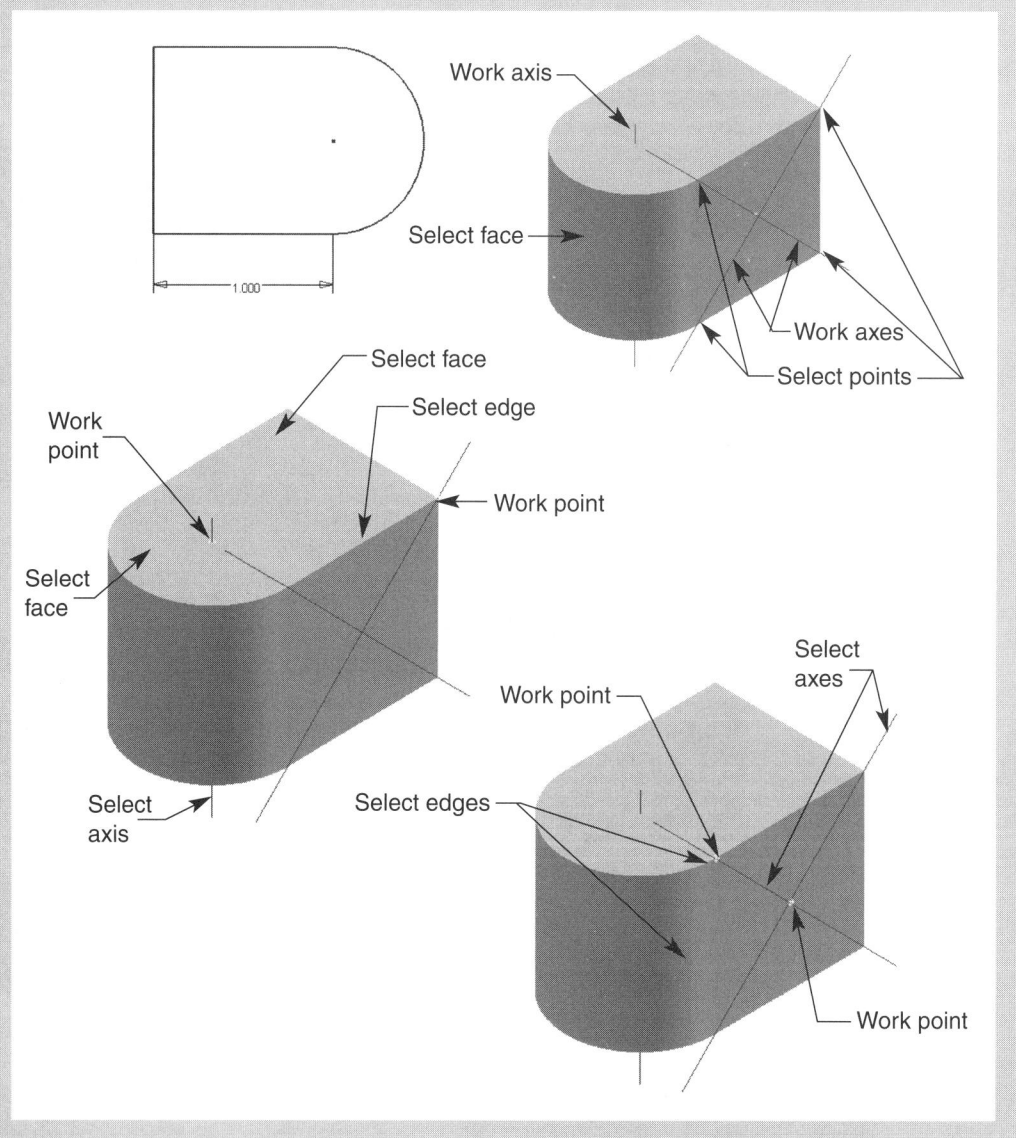

|||||||||||| CHAPTER TEST

1. What are work features?

2. When can work features be created?

3. Identify the only way that work features can be created without existing geometry.

4. What is the purpose of work feature tools?

5. How do you access work feature tools?

6. By default, the panel bar is displayed in learning mode. How do you change it to expert mode?

7. Once you are finished using a work feature, how do you hide it from view?

8. What is a work plane?

9. What does it mean when it is stated that work planes are infinite?

10. List at least three uses for work planes.

11. How do you move a work plane?

12. Moving or resizing the work plane display does not change the specified location, constraints, or area of the work feature. How do you physically relocate the work plane in space?

Questions 13 through 27 relate to creating a work plane using the default reference work features located in the Origin folder of the Browser:

13. Selecting one of the YZ Plane, XZ Plane, or XY Plane default planes displays a preview of the plane, and the drag icon. How do you locate the plane and access the Offset dialog box?

14. Explain how to use the Offset dialog box to create the work plane at its current location.

15. Discuss how to create a work plane by selecting any pair of default reference work axes.

16. Explain how to create a work plane by selecting the Center Point and one of the available axes or planes.

17. Discuss how to create a work plane parallel to and a specified distance from an existing feature face.

18. Explain how to create a work plane from two feature edges or axes.

19. Discuss how to create a work plane using 3 existing points.

20. Explain how to create a work plane perpendicular to the selected edge, through a specified axis or point.

21. Discuss how to create a work plane from a face and a point.

22. Explain how to create a work plane using a tangent face, and an edge or axis.

23. Discuss how to create a work plane using a tangent face and a planar face.

24. Explain how to create a work plane that is angled from a planar edge or axis and a face, and the function of the Angle edit box.

25. Discuss how to create a work plane that is tangent to a cylinder.

26. How do you hide sketch geometry when finished?

27. Once you have added any of the work planes, how do you offset another work plane?

28. What is a work axis?

29. List at least three applications for work axes.

30. What do work axes look like?

31. How do you turn off the visibility of, or hide, a work axis?

Questions 32 through 38 relate to creating a work axis using the Work Axis tool:

32. Explain how to create a work axis using default reference work features.

33. Discuss how to create a work axis from two existing points.

34. Explain how to create a work axis using two intersecting work planes or faces.

35. Discuss how to create a work axis using a revolved or curved feature.

36. Explain how to create a work axis from a feature edge.

37. Discuss how to create a work axis from a face or plane and a point.

38. Explain how to create a work axis using a planar edge and a face.

39. What is a work point?

40. Identify at least four uses for work points.

41. How are work points displayed?

42. What do you do if you want to turn off the visibility, or hide, a work point?

Questions 43 through 48 relate to creating a work point using the Work Point tool:

43. Explain how to create a work point using default reference work features.

44. Discuss how to create a work point at existing feature vertices.

45. Explain how to create a work point using three intersecting work planes or faces.

46. Discuss how to create a work point from an edge or axis and a face or plane.

47. Explain how to create a work point from two feature edges or axes.

48. Discuss how to create a work point at the intersection of two projected coplanar edges.

|||||||||| PROJECTS

Follow the specific instructions provided for each project.

1. Name: Hub

   Units: Inch

   Specific Instructions:

   - Open P6.7, and Save Copy As P7.1.

   - Close P6.7 without saving, and Open P7.1.

   - Offset a work plane from the **XY Plane,** 2.25″.

   - Open a sketch on the work plane, and sketch the geometry shown.

   - Extrude the sketch in a negative direction, using the **To Next** termination option.

   - Add a .5″ fillet to the edges shown. *Note:* The work plane's visibility has been turned off for clarity.

   - Open a sketch on the **YZ Plane,** and sketch the geometry shown.

   - Use the sketch to create a .25″ thick rib, as shown.

   - Use the **Circular Pattern** tool to pattern the extrusion, using the following information:

     Rotation Axis: **Y Axis**

Count: 3

Positioning Method: Fitted

Angle: 360°

- The final part should look like the part shown.
- Save As: P7.1.

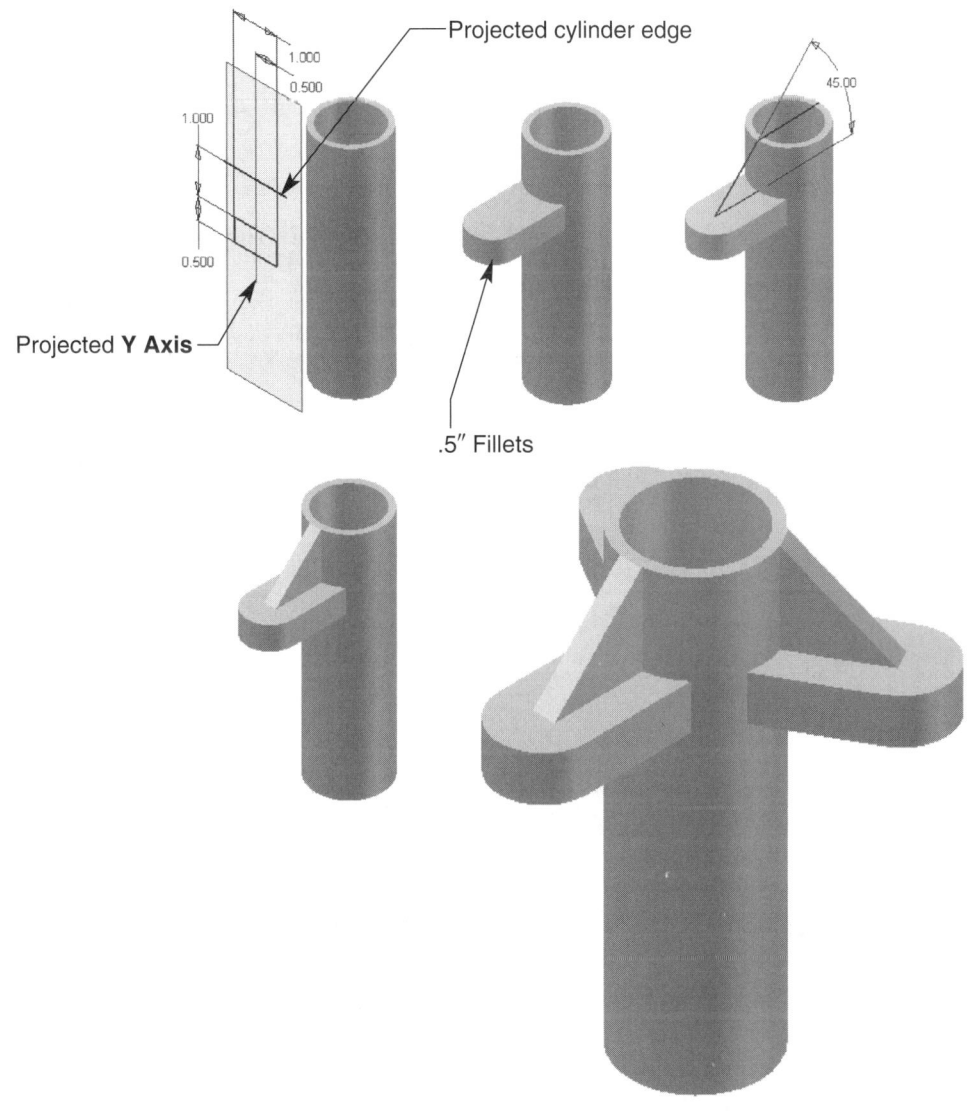

2. Name: 45° Elbow

   Units: Metric

   Specific Instructions:

   ■ Open P6.9, and Save Copy As P7.2.

   ■ Close P6.9 without saving, and Open P7.2.

   ■ Create a work plane by selecting the **XY plane** and **Extrusion1,** as shown.

   ■ Open a sketch on the work plane, and sketch the geometry shown.

   ■ Extrude the sketch in a positive direction, using the **To Next** termination option.

   ■ Create a ⌀50mm circle on face of **Extrusion3,** and extrude 8mm in a positive direction, as shown.

   ■ Add a ⌀25mm hole terminating at the inside face of the 45° elbow.

   ■ Add the specified hole, 19mm from the center of **Extrusion4.**

   ■ Create a work axis by selecting **Hole3.**

   ■ Use the **Circular Pattern** tool to pattern the extrusion, using the following information:

   Rotation Axis: **Work Axis1** in the Browser

   Count: 6

   Positioning Method: Fitted

   Angle: 360°

   ■ Add a 5mm fillet as shown.

   ■ The final part should look like the part shown.

   ■ Save As: P7.2.

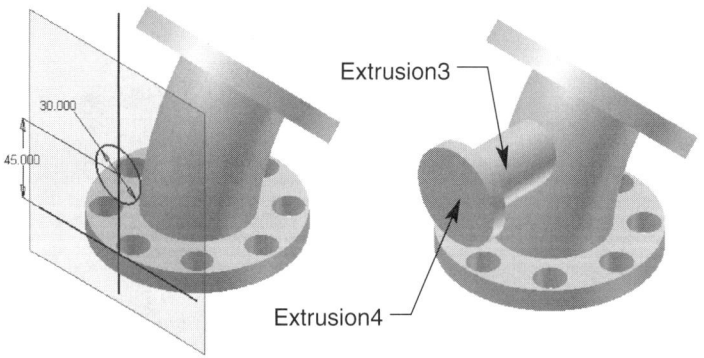

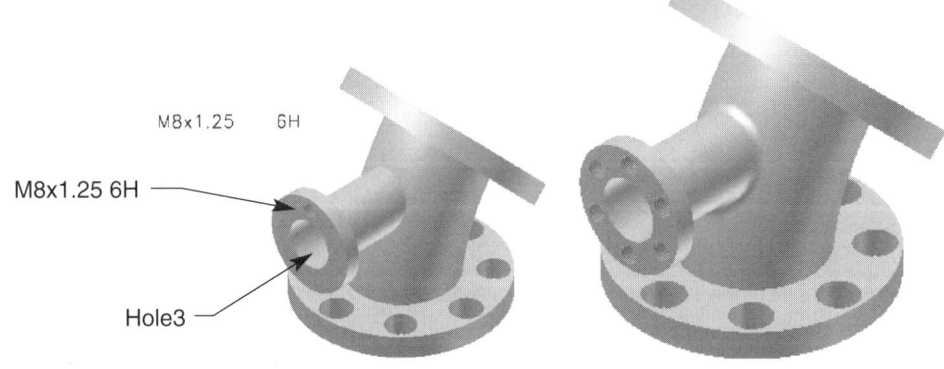

3. Name: Base

Units: Inch

Specific Instructions:

- Open a new part file.

- Create the sketch shown on the **XZ Plane.**

- Extrude the sketch 8″ in both directions.

- Shell the extrusion .125″, removing the bottom face as shown.

- Offset a work plane −.25″ from the bottom face as shown.

- Open a sketch on the work plane, and sketch the lines shown.

- Use the **Rib** tool to create the rib network shown.

- Save As: P7.3.

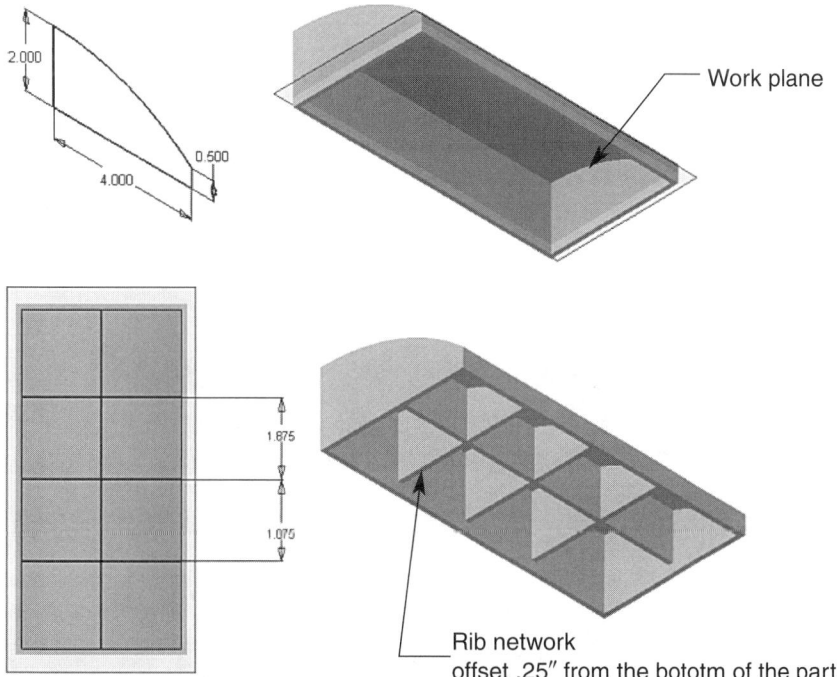

4. Name: Shaft

   Units: Metric

   Specific Instructions:

   ■ Open a new part file.

   ■ Sketch a ∅20mm circle on the **XZ Plane.**

   ■ Extrude the sketch 60mm in both directions.

   ■ Sketch the geometry shown, on the left face of the cylinder.

   ■ Share the sketch.

   ■ Create work plane1 by picking the specified line and point.

   ■ Sketch the geometry shown on the work plane.

   ■ Cut extrude the sketch 5mm.

   ■ Create work plane2 by picking the specified line and point.

   ■ Sketch the geometry shown on the work plane.

   ■ Extrude the sketch 5mm.

   ■ The final part should look like the part shown with all work features, and **Sketch2** hidden.

   ■ Save As: P7.4.

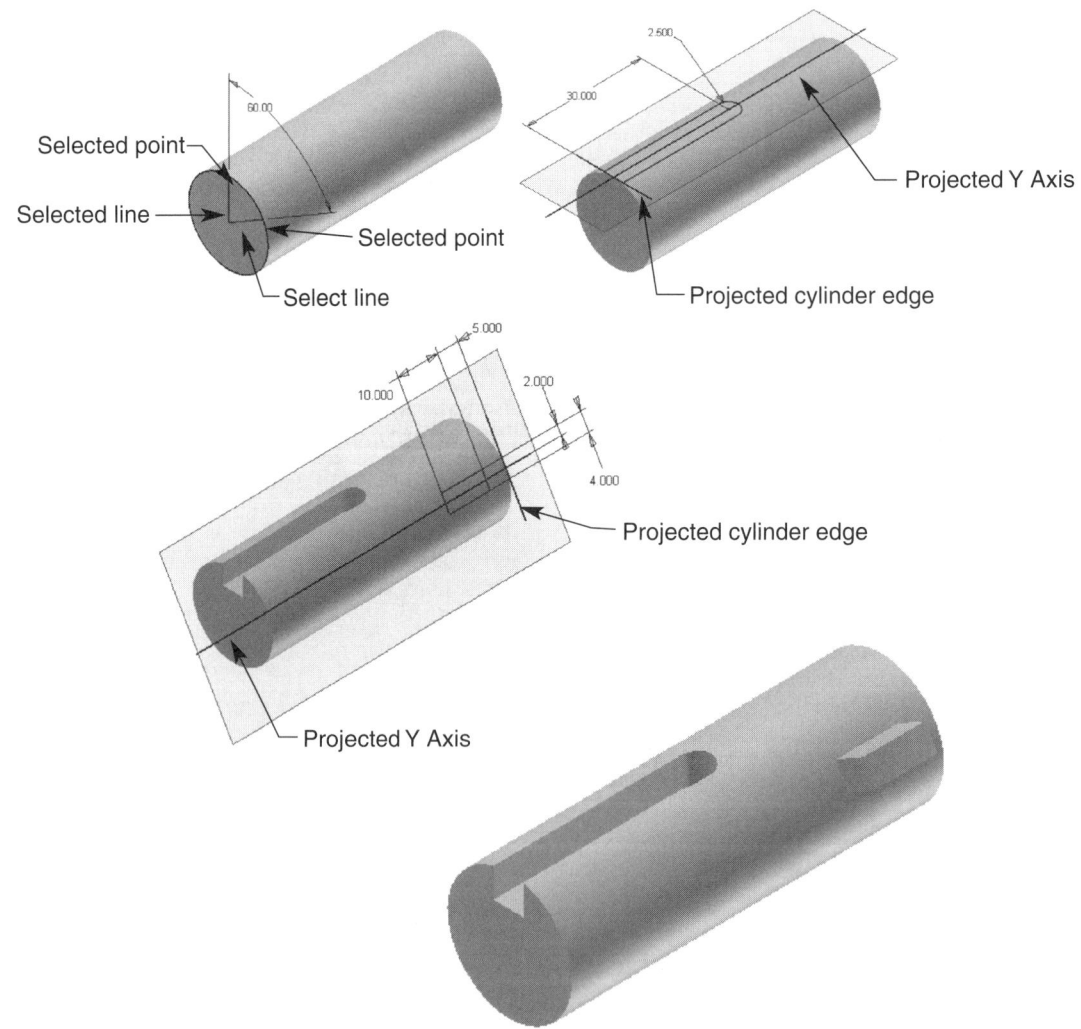

5. Name: Box

Units: Inch

Specific Instructions:

- Open a new part file.

- Sketch a 12″×12″ rectangle on the **XY Plane.**

- Extrude the sketch 12″ in a positive direction.

- Shell the box .25″ inside, removing the top face.

- Create the work planes shown using the techniques discussed.

- Sketch the geometry shown, on one of the box faces.

- Cut extrude the sketch using the **To Next** termination option.

- Use the Rectangular pattern tool to pattern the **Extrusion2** as shown, using the following information:

    Direction 1: Along the X axis (horizontal rows)

    Count: 3

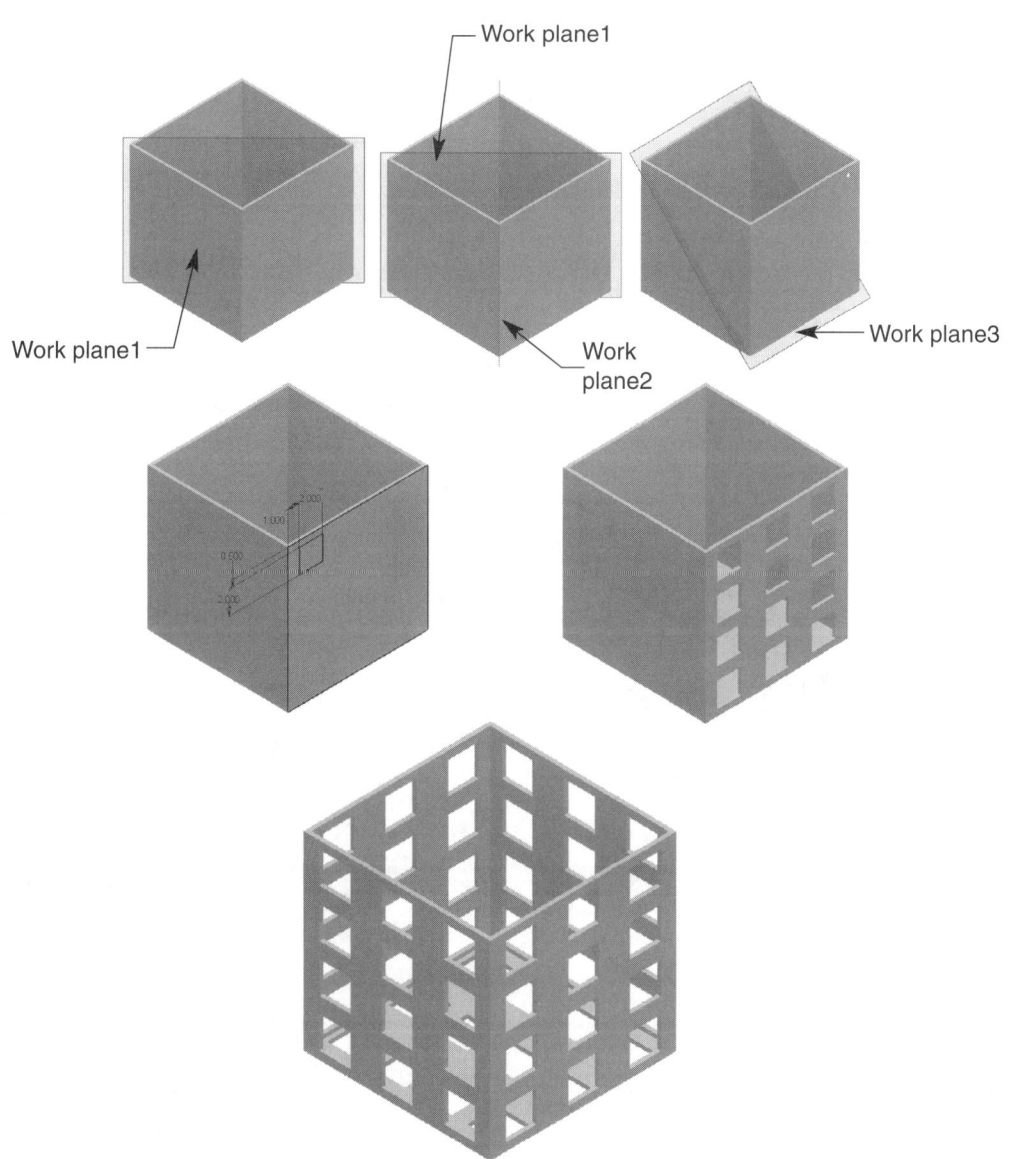

Spacing: 4″

Direction 2: Along the Y axis (vertical columns)

Count: 4

Spacing: 3″

- Mirror **Rectangular Pattern1,** across **Work Plane1.**

- Mirror **Rectangular Pattern1,** across **Work Plane2.**

- Mirror **Rectangular Pattern1,** across **Work Plane3.**

- The final part should look like the part shown with all work features hidden.

- Save As: P7.5.

6. Name: C-Clamp Screw

Units: Metric

Specific Instructions:

- Open P5.3, and Save Copy As P7.6.

- Close P5.3 without saving, and Open P7.6.

- Create a work plane tangent to the screw head, by selecting the tangent face, and the default XZ Plane, as shown.

- Open a sketch on the work plane, and sketch the geometry shown.

- Place a 4.85mm through hole at the sketched Hole, Center Point.

- The final part should look like the part shown.

- Save As: P7.6.

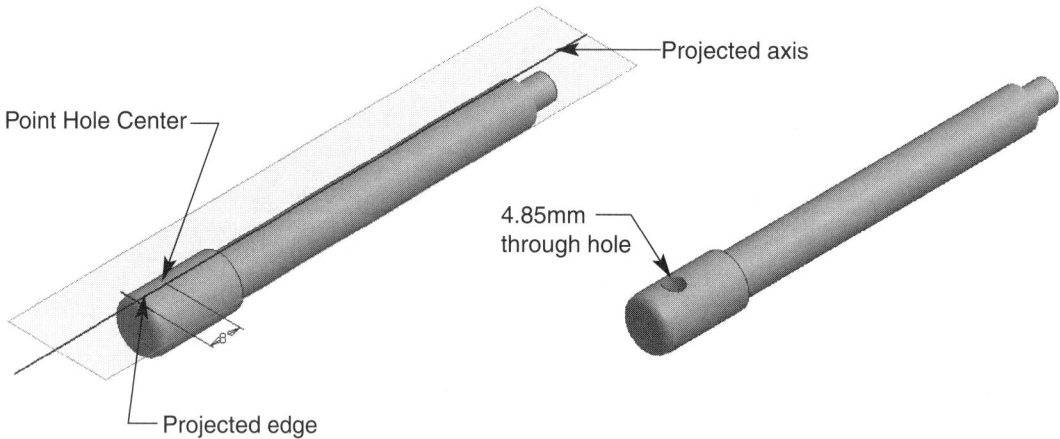

|||||||||||||||||||||||||||||||||||||||||||||||||||||||||||||||||||||||||||||||||||||||||||||||||||||||||||||

CHAPTER 8

Developing Advanced Part Model Features

LEARNING GOALS

After completing this chapter, you will be able to:

◎   Create advance part model features.

◎   Develop **Loft** features.

◎   Produce basic and advanced **Sweep** features.

◎   Work in the 3D sketch environment and interface.

◎   Use 3D sketching tools.

◎   Create **Split** features.

Advanced Part Model Features

Many part models you develop may require complex features that tools such as extrude, revolve, or fillet cannot create, or at least cannot create in one step, using one tool. As a result, Autodesk Inventor provides you with a few other feature tools that may help you develop certain parts. The additional tools allow you to generate lofts, sweeps, and split features. These commands have many applications and are discussed throughout this chapter.

Field Notes

All the interface and work environment options and applications discussed in previous chapters apply to working with lofts, sweeps, and splits.

Using Loft, Sweep, and Split Feature Tools

The **Loft, Sweep,** and **Split** tools are used to create advanced part model features. These tools are accessed from the **Features** toolbar, shown in Figure 8.1, or from the **Features** panel bar. When a feature is present and you are in the feature work environment, the panel bar is in features mode. See Figure 8.2. By default, the panel bar is displayed in learning mode, as seen in Figure 8.3. To change from learning to expert mode, seen in Figure 8.3 and discussed in Chapter 2, right-click any of the panel bar tools or left-click the panel bar title and select the **Expert** option.

FIGURE 8.1 The **Features** toolbar.

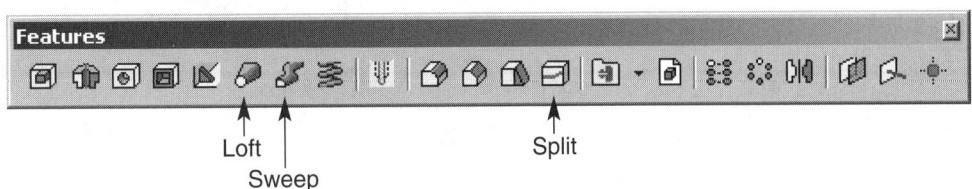

Loft   Sweep   Split

FIGURE 8.2   The panel bar, in features (learning) mode.

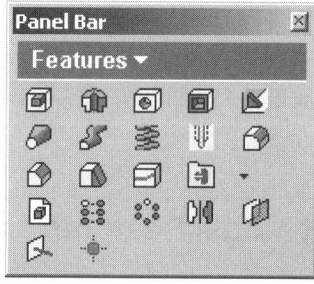

FIGURE 8.3   The panel bar, in features (expert) mode.

Creating Loft Features

A *loft* references two or more sketch profiles, located on different planes, and blends the profiles, or sections, together to make a feature. Lofts are very similar to extrusions, but allow you to create features that are geometrically more complex. See Figure 8.4.

Before you can create a loft feature, you must create two or more sketches. Sketches used to create loft features are referred to as *sections*. The sketches can be independent of any existing geometry, as seen in Figure 8.4. Sketches can also be placed on a planar feature face, allowing you to loft from an existing feature. See Figure 8.5. If you want to use existing feature edges as a loft section, use the **Autoproject edges during curve creation** and/or the **Automatic reference edges for new sketch** tools discussed in Chapter 3.

FIGURE 8.4   A feature created by lofting the sketch of a circle and the offset sketch of a triangle.

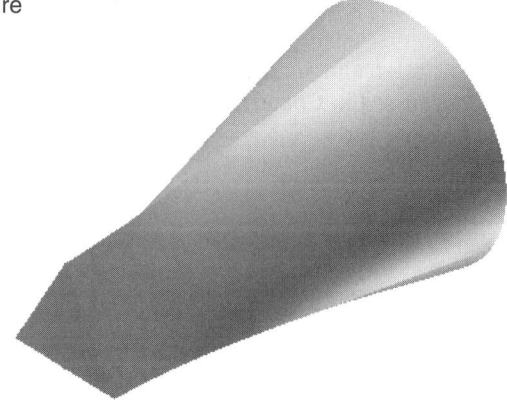

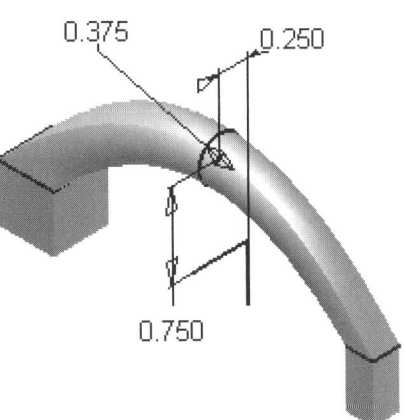

FIGURE 8.5 A handle created by lofting two existing feature face sketches, and an additional offset sketch of a circle. (Sketches are visible for clarity.)

Typically, the additional sketches required for lofting are created on work planes. For example, the second sketch used to create the lofted feature in Figure 8.4 was placed on an offset work plane from the **XZ Plane.** The circle sketched between the existing extrusions in Figure 8.5 was placed on a work plane offset from one of the extrusion faces.

 Field Notes

Section sketches must be closed loop unless you want to create a surface loft. There is no limit to the number of sketch sections, or profiles, used to create a loft feature.

You can create loft sections at any angle and anywhere in space from another section. They do not have to be parallel.

Loft features are created using the **Loft** tool. Access the **Loft** command using one of the following techniques:

✓ Pick the **Loft** button on the panel bar.

✓ Pick the **Loft** button on the **Feature** toolbar.

Assuming you have at least two unconsumed sketches, as previously discussed, the **Loft** dialog box is displayed when you access the **Loft** command. See Figure 8.6.

To create a loft feature, pick the sketch profiles you want to use. The **Sections** list should be active, allowing you to choose sketches. If the **Sections** list is not active, you may need to pick the **Click here to add** button in order to pick sections. For simple lofts, such as those that contain only two sketches, it is not

FIGURE 8.6 The **Loft** dialog box.

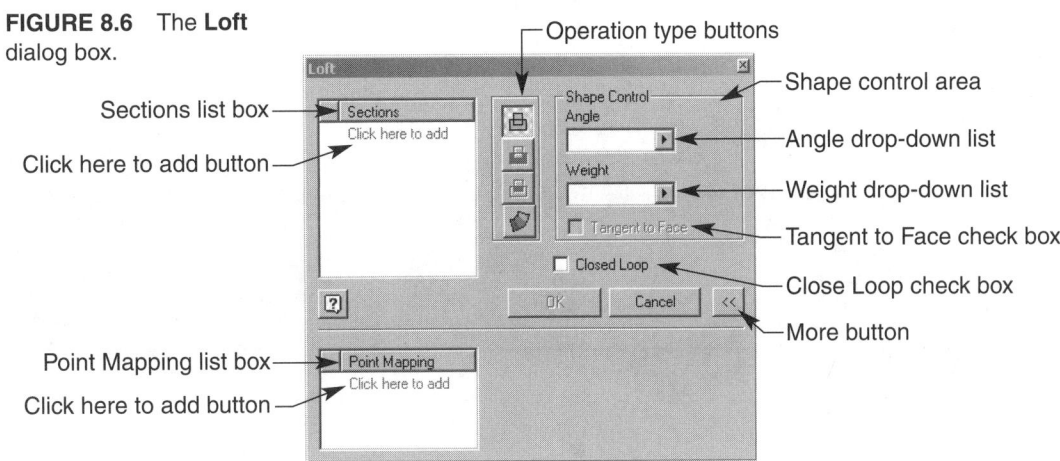

Sections list box

Click here to add button

Operation type buttons

Shape control area

Angle drop-down list

Weight drop-down list

Tangent to Face check box

Close Loop check box

More button

Point Mapping list box

Click here to add button

as critical which sketch you select first. For more complicated lofts, select sketches in a logical, sequential order. Once section profiles are selected, you will notice the specified sketches are listed in the **Sections** list.

Field Notes

> For clarity, you may want to select the desired sketches from the Browser. A preview arrow identifying the direction of the loft will be shown once you have selected the desired sections.

Once you have selected all the sketch sections required to generate the loft, pick the desired operation type button. For simple base features, only the **Join** and **Surface** buttons are available. However, if other features are present in the part model, all the operation type buttons are active, and you can choose to join, cut, intersect, or create a surface, depending on the application.

Field Notes

> The loft operation type buttons function exactly the same as for other feature tools, such as **Extrude.** Refer to Chapter 4 for specific information regarding operation type buttons.

The next step in creating a loft feature is to adjust the desired shape controls in the **Shape Control** area. This area allows you to specify options associated with the appearance of the loft. The following options are available inside the **Shape Control** area of the **Loft** dialog box:

- **Angle**   This drop-down list allows you to specify the angle between the faces that the loft creates and the sketch profiles. See Figure 8.7.

Field Notes

> The default loft angle is 90°. You can modify the angle only when the weight value is greater than 0.

- **Weight**   This drop-down list allows you to specify a unitless loft weight. The weight defines the appearance of the loft. The desired weight is relative to the size of the loft feature

FIGURE 8.7   An example of (A) a loft with the default 90° angle and (B) a loft with a 30° angle.

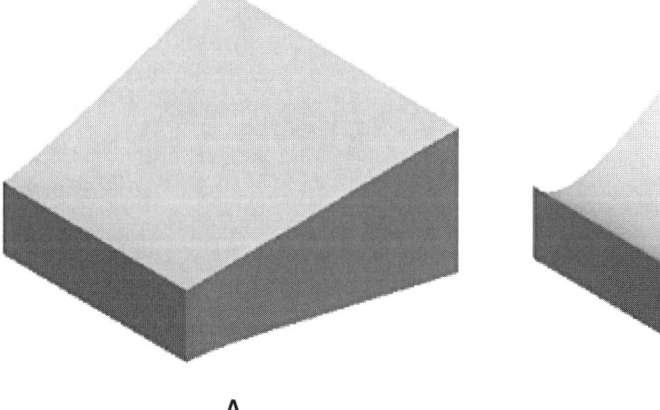

A   B

FIGURE 8.8   An
example of a loft with
(A) the default 0 ul
(smallest) weight and
(B) a large-value weight
(25 ul for this particular
part).

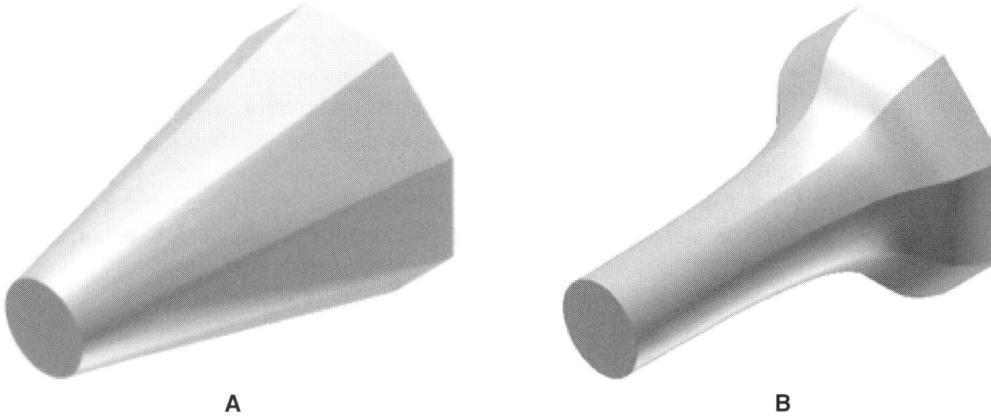

A   B

and can range from 0 to infinite. However, weights that are too high may result in an unde-sirable appearance. When you enter a small value for the weight, a more direct transition occurs between sketch profiles. See Figure 8.8A. When you enter a larger number, a more indirect, meandering transition occurs between sketch profiles, depending on the specified weight. See Figure 8.8B.

- **Tangent to Face**   This check box is available when you create a sketch profile on an exist-ing planar face. The **Tangent to Face** check box functions similarly to the **Angle** drop-down list. Select this check box to place a tangent constraint between the sketch and the planar face on which the sketch is created.

Field Notes

If the **Tangent to Face** check box is selected, the **Angle** drop-down list is not accessible.

The next step in creating a loft feature is to pick the **Closed Loop** check box, if desired, and apply point mapping if necessary. These loft components are described as follows:

- **Closed Loop**   When this check box is selected, a closed loop control is placed on the loft that contains three or more sections. This operation fixes the first and last sketch sections together.

- **Point Mapping**   Often when you are creating a loft that involves several sections that are arranged in a more complex fashion, the loft twists, which results in an inaccurate part. See Figure 8.9. Fortunately, point mapping usually takes care of this problem by allowing you to

FIGURE 8.9   An
example of a tiller blade
created using the loft
tool without point
mapping. Notice the
part has been twisted
into an undesirable
appearance.

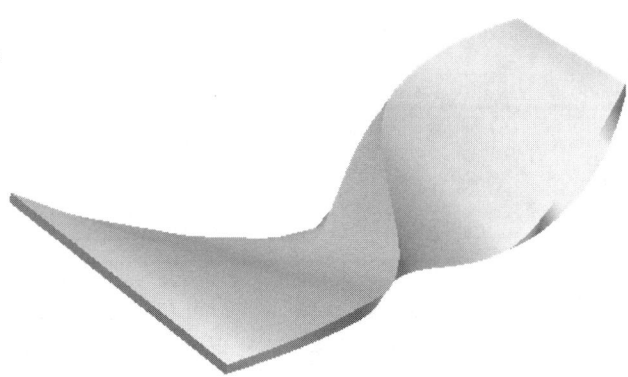

align sketch sections and correctly specify the path of the loft using selected points on each section. Access the **Point Mapping** list box by picking the **More** button in the **Loft** dialog box. Then pick the **Click here to add** button, and select points on each of the sketch section in sequential order. Usually, you can pick any available point on a sketch, but it is very important that you choose corresponding points on each sketch. See Figure 8.10.

Once you have identified the sections, adjusted the desired shape controls, selected the appropriate operation type, and added any necessary point mapping, pick the **OK** button to generate the loft feature.

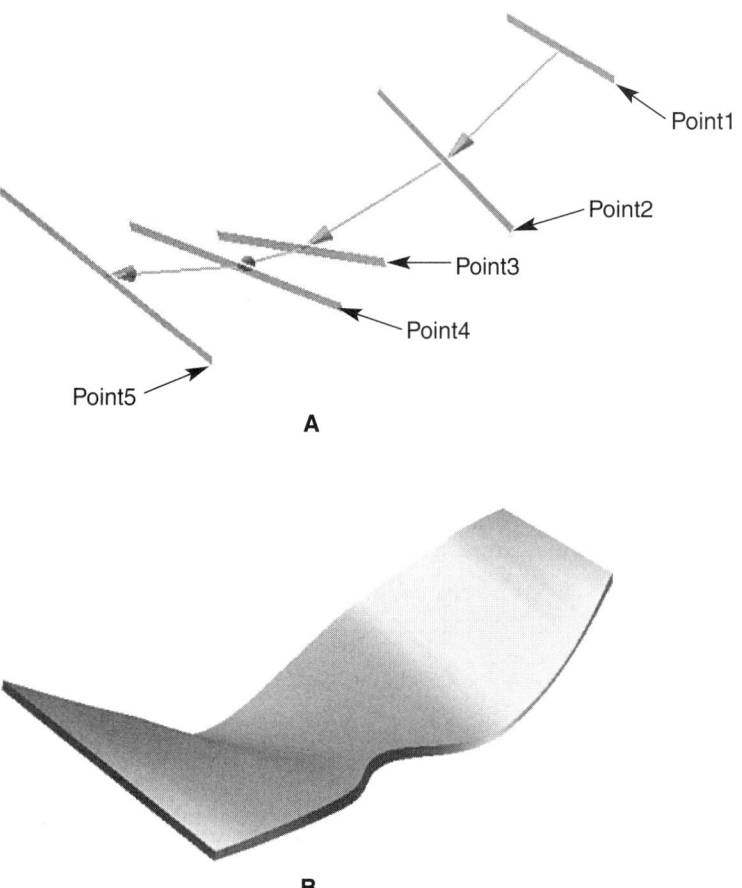

FIGURE 8.10   (A) An example of the points selected for a tiller blade and (B) the lofted tiller blade, both created using the loft tool with point mapping.

Exercise 8.1

1. Launch Autodesk Inventor.
2. Open a new part file.
3. Open a sketch on the **XZ Plane,** and sketch a 20mm×10mm rectangle.
4. Offset a work plane 10mm from the **XZ Plane.**
5. Sketch a circle on the work plane as shown.
6. Use the **Loft** tool to create a loft feature using the following information:

   Sections: Sketch1 and Sketch2
   Operation Type: Join
   Weight: 0 ul

7. The part should look like the part shown.
8. Save the part as EX8.1.
9. Exit Autodesk Inventor or continue working with the program.

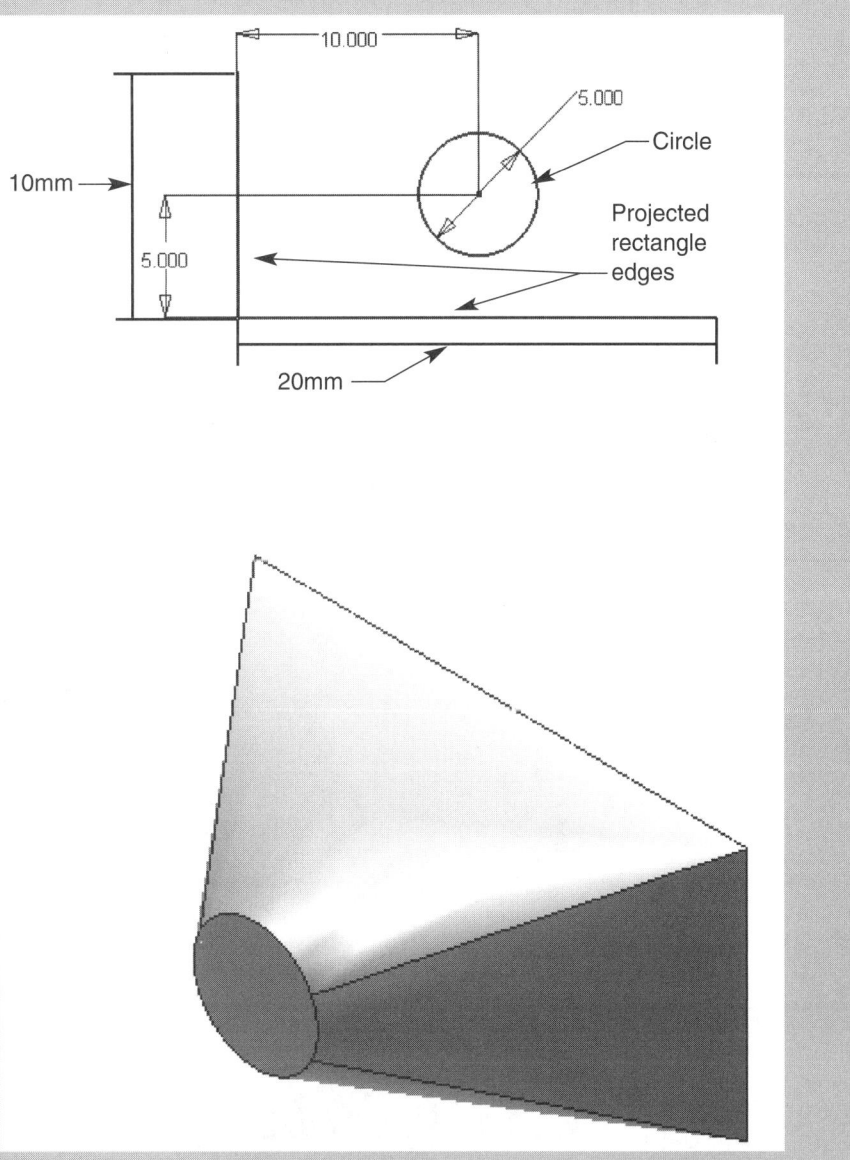

Exercise 8.2

1. Continue from Exercise 8.1, or launch Autodesk Inventor.
2. Open a new part file.
3. Open a sketch on the **XZ Plane,** and sketch the geometry shown in **Sketch1.**
4. Offset a work plane 1.5″ from the **XZ Plane.**
5. Offset another work plane, 1.5″ from **Work Plane1.**
6. Open a sketch on **Work Plane1,** and create **Sketch2.**
7. Open a sketch on **Work Plane2,** and create **Sketch3.**
8. Use the **Loft** tool to create a loft feature using the following information:

   Sections: **Sketch1, Sketch2,** and **Sketch3**
   Operation Type: Join
   Weight: 0 ul
   Point map using the specified points

9. The part should look like the part shown.
10. Save the part as EX8.2.
11. Exit Autodesk Inventor or continue working with the program.

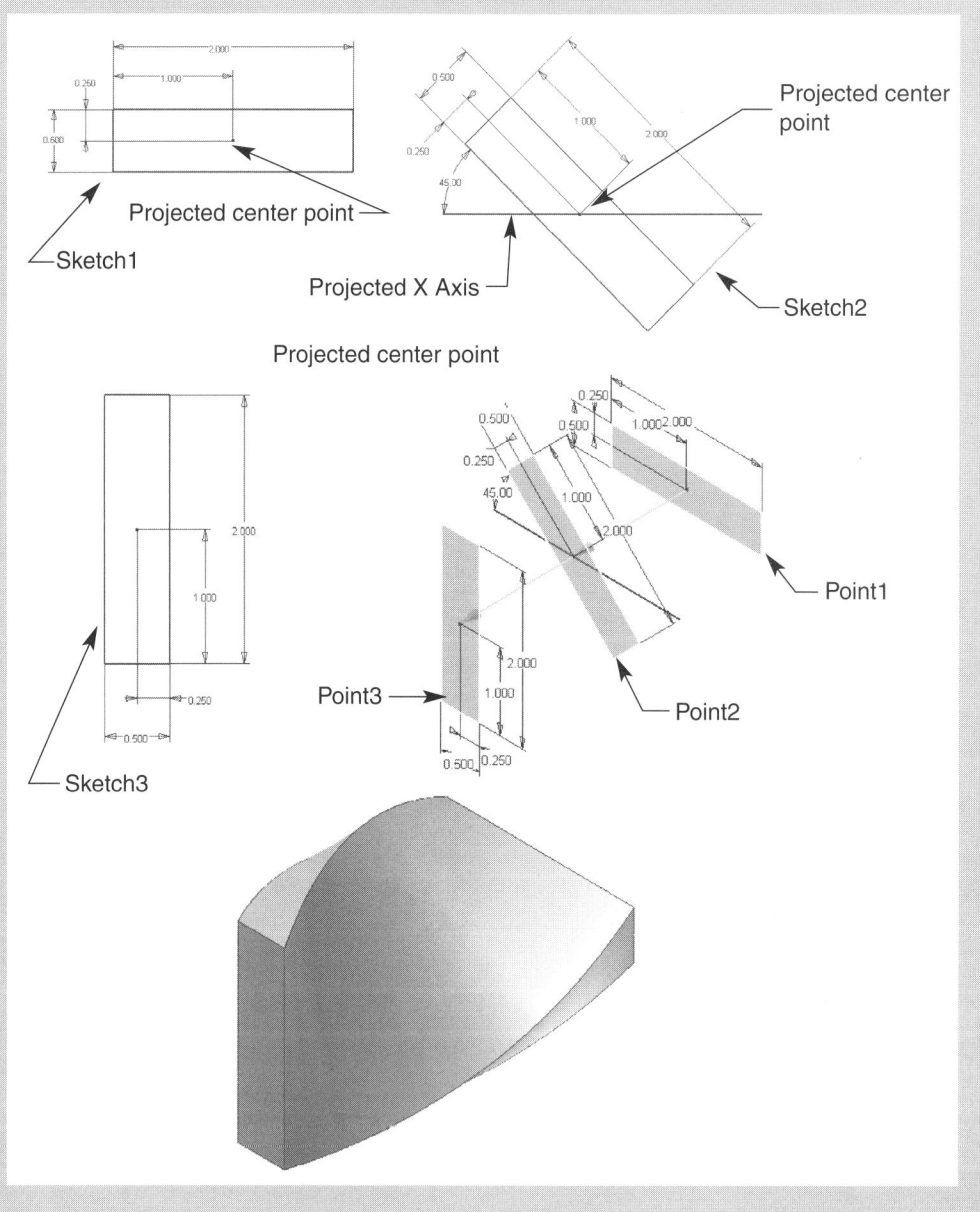

Creating Basic Sweep Features

A *sweep* is a feature that is created by referencing a sketch profile and a sketch path. The sketch profile follows or is swept along the sketch path. Sweeps are similar to extrusions but are geometrically more complex. They allow you to "extrude" a sketch along an irregular, nonlinear path. See Figure 8.11.

Before you can create a sweep feature, you must create two or more sketches. The following information is important for understanding how to create sweep sketches:

- One of the sketches must be the profile that you want to sweep, and another sketch must be the sweep path. See Figure 8.11A.

- The profile and path must be located on intersecting planes. For example, if you create a profile on an XY plane, you must create the path on a XZ or YZ plane. Though the profile and path sketches must be created on intersecting plane, the profile and path sketch geometry do not need to be perpendicular.

- The profile does not have to intersect the path. See Figure 8.12A. Still, it is better to have a profile that intersects the path, for dimensioning and constraining purposes, and to ensure the sweep is accurately generated. See Figure 8.12B. Typically, a **Coincident** constraint is used to establish an intersecting profile and path.

- Sweep sketches can be independent of any existing geometry, as seen in Figure 8.12, or they can be placed on a planar feature face, allowing you to sweep on an existing feature. See

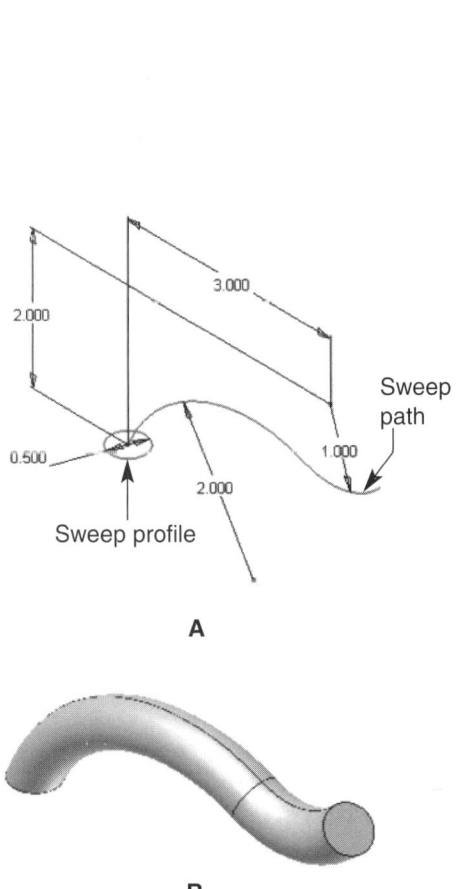

FIGURE 8.11 (A) The sketches used to create a simple sweep feature. (B) The completed sweep.

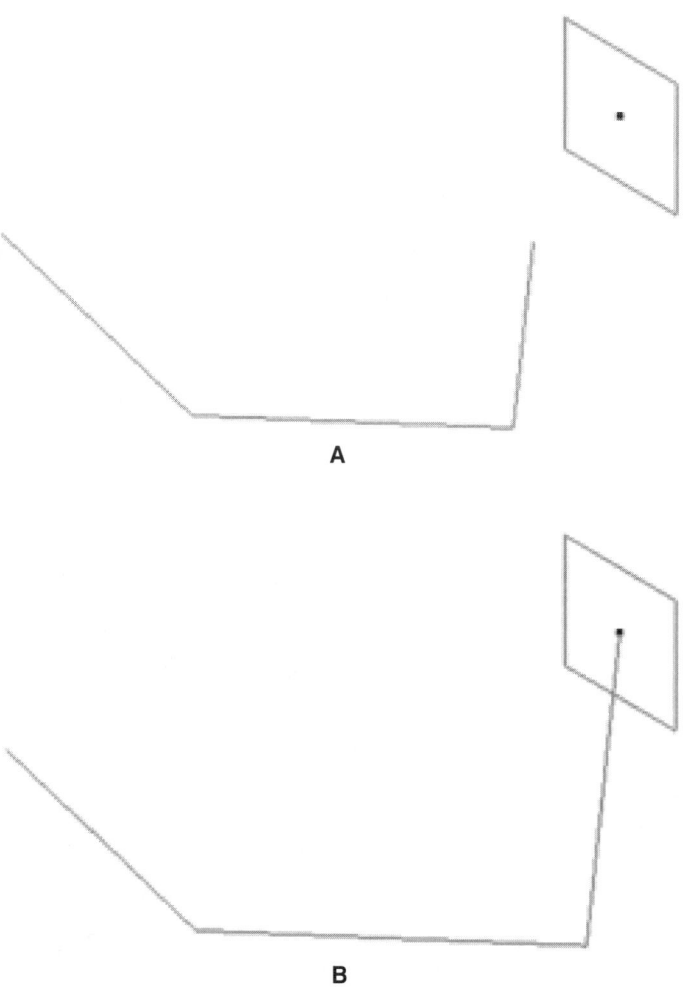

FIGURE 8.12 An example of (A) a sweep profile that does not intersect the path and (B) a sweep profile that does intersect the path.

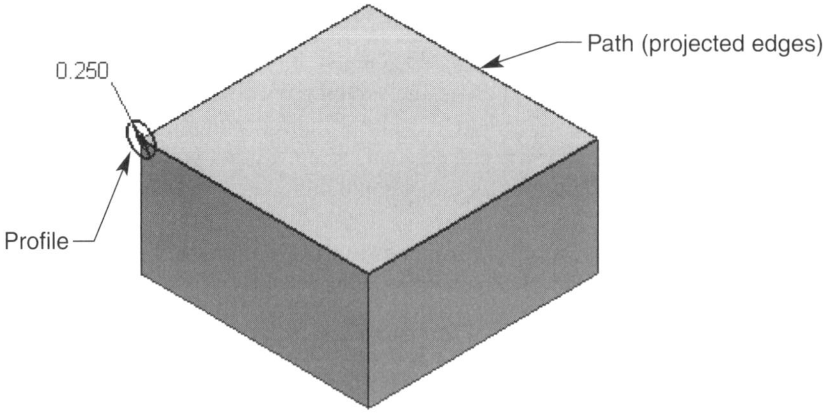

FIGURE 8.13
Creating a sweep
profile and path by
referencing existing
geometry. The profile is
coincident to the corner
point of the extrusion,
and the path is
automatically projected
from the top face.

Figure 8.13. If you want to use existing feature edges as a sweep profile or path, use the **Autoproject edges during curve creation** and/or the **Automatic reference edges for new sketch** tools discussed in Chapter 3.

Field Notes

Profile sketches must be closed loop unless you want to create a surface sweep. Surface features were discussed in previous chapters.

Sweep features are created using the **Sweep** tool. Access the **Sweep** command using one of the following techniques:

✓ Pick the **Sweep** button on the panel bar.

✓ Pick the **sweep** button on the **Features** toolbar.

Assuming you have at least two unconsumed sketches, as previously discussed, the **Sweep** dialog box is displayed when you access the **Sweep** command. See Figure 8.14.

To create a sweep feature, pick the **Profile** button if it is not currently active, and pick the sweep profile. Then pick the **Path** button, and select the sweep path.

Field Notes

If you are creating a very simple sweep feature, the profile and path may be automatically selected when you access the **Sweep** command. Once the profile and the path are selected, you may see a preview of the sweep.

FIGURE 8.14 The **Sweep** dialog box.

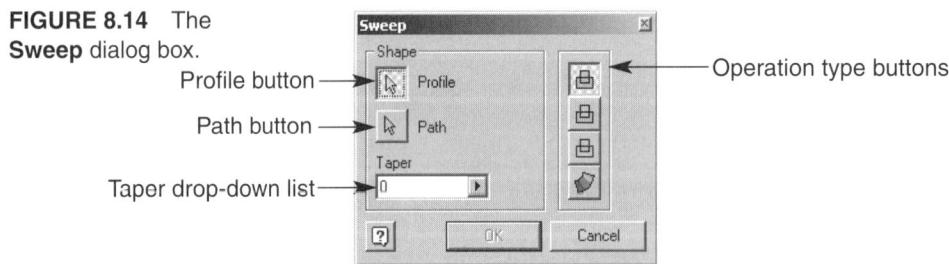

Once you have selected the sketch profile and sketch path, you may want to specify a taper angle using the **Taper** drop-down list. The taper angle defines the increasing or decreasing taper of the sweep profile as it follows the sweep path. The default angle of 0 applies no taper to the sweep, resulting in a feature that is completely controlled by the sketch profile. A positive taper angle constantly increases the size of the swept profile. See Figure 8.15. A negative taper angle constantly decreases the size of the swept profile. See Figure 8.16.

The next step in creating a sweep feature is to pick the desired operation type button. For simple base features, only the **Join** and **Surface** buttons are available. However, if other features are present in the part model, all the operation type buttons are active, and you can choose to join, cut, intersect, or create a surface, depending on the application.

Field Notes

The sweep operation type buttons function exactly the same as for other feature tools, such as **Extrude.** Refer to Chapter 4 for specific information regarding operation type buttons.

Once you have identified the profile and path, specified a taper if necessary, and selected the appropriate operation type, pick the **OK** button to generate the sweep feature.

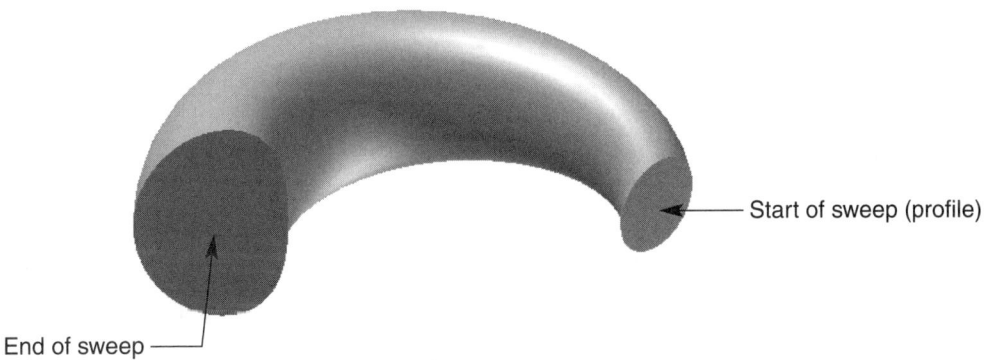

FIGURE 8.15   An example of the effects of a positive taper angle.

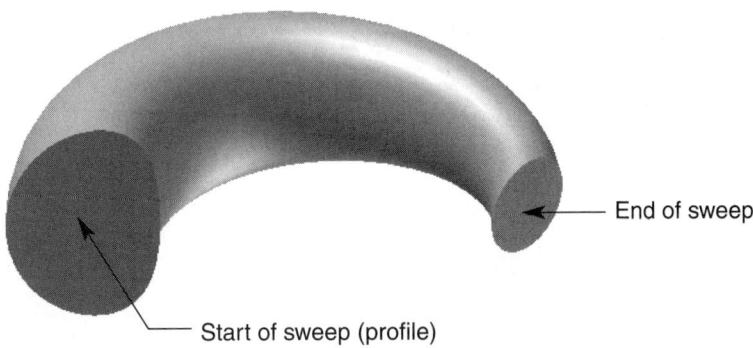

FIGURE 8.16   An example of the effects of a negative taper angle.

Exercise 8.3

1. Continue from Exercise 8.2, or launch Autodesk Inventor.
2. Open a new part file.
3. Open a sketch on the **XY Plane,** and sketch the geometry shown.
4. Open a new sketch on the **XZ Plane,** and sketch the geometry shown.
5. Use the **Sweep** tool to create a sweep feature using the following information:

   Operation Type: Join
   Taper: 0
   The profile and path should automatically be selected. If not, use **Sketch1** as the profile, and **Sketch2** as the path.

6. Open a new sketch on Face1, and create a ⌀.5″ circle at one of the corners, as shown.
7. Open a new sketch on Face2.
8. Using the **Autoproject edges during curve creation** and/or the **Automatic reference edges for new sketch,** project the Face2 sweep feature edges.
9. Use the **Sweep** tool to create a sweep feature using the following information:

   Operation Type: Cut
   Taper: 0
   Profile: **Sketch3**
   Path: **Sketch4**

10. The part should look like the part shown.
11. Save the part as EX8.3.
12. Exit Autodesk Inventor or continue working with the program.

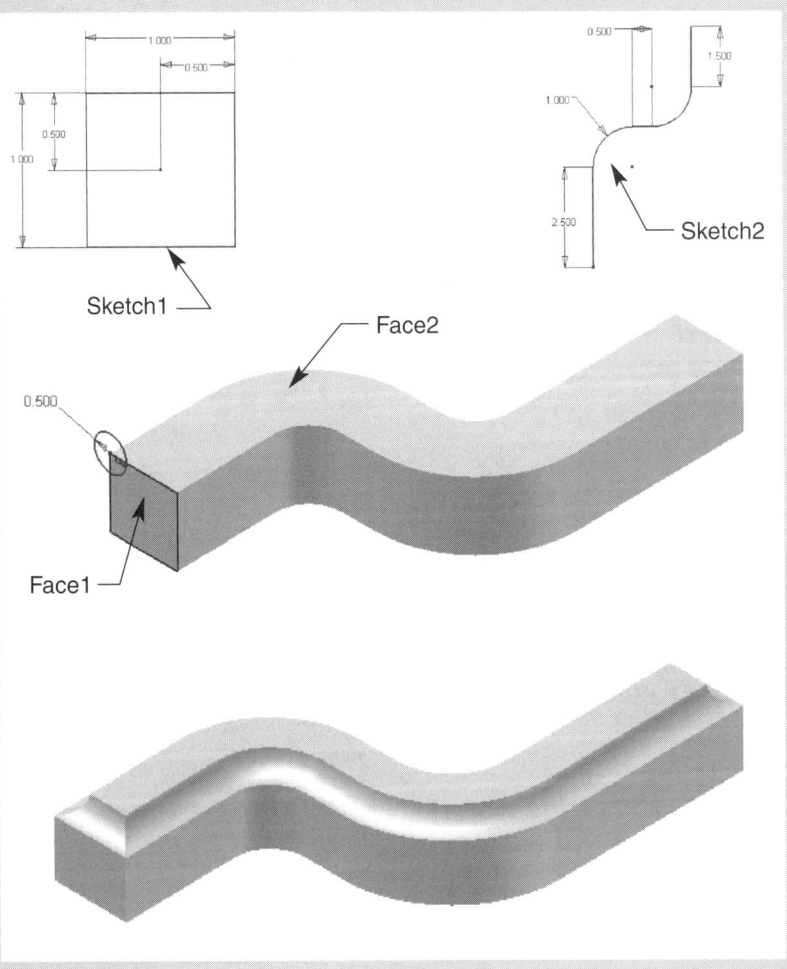

Creating Advanced Sweep Features

A basic sweep feature is generated by creating a two-dimensional sketch profile and a two-dimensional sketch path. An advanced sweep generates a sweep feature that follows a three-dimensional path instead of a two-dimensional path. Using a three-dimensional sketch allows you to create more complex sweeps. Typically a three-dimensional sweep feature is used to produce routed wires or pipes around existing features. Three-dimensional sweeps are usually placed in assemblies, but can also be constructed in parts. See Figure 8.17.

🖉 Field Notes

> The process for creating a three-dimensional sweep feature is the same whether you add the sweep to an assembly or build it as a part.

A three-dimensional sweep feature is very similar to a two-dimensional sweep feature. Once you have sketched a profile and path, the process for generating the sweep is exactly the same as the techniques previously discussed. The main difference between a two- and three-dimensional sweep feature is the path. A three-dimensional sweep feature is developed using a three-dimensional path, which is created using **3D Sketch** tools. Typically, you may want to use the following techniques to sketch a sweep profile and a 3D sweep path:

1. Open a new two-dimensional sketch and create a sweep profile.

2. Finish the 2D profile sketch.

3. Open a new three-dimensional sketch and create a 3D sketch path.

4. Finish the 3D path sketch.

5. Access the **Sweep** tool and generate the sweep feature.

The process for sketching a 2D sweep profile is exactly the same as previously discussed. Once the profile is complete, open a new **3D Sketch** using one of the following techniques:

✓ Right-click in the graphics window or the **Browser,** and select the **New 3D Sketch** menu option.

✓ Pick the **New 3D Sketch** flyout button on the **Command** bar.

When you have opened a new 3D sketch and are in 3D sketch mode, some specific 3D sketch environment and interface options are available.

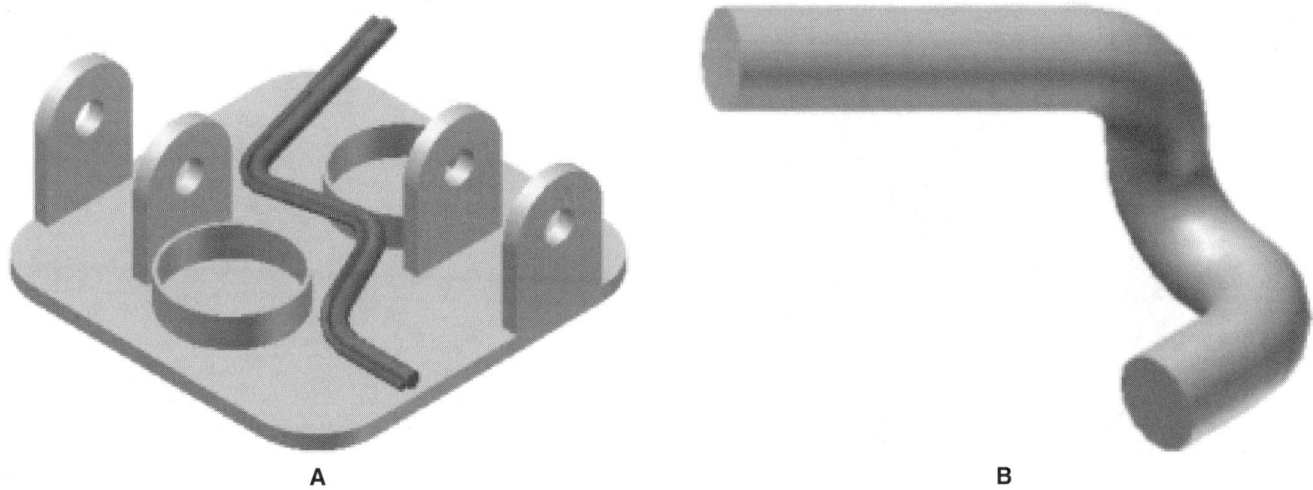

A   **B**

FIGURE 8.17   An example of a three-dimensional sweep feature (A) added to an assembly and (B) developed in the part environment.

Working in the 3D Sketch Environment and Interface

As previously discussed, the Autodesk Inventor interface is different for each stage of the design process. Some of the tools and interface components remain constant or are similar throughout the entire design. However, there are some specific 3D sketch environment components, discussed as follows:

- **Pull-Down Menus**   Most of the pull-down menus discussed in Chapter 2 apply to the 3D sketch environment. However, there are a few specific options that only relate to 3D sketches:

    - **Document Settings...**   Located in the **Tools** pull-down menu, the **Document Settings...** option opens the **Document Settings** dialog box. The only functions specific to the 3D sketch environment are the **3D Sketch** area of the **Sketch** tab, and the **Modeling** tab.

      The **Sketch** tab, shown in Figure 8.18, is specific to the sketch environment, and allows you to adjust the **Auto-Bend Radius** for a 3D sketch in the **3D Sketch** area. When you create a 3D line, using the **3D Line** command, you can choose to automatically place radii, or fillets, at the intersecting corners of two lines by specifying a radius greater than 0 in the **Auto-Bend Radius** edit box.

      The **Modeling** tab, shown in Figure 8.19, is specific to the 3D sketch environment and allows you to set adaptivity and 3D snap spacing. The **Adaptive used in assembly** check box is only available when working in an assembly, and the active part is adaptive. Unselecting the **Adaptive used in assembly** check box removes the adaptive indicator. Typically this check box should be unselected only if the part is not used in the current assembly. Pick the **Compact Model History** check box to purge the file when saved. The **3D Snap Spacing** edit box allows you to set the snap spacing for 3D sketching.

    - **Application Options...**   Located in the **Tools** pull-down menu, the **Application Options...** selection opens the **Options** dialog box. The only function specific to the 3D sketch environment is the **3D Sketch** area of the **Sketch** tab. The **Auto-Bend with 3D Line Creation** check box is located in the **3D Sketch** area. When creating lines in a 3D sketch environment, selecting this check box creates the radius, or corner bend, previously discussed, at the corners of lines without you having to place corner bends manually.

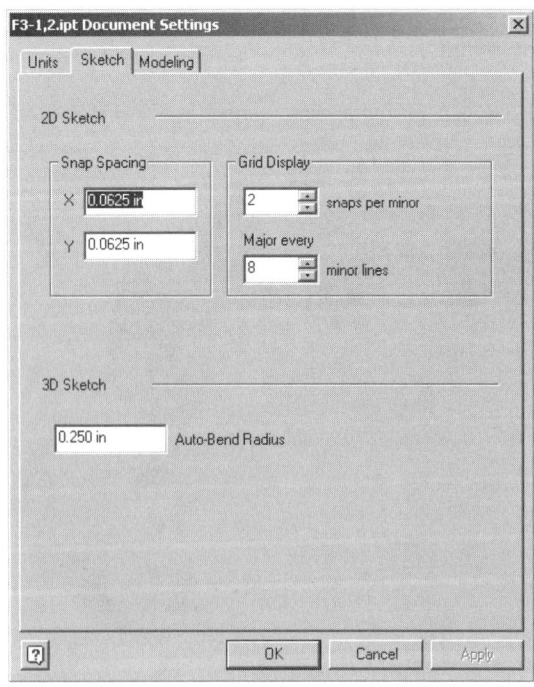

FIGURE 8.18   The **Sketch** tab of the **Document Settings** dialog box.

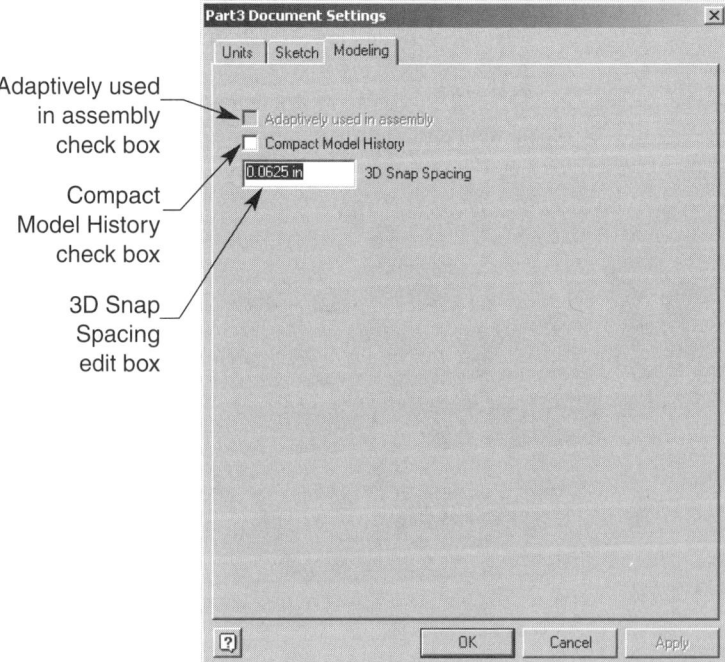

FIGURE 8.19   The **Modeling** tab of the **Document Settings** dialog box.

Using 3D Sketching Tools

3D sketching tools are used to create 3D sweep feature paths. These tools are accessed from the **3D Sketch** toolbar, shown in Figure 8.20, or from the **Sketch** panel bar. When a sketch is active, the panel bar is in sketch mode. See Figure 8.21. By default, the panel bar is displayed in learning mode, as seen in Figure 8.21. To change from learning to expert mode, shown in Figure 8.22 and discussed in Chapter 2, right-click any of the panel bar tools or left-click the panel bar title, and select the **Expert** option.

Creating 3D Lines

Three-dimensional sketches are created in 3D space and are not placed on a plane, or planar face, like a two-dimensional sketch. The only way to sketch a three-dimensional line is to reference work points, or existing feature edge endpoints. Consequently, before you can make a 3D line, two or more work points, or edge endpoints must be available. The **Work Plane, Work Axes,** and **Work Points** tools function exactly the same as those discussed in Chapter 7.

Field Notes

> Refer to Chapter 7 for information about generating work points.
>   While using the **Work Point** tool, right-click and select the **Create Plane** or **Create Axis** menu options to plane a work plane or work axis.

Once you have placed the desired work points or are able to access edge endpoints, access the **Line** command and use one of the following techniques:

✓ Pick the **Line** button on the panel bar.

✓ Pick the **Line** button on the **3D Sketch** toolbar.

✓ When no other commands are active, right-click inside the graphics window and select the **Create 3D Line** menu option.

To sketch a 3D line, pick the work points or endpoints in sequential order. See Figure 8.23. If you want to continue the line, pick additional points. If you do not want to continue the line, press the **Esc** key on your keyboard or right-click and pick the **Done** option.

If the **Auto-Bend with 3D Line Creation** check box, previously discussed, is activated, bends are automatically placed at the line corners. The radius of the bend depends on the value specified in the **Auto-Bend Radius** edit box, previously discussed. An alternative method for activating or deactivating the **Auto-Bend with 3D Line Creation** tool is to access the **Line** command, right-click, and select the **Auto-bend** menu option.

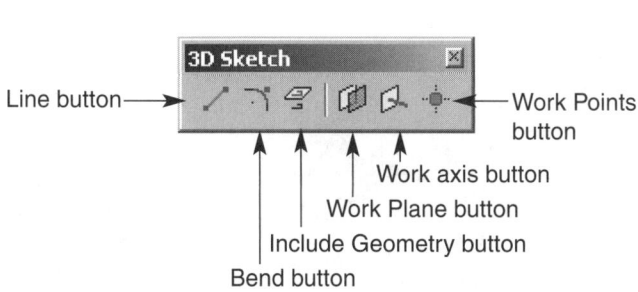

Line button → 3D Sketch — Work Points button
Work axis button
Work Plane button
Include Geometry button
Bend button

FIGURE 8.20   The **3D Sketch** toolbar.

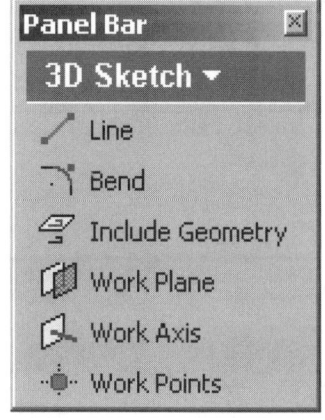

FIGURE 8.21   The panel bar in 3D sketch (learning) mode.

FIGURE 8.22   The panel bar in 3D sketch (expert) mode.

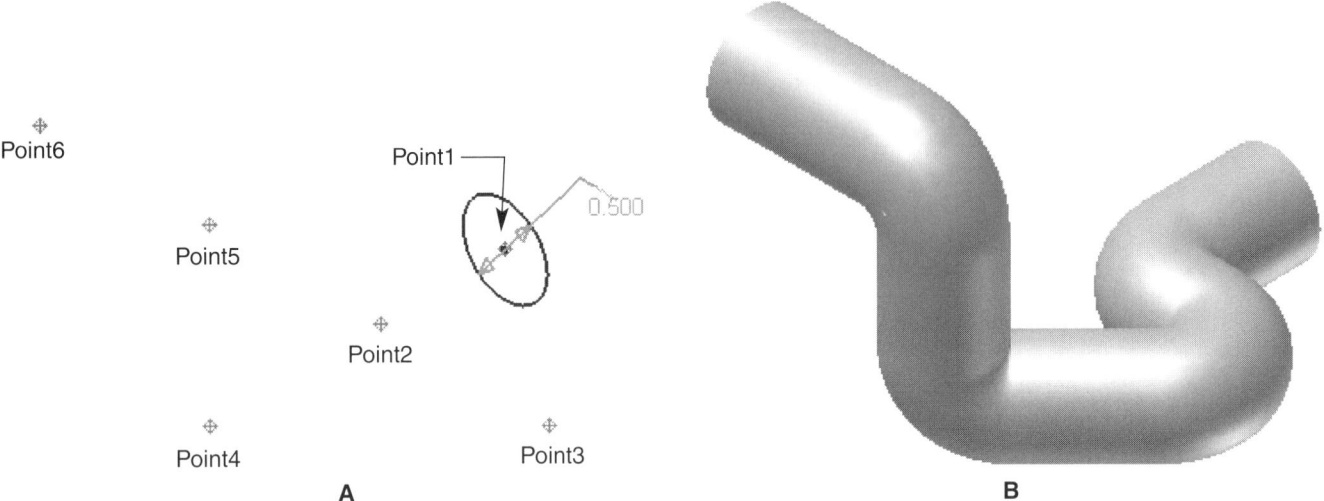

FIGURE 8.23   (A) Picking points to create a 3D sketch. (B) The resulting 3D sweep feature.

If the **Auto-Bend with 3D Line Creation** tool is not activated, no fillet is placed at the 3D line corners. If you would like to add bends to corners that do not have them, access the **Bend** tool, using one of the following options:

✓ Pick the **Bend** button on the panel bar.

✓ Pick the **Bend** button on the **3D Sketch** toolbar.

Accessing the **Bend** tool opens the **3D Sketch Bend** dialog box, shown in Figure 8.24. To create a 3D sketch bend without using the **Auto-Bend with 3D Line Creation** option, first specify a radius in the drop-down list by typing in a radius or picking a radius from the list. Then select the **Equal** button if you would like to create bends with equal radii, without adding additional dimensions. Next you can see a preview of the bend by moving your cursor to the corner of the lines, or by selecting one line and moving your cursor over the next line. If the fillet looks acceptable in the preview, select the intersecting corners of the lines or the two lines that you would like to fillet. See Figure 8.25.

If you want to use existing 2D sketch or model elements and convert them into 3D model geometry, access the **Include Geometry** tool using one of the following options:

✓ Pick the **Include Geometry** button on the panel bar.

✓ Pick the **Include Geometry** button on the **3D Sketch** toolbar.

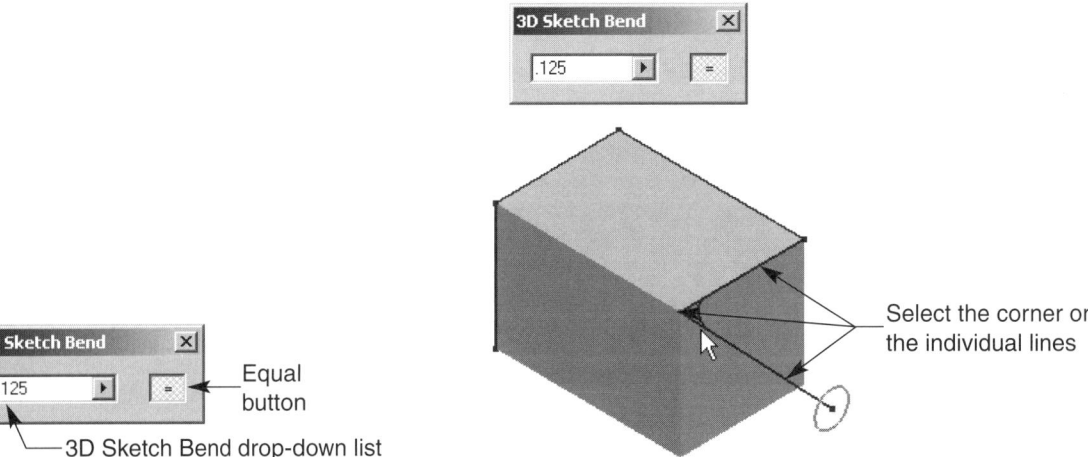

FIGURE 8.24   The **3D Sketch Bend** dialog box.

FIGURE 8.25   Creating a 3D sketch bend.

Then pick the 2D sketch geometry and/or the model edges you want to add to the 3D sketch. When you have specified all the included geometry, press the **Esc** key on your keyboard, or right-click and pick the **Done** option.

Once you have generated a 2D sketch profile and a 3D sketch path, right click and select the **Finish 3D Sketch** menu option, or pick the **Return** button on the **Command** bar. Then, use the techniques previously discussed to generate the 3D sweep feature.

Exercise 8.4

1. Continue from Exercise 8.3, or launch Autodesk Inventor.
2. Open a new part file.
3. Open a 2D sketch on the **YZ Plane.**
4. Sketch a ∅15mm circle around the projected **Center Point.**
5. Finish the sketch.
6. Open a 3D sketch.
7. Offset a work plane 30mm from the **YZ Plane.**
8. Offset a work plane 45mm from the **XY Plane.**
9. Offset a work plane 45mm from the **XZ Plane.**
10. Offset a work plane −60mm from the **XZ Plane.**
11. Offset a work plane 45mm from the **YZ Plane.**
12. Place a work point at the center of the 2D sketch circle.
13. Place a work point by selecting **Work Plane1** and the **X Axis.**
14. Place a work point by selecting **Work Plane2, Work Plane3,** and the **YZ Plane.**
15. Place a work point by selecting **Work Plane1, Work Plane2,** and **Work Plane4.**
16. Place a work point by selecting **Work Plane2, Work Plane5,** and the **XZ Plane.**
17. Ensure the **Auto-Bend with 3D Line Creation** option is activated, and specify a radius of 10mm in the **Auto-Bend Radius.**
18. Use the Line tool to create a line between **Work Points 1, 2, 3, 4, and 5.**
19. Finish the 3D sketch.
20. Use the **Sweep** tool to create a sweep feature using the following information:

    Operation Type: Join
    Taper: 0
    Profile: 2D **Sketch1**
    Path: 3D **Sketch1**

21. The part should look similar to the part shown.
22. Save the part as EX8.4.
23. Exit Autodesk Inventor or continue working with the program.

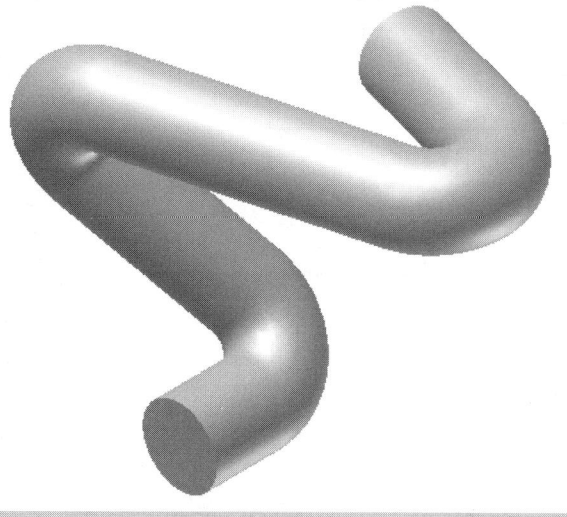

Exercise 8.5

1. Continue from Exercise 8.4, or launch Autodesk Inventor.
2. Open a new part file.
3. Open a 2D sketch on the **XY Plane.**
4. Sketch a 50mm×100mm rectangle.
5. Finish the sketch.
6. Extrude the sketch 75mm in a positive direction.
7. Place an R 25mm fillet at the corner shown.
8. Offset a work plane 150mm from the **YZ Plane.**
9. Offset a work plane −50mm from the **XY Plane.**
10. Open a 2D sketch on **Work Plane1,** and sketch the circle shown.
11. Open a 3D sketch.
12. Place work points at the locations shown.
13. Ensure the **Auto-Bend with 3D Line Creation** option is not activated.
14. Use the **Include Geometry** tool and the **Line** tool to create a line between the points shown.
15. Use the **3D Sketch Bend** tool to place bends at the locations shown using the specified radii.
16. Finish the 3D sketch.
17. Use the **Sweep** tool to create a sweep feature using the following information:

    Operation Type: Join
    Taper: 0
    Profile: 2D **Sketch1**
    Path: 3D **Sketch1**

18. The part should look similar to the part shown.
19. Save the part as EX8.5.
20. Exit Autodesk Inventor or continue working with the program.

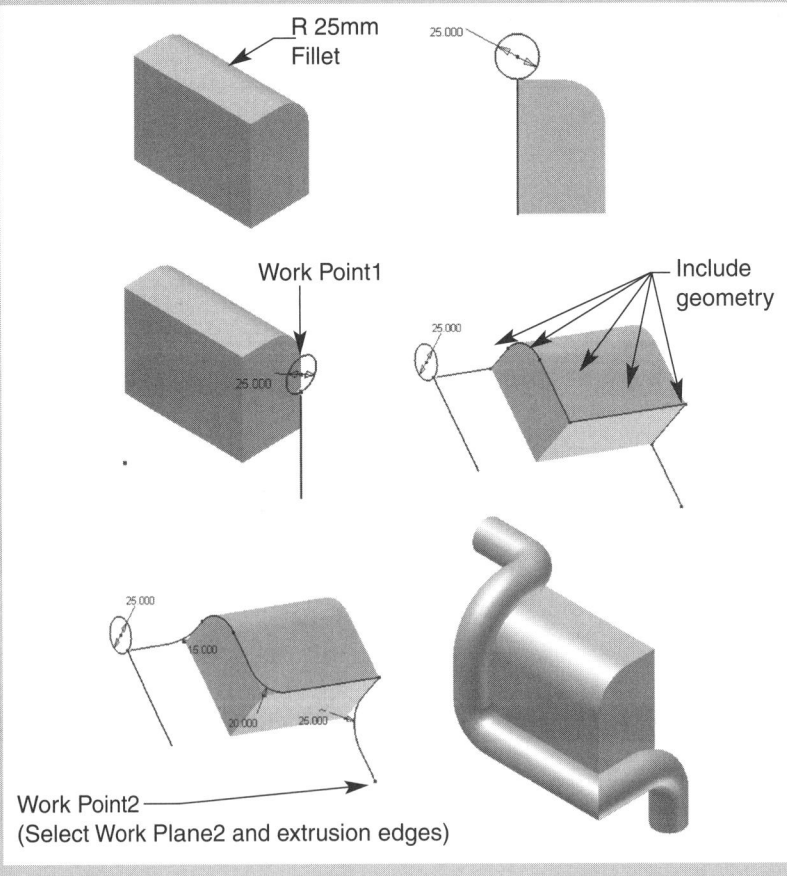

Field Notes

You may want to create an advanced sweep around an existing feature in a part file, as shown in Exercise 8.5. However, typically the sweep is created as a separate part file, while in an assembly.

Creating Split Features

As the name implies, a *split* divides a feature into two separate pieces. Splits reference an existing feature and an additional separation sketch or plane. You may find a number of uses for the split tool, but usually splits are used for casting and forging applications. Autodesk Inventor allows you to split a feature two different ways depending on the purpose. One option is to keep existing feature geometry, and split feature faces. See Figure 8.26. Another option is to remove a section of an existing feature. See Figure 8.27.

Regardless of the technique you use to split a feature, you must create an additional sketch or face (typically a work plane), which acts as the separation line or surface. Use one of the following options to create a parting line or plane for a split:

- Open a 2D sketch on a face or work plane, and sketch geometry using any of the two-dimensional tools. See Figure 8.28.

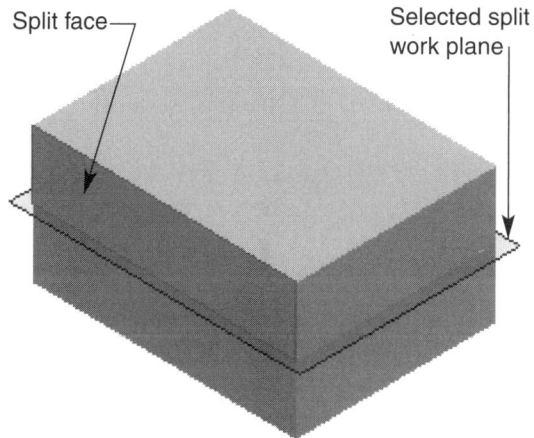

FIGURE 8.26 Splitting feature faces. The original feature, an extrusion, does not actually change shape, but the selected faces are split.

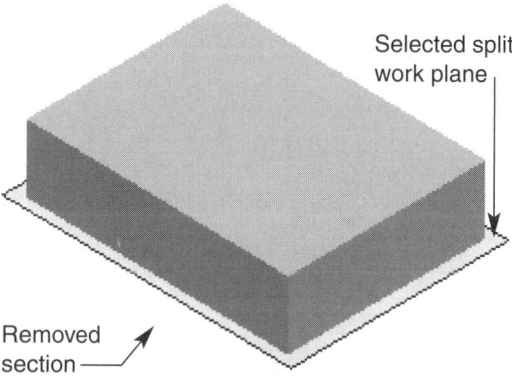

FIGURE 8.27 Splitting a feature. The bottom half of the original feature, an extrusion, is removed.

FIGURE 8.28
Creating a split parting line using a 2D sketch and 2D sketch tools (**Line** tool).

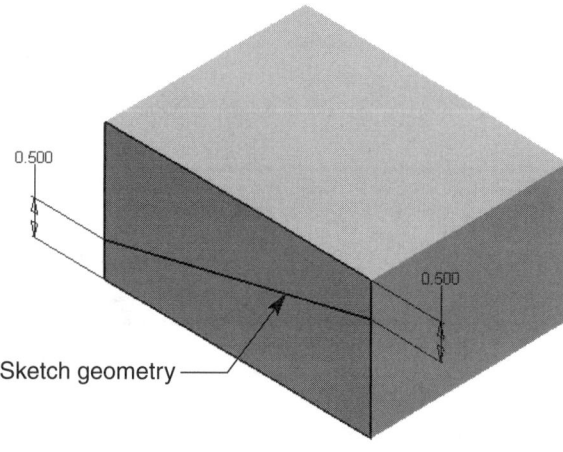

Field Notes

When using the **Spline** tool to create a splitting line, you may want to place and constrain points using the **Point, Hole Center** tool. Then access the **Spline** tool, and connect the points.

- Open a 2D sketch on a face or work plane, and sketch geometry using any of the two-dimensional tools. Then extrude the sketch using the **Surface** operation type option to create a plane. See Figure 8.29.

- Add a work plane to the feature. See Figure 8.30.

Split features are created using the **Split** tool. Access the **Split** command using one of the following techniques:

✓ Pick the **Split** button on the panel bar.

✓ Pick the **Split** button on the **Features** toolbar.

When you access the **Split** command, the **Split** dialog box is displayed. See Figure 8.31.

By default, when you access the **Split** dialog box, the **Split Face** button is active, which allows you to split feature faces without removing a section of the feature. To use the **Split Face** option, pick the **Split Tool** button if it is not already activated, and select the parting line or plane. The next step is to specify the faces you want to split. By default, the **Select** button is active, which allows you to select the specific faces you want to split. This tool is useful for certain applications, such as when you want to apply different draft angles to various split faces. If you only want to split certain faces, ensure the **Select** button and **Faces to Split** buttons are active. Then pick the faces you want to split, followed by the **OK** button. If you want to split all the faces around the entire feature, select the **All** button, followed by the **OK** button.

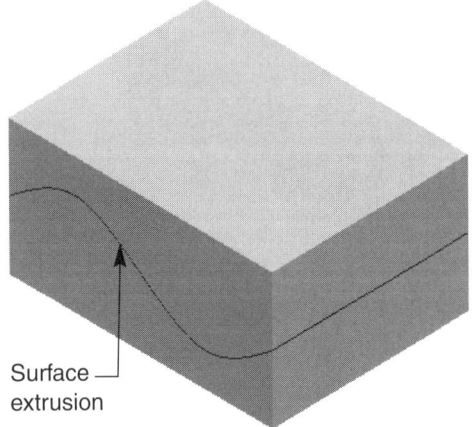

FIGURE 8.29 Creating a split parting line by surface-extruding 2D sketch geometry (spline).

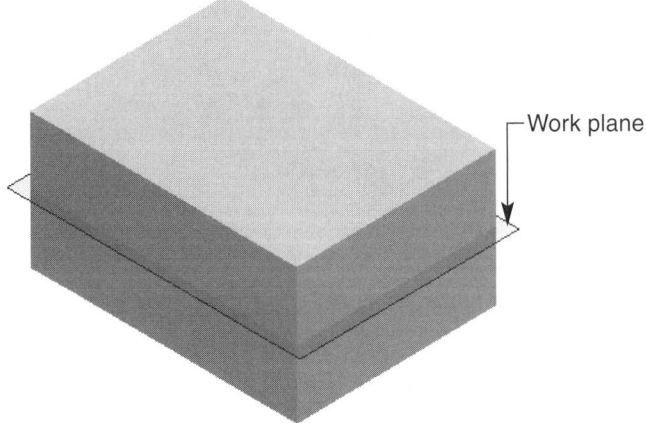

FIGURE 8.30 Creating a split parting line by adding a work plane.

FIGURE 8.31 The **Split** dialog box (**Split Face** method mode).

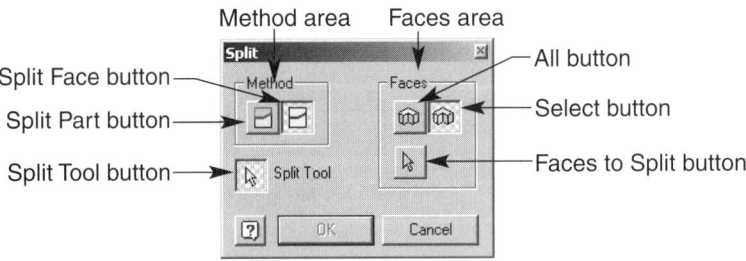

Field Notes

You can only directly access split faces using the **Face Draft** tool. In order to use the split faces for other purposes, such as developing sketch geometry, you must open a new sketch on the face, and project the split edges.

If you want to remove a section of the feature using the **Split** tool, pick the **Split Part** button of the **Split** dialog box. When you select the **Split Part** button, the **Split** dialog box changes. See Figure 8.32.

To use the **Split Part** option, pick the **Split Tool** button if it is not already activated. Then, select the parting line or plane. The next step is to specify which section of the part you want to remove by picking one of the **Side to Remove** buttons. Finally, pick the **OK** button to remove the section of the feature.

Typically, you may want to develop two separate parts when using the **Split Part** option. For example, you may want to create a top half and a bottom half. The two halves should be separate part files, such as Top.ipt and Bottom.ipt. You can then place both parts into an assembly file to make an assembly. To create two part files, use the following steps before you use the split tool:

1. Save the current part with both sections of the part in one piece, and the parting line created. The name of the file should be Top.ipt, for example.

2. Using the **Split** tool, remove a section of the part. Using the previous example, you would want to remove the bottom section.

3. Select **Save Copy As...** from the **File** pull-down menu and save a copy of the current part with one of the sections removed by the **Split** tool. Using the previous example, you would want to save the copy as Bottom.ipt.

4. The initial file, in this example, Top.ipt, should currently be opened. Save and close the active file when you are finished creating the part.

5. Open the saved copy of the original part. Using the previous example, you would want to open Bottom.ipt.

6. Right click the split in the **Browser** bar, and select the **Edit Feature** menu option.

7. Select the opposite **Side to Remove** button to flip the section to be removed, and pick the **OK** button.

8. Save and close the active file when you are finished creating the part.

Field Notes

The steps previously discussed for developing two separate "halves" of a part represent one possible technique. You may develop your own techniques depending on your applications.

FIGURE 8.32 The **Split** dialog box (**Split Part** method mode).

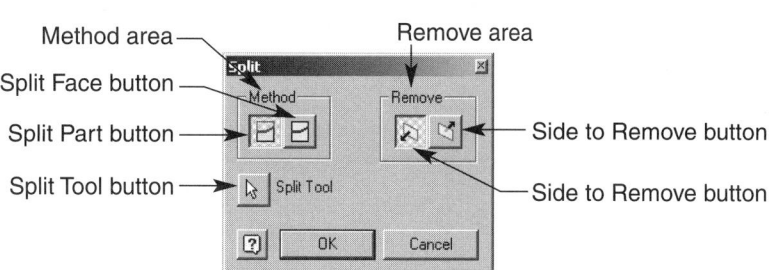

Exercise 8.6

1. Continue from Exercise 8.5, or launch Autodesk Inventor.
2. Open a new part file.
3. Open a sketch on the **XY Plane,** and sketch the geometry shown.
4. Extrude the Sketch 2″ in appositive direction.
5. Offset a work plane −1″ from the bottom extrusion face.
6. Use the **Split** tool to split the extrusion using the following information:

    Method: Split Face
    Split Tool: Work Plane1
    Faces: All

7. Each of the Z Axis faces should be split in half as shown.
8. Save the part as EX8.6.
9. Exit Autodesk Inventor or continue working with the program.

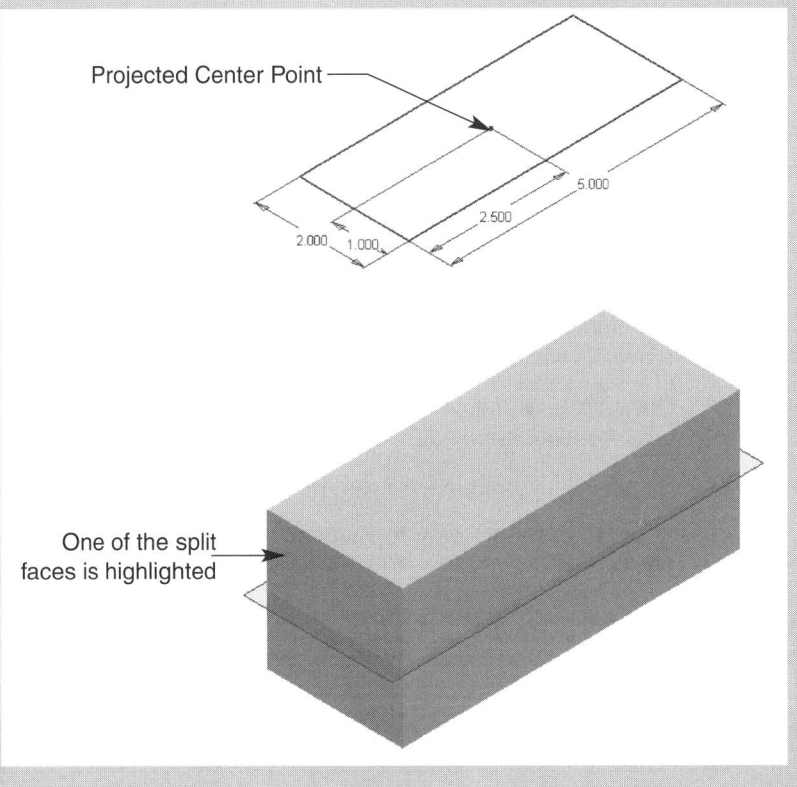

Exercise 8.7

1. Continue from Exercise 8.5, or launch Autodesk Inventor.
2. Open a new part file.
3. Open a sketch on the **XY Plane,** and sketch the geometry shown.
4. Extrude the Sketch 2″ in appositive direction.
5. Open a new sketch on the face shown and place the specified points.
6. Connect the points together using the **Spline** tool.
7. Finish the sketch.
8. Use the **Split** tool to split the extrusion using the following information:

   Method: Split Part
   Split Tool: Sketch2
   Remove: Top section

9. The part should look like the part shown.
10. Save the part as EX8.7.
11. Exit Autodesk Inventor or continue working with the program.

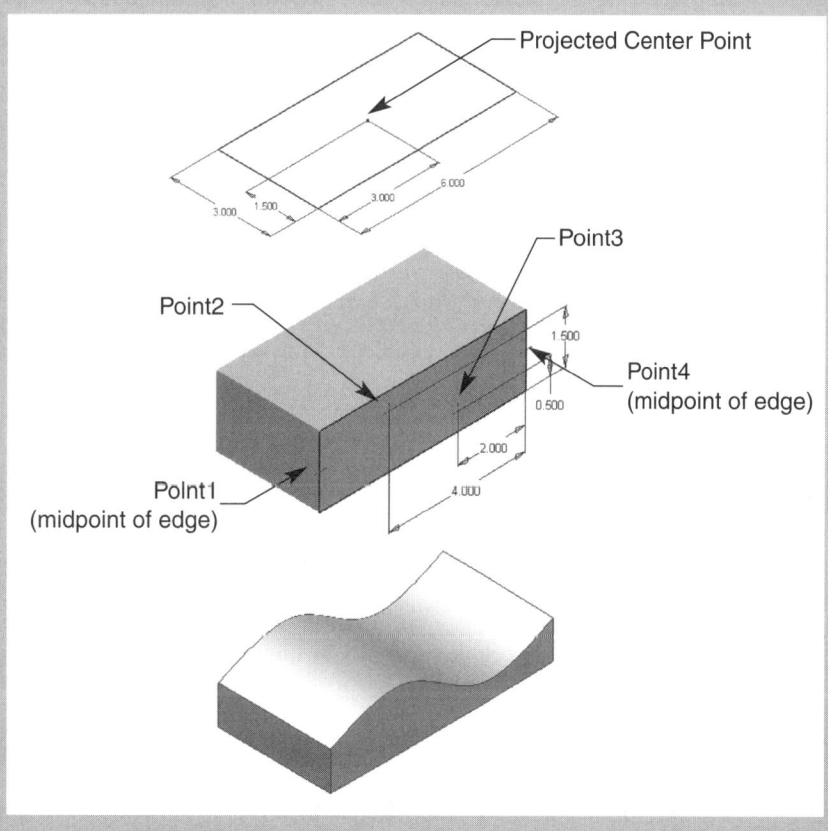

|||||||||||| CHAPTER TEST

1. By default, the panel bar is displayed in learning mode. How do you change it to expert mode?

2. What is a loft?

3. What must you do before you can create a loft feature?

4. How are sketches used to create loft features referred to?

5. Give the first step used to create a loft feature after sketches have been created.

6. What do you do if the Sections list is not active?

7. When creating a loft, which sketches do you select first?

8. From where do you select the desired sketches for a loft if you want a preview arrow identifying the direction of the loft to be shown?

9. The next step in creating a loft feature is to adjust the desired shape controls in the Shape Control area. What does this area allow you to do?

10. Give the function of the Angle drop-down list in the Shape Control area of the Loft dialog box.

11. What is the default loft angle and when can it be modified?

12. Discuss the function of the Weight drop-down list in the Shape Control area of the Loft dialog box.

13. Explain the purpose and function of loft weights.

14. Discuss the function of the Tangent to Face check box in the Shape Control area of the Loft dialog box.

15. Give the function of the Closed Loop check box.

16. Explain the function of Point Mapping.

17. How do you access the Point Mapping list box and how do you select points?

18. What is a sweep and what do sweeps allow you to do?

19. Before you can create a sweep feature, you must create two or more sketches. Identify at least three conditions that must be maintained when creating sweep sketches.

20. Once you have selected the sketch profile and sketch path, you may want to specify a taper angle using the Taper drop-down list. Give the function of the taper angle.

21. What is the difference between basic and advanced sweeps?

22. Give the typical techniques that you may want to use to sketch a sweep profile and a 3D sweep path.

23. Once the profile is complete, identify two options that can be used to open a new 3D Sketch.

24. Three-dimensional sketches are created in 3D space and are not placed on a plane, or planar face, like a two-dimensional sketch. Identify the only way to sketch a three-dimensional line.

25. How do you create a 3D sketch bend without using the Auto-Bend with 3D Line Creation option?

26. What do you do if you want to use existing 2D sketch or model elements and convert them into 3D model geometry?

27. What is a split feature?

28. Identify at least two options for creating a parting line or plane for a split.

29. By default, when you access the Split dialog box, the Split Face button is active. What does this button allow you to do and how is it used?

30. What do you do if you want to remove a section of the feature, using the Split tool?

31. Typically, you may want to develop two separate parts when using the Split Part option. You can then place both parts into an assembly file to make an assembly. Give the steps used to create two part files.

|||||||||||| PROJECTS

Follow the specific instructions provided for each project.

1. Name: Nozzle

   Units: Metric

   Specific Instructions:

   ■ Open EX8.1 and Save Copy As P8.1.

   ■ Close EX8.1 without saving and Open P8.1.

   ■ Open a new sketch on Face1, and offset the projected edges 50mm.

   ■ Open a new sketch on Face2, and offset the projected edges 15mm.

   ■ Use the **Loft** tool to create a loft feature using the following information:

   Sections: **Sketch3** and **Sketch4**

   Operation Type: Cut

   Angle: 90°

   Weight: 100 ul

   ■ Fillet all fillets and all rounds R 5mm.

   ■ The final part should look like the part shown.

   ■ Save As: P8.1.

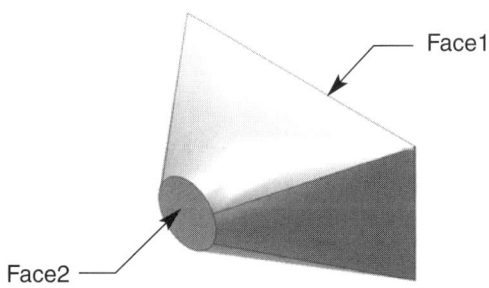

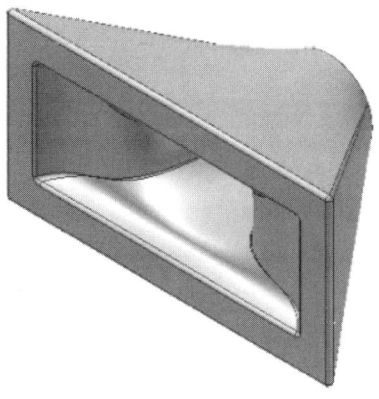

2. Name: Tiller Blade

   Units: Inch or Metric

   Specific Instructions:

   ■ Open a new part file.

   ■ Create a tiller blade similar to the one shown using the following techniques:

   1. Place three equally spaced work planes on the same plane.

   2. Open sketches on the work planes, and sketch rectangles at different angles.

   3. Loft the sketches using point mapping.

   4. Fillet the sharp end corners.

   5. Extrude a center support.

   6. Place a hole in the center of the center support.

   7. Circular pattern the loft feature.

   ■ The final part should look like the part shown.

   ■ Save As: P8.2.

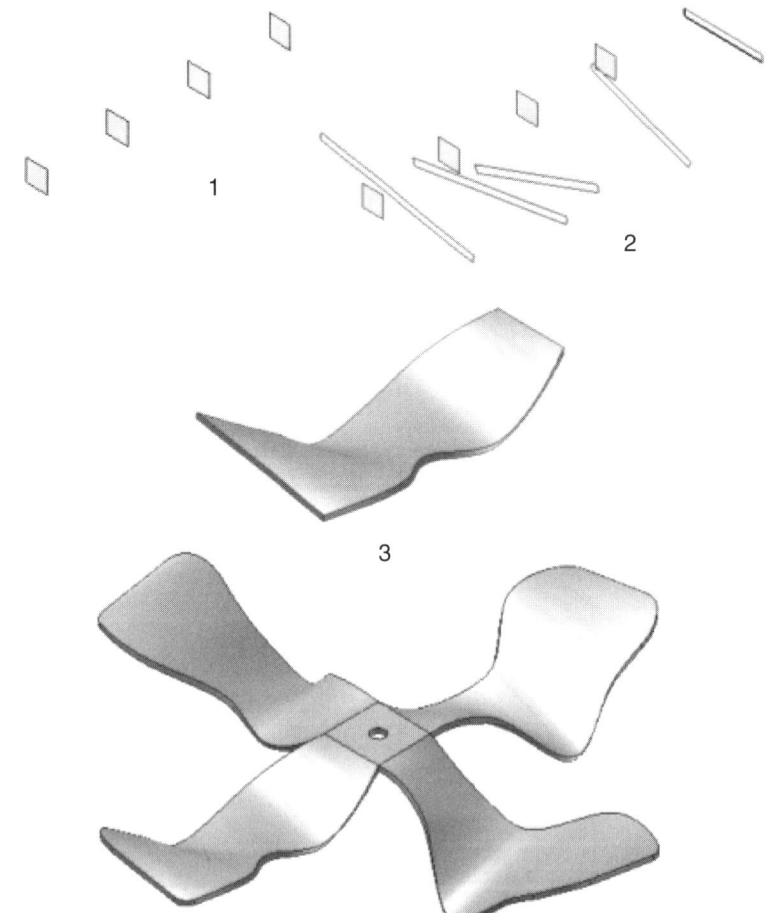

3. Name: Handle

Units: Inch

Specific Instructions:

- Open a new part file.

- Create the sketch shown on the **XY Plane.**

- Extrude the sketch .5″ in a positive direction.

- Offset a work plane, 1.75″ from the **YZ Plane.**

- Sketch the geometry shown on **Work Plane1.**

- Open a new sketch on the top face of one of the extruded rectangles, and project the top face edges.

- Open another new sketch on the top face of the other extruded rectangle, and project the top face edges.

- Use the **Loft** tool to create a loft feature shown using the following information:

  Sections: **Sketch3,** followed by **Sketch2,** followed by **Sketch4.**

  Operation Type: Join

  Weight: 0 ul

- Place the specified holes in the centers of the extrusion bases, as shown.

- The final part should look like the part shown.

- Save As: P8.3.

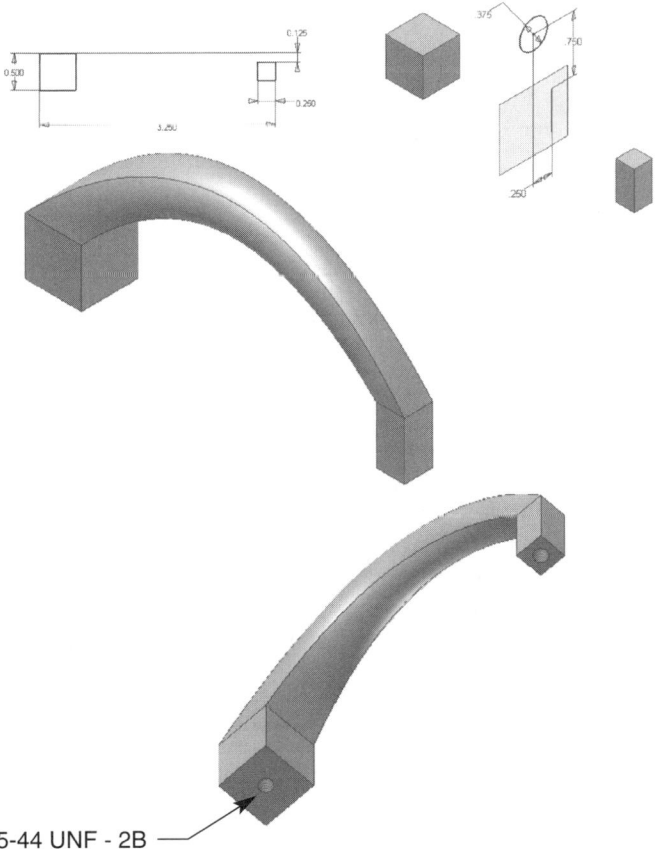

5-44 UNF - 2B

4. Name: 90° Reducing Elbow

Units: Inch

Specific Instructions:

- Open a new part file.

- Sketch a ∅2″ circle around the projected **Center Point** on the **XY Plane.**

- Open a new sketch on the **XZ Plane**, and sketch the arc shown.

- Use the **Sweep** tool to create a sweep feature using the following information:

  Operation Type: Join

  Taper: −6°

  Profile: **Sketch1**

  Path: **Sketch2**

- Shell the sweep feature to create .125″ thick walls, and remove the end faces.

- Add .1″ chamfers to the end edges.

- The final part should look like the part shown.

- Save As: P8.4.

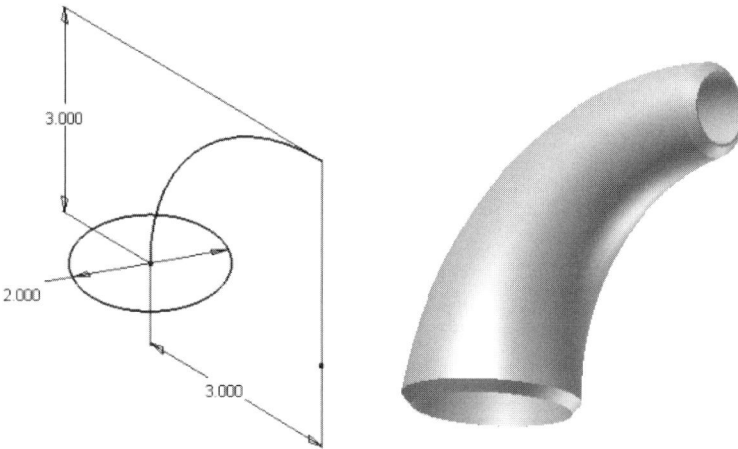

5. Name: Fork Body

Units: Metric

Specific Instructions:

■ Open a new part file.

■ Sketch a 4mm×4mm rectangle, centered around the projected **Center Point,** on the **YZ Plane.**

■ Open a new sketch on the **XY Plane,** and sketch the geometry shown.

■ Use the **Sweep** tool to create a sweep feature using the following information:

Operation Type: Join

Taper: 0

Profile: **Sketch1**

Path: **Sketch2**

■ Open a sketch of Face1, and project the edges.

■ Extrude **Sketch3,** 150mm in a positive direction.

■ Place 2.5mm×15mm chamfers of the front edges of the fork, as shown.

■ Fillet all fillets 2mm, and all rounds .5mm.

■ The final part should look like the part shown.

■ Save As: P8.5.

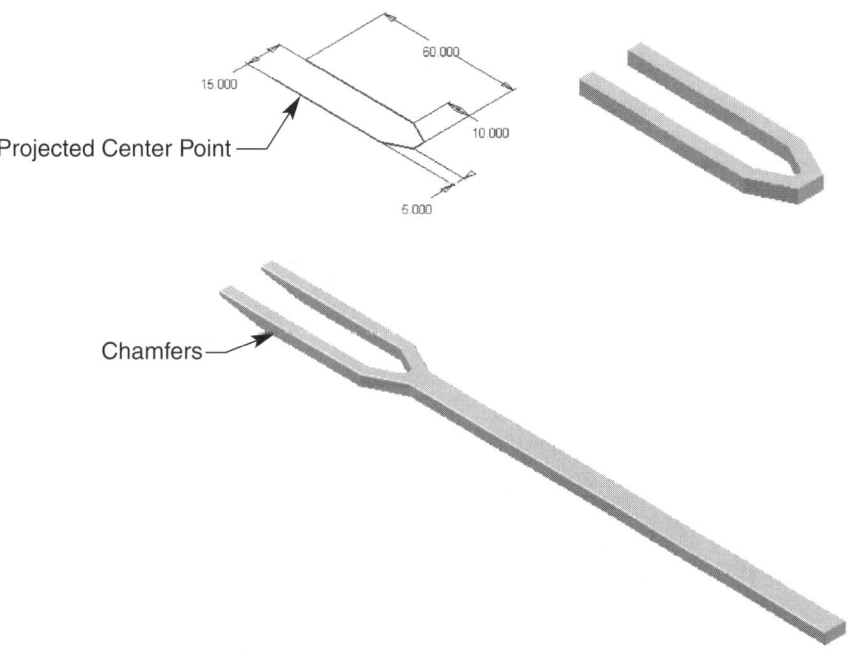

6. Name: Wire

Units: Inch or Metric

Specific Instructions:

■ Open a new part file.

■ Create an extrusion similar to the extrusion shown.

■ Using the 3D sweep techniques discussed and the following steps, create a wire similar to the wire shown.

    1. Open a 2D sketch and sketch the wire profile, a circle.

    2. Finish the 2D sketch.

    3. Open a 3D sketch, and offset the required work planes.

    4. Use the work planes to place work points, similar to the work planes and points shown.

    5. Connect the work points with a line and bends. This creates the sweep path.

    6. Finish the 3D sketch.

    7. Access the **Sweep** tool, and generate the sweep feature.

■ The final part should look like the part shown.

■ Save As: P8.6.

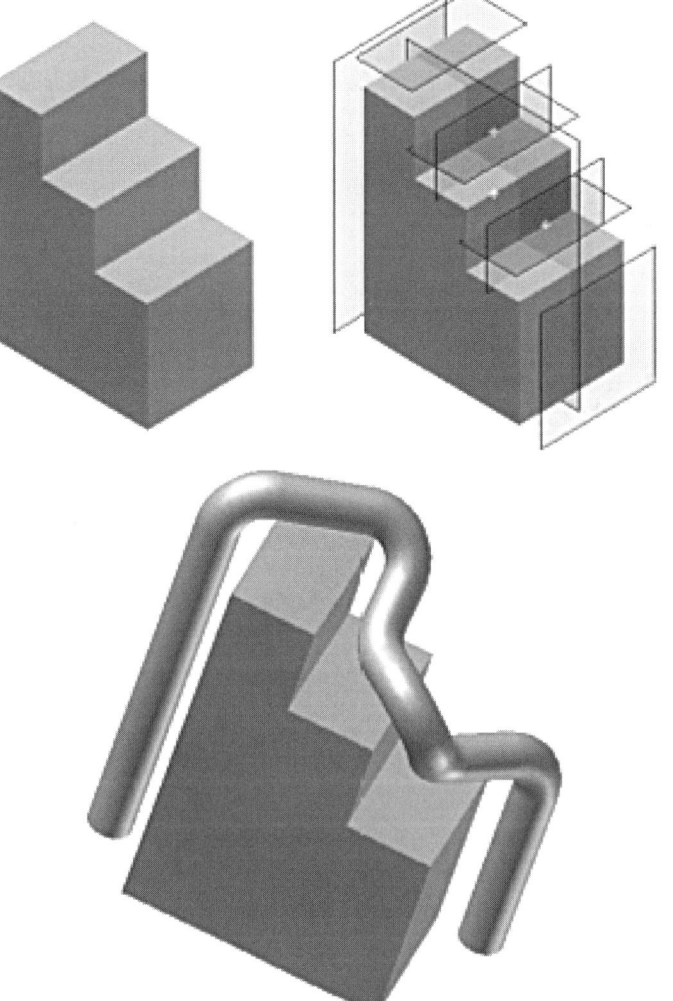

7. Name: Casting

Units: Inch

Specific Instructions:

- Open EX8.6, and Save Copy As P8.7.

- Close EX8.6 without saving, and Open P8.7.

- Apply 2.5° of face draft to the top four split faces.

- Apply 5° of face draft to the bottom four split faces.

- The final part should look like the part shown.

- Save As: P8.7.

8. Name: Soap Pattern Top

Units: Inch

Specific Instructions:

- Open a new part file.

- Sketch a 4″×6″ rectangle on the **XY Plane.**

- Extrude the sketch 3″ in a positive direction.

- Offset a work plane −1.5″ from the top face of the extrusion.

- Use the **Split** tool to split the extrusion using the following information:

  Method: Split Part

  Split Tool: Work Plane1

  Remove: Bottom section

- Open a new sketch on the bottom face of the split feature.

- Offset the projected edges of the rectangle .5″ as shown.

- Cut extrude the sketch 1″ in a positive direction.

- Fillet All Fillets with a R .5″.

- The final part should look like the part shown.

- Save As: P8.8.

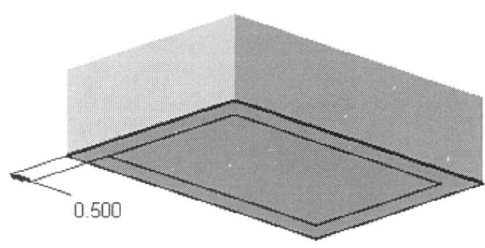

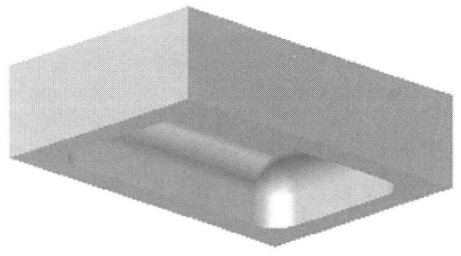

9. Name: Soap Pattern Bottom

Units: Inch

Specific Instructions:

- Open P8.8, and Save Copy As P8.9.
- Save and Close P8.8, and Open P8.9.
- Delete **Extrusion2** and **Fillet1.**
- Right-click **Split1,** and select the **Edit Feature** menu option.
- Pick the opposite **Side to Remove "Flip"** button, and remove the top half of the extrusion.
- Open a new sketch on the top face of the split feature.
- Offset the projected edges of the rectangle .5″ as shown.
- Cut extrude the sketch 1″ in a negative direction.
- Fillet All Fillets with an R .75″.
- Save As: P8.9.

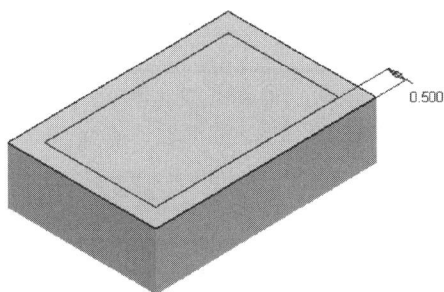

10. Name: Calculator

Units: Inch or Metric

Specific Instructions:

■ Create a calculator face and base similar to the face and base shown.

■ Generate the parts using the techniques discussed in this chapter and previous chapters.

■ Although you may develop your own methods for creating the calculator parts, you may want to refer the following list for assistance:

1. Extrude the entire calculator (face and base intact).

2. Place fillets.

3. Add work planes, or a sketch, to define the split parting line.

4. Save the part as Face.ipt.

5. Split the feature, and remove the bottom (base) section.

6. Save a copy of Face.ipt, as Base.ipt.

7. Create button slots by cut extruding rectangles.

8. Rectangular pattern certain button slots.

9. Place holes, and add fillets.

10. Save and close the active file (Face.ipt).

11. Open Base.ipt, and edit the split to remove the top (face) section.

12. Extrude base feet.

13. Save and close the active file (Base.ipt).

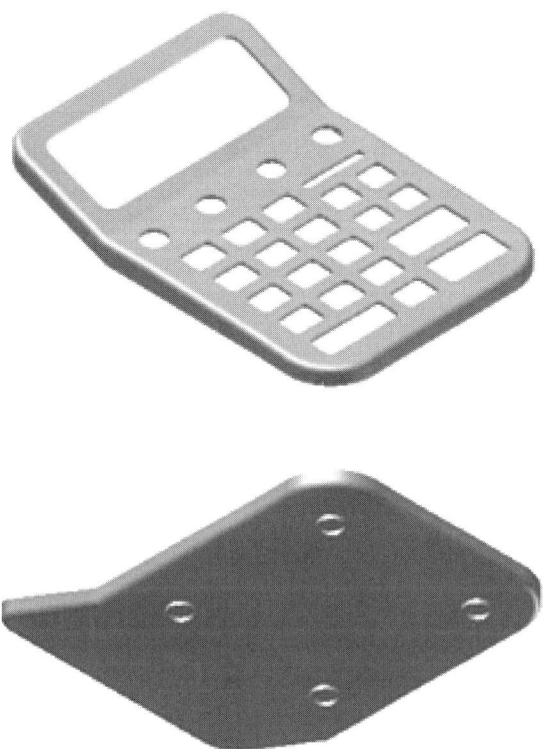

|||||||||||||||||||||||||||||||||||||||||||||||||||||||||||||||||||||||||||||||||||||||||||||||||||||

CHAPTER 9

Creating and Using Catalog Features

LEARNING GOALS

After completing this chapter, you will be able to:

◎ Use design element (iFeature) tools.
◎ View and manage iFeature files.
◎ Create iFeatures.
◎ Insert iFeatures.

Creating and Using Catalog Features

Catalog features are similar to placed features, because you can place them onto an existing part or feature. However, catalog features can be far more complex than a standard shell, fillet, chamfer, or face draft, and consist of several previously created features. See Figure 9.1. Catalog features, also known as *design elements* or *iFeatures*, are existing features you create, such as the three holes and the fillet, shown in Figure 9.1. The features are saved and stored as a single design element, in a catalog, to be used in other models. When inserted into other models, the iFeatures can retain their original size and position values, or can be modified to fit a different design application.

FIGURE 9.1 An example of a simple design element (iFeature or catalog feature).

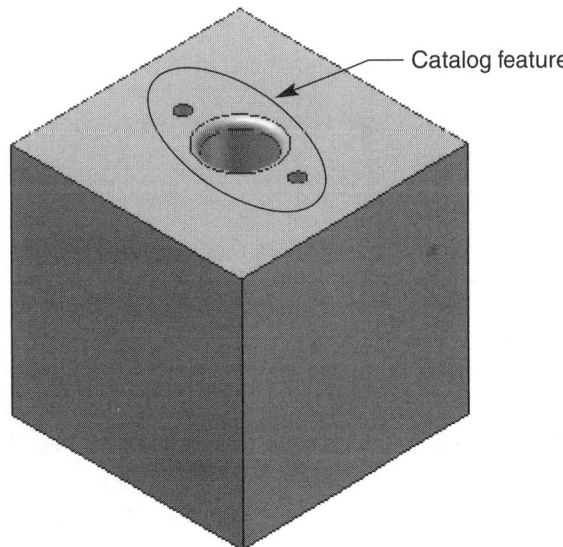

Catalog feature

Field Notes

All the interface and work environment options and applications discussed in Chapter 4 apply to working with iFeatures.

Using Design Element Tools

Design element tools are used to create, store, view, manage, and insert iFeatures. These tools are accessed from the **Features** toolbar, shown in Figure 9.2, or from the **Features** panel bar. When a feature is present and you are in the feature work environment, the panel bar is in features mode. See Figure 9.3. By default, the panel bar is displayed in learning mode, as seen in Figure 9.3. To change from learning to expert mode, seen in Figure 9.4 and discussed in Chapter 2, right-click any of the panel bar tools or left-click the panel bar title, and select the **Expert** option.

Design element tools allow you to view existing iFeatures, create new iFeatures, and insert existing iFeatures. All three tools are contained in a design element flyout button. See Figures 9.2 through 9.4. By default, the **View Catalog** button is active. If you want to create an iFeature, select the arrow next to the active button and choose the **Create iFeature** button. Similarly, if you want to insert an existing design element, select the arrow next to the active button and choose the **Insert iFeature** button.

FIGURE 9.2   The **Features** toolbar.

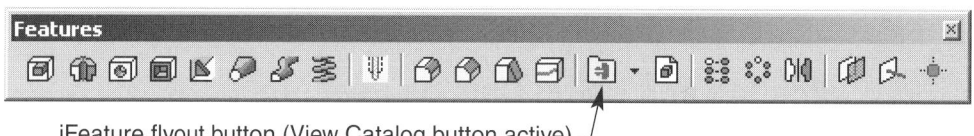

iFeature flyout button (View Catalog button active)

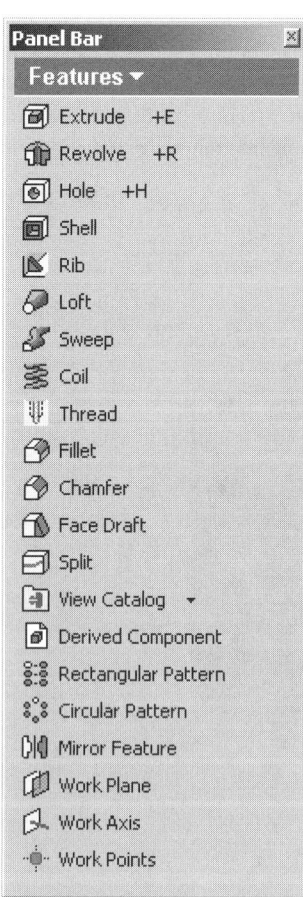

FIGURE 9.3   The panel bar in features (learning) mode.

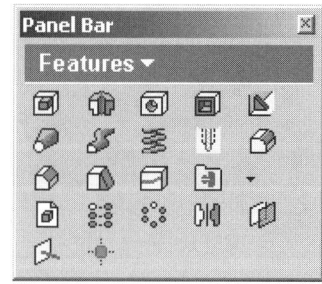

FIGURE 9.4   The panel bar in features (expert) mode.

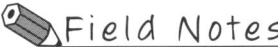Field Notes

When you pick one of the design element flyout options, the selected button becomes active and is displayed on the toolbar or the panel bar, depending on how the tool is accessed. The other two buttons then become invisible.

Viewing and Managing Catalog Features

Design elements are created as iFeature files (.ide) and placed in a desired location. The typical default location for catalog features is the **Catalog** folder, located in the path: C:\Program Files\Autodesk\Inventor 5\Catalog, depending on your settings. This is also the location of several preexisting Autodesk Inventor design elements. To display and manage the iFeatures located in the default path previously discussed, access the **View Catalog** command using one of the following techniques:

✓ Pick the **View Catalog** button on the panel bar.

✓ Pick the **View Catalog** button on the **Features** toolbar.

The **Catalog** folder window is displayed when you access the **View Catalog** command. See Figure 9.5. Using the specified iFeature viewer, depending on your operating system, you can navigate through your folders and files to locate, view, and manage design elements. However, if you want to modify the

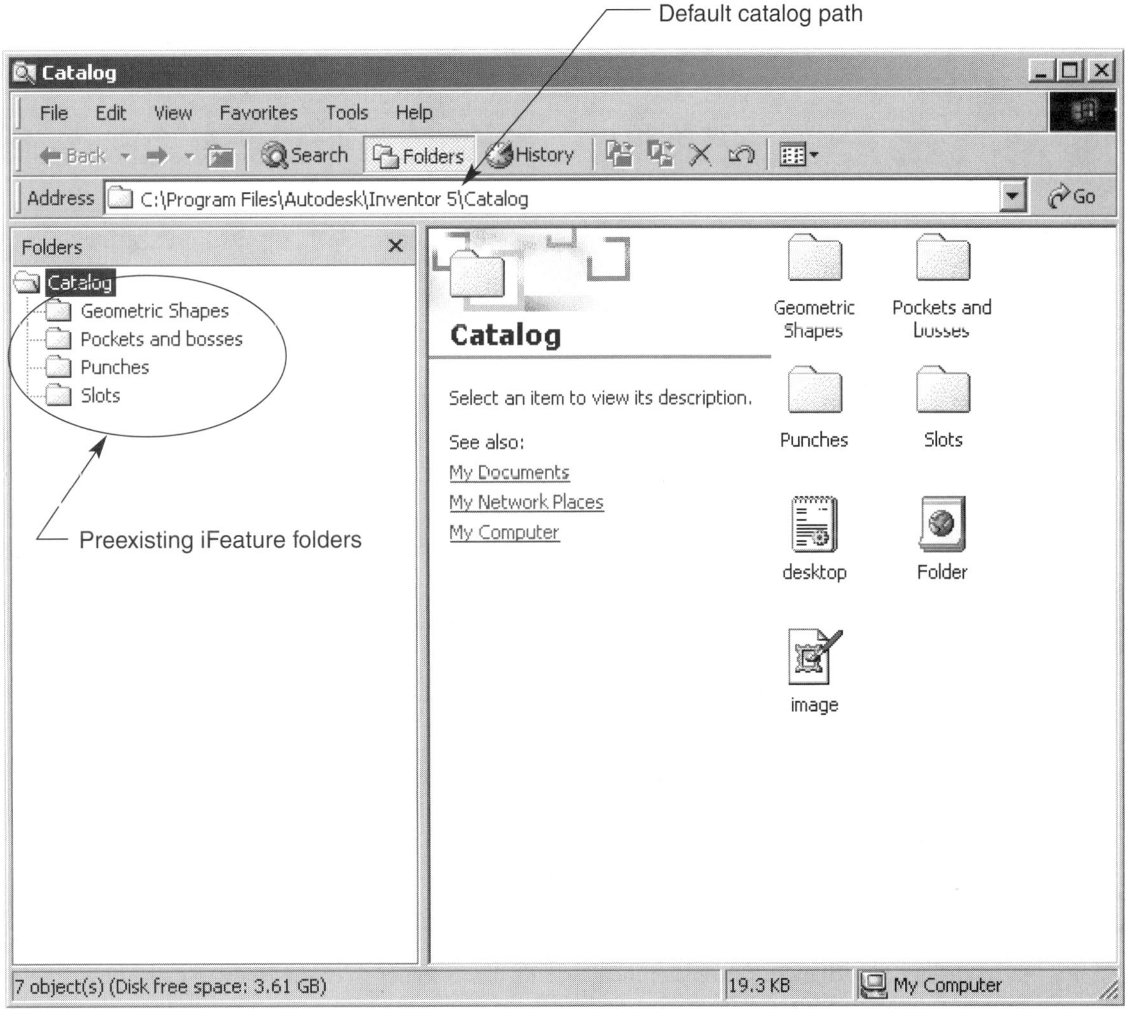

Default catalog path

FIGURE 9.5 The Autodesk Inventor **Catalog** folder (displayed using Windows Explorer).

FIGURE 9.6 The
iFeature tab of the
Options dialog box.

iFeature Viewer edit box ———

iFeature Viewer Argument String edit box ———

iFeature Root edit box ———

iFeature User Root edit box ———

Sheet Metal Punches Root edit box ———

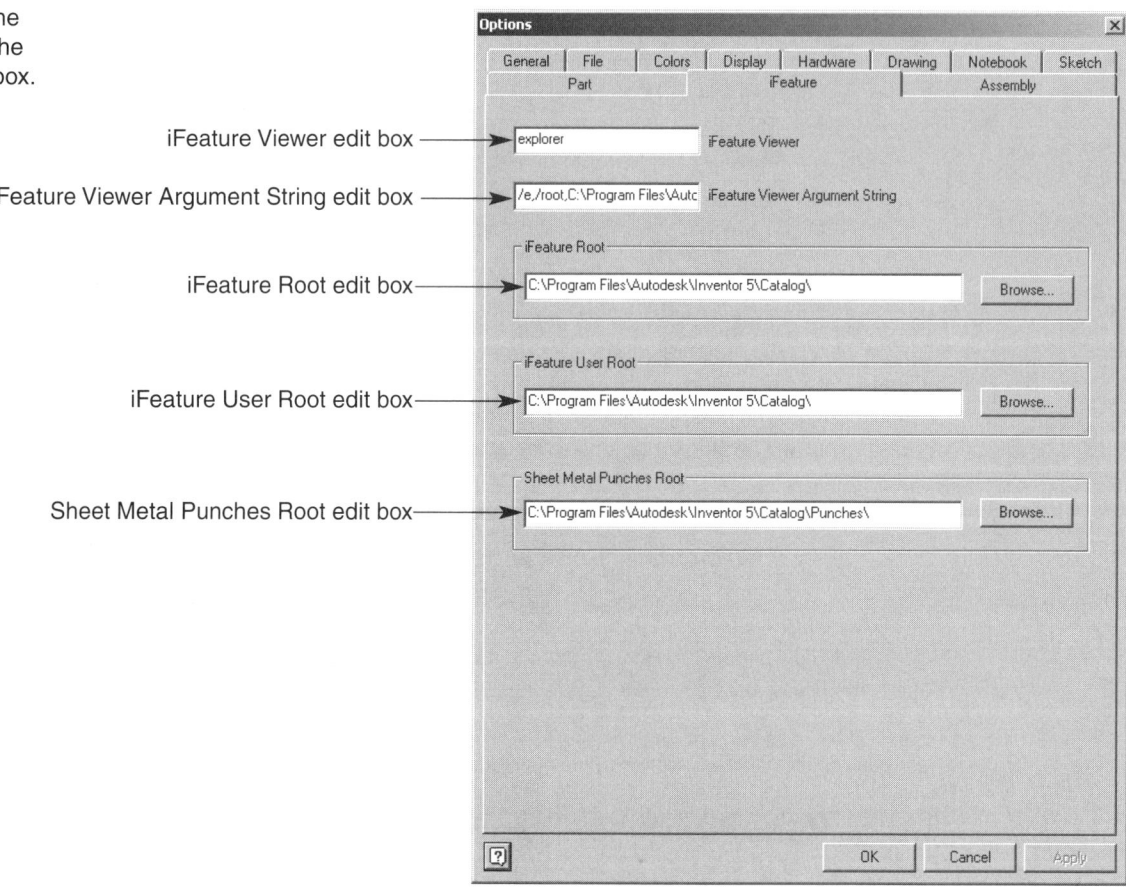

default viewer and default design element locations, select **Application Options...** from the **Tools** pull-down menu, and pick the **iFeature** tab of the **Options** dialog box. See Figure 9.6.

Field Notes

> When you open a design element .ide file, very few tools are available. The purpose of opening an .ide file is to view the design element. You must modify the design element in the part environment, and then redefine the iFeature.

The following options are located in the **iFeature** tab of the **Options** dialog box:

- **iFeature Viewer**   This edit box allows you to enter a viewer application for your operating system, such as Windows Explorer. The specified viewer is used to locate and manage iFeature files.

- **iFeature Viewer Argument String**   This edit box allows you to define the viewer argument string for the specified viewer. The folder listed in the edit box is automatically opened when you access the **View Catalog** command.

- **iFeature Root**   This edit box allows you to specify the iFeature files displayed in the **View Catalog** dialog box by typing the file name in the iFeature **Root** edit box, or selecting the **Browse** button and locating a different folder from the **Browse For Folder** dialog box.

- **iFeature User Root**   This edit box allows you to specify the iFeature files used by the **Create** iFeature and **Insert** iFeature dialog boxes discussed later in this chapter. To change the files, type the file name in the iFeature **User Root** edit box, or select the **Browse** button and locate a different folder from the **Browse For Folder** dialog box.

■ **Sheet Metal Punches Root** This edit box allows you to specify the iFeature files used by the sheet metal **PunchTool** dialog box by typing the file name in the **Sheet Metal Punches Root** edit box, or selecting the **Browse** button and locating a different folder from the **Browse For Folder** dialog box.

Field Notes

Modify the settings in the **iFeature** tab of the **Options** dialog box to fit your needs. The default paths refer to the **Catalog** folder that is part of the Autodesk Inventor program. You can change the location of your iFeature files as desired, such as placing the iFeatures in a network environment, to allow all members of the design team access to commonly used design elements.

Exercise 9.1

1. Launch Autodesk Inventor.
2. Open a new Autodesk Inventor part file, and explore the iFeature viewing interface options previously discussed.
3. Select **Application Options...** from the **Tools** pull-down menu, and pick the **iFeature** tab. Explore the **iFeature** tab, and modify your settings if desired.
4. Close the **Options** dialog box.
5. Access the View Catalog command, and explore the default **Catalog** folder.
6. Pick the **Geometric Shapes** folder, and open the **Cone.ipe** file.
7. Close the design element.
8. Close the iFeature viewer, and close the part file without saving.
9. Exit Autodesk Inventor or continue working with the program.

Creating Catalog Features

Design elements are features that are copied from a part file and placed in an iFeature file (.ide). You can then place the iFeature into another part and continually use the features you created as design elements. In order to create an iFeature, you must open an existing part file that contains features. Typically, iFeatures are created from sketched features, such as extrusions, revolutions, holes, ribs, lofts, sweeps, and coils. You cannot create an iFeature from a placed feature, such as a shell, fillet, chamfer, or face draft. However, if a placed feature is located on, and references, a sketched feature, the placed feature can be part of the design element. See Figure 9.7.

FIGURE 9.7 The iFeature consists of three holes and a fillet. The extruded box is not part of the iFeature.

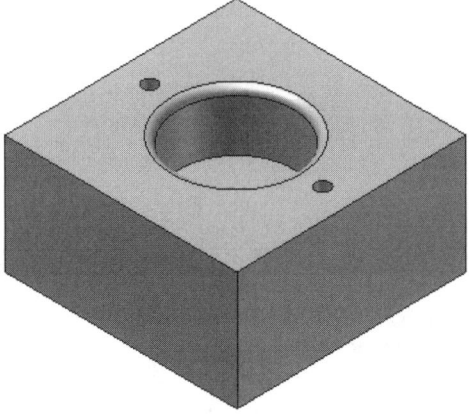

Once you have generated the features you want to extract into design elements, access the **Create iFeature** command using one of the following techniques:

✓ Pick the **Create iFeature** button on the panel bar.

✓ Pick the **Create iFeature** button on the **Features** toolbar.

The **Create iFeature** dialog box is displayed when you access the **Create iFeature** command. See Figure 9.8.

The first step in creating an iFeature is selecting the features that you want to use. You can select the actual features in the graphics window, or pick the features in the Browser Bar. When you select the desired features, the features and all the geometric information are displayed in the **Selected Features** area. The first listed feature is the base feature, and positions the iFeature when it is inserted. Geometrically dependent features, such as fillet on a hole edge, are also displayed. If you do not want the geometrically dependent feature to be associated with the iFeature, right click the undesired feature and select the **Remove Feature** menu option.

Field Notes

> You may want to rename the title of iFeature. For example, iFeature1 could receive a more descriptive name, such as "Bushing Base." To rename the iFeature, select the current name, and then select the name again, or right-click and select the **Rename** option.

When you insert the iFeature, you can select where and how to position, insert, and size the design element. However, you must set specific information when you create the iFeature to ensure the design element will insert and function correctly. As a result, the next step in creating an iFeature is to adjust specific size parameters. As previously mentioned, all the geometric information or parameters associated with the selected features are listed under the features in the **Selected Features** area of the **Create iFeature** dialog box. See Figure 9.9.

Field Notes

> The default values indicated as d1, d2, d3, etc. define which parameter is created first, second, third, etc. For example, if you create an extruded cylinder, the circle diameter is d1, and the extrusion height is d2. You may want to modify the names of the parameters in the **Parameters** dialog box to better understand the various parameters.

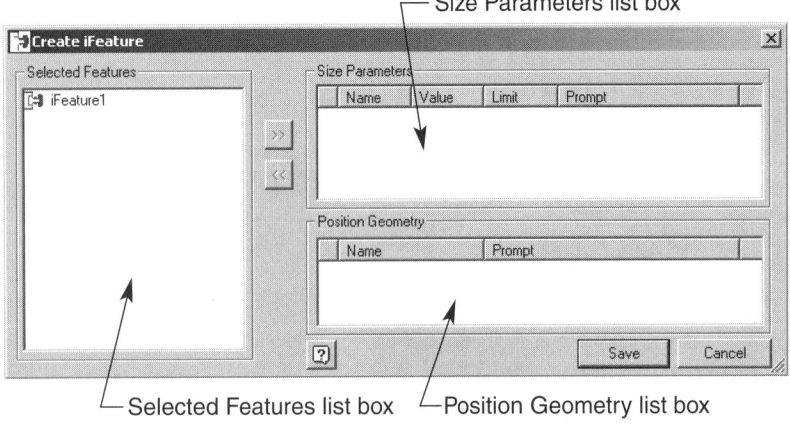

FIGURE 9.8   The Autodesk Inventor **Create iFeature** dialog box.

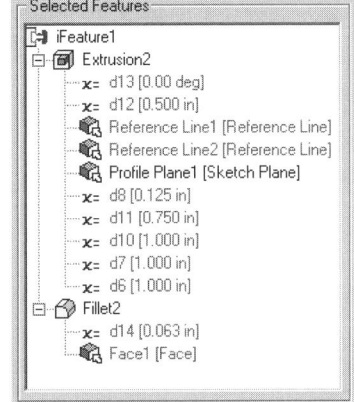

FIGURE 9.9   The **Selected Features** area of the Autodesk Inventor **Create iFeature** dialog box.

When you insert the iFeature, the design element geometry, or parameters, can be modified if you place the parameters into the **Size Parameters** area of the **Create iFeature** dialog box, using one of the following techniques:

✓ Pick the parameter and select the **Add Parameter** button.

✓ Double-click the parameter inside the **Selected Features** area.

✓ Right-click parameter and select the **Add Parameter** menu option.

Similarly, if you want to remove the parameter from the **Size Parameters** area, use one of the following techniques:

✓ Pick the parameter and select the **Remove Parameter** button.

✓ Double-click the parameter for the **Selected Features** area.

✓ Right-click the parameter and select the **Remove Parameter** menu option.

Once you insert the parameters into the **Size Parameters** area of the **Create iFeature** dialog box, you can adjust the parameter name, value, limit, and prompt. See Figure 9.10.
Each of these options is discussed as follows:

■ **Name** You may want to rename the parameter to reflect a more understandable name, such as "Major Hole Diameter." Edit the name of the parameter by selecting the default name, d12, for example, and typing a new name. When finished, pick the pencil icon, or anywhere outside of the **Size Parameters** area.

Field Notes

If you change the name of the parameter in the **Parameters** dialog box, as previously noted, the new name will be displayed in the **Create iFeature** dialog box.

■ **Value** The value displayed in the **Size Parameters** area reflects the size of the parameter when it was created. Edit the parameter value by selecting the default value—25mm, for example—and typing a new value. When finished, pick the pencil icon or anywhere outside of the **Size Parameters** area.

Field Notes

Changing the parameters value in the **Size Parameters** area does not modify the parameter value in the part file.

FIGURE 9.10
Examples of some parameters listed in the **Size Parameters** area of the **Create iFeature** dialog box.

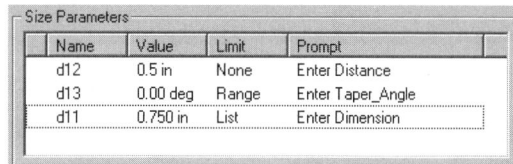

Name	Value	Limit	Prompt
d12	0.5 in	None	Enter Distance
d13	0.00 deg	Range	Enter Taper_Angle
d11	0.750 in	List	Enter Dimension

■ **Limit**   This option allows you to place limits on the amount of size changes that can occur when an iFeature is inserted into a part. You can specify no limit by selecting the **None** option, which is default. You can also specify a **Range** or **List** limit. These three options are found in a **Limit** drop-down list.

To create a limited size range, pick the current limit option, select the arrow, and choose the **Range** option. Selecting **Range** opens the **Specify Range** dialog box for the selected parameter. See Figure 9.11.

The **Specify Range** dialog box allows you to define a minimum limit, less than a specified value, less than or equal to a specified value, or infinitely less than the default value. You can also define a maximum limit, greater than a specified value, greater than or equal to a specified value, or infinitely greater than the default value. Finally you can modify the default value in the **Default** edit box, which is the same as changing the **Value,** previously discussed.

To create a limited size list, pick the current limit option, select the arrow, and choose the **List** option. Selecting **List** opens the **List Values** dialog box for the selected parameter. See Figure 9.12. The **List Values** dialog box allows you to define a certain number of size values. The values you specify in this edit box will be available for selection when the design element is inserted. The current default value is automatically listed in the **List** dialog box. To add another optional value, select the **Click here to add value** button and define the value. You can then choose the new value from the Default drop down list, which is the same as changing the **Value**, previously discussed.

■ **Prompt**   When an iFeature is inserted, a prompt instructs you on the parameter value operation, such as "Enter distance." The **Prompt** option allows you to modify the default prompt to a more understandable description. For example, you may want to change the default hole depth prompt of "Enter Distance," to "Enter the depth of the hole."

In addition to physical feature parameters, reference geometry parameters are also listed in the **Selected Features** area of the **Create iFeature** dialog box. These parameters are indicated by a reference icon. The next step in creating an iFeature is to specify how you want to position the iFeature geometry when the iFeature is inserted using the **Position Geometry** area of the **Create iFeature** dialog box. When

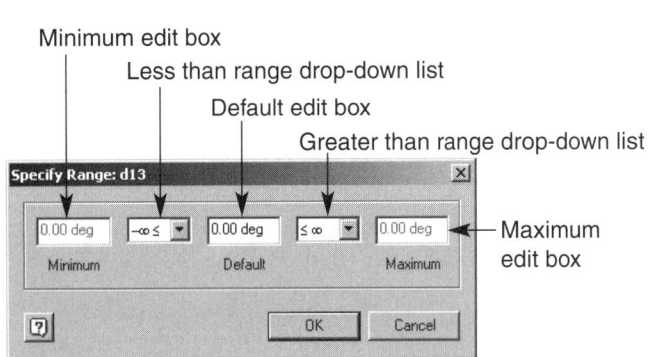

FIGURE 9.11   The **Specify Range** dialog box.

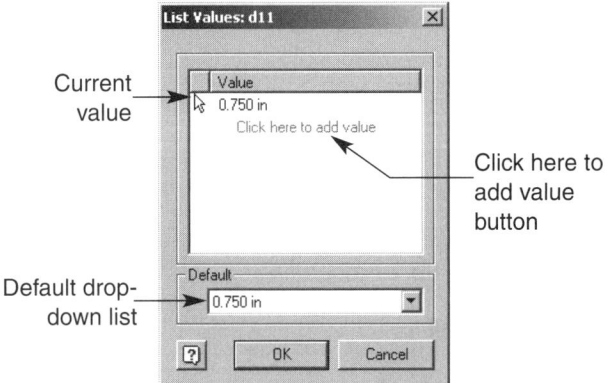

FIGURE 9.12   The **List Values** dialog box.

you insert the iFeature, the design element position can only reference the additional reference parameters you place into the **Position Geometry** area of the **Create iFeature** dialog box. For example, if the only position geometry listed in the **Position Geometry** area is "Profile Plane1" you will only be able to select, and reference, a profile plane on the part the iFeature is inserted into. To add additional position geometry to the **Position Geometry** list box, such as a reference edge ("Reference Line1"), use one of the following techniques:

✓ Pick the position geometry and select the **Add Geometry** button.

✓ Double-click the position geometry inside the **Selected Features** area.

✓ Right-click the position geometry and select the **Expose Geometry** menu option.

Similarly, if you want to remove position geometry from the **Position Geometry** area, use one of the following techniques:

✓ Pick the position geometry and select the **Remove Geometry** button.

✓ Double-click the position geometry the **Selected Features** area.

✓ Right-click position geometry and select the **Remove Geometry** menu option.

Once you insert the position geometry into the **Position Geometry** area of the **Create iFeature** dialog box, you can adjust the position geometry name and prompt the same way you adjust the name and prompt of the parameters previously discussed. In addition, when you right-click certain position geometry listed in the **Position Geometry** area of the **Create iFeature** dialog box, the following menu options are available in addition to the **Remove Geometry** button, previously discussed:

■ **Make Independent** This option allows you to divide shared position geometry. For example, if an extrusion and a fillet share the same position geometry, you are able to make the extrusion and fillet position geometry independent of each other. The **Make Independent** option is only available when two or more features are selected.

■ **Combine Geometry** This option is opposite of the **Make Independent** option previously discussed, and allows you to combine shared position geometry. After you choose the **Combine Geometry** option, select the additional geometry you want to combine with the selected geometry.

Once you have specified all the options in the **Create iFeature** dialog box, select the **Save** button to access the **Save As** dialog box. Saving an iFeature is similar to saving other files, as discussed in Chapter 2. You may notice how the default save location corresponds to the iFeature **User Root,** discussed earlier in this chapter.

 Field Notes

> For the most efficient use of iFeatures, you may want to adjust the options located in the **iFeature** tab of the **Options** dialog box and create an iFeature library or set of libraries. You can then save your design elements in these specific locations.

Exercise 9.2

1. Continue from Exercise 9.1, or launch Autodesk Inventor.
2. Open a new inch part file.
3. Open a sketch on the **XY Plane.**
4. Sketch a ∅1″ circle around the projected **Center Point.**
5. Extrude the sketch 1″ in a positive direction.
6. Open a new sketch on the top face of the cylinder, and create a ∅.5 × .5 deep hole.
7. Place a .125″ fillet around the top edge of the cylinder.
8. Though this exercise uses default parameter names, you may want to change the names of the parameters for clarity.
9. Access the **Create iFeature** command to open the **Create iFeature** dialog box.
10. Rename the default iFeature name (iFeature1), to Base.
11. Select **Extrusion1** in order to enter **Extrusion1, Hole1,** and **Fillet1** into the **Selected Features** area.
12. Add the extrusion diameter parameter (d1), and the extrusion height parameter (d2), to the **Size Parameters** list.
13. Modify the size parameters according to the following specifications:

    Extrusion diameter (d1):
    New Name: Base_Diameter
    New Value: unchanged
    Limit: Minimum range of ≤ 1″, maximum range of ≥ 3″.
    New Prompt: Enter the diameter of the base

    Extrusion height (d2):
    New Name: Base_Height
    New Value: unchanged
    Limit: List of 1″, 1.5″, and 2″.
    New Prompt: Enter the Height of the base

14. Modify the position geometry parameters according to the following specifications:

    Profile plane1 [Sketch Plane]:
    New Name: Base Bottom Face
    New Prompt: Pick the bottom of the base

15. Save the iFeature as Base.ipe, in a location of your choosing.
16. Exit Autodesk Inventor or continue working with the program.

Inserting Catalog Features

Once you have created design elements, you are ready to insert them into a part using the **Insert iFeature** command. Access the **Insert iFeature** command using one of the following techniques:

✓ Pick the **Insert** iFeature button on the panel bar.

✓ Pick the **Insert** iFeature button on the **Features** toolbar.

The **Insert iFeature** dialog box is displayed when you access the **Insert iFeature** command. See Figure 9.13.

The first step in inserting an iFeature is to select the desired iFeature by typing a path in the **File Name:** edit box, or picking the **Browse** button to access the **Open** dialog box, discussed in Chapter 2. The **Open** dialog box allows you to locate an iFeature anywhere on your computer or the network.

Once you select an iFeature, the position window is activated, as shown in Figure 9.14, and a preview of the design element is displayed. The prompt in the position window of the **Insert** iFeature dialog box guides you through the positioning process of inserting the iFeature. Typically, you need to pick the face of an existing feature or the work feature where you want to place the iFeature. You then have the option of

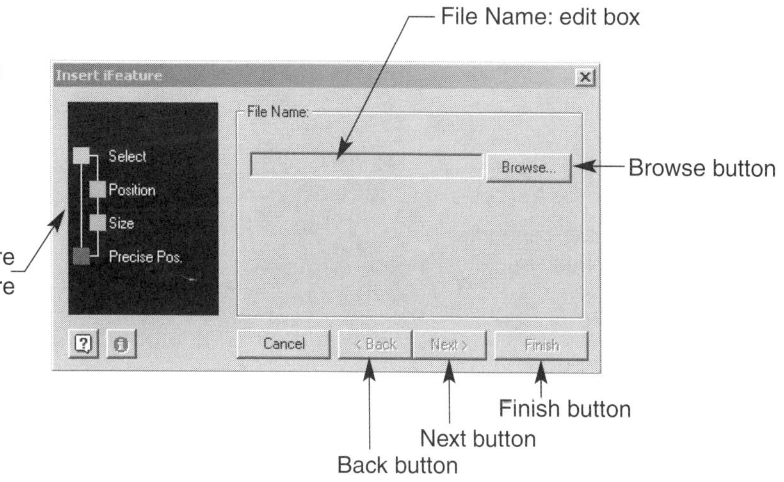

FIGURE 9.13 The Autodesk Inventor **Insert iFeature** dialog box (Select window active).

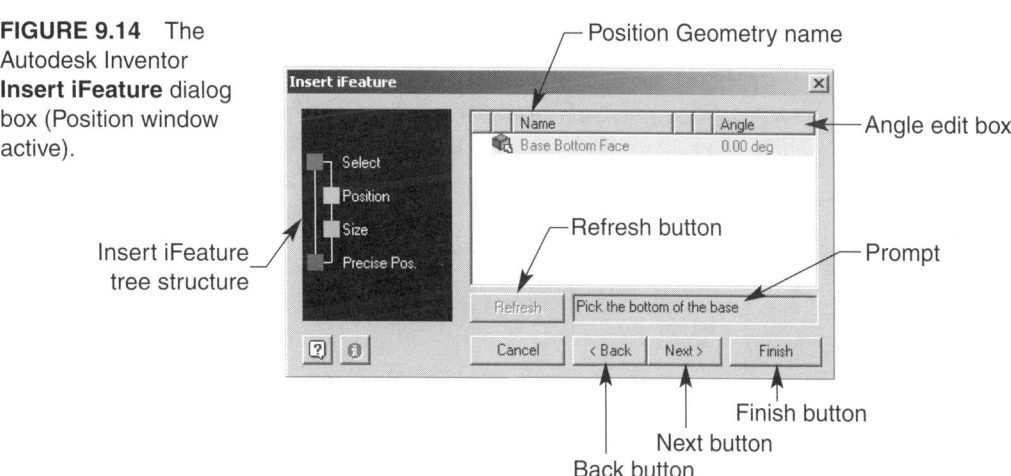

File Name: edit box

Insert iFeature tree structure

Browse button

Finish button

Next button

Back button

FIGURE 9.14 The Autodesk Inventor **Insert iFeature** dialog box (Position window active).

Position Geometry name

Insert iFeature tree structure

Angle edit box

Refresh button

Prompt

Finish button

Next button

Back button

rotating the iFeature, by selecting the default angle in the list box, and entering the desired angle, or by picking the **Rotate** icon and rotating the iFeature. See Figure 9.15. Similarly, you can move the iFeature by selecting the **Move** icon, as shown in Figure 9.15, and move the iFeature to the desired location.

After you position the iFeature, pick the **Next** button, or select the **Size** option on the tree structure to activate the **Size** window. See Figure 9.16. If you specified specific size parameter information when the

FIGURE 9.15 The **Rotate** and **Move** icons.

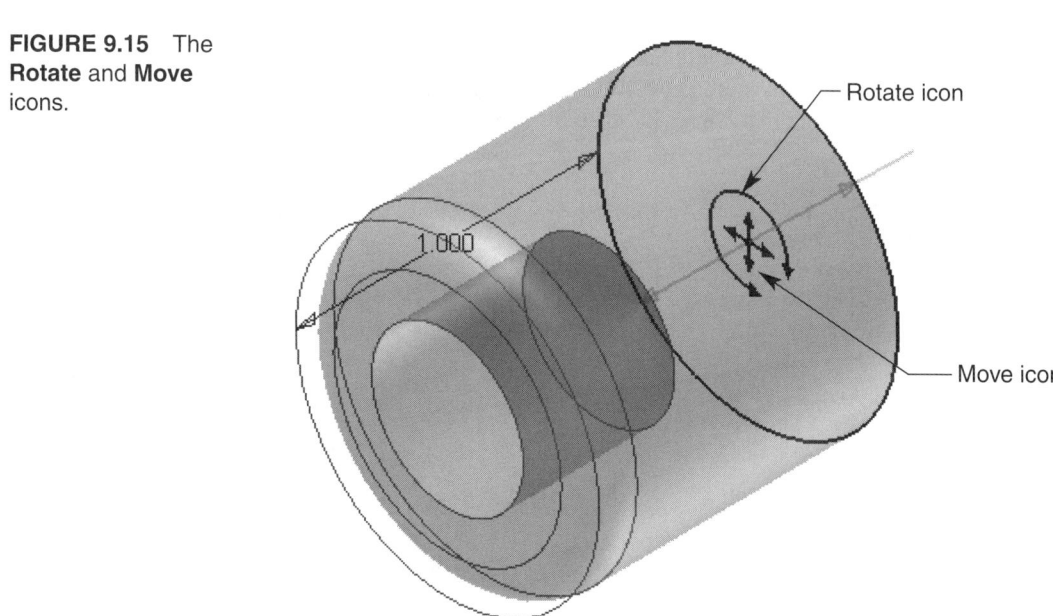

Rotate icon

Move icon

FIGURE 9.16 The
Autodesk Inventor
Insert iFeature dialog
box (Size window
active).

Insert iFeature
tree structure

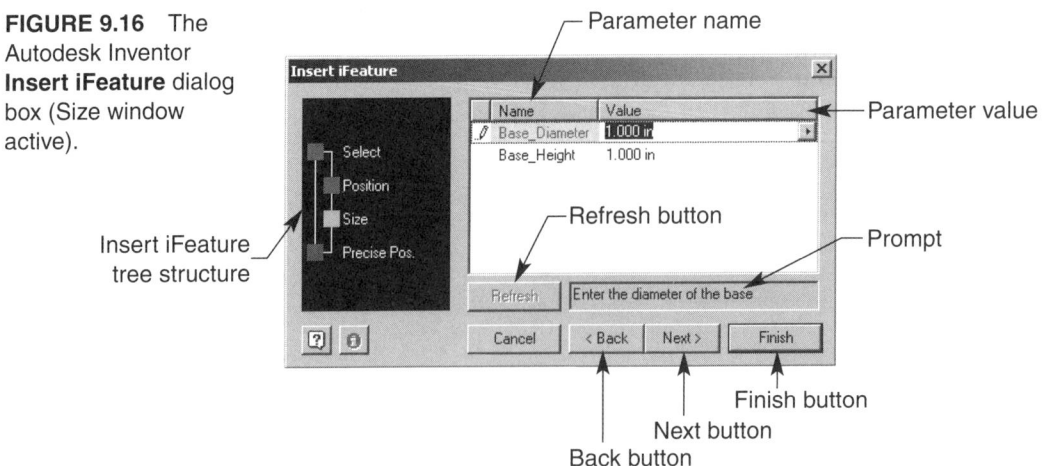

iFeature was created, the size parameters are listed in the **Size** list box. The prompts guide you through the sizing process of inserting the iFeature. If no size limits were placed on the iFeature when it was created, you can enter any value. However, if a range limit was placed on the iFeature when created, you will only be able to enter a specific range of values, or if a list limit was placed on the iFeature when created, you will only be able to choose values from the **Value** drop-down list.

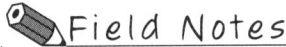 Field Notes

Pick the **Refresh** button in the **Insert iFeature** dialog box anytime you resize or reposition the iFeature, to preview the changes.

Once you make size modifications to the iFeature, pick the **Next** button, or select the **Precise Pos.** option on the tree structure to activate the **Precise Position** window. See Figure 9.17. This window contains the following two radio buttons, which allow you to specify whether or not you want to further position the iFeature with sketch constraints and dimensions:

■ **Activate Sketch Edit Immediately**   When this radio button is selected, the sketch of the iFeature is automatically opened, allowing you to precisely position the iFeature in reference to space or other features using constraint and dimension sketching tools.

■ **Do Not Activate Sketch Edit**   When this radio button is selected, the sketch of the iFeature does not automatically open. Consequently, to precisely position the iFeature using constraint and dimension sketching tools, you must first activate the iFeature sketch by editing.

FIGURE 9.17 The
Autodesk Inventor
Insert iFeature dialog
box (Precise Position
window active).

Insert iFeature
tree structure

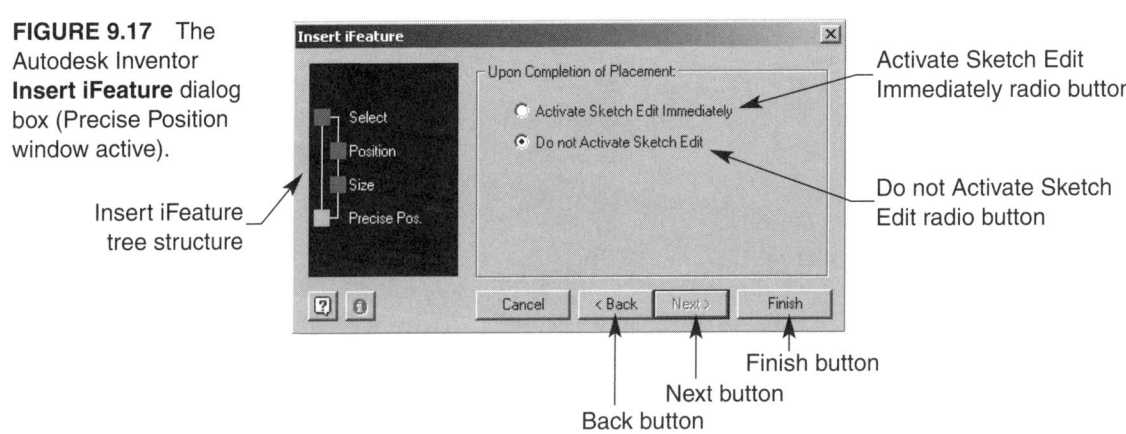

Once you have selected, positioned, and sized the iFeature you want to insert, and specified the desired precise position option, the final step to insert an iFeature is to select the **Finish** button. However, if you at least select and position the iFeature, you can select the **Finish** button and bypass the other insertion options.

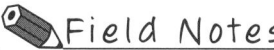Field Notes

Pick the **Back** button in the **Insert iFeature** dialog box anytime you return to the previous window to modify an insertion option. You can also choose any of the options on the tree structure and return to modify an insertion option.

Exercise 9.3

1. Continue from Exercise 9.2, or launch Autodesk Inventor.
2. Open a new inch part file.
3. Open a sketch on the **XY Plane.**
4. Sketch a 2″×4″ rectangle around the projected **Center Point.**
5. Extrude the sketch 1″ in a positive direction.
6. Insert the following iFeature: C:\Program Files\Autodesk\Inventor 5\Catalog\Pockets and bosses\Boss_obround_2_fillets.
7. Position the iFeature on the top of the extruded face, and rotate the iFeature 90°.
8. Change the Length of the iFeature from 1.5″ to 2.5″.
9. Select the **Activate Sketch Edit Immediately** radio button.
10. Use sketch dimensions to center the iFeature sketch in the center of the extrusion top face as shown.
11. The final part should look like the part shown.
12. Save the part as EX9.3.
13. Exit Autodesk Inventor or continue working with the program.

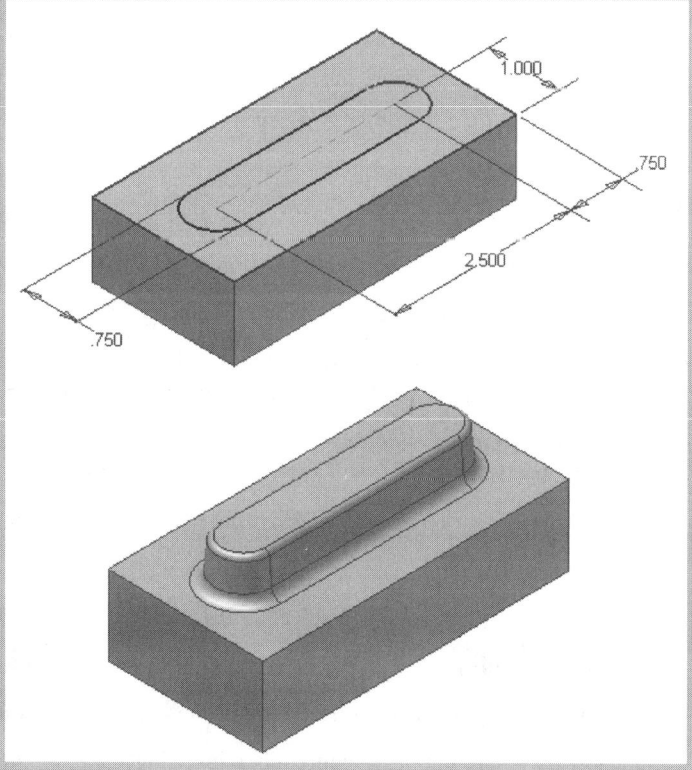

|||||||||||| CHAPTER TEST

1. Give two other names for catalog features.

2. By default, the panel bar is displayed in learning mode. How do you change to expert mode?

3. What do design element tools allow you to do?

4. By default, the View Catalog button is active. What button do you select if you want to create an iFeature?

5. What button do you select if you want to insert an existing design element?

6. Design elements are created as iFeature files. What is the file extension?

7. Where is the typical default folder location for catalog features?

8. Give the function of the iFeature Viewer in the Options dialog box.

9. Explain the function of the iFeature Viewer Argument String in the Options dialog box.

10. Discuss the function of the iFeature Root in the Options dialog box.

11. Explain the function of the iFeature User Root in the Options dialog box.

12. Discuss the function of the Sheet Metal Punches Root in the Options dialog box.

13. Give the first step used in creating an iFeature.

14. What do you do if you do not want a geometrically dependent feature to be associated with the iFeature that you are creating?

15. What must you do before you can insert the iFeature and select where and how to position, insert, and size the design element?

16. Give the function of the Name option in the Size Parameters area of the Create iFeature dialog box.

17. Discuss the function of the Value option in the Size Parameters area of the Create iFeature dialog box.

18. Explain the function of the Limit option in the Size Parameters area of the Create iFeature dialog box.

19. Discuss the function of the Specify Range dialog box.

20. How do you create a limited size list?

21. Explain the function of the List Values dialog box.

22. Give the function of the Prompt option.

23. Discuss the function of the Position Geometry area of the Create iFeature dialog box.

24. Give the function of the Make Independent option when you right-click certain position geometry listed in the Position Geometry area of the Create iFeature dialog box.

25. Explain the function of the Combine Geometry option when you right-click certain position geometry listed in the Position Geometry area of the Create iFeature dialog box.

26. How do you save once you have specified all the options in the Create iFeature dialog box?

27. Give the first step in inserting an iFeature, which involves selecting the desired iFeature.

28. What do you see when you select an iFeature?

29. What does the prompt in the position window of the Insert iFeature dialog box provide?

30. What are the typical steps used to place an iFeature after one has been selected?

31. How do you rotate the iFeature when inserting?

32. How do you move the iFeature when inserting?

33. After you position the iFeature, how do you activate the Size window?

34. Discuss the sizing options and process if you specified specific size parameter information when the iFeature was created, and if no size limits were placed on the iFeature when it was created.

35. How do you preview the changes anytime you resize or reposition the iFeature?

36. Explain the function of the Activate Sketch Edit Immediately in the Precise Position window.

37. Discuss the function of the Do Not Activate Sketch Edit Immediately in the Precise Position window.

38. Once you have selected, positioned, and sized the iFeature you want to insert and specified the desired precise position option, what is the final step to insert an iFeature?

|||||||||||| PROJECTS

Follow the specific instructions provided for each project.

1. Name: Bearing Pocket

   Units: Inch

   Specific Instructions:

   - Open a new part file.

   - Create a 1″×2″×2″ extruded rectangle.

   - Open a sketch on the top face of the extrusion, and place the hole centers as shown.

   - Create the two smaller holes as specified.

   - Share **Sketch2.**

   - Create the larger middle hole as specified.

   - Use the **Parameters** dialog box to rename the internal thread diameter (**Hole1**) to Internal_Threads, and the pocket diameter (**Hole2**) to Pocket.

   - Create an iFeature using the following information:

     Name: Bearing_Pocket

     Selected Features: **Hole1** and **Hole2.**

     Limit the value of "Internal_Threads" to a range of .125<.25>.5.

     Change the prompt of "Internal_Threads" to "Enter Diameter of Internal Threads."

     Limit the value of "Pocket" to a list of .75″, 1″, and 1.25″.

     Change the prompt of "Pocket" to "Enter Pocket Diameter."

     Save Bearing_Pocket as P9.1.ide.

   - Access the View Catalog command, and open P9.1.

   - The iFeature P9.1 should look like the feature shown.

   - Save As: P9.1.

PROJECT 9.1

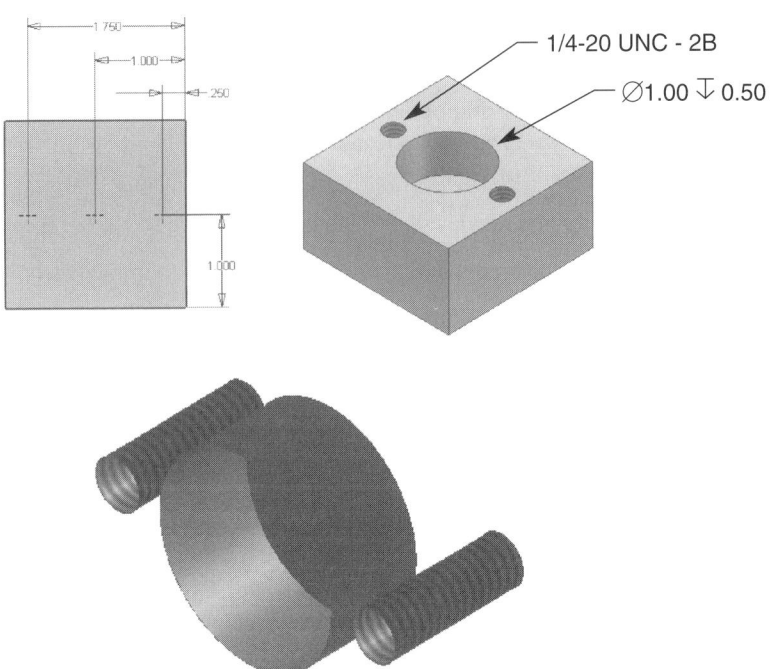

2. Name: Bearing Pocket Base

Units: Inch

Specific Instructions:

- Open a new part file.

- Create a 1″×3″×3″ extruded rectangle.

- Insert the iFeature, P9.1.ide, on the top face of the extrusion.

- Attempt to enter a value of .5″ in the Internal_Threads Value edit box.

- Select the **Activate Sketch Edit Immediately** radio button.

- Use sketch dimensions, center the iFeature in the middle of the extrusion, and finish the sketch.

- Right-click **Bearing_Pocket1** in the browser and select the **Edit iFeature...** menu option.

- Change the bearing pocket diameter to .75″.

- Place a .25″ chamfer around the top edge of the extrusion.

- The final part should look like the part shown.

- Save As: P9.2.

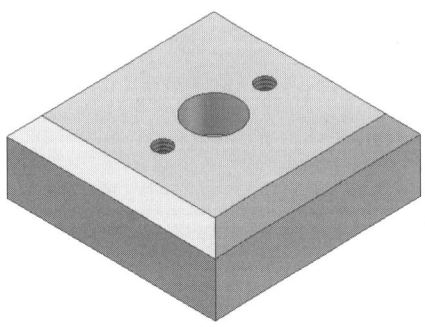

II

Developing Sheet Metal Parts

LEARNING GOALS

After completing this chapter, you will be able to:

- ◎ Create Autodesk Inventor sheet metal parts.
- ◎ Work in the sheet metal part environment.
- ◎ Use sheet metal feature tools.
- ◎ Create sheet metal sketched features including faces, contour flanges, cuts, and folds.
- ◎ Develop sheet metal placed features including flanges, hems, corner seems, bends, corner rounds, and corner chamfers.
- ◎ Use the sheet metal **PunchTool.**
- ◎ Use the **Flat Pattern** tool.

Developing Sheet Metal Parts

Sheet metal parts are created by forming flat metal into a particular shape. Typically, a sheet metal drawing consists of a flat pattern, which represents the final, unfolded part. The sheet metal pattern is then transferred to a flat sheet of metal, which is then sheered and bent to the correct specifications. Sheet metal drafting and sheet metal parts are utilized in a number of industries, including HVAC (heating, ventilating, and air conditioning), automotive, and product chassis. See Figure 10.1.

The process for creating sheet metal parts is very similar to the process for creating other types of parts. Usually, you begin with a sketched feature and then add additional features to the part. However, Autodesk Inventor provides you with specific tools that allow you to develop sheet metal parts. For most sheet metal applications, the metal thickness, material, bend radii, and relief sizes remain the same throughout part constructions. As a result, you should define the sheet metal properties first, so the properties are applied to the part while you develop your design.

FIGURE 10.1 An example of an Autodesk Inventor sheet metal part.

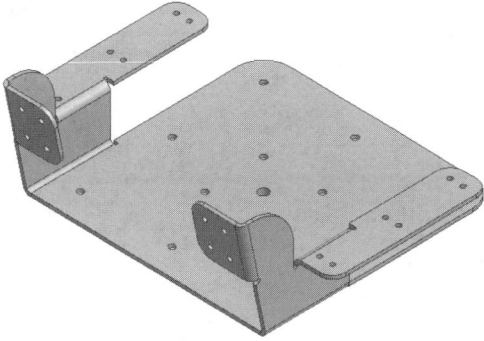

Working in the Sheet Metal Part Environment

Specific work environment and interface options are available when working through each stage of the sheet metal design process. All the sketching tools discussed in Chapter 3 apply to sketching in a sheet metal part. However, some aspects of the sheet metal feature interface and work environment are different.

Typically, when you begin a new design and open a new part file, you specify whether the part is for a modeling application or a sheet metal application, depending on the type of part you want to create. As discussed in Chapter 1, both modeling part files and sheet metal part files carry the extension .ipt, which stands for Inventor part. However, specific sheet metal part templates are available. By default these templates are listed as Sheet Metal.ipt. Sheet metal part templates are similar to other part templates. However, opening a new Sheet Metal.ipt file from the **New File** menu creates a part as a sheet metal subfile and opens the sheet metal tools necessary to create sheet metal parts when required.

To enter the sheet metal feature interface and work environment after you have created a sketch, use one of the following options:

✓ Right-click and select the **Finish Sketch** option from the shortcut menu.

✓ Use the **+S** accelerator key option by pressing the **+** and **S** keys on your keyboard.

✓ Pick the available sheet metal tool from the **Sheet Metal** panel bar or **Sheet Metal** toolbar.

✓ Select the **Return** button on the **Command** bar.

Field Notes

> When beginning a new sheet metal design you should use a sheet metal template. However, if you begin a new sheet metal part in a modeling application and want to access the sheet metal tools, pick the **Sheet metal** menu option from the **Applications** pull-down menu. This menu allows you to change from a **Modeling** design to a **Sheet metal** design while in a part work environment.
>
> The entire interface options discussed in Chapter 4 apply to working in a sheet metal part environment.

Using Sheet Metal Feature Tools

Sheet metal feature tools are used to create sheet metal features, including faces, contour flanges, cuts, flanges, hems, folds, corner seams, bends, holes, corner rounds, corner chamfers, punches, design elements, work planes, work axes, work points, rectangular patterns, circular patterns, and mirrors. These tools are accessed from the **Sheet Metal** toolbar, shown in Figure 10.2, or from the **Sheet Metal** panel bar.

When a sheet metal sketch or feature is present, and you are in the sheet metal work environment, the panel bar is in sheet metal mode. See Figure 10.3. By default, the panel bar is displayed in learning mode, as seen in Figure 10.3. To change from learning to expert mode, seen in Figure 10.4 and discussed in Chapter 2, right-click any of the panel bar tools or left-click the panel bar title, and select the **Expert** option.

The following tools, available in the sheet metal environment, are used exactly the same as in any other part model environment. These tools are covered in previous chapters and are not discussed in this chapter:

■ **Hole** (Chapter 4)

■ **Circular Pattern** (Chapter 6)

■ **Mirror Feature** (Chapter 6)

FIGURE 10.2 The
Sheet Metal toolbar.

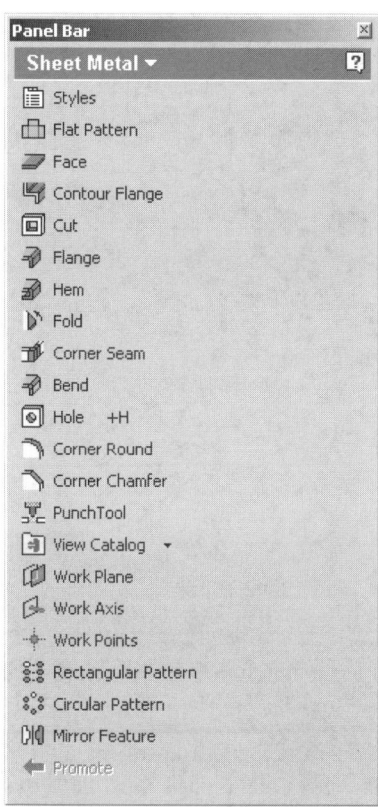

FIGURE 10.3 The panel bar in sheet metal (learning) mode.

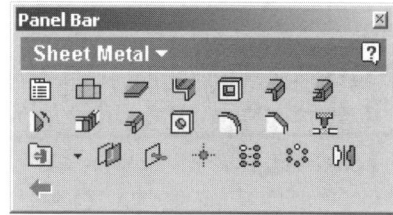

FIGURE 10.4 The panel bar in sheet metal (expert) mode.

- **Rectangular Pattern** (Chapter 6)

- **Work Plane** (Chapter 7)

- **Work Axis** (Chapter 7)

- **Work Point** (Chapter 7)

- **View Catalog** (Chapter 9)

- **Create iFeature** (Chapter 9)

- **Insert iFeature** (Chapter 9)

Creating Sheet Metal Features

As previously mentioned, most of the sheet metal parameters you want to use when creating your sheet metal parts are specified before you begin the design. As a result, the sheet metal parameters are automatically applied to the design as you create the part. You can define the bend radius, unfold options, and bend relief options while you create a sheet metal part. However, you should specify these options and all other parameters, such as the material type, thickness, and corner options, before you begin a design.

To define the sheet metal parameters before you begin, access the **Sheet Metal Styles** command using one of the following techniques:

✓ Pick the **Styles** button on the panel bar.

✓ Pick the **Styles** button on the **Sheet Metal** toolbar.

The **Sheet Metal Styles** dialog box is displayed when you access the **Styles** command. See Figure 10.5. The following options are available inside this dialog box:

FIGURE 10.5 The **Sheet Metal Styles** dialog box (**Sheet** tab active).

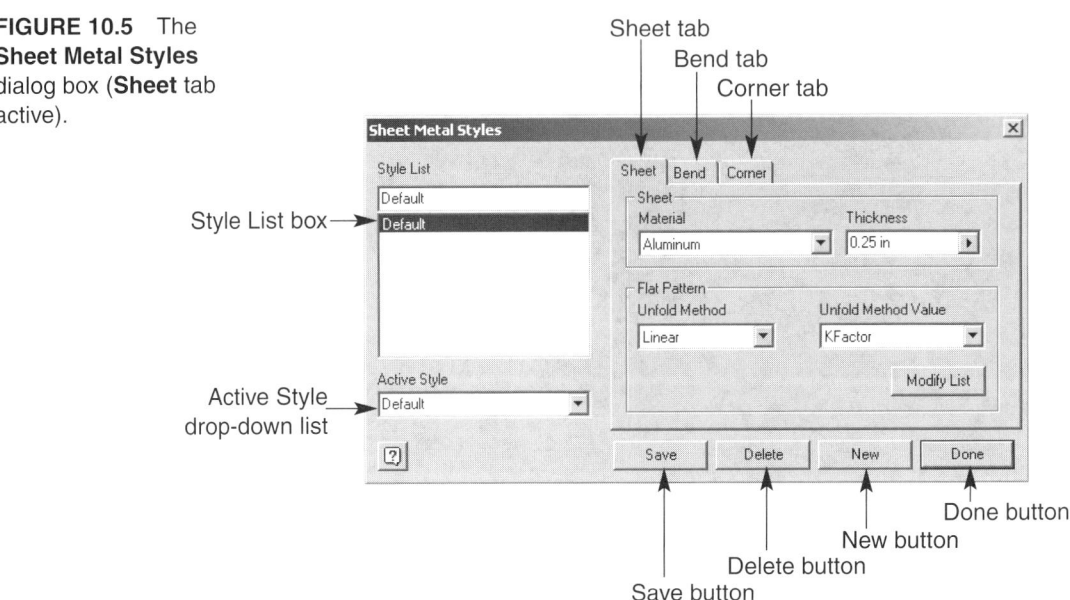

Sheet tab
Bend tab
Corner tab

Style List box

Active Style drop-down list

Done button
New button
Delete button
Save button

- **Style List**   This area lists all the current sheet metal styles. Unless you create additional sheet metal styles, **Default** is the only available style.

- **Active Style**   This drop-down list allows you to activate a sheet metal style. The sheet metal style that you want to use when creating sheet metal parts must be activated. To activate a sheet metal style, pick the drop-down list arrow and select the desired style.

- **Sheet**   This tab, shown in Figure 10.5, allows you to specify the sheet and flat pattern unfold parameters for the sheet metal part. The following are located in the **Sheet** tab:

  - **Sheet**   This area allows you to select the type of material you want to use for the sheet metal part by selecting the **Material** drop-down arrow and choosing the desired material. You can also define the material thickness by entering a value in the **Thickness** edit box, or selecting a value from the drop-down list.

✎ Field Notes

The type of material can also be defined in the **Physical** tab of the **Properties** dialog box, discussed in Chapter 2.

  - **Flat Pattern**   This area allows you to define the flat pattern unfold options for the sheet metal part. Select either the **Linear** or **Bend Table** unfold method option from the **Unfold Method** drop-down list.
    When you select the **Linear** option, you can specify an unfold method value from the **Unfold Method Value** drop-down list. When you select the **Bend Table** option, the **Open** dialog box appears, allowing you to search for an existing bend table, which is a text file.

✎ Field Notes

To create your own bend tables or customize a bend table, use a spread sheet. Then save the spread sheet as a .txt file for use in the sheet metal environment.

FIGURE 10.6 The **Unfold Method List** dialog box.

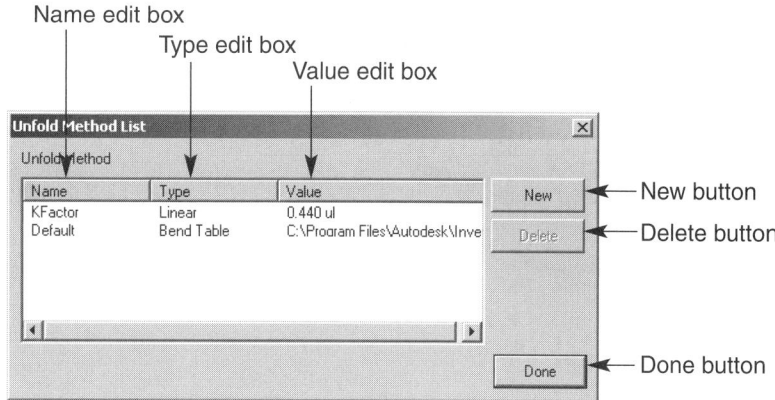

If you want to change, add, or remove an unfold method value, pick the **Modify List** button to open the **Unfold Method List** dialog box. See Figure 10.6. To change the name of an unfold method value, pick the current name and enter a new value. Change the type of unfolding method by selecting the current type, and picking either **Linear** or **Bend Table** option from the **Type** drop-down list. To modify the value, pick the current value and enter the desired value between 0 and 1, or select a value from the **Value** drop-down list. To create a new unfold method value, pick the button and create a new unfold method value with the desired name, type, and value, using the options previously discussed. You can also remove unfold method values by selecting the name you want to remove, and picking the **Delete** button.

When you are finished defining the sheet parameters, select another sheet metal style tab to continue defining the sheet metal style, or pick the **Save** button to save changes to the style, and select the **Done** button to exit the **Sheet Metal Styles** dialog box.

- **Bend** This tab, shown in Figure 10.7, allows you to specify the sheet metal part, bend parameters. The following areas are located in the **Bend** tab:

  - **Radius** This drop-down list allows you to specify the bend radius. The default bend radius is equal to the thickness of the material. You can also change the bend radius setting for certain features when they are created.

  - **Relief Shape** This drop-down list allows you to select a **Straight** or a **Round** bend relief shape. See Figure 10.8.

  - **Minimum Remnant** This drop-down list allows you to specify the amount of sheet metal material between the inside relief edge and the feature edge. The default minimum remnant is twice the material thickness. If the size of the remnant is less than the value specified, the remnant is removed. See Figure 10.9.

  - **Relief Width** This drop-down list allows you to define the width of the bend relief. The default relief width is the same as the material thickness. However, you can define a different relief width, with a value greater than 0. See Figure 10.10.

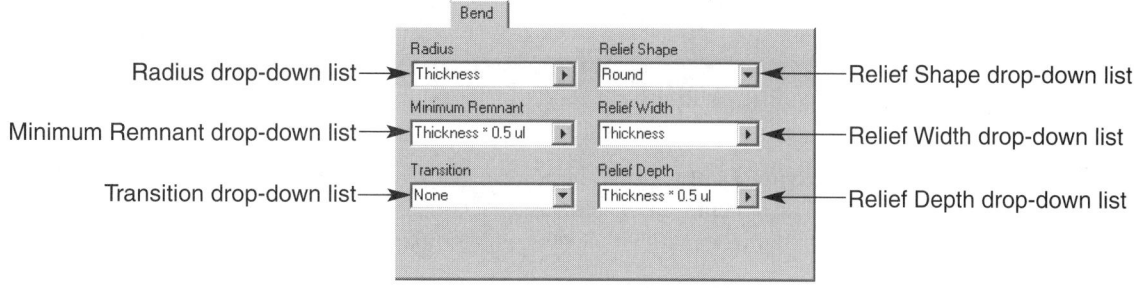

FIGURE 10.7 The **Bend** tab of the **Sheet Metal Styles** dialog box.

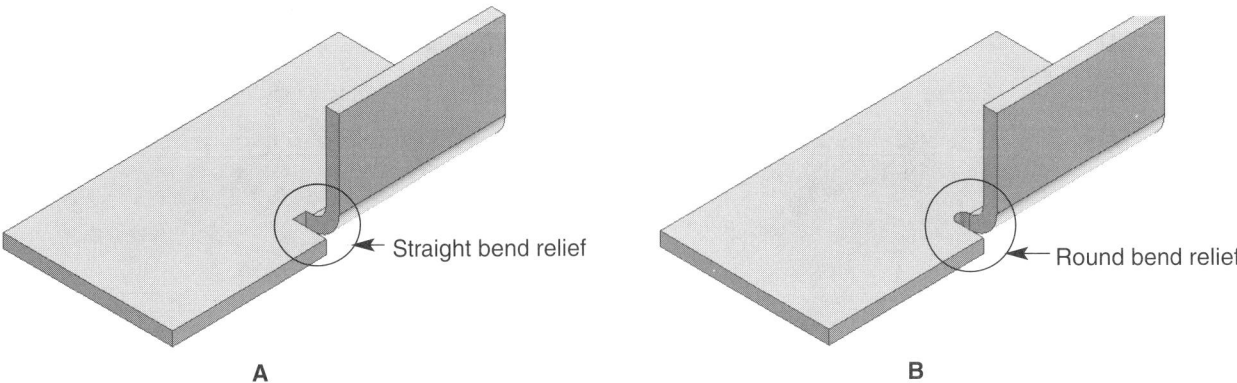

FIGURE 10.8   An example of (A) a straight bend relief and (B) a round bend relief.

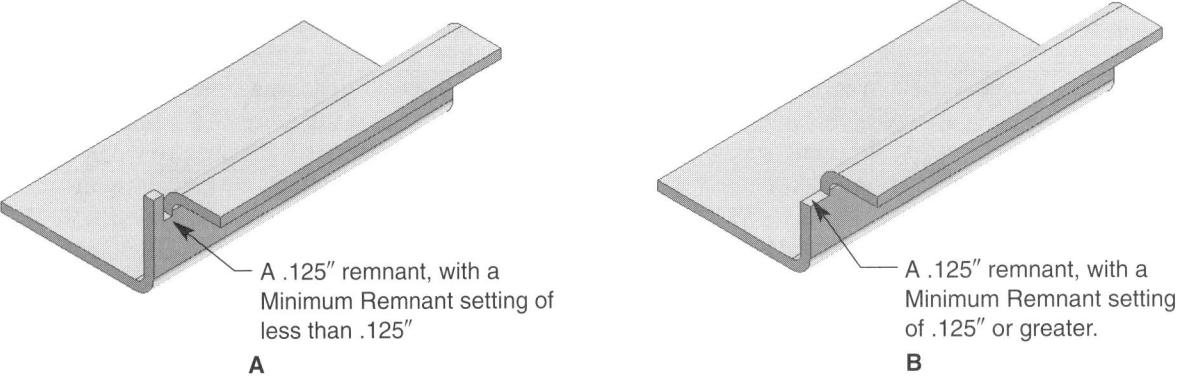

FIGURE 10.9   An example of (A) a remnant and (B) a removed remnant.

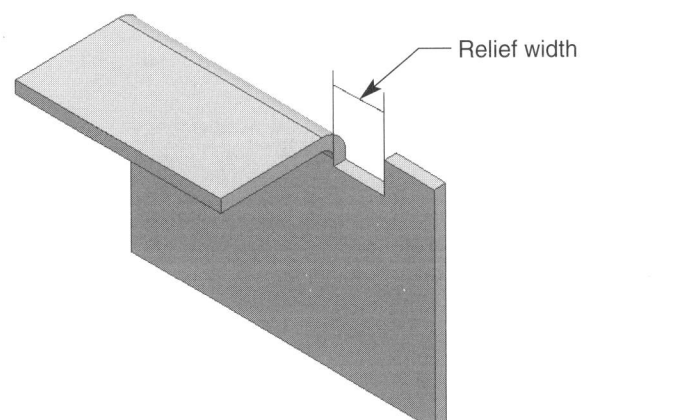

FIGURE 10.10   An example of a relief width greater than the material thickness.

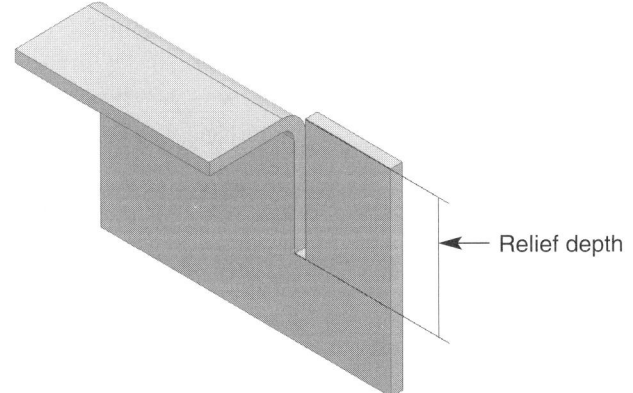

FIGURE 10.11   An example of a relief depth greater than 0.

- **Relief Depth**   This drop-down list allows you to specify the depth of the bend relief. The default relief depth is half the material thickness. However, you can specify a different relief depth, with a value greater than or equal to 0. See Figure 10.11.

- **Transition**   This drop-down list allows you to select one of four different types of bend transitions to use when the sheet metal part is unfolded and when no bend relief is specified.

FIGURE 10.12 The **Corner** tab of the **Sheet Metal Styles** dialog box.

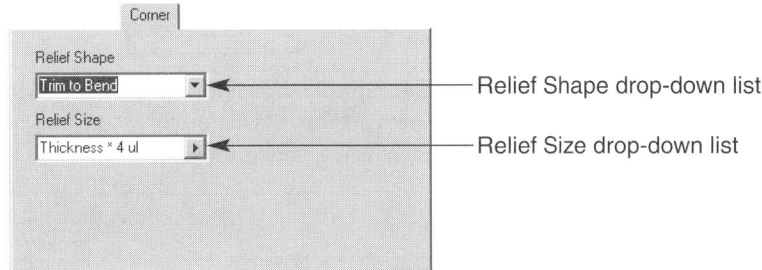

Relief Shape drop-down list

Relief Size drop-down list

- **Corner**   This tab, shown in Figure 10.12, allows you to specify the sheet metal part, corner relief parameters. *Corner relief* is typically added to a sheet metal part when two edges are bent, to relieve stress in the corner of the bend. The following are located in the **Corner** tab:

  - **Relief Shape**   This drop-down list allows you to specify the corner relief shape. You can specify round, square, or tear corner relief. You can also specify no corner relief by choosing the **Trim to Bend** relief shape option. See Figure 10.13.

  - **Relief Size**   This drop-down list allows you to specify the size of the corner relief. The default relief size is four times larger than the specified thickness of the material.

To develop a new sheet metal style, pick the **New** button. A copy of the currently active sheet metal style is displayed in the **Style List** edit box. Change the name of the style by picking the style name and entering the desired name. Then specify all the sheet metal options, using the previously discussed techniques. Once you have fully defined the sheet metal style, pick the **Save** button to save the style. Continue adding new styles as needed. If you want to remove a style from the list, pick the style in the style list and select the **Delete** button. When you finish modifying, adding, or removing sheet metal styles, pick the **Done** button to exit the **Sheet Metal Styles** dialog box.

FIGURE 10.13   An example of (A) no corner relief using the **Trim to Bend** option, (B) round corner relief using the **Round** option, (C) square corner relief using the **Square** option, and (D) torn corner relief using the **Tear** option.

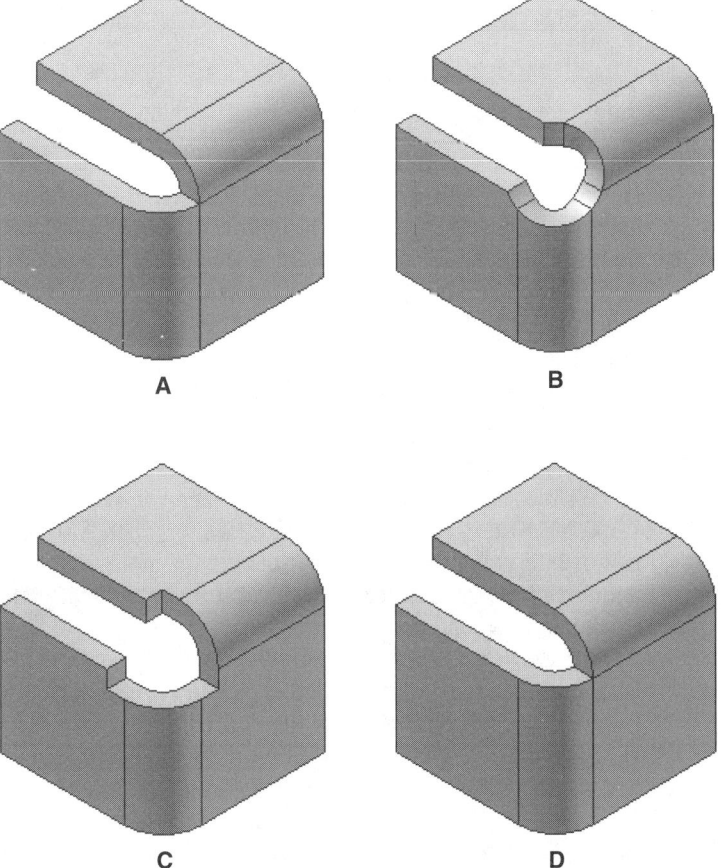

A

B

C

D

Field Notes

Even if you have completed a sheet metal part, you can modify the sheet metal style options, located in the **Sheet Metal Styles** dialog box, and update the part to reflect the changes. Changes to sheet metal parameters can also be made in the **Parameters** dialog box.

Exercise 10.1

1. Launch Autodesk Inventor.
2. Open a new Autodesk Inventor sheet metal part file and explore the feature interface options previously discussed.
3. Open the **Sheet Metal Styles** dialog box, and review the options discussed.
4. Create a new sheet metal style using specifications of your choice.
5. Save the new sheet metal style using a name of your choice.
6. Close the sheet metal part file without saving.
7. Exit Autodesk Inventor or continue working with the program.

Developing Sheet Metal Sketched Features

Once you have created a sketch, you are ready to develop a three-dimensional sheet metal part using Autodesk Inventor sheet metal sketched feature tools. As in most other part applications, usually the first type of sheet metal feature you create is a sketched feature. Typically a face, created using the **Face** tool, is used as the initial base feature. Additional features then add material to, or take material from, the face. Sketches and sheet metal sketched features can be created anytime during sheet metal part model development. Sheet metal sketch features include faces, contour flanges, cuts, and folds.

Creating Sheet Metal Faces

A *sheet metal face* is similar to an extrusion, and is typically used to create a base sheet metal feature. By default, a sheet metal face is created using the parameters specified in the **Sheet Metal Styles** dialog box. However, when you create individual sheet metal faces, you can override the bend, unfold, and bend relief options. All other options must be defined in the **Sheet Metal Styles** dialog box, or updated using the **Parameters** dialog box. An example of a simple sheet metal face is shown in Figure 10.14.

Once you have developed sketch geometry to use for the sheet metal face, access the **Face** command using one of the following techniques:

✓ Pick the **Face** button on the panel bar.

✓ Pick the **Face** button on the **Sheet Metal** toolbar.

The **Face** dialog box is displayed when you access the **Face** command. See Figure 10.15.

When you create a base feature face, typically you need to specify only the sketch profile and offset. However, if you are creating additional faces, more options are available. The tools required to generate an initial base feature face are located in the **Shape** area of the **Face Shape** tab. This tab contains the face shape definition options.

If only one sketch loop exists, the sketch profile is already selected. To create a face from a sketch with more than one profile available, pick the **Profile** button from the **Shape** area of the **Face Shape** tab if it is not already active. Then pick the desired face profile sketch. When the sketch profile is selected, a preview of the face is displayed. See Figure 10.16. The next step is to determine which side of the sketch plane you want the face depth to occur. If you want the face to be created on the opposite side of the current preview, pick the **Offset** button. If you are creating only one face, no other faces are present, and you do not need to define any other sheet metal face options, pick the **OK** button to generate the sheet metal face.

FIGURE 10.14 An example of a simple sheet metal face, created from a sketched rectangle.

FIGURE 10.15 The **Face** dialog box (**Face Shape** tab active).

Face Shape tab

Unfold Options tab

Bend Relief Options

Shape area

Profile button

Offset button

Double Bend area

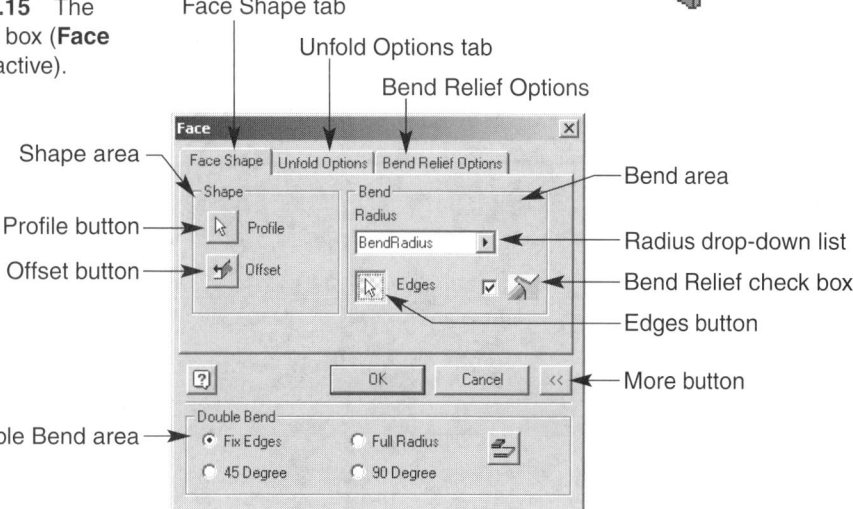

Bend area

Radius drop-down list

Bend Relief check box

Edges button

More button

FIGURE 10.16 A sheet metal face preview.

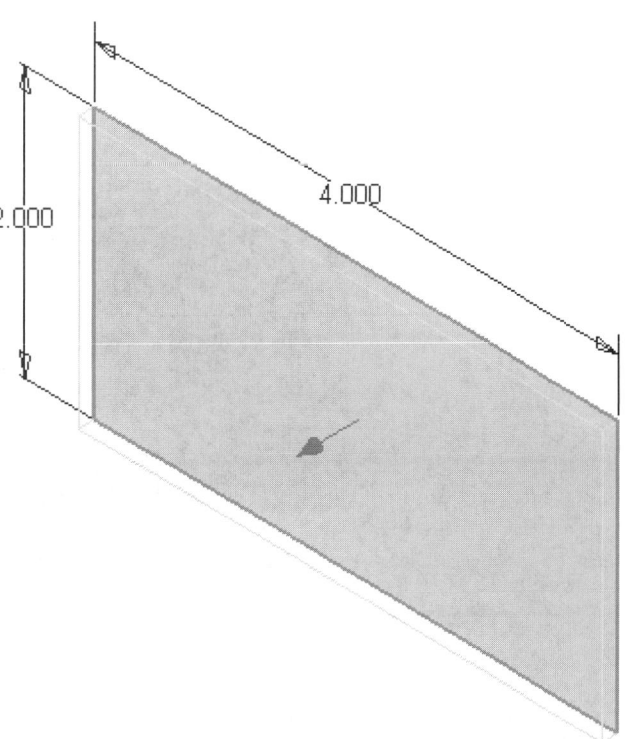

As previously discussed, more options are available in the **Face** dialog box when you add additional faces to the base feature face. The additional face shape options are found in the **Bend** area of the **Face Shape** tab, the **Unfold Options** tab, and the **Bend Relief Options** tab. To add a face to an existing face, select the sketch profile as shown in Figure 10.17 and choose the desired offset. Then define the face bend using the following options found in the **Bend** area of the **Face Shape** tab:

- **Radius**   This drop-down list allows you to define the bend radius. The default bend radius specified in the **Sheet Metal Styles** dialog box is described as "BendRadius" in the **Radius** drop-down list. However, you can override the default by entering any appropriate bend radius.

- **Bend Relief**   This check box allows you to specify whether you want to have a bend relief. If you do not want a bend relief, as shown in Figure 10.18, deselect the checkbox.

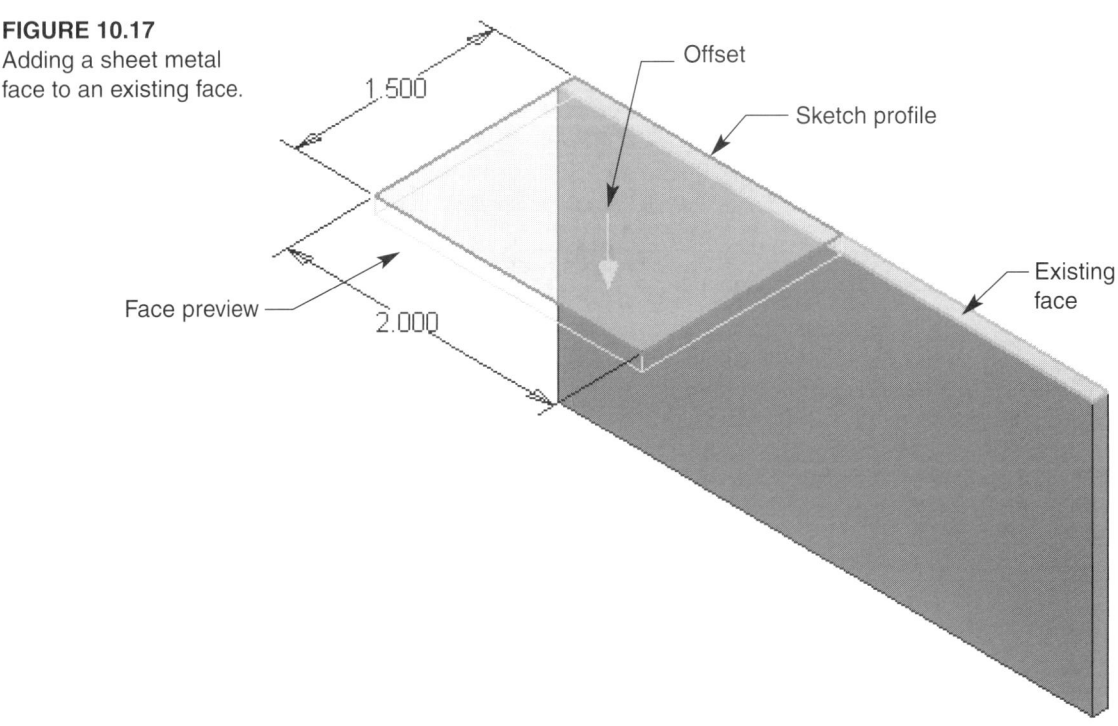

FIGURE 10.17
Adding a sheet metal
face to an existing face.

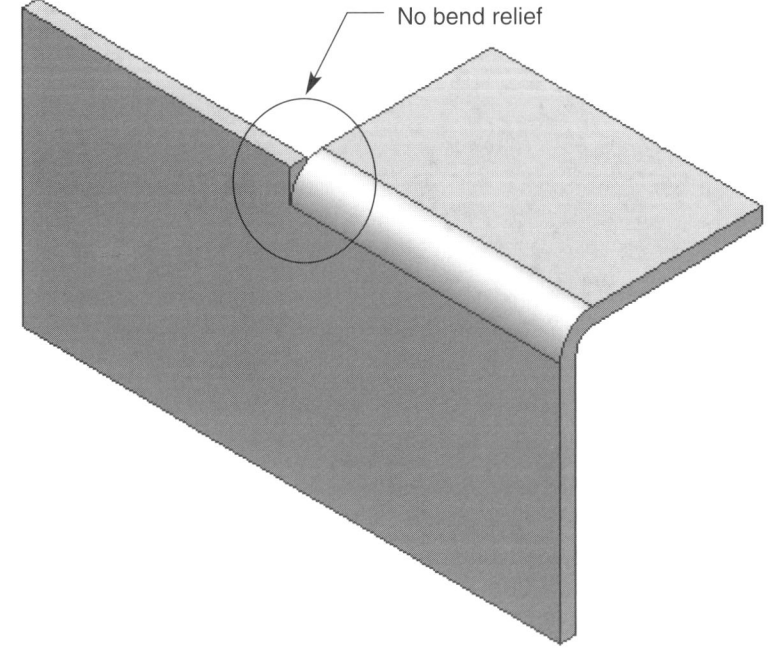

FIGURE 10.18   An
example of no bend
relief.

■ **Edges**   This button is used to select edges when defining a double bend. A *double bend* allows you to generate a bend between two parallel, noncoplanar faces. See Figure 10.19. To create a double bend, pick the **Edge** button, and select the edge you want to bend from. See Figure 10.20. Once you select the edge you want to bend from, pick the **More** button to access the **Double Bend** area of the **Face** dialog box. Then select the type of double bend you want to create. A **Fix Edges** double bend is shown in Figure 10.20; all other double bends are shown in Figure 10.21. If the double edge preview does not look correct, you may need to need to modify the selected edges using the **Flip Fixed Edge** button. See Figure 10.22.

The next step in creating a sheet metal face is to modify the unfold and bend relief options located in the **Unfold Options** and **Bend Relief Options** tabs of the **Face** dialog box, if necessary. See Figure 10.23. Each of the options available inside are the same as those located in the **Sheet Metal Styles** dialog box. The preferences in **Unfold Options** and **Bend Relief Options** tabs allow you to override the default options specified in the **Sheet Metal Styles** dialog box. However, if you want to use the sheet metal settings defined in the **Sheet Metal Styles** dialog box, select the **Use Default Settings** check box.

When you have specified all the sheet metal face options, pick the **OK** button to complete the face.

FIGURE 10.19   Two parallel, noncoplanar sheet metal faces.

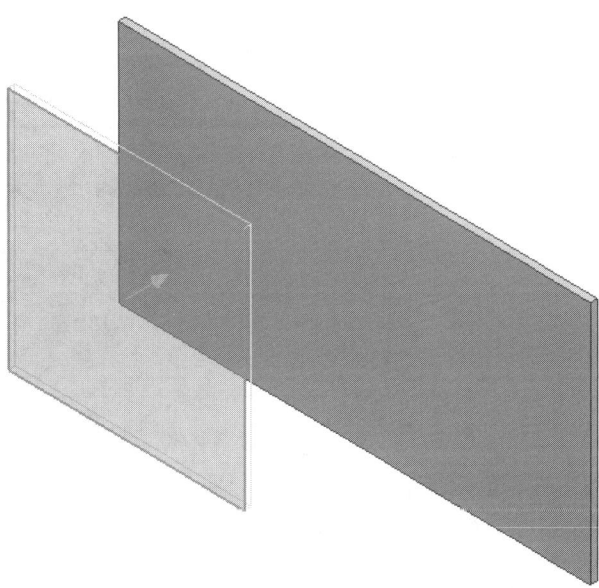

FIGURE 10.20
Creating a double bend face (**Fix Edges** option).

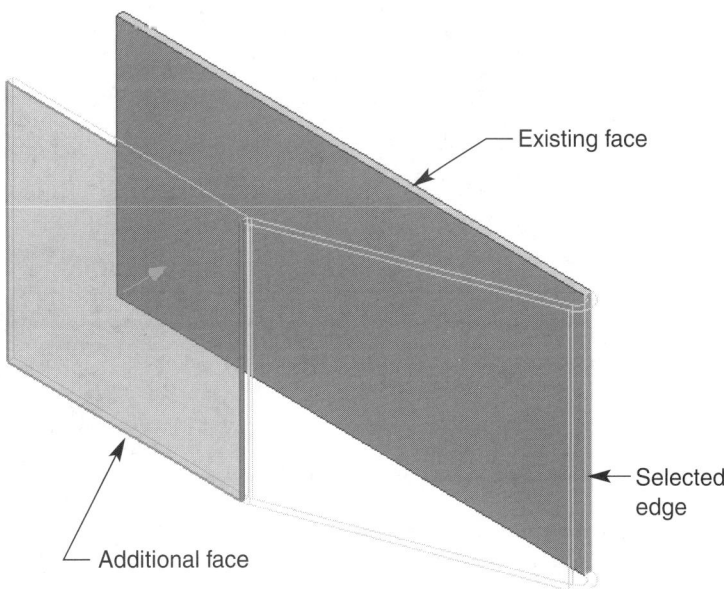

Existing face

Selected edge

Additional face

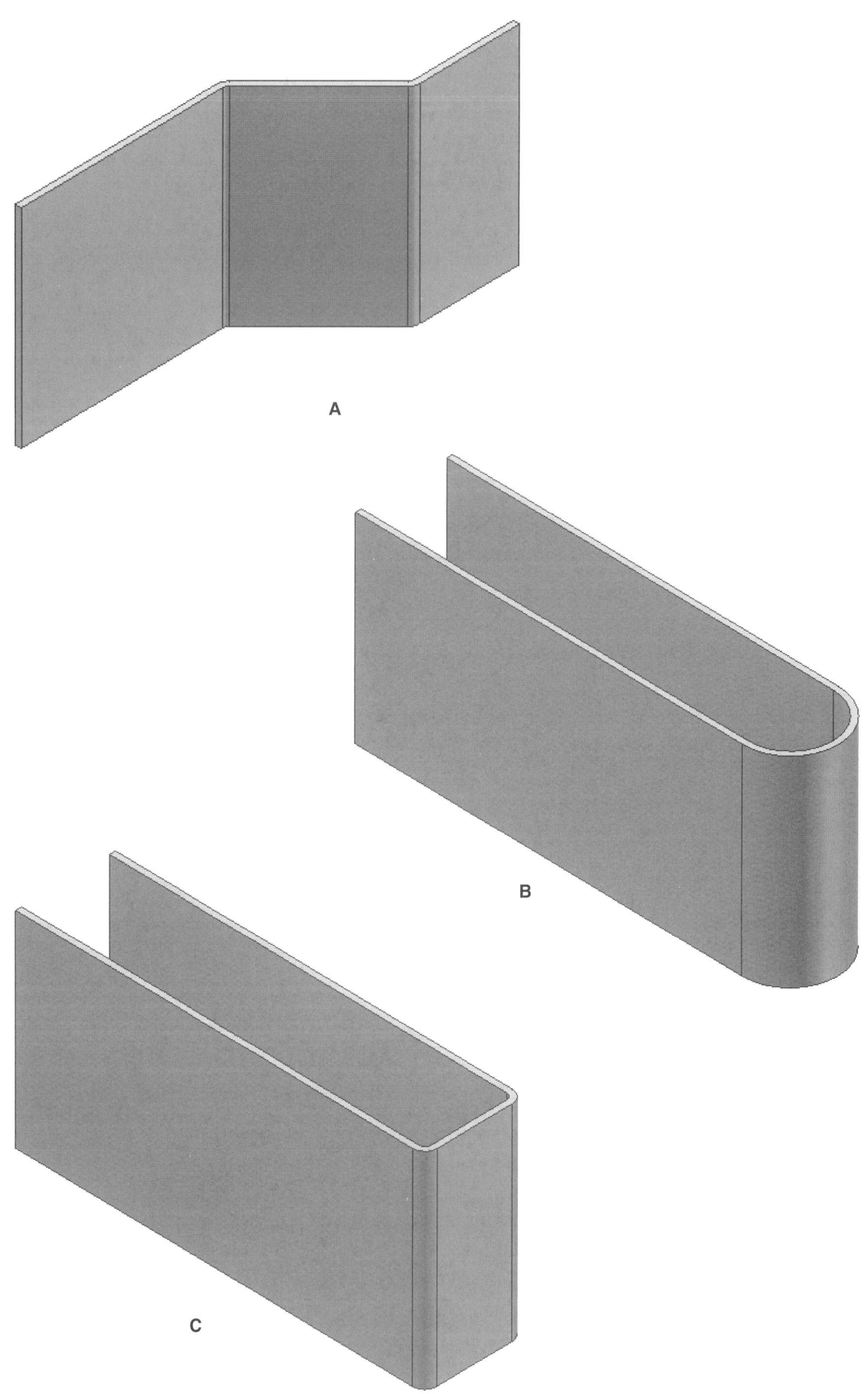

FIGURE 10.21 An example of (A) a 45 Degree double bend, (B) a Full Radius double bend, and (C) a 90 Degree double bend.

FIGURE 10.22   (A) A double bend created at the selected edge. The selected edge is the fixed edge. (B) An example of the effects of flipping the fixed edges.

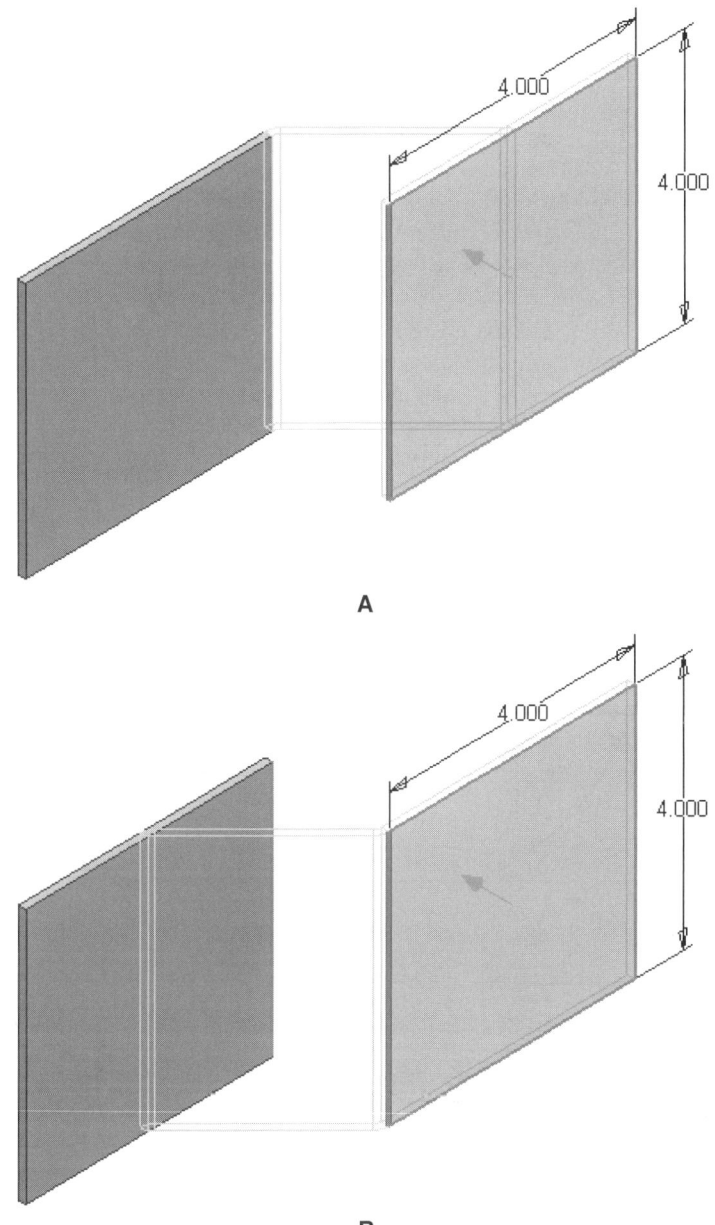

FIGURE 10.23   (A) The **Unfold Options** tab and (B) the **Bend Relief Options** tab of the **Face** dialog box.

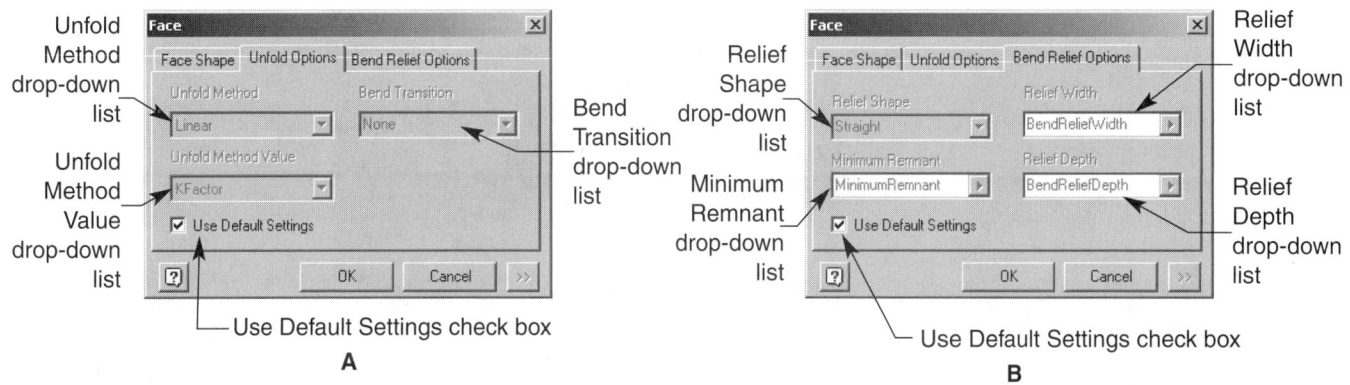

Exercise 10.2

1. Continue from Exercse 10.1, or launch Autodesk Inventor.
2. Open a new sheet metal part file.
3. Open a sketch on the **XZ Plane.**
4. Use the default sheet metal style.
5. Sketch a 12″×16″ rectangle, with the projected **Center Point** to fix the corner.
6. Use the **Face** tool to create a sheet metal face. Offset the face in a positive direction.
7. Offset a work plane 8″ from the **XZ Plane.**
8. Sketch an 8″×12″ rectangle on the work plane, with the projected **Center Point** to fix the corner, as shown.
9. Use the **Face** tool to create a sheet metal face. Offset the face in a negative direction. Create a 90 Degree bend between Face2 and Face1, by selecting the specified edge.
10. The final part should look like the part shown.
11. Save the file as EX10.2.
12. Exit Autodesk Inventor or continue working with the program.

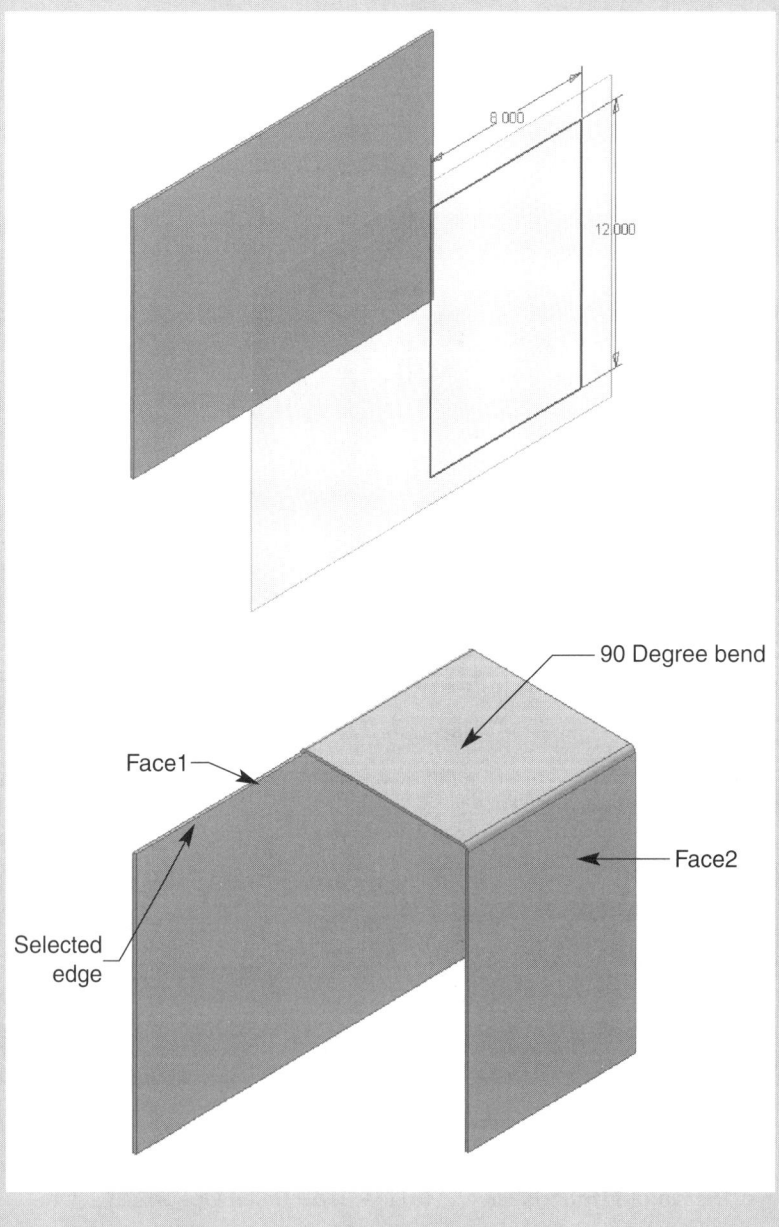

Creating a Sheet Metal Contour Flange

A sheet metal contour flange is similar to a face, but is typically created using an open sketch profile, such as the sketch shown in Figure 10.24. By default, a sheet metal contour flange is created using the parameters specified in the **Sheet Metal Styles** dialog box. However, when you create individual contour flanges, you can override the bend, unfold, and bend relief options. All other options must be defined in the **Sheet Metal Styles** dialog box or updated using the **Parameters** dialog box. Contour flanges are often a quick and easy way to generate extruded sheet metal parts such as a roof gutter. See Figure 10.25.

Once you have developed sketch geometry to use for the sheet metal contour flange, access the **Contour Flange** command using one of the following techniques:

✓ Pick the **Contour Flange** button on the panel bar.

✓ Pick the **Contour Flange** button on the **Sheet Metal** toolbar.

The **Contour Flange** dialog box is displayed when you access the **Contour Flange** command. See Figure 10.26.

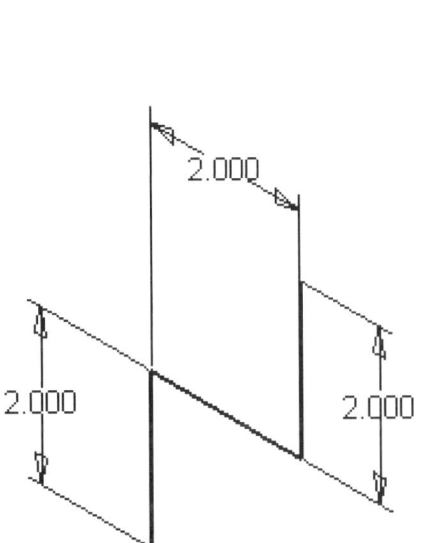

FIGURE 10.24 An example of a simple sketch used to generate a contour flange.

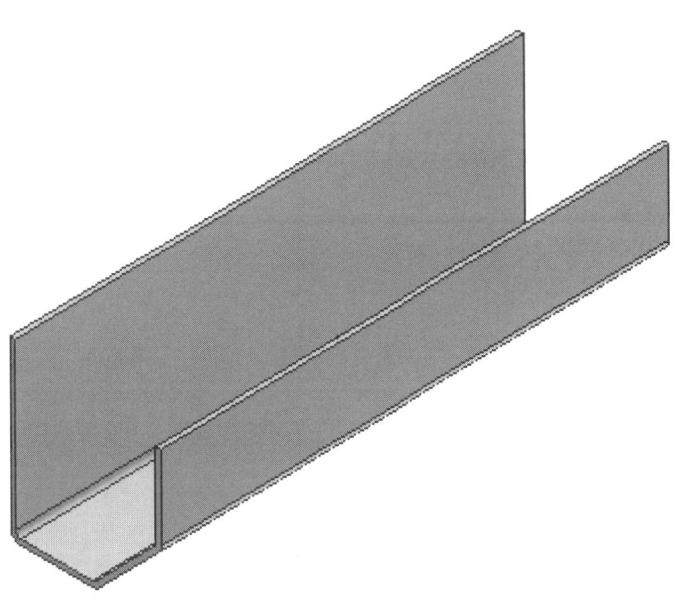

FIGURE 10.25 An example of a sheet metal contour flange, created from an open sketch profile.

FIGURE 10.26 The **Contour Flange** dialog box (**Contour Flange Shape** tab active).

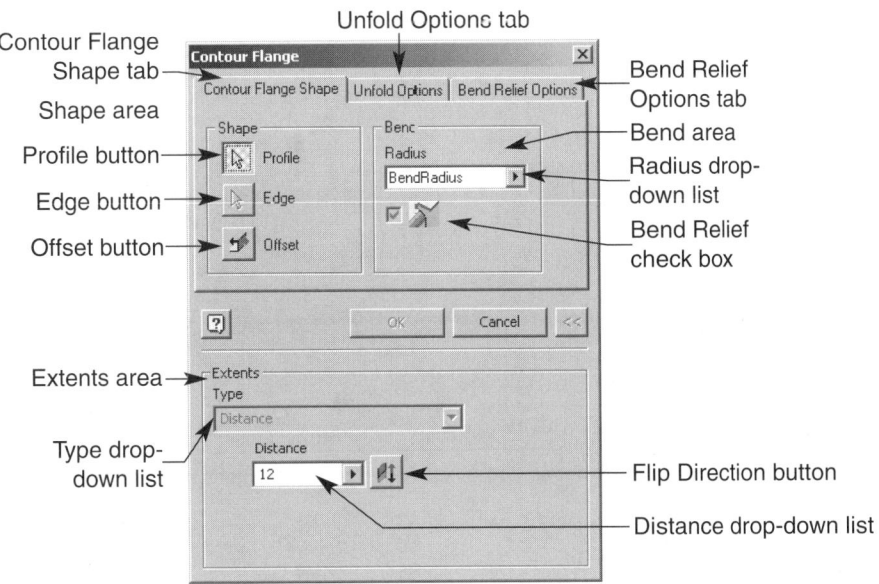

When you create a base feature contour flange, typically you need to specify only the sketch profile and length of the feature (**Distance** extents option). However, if you are creating additional contour flanges or other sheet metal features are present, more options are available.

To create a contour flange, first specify the options in the **Contour Flange Shape** tab of the **Contour Flange** dialog box. Initially the **Profile** button should be selected. If it is not already selected, select the **Profile** button and pick the sketch profile used to create the contour flange. When the sketch profile is selected, a preview of the face is displayed. See Figure 10.27.

Next, define the offset of the contour flange by selecting the **Offset** button, if necessary. The offset determines whether the depth of the sheet metal thickness is applied to the inside or the outside of the sketch profile. Once you determine the contour flange offset, define the flange bends using the following options found in the **Bend** area of the **Contour Flange Shape** tab:

- **Radius**   This drop-down list allows you to define the bend radius. The default bend radius specified in the **Sheet Metal Styles** dialog box is described as "BendRadius" in the **Radius** drop-down list. However, you can override the default by entering any appropriate bend radius.

- **Bend Relief**   This check box allows you to specify whether you want to have a bend relief. If you do not want a bend relief, deselect the checkbox.

The next step in creating a sheet metal contour flange is to modify the unfold and bend relief options located in the **Unfold Options** and **Bend Relief Options** tabs of the **Contour Flange** dialog box, if necessary. See Figure 10.28. Each of the options available are the same as those located in the **Sheet Metal Styles** dialog box. The preferences in **Unfold Options** and **Bend Relief Options** tabs allow you to override the default options specified in the **Sheet Metal Styles** dialog box. However, if you want to use the sheet metal settings defined in the **Sheet Metal Styles** dialog box, select the **Use Default Settings** check box.

As previously mentioned, when you are creating a base feature and no other sheet metal features are present, the only way to define the length, or depth, of the contour flange is with the **Distance** extents option. However, you can use the **Distance** extents option anytime to specify the length of the flange. To define the extents of the flange using the **Distance** extents option, access the **Extents** area by selecting the

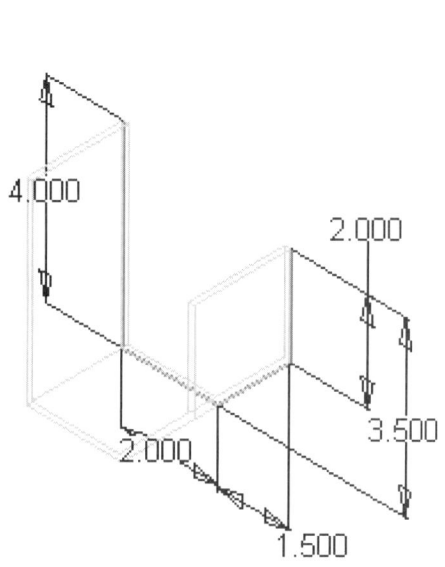

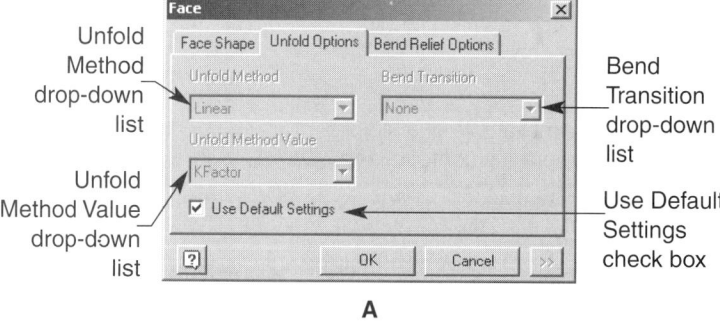

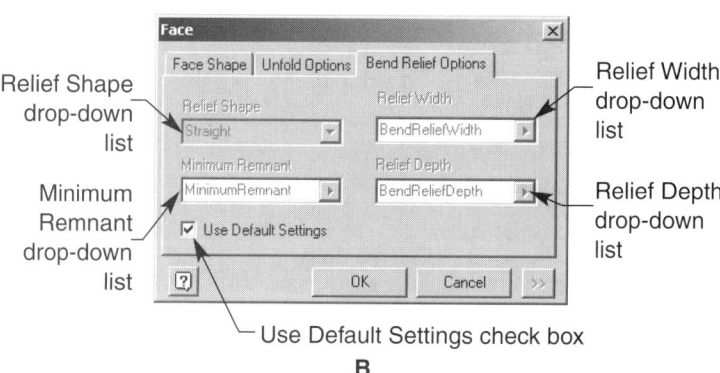

FIGURE 10.27   A sheet metal contour flange preview.

FIGURE 10.28   (A) The **Unfold Options** tab and (B) the **Bend Relief Options** tab of the **Contour Flange** dialog box.

More button, if it is not already selected. Then choose the **Distance** option from the **Type** drop-down list. Next enter the flange length, or choose a length from the Distance drop-down list. Finally, if you want the depth of the flange to occur on the opposite side of the plane, pick the **Flip Direction** button. An example of a distance extents contour flange is shown in Figure 10.29.

In addition to the **Distance** extents option, the following extents specifications are available when additional sheet metal contour features are present, such as adding a contour flange to a face:

■ **Edge**   This extents option allows you to specify an existing sheet metal feature edge as the length of the contour flange. To use the **Edge** extents option, first select the sketch profile. Then pick the **Edge** button if it is not already active. Finally, select the desired existing feature edge. As shown in Figure 10.30, selectable edges are perpendicular to the sketch plane.

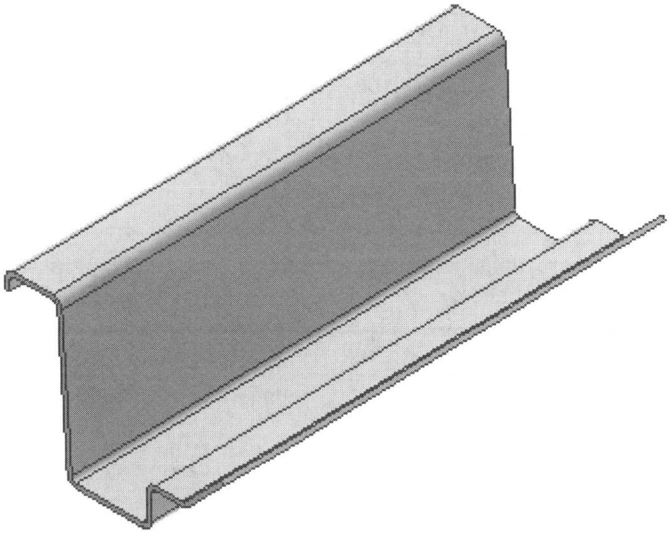

FIGURE 10.29   A 12″ contour flange created using the **Distance** extents option.

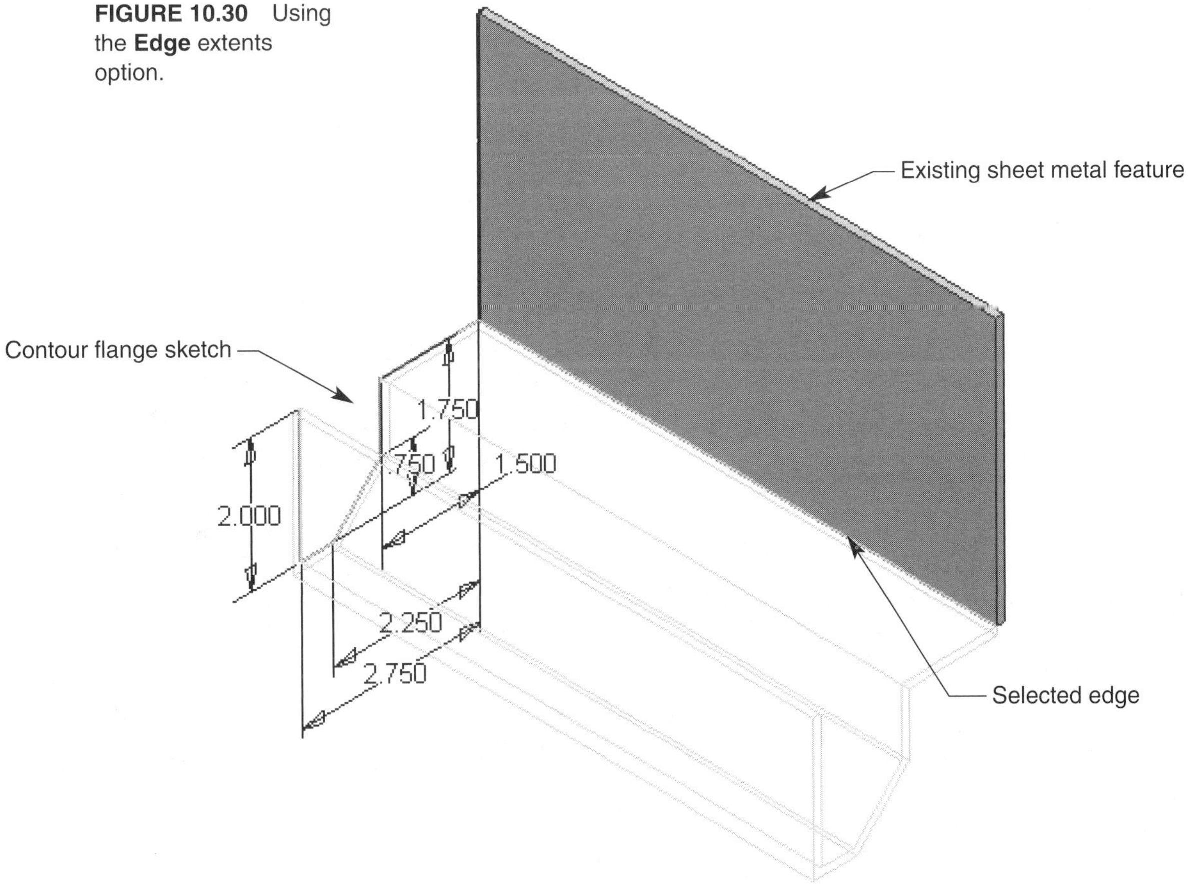

FIGURE 10.30   Using the **Edge** extents option.

Existing sheet metal feature

Contour flange sketch

1.750

.750

1.500

2.000

2.250

2.750

Selected edge

■ **Width**   This extents option allows you to specify the width of the contour flange and an offset distance from a selected point. To use the **Width** extents option, select the sketch profile, pick the **Edge** button, if it is not already active, and select the desired existing feature edge. Then pick the **More** button to access the **Extents** area of the **Contour Flange** dialog box, and select the **Width** option from the **Type** drop-down list. See Figure 10.31.

Next pick the **Select Start Point** button if it is not active, and pick an endpoint on the existing feature. The selected point identifies where the contour flange is offset from. Enter an offset value or select an offset value from the **Offset** drop-down list. If the offset occurs on the wrong side, pick the **Flip Direction** button to flip the offset. Finally, specify the width of the contour flange by entering a value or selecting a value from the width drop-down list. See Figure 10.32.

■ **Offset**   This extents option allows you to specify the width of the contour flange, by defining the offset distance from two selected points. To use the **Offset** extents option, select the sketch profile, pick the **Edge** button if it is not already active, and select the desired existing feature edge. Then pick the **More** button to access the **Extents** area of the **Contour Flange** dialog box, and select the **Offset** option from the **Type** drop-down list. See Figure 10.33. Next pick the **Select Start Point** button if it is not active, and pick an endpoint on the exist-

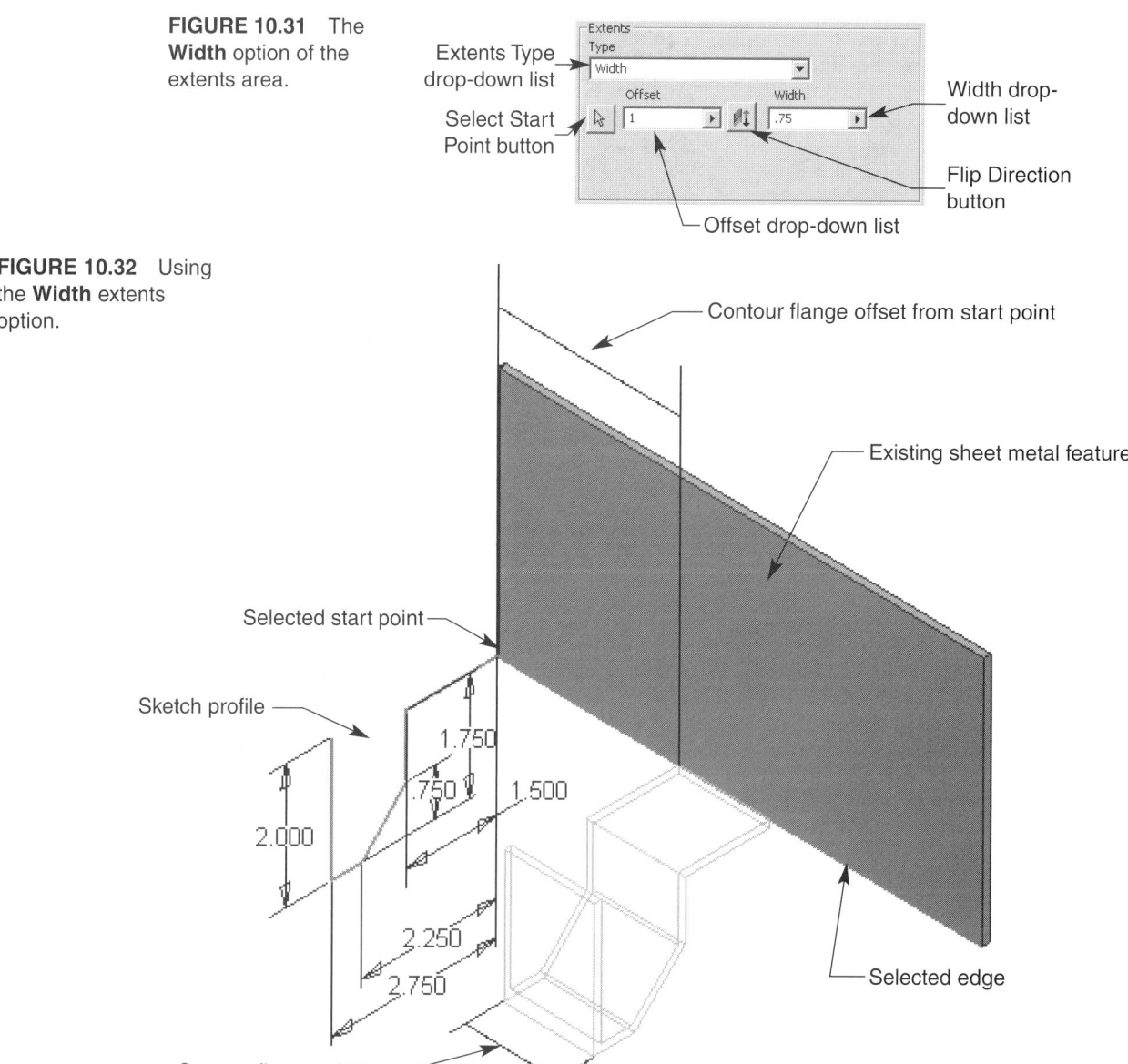

FIGURE 10.31   The **Width** option of the extents area.

Extents Type drop-down list

Select Start Point button

Width drop-down list

Flip Direction button

Offset drop-down list

FIGURE 10.32   Using the **Width** extents option.

Contour flange offset from start point

Existing sheet metal feature

Selected start point

Sketch profile

Selected edge

Contour flange width

FIGURE 10.33 The **Offset** option of the extents area.

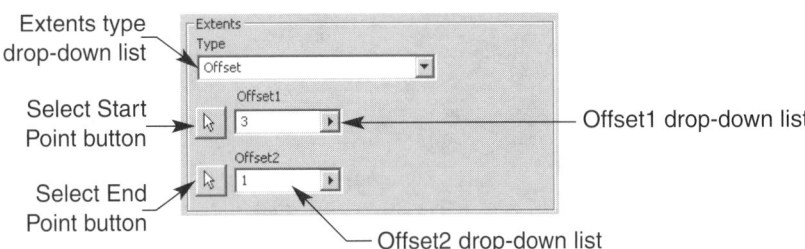

Extents type drop-down list

Select Start Point button

Select End Point button

Offset1 drop-down list

Offset2 drop-down list

ing feature where you want the contour flange offset to begin. Once the start point is selected, the **Select End Point** button should be come active. If not, select the **Select End Point** button, and pick an endpoint on the existing feature where you want the contour flange offset to end. Finally, enter an offset value or select a value from the **Offset1** and **Offset2** drop-down lists. See Figure 10.34.

When you have specified all the sheet metal contour feature options, pick the **OK** button to complete the contour feature.

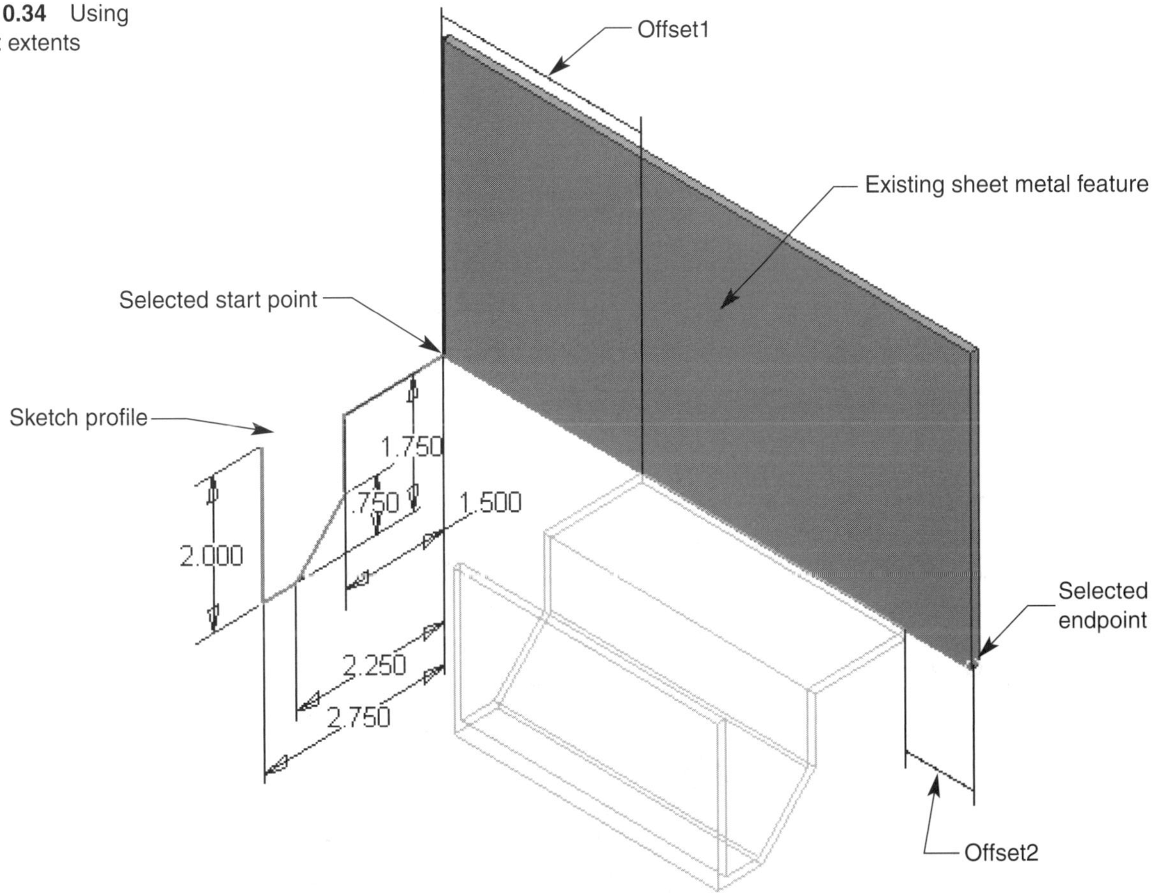

FIGURE 10.34 Using the **Offset** extents option.

Offset1

Existing sheet metal feature

Selected start point

Sketch profile

1.750

.750

1.500

2.000

2.250

2.750

Selected endpoint

Offset2

Exercise 10.3

1. Continue from Exercse 10.2, or launch Autodesk Inventor.
2. Open a new sheet metal part file.
3. Open a sketch on the **YZ Plane.**
4. Use the default sheet metal style.
5. Create the sketch shown, using the projected **Center Point** to fix the sketch.
6. Use the **Contour Flange** tool to create a sheet metal contour flange. Offset the flange toward the inside of the sketch, 30mm in a positive direction.
7. Open a new sketch on the **YZ Plane.**
8. Create the sketch shown, using the corner of **Contour Flange1** to fix the sketch.
9. Offset a work plane 8″ from the **XZ Plane.**
10. Use the **Contour Flange** tool to create a sheet metal contour flange. Use the Edge extents option to create the flange along the specified edge. Offset the flange toward the inside of the sketch.
11. The final part should look like the part shown.
12. Save the file as EX10.3.
13. Exit Autodesk Inventor or continue working with the program.

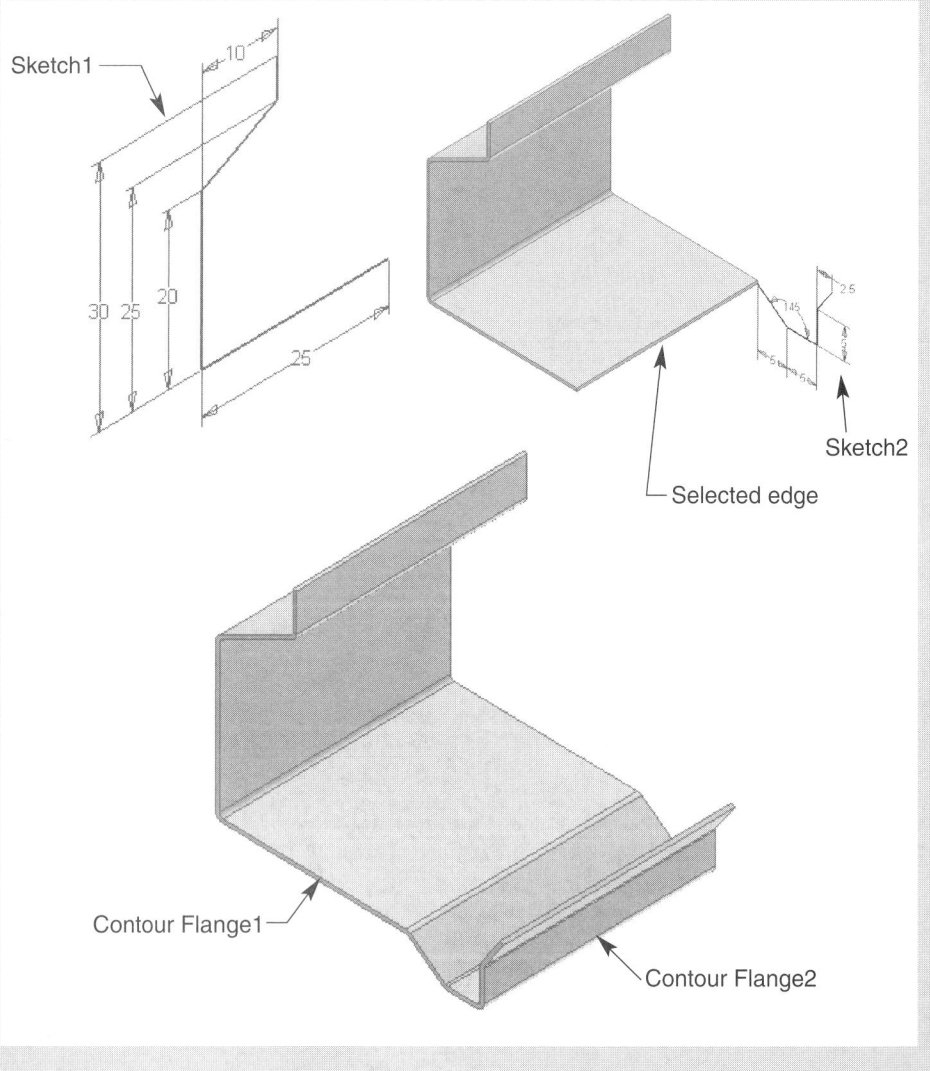

Using the Sheet Metal Cut Tool

The sheet metal cut tool is used to remove material through a selected sketch profile on an exiting sheet metal feature. Using the **Cut** tool is similar to using the **Extrusion** tool with the cut operation. See Figure 10.35.

Once you have developed sketch geometry to use for the sheet metal cut, access the **Cut** command using one of the following techniques:

✓ Pick the **Cut** button on the panel bar.

✓ Pick the **Cut** button on the **Sheet Metal** toolbar.

The **Cut** dialog box is displayed when you access the **Cut** command. See Figure 10.36.

To use the **Cut** tool, pick the sketch profile you want to use for the cut. Then specify the cut depth using one of the following techniques:

■ **Cut Across Bend** This check box, located in the **Shape** area of the **Cut** dialog box, allows you to cut through the thickness of the sheet metal and the full length on the bend. See Figure 10.37. Other extent operations are not available when the **Cut Across Bend** check box is selected.

■ **Distance** This option, located in the **Extents** area of the **Cut** dialog box, allows you to define the amount of material the cut removes, using a specified distance. See Figure 10.38. To use the **Distance** extent option, pick **Distance** from the **Extent Type** drop-down list, and specify the distance in the **Distance** drop-down list. The default distance is "Thickness," which is the thickness of the material specified in the **Sheet Metal Styles** dialog box. To modify the distance, enter a new value or select a value from the **Distance** drop-down list. Finally, select the direction you want the cut to occur, and pick the **OK** button to create the cut.

FIGURE 10.35 An example of a simple cut through a sheet metal face.

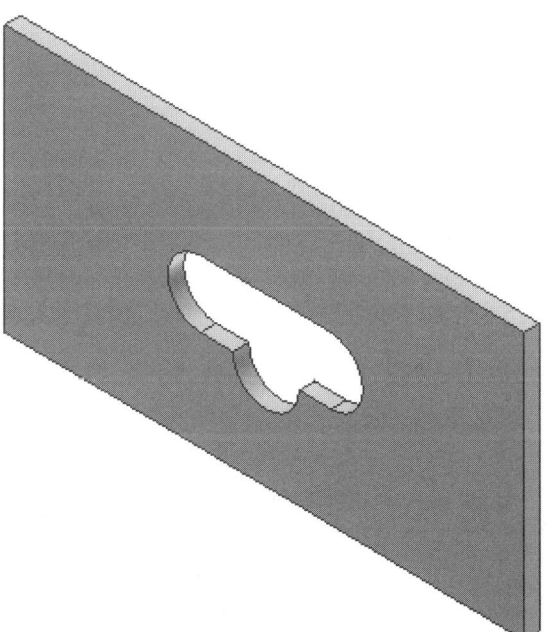

FIGURE 10.36 The **Cut** dialog box (**Distance** extents option active).

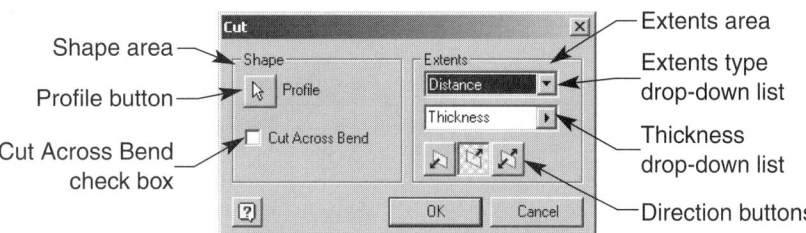

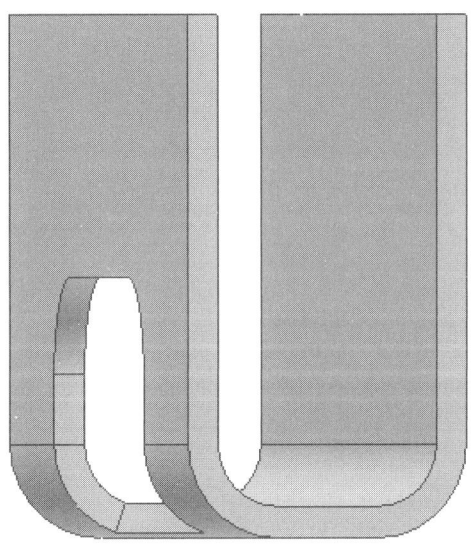

FIGURE 10.37   An example of a cut created
using the **Cut Across Bend** option.

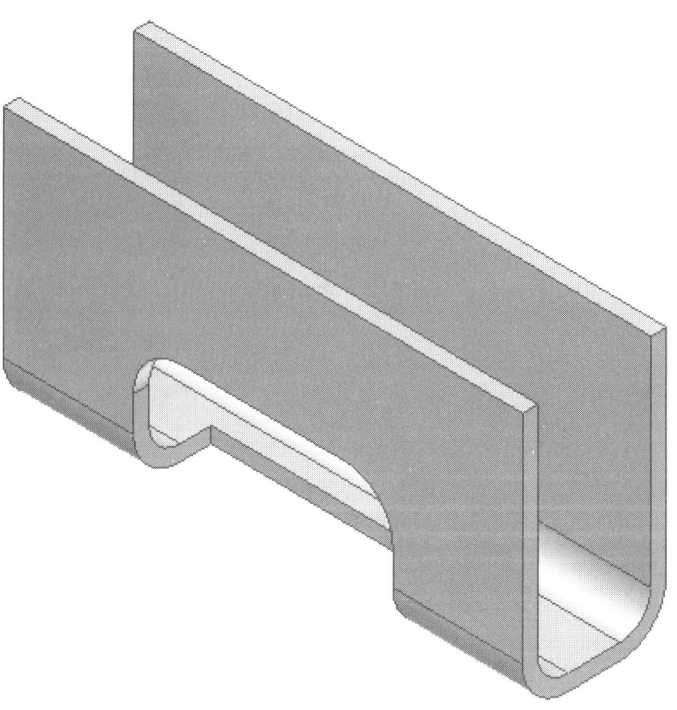

FIGURE 10.38   An example of a cut created using the
Distance extent option five times the material thickness.

■ **To Next**   Selecting this option, located in the **Extents** area of the **Cut** dialog box, enables
you to cut a sketch profile to the next possible face or plane. See Figure 10.39. To cut a
sketch profile to the next face or plane, pick the sketch profile, and select the direction you
want the cut to occur. Then, pick the **OK** button to create the cut.

Field Notes

There is no equal side, or midplane, direction option available when cutting **To Next** face or plane.
If a face or plane does not exist that can end the extrusion, the feature cannot be created.

FIGURE 10.39
Cutting through sheet
metal using the **To Next**
extents option.

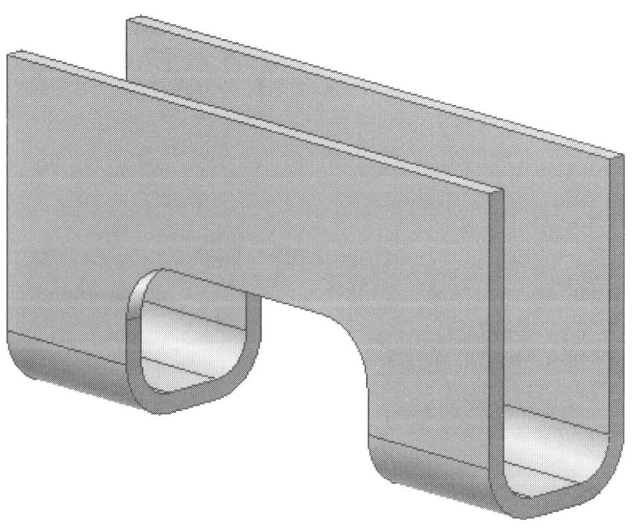

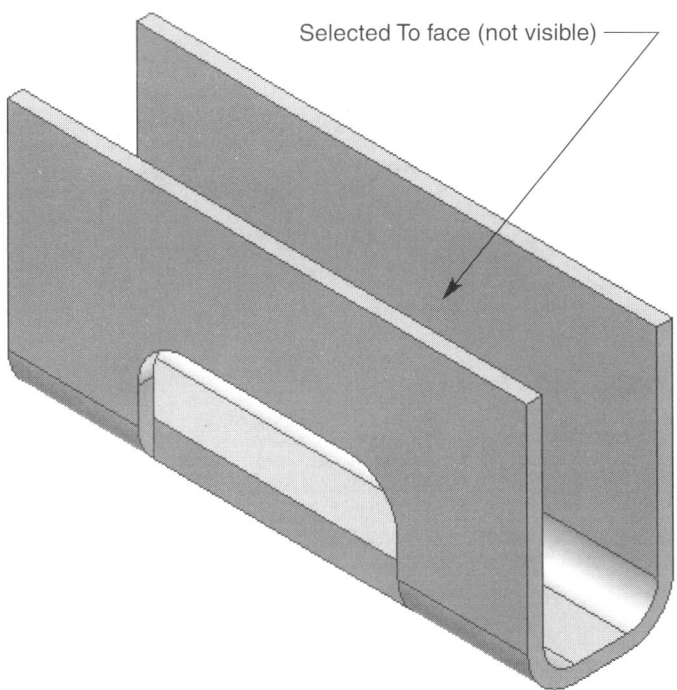

Selected To face (not visible)

FIGURE 10.40   Creating an extrusion using the **To** extents option.

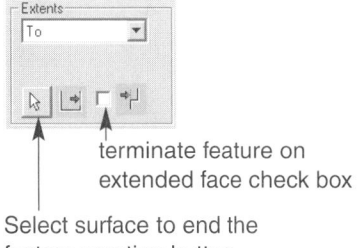

terminate feature on
extended face check box

Select surface to end the
feature creation button

FIGURE 10.41   The **Extents** area
of the **Cut** dialog box (**To** option).

- ■ **To**   This option is similar to the **To Next** option previously described. However, selecting **To** allows you to choose a face or plane to end the cut, instead of the cut ending at the next possible face or plane. See Figure 10.40. To terminate a cut at a specified face or plane using the **To** option, pick the face or plane you want to extrude the feature to. You may need to choose the **Select surface to end the feature creation** button if it is not already activated. See Figure 10.41. In addition, if the face or plane you choose does not intersect the extrusion path, you need to check the **terminate feature on extended face** check box.

- ■ **From To**   This option is similar to the **To** option previously described. However, selecting **From To** allows you to choose a face or plane to begin and end the cut, instead of just the ending surface. See Figure 10.42. To begin and terminate a cut at a specified surface using the **From To** option, pick the face or plane you want to begin the cut from. You may need to choose the **Select surface to start the feature creation** button if it is not already activated. Next, pick the face or plane where you want to end the extrusion. Again, you may need to choose the **Select surface to end the feature creation** button if it is not already activated. If the surfaces you choose to start and end the extrusion do not intersect the extrusion path, you need to check the **terminate feature on extended face** check boxes. See Figure 10.43.

- ■ **All**   Selecting this option enables you to cut a sketch profile directly through all other features. If you know for sure you want to cut the entire way through existing features, use the **All** option. It does not require any distances, or surface selections, and may save some time. To cut a feature through all existing features, pick the sketch profile and select the direction you want the cut to occur. Then pick the **OK** button to create the cut.

When you have specified all the sheet metal cut options, pick the **OK** button to complete the cut.

FIGURE 10.42
Cutting through sheet
metal using the **From
To** extents option.

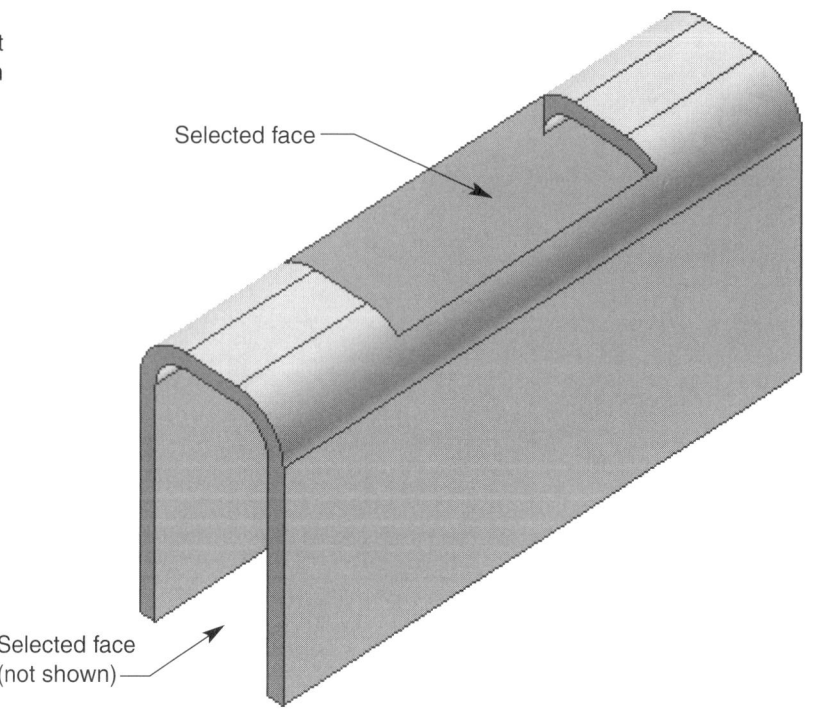

Selected face

Selected face
(not shown)

FIGURE 10.43   The
Extents area of the **Cut**
dialog box (**From To**
option).

Select surface to start the
feature creation button

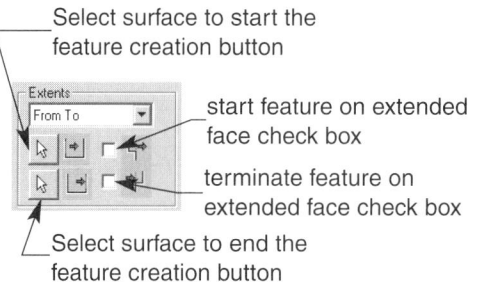

start feature on extended
face check box

terminate feature on
extended face check box

Select surface to end the
feature creation button

Exercise 10.4

1. Continue from Exercse 10.3, or launch Autodesk Inventor.
2. Open EX10.2.
3. Save a copy of EX10.2 as EX10.4.
4. Close EX10.2 without saving, and open EX10.4.
5. Open a sketch on **Face1,** and sketch the geometry shown.
6. Use the **Cut** tool to create a sheet metal cut of the sketch. Specify the **Distance** extents option with a distance of the material thickness.
7. Open a sketch on **Face2,** and sketch the geometry shown.
8. Use the **Cut** tool to create a sheet metal cut of the sketch. Specify the **All** extents option.
9. Resave EX10.4.
10. Exit Autodesk Inventor or continue working with the program.

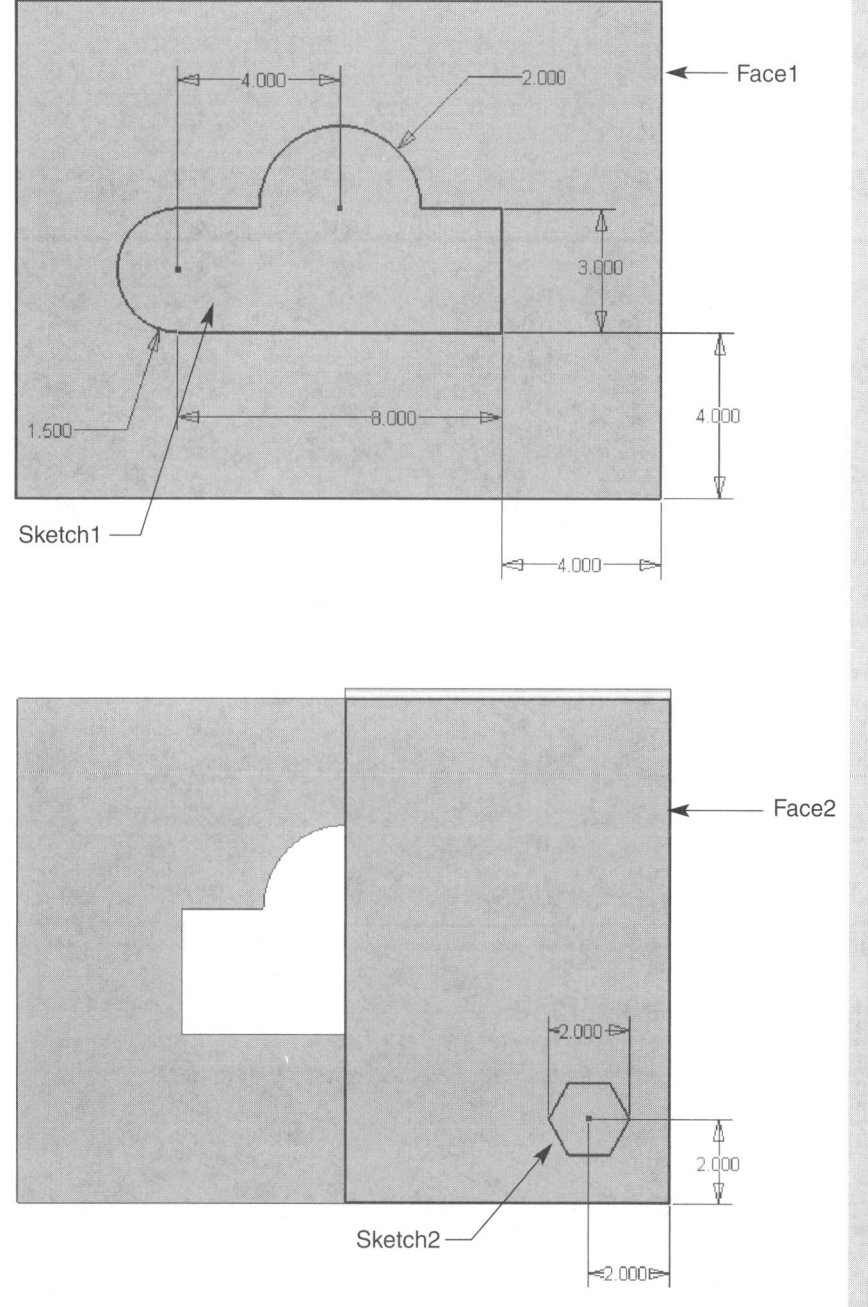

Creating Sheet Metal Folds

Sheet metal *folds* are created by folding a section of another sheet metal feature, such as a face or a contour flange, along a sketched line. See Figure 10.44.

The first step in creating a fold is to make a sketch. Most of the time, a single sketched line is used to define the edge the material is folded over. See Figure 10.45.

Once you have developed a sketch, access the **Fold** command using one of the following techniques:

✓ Pick the **Fold** button on the panel bar.

✓ Pick the **Fold** button on the **Features** toolbar.

The **Fold** dialog box is displayed when you access the **Fold** command; it is used to specify the fold options and parameters. See Figure 10.46. When you initially open the **Fold** dialog box, the **Fold Shape** tab is active and contains the following options:

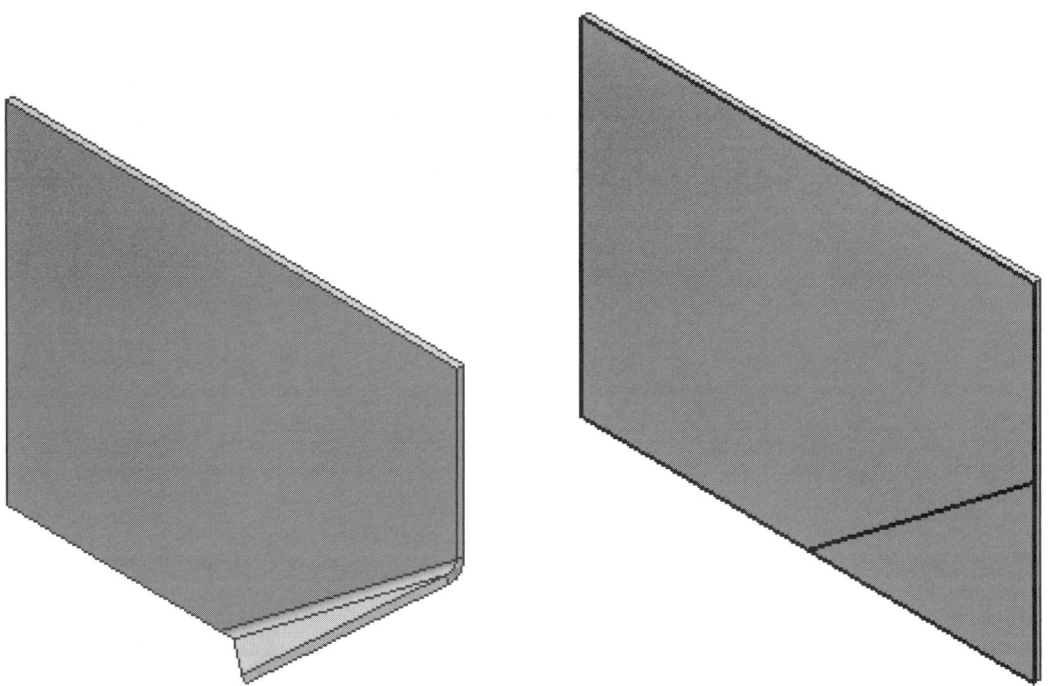

FIGURE 10.44   An example of a simple sheet metal fold.

FIGURE 10.45   A sketched line used as a fold edge.

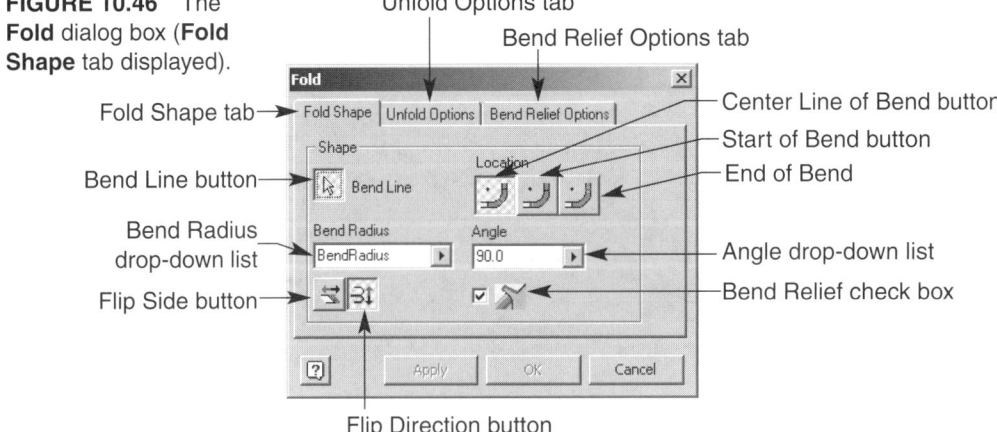

FIGURE 10.46   The **Fold** dialog box (**Fold Shape** tab displayed).

Unfold Options tab

Bend Relief Options tab

Fold Shape tab

Bend Line button

Bend Radius drop-down list

Flip Side button

Center Line of Bend button

Start of Bend button

End of Bend

Angle drop-down list

Bend Relief check box

Flip Direction button

- **Bend Line**   This button allows you to pick the sketched line where you want the fold to occur.

- **Bend Radius**   This drop-down list allows you to define the bend radius. The default bend radius specified in the **Sheet Metal Styles** dialog box is described as "BendRadius" in the **Radius** drop-down list. However, you can override the default by entering any appropriate bend radius.

- **Location**   The following three buttons control where the sheet metal material is folded in reference to the sketched line:

  - **Center Line of Bend**   This button uses the sketched line as a center line and bends the sheet metal material equally between the sketch lines.

  - **Start of Bend**   This button begins the bend at the sketched line and does not alter the material that is not bent.

  - **End of Bend**   This button ends the bend at the sketched line and does not alter the material that is bent.

Field Notes

The displays on the three Location buttons, and the preview of the operation on the part, help you determine the desired bend location.

- **Angle**   This drop-down list allows you to specify the amount of bend applied to the section of material being bent. The default bend is 90°, which results in a right-angle bend. However, you can enter any appropriate angle or select an angle from the **Angle** drop-down list. See Figure 10.47.

- **Flip Side**   This button controls where the angle of the fold is applied. See Figure 10.48.

FIGURE 10.47   An example of (A) a 90° fold and (B) a 20° fold.

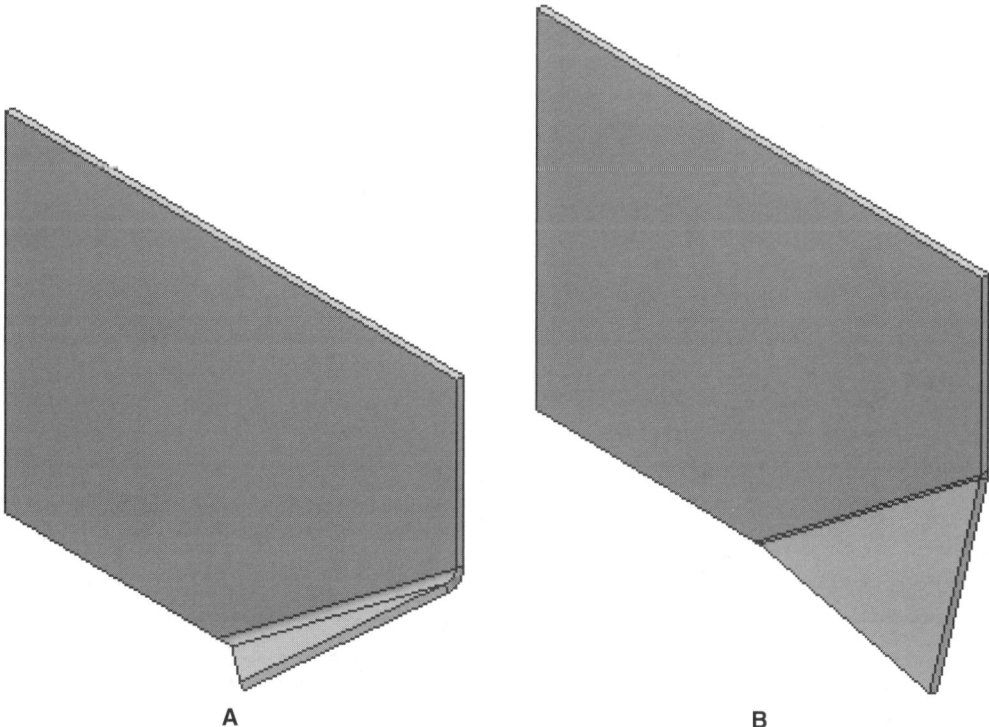

A   B

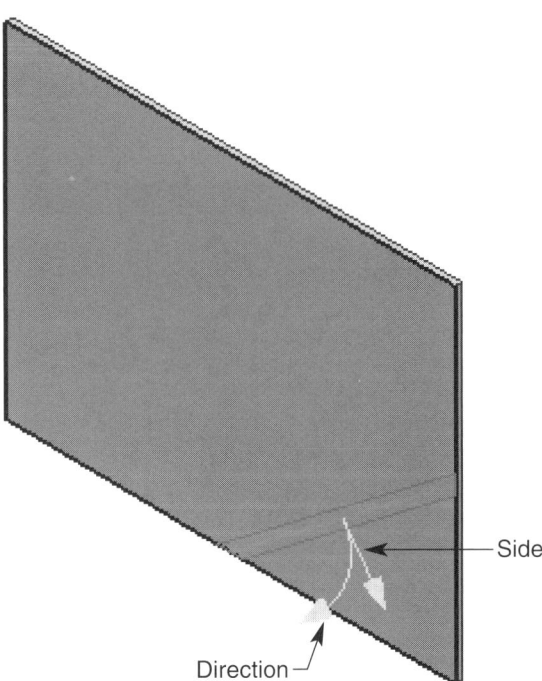

FIGURE 10.48 The preview of side and direction options.

Side

Direction

- **Bend Direction** This button controls which side of the sheet metal feature the fold occurs. See Figure 10.48.

- **Bend Relief** This check box allows you to specify whether or not you want to have a bend relief. If you do not want a bend relief, deselect the checkbox.

Once you have defined all the fold shape options, the next step in creating a sheet metal fold is to modify the unfold and bend relief options located in the **Unfold Options** and **Bend Relief Options** tabs of the **Fold** dialog box, if necessary. See Figure 10.49. Each of the options available inside these tabs are the same as those located in the **Sheet Metal Styles** dialog box. The preferences in these tabs allow you to override the default options specified in the **Sheet Metal Styles** dialog box. However, if you want to use the sheet metal settings defined in the **Sheet Metal Styles** dialog box, select the **Use Default Settings** check box.

When you have specified all the sheet metal fold options, pick the **OK** button if you are generating only one fold. If you want to use the **Fold** tool again, pick the **Apply** button and continue producing folds as needed. When finished, pick the **OK** button to exit the command.

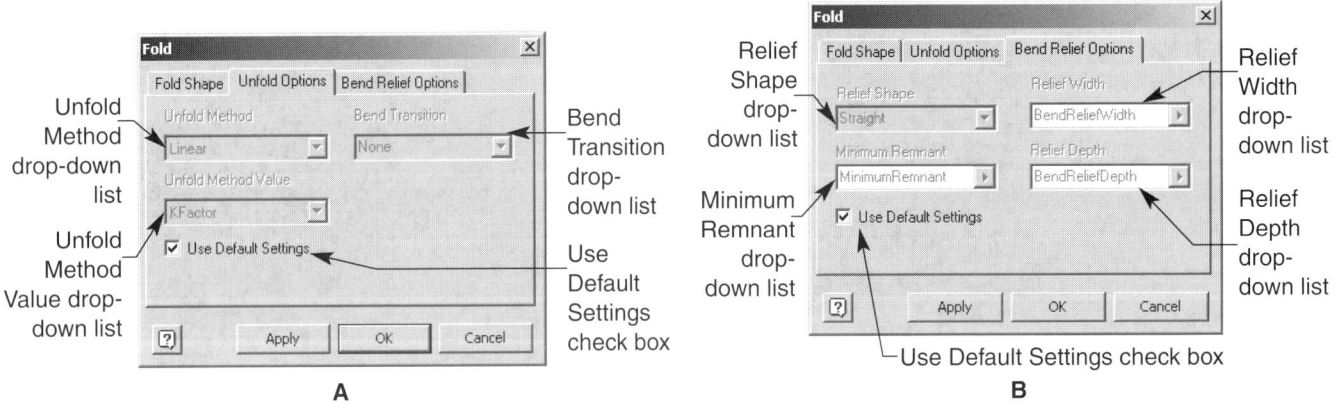

FIGURE 10.49 (A) The **Unfold Options** tab and (B) the **Bend Relief Options** tab of the **Fold** dialog box.

Exercise 10.5

1. Continue from Exercse 10.4, or launch Autodesk Inventor.
2. Open EX10.2.
3. Save a copy of EX10.2 as EX10.5.
4. Close EX10.2 without saving, and open EX10.5.
5. Create sketches on the faces shown using the **Line** tool and projected midpoint references.
6. Create the specified folds as shown using the following information:

Fold1:
Bend Line: Sketch3 line
Bend Location: Centerline of Bend
Angle: 20°
Flip Side if necessary

Fold2:
Bend Line: Sketch4 line
Bend Location: Start of Bend
Angle: 45°
Flip Direction if necessary

Fold3:
Bend Line: Sketch5 line
Bend Location: End of Bend
Angle: 90°
Flip Direction if necessary

7. The final part should look like the part shown.
8. Resave EX10.5.
9. Exit Autodesk Inventor or continue working with the program.

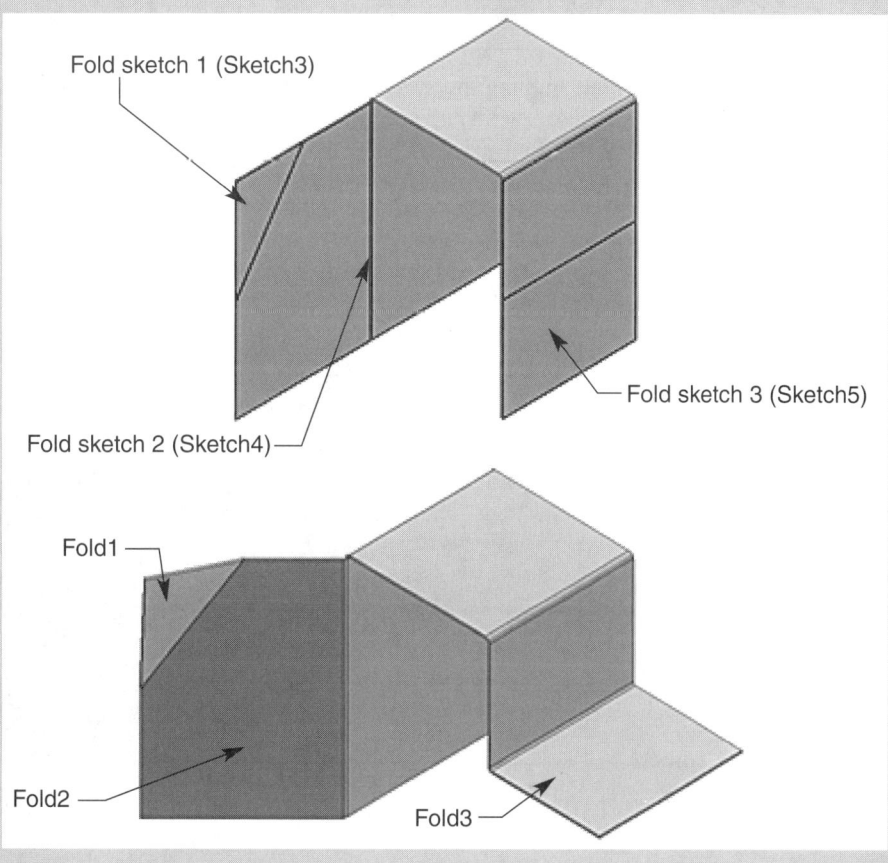

Developing Sheet Metal Placed Features

Once you have generated at least one of the previously discussed sheet metal sketched features, you can further develop the part using Autodesk Inventor sheet metal placed feature tools. ***Placed features*** can be added to the most basic feature, such as a face or cut, or to very complex, multifeature sheet metal parts. As the name implies, placed features are placed on existing features. For example, a corner round is placed on the corner of a contour flange.

In contrast to the sketched features previously discussed, placed features do not require a sketch to be created. However, they do require an acceptable sketched, placed, catalog, or pattern feature to be present. Placed features can be created anytime during sheet metal part model development, as long as one of these suitable features exists. Sheet metal placed features include flanges, hems, corner seams, bend, corner rounds, and corner chamfers.

Adding Sheet Metal Flanges

A *flange* is a type of sheet metal face that can be added to an existing feature such as a face. See Figure 10.50. In contrast to a sheet metal face, flanges do not require a sketch and are typically much less geometrically complex. Flanges are added to existing sheet metal feature edges and create specified shape and bend preferences between the existing feature and the placed flange.

Flanges are created using the **Flange** tool, which is accessed using one of the following techniques:

✓ Pick the **Flange** button on the panel bar.

✓ Pick the **Flange** button on the **Sheet Metal** toolbar.

The **Flange** dialog box is displayed when you access the **Flange** command. See Figure 10.51.

To place a flange, first pick the **Select Edge** button, if it is not already selected, and pick the existing feature edge on which the flange will be created. The edge you select will identify the beginning of the flange width and the location of the bend. See Figure 10.52. Once you have selected the flange edge, use the **Flip Offset** button to identify where you want the flange depth to occur. (This is similar to how a particular edge is selected.) The flange offset will identify the beginning of the flange depth and the location of the bend.

Next specify the width of the flange by entering a value or selecting a value from the **Distance** drop-down list. The distance of the flange specifies the flange width from the existing feature. You may need to

FIGURE 10.50 An example of a simple sheet metal flange.

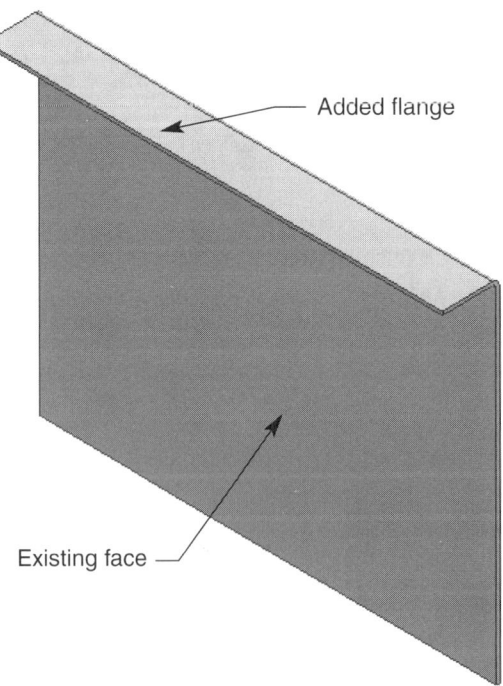

Added flange

Existing face

FIGURE 10.51 The
Flange dialog box
(**Flange Shape** tab
displayed and **Edge**
extents option active).

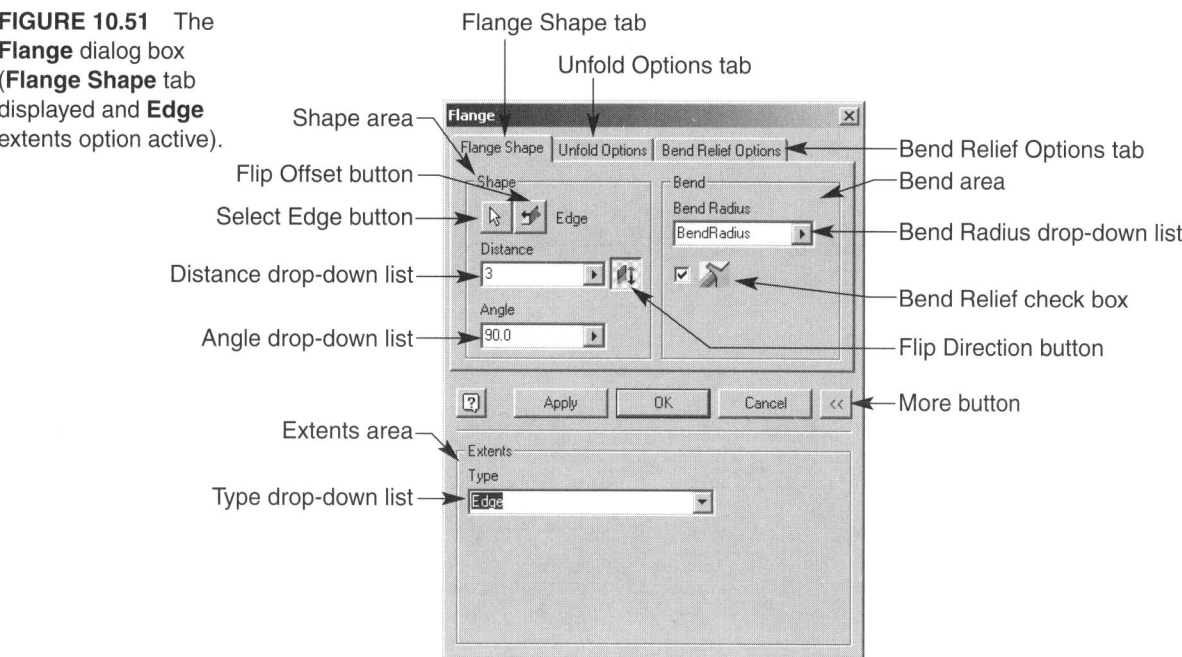

Flange Shape tab
Unfold Options tab
Shape area
Flip Offset button
Select Edge button
Distance drop-down list
Angle drop-down list
Bend Relief Options tab
Bend area
Bend Radius drop-down list
Bend Relief check box
Flip Direction button
More button
Extents area
Type drop-down list

FIGURE 10.52 The
effects of choosing
certain feature edges.

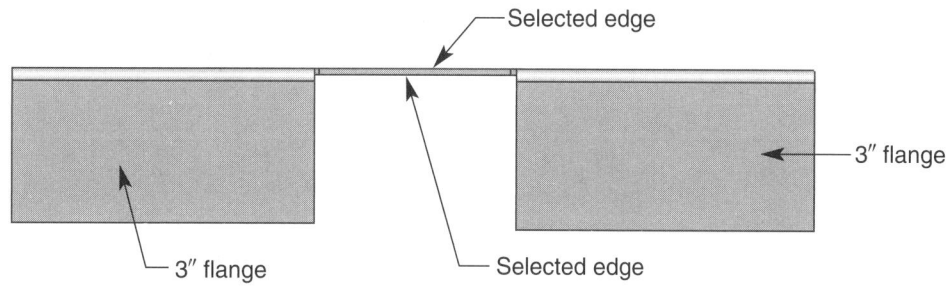

Selected edge
3″ flange
3″ flange
Selected edge

select the **Flip Direction** button to reverse the direction of the flange width. This direction defines the side of the existing feature on which the flange is created and will also influence the beginning of the flange width and the location of the bend.

After you determine the flange distance, specify the angle of the flange by entering a value or selecting a value from the **Angle** drop-down list. Then specify the following options located in the **Bend** area of the **Flange** dialog box.

- **Bend Radius**   This drop-down list allows you to define the bend radius. The default bend radius specified in the **Sheet Metal Styles** dialog box is described as "BendRadius" in the **Radius** drop-down list. However, you can override the default by entering any appropriate bend radius.

- **Bend Relief**   This check box allows you to specify whether or not you want to have a bend relief. If you do not want a bend relief, deselect the checkbox.

Once you have defined all the flange shape options, you may want to modify the unfold and bend relief options located in the **Unfold Options** and **Bend Relief Options** tabs of the **Flange** dialog box, as previously discussed in the sheet metal sketched feature tools.

The final step in creating a sheet metal flange is to define the flange extents, using the **Extents** area of the **Flange** dialog box. This area is accessed by picking the **More** button. By default, the **Edge** extent type is active. This option, as shown in Figure 10.50, creates a flange that extends the entire length of the selected feature edge. In addition to the **Edge** extents option, the following extents specifications are available when placing a sheet metal flange:

- ■ **Width**   This extents option allows you to specify the width of the contour flange and an offset distance from a selected point. To use the **Width** extents option, select **Width** from the **Type** drop-down list. See Figure 10.53. Next, pick the **Select Start Point** button if it is not active, and pick an endpoint on the existing feature. The selected point identifies where the flange is offset from. Then enter an offset value or select an offset value from the **Offset** drop-down list. If the offset occurs on the wrong side, pick the **Flip Direction** button to flip the offset. Finally, specify the width of the flange by entering a value or selecting a value from the width drop-down list. See Figure 10.54.

FIGURE 10.53   The **Width** option of the extents area.

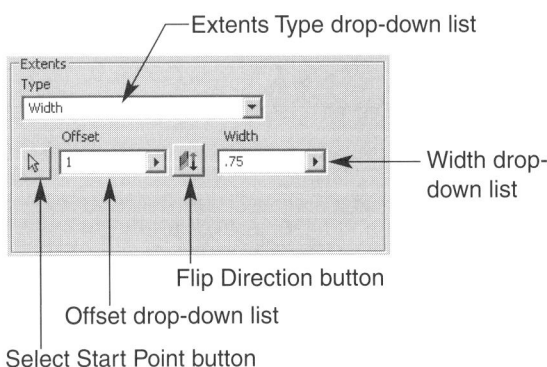

FIGURE 10.54   Using the **Width** extents option.

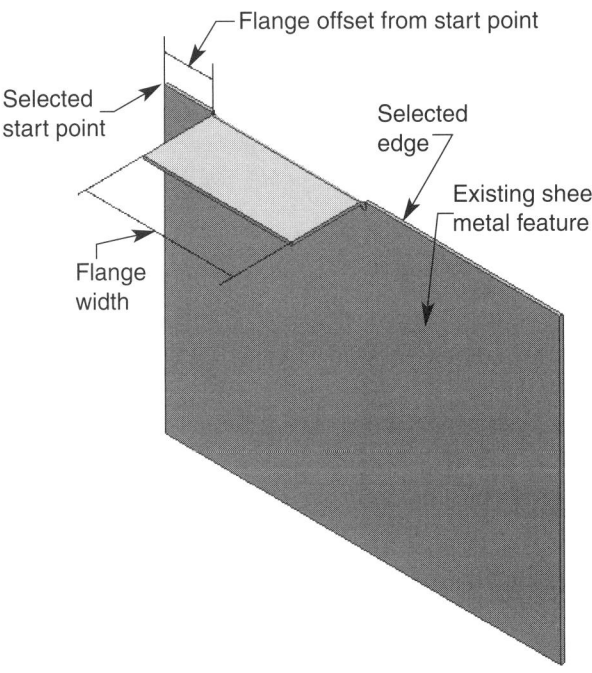

■ **Offset**   This extents option allows you to specify the width of the flange by defining the offset distance from two selected points. To use the **Offset** extents option, select the **Offset** option from the **Type** drop-down list. See Figure 10.55. Next, pick the **Select Start Point** button if it is not active and pick an endpoint on the existing feature where you want the flange offset to begin. Once the start point is selected, the **Select End Point** button should become active. If not, select the **Select End Point** button and pick an endpoint on the existing feature where you want the flange offset to end. Finally, enter an offset value or select a value from the **Offset1** and **Offset2** drop-down lists. See Figure 10.56. When you have specified all the sheet metal flange options, pick the **OK** button if you are generating only one flange. If you want to use the **Flange** tool again, pick the **Apply** button and continue producing flanges as needed. When finished, pick the **OK** button to exit the command.

FIGURE 10.55   The **Offset** option of the extents area.

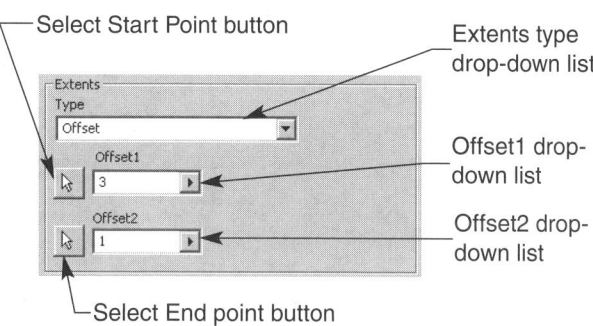

FIGURE 10.56   Using the **Offset** extents option.

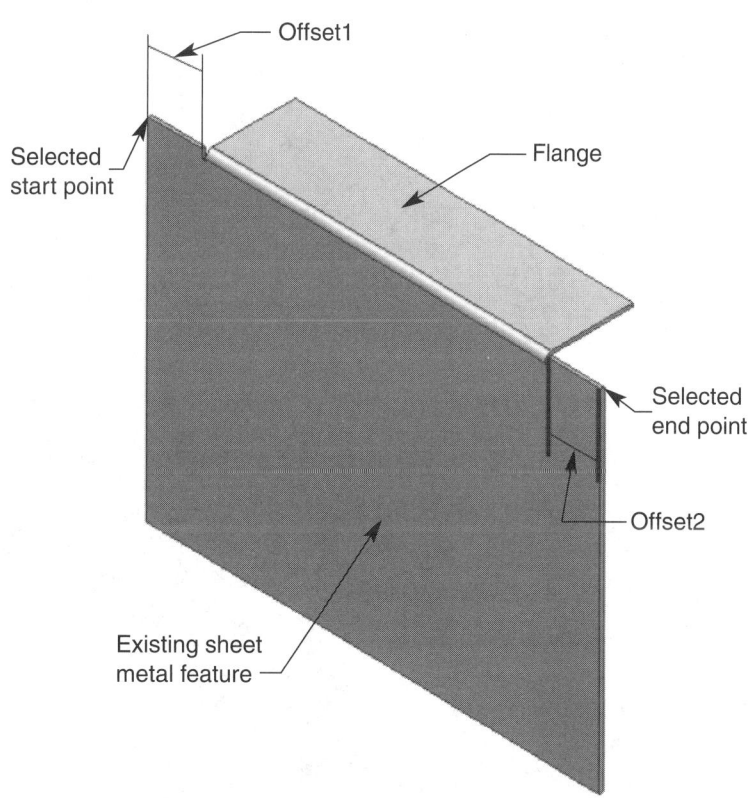

Exercise 10.6

1. Continue from Exercse 10.5, or launch Autodesk Inventor.
2. Open EX10.2.
3. Save a copy of EX10.2 as EX10.6.
4. Close EX10.2 without saving, and open EX10.6.
5. Create the specified flanges as shown using the following information:

Flange1:
Offset into face
Distance: 4″
Angle: 90°
Extents Type: Offset
Offset1: 3″
Offset2: 2″

Flange2:
Offset into face
Distance: 3″
Angle: 90°
Extents Type: Width
Offset: 3″
Width: 5″

6. The final part should look like the part shown.
7. Resave EX10.6.
8. Exit Autodesk Inventor or continue working with the program.

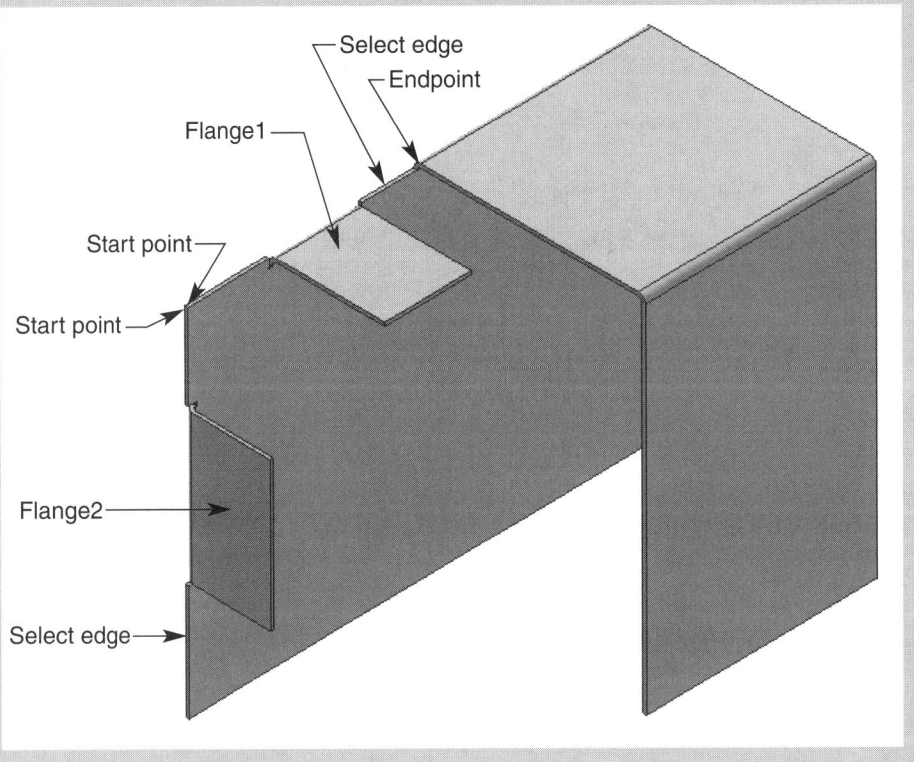

Placing Sheet Metal Hems

A *hem* is a special type of sheet metal flange that is added to an exposed edge for strength, or to relieve sharpness. See Figure 10.57. Like flanges, hems do not require a sketch and can be added to an existing feature such as a face. In contrast to flanges, hems typically create more complex features.

Hems are created using the **Hem** tool, which is accessed using one of the following techniques:

✓ Pick the **Hem** button on the panel bar.

✓ Pick the **Hem** button on the **Sheet Metal** toolbar.

The **Hem** dialog box is displayed when you access the **Hem** command. See Figure 10.58.

To place a hem, first pick the **Select Edge** button, if it is not already selected, and pick the existing feature edge on which the hem will be created. The edge you select will identify the beginning of the hem width and the location of the bend. Once you have selected an edge, specify the type of hem you want to create by picking a hem type from the **Type** drop-down list. A single hem is the default hem type, shown in Figure 10.57. All other hem type options are shown in Figure 10.59.

Once you have selected the feature edge and specified the type of hem you want to create, define the direction and bend relief using the following options in the **Shape** area of the **Hem** dialog box for each of the hem types:

- **Flip Direction**   This button allows you to identify on which side of the existing feature you want the hem to be placed.

- **Bend Relief**   This check box allows you to specify whether you want to have a bend relief. If you do not want a bend relief, deselect the checkbox.

Then, define the additional hem size and shape parameters using the options in the **Shape** area of the **Hem** dialog box for each of the hem types:

- **Single and Double**   When you are creating either a single or a double hem, the **Gap** and the **Length** drop-down lists are available. The gap represents the opening between the inside hem face and the inside feature face. By default, the gap is half of the material thickness

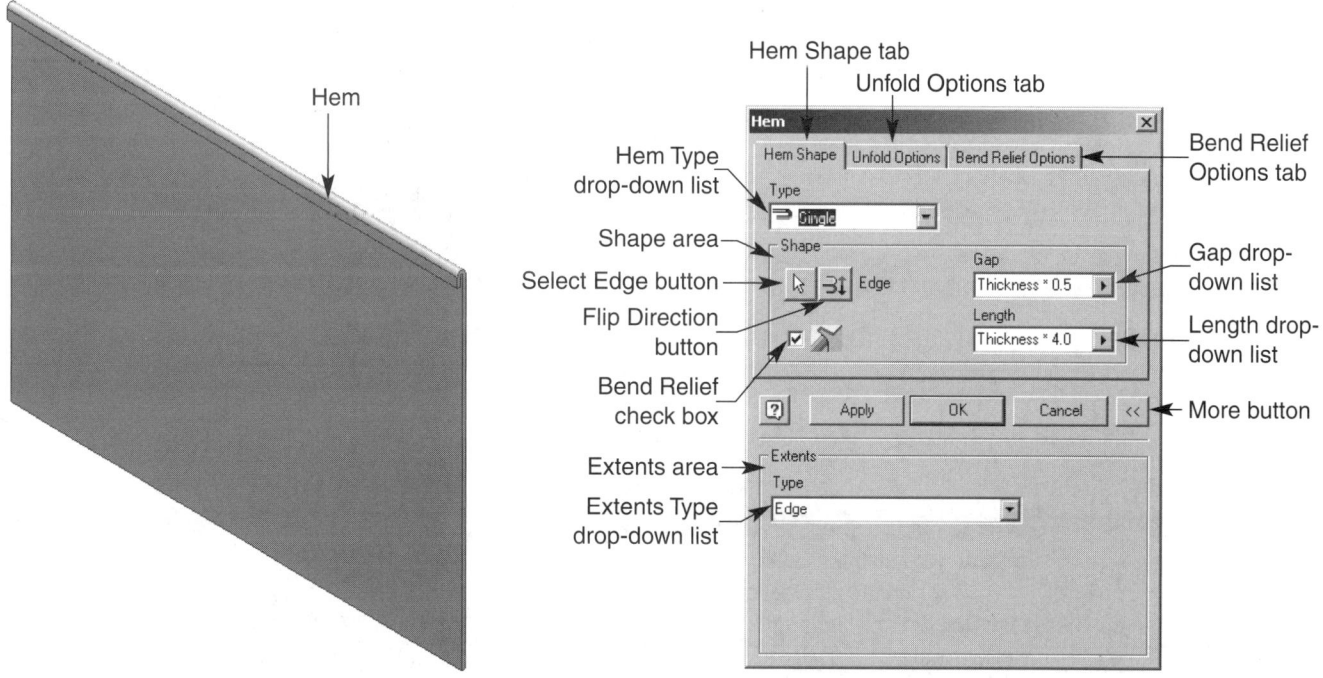

FIGURE 10.57   An example of a simple, single sheet metal hem.

FIGURE 10.58   The **Hem** dialog box (**Hem Shape** tab displayed and **Edge** extents option active).

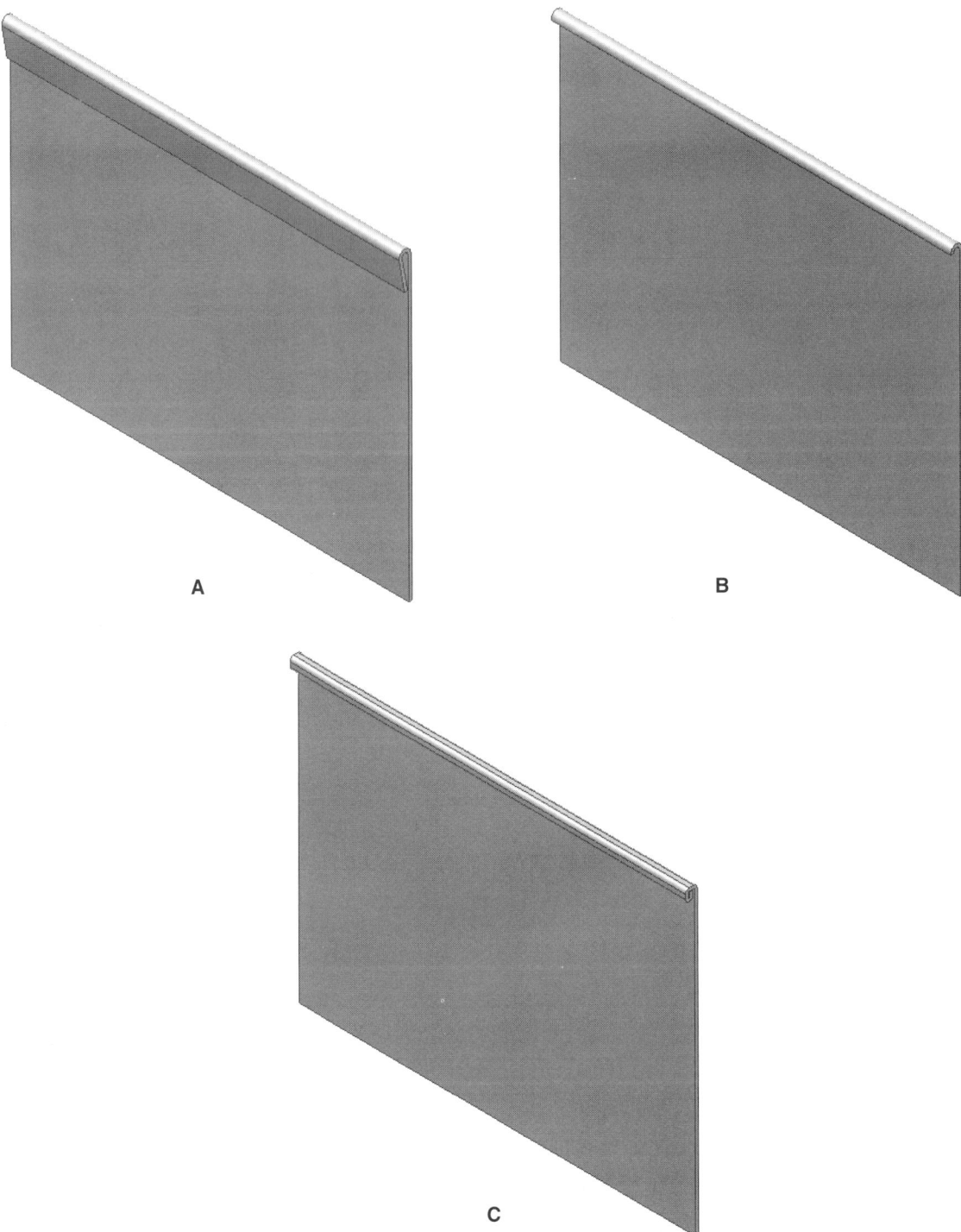

FIGURE 10.59   An example of (A) a teardrop hem, (B) a rolled hem, and (C) a double hem.

specified in the **Sheet Metal Styles** dialog box. However, you can enter a different value or select a value from the **Gap** drop-down list.

The length represents the amount of sheet metal material overlap used to create the hem. By default, the length is four times the material thickness specified in the **Sheet Metal Styles** dialog box. However, you can enter a different value or select a value from the **Length** drop-down list.

■ **Teardrop and Rolled**   When you are creating either a teardrop or a rolled hem, the **Radius** and the **Angle** drop-down lists are available. The radius represents the radius of the bend. By default, the radius is equal to the bend radius specified in the **Sheet Metal Styles** dialog box. However, you can enter a different value or select a value from the **Radius** drop-down list.

The angle represents the amount of sheet metal material overlap used to create the hem and the angle at which the material is bent. By default, the angle for a teardrop hem is 190°. However, you can enter any value greater than 180° and less than 359° or select a value from the **Angle** drop-down list. By default, the angle for a rolled hem is 190°. However, you can enter any value greater than 0° and less than 359° or select a value from the **Angle** drop-down list.

Once you have defined all the hem shape options, you may want to modify the unfold and bend relief options located in the **Unfold Options** and **Bend Relief Options** tabs of the **Hem** dialog box, as previously discussed.

The final step in creating a sheet metal hem is to define the hem extents, using the **Extents** area of the **Hem** dialog box. This area is accessed by picking the **More** button. By default, the **Edge** extent type is active. This option, as shown in Figures 10.57–10.59, creates a hem that extends the entire length of the selected feature edge. In addition to the **Edge** extents option, the **Width** and **Offset** extents options are available in the **Hem** dialog box. These extents preferences function exactly the same as the **Width** and **Offset** extents options, available in the **Flange** dialog box, as previously discussed. See Figure 10.60. When you have specified all the sheet metal hem options, pick the **OK** button if you are generating only one hem. If you want to use the **Hem** tool again, pick the **Apply** button and continue producing hems as needed. When finished, pick the **OK** button to exit the command.

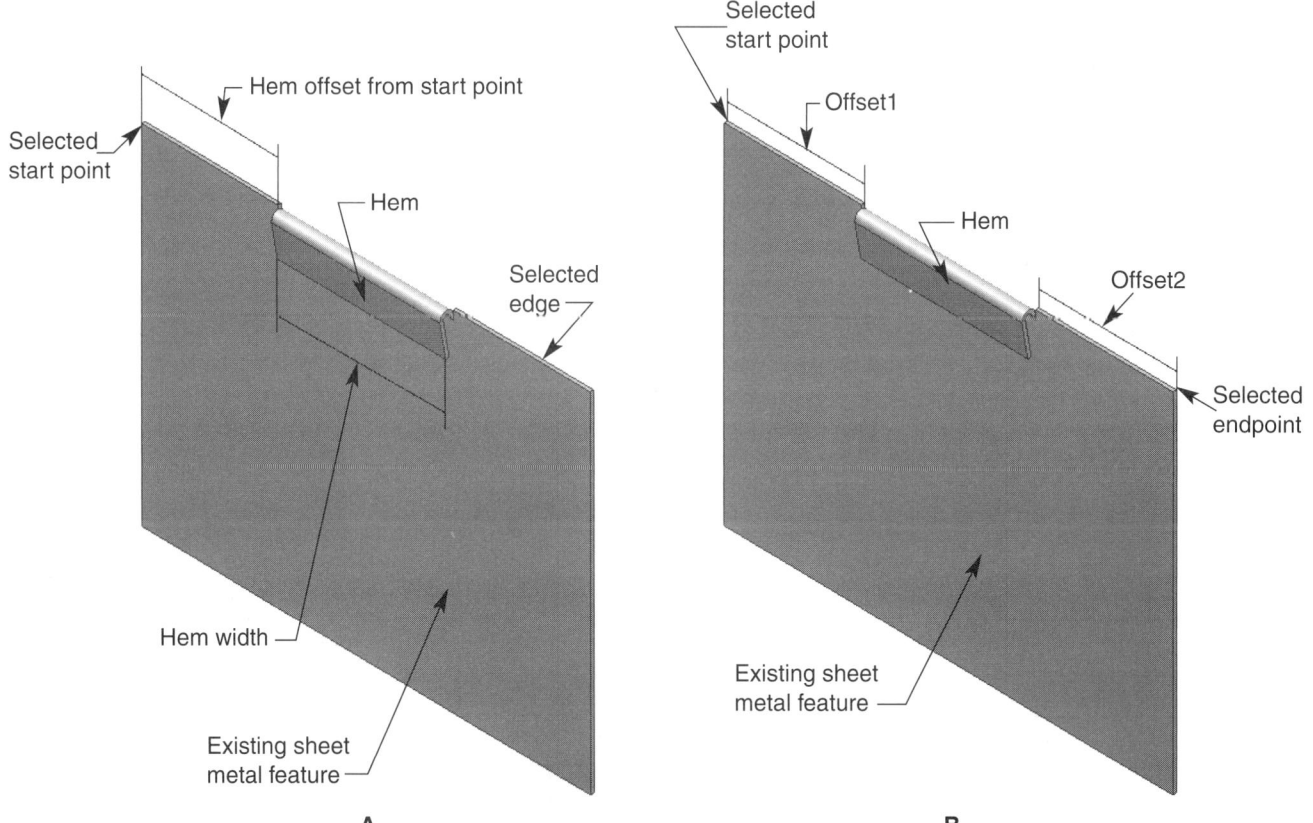

FIGURE 10.60   An example of (A) a hem using the width extents option and (B) a hem using the offset extents option.

Exercise 10.7

1. Continue from Exercse 10.6, or launch Autodesk Inventor.
2. Open EX10.2.
3. Save a copy of EX10.2 as EX10.7.
4. Close EX10.2 without saving, and open EX10.7.
5. Create the specified hems as shown using the following information:

   Hem1:
   Type: Single
   Extents Type: Edge

   Hem2:
   Type: Teardrop
   Extents Type: Offset
   Offset1: 2″
   Offset2: 2″

   Hem3:
   Type: Rolled
   Extents Type: Edge

   Hem4:
   Type: Double
   Extents Type: Width
   Offset: 2″
   Width: 1.5″
   Flip Direction if necessary

6. The final part should look like the part shown.
7. Resave EX10.7.
8. Exit Autodesk Inventor or continue working with the program.

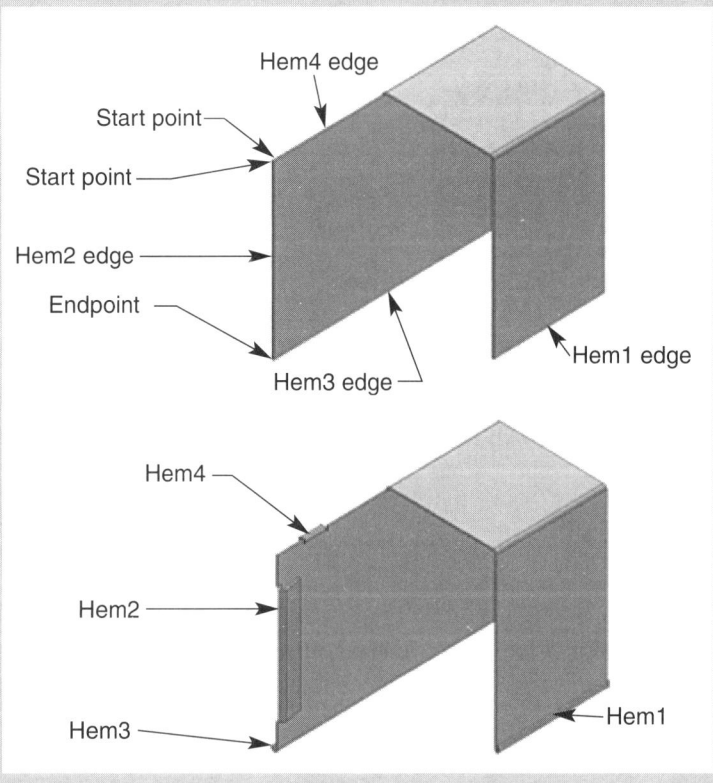

Creating Sheet Metal Corner Seams

Often when creating sheet metal parts, it may be necessary to place seams at intersecting, perpendicular, or coplanar sheet metal faces. *Corner seams* add or remove material at the corners of sheet metal features to create a smooth transition and allow for unfolding. See Figure 10.61. Like other placed sheet metal features, corner seams do not require a sketch and can be added to any intersecting, perpendicular, or coplanar features such as two faces.

Corner seams are created using the **Corner Seam** tool, which is accessed using one of the following techniques:

✓ Pick the **Corner Seam** button on the panel bar.

✓ Pick the **Corner Seam** button on the **Sheet Metal** toolbar.

The **Corner Seam** dialog box is displayed when you access the **Corner Seam** command. See Figure 10.62. The corner seam tool contains options that allow you to create the following corner seams:

FIGURE 10.61 An example of a nonoverlapped corner seam applied to a face and two flanges with a round bend relief.

FIGURE 10.62 The **Corner Seam** dialog box (**Corner Seam Shape** tab and **Seam** area displayed).

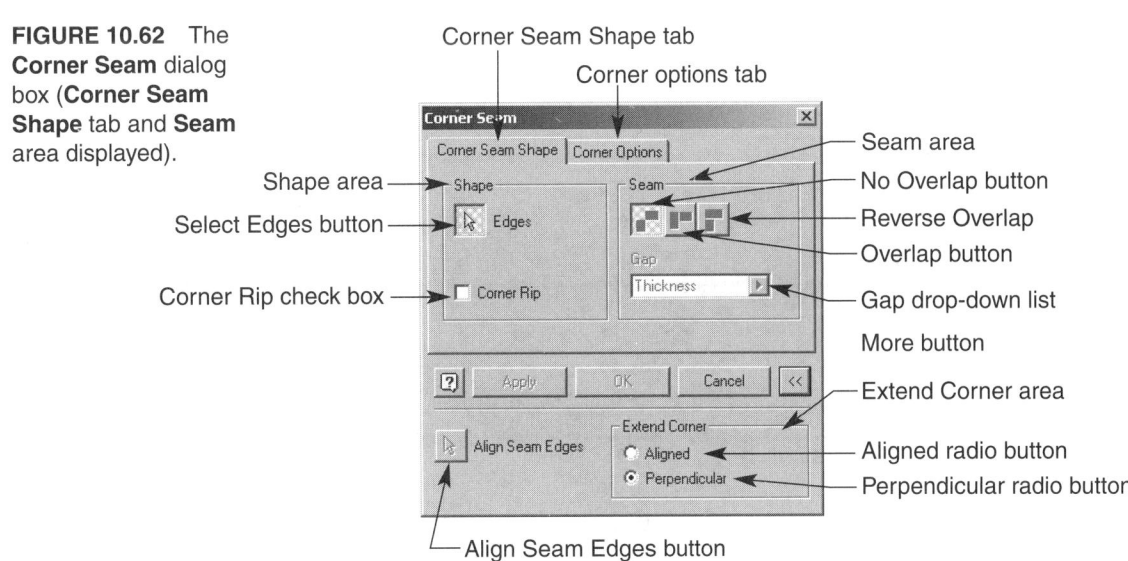

■ A smooth transitioning corner seam between sheet metal faces that are not connected and are perpendicular to each other or intersecting.

To place a corner seam between faces that are not connected and are perpendicular to each other as shown in Figure 10.63, first pick the **Select Edge** button, if it is not already selected. Then pick one of the existing feature edges, followed by the other existing feature edge, to define the seam corner.

Once you have selected the feature edges, specify the type of corner seam you want to create by picking one of the three corner seam type buttons. A **No Overlap** corner seam is the default corner seam type, shown in Figure 10.61. The other two corner seam types are shown in Figure 10.64. The next step in creating a corner seam is to define the gap between the selected sheet metal features by entering a value or selecting a value from the **Gap** drop-down list. The default gap value is the thickness of the material specified in the **Sheet Metal Styles** dialog box.

■ A smooth transitioning corner seam between sheet metal faces that are not connected and are coplanar.

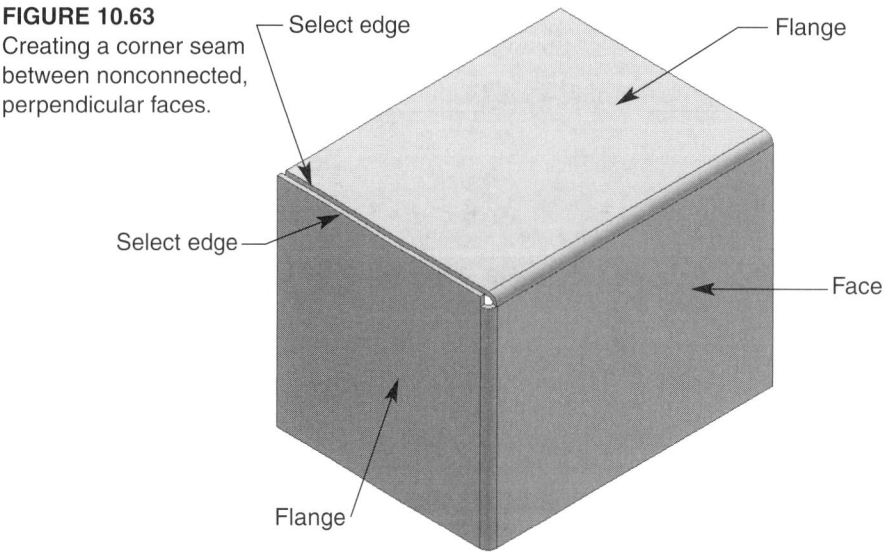

FIGURE 10.63
Creating a corner seam between nonconnected, perpendicular faces.

Select edge

Select edge

Flange

Face

Flange

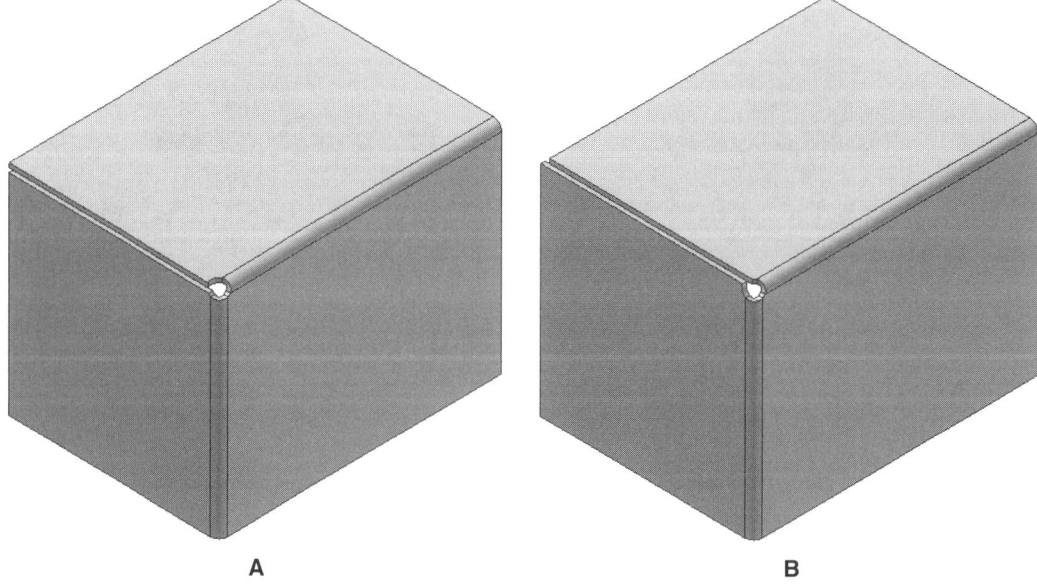

FIGURE 10.64   An example of (A) an overlapped corner seam and (B) a reverse overlap corner seam.

A

B

To place a corner seam between faces that are not connected and are coplanar as shown in Figure 10.65, first pick the **Select Edge** button, if it is not already selected. Then pick an existing feature edge, followed by an edge on the coplanar face, perpendicular to the first edge selected, to define the location of the seam corner.

Once you have selected the feature edges, the **Seam** area of the **Corner Seam** dialog box changes to **Miter.** See Figure 10.66. The **Miter** area allows you to specify the type of corner miter you want to create by picking one of the three corner miter type buttons. The effects of each of these options are shown in Figure 10.67. The next step in creating a corner seam is to define the gap between the selected sheet metal features by entering a value or selecting a value from the **Gap** drop-down list. The default gap value is the thickness of the material specified in the **Sheet Metal Styles** dialog box.

■ A corner seam between sheet metal faces that are square and intersecting.

A corner seam placed between intersecting square sheet metal features is known as a *corner rip.* To add a corner rip, as shown in Figure 10.68, ensure the **Select Edge** button is selected, pick the **Corner Rip** check box, and pick an existing square feature edge.

Once you have selected the feature edge or edges, specify the type of corner rip you want to create by picking one of the three corner seam type buttons. A **No Overlap** corner rip is the default corner seam type, shown in Figure 10.68. The other two corner rip types are shown in Figure 10.69.

Once you have determined the type of corner seam you are creating and fully specified the corner seam shape options in the **Corner Seam Shape** tab of the **Corner Seam** dialog box, define the corner parameters located in the **Corner Options** tab. See Figure 10.70. The options available inside the **Corner Options** tab are the same as those in the **Sheet Metal Styles** dialog box. The preferences in **Corner Options** tab allow you to override the default options specified in the **Sheet Metal Styles** dialog box. However, if you want to use the sheet metal settings defined in the **Sheet Metal Styles** dialog box, select the **Use Default Settings** check box.

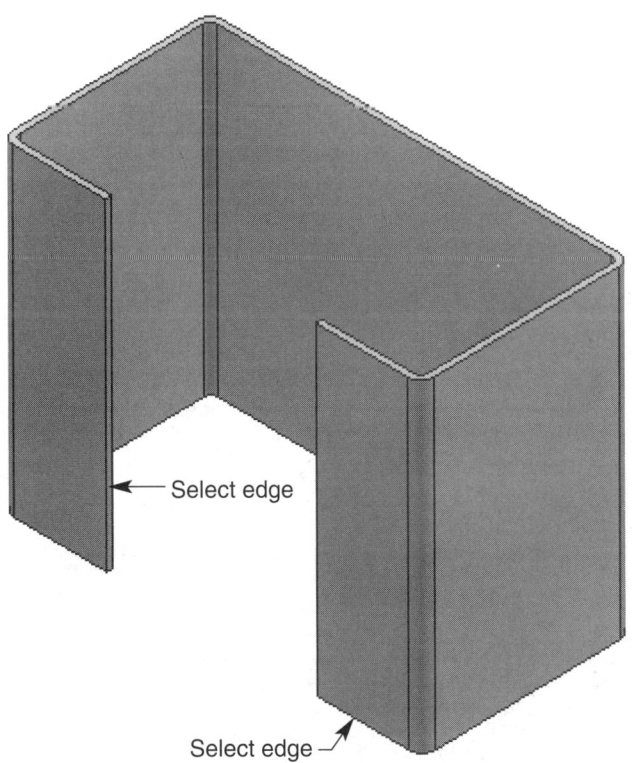

FIGURE 10.65 Creating a corner seam between coplanar faces.

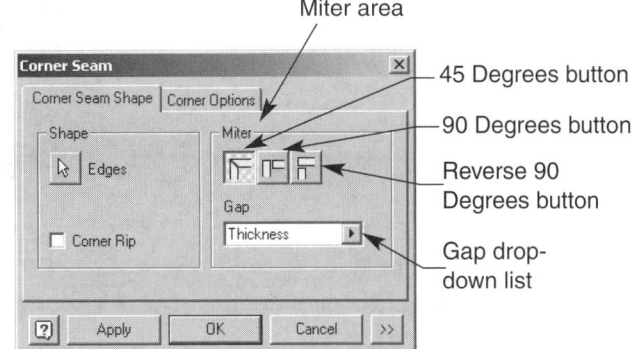

FIGURE 10.66 The **Corner Seam** dialog box (**Corner Seam Shape** tab and **Miter** area displayed).

FIGURE 10.67 An example of (A) a **45 Degrees** corner miter, (B) a **90 Degrees** corner miter, and (C) a **Reverse 90 Degrees** corner miter.

A

B

C

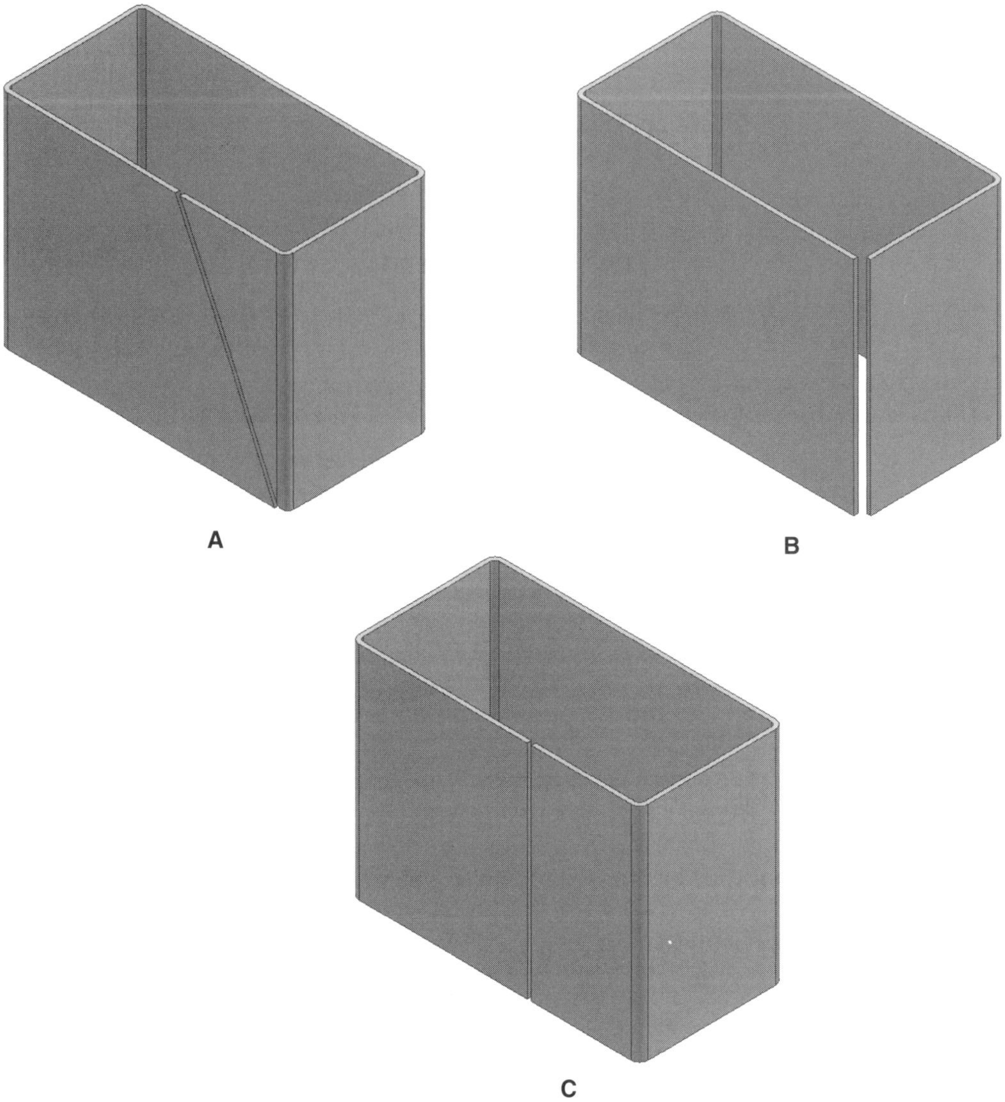

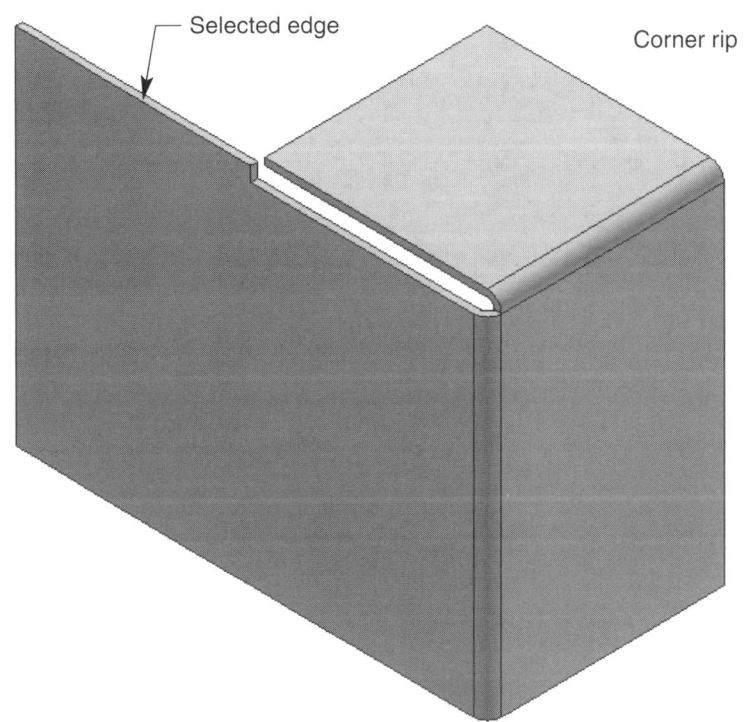

FIGURE 10.68 Creating a corner rip.

Selected edge

Corner rip

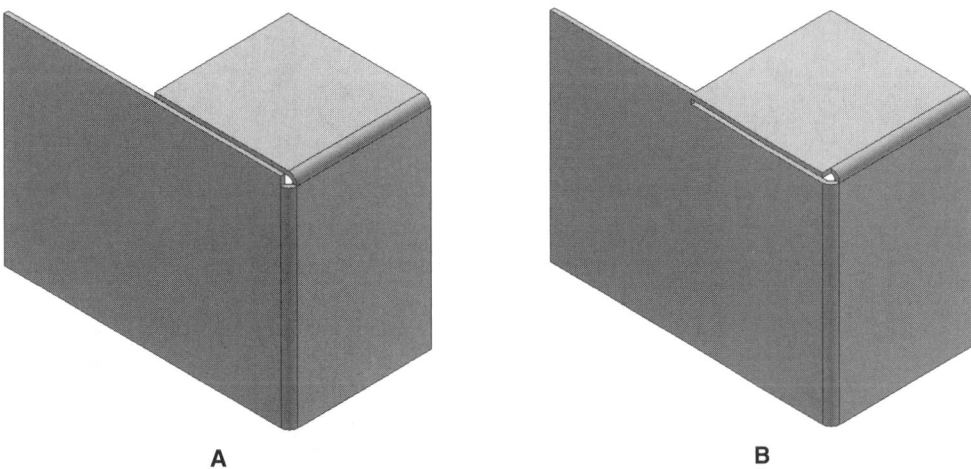

FIGURE 10.69 An example of (A) an overlapped corner rip and (B) a reverse overlap corner rip.

A   B

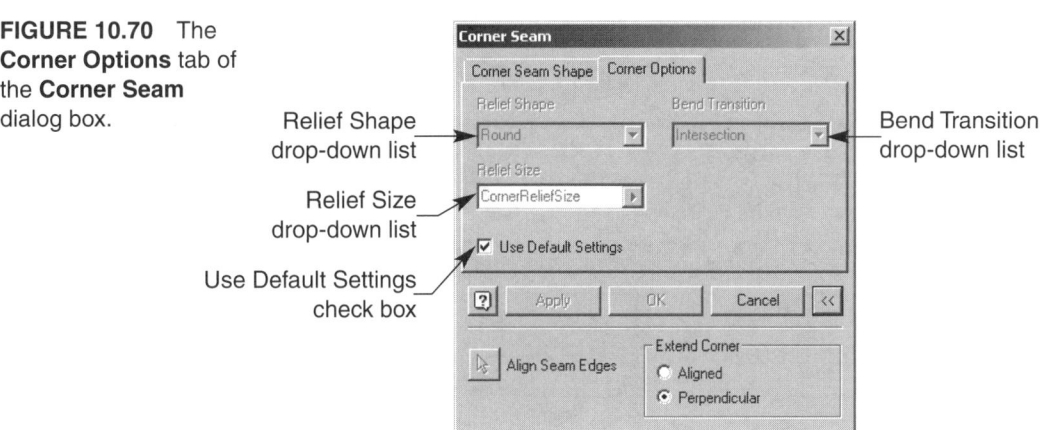

FIGURE 10.70 The **Corner Options** tab of the **Corner Seam** dialog box.

Relief Shape drop-down list

Relief Size drop-down list

Use Default Settings check box

Bend Transition drop-down list

Finally, you may want to further define the corner seam using the following options accessed by picking the **More** button:

- **Align Seam Edges**   This button allows you to select edges to align with the seam.

- **Aligned** extended corner   This radio button aligns the selected face edges with each other by projecting the first face edge to the second face edge.

- **Perpendicular** extended corner   This radio button projects the first selected face edge perpendicular to the second selected face edge.

When you have specified all the sheet metal corner seem options, pick the **OK** button if you are only generating one corner seam. If you want to use the **Corner Seam** tool again, pick the **Apply** button and continue producing corner seams as needed. When finished, pick the **OK** button to exit the command.

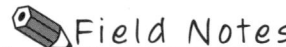 Field Notes

If you want to change a corner seam into a bend, discussed later in the chapter, right-click the corner seam in the **Browser** and select the **Change to Bend** menu option. The corner seam is removed, and the **Bend** dialog box is displayed.

Exercise 10.8

1. Continue from Exercse 10.7, or launch Autodesk Inventor.
2. Open a new metric sheet metal part file.
3. Open a sketch on the XY plane, and sketch a rectangle 50mm×75mm.
4. Using the **Face** tool, create a sheet metal face (**Face1**) from the sketch, with default sheet metal style settings. Face depth should occur in a positive direction.
5. Using the **Flange** tool, create 30mm flanges (**Flange1** and **Flange2**) as shown. Flange depth should occur in towards the face.
6. Open a sketch on "Sketch Face" as shown, and create the specified sketch geometry (**Sketch2**).
7. Using the **Face** tool, create sheet metal faces (**Face2**) from the sketch profiles, with default sheet metal style settings. Face depth should occur in towards **Face1.**
8. Create the following corner seams as shown using the **Corner Seam** tool; use default settings for all other specifications:

Corner1:
Seam: No Overlap
Relief Shape: Round

Corner2:
Corner Rip
Seam: No Overlap

9. When complete, the part should look like the part shown.
10. Save the file as EX10.8.
11. Exit Autodesk Inventor or continue working with the program.

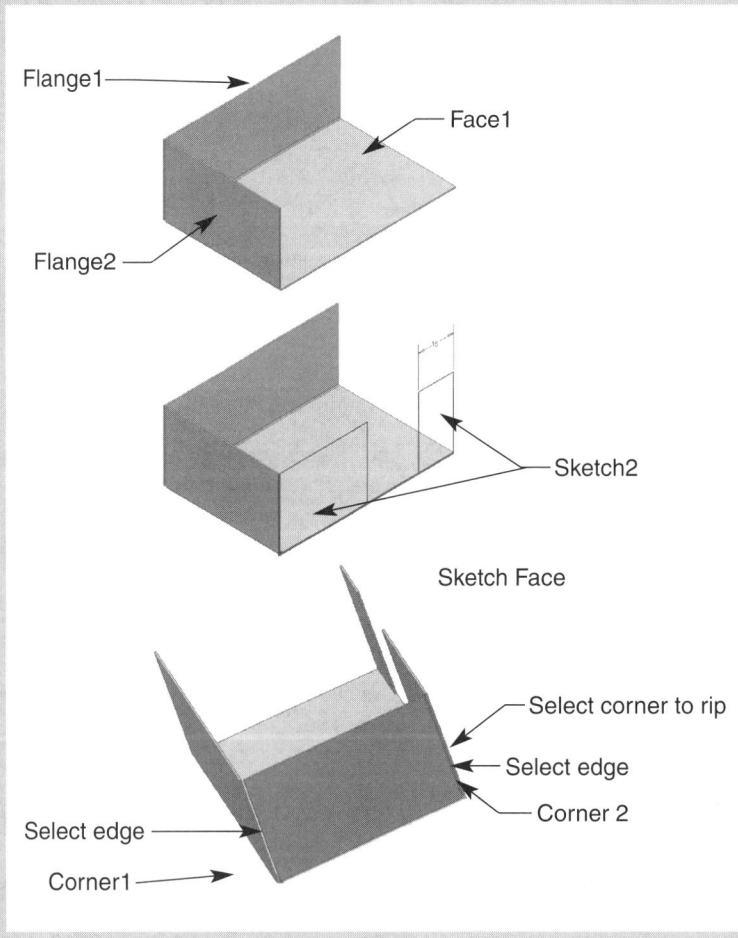

Adding Sheet Metal Bends

A sheet metal ***bend*** is typically created during sheet metal part development using the tools previously discussed. However, sometimes it is necessary to add an additional bend between two perpendicular, or two parallel, sheet metal features. See Figure 10.71.

Once you have created at least two acceptable sheet metal features, access the **Bend** command using one of the following techniques:

✓ Pick the **Bend** button on the panel bar.

✓ Pick the **Bend** button on the **Sheet Metal** toolbar.

The **Bend** dialog box is displayed when you access the **Bend** command. See Figure 10.72.

To create a sheet metal bend, pick the **Select Edges** button from the **Bend** area of the **Bend Shape** tab, if it is not already active. Then pick the edges on both features where you want the bend to be created. When both features are selected, a preview of the bend is displayed.

The next step is to define the face bend using the following options found in the **Bend** area of the **Bend Shape** tab:

FIGURE 10.71   An example of a sheet metal bend between two parallel sketched faces (full radius bend).

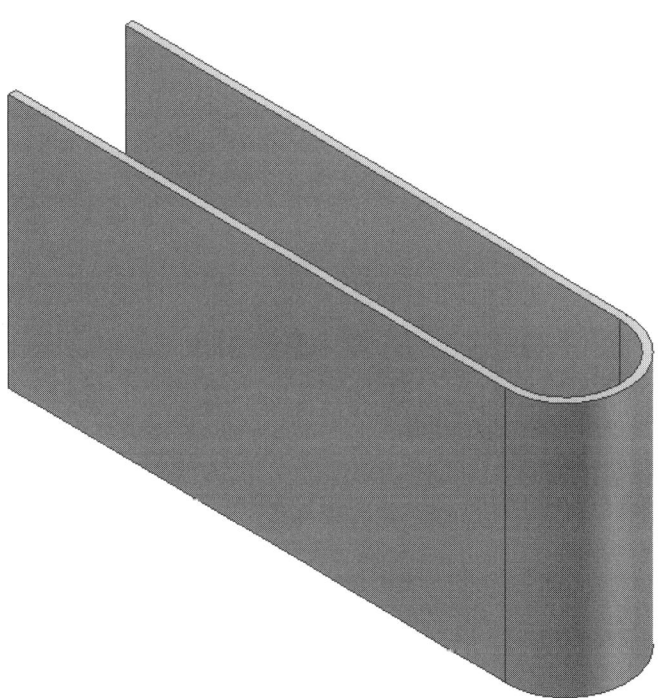

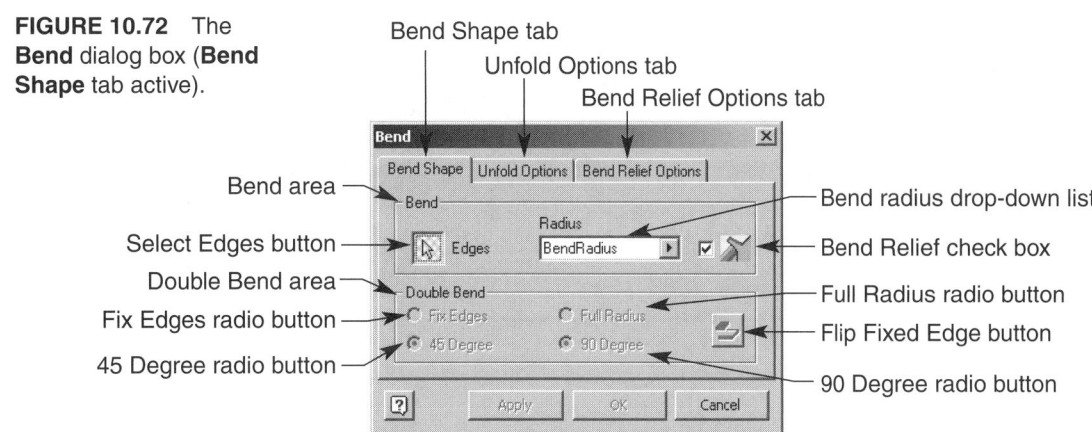

FIGURE 10.72   The **Bend** dialog box (**Bend Shape** tab active).

Bend Shape tab

Unfold Options tab

Bend Relief Options tab

Bend area

Select Edges button

Double Bend area

Fix Edges radio button

45 Degree radio button

Bend radius drop-down list

Bend Relief check box

Full Radius radio button

Flip Fixed Edge button

90 Degree radio button

- **Radius**   This drop-down list allows you to define the bend radius. The default bend radius "BendRadius" is specified in the **Sheet Metal Styles** dialog box. However, you can override the default radius by entering any appropriate bend radius.

- **Bend Relief**   This check box allows you to specify whether you want to have a bend relief. If you do not want a bend relief, deselect the checkbox.

- **Double Bend**   This area allows you to define the type of bend you want to create. These options are exactly the same as those available when developing a face, in the **Face** dialog box, as previously discussed. You can choose to create a fix edges, 45°, full radius, or 90° bend depending on the application. A **Full Radius** double bend is shown in Figure 10.71; all other bends are shown in Figure 10.73.

- **Flip Fixed Edge**   This button allows you to flip the selected edges used to create the bend, without reselecting edges. If the bend preview does not look correct, you may need to modify the selected edges using this button.

Once you have defined all the bend shape options, you may want to modify the unfold and bend relief options located in the **Unfold Options** and **Bend Relief Options** tabs of the **Bend** dialog box, as previously discussed. Finally, when you have specified all the sheet metal bend options, pick the **OK** button if you are generating only one bend. If you want to use the **Bend** tool again, pick the **Apply** button and continue producing bends as needed. When finished, pick the **OK** button to exit the command.

FIGURE 10.73   An example of (A) a **45 Degree** double bend, (B) a **Fix Edges** double bend, and (C) a **90 Degree** double bend.

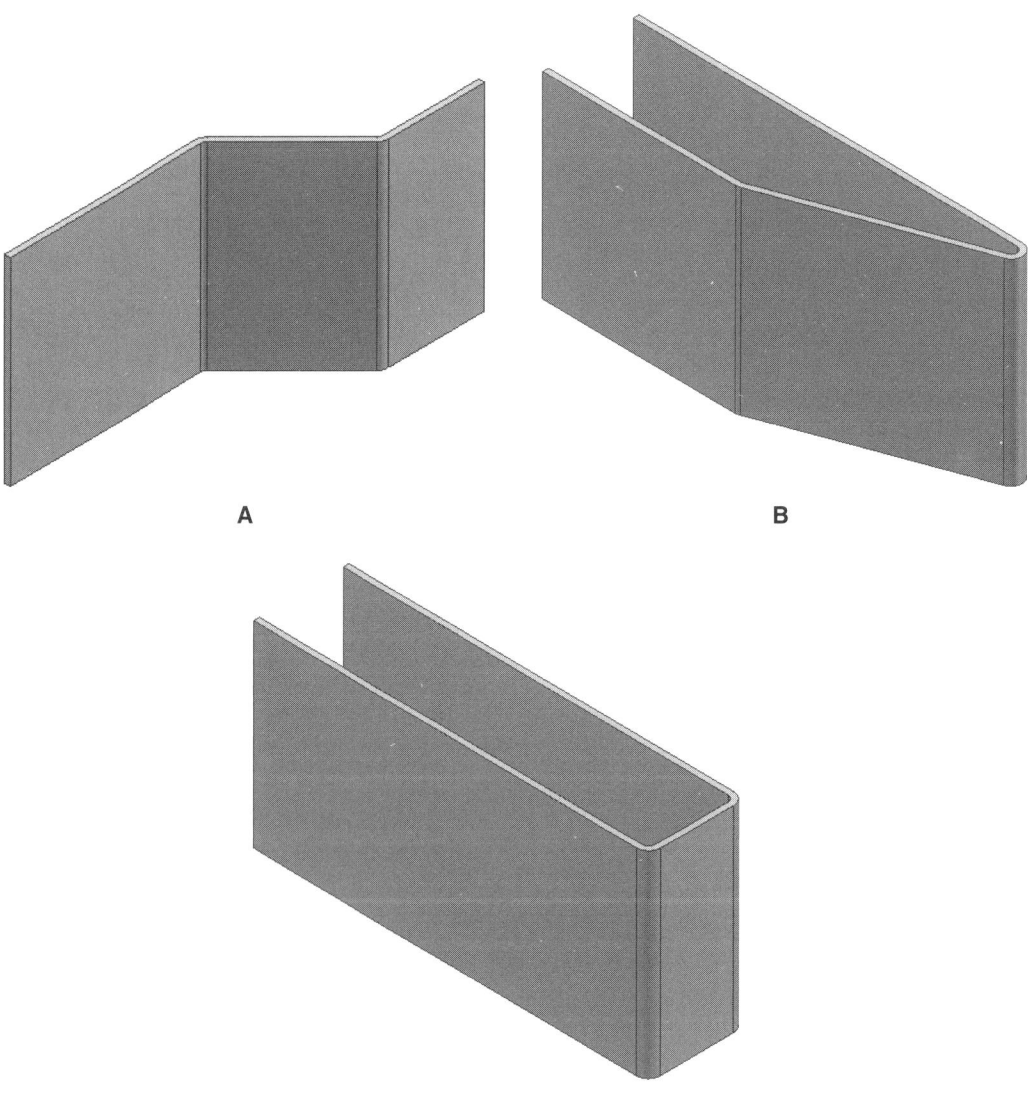

A

B

C

Field Notes

If you want to change a bend into a corner seam, right-click the bend in the **Browser,** and select the **Change to Corner** menu option. The bend is removed, and the **Corner Seam** dialog box is displayed.

Exercise 10.9

1. Continue from Exercse 10.8, or launch Autodesk Inventor.
2. Open EX 10.9.
3. Save a copy of EX10.8 as EX10.9.
4. Close EX10.8 without saving, and open EX10.9.
5. Use the **Bend** tool to create the bends shown, using the following information. Use default settings for all other specifications:

   Bend4:
   Select specified edge.

   Bend5:
   Select specified edge.

   Bend6:
   Select specified edges.
   Double Bend type: Full Radius

6. The final part should look like the part shown.
7. Resave EX10.9.
8. Exit Autodesk Inventor or continue working with the program.

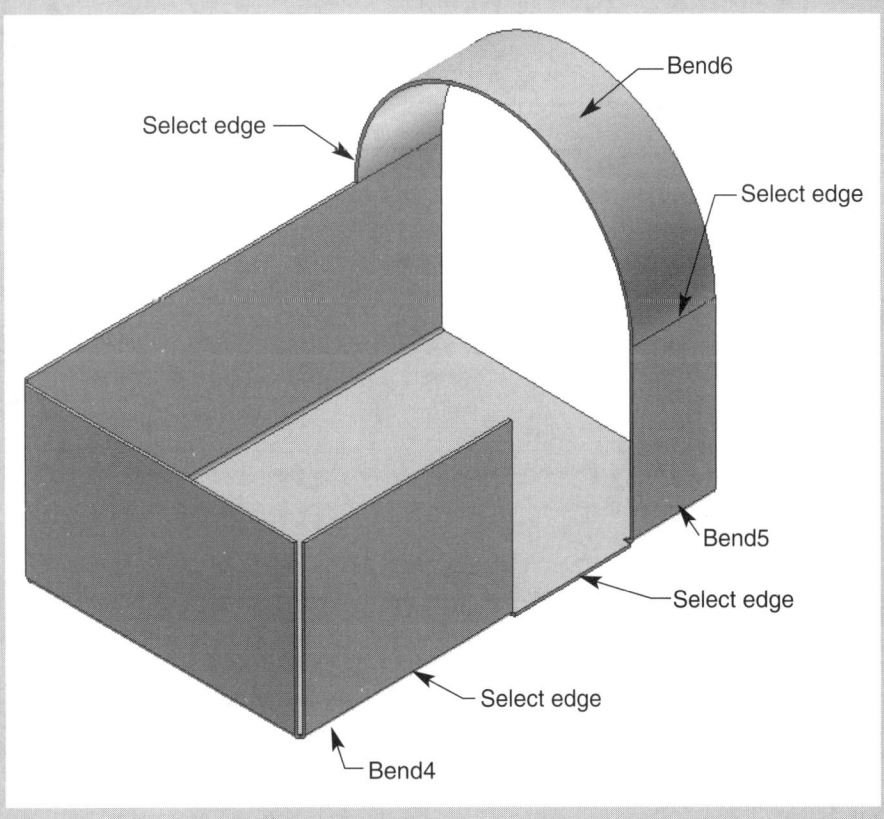

Corner Round

A **_corner round_** in the sheet metal environment is very similar to a fillet and round in the part model environment. Corner rounds are curves placed at the inside and outside sheet metal corner intersections. See Figure 10.74.

Typically, corner rounds are used to remove points or apply a specified radius to corners. Corner rounds can be placed at the corner of any acceptable sheet metal feature previously discussed.

Corner rounds are created using the **Corner Round** tool. Access the **Corner Round** command using one of the following techniques:

✓ Pick the **Corner Round** button on the panel bar.

✓ Pick the **Corner Round** button on the **Sheet Metal** toolbar.

The **Corner Round** dialog box is displayed when you access the **Corner Round** command. See Figure 10.75.

FIGURE 10.74
Examples of sheet
metal corner rounds.

FIGURE 10.75   The
Corner Round dialog
box.

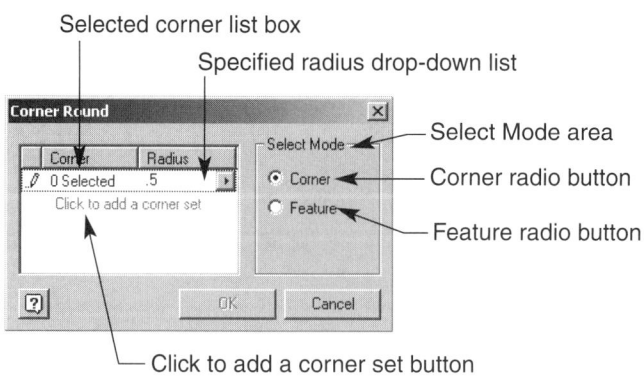

Most of the time, you should be able to apply all the corner rounds needed in the part using a single **Corner Round** command, because the **Corner Round** dialog box allows you apply different corner rounds to corners, using different radii. You do not need to access the **Corner Round** command continually to create a different type of corner round. Figure 10.74 shows an example of a sheet metal part with a variety of corner round radii. All the fillets and rounds were created using a single fillet command, and each of the fillets placed during one operation create a single feature.

 Field Notes

> If you plan to suppress only certain corner rounds or want to have additional corner rounds, you may want to access the **Corner Round** command more than once.

To apply a corner round to a square sheet metal feature corner, first specify how you want to select the corners, by picking either the **Corner** or the **Feature** radio button. By default, the **Corner** radio button is selected. This option allows you to select specific corners individually. If you want to apply corner rounds to all the corners in an entire feature, pick the **Feature** radio button. Once you determine how you want to select corners to apply corner rounds, specify a corner round radius by highlighting the default radius and entering the desired value, or by selecting a value from the drop-down list.

The next step is to pick the corners where you want to apply corner rounds. If you are using the **Corner** select method, identify which corners you want to apply corner rounds to, by selecting the individual corners. If you are using the **Features** select method, identify the features you want to apply corner rounds to, by selecting the entire features. A preview of the selected corners is displayed. If the preview looks correct, and you do not want to place any additional corner rounds with different radii, pick the **OK** button to accept the curves. If you would like to apply other corner rounds with different radii, pick the **Click here to add** button. Then, as previously discussed, enter the corner round radius and select the corner. Continue adding corner rounds as required. When finished, pick the **OK** button.

 Field Notes

> You can deselect any of the selected corners by holding down the **Ctrl** key on your keyboard and picking the object.

 Field Notes

> To delete a corner round corner, listed in the **Corner** list box, pick the corner you want to remove and press the **Delete** key on your keyboard.

Exercise 10.10

1. Continue from Exercse 10.9, or launch Autodesk Inventor.
2. Open EX 10.4.
3. Save a copy of EX10.4 as EX10.10.
4. Close EX10.4 without saving, and open EX10.10.
5. Use the **Corner Round** tool to create the corner rounds shown. You should have to access the **Corner Round** tool only once.
6. The final part should look like the part shown.
7. Resave EX10.10.
8. Exit Autodesk Inventor or continue working with the program.

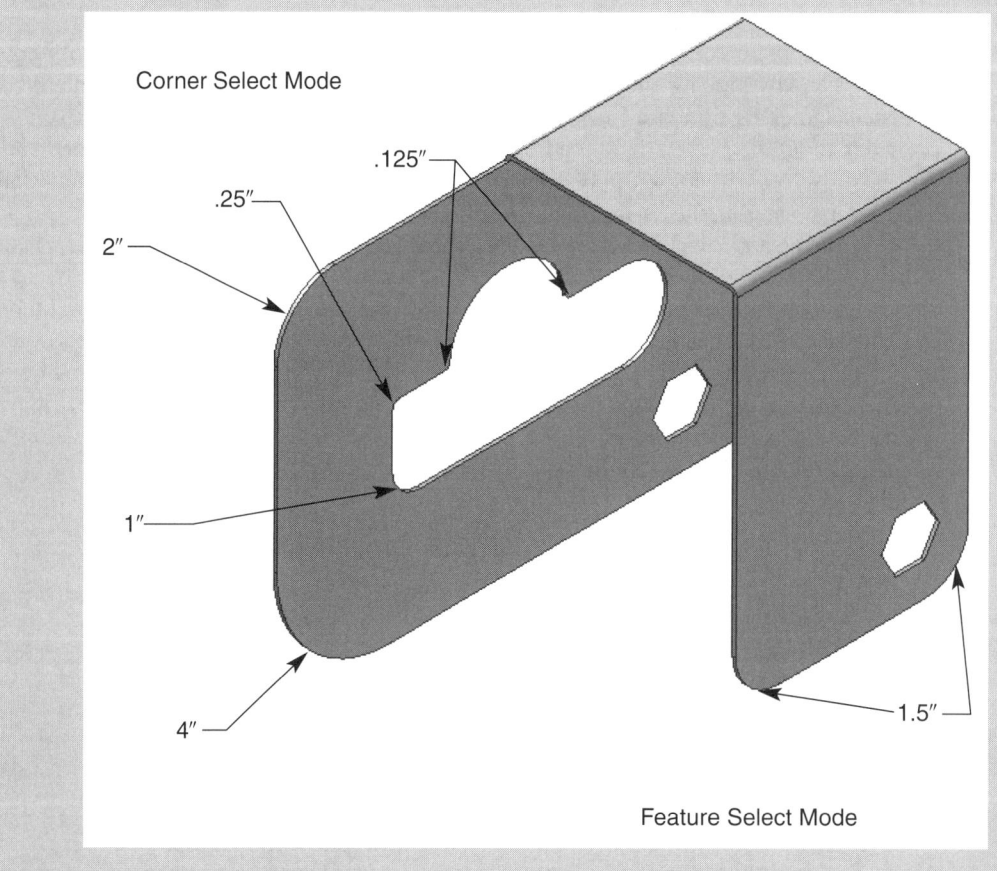

Corner Chamfer

A *corner chamfer* is an angled face, placed on a sheet metal feature corner. See Figure 10.76. Corner chamfers are used to relieve sharp edges or create specific angled sheet metal geometry. Corner chamfers can remove material from an external corner or add material to an internal corner.

Corner chamfers are created using the **Corner Chamfer** tool. Access the **Corner Chamfer** command by using one of the following techniques:

✓ Pick the **Corner Chamfer** button on the panel bar.

✓ Pick the **Corner Chamfer** button on the **Sheet Metal** toolbar.

FIGURE 10.76
Examples of several
chamfers placed on two
parts.

The **Corner Chamfer** dialog box is displayed when you access the **Corner Chamfer** command. See Figure 10.77.

Field Notes

Unlike the **Corner Round** tool, you need to access the **Chamfer** command every time you want to create a different-size chamfer. However, you can fully chamfer a sheet metal part using only one **Corner Chamfer** command if the distances are constant for each corner. Still, each of the corner chamfers placed during one operation create a single feature.

To create a corner chamfer, first select one of the following corner chamfer method buttons: **One Distance, Distance and Angle,** and **Two Distances.** The techniques used by each of these buttons to create chamfers are shown in Figure 10.78. The method button selected by default is **One Distance;** it is the simplest option to use. The **Distance** option allows you to create a corner chamfer with equal-distance sides from the selected corner, or a 45° angle. To place a corner chamfer using the **One Distance** option, you must first be sure the **One Distance** button is selected. Then pick the edges you want to chamfer, and enter

FIGURE 10.77   The
Corner Chamfer dialog
box (**Distance** option).

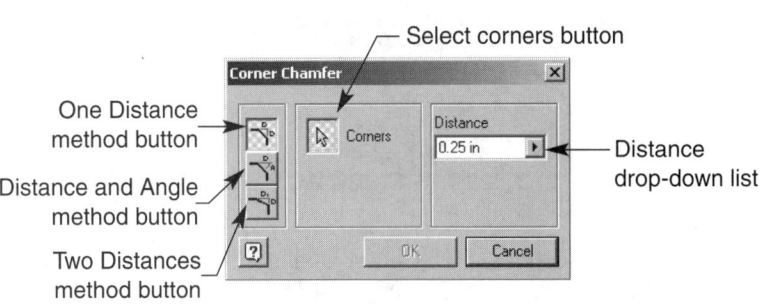

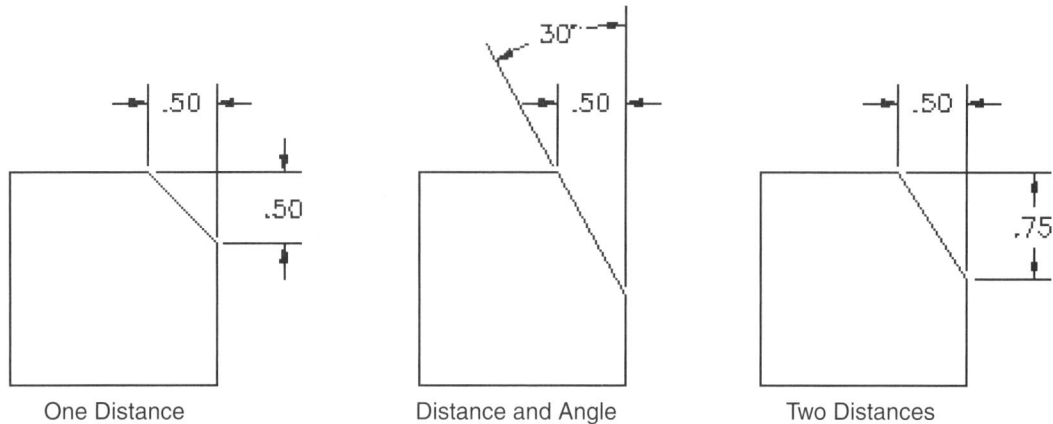

FIGURE 10.78   Examples of the effects of each type of corner chamfer creation method.

or select a corner chamfer distance from the **Distance** drop-down list. If the chamfer preview looks correct, pick the **OK** button to create the chamfer.

Field Notes

> You can deselect any of the selected objects in any of the **Chamfer** dialog box drop-down lists by holding down the **Ctrl** key on your keyboard and picking the object.

If you want to specify one corner chamfer side as a distance from the corner, and the other side as an angle from the corner, pick the **Distance and Angle** method button. Figure 10.79 shows the **Chamfer** dialog box with the **Distance and Angle** button selected. Initially, the **Edge** button should be active. The **Edge** button allows you to select the face you want the distance from the corner to be measured. Once you have defined the face, the **Corners** button should become active. The **Corners** button allows you to select the corners you want the angle to be measured from. To change the face, pick the **Edge** button again, and then reselect the corners.

Once you have established the face and corners you want to chamfer, enter or select a corner chamfer distance from the **Distance** drop-down list, and enter or select a corner chamfer angle from the **Angle** drop-down list. If the chamfer preview looks correct, pick the **OK** button to create the chamfer.

If you want to specify one corner chamfer side with a distance from the edge, and the other side with a different distance from the edge, pick the **Two Distances** method button. Figure 10.80 shows the **Corner Chamfer** dialog box with the **Two Distances** button selected. To place a corner chamfer using the **Two Distances** method, with the **Corner** button selected, pick the corners you want to chamfer. Then enter or select a chamfer distance from the **Distance1** drop-down list, and enter or select a chamfer distance from the **Distance2** drop-down list. If the corner chamfer preview looks correct, pick the **OK** button to create the corner chamfer. However, if the distances you enter need to be reversed, pick the **Flip** button to redefine the direction of the corner chamfer distances from the edge, and pick the **OK** button to create the corner chamfer.

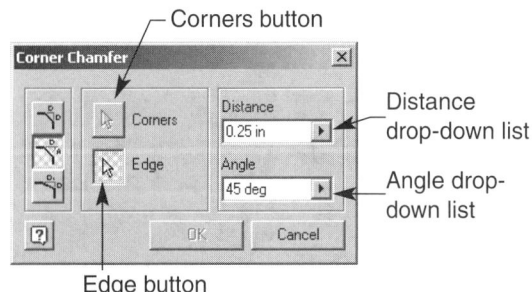

FIGURE 10.79   The **Corner Chamfer** dialog box (**Distance and Angle** option).

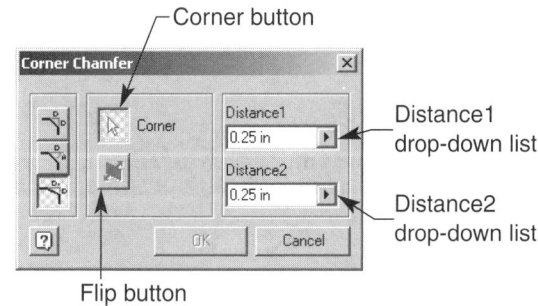

FIGURE 10.80   The **Corner Chamfer** dialog box (**Two Distances** option).

Exercise 10.11

1. Continue from Exercse 10.10, or launch Autodesk Inventor.
2. Open EX 10.4.
3. Save a copy of EX10.4 as EX10.11.
4. Close EX10.4 without saving, and open EX10.11.
5. Use the **Corner Chamfer** tool to create the corner chamfers shown.
6. The final part should look like the part shown.
7. Resave EX10.11.
8. Exit Autodesk Inventor or continue working with the program.

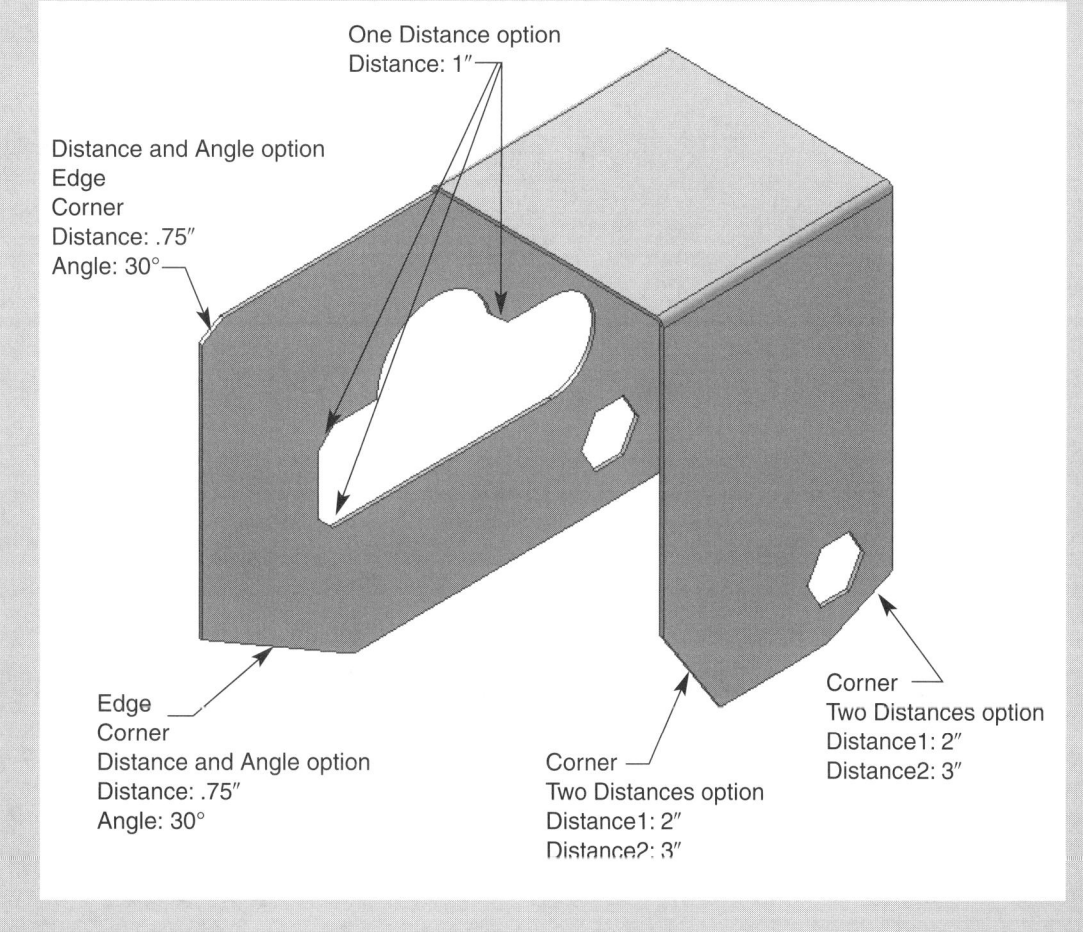

One Distance option
Distance: 1″

Distance and Angle option
Edge
Corner
Distance: .75″
Angle: 30°

Edge
Corner
Distance and Angle option
Distance: .75″
Angle: 30°

Corner
Two Distances option
Distance1: 2″
Distance2: 3″

Corner
Two Distances option
Distance1: 2″
Distance2: 3″

Using the Sheet Metal Punch Tool

The Autodesk Inventor sheet metal **PunchTool** is used to place special sheet metal catalog iFeatures on a sheet metal part. Autodesk Inventor iFeatures are discussed in Chapter 9. Using the **PunchTool** is very similar to inserting an iFeature, as discussed in Chapter 9. However, the base feature sketch used to create the iFeature must contain a **Hole Center,** because a **PunchTool** iFeature must be inserted on a sketched **Hole Center.** See Figure 10.81.

Once you have created an iFeature with a sketched **Hole Center** for positioning and have placed **Hole Centers** at the desired locations on the sheet metal feature, you are ready to use the sheet metal **PunchTool** to insert or "punch" features.

FIGURE 10.81 (A) The base feature sketch of the **PunchTool** iFeature must contain a **Hole Center.** (B) The sheet metal feature a **PunchTool** iFeature is inserted on must contain a **Hole Center.**

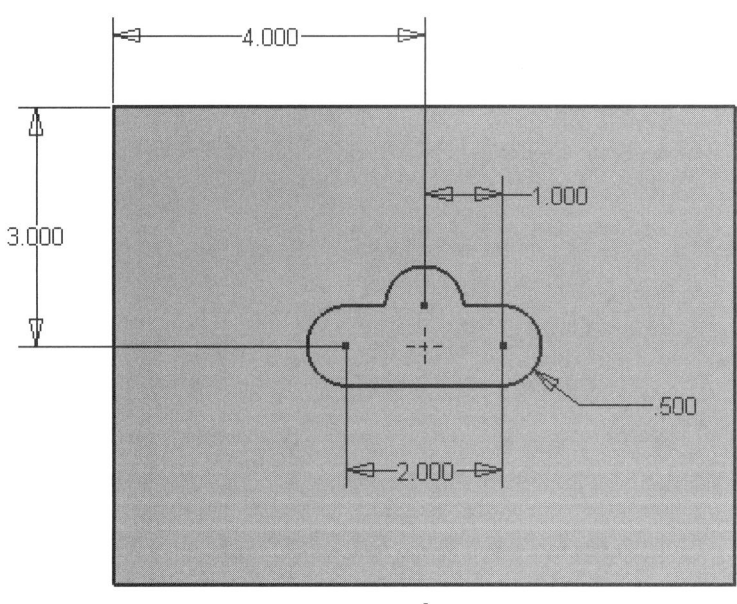

A

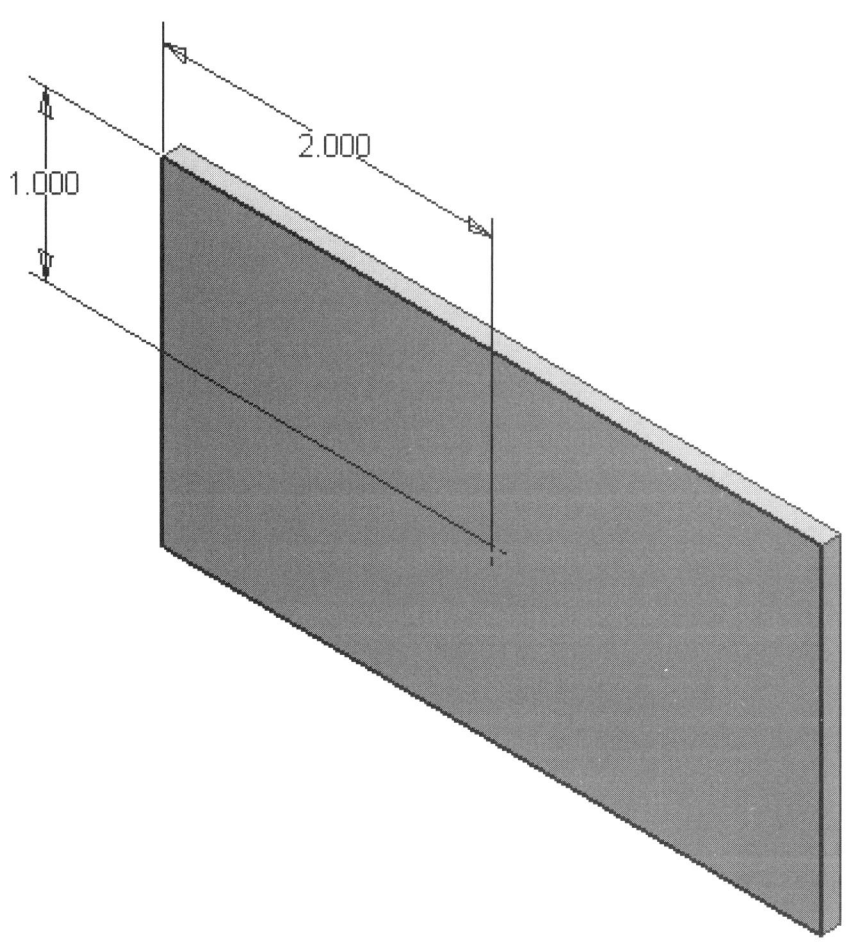

B

Field Notes

You may want to save all your **PunchTool** iFeatures in a specific location. The Autodesk Inventor **PunchTool** iFeatures are located in the following path: Program Files/Autodesk/Inventor 5/Catalog/Punches.

Access the **PunchTool** command by using one of the following techniques:

✓ Pick the **PunchTool** button on the panel bar.

✓ Pick the **PunchTool** button on the **Sheet Metal** toolbar.

The **PunchTool** dialog box is displayed when you access the **PunchTool** command. See Figure 10.82.

When you initially open the **PunchTool** dialog box, the select file window is displayed, and the contents of the **Punches** folder is shown in the file list box. If you want to use one of the iFeatures found in the **Punches** folder, select the file in the file list box and pick the **Next** button. If you want to locate a different folder that contains other **PunchTool** iFeatures, pick the **Browse...** button to access the **Browse For Folder** dialog box. Here you can locate and select any folder on your computer or the network. When you find the desired folder, pick the **OK** button. The folder path should be displayed in the **File Name** area, and the files contained in the folder should be listed in the file list box. Select the iFeature file you want to use, and pick the **OK** button.

When you select the **Next** button, the specify geometry window should be displayed. See Figure 10.83. In addition, the selected iFeature is automatically positioned on the sketched **Hole Centers,** and a preview of the design element is displayed. If you need to select additional hole centers, pick the **Select Centers** button, if it is not already active, and pick the hole centers. However, if you want to deselect certain hole centers, and not place the iFeature at those locations, assure the **Select Centers** button is selected, hold down the **Ctrl** key on your keyboard, and pick the hole centers you want to deselect. The next step is to define the angle of the punches by entering a value, or selecting a value from the **Angle** drop-down list.

The final step in positioning the iFeature is to satisfy any dangling geometry references. Dangling geometry results when additional positioning information is required in order for iFeature insertion to occur, primarily due to initial iFeature sketch, and feature geometry. Dangling geometry references may include faces, edges, angles, or other reference points, and are displayed in the **Dangling Geometry** list box. To satisfy dangling geometry, pick the name in the list box, and follow the prompt in the position

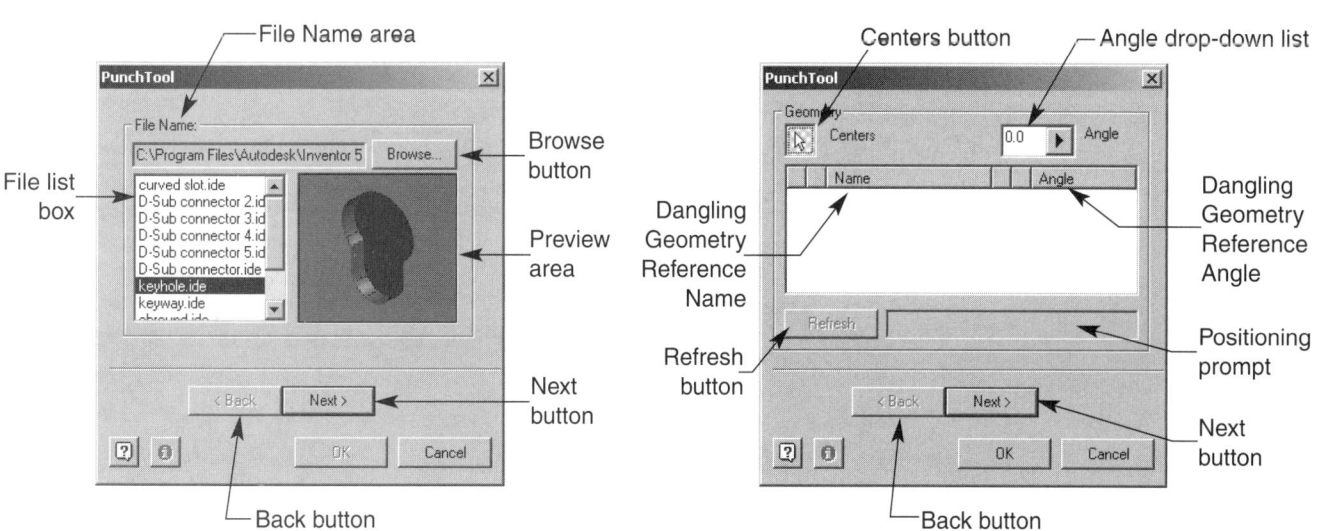

FIGURE 10.82 The **PunchTool** dialog box (select file window).

FIGURE 10.83 The **PunchTool** dialog box (position geometry window).

FIGURE 10.84 The
PunchTool dialog box
(modify size parameters
window).

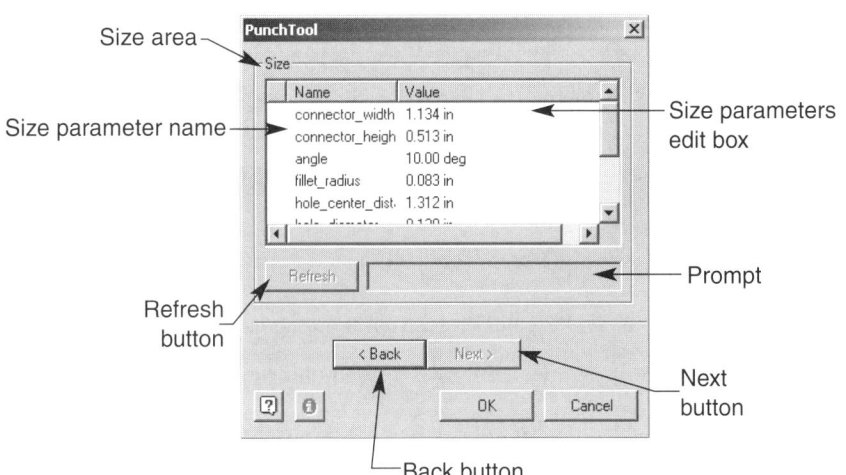

geometry window of the **PunchTool** dialog box. After you position the iFeature, pick the **Next** button to activate the **Modify Size Parameters** window. See Figure 10.84.

If you specified specific size parameter information when the iFeature was created, the size parameters are listed in the **Size** list box; otherwise the area is blank. The prompts guide you through the sizing process of inserting the iFeature. If no size limits were placed on the iFeature when it was created, you can enter any value. However, if a range limit was placed on the iFeature when created, you will be able to enter only a specific range of values, or if a list limit was placed on the iFeature when created, you will be able to choose only values from the **Value** drop-down list.

Field Notes

> Pick the **Refresh** button in the **Insert** iFeature dialog box anytime you resize or reposition the iFeature, to preview the changes.

Once you have selected, positioned, and sized the **PunchTool** iFeature you want to insert, pick the OK button to insert the iFeature. However, if you at least selected and positioned the iFeature, you can select the **OK** button and bypass the size options.

Field Notes

> Pick the **Back** button in the **PunchTool** dialog box anytime to return to the previous window to modify an insertion option.

Exercise 10.12

1. Continue from Exercse 10.11, or launch Autodesk Inventor.
2. Open a new inch sheet metal part file.
3. Open a sketch on the **XY Plane.**
4. Sketch a 12″×16″ rectangle, using the projected **Center Point** to fix the sketch in place.
5. Using default sheet metal style settings, create a face from the sketch.
6. Open a sketch on the top face of **Face1,** and sketch the **Hole Center** points as shown.
7. Access the **PunchTool** command.
8. The default folder should be Punches, located in the directory, Program Files/Autodesk/ Inventor 5/Catalog/Punches, and should be displayed. If this is not the currently selected folder, pick the **Browse...** button to access the **Browse For Folder** dialog box, and locate the **Punches** folder.
9. Select the file: Square Emboss.ipt, and pick the **Next** button.
10. The iFeature should automatically be positioned at all the hole centers. If it is not, select the sketched hole centers on the sheet metal face.
11. Pick the **Next** button.
12. Change the default height to .120″, the default length to 2″, and the default width to 2″.
13. Pick the **OK** button.
14. Save the sheet metal part file as EX10.12.
15. Exit Autodesk Inventor or continue working with the program.

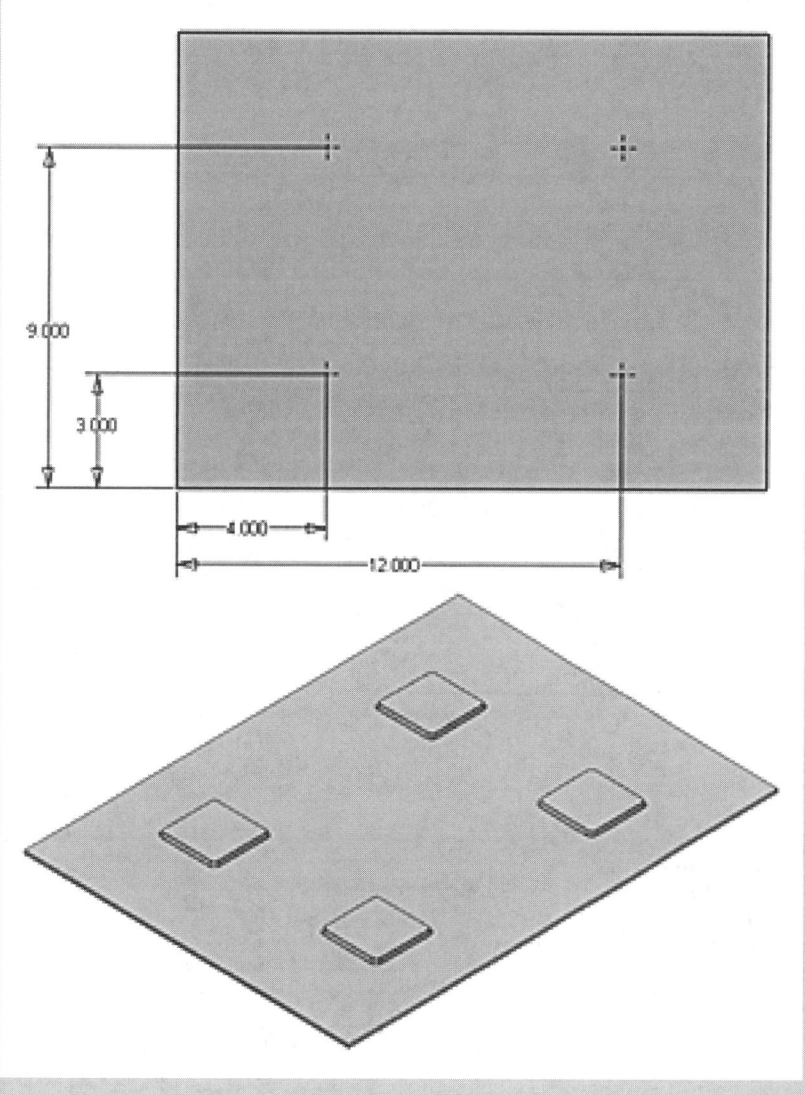

Using the Flat Pattern Tool

Once you have created a sheet metal part or anytime during sheet metal part development, you can "unfold" the entire part and create a *flat pattern.* A flat pattern is the flat layout of the sheet metal part. The flat pattern can eventually be transferred to a flat piece of sheet metal and formed into the sheet metal part. See Figure 10.85.

The **Flat Pattern** tool is used to create a flat pattern from an Autodesk Inventor sheet metal part. Access the **Flat Pattern** command by using one of the following techniques:

✓ Pick the **Flat Pattern** button on the panel bar.

✓ Pick the **Flat Pattern** button on the **Sheet Metal** toolbar.

You can create a flat pattern from any sheet metal part. If the sheet metal part contains at least one bend, when you access the **Flat Pattern** command the part is automatically flattened. However, if the part does not contain any bends, such as a cylindrical or conical sheet metal feature, you must first select a face. You can also select any face, on any part, to define that face as the base for the flat pattern.

The **Flat Pattern** tool calculates all the features and bends in the part and creates a flat pattern. When you generate a flat pattern, you will notice the following things happen:

■ The **FlatPattern** specification and icon are displayed in the **Browser.** See Figure 10.86.

■ A new window is automatically created with the flat pattern displayed in the graphics window. You cannot modify the sheet metal part in the flat pattern window. However, changes made to the sheet metal part, in the ipt. file, are updated in the flat pattern, unless the change made in the sheet metal part cannot be flattened.

FIGURE 10.85   An example of a sheet metal flat pattern.

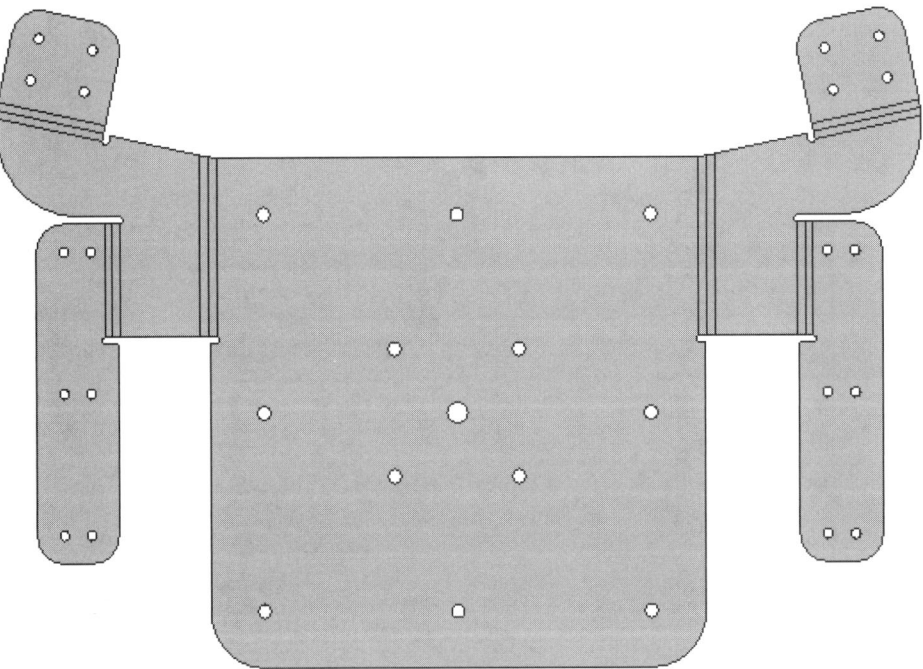

FIGURE 10.86   The **FlatPattern** specification and icon in the **Browser.**

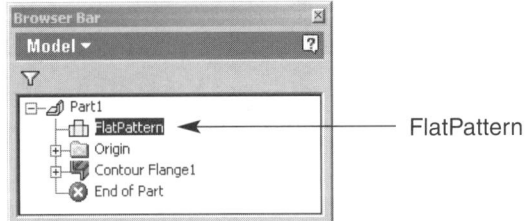

FlatPattern

Field Notes

Typically the parent part file is hidden by the new flat pattern, but is not closed. You can arrange the part file and flat pattern windows just like any other windows, discussed in Chapter 2.

- The flat pattern is differentiated from the parent sheet metal part as **FlatPattern:,** followed by the name of the part.

Once you create a flat pattern, the following menu options specific to flat patterns are available when you right click on **FlatPattern,** or the flat pattern icon in the **Browser:**

- **Extents...**   This option opens the **Flat Pattern Extents** display box, which allows you to view the overall length and width of the piece of sheet metal material needed to create the part. See Figure 10.87.

Field Notes

You cannot edit the length and width in the **Flat Pattern Extents** display box.

- **Save Copy As...**   This option opens the **Save Copy As** dialog box, shown in Figure 10.88. This dialog box allows you to save a copy of the flat pattern as an SAT, DWG, or DXF file, anywhere on your computer or the network.
- **Show Window**   If the flat pattern graphics window is not active and is hidden by another window, pick this option to display the flat pattern graphics window.
- **Close Window**   Use this option to close the flat pattern graphics window.
- **Open Window**   This option allows you to reopen the flat pattern graphics window and is only available when the flat pattern window has been closed.

FIGURE 10.87   The **Flat Pattern Extents** display box.

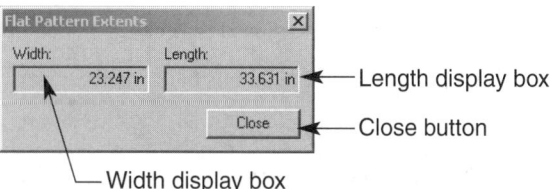

FIGURE 10.88   The **Save Copy As** dialog box.

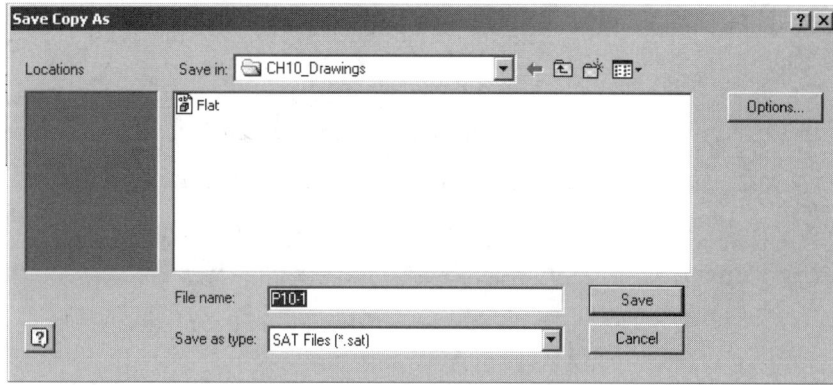

- **Align <u>H</u>orizontal**   Use this option to horizontally align the flat pattern window.

- **Align <u>V</u>ertical**   Use this option to vertically align the flat pattern window.

- **Select <u>B</u>ase Face**   This option allows you to reselect the flat pattern base feature from the sheet metal part. To select a base feature, pick the **Select <u>B</u>ase Face** menu option and choose the feature you want to specify as the base feature.

Field Notes

> Often complex features, such as embossed iFeatures, cannot be flattened, and remain three-dimensional when other features are flattened.

‖‖‖‖‖‖‖‖‖‖ CHAPTER TEST

Answer the following questions on a separate sheet of paper.

1. Briefly describe sheet metal parts and sheet metal drafting.

2. What do you get when you open a new Sheet Metal.ipt file from the New File menu?

3. If you begin a new sheet metal part in a modeling application, what do you do to access the sheet metal tools?

4. Name the command used to define the material type, thickness, and corner options when creating sheet metal parts.

5. Briefly explain the function of the Sheet Metal Styles dialog box.

6. Give the function of the Unfold Method List dialog box.

7. Define corner relief.

8. Describe a sheet metal face and its use.

9. Name the command that is used to access the Face dialog box.

10. What is a sheet metal contour flange?

11. Name the command that is used to access the Contour Flange dialog box.

12. What does the offset of a contour flange determine?

13. What is the purpose of the sheet metal cut tool?

14. Name the command for accessing the Cut dialog box.

15. How do you cut a sketch profile to the next face or plane?

16. How are sheet metal folds created?

17. Name the command and dialog box that is used to specify the fold options and parameters.

18. Define placed features.

19. What is required before a placed feature can be created?

20. When can placed features be created?

21. Name at least four types of placed features.

22. Define flange.

23. Name the command that is used to access the Flange dialog box.

24. Define hem.

25. Name the command that is used to access the Hem dialog box.

26. Explain the purpose of corner seams.

27. Define corner rip.

28. Define corner rounds.

29. Define corner chamfer

30. Briefly explain the function of the Autodesk Inventor sheet metal PunchTool.

31. Define flat pattern.

32. When can you create a flat pattern from any sheet metal part?

33. When is the part automatically flattened?

34. When creating a flat pattern, what do you have to do if the part does not contain any bends, such as a cylindrical or conical sheet metal feature?

35. Give the function of the Save Copy As dialog box when saving a flat pattern.

|||||||||||| PROJECTS

Follow the specific instructions for each project.

1. Name: Seat Base

   Units: Inch or Metric

   Specific Instructions:

   ■ Using the part modeling techniques discussed in previous chapters and the sheet metal part modeling techniques discussed in this chapter, create a seat base similar to the seat base shown.

   ■ When complete, use the flat pattern tool to view the seat base flat pattern.

   ■ Save the sheet metal part as: P10.1.ipt

   ■ Save the flat pattern as: P10.1.sat

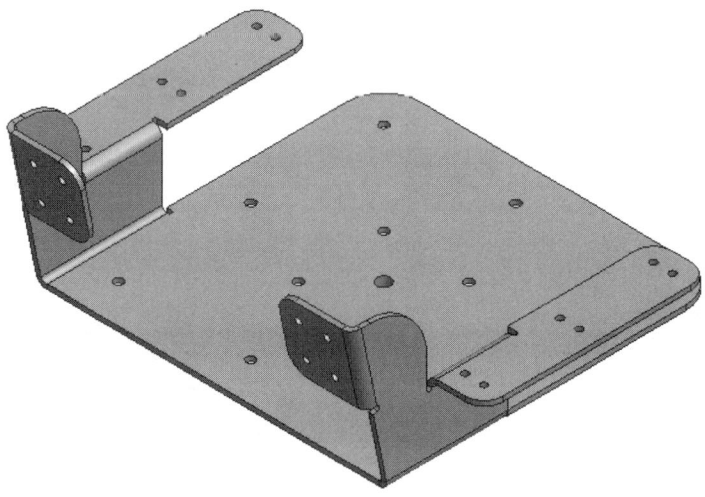

2. Name: Housing

Units: Inch

Specific Instructions:

- Open a new inch sheet metal part file.

- Create a 4″×6″ face as shown.

- Add 2″ flanges around the face as shown. The flanges should offset toward the face and occur in a positive direction. Use default sheet metal style settings.

- Add two 1″ flanges as shown. The flanges should offset toward the base flanges. Use default sheet metal style settings.

- Place four .25″ holes on the flanges as shown.

- Modify each of the corners with a **No Overlap** seam and a round bend relief as shown.

- Using the dimensions shown, cut a rectangle through the specified flange.

- Using the dimensions shown, cut the circle through the specified flange.

- The final part should look like the part shown.

- Save As: P10.2.

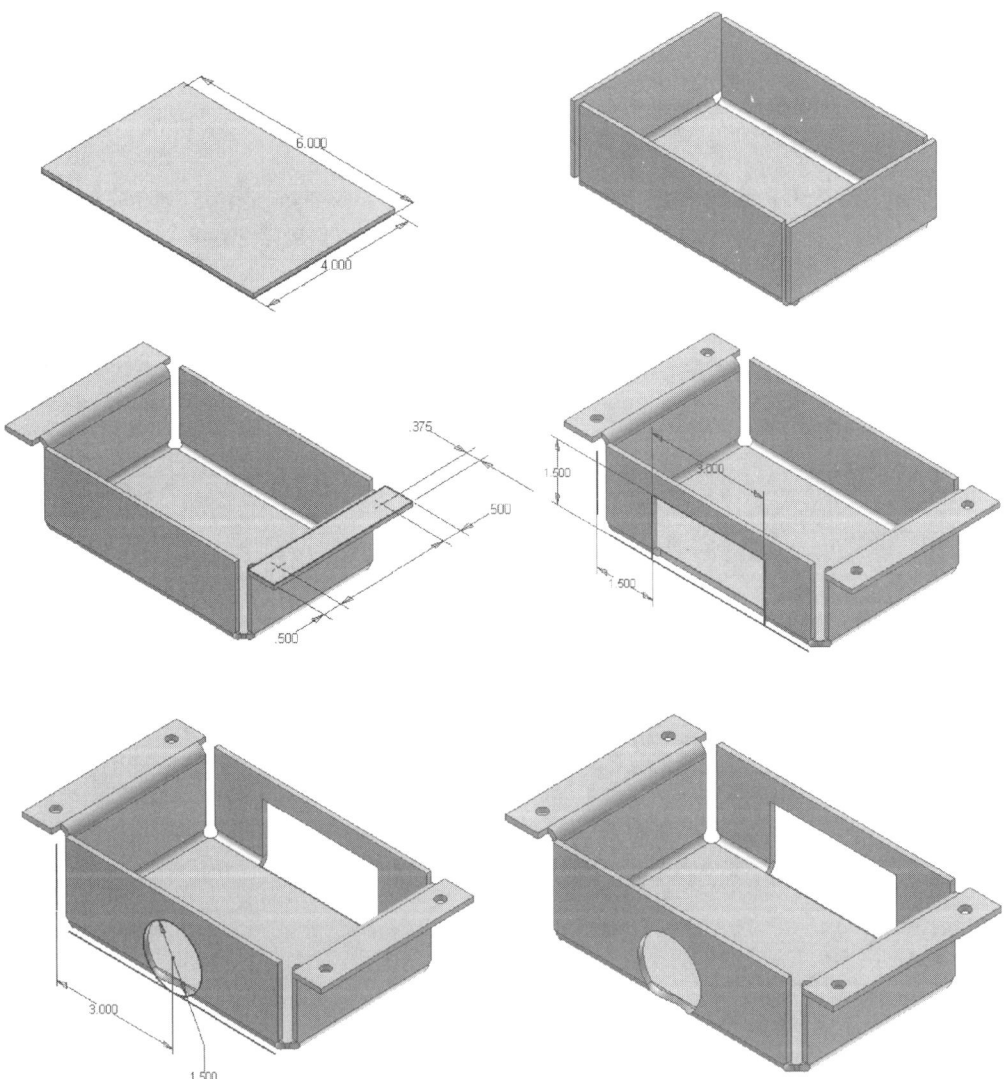

3. Name: Louver

   Units: Metric

   Specific Instructions:

- Open a new metric sheet metal part file.

- Create a 50mm×100mm face as shown.

- Sketch the rectangle and **Hole Center** on the face as shown for reference.

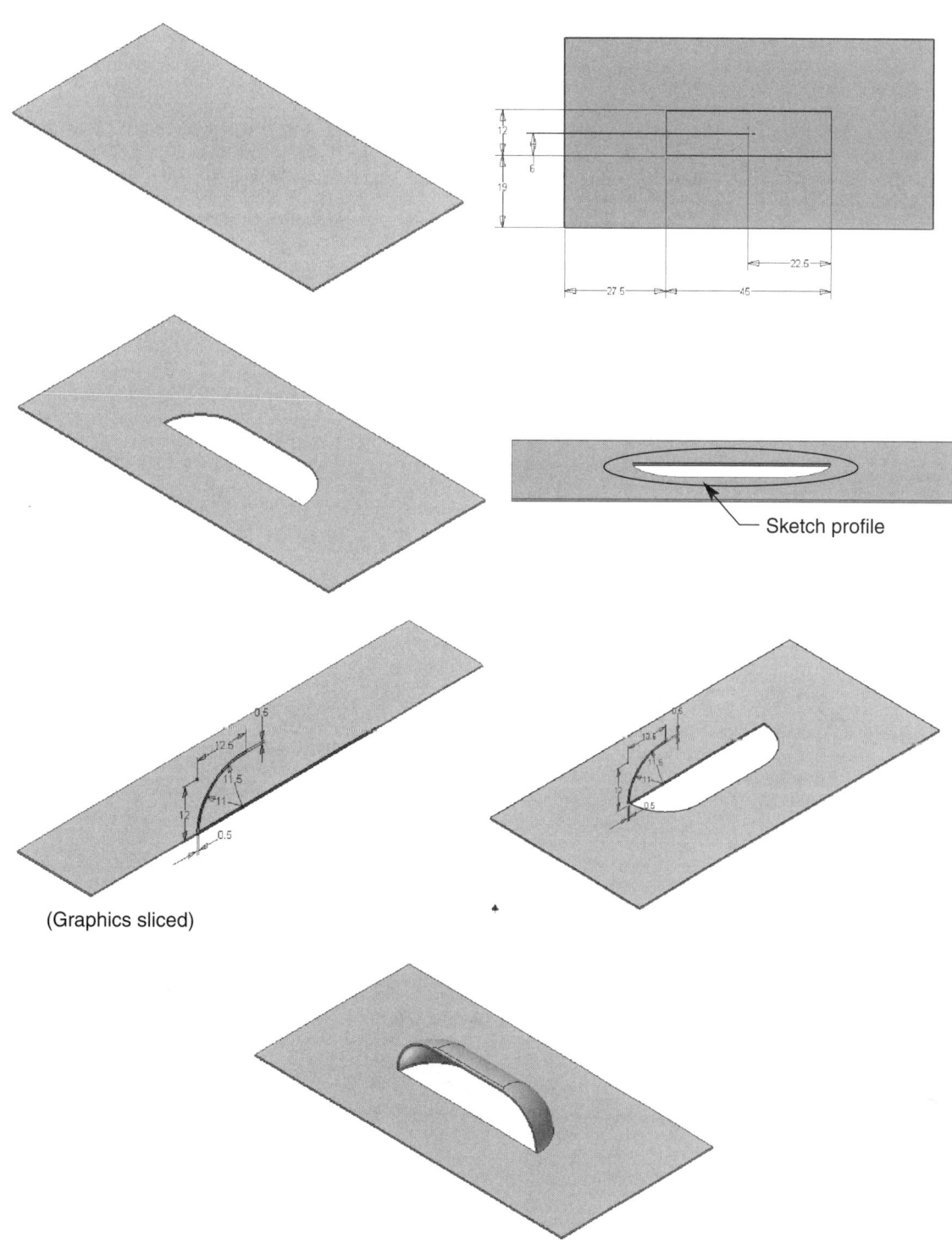

Sketch profile

(Graphics sliced)

- Cut the rectangle through the face.

- Place R 12mm corner rounds as shown.

- Open a sketch on the inside of the cut and sketch the geometry shown. You may want to Slice Graphics for clarity. This sketch is used as a sweep profile.

- Open a new sketch on the top of the face and project the geometry shown. This sketch is used as a sweep path.

- Access the Features tools, and create a sweep feature using Sketch3 as the profile and Sketch4 as the path. The sweep should look like the feature shown.

- Add 2″ flanges around the face as shown. The flanges should offset towards the face and occur in a positive direction.

- The final part should look like the part shown.

- Save As: P10.3.

- Using the Create iFeature dialog box, create an iFeature from the cut, corner round, and sweep. Name the iFeature Louver, and specify thickness, length, and width size parameters.

- Save the iFeature in a location of your choice for future access.

4. Name: Louvered Cover

Units: Metric

Specific Instructions:

- Open a new metric sheet metal part file.
- Change the default sheet metal style thickness to 1.5mm
- Create a 300mm×400mm face as shown.
- Add 25mm flanges to the 300mm edge as shown.
- Add single hems to the 400mm edge as shown.
- Sketch a **Hole Center** on the top of the face as shown.
- Insert the louver created in P10.3 on the **Hole Center.**
- Use the **Rectangular Pattern** tool, and equally space louvers as shown.
- Save As: P10.4

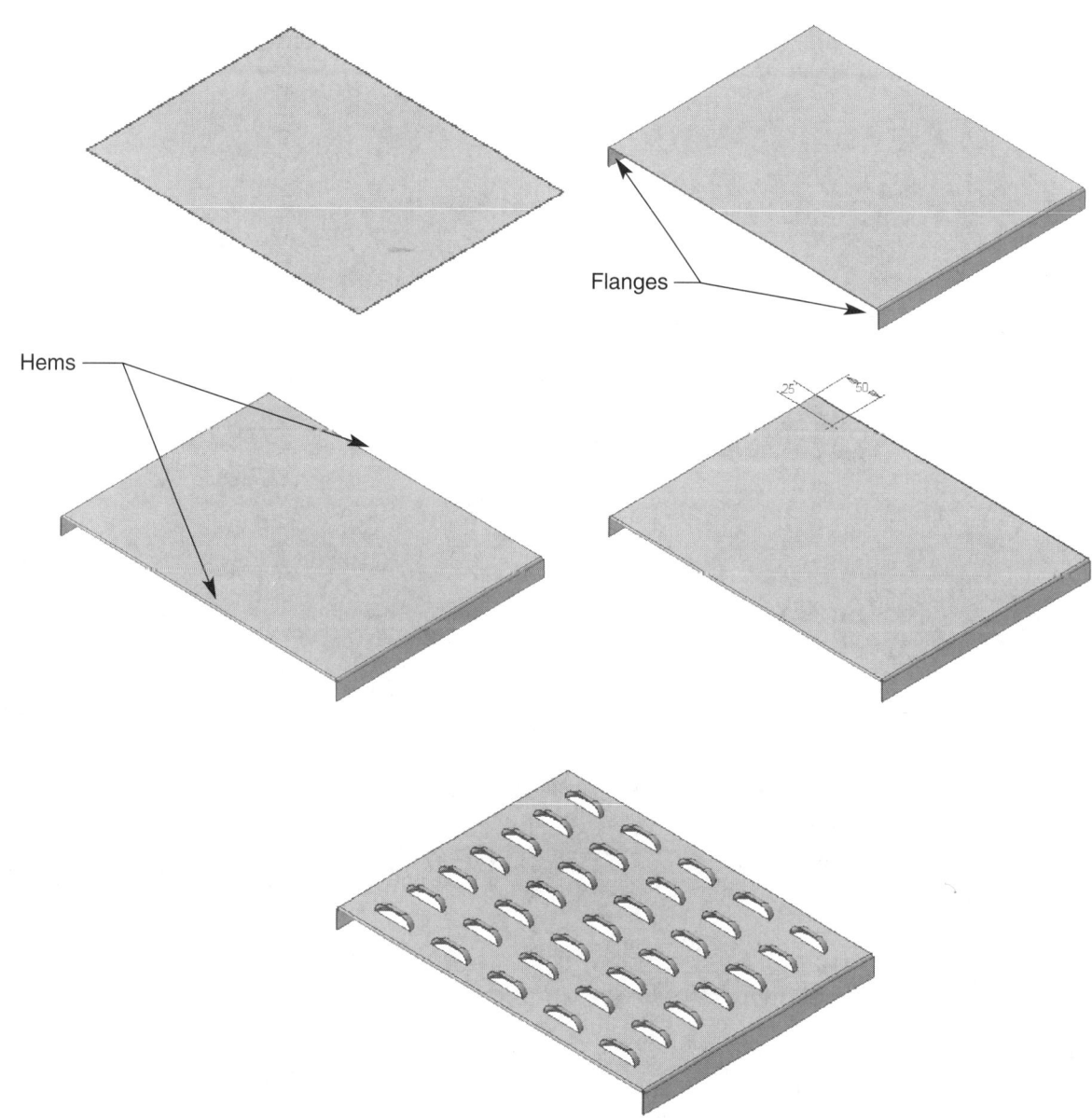

CHAPTER **11**

||||||||||||||||||||||||||||||||||||||||||||||||||||||||||||||||||||||||||||||||||||||||||||||||||||

Working with Solids

LEARNING GOALS

After completing this chapter, you will be able to:

◎   Use base solids from other CAD programs or those created using Autodesk Inventor.

◎   Operate in the solids work environment.

◎   Use solids tools.

◎   Move solid faces.

◎   Extending and contracting solid bodies.

Working with Solids

As discussed in previous chapters, Autodesk Inventor part models are fully parametric and contain a great deal of information. These types of part models are known as *intelligent solids.* In contrast, many CAD programs allow you to generate *basic* or *dumb solids,* and you can even save an Autodesk Inventor file as a dumb solid. Dumb solids are created using basic Boolean methods and do not contain any information about specific parameters, dimensions, constraints, part history, or features (not including work features). Only fundamental volume information, such as length, width, and height, are stored in dumb solid parts.

When using Autodesk Inventor, you can open any solid saved as an .iges, .step, or .sat file in another CAD program or Autodesk Inventor. You can open these types of files the same way you open any other file, and you can choose to use one of the following access techniques:

✓ After launching Autodesk Inventor, access the **Open** dialog box and open the desired file.

✓ While in a part file, select the **Import...** option from the **Insert** pull-down menu, and import the desired file.

✓ While in an assembly file, discussed in future chapters, select the **Existing Component...** option from the **Insert** pull-down menu, and import the desired file.

Working in the Solids Environment

Though solids may look the same as other parts or features, the solid feature interface and work environment is quite different. Solids are opened as base features and cannot be placed in different areas in space. When opened, the 0,0,0 corner of the solid coincides with the **Center Point.** The browser displays the solid base feature as **Base1** next to the solid icon. See Figure 11.1. You will notice no sketch, or other parametric, information is available inside a base solid, because no dimension, constraint, sketch, or feature (not including work feature) information is contained in a dumb solid.

401

FIGURE 11.1   A base
solid in the **Browser**
bar.

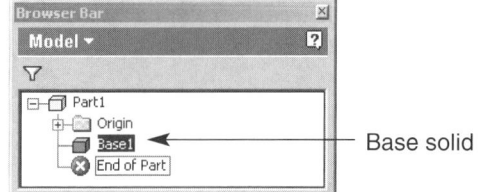

Base solid

Field Notes

When a base feature already exists, .iges, .step, and .sat files are inserted as surfaces, not solids.

Using Solids Tools

The solids tools are used to edit the basic volume parameters of the solid, such as length, width, and height. These tools allow you to move faces and extend or contract solid bodies. They also allow you to create work planes, work axes, and work points. Unlike other parts, in order to activate the solids tools, you must use one of the following techniques:

✓ Double-click the base solid in the browser.

✓ Right-click the base solid in the browser and select the **Edit Solid** menu option.

Once you activate the solids tools by editing the solid and you are in the solids work environment, the panel bar is in **Solids** mode. See Figure 11.2. By default, the panel bar is displayed in learning mode, as seen in Figure 11.2. To change from learning to expert mode, seen in Figure 11.3 and discussed in Chapter 2, right-click any of the panel bar tools or left-click the panel bar title, and select the **Expert** option. You can also access the solids tools from the **Solids** toolbar, shown in Figure 11.4.

The following tools, available in the solids environment, are used exactly the same as in any other part model environment. These tools are covered in previous chapters and are not discussed in this chapter:

■ **Work Plane** (Chapter 7)

■ **Work Axis** (Chapter 7)

■ **Work Point** (Chapter 7)

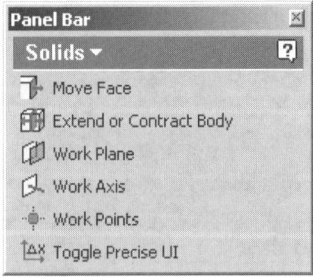

FIGURE 11.2   The panel bar
in solids (learning) mode.

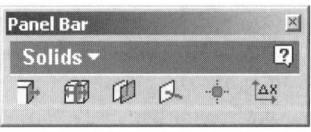

FIGURE 11.3   The panel bar
in solids (expert) mode.

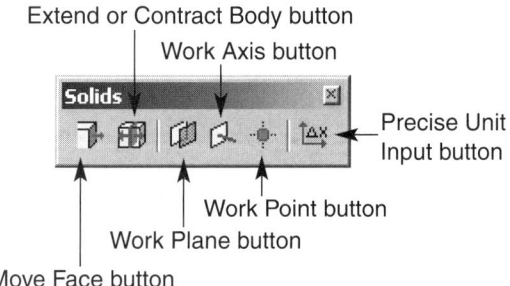

Extend or Contract Body button
Work Axis button
Precise Unit Input button
Work Point button
Work Plane button
Move Face button

FIGURE 11.4   The **Solids** toolbar.

Exercise 11.1

1. Launch Autodesk Inventor.
2. Open a new inch part file.
3. Open a sketch on the **XY Plane,** and sketch a 2″×4″ rectangle, using the projected **Center Point** to fix the corner of the sketch.
4. Extrude the sketch 1″ in a positive direction.
5. Use the **Save Copy As** dialog box to save the current file as EX11.1, in an .iges, .step, or .sat file.
6. Close the active file (Part1) without saving.
7. Open EX11.1.iges, .step, or .sat.
8. Explore the solid work environment and interface options previously discussed.
9. Exit Autodesk Inventor or continue working with the program.

Moving Solid Faces

Though you cannot change the parameters of a solid using sketch dimensions or constraints, you can move base solid faces nonparametrically. You can move a single face or a group of faces, as shown in Figure 11.5. Once you have opened or inserted a base solid, access the **Move Face** command using one of the following techniques:

✓ Pick the **Move Face** button on the panel bar.

✓ Pick the **Move Face** button on the **Solids** toolbar.

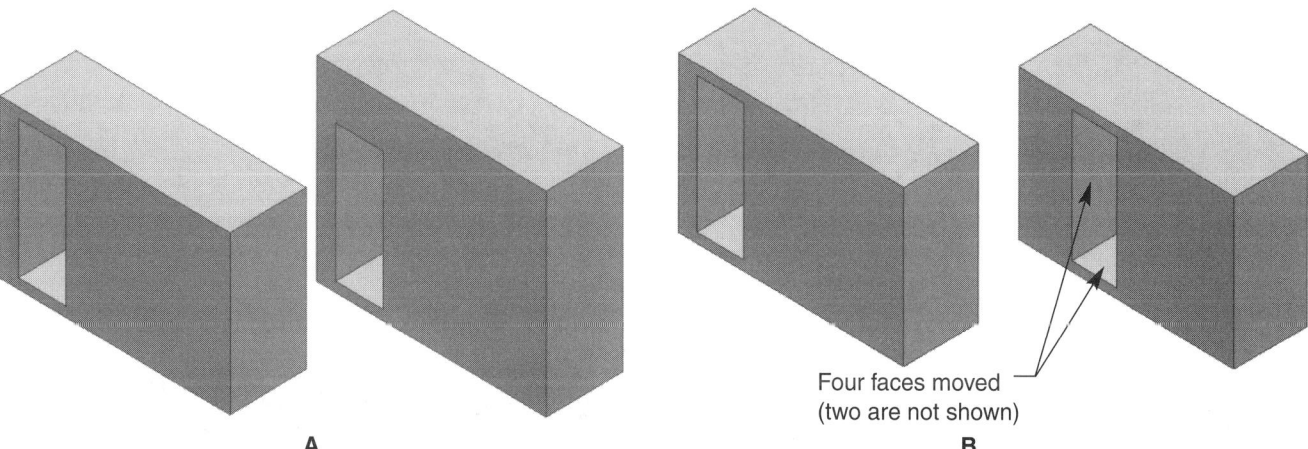

Four faces moved
(two are not shown)

A   B

FIGURE 11.5   An example of moving (A) a single solid face and (B) a group of solid faces.

The **Move Face** dialog box is displayed when you access the **Move Face** command. See Figure 11.6.

You can choose to move a face using the **Distance and Direction** option or the **Planar Move** option. By default, the **Distance and Direction** option is active in the **Move Face** dialog box. This method allows you to move a face or group of faces a specified distance, along a linear edge. To move a face using this technique, first pick the **Select Faces** button if it is not already selected. Then pick the faces that you want to move. Once you have selected one or more solid faces to move, pick the **Select Edge** button located in the **Distance & Direction** area, and pick the desired edge. The edge defines where the face or faces are moved. For example, if you select a horizontal edge, the face moves along a horizontal plane. When you select an edge, a preview arrow displays the direction and intended effect of the move. See Figure 11.7.

If the direction of the move is not correct, pick the **Flip Direction** button to move the face in the opposite direction. Finally, specify how far you want the face to move by entering a value, or selecting a value from the **Distance** drop-down list. When complete, pick the **OK** button to move the selected faces. See Figure 11.8.

You can also move a face using the **Planar Move** method, by picking the **Planar Move** button, to access the **Plane** and **Points** buttons. See Figure 11.9. The **Planar Move** option allows you to move one or more faces from one point to another on a specified plane. To use the **Planar Move** technique, first pick the **Select Faces** button if it is not already selected. Then pick the faces that you want to move. Once you have selected one or more solid faces to move, pick the **Select Plane** button located in the **Direction & Distance** area, and pick the desired plane. The plane defines where the face or faces are moved. Typically, a plane that is not parallel to the specified faces is selected.

The next step is to pick the **Select Points** button to define the move distance by specifying points. You can select any available points on the solid, such as endpoints and midpoints, or work points. See Figure 11.10. You can also define the points using precise input techniques. To use this method, after selecting the desired plane, pick the **Toggle Precise UI** button on the panel bar, or the **Solids** toolbar, to access the **Precise Input** toolbar. Then pick the **Select Points** button to activate the precise input tools. Specify the start point of the move by entering precise input values and pressing **Enter** on your keyboard. Specify the endpoint of the move by entering precise input values and pressing **Enter** on your keyboard. Finally pick the **OK** button to move the selected faces. See Figure 11.11.

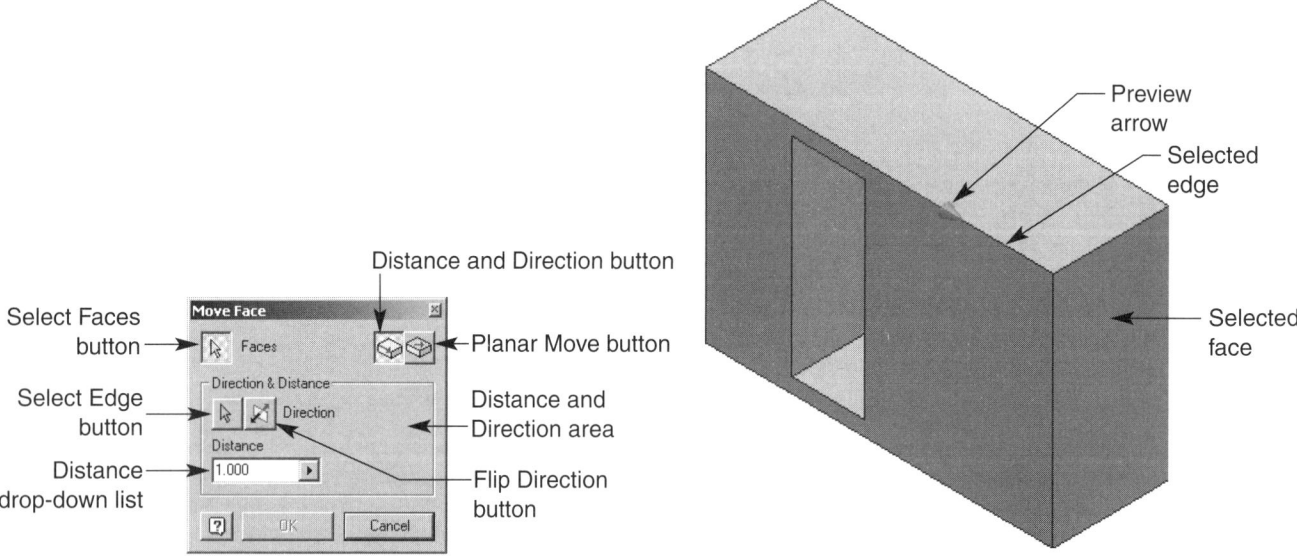

FIGURE 11.6   The **Move Face** dialog box (**Direction and Distance** move option).

FIGURE 11.7   Moving a solid face.

FIGURE 11.8 An example of a solid face moved 1″.

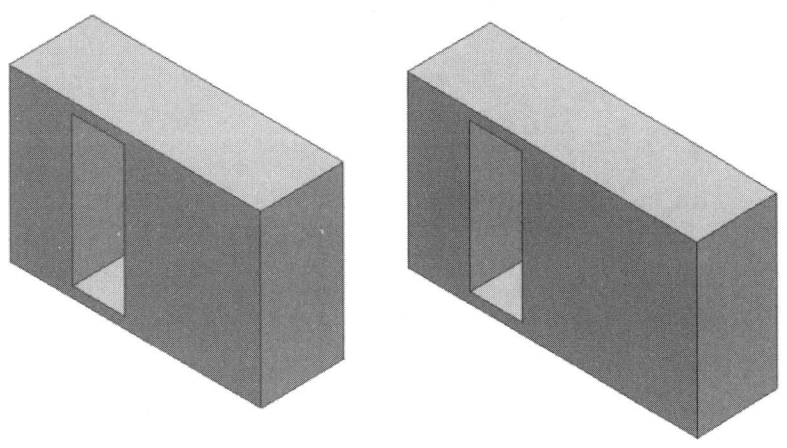

FIGURE 11.9 The **Move Face** dialog box (**Planar Move** option).

Direction and Distance button

Select Faces button →

Planar Move button

Select Plane button →

Direction & Distance area

Select Points button →

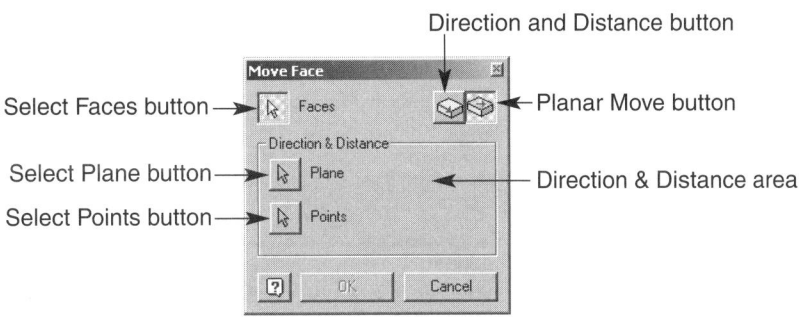

FIGURE 11.10 Moving a solid face using points on the solid.

Selected endpoint

Selected midpoint

Four selected faces

Selected plane

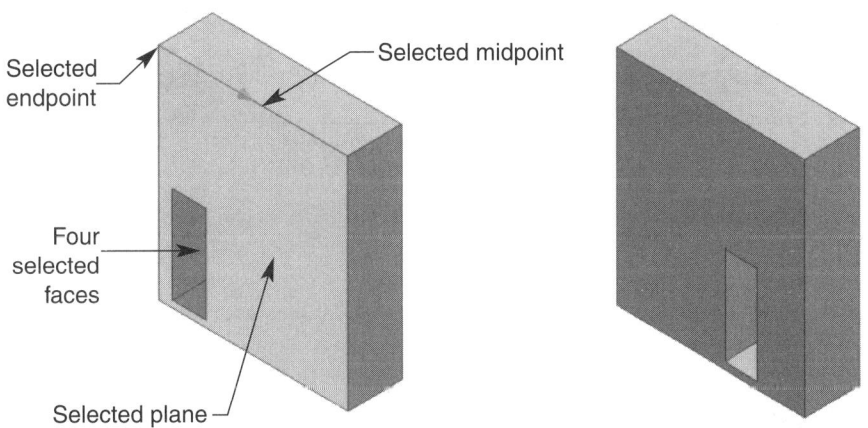

FIGURE 11.11 Moving a solid face using precise input points.

Preview vector

Four selected faces

Selected plane

Start point (1.5,1.5)

Endpoint (3,3)

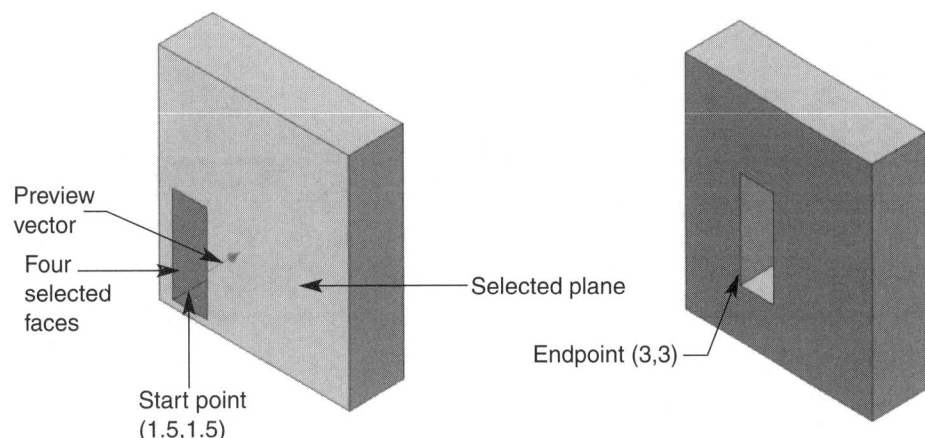

Exercise 11.2

1. Continue from Exercise 11.1, or launch Autodesk Inventor.
2. Open EX11.1.
3. Save a copy of EX11.1 as EX11.2.ipt.
4. Close EX11.1 without saving, and open EX11.2.
5. Double-click **Base1** in the **Browser** or right-click **Base1** and select the **Edit Solid** menu option, to activate the **Solids** tools.
6. Access the **Move Face** dialog box.
7. Using the **Direction Distance** option, move the top face up, 1.5″ as shown.
8. Access the **Move Face** dialog box again.
9. Using the **Planar Move** option, move the specified face, using the plane and points shown.
10. Resave EX11.2.
11. Exit Autodesk Inventor or continue working with the program.

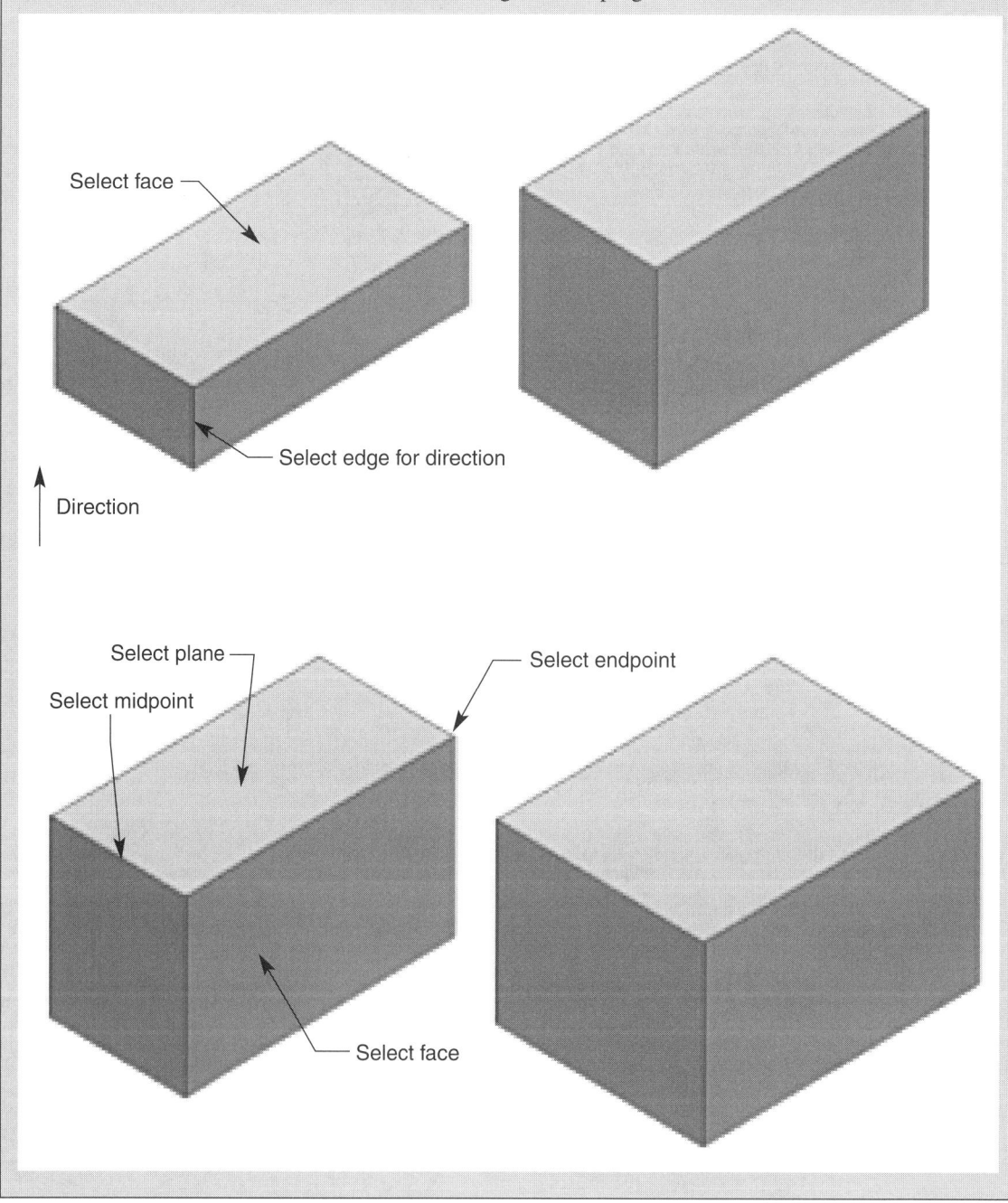

Exercise 11.3

1. Continue from Exercise 11.2, or launch Autodesk Inventor.
2. Open EX11.1.
3. Save a copy of EX11.1 as EX11.3.ipt.
4. Close EX11.1 without saving, and open EX11.3.
5. Double-click **Base1** in the **Browser** or right-click **Base1** and select the **Edit Solid** menu option, to activate the **Solids** tools.
6. Access the **Move Face** dialog box.
7. Using the **Planar Move** option, move the specified face, using the plane shown.
8. To define the points, pick the Points button. Then pick the **Toggle Precise UI** button from the **Solids** toolbar or **Panel** bar, to access the **Precise Input** toolbar.
9. Define the relative origin, and the relative orientation at the locations shown.
10. Using the XY input technique, locate the first point as 1.5″, 1.5″ and the second point as 2.5″, 2.5″.
11. Resave EX11.3.
12. Exit Autodesk Inventor or continue working with the program.

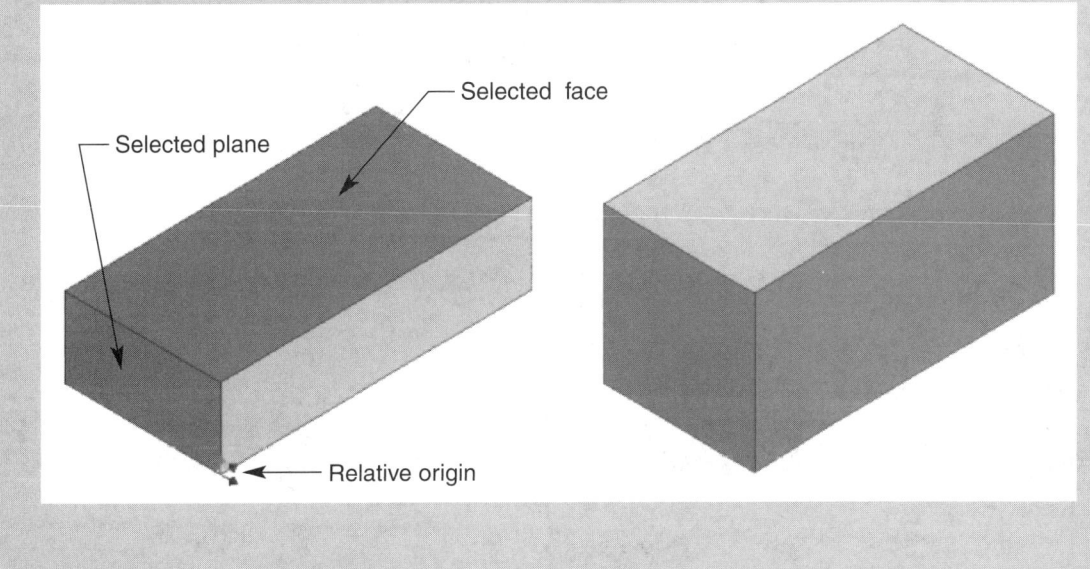

Extending and Contracting Solid Bodies

Though you cannot change the parameters of a solid using sketch dimensions or constraints, you can extend or contract the body of a solid by selecting a single plane. See Figure 11.12. Once you have opened or inserted a base solid, access the **Extend or Contract Body** command using one of the following techniques:

✓ Pick the **Extend or Contract Body** button on the panel bar.

✓ Pick the **Extend or Contract Body** button on the **Solids** toolbar.

The **Extend or Contract Body** dialog box is displayed when you access the **Extend or Contract Body** command. See Figure 11.13.

You can either extend or contract a base solid body using the **Extend or Contract Body** dialog box. Whether you want to extend or contract the body, the first step is to pick the **Plane** button, if it is not currently active, and pick the plane from where you want to expand or contract the solid body. Then, if you want to extend the body, pick the **Expand** button, or pick the **Contract** button if you want to contract the body. Finally, specify how much expansion or contraction you want to occur by entering a value or selecting a value from the **Distance** drop-down list. See Figure 11.14. Once you have selected a plane, picked the **Expand** or **Contract** button, and entered the desired distance, pick the **OK** button to complete the command.

FIGURE 11.12   An example of (A) extending a solid body and (B) contracting a solid body.

FIGURE 11.13   The **Extend or Contract Body** dialog box.

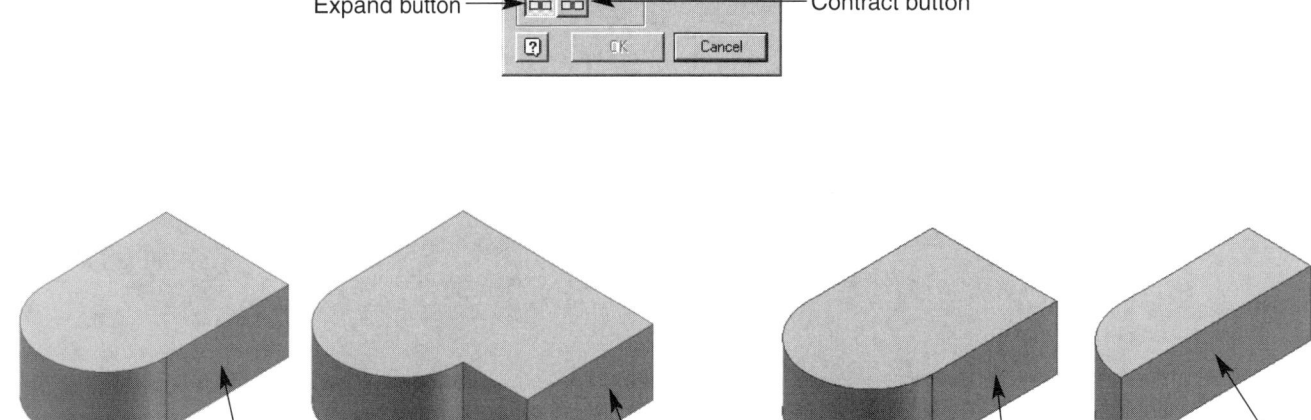

FIGURE 11.14   (A) Extending a base solid face. (B) Contracting a base solid face.

 Field Notes

You can select any face on the base solid or any existing work plane as the plane from where the extension or contraction occurs.

Exercise 11.4

1. Continue from Exercise 11.3, or launch Autodesk Inventor.
2. Open EX11.1.
3. Save a copy of EX11.1 as EX11.4.ipt.
4. Close EX11.1 without saving, and open EX11.4.
5. Double-click **Base1** in the **Browser** or right-click **Base1** and select the **Edit Solid** menu option to activate the **Solids** tools.
6. Access the **Extend or Contract Body** dialog box.
7. Pick the specified plane.
8. Select the **Expand** button.
9. Specify a distance of 2″
10. The final solid should look like the solid shown.
11. Resave EX11.4.
12. Exit Autodesk Inventor or continue working with the program.

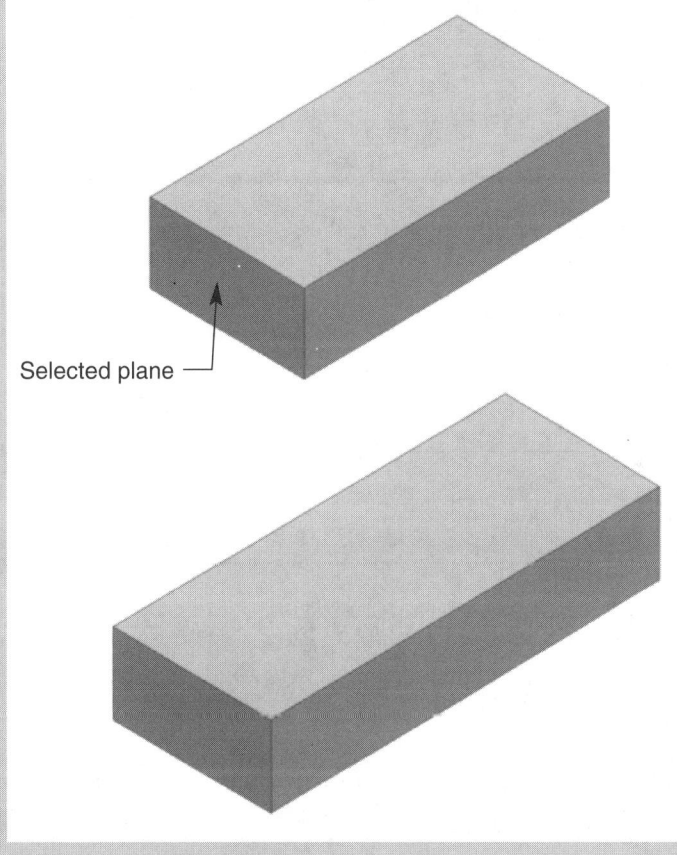

Selected plane

Exercise 11.5

1. Continue from Exercise 11.4, or launch Autodesk Inventor.
2. Open EX11.1.
3. Save a copy of EX11.1 as EX11.5.ipt.
4. Close EX11.1 without saving, and open EX11.5.
5. Double click **Base1** in the **Browser** or right-click **Base1** and select the **Edit Solid** menu option, to activate the **Solids** tools.
6. Access the **Extend or Contract Body** dialog box.
7. Pick the specified plane.
8. Select the **Contract** button.
9. Specify a distance of 1.5″
10. The final solid should look like the solid shown.
11. Resave EX11.5.
12. Exit Autodesk Inventor or continue working with the program.

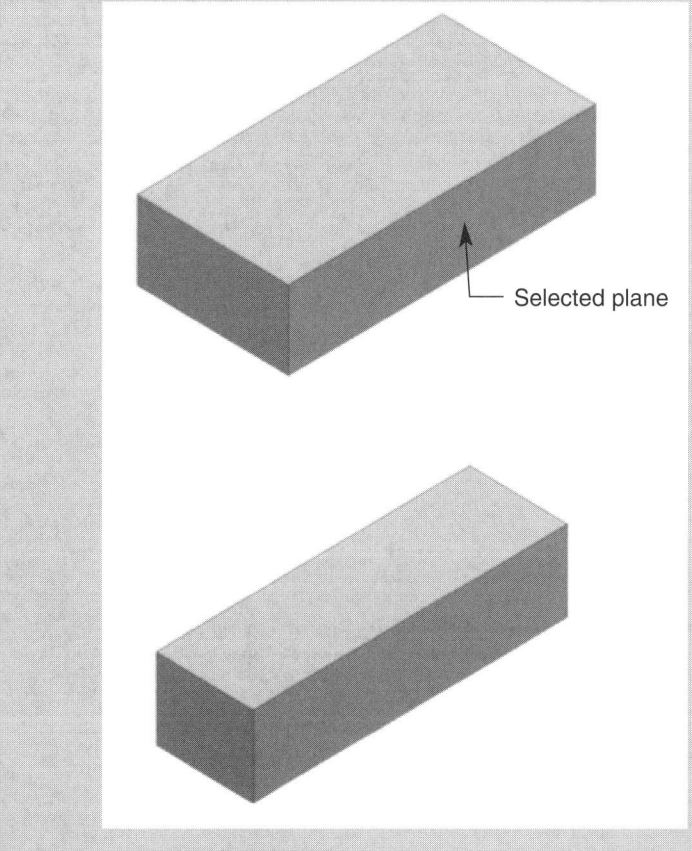

Selected plane

|||||||||||| CHAPTER TEST

Answer the following questions on a separate sheet of paper.

1. What are intelligent solids?

2. What are dumb solids?

3. Discuss the general function of solids tools.

4. When you open a solid, with what does the 0,0,0 corner coincide?

5. Name the dialog box where you can move a single face or a group of faces.

6. What does the Distance and Direction option allow you to do?

7. What does the Planar Move option allow you to do?

8. Name the dialog box where you can extend or contract the body of a solid by selecting a single plane.

||||||||||| PROJECTS

Follow the specific instructions provided for each project.

1. Name: Support Block

   Units: Metric

   Specific Instructions:

   ■ Open a new metric part file.

   ■ Create a 25mm×50mm extruded rectangle as shown.

   ■ Cut a 5mm×28mm slot in the middle of the extrusion as shown.

   ■ Save a copy of the part as P11.1.sat.

   ■ Close the original part file without saving, and open P11.1.sat.

   ■ Using the **Move Face** tool, move the specified faces out 5mm.

   ■ Using the **Extend** tool, extend the specified plane 15mm.

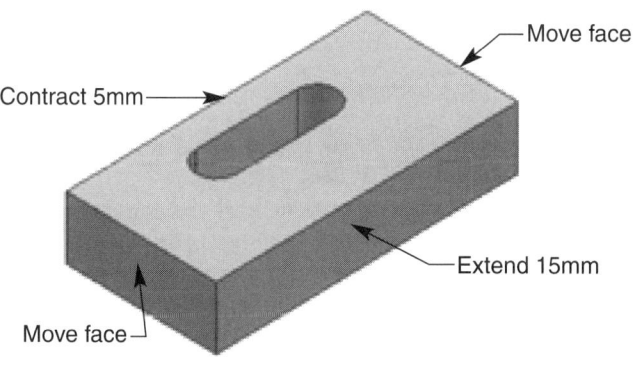

Move face
Contract 5mm
Extend 15mm
Move face

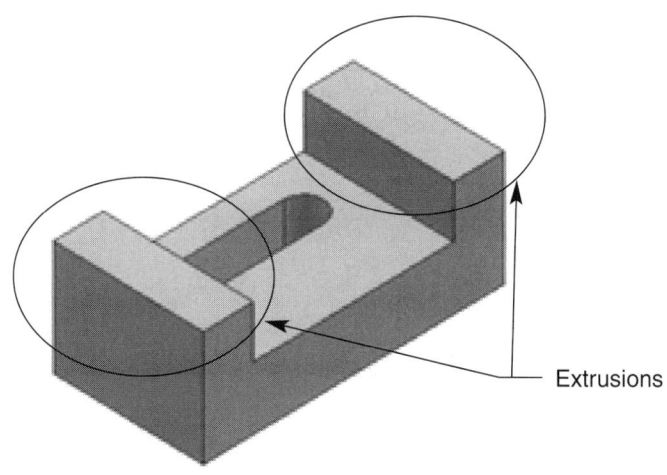

Extrusions

- Using the **Contract** tool, contract the specified plane 5mm.

- Open a sketch on the top of the base solid.

- Sketch 10mm×30mm rectangles, and extrude the sketches 10mm in a positive direction as shown.

- Resave P11.1.

2. Name: Base Solid

   Units: Inch or Metric

   Specific Instructions:

   - Open an .iges, .step, or .sat file from another CAD program, into Autodesk Inventor, if available.

   - Using the techniques discussed in this chapter, manipulate the imported base solid.

   - Add additional features to the base solid if desired.

   - Save As: P11.2.

||||||||||||||||||||||||||||||||||||||||||||||||||||||||||||||||||||||||||||||||||||||

CHAPTER 12

Creating Part Drawings

LEARNING GOALS

After completing this chapter, you will be able to:

◎ Work in the part drawing environment.

◎ Define common, sheet, and terminator drafting standards.

◎ Specify drawing sheet parameters.

◎ Use default and create your own drawing borders.

◎ Use default and create your own title blocks.

◎ Create mutiview drawings.

◎ Use the **Create View, Project View, Auxiliary View, Section View, Detail View,** and **Broken View** tools.

◎ Sketch two-dimensional drawing views.

◎ Use sheet formats.

Working with Part Drawings

Unlike traditional CAD programs, Autodesk Inventor combines three dimensional (3D) solid modeling with powerful, parametric two-dimensional (2D) drawing capabilities. See Figure 12.1. Part drawings are created quickly and easily from existing part models. When you edit the part model, the corresponding part drawing modifies to reflect the new design. Similarly, you can edit a part model by modifying the parametric model dimensions inside the part drawing. Using existing part model geometry, you can create a *monodetail* drawing, which is a drawing of a single part on one sheet, or a *multidetail* drawing, which is a drawing of several parts on one sheet. You can also develop 2D part drawings using sketch tools, or by inserting an AutoCAD drawing into an Autodesk Inventor drawing file with the extension .idw. Once you have inserted a part model view, you can create additional views, including projected, isometric, auxiliary, section detail, and broken views. You can also fully define the drawing with a variety of dimensions, notes, and text.

There are a number of methods for creating drawings using Autodesk Inventor. For example, you may want to initially place drawing views, then add annotations, and finally define a border and a title block. Or you may want to place drawing views, then specify a border and title bock, and add annotations toward the end of the drawing process. Any of these techniques can be effective. However, this chapter approaches the creation of part drawings in the following order:

1. Identify the sheet parameters.

2. Define the border options.

3. Specify the title block options.

4. Place and create drawing views.

The next step is to add additional drawing annotations, which is discussed in Chapter 13.

FIGURE 12.1 An example of a 2D part drawing created using Autodesk Inventor.

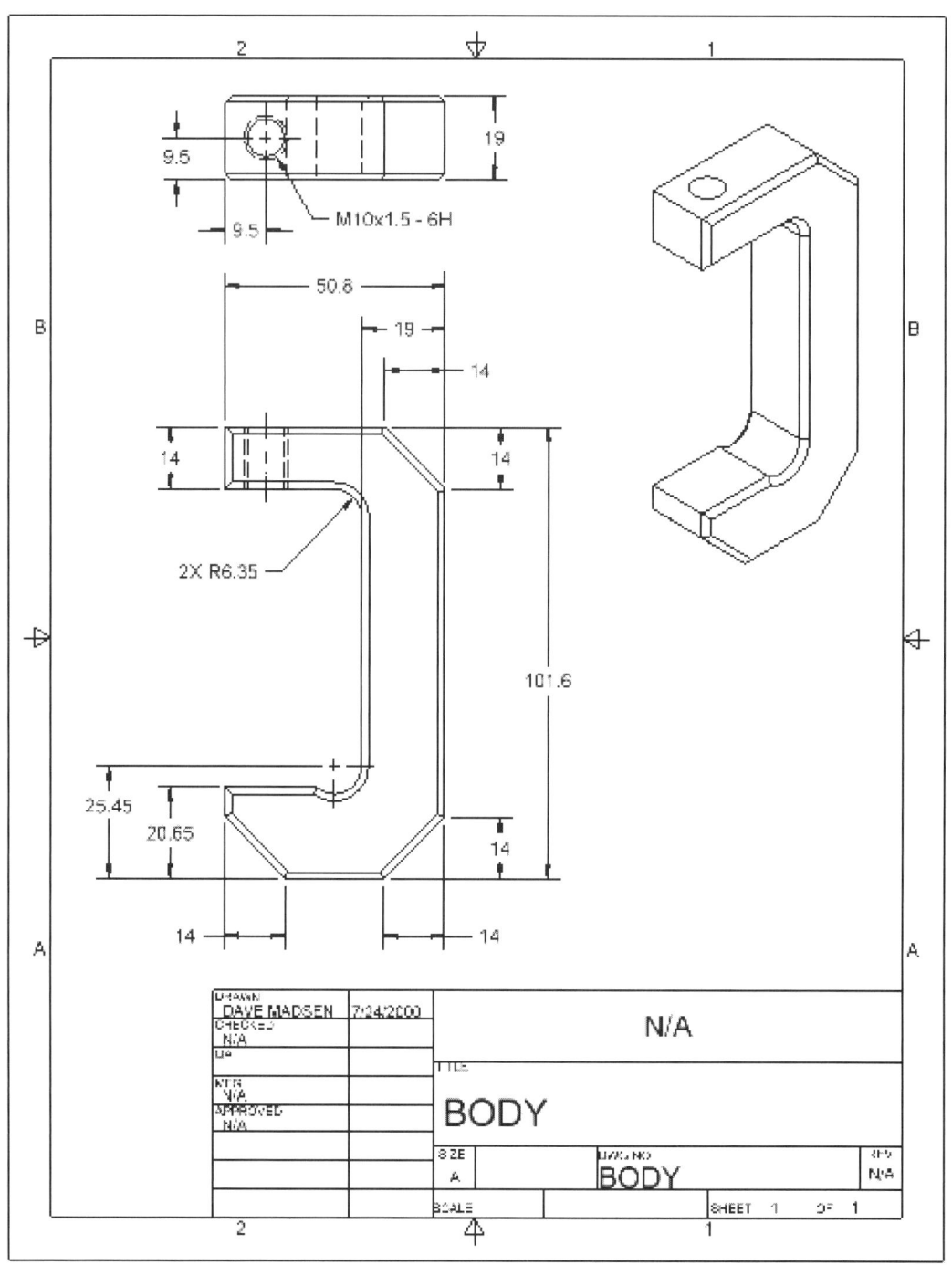

> **Field Notes**
>
> This chapter explores only the creation of part drawings. All the tools and options discussed in this chapter are required for creating additional part drawings and assembly drawings. For specific information regarding assembly drawings, multiple drawing sheets, and parts lists, refer to Chapter 15.

Working in the Part Drawing Environment

The work environment and interface are different for each stage of the design process. Some of the tools and interface components remain constant, or are similar throughout the entire design process. However, the part drawing environment and many of the drawing interface components are specific to working with drawings.

By default, when you open a new Autodesk Inventor drawing the work environment should look like the display in Figure 12.2. A sheet, border, and title block are automatically inserted into the drawing. However, with the drawing interface options, you have full control over the display of the drawing and can modify any drawing information as required. Throughout this chapter you will explore specific tools, commands, and options that allow you to develop a part drawing, complete with a border, title block, and multiple views.

Using the Command Bar for Drawings

In previous chapters you have become familiar with the Autodesk Inventor command bar, shown in Figure 12.3. Some of the command bar options are the same, or similar, when working in drawings as they are when working with parts. However, there are also some differences. Each of the command bar tools available while in the drawing environment is discussed as follows:

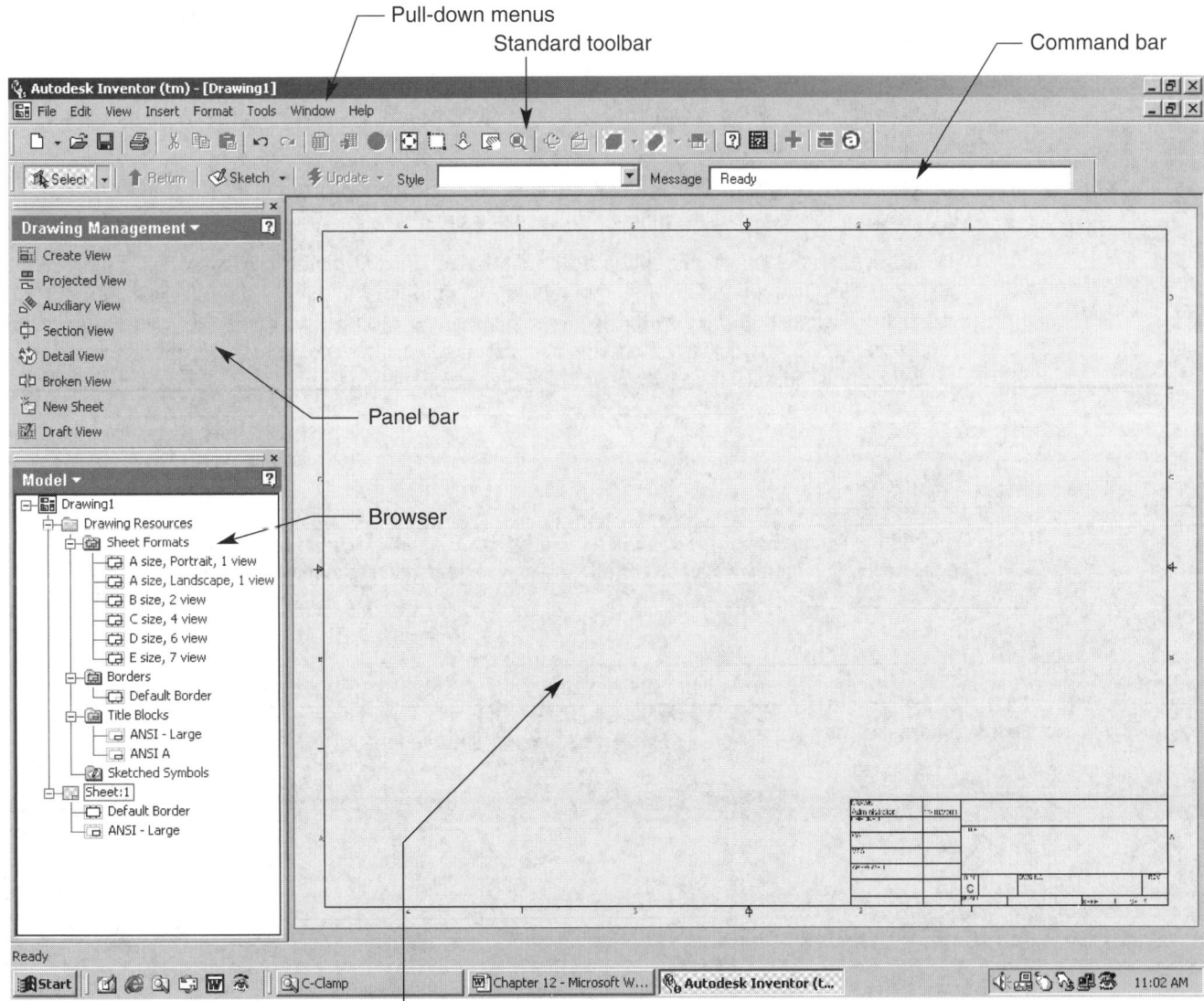

FIGURE 12.2   The default Autodesk Inventor drawing work environment.

Return button   Update button   Style drop-down list   Message display box

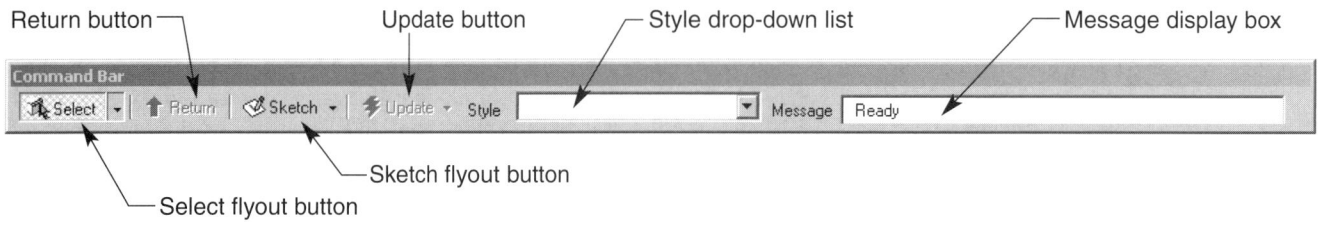

Sketch flyout button

Select flyout button

FIGURE 12.3   The command bar.

- **Select**   This flyout button allows you to specify which objects you want to select when working in a drawing. By default, select priority is set to **Edge,** which allows you to pick part edges. You can choose the **Feature** option in order to select features, or **Part** to select entire parts.
  In addition, the drawing **Select** flyout button contains filter options. Filters allow you to further specify which drawing components are selectable. The following filter options are available from the **Select** flyout button:

  - **No Filters**   When you choose this option, no filters are used, and you will be able to select any elements in the drawing environment.

  - **Layout**   When you choose this option, you will be able to select only layout information, including sheets and views.

  - **Detail**   This option will allow you to select detail information such as lines, dimensions, and text. However, you will not be able to select other drawing components such as symbols, sheets, and views.

  - **Custom**   This option allows you to select the elements of a drawing specified in the **Custom Filters** set of the **Select Filters** dialog box.

  - **Filters...**   Selecting this option opens the **Select Filters** dialog box, shown in Figure 12.4. This dialog box allows you to define specific filter options for the filter sets previously discussed. To modify the specified filter sets choose the desired filter set from the **Filter Set** drop-down list. Then select or deselect the filters you want to use or remove. When finished, pick the **OK** button to apply the filter options.

Field Notes

Typically the default filters specified in the **No Filters, Layout, Detail,** and **Custom** filter sets of the **Select** flyout button are adequate. Still, you may want to adjust them to fit your specific needs.

FIGURE 12.4   The **Select Filters** dialog box.

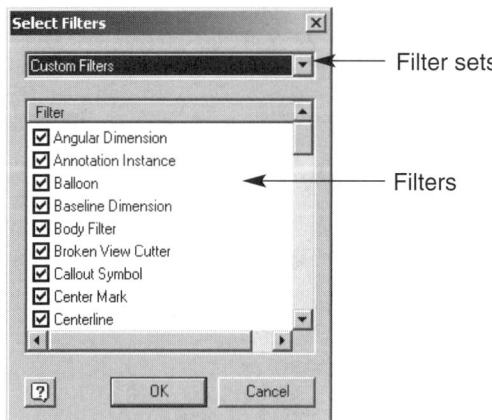

Filter sets

Filters

- **Return**　Picking this button finishes a sketch and takes you from sketch mode to drawing management mode.

- **Sketch**　Select this button to activate a new sketch or draft as it is referred to in the drawing environment, and enter sketch mode. When complete pick the sketch flyout button to finish the sketch.

- **Update**　The Update allows you to regenerate the current drawing and its dependent children.

- **Style**　Use this drop-down list to change the line style of sketch geometry or annotation style. You may choose to specify a line or annotation style and then draw objects and create annotations, or select existing objects and pick the style you would like to change.

- **Message**　The message display box is similar to the status bar discussed in Chapter 2, and provides you with prompts and drawing information.

Using the Browser in a Drawing

The browser for a typical part drawing displays the drawing icon and drawing name, followed by the **Drawing Resources** folder and then each of the sheets. Figure 12.5 shows the browser for a c-clamp part drawing named BODY.ipt.

A *sheet* is the base of a drawing and represents the physical limits of the drawing area or the paper size. Think of the sheet as the piece of paper on which you are drawing. Drawing sheets contain all the drawing information and components, including a border, a title bock, and one or more drawing views. As mentioned, a monodetail drawing will have only one part drawing on one sheet, while a multidetail drawing will have numerous part drawings on a single sheet. In addition, special symbols, dimensions, and other annotations displayed on the drawing are located on the **Sheet.**

By default, there is only one sheet in a drawing file. If you are working with a single part, you may need only one sheet. However, when you generate more than one part, each of which is a separate assembly component, you may choose to create several sheets in a single drawing file. For example, a drawing file may contain a sheet with an assembly drawing, and number of sheets that contain the individual part details.

FIGURE 12.5　The browser display for a typical part drawing.

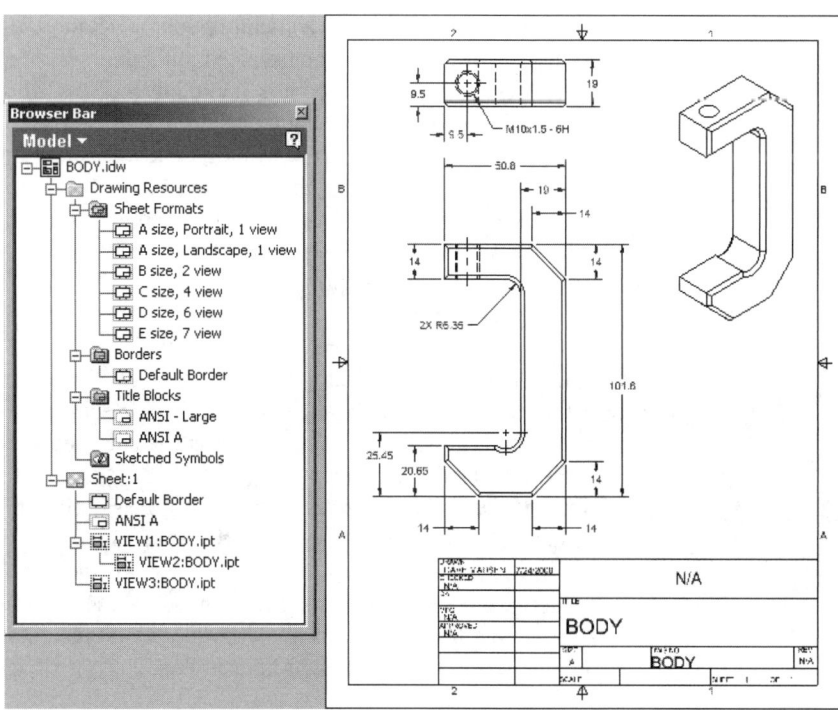

> For specific information regarding assembly drawings, multiple drawing sheets, and parts lists, refer to Chapter 17
>
>   Additional part drawing browser options are discussed throughout this chapter.

Exercise 12.1

1. Launch Autodesk Inventor.
2. Open a new inch or metric drawing file.
3. Explore the basic drawing file interface options and work environment as previously discussed.
4. Exit Autodesk Inventor without saving, or continue working with the program.

Defining Drafting Standards

Drafting standards allow you to control every aspect of the drawing display including dimension style, text style, sheet display, and multiple other drafting elements. You can modify the drafting standards at any time during the drawing creation process. However, before you begin a drawing, you may want to become familiar with the drafting standard options, or adjust the drafting standards to fit your specific application.

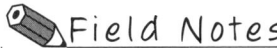

> Drafting standards are very important in order to make a drawing look exactly the way you want it to look and to capture design intent. However, if you are not concerned about defining drafting standards at this time, you may choose to skip ahead to the next section of this chapter.

To explore the drafting standard options, modify existing drafting standards, or create new drafting standards, access the **Drafting Standards** command by selecting **Standards...** from the **Format** pull-down menu. Selecting the **Standards...** menu option opens the **Drafting Standards** dialog box, shown in Figure 12.6. When the **More** button is not selected, only the **Select Standard** area is displayed. This area contains a number of default drafting standards including ANSI, BSI, DIN, GB, ISO, JIS. Each of these standards includes a multiple number of default drafting specifications, set according to the particular standard.

To select and use one of the default standards, pick the radio button next to the standard. When the **More** button is selected, displaying the standard tabs, as shown in Figure 12.6, the options in the tabs will change when you select a different default standard. You can modify any of the specifications in the tabs, but if you want to revert to the default settings, right-click on the standard in the **Select Standard** area, and pick the **Set Defaults** menu option.

If you plan on changing a number of the default drafting standard settings or want to specify your own standard with a particular name, pick the **Click to add new standard** button. When you select this button, the **New Standard** dialog box appears. See Figure 12.7. To create a new standard, enter the desired name of the standard in the **Name** edit box, then select the standard you want to base your standard on from the **Based On** drop-down list. The standards in the **Based On** drop-down list correspond to the standards in the **Select Standard** area. As a result, initially you are creating a copy of the selected standard. Once you enter the desired name and select the base standard, pick the **OK** button to create the new standard. The new standard should be displayed in the **Standard** list and be currently active. You are now ready to modify the settings in the standard tabs to meet your specifications.

You can modify the information in the **Drafting Standards** dialog box at any time. You may want to adjust the setting before beginning a drawing, or to create a template. Tabs that relate to specific tools and drawing applications are discussed later in this chapter, Chapter 13, and Chapter 15. However, the options

FIGURE 12.6 The **Drafting Standards** dialog box (**Common** tab displayed).

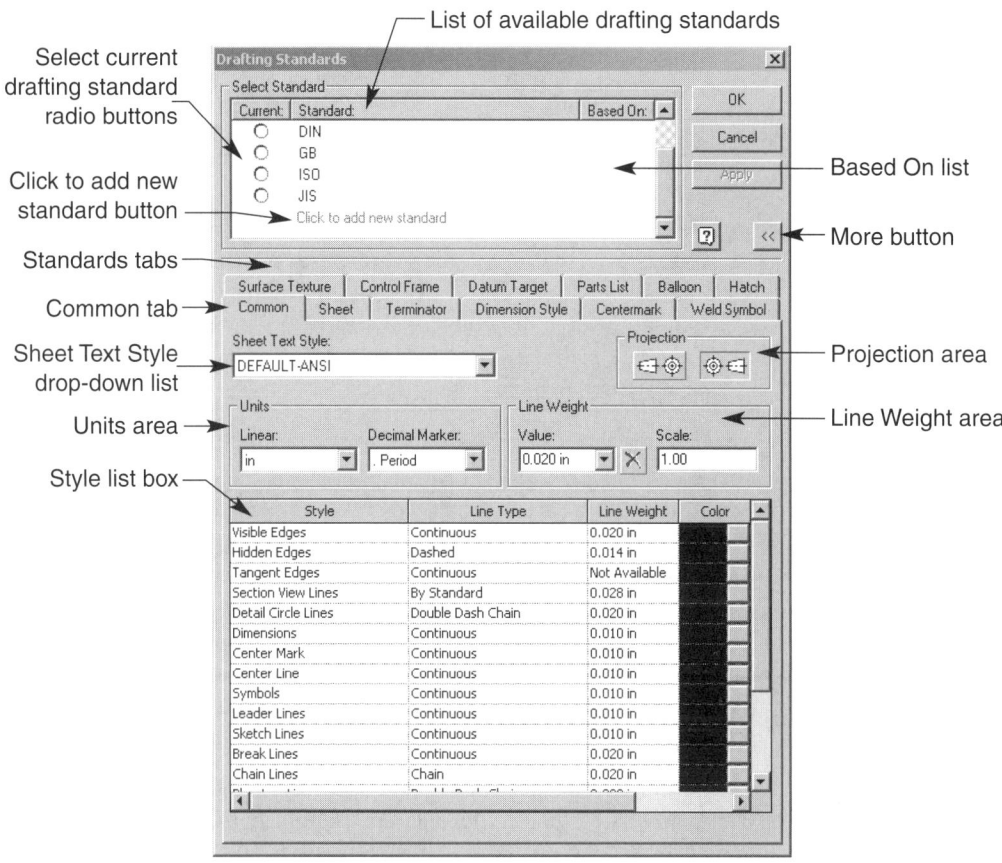

FIGURE 12.7 The **New Standard** dialog box.

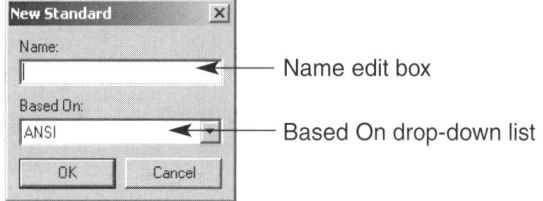

available in the following **Drafting Standards** dialog box tabs are more general than the options located in the tabs that control specific tool or application functions:

- **Common**　This tab, shown in Figure 12.6, contains the following general drawing settings:

  - **Sheet Text Style**　Use this drop-down list to set the default text style for the sheet. The default text style is not related to dimension text styles, and can be overridden when you generate text. You can create your own text style using the **Text Styles** dialog box, discussed later in this chapter.

  - **Projection**　This area allows you to specify drawing view projection angles. Pick the **First angle of projection** button to project a first angle, or pick the **Third angle of projection** button to project the third angle.

  - **Units**　Use this area to set the units for the drawing. Specify the type of unit measurement, such as inch or mm, from the **Linear** drop-down list. You can also define the type of decimal marker, such as a decimal point or comma, from the **Decimal Marker** drop-down list.

  - **Line Weight**　This area is used to define the line weights available for line styles and set the line scale factor. Use the value drop-down list to display the line weights currently available, or enter a new line weight to be used in the drawing. If you want to delete a line weight, select the one you want to remove and pick the **Delete** button. Use the **Scale** edit box to define a new line scale factor. This scale is applied to all existing line styles.

■ **Styles**   This area contains all the available line styles in the drawing, and allows you to modify line style parameters. You can change the line type by selecting the default line type in the **Line Type** list box and choosing a different line type from the drop-down list. Similarly, you can modify the line weight by selecting the default line weight in the **Line Weight** list box, and choosing a different line weight from the drop-down list. The line weights available in the drop-down lists correspond to those in the **Line Weight Value** drop-down list previously discussed. By default, all of the line styles are black in color. You can change the color of line style by picking the **Color** button next to the color. When you pick the **Color** button, the **Color** dialog box appears, allowing you to specify the desired line color.

■ **Sheet**   This tab, shown in Figure 12.8, allows you to adjust the sheet labels and display colors. *Sheet labels* define the names of the drawing elements displayed in the **Browser** bar. By default, the sheet label is **Sheet,** the view label is **VIEW,** and the draft view label is **DRAFT.** The **Colors** area allows you to define the colors used in the drawing environment. Sheet sets the color of the sheet, or "paper"; **Sheet Outline** specifies the color of the sheet edges; and **Highlight** sets the color of drawing elements that can be selected (this color is visible when you move your cursor over selectable objects). Finally, **Selection** specifies the color of drawing elements that have been selected and are currently active. Each of the drawing element colors can be modified by picking the color button next to the element. When you pick the **Color** button, the **Color** dialog box appears, allowing you to specify the desired line color.

■ **Terminator**   This tab, shown in Figure 12.9, allows you to adjust the terminator and arrowhead options for certain leaders and datum identifier terminators. The following options define leader terminator specifications used in creating hole/thread notes, surface texture symbols, weld symbols, feature control frames, feature identifier symbols, datum identifier symbols, text leaders, datum target leaders, and balloon leaders only:

■ **General**   Use this drop-down list to define the type of arrowhead you want to use.

■ **Size**   Enter a value in this edit box to define the size of the terminator. For arrowheads, the size represents the arrowhead length. For circles, the size represents the circle diameter.

■ **Aspect**   Enter a value in this edit box to define the aspect, or width, in reference to the size of the arrowhead. For example, to create a terminator that is three times as long as it is wide, the aspect should be one third the value of the size.

■ **Datum**   Use this drop-down list to define the datum identifier terminator style.

FIGURE 12.8   The **Sheet** tab of the **Drafting Standards** dialog box.

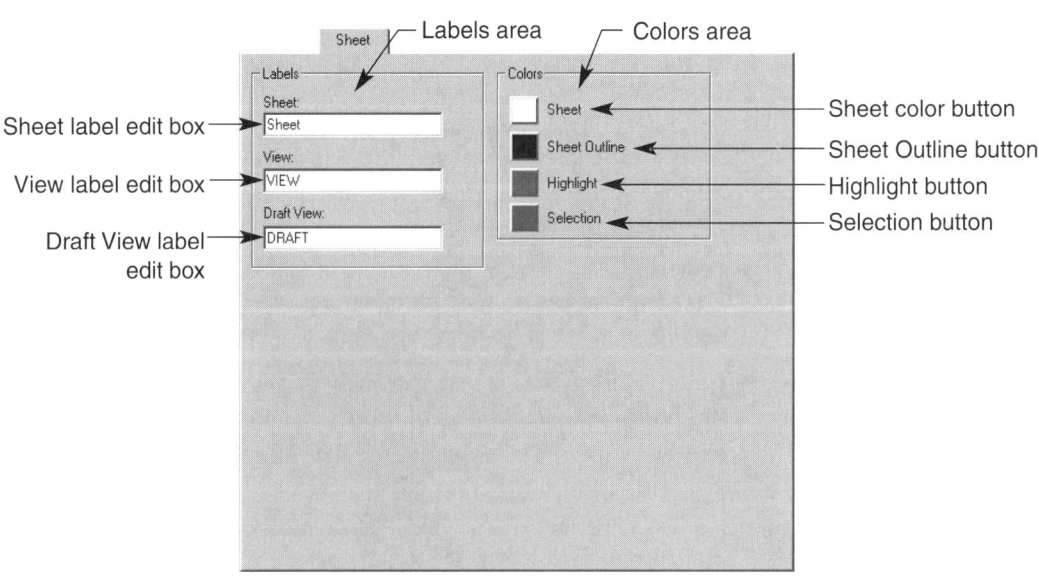

FIGURE 12.9 The **Terminator** tab of the **Drafting Standards** dialog box.

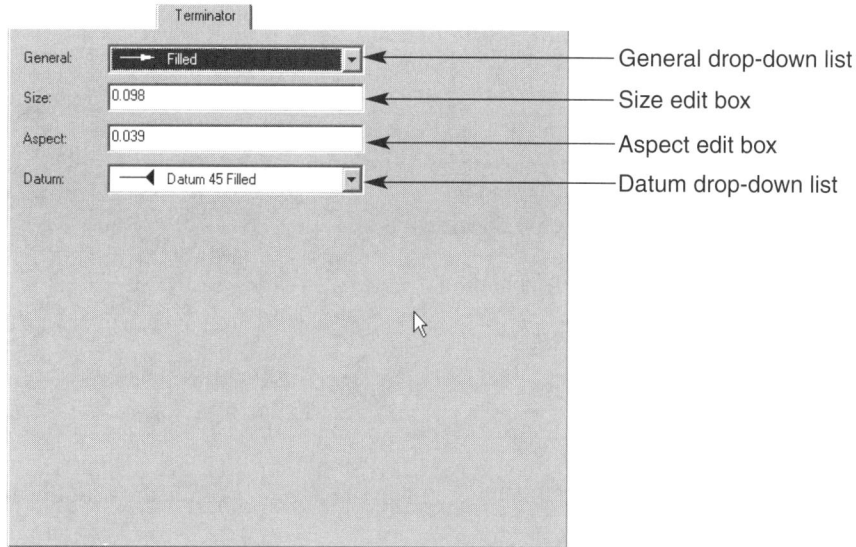

Terminator

General: ▬► Filled ▼ ◄──── General drop-down list

Size: 0.098 ◄──── Size edit box

Aspect: 0.039 ◄──── Aspect edit box

Datum: ▬◄ Datum 45 Filled ▼ ◄──── Datum drop-down list

✏ Field Notes

Linear, angular, aligned, diameter, radial, and ordinate dimension terminators, which are controlled by dimension styles, are not affected by the changes in the **Terminator** tab. Dimension styles and the **Dimension Styles** dialog box are discussed later in this chapter.

Exercise 12.2

1. Continue from EX12.1, or launch Autodesk Inventor.
2. Open a new inch drawing file.
3. This exercise begins the process of making a custom drawing template. For this exercise, save the file as EX12.2. When in a work setting, you will want to save the file in the **Templates** folder located in the following directory: Program Files/Autodesk/Inventor 5/Templates/Metric. You will want to name the file—"ANSI (mm)," for example.
4. From the **Format** pull-down menu, pick the **Standards...** option to access the **Drafting Standards** dialog box.
5. Review the options and tabs available in the **Drafting Standards** dialog box as discussed.
6. Pick the **Click to add new standard** button, and name the standard **ANSI (mm)** based on the **ANSI** standard.
7. In the **Common** tab, change the linear units to mm.
8. In the **Sheet** tab, change the sheet color to white.
9. In the **Terminator** tab, change the size of the terminator to 3 and the aspect to 1.
10. Resave EX12.2.
11. Exit Autodesk Inventor or continue working with the program.

Specifying Drawing Sheet Parameters

Depending on your approach, the first step in developing a part drawing is to define the sheet parameters, including the sheet name, size, and layout. The sheet size is the physical limits of the sheet, or the size of the paper. When you open a new Autodesk Inventor drawing file for the first time, a sheet named **Sheet:1** is available. The size of the sheet depends on the template you open. For example, if you open ANSI [in].idw the default sheet size is C, or 11″×17″. If you want to use the default sheet parameters, you do not have to make any modifications and can move on to the next step of drawing development. However, if you want to redefine the current sheet information, right-click the sheet in the **Browser** and select the **Edit Sheet...** menu option to access the **Edit Sheet** dialog box. See Figure 12.10.

FIGURE 12.10   The
Edit Sheet dialog box.

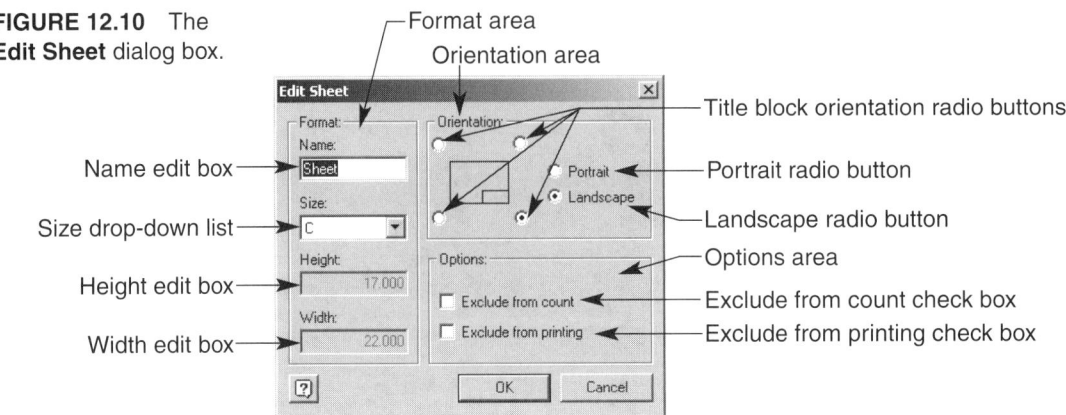

The following options are available inside the **Edit Sheet** dialog box:

- **Name**   This edit box allows you to rename the sheet by highlighting the current name and typing a new name. The name you enter is followed by a colon and the sheet number. For example, if the sheet you are editing is the first sheet in the drawing, the name is followed by **:1.**

- **Size**   This drop-down list allows you to specify the sheet size. There are a number of standard sheet sizes available such as A, B, C, D, E, F, or you can define your own sheet size by selecting the **Custom Size** option for the desired units (**Custom Size [inches]** or **Custom Size [mm]**). Once you choose a custom size option, the **Height** and **Width** edit boxes become active, allowing you enter the desired height and width of the custom sheet size.

- **Title Block Orientation**   Title block orientation is located in the **Orientation** area and consists of four radio buttons. The radio buttons correspond to the corner where the title block is placed. To change the orientation of the title block, pick the desired radio button. A preview of the action is displayed.

Field Notes

The title block orientation radio buttons are available only when there is an existing title block on the sheet.

- **Portrait**   This radio button allows you to apply a portrait, or vertical sheet, orientation.

- **Landscape**   This radio button allows you to apply a landscape, or horizontal, sheet orientation.

- **Exclude from count**   When this check box is selected, the sheet is excluded from the count of sheets in the drawing. A sheet that is excluded from the count does not have a number attached to its name.

Field Notes

If only one sheet is present, the sheet number is always 1, unless you exclude the sheet from the count.
   If more than one sheet is present in the drawing you can drag sheets above or below other sheets to modify the sheet number.

- **Exclude from printing**   When this check box selected, the sheet is excluded from the printing process and is not printed with the other drawing sheets.

Once you have modified the desired options, pick the **OK** button to complete the sheet edit.

Exercise 12.3

1. Continue from EX12.2, or launch Autodesk Inventor.
2. Open EX12.2.
3. Save a copy of EX12.2 as EX12.3.idw.
4. Close EX12.2 without saving, and open EX12.3.
5. Right-click on **Sheet:1** in the **Browser** and select the **Edit Sheet...** menu option.
6. Change the sheet name to **BODY.**
7. Change the sheet size to **A4.**
8. Select the **Portrait** orientation radio button.
9. The final drawing should look like the drawing shown.
10. Resave EX12.3.
11. Exit Autodesk Inventor or continue working with the program.

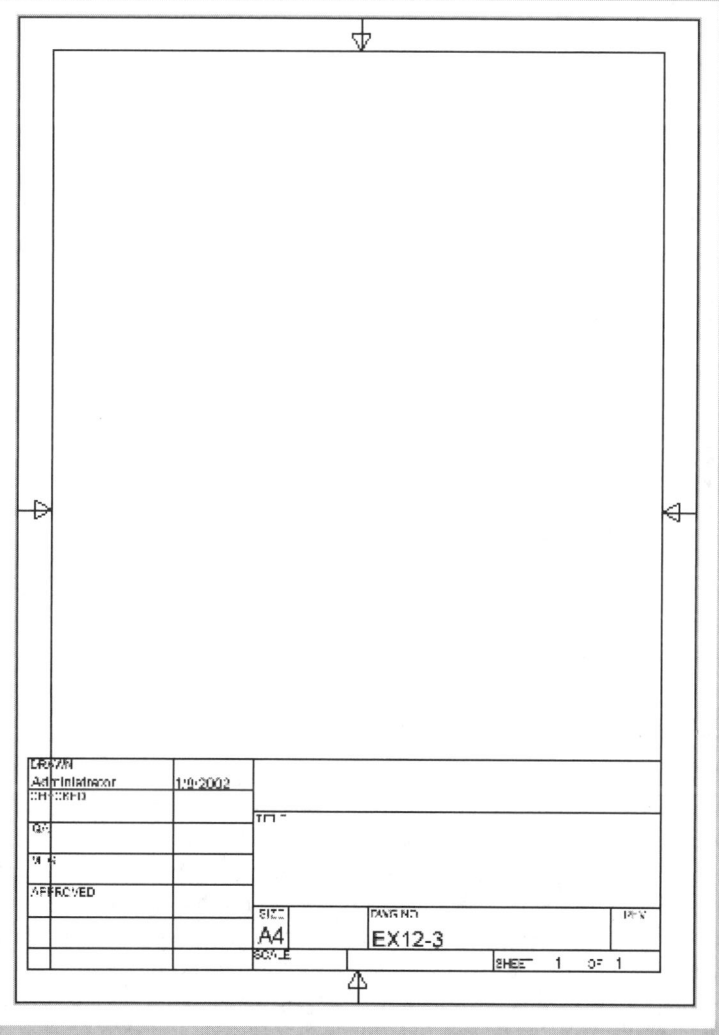

Exercise 12.4

1. Continue from EX12.3, or launch Autodesk Inventor.
2. Open a new inch drawing file.
3. Right-click on **Sheet:1** in the **Browser** and select the **Edit Sheet...** menu option.
4. Change the sheet name to **D SIZE.**
5. Change the sheet size to **D.**
6. The final drawing should look like the drawing shown.
7. Save the drawing as EX12.4.
8. Exit Autodesk Inventor or continue working with the program.

Using Default Drawing Borders

Once you have defined the sheet name, size, orientation, and other parameters, depending on your approach, the next step in developing a part drawing is to define a border. A border can be as simple as a rectangle inside the sheet limits, or it may include zone numbers and center marks. See Figure 12.11. When you open an Autodesk Inventor drawing file for the first time, a border is already placed on the sheet. If you want to use the default sheet border, you do not have to make any modifications and can move on to the next step of drawing development. However, if you want to redefine the current border on **Sheet:1,** you must first remove the existing border by right-clicking the default border and selecting the **Delete** menu option, or by picking the default border and pressing the delete key on your keyboard. Then, right-click the **Default Border** located in the **Borders** folder of the **Drawing Resources** folder, and select the **Insert Drawing Border...** option to access the **Default Drawing Border Parameters** dialog box. See Figure 12.12. The **Default Drawing Border Parameters** dialog box contains the following options that allow you to create a specific border (refer to Figure 12.13 for further information regarding default border options):

- **Horizontal Zones**   This area allows you to define the horizontal border zones applied to the top and bottom of the border. Specify the number of zones you want to create in the **Number of Zones** edit box. Then pick the type of zone label by selecting the **Alphabetical** radio button to use letters, or the **Numerical** radio button to use numbers. You can also suppress the horizontal zones by choosing the **None** radio button.

FIGURE 12.11 An example of a drawing border.

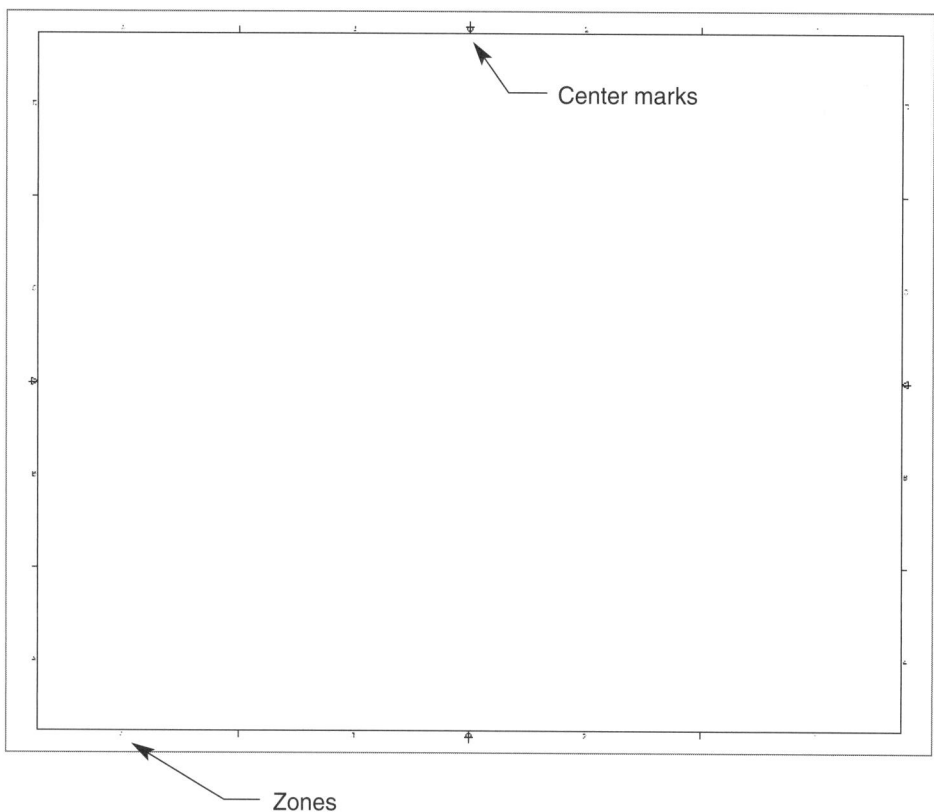

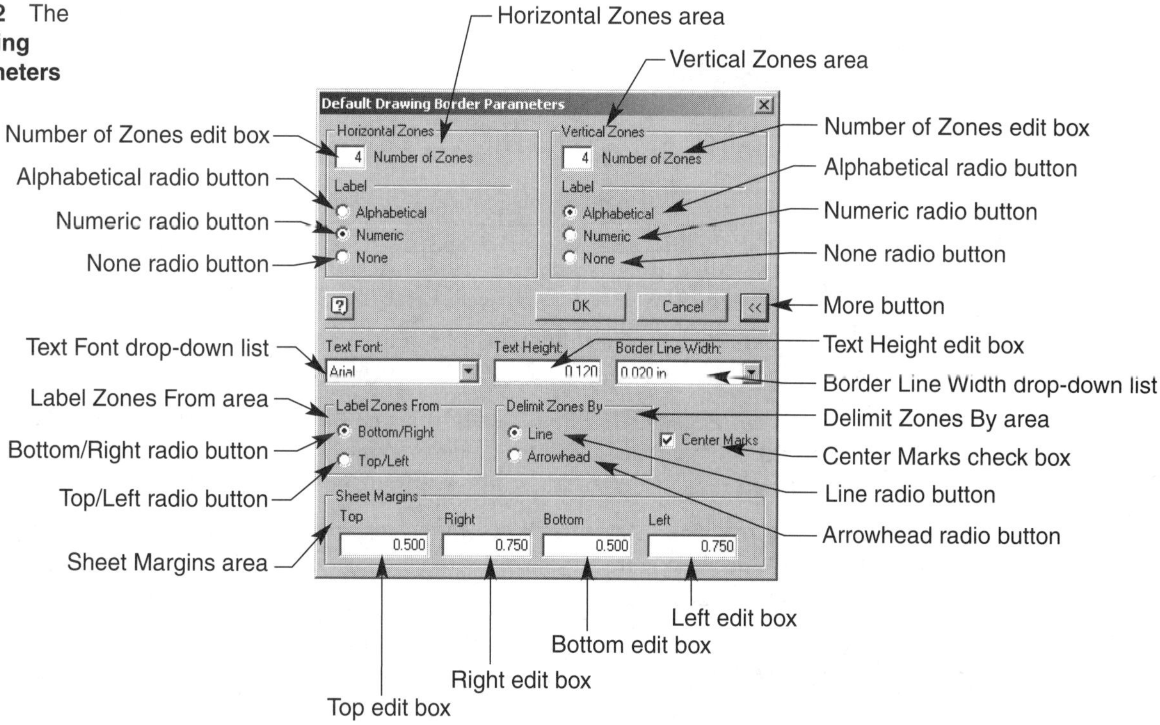

FIGURE 12.12 The **Default Drawing Border Parameters** dialog box.

- **Vertical Zones**   This area allows you to define the vertical border zones applied to the left and right sides of the border. Specify the number of zones you want to create in the **Number of Zones** edit box. Then pick the type of zone label by selecting the **Alphabetical** radio button to use letters, or the **Numerical** radio button to use numbers. You can also suppress the vertical zones by choosing the **None** radio button.

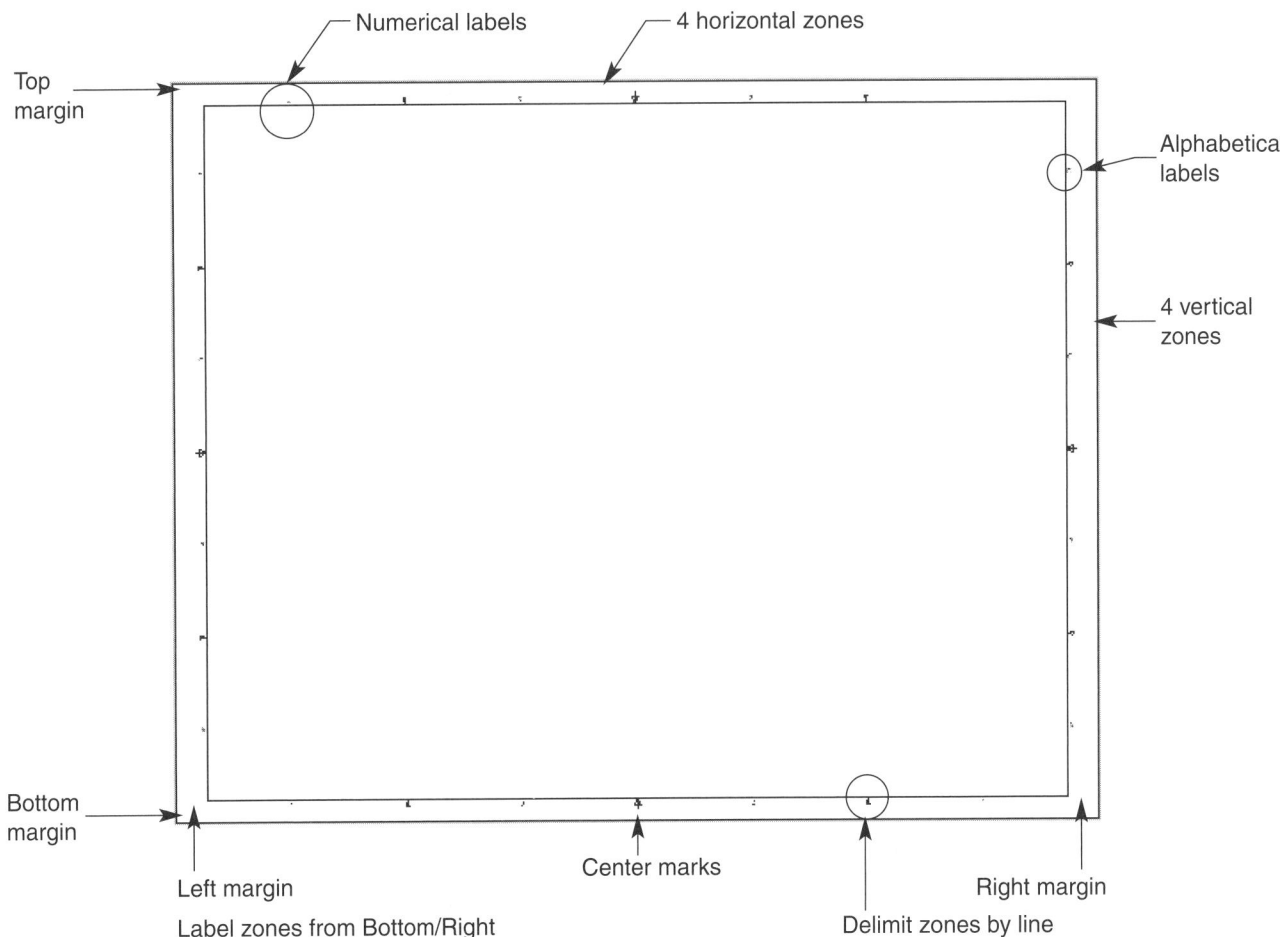

FIGURE 12.13 Default border parameters.

- **Text Font**   Use this drop-down list to specify the zone label text font type.

- **Text Height**   Use this edit box to define the height of the zone label text.

- **Border Line Width**   This drop-down list allows you to set the width, or thickness, of the border line.

- **Label Zones From**   This area is used to define the zone label justification. Pick the **Bottom/Right** radio button to label the zones from the bottom/right corner of the border, or pick the **Top/Left** radio button to label the zones from the top/left corner of the border.

- **Delimit Zones By**   This area allows you to specify how you want to mark the zone limits. Pick the **Line** radio button to mark the zone limits with small lines, or pick the **Arrowhead** radio button to mark the zone limits with an arrowhead.

- **Center Marks**   By default, this check box is selected. If you to not want to display center marks deselect the **Center Marks** check box.

- **Sheet Margins**   This area is used to define the top, right, bottom, and left border margins. The margins define the placement of the border in reference to the sheet edges.

Once you have fully defined the default border parameters, pick the **OK** button to insert the border onto the current sheet.

Exercise 12.5

1. Continue from EX12.4, or launch Autodesk Inventor.
2. Open EX12.4.
3. Save a copy of EX12.4 as EX12.5.idw.
4. Close EX12.4 without saving, and open EX12.5.
5. Right-click on **Default Border** in the **Browser** and select the **Delete** menu option.
6. Right-click on **ANSI-Large** in the **Browser** and select the **Delete** menu option.
7. Access the **Default Drawing Border Parameters** dialog box by right-clicking **Default Border** in the **Borders** folder of the **Drawing Resources** folder, and selecting the **Insert Drawing Border...** menu option.
8. Change the following border parameters:

   Number of Horizontal Zones: 12
   Number of Vertical Zones: 8
   Text Font: RomanS
   Text Height: .24

9. The final drawing should look like the drawing shown.
10. Resave EX12.5.
11. Exit Autodesk Inventor or continue working with the program.

Creating Your Own Drawing Border

The default borders created using your own specific parameters can be useful for many applications. Still, you may want to create your own borders using sketching tools. To develop a custom border, you may first want to delete any existing borders currently in the sheet. Then, select the **Define New Border** option from the **Format** pull-down menu. When you pick the **Define New Border** menu option, the panel bar automatically enters **Sketch** mode and the tools in **Sketch** toolbar become active. See Figure 12.14.

Sketch tools allow you to generate your own border design, title block, or drawing view. Most of the sketch tools available in a drawing file, such as line, circle, and mirror, function exactly the same as those in a part file. For information regarding these tools refer to Chapter 3. However, the following sketch tools are available only for use in drawings:

FIGURE 12.14 (A)
The **Panel** bar in sketch
(learning) mode. (B)
The **Sketch** toolbar.

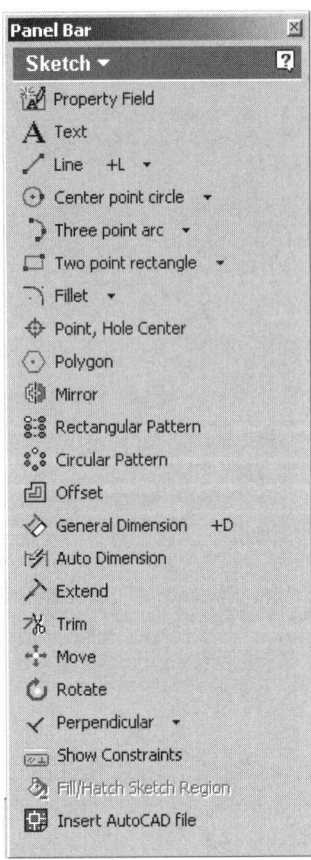

A

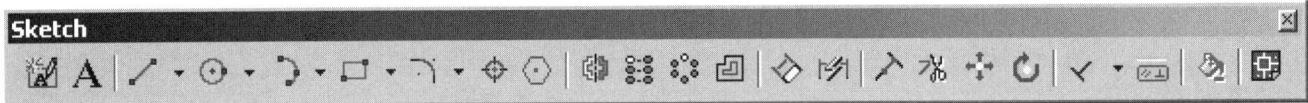

B

- **Property Field**   This tool allows you to add a property, corresponding to one of the properties in the **Properties** dialog box, a static value, or a prompted text entry, to the border. You can also add property fields to custom title blocks or symbols. When the border, or other drawing element, that contains a property field is inserted on a sheet, the property or prompted text created using the **Property Field** tool is also inserted.

    Property fields consist of insertable text and text symbols, controlled by text styles. You can use the default text styles and override some of the format options in the **Property Field** dialog box. However, you may want to create your own text styles to use when developing property fields, drawing text, and drawings dimensions. To create a new text style or modify an existing text style, select the **Text Styles...** option from the **Format** pull-down menu. When you select the **Text Styles...** menu option, the **Text Styles** dialog is displayed. See Figure 12.15.

    To modify an existing text style, pick the text standard you want to modify from the **Standard** pull-down menu. Then select the style you want to change from the **Style Name** list box. If you select the default text style you will not be able to modify any of the text parameters. However, you can change the text options of the current text style, which is a copy of the default text style. You can also define your own text style by picking the **New** button. When you select the **New** button, a copy of the active text style is created, and the new name, Copy of Default-ANSI, for example, is highlighted. When the name is highlighted you can enter a different name and press the **Enter** key on your keyboard to accept the new style name.

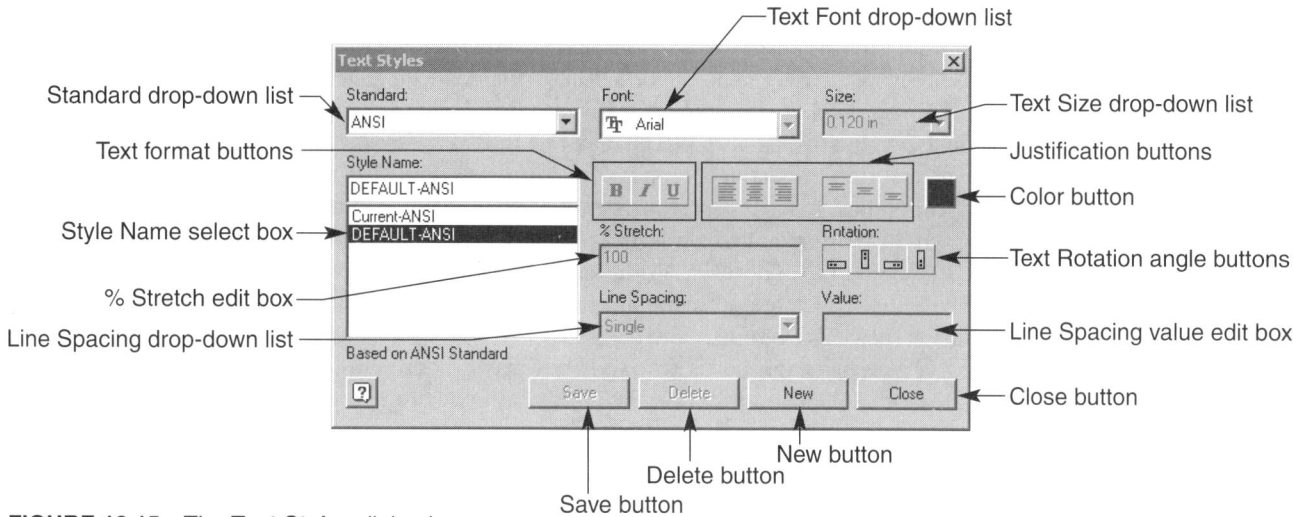

FIGURE 12.15   The **Text Styles** dialog box.

Once the new style is selected, use the following properties to create a custom text style:

- **Font**   Use this drop-down list to define the font type for the new text style.

- **Size**   This drop-down list allows you to select or enter a value for the size, or height, of the text.

- **Format**   These buttons define the display options of the text. Pick the **Bold** button to create bold text, the **Italic** button to create italicized text, and the **Underlined** button to underline text.

- **Justification**   These buttons allow you to control the left, center, right, top, middle, and bottom justification, or placement of the text.

- **Color**   Pick this button to display the **Color** dialog box and define the color of the text.

- **% Stretch**   This edit box allows you to define the amount of stretch, or width, of the text. To create a narrower text, enter a percent stretch less than 100, and to create a wider text, enter a percent stretch greater than 100.

- **Rotation Angle**   Use these buttons to define the rotation of the text. A 0° rotation generates text horizontally from left to right, a 270° rotation generates text vertically from top to bottom, a 180° rotation generates text horizontally from right to left, and a 90° rotation generates text vertically from bottom to top.

- **Line Spacing**   This drop-down list allows you to specify the distance between multiple lines of text. When you select the **Multiple** option, the **Value** edit box becomes active, allowing you to define a smaller or greater spacing ratio, such as triple, or half. When you select the **Exactly** option, the **Value** edit box becomes available, allowing you to enter a specific distance between multiple lines of text.

When you have fully defined the text parameters for a new or existing text style, pick the **Save** button to save the changes. You can also remove text styles, by picking the style, and selecting the **Delete** button. When you are finished, pick the **Close** button to exit the **Text Styles** dialog box.

To create a property field, access the **Property Field** command using one of the following techniques:

✓ Pick the **Property Field** button on the panel bar.

✓ Pick the **Property Field** button on the **Sketch** toolbar.

✓ Right-click an existing, unsaved property field and select the **Edit** menu option.

Once you access the **Property Field** command, use the positioning cursor to select a point, or hold down the left mouse button and drag the cursor to create a window on the sheet. See Figure 12.16. The point you select, or window you create, defines the location and extents of the property field. When you have located the property field on the sheet, the **Format Field Text** dialog box is displayed. See Figure 12.17.

The first step in creating a property field is to select the text style you want to use from the **Style** drop-down list. If you have previously created a custom text style or modified an existing style you may not need to change any of the text style parameters in the **Format Field Text** dialog box. However, all the text parameter options previously discussed in the **Text Styles** dialog box are available in the **Format Field Text** dialog box, allowing you to override the selected text style properties.

Once you have selected a text style and have modified the desired text properties, select the type of field you want to create from the **Type** drop-down list. The following field property types are available:

- **Model Properties**   These properties correspond to those you specify in the **Properties** dialog box, while in the model file. For example, when you insert a part model onto the sheet, those properties you specified while in the part file are available to be used in a property field. When this option is selected, the design properties available from the model are listed in the **Property** drop-down list.

- **Design Properties**   These properties correspond to those you specify in the **Properties** dialog box, for the current drawing file. When this option is selected, the design properties available in the drawing are listed in the **Property** drop-down list.

FIGURE 12.16
Selecting a property field location.

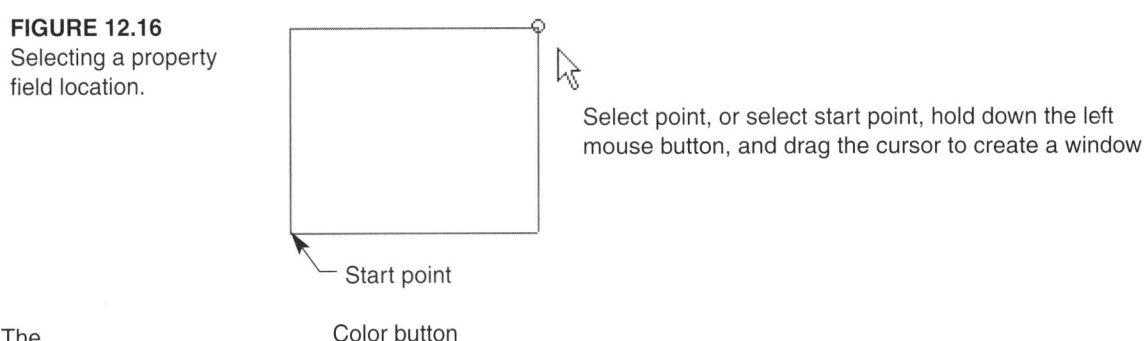

Select point, or select start point, hold down the left mouse button, and drag the cursor to create a window

Start point

FIGURE 12.17   The **Format Field Text** dialog box.

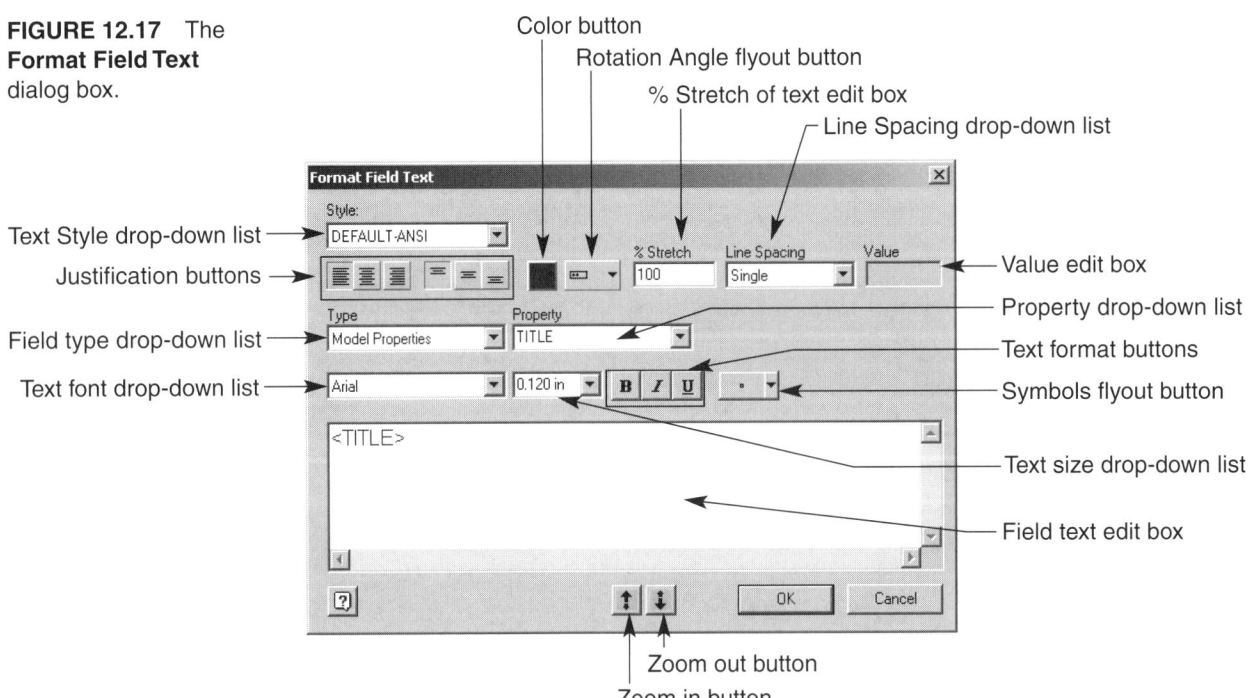

- **Custom Properties**   This option is only available when you have created a custom property in the **Properties** dialog box, for the current drawing file. If a custom property is available and you select the **Custom Properties** option, the custom design properties available in the drawing are listed in the **Property** drop-down list.

- **Drawing Properties**   These properties relate to elements of the current drawing, such as the number of sheets. When inserted, the value of the property field reflects the sheet number it is inserted on. For example, if you insert this property on sheet 1, the value of the property is 1.

- **Sheet Properties**   These properties are similar to drawing properties, because they relate to current drawing elements, such as sheet number and sheet size. When inserted, the value of the property field reflects the sheet number or sheet size it is inserted on. For example, if you insert this property on a C-size sheet, the value of the property is C.

- **Static Value**   This property is similar to placing text on a sheet and is typically used when you want to repeatedly insert the same piece of text or a static value. The text you place in the **Field Text** edit box is displayed when inserted.

- **Prompted Entry**   This option is used to insert a prompted text value on a sheet. Use this technique when you insert the property and want to be prompted for a particular value such as "What is the date?" To create a prompted entry, type the prompt in the **Field Text** edit box. When inserted, the **Field Entry** dialog box is displayed. See Figure 12.18. The **Field Entry** dialog box prompts you with the prompt you specified. **Enter** the desired value in the edit box and pick the **Check Mark** button to insert the value, or pick the **Delete** button if you do not want to insert a particular prompted value, but do want to insert the border, title block, or symbol.

Additional tools available in the **Format Field Text** dialog box include the **Symbols** flyout button and the **Zoom** buttons. To enter a text symbol in the property field, select the desired symbol from the **Symbols** flyout button. To zoom in on the text in the **Text** edit box, pick the **Zoom In** button. Similarly, to zoom out on the text, pick the **Zoom Out** button.

Field Notes

> Zooming in or out on the text in the text edit box does not change the physical size of the text or symbols.

Once you have fully defined the property field, pick the **OK** button to exit the **Format Field Text** dialog box. Continue placing property fields, or exit the **Property Field** command by right-clicking and selecting the **Done [ESC]** menu option, pressing the **ESC** key on your keyboard, or accessing a different command.

Field Notes

> Once you have generated a property field and before you save the border, title block, or symbol, you can access the **Format Field Text** dialog box and edit the property field by right-clicking the property field and selecting the **Edit** menu option.

FIGURE 12.18   The **Field Entry** dialog box.

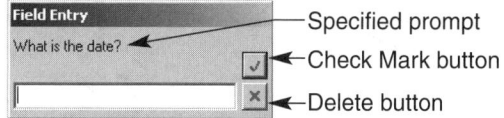

■ **Text**   This tool allows you to add text, such as drawing notes, to the border, which is not controlled by properties or insertable values. You can also add text to custom title blocks, symbols, or sketched drawing views.

Text properties are controlled by the text styles previously discussed. You can use the default text styles and override some of the format options in the **Format Text** dialog box. However, you may want to create your own text styles to use when placing text. Refer to the previous discussion about the **Text Styles...** command and the **Text Styles** dialog box for information regarding text styles.

To create text, access the **Text** command using one of the following techniques:

✓ Pick the **Text** button on the panel bar.

✓ Pick the **Text** button on the **Sketch** toolbar.

✓ Right-click existing, unsaved text and select the **Edit** menu option.

Once you access the **Text** command, use the positioning cursor to select a point, or hold down the left mouse button and drag the cursor to create a window on the sheet. See Figure 12.19. The point you select, or window you create, defines the location and extents of the text. When you have located where you want to place the text, the **Format Text** dialog box is displayed. See Figure 12.20.

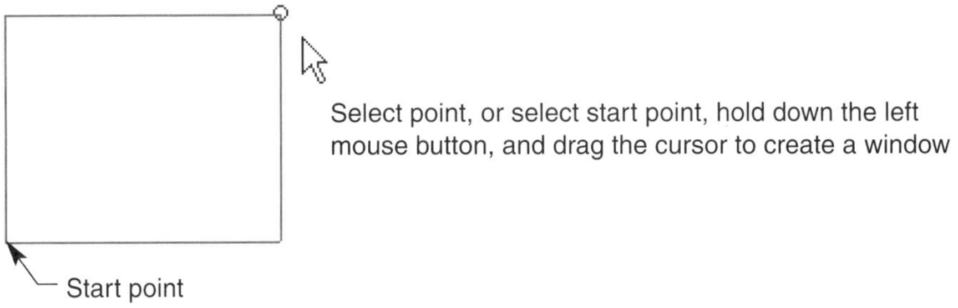

FIGURE 12.19
Selecting a text location.

Select point, or select start point, hold down the left mouse button, and drag the cursor to create a window

└ Start point

FIGURE 12.20   The **Format Text** dialog box.

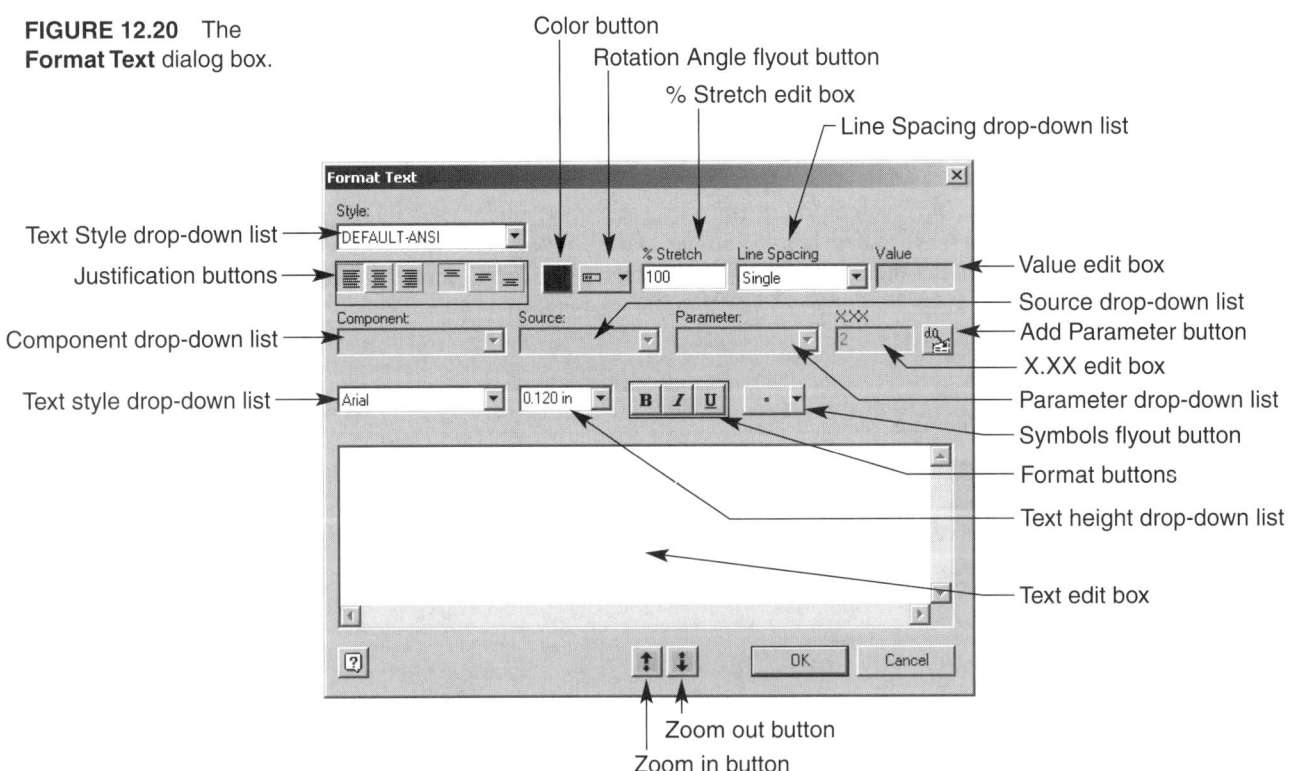

The first step in creating text is to select the text style you want to use from the **Style** drop-down list. If you have previously created a custom text style or modified an existing style you may not need to change any of the text style parameters in the **Format Text** dialog box. However, all the text properties previously discussed in the **Text Styles** dialog box are available in the **Format Text** dialog box, allowing you to override the selected text style properties.

Once you have selected a text style and have modified the desired text properties, enter the desired text in the **Text** edit box. You can specify a text symbol by selecting the desired symbol from the **Symbols** flyout button. You can also zoom in or out on the text in the **Text** edit box by picking the **Zoom In** or **Zoom Out** buttons. After you have fully defined the text, pick the **OK** button to exit the **Format Text** dialog box. Continue placing text, or exit the **Text** command by right-clicking and selecting the **Done [ESC]** menu option, pressing the **ESC** key on your keyboard, or accessing a different command.

Field Notes

Zooming in or out on the text in the text edit box does not change the physical size of the text or symbols.

You may have noticed the following options in the **Format Text** dialog box, which are available only when adding or editing specific model parameter text, such as dimensions or model notes:

- **Component**   This drop-down list contains the names of all the components, or models, currently inserted into the drawing. Only one component is typically available when creating a single part drawing.

- **Source**   This drop-down list specifies the type or source of parameters to use. **Model Parameters** are specific parameters that relate to the model. These parameters are added when you insert a model view or add additional model information, such as dimensions to the drawing. **User Parameters** are additional parameters added to the model. If you do not create any user parameters, you cannot use the **User Parameters** source option.

- **Parameter**   This drop-down list contains all the parameters available in the selected source. The parameters correspond to the parametric model or user parameters used to define model geometry.

- **X.XX**   Use this edit box to define the precision of the displayed value by entering the number of decimal places.

- **Add Parameter**   Once you have selected a parameter and fully defined the parameter options, pick this button to add the parameter to the text edit box.

After you have fully defined the text and parameters, pick the **OK** button to exit the **Format Text** dialog box. Continue placing text, or exit the **Text** command by right-clicking and selecting the **Done [ESC]** menu option, pressing the **ESC** key on your keyboard, or accessing a different command.

- **Fill/Hatch Sketch Region**   This tool allows you to add a solid color fill or a hatch pattern to a sketched region. See Figure 12.21. A sketch region must be available before you can access the **Hatch/Color Fill** command. A sketch region can consist of any closed-loop piece of geometry, such as a circle or a rectangle.

Before you fill or hatch a sketched region, you may want to adjust some of the hatch properties in the **Drafting Standards** dialog box. Access the **Drafting Standards** dialog box by picking the **Standards...** menu option from the **Format** pull-down menu. As shown

FIGURE 12.21 Filling or hatching a sketch region.

Filled sketched region

Hatched sketched region

FIGURE 12.22 The **Hatch** tab of the **Drafting Standards** dialog box.

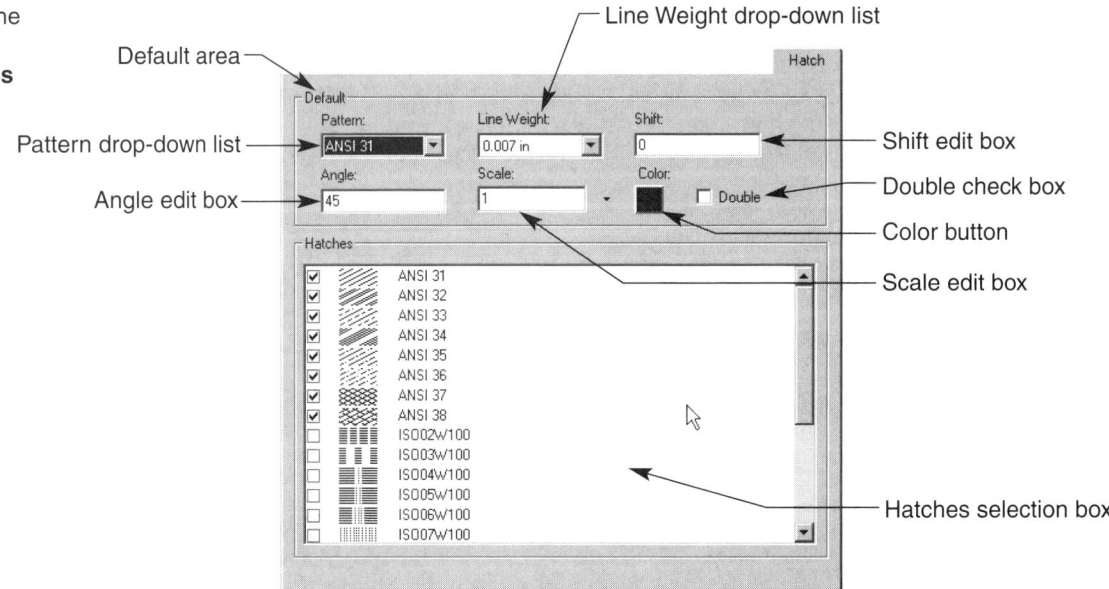

in Figure 12.22, hatch properties are located in the **Hatch** tab of the **Drafting Standards** dialog box. The following options in the **Hatch** tab control the default hatch pattern options when hatching a sketched region or a section view:

- **Hatches**   This selection box contains all the available hatch patterns. Hatch patterns that are selected are available when hatching a sketch region, or section, and can be modified. Select an additional hatch pattern for use by picking the check box next to the hatch name. Make a hatch pattern unusable by deselecting the check box.

- **Default**   This area allows you to define the default properties of the selected hatch pattern and contains the following options:

  - **Pattern**   Pick the hatch pattern you want to modify from this pull-down menu. Only hatch patterns selected in the **Hatch** area are available.

  - **Line Weight**   This drop-down list allows you to select a hatch pattern line weight.

  - **Shift**   Enter a value in this edit box to shift, or offset the hatch pattern, when hatching next to another hatched region. A shift does not align the hatch patterns with each other.

  - **Angle**   Use this edit box to define the angle of the hatch. For example, a hatch pattern that consists of horizontal lines has an angle of 0° or 180°.

  - **Scale**   This edit box defines the dimensions of hatch pattern. When you increase the scale, the density of hatch lines decreases. When you decrease the scale, the density of hatch lines increases. Enter the hatch pattern scale in the edit box, or pick the flyout button, and select a scale.

  - **Color**   This button opens the Color dialog box and allows you to specify the color of the hatch pattern.

  - **Double**   Select this check box to create a double, or mirrored, hatch pattern. See Figure 12.23.

FIGURE 12.23 (A) A single hatch pattern. (B) The same hatch pattern shown in (A) using the double option.

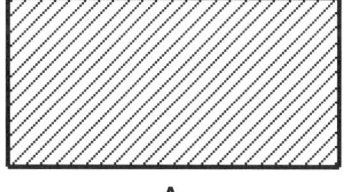

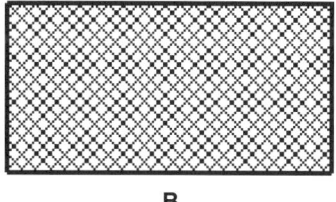

A   B

> The hatch properties defined in the Hatch tab of the **Drafting Standards** dialog box are default and can be overridden when you place hatching.

Once you have defined the default options in the **Hatch** tab of the **Drafting Standards** dialog you can use the **Fill/Hatch Sketch Region** tool. Access the **Hatch/Color Fill** command using one of the following techniques:

✓ Pick the **Fill/Hatch Sketch Region** button on the panel bar.

✓ Pick the **Fill/Hatch Sketch Region** button on the **Sketch** toolbar.

When you access the **Hatch/Color Fill** command, use the cursor to select a sketch region to define the area to fill or hatch. See Figure 12.24. Once you have selected appropriate sketch geometry, the **Hatch/Color Fill** dialog box is displayed. See Figure 12.25.

If you want to add hatching to the sketched region, pick the **Enable** check box in the **Hatch** area. Then pick the **Hatch** pattern you want to use from the **Pattern** drop-down list. Only hatch patterns selected from the **Hatch** tab of the **Drafting Standards** dialog box are available. You can override any of the default properties in the **Hatch** area, and the hatch color. These properties correspond to those previously discussed in the **Hatch** tab of the **Drafting Standards** dialog box. Changes made to the hatch properties are previewed in the

FIGURE 12.24
Selecting a fill or hatch sketched region.

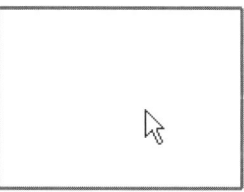

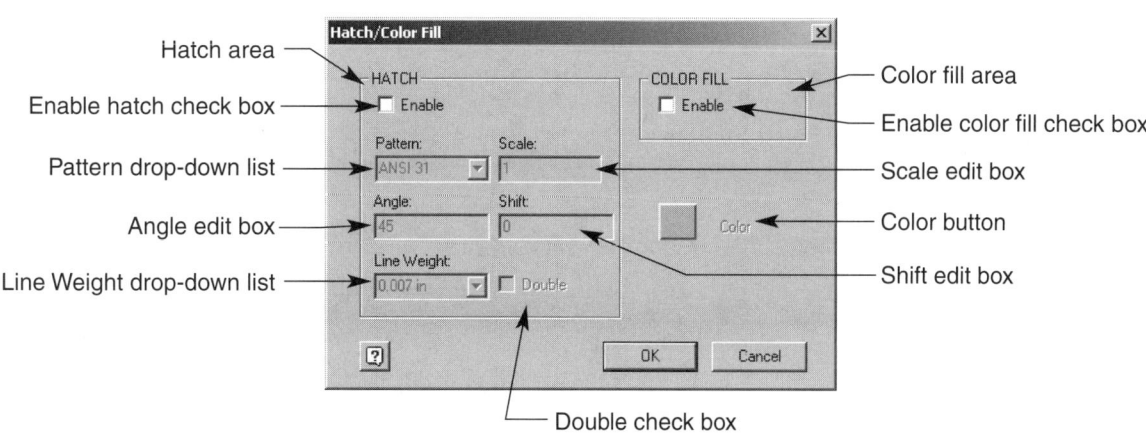

FIGURE 12.25 The **Hatch/Color Fill** dialog box.

selected sketch region. When you are satisfied with the hatch preview pick the **OK** button to create the hatching.

If you want to add a solid color to the sketched region, pick the **Enable** check box in the **Color Fill** area. Then specify the color of the fill by selecting the **Color** button and choosing a color from the **Color** dialog box. When you are satisfied with the hatch preview in the selected sketched region, pick the **OK** button to create the fill.

 Field Notes

> Use the **Insert AutoCAD file** tool in the sketch environment to create a border from an existing AutoCAD file.

The final step in creating your own border is to define the location where you want the corner of the title block to be inserted. To specify the insertion point, you must first pick the point you want to use, which is typically the corner of the border. Then select the **Insert Point** option from the **Style** drop-down list in the **Command** bar. With a specified insertion point, the title block will be placed in the correct location.

After you have sketched your border using sketching tools and defined the title block insertion point, you are ready to generate the border. To exit the sketch environment use one of the following options:

✓ Right-click and select the <u>S</u>ave Border menu option.

✓ Pick the **Return** button on the **Command** bar.

✓ Select the <u>S</u>ave Border option from the **F<u>o</u>rmat** pull-down menu.

When you exit the sketch environment, the **Border** dialog box is displayed, which allows you to save the border. See Figure 12.26. Type the name of the border in the **Name** edit box and pick the Save button to create the border. If you pick the **Discard** button, the sketch is deleted, and the border is not created.

The border you create as a sketch is transformed into an insertable feature. Your custom border, along with any other borders currently in the drawing, are contained in the **Borders** folder of the **Drawing Resources** folder in the **Browser.** See Figure 12.27. A border contained in the **Borders** folder can be inserted into the current sheet and future sheets by double-clicking the border name, or right-clicking the border name and selecting the **Insert** menu option. To edit a custom border, right-click the border name in the **Borders** folder and pick the **Edit** menu option.

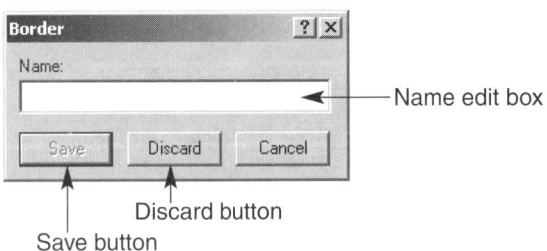

FIGURE 12.26   The **Border** dialog box.

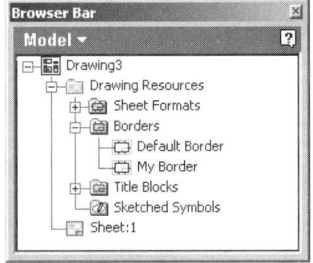

FIGURE 12.27   The available drawing borders.

Exercise 12.6

1. Continue from EX12.5, or launch Autodesk Inventor.
2. Open a new inch drawing file.
3. Delete the default border and title block in **Sheet:1.**
4. Right-click on **Sheet:1** in the **Browser** and select the **Edit Sheet...** menu option.
5. Change the sheet name to **B SIZE.**
6. Change the sheet size to **B.**
7. Access the **Define New Border** command.
8. Using the **Precise Input** toolbar, create the rectangle shown starting at point .62, .38 and ending at 16.38, 10.62.
9. Ground the rectangle in space using the **Fix** constraint.
10. Create the microfilm arrowheads shown, using lines and the **Color Fill** option of the **Hatch** command. The arrowheads are fixed to the midpoints of the rectangle lines.
11. Using the specified dimensions, sketch the margin drawing number block shown, in the bottom left corner of the sheet.
12. Using the **Text** command and the following specifications, enter the text shown in the margin drawing number block.

   Style: DEFAULT-ANSI
   Rotation Angle: 90°
   Font: Arial
   Text Height: .120

13. Use the **Property Field** tool to create the property fields shown in the margin drawing number block. The following specifications are used:

   Part Number
   Center Justification
   Rotation: 90°
   Property Type: Design Properties
   Property: Part Name
   Text Font: Arial
   Text Height: .12

   Sheet Number
   Center Justification
   Rotation: 90°
   Property Type: Design Properties
   Property: Sheet number
   Text Font: Arial
   Text Height: .12

   Revision Number
   Center Justification
   Rotation: 90°
   Property Type: Design Properties
   Property: REVISION NUMBER
   Text Font: Arial
   Text Height: .12

14. Once you have finished creating the property fields, pick the point shown and change its style to **Insert Point.**
15. Save the border as BORDER.
16. In the **Properties** dialog box, enter EX12.6 in the **Part Name** edit box, and 1 in the **Revision Number** edit box.

Exercise 12.6 continued on next page

Exercise 12.6 continued

17. Close the **Properties** dialog box, and insert BORDER onto the sheet.
18. The final drawing should look like the drawing shown.
19. Save the drawing as EX12.6.
20. Exit Autodesk Inventor or continue working with the program.

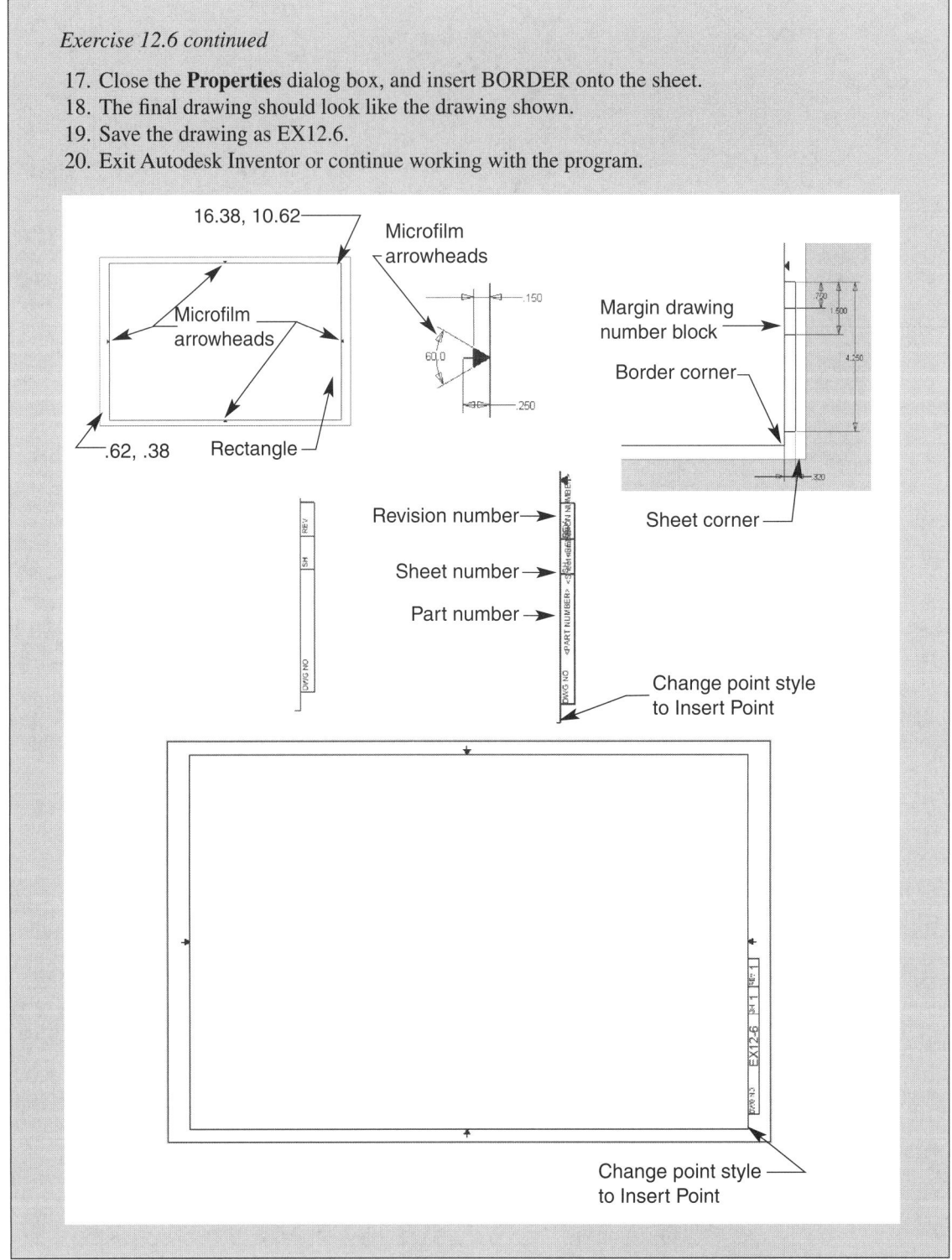

Using Default Drawing Title Blocks

After you have inserted a border onto the drawing sheet, depending on your approach, the next step in developing a part drawing is to define a title block. Title blocks contain information about the model, company, drafter, tolerances, and other design information. A title block can be as simple as a rectangle inside the drawing border, with a couple lines of text, or may include several property areas. See Figure 12.28.

When you open an Autodesk Inventor drawing file for the first time, a title block is already placed on the sheet, along with a border. If you want to use the default sheet title block and do not have to make any modifications to the title block design, you can begin to add information to the title block. All the text infor-

FIGURE 12.28 An example of a drawing title block.

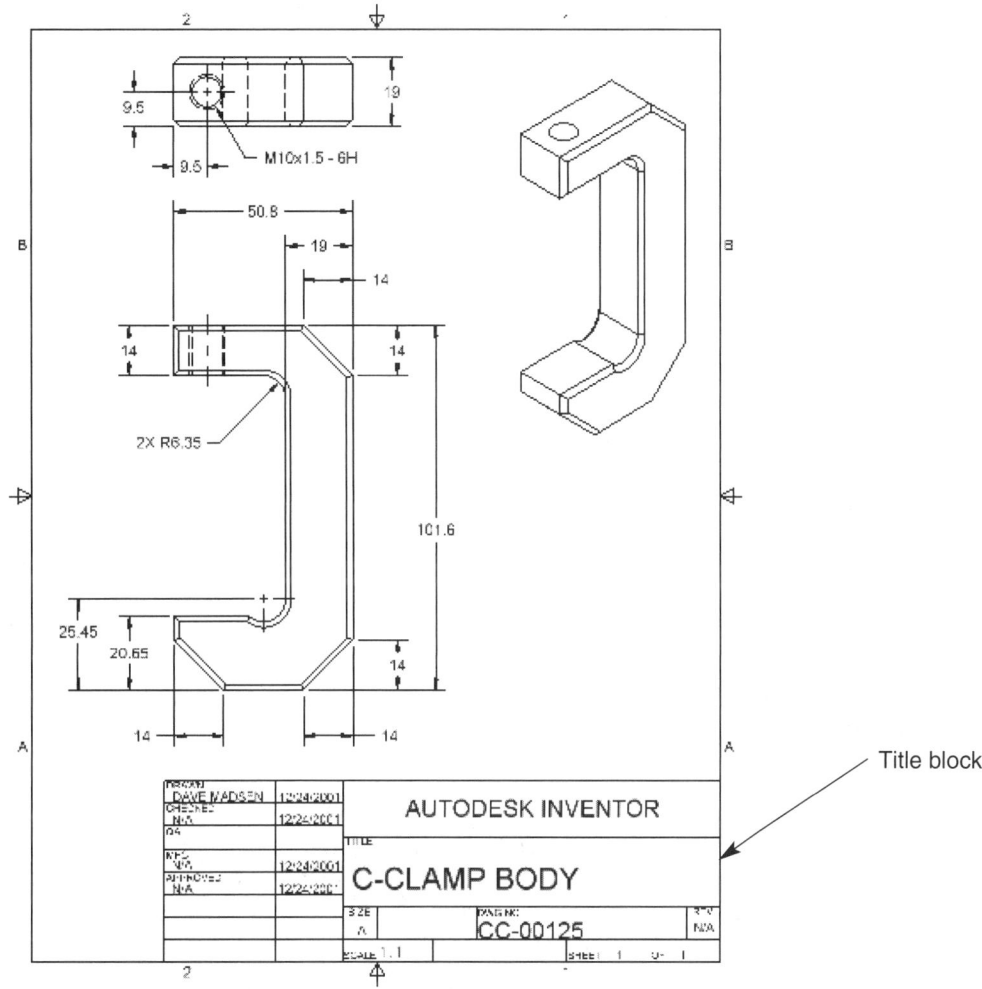

Title block

mation on the default title blocks is created using **Property Fields.** See Figure 12.29. All the property fields correspond to entries in the **Properties** dialog box. See Figure 12.30. To add information to a default title block, access the **Properties** dialog box from the **File** pull-down menu and enter information in various edit boxes. The information you add is automatically displayed in the specified areas of the title block.

Notice in Figure 12.29 that the default large title block contains a scale property field. The scale property field is custom, because there is not a scale edit box in the **Properties** dialog box. To enter a scale in the title block, pick the **Custom** tab of the **Properties** dialog box. See Figure 12.31. Then enter the name of the property in the **Name** edit box, or select a name if it is available from the **Name** drop-down list. You must enter the name exactly as it is named when it was created. For example, the scale property is named SCALE. If you enter scale as the name, the value will not be applied to the property field. The next step is to define the type of property you are creating from the **Type** drop-down list. A property type can be text, a date, numbers, or Yes or No. Then enter a value for the property, such as 1:1 for scale, in the **Value** edit box, and pick the **Add** button to add the custom property to the **Property** list. If you want to delete a custom property, pick the property you want to remove and select the **Delete** button. Finally, pick the **Apply** button to generate the changes.

FIGURE 12.29
Property fields in a default title block.

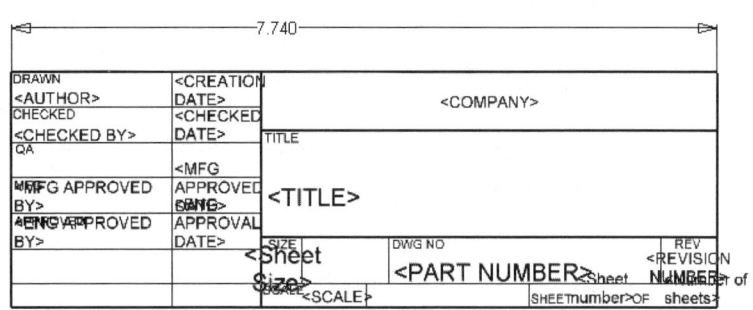

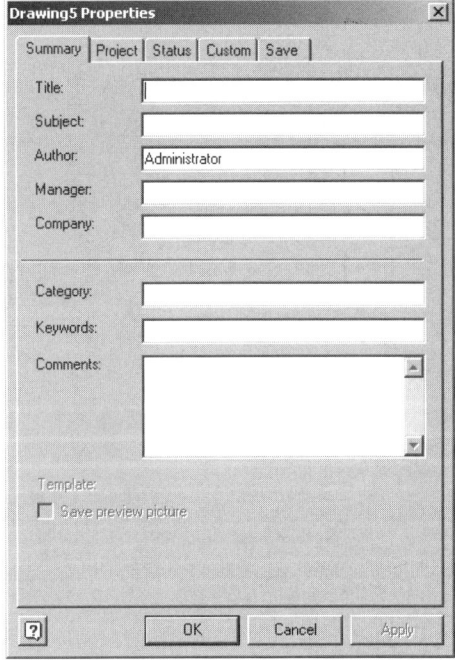

FIGURE 12.30   The **Properties** dialog box (**Summary** tab shown).

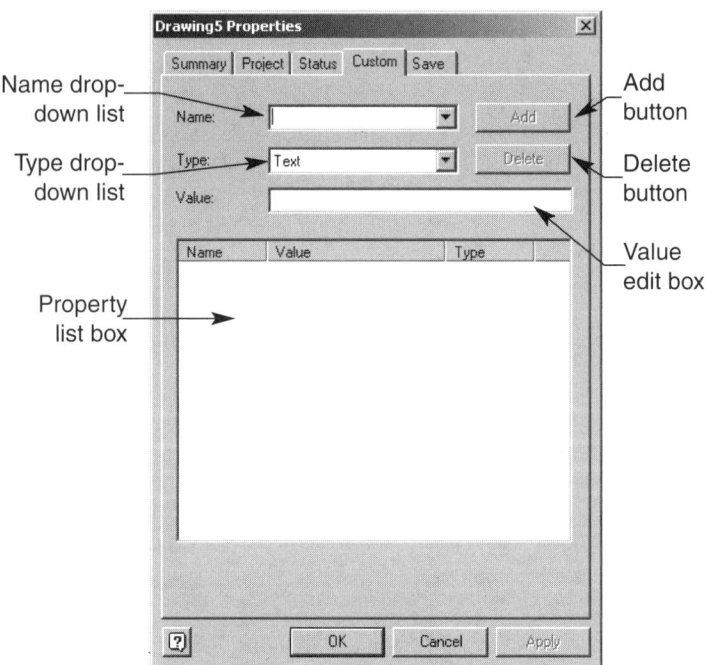

FIGURE 12.31   The **Custom** tab of the **Properties** dialog box.

If you can fully define the default title block using the **Properties** dialog box, you are ready to move on to the next step of drawing development. However, if you want to redefine the size, property fields, text, or any other aspect of the current title block on **Sheet:1,** right-click the default title block in the **Title Blocks** folder of **Drawing Resources** folder in the **Browser** and select the **Edit** menu option to access the sketched title block. Use sketching tools to modify the desired information.

Once you have fully defined the default title block as needed, use one of the following techniques to exit the sketch environment:

✓ Right-click and select the **Save Title Block** menu option.

✓ Pick the **Return** button on the **Command** bar.

✓ Select the **Save Title Block** option from the **Format** pull-down menu.

You will be prompted to save edits, as shown in Figure 12.32. If you want to save the edits to the original title block, pick the **Save** button. However, if you want to create a copy of the original title block and save the edits to the copy, pick the **Save As...** button to access the **Title Block** dialog box. See Figure 12.33. Type the name of the title block in the **Name** edit box and pick the **Save** button to create the title block. If you pick the **Discard** button, the sketch changes are deleted, and the title block is not modified.

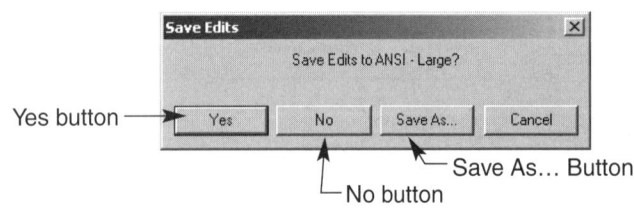

FIGURE 12.32   The **Save Edits** dialog box.

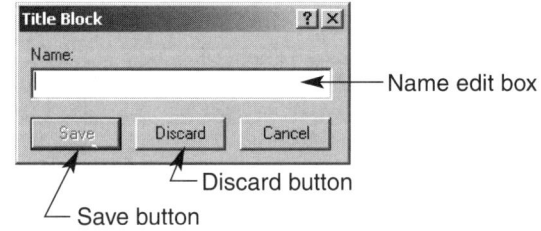

FIGURE 12.33   The **Title Block** dialog box.

Exercise 12.7

1. Continue from EX12.6, or launch Autodesk Inventor.
2. Open EX12.3.
3. Save a copy of EX12.3 as EX12.7.idw.
4. Close EX12.3 without saving, and open EX12.7.
5. Right-click on **ANSI-Large** in the **Browser** and select the **Delete** menu option.
6. Insert the ANSI A title block from the **Title Blocks** folder of the **Drawing Resources** folder.
7. The final drawing should look like the drawing shown.
8. Resave EX12.7.
9. Exit Autodesk Inventor or continue working with the program.

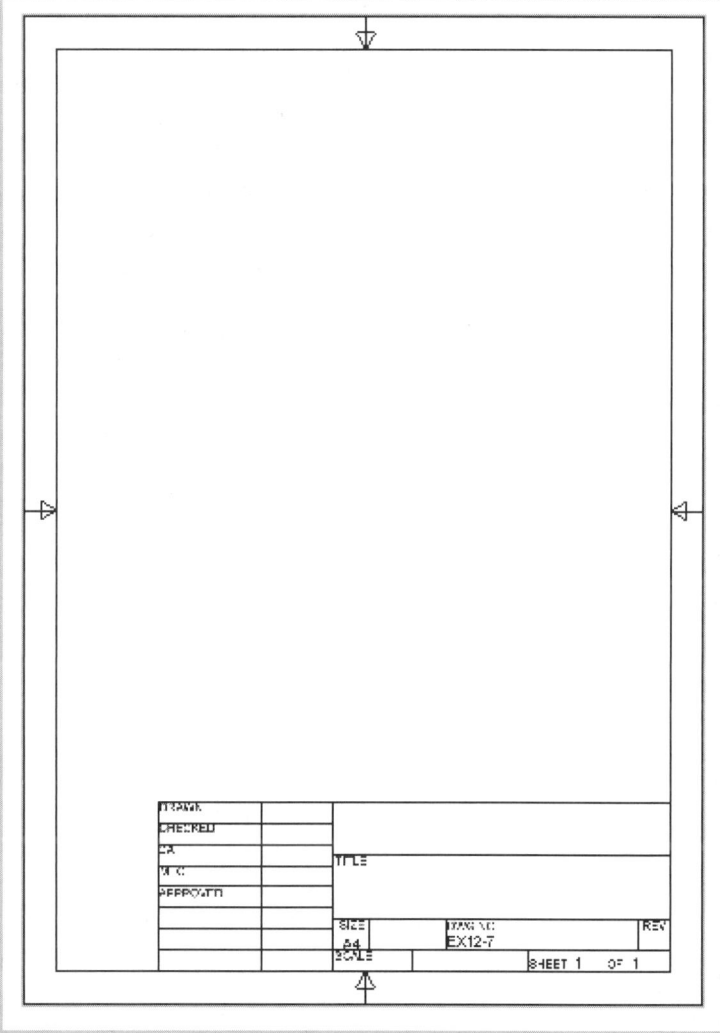

Exercise 12.8

1. Continue from EX12.7, or launch Autodesk Inventor.
2. Open EX12.4.
3. Save a copy of EX12.4 as EX12.8.idw.
4. Close EX12.4 without saving, and open EX12.8.
5. Select **Properties...** from the **File** pull-down menu.
6. Enter as much information as possible in the edit boxes of the **Properties** dialog box, and pick the **Apply** button.
7. Close the **Properties** dialog box, and notice the property displays in the title block.
8. Open the **Properties** dialog box again.
9. Specify the following options the **Custom** tab:

   Name: SCALE
   Type: Text
   Value: 1:1

10. Pick the add button to add the custom property to the **Custom Property** list box.
11. Pick the **Apply** button, followed by the **Close** button.
12. You should notice the 1:1 scale factor in the Scale area of the title block once the title block is edited.
13. The final drawing should look like the drawing shown.
14. Resave EX12.8.
15. Exit Autodesk Inventor or continue working with the program.

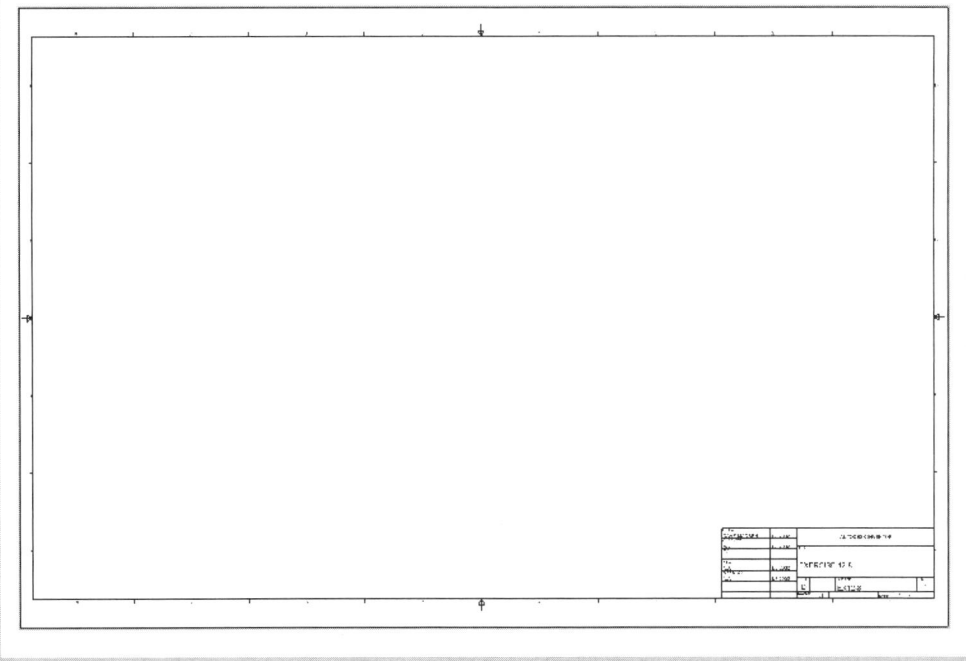

Creating Your Own Drawing Title Block

The default title blocks and the ability to modify the default title blocks is useful for many applications. Still, you may want to create your own title blocks using sketching tools. Creating your own title block is very similar to editing an existing title block, as previously discussed. To develop a custom title block, you may first want to delete any existing title blocks currently in the sheet. Then, select the **Define Title Block** option from the **Format** pull-down menu. When you pick the **Define Title Block** menu option, the panel bar automatically enters **Sketch** mode and the tools in **Sketch** toolbar become active.

The drawing sketch environment is exactly the same when sketching a title block as it is when sketching a border. Refer to the previous discussion on creating your own border for information regarding custom title block development.

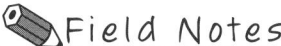 Field Notes

> Use the **Insert AutoCAD file** tool in the sketch environment to create a title block from an existing AutoCAD file.

After you have sketched your title block using sketching tools, you are ready to generate the title block. To exit the sketch environment use one of the following options:

✓ Right-click and select the **Save Title Block** menu option.

✓ Pick the **Return** button on the **Command** bar.

✓ Select the **Save Title Block** option from the **Format** pull-down menu.

When you exit the sketch environment, the **Title Block** dialog box, previously discussed, is displayed, which allows you to save the title block with the desired name. The title block you create as a sketch is transformed into an insertable feature. Your custom title blocks, along with any other title blocks currently in the drawing, are contained in the **Title Blocks** folder of the **Drawing Resources** folder in the **Browser.** See Figure 12.34. A title block contained in the **Title Blocks** folder can be inserted into the current sheet and future sheets by double-clicking the title block name, or right-clicking the title block name and selecting the **Insert** menu option. The insertion point of the title block is defined by the extents of the border and the orientation options in reference to the sheet layout. To adjust the location of the title block, enter the desired orientation in the **Edit Sheet Dialog** box, previously discussed. You can also define the location of the title block before you insert the title block, by selecting **Application Options** from the **Tools** pull-down menu and picking the **Drawing** tab. See Figure 12.35. In the **Alternative Title Block Alignment** area, select the radio button that corresponds to the corner where you want the title block to be inserted. Then pick the **Apply** button to apply the changes, and the **OK** button to exit the dialog box.

To edit a custom title block, right-click the title block name in the **Title Blocks** folder and pick the **Edit** menu option.

FIGURE 12.34 The available drawing title blocks.

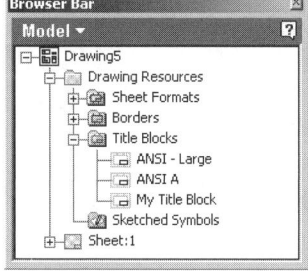

FIGURE 12.35   The
Drawing tab of the
Options dialog box.

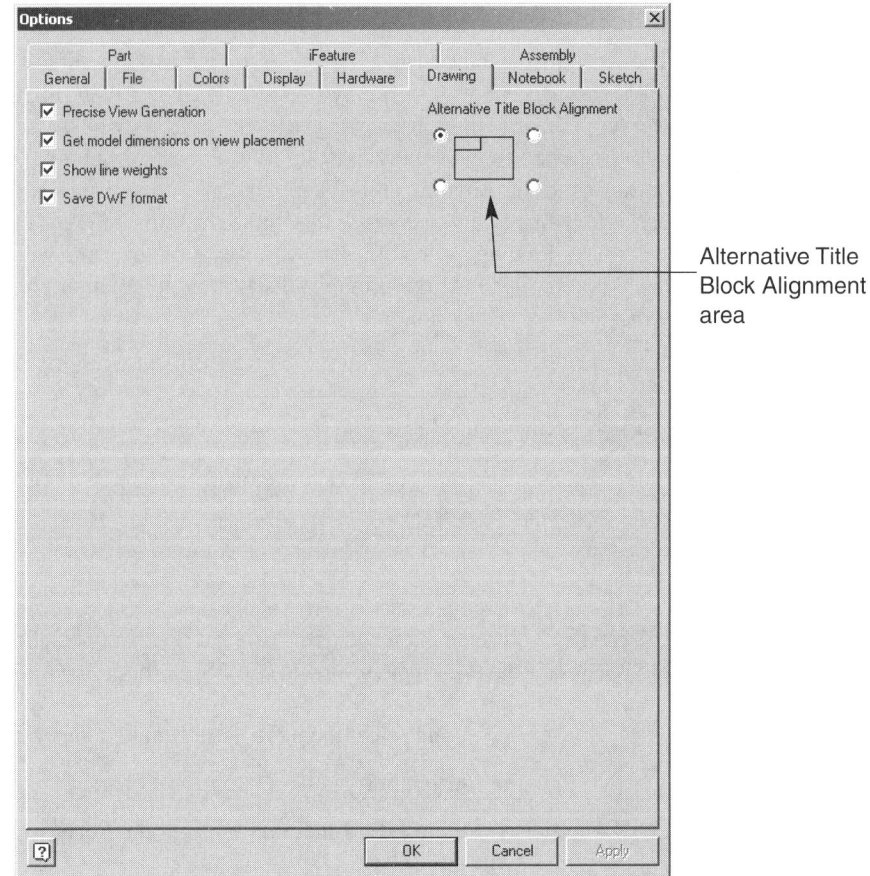

Alternative Title
Block Alignment
area

Exercise 12.9

1. Continue from EX12.8, or launch Autodesk Inventor.
2. Open EX12.6.
3. Save a copy of EX12.6 as EX12.9.idw.
4. Close EX12.6 without saving, and open EX12.9.
5. Open the **Properties** dialog box and create a custom property named Scale with a value of 1:1.
6. Access the **Define New Title Block** command.
7. Using sketching tools, create the title block shown.
8. Save the title block as TITLE BLOCK.
9. Insert TITLE BLOCK onto the sheet.
10. If you changed the corner point style to **Insert Point** when you created the border and have specified the bottom-right-corner title block orientation, the title block should automatically be inserted correctly.
11. The final drawing should look like the drawing shown.
12. Save the drawing as EX12.9.
13. Exit Autodesk Inventor or continue working with the program.

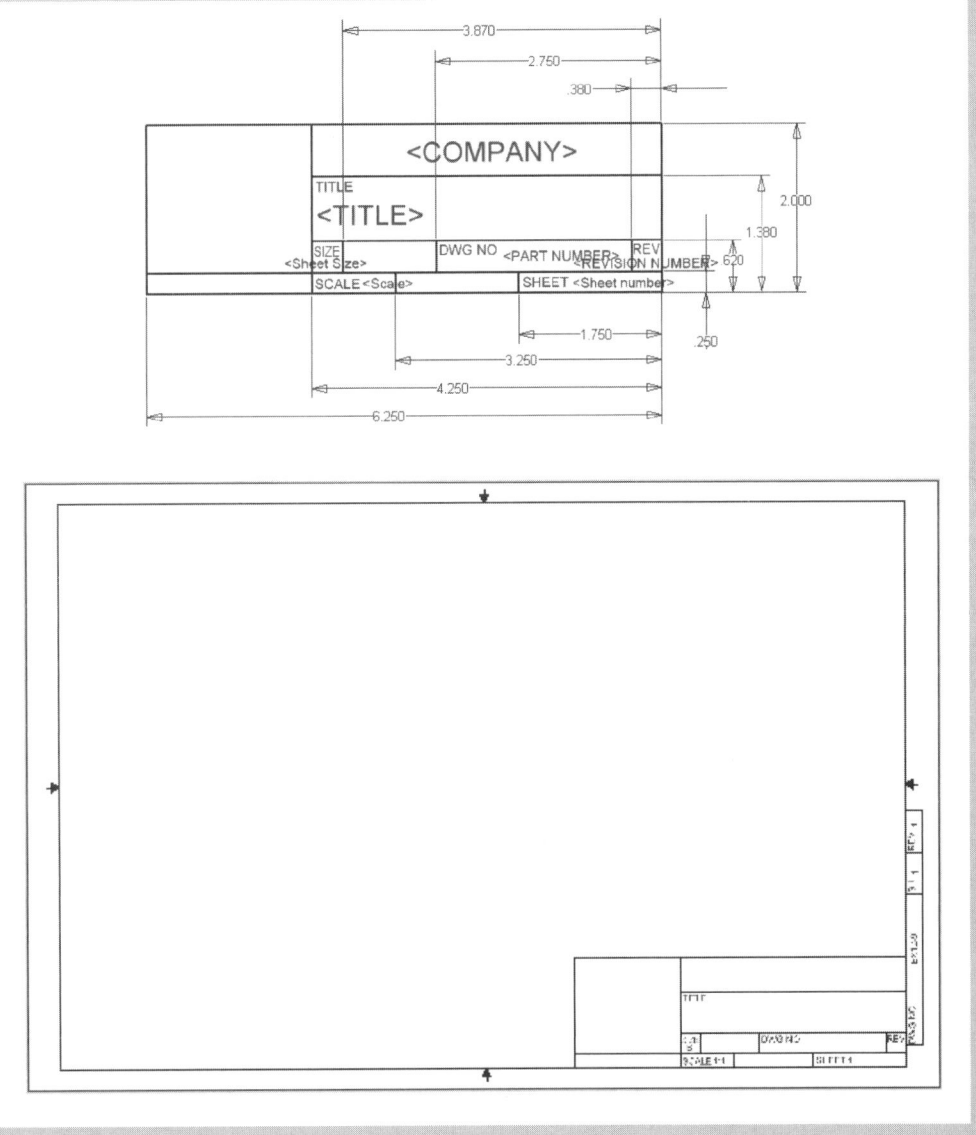

Creating Drawing Views

At some point in the creation of the drawing, depending on the sequence you choose, you want to develop drawing views. As previously discussed, you can create multiple views of a part, including projected, isometric, auxiliary, section, detail, and broken views. However, before you can project additional views, or create auxiliary, section, detail, and broken views, you must have an inserted view, such as a front view. See Figure 12.36.

You can use one or a combination of the following options to create drawing views in Autodesk Inventor:

✓ Place model views using drawing management tools.

✓ Sketch drawing views using sketching tools.

✓ Insert views automatically into a sheet, using the sheet formats located in the **Sheet Formats** folder of the **Drawing Resources** folder in the **Browser** bar.

Using Drawing Management Tools

Drawing management tools are used to create drawing views of existing part models, new sheets, and draft views. These tools are accessed from the **Drawing Management** toolbar, the **Drawing Management** panel bar, or the **Insert** pull-down menu. See Figure 12.37.

You can use the drawing management tools with the default settings applied to the drawing. However, you may want to review or adjust some of the following settings in the **Drawing** tab of the **Options** dialog box, as shown in Figure 12.38, by picking **Application Options...** from the **Tools** pull-down menu:

■ **Precise View Generation** When this check box is selected, drawing views are placed to the exact specifications and precisions as they were created in the model. When this check box is deselected, the view is created and can be modified more quickly, but geometry is not precisely generated.

■ **Get model dimensions on view placement** If this check box is selected when you insert the view, model dimensions are automatically placed. You can override this setting when you create a view using the **Create View** tool.

FIGURE 12.36 An initial drawing view (front view).

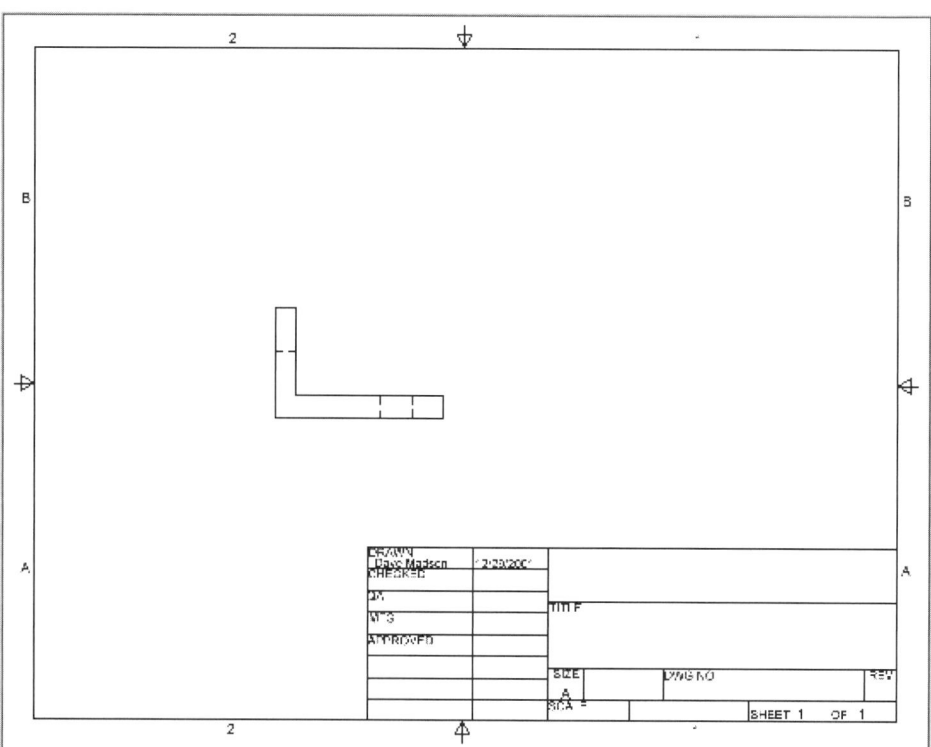

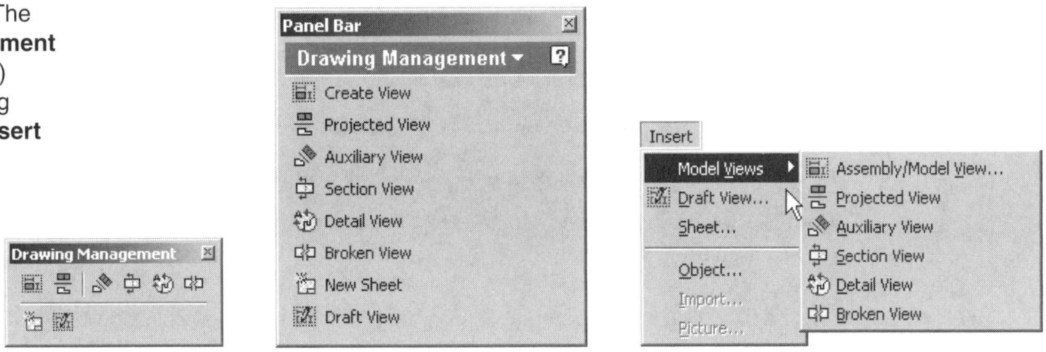

FIGURE 12.37 The **Drawing Management** (A) toolbar and (B) panel bar (learning mode). (C) The **Insert** pull-down menu.

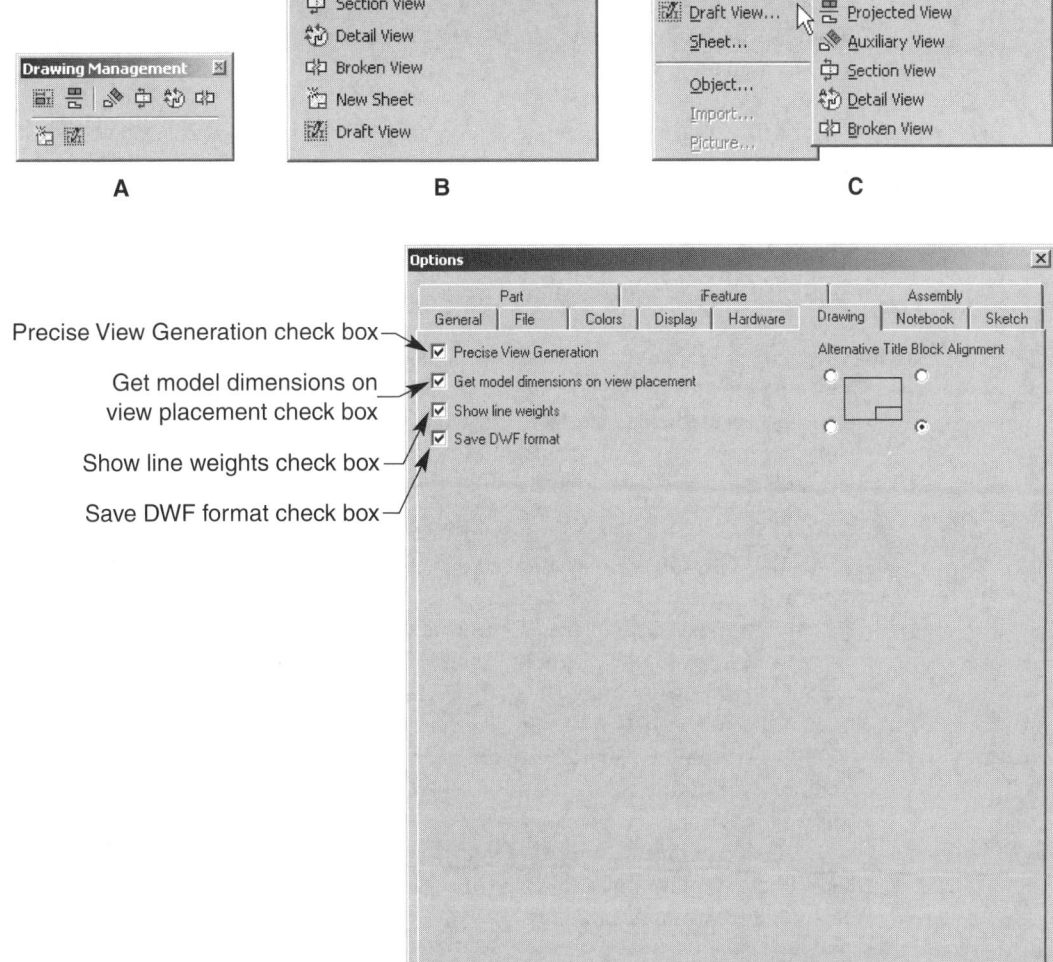

FIGURE 12.38 The **Drawing** tab of the **Options** dialog box.

- **Show line weights** If this check box is selected when you insert a view, line weights specified by drafting standards are displayed. When you deselect this check box, all lines are displayed with the same line weight.

- **Save DWF format** When you select this check box, drawing information is saved in a DFW format for use with the WHIP! Viewer.

Using the Create View Tool

The **Create View** tool allows you to insert a single view, referenced by an existing model, into the drawing. You can use this tool anytime during drawing development and generate a wide variety of model views. Often the **Create View** tool is used to place the initial drawing view, or base view, to which projected, isometric, auxiliary, section, detail, and broken views are added.

Access the **Create View** command using one of the following techniques:

✓ Pick the **Create View** button on the panel bar.

✓ Pick the **Create View** button on the **Drawing Management** toolbar.

✓ Right-click on a sheet in the graphics window and select the **Create View...** menu option.

✓ From the **Insert** pull-down menu, select the **Assembly/Model View...** option of the **Model Views** cascading submenu.

When you access the **Create View** command, the **Create View** dialog box is displayed. See Figure 12.39.

The first step in creating a drawing view is to select the existing part model file from which you want to reference the drawing view. Pick the **Explore** button to access the **Open** dialog box. Using the **Open** dialog box, locate the file you want to use, and pick the Open button. The file and its directory should be displayed in the **File** drop-down list.

Then specify the following view options located in the **View** area of the **Create View** dialog box:

■ **View**   This selection box contains a number of views to choose from. When you pick a view, the preview of the drawing view changes to reflect the selection. You can select any of the views available, such as Iso Top Right, and create the drawing view. However, for most part drawings, the initial view should be front, top, or side, because these allow you to project additional views, including isometrics.

■ **Custom View**   This button opens the **Custom View** window of Autodesk Inventor. See Figure 12.40. The **Custom View** window allows you to develop a custom drawing view for applications when none of the views available in the **View** selection box is appropriate. Many of the typical Autodesk Inventor viewing tools are available when working in the custom view environment. **Rotate, Look At, Zoom, Zoom Window, Zoom All, Zoom Select, WireFrame Display, Hidden Edge Display, Shaded Display,** and **Pan** are accessible from the **View** pull-down menu or from the **Custom View** toolbar. In addition, the **Rotate at Angle** tool is available.

The zoom, pan, and display tools are used for display purposes only, while the **Rotate, Look At,** and **Rotate at Angle** tools are used to actually define the custom view. For example, if you rotate the front view of an object 10°, a 10° custom view, as shown in Figure 12.41, is created.

At this point, the only tool you may not be familiar with is **Rotate at Angle,** which allows you to rotate a view at a specified angle. Access the **Rotate at Angle** command using one of the following techniques:

FIGURE 12.39 The **Create View** dialog box.

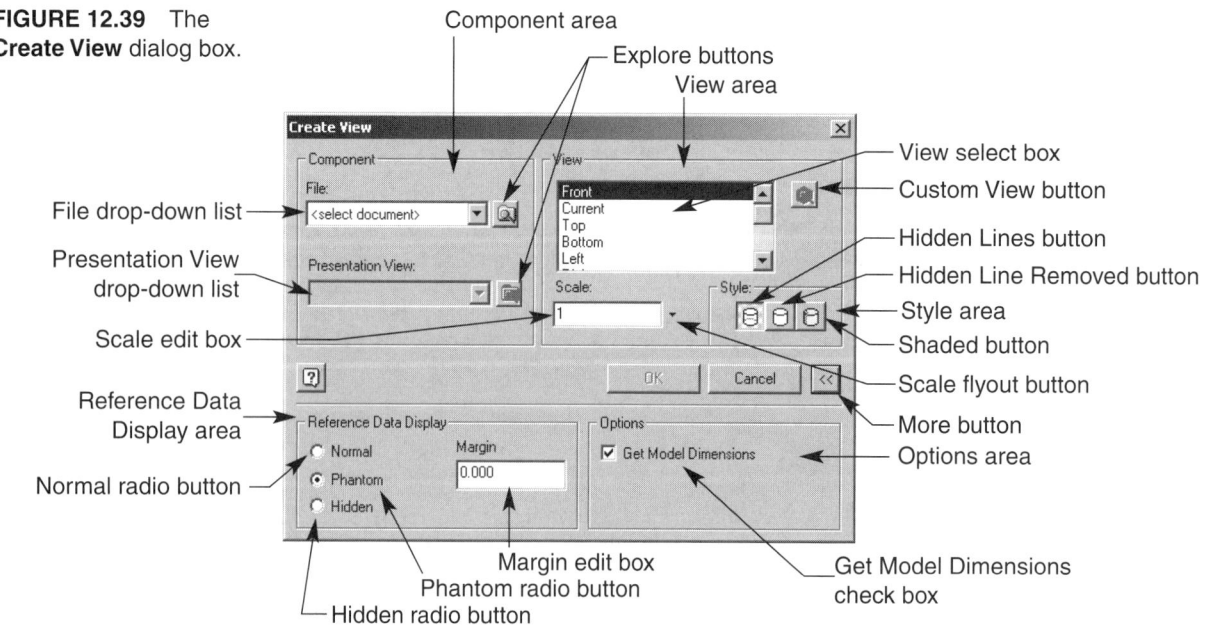

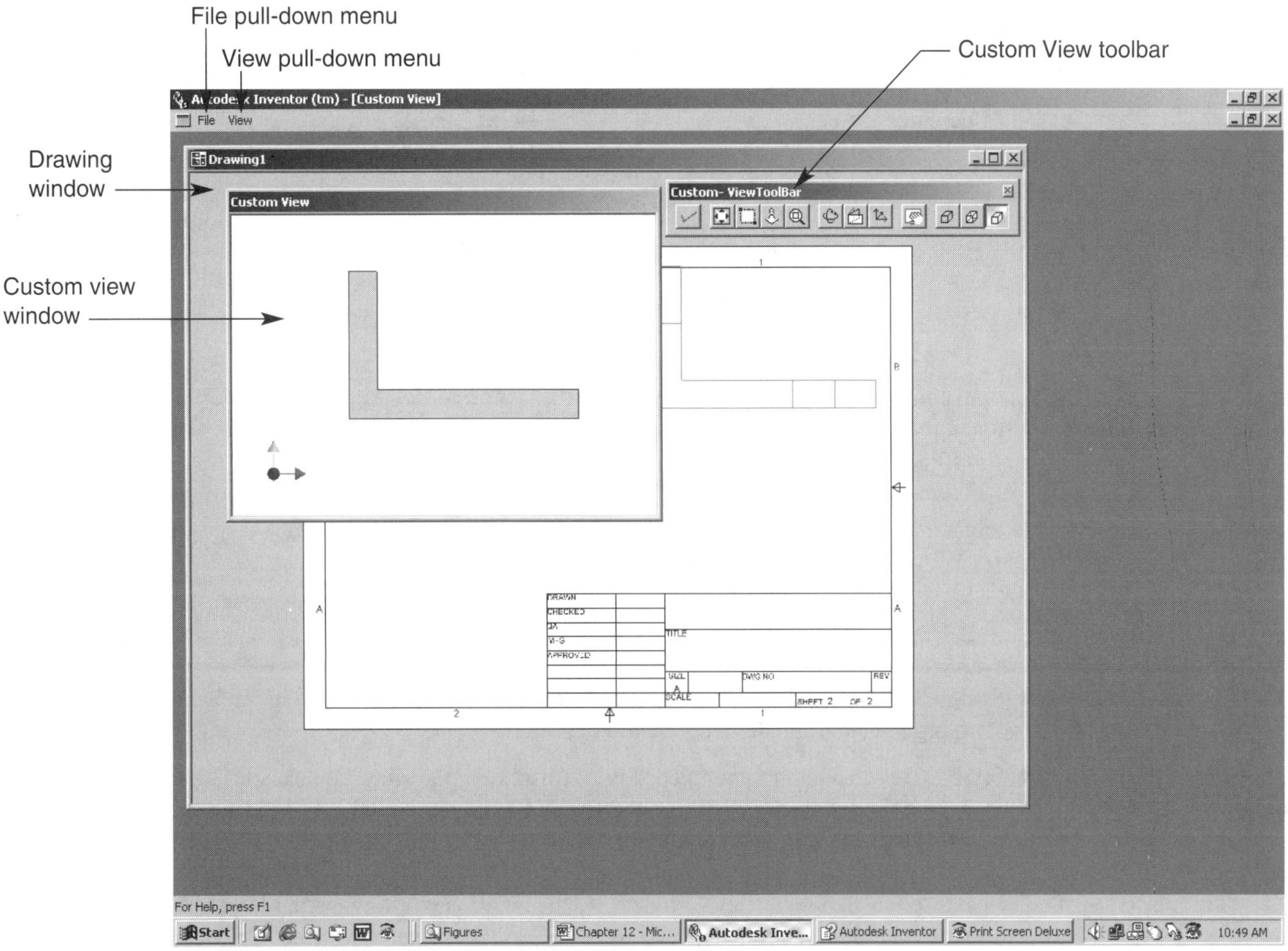

File pull-down menu

View pull-down menu

Custom View toolbar

Drawing window

Custom view window

FIGURE 12.40   The **Custom View** window.

FIGURE 12.41   A 10° custom view.

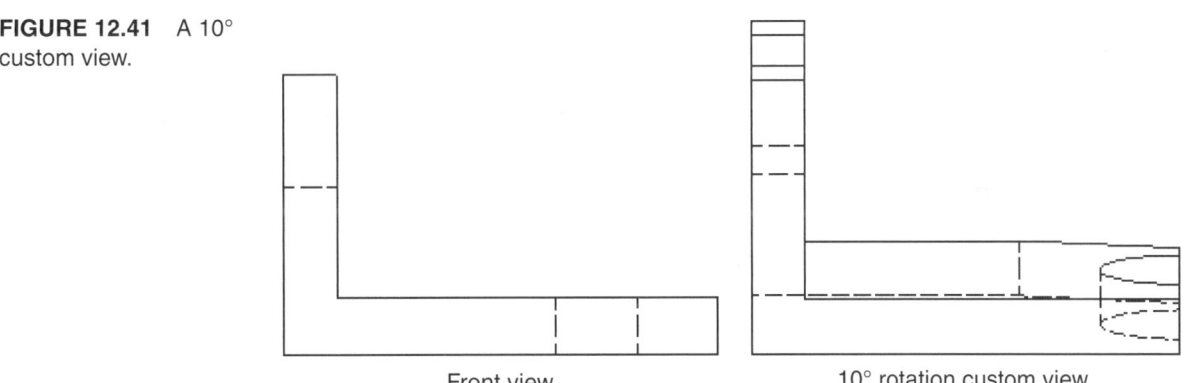

Front view

10° rotation custom view

✓ Pick the **Rotate at Angle** button on the **Custom View** toolbar.

✓ Select the **Rotate At Angle** option from the **Rotate** cascading submenu of the **View** pull-down menu.

When you access the **Rotate at Angle** command, the **Incremental View Rotate** dialog box is displayed. See Figure 12.42.

To use the **Incremental View Rotate** dialog box, enter the desired amount of rotation in the **Increment** edit box. Then use the rotate buttons to rotate the view to the desired location. Each time you select one of the rotate buttons, the view rotates in the specified direc-

FIGURE 12.42 The
**Incremental View
Rotate** dialog box.

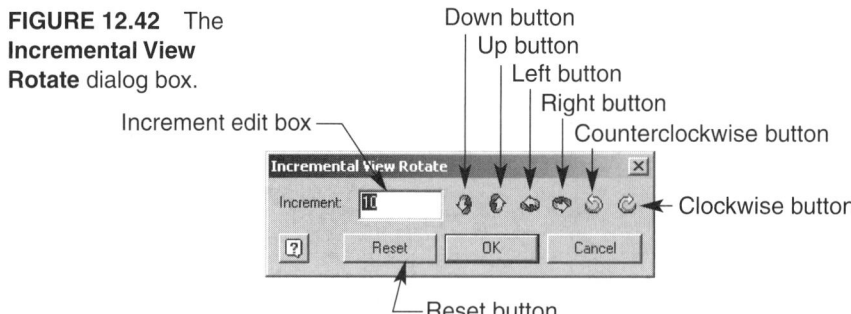

tion at the increment to define. If you want to return to the original view, pick the **Reset** button. Finally, when you have rotated the view as desired, pick the **OK** button to exit the
Rotate at Angle command.

Once you are satisfied with the display in the **Custom View** window, use one of the
following options to exit the custom view environment:

✓ Pick the **Exit Custom View** button on the **Custom View** toolbar.

✓ Select the **Exit Custom** option from the **File** pull-down menu.

■ **Scale**   Here you can define the scale of the view by entering a scale in the **Scale** edit box or
selecting a scale from the flyout button options.

■ **Style**   This area allows you to specify the display of the view. You can choose to display the
view with hidden lines by picking the **Hidden Line** button, without hidden lines by picking
the **Hidden Line Removed** button, or as shaded by picking the **Shaded** button.

The next step in creating a view is to specify the additional settings in the **Create View** dialog box by
picking the **More** button. The following two areas are available when you select the **More** button:

■ **Reference Data Display**   This area allows you to specify the display characteristics of reference data in the view. If you select the **Normal** radio button, reference data is not set apart
from other geometry. If you select the **Phantom** radio button, special characteristics are
applied to reference data for specific display applications. If you choose the **Hidden** radio
button, reference data is not shown in the view. When reference data extends past the margins of the view, specify a margin in the **Margin** edit box to increase the view area, and display all reference data.

■ **Options**   This area contains the **Get Model Dimension** check box. This check box will be
selected or deselected if the **Get Model Dimension** check box in the **Options** dialog box is
selected or deselected. If this check box is selected, when you insert the view, model dimensions are automatically placed.

✎ Field Notes

Additional options may be available inside the **Create View** dialog box depending on the application. The **Design View** option is available when creating assembly drawings and is used to
specify an assembly design view. The **Presentation View** option is available when creating
sheet metal drawings, where it is used to specify a folded or flat view, and in presentation drawings, where it is used to specify a presentation view.

Once you have selected a component to insert into the drawing sheet, specified the desired view, and
defined any other view options, pick the location on the sheet where you want to place the view. See Figure
12.43. Selecting the view location ends the **Create View** command.

FIGURE 12.43
Placing the view on the sheet.

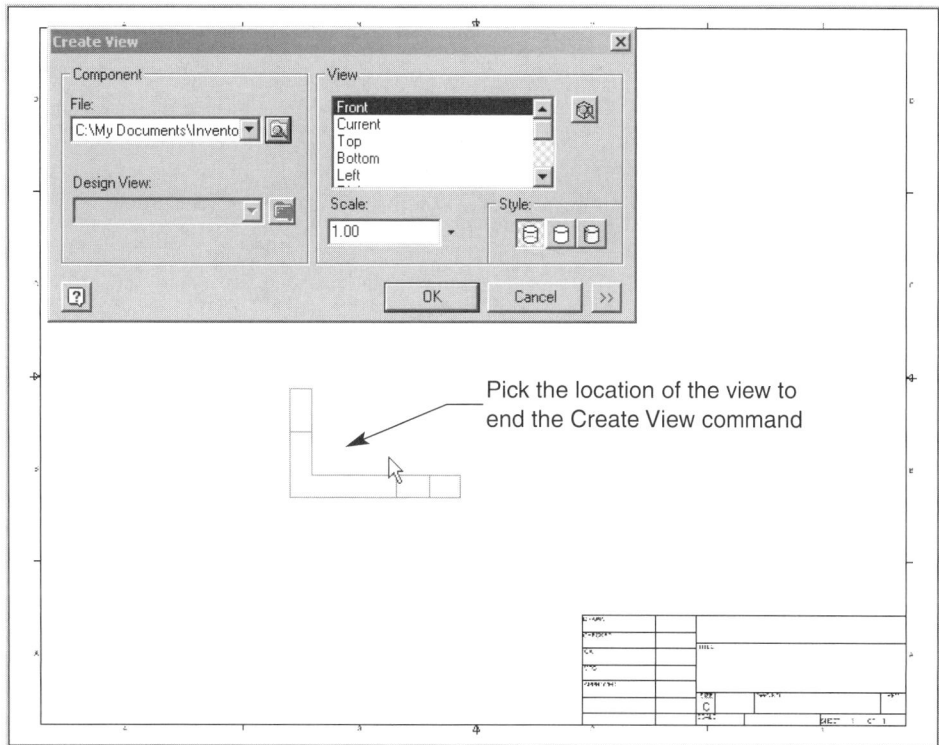

After you create a view, you can move the view by picking inside the view boundaries and dragging the view to the desired location. You can also manipulate many of the view characteristics by right-clicking inside the view boundaries, or the view name in the **Browser,** and selecting one of the following options:

- **Edit View...** Choose this option to access the **Edit View** dialog box, as shown in Figure 12.44. The **Edit View** dialog box allows you to edit many of the view options previously discussed when creating a view. The following additional options are also available:

  - **Label** This area allows you to modify the view label. You can edit the default name of the label in the **Label** edit box and pick the **Show Label** check box if you want the label displayed in the view. See Figure 12.45.

  - **Tangent Edges** Pick this check box to display tangent edges in the drawing view.

FIGURE 12.44 The **Edit View** dialog box.

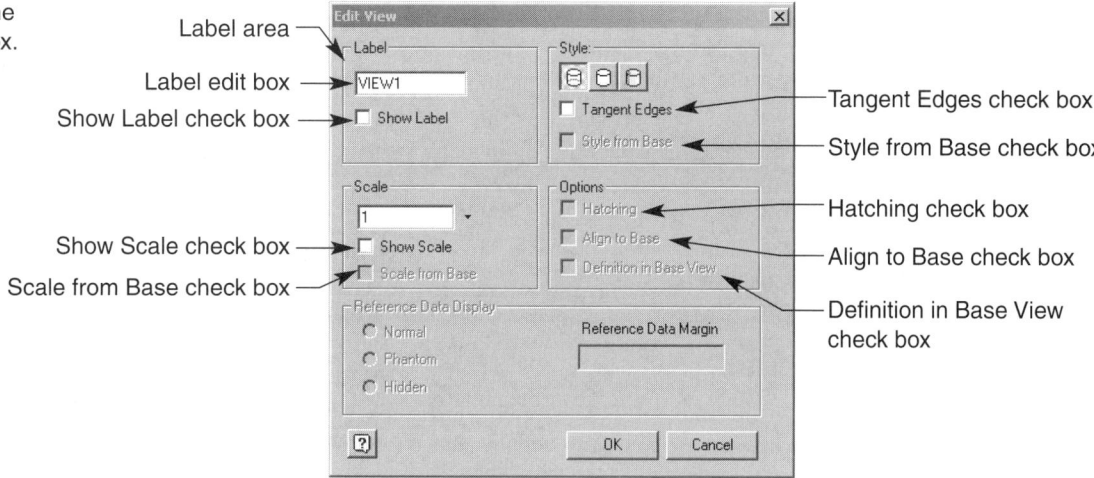

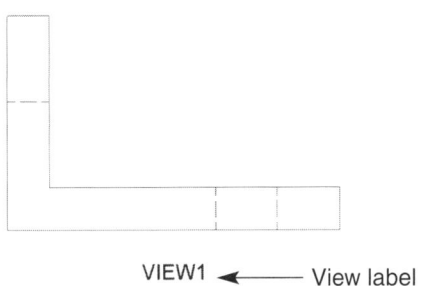

VIEW1 ◄——— View label

FIGURE 12.45   An example of a
shown view label

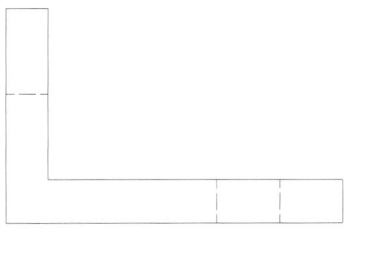

SCALE 1 : 1 ◄——— Visible scale

FIGURE 12.46   An example of a visible scale.

- **Style from Base**   Typically the first view you create in a drawing is the base view. Additional, dependent views are created from the base view. Select this check box to set the style characteristics of the selected view, the same as the base view.

- **Show Scale**   Select this check box if you want the view scale to be displayed in the view. See Figure 12.46.

- **Scale from Base**   Pick this check box to set the scale of the selected view, equal to the scale of the base view.

- **Options**   This area controls additional view display characteristics. The **Hatching** check box sets the display of hatching in section view. Deselect this check box to hide section hatching. The **Align to Base** check box sets view alignment for views where alignment constraints exist. If this check box is deselected, the specified view is not constrained, or aligned, with the base view. The **Definition in Base View** check box controls the display of view projection lines. Pick this check box to show projection lines.

Field Notes

> **Style from Base** and **Scale from Base** check boxes are only available when editing a dependent view, such as a view projected from the base view.
>
> When you use the **Style from Base** and **Scale from Base** options, the selected view style and scale cannot be modified.
>
> You can modify the displayed label and scale by right-clicking the text and selecting the **Edit View Label...** menu option.

- **Rotate**   Choose this option to access the **Rotate View** dialog box, as shown in Figure 12.47. The **Rotate View** dialog box allows you to rotate the selected view. The first step is to specify how you want to rotate the view by selecting **Edge** or **Angle** from the **By** drop-down list.

    The **Edge** option rotates the view by aligning a selected edge horizontally or vertically. To use the **Edge** rotation technique, pick the edge you want to align, then pick the **Horizon-**

FIGURE 12.47   The
Rotate View dialog box
(rotate by **Edge** option).

By drop-down list

Horizontal radio button

Vertical radio button

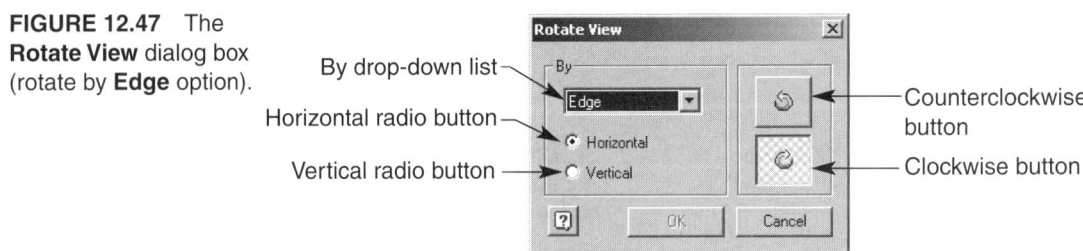

Counterclockwise button

Clockwise button

tal or **Vertical** radio button depending on the application. For example, if you want an edge that is currently horizontal to become vertical, pick the horizontal edge and the **Vertical** radio button. Next, select the **Counterclockwise** button to rotate the view counterclockwise, or the **Clockwise** button to rotate the view clockwise. When you are satisfied with the view rotation, pick the **OK** button to exit the command.

The **Angle** option rotates the view by a specified angle. To use the **Angle** rotation technique, enter the rotation angle in the **Angle** edit box, as shown in Figure 12.48. Next, select the **Counterclockwise** button to rotate the view counterclockwise, or the **Clockwise** button to rotate the view clockwise. When you are satisfied with the view rotation, pick the **OK** button to exit the command.

Field Notes

> The position of annotations, such as dimensions, is automatically modified to remain associated to the view geometry.

- **Hide Dimensions**   The **Hide Dimensions** cascading submenu contains the **Hide Model Dimensions** and **Hide Drawing Dimensions** options. Model dimensions are parametrically controlled dimensions related directly to the model; drawing dimensions are additional dimensions specified after a view is created. Select the **Hide Model Dimensions** option to conceal the model dimensions in the specified view, and pick the **Hide Drawing Dimensions** option to conceal the drawing dimensions in the specified view.

- **Get Model Dimensions**   Pick this option to display model dimensions for the specified view. If you select the **Get model dimension on view placement** radio button in the **Option** dialog box, model dimensions are automatically displayed.

- **Show Contents**   When you choose this menu option, all the components and contents of the model are shown in the Browser, inside the view. See Figure 12.49.

Field Notes

> You can edit the properties of specific drawing features in a view by right-clicking the component in the **Browser**, if the contents are shown, and picking the **Properties...** menu option. You can also right-click on the actual drawing feature in the view and pick the **Properties...** menu option.

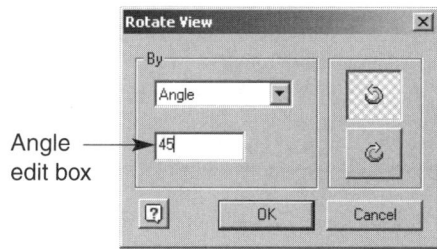

Angle edit box

FIGURE 12.48 The **Rotate View** dialog box (rotate by **Angle** option).

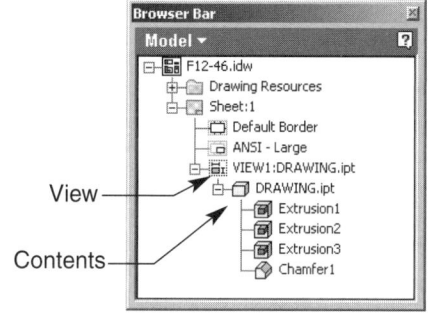

View

Contents

FIGURE 12.49 An example of shown model contents in a drawing view.

Exercise 12.10

1. Continue from EX12.9, or launch Autodesk Inventor.
2. Open EX12.9.
3. Save a copy of EX12.9 as EX12.10.idw.
4. Close EX12.9 without saving, and open EX12.10.
5. Select **Application Options...** from the **Tools** pull-down menu.
6. Ensure each of the check boxes in the **Drawing** tab of the **Options** dialog box is selected.
7. Access the **Create View** command.
8. Specify the following information in the **Create View** dialog box:

   File: EX4.4
   View: Right
   Scale: 1:1
   Style: Hidden Line
   Deselect the **Get Model Dimensions** check box.

9. Pick a location for the drawing view similar to the location shown.
10. When you created the title block, you used model property fields. As a result, the properties specified in the **Properties** dialog box for EX4.4 are displayed.
11. Resave EX12.10.
12. Exit Autodesk Inventor or continue working with the program.

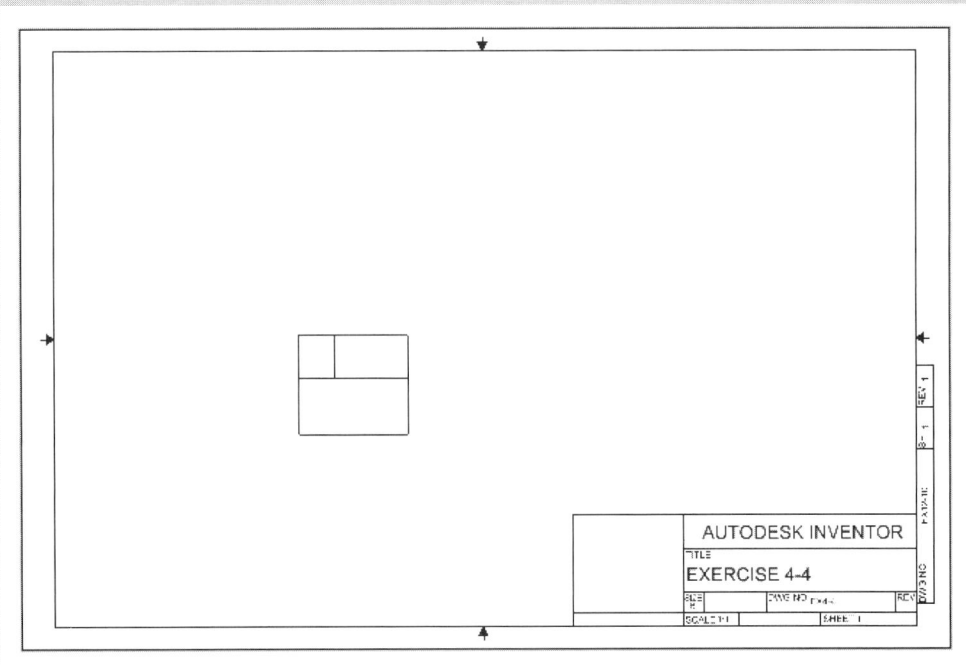

Exercise 12.11

1. Continue from EX12.10, or launch Autodesk Inventor.
2. Open EX12.9.
3. Save a copy of EX12.9 as EX12.11.idw.
4. Close EX12.9 without saving, and open EX12.11.
5. Access the **Create View** command.
6. Specify the following information in the **Create View** dialog box:

   File: P5.10
   View: Bottom
   Scale: 1:1
   Style: Hidden Line
   Deselect the **Get Model Dimensions** check box.

7. Pick a location for the drawing view similar to the location shown.
8. Resave EX12.11.
9. Exit Autodesk Inventor or continue working with the program.

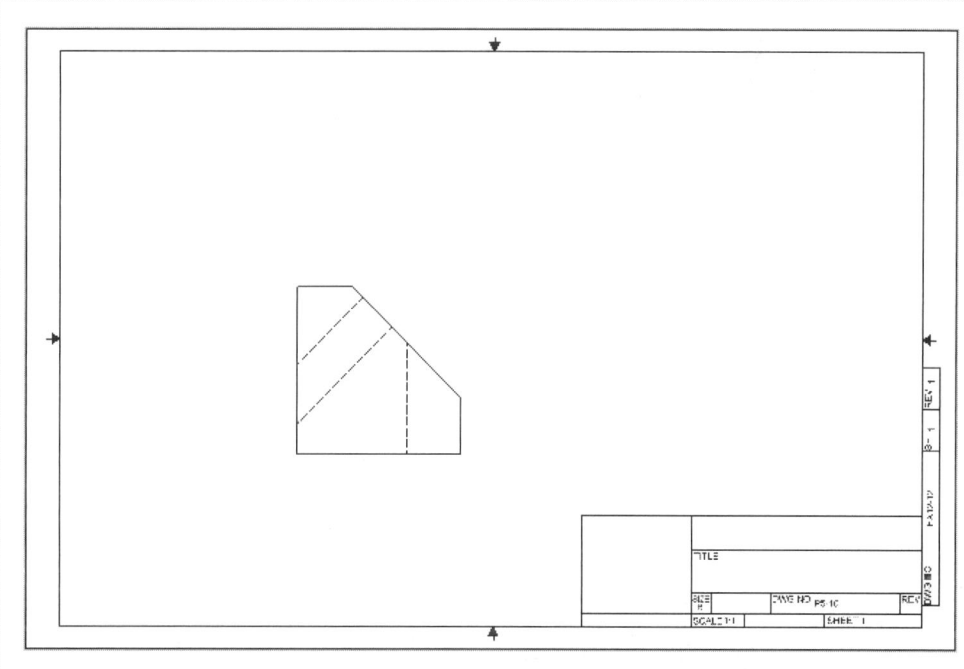

Projecting Views

Once you have created a base drawing view, such as a front view, using the **Create View** tool, the next step in creating a multiview drawing, depending on the application, is to project additional views, such as a top and side view. Projected views are generated quickly and easily from existing base views or dependent views, because model information is already stored in the drawing. The **Projected View** tool allows you to project as many views as you want from base or dependent views. Assuming at least one view referencing a model exists in the drawing, you can use the **Projected View** tool anytime during drawing development and generate a wide variety of additional model views.

Access the **Projected View** command using one of the following techniques:

✓ Pick the **Projected View** button on the panel bar.

✓ Pick the **Projected View** button on the **Drawing Management** toolbar.

✓ Right-click an existing parametric drawing view in the graphics window, and select **Project-ed** from the **Create View** cascading submenu.

✓ Right-click an existing parametric drawing view name in the **Browser,** and select **Projected** from the **Create View** cascading submenu.

✓ From the **Insert** pull-down menu, select the **Projected View** option of the **Model Views** cascading submenu.

Field Notes

If you access the **Projected View** command from the panel bar, **Drawing Management** tool-bar, or the **Insert** pull-down menu, you must select the view you want to project from.

When you access the **Projected View** command and a view is selected, use your mouse to move the projected view in place. See Figure 12.50. When you are satisfied with the location of the projected view select the location by pressing the left mouse button. Then continue placing views as desired using the techniques discussed. By accessing the **Projected View** tool once, you can create multiple views. See Figure 12.51.

FIGURE 12.50
Projecting views.

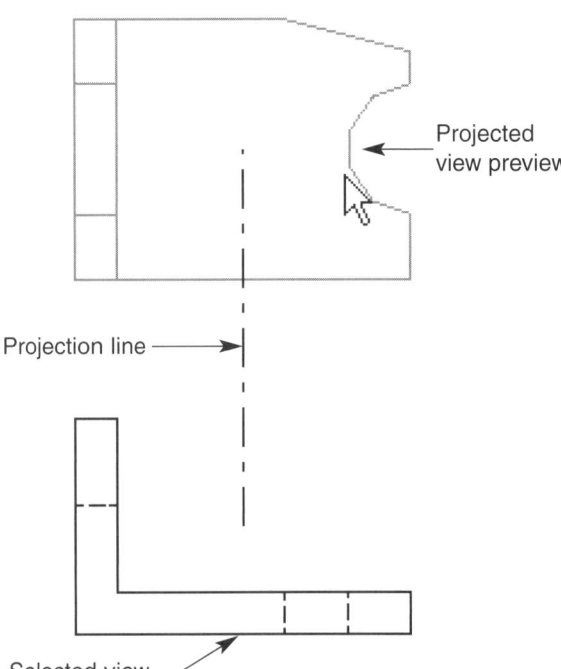

FIGURE 12.51
Creating multiple
projected views using a
single **Projected View**
command.

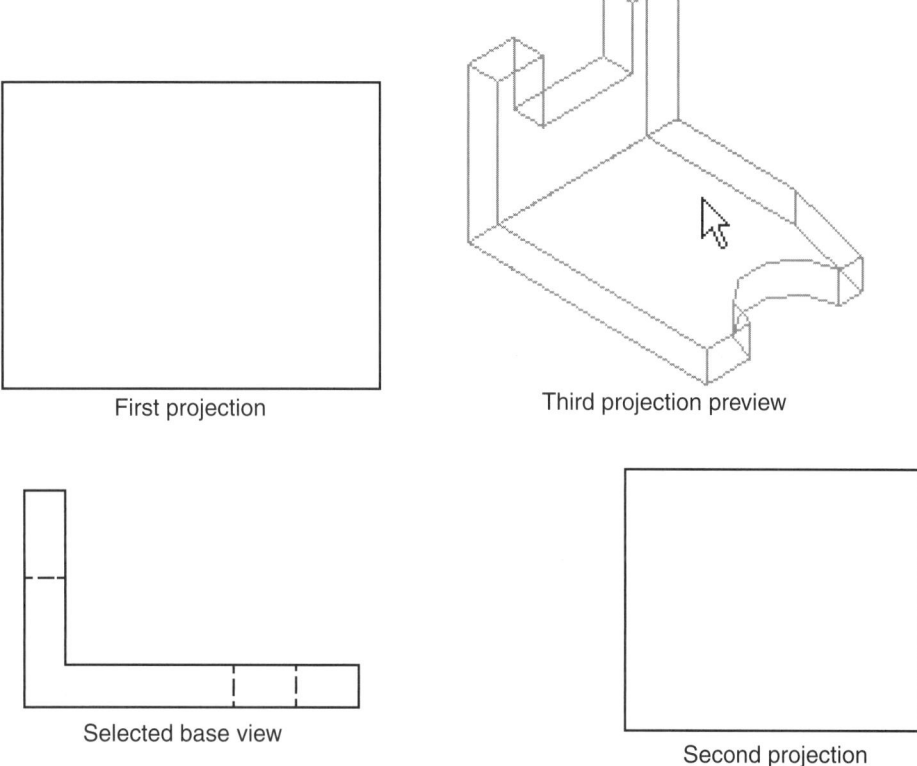

First projection

Third projection preview

Selected base view

Second projection

Once you are satisfied with the location and number of projected views, right-click and select the **Create** menu option to generate the projected views.

> ✏ Field Notes
>
> Each of the view editing options previously discussed in the **Create View** section apply to editing projected views.
>
> To display model dimensions in a projected view, you must right-click inside the view boundaries or the view name in the **Browser** and select the **Get Model Dimensions** menu option.

After you create a projected view, you can move the view by picking inside the view boundaries and dragging the view to the desired location. However, you do not have total control over the movement of a projected view because of alignment constraints. Similarly, when you move the base view, the projected view also moves because of alignment constraints. Typically the alignment constraints associated with drawings ensure valid relationships between views and do not need to be edited. However, you can modify the alignment of views by accessing alignment commands, using one of the following techniques:

✓ Right-click inside the view boundaries and select one of the alignment options from the **Alignment** cascading submenu.

✓ Right-click on the view name in the **Browser,** and select one of the alignment options from the **Alignment** cascading submenu.

✓ Pick the view you want to align, then from the **Tools** pull-down menu, select an alignment option from the **Align Views** cascading submenu.

The following alignment options are available:

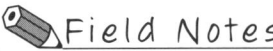

> If an alignment constraint already exists, you will not be able to modify alignment of the view, unless the constraint is removed.

- **Horizontal**   Select this option to create a horizontal alignment constraint, which allows you to move the view only horizontally, or along the X axis.

- **Vertical**   Choose this option to create a vertical alignment constraint, which allows you to move the view only vertically, or along the Y axis.

- **In Position**   Select this option to create a constraint that is neither vertical nor horizontal. For example, this type of alignment is typically used for projected isometric views, which are not constrained to the X or Y axis of a base view.

- **Break**   Pick this option to remove alignment constraints. A view with a broken alignment can be moved anywhere, without being controlled by any other view. You must first break the existing alignment constraints of a view, before the alignment can be modified.

Once the view you want to align is selected and you have chosen the type of alignment you want to create (horizontal, vertical, or in position), the final step is to pick the view from which you want the alignment to be referenced, which is typically the base view.

> To delete a base view but not remove dependent views, use the **Delete View** dialog box to select the dependent views that you do not want to delete.

Exercise 12.12

1. Continue from EX12.11, or launch Autodesk Inventor.
2. Open EX12.10.
3. Save a copy of EX12.10 as EX12.12.idw.
4. Close EX12.10 without saving, and open EX12.12.
5. Select **Standards…** from the **Format** pull-down menu.
6. Ensure the Third angle of projection button in the **Common** tab is selected.
7. Access the **Projected View** command.
8. Project the views shown, from View:1.
9. To explore the alignment characteristics of projected views, move the base view (View:1), down and to the left. All the views except the isometric view should move. Next, move the top view up and the right side view closer to the base view. You should be able to move the top view only along the Y axis, and the side view along the X axis. Finally, move the isometric view to observe a broken alignment.
10. The final drawing should look like the one shown.
11. Resave EX12.12.
12. Exit Autodesk Inventor or continue working with the program.

Exercise 12.12 continued on next page

Exercise 12.12 continued

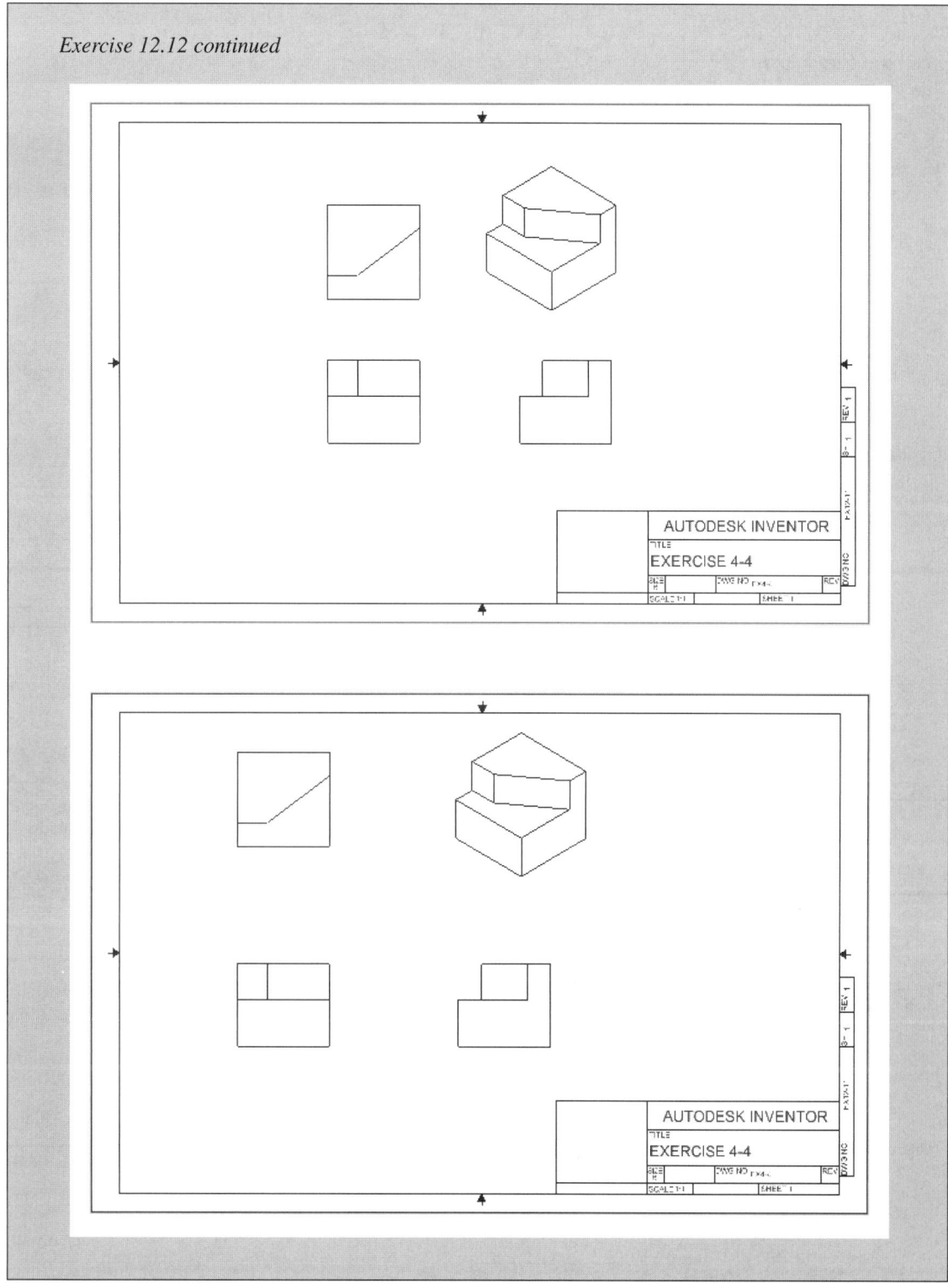

Creating Auxiliary Views

Often drawing views are needed that show the actual size and geometry of a surface that is not parallel to any of the projected views, including the front, top, bottom, left and right side, and back views. In order to show the true size and shape of an inclined surface, ***auxiliary views*** are required. See Figure 12.52.

FIGURE 12.52   An example of an auxiliary view (partial auxiliary view).

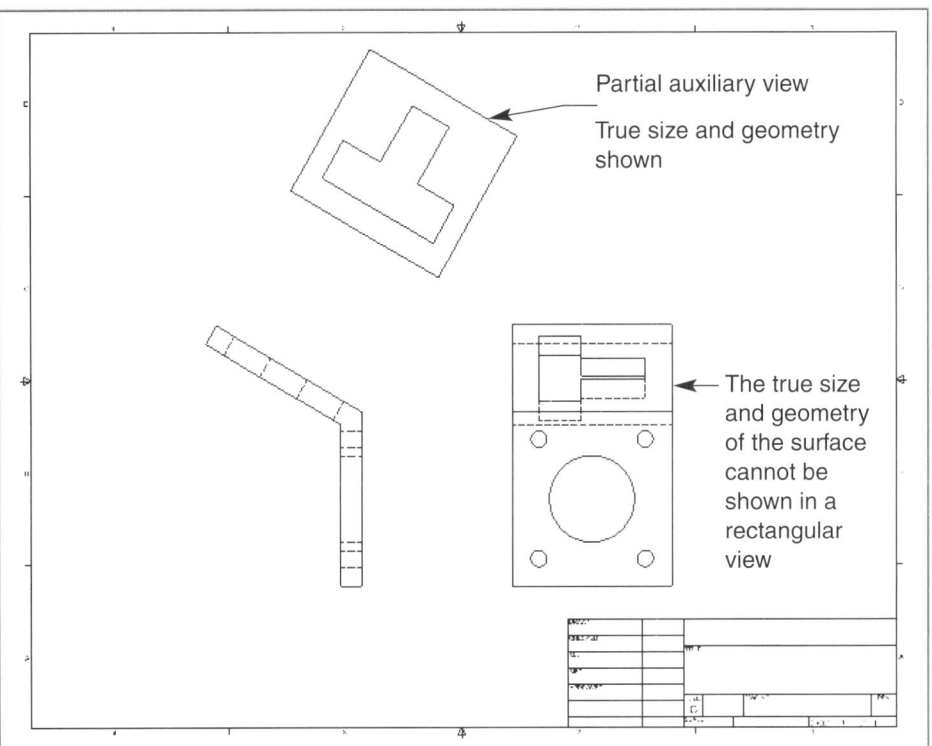

Partial auxiliary view

True size and geometry shown

The true size and geometry of the surface cannot be shown in a rectangular view

Once you have created a base drawing view, such as a front view, using the **Create View** tool, you can create an auxiliary view for drawing applications where an auxiliary view is needed. Like projected views, auxiliary views are generated from existing base views or dependent views, because model information is already stored in the drawing. The **Auxiliary View** tool allows you to create auxiliary views, which are projected from existing inclined surfaces onto the auxiliary plane. Assuming at least one view referencing a model exists in the drawing, you can use the **Auxiliary View** tool anytime during drawing development.

Access the **Auxiliary View** command using one of the following techniques:

✓ Pick the **Auxiliary View** button on the panel bar.

✓ Pick the **Auxiliary View** button on the **Drawing Management** toolbar.

✓ Right-click an existing parametric drawing view in the graphics window, and select **Auxiliary** from the **Create View** cascading submenu.

✓ Right-click an existing parametric drawing view name in the **Browser,** and select **Auxiliary** from the **Create View** cascading submenu.

✓ From the **Insert** pull-down menu, select the **Auxiliary View** option of the **Model Views** cascading submenu.

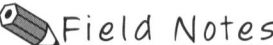 Field Notes

If you access the **Auxiliary View** command from the panel bar, **Drawing Management** toolbar, or the **Insert** pull-down menu, you must select the view you want to project from.

When you access the **Auxiliary View** command and a view is selected, the **Auxiliary View** dialog box is displayed. See Figure 12.53.

Use the following options in the **Auxiliary View** dialog box to define the auxiliary view:

FIGURE 12.53 The **Auxiliary View** dialog box.

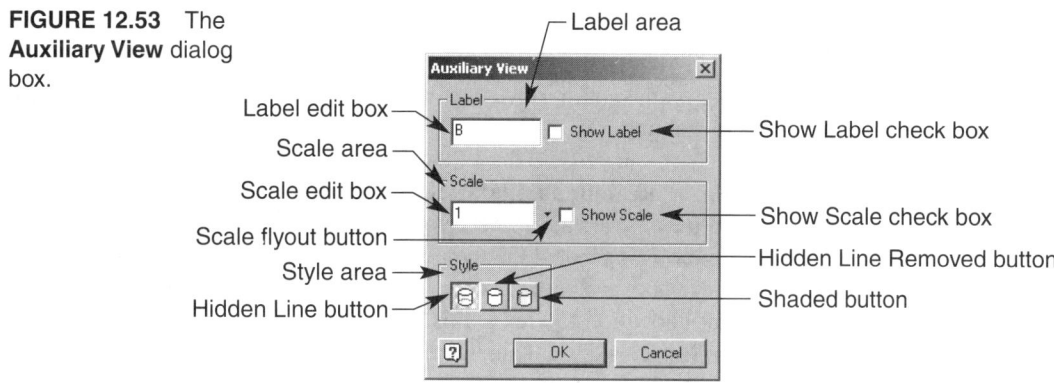

- **Label** This area allows you to modify the view label. You can edit the default name of the label in the **Label** edit box and pick the **Show Label** check box if you want the label displayed in the view.

- **Scale** Here you can define the scale of the view by entering a scale in the **Scale** edit box, or select a scale from the flyout button options. If you want the view scale to be displayed in the view, select **Show Scale** check box.

- **Style** This area allows you to specify the display of the view. You can choose to display the view with hidden lines by picking the **Hidden Line** button, without hidden lines by picking the **Hidden Line Removed** button, or as shaded by picking the **Shaded** button.

Once you have specified the label, scale, and display options in the **Auxiliary View** dialog box, pick the edge of the surface from which you want the auxiliary view to be projected. The selected edge can be parallel or perpendicular to the desired auxiliary view. Next, drag the preview of the auxiliary view to the desired location. See Figure 12.54.

Finally, generate the auxiliary view by selecting the location where you want the view to be displayed, or by picking the **OK** button in the **Auxiliary View** dialog box.

If the auxiliary view you create is not acceptable, you can redefine the surface edge by:

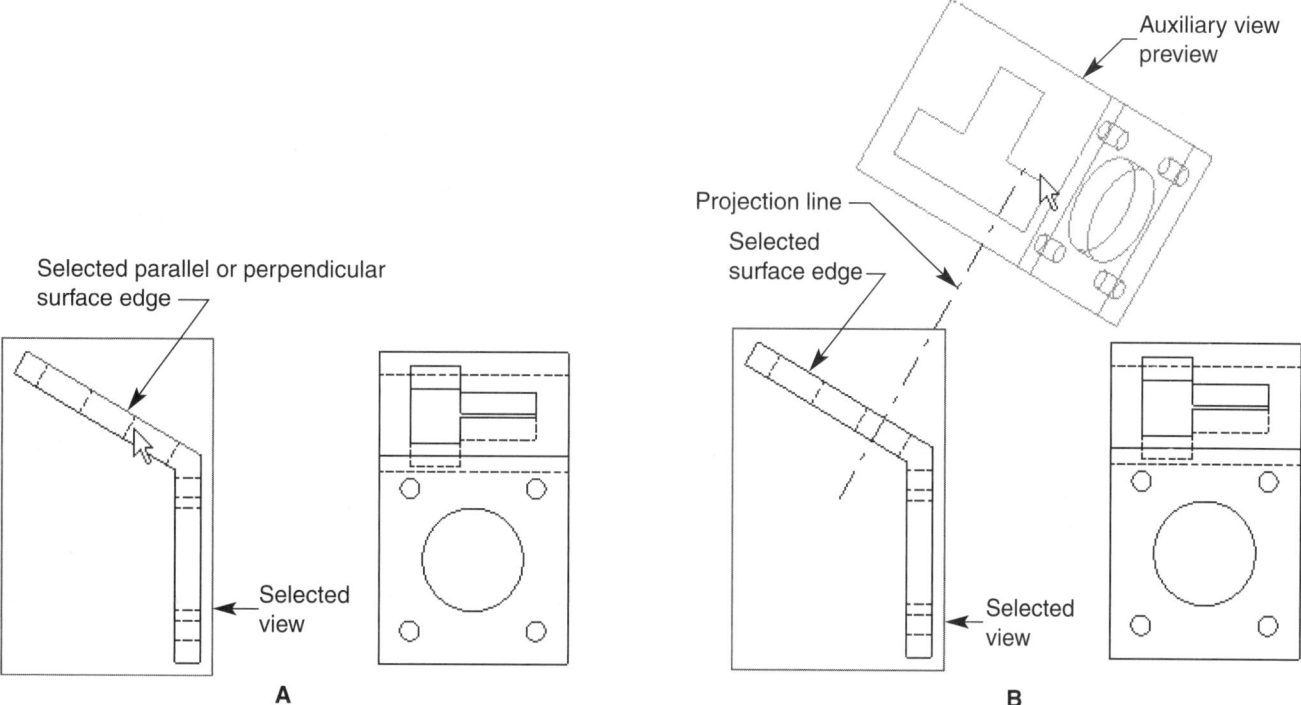

FIGURE 12.54 (A) An example of the selected auxiliary view edge. (B) The auxiliary view preview.

✓ Right-clicking inside the auxiliary view boundaries and selecting the **Realign auxiliary views** menu option.

✓ Right-clicking on the auxiliary view name in the **Browser,** and selecting the **Realign auxiliary views** menu option.

Once you access the **Realign auxiliary views** command, select the edge you want to use to redefine the auxiliary view, and pick the location where you want the view to be displayed.

Often auxiliary views are created that do not contain all the projected geometry from a selected view. These types of auxiliary views are known as partial auxiliary views. See Figure 12.55. There are no tools in Autodesk Inventor that allow you to create a partial auxiliary view directly, unless you want to create an auxiliary view section. However, partial auxiliary views can be easily created by turning off the visibility of specific geometry. To disable geometry visibility right-click the geometry you want to hide and deselect the **Visibility** menu option. See Figure 12.56. Continue tuning off the visibility of geometry until the partial view is created.

FIGURE 12.55 An example of a partial auxiliary view.

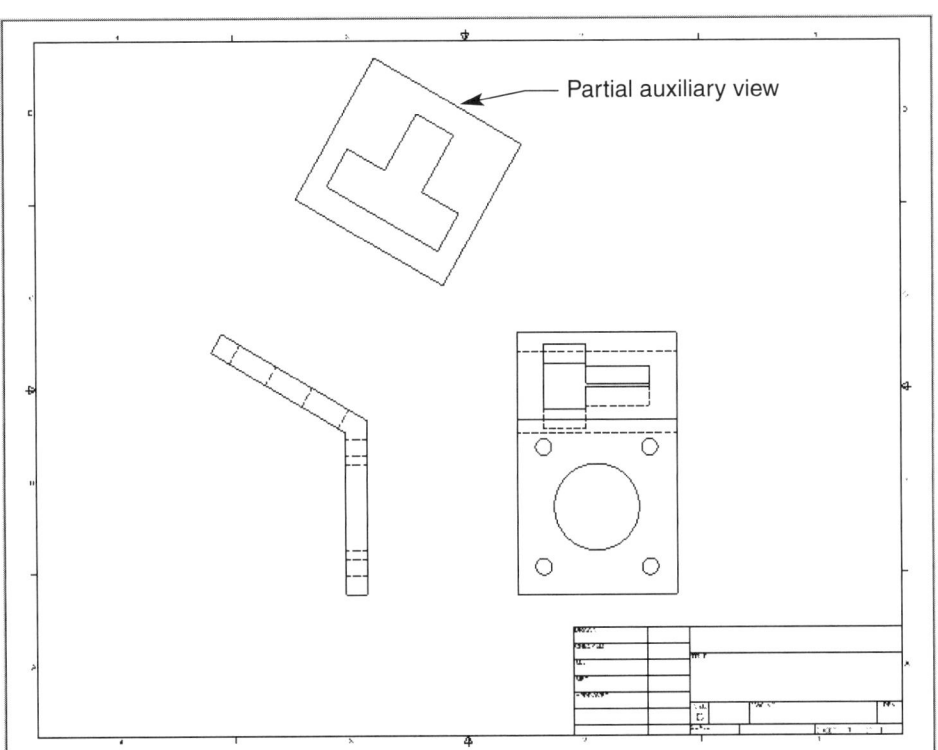

Partial auxiliary view

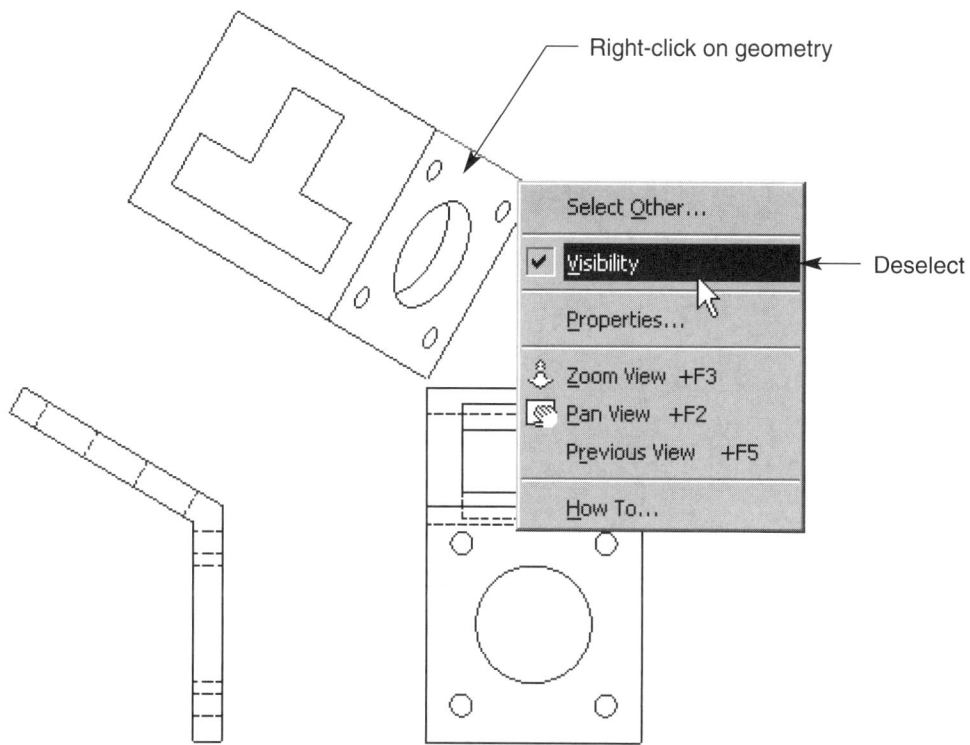

Right-click on geometry

Deselect

FIGURE 12.56 Creating a partial auxiliary view.

Field Notes

You can disable the visibility of most geometry in all types of views, not just auxiliary views.

Exercise 12.13

1. Continue from EX12.12, or launch Autodesk Inventor.
2. Open EX12.11.
3. Save a copy of EX12.11 as EX12.13.idw.
4. Close EX12.11 without saving, and open EX12.13.
5. Access the **Auxiliary View** command.
6. Select View:1 (the base view).
7. Specify the following information in the **Auxiliary View** dialog box:

   Label: A (do not show label)
   Scale: 1:1 (do not show scale)
   Style: Hidden Line

8. Select one of the specified edges.
9. Pick a location for the auxiliary view, similar to the location shown.
10. Turn off the visibility of the geometry shown to create a partial auxiliary view.
11. The final drawing should look like the drawing shown.
12. Resave EX12.13.
13. Exit Autodesk Inventor or continue working with the program.

Exercise 12.13 continued on next page

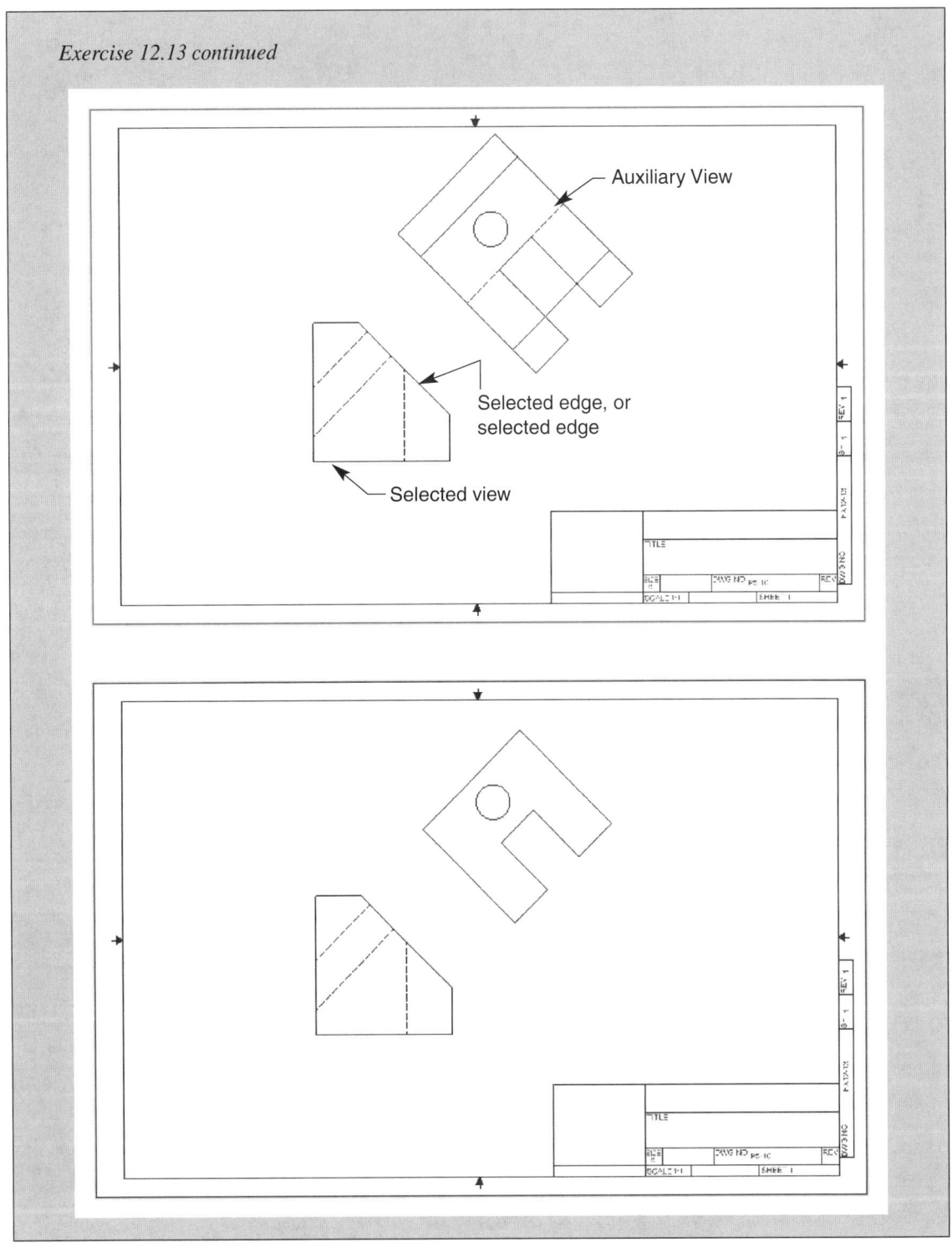

Developing Section Views

Many part models have complex interior geometry that is difficult to visualize and display in a drawing. To overcome this problem, *section views,* also known as *sections,* are used to expose the interior features of a part. Hidden lines are omitted in sections, so the geometry shown using sections can be dimensioned and provides a much clearer representation of the part. See Figure 12.57.

FIGURE 12.57 An
example of section
views (full sections).

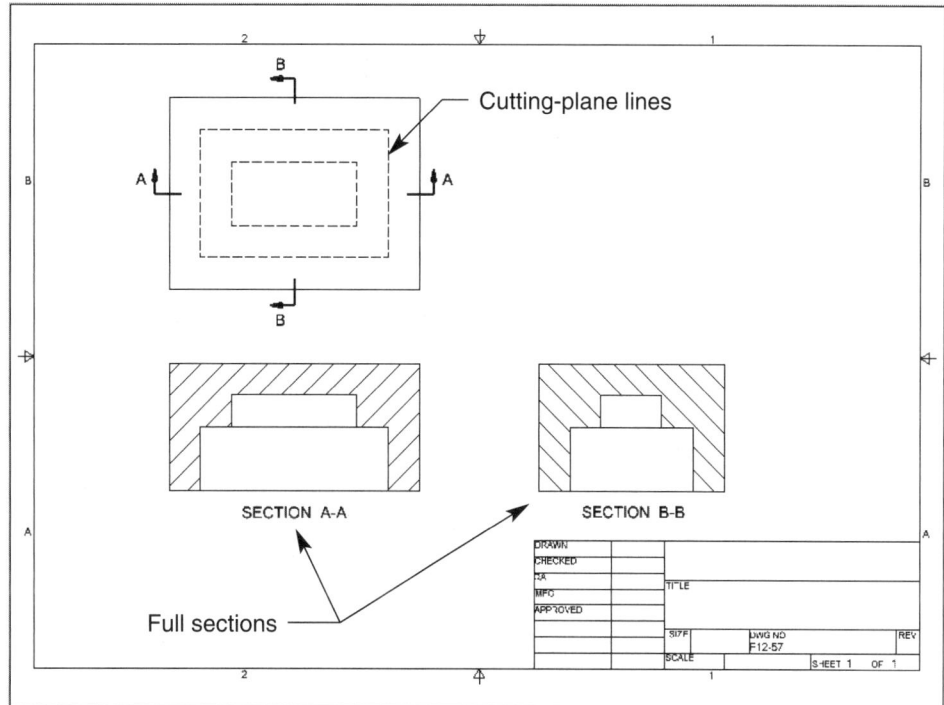

Once you have created a base drawing view, such as a front or top view, using the **Create View** tool,
you can create a section view for drawing applications where a section is needed. Like projected views, section views are generated from existing base views or dependent views, because model information is
already stored in the drawing. The **Section View** tool allows you to create section views, which are projected from a specified cutting plane, through an existing surface. Assuming at least one view, referencing a
model, exists in the drawing, you can use the **Section View** tool any time during drawing development.

Access the **Section View** command using one of the following techniques:

✓ Pick the **Section View** button on the panel bar.

✓ Pick the **Section View** button on the **Drawing Management** toolbar.

✓ Right-click an existing parametric drawing view in the graphics window, and select **Section**
from the **Create View** cascading submenu.

✓ Right-click an existing parametric drawing view name in the **Browser,** and select **Section**
from the **Create View** cascading submenu.

✓ From the **Insert** pull-down menu, select the **Section View** option of the **Model Views** cascading submenu.

 Field Notes

> If you access the **Section View** command from the panel bar, **Drawing Management** toolbar,
> or the **Insert** pull-down menu, you must select the view from which you want to project the
> section.

When you access the **Section View** command and a view is selected, the next step in creating a section view is to define a cutting-plane line. The ***cutting-plane line*** represents the cutting plane of the section,

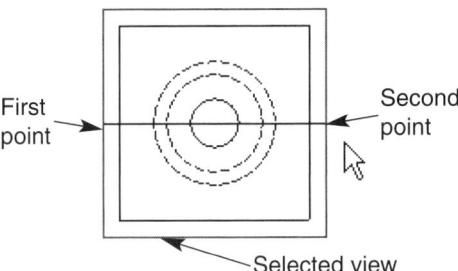

FIGURE 12.58
Sketching a cutting-plane line (full section).

or the location where the view is sliced. Placing a cutting-plane line is just like sketching a line in the Autodesk Inventor sketch environment. See Figure 12.58. The cutting-plane line you create defines the type of section that is generated. Figure 12.58 shows a single cutting-plane line that extends the entire length of the view, resulting in a full section. If you want to show a half, offset, or aligned section, for example, you must sketch a cutting-plane line that will result in the desired section. See Figure 12.59.

Once you have sketched the cutting-plane line, right-click and select the **Continue** menu option. When you select **Continue,** the cutting-plane line is created, and the **Section View** dialog box is displayed. See Figure 12.60. Use the following options in the **Section View** dialog box to define the section view:

- **Label**   This area allows you to modify the view label. You can edit the default name of the label in the **Label** edit box, and pick the **Show Label** check box if you want the label displayed in the view.

- **Scale**   Here you can define the scale of the view by entering a scale in the **Scale** edit box, or select a scale from the flyout button options. If you want the view scale to be displayed in the view, select **Show Scale** check box.

- **Style**   This area allows you to specify the display of the view. You can choose to display the view with hidden lines by picking the **Hidden Line** button, without hidden lines by picking the **Hidden Line Removed** button, or as shaded by picking the **Shaded** button.

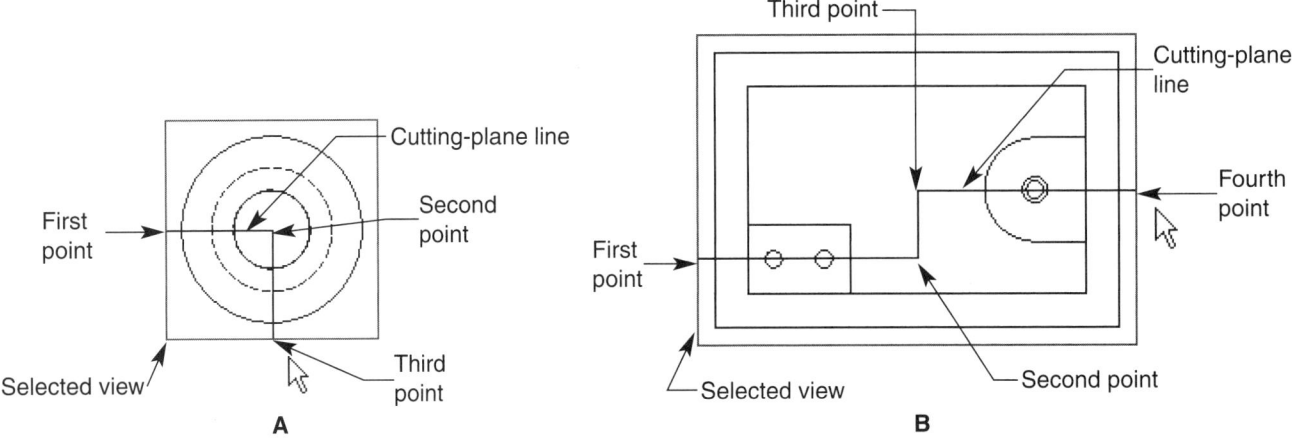

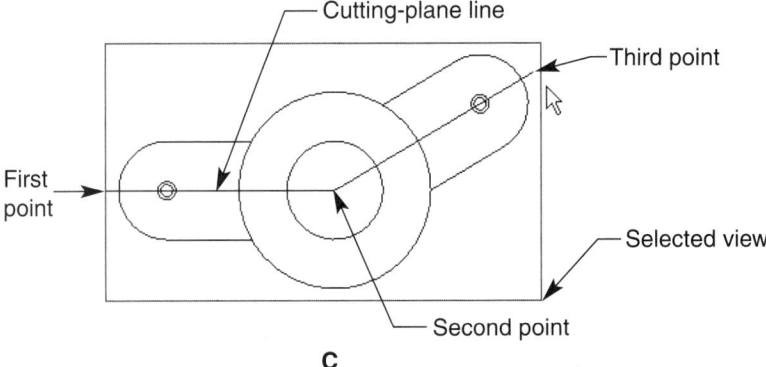

FIGURE 12.59 Sketching cutting-plane lines for (A) a half section, (B) an offset section, and (C) an aligned section.

FIGURE 12.60 The **Section View** dialog box.

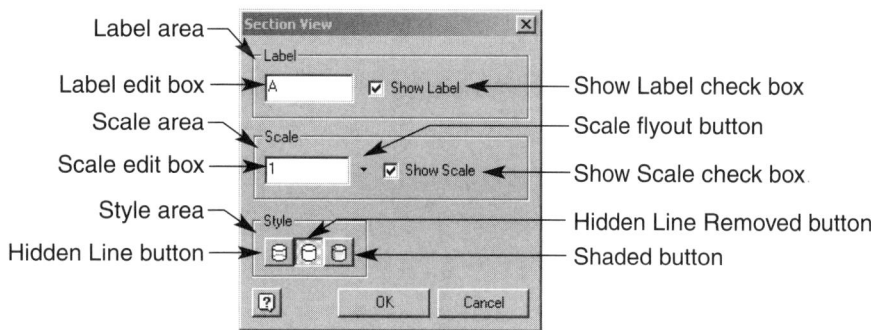

Once you have specified the label, scale, and display options in the **Section View** dialog box, drag the preview of the section view to the desired location. See Figure 12.61.

Finally, generate the section view by selecting the location where you want the view to be displayed or by picking the **OK** button in the **Section View** dialog box.

You will notice that the section lines, cutting plane lines, and labels are automatically created based on your specifications. However, you can modify any of the following section view components, using the techniques discussed:

■ **Section lines** *Section lines,* or hatching, show where the section has cut through the material. The default section lines displayed when you create a section view correspond to the settings in the **Hatch** tab of the **Drafting Standards** dialog box, previously discussed.

To change the display of the section lines in a section view, right-click on the hatching, and pick the **Modify Hatch…** menu option to open the **Modify Hatch Pattern** dialog box. See Figure 12.62. Refer to the previous discussion about the **Fill/Hatch Sketch Region** tool for information regarding the options in the **Modify Hatch Pattern** dialog box.

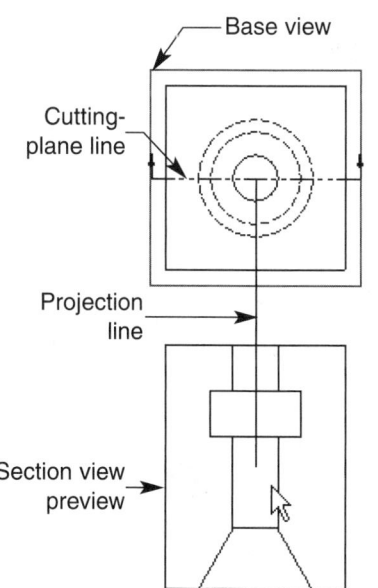

FIGURE 12.61 The section view preview.

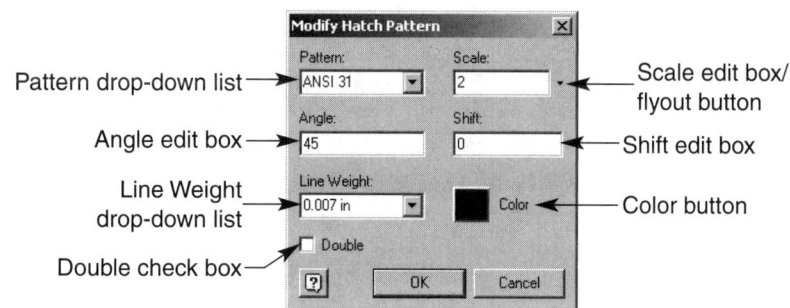

FIGURE 12.62 The **Modify Hatch Pattern** dialog box.

FIGURE 12.63 (A) A full-length cutting-plane line. (B) Cutting-plane line ends.

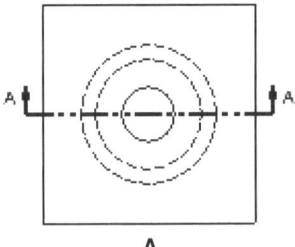
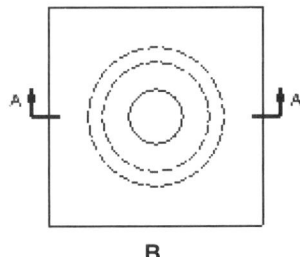

A   **B**

- **Cutting-plane lines**   There are a number of options for modifying cutting-plane lines. To change the size or location of the lines, right-click on the cutting-plane lines or the sketch listed in the **Browser,** and select the **Edit +S** menu option to enter the sketch environment. Cutting plane lines are sketched features, and are edited like any other sketched lines. Once you have changed the size or location of the sketch, exit or finish the sketch environment to complete the modification.

    In addition to changing the actual physical size or location of the cutting plane lines, you can reverse the direction of the cutting-plane direction arrows by right-clicking on the cutting-plane line and selecting the **Reverse Direction** menu option. Also, many sections where the extent of the cutting-plane is obvious do not contain full-length cutting-plane lines, as shown in Figure 12.63. To remove the inner section of the cutting-plane line, right-click on the cutting-plane line and deselect the **Show Entire Line** menu option.

- **Label**   You can edit the default label parameters by right-clicking inside the section view boundaries or the section view name in the **Browser** and selecting the **Edit View...** menu option. However, you can modify and add information to the label text by right-clicking the label and selecting the **Edit View Label...** menu option to display the **Format Text** dialog box previously discussed.

Exercise 12.14

1. Continue from EX12.13, or launch Autodesk Inventor.
2. Open a new inch drawing file.
3. Delete the existing border and title block.
4. Change the sheet size to C, if it is not currently.
5. Access the **Create View** command.
6. Specify the following information in the **Create View** dialog box:

    File: P6.6
    View: Front
    Scale: 1:1
    Style: Hidden Line
    Deselect the **Get Model Dimensions** check box.

7. Pick a location for the drawing view similar to the location shown.
8. Access the **Section View** command.
9. Select View:1 (the base view).
10. Using the center of the part for reference, sketch the line shown.
11. Right-click and select the **Continue** menu option.
12. Specify the following information in the **Section View** dialog box:

Exercise 12.14 continued on next page

Exercise 12.14 continued

Label: A (show label)
Scale: 1:1 (do not show scale)
Style: Hidden Line Removed

13. Pick a location for the section view, similar to the location shown.
14. Change the scale of the section lines to 2:1.
15. The final drawing should look like the drawing shown.
16. Save the drawing as EX12.14.
17. Exit Autodesk Inventor or continue working with the program.

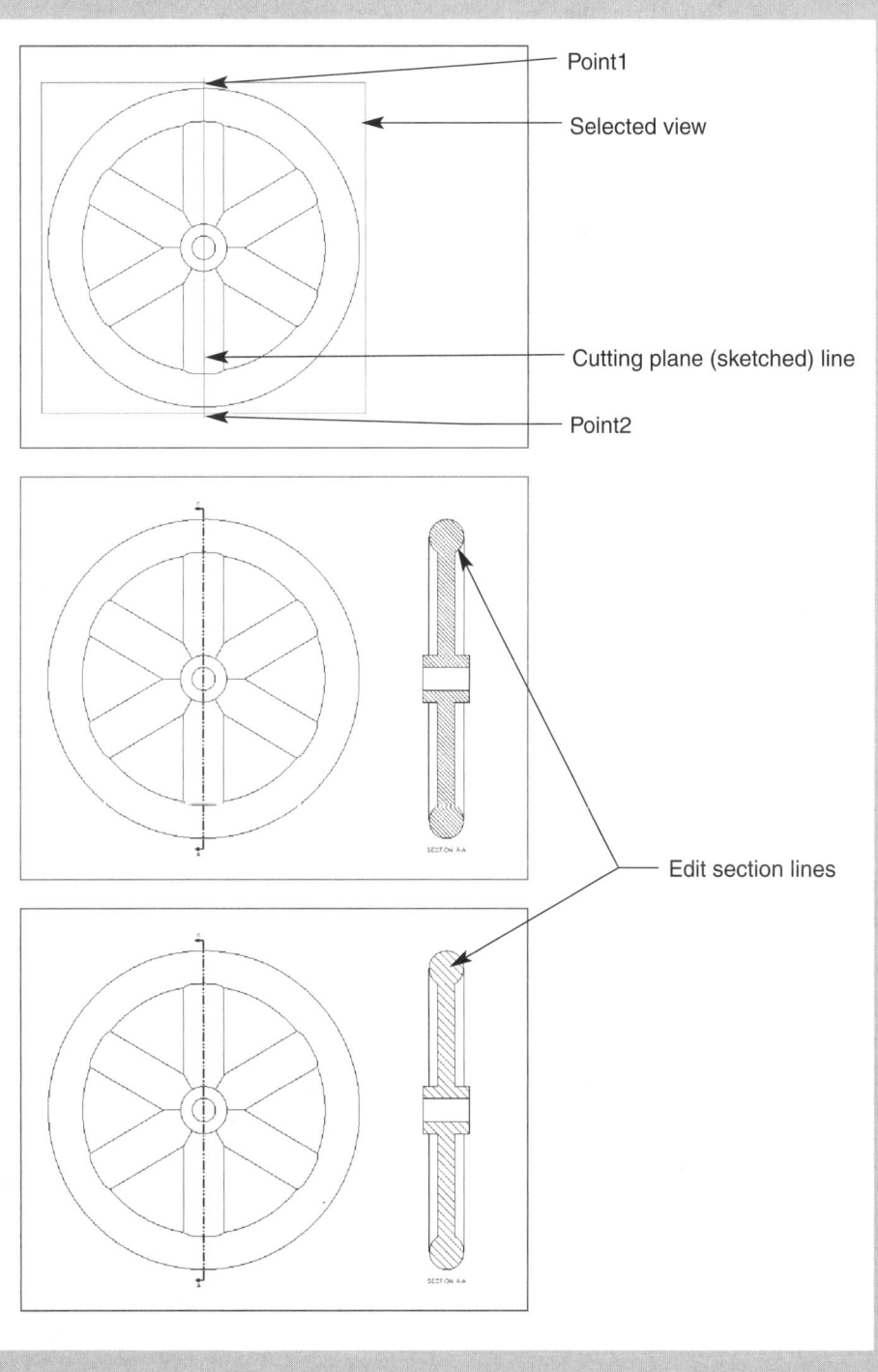

Point1

Selected view

Cutting plane (sketched) line

Point2

Edit section lines

Creating Detail Views

Often part models contain complex geometry that is small in relation to the overall size of the part. This type of small geometry is typically difficult to see and dimension in a drawing. To overcome this problem, *detail views* are used to magnify the size of specific features of a part. For many parts, the use of detail views creates a cleaner, easier to understand drawing, and may even be necessary for dimensioning purposes. See Figure 12.64.

Once you have created a base drawing view, such as a front or top view, using the **Create View** tool, you can create a detail view for drawing applications where a detail view is needed. Like projected views, detail views are generated from existing base views or dependent views because model information is already stored in the drawing. The **Detail View** tool allows you to create detail views, which are projected and magnified from specified boundaries on an existing surface. Assuming at least one view referencing a model exists in the drawing, you can use the **Detail View** tool any time during drawing development.

Access the **Detail View** command using one of the following techniques:

✓ Pick the **Detail View** button on the panel bar.

✓ Pick the **Detail View** button on the **Drawing Management** toolbar.

✓ Right-click an existing parametric drawing view in the graphics window, and select **Detail** from the **Create View** cascading submenu.

✓ Right-click an existing parametric drawing view name in the **Browser,** and select **Detail** from the **Create View** cascading submenu.

✓ From the **Insert** pull-down menu, select the **Detail View** option of the **Model Views** cascading submenu.

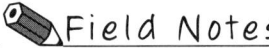 Field Notes

> If you access the **Detail View** command from the panel bar, **Drawing Management** toolbar, or the **Insert** pull-down menu, you must select the view from which you want to create the detail.

When you access the **Detail View** command and a view is selected, the **Detail View** dialog box is displayed. See Figure 12.65.

FIGURE 12.64 An example of a detail view.

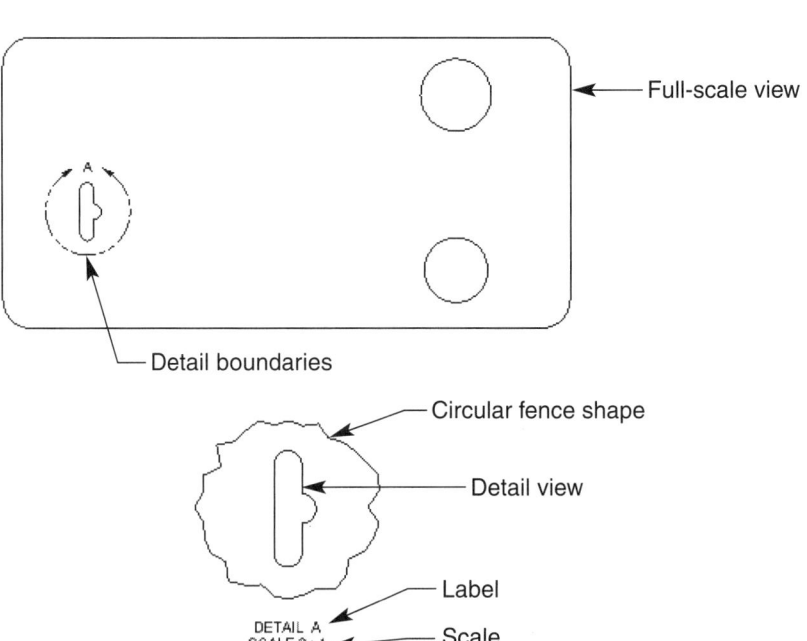

FIGURE 12.65 The **Detail View** dialog box.

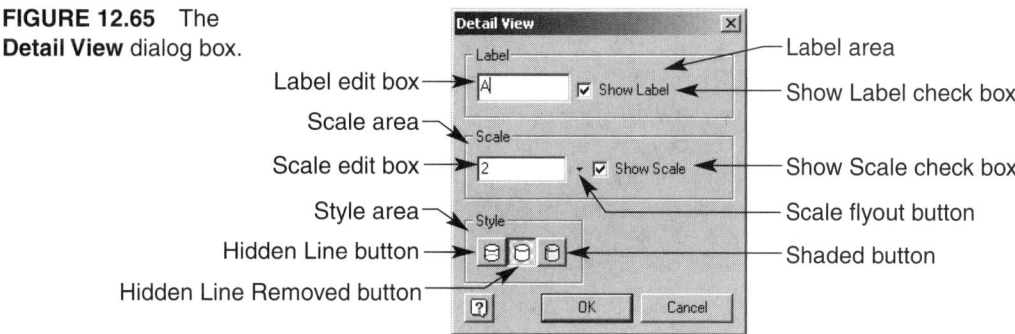

Use the following options in the **Detail View** dialog box to define the detail view options:

- **Label**   This area allows you to modify the view label. You can edit the default name of the label in the **Label** edit box and pick the **Show Label** check box if you want the label displayed in the view.

- **Scale**   Here you can define the scale of the view by entering a scale in the **Scale** edit box, or select a scale from the flyout button options. If you want the view scale to be displayed in the view, select **Show Scale** check box.

- **Style**   This area allows you to specify the display of the view. You can choose to display the view with hidden lines by picking the **Hidden Line** button, without hidden lines by picking the **Hidden Line Removed** button, or as shaded by picking the **Shaded** button.

Once you have specified the label, scale, and display options in the **Detail View** dialog box, the next step in creating a detail view is to define a boundary. The boundary consists of a phantom line style circle, which represents the detail area being magnified. You can create a detail view with a circular or a rectangular fence, as shown in Figure 12.66. Placing a detail boundary is just like sketching a circle or a four-sided polygon in the Autodesk Inventor sketch environment.

To create a detail view with a circular fence, select the boundary center, followed by the boundary extents, or diameter of the circle. See Figure 12.67.

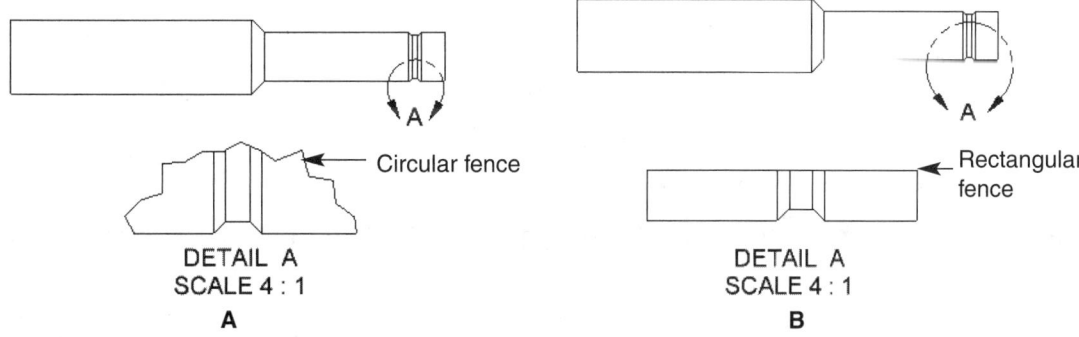

FIGURE 12.66   A detail view with (A) a circular fence and (B) a rectangular fence.

FIGURE 12.67   Sketching a detail view boundary circle.

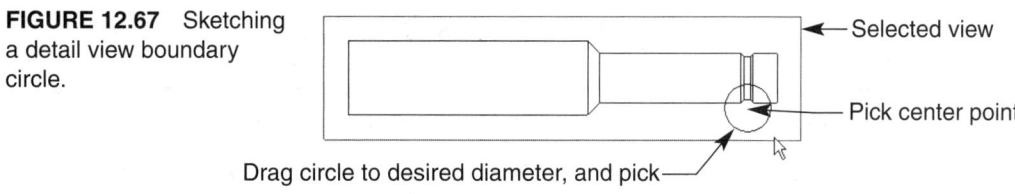

To create a detail view with a rectangular fence, you must first right-click inside the graphics window and select the **Rectangular Fence** menu option, because the circular fence is default. Then select the boundary center, followed by the boundary extents or edge of the rectangle. See Figure 12.68.

Finally, generate the detail view by selecting the location where you want the view to be displayed, or by picking the **OK** button in the **Detail View** dialog box.

You will notice that the detail view boundary and labels are automatically created based on your specifications. Occasionally, as shown in Figure 12.69, the detail display is not acceptable. You can modify the following detail view components, using the techniques discussed:

- **Boundaries** Though you cannot directly edit the sketch of a boundary using sketch tools, you can change the size and location of the boundary and the location of the boundary arrows and detail label. To change the location of the boundary by moving the center point, move your cursor over the boundary, or select the boundary until the boundary is highlighted and you see green dots. Then, pick the green dot in the center and drag the boundary to the desired location. See Figure 12.70. Similarly, to increase or decrease the boundary size, move your cursor over the boundary or select the boundary until the boundary is highlighted and you see green dots. Then, pick one of the three green dots located around the boundary circumference and drag the boundary edge to the desired location. See Figure 12.71.

Field Notes

> When you adjust the size or location of the detail boundary, the detail view automatically updates to reflect the changes.

- **Boundary label** To change the location of the boundary label, move your cursor over the boundary label until you see the A icon. Then select and drag the label to the desired location by rotating the boundary display. See Figure 12.72.

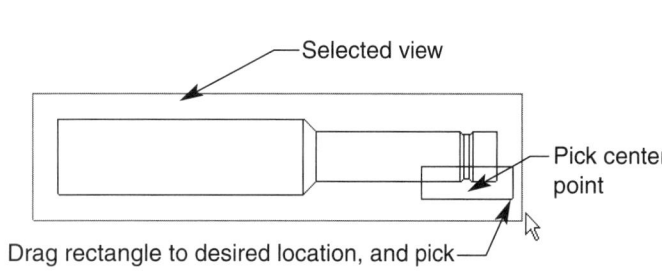

FIGURE 12.68 Sketching a detail view boundary rectangle.

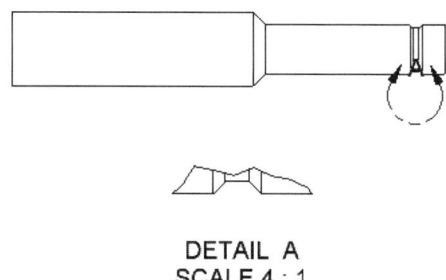

FIGURE 12.69 An unsatisfactory detail display.

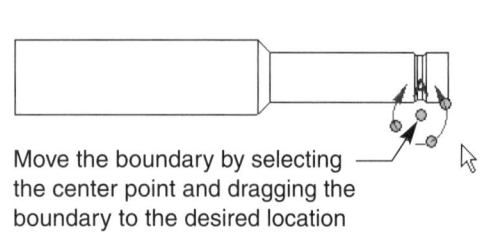

FIGURE 12.70 Moving a detail boundary.

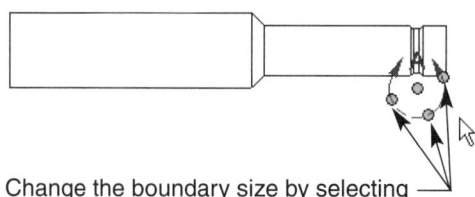

FIGURE 12.71 Changing the size of a detail boundary.

FIGURE 12.72
Rotating the boundary
and the boundary label.

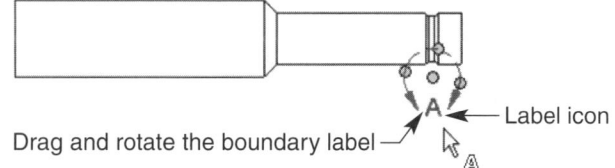

Drag and rotate the boundary label ⟋   Ⓐ

■ **Label**   You can edit the default label parameters by right-clicking inside the detail view
boundaries or the detail view name in the **Browser,** and selecting the **Edit View...** menu
option. However, you can modify and add information to the label text by right-clicking the
label and selecting the **Edit View Label...** menu option to display the **Format Text** dialog
box previously discussed.

Field Notes

Detail views are not bound by any alignment constraints and can be freely moved without
affecting other views.

Exercise 12.15

1. Continue from EX12.14, or launch Autodesk Inventor.
2. Open EX12.14.
3. Save a copy of EX12.14 as EX12.15.idw.
4. Close EX12.14 without saving, and open EX12.15.
5. Change the sheet size to D.
6. Move the existing views to the locations shown.
7. Access the **Detail View** command.
8. Select the section view.
9. Specify the following information in the **Detail View** dialog box:

   Label: B (show label)
   Scale: 2:1 (show scale)
   Style: Hidden Line Removed

10. Right-click and select the **Rectangular Fence** menu option.
11. Select the center of the fence shown, and drag the fence edge to a location similar to the
    location shown.
12. Pick a location for the detail view, similar to the location shown.
13. Rotate the boundary label as shown.
14. Access the **Detail View** command.
15. Select the section view.
16. Specify the following information in the **Detail View** dialog box:

    Label: C (show label)
    Scale: 2:1 (show scale)
    Style: Hidden Line Removed

17. Ensure the **Circular Fence** menu option is selected.
18. Select the center of the fence shown, and drag the fence circle edge to a location similar to
    the location shown.
19. Pick a location for the detail view, similar to the location shown.
20. Rotate the boundary label as shown.
21. The final drawing should look like the drawing shown.

Exercise 12.15 continued on next page

Exercise 12.15 continued

22. Resave EX12.15.
23. Exit Autodesk Inventor or continue working with the program.

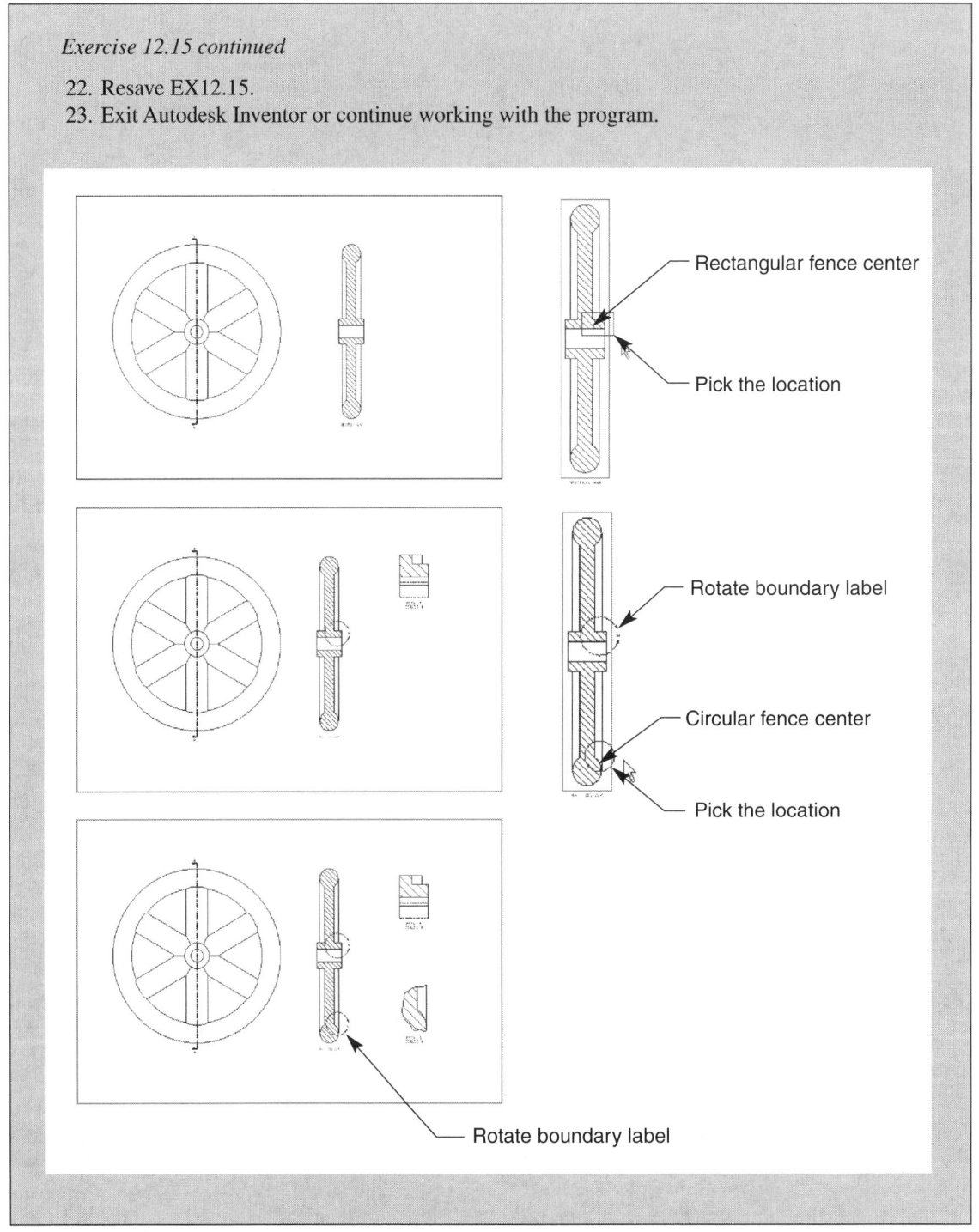

Rectangular fence center

Pick the location

Rotate boundary label

Circular fence center

Pick the location

Rotate boundary label

Creating Broken Views

When you are creating a drawing of a long, constant shape part, the view may not fit on the desired paper size, or may look strange in reference to the rest of the drawing. See Figure 12.73.

To overcome this problem, **broken views** are used to shorten the view. For many long, constant shape parts, the use of breaks is required to display views, or increase the view scale, without increasing the paper size. See Figure 12.74.

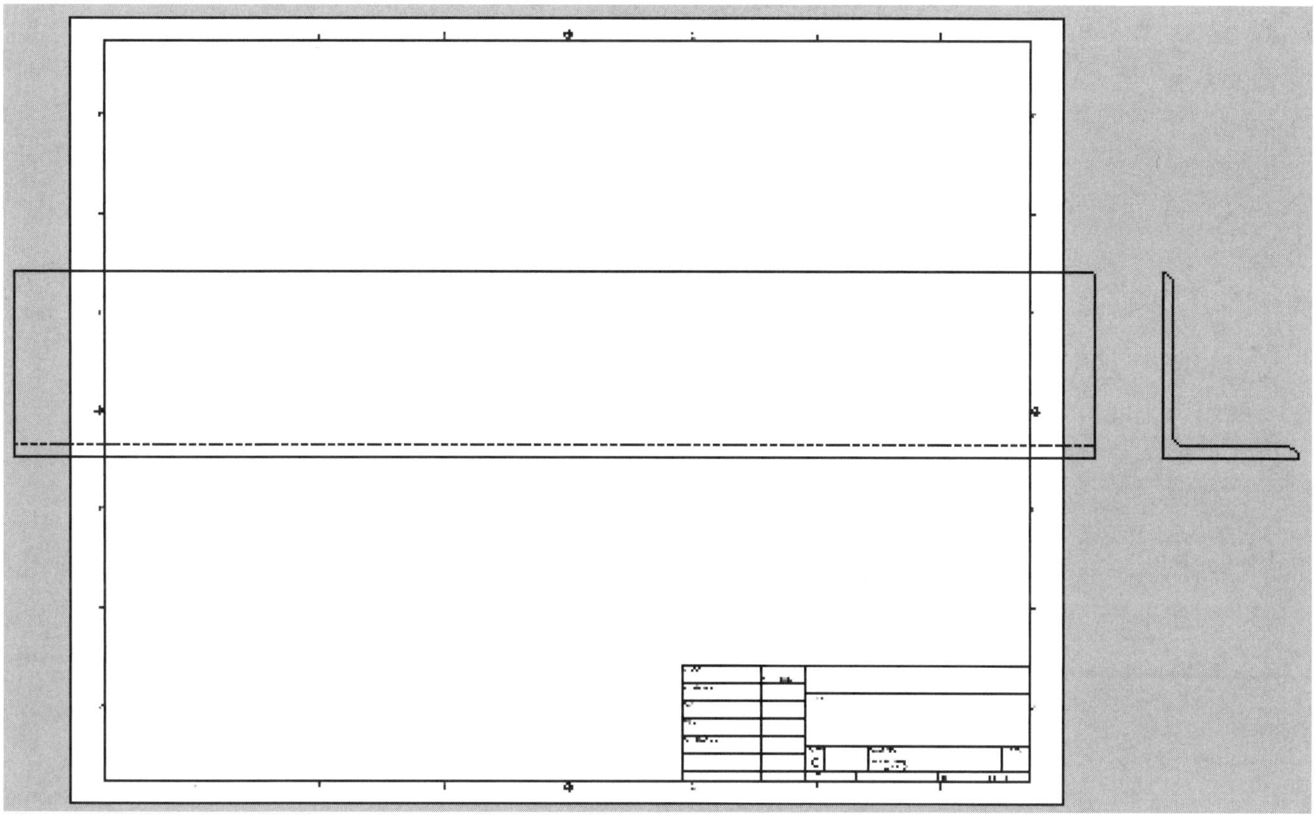

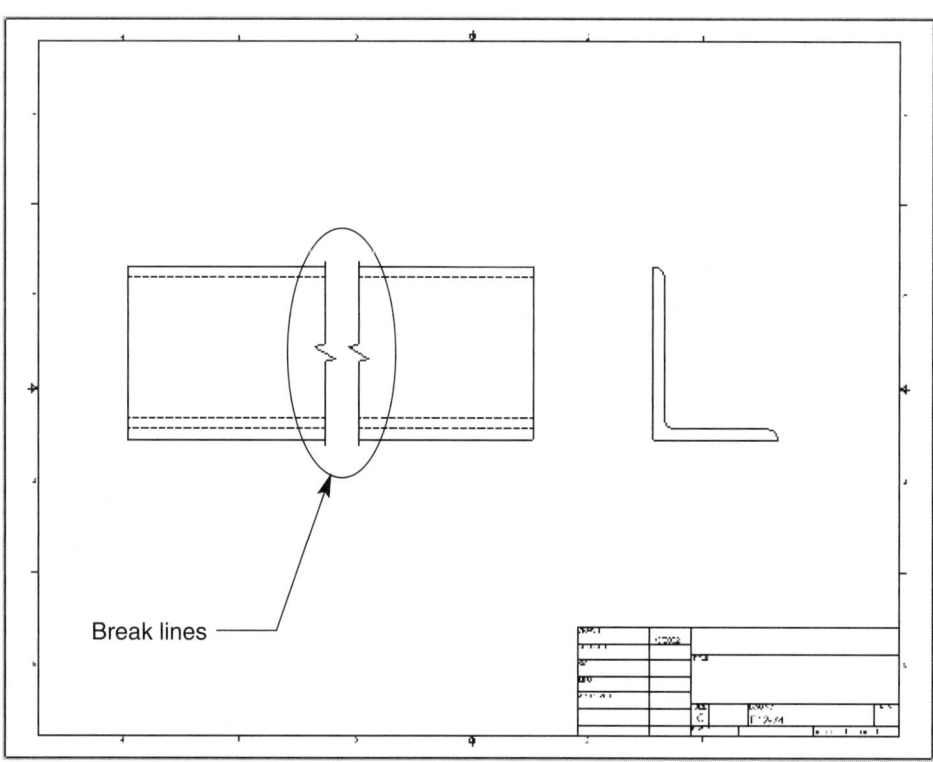

FIGURE 12.73   An example of a long, constant shape, unbroken part view.

FIGURE 12.74
Breaking the view
shown in Figure 12.73
to fit the desired paper
size without reducing
the view scale.

Break lines

**Field Notes**

> Breaking a drawing view does not change the actual parametric size, shape, or dimensions of the part.

Once you have created a base drawing view, such as a front or top view, using the **Create View** tool, you can create a broken view for drawing applications where a broken view is needed. Like projected views, broken views are generated from existing base views or dependent views, because model information is already stored in the drawing. However, breaking a view modifies the existing view and does not create a new view.

Broken views are created using the **Broken View** tool, which can be used anytime during drawing development. Access the **Broken View** command using one of the following techniques:

✓ Pick the **Broken View** button on the panel bar.

✓ Pick the **Broken View** button on the **Drawing Management** toolbar.

✓ Right-click an existing parametric drawing view in the graphics window, and select **Broken** from the **Create View** cascading submenu.

✓ Right-click an existing parametric drawing view name in the **Browser,** and select **Broken** from the **Create View** cascading submenu.

✓ From the **Insert** pull-down menu, select the **Broken View** option of the **Model Views** cascading submenu.

Field Notes

> If you access the **Broken View** command from the panel bar, **Drawing Management** toolbar, or the **Insert** pull-down menu, you must select the view you want to break.

When you access the **Broken View** command and a view is selected, the **Broken View** dialog box is displayed. See Figure 12.75.

Use the following options in the **Broken View** dialog box to define the broken view display:

■ **Style**   This area contains the **Rectangular Style** and **Structural Style** buttons. Pick the type of break style you want to use for your particular application. See Figure 12.76.

■ **Orientation**   This area defines the orientation and direction of the break. Pick the **Horizontal Orientation** button when breaking a horizontal view, and pick the **Vertical Orientation** button when breaking a vertical view. See Figure 12.77.

FIGURE 12.75   The **Broken View** dialog box.

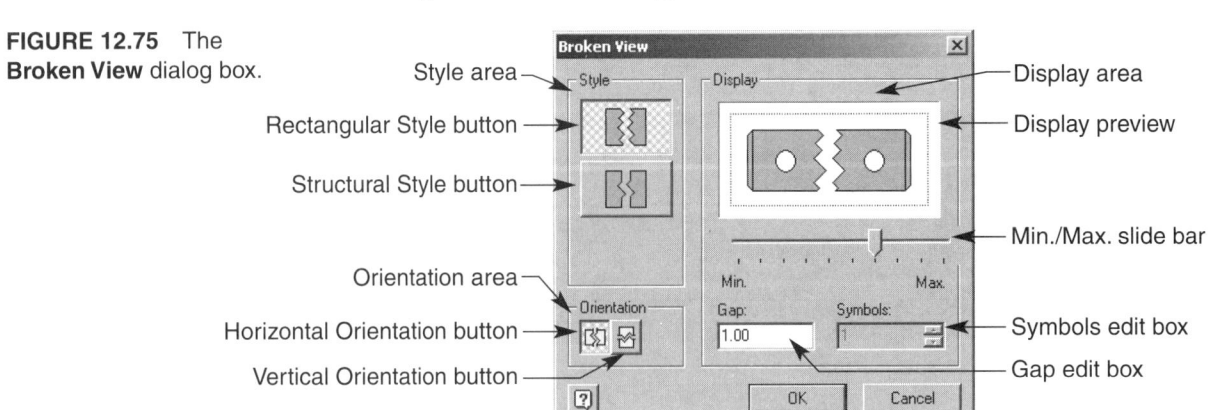

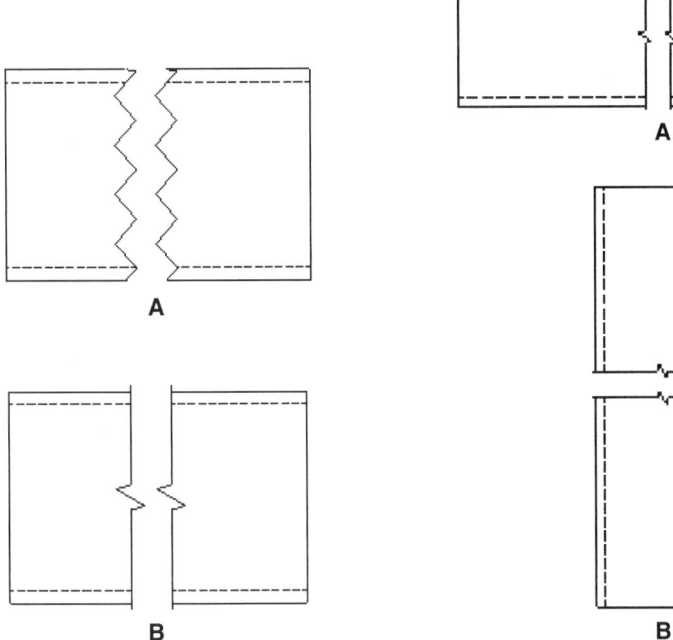

FIGURE 12.76 An example of (A) a rectangular break style and (B) a structural break style (one break symbol).

FIGURE 12.77 An example of (A) a horizontal orientation broken view and (B) a vertical orientation broken view.

Field Notes

Typically, the correct orientation is automatically chosen, depending on the view you select.

- **Min./Max.** This slide bar allows you to adjust the size and number of the break edges when creating a rectangular style break, or the size of break symbols when creating a structural style break. See Figure 12.78.

FIGURE 12.78 (A) A rectangular break style with a minimum edge size (large number of edges) and small gap value. (B) A structural break style with a maximum break symbol size and large gap value.

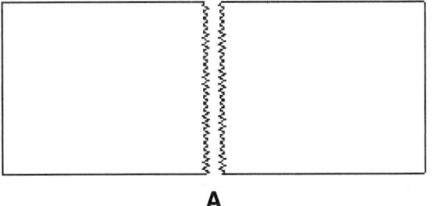

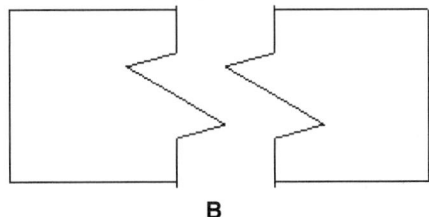

- **Gap**   Use this edit box to enter the gap or distance between break lines. See Figure 12.78. The units specified in this edit box correspond to those in the current drawing.

- **Symbols**   This edit box is available only when creating a structural style broken view and allows you to adjust the number of break symbols along the break lines.

Once you have specified all the broken view display options in the **Broken View** dialog box, the next step in creating a broken view is to define the break. The break is the section of view that you want to remove and define as the broken length of the part. To specify the break, pick the location of the first break line, then select the location for the second break line. See Figure 12.79. The area between the first and second selections is removed by the break. Once you pick the location of the second break line, the broken view is generated.

To edit the break information you specified in the **Broken View** dialog box, right-click on the break lines and select the **Edit Break...** menu option. To change the location of the break lines without shortening or expanding the broken view, move your cursor over the break lines, or select the break lines until they are highlighted and you see the green dot. Then pick the green dot and drag the break lines to the desired location. If you want to shorten the broken view, select one of the break lines and drag it away from the center of the break lines. Similarly, to extend the broken view, drag one of the break lines toward and past the center of the break lines. See Figure 12.80.

Field Notes

Autodesk Inventor does not currently provide options for creating the ASME preferred appearance for cylindrical and tubular breaks.

FIGURE 12.79
Selecting break line locations.

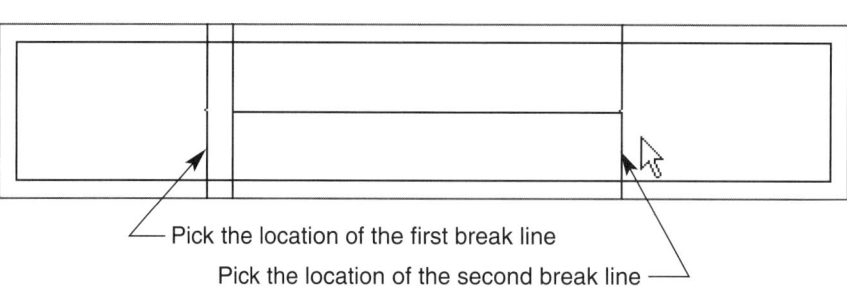

Pick the location of the first break line

Pick the location of the second break line

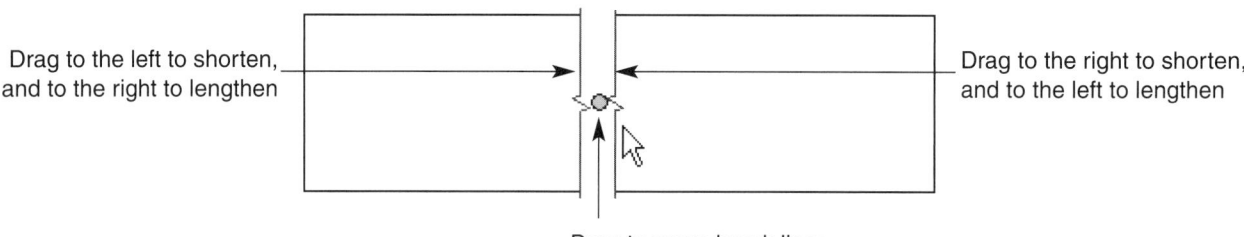

Drag to the left to shorten, and to the right to lengthen

Drag to the right to shorten, and to the left to lengthen

Drag to move break lines

FIGURE 12.80   Moving break lines and shortening or lengthening the broken view.

Exercise 12.16

1. Continue from EX12.15, or launch Autodesk Inventor.
2. Open EX12.7.
3. Save a copy of EX12.7 as EX12.16.idw.
4. Close EX12.7 without saving, and open EX12.16.
5. Access the **Create View** command.
6. Specify the following information in the **Create View** dialog box:

   File: P6.7
   View: Custom view as shown
   Scale: 2:1
   Style: Hidden Line
   Deselect the **Get Model Dimensions** check box.

7. Pick a location for the drawing view similar to the location shown.
8. Access the **Broken View** command.
9. Select View:1 (the base view).
10. Specify the following information in the **Broken View** dialog box:

    Style: Rectangular
    Min./Max. slide bar: Half way between Min. and Max.
    Gap: 1
    Orientation: Horizontal

11. Select the location of the break lines as shown.
12. Access the **Broken View** command.
13. Select View:1 (the base view).
14. Specify the following information in the **Broken View** dialog box:

    Style: Structural
    Min./Max. slide bar: Max. setting
    Gap: 1
    Symbols: 1
    Orientation: Horizontal

15. Select the location of the break lines as shown.
16. The final drawing should look similar to the drawing shown.
17. Resave EX12.16.
18. Exit Autodesk Inventor or continue working with the program.

Exercise 12.16 continued on next page

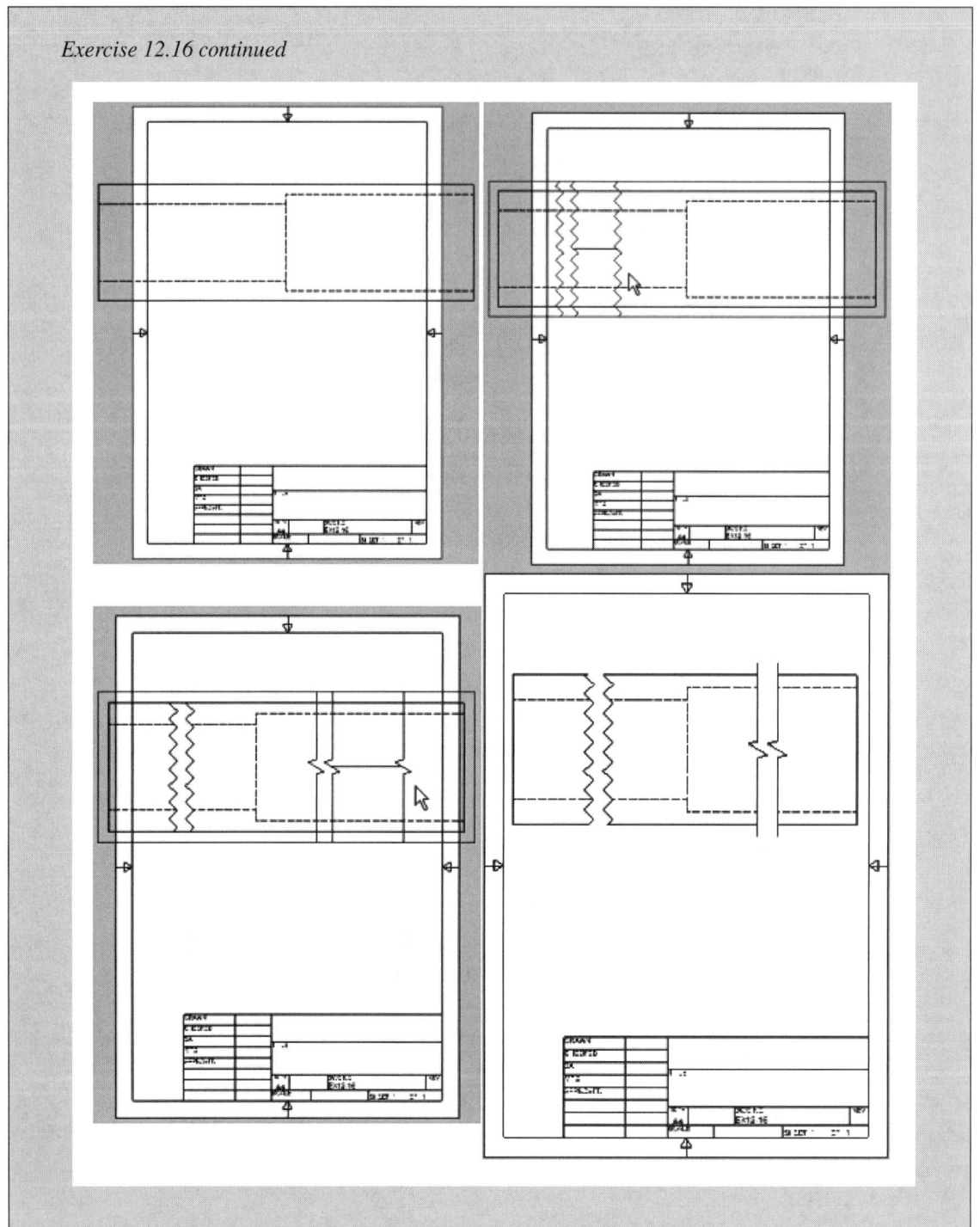

Exercise 12.16 continued

Sketching Views

As you have learned in this chapter, the Autodesk Inventor drawing environment is primarily used to create two-dimensional drawings that reference three-dimensional models. As a result, both the drawings and the models are parametrically controlled and change to reflect design intent. The ability to generate a two-dimensional drawing quickly and easily from a three-dimensional model is very powerful. However, if desired, you can produce drawing views by using sketching tools.

To sketch a drawing view, access the **Draft View** command using one of the following techniques:

✓ Pick the **Draft View** button on the panel bar.

FIGURE 12.81 The
Draft View dialog box.

✓ Pick the **Draft View** button on the **Drawing Management** toolbar.

✓ From the **Insert** pull-down menu, select the **Draft View...** option.

Once you access the **Draft View** command, the **Draft View** dialog box is displayed. See Figure 12.81. The **Draft View** dialog box allows you to change the default label in the **Label** edit box and adjust the scale of the view using the **Scale** edit box and flyout button. When you have specified the view label and scale, pick the **OK** button to enter the sketch environment. Sketching in a drawing file is the same as sketching in any other Autodesk Inventor file, such as a part. Refer to Chapter 3 and the discussions on specific drawing environment sketch tools for more information regarding sketches.

To edit the information you specified in the **Draft View** dialog box, right-click inside the draft view boundaries or the draft view name in the **Browser,** and select the **Edit View...** menu option.

Field Notes

When you are finished drafting the view and have exited the sketch environment, the dimensions you used to define the sketch are converted into view dimensions. The dimension style and specifications correspond to the current dimension settings.

You can also place a sketch on a drawing sheet by accessing the sketch command and using sketching tools to create geometry. If a view is selected, when you activate a sketch, the new sketch is connected to the selected view.

Exercise 12.17

1. Continue from EX12.16, or launch Autodesk Inventor.
2. Open a new metric drawing file.
3. Delete the existing border and title block.
4. Change the sheet size to A4, if it is not currently.
5. Access the **Draft View** command.
6. Specify the following information in the **Draft View** dialog box:

   Label: DRAFT
   Scale: 1:1

7. Sketch the geometry shown.
8. Exit the sketch environment.

Exercise 12.17 continued on next page

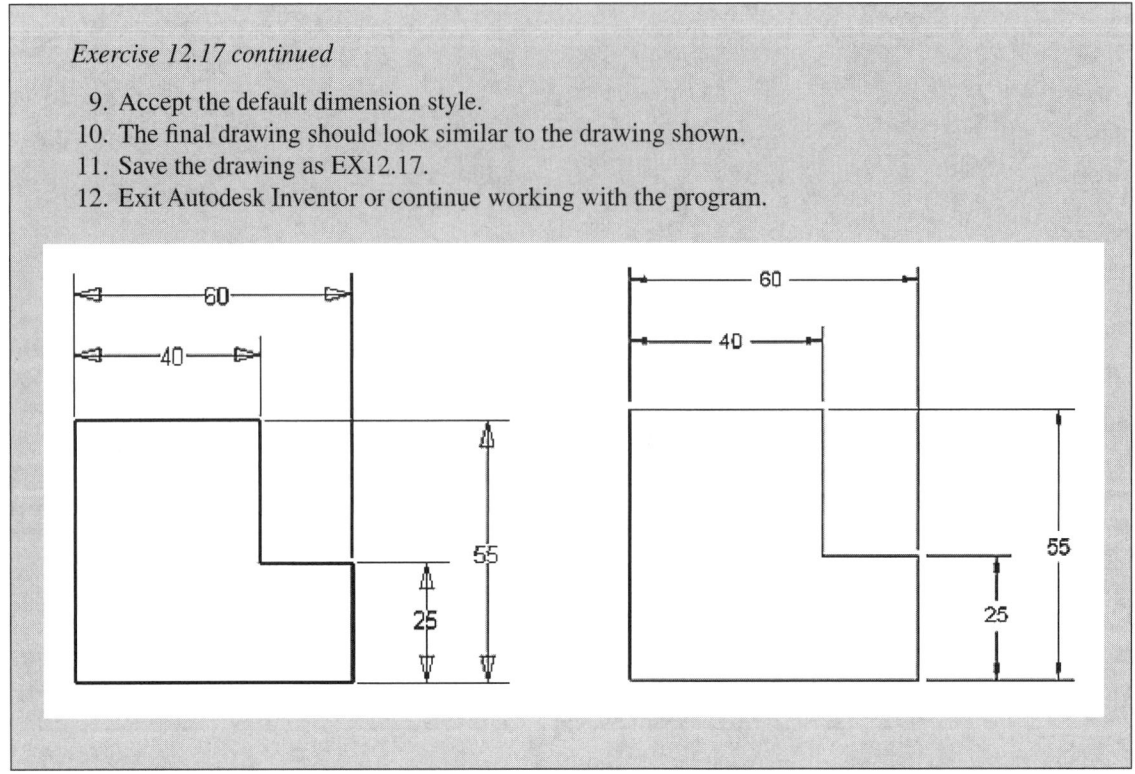

Exercise 12.17 continued

9. Accept the default dimension style.
10. The final drawing should look similar to the drawing shown.
11. Save the drawing as EX12.17.
12. Exit Autodesk Inventor or continue working with the program.

Using Sheet Formats

Sheet formats are templates stored in the drawing file. When you use a sheet format, a new sheet is created, and views from a specified model are inserted onto the sheet, along with a border and title block if they are part of the sheet format. Sheet formats are located in the **Sheet Formats** folder of the **Drawing Resources** folder in the **Browser** bar. See Figure 12.82.

By default, several sheet formats exist in a drawing file. To use one of these sheets, double-click the sheet format, or right-click the sheet format and select the **New Sheet** menu option. Once the sheet format is selected, the **Select Component** dialog box appears. See Figure 12.83. If you have previously inserted a component, the model is available in the **Document Name** drop-down list. Otherwise, you need to pick the **Browse** button to display the **Open** dialog box and locate the model you want to insert. Then pick the **OK** button to create the sheet complete with a border, title block, and the specified views.

To create your own sheet format, generate a border and title block, and insert views from any model. The views you insert are not displayed when you create a new sheet and act only as a view reference. Then right-click in the sheet the **Browser** and select the **Create Sheet Format...** menu option. This displays the **Create Sheet Format** dialog box. See Figure 12.84. Enter the desired name in the **Format Name** edit box, and pick the **OK** button to create the sheet format.

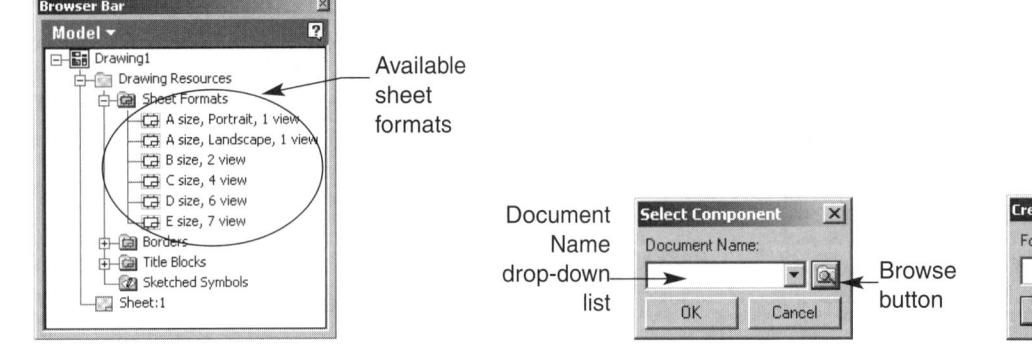

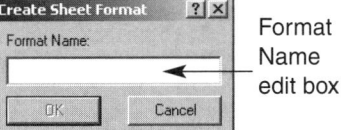

FIGURE 12.82 The **Sheet Formats** folder.

FIGURE 12.83 The **Select Component** dialog box.

FIGURE 12.84 The **Create Sheet Format** dialog box.

Exercise 12.18

1. Continue from EX12.17, or launch Autodesk Inventor.
2. Open EX12.12.
3. Save a copy of EX12.12 as EX12.18.idw.
4. Close EX12.12 without saving, and open EX12.18.
5. Right-click on the sheet (B SIZE:1) in the Browser, and select the **Create Sheet Format...** menu option.
6. Enter B size, 3 view in the **Format Name** edit box of the **Create Sheet Format** dialog box.
7. Delete the current sheet (B SIZE:1), by right-clicking on B SIZE:1 and selecting the **Delete** menu option.
8. Access the sheet format **B size, 3 view** in the **Sheet Formats** folder of the **Drawing Resources** folder.
9. Using the **Explore** button in the **Select Component** dialog box, open EX5.5 and pick the **OK** button.
10. Delete Sheet:1.
11. The drawing (Sheet:1) should look like the drawing shown.
12. Resave EX12.18.
13. Exit Autodesk Inventor or continue working with the program.

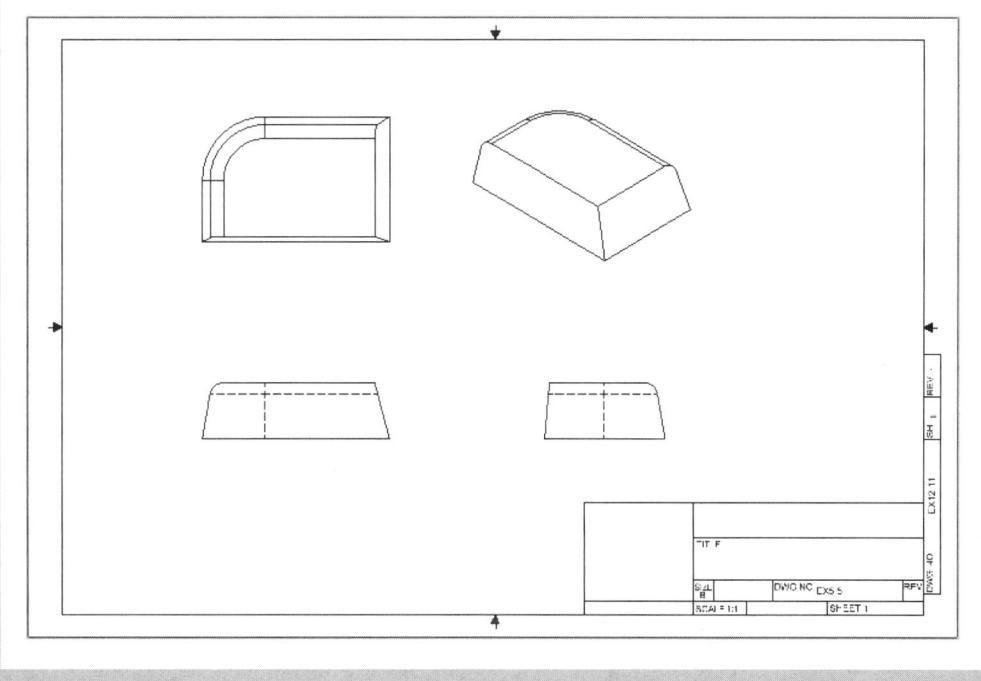

|||||||||||| CHAPTER TEST

Answer the following questions on a separate sheet of paper.

1. Explain the function of the Select flyout button and its options.

2. Briefly discuss the function of the drawing Select flyout button filter options.

3. What is a sheet?

4. How do you open the Drafting Standards dialog box?.

5. How do you access the Edit Sheet dialog box and what is its basic function?.

6. How do you access the Default Drawing Border Parameters dialog box, and what is its basic function?

7. Explain the function of the Property Field sketch tool.

8. How do you access the Text Styles dialog box?

9. Give the procedures used to access the Format Field Text dialog box.

10. Briefly outline the steps used to create text after you have accessed the Format Text dialog box.

11. Name the tool that allows you to add a solid color fill or a hatch pattern to a sketched region.

12. Name the dialog box that allows you to adjust the hatch properties.

13. What must be available before you can access the Hatch/Color Fill command?

14. Give the steps used to add hatching to the sketched region after you have entered the Hatch/Color Fill dialog.

15. What do you do if you want to add a solid color to the sketched region?

16. Give the steps used to enter a scale in the title block.

17. How do you delete a custom property?

18. What do you do if you want to create a copy of the original title block and save the edits to the copy?

19. How do you insert and adjust the location of a title block contained in the Title Blocks folder into the current sheet, and future sheets?

20. Discuss the function of the Precise View Generation check box.

21. Explain the function of the Get model dimensions on view placement check box.

22. Give the function of the Show line weights check box.

23. What does the Create View tool allow you to do, and when can you use this tool?

24. Discuss the function of the View selection box in the Create View dialog box.

25. What is the purpose of the Custom View window?

26. Give the function of the Rotate at Angle tool.

27. What does the Edit View dialog box allow you to do?

28. Give the purpose of the Rotate View dialog box.

29. Why can projected views be generated quickly and easily from existing base views or dependent views?

30. How many views can you project from the base view?

31. When can you use the Projected View tool?

32. When you access the Projected View command and a view is selected, how do you move the view in place and select the location for the view?

33. After you create a projected view, how can you move it?

34. What restrictions does view alignment constraints have on your ability to move a view?

35. Why are there alignment constraints?

36. What is an auxiliary view?

37. Give the function of the Label edit box in the Auxiliary View dialog box.

38. Briefly discuss the function of the Style area in the Auxiliary View dialog box.

39. How do you create a partial auxiliary view?

40. What is a cutting plane line?

41. How do you display the Section View dialog box?

42. How do you change the display of section lines?

43. How do you modify cutting plane lines?

44. Give two ways to edit the cutting plane line labels.

45. Give the function of the Detail View tool.

46. What does the detail boundary represent? Name the boundary shapes that can be created.

47. How do you create a detail view with a circular fence?

48. Once you have specified all the broken view display options in the Broken View dialog box, how do you create the broken view?

49. Explain the function of the Draft View dialog box.

50. By default, several sheet formats exist in a drawing file. How do you use one of these sheets?

||||||||||| PROJECTS

Instructions:

- Open a new drawing file for each of the projects.

- Use the information provided to create the drawings shown.

- Define the drawing properties in the **Properties** dialog box, so the properties are displayed in the title block.

- Do not add any dimensions or annotations to the drawings.

1. Name: Bracket

   Units: Metric

   Sheet Size: A2

   Drawing View Scale (unless otherwise noted): 1:1

   Border: Default or custom

   Title Block: Default or custom

   Base View File: P4.10

   Special Instructions:

   1. Place the base view first using the **Create View** tool.

   2. Project the three additional views using the **Project View** tool.

PROJECT 12.1

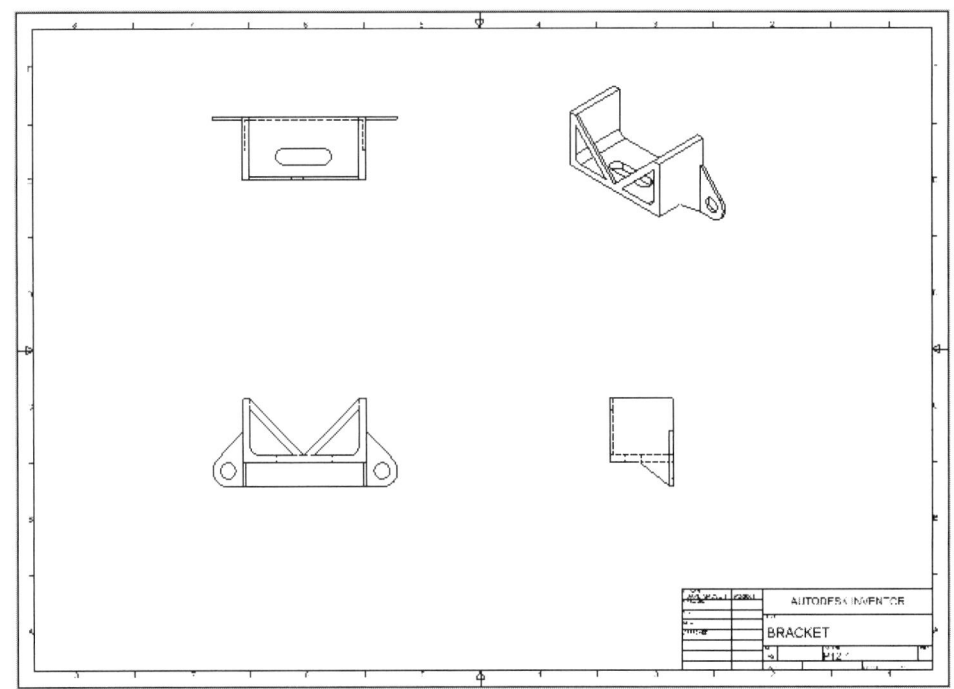

Save As: P12.1

2. Name: 45° Elbow

Units: Metric

Sheet Size: A2

Drawing View Scale (unless otherwise noted): 1:1

Border: Default or custom

Title Block: Default or custom

Base View File: P7.2

PROJECT 12.2

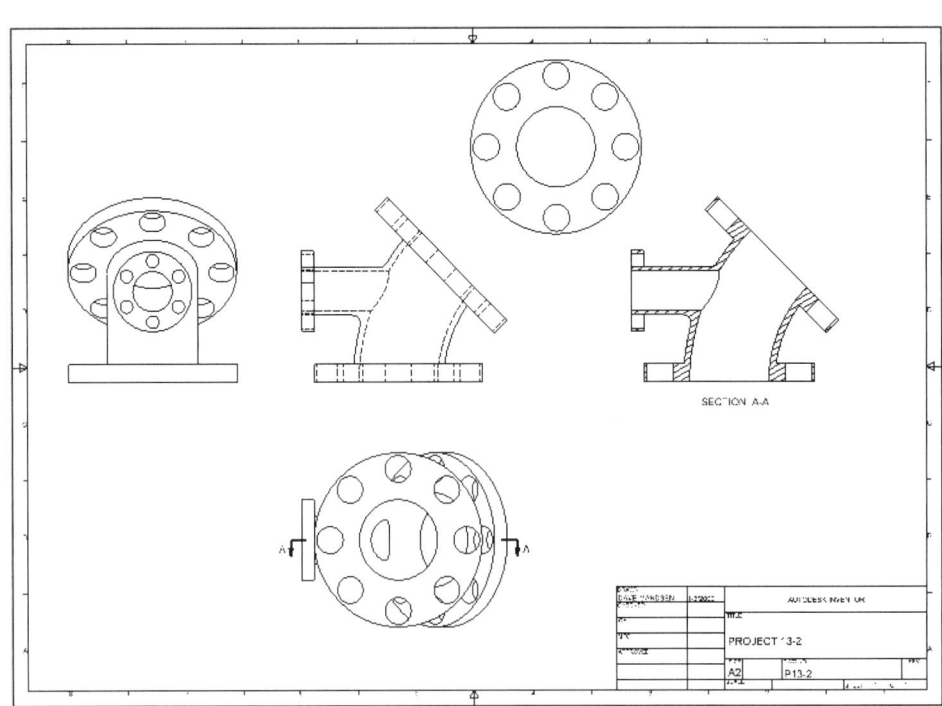

Special Instructions:

1. Place the front view first using the **Create View** tool.

2. Project the side and bottom views using the **Project View** tool.

3. Edit the side and bottom views by removing the style from base view constraint and selecting the **Hidden Line Removed** button.

4. Create the auxiliary view using the **Auxiliary View** tool.

5. Create the section view using the **Section View** tool.

6. Align the section view with the front view.

Save As: P12.2

3. Name: Funnel

Units: Inch

Sheet Size: B

Drawing View Scale (unless otherwise noted): 1:1

Border: Default or custom

Title Block: Default or custom

PROJECT 12.3

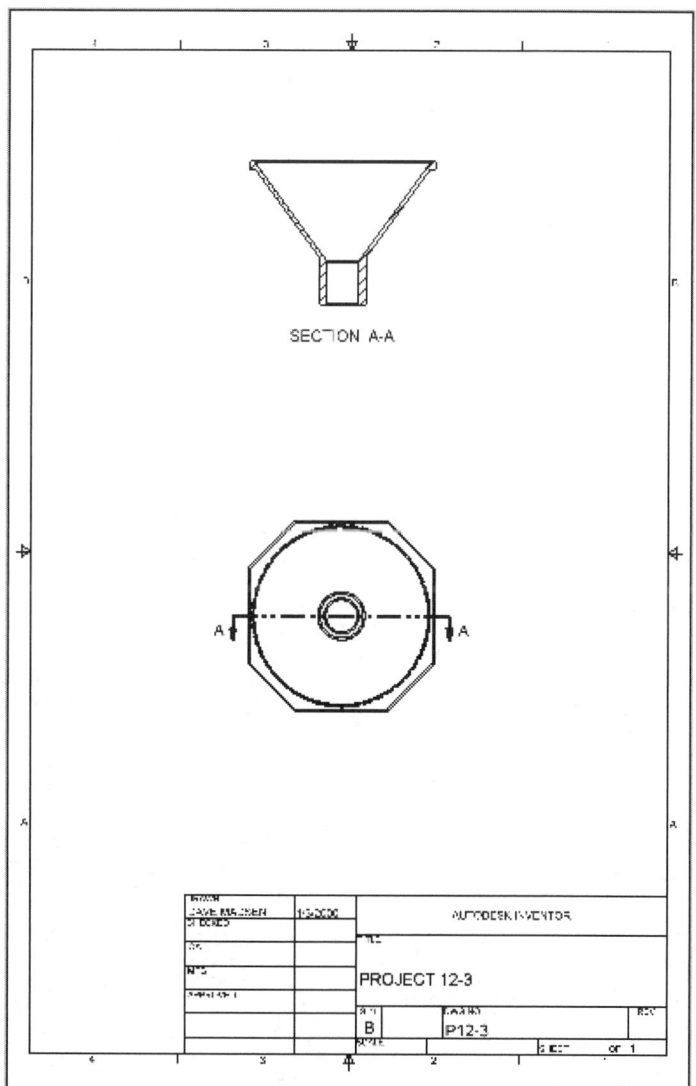

Base View File: P5.5

Special Instructions:

1. Place the top view first using the **Create View** tool.

2. Create the section view using the **Section View** tool.

Save As: P12.3

4. Name: C-Clamp Swivel

Units: Metric

Sheet Size: A4

Drawing View Scale (unless otherwise noted): 2:1

Border: Default or custom

Title Block: Default or custom

Base View File: P5.2

Special Instructions:

1. Place the top view first using the **Create View** tool.

2. Create the section view using the **Section View** tool.

Save As: P12.4

PROJECT 12.4

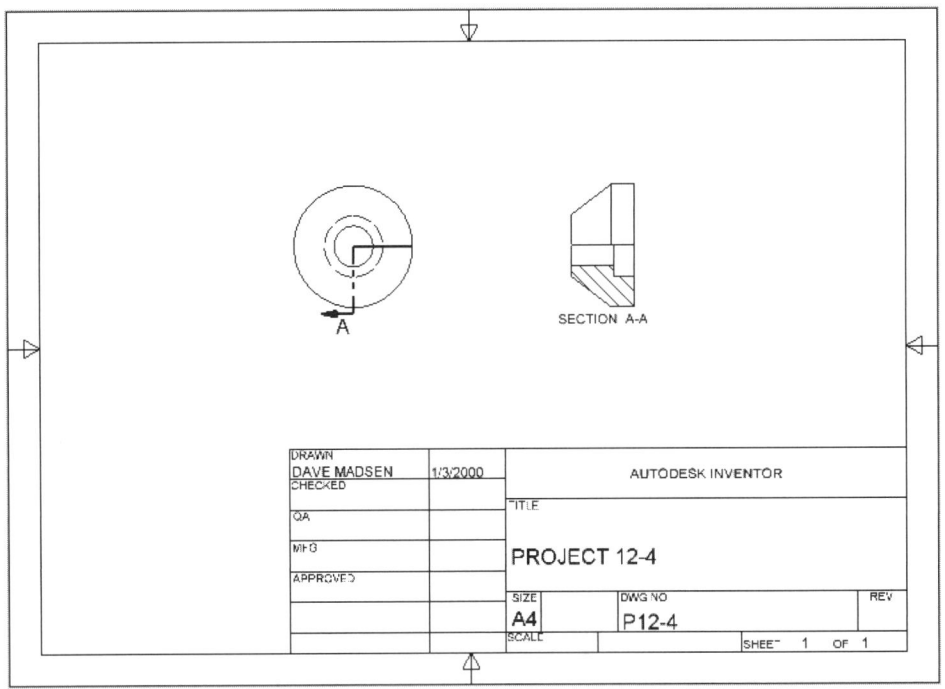

5. Name: C-Clamp Screw

Units: Metric

Sheet Size: A4

Drawing View Scale (unless otherwise noted): 2:1

Border: Default or custom

Title Block: Default or custom

Base View File: P4.3

Special Instructions:

1. Place the side view first using the **Create View** tool.

2. Break the side view using the **Broken View** tool.

3. Create the front and back views using the **Project View** tool.

Save As: P12.5

PROJECT 12.5

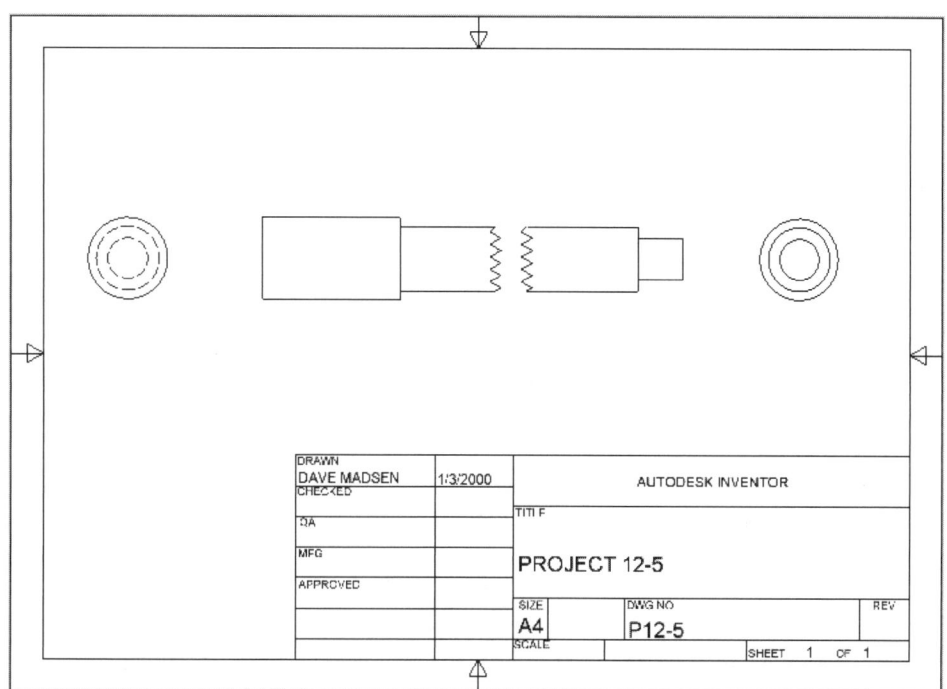

DRAWN DAVE MADSEN	1/3/2000	AUTODESK INVENTOR		
CHECKED				
QA		TITLE		
MFG		PROJECT 12-5		
APPROVED				
		SIZE A4	DWG NO P12-5	REV
		SCALE		SHEET 1 OF 1

6. Name: Support Bracket

Units: Inch

Sheet Size: B

Drawing View Scale (unless otherwise noted): 3:1

Border: Default or custom

Title Block: Default or custom

Base View File: P5.4

Special Instructions:

1. Place the top view first using the **Create View** tool.

2. Create the section view using the **Section View** tool.

Save As: P12.6

PROJECT 12.6

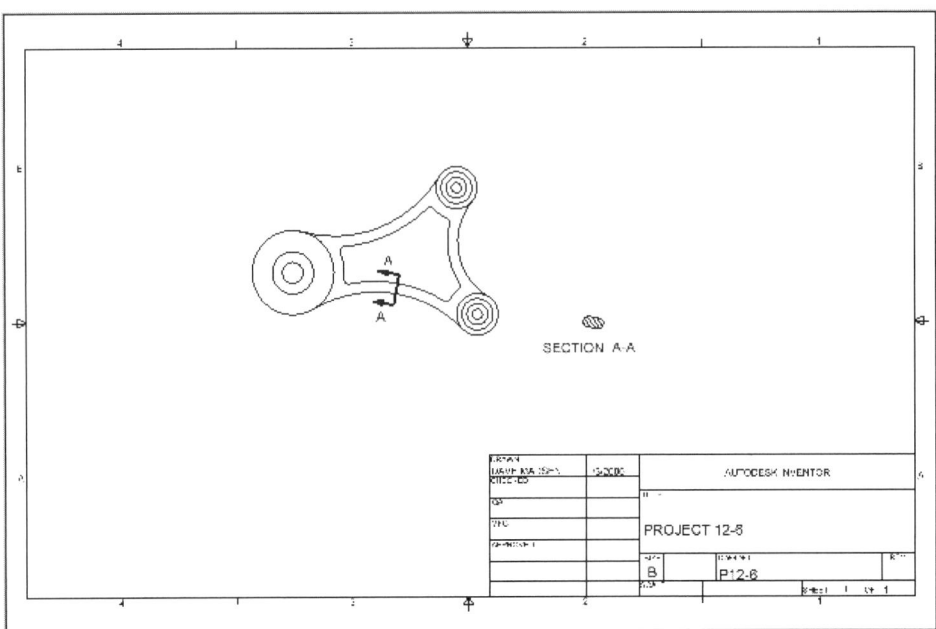

7. Name: Slide Bar Hinge Connector

   Units: Inch

   Sheet Size: B

   Drawing View Scale (unless otherwise noted): 1:1

   Border: Default or custom

   Title Block: Default or custom

   Base View File: P6.4

   Special Instructions:

       1. Place the front view first using the **Create View** tool.

       2. Create the additional views using the **Project View** tool.

   Save As: P12.7

PROJECT 12.7

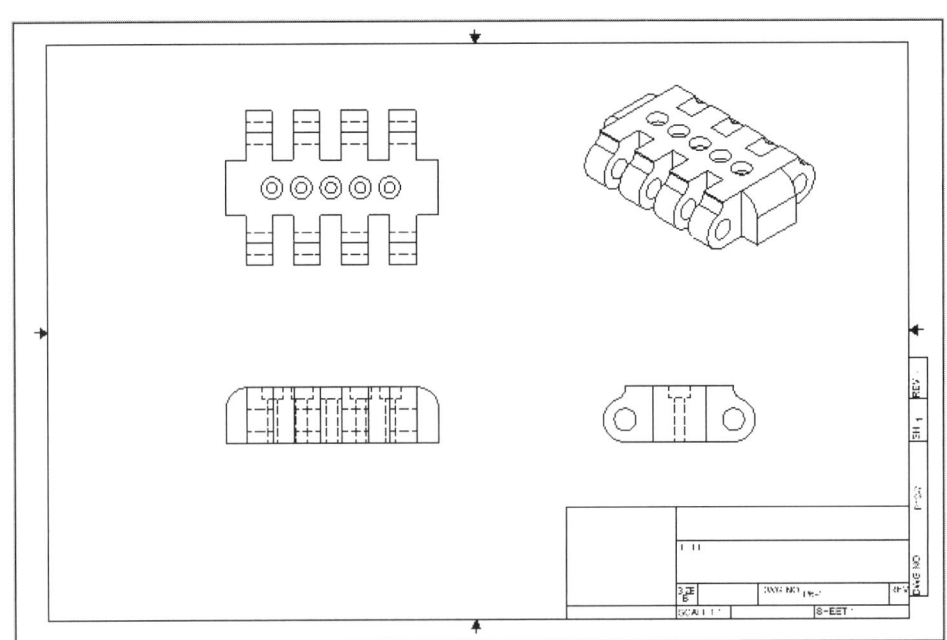

8. Name: Selector Bracket

Units: Metric

Sheet Size: A2

Drawing View Scale (unless otherwise noted): 1:1

Border: Default or custom

Title Block: Default or custom

Base View File: P6.3

Special Instructions:

1. Place the front view first using the **Create View** tool.

2. Create the additional views using the **Project View** tool.

Save As: P12.8

PROJECT 12.8

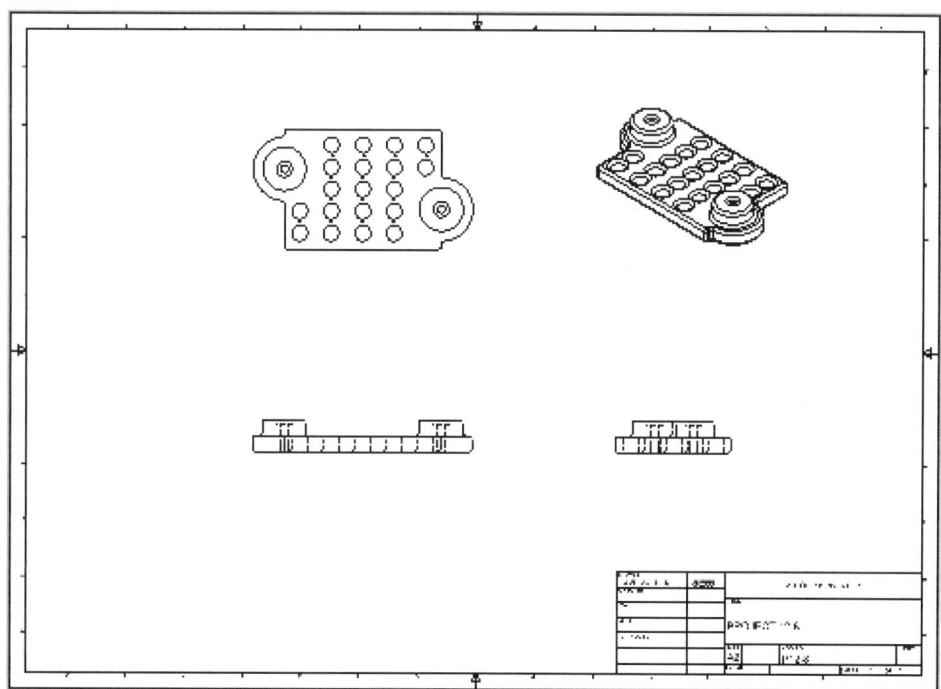

9. Name: Cover Plate

Units: Metric

Sheet Size: A4

Drawing View Scale (unless otherwise noted): 1:1

Border: None

Title Block: None

Base View File: EX6.2

Special Instructions:

1. Place the front view first using the **Create View** tool.

2. Create the additional view using the **Project View** tool.

Save As: P12.9

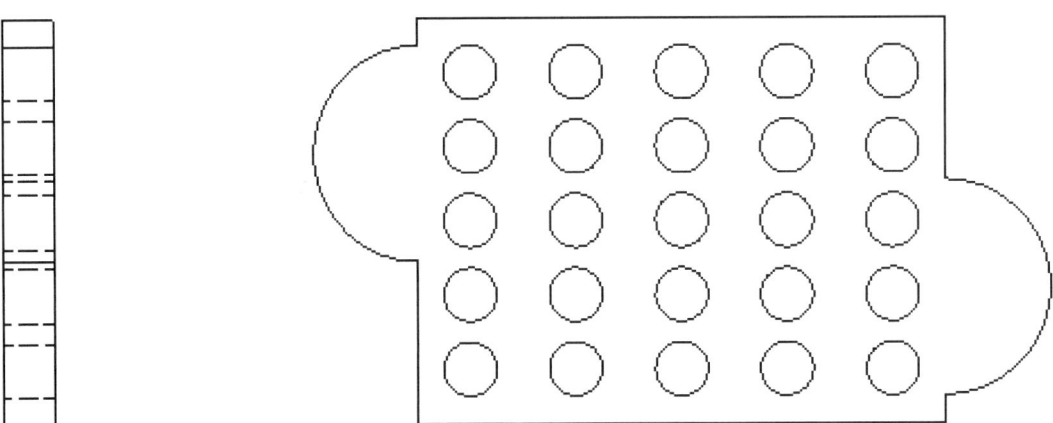

|||||||||||||||||||||||||||||||||||||||||||||||||||||||||||||||||||||||||||||||||||||||||||||||||||||||||||||||||||||||

CHAPTER 13

Dimensioning Drawings

LEARNING GOALS

After completing this chapter, you will be able to:

◎ Fully dimension part drawings.

◎ Specify dimension styles and define additional drafting standards.

◎ Work with, and differentiate between, Model and drawing dimensions.

◎ Add centerlines and centermarks to drawing view geometry.

◎ Use the **General Dimension,** and **Baseline Dimension** tools.

◎ Add text and leader text.

◎ Use ordinate and tabular dimension tools.

◎ Add hole and thread notes.

◎ Place surface texture and weld symbols.

◎ Use geometric dimensioning and tolerancing tools to add feature control frames, feature identifier symbols, datum feature symbols, and datum target symbols.

◎ Modify dimensions and override specific dimension style options.

Dimensioning Drawings

Once you have placed views in a drawing, the next step is to add dimensions to the view geometry. ***Dimensions*** are used to define the size and shape of features of an object so the part can be manufactured. Dimensions, along with other notes and text, also specify the location and characteristics of geometry, surface texture, and weld information.

As discussed in Chapter 12, most part drawings are generated from existing 3D parametric solid models. As a result, there are both model and drawing dimensions in Autodesk Inventor. ***Model dimensions*** are specified when you add dimensions to the model sketch or define the size of an object, such as an extrusion depth. When you insert a model into a drawing, the model dimensions can be displayed automatically. Model dimensions are parametric, and when edited, change the physical size and shape of the corresponding model. Similarly, if you modify a model dimension while in a part file, the model dimension and the size and shape of the object in the corresponding drawing also change.

Drawing dimensions are additional dimensions added to drawing view geometry to further document the design for manufacturing purposes. Unlike model dimensions, drawing dimensions do not control any model parameters. When you edit a model while in a part file, the drawing dimensions in the drawing file change to reflect the new values. However, you cannot change the size or shape of a part model by manipulating drawing dimensions.

Autodesk Inventor provides you with multiple tools that allow you to create a wide variety of drawing dimensions. These tools are referred to as drawing annotation tools. Throughout this chapter, you will explore specific tools, commands, and options that allow you to dimension a part drawing fully.

Field Notes

The **Balloon, Balloon All,** and **Parts List** drawing annotation tools correspond to assembly drawings and are discussed in Chapter 15.

Specifying Dimension Styles

Dimension styles allow you to control most of the dimension display characteristics including units, text, line display, and other dimension features. You can select a new dimension style or modify a dimension style at any time during the drawing creation process. However, before you begin to dimension drawing views, you may want to become familiar with the dimension style options or adjust a dimension style to fit your specific application. Typically, dimension styles are defined in a drawing template.

Field Notes

Dimension styles are very important for dimensioning a drawing exactly the way you want it to be dimensioned and capturing design intent. However, if you are not concerned about defining dimension styles at this time, you may choose to skip ahead to the next section of this chapter.

To explore the dimension styles options, modify existing dimension styles, or create new dimension styles, access the **Dimension Styles** command by selecting **Dimension Styles...** from the **Format** pull-down menu. Selecting the **Dimension Styles...** menu option opens the **Dimension Styles** dialog box, shown in Figure 13.1. The **Dimension Styles** dialog box contains a number of default dimension styles based on ANSI, BSI, DIN, GB, ISO, and JIS drafting standards. To access the default dimension styles for a specific standard, select the desired standard from the **Standard** drop-down list. When you select a different standard from the **Standard** drop-down list, the available default dimensions styles are displayed in the **Style Name** selection box.

You can pick any of the dimension styles available in the **Style Name** selection box and use the style to dimension your drawing. Notice when you choose different dimension styles, the options in each of the

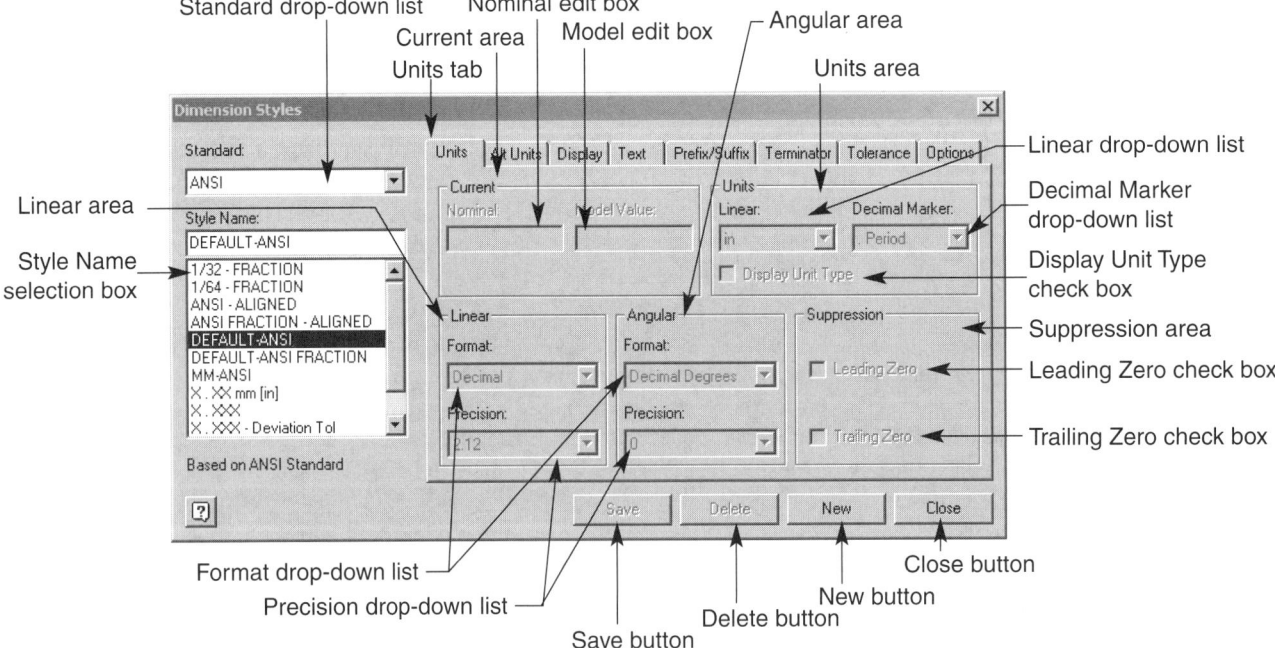

FIGURE 13.1   The **Dimension Styles** dialog box (**Units** tab displayed).

Dimension Styles dialog box tabs change to reflect the particular standards settings. You can also modify the characteristics of the selected dimension style using the options in the various tabs. However, you may want to create a new dimension style for applications that require significant changes to an existing style.

Field Notes

Default dimension styles, which are identified by DEFAULT-, cannot be modified or removed.

To create a new dimension style, first pick an existing dimension style that is similar to the dimension style you want to create. For example, if you are creating an inch drawing using ANSI standards, you may want to select DEFAULT-ANSI. Then pick the **New** button. As shown in Figure 13.2, a new style is placed in the **Style Name** edit box, named Copy of DEFAULT-ANSI. You can use the default name or enter a different name in the **Style Name** edit box. Once you have specified the desired dimension style name, pick the **Save** button to create the style. The new style should now be listed in the **Style Name** selection box and can be removed by selecting the style and picking the **Delete** button.

Once you access the **Dimension Styles** dialog box and have selected the desired standard and dimension style, use the following tabs to control the features and attributes of dimensions:

- **Units** This tab, shown in Figure 13.1, controls a variety of unit and unit format specifications. The following options are available inside the **Units** tab:

  - **Current** This area may allow you to define the current dimension value displayed. Use the **Nominal** edit box to specify the value shown in the drawing, but not to change the value. Although model dimensions cannot be changed, the Model Value box may display the value of the current dimension.

  - **Units** Use this area to set the units for the dimension style. Specify the type of unit measurement, such as inch or mm, from the **Linear** drop-down list. You can also define the type of decimal marker, such as a decimal point or comma, from the **Decimal Marker** drop-down list. If you want the unit type to be displayed along with the dimension value, such as 12mm, pick the **Display Unit Type** check box.

  - **Linear** This area allows you to set the desired unit format and precision for linear or nonangular dimensions. You can choose either a Decimal or Fractional unit format from the **Format** drop-down list, depending on the application. Then specify the linear unit precision by selecting a value from the **Precision** drop-down list.
      Precision identifies the accuracy of the dimension by placing more or fewer numbers after the decimal point when using a decimal format, or a higher or lower denominator when using fractional format. For example, .125in is higher precision than .13in, and 1/16in is higher precision than 1/8in.

  - **Angular** This area allows you to set the desired unit format and precision for angular or nonlinear dimensions. You can choose either a Decimal Degrees or Deg-Min-Sec format from the **Format** drop-down list, depending on the application. Then specify the angular unit precision by selecting a value from the **Precision** drop-down list.

FIGURE 13.2
Creating a new
dimension style.

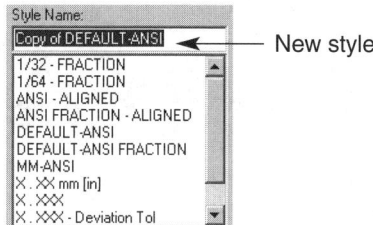

New style

Again, precision identifies the accuracy of the dimension by placing more or fewer numbers after the decimal point when using a decimal degree format, or degrees, minutes, and seconds when using deg-min-sec format. For example, 45.5° is higher precision than 45°, and 36°29′10″ is higher precision than 36°29′ or 36°.

- ■ **Suppression**   Use this area to control the display, or suppression, of leading and trailing zeros. Select the **Leading Zero** check box to suppress the leading zero, which changes the display of 0.5 to .5 for example. Select the **Trailing Zero** check box to suppress the trailing zero, which changes the display of 45.0 to 45 for example.

■ **Alt Units**   This tab, shown in Figure 13.3, contains options that allow you to display alternate dimension units in addition to the units specified in the **Units** tab. If you want to display alternate units in your dimensions, pick the **Display Alternate Units** check box, located in the **Alternate Units** area of the **Alt Units** tab. When you select the **Display Alternate Units** check box, the following areas and options become available in the **Alt Units** tab:

- ■ **Alternate Units**   This area allows you to activate alternate unit display and modify the alternate units. Specify the type of alternate unit measurement, such as inch or mm, from the **Linear** drop-down list, and the type of decimal marker, such as a decimal point or comma, from the **Decimal Marker** drop-down list. The purpose of alternate units is to show two types of dimension values, such as "in" and "mm." Consequently, if you select "in" as the primary units in the **Units** tab, you should pick "mm," for example, in the **Alt Units** tab. Then if you want the alternate unit type to be displayed along with the dimension value, such as 12mm, pick the **Display Unit Type** check box.

- ■ **Linear**   This area allows you to set the desired alternate unit format and precision for linear or nonangular dimensions. You can choose either a Decimal or Fractional unit format from the **Format** drop-down list, depending on the application. Then specify the linear unit precision by selecting a value from the **Precision** drop-down list.

- ■ **Display Format**   This area is used to define the display characteristics of alternate units in reference to primary units. Select the type of unit display you want to use from **Style** drop-down list.

- ■ **Suppression**   Use this area to control the display, or suppression, of leading and trailing zeros for alternate units. Select the **Leading Zero** check box to suppress the leading zero, and the **Trailing Zero** check box to suppress the trailing zero.

■ **Display**   This tab, shown in Figure 13.4, contains options that allow you to control a number of physical dimension display options. Use the following options to define the appear-

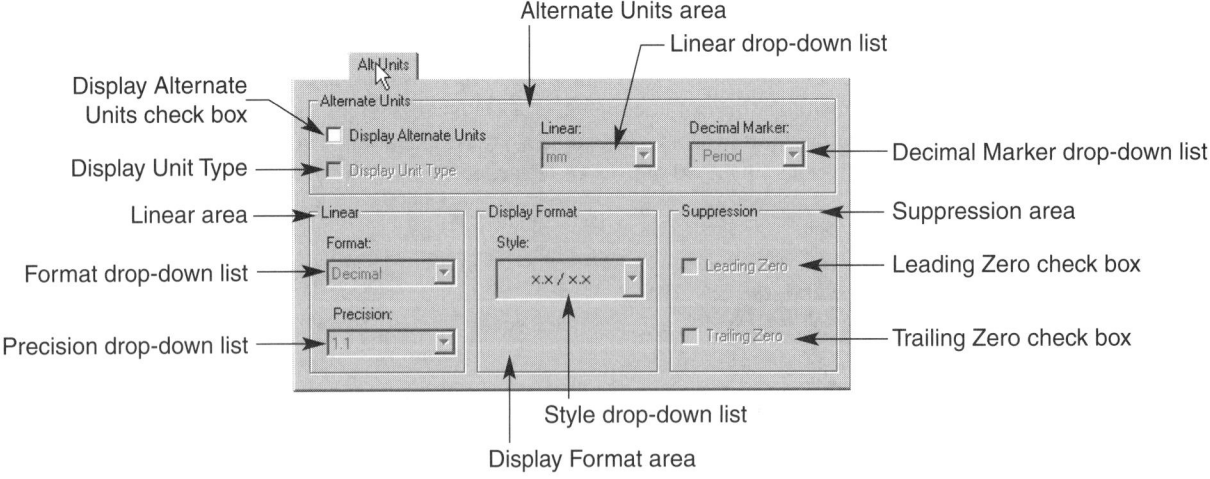

FIGURE 13.3   The **Alt Units** tab of the **Dimension Styles** dialog box.

FIGURE 13.4 The **Display** tab of the **Dimension Styles** dialog box.

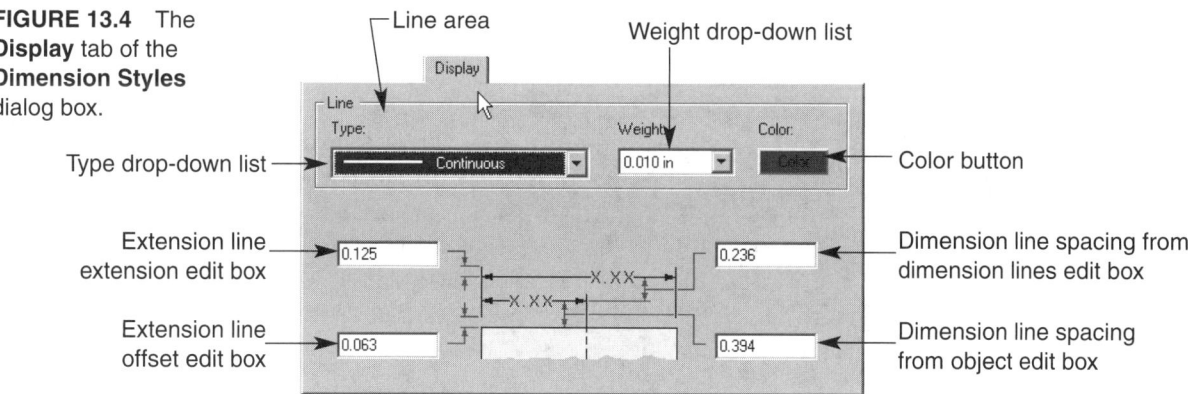

ance of dimension and extension lines, the length of the extension line extension, dimension line spacing, and the extension line offset:

- **Line**   Use this area to define the appearance of the dimensioning lines. Select the line type you want to use from the **Type** drop-down list, the line weight from the **Weight** drop-down list, and the line color by selecting the **Color** button.

- **Extension line extension**   Enter a value in this edit box to specify the length of the extension line extension from the dimension line.

- **Extension line offset**   Specify a value in the edit box to define the extension line offset, or gap, between the extension line and the object.

- **Dimension line spacing from dimension lines**   Enter a value in this edit box to specify the distance, or spacing, between dimension lines.

- **Dimension line spacing from object**   Enter a value in this edit box to define the distance, or spacing, between the object and the first dimension line of dimension lines that are parallel to the object.

Field Notes

> Dimension line spacing does not automatically occur. As discussed later in this chapter, when you place or move a dimension, it becomes highlighted at the spacing locations specified in the **Display** tab. You must select the highlighted location.

- **Text**   This tab, shown in Figure 13.5, contains options that allow you to manage dimension text display preferences. Use the following options to define various types of dimension text appearance information:

  - **Text Style**   Use this drop-down list to select a specific text style to use for dimensioning. Text styles and the **Text Style** dialog box are discussed in Chapter 12.

  - **Font**   Use this drop-down list to define the font type for the new text style.

  - **Size**   This drop-down list allows you to select or enter a value for the size, or height, of the text.

  - **Format**   These buttons define the display text options. Pick **Bold** to create bold text, **Italic** to create italicized text, and **Underlined** to underline text.

  - **Justification**   These buttons allow you to control the left, center, right, top, middle, and bottom justification, or placement, of the text.

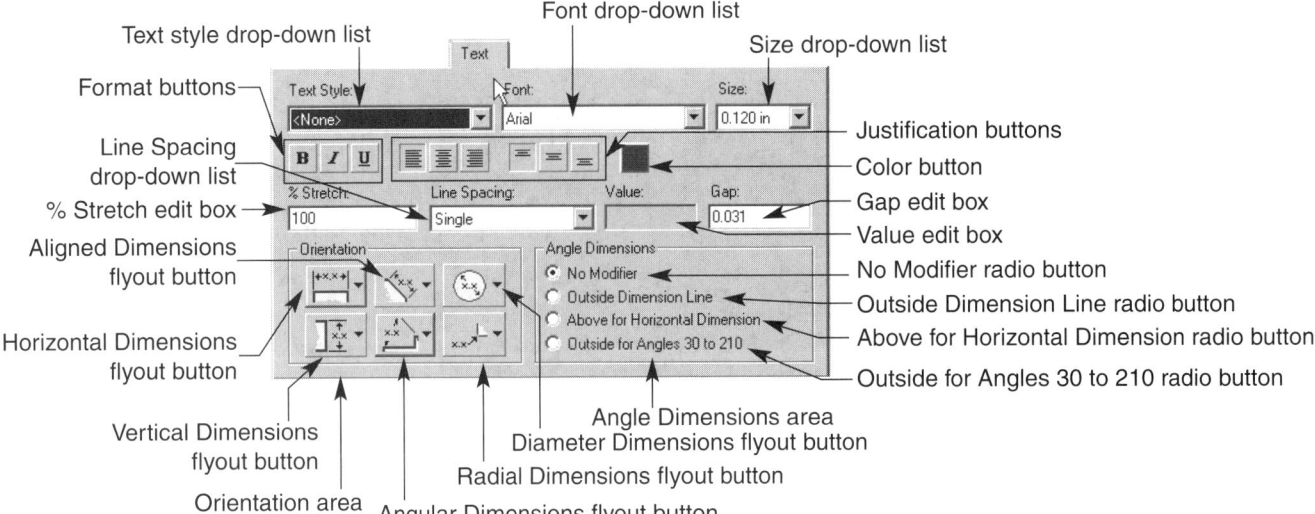

FIGURE 13.5 The **Text** tab of the **Dimension Styles** dialog box.

- **Color** Pick this button to display the **Color** dialog box and define the color of the text.

- **% Stretch** This edit box allows you to define the amount of stretch, or width of the text. To create a narrower text, enter a percent less than 100, and to create a wider text, enter a percent greater than 100.

- **Line Spacing** This drop-down list allows you to specify the distance between multiple lines of text. When you select the **Multiple** option, the **Value** edit box becomes active, allowing you to define a smaller or greater spacing ratio, such as triple, or half. When you select the **Exactly** option, the **Value** edit box becomes available, allowing you to enter a specific distance between multiple lines of text.

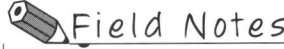Field Notes

Text font, size, format, justification, color, % stretch, and line spacing can only be modified inside the Text tab when the test style is set to <None>.

- **Gap** Use this edit box to specify the distance, or gap, between the text and the dimension line. See Figure 13.6.

- **Orientation** This area contains several flyout buttons that allow you to control the orientation and location of text in reference to various types of geometry. Graphics on the buttons help identify the desired orientation; holding the cursor over the flyout button gives the button name. The following flyout buttons are available:

  - **Horizontal Dimensions** This flyout button contains two orientation options. Pick the aligned button to place horizontal dimension text that is in line with dimension lines, or pick the above button to place horizontal dimension text that is aligned with and above the dimension line.

FIGURE 13.6 The dimension text gap.

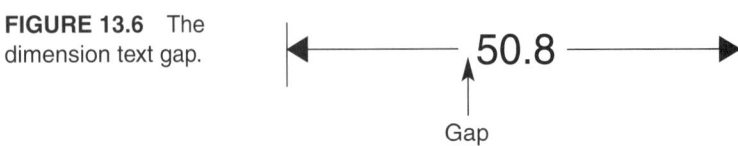

- **Vertical Dimensions**   This flyout button contains three orientation options. Pick the unidirectional button to place unidirectional vertical dimension text, which is perpendicular to the dimension line. *Unidirectional dimensions* are dimensions that read horizontally from the bottom of the sheet for both horizontal and vertically placed dimensions. Select the aligned button to place vertical dimension text that is in line with dimension lines, or pick the above button to place vertical dimension text that is aligned with and above the dimension line.

- **Aligned Dimensions**   This flyout button contains three orientation options. Pick the unidirectional button to place unidirectional aligned dimension text, which is horizontal no matter what the angle of the dimension line. Select the aligned button to place aligned dimension text that is in line with dimension lines, or pick the above button to place aligned dimension text that is aligned with and above the dimension line.

- **Angular Dimensions**   This flyout button contains four orientation options. Pick the unidirectional/inside button to place unidirectional angular dimension text that is inside the dimension line. Select the aligned button to place angular dimension text that is in line with dimension lines, or pick the above button to place angular dimension text that is aligned with and above the dimension line. You can also choose the unidirectional/outside button to place unidirectional angular dimension text that is outside the dimension line.

- **Diameter Dimensions**   This flyout button contains three orientation options. Pick the unidirectional button to place unidirectional diameter dimension text. Select the aligned button to place diameter dimension text that is in line with dimension lines, or pick the above button to place diameter dimension text that is aligned with and above the dimension line.

- **Radial Dimensions**   This flyout button contains four orientation options. Pick the unidirectional/centered button to place unidirectional radial dimension text that is horizontal and in line with the leader shoulder. Select the unidirectional/above button to place unidirectional radial dimension text that is horizontal and above the leader shoulder. Select the aligned/centered button to place radial dimension text that is in line with, and centered on, leader lines, or pick the aligned/above button to place radial dimension text that is aligned with and above the leader line.

- **Angle Dimensions**   This area further specifies the placement of text for specific applications when using angular dimensions. Pick the **Outside Dimension Line** radio button to place angular dimension text outside the dimension line. Choose the **Above for Horizontal Dimension** radio button to place angular dimension text above the extension line, when text is unidirectional. Select the Outside for Angles 30 to 210 radio button to place angular dimension text outside the dimension line when the object being dimensioned is between 30 and 210°.

- **Prefix/Suffix**   This tab, shown in Figure 13.7, is used to add a prefix and a suffix to the dimension text. A *prefix* is a word, character, or symbol added before or on top of the dimension value, while a *suffix* is a word, character, or symbol located after or below the dimension value. Prefixes and suffixes are automatically added to the dimension text when a dimension is created.

  If you want to add a prefix or a suffix to the dimension value, first select the order in which you want the prefix and suffix to be placed. As the graphics illustrate, you can choose to place the prefix before the dimension value and the suffix after the dimension value by selecting the **Before/After** radio button. Or you can place the prefix above the dimension value and the suffix below the dimension value by selecting the **Above/Below** radio button.

  The next step is to enter the prefix in the **Prefix** edit box, and the suffix in the **Suffix** edit box. You may also choose to add symbols to the **Prefix** and **Suffix** edit boxes by selecting the desired symbol from the **Symbol** flyout button.

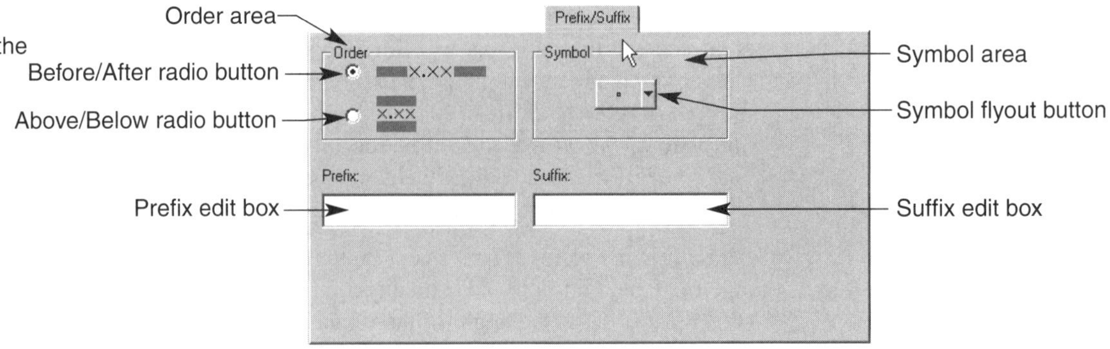

FIGURE 13.7 The **Prefix/Suffix** tab of the **Dimension Styles** dialog box.

Order area — Before/After radio button — Above/Below radio button — Prefix edit box — Symbol area — Symbol flyout button — Suffix edit box

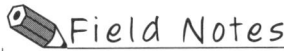

Field Notes

If you do not want to add a prefix or a suffix to the dimension text, leave the edit boxes clear.

■ **Terminator**   This tab, shown in Figure 13.8, allows you to define the style, size, and aspect of the dimension line terminator, such as an arrowhead, depending on the application. The options in the **Terminator** tab control the style, size, and aspect of linear, aligned, angular, diameter, and radial dimensions only.

To define the terminator style, select the desired arrowhead from the **General** drop-down list. Then enter the size and aspect of the arrowhead in the **Size** and **Aspect** edit boxes. *Size* defines the length of the arrowhead from the point to the end, while *aspect* specifies the width or thickness of the arrowhead in reference to the size. See Figure 13.9. For example, most arrowheads are drawn at a 3:1 ratio. As a result, the size of the arrowhead would be 3mm, or .125in, while the aspect would be 1mm, or .042in.

FIGURE 13.8 The **Terminator** tab of the **Dimension Styles** dialog box.

General drop-down list — Size edit box — Aspect edit box

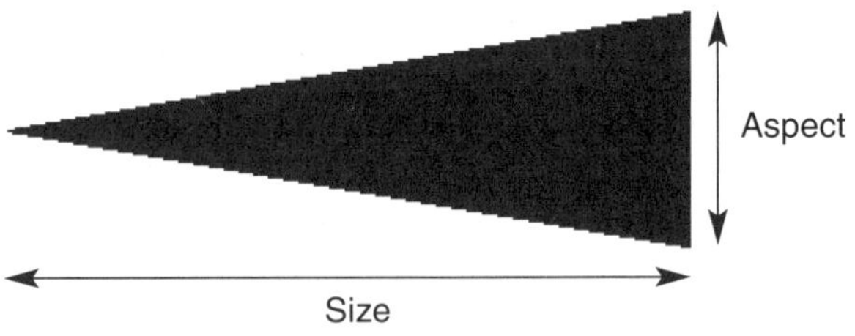

FIGURE 13.9 Understanding arrowhead size and aspect.

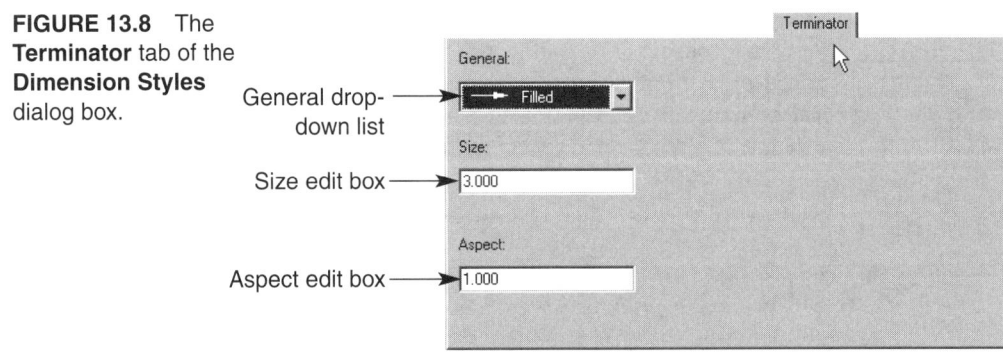

Aspect

Size

Field Notes

The terminator options for hole/thread notes, surface texture symbols, weld symbols, feature control frames, feature identifier symbols, datum identifier symbols, text leaders, datum target leaders, and balloon leaders are defined in the **Terminator** tab of the **Drafting Standards** dialog box, discussed later in this chapter.

■ **Tolerance**   This tab, shown in Figure 13.10, allows you to specify tolerancing options for drawing applications that require dimension tolerances. A *tolerance* is the total variation permissible in a size or location dimension. To define a dimension style with tolerances, first select the tolerance type or method from the **Method** drop-down list. Options in the **Tolerance** tab become active depending on the method you choose. If you select the **Default** tolerance method, no tolerance is applied to the dimension style.

  Once you select the desired tolerance method, use the following options to fully define the tolerance specifications:

■ **Linear Precision**   Use this drop-down list to set the desired unit precision for linear tolerance dimensions.

  ■ **Alt. Unit Precision**   Use this drop-down list to set the desired unit precision for alternate unit tolerance dimensions. This drop-down list is available only when you activate alternative units in the **Alt Units** tab of the **Dimension Styles** dialog box.

  ■ **Text Size**   This drop-down list is used to define the text size, or height, of the tolerance text, which can be larger or smaller than the nominal or specified dimension text. The tolerance height is normally the same as the nominal text height.

  ■ **Angular Precision**   Use this drop-down list to set the desired unit precision for angular tolerance dimensions.

  ■ **Suppression**   Use this area to control the display, or suppression, of leading and trailing zeros for tolerance units. Select the **Leading Zero** check box to suppress the leading zero, and the **Trailing Zero** check box to suppress the trailing zero.

  ■ **Tolerance**   This area allows you to set tolerance values for certain applications. Again, the options in the **Tolerance** area are available only when a specific tolerance method is selected.

    The **Lower** and/or **Upper** edit boxes are available when you define limit tolerance methods such as Symmetrical, Deviation, Limits - Stacked, and Limits - Linear. Enter a

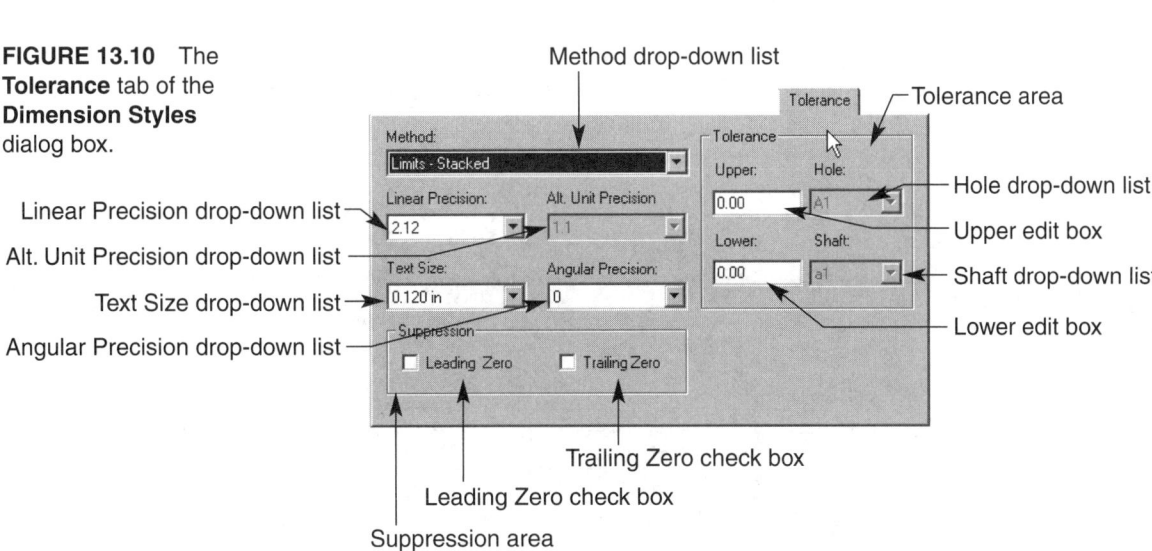

FIGURE 13.10   The **Tolerance** tab of the **Dimension Styles** dialog box.

value in the **Upper** edit box to set an upper value for a tolerance with a value range. For example, an object that is 8in long, with an upper range of .05in, has an upper tolerance limit of 8.05in. Similarly, you can enter a value in the **Lower** edit box to set a lower value for a tolerance with a value range. Using the previous example, an object with a lower range of .05in has a lower tolerance limit of 7.95in. The tolerance for this dimension reads 8±.05 for a Symmetric tolerance Method option.

When using the Deviation tolerance method option, both the **Upper** and **Lower** edit boxes are available. The normal Deviation application is used to set up a bilateral tolerance such as .625±.005 or .625+.005/−.002. To accomplish this, you need to enter a minus (−) symbol in front of the value in the **Lower** edit box. A positive symbol is automatically applied to a positive number, without adding a plus (+) sign.

When using limits and fits tolerance methods for dimensioning holes and shafts, the **Hole** and **Shaft** drop-down lists are available. The hole values—A1 or B2, for example— are available in the **Hole** drop-down list corresponding to specific hole tolerances established in ANSI size and fit tables. For example, a hole dimensioned 20 H9 takes the place of 20+0.05/0. In the same respect as the **Hole** drop-down list, the shaft values—a1 or b2, for example—are available in the **Shaft** drop-down list corresponding to specific hole tolerances established in ANSI size and fit tables. For example, a shaft dimensioned 10d9 takes the place of 9.95 0/−0.05.

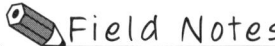

When you select a value from the **Hole** drop-down list, the **Shaft** value is N/A. Similarly, when you select a value from the **Shaft** drop-down list, the **Hole** value is N/A, because the ANSI size and fit tolerances apply to either the hole or shaft.

■ **Options**   This tab, shown in Figure 13.11, allows you to define additional dimension style options for dimensioning specific geometry. To specify particular dimension style options, first pick the desired dimension method radio button from the **Method** area. Dimensioning options for each of the available methods are displayed in the **Type** and **Options** areas. For example, if you select the **Line** radio button, only dimension options that pertain to dimensioning lines are available, or if you select the **Circle** button, only dimension options that apply to dimensioning circles are available.

FIGURE 13.11   The **Options** tab of the **Dimension Styles** dialog box (**Line** radio button selected).

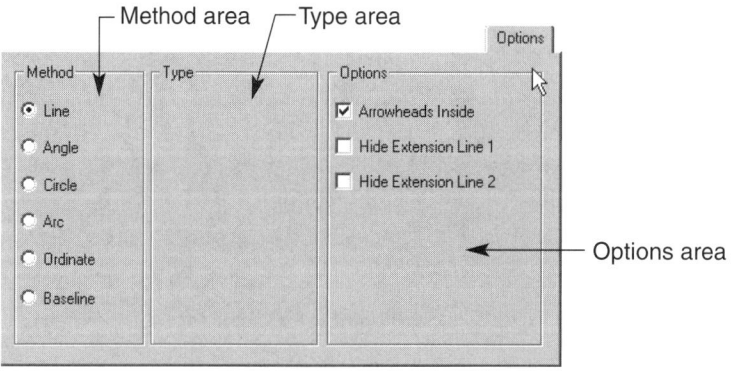

Once you select the dimension method, radio buttons and check boxes that correspond to the method are available in the **Type** and **Options** area. Use the radio buttons and check boxes to define the dimension specifications for the selected method. Then choose another method, if needed, and define the dimensioning options for the selected method. The selected method determines only which options are displayed and does not control your ability to dimension other objects. For example, if the **Line** radio button is selected, you can still dimension circles or arcs. You can think of the method radio buttons as separate dimension style tabs. The **Type** and **Options** areas have options that relate to specific drafting practices and applications. A word or description is provided for each option, giving you an understanding of the option results.

Exercise 13.1

1. Launch Autodesk Inventor.
2. Open a new inch drawing file.
3. Delete the current border and title block.
4. From the **Format** pull-down menu, pick the **Dimension Styles...** option to access the **Dimension Styles** dialog box.
5. Review the options and tabs available in the **Dimension Styles** dialog box as discussed.
6. Ensure the **ANSI** standard is active, and delete all removable styles from the **Style Name:** list box.
7. Pick the **New** button, and create a new dimension style named **ANSI-INCH** based on the default ANSI standard.
8. Specify the following options in the **Dimension Styles** dialog (retain default settings, or use desired specifications for options not listed):

   Linear Units / Decimal Marker: in. / . Period
   Linear Format / Precision: Decimal / 3.123
   Angular Format / Precision: Decimal Degrees / 0
   Suppression: Leading Zero
   Line Type / Line Weight: Continuous / .01 in
   Extension Line Extension: .12
   Extension Line Offset: .0625
   Dimension Line Spacing From Dimension: .45
   Dimension Line Spacing From Object: .625
   Text Font / Size / % Stretch / Gap: Arial / .120 / 100 / .0625
   Terminator Style / Size / Aspect: Closed Arrowhead / .125 / .042

9. Pick the **Save** button.
10. Pick the **Close** button.
11. Save As EX13.1.
12. Exit Autodesk Inventor or continue working with the program.

Field Notes

As discussed earlier in this chapter, dimension styles are typically specified in a template file and are available even before you add drawing views. However, the dimension styles you created in Exercise 13.1 will be inserted into future inch exercises and projects.

Exercise 13.2

1. Launch Autodesk Inventor.
2. Open a new metric drawing file.
3. Delete the current border and title block.
4. From the **Format** pull-down menu, pick the **Dimension Styles...** option to access the **Dimension Styles** dialog box.
5. Ensure the **ANSI** standard is active, and delete all removable styles from the **Style Name:** list box.
6. Pick the **New** button, and create a new dimension style named **ANSI-METRIC** based on the default ANSI standard.
7. Specify the following options in the **Dimension Styles** dialog (retain default settings, or use desired specifications for options not listed):

   Linear Units / Decimal Marker: mm / . Period
   Linear Precision: 2.12
   Angular Format / Precision: Decimal Degrees / 0
   Suppression: Trailing Zero
   Line Type / Line Weight: Continuous / .25mm
   Extension Line Extension: 3.175
   Extension Line Offset: 1.588
   Dimension Line Spacing From Dimension: 9.525
   Dimension Line Spacing From Object: 12.7
   Text Font / Size / % Stretch / Gap: Arial / 3.17 / 100 / .8
   Terminator Style / Size / Aspect: Closed Arrowhead / 3 / 1

8. Pick the **Save** button.
9. Pick the **Close** button.
10. Save As EX13.2.
11. Exit Autodesk Inventor or continue working with the program.

Field Notes

As discussed earlier in this chapter, dimension styles are typically specified in a template file and are available even before you add drawing views. However, the dimension styles you created in Exercise 13.2 will be inserted into future metric exercises and projects.

Defining Additional Drafting Standards

As discussed in Chapter 12, drafting standards allow you to control every aspect of the drawing display including dimension style, text style, sheet display, and multiple other drafting elements. You can modify the drafting standards at any time during the drawing creation process by selecting **Standards...** from the **Format** pull-down menu to access the **Drafting Standards** dialog box. However, before you begin a drawing or dimension an object, you may want to become familiar with the drafting standard options, or adjust the drafting standards to fit your specific applications. Chapter 12 explains how to use the options in the **Common, Sheet, Terminator,** and **Hatch** tabs of the **Drafting Standards.** The additional tabs allow you to define dimensioning and annotation standards.

Field Notes

Drafting standards are very important in order to make a drawing look exactly the way you want it to look, and capture design intent. However, if you are not concerned about defining drafting standards at this time, you may choose to skip ahead to the next section of this chapter.

As previously mentioned, you may want to adjust some of the settings in the **Drafting Standards** dialog box before dimensioning a drawing, or to create a template complete with all the desired settings. This chapter explores only the drafting standards tabs that relate to part drawing dimensions. The **Parts List** and **Balloon** tabs are discussed in Chapter 17. Tabs that relate to specific tools and dimensioning applications are discussed later in this chapter. However, the options available in the following dimensioning **Drafting Standards** dialog box tabs are more general than the options located in the tabs that control specific tool or application functions:

- **Terminator** This tab, shown in Figure 13.12, allows you to adjust the terminator and arrowhead options for certain leaders and datum identifier terminators. As previously discussed, terminator options for linear, aligned, angular, diameter, and radial dimensions are controlled in the **Terminator** tab of the **Dimension Styles** dialog box. The following options located in the **Terminator** tab of the **Drafting Standards** dialog box define leader terminator specifications used in creating hole/thread notes, surface texture symbols, weld symbols, feature control frames, feature identifier symbols, datum identifier symbols, text leaders, datum target leaders, and balloon leaders only:

  - **General** Use this drop-down list to define the type of arrowhead you want to use.

  - **Size** Enter a value in this edit box to define the size of the terminator. For arrowheads, the size represents the arrowhead length. For circles, the size represents the circle diameter.

  - **Aspect** Enter a value in this edit box to define the aspect or width in reference to size of the arrowhead. For example, to create a terminator that is three times as long as it is wide, the aspect should be one-third the value of the size.

  - **Datum** Use this drop-down list to define the datum identifier terminator style.

- **Dimension Style** This tab, shown in Figure 13.13, displays the dimension styles and characters currently available in the drawing. It also allows you to activate a different dimension style. Pick the dimension style you want to use to dimension the drawing from the **Active**

FIGURE 13.12 The **Terminator** tab of the **Drafting Standards** dialog box.

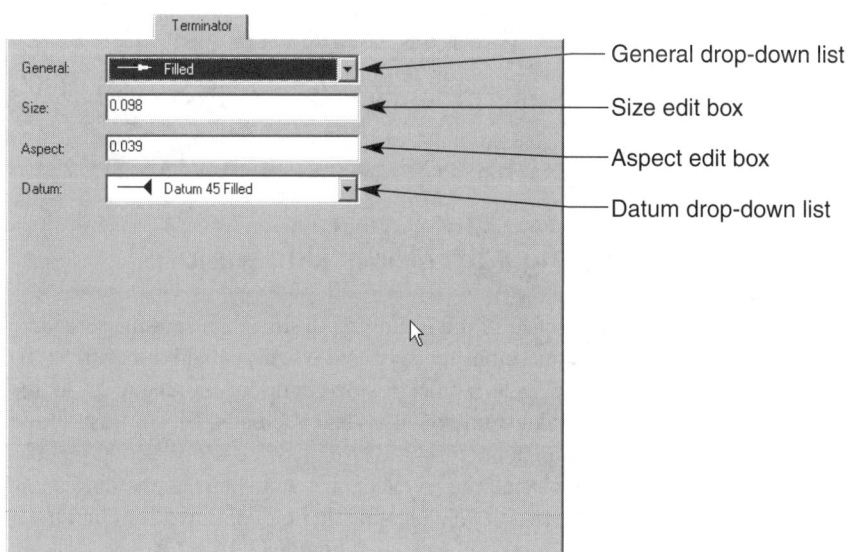

FIGURE 13.13   The
Dimension Style tab of
the **Drafting Standards**
dialog box.

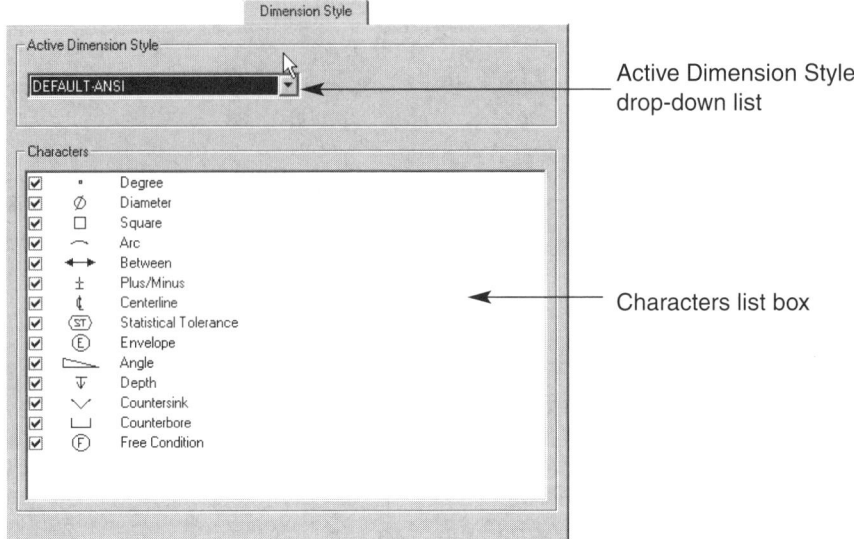

Active Dimension Style
drop-down list

Characters list box

Dimension Style drop-down list. Typically, the dimension style that is highlighted in the
Dimension Style dialog box is active. Still, although a certain dimension style may be dis-
played in the **Dimension Style** dialog box, you must activate the style in the **Dimension
Style** tab of the **Drafting Standards** dialog box, if you want to use the style in the drawing.

The available dimension characters are displayed in the **Characters** list box. When a
characters check box is selected, the character is usable. Deselect the check box to make a
character unusable.

 Field Notes

Dimension styles define most of the dimension specification for most dimensions. Dimension
styles and the **Dimension Styles** dialog box are discussed later in this chapter.

Exercise 13.3

1. Continue from EX13.2, or launch Autodesk Inventor.
2. Open a new inch drawing file.
3. Explore the options and tabs available in the **Drafting Standards** dialog box as discussed.
4. Close the drawing file without saving.
5. Exit Autodesk Inventor or continue working with the program.

Working with Model Dimensions

As mentioned in the beginning of this chapter, Autodesk Inventor contains model and drawing dimensions.
Model dimensions are specified when you add dimensions to the model sketch or define the size or shape of
an object, such as an extrusion depth or hole diameter. As the name implies, model dimensions correspond
to the model from which a view is created. In contrast, drawing dimensions are additional dimensions you
add that do not control the size or shape of the actual model. Model dimensions are parametric; when edited, they change the physical size and shape of the corresponding model. Similarly, if you modify a model
dimension while in a part file, the model dimension and the size and shape of the object in the corresponding drawing also change. See Figure 13.14.

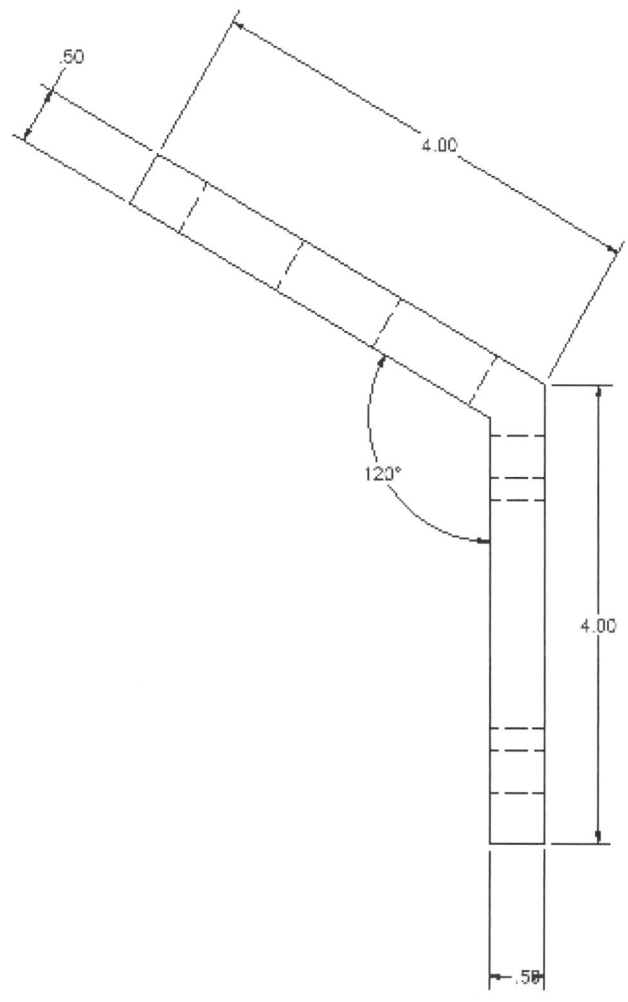

FIGURE 13.14 An example of model dimensions in a drawing view. Often it may be difficult to differentiate between model and drawing dimensions.

Only model dimensions that are planar to the drawing view are shown in a view. Use one of the following techniques to display model dimensions:

✓ Before you create a base view, access the **Options** dialog box by selecting **Application Options...** from the **Tools** pull-down menu. Then choose the **Get model dimensions on view placement** check box. Now when you insert a drawing view the model dimensions should automatically show.

✓ Right-click inside an existing drawing view and select the **Get Model Dimensions** menu option.

✓ Right-click on an existing drawing view name in the **Browser,** and select the **Get Model Dimensions** menu option.

Typically model dimensions are placed directly over view geometry. You can move model dimensions by selecting the dimension text and dragging the entire dimension to the desired location. You can never actually remove a model dimension from a drawing, unless you delete the parameter from the model in the part file. However, you can hide individual model dimensions by right-clicking on the dimension you want to hide and selecting the **Delete** menu option. You can also hide all the model dimensions in a view, by right-clicking inside a view and selecting the **Hide Model Dimensions** option of the **Hide Dimension** cascading submenu.

If you want to change the size or shape of a model while working in the drawing environment, right-click the model dimension of the geometry you want to change and select the **Edit Model Dimension** menu option. When you pick the **Edit Model Dimension** menu option, the **Edit Dimension** dialog box for the selected parameter is displayed. See Figure 13.15. Enter the new parameter value in the edit box and

FIGURE 13.15 The
Edit Dimension dialog
box.

 ← OK button

pick the **OK** button. It may take some time for the changes to occur, because the geometry of the model is also modified to reflect the new value. Additional information regarding dimension modification and dimension style override is discussed later in this chapter.

Exercise 13.4

1. Continue from EX13.3, or launch Autodesk Inventor.
2. Open a new inch drawing file.
3. Delete the default border and title block.
4. Access the **Options** dialog box by selecting **Application Options...** from the **Tools** pull-down menu, and ensure the **Get model dimensions on view placement** check box is selected.
5. Use the **Create View** tool to place the top view of EX6.3, as shown.
6. Move the model dimensions as shown.
7. Right-click on the ∅5.00 dimension and select the **Edit Model Dimension...** menu option.
8. Change the dimension to ∅7.00.
9. While still in the drawing file, open EX6.3 and notice the changes to the model.
10. Pick the **Undo** button to undo the model dimension edit. You should clearly see the connection between model dimensions in a drawing and part model geometry.
11. Close EX6.3 without saving.
12. Save the drawing file as EX13.4.
13. Exit Autodesk Inventor or continue working with the program.

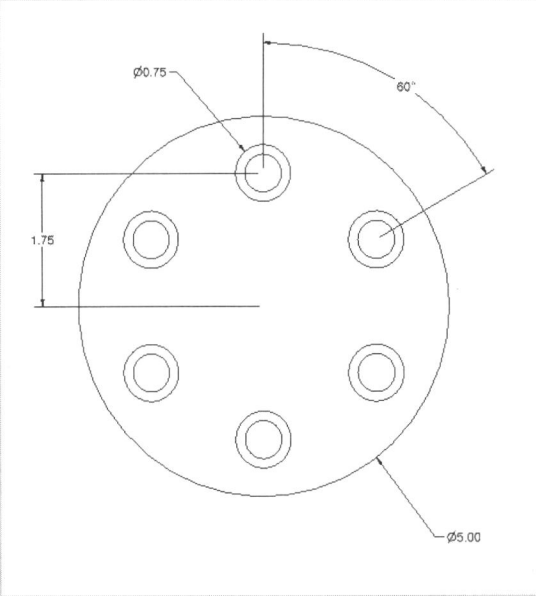

Working with Drawing Dimensions

Often model dimensions do not fully define part drawing design intention. As a result, drawing dimensions are used to place additional dimensions on drawing view geometry to further document the design for manufacturing purposes. As the name implies, drawing dimensions relate only to the drawing in which they are placed. Unlike model dimensions, drawing dimensions do not control any model parameters. When you edit a model while in a part file, the drawing dimensions in the drawing file change to reflect the new values. However, you cannot change the size or shape of a part model by manipulating drawing dimensions.

Autodesk Inventor provides you with multiple tools that allow you create a wide variety of drawing dimensions. These tools are referred to as drawing annotation tools. Throughout this chapter, you will explore specific tools, commands, and options that allow you to apply additional dimensions and notes to a part drawing.

Field Notes

> The **Balloon, Balloon All,** and **Parts List** drawing annotation tools correspond to assembly drawings and are discussed in Chapter 15.

Drawing annotation tools are used to create a wide variety of drawing dimensions and notes including linear, angular, radial, diameter, and ordinate dimensions, hole and thread notes, and weld and surface texture symbols, just to name a few. These tools are accessed from the **Drawing Annotation** toolbar, shown in Figure 13.16, or from the **Drawing Annotation** panel bar.

Once a drawing view is created, you can enter the drawing annotation work environment by right-clicking on the panel bar and selecting the **Drawing Annotation** menu option. Figure 13.17 shows the drawing annotation tools available in the panel bar. By default, the panel bar is displayed in learning mode, as shown in Figure 13.17. To change from learning to expert mode, as shown in Figure 13.18 and discussed in Chapter 2, right-click any of the panel bar tools or left-click the panel bar title and select the **Expert** option.

FIGURE 13.16 The **Drawing Annotation** toolbar.

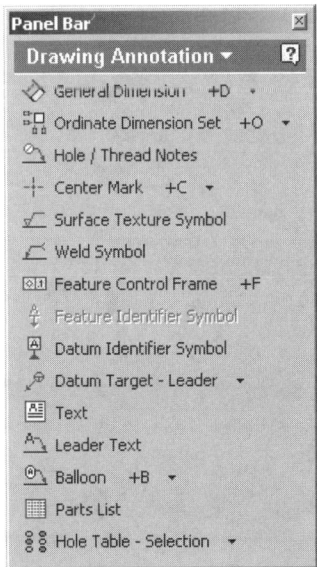

FIGURE 13.17 The panel bar in **Drawing Annotation** (learning) mode.

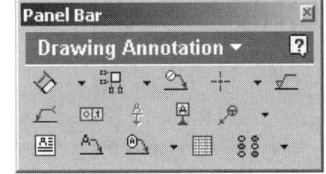

FIGURE 13.18 The panel bar in **Drawing Annotation** (expert) mode.

Adding Centerlines and Centermarks

Drawings that contain circles, arcs, other round geometry, or symmetrical features, require centermarks or centerlines before they can be dimensioned. Autodesk Inventor provides you with four different tools that allow you to place centermarks on single circles and arcs, centermarks on linear and circular groups of circles and arcs, and centerline bisectors. See Figure 13.19.

Before you place centermarks and centerlines, you may want to adjust some of the centermark and centerline properties in the **Drafting Standards** dialog box. Access the **Drafting Standards** dialog box by picking the **Standards...** menu option from the **Format** pull-down menu. As shown in Figure 13.20, centermark and centerline properties are located in the **Centermark** tab of the **Drafting Standards** dialog box.

This tab allows you to adjust the settings for creating centermarks, centerlines, centerline bisectors, and center patterns. Centerlines, also called centermarks, are used to show the center of a circle or arc and the axis or center plane of a symmetrical feature. The centerline is made up of alternating short and long dashes, with a gap between the dashes. Short dashes typically cross at the center of a circle or arc. The centermark elements available in the tab correspond to the centermark options description area next to the edit boxes. The top description displays the effects on a circle, while the bottom description displays the effects on a center line bisector. The centermark components follow:

- **Mark**   Defines length of the short dashes in the centermark indicator size.

- **Gap**   Sets the gap, which is the space between the short and long dashes of the centerlines.

- **Extension**   Establishes the distance a symmetrical centerline extends past the feature, and controls the length relationship between short and long centerline dashes.

- **Overshoot**   Sets the distance centerlines extend past a circle or arc perimeter.

- **Fictitious Diameter**   This option identifies the centermark size for a suppressed feature in a pattern of features.

FIGURE 13.19
Examples of the types of centermarks and centerlines you can create.

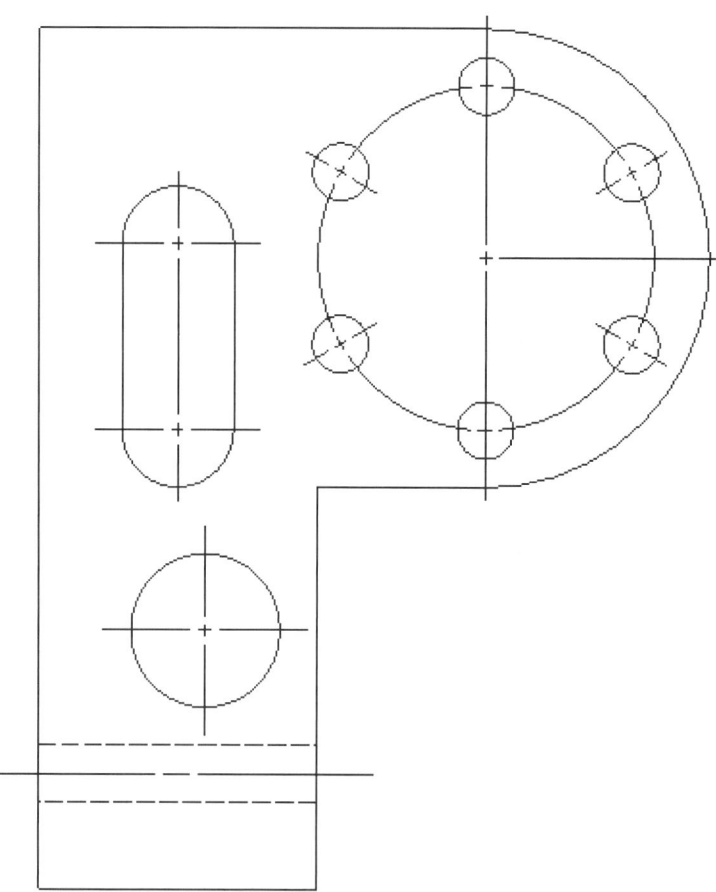

FIGURE 13.20 The **Centermark** tab of the **Drafting Standards** dialog box.

Mark edit box →

Gap edit box →

Extension edit box →

Overshoot edit box →

Fictitious Diameter edit box →

Centermark options description area

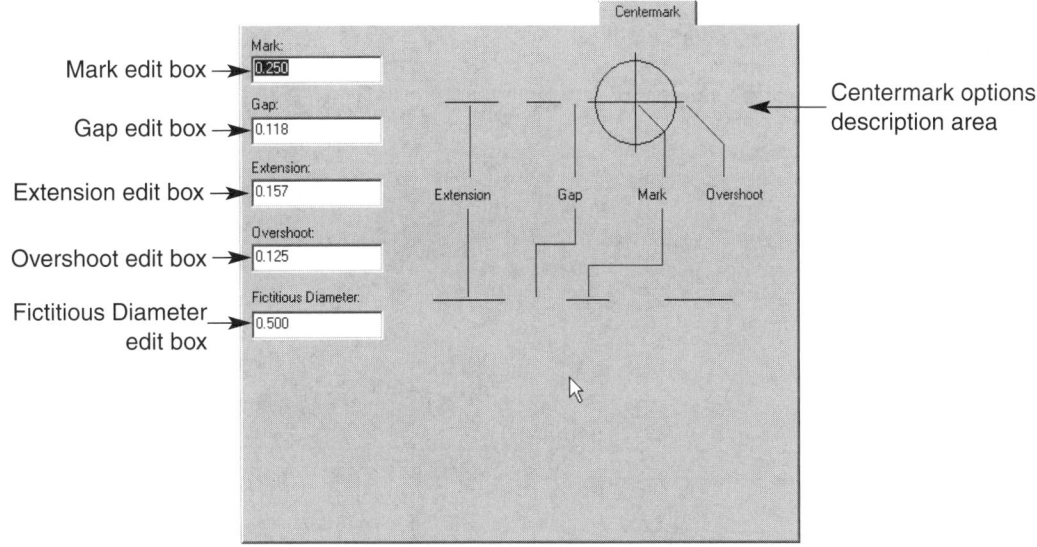

You can modify each of the centermark components by entering values in the edit boxes.

Once you have defined the centermark and centerline display options in the **Centermark** tab, you are ready to place centermarks and centerline bisectors on the drawing. Use the following tools and techniques to add the desired centermarks to your drawings:

 Field Notes

The **Center Mark, Center Line, Center line bisector,** and **Centered pattern** tools are all located in a flyout button.

■ **Center Mark** This tool is used to add centermarks to a variety of curved geometric shapes, such as circles, ellipses, and arcs. See Figure 13.21. Access the **Center Mark** command using one of the following techniques:

✓ Pick the **Center Mark** button on the panel bar.

✓ Pick the **Center Mark** button on the **Drawing Annotation** toolbar.

✓ Type the **+** and **C** keys on your keyboard.

To use the **Center Mark** tool, pick the geometry or the center of the geometry on which you want to place a centermark. Continue placing centermarks as required. When finished, press the **Esc** key on your keyboard, or right-click and select the **Done [Esc]** menu option.

FIGURE 13.21 Example of various centermarks created using the **Center Mark** tool.

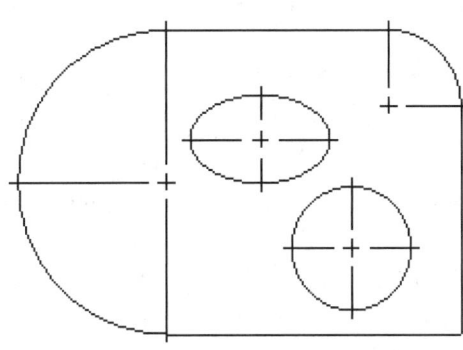

FIGURE 13.22
Example of various centermarks created using the **Center Line** tool.

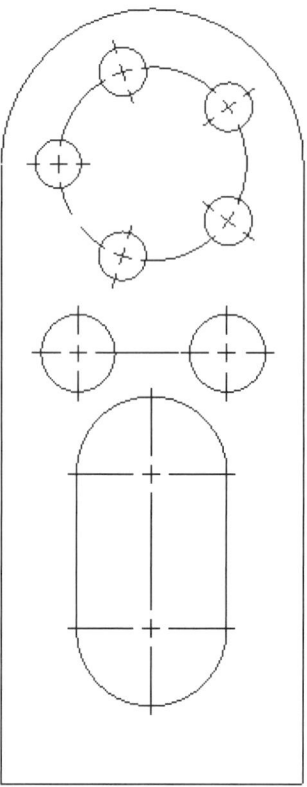

- **Center Line**　This tool is used to add centermarks to two or more round objects with connected centerlines. See Figure 13.22.

  Access the **Center Line** command using one of the following techniques:

  ✓ Pick the **Center Line** button on the panel bar.

  ✓ Pick the **Center Line** button on the **Drawing Annotation** toolbar.

  To use the **Center Line** tool, pick the geometry or the center of the geometry on which you want to place the first centermark. Continue picking additional objects as required. Then, generate the centermarks by right-clicking and selecting the **Create** menu option. Continue placing centermarks as needed. When finished, press the **Esc** key on your keyboard, or right-click and select the **Done [Esc]** menu option.

- **Centered pattern**　This tool is similar to the **Center Line** tool previously discussed, but is primarily used to add center marks to multiple round objects and the bolt circle when applying polar coordinate dimensions. See Figure 13.23. Access the **Centered pattern** command using one of the following techniques:

  ✓ Pick the **Centered pattern** button on the panel bar.

  ✓ Pick the **Centered pattern** button on the **Drawing Annotation** toolbar.

  To use the **Centered pattern** tool, first select the outside diameter or radius of the part. Then, in a clockwise or counterclockwise fashion, select the objects in the pattern. See Figure 13.23. To generate the centermarks, right-click and select the **Create** menu option. Continue placing centermarks on other objects as needed. When finished, press the **Esc** key on your keyboard, or right-click and select the **Done [Esc]** menu option.

- **Center line bisector**　This tool allows you to place a centerline that bisects two symmetrical objects, such as a hidden or section view of a hole. See Figure 13.24. Access the **Center line bisector** command using one of the following techniques:

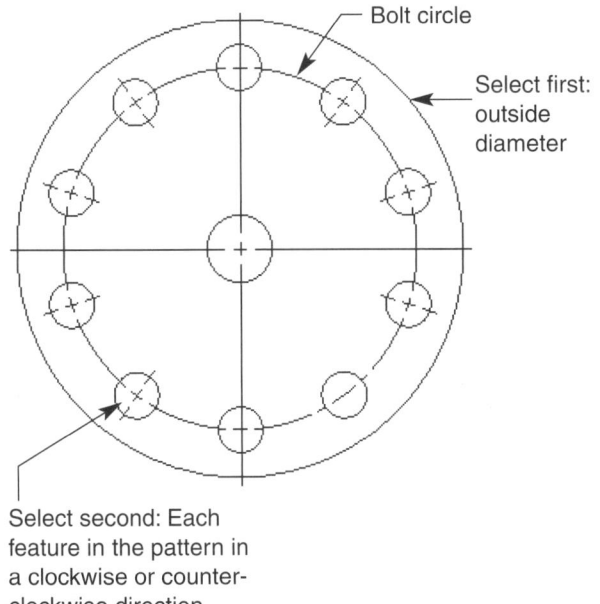

FIGURE 13.23 An example of the centermarks created using the **Centered Pattern** tool.

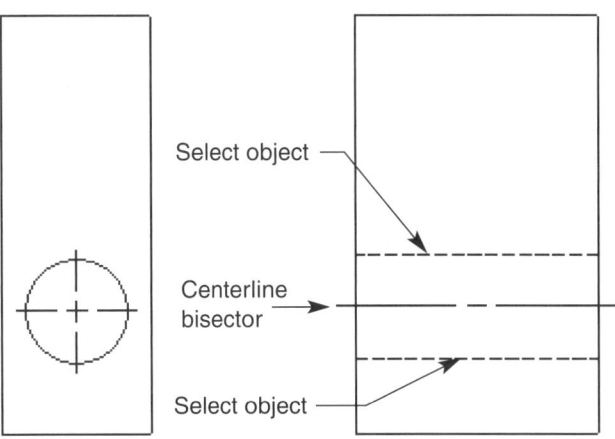

FIGURE 13.24 An example of a centerline bisector created using the **Center line bisector** tool.

✓ Pick the **Center line bisector** button on the panel bar.

✓ Pick the **Center line bisector** button on the **Drawing Annotation** toolbar.

To use the **Center line bisector** tool, select the first side of the symmetrical object, followed by the opposite side. See Figure 13.24. Then, in a clockwise or counterclockwise fashion, select the objects in the pattern. See Figure 13.23. To generate the centerline, right-click and select the **Create** menu option. Continue placing centerline bisectors on other objects as needed. When finished, press the **Esc** key on your keyboard, or right-click and select the **Done [Esc]** menu option.

Exercise 13.5

1. Continue from EX13.4, or launch Autodesk Inventor.
2. Open a new inch drawing file.
3. Access the **Centermark** tab of the **Drafting Standards** dialog box by picking **Standards...** from the **Format** pull-down menu.
4. Enter the specified values in the following edit boxes:

   Mark: .125
   Gap: .0625
   Extension: .25
   Overshoot: .125
   Fictitious Diameter: .125

5. Select **Draft View...** from the **Insert** pull-down menu to open a draft view with a 1:1 scale.
6. Sketch geometry similar to the drawing shown in Figure 13.19.
7. Finish the sketch.

Exercise 13.5 continued on next page

Exercise 13.5 continued

8. Activate the panel bar **Drawing Annotation** mode, or access the **Drawing Annotation** toolbar.
9. Using the information discussed in this chapter and the specified tools, add centermarks and centerlines to the drawing as shown.
10. Save the file as EX13.4.
11. Exit Autodesk Inventor or continue working with the program.

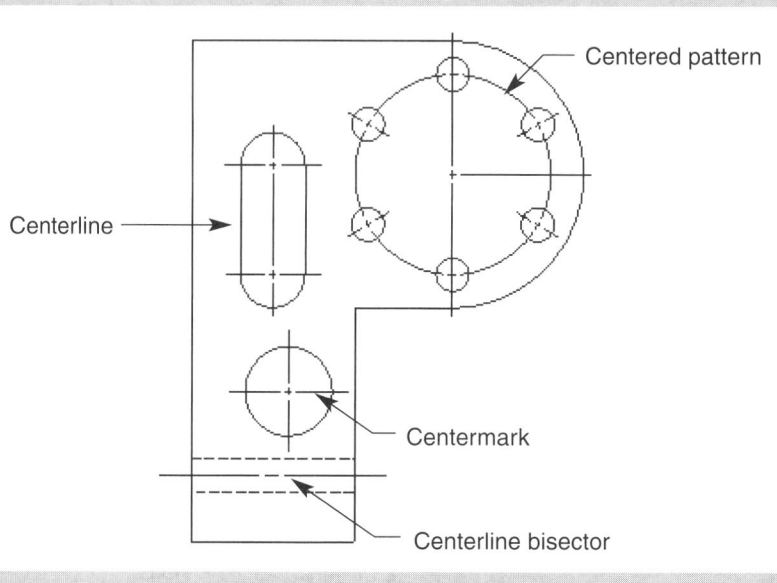

Exercise 13.6

1. Continue from EX13.5, or launch Autodesk Inventor.
2. Open EX12.13.
3. Save a copy of EX12.13 as EX13.6.idw.
4. Close EX12.13 without saving, and open EX13.5.
5. Right-click on **B Size:1** in the **Browser** and select the **Edit Sheet...** menu option.
6. Change the sheet name to **C Size.**
7. Change the sheet size to **C,** and pick the **OK** button.
8. Delete the current border and insert a default C-size border.
9. Access the **Project View** tool, and project a bottom view from the front view.
10. Reposition views as needed.
11. Activate the panel bar **Drawing Annotation** mode, or access the **Drawing Annotation** toolbar.
12. Access the **Center Mark** tool and place a centermark at the hole shown.
13. Right-click the centermark you just created, and select the **Align to Edge** option from the **Edit** cascading submenu.
14. Select any of the auxiliary view edges to correctly align the centermark.
15. Access the **Center line bisector** tool and place a center line as shown.
16. The final drawing should look like the drawing shown.
17. Resave EX13.5.
18. Exit Autodesk Inventor or continue working with the program.

Exercise 13.6 continued on next page

Exercise 13.6 continued

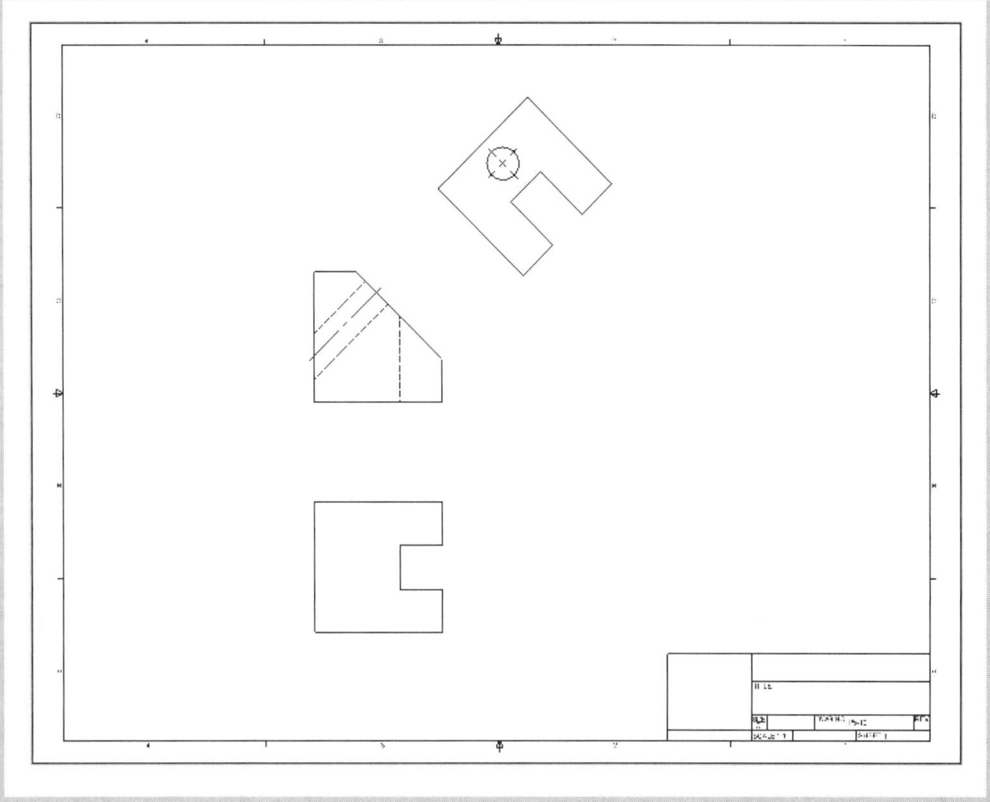

Using the General Dimension Tool

The *general dimension* tool allows you to create a number of different drawing dimensions including linear, aligned, radial, diameter, and angular dimensions. See Figure 13.25. General dimensions are placed on a drawing using slightly different techniques, depending on the type of geometry you are dimensioning. However, each of the dimensions previously discussed are created using the **General Dimension** tool, which is accessed using one of the following techniques:

FIGURE 13.25
Examples of dimensions created using the general dimension tool.

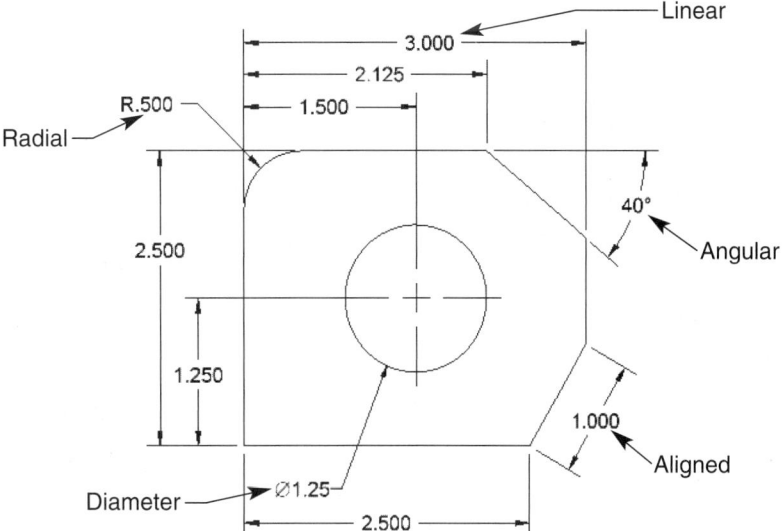

✓ Pick the **General Dimension** button on the panel bar.

✓ Pick the **General Dimension** button on the **Drawing Annotation** toolbar.

✓ Type the **+** and **D** keys on your keyboard.

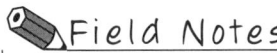Field Notes

The **General Dimension** tool is located in a flyout button that also contains the **Baseline Dimension** tool.

Using the techniques discussed, you can create the following dimensions with the **General Dimension** tool:

■ **Linear**   Use *linear* dimensions to define and locate horizontal and vertical geometry. To create a linear dimension, select the object you want to dimension, the endpoints of the objects, or an object and an endpoint. The selection method depends on the dimension you want to create. See Figure 13.26. If you select two endpoints, or an object and an endpoint, you can specify the dimension type by moving the dimension text, or by selecting either the **Horizontal, Vertical, Aligned, Linear Diameter,** or **Linear Symmetric** option from the **Dimension Type** cascading submenu, depending on the application. To create the dimension, drag the dimension line to the desired location and pick the position. The dimension will become dashed when you drag the dimension over the location designated in the **Dimension line spacing from dimension lines** and **Dimension line spacing from object** edit boxes of the **Dimension Styles** dialog box. When the dimension becomes dashed, pick the location to accept the specified offset. See Figure 13.27.

■ **Aligned**   Typically *aligned* dimensions refer to a style of dimensioning in which the dimension text is aligned with the dimension line. However, Autodesk Inventor describes an aligned dimension as a dimension in which the dimension line is parallel to an inclined angular surface. See Figure 13.28. To create an aligned dimension, first select an angled

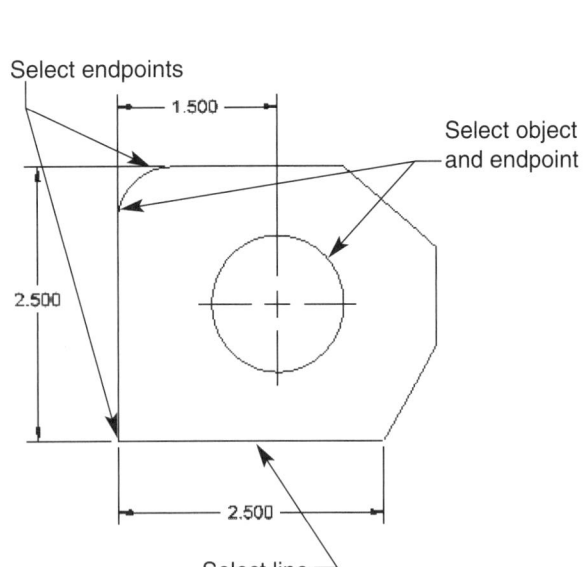

FIGURE 13.26 Selecting objects to apply general dimensions.

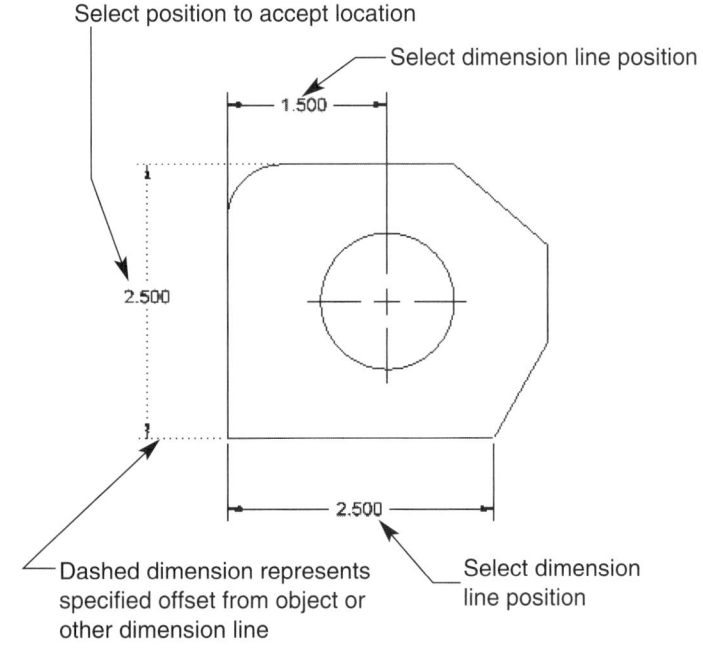

FIGURE 13.27 Selecting dimension line positions and offsets specified in the **Dimension Styles** dialog box.

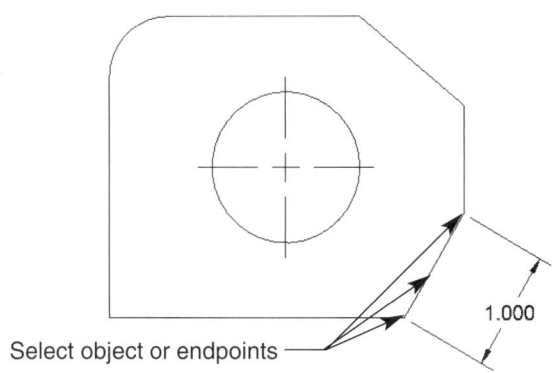

FIGURE 13.28 Aligned dimensioning.

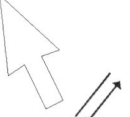

FIGURE 13.29 The aligned dimension icon.

object or the endpoints of the object. Then pick again, once you see the aligned dimension icon, shown in Figure 13.29. An alternative method is to right-click after you initially pick the object or endpoints, and select the **Aligned** option from the **Dimension Type** cascading submenu. To generate the dimension, drag the dimension line to the desired location and pick the position. Again, the dimension will become dashed when you drag the dimension over the location designated in the **Dimension line spacing from dimension lines** and **Dimension line spacing from object** edit boxes of the **Dimension Styles** dialog box. When the dimension becomes dashed, pick the location to accept the specified offset.

- **Radial** *Radial* dimensions are used to define the radius of an arc, such as a half circle or a fillet, and are identified by an R, followed by the radius value. See Figure 13.30. To create a radial dimension, as shown in Figure 13.30, select the arc you want to dimension. Then drag the dimension text to the desired location and pick the position.

    If you want to override the radial dimension style, right-click before you pick the dimension text position and select **Arrowheads Inside** or **Jogged** from the **Options** cascading submenu. You can also specify a different dimension type by selecting **Diameter, Angle, Arc Length,** or **Chord Length** from the **Dimension Type** cascading submenu, depending on the application.

- **Diameter** *Diameter* dimensions are used to define the diameter of a circle, such as a hole, and are identified by $\varnothing$, followed by the diameter value. See Figure 13.31. Placing a diameter dimension is similar to placing an angular dimension. First, select the circle you want to dimension. Then drag the dimension text to the desired location and pick the position. See Figure 13.31.

    You can also override the current diameter dimension options, by right-clicking before you pick the dimension text position, and select **Leader From Center** from the **Options** cascading submenu or Radius from the **Dimension Type** cascading submenu, depending on the application.

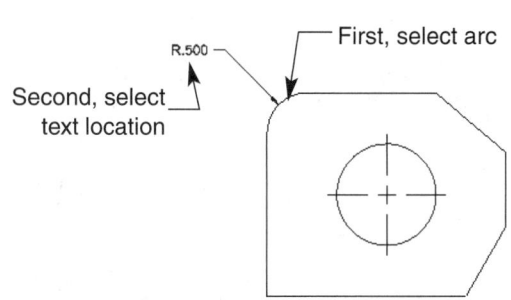

FIGURE 13.30 Radial dimensioning.

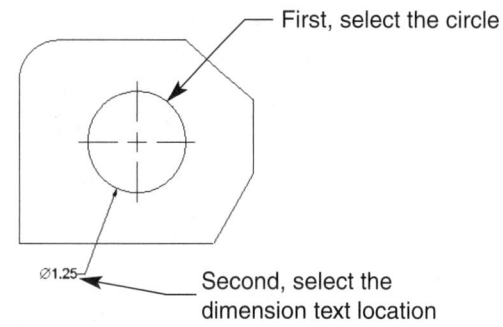

FIGURE 13.31 Diameter dimensioning.

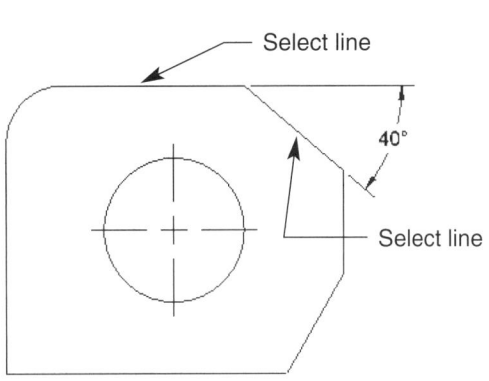

FIGURE 13.32   Angular dimensioning.

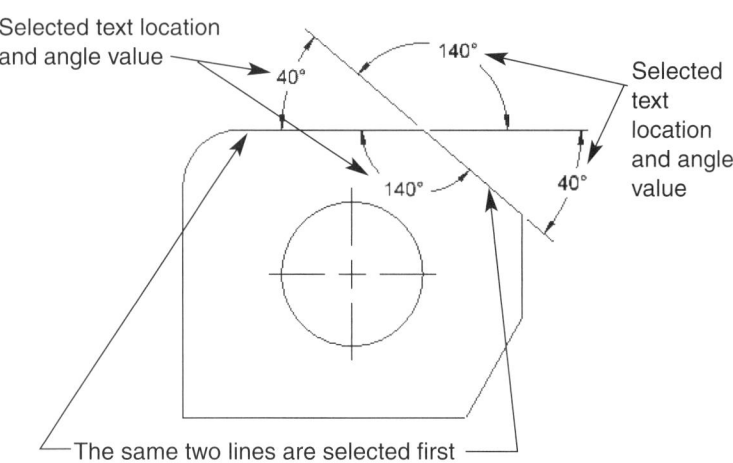

FIGURE 13.33   Defining an angular dimension.

- **Angular**   *Angular* dimensions are used to specify the incline of an angular surface, such as a chamfer. See Figure 13.32. Typically angular dimension values are set to degrees—for example, 45°—but an angle may also be represented by degrees, minutes, and seconds, such as 25°12′36″. To create an angular dimension, select one of the lines that define the angled surface, followed by the second line that defines the angled surface. See Figure 13.32. To generate the dimension, drag the dimension text to the desired location and pick the position. The location you select defines the specified angle, as shown in Figure 13.33.

  Before you select the dimension text position, you can toggle between the currently displayed dimension and the opposite angle by right-clicking and selecting the **Opposite Angle** option from the **Options** cascading submenu. As with linear and aligned dimensions, angular dimensions will become dashed when you drag the dimension over the location designated in the **Dimension line spacing from dimension lines** and **Dimension line spacing from object** edit boxes of the **Dimension Styles** dialog box. When the dimension becomes dashed, pick the location to accept the specified offset.

When you are finished placing general dimensions, right-click and select the **Done [Esc]** menu option, press the **Esc** key on your keyboard, or access a different tool.

Exercise 13.7

1. Continue from EX13.6, or launch Autodesk Inventor.
2. Open EX13.6.
3. Save a copy of EX13.6 as EX13.7.idw.
4. Close EX13.6 without saving, and open EX13.7.
5. Select **Organizer…** from the **Format** pull-down menu to access the **Drawing Organizer** dialog box.
6. Pick the **Browse** button, and open the file EX13.1.
7. Select ANSI-INCH in the **Source Document:** list box, and pick the **Copy** button.
8. The dimension style ANSI-INCH is now in the current drawing file.
9. Access the **Dimension Style** tab of the **Drafting Standards** dialog box, and activate the ANSI-INCH drafting standard.
10. Pick the **Apply** button, followed by the **OK** button to exit the **Drafting Standards** dialog box.

Exercise 13.7 continued on next page

Exercise 13.7 continued

11. Activate the panel bar **Drawing Annotation** mode, or access the **Drawing Annotation** toolbar.
12. Use the **General Dimension** tool to dimension the drawing views as shown.
13. Resave EX13.7.
14. Exit Autodesk Inventor or continue working with the program.

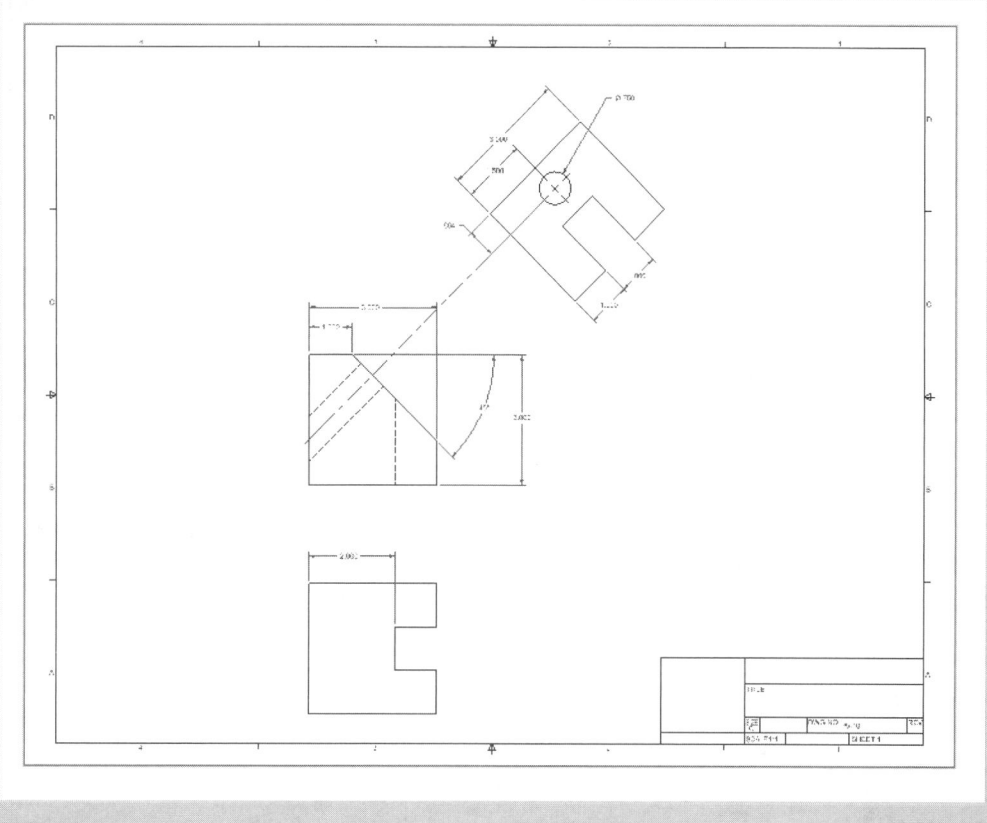

Field Notes

Although using the **Drawing Organizer** dialog box to access existing dimension styles is an acceptable practice, this exercise identifies the importance of defining dimension styles in a template.

For more information regarding the **Drawing Organizer** dialog box, refer to Chapter 2.

Creating Datum Dimensions

Datum dimensioning, also referred to as ***Baseline dimensioning*** in Autodesk Inventor, is a way to dimension drawing geometry by defining the size and location of each feature in reference to an origin, also called a datum. Adjacent dimension values are staggered to reduce dimension crowding and improve clarity. See Figure 13.34.

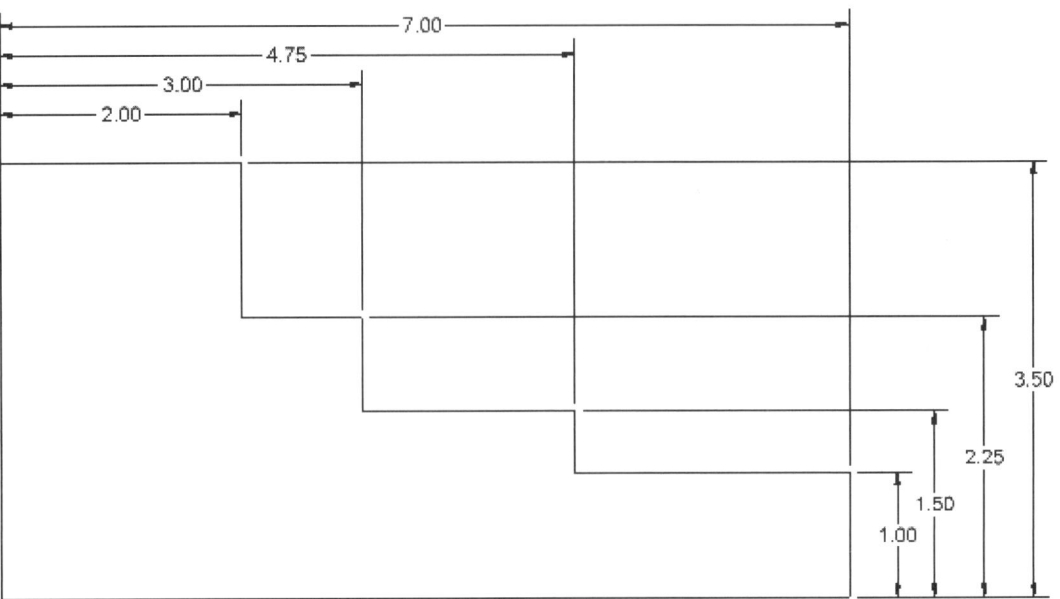

FIGURE 13.34   Datum or baseline dimensions (one set of datum dimensions for the horizontal features, and one set for the vertical features).

Datum dimensioning can be accomplished using the **General Dimension** tool, previously discussed. However, Autodesk Inventor provides the **Baseline Dimension** tool for quickly and easily placing multiple dimensions in a datum dimensioning fashion. Access the **Baseline Dimension** tool using one of the following techniques:

✓ Pick the **Baseline Dimension** button on the panel bar.

✓ Pick the **Baseline Dimension** button on the **Drawing Annotation** toolbar.

✓ Type the **+** and **A** keys on your keyboard.

Field Notes

The **Baseline Dimension** tool is located in a flyout button that also contains the **General Dimension** tool.

After you access the **Baseline Dimension** tool, create datum dimensions by first selecting the common surface, which is the object that each of the dimensions originate. Then, pick each of the additional objects you want to include in the dimension set. Typically the order in which you select additional features does not matter. However, the common surface must be selected first. See Figure 13.35.

Once you select the common surface and all the geometry you want to include in the dimension set, right-click and select the **Continue** menu option. Then locate the group of dimensions by picking the desired location of the initial dimension line. As in other dimensions, the dimension will become dashed when you drag the dimension over the location designated in the **Dimension line spacing from dimension lines** and **Dimension line spacing from object** edit boxes of the **Dimension Styles** dialog box. When the dimension becomes dashed, pick the location to accept the specified offset. See Figure 13.36.

After you generate a group of datum dimensions, you can create another set of dimensions by selecting additional features. When you are finished using the **Baseline Dimension** tool, right-click and select the **Done [Esc]** menu option, press the **Esc** key on your keyboard, or access a different tool.

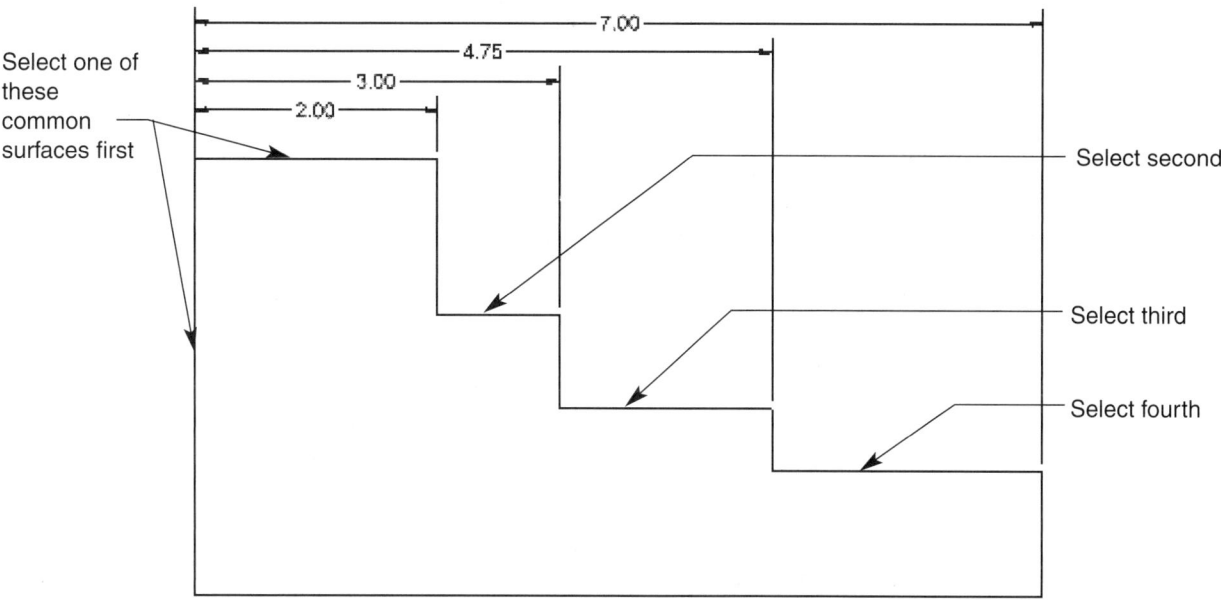

FIGURE 13.35 Creating datum, or baseline dimensions.

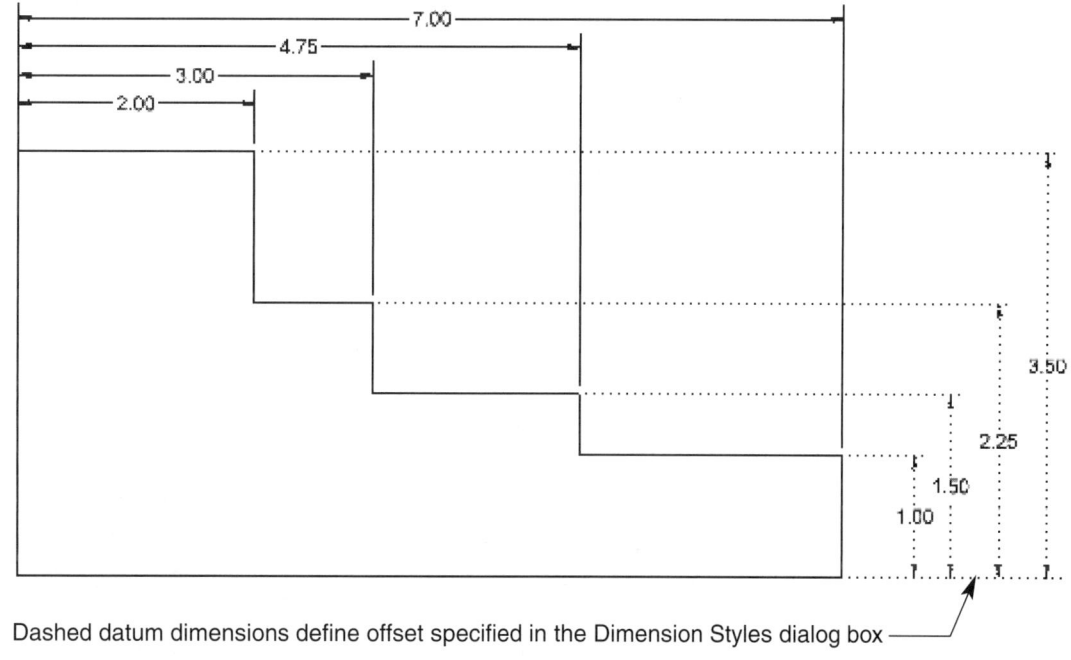

Dashed datum dimensions define offset specified in the Dimension Styles dialog box ——

FIGURE 13.36 Selecting datum dimension line positions and offsets specified in the **Dimension Styles** dialog box.

Exercise 13.8

1. Continue from EX13.7, or launch Autodesk Inventor.
2. Open EX12.12.
3. Save a copy of EX12.12 as EX13.8.idw.
4. Close EX12.12 without saving, and open EX13.8.
5. Select **Organizer...** from the **Format** pull-down menu to access the **Drawing Organizer** dialog box.
6. Pick the **Browse** button, and open the file EX13.1.
7. Select ANSI-INCH in the **Source Document:** list box, and pick the **Copy** button.
8. The dimension style ANSI-INCH is now in the current drawing file.
9. Access the **Dimension Style** tab of the **Drafting Standards** dialog box and activate the ANSI-INCH drafting standard.
10. Pick the **Apply** button, followed by the **OK** button to exit the **Drafting Standards** dialog box.
11. Activate the panel bar **Drawing Annotation** mode, or access the **Drawing Annotation** toolbar.
12. Use the **Baseline Dimension** tool to dimension the drawing views as shown.
13. Resave EX13.8.
14. Exit Autodesk Inventor or continue working with the program.

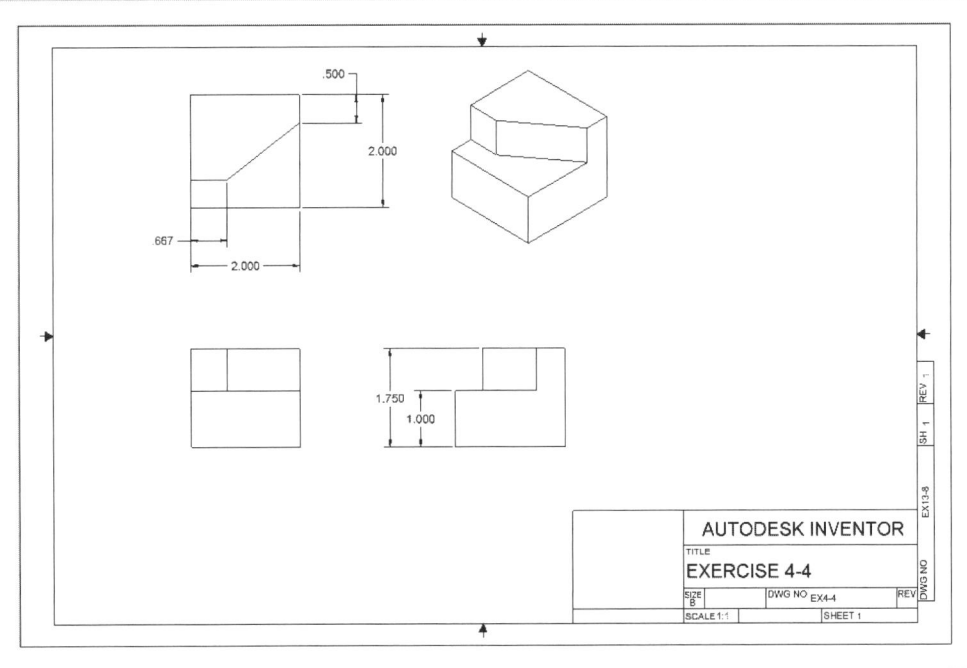

Field Notes

Although using the **Drawing Organizer** dialog box to access existing dimension styles is an acceptable practice, this exercise identifies the importance of defining dimension styles in a template.

For more information regarding the **Drawing Organizer** dialog box, refer to Chapter 2.

Adding Text

Often you may need to apply specific or general notes or some type of additional text to a drawing, without opening a new sketch or adding a draft view. *Specific notes,* also called *local notes,* are notes that apply to a specific feature or features on the drawing. *General notes* apply to the entire drawing and are usually placed together in a location on the drawing such as the lower left corner, upper right corner, or over or adjacent to the title block. To place text in these types of situations, Autodesk Inventor includes a **Text** tool as part of the drawing annotation commands. The **Text** tool available in the drawing annotation environment functions exactly the same as the **Text** tool available in the drawing sketch environment, discussed in Chapter 12. Many of the text properties are controlled by text styles, also discussed in Chapter 12. You can use the default text styles and override some format options when you place text, but you may want to create your own text styles. Refer to the discussion on the **Text Styles...** command and the **Text Styles** dialog box in Chapter 12 for information regarding text styles.

To create text, access the **Text** command using one of the following techniques:

✓ Pick the **Text** button on the **Drawing Annotation** panel bar.

✓ Pick the **Text** button on the **Drawing Annotation** toolbar.

Once you access the **Text** command, use the positioning cursor to select a point, or hold-down the left mouse button and drag the cursor to create a window on the sheet. See Figure 13.37. The point you select, or window you create, define the location and extents of the text. When you have located where you want to place the text, the **Format Text** dialog box is displayed. See Figure 13.38. The **Format Text** dialog box functions exactly the same in the drawing annotation environment as it does in the sketch environment. For more information regarding text and the **Format Text** dialog box, refer to Chapter 12. When you have finished defining options in the **Format Text** dialog box, pick the **OK** button to generate the text. The **Text** tool will

FIGURE 13.37
Selecting a text location

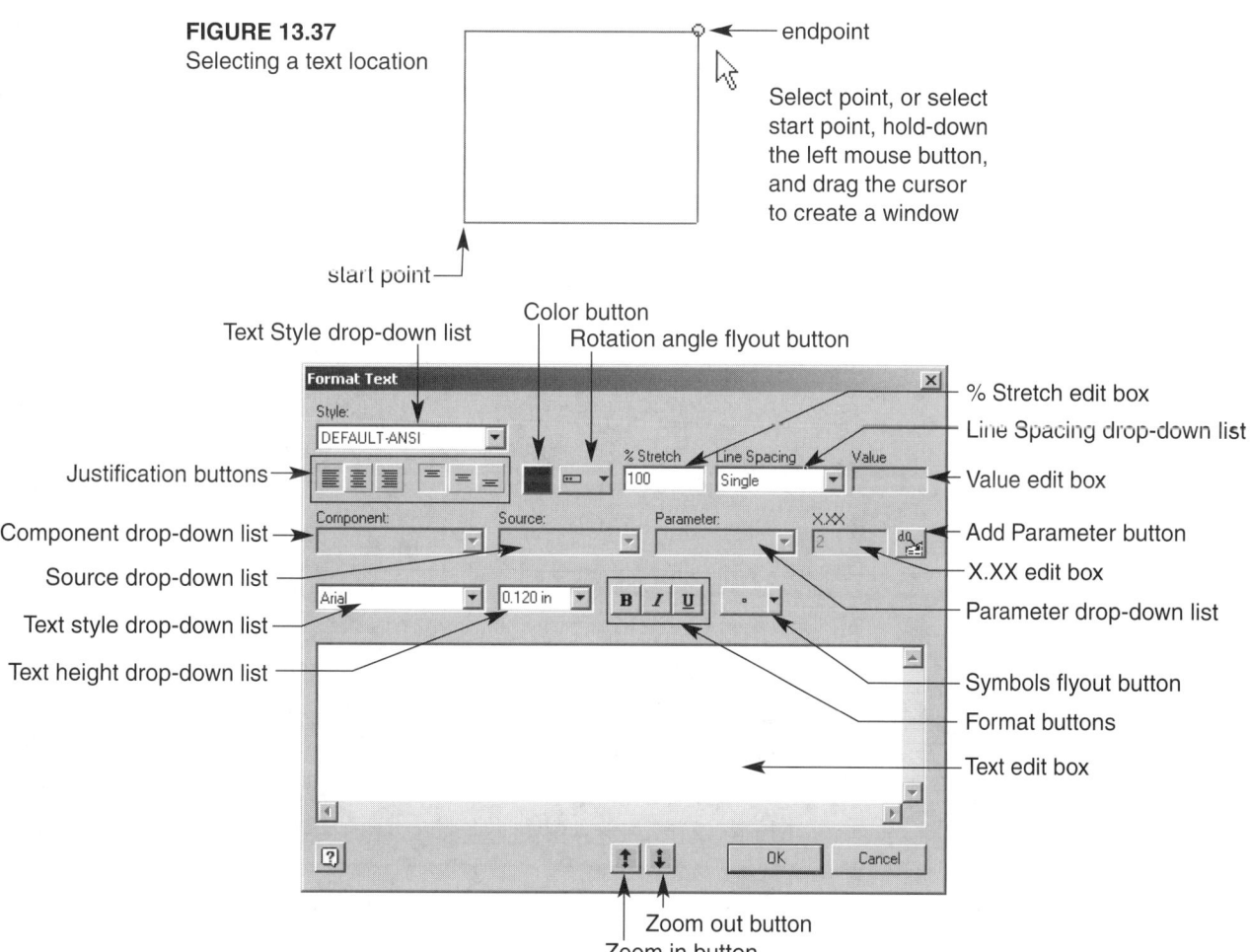

FIGURE 13.38   The **Format Text** dialog box.

still be active, allowing you to continue placing text on the drawing. To exit the **Text** command, right-click and select the **Done [Esc]** menu option, press the **Esc** key on your keyboard, or access a different tool.

Exercise 13.9

1. Continue from EX13.8, or launch Autodesk Inventor.
2. Open EX13.6.
3. Save a copy of EX13.6 as EX13.9.idw.
4. Close EX13.6 without saving, and open EX13.9.
5. Activate the panel bar **Drawing Annotation** mode, or access the **Drawing Annotation** toolbar.
6. Select the **Text** tool, and pick a position near the bottom left edge of the drawing sheet.
7. Use the Default-ANSI text style and type the following text:

   NOTES:
   1. INTERPRET DIMENSIONS AND TOLERANCES PER ASME Y14.5M-1994.
   2. REMOVE ALL BURRS AND SHARP EDGES.

8. Highlight the word **NOTES:** and change the text height to .24in.
9. Ensure the additional text is .12in in height.
10. The final exercise should look similar to the drawing shown.
11. Resave EX13.9.
12. Exit Autodesk Inventor or continue working with the program.

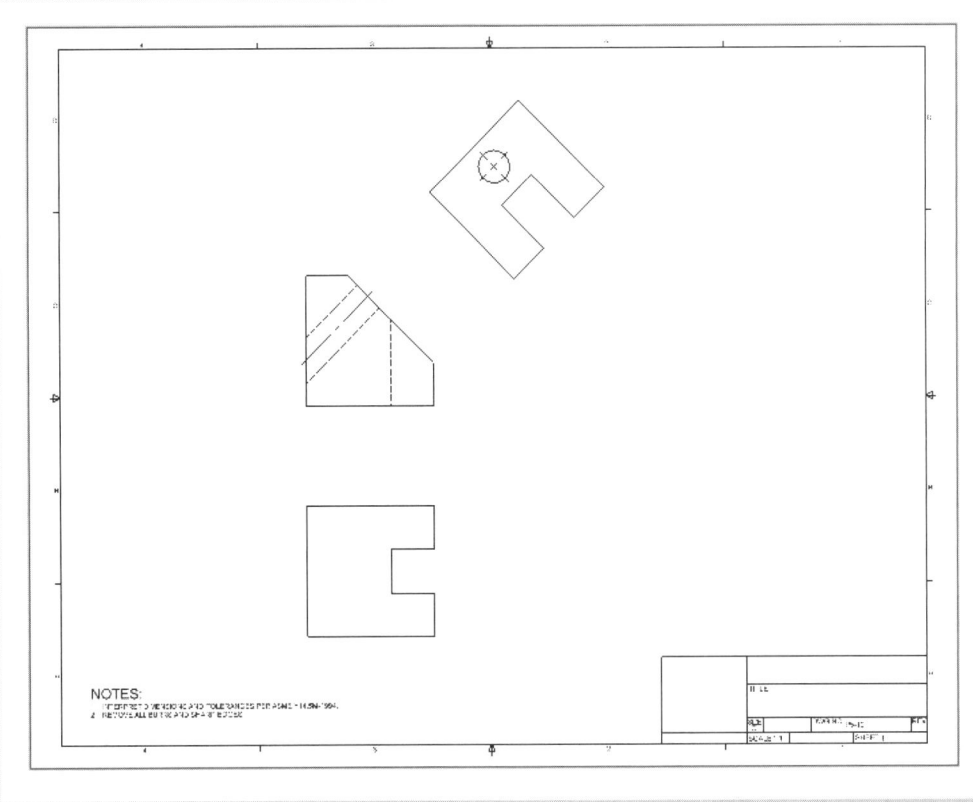

Placing Leader Notes

Specific or *local notes* are notes that apply to specific features on the drawing. These notes are often connected to the feature using leader lines, or leaders as they are typically called. A leader connects to the beginning or the end of a note with a horizontal shoulder and is drawn at an angle where it is capped on the end with an arrowhead where it touches the desired feature. The **General Dimension** tool, previously discussed, allows you to dimension circles and arcs with leaders, and the **Hole/Thread Notes** tool, discussed later in this chapter, is used to dimension holes and threads. However, you can dimension all other types of specific geometry with leaders, using the **Leader Text** command. The **Leader Text** command is similar to the text command discussed earlier, but includes a leader. See Figure 13.39.

Access the **Leader Text** command using one of the following techniques:

✓ Pick the **Leader Text** button on the panel bar.

✓ Pick the **Leader Text** button on the **Drawing Annotation** toolbar.

Once you access the **Leader Text** command, use the cursor to select the feature, or point, to which you want to add a leader and text, or pick a point in the drawing sheet. Then drag the leader shoulder to the desired location. See Figure 13.40.

After you select appropriate drawing geometry or a point on the drawing sheet and locate the text leader shoulder, press the enter key on your keyboard, or right-click and select the **Continue** menu option to display the **Format Text** dialog box. See Figure 13.41. The **Format Text** dialog box displayed when

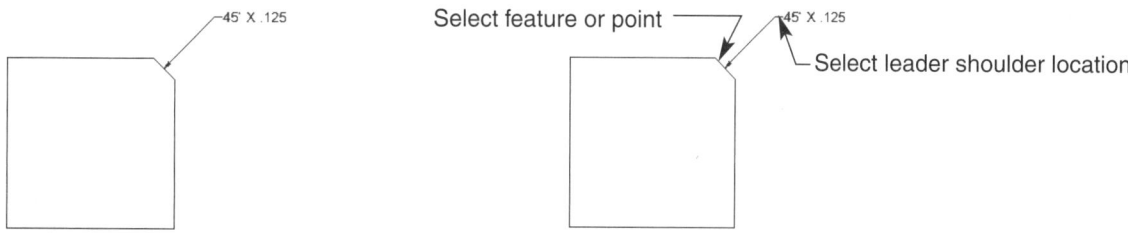

FIGURE 13.39   An example of a leader text dimension.

FIGURE 13.40   Placing a text leader.

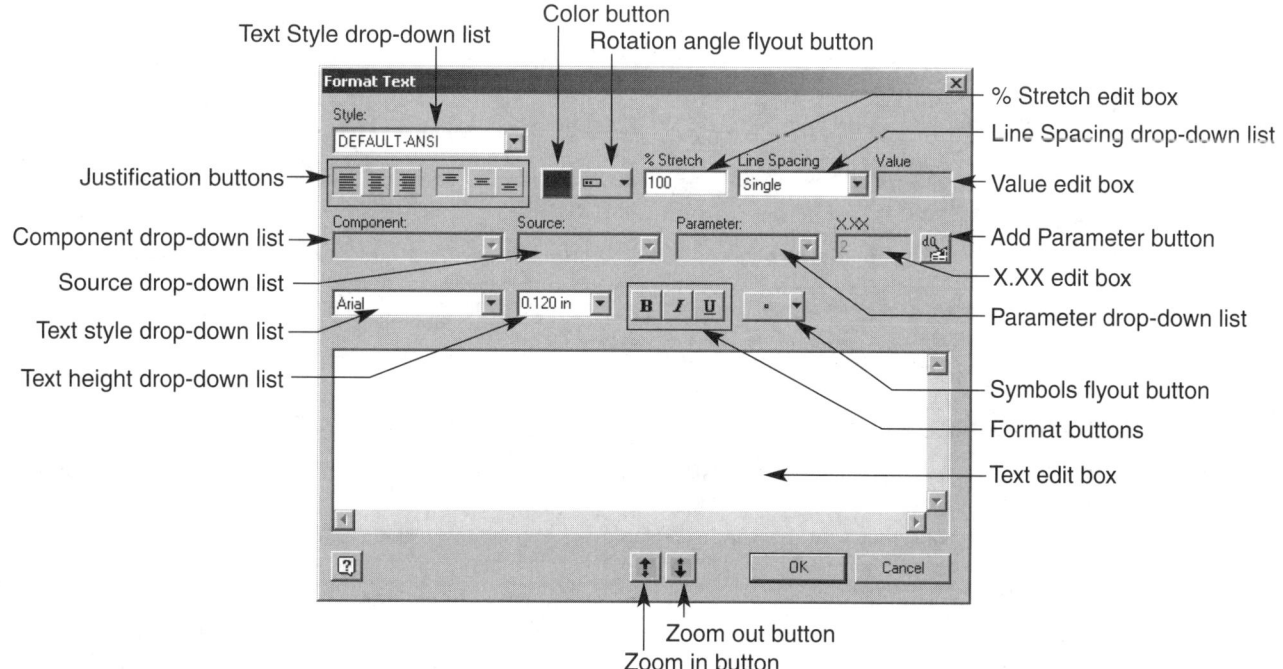

FIGURE 13.41   The **Format Text** dialog box.

placing leader text is exactly the same as the **Format Text** dialog box displayed when placing text without a leader. Again, for more information regarding text and the **Format Text** dialog box, refer to Chapter 12. When you have finished defining options in the **Format Text** dialog box, pick the **OK** button to generate the leader and the text. The **Leader Text** tool will still be active, allowing you to continue placing leaders on the drawing. To exit the **Leader Text** command, right-click and select the **Done [Esc]** menu option, press the **Esc** key on your keyboard, or access a different tool.

Exercise 13.10

1. Continue from EX13.9, or launch Autodesk Inventor.
2. Open a new metric drawing file.
3. Delete the default border and title block.
4. Select **Draft View...** from the **Insert** pull-down menu to open a draft view with a 1:1 scale.
5. Sketch the geometry shown using sketch tools.
6. Finish the sketch and remove unnecessary dimensions (see final drawing).
7. Explore the **Terminator** tab of the **Drafting Standards** dialog box.
8. Activate the panel bar **Drawing Annotation** mode or access the **Drawing Annotation** toolbar.
9. Use the **Leader Text** tool to add the leader note as shown.
10. Save the exercise as EX13.10.
11. Exit Autodesk Inventor or continue working with the program.

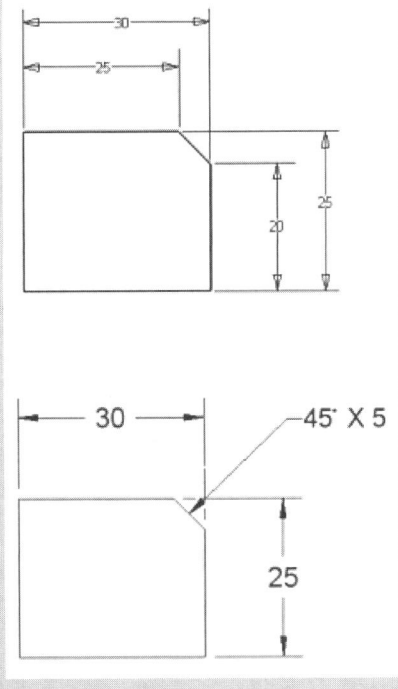

Ordinate Dimensioning

Ordinate dimensioning is also known as *arrowless dimensioning,* because traditional dimension lines and arrowheads are not used. With ordinate dimensioning, dimension values are placed at the end of extension lines, providing coordinates from established datums that are often the corner of the part or the axis of a feature. See Figure 13.42.

Using ordinate dimensioning techniques, features such as holes and radii can be dimensioned in a traditional manner, or letters can be placed on the drawing identifying each feature and coordinating the letter to a table where the dimensional value is given. For example, all holes of one diameter are labeled (A), and holes of another diameter are labeled (B), and so on.

FIGURE 13.42 An example of ordinate dimensioning.

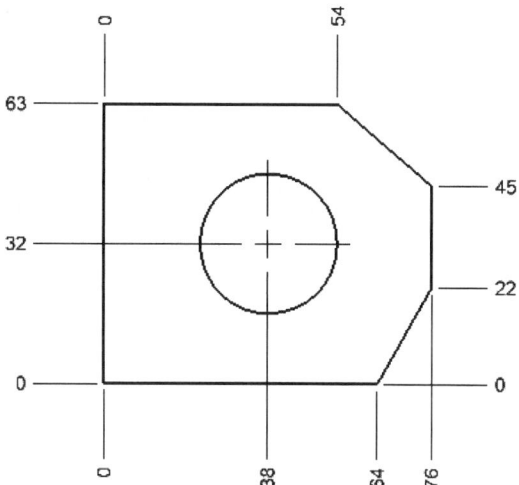

Autodesk Inventor provides two tools, **Ordinate Dimension Set** and **Ordinate Dimension,** for adding ordinate dimensions to drawing view geometry. The **Ordinate Dimension Set** tool allows you to place a group of ordinate dimensions along the same plane, as shown in Figure 13.43A. Dimensions created using the **Ordinate Dimension Set** tool are linked together. You cannot delete or modify individual dimension values. The **Ordinate Dimension** tool allows you to fully dimension one or more drawing views by placing several groups of ordinate dimensions along various planes, as shown in Figure 13.43B. Unlike the **Ordinate Dimension Set** tool, you have to access the **Ordinate Dimension** tool only once to fully define view geometry, and the dimensions you create are not linked together. You can delete and modify individual dimension values.

To use the **Ordinate Dimension Set** tool, access the **Ordinate Dimension Set** command using one of the following techniques:

✓ Pick the **Ordinate Dimension Set** button on the panel bar.

✓ Pick the **Ordinate Dimension Set** button on the **Drawing Annotation** toolbar.

✓ Type the **+** and **O** keys on your keyboard.

Field Notes

The **Ordinate Dimension Set** tool is located in a flyout button that also contains the **Ordinate Dimension** tool.

FIGURE 13.43 (A) A set of ordinate dimensions created by accessing the **Ordinate Dimension Set** tool once. (B) Dimensioning a drawing view by accessing the **Ordinate Dimension** tool once.

Ordinate dimension set

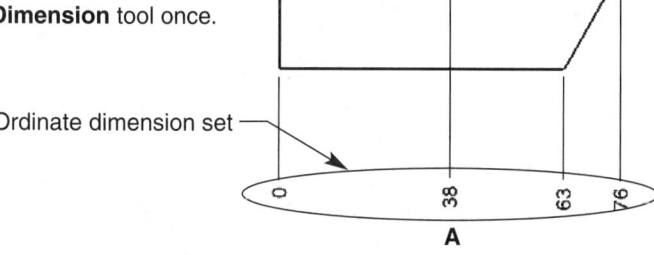

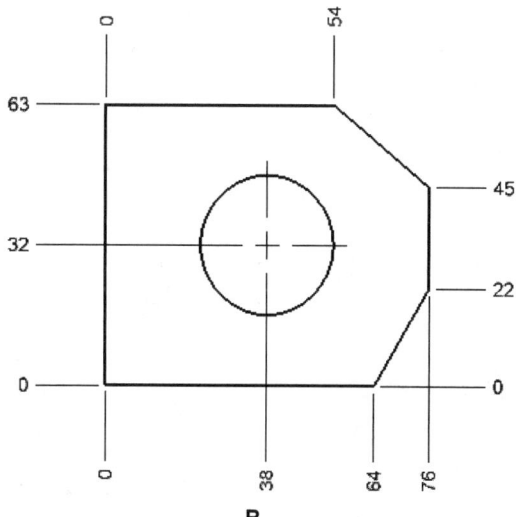

Once you access the **Ordinate Dimension Set** command, the first step in creating a set of ordinate dimensions is to select the origin. The origin establishes the 0,0 point, or datum, which is typically the corner of the part, or the axis of a feature. Next, drag the initial dimension value, which should be 0, to the desired location, and select the position. Continue selecting additional object endpoints to dimension. The placement of the additional dimension values is automatically defined by the selected location of the initial 0 value. See Figure 13.44. While placing ordinate dimensions, you can right-click and select any of the choices in the **Options** cascading submenu to apply the options, depending on your application. Once you have selected all the geometry you want to dimension and specified any required options, right-click and select the **Create** menu option to generate the ordinate dimensions, and exit the command.

To use the **Ordinate Dimension** tool, access the **Ordinate Dimension** command using one of the following techniques:

✓ Pick the **Ordinate Dimension** button on the panel bar.

✓ Pick the **Ordinate Dimension** button on the **Drawing Annotation** toolbar.

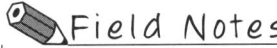

Field Notes

The **Ordinate Dimension** tool is located in a flyout button that also contains the **Ordinate Dimension Set** tool.

Once you access the **Ordinate Dimension** command, the first step in creating a set of ordinate dimensions is to select the view you want to dimension. Then, as with the **Ordinate Dimension Set** tool, select the origin. Again, the origin establishes the 0,0 point, or datum, which is typically the corner of the part, or the axis of a feature. Next, drag the initial dimension value, which should be 0, to the desired location, and select the position. Continue selecting additional object endpoints to dimension. The placement of the additional dimension values is automatically defined by the selected location of the initial 0 value. See Figure 13.45. Once you have selected all the geometry you want to dimension along a specified plane, right-click and select the **Create** menu option to generate the ordinate dimensions. You will notice after you create the dimensions, the **Ordinate Dimension** command is still active. You can repeat the process of selecting the view, followed by the datum, and then the other view geometry to place additional view

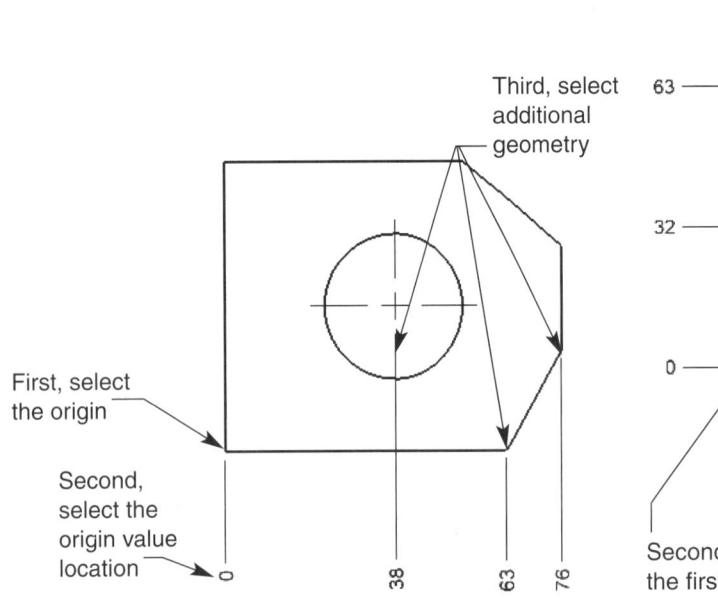

FIGURE 13.44 Creating ordinate dimensions using the **Ordinate Dimension Set** tool.

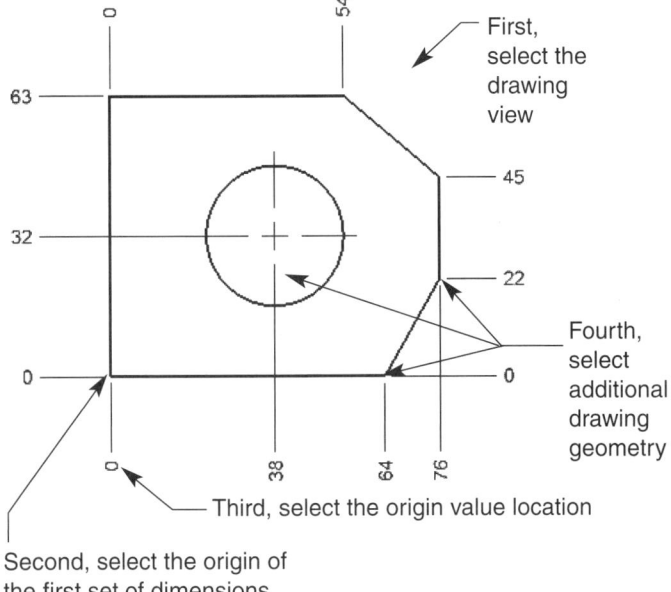

FIGURE 13.45 Creating ordinate dimensions using the **Ordinate Dimension** tool.

dimensions. You may also choose to exit the command by right-clicking and selecting the **Done [Esc]** menu option or selecting the **Esc** key on your keyboard.

Exercise 13.11

1. Continue from EX13.10, or launch Autodesk Inventor.
2. Open P12.9.
3. Save a copy of P12.9 as EX13.11.idw.
4. Close P12.9 without saving, and open EX13.11.
5. Copy the ANSI-METRIC dimension style from EX13.1, using **Drawing Organizer** dialog box.
6. Activate the ANSI-METRIC drafting standard.
7. Activate the panel bar **Drawing Annotation** mode, or access the **Drawing Annotation** toolbar.
8. Use the **Center Mark** and **Center Line Bisector** tools to place the centermarks and centerlines shown.
9. Add the radial dimension shown using the **General Dimension** tool.
10. Use both the **Ordinate Dimension Set** and **Ordinate Dimension** tools to dimension the drawing views as shown.
11. Resave EX13.11.
12. Exit Autodesk Inventor or continue working with the program.

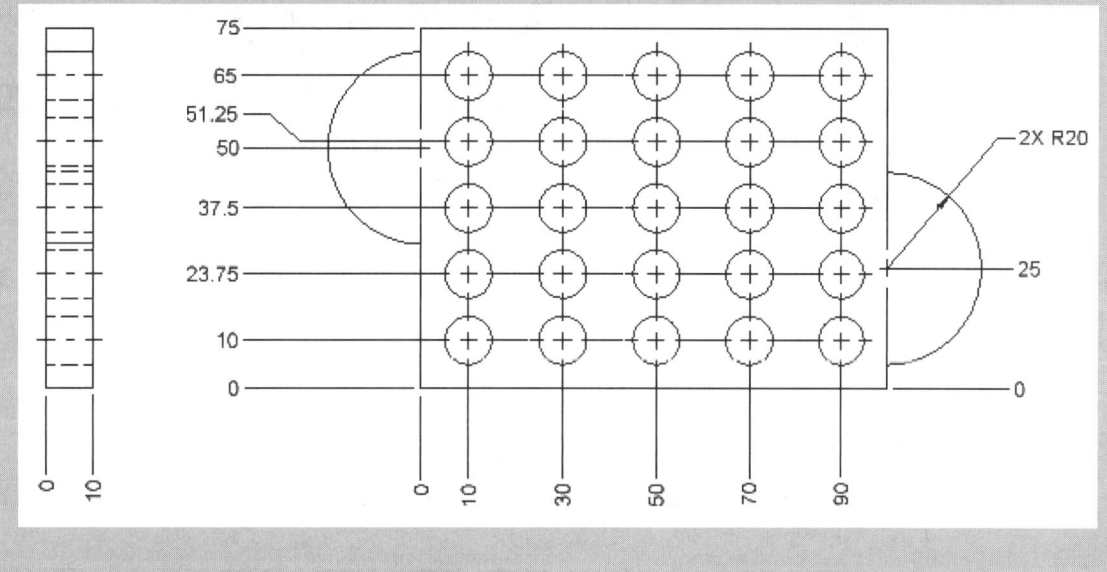

 Field Notes

Although using the **Drawing Organizer** dialog box to access existing dimension styles is an acceptable practice, this exercise identifies the importance of defining dimension styles in a template.

For more information regarding the **Drawing Organizer** dialog box, refer to Chapter 2.

Tabular Dimensioning

Tabular dimensioning is a type of arrowless, ordinate dimensioning. For more information regarding ordinate dimensioning, refer to the previous section of this chapter. Unlike standard ordinate dimensioning, tabular dimensioning involves a system in which coordinate dimensions and size dimensions are given in a table, correlating features on the drawing. See Figure 13.46.

FIGURE 13.46 An example of tabular dimensioning.

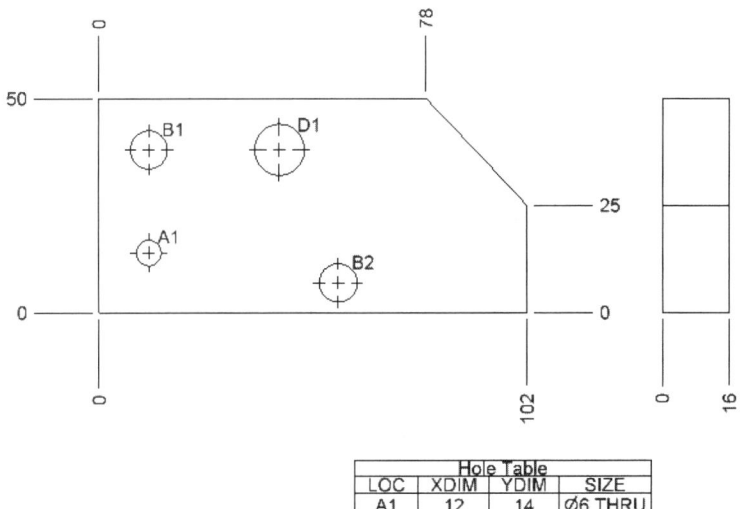

Hole Table			
LOC	XDIM	YDIM	SIZE
A1	12	14	Ø6 THRU
B1	12	38	Ø9 ⊽ 9
B2	57	7	Ø9 ⊽ 12
D1	43	38	Ø12 THRU

Autodesk Inventor provides two tools: **Hole Table – Selection** and **Hole Table – View,** for placing tabular dimensions. The **Hole Table – Selection** tool allows you to create a dimension table by selecting specific drawing view features, while the **Hole Table – View** allows you to create a table by simply picking a drawing view. Like other dimensioning tools, the **Hole Table – Selection** and **Hole Table – View** tools do not fully dimension a feature. For example, you must still use the Center Mark tool to place hole center-marks, and the **Ordinate Dimension Set** or **Ordinate Dimension** tools to place ordinate dimensions, as shown in Figure 13.46. Figure 13.47 displays the results of using the **Hole Table – Selection** or **Hole Table – View** tools without adding any other dimension information.

To use the **Hole Table – Selection** tool, access the **Hole Table – Selection** command using one of the following techniques:

✓ Pick the **Hole Table – Selection** button on the panel bar.

✓ Pick the **Hole Table – Selection** button on the **Drawing Annotation** toolbar.

Field Notes

The **Hole Table – Selection** tool is located in a flyout button that also contains the **Hole Table – View** tool.

FIGURE 13.47 The effects of using the **Hole Table – Selection** and **Hole Table – View** tools.

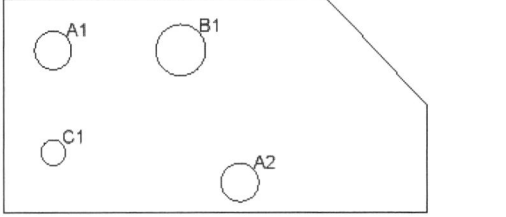

Hole Table			
LOC	XDIM	YDIM	SIZE
A1	12	38	Ø9 ⊽ 9
B1	43	38	Ø12 THRU
C1	12	14	Ø6 THRU
A2	57	7	Ø9 ⊽ 12

FIGURE 13.48
Tabular dimensioning
using the **Hole Table –
Selection** tool.

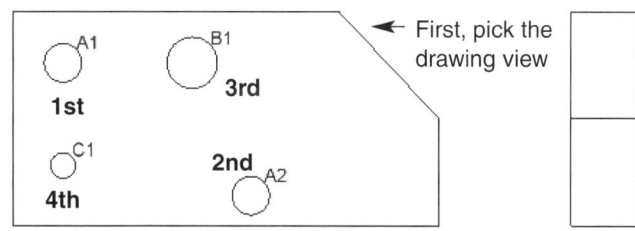

Second, select
the desired
holes in the
given order:

	Hole Table		
LOC	XDIM	YDIM	SIZE
A1	12	38	Ø9 ⊽ 9
B1	43	38	Ø12 THRU
C1	12	14	Ø6 THRU
A2	57	7	Ø9 ⊽ 12

Once you access the **Hole Table – Selection** command, the first step in creating a hole table is to select the drawing view. Next, select each of the holes you want to include in the table. The order in which you select holes designates the hole tag specified in the hole table. For example, the first hole you select is labeled A1, and the second hole you select is labeled B1. Once you select each of the holes that you want to display in the table, right-click and pick the **Create** menu option. Then drag the table to the desired location and pick the position. See Figure 13.48.

To use the **Hole Table – View** tool, access the **Hole Table – View** command using one of the following techniques:

✓ Pick the **Hole Table – View** button on the panel bar.

✓ Pick the **Hole Table – View** button on the **Drawing Annotation** toolbar.

Field Notes

The **Hole Table – View** tool is located in a flyout button that also contains the **Hole Table – Selection** tool.

Once you access the **Hole Table – View** command, dimensioning a drawing view is a two-step process. The first step is to select the drawing view to select automatically all the available holes in the view. The second step is to drag the table to the desired location and pick the position. See Figure 13.49.

FIGURE 13.49
Tabular dimensioning
using the **Hole Table –
View** tool.

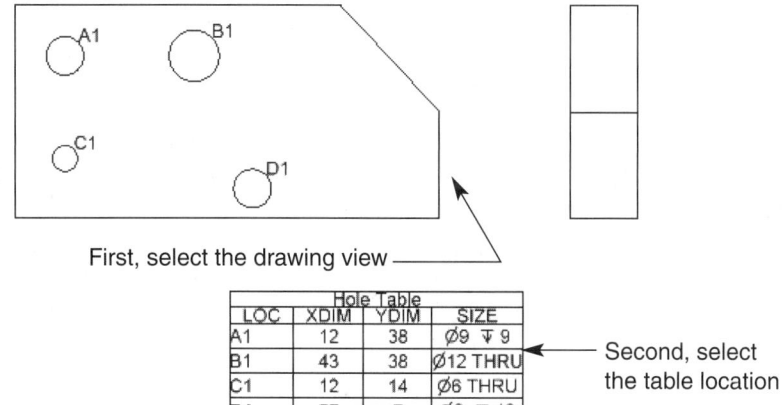

First, select the drawing view

	Hole Table		
LOC	XDIM	YDIM	SIZE
A1	12	38	Ø9 ⊽ 9
B1	43	38	Ø12 THRU
C1	12	14	Ø6 THRU
D1	57	7	Ø9 ⊽ 12

Second, select
the table location

Exercise 13.12

1. Continue from EX13.11, or launch Autodesk Inventor.
2. Open a new inch drawing file.
3. Delete the existing tile block and border.
4. Use the **Create View** tool to place a 1:1 front view of EX6.3 as shown.
5. Project the side view as shown, using the **Project View** tool.
6. Activate the DEFAULT-ANSI drafting standard.
7. Activate the panel bar **Drawing Annotation** mode, or access the **Drawing Annotation** toolbar.
8. Use the **Hole Table – Selection** tool to dimension the drawing view as shown.
9. To define the datum at the circle center, pick the circle edge. Then, in a clockwise fashion select each of the outside hole circles starting with the top hole.
10. The final exercise should look like the drawing shown.
11. Save the file as EX13.12.
12. Exit Autodesk Inventor or continue working with the program.

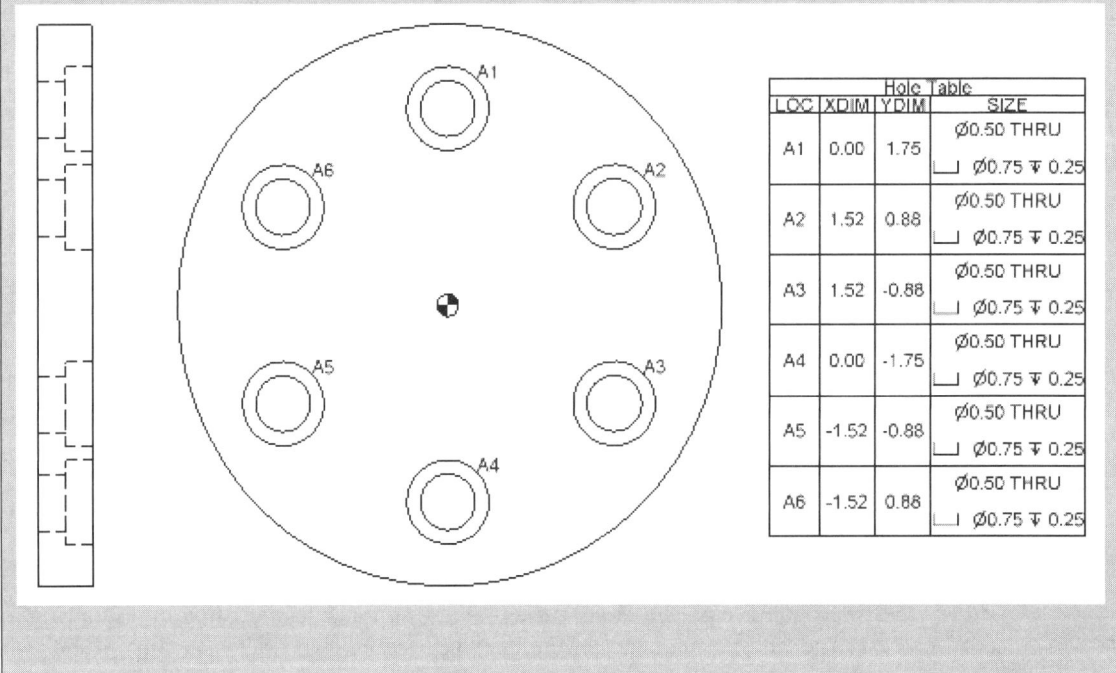

Adding Hole and Thread Notes

Hole and thread notes are normally dimensioned to the feature using a leader line. Holes are dimensioned using diameter dimensions with the diameter symbol preceding the dimension value as in ⌀1.625. Thread notes are commonly connected to the thread representation using a leader line and always follow the same format. For Unified and American National Threads, the thread note format is: major diameter, dash (-), number of threads per inch, thread series, dash (-), class of fit, external (A) or internal (B), right-hand thread is assumed or left-hand (LH), such as ½ - 13UNC-2A. Metric thread notes also follow a specific format: (M) denotes metric, nominal diameter in millimeters, (X) symbol, pitch in millimeters, dash (-), tolerance grade, such as M10X1.5-6H.

Autodesk inventor provides you with the **Hole/Thread Notes** tool, which allows you to fully dimension holes, holes with threads, and threaded shafts. The **Hole/Thread Notes** tool automatically provides information such as countersink, counter bore, depth, tap, thru, and any other hole and thread information that corresponds to the design. See Figure 13.50.

Access the **Hole/Thread Notes** tool using one of the following techniques:

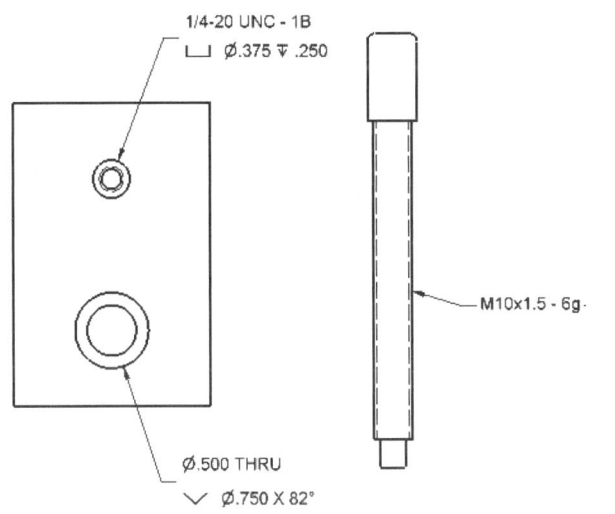

FIGURE 13.50 Examples of holes and threads dimensioned using the **Hole/Thread Notes** tool.

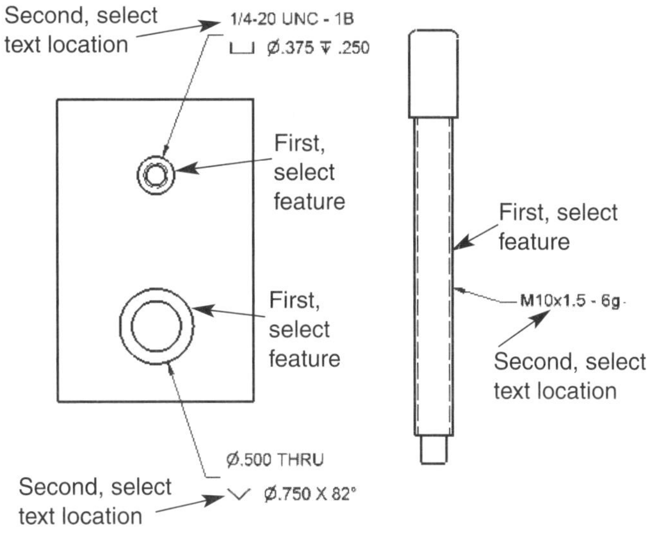

FIGURE 13.51 Using the **Hole/Thread Notes** tool.

✓ Pick the **Hole/Thread Notes** button on the panel bar.

✓ Pick the **Hole/Thread Notes** button on the **Drawing Annotation** toolbar.

After you access the **Hole/Thread Notes** tool, dimension a hole or threaded stud by first selecting the feature. Then drag the leader and dimension text to the desired location and pick the position. Continue adding hole and thread notes as needed. See Figure 13.51. When you are finished, right-click and select the **Done [Esc]** menu option, press the **Esc** key on your keyboard, or access a different tool.

Exercise 13.13

1. Continue from EX13.12, or launch Autodesk Inventor.
2. Open a new metric drawing file.
3. Delete the default border and title block in **Sheet:1.**
4. Use the **Create View** command to place the view shown, using the following information in the **Create View** dialog box:

   File: P4.2
   Scale: 2:1
   Style: Hidden Line
   Deselect the **Get Model Dimensions** check box.

5. Pick a location for the drawing view similar to the location shown.
6. Copy the ANSI-METRIC dimension style from EX13.1, using **Drawing Organizer** dialog box.
7. Activate the ANSI-METRIC drafting standard.
8. Activate the panel bar **Drawing Annotation** mode, or access the **Drawing Annotation** toolbar.
9. Select the **Hole/Thread Notes** tool, and select the drawing view hole.
10. Pick the leader shoulder position, right-click and select the <u>C</u>**ontinue** menu option.
11. The final exercise should look similar to the drawing shown.
12. Save the exercise as EX13.13.
13. Exit Autodesk Inventor or continue working with the program.

Exercise 13.13 continued on next page

Exercise 13.13 continued

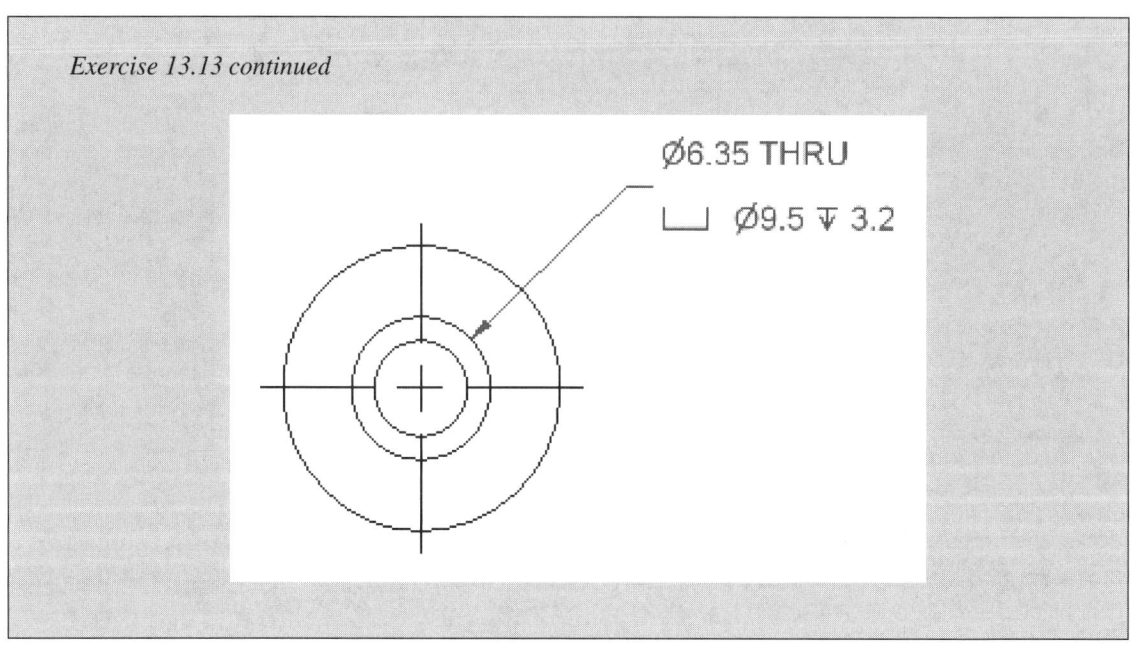

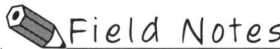**Field Notes**

Although using the **Drawing Organizer** dialog box to access existing dimension styles is an acceptable practice, this exercise identifies the importance of defining dimension styles in a template.

For more information regarding the **Drawing Organizer** dialog box, refer to Chapter 2.

Placing Surface Texture Symbols

Surface finish refers to the allowable roughness, waviness, law, and flaws of a surface, obtained by machining, grinding, honing, or lapping. The surface finish symbol is a "V" shape that contains values related to the finish characteristics. Using the **Surface Texture Symbol** tool, Autodesk Inventor allows you to place the specified surface finish characteristic values and surface finish symbol at the desired location or using a leader line. See Figure 13.52.

Before you add surface texture information and use the **Surface Texture Symbol** tool, you may want to adjust some of the surface texture properties in the **Drafting Standards** dialog box. Access the **Drafting Standards** dialog box by picking the **Standards...** menu option from the **Format** pull-down menu. As shown in Figure 13.53, surface texture properties are located in the **Surface Texture** tab.

Depending on the specified drafting standard, the following options in the **Surface Texture** tab control the default surface texture options when adding surface finish definitions:

FIGURE 13.52
Examples of surface
finish definitions.

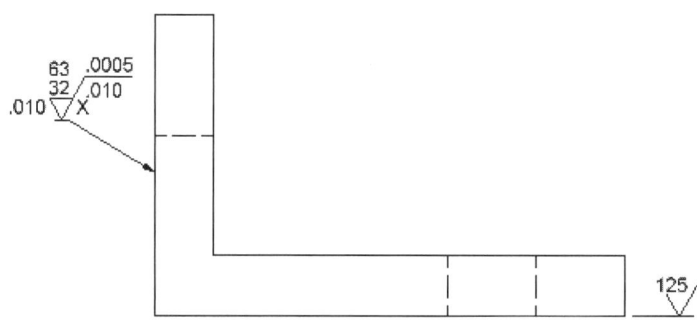

FIGURE 13.53 The
Surface Texture tab of
the **Drafting Standards**
dialog box.

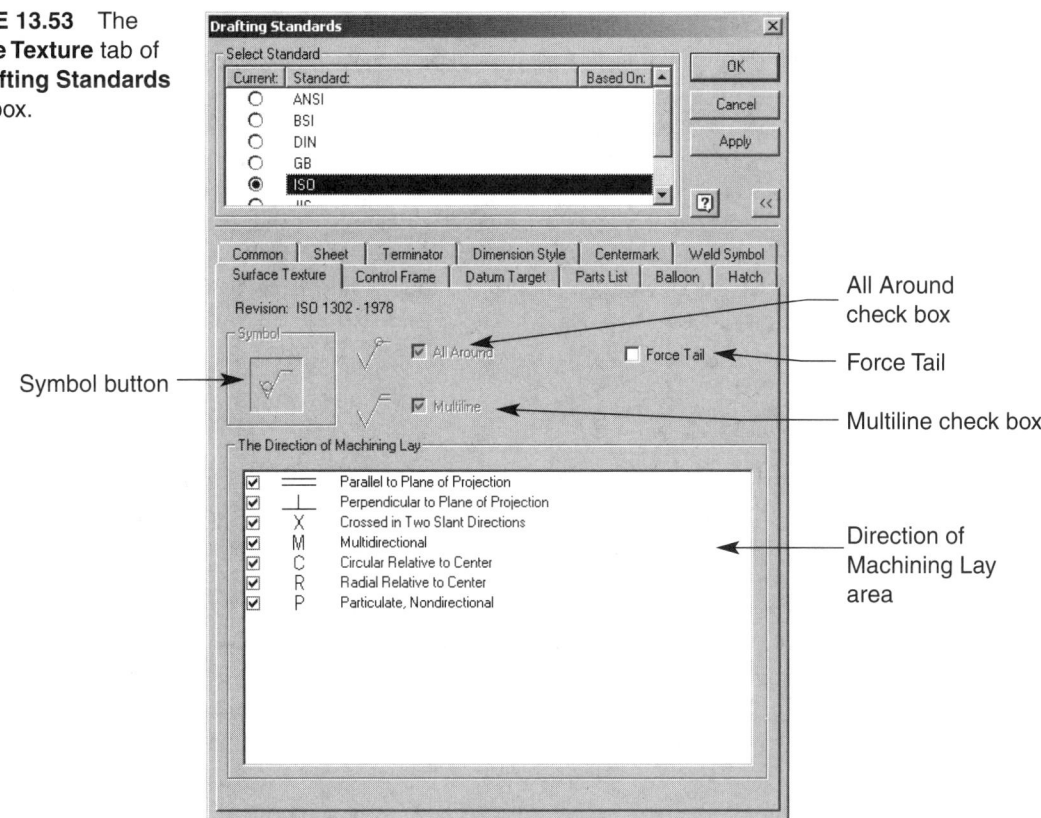

- **Symbol**   This button allows you to set the machining prohibited symbol size when creating or modifying a custom drafting standard.

- **All Around**   This check box is only available when creating or modifying DIN, ISO, JIS, or custom drafting standards. Select this check box to add the **All Around** symbol button to the **Surface Texture Symbol** dialog box.

- **Multiline**   This check box is available only when creating or modifying DIN, ISO, or custom drafting standards. Select this check box to place two line notes.

- **Force Tail**   Choose this check box if you want the **Force Tail** button in the **Surface Texture Symbol** dialog box to be selected automatically.

- **Direction of Machining Lay**   This area contains a number of check boxes, which control the machining lay symbols available to use when placing surface finish definitions. If you do not want to have access to a certain symbol in the **Surface Texture Symbol** dialog box, select the desired check box.

Field Notes

> The surface texture properties defined in the **Surface Texture** tab of the **Drafting Standards** dialog box are default. Some of the options can be overridden when you place surface texture definitions.

After you define the default options in the **Surface Texture** tab of the **Drafting Standards** dialog box, if needed, you are ready to use the **Surface Texture Symbol** tool. Access the **Surface Texture Symbol** command using one of the following techniques:

✓ Pick the **Surface Texture Symbol** button on the panel bar.

✓ Pick the **Surface Texture Symbol** button on the **Drawing Annotation** toolbar.

Once you access the **Surface Texture Symbol** command, use the cursor to select the feature, or point, to which you want to add surface texture information. Then drag the surface texture symbol to the desired location. See Figure 13.54.

After you select appropriate drawing geometry and locate the surface texture symbol, press the enter key on your keyboard, or right-click and select the **Continue** menu option, to display the **Surface Texture** dialog box. See Figure 13.55. Use the following options in the **Surface Texture** dialog box to define the finish of a part surface fully:

Field Notes

As previously discussed, the options in the **Surface Texture** dialog box will vary depending on the settings you specify in the **Surface Texture** tab of the **Drafting Standards** dialog box.

■ **Surface Type**   This area contains buttons that allow you to define the type of surface finish symbol to use. You can select the **Standard Surface Finish Symbol** button, **Removal of Material Required** button, or the **Removal of Material Prohibited** button, depending on the application.

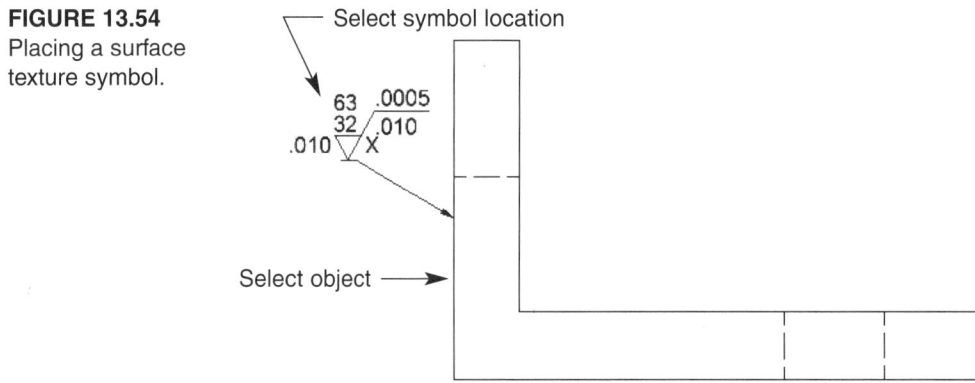

FIGURE 13.54 Placing a surface texture symbol.

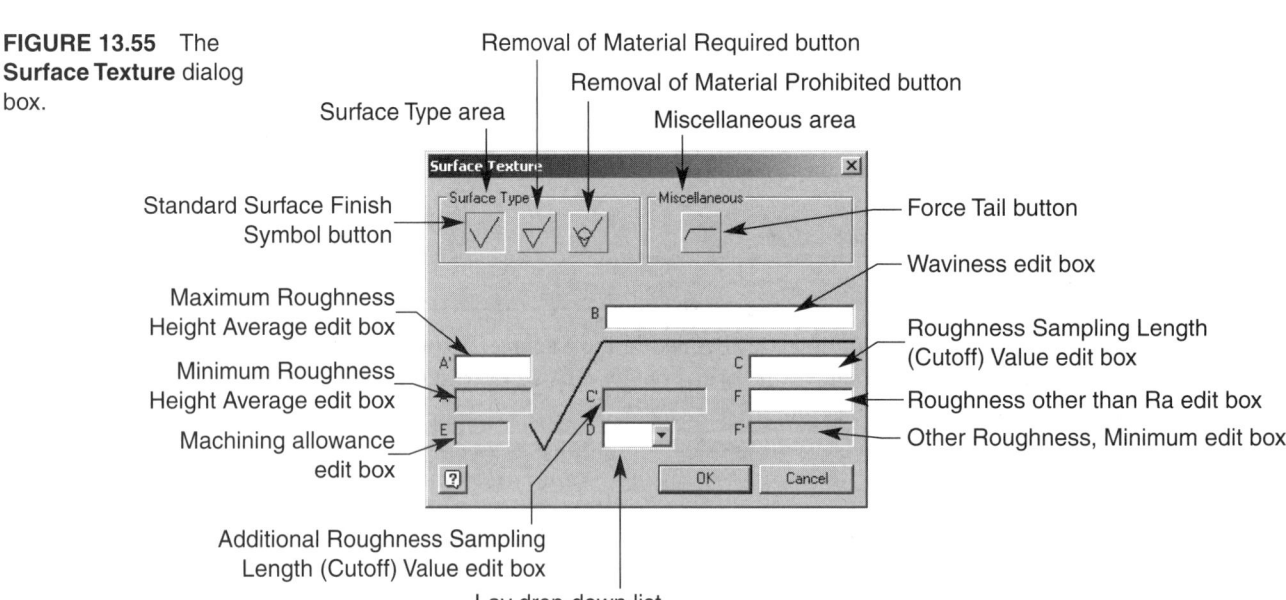

FIGURE 13.55 The **Surface Texture** dialog box.

■ **Miscellaneous**   This area contains the Force Tail button, which allows you to force a tail on the surface texture symbol. Other buttons specified in the **Surface Texture** tab of the **Drafting Standards** dialog box are also located in the **Miscellaneous** area.

■ **A′**   This edit box allows you to define the maximum roughness height, or if you do not specify a minimum roughness height, use this edit box to define the roughness average or Ra.

■ **A**   This edit box allows you to define the minimum roughness height, when a maximum roughness height is set in the **A′** edit box.

■ **B**   Use this edit box to enter the waviness of the surface texture.

■ **C′**   Use this edit box to enter the roughness sampling length, or cutoff value.

■ **C**   This edit box is used to also define the roughness sampling length, or cutoff value, or an additional value.

■ **D**   Use this drop-down list to select the desired machining lay. You can choose from parallel (5) or perpendicular (b) to edge, angular in both directions (X), multidirectional (M), circular relative to center (C), radial relative to center (R), or nondirectional particulate (P).

■ **F**   This edit box is used to specify a roughness other than Ra, which is defined in the **A′** edit box

■ **F′**   This edit box is used to specify a minimum roughness other than Ra, when a value is entered in the **F** edit box.

■ **E**   Use this edit box to specify the machining allowance, which specifies the amount of stock to be removed by machining.

When you have fully defined the surface texture specifications in the **Surface Texture** dialog box, pick the **OK** button to create the surface texture symbol, and exit the command.

Exercise 13.14

1. Continue from EX13.13, or launch Autodesk Inventor.
2. Open P12.13.
3. Save a copy of P12.13 as EX13.14.idw.
4. Close P12.13 without saving, and open EX13.14.
5. Explore the options in the **Surface Texture** tab of the **Drafting Standards** dialog box as discussed.
6. Activate the panel bar **Drawing Annotation** mode, or access the **Drawing Annotation** toolbar.
7. Use the **Surface Texture Symbol** tool to place surface texture symbols as shown.
8. To generate surface texture symbols that include an extension line, move your cursor over the surface texture symbol you want to move or pick the surface texture symbol. Then move your cursor over the green dot that intersects the leader line and leader shoulder. When you see the **Move** icon, pick the green dot and drag the symbol to the desired location. The leader should change into an extension line.
9. Save the exercise as EX13.14.
10. Exit Autodesk Inventor or continue working with the program.

Exercise 13.14 continued on next page

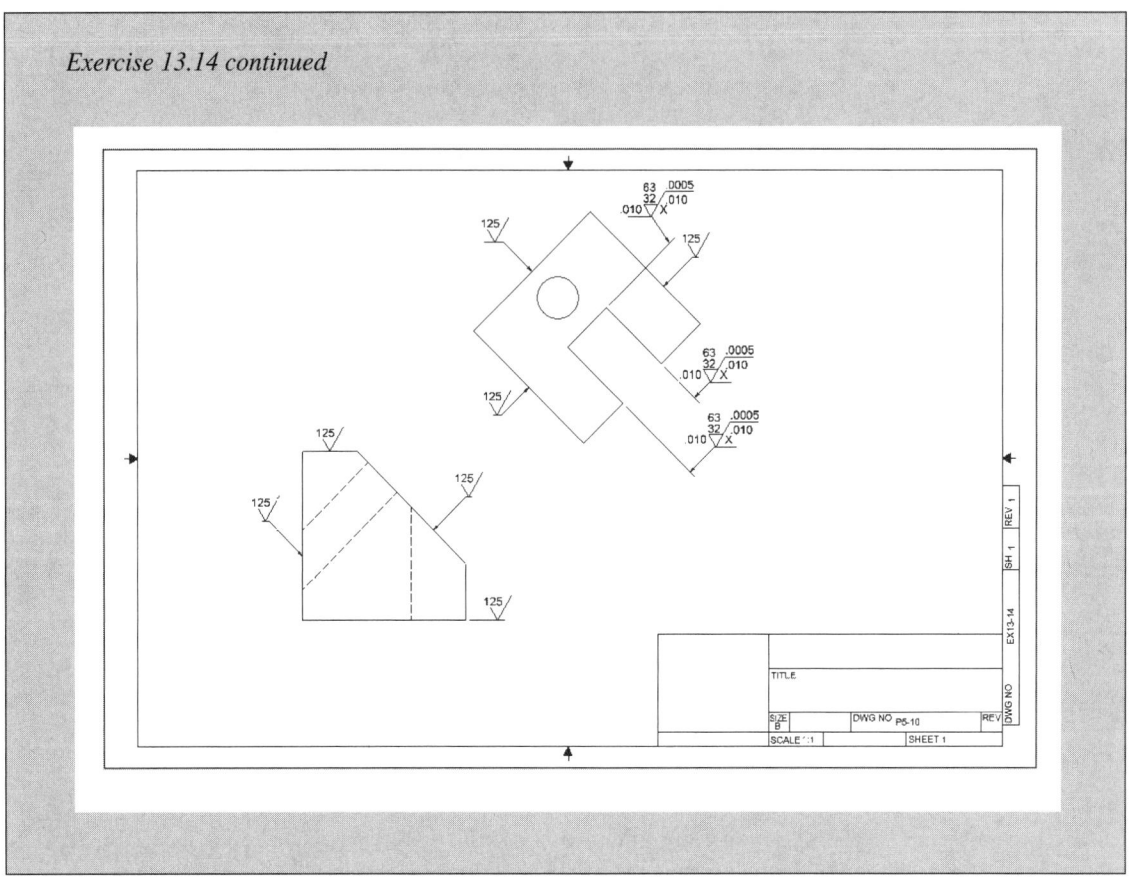

Exercise 13.14 continued

Placing Weld Symbols

Welding is a process of joining two or more pieces of like metals by heating the material to a temperature that is high enough to cause softening or melting so that the pieces combine grain structure and become joined together. *Welding,* or *weld, symbols* are used to display the complete information about the desired weld. The components of the weld symbol are the reference line, tail, leader, and weld symbol and specifications. You can create welding symbols and display weld information in Autodesk Inventor using the **Weld Symbol** tool. See Figure 13.56.

Before you add welding symbols and use the **Weld Symbol** tool, you may want to adjust some of the weld symbol properties in the **Drafting Standards** dialog box. Access the **Drafting Standards** dialog box by picking the **Standards...** menu option from the **Format** pull-down menu. As shown in Figure 13.57, weld symbol properties are located in the **Weld Symbol** tab of the **Drafting Standards** dialog box.

Depending on the specified drafting standard, the following options in the **Weld Symbol** tab control the default welding symbol options when adding weld information to a drawing view:

FIGURE 13.56 An example of a weld symbol and specifications.

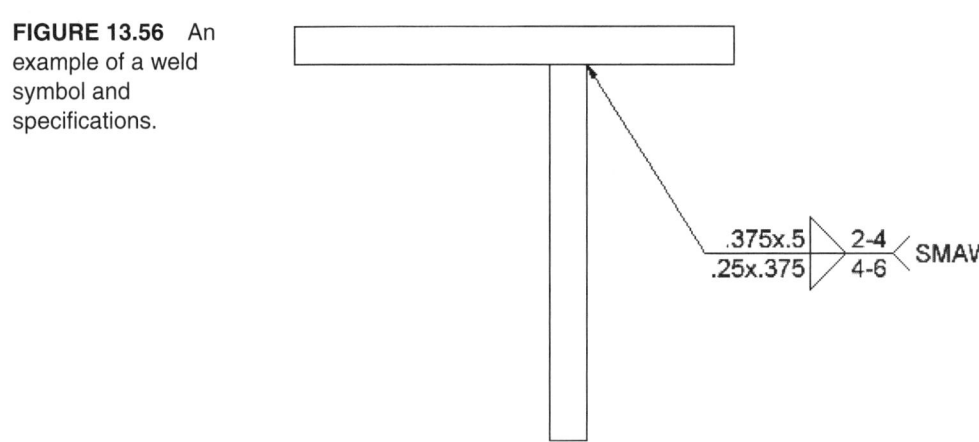

FIGURE 13.57 The **Weld Symbol** tab of the **Drafting Standards** dialog box.

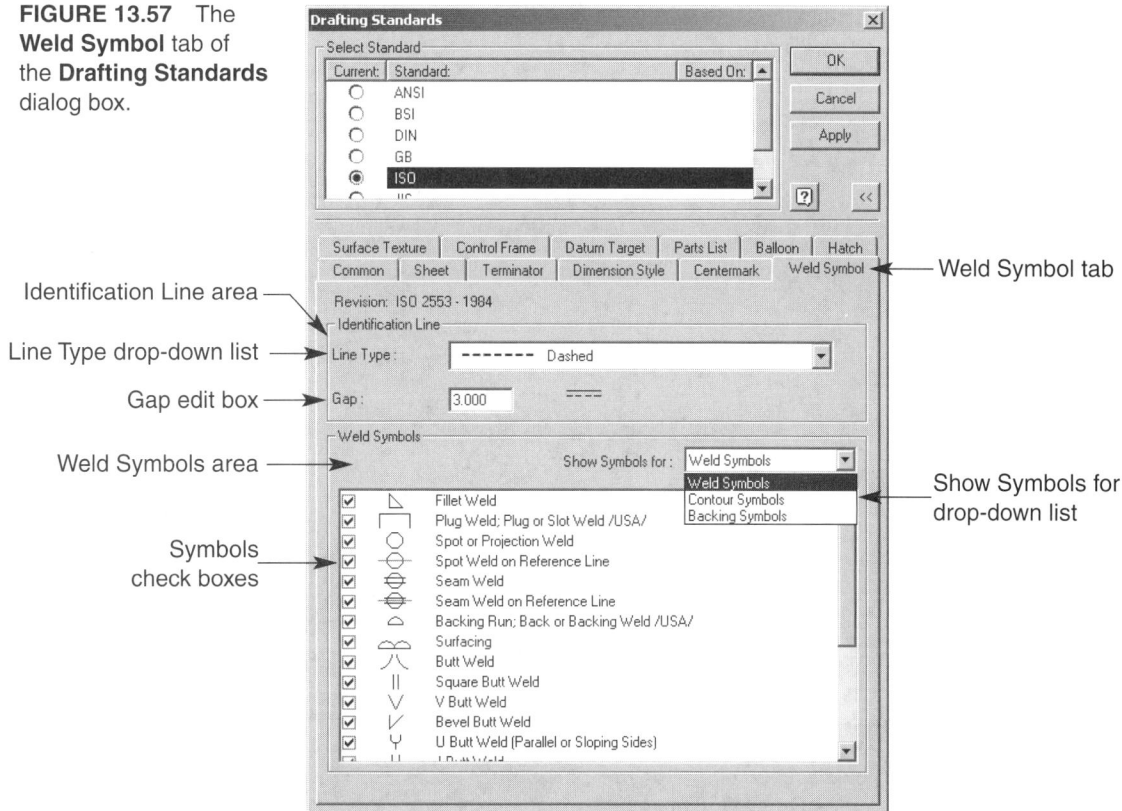

Identification Line area

Line Type drop-down list

Gap edit box

Weld Symbols area

Symbols check boxes

Weld Symbol tab

Show Symbols for drop-down list

- **Identification Line** This area allows you to define the display options for welding symbol identification lines and is not available when creating or modifying ANSI or JIS drafting standards. To change the welding symbol identification line type, select a line type from the **Line Type** drop-down list. You can also change the amount of space between the welding symbol identification line and the welding symbol by entering the desired value in the **Gap** edit box.

- **Weld Symbols** This area contains a drop-down list and a number of check boxes, which control the symbols available to use when placing weld definitions. If you want to have access to only certain weld, contour, or backing symbols, select **Weld Symbols, Contour Symbols,** or **Backing Symbols** from the **Show Symbols for:** drop-down list. Then deselect the check box of the symbol you do not want to have available for use.

After you define the default options in the **Weld Symbol** tab of the **Drafting Standards** dialog box, if needed, you are ready to use the **Weld Symbol** tool. Access the **Weld Symbol** command using one of the following techniques:

✓ Pick the **Weld Symbol** button on the panel bar.

✓ Pick the **Weld Symbol** button on the **Drawing Annotation** toolbar.

Once you access the **Weld Symbol** command, use the cursor to select the feature or point where you want to add welding information. Then drag the weld symbol to the desired location. See Figure 13.58.

After you select appropriate drawing geometry and locate the weld symbol, press the enter key on your keyboard, or right-click and select the **Continue** menu option, to display the **Weld Symbol** dialog box. See Figure 13.59.

FIGURE 13.58
Placing a welding
symbol.

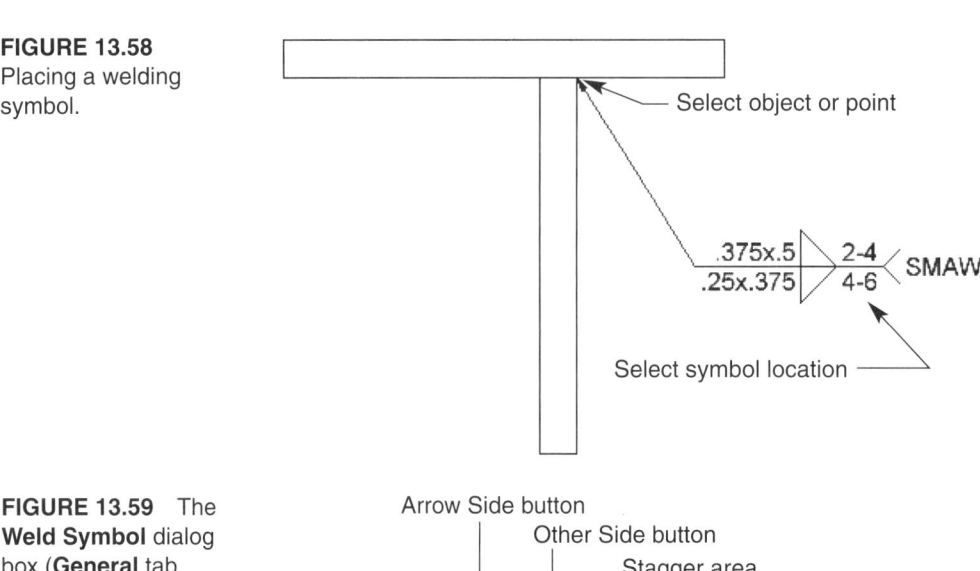

FIGURE 13.59   The
Weld Symbol dialog
box (**General** tab
selected).

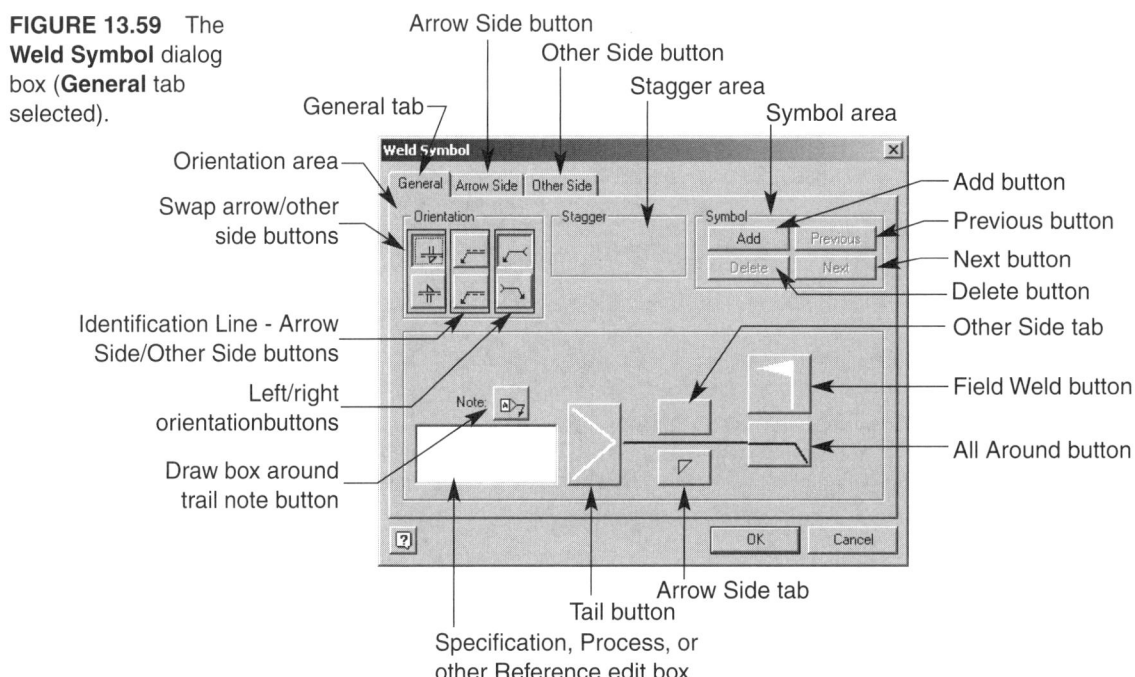

Field Notes

As previously discussed, the options in the **Weld Symbol** dialog box will vary depending on the settings you specify in the **Weld Symbol** tab of the **Drafting Standards** dialog box.

You can fully define weld symbols using the options in the **General** tab of the **Weld Symbol** dialog box because when you select the **Arrow Side** button, the **Arrow Side** tab becomes active, and when you select the **Other Side** button, the **Other Side** tab becomes active. However, you may choose to define some of the options in the **General** tab, then select the **Other Side** and **Arrow Side** tabs to specify additional options. Consequently, you may develop your own approach to creating weld symbols. The tabs are available in the **Weld Symbol** dialog box:

- **General**   As previously mentioned, this tab can be used to fully define a weld symbol and contains the following options:

  - **Orientation**   This area contains the following buttons that allow you to adjust the orientation of the weld symbol elements:

- **Swap Arrow/Other Side**   These buttons allow you to define the side of the feature to be welded. Select the button that corresponds with geometry to be welded, and the side you want to weld. The specified side is shown in the display area of the **Weld Symbol** dialog box and on the buttons themselves.

- **Identification Line - Arrow Side/Other Side**   These bottoms are used to control the placement of the identification line, if used. Select the **Arrow Side** or **Other Side** button, depending on the application. The effects of the button you select are displayed in drawing view and the buttons themselves, not the preview area of the **Weld Symbol** dialog box.

- **Left/Right orientation**   Use these buttons to specify the orientation of weld symbols. Select the **Left Orientation** button or the **Right Orientation** button depending on the application. Again, the effects of the button you select are displayed in the drawing view and the buttons themselves.

■ **Stagger**   The area contains buttons that allow you to adjust the stagger of welds on both sides of a feature. Buttons in the **Stagger** area are available only when you are specifying a fillet weld symbol for welding both the arrow and the other side of a feature.

■ **Symbol**   This area is used to add an additional symbol reference line to the weld symbol arrow. Typically another reference line is added after the initial line has been fully defined. However, you can add a symbol at any time by selecting the **Add** button. Then, pick the **Next** button to activate the added symbol. Once activated, you can adjust the settings of the added reference line. You can also return to a previous reference line by selecting the **Previous** button, or delete a reference line by picking the **Delete** button.

■ **Specification, Process, or other Reference**   This edit box is used to add a weld note to the symbol. To display a note, type the required information in the edit box. When you add information to the weld symbol, the **Tail** button is automatically selected, and a tail is added to the end of the symbol reference line. Depending on the specified drafting standard, you can choose the **Draw Box Around Tail** Note button to add a box around the weld note.

■ **Arrow Side button**   Depending on the specified orientation, this button will be above or below the symbol reference line. Pick this button to open the **Arrow Side** tab of the **Weld Symbol** edit box automatically.

■ **Other Side button**   As with the **Arrow Side Button,** depending on the specified orientation this button will be above or below the symbol reference line. Pick this button to open the **Other Side** tab of the **Weld Symbol** edit box automatically.

■ **Field Weld**   Pick this button to add a field weld symbol for field welding applications.

■ **All Around**   Select this button to add a weld-all-around symbol for applications that require all-around welding.

■ **Arrow Side**   As previously mentioned, you can access this tab, shown in Figure 13.60, by selecting the **Arrow Side** button in the **General** tab, or by picking the **Arrow Side** tab. The first symbol selected in the **Weld Symbol** tab of the **Drafting Standards** dialog box, fillet by default, is automatically applied to the arrow side of the weld symbol. To change the default symbol, select the **Weld Symbol Palette** button and choose the desired symbol. Then enter a value in each of the weld specification edit boxes, and select a contour from the **Contour** drop-down list if required.

■ **Other Side**   As previously mentioned, you can access this tab, shown in Figure 13.61, by selecting the **Other Side** button in the **General** tab, or by picking the **Other Side** tab. No symbol is specified for the other side of a weld. As a result, when you initially access the **Other Side** tab, the **Symbol Pallet** is automatically displayed. See Figure 13.62. Once you select the desired symbol, the weld specification edit boxes, **Weld Symbol Palette** button, and **Contour** drop-down list are available as shown in Figure 13.61. To change the default

FIGURE 13.60   The **Weld Symbol** dialog box (**Arrow Side** tab selected).

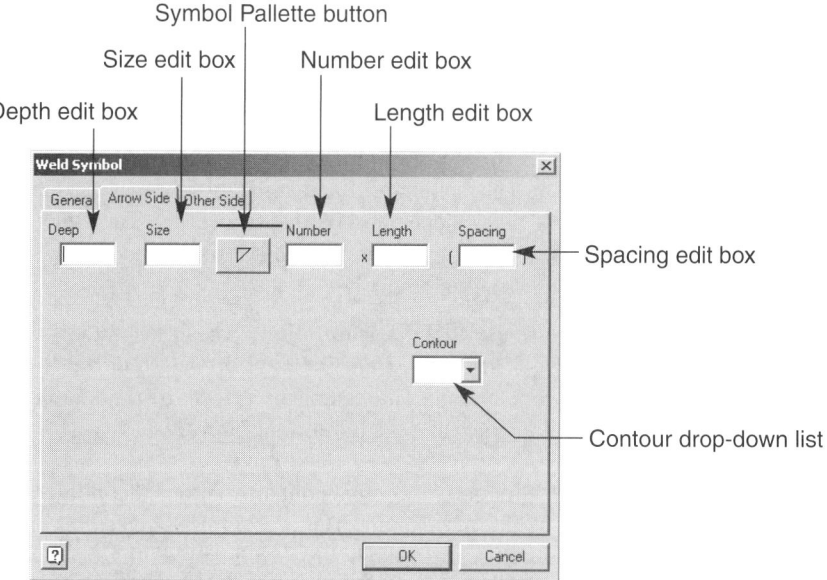

FIGURE 13.61   The **Weld Symbol** dialog box (**Other Side** tab selected).

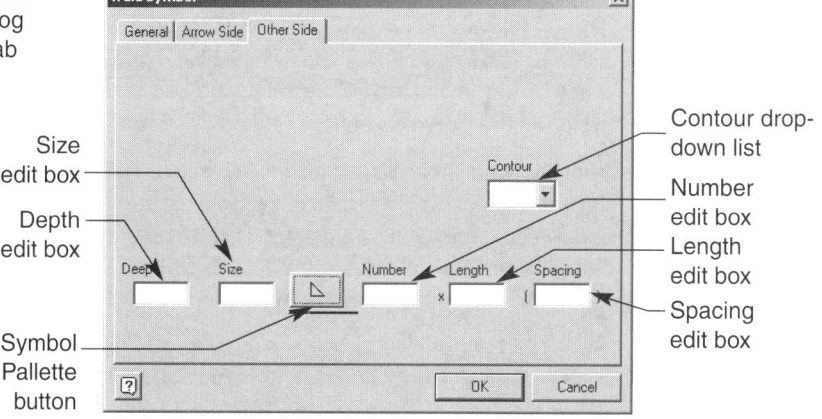

FIGURE 13.62   The **Symbol Palette** displayed in the **Other Side** tab of the **Weld Symbol** dialog box.

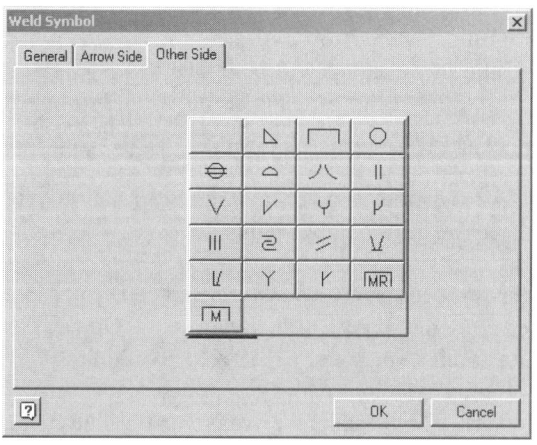

symbol, select the **Weld Symbol Palette** button and choose the desired symbol. Then enter a value in each of the weld specification edit boxes, and select a contour from the **Contour** drop-down list if required.

When you have fully defined the weld symbol specifications in the **Weld Symbol** dialog box, pick the **OK** button to create the weld symbol. You will notice the **Weld Symbol** tool is still active. Continue placing weld symbols as required or right-click and select the **Done [Esc]** menu option, press the **Esc** key on your keyboard, or access a different tool.

Exercise 13.15

1. Continue from EX13.14, or launch Autodesk Inventor.
2. Open a new inch drawing file.
3. Delete the current title block and border.
4. Explore the options in the **Weld Symbol** tab of the **Drafting Standards** dialog box as discussed.
5. Activate the panel bar **Drawing Annotation** mode, or access the **Drawing Annotation** toolbar.
6. Use the **Weld Symbol** tool to place the weld symbols shown.
7. Save the exercise as EX13.15.
8. Exit Autodesk Inventor or continue working with the program.

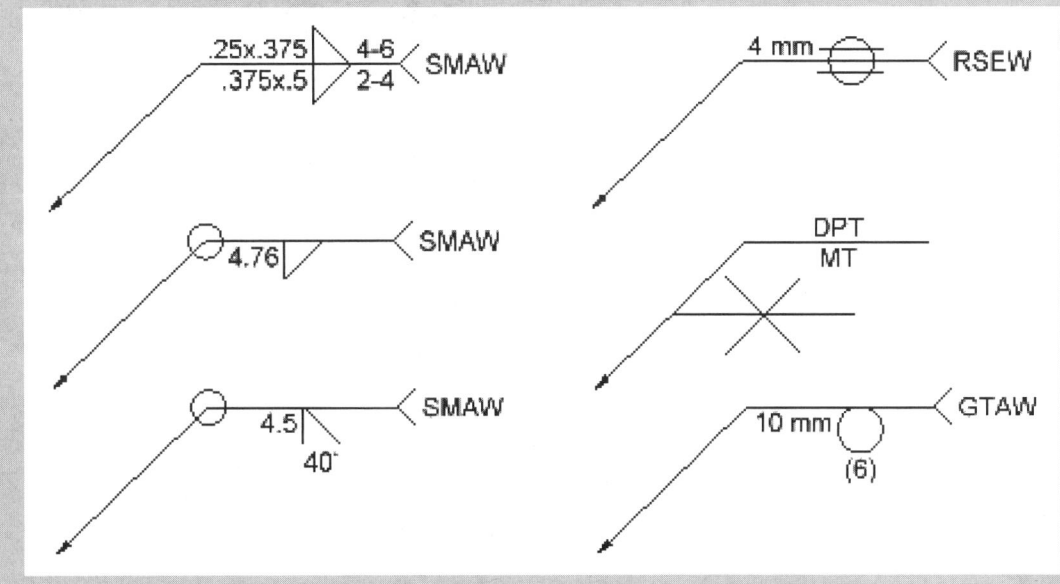

Geometric Dimensioning and Tolerancing

Geometric dimensioning and tolerancing (GD&T) is the dimensioning and tolerancing of individual features of a part where the permissible variations relate to characteristics of form, profile, orientation, runout, or the location of features. GD&T uses symbols and related tolerances to display the desired specifications on the drawing. Autodesk Inventor provides you with tools that allow you to generate control frames, feature identifier symbols, datum identification symbols, and datum target symbols.

Adding Feature Control Frames

A *feature control frame* contains a geometric characteristic symbol, geometric tolerance, material condition symbol when used, and datum reference when used. The feature control frame is divided into compartments containing the geometric characteristic symbol in the first compartment, followed by the geometric tolerance in a compartment and each datum reference in a separate compartment. The geometric tolerance is preceded by a diameter symbol if the geometric tolerance zone is cylindrical. The feature control frame is generally placed with a related dimension on the drawing.

Feature control frames can be added to drawings by using the **Feature Control Frame** tool. See Figure 13.63. Before you create feature control frames and use the **Feature Control Frame** tool, you may want to adjust some of the feature control frame properties in the **Drafting Standards** dialog box. Access the dialog box by picking the **Standards...** menu option from the **Format** pull-down menu. As shown in Figure 13.64, feature control frame properties are located in the **Control Frame** tab.

FIGURE 13.63
Examples feature
control frames created
using Autodesk
Inventor.

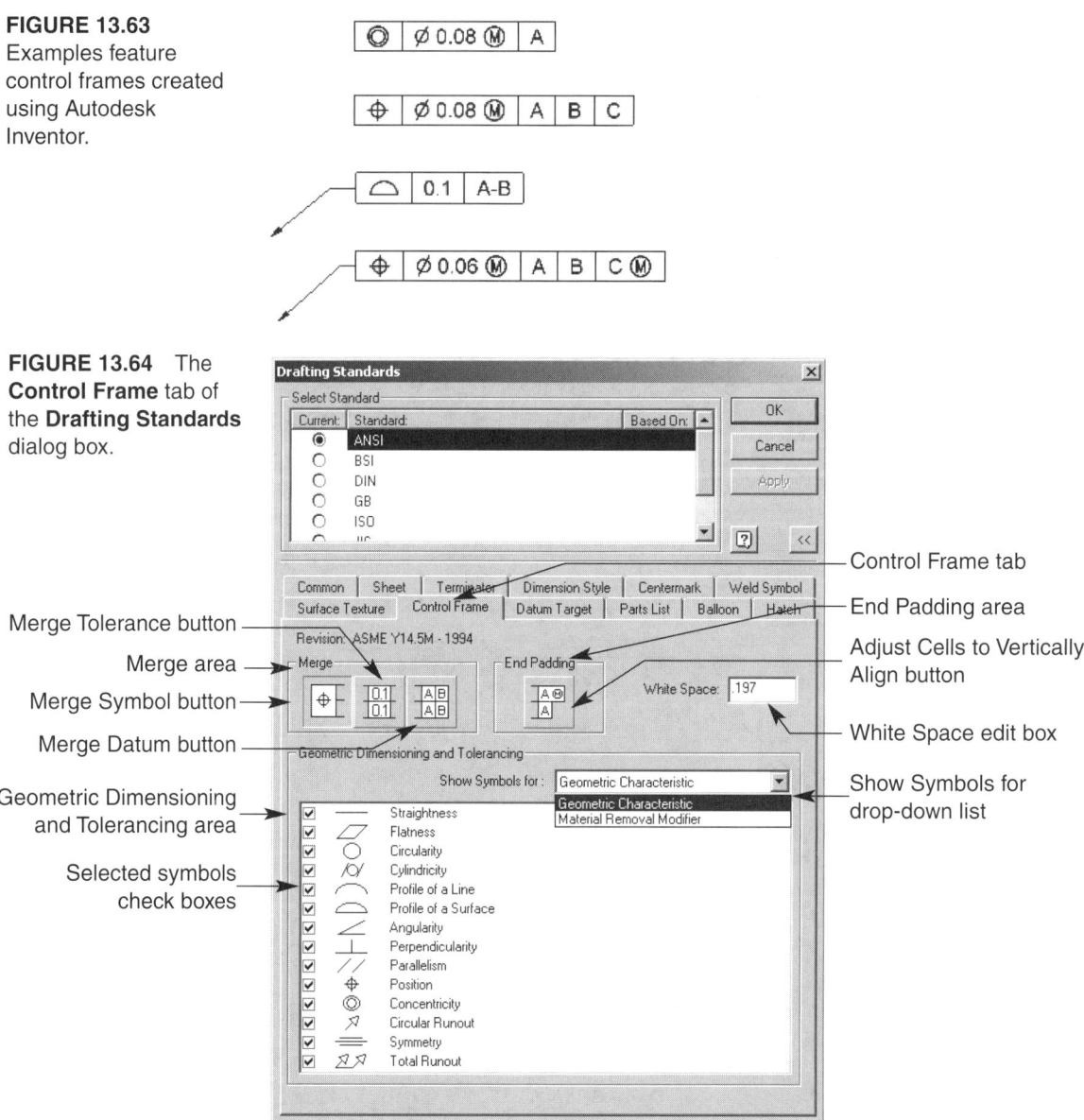

FIGURE 13.64 The
Control Frame tab of
the **Drafting Standards**
dialog box.

Depending on the specified drafting standard, the options in the **Control Frame** tab are displayed slightly differently. However, the following properties are available for all drafting standards and define the default feature control frame options when adding feature control frames:

- **Merge** This area contains three buttons that allow you to adjust the display of composite position and multiple single position tolerances. Pick the **Merge Symbol** button to create a composite position tolerance, as shown in Figure 13.65A. Select the **Merge Tolerance** button to combine like tolerances, as shown in Figure 13.65B. Pick the **Merge Datum** button to combine like datums, as shown in Figure 13.65C.

- **End Padding** This area contains the **Adjust Cells to Vertically Align** button, which as the name implies, vertically aligns composite position and multiple single position tolerances cells. See Figure 13.66.

- **White Space** This edit box allows you to define the width of symbol, tolerance, and datum cells in reference to the amount of text in each cell. Increase or decrease the number in the edit box to create more or less white space, which widens or narrows the cells in reference to the text.

FIGURE 13.65 The effects of merging like symbols, tolerances, and datums.

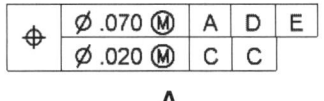

FIGURE 13.66 (A) Unaligned and (B) aligned vertical cells.

- **Geometric Dimensioning and Tolerancing** This area contains a number of check boxes, which control the geometric dimensioning and tolerancing symbols available to use when placing feature control frames. If you do not want to have access to a certain symbol in the **Feature Control Frame** dialog box, select the desired check box. Pick the **Geometric Characteristic** option from the **Show Symbols for:** drop-down list, to display geometric characteristic check boxes. Similarly, pick the **Material Removal Modifier** option from the **Show Symbols for:** drop-down list, to material removal modifier check boxes.

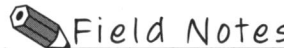Field Notes

The feature control frame properties defined in the **Control Frame** tab of the **Drafting Standards** dialog box cannot be overridden when you place feature control frames.

After you define the default options in the **Control Frame** tab of the **Drafting Standards** dialog box, if needed, you are ready to use the **Feature Control Frame** tool. Access the **Feature Control Frame** command using one of the following techniques:

✓ Pick the **Feature Control Frame** button on the panel bar.

✓ Pick the **Feature Control Frame** button on the **Drawing Annotation** toolbar.

✓ Type the **+** and **F** keys on your keyboard.

Once you access the **Feature Control Frame** command, use the cursor to select the feature, or point, to which you want to add feature control frame information. Then drag the feature control frame symbol to the desired location, and pick the point. See Figure 13.67.

FIGURE 13.67 Placing a feature control frame.

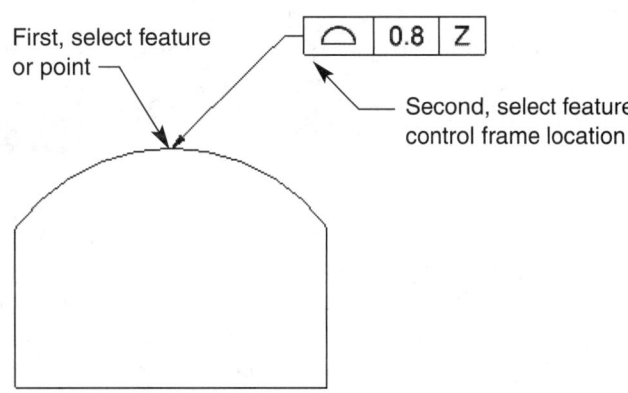

Field Notes

If you do not want to create a leader from the feature control frame to the feature, do not pick the position of the leader shoulder. After you select the first point, press the enter key on your keyboard, or right-click and select the **Continue** menu option.

After you select appropriate drawing geometry or a point on the sheet and locate the feature control frame, press the enter key on your keyboard, or right-click and select the **Continue** menu option, to display the **Feature Control Frame** dialog box. See Figure 13.68. Use the following options in the **Feature Control Frame** dialog box to fully define a feature control frame:

- **Symbol**   These buttons allow you to define the geometric characteristic symbols for the first and second lines of feature control frame information. Select the button to access the symbols palette, and pick the desired symbol.

- **Tolerance**   These edit boxes are used to specify geometric tolerances.

- **Datum**   These edit boxes allow you to enter primary, secondary, and tertiary datums in the feature control frame. The **Datum 2** edit box is available only when a value is entered in the **Datum 1** edit box, and the **Datum 3** edit box is available only when a value is entered in the **Datum 2** edit box.

- **Insert Modifier**   These buttons are located towards the bottom of the dialog box and control multiple geometric dimensioning and tolerancing information such as material condition and tolerance zone descriptors. To use these buttons, pick a location inside the desired edit box, and pick the required button. The symbol should appear in the selected edit box.

- **All Around**   Select this check box to add an all around symbol on the leader, next to the feature control frame.

When you have fully defined the feature control frame specifications in the **Feature Control Frame** dialog box, pick the **OK** button to create the feature control frame. You will notice the **Feature Control Frame** tool is still active. Continue placing feature control frames as required or right-click and select the **Done [Esc]** menu option, press the **Esc** key on your keyboard, or access a different tool, to exit the command.

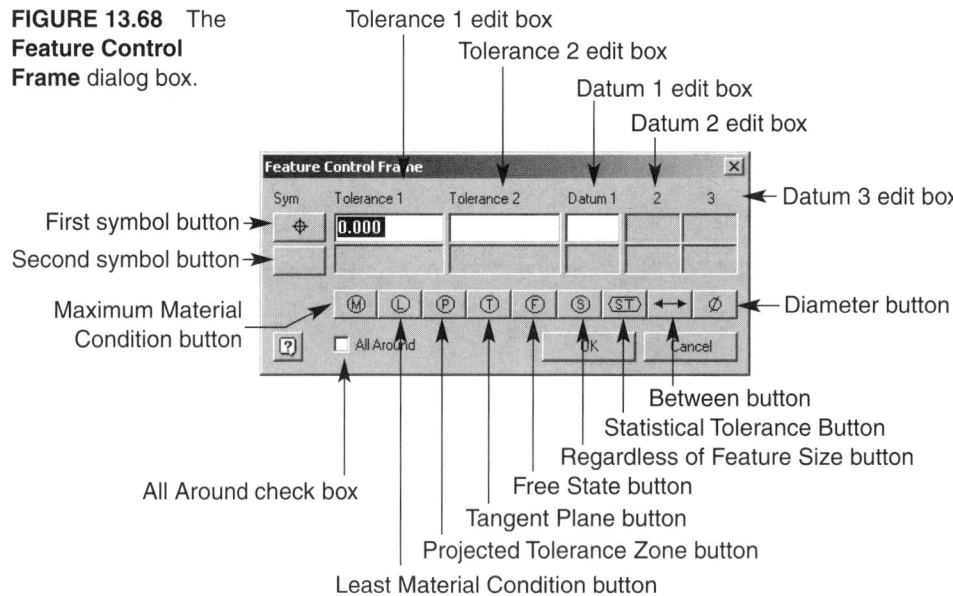

FIGURE 13.68   The **Feature Control Frame** dialog box.

Placing Feature Identifier Symbols

The *feature identifier* symbol is not used in the ASME standard. It is a symbol that is used to correlate certain features to specifications in the ISO, DIN, and JIS standards. See Figure 13.69.

Feature identifier symbols are placed using the **Feature Identifier Symbol** command. Access the **Feature Identifier Symbol** command using one of the following techniques:

✓ Pick the **Feature Identifier Symbol** button on the panel bar.

✓ Pick the **Feature Identifier Symbol** button on the **Drawing Annotation** toolbar.

Once you access the **Feature Identifier Symbol** command, use the cursor to select the feature, or point, to which you want to add a feature identifier symbol, or pick a point in the drawing sheet. Then drag the symbol to the desired location and pick the position to specify the leader length. See Figure 13.70.

After you select appropriate drawing geometry or a point on the drawing sheet, and locate the feature identifier symbol, press the enter key on your keyboard, or right-click and select the **Continue** menu option to display the **Format Text** dialog box, discussed earlier in this chapter. The default feature identifier value is displayed in the **Format Text** dialog box and can be modified. For more information regarding the **Format Text** dialog box, refer to Chapter 12.

 Field Notes

> If you select two points in addition to the initial feature or point to define the location of the feature identifier symbol, the **Format Text** dialog box automatically opens.

When you are satisfied with the value of the feature identifier symbol and have finished defining options in the **Format Text** dialog box, pick the **OK** button to generate the feature identifier symbol. The **Feature Identifier Symbol** tool will still be active, allowing you to continue placing feature identifier symbols. To exit the **Feature Identifier Symbol** command, right-click and select the **Done [Esc]** menu option, press the **Esc** key on your keyboard, or access a different tool.

Placing Datum Feature Symbols

The *datum feature symbol* is also referred to as a datum identification symbol. Datums are considered theoretically perfect surfaces, planes, points, or axes. The datum feature symbol is used to identify datums on the drawing. The datum feature symbol is a square box that contains the datum reference letter. The box is connected to a triangular datum arrow that contacts the datum surface or related application. Datum feature symbols are attached to the surface with a leader or off of an extension line, with a diameter or center plane dimension, or attached to a feature control frame. See Figure 13.71.

Datum feature symbols are placed using the **Datum Identifier Symbol** command. Access the **Datum Identifier Symbol** command using one of the following techniques:

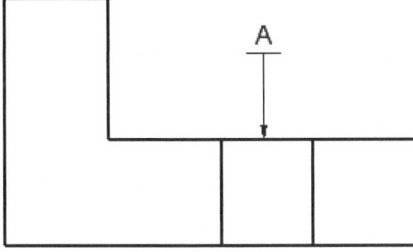

FIGURE 13.69 An example of a feature identifier symbol.

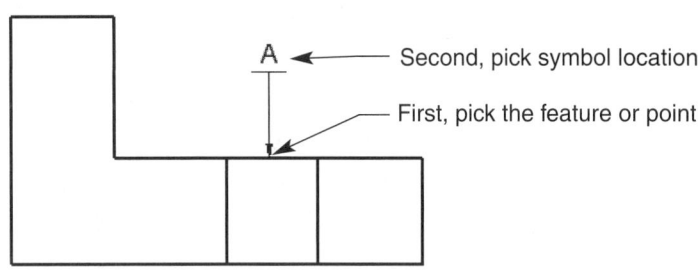

FIGURE 13.70 Placing a feature identifier symbol.

FIGURE 13.71
Examples of various
datum feature symbols.

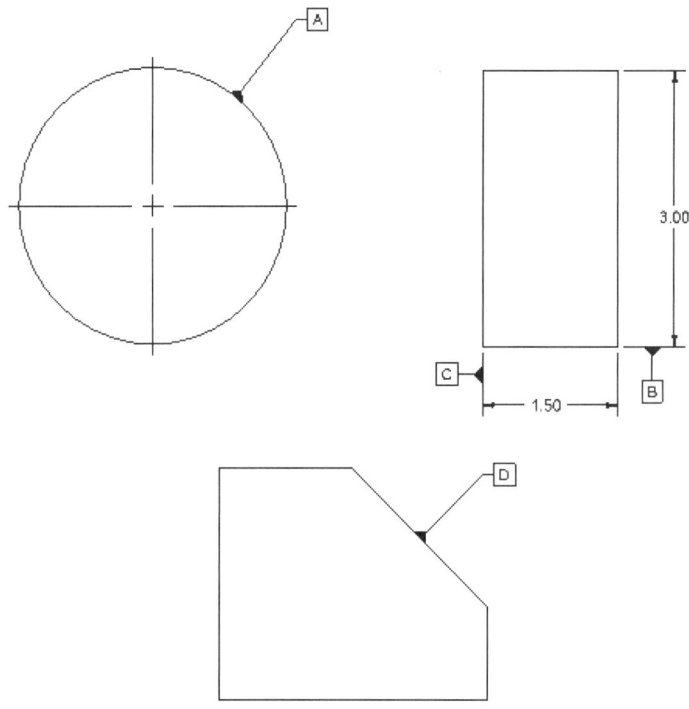

✓ Pick the **Datum Identifier Symbol** button on the panel bar.

✓ Pick the **Datum Identifier Symbol** button on the **Drawing Annotation** toolbar.

Once you access the **Datum Identifier Symbol** command, use the cursor to select the feature, or point, to which you want to add a datum identifier symbol, or pick a point in the drawing sheet. Then drag the symbol to the desired location, and pick the position to specify the leader length. See Figure 13.72.

After you select appropriate drawing geometry or a point on the drawing sheet, and locate the datum identifier symbol, press the enter key on your keyboard, or right-click and select the **Continue** menu option to display the **Format Text** dialog box, discussed earlier in this chapter. The default datum identifier value is displayed in the **Format Text** dialog box and can be modified. For more information regarding the **Format Text** dialog box, refer to Chapter 12.

Field Notes

If you select two points, in addition to the initial feature or point, to define the location of the datum identifier symbol, the **Format Text** dialog box automatically opens.

FIGURE 13.72
Placing a datum
identifier symbol.

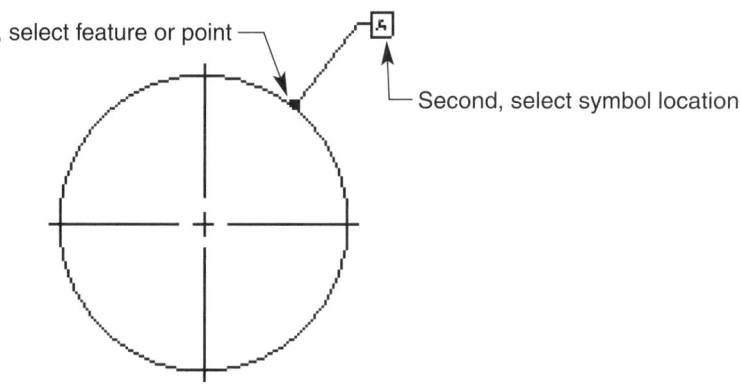

When you are satisfied with the value of the datum identifier symbol and have finished defining options in the **Format Text** dialog box, pick the **OK** button to generate the datum identifier symbol. The **Datum Identifier Symbol** tool will still be active, allowing you to continue placing datum identifier symbols. To exit the **Datum Identifier Symbol** command, right-click and select the **Done [Esc]** menu option, press the **Esc** key on your keyboard, or access a different tool.

Adding Datum Target Symbols

Datum targets are used to specify points, lines, or areas of contact on a part that establishes datums when it is not possible or appropriate to use a surface. The datum target symbol is drawn as a circle with a horizontal line through the center. The top half of the circle is left blank unless the datum target symbol refers to a datum target area. In this case, the size of the datum target area is provided in the top half of the datum target symbol. The bottom half of the datum target symbol is used to identify the related datum with the datum reference letter and datum target number assigned sequentially to the datum, such as A1, A2, and A3. The datum target symbol is connected to the related datum target point, line or area with a leader. There is no arrowhead at the end of the leader. See Figure 13.73.

Autodesk Inventor provides five separate tools, **Datum Target – Leader, Datum Target – Circle, Datum Target – Line, Datum Target – Point,** and **Datum Target – Rectangle,** for adding datum target symbols for various applications. The **Datum Target – Leader** tool allows you to place a datum target with a leader attached to the specified feature or point, as shown in Figure 13.73A. The **Datum Target – Circle** tool allows you to place a datum target with a leader attached to a circular datum target area, as shown in Figure 13.73B. The **Datum Target – Line** tool allows you to place a datum target with a leader attached to a specified line, with points and the line ends, as shown in Figure 13.73C. The **Datum Target – Point** tool allows you to place a datum target with a leader attached to a point, as shown in Figure 13.73D. The **Datum Target – Rectangle** tool allows you to place a datum target with a leader attached to a rectangular datum target area, as shown in Figure 13.73E.

Before you create datum targets and use the various datum target tools, you may want to adjust some of the datum target properties in the **Drafting Standards** dialog box. Access the **Drafting Standards** dialog box by picking the **Standards...** menu option from the **Format** pull-down menu. As shown in Figure 13.74, datum target properties are located in the **Datum Target** tab of the **Drafting Standards** dialog box.

Depending on the specified drafting standard, the default values in the **Control Frame** tab are displayed slightly differently. However, the following properties are available for all drafting standards and define the default datum target options:

- **Target Point**   This area allows you to specify the display options for point and line datum targets. Enter the desired point size in the **Size** edit box, and select a point color by picking the **Color** button.

- **Area Hatch**   This area allows you to specify the hatch display options of circular and rectangular datum target areas. Enter the scale, or the distance between hatch lines, in the **Distance** edit box, and enter the hatch line angle in the **Angle** edit box.

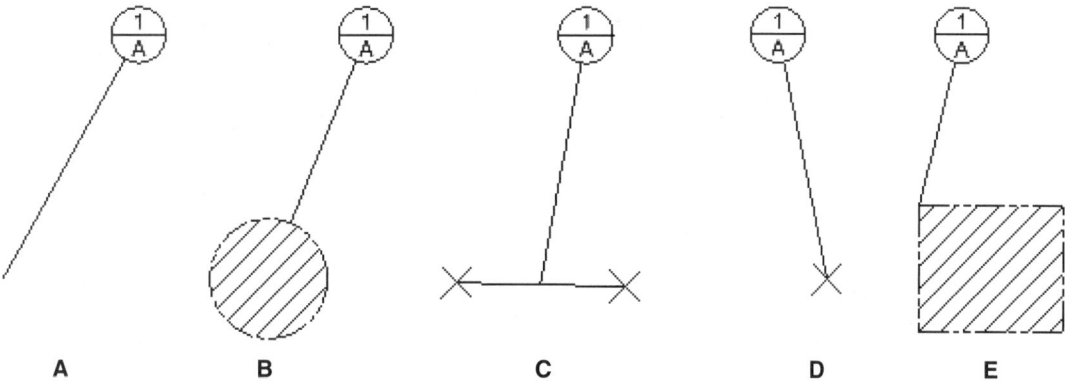

A   B   C   D   E

FIGURE 13.73   Examples of various types of datum targets.

FIGURE 13.74 The **Datum Target** tab of the **Drafting Standards** dialog box.

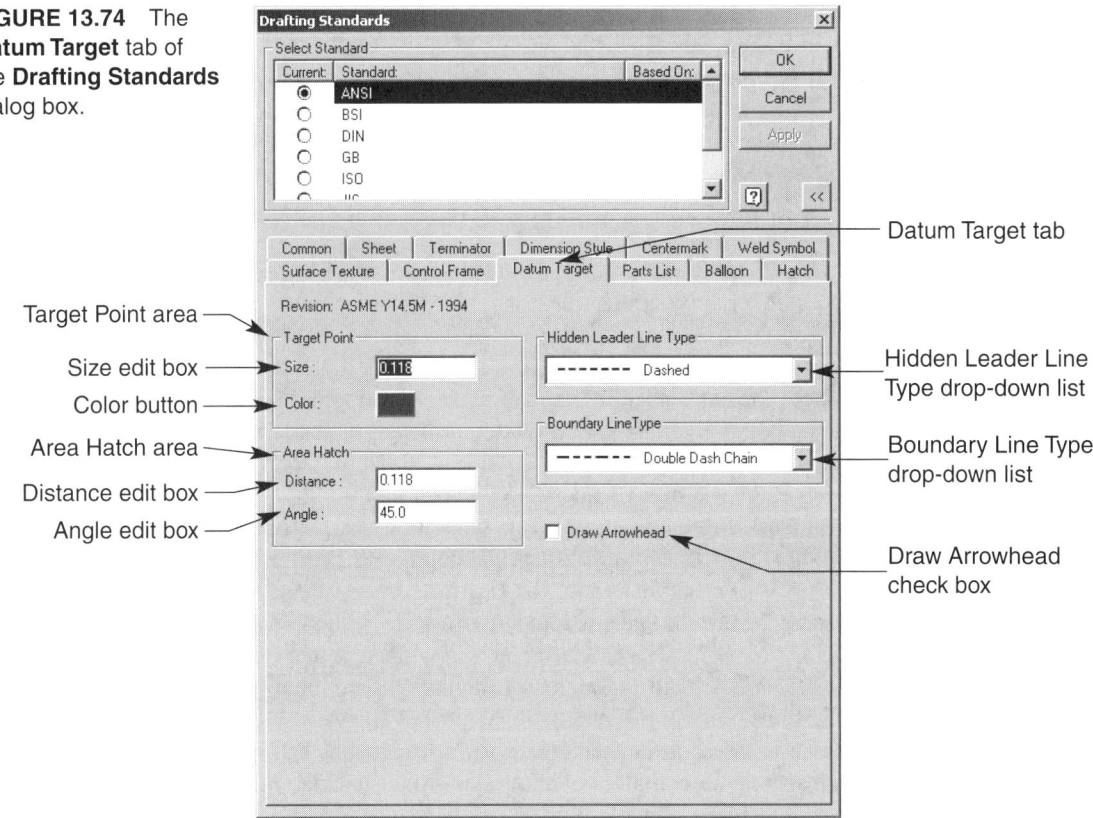

- **Hidden Leader Line Type**   Use this drop-down list to define the type of leader line to use for far side datum target.

- **Boundary Line Type**   Use this drop-down list to specify the type of line to use for the boundary, or outline, of circular and rectangular datum target areas.

- **Draw Arrowhead**   Select this check box if you want to display an arrowhead at the end of the datum target leader.

Field Notes

The datum target properties defined in the **Datum Target** tab of the **Drafting Standards** dialog box cannot be overridden when you place datum targets in a drawing.

Each of the different types of datum targets are contained in a flyout button and can be accessed using one of the following techniques:

✓ Pick the desired datum target tool button from the **Datum Target** flyout button on the panel bar.

✓ Pick the desired datum target tool button from the **Datum Target** flyout button on the **Drawing Annotation** toolbar.

Once you access one of the available datum target tools, use the following information to create the datum target:

- **Datum Target – Leader**   First, use the cursor to select the feature, or point, to which you want to add a datum target, or pick a point in the drawing sheet. Then drag the datum target

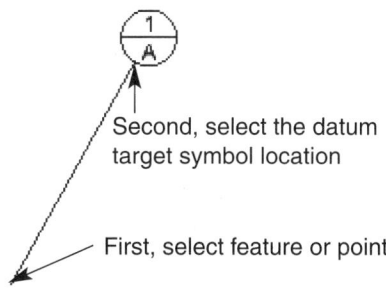

Second, select the datum
target symbol location

First, select feature or point

FIGURE 13.75 Placing a leader
datum target.

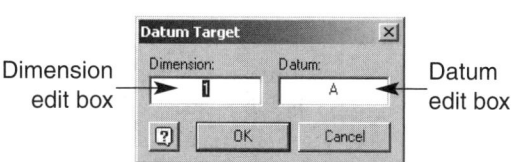

Dimension
edit box

Datum
edit box

FIGURE 13.76 The **Datum Target** dialog box.

symbol to the desired location, and pick the position to specify the leader length. See Figure
13.75. After you select appropriate drawing geometry or a point on the drawing sheet, and
locate the datum target symbol, press the enter key on your keyboard, or right-click and
select the **Continue** menu option to display the **Datum Target** dialog box. See Figure 13.76.
Enter the dimension in the **Dimension** edit box, and the datum in the **Datum** edit box. Then
pick the **OK** button to generate the datum target. The **Datum Target – Leader** tool will still
be active, allowing you to continue placing leader datum targets. To exit the **Datum Target –
Leader** command, right-click and select the **Done [Esc]** menu option, press the **Esc** key on
your keyboard, or access a different tool.

■ **Datum Target – Circle**   First, use the cursor to select the center point of the circular datum
target area. Then drag the datum target area circle edge to the desired point, and pick the
position to specify the area diameter length. Next, drag the datum target symbol to the
desired location, and pick the position to specify the leader length. See Figure 13.77. After
you select appropriate drawing geometry or a point on the drawing sheet, and locate the
datum target symbol, press the enter key on your keyboard, or right-click and select the
Continue menu option to display the **Datum Target** dialog box, shown in Figure 13.76.
Enter the dimension in the **Dimension** edit box, and the datum in the **Datum** edit box. Then
pick the **OK** button to generate the datum target. The **Datum Target – Circle** tool will still
be active, allowing you to continue placing leader datum targets. To exit the **Datum Target –
Circle** command, right-click and select the **Done [Esc]** menu option, press the **Esc** key on
your keyboard, or access a different tool.

■ **Datum Target – Line**   First, use the cursor to select the start point of the datum target line.
Then drag the datum target line to the desired location, and pick the position to specify the
length of the line. Next, drag the datum target symbol to the desired location, and pick the
position to specify the leader length. See Figure 13.78. After you select appropriate drawing
geometry or a point on the drawing sheet, and locate the datum target symbol, press the enter
key on your keyboard, or right-click and select the **Continue** menu option to display the
Datum Target dialog box, shown in Figure 13.76. Enter the dimension in the **Dimension** edit

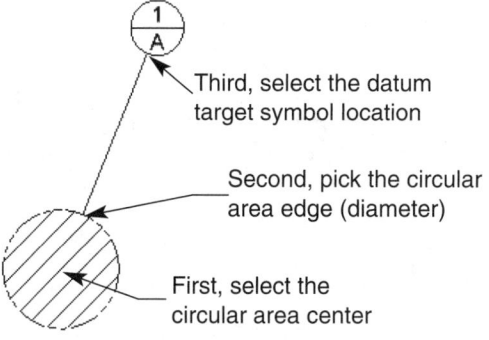

Third, select the datum
target symbol location

Second, pick the circular
area edge (diameter)

First, select the
circular area center

FIGURE 13.77 Placing a circle datum target.

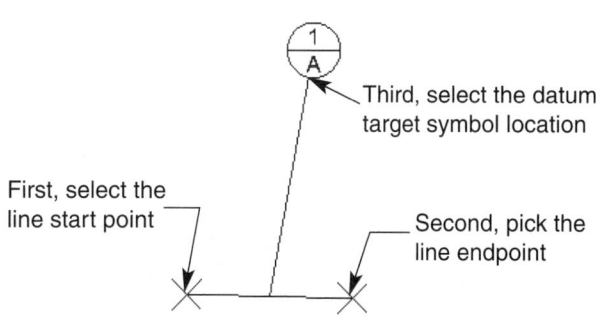

Third, select the datum
target symbol location

First, select the
line start point

Second, pick the
line endpoint

FIGURE 13.78 Placing a line datum target.

box, and the datum in the **Datum** edit box. Then pick the **OK** button to generate the datum target. The **Datum Target – Line** tool will still be active, allowing you to continue placing leader datum targets. To exit the **Datum Target – Line** command, right-click and select the **Done [Esc]** menu option, press the **Esc** key on your keyboard, or access a different tool.

- **Datum Target – Point**   First, use the cursor to select the feature, or point, to which you want to add a datum target, or pick a point in the drawing sheet. Then drag the datum target symbol to the desired location, and pick the position to specify the leader length. See Figure 13.79. After you select appropriate drawing geometry or a point on the drawing sheet, and locate the datum target symbol, press the enter key on your keyboard, or right-click and select the **Continue** menu option to display the **Datum Target** dialog box, shown in Figure 13.76. Enter the dimension in the **Dimension** edit box, and the datum in the **Datum** edit box. Then pick the **OK** button to generate the datum target. The **Datum Target – Point** tool will still be active, allowing you to continue placing leader datum targets. To exit the **Datum Target – Point** command, right-click and select the **Done [Esc]** menu option, press the **Esc** key on your keyboard, or access a different tool.

- **Datum Target – Rectangle**   First, use the cursor to select the center point of the rectangular datum target area. Then drag the datum target area rectangle corner to the desired point, and pick the position to specify the rectangle width and length. Next, drag the datum target symbol to the desired location, and pick the position to specify the leader length. See Figure 13.80. After you select appropriate drawing geometry or a point on the drawing sheet, and locate the datum target symbol, press the enter key on your keyboard, or right-click and select the **Continue** menu option to display the **Datum Target** dialog box, shown in Figure 13.76. Enter the dimension in the **Dimension** edit box, and the datum in the **Datum** edit box. Then pick the **OK** button to generate the datum target. The **Datum Target – Rectangle** tool will still be active, allowing you to continue placing leader datum targets. To exit the **Datum Target – Rectangle** command, right-click and select the **Done [Esc]** menu option, press the **Esc** key on your keyboard, or access a different tool.

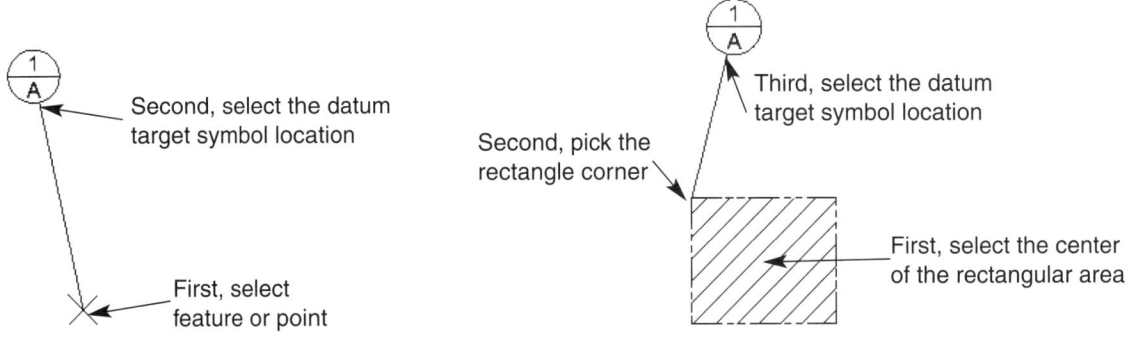

FIGURE 13.79   Placing a point datum target.

FIGURE 13.80   Placing a rectangle datum target.

Exercise 13.16

1. Continue from EX13.15, or launch Autodesk Inventor.
2. Open EX13.6.
3. Save a copy of EX13.6 as EX13.16.idw.
4. Close EX13.6 without saving, and open EX13.16.
5. Explore the options in the **Control Frame** and **Datum Target** tabs of the **Drafting Standards** dialog box as discussed.
6. Activate the panel bar **Drawing Annotation** mode, or access the **Drawing Annotation** toolbar.
7. Use the various geometric dimensioning and tolerancing tool to dimension the drawing views as shown.
8. Resave EX13.16.
9. Exit Autodesk Inventor or continue working with the program.

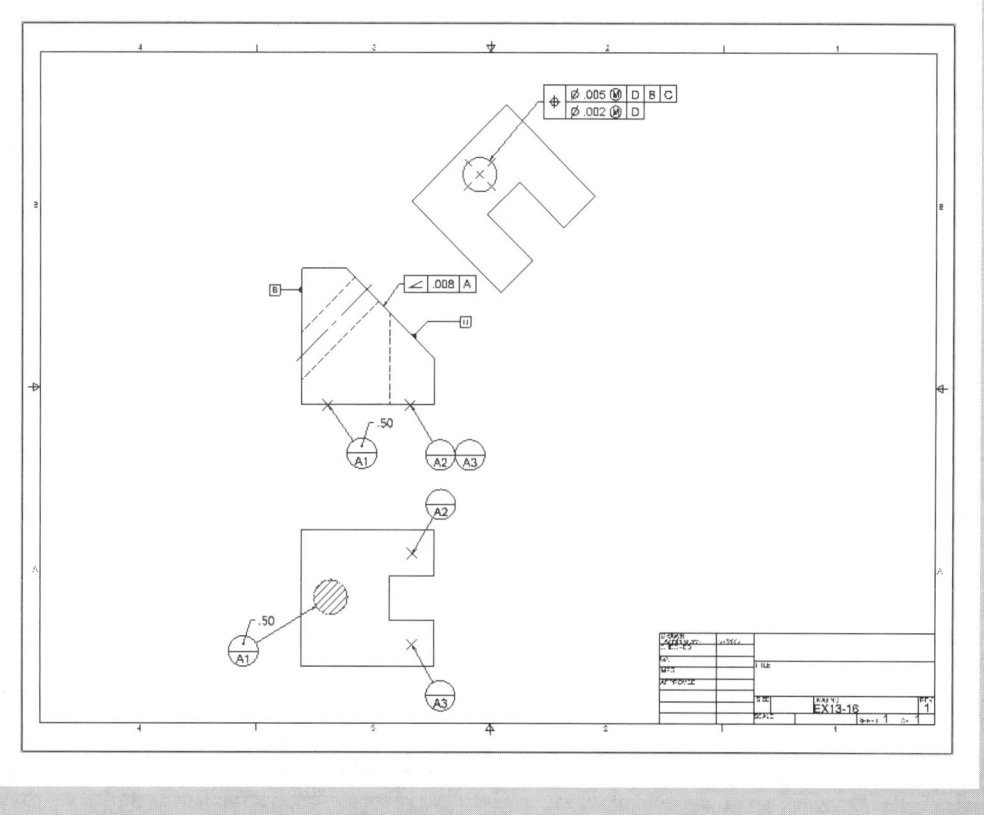

Modifying Dimensions and Overriding Dimension Styles

Most dimensions can be manipulated by changing display characteristics or overriding specific dimension style standards after they have been created. For most dimensions, the dimension style specification you establish should be appropriate for the specific application. In addition, to change specific object values you should modify the actual part, not just the dimensions. However, there are times when you must change the display of dimensions or add data to dimension values. The following list represents several options for modifying existing dimensions:

- Stretch extension lines to offset a dimension further from an object, or create extension lines that are not perpendicular to the object.

- Move a leader arrowhead, or stretch a leader shoulder.

- Move dimension text to a new location.

- Extend or shorten centerlines and centermarks.

- Change dimension value specifications, such as the default tolerance.

- Edit dimension value information, such as the parameters in a feature control frame.

Dimension modification options are specific to the dimensions you create. For example, you can choose to reverse the direction of ordinate dimensions, or change the default leader arrowhead.

To modify physical dimension properties, such as a dimension line offset, pick the dimension you want to change. You will notice several green dots, which designate the movable dimension features. See Figure 13.81. Once you see the move icon, as shown in Figure 13.82, pick one of the points or the dimension text and drag the dimension information to the desired location.

If you move the cursor over the green dot and the arrow icon is displayed, double-click the dot to open the dialog box for the specific dimension feature. For example, if you double-click on a leader arrowhead, the Change Arrowhead dialog box is displayed, which allows you to override the arrowhead style. You can also modify, or override, dimension characteristics by right-clicking on a dimension and selecting a menu option. For example, Figure 13.83 shows the menu options available when you right-click on a horizontal linear dimension, created using the **General Dimension** tool.

FIGURE 13.81
Examples of dimension
alteration points.

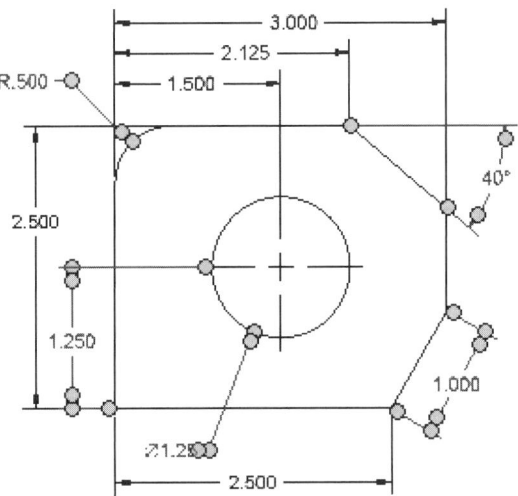

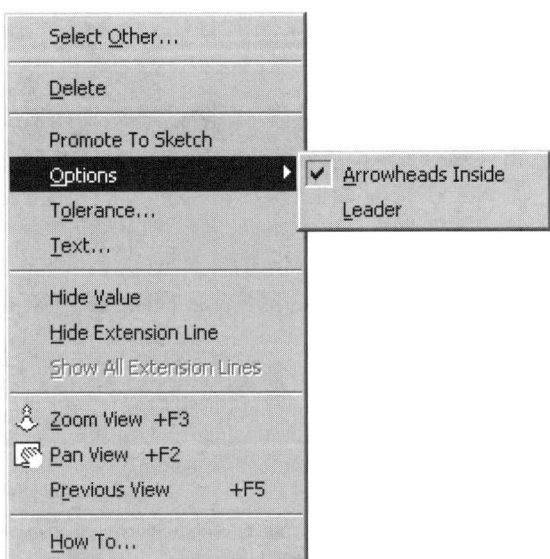

FIGURE 13.82   The move icon   **FIGURE 13.83**   Dimensioning Options cascading submenu.

|||||||||||| CHAPTER TEST

Answer the following questions on a separate sheet of paper.

1. What are dimensions?

2. What are model dimensions?

3. What are drawing dimensions?

4. What are dimension styles?

5. How do you select a dimension style based on an existing standard?

6. How do you start the process of creating a new dimension style?

7. Give the settings that you can use to establish a dimension style with horizontal dimension text placed horizontally in a space provided in the dimension line, and vertical dimension text placed in a space in the dimension line and reading horizontally from the bottom of the sheet for unidirectional dimensioning.

8. What is a terminator?

9. Define tolerance.

10. How do you set a symmetrical bilateral tolerance of $\pm.05$?

11. What does it mean when it is stated that model dimensions are parametric?

12. How do you move model dimensions?

13. Explain if and how model dimensions can be removed from the drawing.

14. How do you hide individual model dimensions?

15. How do you hide all the model dimensions in a view?

16. How do you change the size and shape of a model while working in the drawing environment?

17. Why is it necessary to place drawing dimensions when model dimensions are parametric?

18. Once a drawing view is created, how do you enter the drawing annotation work environment?

19. Describe the appearance of a centerline.

20. How do you place centermarks?

21. How do you place centermarks on a group of features in a circular pattern?

22. How do you draw a linear dimension using the General Dimension tool?

23. How do you place a diameter dimension on a circle?

24. How do you override a current diameter dimension option?

25. Define datum dimensioning and give another name for it.

26. Why should adjacent dimension values be staggered?

27. Give the steps used to place datum dimensions.

28. Explain the difference between specific and general notes.

29. Name the command that allows you to place a specific note with a leader.

30. Define ordinate dimensioning.

31. Explain the difference between the Ordinate Dimension Set and Ordinate Dimension tools.

32. Give the steps used to place a set of ordinate dimensions.

33. Discuss the difference between the Hole Table – Selection tool and the Hole Table – View.

34. Give the steps for using the Hole Table – Selection command.

35. What does the Hole/Thread Notes tool automatically provide?

36. Give the steps used to create a hole or thread note.

37. Define surface finish and describe the appearance of the surface finish symbol.

38. How do you display the Surface Texture dialog box?

39. Define welding.

40. Before you add welding symbols and use the Weld Symbol tool, you may want to adjust some of the weld symbol properties. Where is this done?

41. Give the steps used to add weld symbols to the drawing.

42. Briefly define geometric dimensioning and tolerancing.

43. Before you create feature control frames and use the Feature Control Frame tool, you may want to adjust some of the feature control frame properties. Where is this done?

44. How do you place a feature control frame in your drawing?

45. Define datums.

46. Explain the function of the datum feature symbol and discuss its appearance.

47. Give the steps used to place a datum feature symbol.

48. Give the steps used to draw a datum target point.

49. To modify physical dimension properties, such as a dimension line offset, pick the dimension you want to change. You will notice several green dots. What do these dots represent?

50. When editing dimensions, how do you open the dialog box for the specific dimension feature?

||||||||||| PROJECTS

Instructions:

■ Open the indicated Chapter 12 drawing file.

■ Save a copy of the file using the specified name.

■ Close the Chapter 12 file, and open the Chapter 13 copy.

■ Use the information provided to complete the drawings shown.

■ Drawing view scale is 1:1 unless otherwise noted.

1. Name: Bracket

   Units: Metric

   Sheet Size: A2

   Border: Default or custom

   Title Block: Default or custom

   Base View File: P12.1

   Dimension Style: ANSI-METRIC or custom

   Save As: P13.1

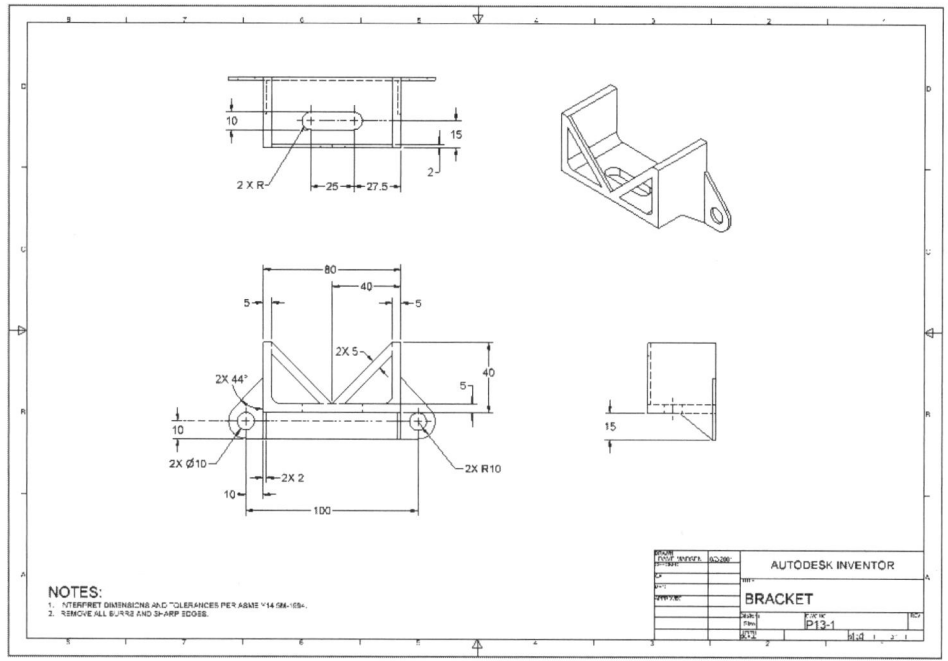

2. Name: Angle Block

   Units: Inch

   Sheet Size: B

   Border: Default or custom

   Title Block: Default or custom

   Base View File: EX12.12

   Dimension Style: ANSI-INCH or custom

   Save As: P13.2

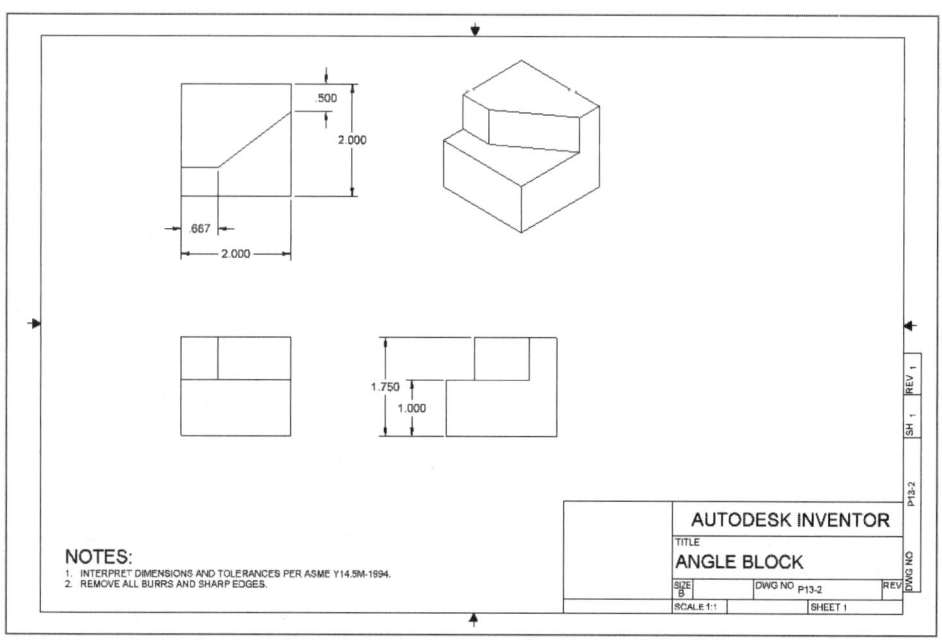

3. Name: Angle Bracket

   Units: Inch

   Sheet Size: C

   Border: Default or custom

   Title Block: Default or custom

   Base View File: EX13.7

   Dimension Style: ANSI-INCH or custom

   Save As: P13.3

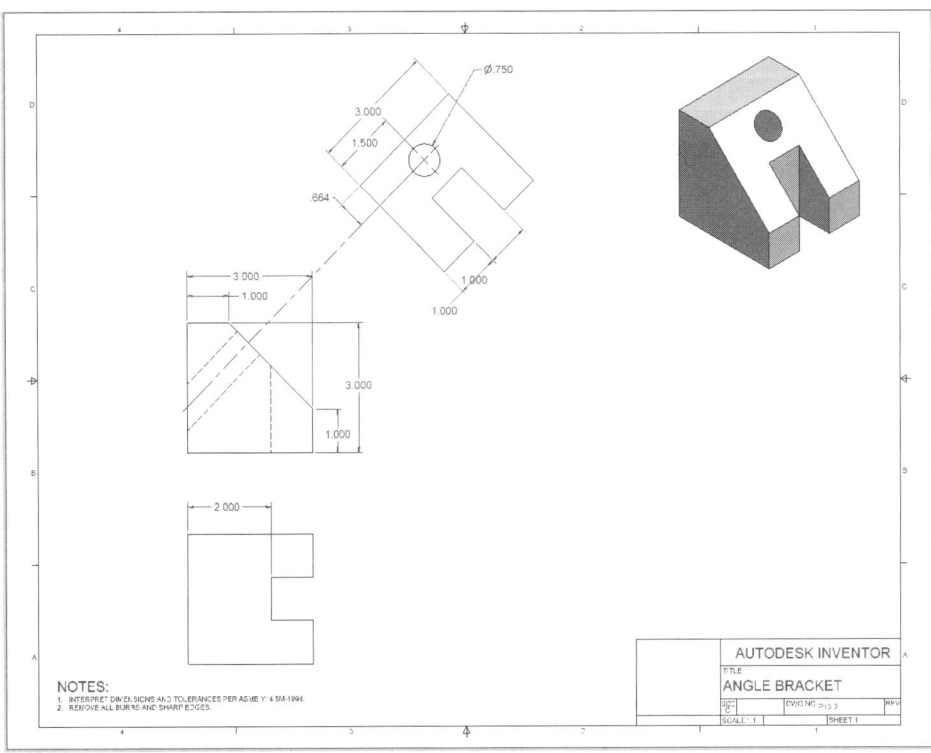

4. Name: Shim

   Units: Inch

   Sheet Size: B

   Border: Default or custom

   Title Block: Default or custom

   Base View File: EX12.18

   Dimension Style: ANSI-INCH or custom

   Save As: P13.4

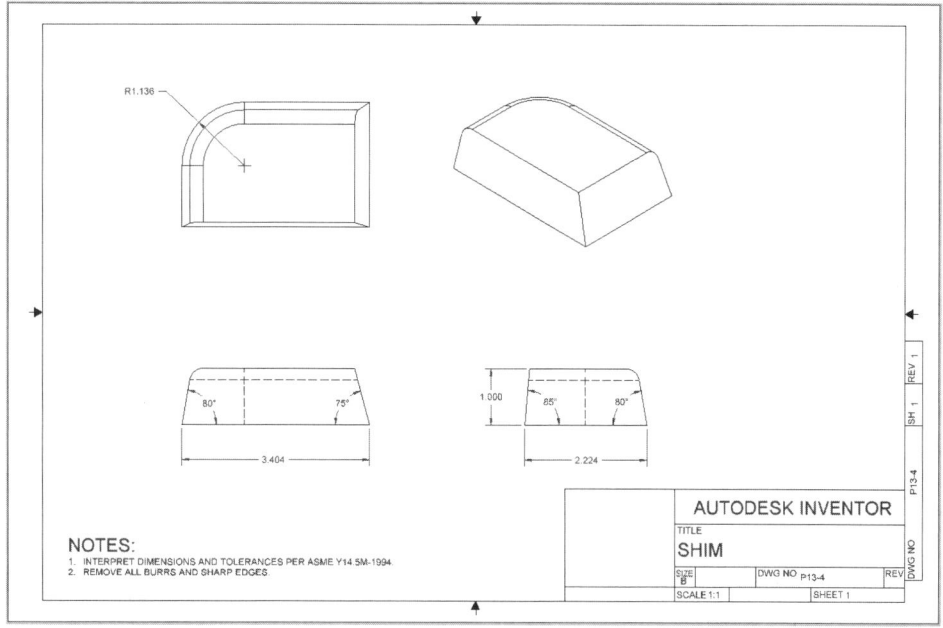

5. Name: Cover Plate

   Units: Inch

   Sheet Size: A2

   Border: Default or custom

   Title Block: Default or custom

   Base View File: EX13.12

   Dimension Style: ANSI-INCH or custom

   Save As: P13.5

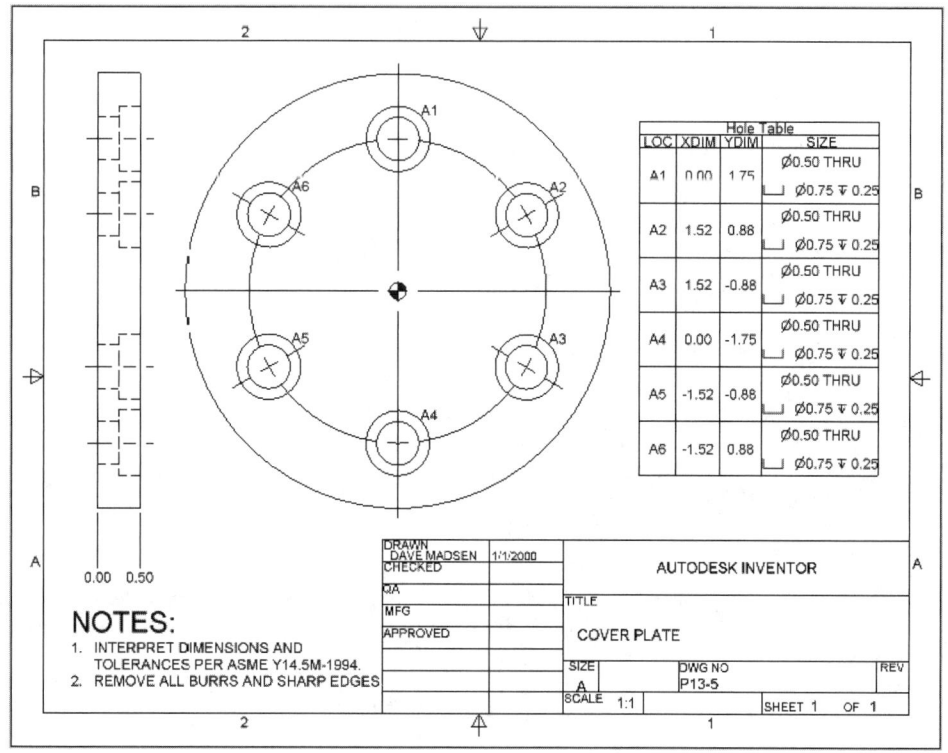

6. Name: Wheel

    Units: Inch

    Sheet Size: D

    Border: Default or custom

    Title Block: Default or custom

    Base View File: EX12.14

    Dimension Style: ANSI-INCH or custom

    Save As: P13.6

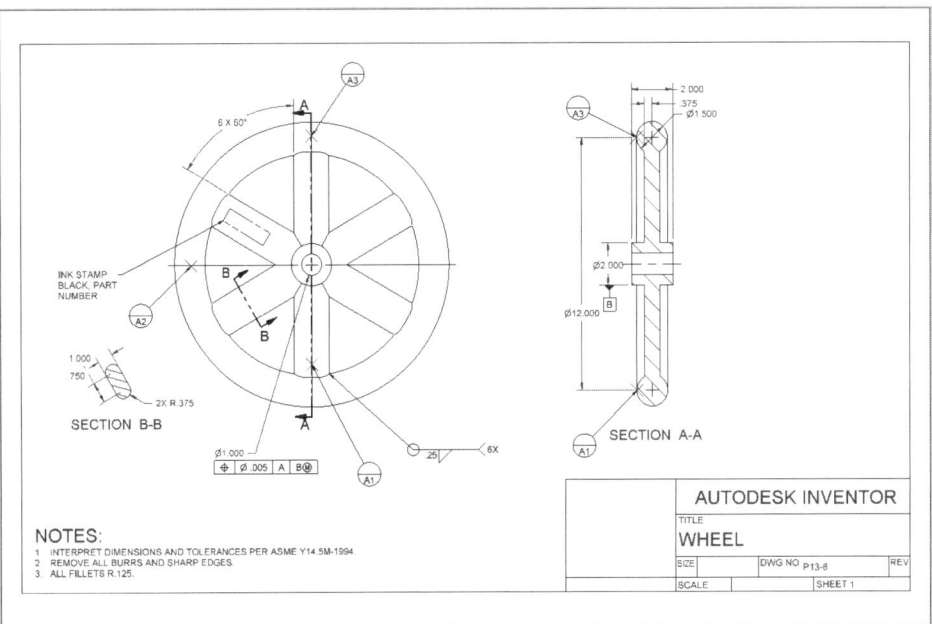

7. Name: Angle Bracket

    Units: Inch

    Sheet Size: C

    Border: Default or custom

    Title Block: Default or custom

    Base View File: EX13.9

    Dimension Style: ANSI-INCH or custom

    Save As: P13.7

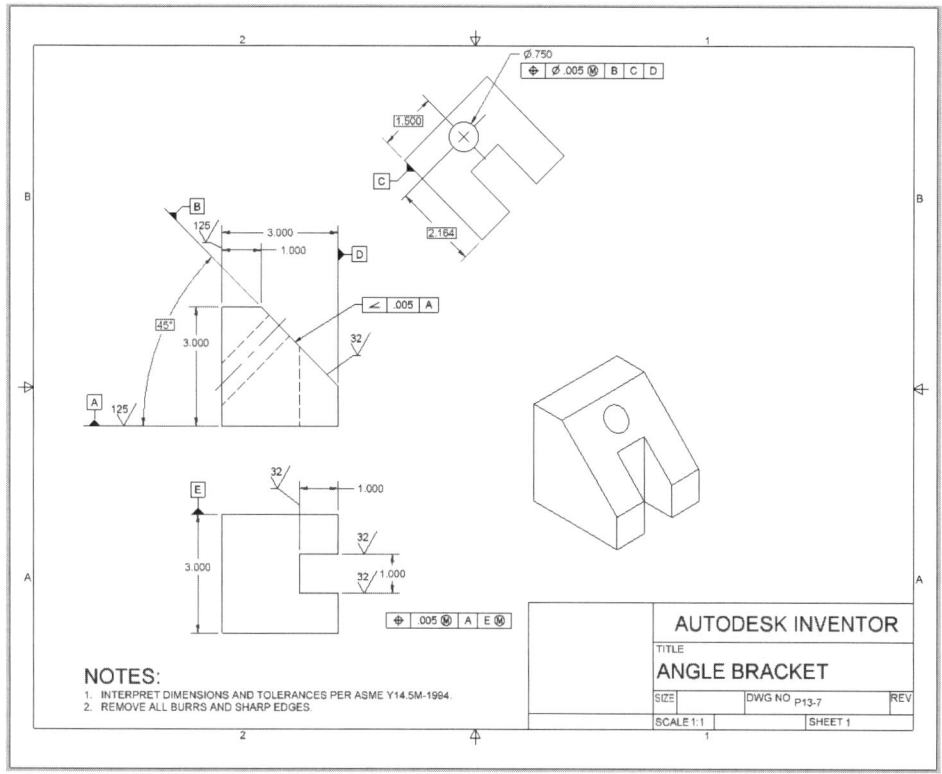

8. Name: Slide Bar Hinge Connector

Units: Inch

Sheet Size: B

Border: Default or custom

Title Block: Default or custom

Base View File: P12.7

Dimension Style: ANSI-INCH or custom

Save As: P13.8

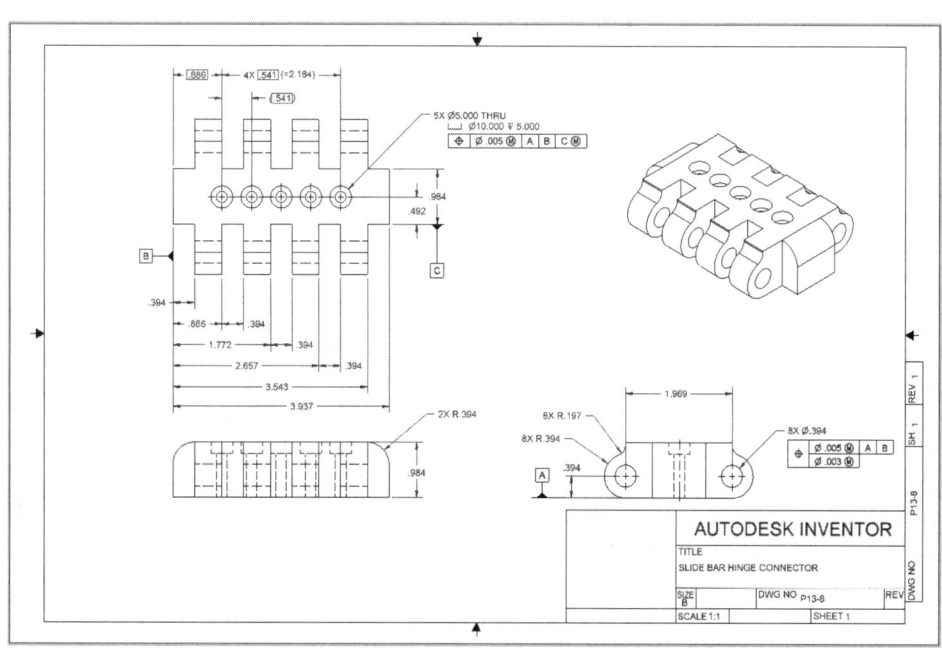

|||||||||||||||||||||||||||||||||||||||||||||||||||||||||||||||||||||||||||||||||||||||||||||

CHAPTER 14

Working with Assemblies

LEARNING GOALS

After completing this chapter, you will be able to:

◎　Work in the assembly environment.

◎　Create assemblies from existing part and subassembly files.

◎　Develop assembly components while in an assembly file.

◎　Apply, edit, and drive assembly component constraints.

◎　Pattern assembly components.

◎　Manipulate assembly components.

◎　Use assembly section views.

◎　Work with design views.

Working with Assemblies

An *assembly* is a grouping of one or more design components. See Figure 14.1. Components may include part models and smaller assemblies, called subassemblies. *Subassemblies* are groups of parts that are typically standard product components and may be repeated several times in various assemblies. For example,

FIGURE 14.1 An example of a c-clamp assembly, which consists of four parts.

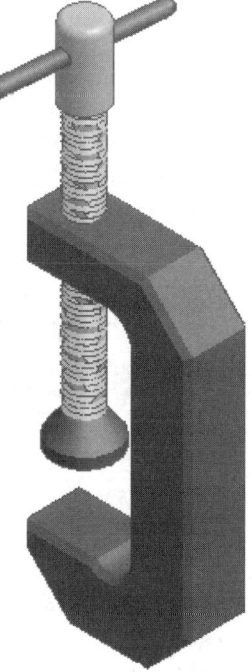

subassemblies may include typical items like switches, or spring assemblies, which are composed of two or more parts, and may be used multiple times in the final assembly product. Although some designs and products may consist of only one part, such as a bolt, most products usually consist of at least two parts, such as a bolt and a nut. In an assembly file, multiple parts and subassemblies are brought together and constrained to form an assembly. As discussed in Chapter 3, constraints define relationships between sketch geometry when creating parts. Constraints are used similarly in the assembly environment, to link individual components together.

As discussed in Chapter 1, Autodesk Inventor assembly files are separate files and carry the extension .iam. Assembly template files are used to create assemblies and subassemblies by inserting parts and subassemblies into an assembly template file or assembly file. You can also create multiple individual components in the assembly template file, and build the assembly without inserting existing components. For this method, you may find it helpful to think of an assembly file as a part file that also contains assembly tools. A combination of assembly creation techniques can also be used. Although either option is effective, it is often faster, easier, and more productive to create parts within an assembly template file or assembly file, because building parts inside an assembly file creates both an assembly and a separate part file for each of the parts. Both of these techniques are discussed in this chapter.

Working in the Assembly Environment

The work environment and interface is different for each stage of the design process. Some of the tools and interface components remain constant or are similar throughout the entire design process. As previously discussed, parts can be created in the assembly environment, which means that all the part model interface options are available in the assembly environment. However, the assembly environment and many of the assembly interface components are specific to working with assemblies.

By default, when you open a new Autodesk Inventor assembly the work environment should look like the display in Figure 14.2. When you open an assembly template, depending on your approach, you are ready to insert individual parts or subassemblies and create parts to generate an assembly. Throughout this chapter you will explore specific tools, commands, and options that allow you to develop Autodesk Inventor assemblies.

Using the Command Bar for Assemblies

In previous chapters you have become familiar with the Autodesk Inventor command bar, shown in Figure 14.3. Some of the command bar options are the same, or similar, when working in assemblies, as they are when working with parts and drawings. However, there are also some differences. The following command bar tools are available while in the drawing environment:

- **Select**   This flyout button allows you to specify which objects you want to select when working in an assembly. By default, select priority is set to **Component Priority,** which allows you to pick individual assembly components such as parts or subassemblies. You can also choose the **Leaf Part Priority** option to select parts only, not subassemblies. In addition, the **Feature, Face,** and **Sketch** priority options are available when creating assembly components.

- **Return**   Picking this button finishes a sketch or component creation and takes you from sketch mode to feature mode, and from feature mode to assembly mode.

- **Sketch**   Select either the 2D Sketch button or 3D Sketch button to activate a new sketch, and enter sketch mode.

- **Update**   These buttons allow you to update the assembly or the assembly and its dependent children.

- **Style**   Use this drop-down list to change the color of specific assembly components, or various styles in the sketch or feature environments.

- **Message**   The message display box is similar to the status bar discussed in Chapter 2 and provides you with prompts and assembly information.

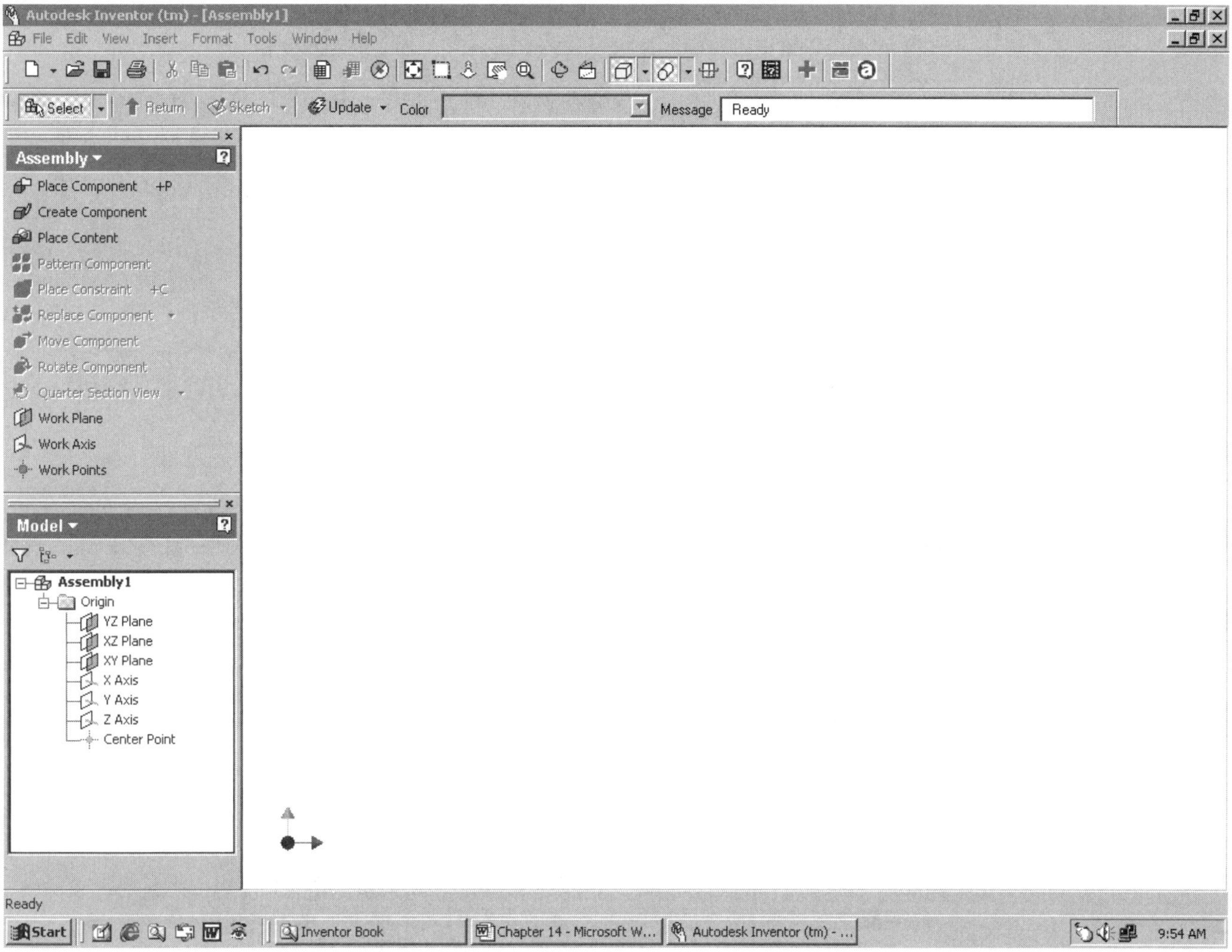

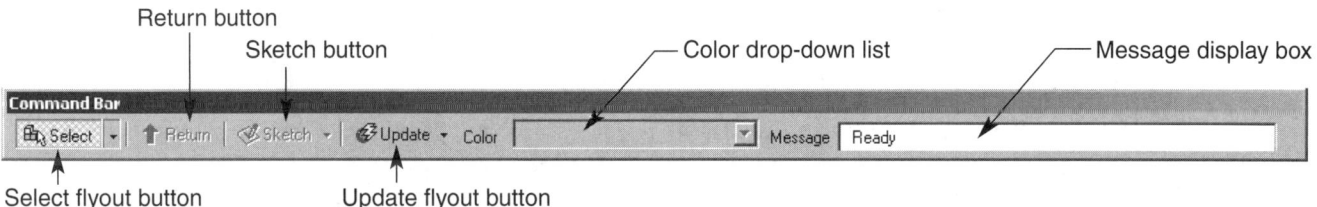

FIGURE 14.2 The default Autodesk Inventor assembly work environment.

Return button

Sketch button

Color drop-down list

Message display box

Command Bar

Select flyout button

Update flyout button

FIGURE 14.3 The command bar.

Using the Browser in an Assembly

The browser for a typical assembly displays the assembly icon and name, followed by the **Origin** folder, and then each of the assembly components. Figure 14.4 shows the browser for a c-clamp assembly named F14.1.iam. As in Autodesk Inventor part and presentation files, the assembly browser contains a **Browser Filters** button. The following Browser options are specific to the assembly environment. For information regarding other Browser options available in both the assembly and part Browsers, refer to Chapter 4.

- **Assembly Tasks** Select this option to display the assembly tasks associated with the various components listed in the Browser. For example, when this option is active, rotation and component constraint information is displayed for each of the components.

FIGURE 14.4   The browser display for a typical assembly.

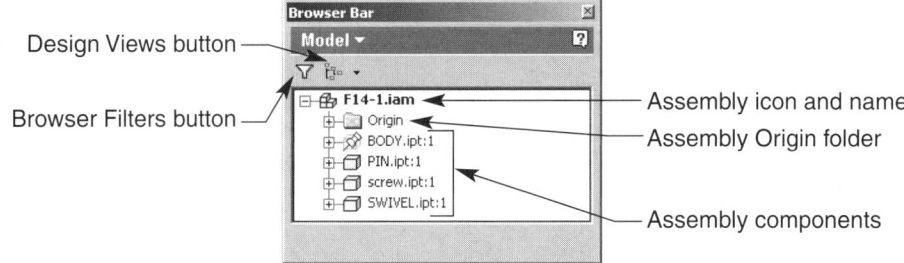

Design Views button

Browser Filters button

Assembly icon and name
Assembly Origin folder

Assembly components

- **Modeling Tasks**   Select this option to display the modeling tasks associated with the specific components listed in the Browser. For example, when this option is active, features like extrusions and holes are displayed for each of the components.

- **Show Children Only**   Select this option to display only certain elements of subassemblies.

In addition to the **Browser Filters** button, you may want to become familiar with the following options, available only in an assembly file when you right-click a component in the Browser or the actual component in the graphics window:

- **Edit**   This option activates the specified component and shows the modeling tasks, allowing you to modify the part or subassembly.

- **Open**   This option opens the actual component (part or assembly) file, allowing you to work in the part or assembly environment.

- **Visibility**   When unselected, this option hides the specified component in the graphics window, but does not actually remove the component. Use this selection for clarity and program operating speed when working with complex assemblies.

- **Enabled**   Unselect this option if you do not want to be able to select certain components. Components that are not enabled are similar to those that are not visible; they allow quicker program operating speed and help clarity when working with large assemblies.

- **Grounded**   Grounded components are discussed in depth throughout this chapter.

 Field Notes

Assembly components can be moved up and down in the Browser hierarchy, just like features in a part model Browser. Additional assembly browser options are discussed throughout this chapter.

Specific Assembly Pull-Down Menu System Options

All the pull-down menu options that relate to working with parts are available inside the assembly environment, because you are able to develop parts while working in an assembly file. For information on these options, refer to Chapters 3 and 4. There are also some specific pull-down menu options available only while in the assembly environment. Not all the assembly file pull-down menu options are discussed here. Many are explained in the section where they apply. Still, you may want to become familiar with these selections for future reference:

- **View**   As discussed in Chapter 2, the **View** pull-down menu contains a number of tools that allow you to manipulate the display of objects in the graphics window.

  - **Refresh**   Select this option to update component and assembly files that have been modified.

■ **Degrees of Freedom**   Use this option to analyze component freedom of movement in an assembly. As shown in Figure 14.5, components that are still free to move are indicated by the degrees of freedom icon and arrows.

■ <u>**Tools**</u>   This pull-down menu contains the following options for working with assemblies.

■ **Analyze Interference**   This tool is used to evaluate any interference between selected assembly components. When you pick this option, the **Interference Analysis** dialog box is displayed. See Figure 14.6. You can analyze one or two sets, or groups, of assembly components. To analyze one set, pick the **Define Set #1** button, and select one or more components, such as a hole and a shaft. If you want to analyze an additional set of components, pick the **Define Set #2** button, and select one or more other components in the assembly. Finally, pick the OK button to activate the analysis.

If there is no interference between the components, a corresponding message will show. However, if interference is found, the **Interference Detected** message and table are displayed, allowing you to view the interference analysis. See Figure 14.7.

■ <u>**Bill of Materials…**</u>   Pick this option to access the *assembly bill of materials,* which identifies the quantity and properties of the assembly components. See Figure 14.8. By default, item number, quantity, part number, and description are displayed in the bill of materials. The item numbers are set based on the listed browser hierarchy; the quantity of each part is defined by the number of each component in the assembly; and the part number and description correspond to the values entered in the **Properties** dialog box of each of the components.

To modify property columns, pick a column or a column cell to display the column properties in the **Column Properties** area. Then edit the name by entering a value in the **Name** edit box, and change the column width by entering a value in the **Width** edit box. The column width can also be altered by picking and dragging the column edge to the desired location. To modify the justification of the column name, select the desired **Name**

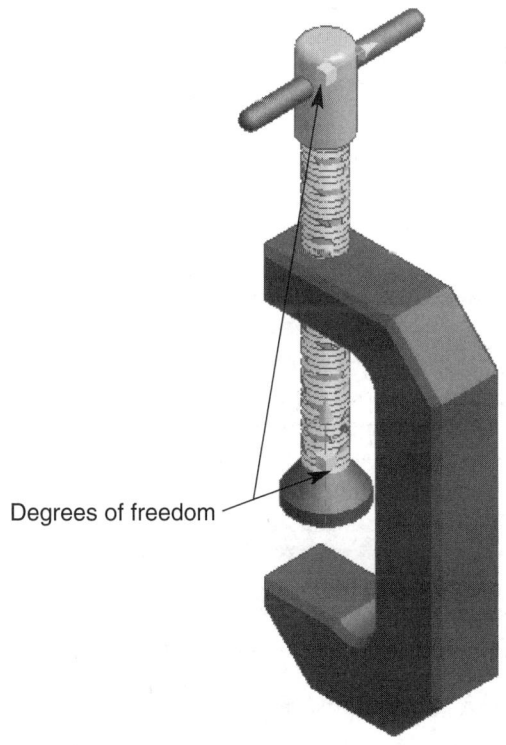

Degrees of freedom

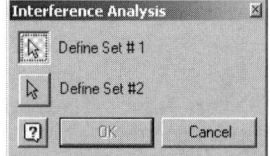

FIGURE 14.5   Degrees of freedom identification.

FIGURE 14.6   Interference Analysis dialog box.

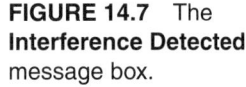

FIGURE 14.7   The **Interference Detected** message box.

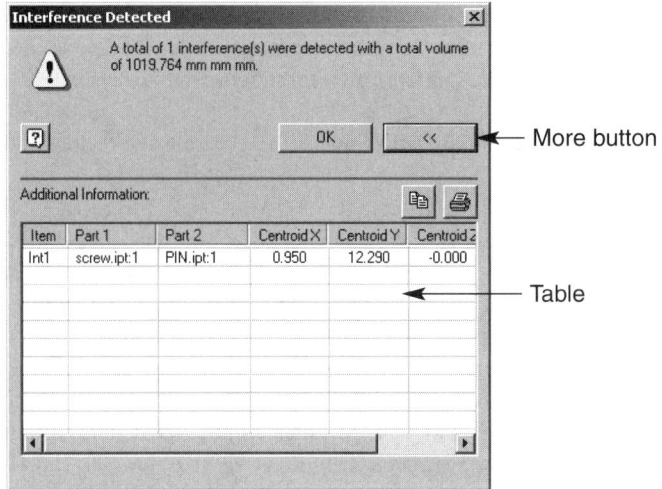

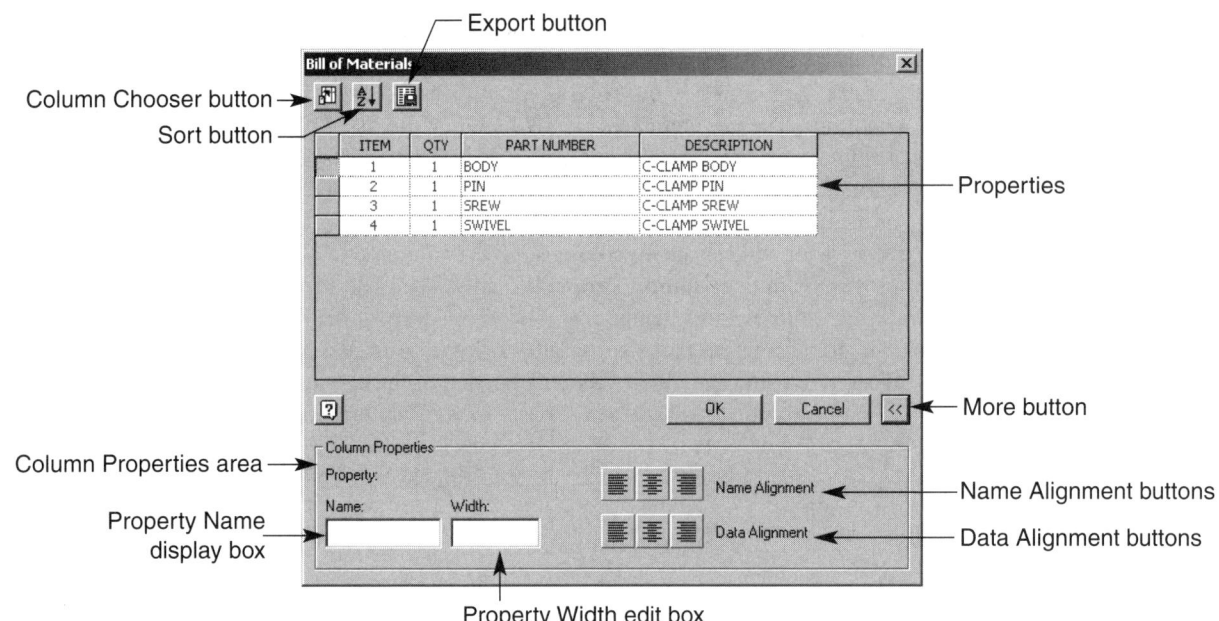

FIGURE 14.8   An example of an assembly bill of materials.

Alignment button. Similarly, to adjust the justification of the column data, select the desired **Data Alignment** button.

Property columns are placed in a default order. If you want to change the order, or sorting, of the property columns, pick the **Sort** button to display the **Sort Bill Of Material** dialog box. See Figure 14.9. Select the first column from the **Sort by** drop-down list, followed by the second column from the **Then by** drop-down list, and finally the third column from the **Then by** drop-down list. For each column, pick the **Ascending** or **Descending** radio button to sort either by the lowest value to the highest value, or the highest value to the lowest value.

You can add properties or columns to the bill of materials by selecting the **Column Chooser** button to display the **Bill Of Material Column Chooser** dialog box. See Figure 14.10. To add a property column, activate the desired field of properties by selecting an option from the **Select available fields from:** drop-down list. Then choose the property you want to add to the bill of materials from the **Available Properties** list box, and pick the **Add** button. The **Bill Of Material Column Chooser** dialog box also allows you to change the order the property columns are displayed from left to right, by highlighting the property you want to move in the **Selected Properties** list box, and picking either the

FIGURE 14.9 The
Sort Bill Of Material
dialog box.

Sort by (first column) area

Then by (second column) area

Then by (third column) area

Ascending radio button
Descending radio button
Ascending radio button
Descending radio button
Ascending radio button
Descending radio button

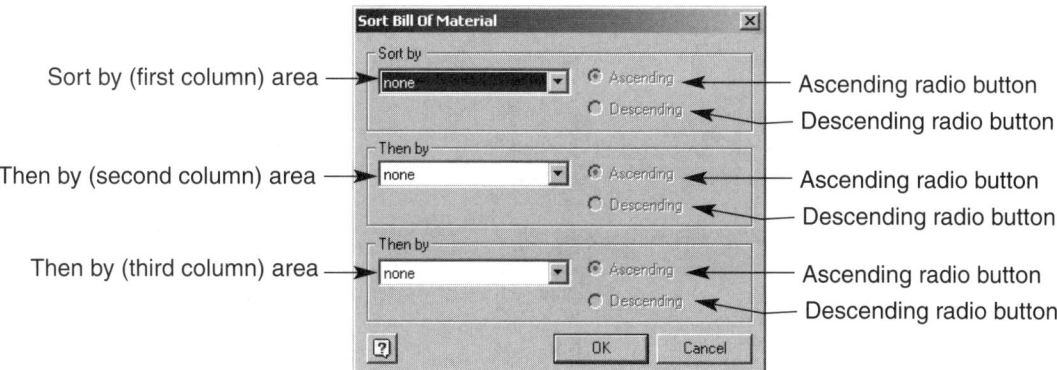

FIGURE 14.10 The
**Bill Of Material
Column Chooser**
dialog box.

Select available fields from: drop-down list
Add to Selected
Properties button

Available Properties
list box

Selected Properties
list box

Move Up button

Move Down button

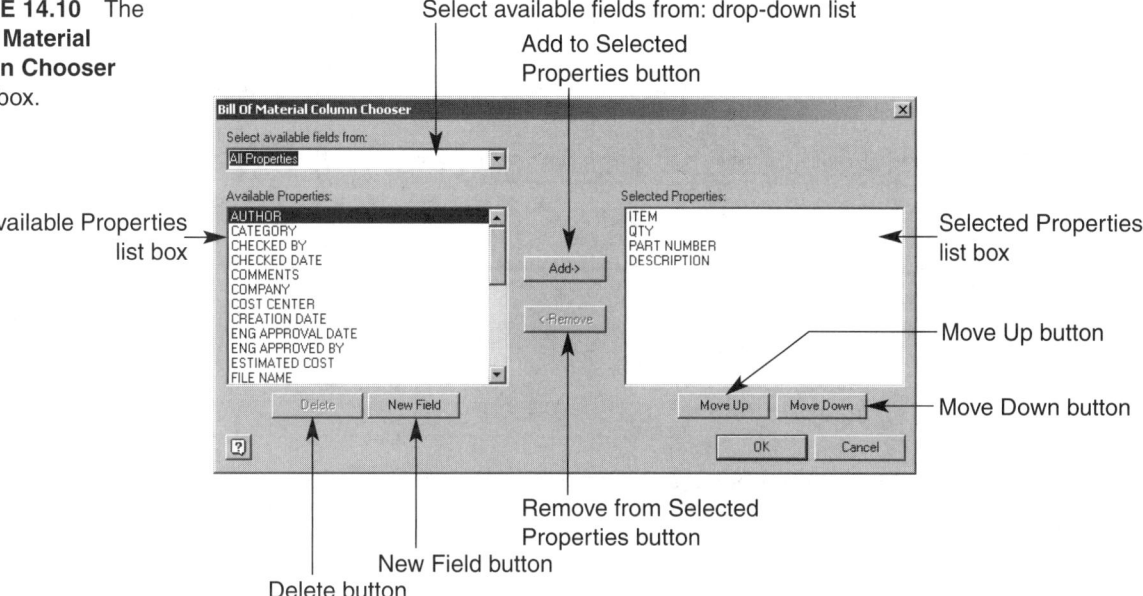

Remove from Selected
Properties button

New Field button

Delete button

Move Up or **Move Down** button. In addition, you can remove a selected property column by highlighting the property you want to remove, and picking the **Remove** button. To create a custom field property in this dialog box, not related to any of the available fields, pick the **New Field** button. Then enter the name of the field property in the **Define New Field** dialog box, and pick the **OK** button. A custom field must be placed in the **Available Properties** list box before it can be completely removed using the **Delete** button.

■ **Application Options...** As discussed in previous chapters, this selection opens the **Options** dialog box. The **Assembly** tab, shown in Figure 14.11, relates exclusively to the assembly environment and contains the following preferences:

■ **Defer Update** When you select this check box, you must choose the **Update** button after you modify an assembly component to update the assembly. If you do not select this check box, the assembly is automatically updated when a component is modified.

■ **Delete Component Pattern Source(s)** When you select this check box and delete a pattern of components, the original pattern source is also deleted. Do not pick this check box if you want to delete a component pattern and retain the source component.

■ **Part Feature Adaptivity** This area is used to define the Adaptivity of assembly components you create. Pick the **Features are initially adaptive** radio button if you want to automatically make new component features adaptive, or select the **Features are initially non-adaptive** radio button if you want to manually control the Adaptivity of component features.

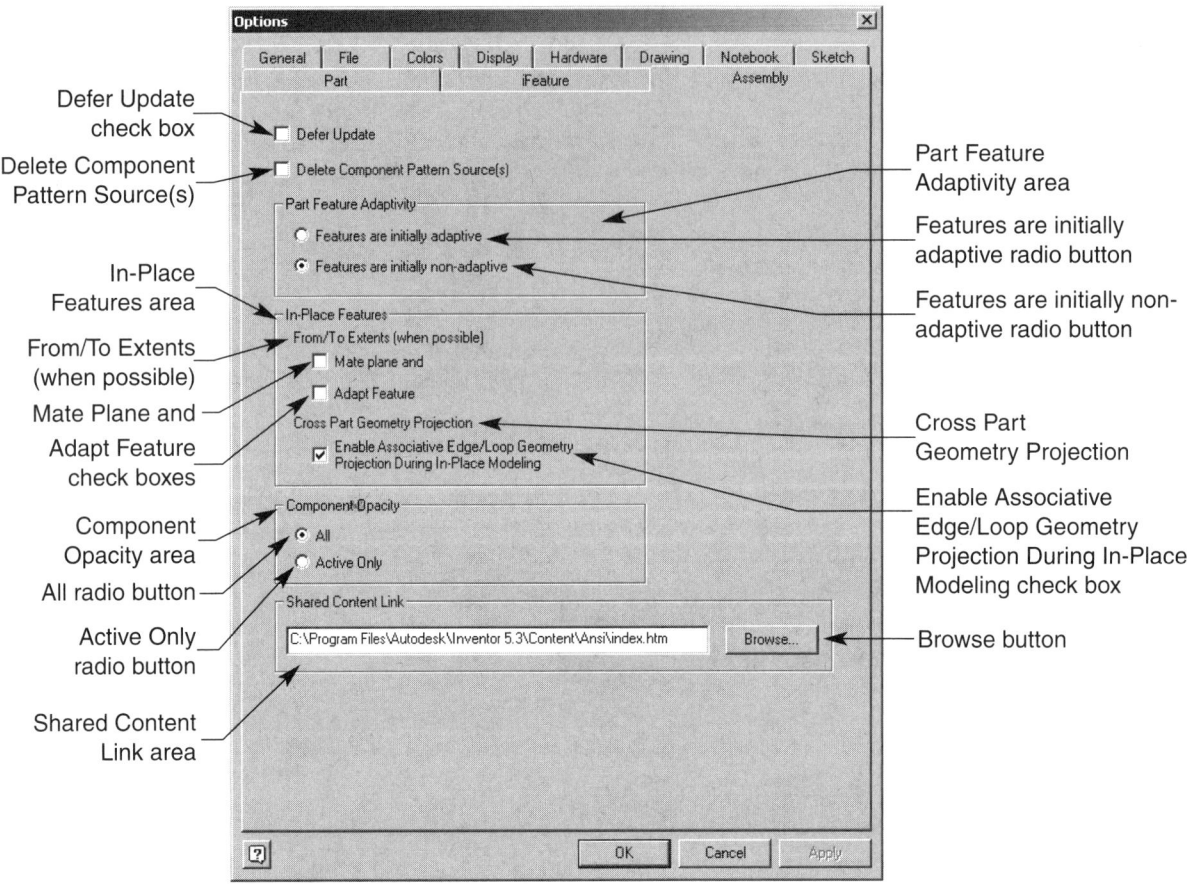

FIGURE 14.11　The **Assembly** tab of the **Options** dialog box.

- **In-Place Features**　This area allows you to specify preferences for creating components inside an assembly file. Select the **Mate Plane** check box if you do not want features to be adaptive but do want a mate constraint to occur between the feature and the plane. Pick the **Adapt Feature** check box to automatically adapt features to changing geometric places. Choose the **Enable Associative Edge/Loop Geometry Projection During In-Place Modeling** check box to generate a sketch from selected geometry when creating a new feature.

- **Component Opacity**　This area allows you to define the opacity of components when sectioning an assembly. Pick the **All** radio button to show all assembly components in an opaque style, or pick the **Active Only** radio button to show only the active components in an opaque style. You can also adjust this preference by picking the **Component Opacity** button in the **Standard** toolbar.

- **Shared Content Link**　Use this area to define a location for shared, third-party, I-drop component files. The path you specify is available when using the **Shared Content/Place Component** tool.

 Field Notes

Additional assembly pull-down menu system options and those not previously discussed are identified throughout this chapter in the sections where they apply.

Using Assembly Tools

Assembly tools are used to develop assemblies by inserting parts and subassemblies, creating parts, placing assembly constraints, patterning components, and other processes. These tools are accessed from the **Assembly** toolbar, shown in Figure 14.12, or from the **Assembly** panel bar. In addition, many of the tools can be accessed from the **Insert** pull-down menu. When you are in the assembly work environment, the panel bar is in assembly mode. See Figure 14.13. By default, the panel bar is displayed in learning mode, as seen in Figure 5.3. To change from learning to expert mode, as seen in Figure 14.14 and discussed in Chapter 2, right-click any of the panel bar tools or left-click the panel bar title, and select the **Expert** option. The **Work Plane, Work Axis,** and **Work Point** tools discussed in previous chapters are also available in the **Assembly** environment. These tools are not specifically addressed in this chapter. However, they can be used for any required application, including creating model geometry, locating and constraining assembly components, sectioning assemblies, and whenever a work feature is needed.

FIGURE 14.12 The **Assembly** toolbar.

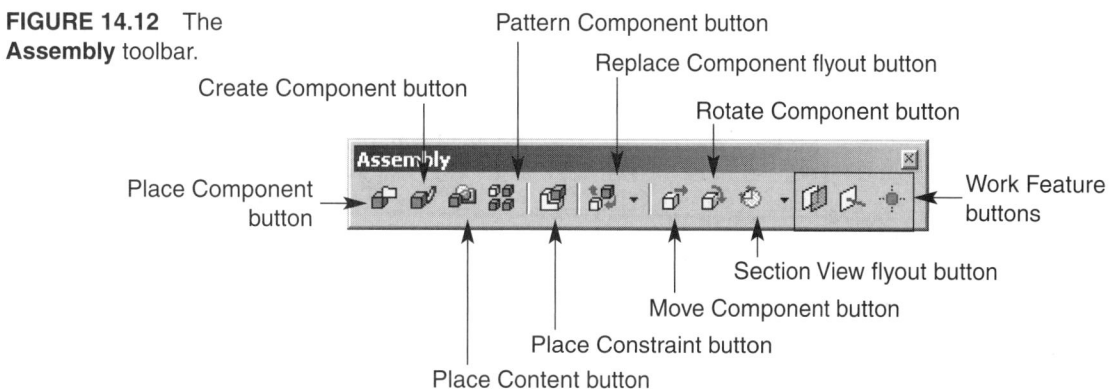

FIGURE 14.13 The panel bar in feature (learning) mode.

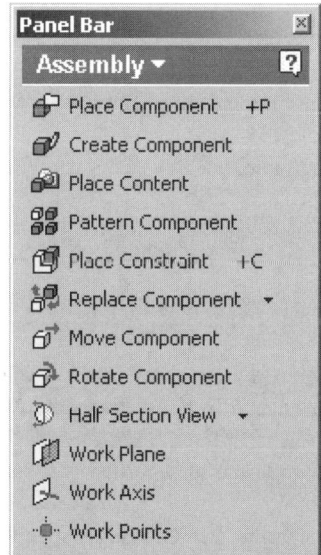

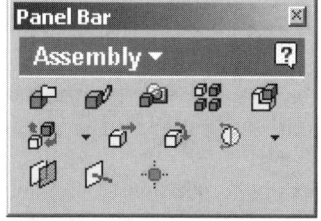

FIGURE 14.14 The panel bar, in assembly (expert) mode.

Exercise 14.1

1. Launch Autodesk Inventor.
2. Open a new inch or metric assembly file.
3. Explore the basic assembly file interface options and work environment, as previously discussed.
4. Exit Autodesk Inventor without saving, or continue working with the program.

Creating Assemblies from Existing Components

In previous chapters you learned how to develop individual part models using separate Autodesk Inventor part files. As previously discussed, one option for generating an assembly is to insert existing parts or subassemblies into an assembly file. Typically the first component you insert should be a major part or subassembly, such as the assembly body or a main support base. The initial part or subassembly, which is the first component listed in the Browser, is similar to the base feature of a part. However, defining the first component is usually not as critical and specifying a base feature. When you insert the initial assembly component, it automatically becomes grounded. As the name implies, a ***grounded*** part or assembly cannot be moved or shifted because it is fastened to a point, or to space. The initial component is grounded to the assembly origin, or **Center Point,** located in the Browser **Origin** folder. Each component you insert after the first component is not grounded and can be placed anywhere in graphics window.

Field Notes

> A ground constraint can be removed by right-clicking the component name in the Browser, or the component itself in the graphics window, and selecting the **Grounded** menu option.

If you choose to apply the method of inserting existing components for creating an assembly, the first tool you will use is the **Place Component** tool, which inserts parts or subassemblies into an Autodesk Inventor assembly file. Access the **Place Component** tool using one of the following techniques:

✓ Pick the **Place Component** button on the **Assembly** panel bar.

✓ Pick the **Place Component** button on the **Assembly** toolbar.

✓ Type the **+** and **P** keys on your keyboard.

✓ Select the **Existing Component…** option of the **Insert** pull-down menu.

When you access the **Place Component** tool, the **Open** dialog box, shown in Figure 14.15, is displayed. As discussed in previous chapters, the **Open** dialog box allows you to locate the components you want to insert in the assembly file. The following options represent the main differences between the **Open** dialog box displayed when placing an assembly component, and other **Open** dialog boxes you have used:

■ **Files of type**   In addition to part and assembly files, you can choose to specify component files, which display both part and assembly files located in the specified path. You can also choose to display components created using other cad files or those saved as STEP or SAT files.

Field Notes

Imported component files such as STEP or SAT files are not parametrically controlled.

FIGURE 14.15   The
Open dialog box.

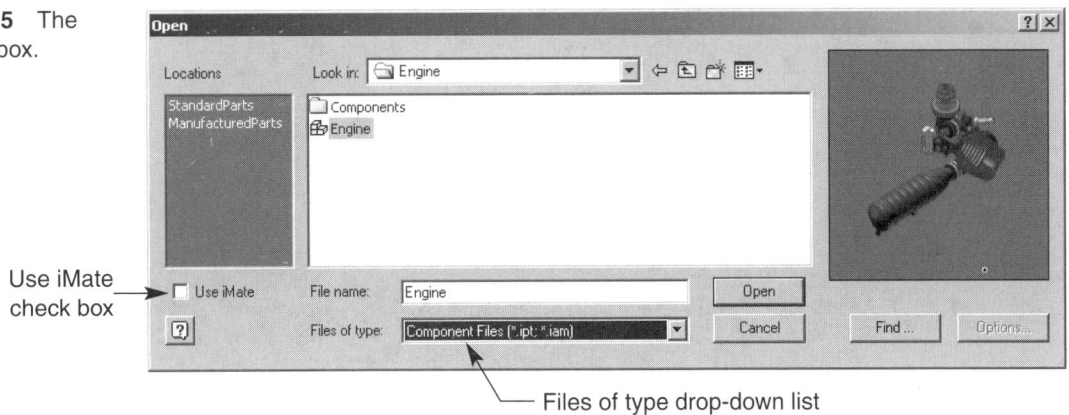

Use iMate
check box

Files of type drop-down list

- **Use iMate**   Select this check box if you want to automatically place a constraint between components with iMates when you place the components. iMates are discussed later in this chapter.

As mentioned earlier, the first component you place in an assembly is automatically grounded and located at the **Center Point.** See Figure 14.16. You can automatically place additional base components by picking a location in space to place the component. If you want only one component, or after you have placed all the desired components, right-click and select the **Done** menu option, or press the **Esc** key on your keyboard. After the initial component is added to the assembly file, all other components, including those that are identical to the first component, must be located in space and are not automatically grounded or constrained unless they contain iMates.

You may choose to continue inserting parts and subassemblies until all the required components are placed in the assembly file. See Figure 14.17. The next step in assembly development is to establish links between individual components using assembly constraints, discussed later in this chapter. Depending on your approach to inserting assembly components and the complexity of the assembly, you may choose to insert and constrain one or two components at a time.

If you receive the **Resolve Link** dialog box when you try to insert an existing component, most likely the component file is not located in the specified project folder; is not available because of a network problem if the file is contained in a local network; or the Internet or Intranet server where the file is located is not functioning correctly. To ensure you do not receive this dialog box, identify and solve the problem. However, you can pick the **Open** button in an attempt to locate the file elsewhere, or skip the entire process by picking the **Skip** button.

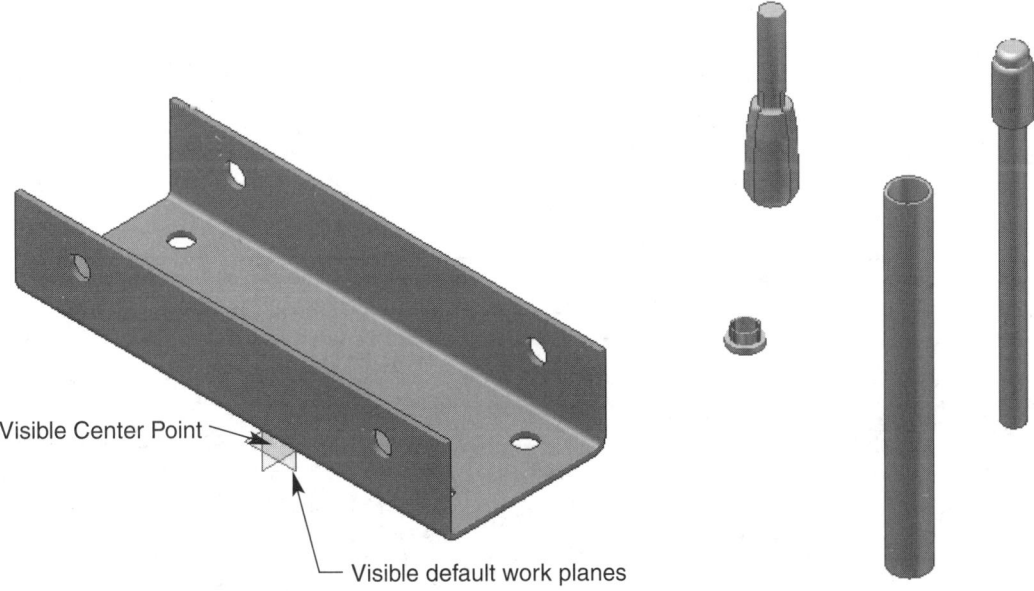

Visible Center Point

Visible default work planes

FIGURE 14.16   An initially placed assembly component.

FIGURE 14.17   Inserting
assembly components.

Exercise 14.2

1. Continue from EX14.1, or launch Autodesk Inventor.
2. Open a new metric assembly file.
3. Access the **Place Component** tool.
4. Locate P5.6.ipt, and pick the **Open** button to automatically place the base component. P5.6.ipt should be grounded as indicated by the **Ground** icon displayed over the part icon in the Browser.
5. Test the ground of the component by selecting the part, holding down the left mouse button, and attempting to drag the component. The **Ground** icon should be displayed in the graphics window.
6. Access the **Place Component** tool.
7. Locate P7.6.ipt, and pick the **Open** button.
8. Pick a location for the screw similar to the location shown, and exit the command to place only one copy.
9. Save the assembly as EX14.2. You may be asked to save the assembly file and its dependents. Dependents represent the separate component files—in this case, P5.6.ipt. Dependents are saved in addition to the assembly to ensure parametric relationships.
10. Exit Autodesk Inventor or continue working with the program.

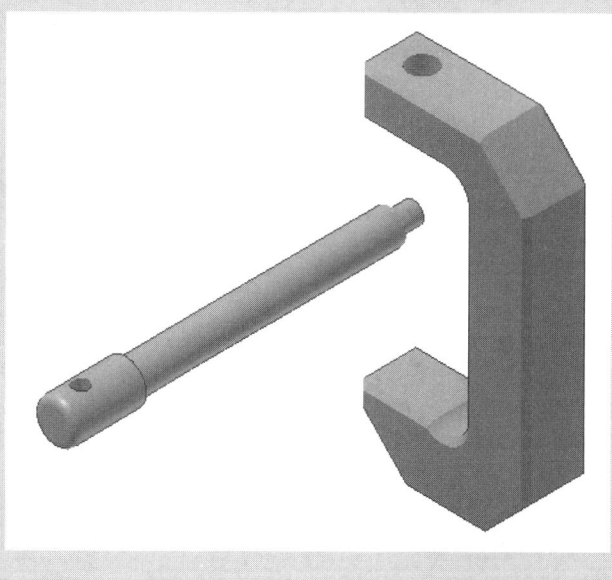

Alternative Methods for Inserting Assembly Components

Although the typical method for inserting assembly parts and subassemblies involves using the **Place Component** tool, there are alternative techniques for placing assembly components. You may choose to use one of the following methods depending on your preference, or the type of assembly you create:

■ Inserting components from one open window to another. Open an assembly file that will be used to receive the components, and open one or more components (part or assembly files) that you want to insert into the designated assembly file. Then arrange the file windows so they are all visible, as discussed in Chapter 2. See Figure 14.18. Next, select the window of the component from which you want to place a part or subassembly into a different assembly file. Finally, drag the component icon and name from the browser of the selected component window, and drop the component into the assembly window.

FIGURE 14.18
IInserting components
from one open window
to another.

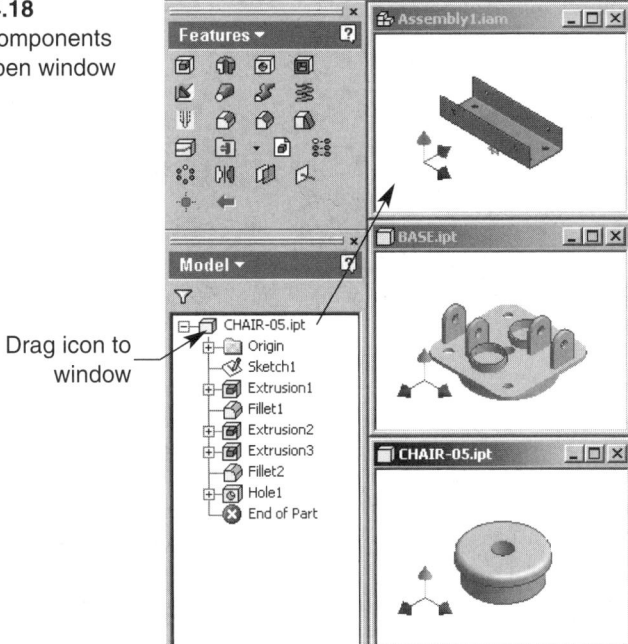

Drag icon to
window

Active window

Field Notes

You can only drag and drop one component at a time.

■ Using Windows Explorer and an open assembly file. Open an assembly file that will be used to receive the components. The assembly file can be default such as Assembly1, a saved file, or a saved file with existing components. Then access Windows Explorer, and locate the part or subassembly files you want to insert into the assembly. Next, select all the components you want to insert, drag them to the assembly button on the taskbar, and hold them there until the Autodesk Inventor assembly window opens, or drag them into the opened Autodesk Inventor assembly window, if it is displayed. Finally, continue dragging the components into the graphics window, and drop them into place.

Exercise 14.3

1. Continue from EX14.2, or launch Autodesk Inventor.
2. Open EX14.2.
3. Save a copy of EX14.2 as EX14.3.
4. Close EX14.2 without saving, and open EX14.3.
5. Open P5.1.ipt.
6. Pick the **Arrange All** option from the **Windows** pull-down menu.
7. Activate the P5.1.ipt window by selecting inside the graphics window.
8. Drag the part, P5.1.ipt from the Browser, and drop it into the assembly window. P5.1.ipt should now be a component in the assembly file.
9. Close P5.1.ipt without saving.
10. Open Windows Explorer, and locate P5.2.ipt.

Exercise 14.3 continued on next page

Exercise 14.3 continued

11. Drag P5.2.ipt from its location in Windows Explorer to the assembly button on the taskbar. Continue holding the left mouse button, with P5.2.ipt hovering over the assembly button, until the Autodesk Inventor window opens. Then drag and drop the part file into the assembly window.
12. The final exercise should look like the assembly shown.
13. Resave EX14.3.
14. Exit Autodesk Inventor or continue working with the program.

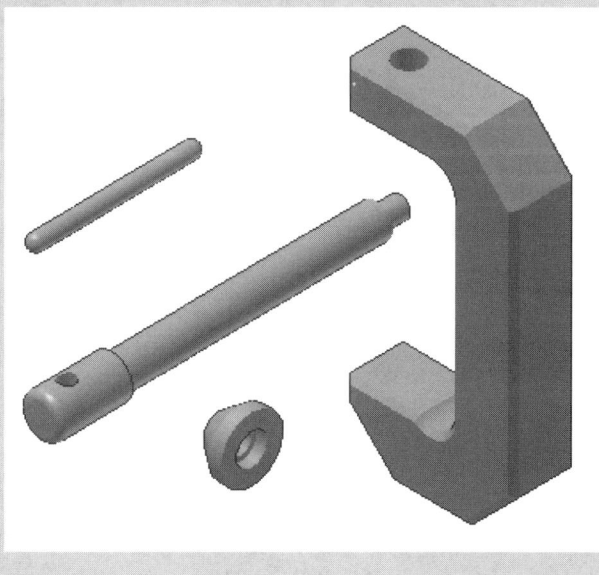

Inserting Shared Content

While you may create most assemblies using parts and subassemblies located on your computer or a local network, you can also access part files (.ipt) and SAT files using the Internet and Intranet. ***Shared content,*** also known as ***third-party content,*** consists of files available on the Internet, such as bolts from a bolt manufacturer or supplier, or parts accessible on an Intranet system, such as standard parts a company creates, and then used for developing assemblies. The shared content capabilities of Autodesk Inventor are typically used to generate image part libraries accessible to multiple users, to develop assemblies that require components available through the Internet or Intranet, and for companies located in multiple areas to share design information.

Shared, or third-party content, is also referred to as ***I-drop,*** because of the ability to locate .ipt and SAT files through an Internet or Intranet connection, then drag and drop the file into an assembly. There are a couple different options for using I-drop. One option is to open the assembly file into which you want to place the part. Then, using your Web browser, locate the Website that contains the .ipt or SAT files that you want to insert. Move your cursor over the bit map image of the desired part file and hold down the left mouse button to "fill" the I-drop icon. Now drag the file to the assembly button on the taskbar, and hold it there until the Autodesk Inventor assembly window opens, or drag the file into the opened Autodesk Inventor assembly window, if it is displayed. Finally, continue dragging the components into the graphics window, and drop them into place.

Another I-drop method is to open the assembly file into which you want to place the part. Then specify the shared content link, which is the location of the third-party files, in the **Shared Content Link** area of the **Assembly** tab in the **Options** dialog box, discussed earlier. The next step is to access the Place Content tool, using one of the following options:

✓ Pick the **Place Content** button on the **Assembly** panel bar.

✓ Pick the **Place Content** button on the **Assembly** toolbar.

✓ Select the **Shared Content...** option of the **Insert** pull-down menu.

When you select this tool, your Web browser automatically opens, and the Website specified in the **Shared Content Link** area is displayed. Next, located the desired .ipt or SAT files, move your cursor over the bit map image, and hold-down the left mouse button, to "fill" the I-drop icon. Now drag the file to the assembly button on the taskbar, and hold it there until the Autodesk Inventor assembly window opens, or drag the file into the opened Autodesk Inventor assembly window, if it is displayed. Finally, continue dragging the components into the graphics window, and drop them into place.

Exercise 14.4

1. Continue from EX14.3, or launch Autodesk Inventor.
2. Open a new inch assembly file.
3. Ensure "C:\Program Files\Autodesk\Inventor 5.3\Content\Ansi\index.htm" is the specified **Shared Content Link** in the **Assembly** tab of the **Options** dialog box.
4. Access the **Place Content** tool.
5. Reduce the size of the page to display the both the Inventor Content page and the assembly file.
6. Locate, and select "Shoulder Pattern 2-A – UNC (Regular Thread – Inch)" from the following directory: Fasteners/Screws and Threaded Bolts/Specialty Head Types/Eye Bolts.
7. Specify a nominal diameter of ¼ -20″, and a nominal length of 1″.
8. Move your cursor over the bit map image.
9. Hold down the left mouse button, to fill the I-drop icon.
10. Drag the file into the opened Autodesk Inventor assembly window. It should become grounded
11. Repeat the process using the following content files (you must select a location for the parts):

    (1) Fasteners/Nuts/Hex Nuts/Hex Nut – UNC (Regular Thread Inch) with a nominal diameter of Eye Bolts.
    (1) Fasteners/Washers/Plain/Type A – Plain Washer (Inch) with a nominal diameter of ¼″.
    (1) Fasteners/ Washers/Plain/Regular Helical Spring Lock Washer (Inch) with a nominal diameter of ¼″.

12. Close the Inventor Content page.
13. The final exercise should look like the assembly shown.
14. Save the assembly as EX14.4.
15. Exit Autodesk Inventor or continue working with the program.

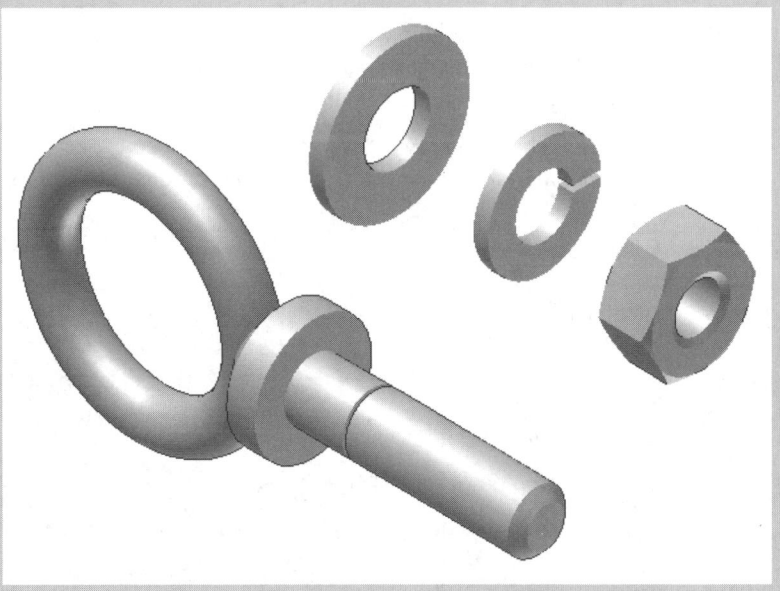

Developing Components While in an Assembly File

Previously in this chapter, you learned how to generate assemblies by inserting existing subassemblies (assembly files) and parts (part files) into a specified assembly file. As discussed, the other option for creating assemblies involves building parts inside the assembly file, also known as *in-place*. However, usually both assembly creation methods are used to develop more complex assemblies. Often standard base parts or subassemblies are individually built, then inserted into an assembly file. Other times entire, usually less complex assemblies are generated by developing all components inside a single assembly file. The advantage to creating components in-place is that you can quickly develop an assembly and analyze design intent as you create a product. In addition, when you build a part component inside an assembly file, a separate, fully parametric, individual part file (.ipt) is also generated.

In previous chapters you learned how to develop individual part models, using separate Autodesk Inventor part files. The process for creating parts in an assembly file is identical to creating parts in a part file, except for a few processes, discussed in this chapter. All tools and commands available in a part file are found in an assembly file, including the sketch, 3D sketch, features solids, and sheet metal tools. For more information regarding the part model tools and commands, refer to Chapters 1 through 11.

Using the Create Component Tool

If you choose to create assembly components while in an assembly file, the first tool you will use is the **Create Component** tool. When you develop a component in-place, the component is created in the assembly file, but also in a separate part or assembly file. As a result, the **Create Component** tool allows you to define the component and component file name, the type of component and component file (part or subassembly), the location of the new component file, and the template file used for creating the component. In addition, this command automatically opens the 2D sketch environment when developing a part, or the assembly work environment when generating subassemblies.

Access the **Create Component** tool using one of the following techniques:

✓ Pick the **Create Component** button on the **Assembly** panel bar.

✓ Pick the **Create Component** button on the **Assembly** toolbar.

✓ Select the **New Component...** option of the **Insert** pull-down menu.

When you access the **Create Component** tool, the **Create In-Place Component** dialog box, shown in Figure 14.19, is displayed. Use the following options in the **Create In-Place Component** dialog box to define the new component:

■ **New File Name**   Use this edit box to enter the name of the new component and the name of the additional component file. A default name is specified depending on the type of file you create, if you do not enter a different name.

■ **File Type**   This drop-down list allows you to specify the type of component you want to create, which corresponds to the type of new component file. Select Part to generate a part component and a part file, or Assembly to generate a subassembly component and assembly file.

FIGURE 14.19 The **Create In-Place Component** dialog box.

New File Name edit box

New File Location edit box

Template drop-down list

Constrain sketch plane to selected face or plane check box

File Type drop-down list

Browse buttons

OK button

■ **New File Location**   Use this edit box and the **Browse** button to specify the location of the new, separate component file.

■ **Template**   Use this drop-down list and the **Browse** button to identify the desired template file to use for the component file.

■ **Constrain sketch plane to selected face or plane**   When this check box is selected, you have the option of establishing a constraint between the sketch plane and the selected existing feature plane. This option applies only when creating a new component in an assembly file that already contains features. Clear this check box if you do not want to specify a constraint between an existing feature plane and the new component.

Once you have fully defined the options in the **Create In-Place Component** dialog box, pick the **OK** button to begin component creation.

Creating a Part In-Place

To create a part component, you must specify Part as the file type in the **Create In-Place Component** dialog box. If there are no other components in the assembly file, a new sketch is automatically opened on the plane specified in the **Part** tab of the **Options** dialog box, and you enter the 2D sketch environment. As with inserting components, the initial component is grounded. If there are other components present in the assembly file, you must pick a sketch location. You can choose to locate the new component sketch in relation to existing geometry by selecting a feature plane or a work plane. You can also select any of the default work planes available in the **Origin** folder, or choose to not relate the new sketch to existing features, by picking a location anywhere in the graphics window. Once you pick the desired sketch plane, the 2D sketch environment becomes active.

Field Notes

> If you position a sketch and part by selecting an existing component plane, a flush constraint is automatically placed between the new and existing components. If you do not want to automatically establish a constraint, pick the **Constrain sketch plane to selected face or plane** check box, previously discussed, or pick in a location in the graphics window, not related to any features, to identify the sketch plane.

From this point on, creating parts in an assembly file is almost identical to generating parts in a part file, as shown with the Browser display in Figure 14.20. Use the tools and commands discussed in Chapters 2 and 3 to develop your part sketch. When complete, finish the sketch to exit the sketch environment. Then use the tools and commands discussed in Chapters 4 through 11 to fully develop your part.

Field Notes

> If you want to reference existing features and components when creating a sketch for a new part, use the projection techniques discussed in Chapter 4.

When you create a component in-place, you are actually editing the existing assembly file by placing additional content in the form of assembly components. Once you have created the part, you must finish the assembly edit and reenter the assembly work environment using one of the following options:

✓ Right-click and select the **Finish Edit** option from the shortcut menu.

✓ Select the **Return** button on the **Command** bar.

✓ Double-click the parent assembly name in the Browser.

Once you reenter the assembly work environment you can insert or create additional components, establish constraints between components, or use any of the other tools available.

FIGURE 14.20 The
Browser display in an
assembly file named
Assembly8, and a new
part named BLOCK.

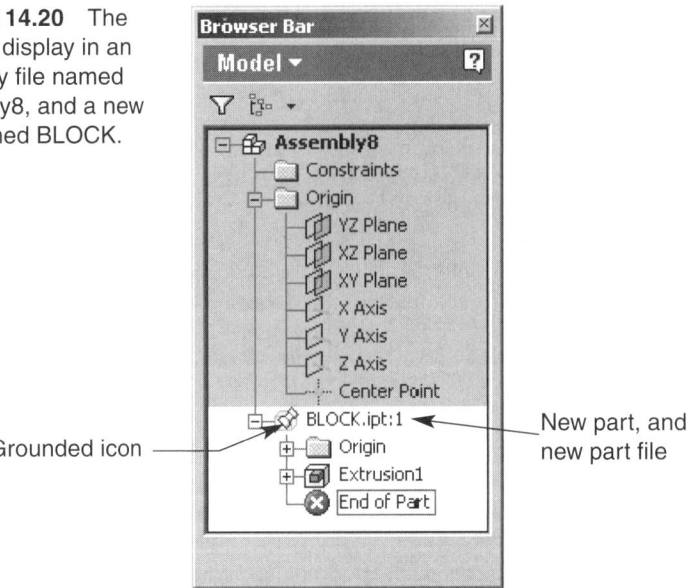

Grounded icon

New part, and
new part file

Exercise 14.5

1. Continue from EX14.4, or launch Autodesk Inventor.
2. Open a new inch assembly file.
3. Access the **Create Component** tool.
4. Specify a new file name of EX14.5.ipt, a part file type, and a location and template of your choice.
5. Pick the **OK** button.
6. Project the Center Point, and sketch a 6″×8″ rectangle on the XY Plane.
7. Extrude the sketch .5″ in a positive direction.
8. Open a sketch on the face shown, and sketch the Hole, Center Points.
9. Place .25″ through holes at the Hole, Center Points.
10. Add .5″ fillets at the edges shown.
11. Finish the edit.
12. The final exercise should look like the assembly shown.
13. Save the assembly as EX14.5.
14. Exit Autodesk Inventor or continue working with the program.

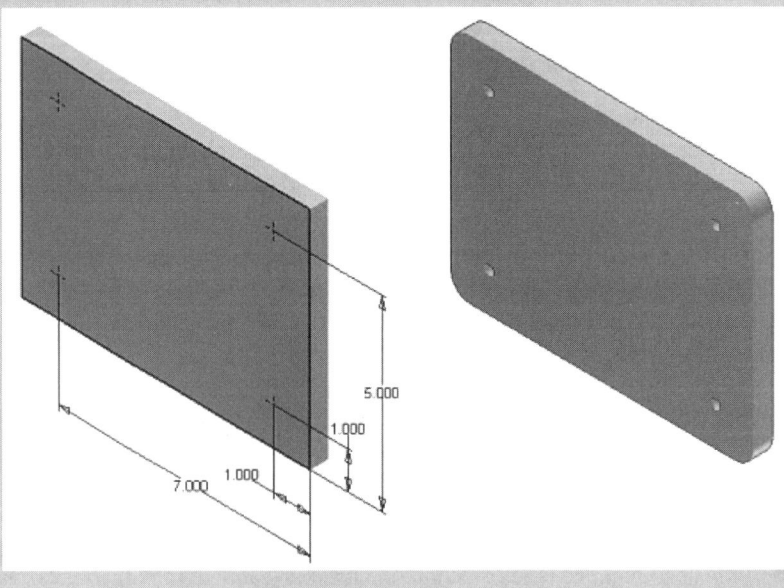

Creating a Subassembly In-Place

A *subassembly* is a group of two or more parts that are linked together to form a portion of a typically larger assembly. As discussed earlier in this chapter, subassemblies may include standard items like switches or spring assemblies, which may be used multiple times in the final assembly product. Like the two assembly creation techniques discussed in this chapter, subassemblies can be created in-place by inserting existing parts or assemblies, or by creating new parts with constraints. To create a subassembly component, you must specify Assembly as the file type in the **Create In-Place Component** dialog box. As shown with the Browser display in Figure 14.21, a secondary or subassembly is embedded into the assembly file and automatically becomes active.

Once you activate a new subassembly using the **Create Component** tool, the process for generating a subassembly is the same as for creating any other type of assembly. You can insert parts or subassemblies using the **Place Component** tool, or develop parts and even additional subassemblies in-place, by accessing the **Create Component** tool again. When you create a component inside a subassembly, or insert a component, as shown in the Browser display of Figure 14.22, the new component is part of the subassembly, not the overall assembly file.

Continue inserting subassembly components or creating subassembly components in-place using the techniques discussed. You must ensure the subassembly remains active while you are inserting and creating subassembly components. However, you can apply constraints, discussed later in this chapter, when the subassembly is active, or the parent assembly file is active. Once you have created the subassembly, you must finish the assembly edit and reenter the parent assembly work environment using one of the following options:

✓ Right-click and select the **Finish Edit** option from the shortcut menu.

✓ Select the **Return** button on the **Command** bar.

✓ Double-click the parent assembly name in the Browser.

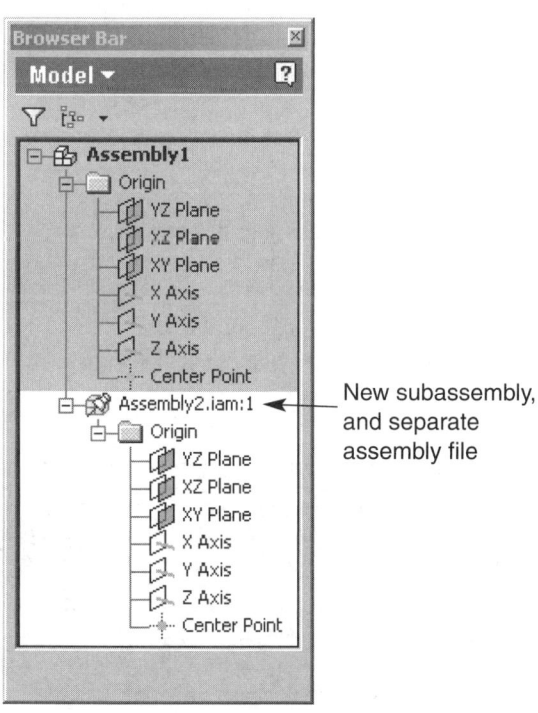

New subassembly, and separate assembly file

FIGURE 14.21   The Browser display in an assembly file named Assembly1, and a new subassembly named Assembly2.

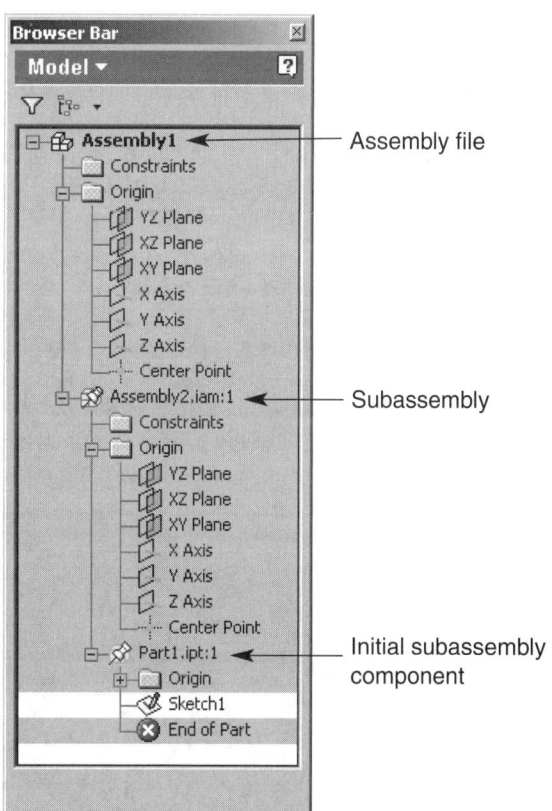

Assembly file

Subassembly

Initial subassembly component

FIGURE 14.22   The Browser display in an assembly file named Assembly1, with a new subassembly named Assembly2, and a part in the subassembly named Part1.

Once you reenter the assembly work environment you can insert or create additional components, establish constraints between components, or use any of the other tools available.

Another option for creating subassemblies is to insert or create components inside the parent assembly file, just like when you are creating an assembly. Then use one of the following techniques to place the components into a separate subassembly file:

✓ In the Browser, right-click the component you want to place into a subassembly, and select the **Dem<u>o</u>te** option from the shortcut menu.

✓ In the graphics window, right-click the actual component you want to place into a subassembly, and select the **Dem<u>o</u>te** option from the shortcut menu.

✓ Select the actual component in the graphics window, or the component name in the Browser, and press the **+** and **Tab** keys on your keyboard.

When you use this method for creating subassemblies, the **Create In-Place Component** dialog box opens, allowing you to specify the subassembly file name and folder location. When you pick **OK** to exit the **Create In-Place Component** dialog box, a subassembly is generated, and the component is automatically placed in the file.

Conversely, components can be removed from subassemblies and placed in the parent assembly using one of the following options:

✓ In the Browser, right-click the component you want to place into a subassembly, and select the **Promote** option from the shortcut menu.

✓ In the graphics window, right-click the actual component you want to place into a subassembly, and select the **Promote** option from the shortcut menu.

✓ Select the actual component in the graphics window, or the component name in the Browser, and press the **Shift, +,** and **Tab** keys on your keyboard.

You can also place components from the parent assembly file into an existing subassembly by dragging the component in the Browser from its current location to a location under the desired subassembly Origin folder. Similarly, place subassembly components into the parent assembly file by dragging the component from the subassembly and inserting it under the parent assembly Origin folder. If you place the component under the Origin folder of a subassembly it will be placed in the subassembly.

Field Notes

Dragging and dropping components into and out of assemblies and subassemblies can become confusing. To confirm the location of components, you may want to collapse all child nodes, then open and close individual subassemblies.

Use the **<u>F</u>ind** option available in the **Edit** pull-down menu, or by pressing the **Ctrl** and **F** buttons of your keyboard, to locate components in the Browser and in the graphics window.

Exercise 14.6

1. Continue from EX14.5, or launch Autodesk Inventor.
2. Open a new metric assembly file.
3. Access the **Create Component** tool.
4. Specify a new file name of EX14.6SUB.iam, an assembly file type, and a location and template of your choice.

Exercise 14.6 continued on next page

Exercise 14.6 continued

5. Pick the **OK** button.
6. Use the Place Component tool to insert the following components. The first part will automatically become grounded; all other parts should be placed in space and not reference other parts.

   P4.7.ipt
   P4.8.ipt
   P5.7.ipt

7. Finish the edit.
8. The final exercise should look like the assembly shown.
9. Save the assembly as EX14.6.
10. Exit Autodesk Inventor or continue working with the program.

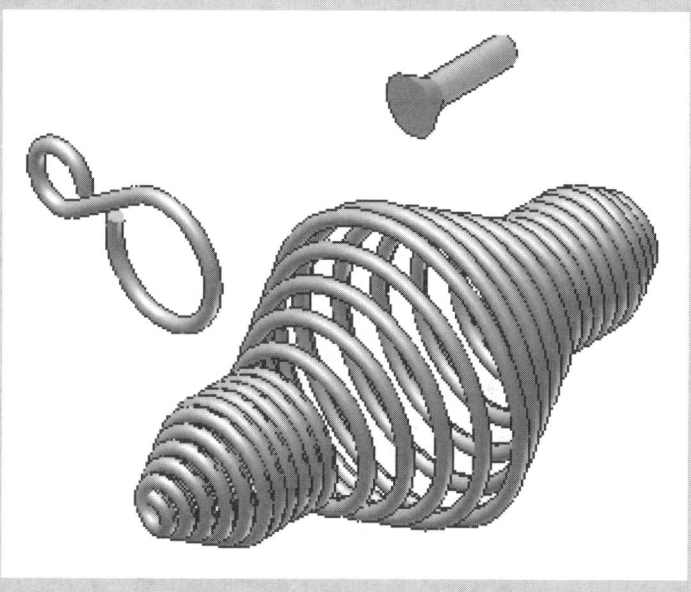

Applying Component Constraints

Once you have inserted or created assembly components, usually the next step in assembly development is to add constraints between components. As previously mentioned, multiple assembly components, including parts and subassemblies, are brought together, or linked, using constraints. In the same way that constraints are used to define relationships between geometric shapes in the part sketch environment, assembly constraints bring together individual and multiple components by parametrically assigning location specifications of each component in reference to other components. Assembly constraints remove a certain amount of component movement freedom. This means that constrained components are no longer free to move and be located anywhere in space. In addition, any movement freedom still existing can be driven, allowing you to examine design concepts.

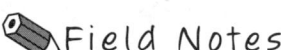Field Notes

> You can observe the degrees of freedom in an assembly by dragging components, or selecting the **Degrees of Freedom** option of the **View** pull-down menu.

All components can be constrained to other components in the assembly, assuming a constraint can be established between certain components and a preexisting relationship does not affect the constraint. Like all Autodesk Inventor tools, the types of constraints you use in an assembly are determined by the design. You have already observed one type of constraint, grounding. Grounded components have no movement freedom, are fixed in position, and cannot be driven. Any component can be grounded, but typically only the first, base component, or components that cannot move in reference to other components, are grounded. In addition to grounding, the following constraints are possible:

- **Assembly**   These constraints are the most widely used and establish geometric relationships and positions between one component face or edge and another component face or edge.

- **Motion**   This type of constraint defines the desired movement between one component and another, using a specified ratio and direction.

- **Transitional**   Use this constraint to identify relationships between the transitioning path of a fixed component and a component moving along the path.

The typical method for constraining most components is to use the **Place Constraint** tool. However, there are other techniques, such as dragging components, identifying iMates, and establishing constraints as you create components in an assembly file. Each of these constraining options is discussed in this chapter.

Placing Constraints

One of the most common techniques for constraining assembly components is to use the **Place Constraint** tool, which uses the typical Autodesk Inventor dialog box approach. The **Place Constraint** tool allows you to establish assembly, motion, and transitional constraints between components. In many situations you can access the **Place Constraint** tool once and fully constrain an entire assembly. However, you can use the tool anytime during assembly development. Access the **Place Constraint** tool using one of the following techniques:

✓ Pick the **Place Constraint** button on the **Assembly** panel bar.

✓ Pick the **Place Constraint** button on the **Assembly** toolbar.

✓ Select the **Constraint ...** option of the **Insert** pull-down menu.

✓ Type the **+** and **C** keys on your keyboard.

When you access the **Place Constraint** tool, the **Place Constraint** dialog box is displayed. See Figure 14.23. The first step in constraining components using the **Place Constraint** dialog box is to select the tab that corresponds to the type of constrains you want to create. As previously discussed you can choose to define assembly, motion, or transitional constraints by picking the corresponding tab. The following describes each of the tabs, and the process of using the various constraining techniques:

FIGURE 14.23 The **Place Constraint** dialog box (**Assembly** tab displayed, **Mate** constraint button selected).

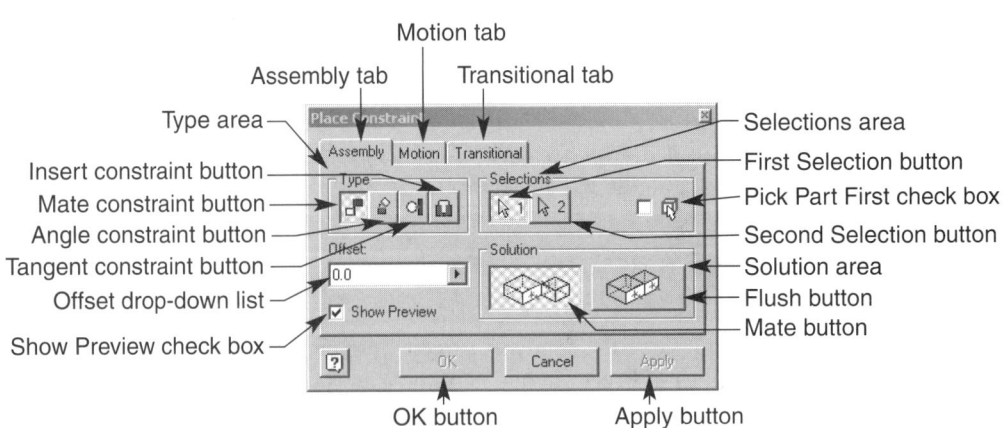

■ **Assembly**   This tab is active by default and allows you to place mate, angle, tangent, and insert constraints. The constraining options available in the **Assembly** tab represent the most commonly used constraints for most assembly applications. Almost all assemblies contain at least one of these types of constraints, but usually several. In addition, typically before you add motion or transitional constraints, you place assembly constraints. Choose one of the following assembly constraint type buttons from the **Type** area of the **Assembly** tab, and use the information provided to create the required constraint:

■ **Mate**   Select this button to establish a mate or flush constraint. *Mate* and *flush constraints* are common in many assemblies and are used to define a mate or flush relationship between the face, plane, axis, edge, or point of one component, and the face, plane, axis, edge, or point of another component. See Figure 14.24. When you pick the **Mate** button, the **Assembly** tab of the **Place Constraint** dialog box, shown in Figure 14.23, displays mate constraint options. Though you may develop your own approach to applying constraints, typically the first step is to identify the mate constraint solution, by selecting the **Mate** or **Flush** button in the **Solution** area, though you can change the solution any time during constraint development. A mate solution places two planes, co-planar to each other, as shown in Figure 14.25A, two axes collinear to each other, as shown in Figure 14.25B, two edges collinear to each other, as shown in Figure 14.25C, or two points matched together, as shown in Figure 14.25D. In contrast, a flush solution positions two faces along the same plane, facing the same direction. See Figure 14.26. Once you determine the mate solution you want to use, and pick the appropriate solution button, you can begin selecting component features to constrain. If you are working on a complex, or detailed assembly, you may find it helpful to pick the **Pick Part First** check box. When selected, the **Pick Part First** check box allows you to pick a part first, isolating the part from other assembly components, and making easier to choose specific geometry on the selected part. Once you select the part or if the **Pick Part First** check box is not active, you can select the required component object, including a face, edge, axis, or point. See Figure 14.27. By default the **First Selection** button is initially active, allowing you to select the first component element. Once you choose the first object on one component, the **Second Selection** button becomes active, allowing you to select the second object on the other component. Again, if the **Pick Part First** check box is selected, you must pick the second component before you can make the second component element selection. The final step in applying a mate constraint is to specify an offset between the selected components by entering or selecting a value from the **Offset** drop-down list. For many appli-

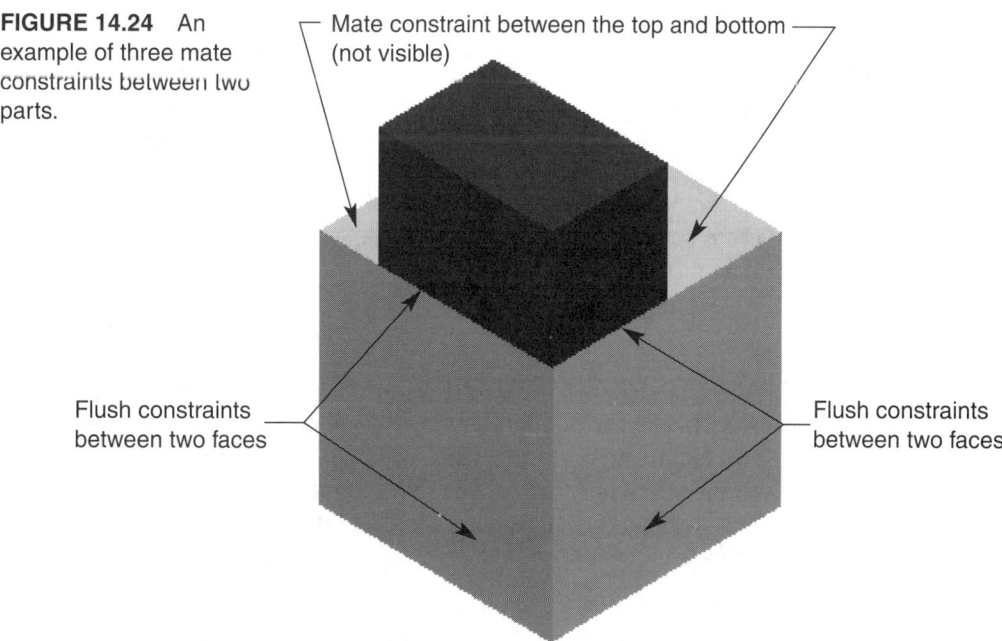

FIGURE 14.24   An example of three mate constraints between two parts.

Mate constraint between the top and bottom (not visible)

Flush constraints between two faces

Flush constraints between two faces

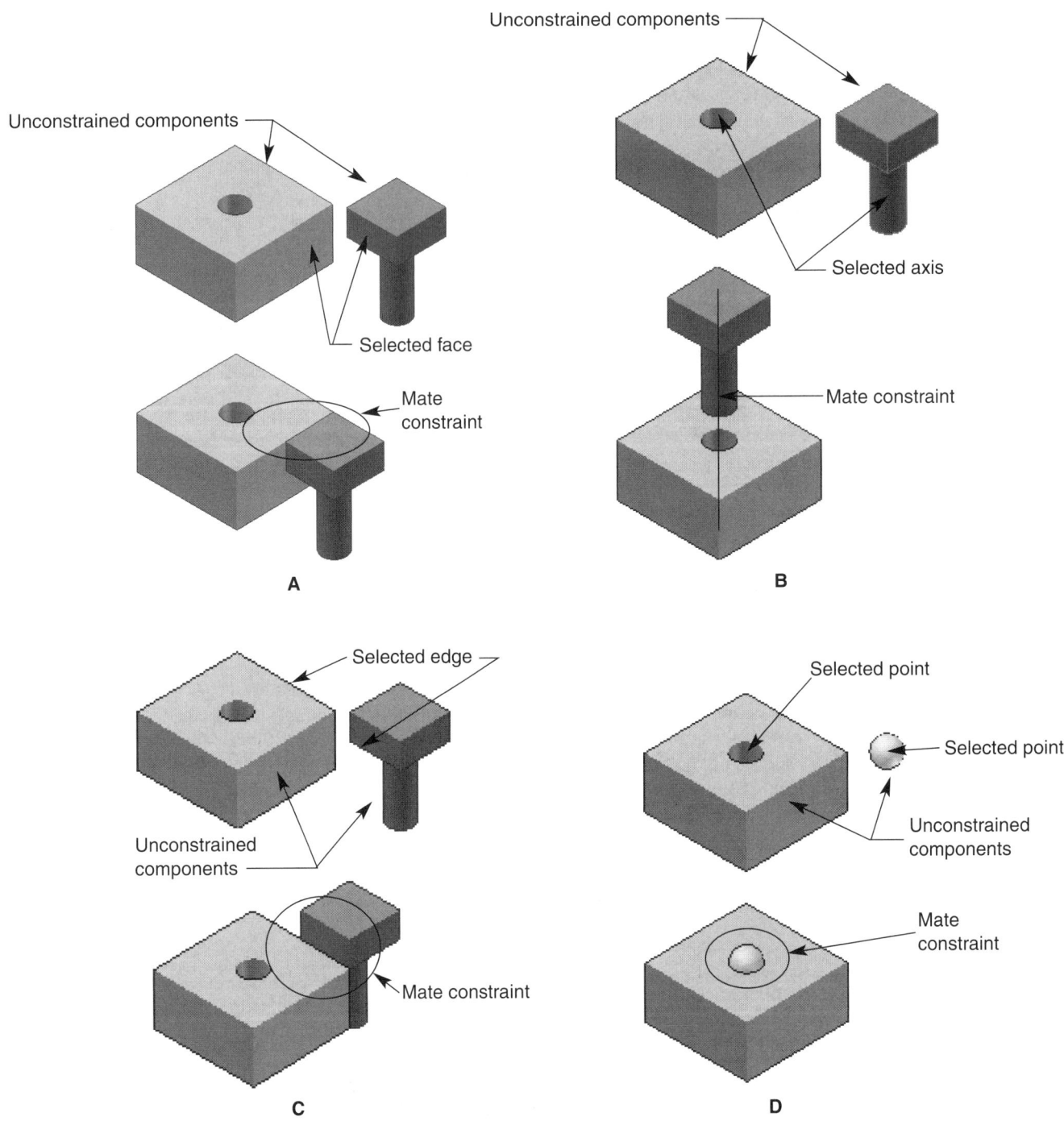

FIGURE 14.25   A mate constraint mate solution between (A) two planes, (B) two axes, (C) two edges, and (D) two points.

cations an offset of 0 is used to specify a contacting fit, with no offset between components. To preview the effects of the constraint, select the **Show Preview** check box. Then, if the constraint looks acceptable, pick the **Apply** button to generate the constraint. When you pick the **Apply** button, you remain in the **Place Constraint** command, allowing you to continue placing constraints. Pick the **OK** button when finished placing constraints, or to generate a constraint, and exit the **Place Constraint** command.

■ **Angle**   As the name implies, this constraint allows you to specify an angle between the face, axes, or edge of one component, and the face, axis or edge of another component.

FIGURE 14.26 A mate constraint flush solution between two faces.

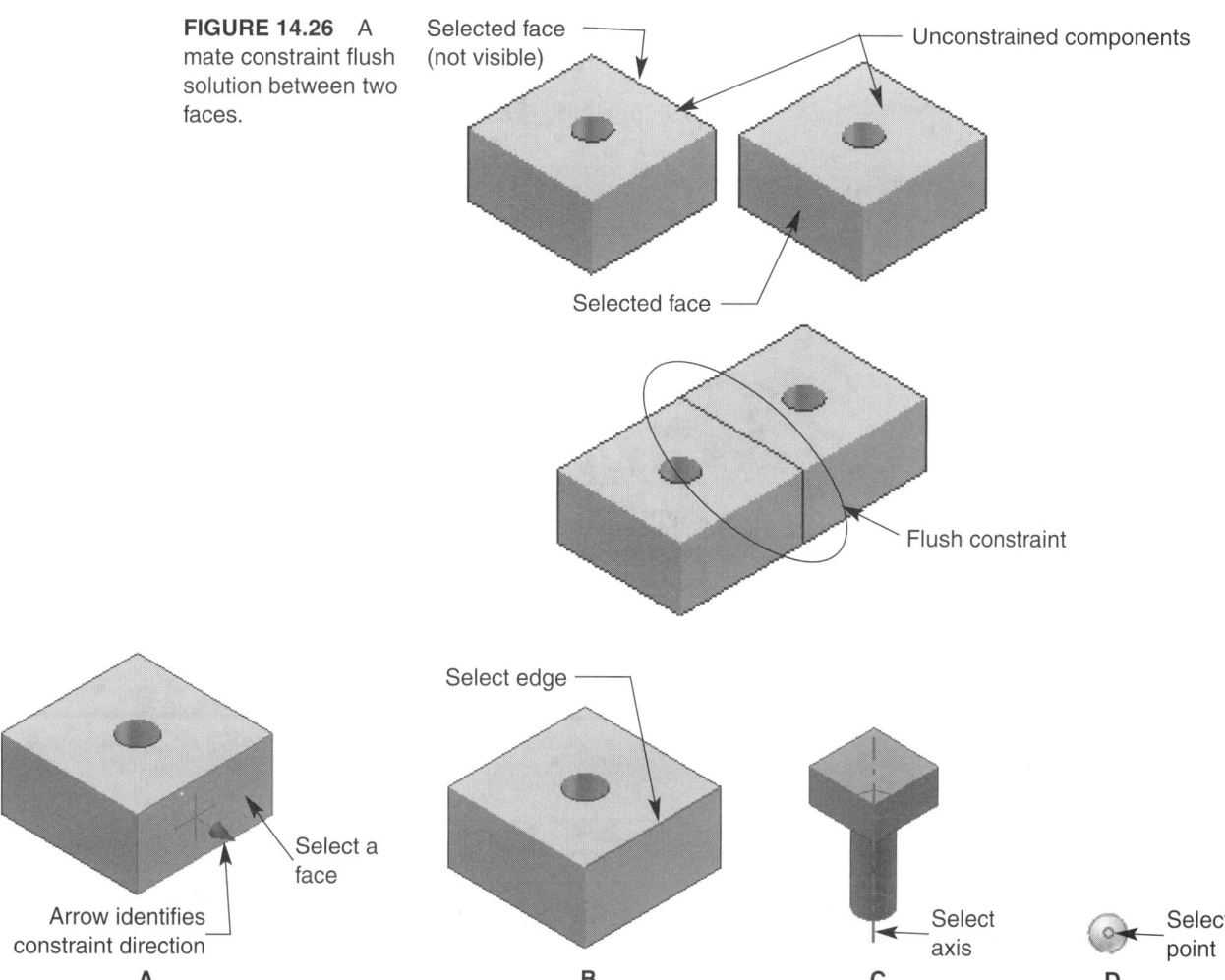

Selected face (not visible)

Unconstrained components

Selected face

Flush constraint

Select edge

Select a face

Arrow identifies constraint direction

Select axis

Select point

A

B

C

D

FIGURE 14.27 Selecting a component (A) face, (B) edge, (C) axis, and (D) point.

See Figure 14.28. When you pick the **Angle** button, the **Assembly** tab of the **Place Constraint** dialog box and angle constraint options are displayed. See Figure 14.29. Typically the first step in defining an angle constraint is to pick component geometry to constrain. If you are working on a complex or detailed assembly, you may find it helpful to pick the **Pick Part First** check box, which isolates the selected part from other assembly components and makes choosing specific features on the specified part easier. Once you select the part, or if the **Pick Part First** check box is not active, you can select component elements including faces, edges, and axes. See Figure 14.30. By default the **First Selection** button is initially active, allowing you to select the first component object. Once you choose the first element on one component, the **Second Selection** button becomes active, allowing you to select the second object on the other component. Again, if the **Pick Part First** check box is selected, you must pick the second component before you can make the second feature selection. Any combination of faces, edges, or axes can be selected depending on design intent. Figure 14.30 shows the effects of creating a 0° angle when you select the two objects shown, while Figure 14.31 displays the effects of creating a 0° angle when typical object combinations are selected. To preview the effects of the constraint, select the **Show Preview** check box. If the constraint does not look acceptable, you may need to rotate the first selection 180° by picking the **Flip the first selection** button in the **Selection** area, and/or the rotate the second selection 180° by picking the **Flip the second selection** button in the **Selection** area. The final step in applying an angle constraint is to specify an angle between the selected components. For some applications an offset of 0° is used, such as for driving purposes, discussed later in this chapter. How-

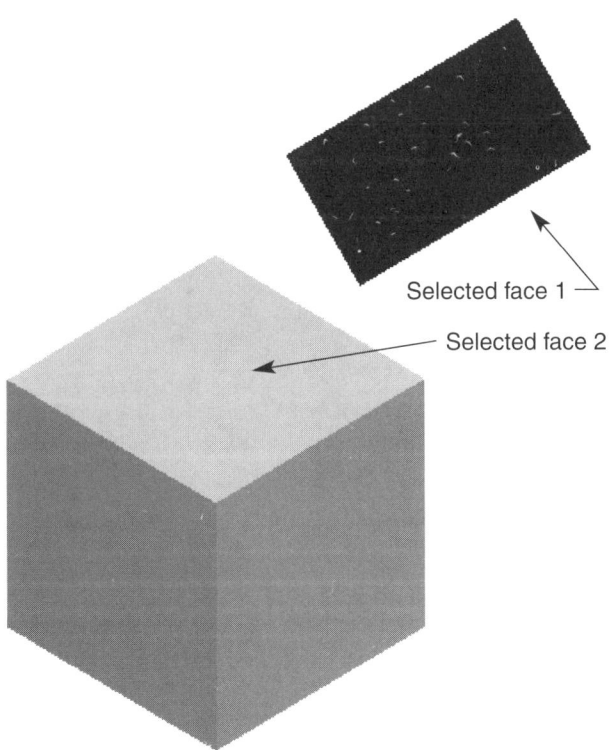

FIGURE 14.28 An example of a 45° angle constraint between two selected faces.

Selected face 1
Selected face 2

FIGURE 14.29 The **Place Constraint** dialog box (**Assembly** tab displayed, **Angle** constraint button selected).

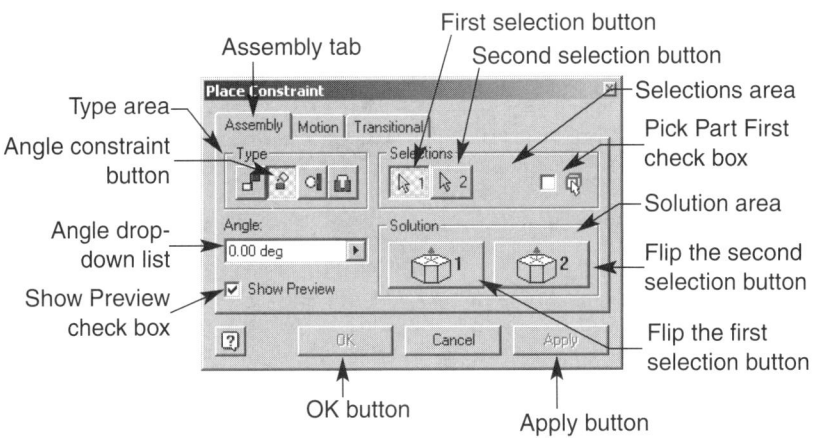

Assembly tab
First selection button
Second selection button
Type area
Selections area
Angle constraint button
Pick Part First check box
Solution area
Angle drop-down list
Flip the second selection button
Show Preview check box
Flip the first selection button
OK button
Apply button

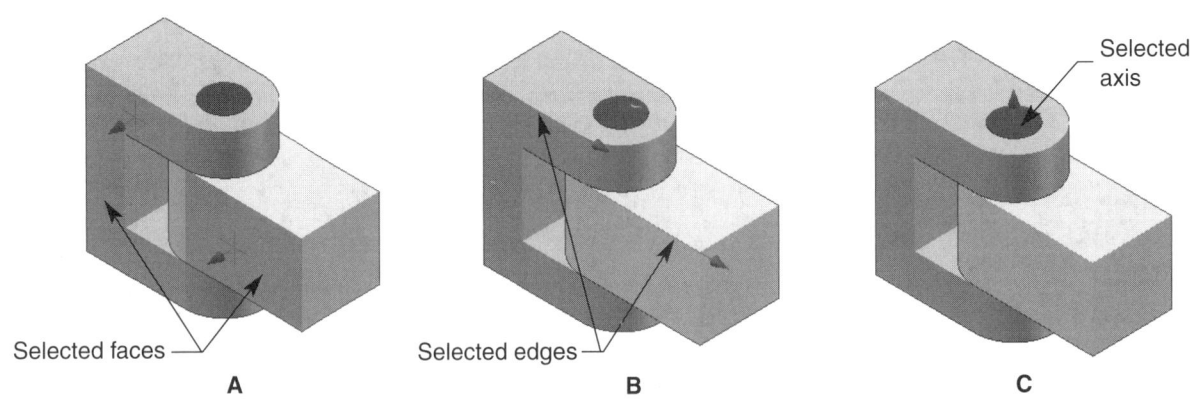

Selected axis
Selected faces
Selected edges
A
B
C

FIGURE 14.30 Selecting component (A) faces, (B) edges, and (C) axes.

FIGURE 14.31 (A) Selecting a component face and a component edge. Notice the preview arrows are pointed in the same direction, identifying an angle of 0°. (B) Selecting component edge and a component axis.

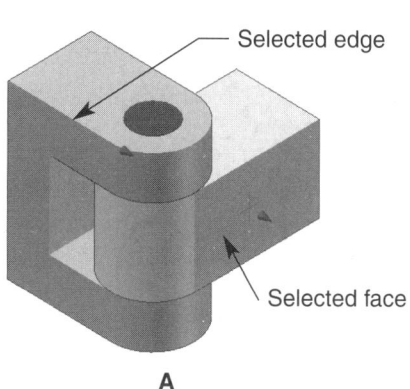
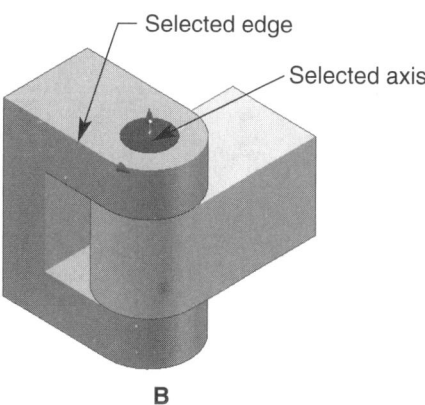

Selected edge

Selected face

Selected edge

Selected axis

A

B

ever, often a specific angle is specified to incorporate design intent. If the constraint looks acceptable, pick the **Apply** button to generate the constraint. When you pick the **Apply** button, you remain in the **Place Constraint** command, allowing you to continue placing constraints. Pick the **OK** button when finished placing constraints, or to generate a constraint, and exit the **Place Constraint** command.

 Field Notes

The **Flip the first selection** and **Flip the second selection** buttons may automatically become active depending on the component elements you select. As in all assembly component constraint selection options, there are multiple ways to identify component objects for constraining. In addition, the available options are multiplied depending on the selected, or unselected, flip buttons. You need to determine which component and button selections are needed depending on the application and design intent.

■ **Tangent** Use this constraint to define a relationship between one component face or edge and the tangency of a curved component feature face, such as a cone, cylinder, sphere, or other round face. You can also establish a constraint between two curved component feature faces. See Figure 14.32. When you pick the **Tangent** button, the **Assembly** tab of the **Place Constraint** dialog box displays tangent constraint options. See Figure 14.33. Typically, the first step in establishing a tangent constraint is to identify the tangent

FIGURE 14.32 An example of an inside and outside tangent constraint.

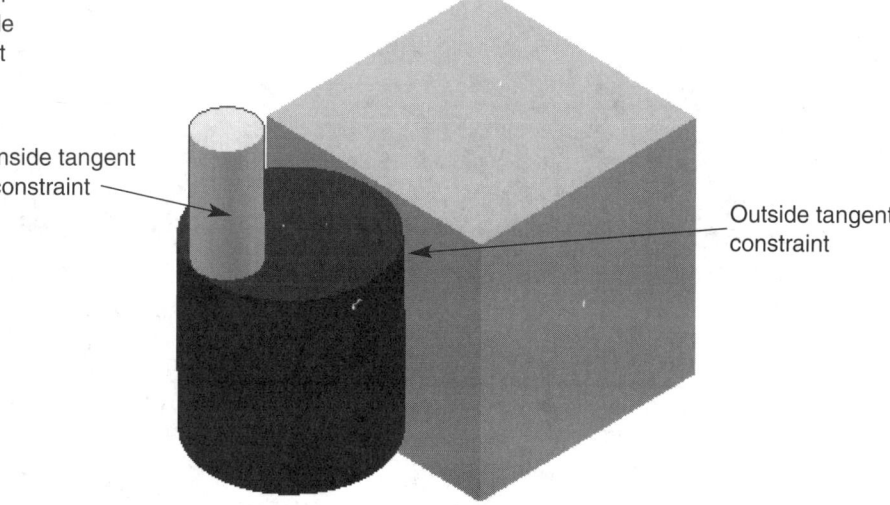

Inside tangent constraint

Outside tangent constraint

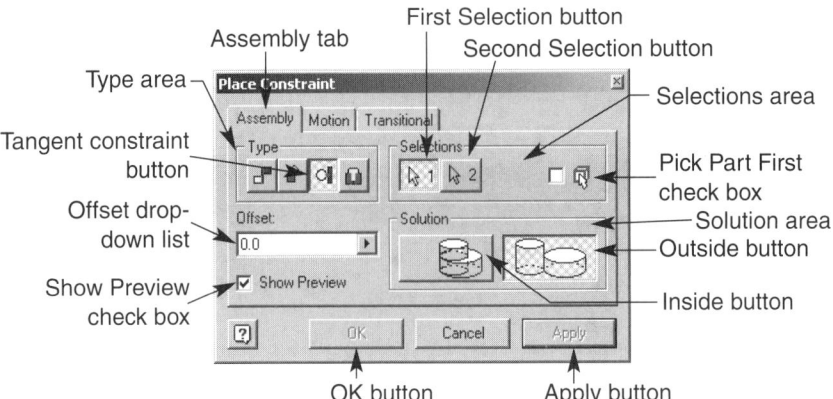

constraint solution by selecting the **Inside** or **Outside** button in the **Solution** area, although you can change the solution any time during constraint development. An inside solution places the component you select first, inside the component you select second, as shown in Figure 14.34A. In contrast, a flush solution positions the first component you select, outside of the second component selected, as shown in Figure 14.34B. Once you determine the tangent solution you want to use and pick the appropriate solution button, you can begin selecting faces or edges to constrain the components. If you are working on a complex or detailed assembly, you may find it helpful to pick the **Pick Part First** check box, which isolates the selected part from other assembly components and makes choosing specific features on the specified part easier. Once you select the part, or if the **Pick Part First** check box is not active, you can select the required component objects. By default the **First Selection** button is initially active, allowing you to select the first component element. You can choose either an edge, planar face, or a curved face for the first selection. If the first selection is an edge or planar face, the second selection must be a tangent face. Similarly, if the second selection is an edge or planar face, the first selection must be a tangent face. See Figure 14.35. Once you choose the first object on one component, the **Second Selection** button becomes active, allowing you to select the second object on the other component. Again, if the **Pick Part First** check box is selected, you must pick the second component before you can make the second component element selection. As previously discussed, the first component you select moves to constrain the second component selected, which does not move. The final step in applying a tangent constraint is to specify an offset between the selected components by entering or selecting a value from the **Offset** drop-down list. For many applications an offset of 0 is

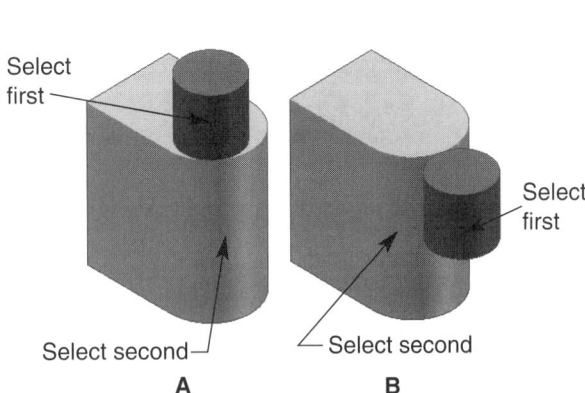

FIGURE 14.34 An example of (A) an inside tangent constraint solution and (B) an outside tangent constraint solution.

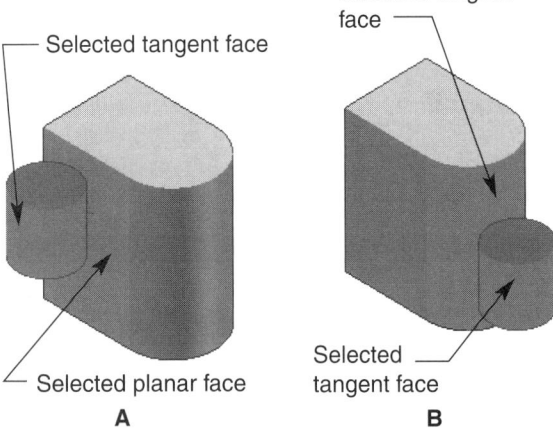

FIGURE 14.35 Selecting (A) a planar face and a tangent edge for an outside tangent constraint and (B) a tangent edge and a tangent edge for an outside tangent constraint.

used to specify a contacting fit, with no offset between components. To preview the effects of the constraint, select the **Show Preview** check box. Then, if the constraint looks acceptable, pick the **Apply** button to generate the constraint. When you do this you remain in the **Place Constraint** command, allowing you to continue placing constraints. Pick the **OK** button when finished placing constraints, or to generate a constraint, and exit the **Place Constraint** command.

■ **Insert**   Use this constraint to establish a relationship between two cylindrical objects, such as a hole and a shaft. An insert constraint is actually like defining both an axis-to-axis and face-to-face mate constraint. See Figure 14.36. Using an insert constraint is an easy way to link two components that may require both an axis-to-axis and face-to-face mate constraint, such as the constraints typically found between a screw and a hole. In this example, a constraint is established between the axis of the screw and the axis of the hole, and the planar face of the part and the bottom of the screw head. When you pick the **Insert** button, the **Assembly** tab of the **Place Constraint** dialog box displays the insert constraint options. See Figure 14.37. Typically the first step in defining an insert constraint is to pick component geometry. If you are working on a complex or detailed assembly, you may find it helpful to pick the **Pick Part First** check box, which isolates the selected part from other assembly components and makes choosing specific features on the specified part easier. Once you select the part, or if the **Pick Part First** check box is not active, you can select components to constrain. See Figure 14.38. By default the **First Selection** button is initially active, allowing you to select the first component. Once you choose the first component, the **Second Selection** button becomes active, allowing you to select the second component. Again, if the **Pick Part First** check box is selected, you must pick the second component before you can make the second selection. An insert selection picks both the axis and the specified edge of a component. As shown in Figure 14.39, you must select the corresponding edges on both components to ensure the proper mate constraint is generated between components. To preview the effects of the constraint, select the **Show Preview** check box. If the constraint does not look acceptable, you may need to flip the direction of the inserted component by picking the **Aligned** button in the **Selection** area. Many insert constraints use an **Opposed** solution, as shown in Figure 14.39. However, other insert constrains require an aligned solution, as shown in Figure 14.36, which aligns the specified edges and axes. The next step in applying

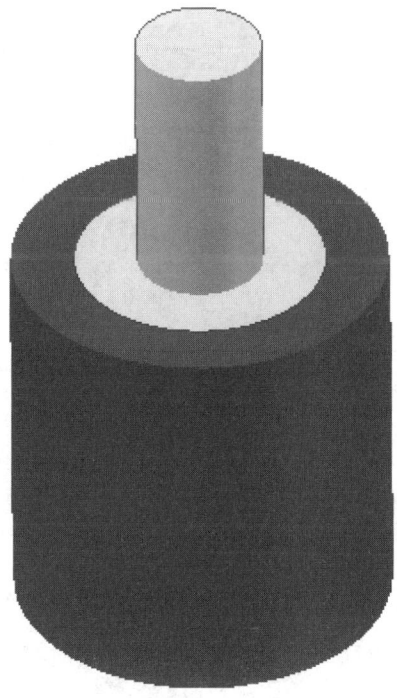

FIGURE 14.36   An example of an insert constraint.

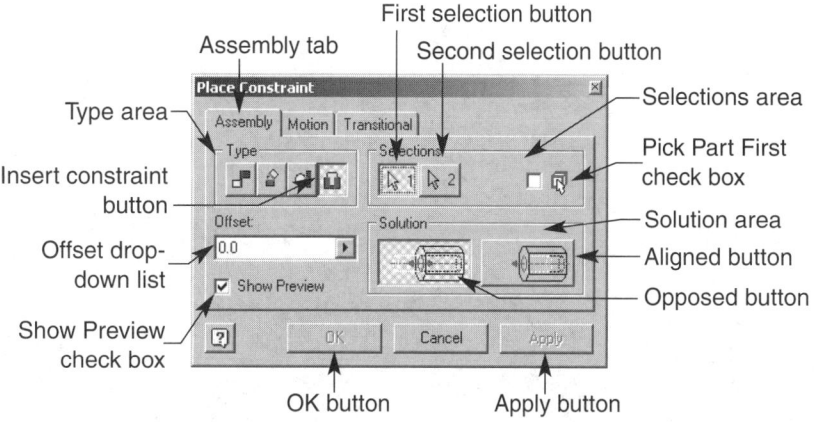

FIGURE 14.37   The **Place Constraint** dialog box (**Assembly** tab displayed, **Insert** constraint button selected).

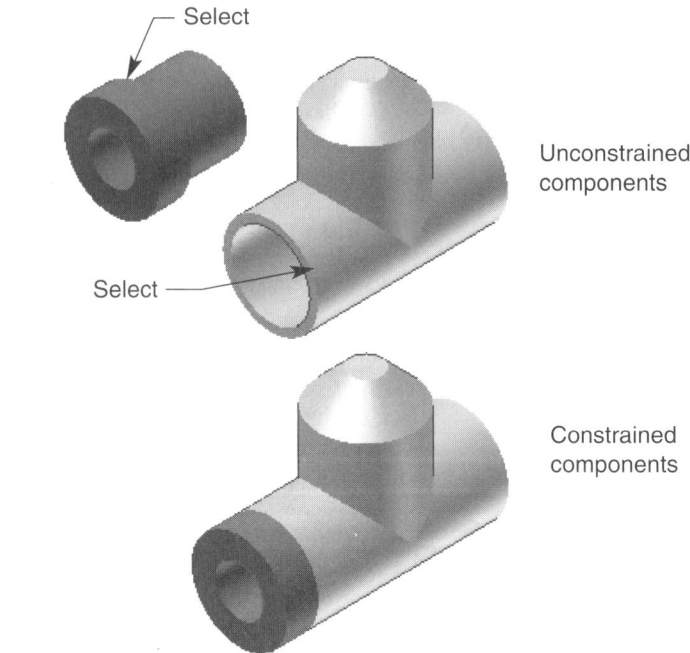

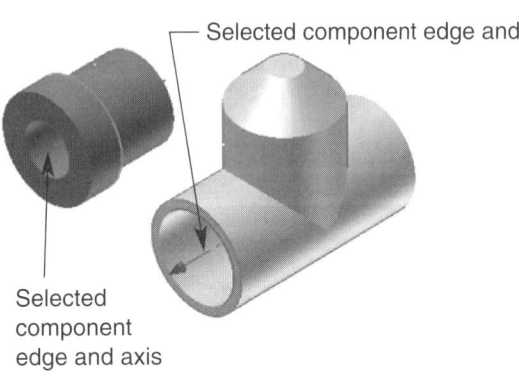

FIGURE 14.38   Selecting components to apply an insert constraint.

FIGURE 14.39   Selecting components to apply an insert constraint.

an insert constraint is to specify an offset between the selected components by entering or selecting a value from the **Offset** drop-down list. For many applications an offset of 0 is used to specify a contacting fit, with no offset between components. Finally, to generate the constraint pick the **Apply** button. When you pick the **Apply** button, you remain in the **Place Constraint** command, allowing you to continue placing constraints. Pick the **OK** button when finished placing constraints, or to generate a constraint, and exit the **Place Constraint** command.

Exercise 14.7

1. Continue from EX14.6, or launch Autodesk Inventor.
2. Open EX14.3.
3. Save a copy of EX14.3 as EX14.7.
4. Close EX14.3 without saving, and open EX14.7.
5. Create the constraints shown using the selections, and specifications:

   (1) Constraint: Assembly
   Type: Mate
   Offset: 25mm
   Solution: Mate

   (2) Constraint: Assembly
   Type: Mate
   Offset: 0mm
   Solution: Mate

   (3) Constraint: Assembly
   Type: Mate
   Offset: 0mm
   Solution: Mate

Exercise 14.7 continued on next page

Exercise 14.7 continued

(4) Constraint: Assembly
Type: Angle
Angle: 0°
Solution: Both solution buttons unselected (see arrows)

(5) Constraint: Assembly
Type: Tangent
Offset: 15mm
Solution: Outside

(6) Constraint: Assembly
Type: Insert
Offset: 0mm
Solution: Opposed

6. The final exercise should look like the assembly shown in 6.
7. Resave EX14.7.
8. Exit Autodesk Inventor or continue working with the program.

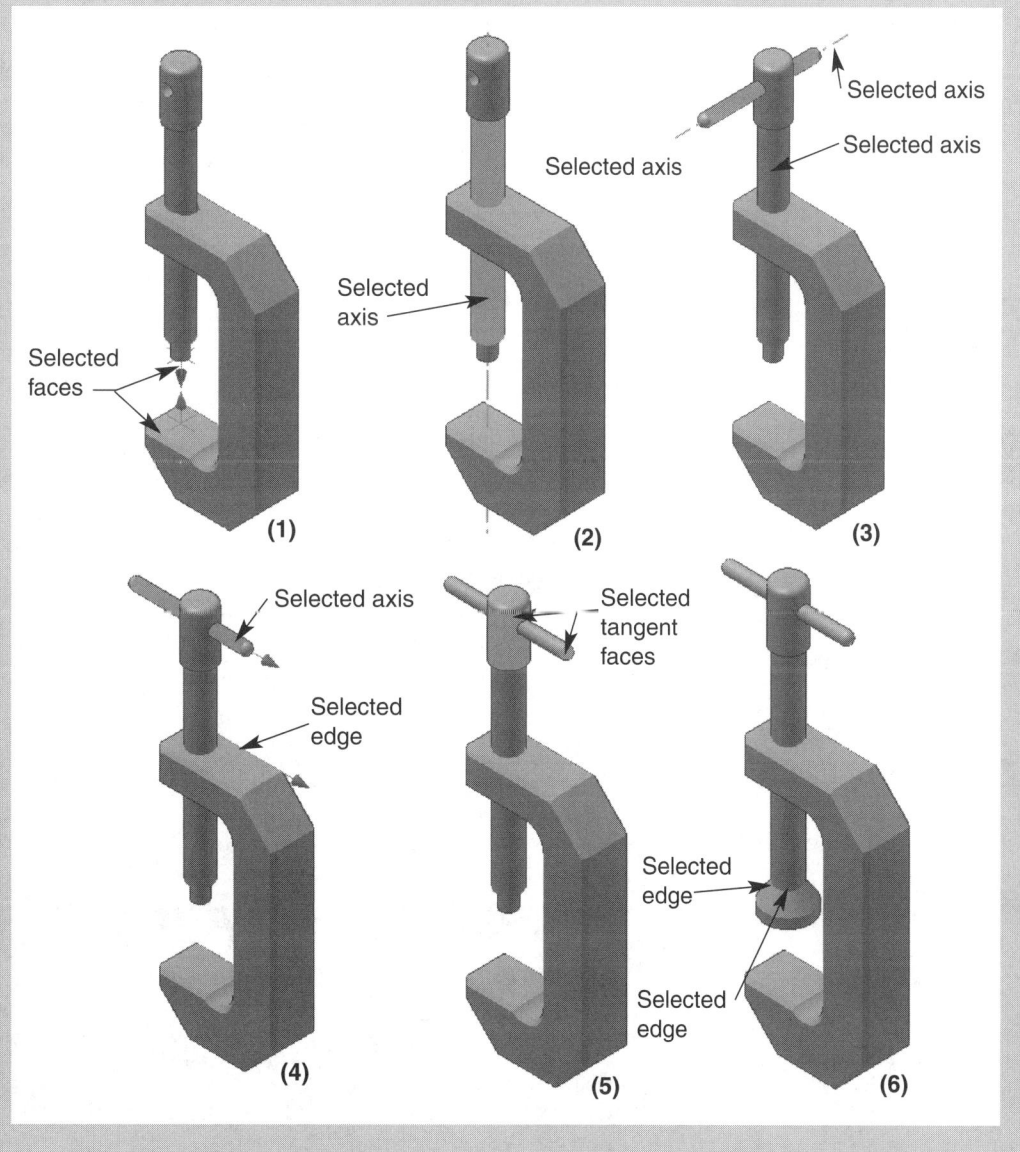

■ **Motion**   This tab, shown in Figure 14.40, contains options that allow you to define the desired movement between one component and another, using a specified ratio and direction. Only components that rotate and move together, such as gears, pulleys, racks and pinions, bearings, and other multiple rotating components, can contain motion constraints. Motion constraints do not define the location or position of components and do not fix components in space. The purpose of motion constraints is to identify how movable components should move in reference to other movable components. For example, a motion constraint can identify whether two gears move in the same direction or opposite directions, and specify the ratio of rotation, such as 1:1. Consequently, you must use assembly constraints, previously discussed, to establish all other types of constraints, such as a flush relationship between the tops of two gears. Assembly constraints are usually added before motion constraints. In addition, you may need to suppress or drive assembly constraints that do not allow for movement in order to observe a motion constraint. Driven components are discussed later in this chapter.

Choose one of the following motion constraint type buttons from the Type area of the Assembly tab, and use the information provided to create the required constraint:

■ **Rotation**   Select this button to establish a direction of rotation and a rotation ratio for two revolving components, such as gears, pulleys, and wheels. When the **Rotation** button is selected, the **Motion** tab of the **Place Constraint** dialog box contains options that allow you to define a rotation motion constraint. See Figure 14.40. As with other constraining options, if you are working on a complex or detailed assembly, you may find it helpful to first pick the **Pick Part First** check box to isolate the selected part from other assembly components. Once you select the part, or if the **Pick Part First** check box is not active, you can select the required component objects including edges, axes, planar faces, and tangent faces. By default the **First Selection** button is initially active, allowing you to select the first component element. Once you choose the first object on one component, the **Second Selection** button becomes active, allowing you to select the second object on the other component. Again, if the **Pick Part First** check box is selected, you must pick the second component before you can make the second component element selection. Selecting certain component elements is not critical when applying motion constraints, because motion constraints define the position of components in reference to other components. As a result, Figure 14.41 shows a couple examples of selection options that create the same constraint between two gears. However, the pattern of selection does define the specified ratio. So, if you specify a 3:1 ratio, the first component you select should be the component that revolves once for every three times the second component rotates. The next step in defining a rotation motion constraint is to specify the rotation solution. By default, the **Forward** solution button is selected, which creates a forward motion between the selected components. A forward motion allows one component to rotate in the same direction as the other component. The arrows shown in Figure 14.41 identify the rotation direction. To reverse the rotation, pick the **Reverse** solution button. A reverse motion allows one component to rotate in the opposite direction of the other component. The final step in applying a rotation motion constraint is to specify a rotation ratio

FIGURE 14.40 The **Place Constraint** dialog box (**Motion** tab displayed, **Rotation** constraint button selected).

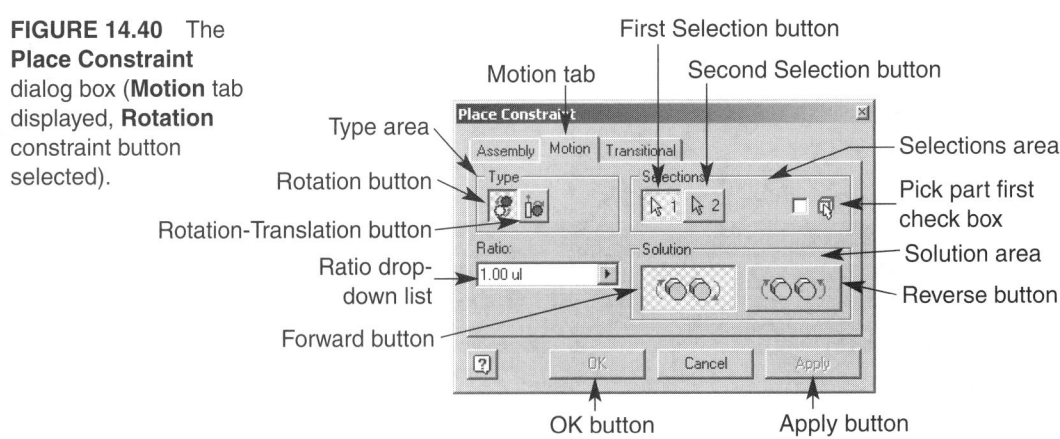

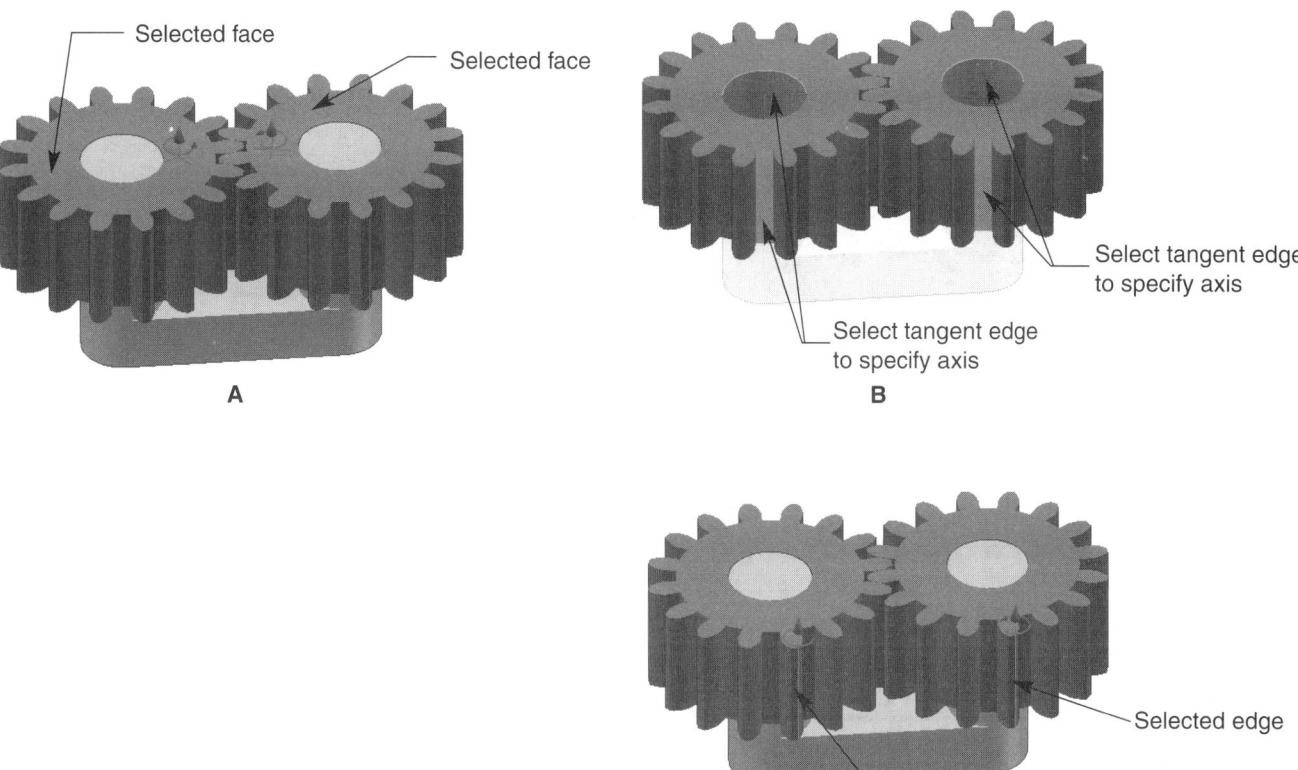

FIGURE 14.41 Selecting (A) two planar faces, (B) two axes (gear bracket enabled for clarity), and (C) two edges.

between the selected components by entering or selecting a value from the **Ratio** drop-down list. The specified ratio defines how much the second selected component rotates in reference to the first component selected. For example, if you apply a rotation constraint between the same-size gears, as shown in Figure 14.41, a ratio of 1:1, or 1 in the **Ratio** drop-down list, is appropriate. If the first selected gear, wheel, or pulley is half the size of the second component and rotates twice for every single rotation of the second component, a ratio of 1:2, or .5 in the **Ratio** drop-down list, is used. To generate the constraint, pick the **Apply** button. Again, when you pick the **Apply** button, you remain in the **Place Constraint** command, allowing you to continue placing constraints. Pick the **OK** button when finished placing constraints, or to generate a constraint, and exit the **Place Constraint** command.

- **Rotation-Translation** Select this button to establish a direction of rotation and a rotation distance between a rotating component and a translating component, such as racks and pinions, or wheels and part faces. When the **Rotation-Translation** button is selected, the **Motion** tab of the **Place Constraint** dialog box contains options that allow you to define a rotation-translation motion constraint. See Figure 14.42. As with other constraining options, if you are working on a complex or detailed assembly, you may find it helpful to first pick the **Pick Part First** check box to isolate the selected part from other assembly components. Once you select the part, or if the **Pick Part First** check box is not active, you can select the required component objects including edges, axes, planar faces, and tangent faces. By default the **First Selection** button is initially active, allowing you to select the first component element. Once you choose the first object on one component, the **Second Selection** button becomes active, allowing you to select the second object on the other component. Again, if the **Pick Part First** check box is selected, you must pick the second component before you can make the second component element selection. Selecting certain component elements is not critical when applying motion constraints, because motion constraints define the position of components in reference to other components. As a result, Figure 14.43 shows two examples of selection options that create the

FIGURE 14.42 The **Place Constraint** dialog box (**Motion** tab displayed, **Rotation-Translation** constraint button selected).

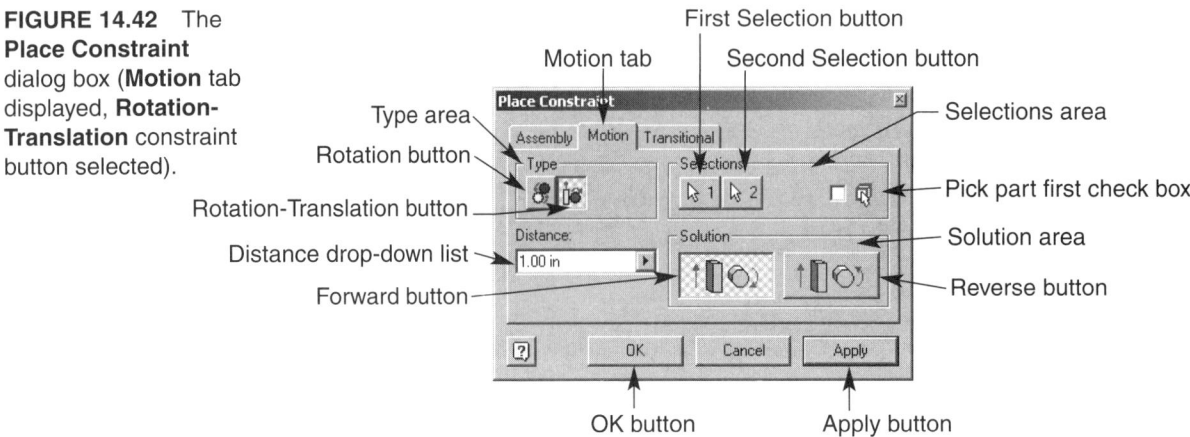

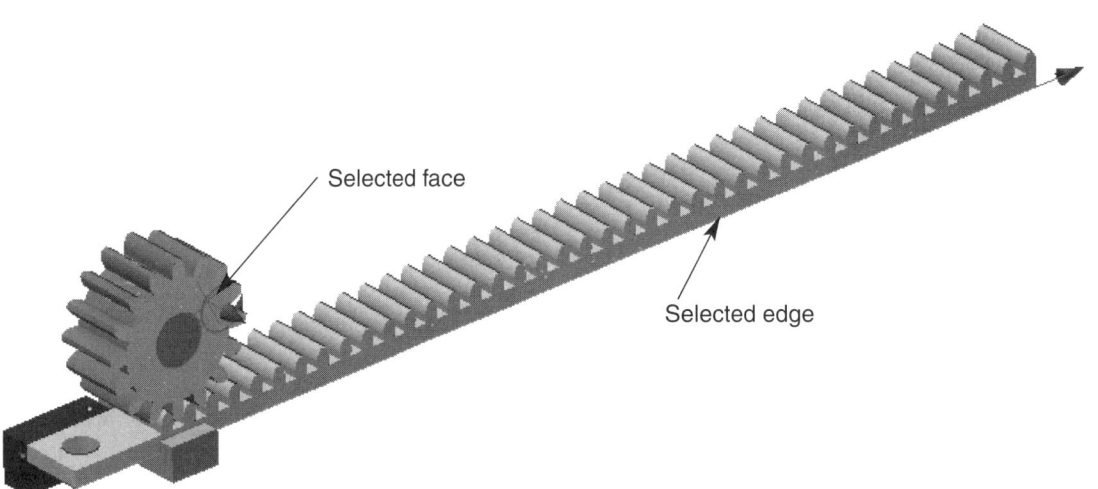

FIGURE 14.43 Selecting a planar face and an edge.

same constraint between rack and pinion assemblies. However, the pattern of selection does define the specified distance. So if you specify a distance of 3″, the first component you select will revolve 360° for every 3″ the second component moves. The next step in defining a rotation-translation motion constraint is to specify the rotation-translation solution. By default, the **Forward** solution button is selected, which creates a forward motion between the selected components. A forward motion allows the rotating component to rotate in the same direction as the sliding component. The arrows shown in Figure 14.43 identify the rotation direction. To change the component motion to a reverse rotation, pick the **Reverse** solution button. A reverse motion allows the rotating component to rotate in the opposite direction of the sliding component. The final step in applying a rotation-translation motion constraint is to specify a rotation-translation distance between the selected components by entering or selecting a value from the **Distance** drop-down list. The specified ratio defines how far the sliding component, second selection, moves in reference to the rotating component, first selection. For example, if you apply a rotation-translation constraint of 3″, as previously mentioned, for every full rotation (360°) the first component revolves, the second component moves 3″. In the same respect a distance of 0 does not allow the second component to move in reference to the revolution of the first component. To generate the constraint, pick the **Apply** button. When you do this, you remain in the **Place Constraint** command, allowing you to continue placing constraints. Pick the **OK** button when finished placing constraints, or to generate a constraint, and exit the **Place Constraint** command.

 Field Notes

Exercises using motion constraints are presented later in this chapter.

■ **Transitional** This tab, shown in Figure 14.44, contains options that allow you to define the desired movement between a moving part and a path. The clearest example of a transitional constraint is between a moving cylinder and a stationary route, such as a roller and a cam. See Figure 14.45. By selecting two planar or tangent edges, a transitional constraint links faces together, applying constant contact between two components. Transitional constraints help define the location or position of components by adding a type of mate constraint between faces, but do not fix components in space. The purpose of a transitional constraint is to identify how a movable component should follow a certain path. Consequently, you must use assembly constraints, previously discussed, to establish all other types of constraints, such as the flush relationship between the top of the roller and the top of the cam in Figure 14.45. Assembly constraints are usually added before transitional constraints and may need to be suppressed or driven in order to show a transitional constraint. Driven components are discussed later in this chapter.

To establish a transitional constraint, as with other constraining options, you may find it helpful to first pick the **Pick Part First** check box. Use this option if you are working on a

FIGURE 14.44 The **Place Constraint** dialog box (**Transitional** tab displayed).

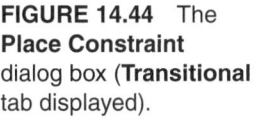
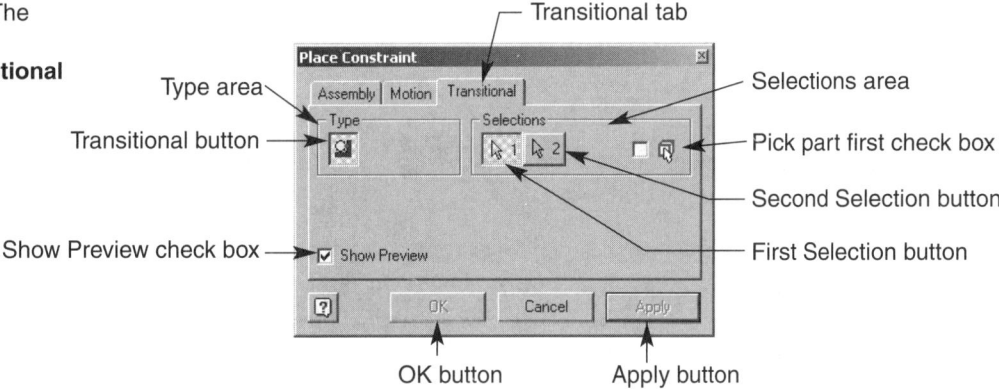

FIGURE 14.45 An example of a transitional constraint application, where the roller moves along the faces of the cam.

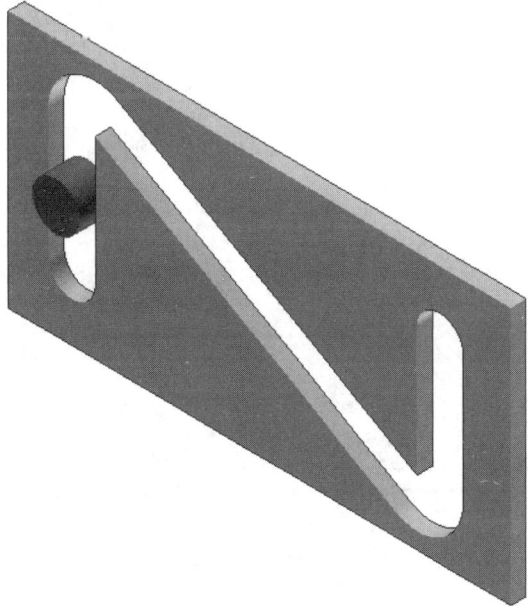

complex or detailed assembly in order to isolate the selected part from other assembly components. Once you select the part, or if the **Pick Part First** check box is not active, you can select the required component objects including planar or tangent faces. By default the **First Selection** button is initially active, allowing you to select the first component element. Once you choose the first object on one component, the **Second Selection** button becomes active, allowing you to select the second object on the other component. Again, if the **Pick Part First** check box is selected, you must pick the second component before you can make the second component face selection. The first component you select moves to constrain the second component selected, which does not move. To preview the effects of the constraint, select the **Show Preview** check box. Then, if the constraint looks acceptable, pick the **Apply** button to generate the constraint. When you do this, you remain in the **Place Constraint** command, allowing you to continue placing constraints. Pick the **OK** button when finished placing constraints, or to generate a constraint, and exit the **Place Constraint** command.

Field Notes

Help locate geometry to constrain, using the wheel mouse selection techniques discussed in previous chapters. You can cycle through possible edges, axes, planes, and other elements for constraint purposes.

Exercise 14.8

1. Continue from EX14.7, or launch Autodesk Inventor.
2. Open a new inch assembly file.
3. Access the **Create Component** tool.
4. Specify a new file name of EX14.8a.ipt, a part file type, and a location and template of your choice.
5. Pick the OK button.
6. Sketch a 4″×8″ inch rectangle using the projected Center Point to fix the sketch.
7. Extrude the rectangle .25″ in a positive direction.
8. Open a new sketch on the extrusion face, and sketch the geometry shown.
9. Cut extrude the sketch through all.
10. Finish the edit.
11. Access the **Create Component** tool.
12. Specify a new file name of EX14.8b.ipt, a part file type, and a location and template of your choice.
13. Pick the **OK** button.
14. Pick a location in space to place the sketch.
15. Sketch a .75″ circle.
16. Extrude the circle .75″ in a positive direction.
17. Finish the edit.
18. Establish the constraints shown using the selections and specifications:

   (1) Constraint: Assembly
   Type: Mate
   Offset: 0
   Solution: Flush

   (2) Constraint: Transitional

Exercise 14.8 continued on next page

Exercise 14.8 continued

19. The final exercise should look like the assembly shown in 2.
20. Drag the roller to observe the transitional constraint.
21. Resave EX14.7.
22. Exit Autodesk Inventor or continue working with the program.

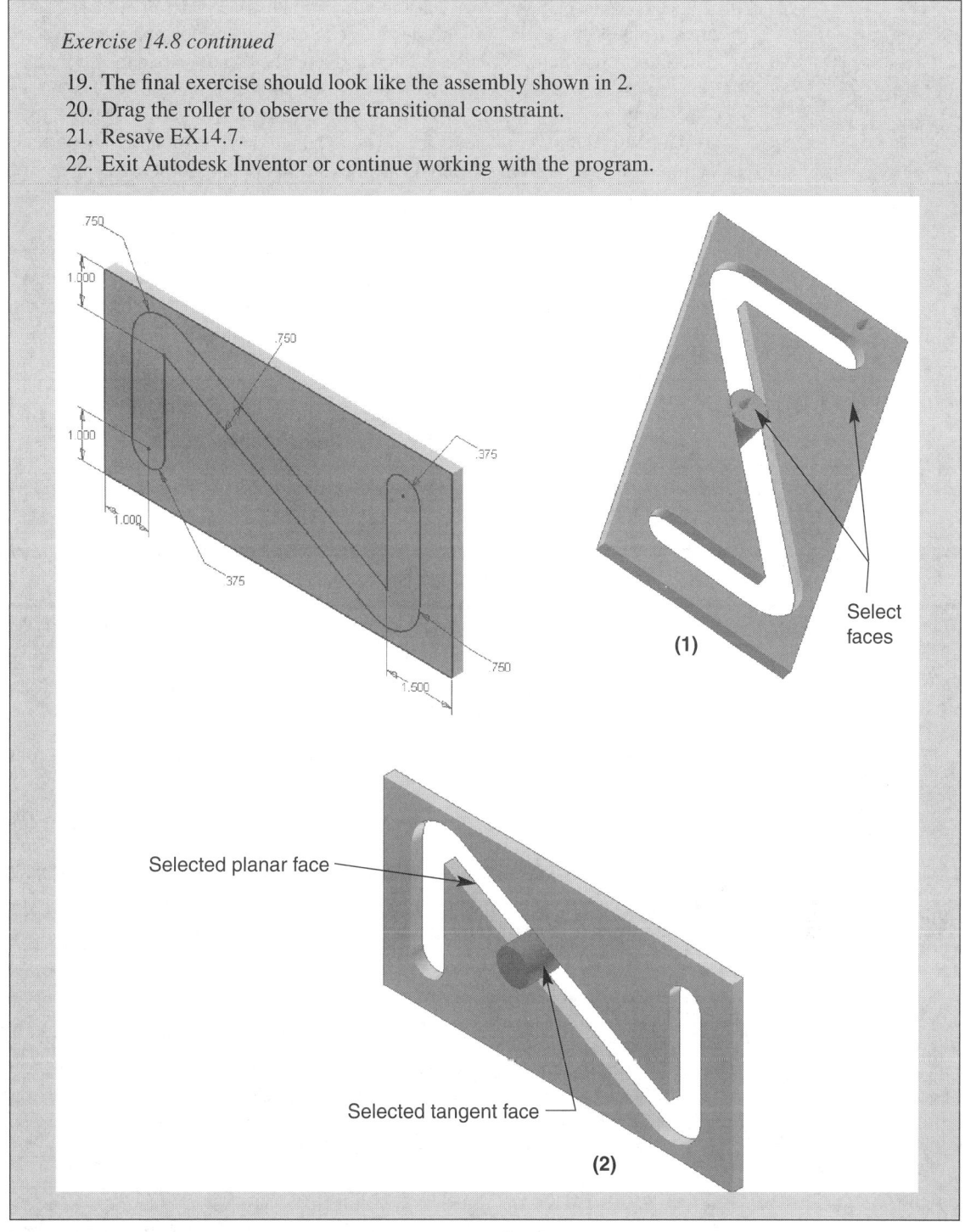

Select faces

(1)

Selected planar face

Selected tangent face

(2)

Alternative Constraining Methods

While using the **Place Constraint** tool and dialog box is one of the most common ways of constraining assembly components, there are other methods for linking components together with constraints. You may use a single constraining method, such as just using the **Place Constraint** tool, or you may incorporate a combination of constraining techniques depending on the application and design process. The alternative constraining procedures are as follows:

- **Dragging**   Use this method, also known as *drag-mate* or *alt-drag,* to establish various assembly constraints, including mate, flush, tangent, and insert constraints, by dragging one component to another component. Like other constraining options, you cannot drag-mate constrain grounded components or those that are already fully constrained. To move an unconstrained component, you can hold down the left mouse button and drag the component to the desired location. Establishing mate constraints between components is a similar process. However, before you pick the component, you must first hold down the **Alt** key on your keyboard. Then, while still pressing the **Alt** key, pick the component face, edge, or point you want to move into place, hold down the left mouse button, and drag the component to the other component face, edge, or point you want to constrain to. As you pass the selected component over and around other components, the constraining options are previewed, along with the possible constraint icon.

Often the automatic constraint option available when using the drag-mate constraint technique is not correct. For these situations, you can toggle between different types of constraints using shortcut keys. To use the shortcut keys, while you are holding down the left mouse button, release the **Alt** key on your keyboard, and select one of the following keys to activate the specified constraint and solution:

- **A** or **2**   Select either key to change the current constraint to an angle constraint. Then press the spacebar to change the angle direction if required.

- **I** or **4**   Select either key to change the current constraint to an insert constraint. Then, if needed, press the spacebar to change the insert direction.

- **M** or **1**   Select either key to change the current constraint to a mate constraint. Then, if required, press the spacebar to change a flush solution.

- **R** or **5**   Select either key to change the current constraint to a rotation motion constraint. Then, if needed, press the spacebar to change the rotation direction.

- **S** or **6**   Select either key to change the current constraint to a translation motion constraint. Then press the spacebar to change the slide direction if needed.

- **T** or **3**   Select either key to change the current constraint to a tangent constraint. Press the spacebar to change to an inside or outside tangent constraint.

- **X** or **8**   Select either key to change the current constraint to a transitional constraint.

Once you are satisfied with the intended constraint, release the left mouse button to generate the constraint. See Figure 14.46. To ensure the constraint is correct, you can see the type of drag-mate constraint defined by looking in the Browser.

Field Notes

Once you begin the drag-mate constraint process, you can use the Help locate geometry to constrain, using the wheel mouse selection techniques discussed in previous chapters. You can cycle through possible edges, axes, planes, and other elements for constraint purposes.

- **Creating components in-place**   As previously discussed, the first step in creating parts and subassemblies in-place is to identify a sketch location. The sketch can be placed anywhere in the modeling space or can be added to an existing component face or plane. When you sketch on an existing component face, a flush constraint is automatically created between the sketch plane of the new component and the face of the existing component. See Figure 14.47. In addition, when you create a component in-place, using existing component geometry, the new component automatically becomes adaptive, as indicated in the Browser. Because of the adaptive nature of the components in Figure 14.47, for example, the shaft is constrained to the hole. If you remove the adaptivity between the shaft and the hole, the only constraint still present is the flush constraint. Consequently, you must specify a mate constraint between the shaft and hole axes in order to keep the shaft positioned in the hole. Adapting parts and assemblies is discussed in Chapter 15.

First, hold down the Alt key on your keyboard

Second, hold down the left mouse button on the desired component face, edge, or point

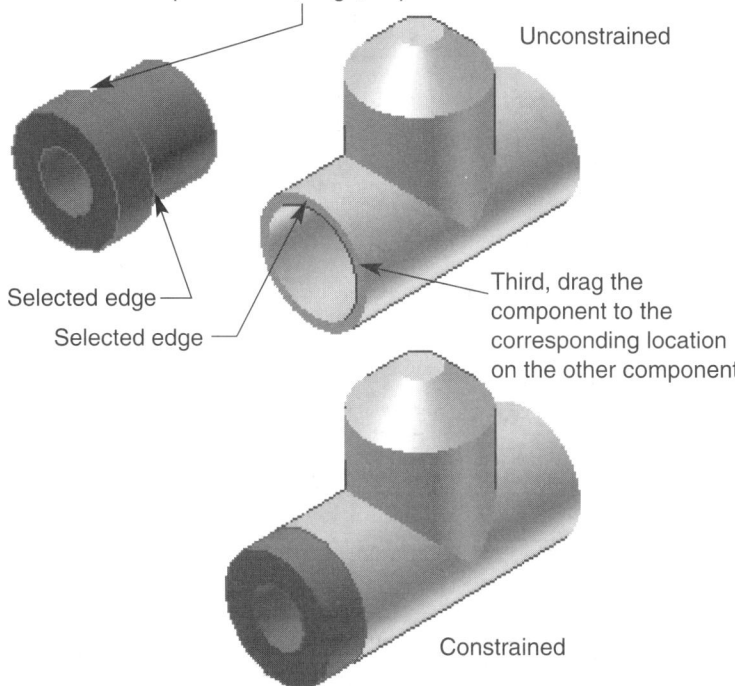

Unconstrained

Selected edge

Selected edge

Third, drag the component to the corresponding location on the other component

Constrained

FIGURE 14.46   An example of a drag-mate constraint.

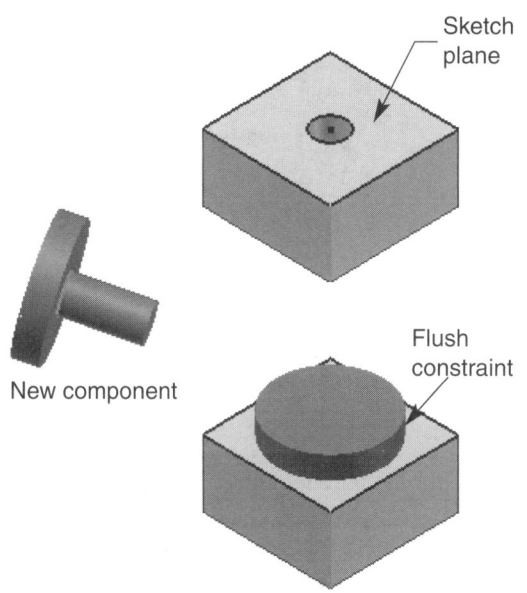

Sketch plane

Flush constraint

New component

FIGURE 14.47   An example of a constraint placed between the sketch plane of a new component and the face of an existing component.

Exercise 14.9

1. Continue from EX14.8, or launch Autodesk Inventor.
2. Open a new inch assembly file.
3. Access the **Create Component** tool.
4. Specify a new file name of EX14.9a.ipt, a part file type, and a location and template of your choice.
5. Pick the **OK** button.
6. Create the sketch shown as A.
7. Extrude the sketch .125″ in a positive direction.
8. Open a new sketch on the extrusion face, and sketch the circles shown in B.
9. Extrude the circles 1″ in a positive direction.
10. Finish the edit.
11. Access the **Create Component** tool.
12. Specify a new file name of EX14.9b.ipt, a part file type, and a location and template of your choice.
13. Pick the **OK** button.
14. Specify the sketch location, by picking the face shown in C.
15. Create the sketch shown in C.
16. Extrude the circle .625″ in a negative direction.
17. Finish the edit.

Exercise 14.9 continued on next page

Exercise 14.9 continued

18. Drag another copy of EX14.9b into the assembly by holding down the left-mouse button on the part icon and EX14.9b name in the Browser, and dragging and dropping the file into the graphics window.
19. Hold down the **Alt** key on you keyboard, select the specified edge of EX14.9b, and drag it to the specified edge to establish an insert constraint shown in D.
20. Double-click on the copy of the EX14.9b in the Browser, to edit the constraint.
21. Select the **Aligned** solution, and pick the **Cancel** button.
22. Establish the constraints shown using the selections, and specifications:

> (E) Constraint: Assembly
> Type: Angle
> Offset: 0
> Solution: **Flip the Second Selection** button selected.

> (F) Constraint: Assembly
> Type: Mate
> Offset: 0
> Solution: Flush

> (G) Constraint: Motion
> Type: Rotation
> Ratio: 0
> Solution: Forward

23. The final exercise should look like the assembly shown in G.
24. Save the assembly file as EX14.9.
25. Exit Autodesk Inventor or continue working with the program.

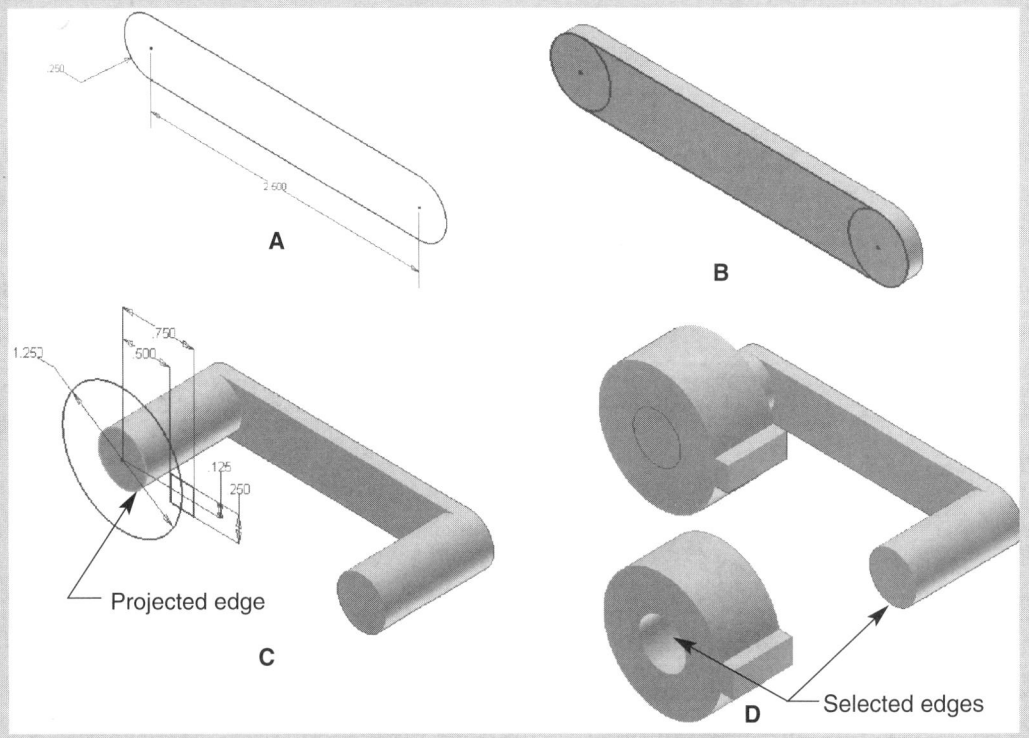

A

B

Projected edge

C

D

Selected edges

Exercise 14.9 continued on next page

Exercise 14.9 continued

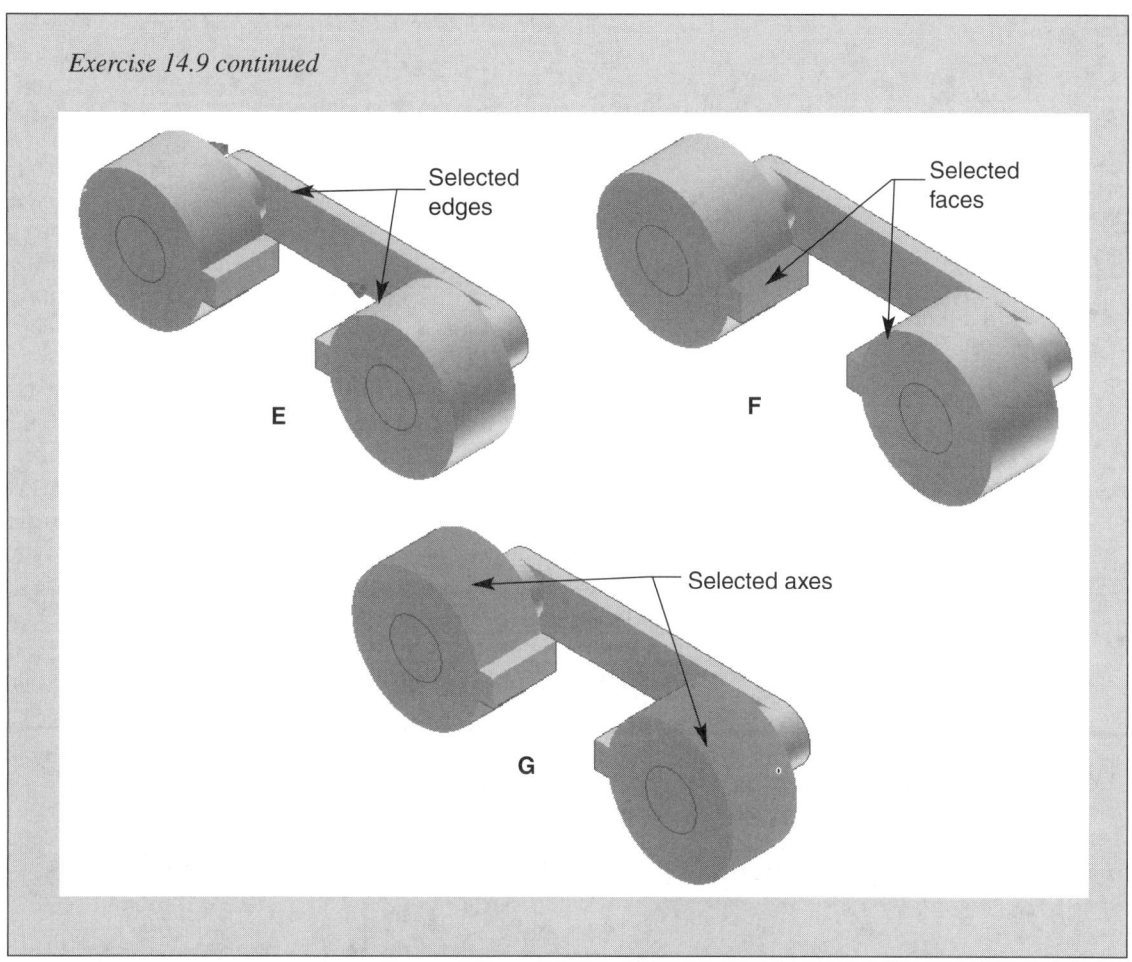

■ **iMates**   As you have seen, most constraints are applied to two components at the same time while in an assembly file. For example, using the **Place Constraint** tool, you establish an insert constraint between a bolt and part with a threaded hole. Another option for identifying constraints is to use iMates. *iMates* allow you to define one or more insertable constraints in the actual component, not the assembly. You may want to think about an individual iMate as half of a constraint that is added to a component when it is created, or edited, not in the assembly work environment. The two constraint halves, iMates, can then be linked together, forming a constraint. Using the previous example, you can place an insert iMate when you create or edit the bolt and an insert iMate when you create or edit the part with the threaded hole. Then, when both the bolt and the part with the threaded bolt are placed in an assembly file, they can be constrained together using the predefined iMates.

Probably the most difficult aspect of working with iMates revolves around planning and identifying iMate placement. If you do not use iMates that correspond to other iMates or do not place iMates in the correct locations in reference to other components, incorrect constraining occurs. The first step to applying iMate relationships is to create iMates for each of the components you want to constrain together. As previously mentioned you can create iMates in a component (part or subassembly) file, or while in the assembly by editing the existing component. Consequently, to create an iMate, you must open a part or subassembly file or edit an existing component in the assembly file by right-clicking on the component name in the Browser, or the actual component in the graphics window, and selecting the **Edit** menu option. Then access the **Create iMate** tool using one of the following options.

✓ Pick the **Create iMate** button on the **Standard** toolbar.

✓ Right-click inside the graphics window, and select the **Create iMate** menu option.

✓ Right-click on an existing iMate folder in the Browser, and select the **Create iMate** menu option.

✓ Select the **Create iMate** option of the **Insert** pull-down menu.

When you access the **Create iMate** tool, the **Create iMate** dialog box is displayed. See Figure 14.48. As you can see, the **Create iMate** dialog box is similar to the **Place Constraint** dialog box, discussed earlier. The only differences between creating an assembly constraint using the **Place Constraint** dialog box and creating an iMate using the **Create iMate** dialog box is that you cannot define a transitional iMate in the **Create iMate** dialog box, and because an iMate is only half of a constraint, only one selection (and selection button) is available in the **Create iMate** dialog box.

The first step in creating an iMate using the **Create iMate** dialog box is to select the tab that corresponds to the type of iMate constraint you want to use. As previously discussed, you can choose to define assembly and motion iMates by picking the corresponding tab. Refer to the discussion on placing constraints and the **Place Constraint** dialog box for information regarding the **Assembly** and **Motion** tabs and the process of using the various iMate techniques. Again, you select only one component in the **Create iMate** dialog box, instead of selecting two components, as in the **Place Constraint** dialog box.

The following information identifies a typical approach to creating iMates for three separate parts. The first part, "imate1.ipt," shown in Figure 14.49, is opened in a part file, and five iMates are added. Figure 14.49 shows that many iMates can be found in one compo-

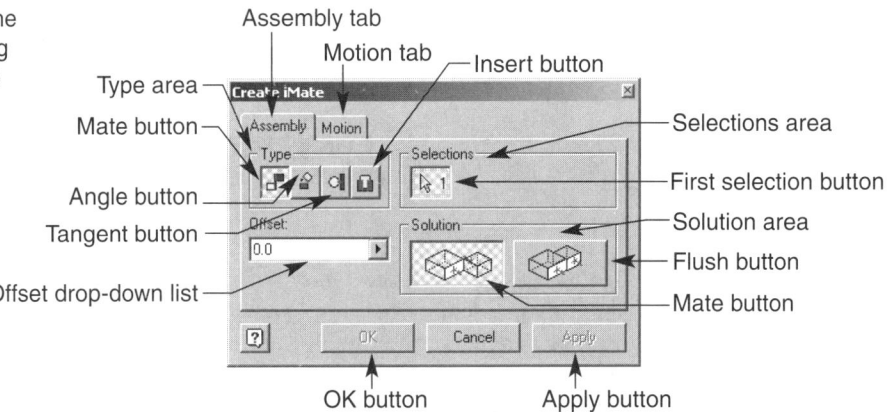

FIGURE 14.48 The **Create iMate** dialog box (**Assembly** tab displayed).

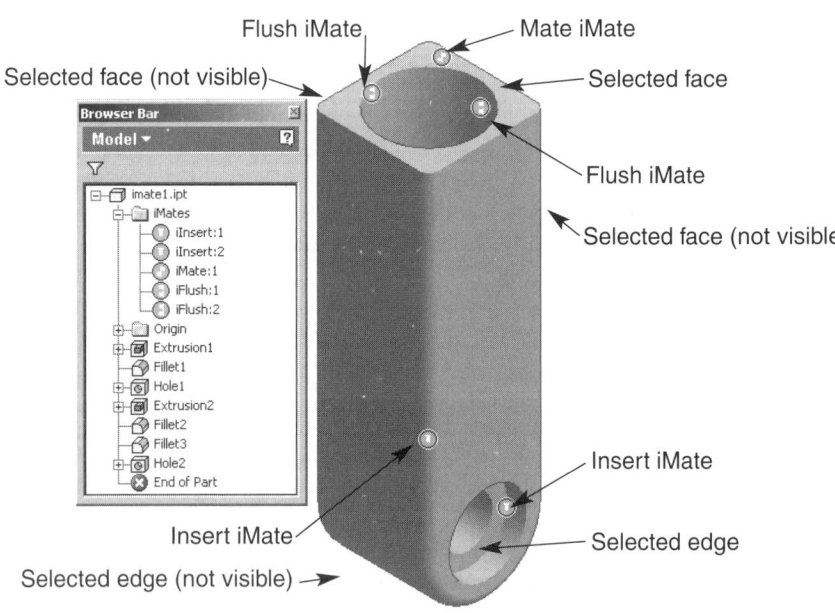

FIGURE 14.49 Creating multiple iMates in a single part.

nent. Two or more iMates in a single component that are used to constrain the same component can be linked together to form a composite iMate. To create a composite iMate, expand the iMate folder in the Browser for the selected component. Then select each of the iMates in the iMate folder that you want to link together, using the **Ctrl** or **Shift** key on your keyboard. Finally, right-click and select the **Create Composite** menu option. See Figure 14.50. The iMate symbols specify the type of iMate created and the iMate status. When a constraint between two iMates is identified, the symbol is no longer visible. Next, "imate1.ipt" is closed and "imate2.ipt," shown in Figure 14.51, is opened, which only contains one iMate. Then, "imate2.ipt" is closed and "imate3.ipt," shown in Figure 14.52, is opened; "imate3.ipt" is another composite iMate and contains three iMates.

Field Notes

Notice only iMates that control constraints between two components are composite. All other iMates remain individual.

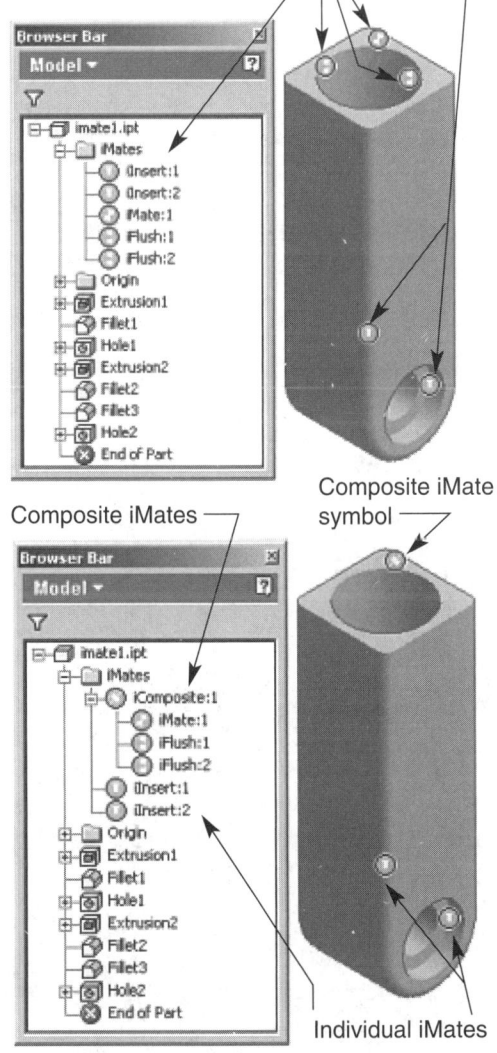

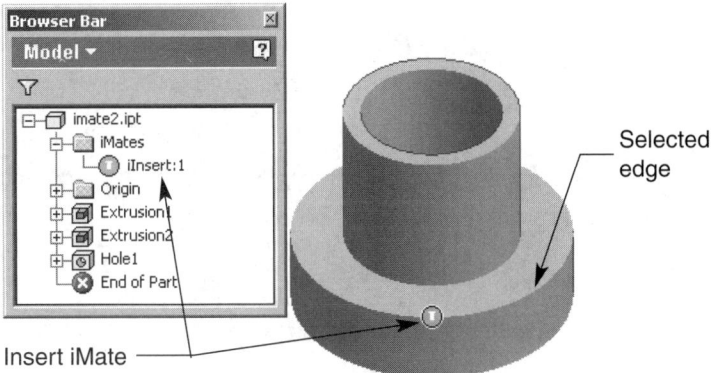

FIGURE 14.50   Creating a composite iMate.   **FIGURE 14.51**   Creating one iMate in a single part.

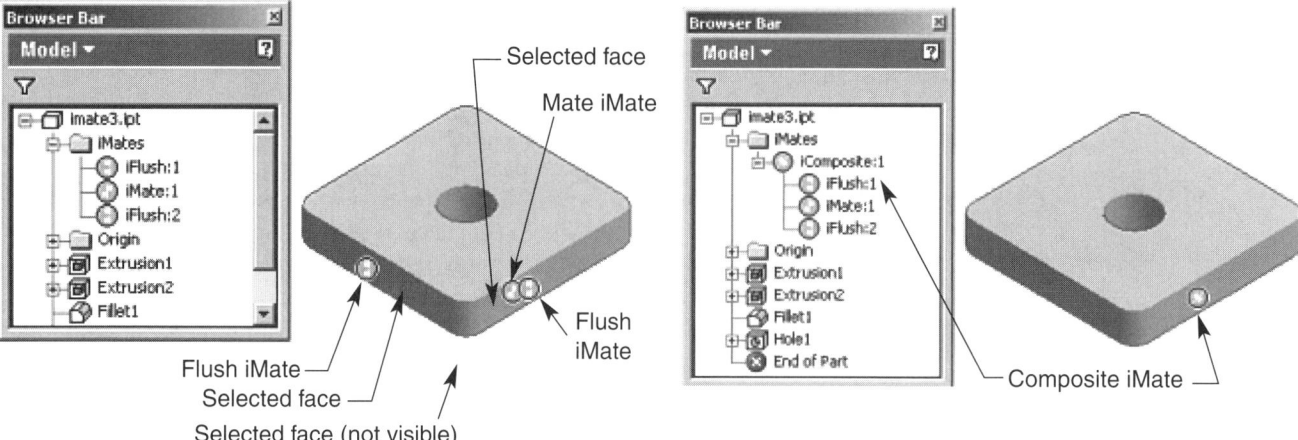

FIGURE 14.52   Creating a composite iMate.

Field Notes

You may find it helpful to rename individual and composite iMates to help identify the intended constraint.

Exercise 14.10

1. Continue from EX14.9, or launch Autodesk Inventor.
2. Open a new inch part file.
3. Sketch a 1″×4″ rectangle on the XY Plane, using the projected Center Point to fix the sketch in space.
4. Extrude the sketched rectangle .25″ in a positive direction.
5. Open a sketch on the extruded face, and sketch the .25″×.5″ rectangle shown in A.
6. Extrude the sketch to the opposite face.
7. Pattern the extrusion as shown in B, with a count of 4, and a spacing of 1″.
8. Establish the iMates shown using the selections and specifications:

> (1) iMate: Assembly
> Type: Mate
> Offset: 0
> Solution: Mate
>
> (2) iMate: Assembly
> Type: Mate
> Offset: 0
> Solution: Flush
>
> (3) iMate: Assembly
> Type: Tangent
> Offset: 0
> Solution: Outside
>
> (4) iMate: Motion
> Type: Rotation-Translation
> Ratio: 1
> Solution: Forward

Exercise 14.10 continued on next page

Exercise 14.10 continued

   (5) iMate: Assembly
   Type: Angle
   Angle: 0
   Solution: Unselected.

 9. Create a composite iMate of iTangent:1 and iRotateTranslate:1.
10. Create a composite iMate of iMate:1 and iAngle:2.
11. Save the part as EX14.10.
12. Exit Autodesk Inventor or continue working with the program.

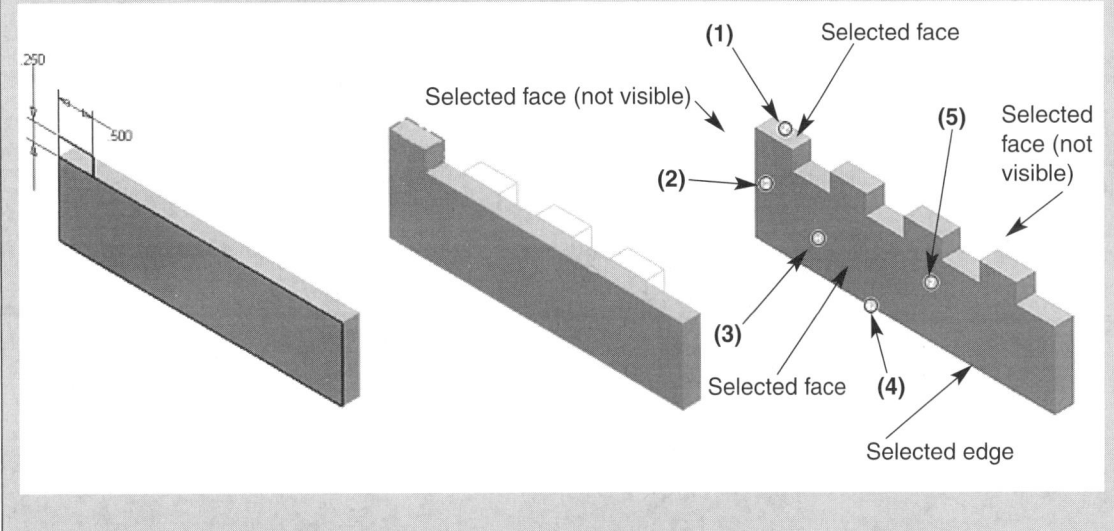

Exercise 14.11

1. Continue from EX14.10, or launch Autodesk Inventor.
2. Open a new inch part file.
3. Sketch a .5″ circle on the XY Plane, using the projected Center Point to fix the sketch in space.
4. Extrude the sketched circle .75″ in a negative direction.
5. Open a sketch on the extruded face, and sketch the .25″×.5″ triangle shown in A.
6. Extrude the sketch .125″ in a positive direction.
7. Open a sketch on the opposite cylinder face and place a hole. Then place a ∅.25″ hole × .5″, with a flat drill point as shown in B.
8. Establish the iMates shown using the selections and specifications:

   (1) iMate: Assembly
   Type: Insert
   Offset: 0
   Solution: Opposed

   (2) iMate: Assembly
   Type: Tangent
   Offset: 0
   Solution: Outside

Exercise 14.11 continued on next page

Exercise 14.11 continued

    (3) iMate: Motion
    Type: Rotation-Translation
    Ratio: 1
    Solution: Forward

9. Create a composite iMate of iTangent:1 and iRotateTranslate:1.
10. Save the part as EX14.11.
11. Exit Autodesk Inventor or continue working with the program.

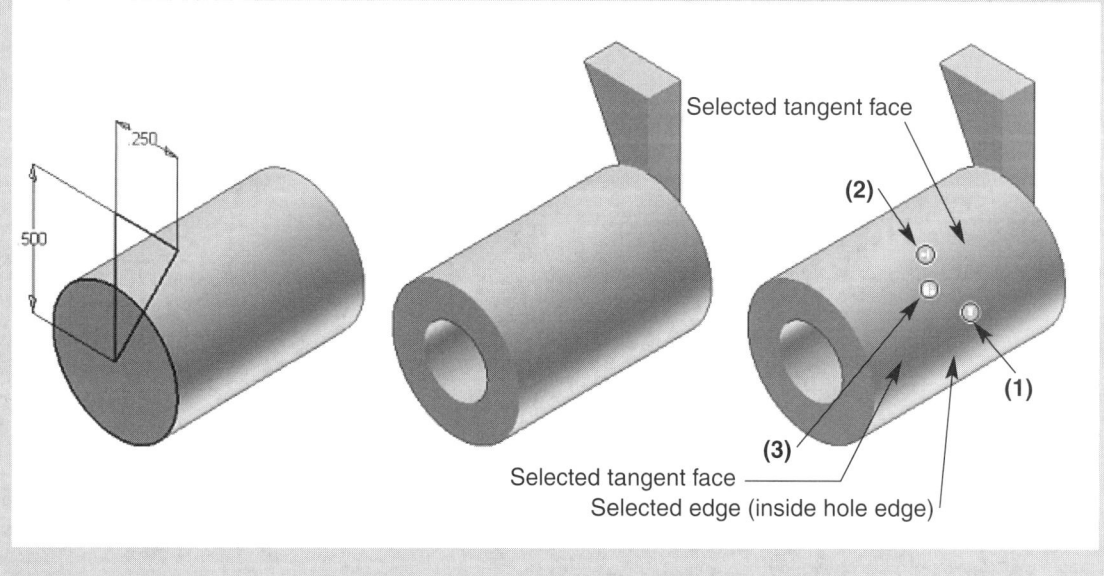

Selected tangent face

Selected tangent face
Selected edge (inside hole edge)

Exercise 14.12

1. Continue from EX14.11, or launch Autodesk Inventor.
2. Open a new inch part file.
3. Sketch a .25″ circle on the XY Plane, using the projected Center Point to fix the sketch in space.
4. Extrude the sketched circle 1″ in a positive direction.
5. Open a sketch on the extruded face, and sketch the geometry shown in A.
6. Extrude the sketch .125″ in a positive direction.
7. Establish the iMates shown using the selections, and specifications:

    (1) iMate: Assembly
    Type: Mate
    Offset: 0
    Solution: Mate

    (2) iMate: Assembly
    Type: Mate
    Offset: 0
    Solution: Flush

Exercise 14.12 continued on next page

Exercise 14.12 continued

(3) iMate: Assembly
Type: Insert
Offset: 0
Solution: Opposed

(4) iMate: Assembly
Type: Angle
Angle: 0
Solution: Unselected.

8. Create a composite iMate of iMate:1 and iAngle:1.
9. Save the part as EX14.12.
10. Exit Autodesk Inventor or continue working with the program.

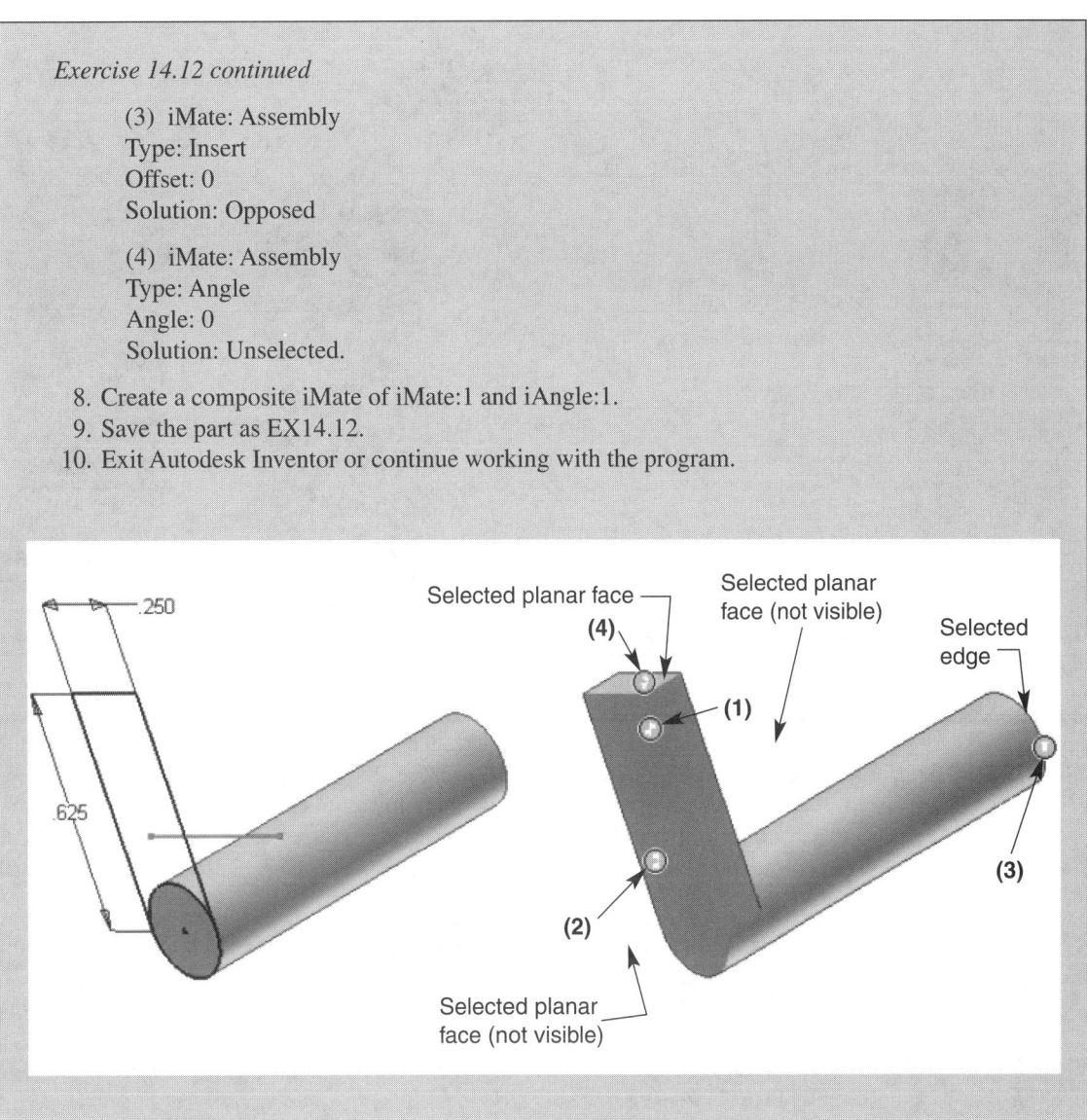

Once you have specified iMates for components, there are a few different techniques to link the iMates together, depending on the application and design process. These methods follow:

■ **Using the Place Component tool** Probably the most effective way of using iMates to create constraints between components is to use the **Place Component** tool, previously discussed. However, this technique does not work if the components are already inserted in the assembly file. When you use the **Place Component** tool to insert components that contain iMates, the inserted components automatically constrain to each other. The first step in applying this technique of assembly creation is to insert the base component, as shown in Figure 14.53. Use the **Place Component** tool to insert the additional components. The key is to ensure the **Use iMate** check box is selected in the **Place Component** dialog box. When you insert components using the **Place Component** tool and the selected **Use iMate** check box, components are automatically constrained to each other, assuming the correct iMates are specified.

■ **Dragging** This method is similar to the dragging technique previously discussed. You can use the alt-drag method when components with iMates are already inserted into the assembly and are not constrained. To drag-mate two iMate components, hold down the

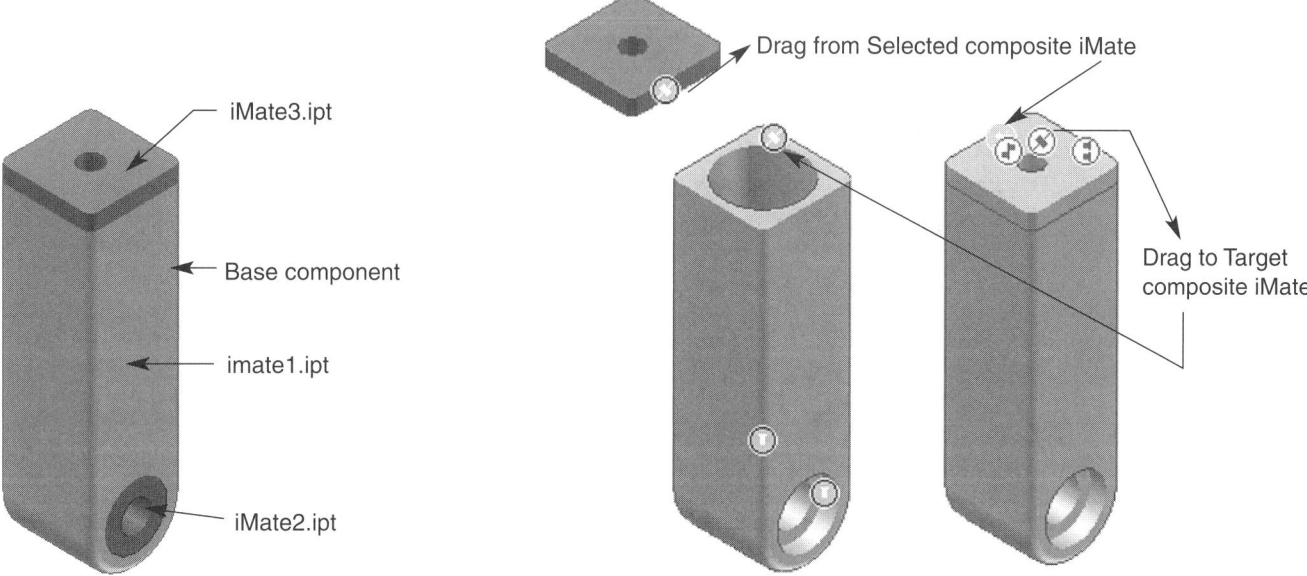

FIGURE 14.53   An example of the constrained iMate assembly.

FIGURE 14.54   An example of a drag-mate iMate constraint.

Alt key on your keyboard. Then, while still pressing the **Alt** key, pick the iMate symbol on the component you want to move into place, hold down the left mouse button, and drag the component iMate symbol to the other component iMate symbol you want to constrain to. Only iMates that can be used for the selected iMate are visible. Once you are satisfied with the intended constraint, release the left mouse button to generate the constraint. See Figure 14.54.

■ **Using the Place Constraint tool**   You can use this technique when components, with iMates, are already inserted into the assembly and are not constrained. The process is similar to using the **Place Constraint** tool to constrain assembly components without iMates. The only difference is the component elements you select. Instead of selecting faces, edges, axes, or points, you select iMates on components to establish constraints between the components. You must select the constraint type that corresponds to the specified iMate.

Field Notes

You cannot select composite iMate symbols when using the **Place Constraint** tool. However, you can expand a composite iMate in the Browser and select individual iMates in the Browser.

Exercise 14.13

1. Continue from EX14.12, or launch Autodesk Inventor.
2. Open a new assembly file.
3. Use the **Place Component** tool to insert EX14.12.ipt, which should automatically become grounded.
4. Use the **Place Component** tool to insert EX14.10.ipt. Ensure the **Use iMate** check box in the **Open** dialog box is selected.
5. Use the **Place Component** tool to insert EX14.11.ipt. Ensure the **Use iMate** check box in the **Open** dialog box is selected.
6. The final exercise should be fully constrained from the iMates and look like the assembly shown.
7. Save the assembly as EX14.13.
8. Exit Autodesk Inventor or continue working with the program.

Editing Constraints

Assembly constraints can be edited much like other Autodesk Inventor elements. Use the following options to modify existing constraints, depending on the required edit:

- **Suppress**   Use this option if you do not want the effects of a constraint to be displayed and do not want to delete the constraint. Suppress a constraint by right-clicking the constraint in the Browser and selecting the **Suppress** menu option.

- **Edit**   Use this option to access the **Edit Constraint** dialog box, shown in Figure 14.55. The **Edit Constraint** dialog box is exactly the same as the **Place Constraint** dialog box, but contains the options used to create the constraint. Access the **Edit Constraint** dialog box using one of the following options.

  ✓ Right-click an existing constraint in the Browser, and select the **Edit** menu option.

  ✓ Double-click an existing constraint in the Browser.

  - **Value edit**   This option allows you to modify the specified constraint offset, angle, ratio, or distance, depending on the selected constraint. Although you can use the **Edit Constraint** dialog box to redefine values, the **Value** drop-down list is often an easier solution. See Figure 14.56. Access the **Value** drop-down list by picking the desired constraint in the Browser. Then enter or select a new value and press the **Enter** key on your keyboard.

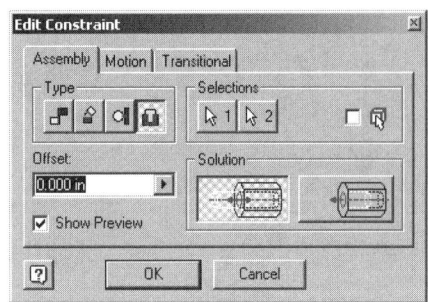

FIGURE 14.55   The **Edit Constraint** dialog box.

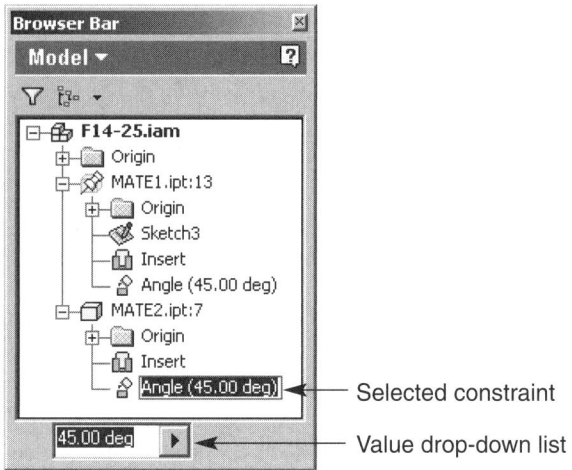

Selected constraint

Value drop-down list

FIGURE 14.56   The **Value** drop-down list.

Driving Constraints

As you have probably already discovered, one way to view component freedom of movement is to drag a component by selecting it, holding down the left mouse button, and rotating, sliding, or offsetting the selected component, depending on the design and existing constraints. Although dragging is an effective way to quickly analyze design progress and intent, driving constraints is also valuable, especially for assembly components that may be difficult to drag. You can *drive* a constraint to show the amount of movement between components, pause movement, drive Adaptivity, and detect collisions between components. A good example of a driven constraint can be seen with the assembly shown in Figure 14.57. In this example, the components can be driven using an angle constraint from a specified angle, such as −110° to another specified angle, such as 95°.

Assuming an acceptable driving constraint is available, typically an angle or mate, you can drive the constraint by right-clicking on the constraint in the Browser and selecting the **Drive Constraint** menu option. When you choose the **Drive Constraint** menu option, the **Drive Constraint** dialog box is displayed with the appropriate units for the selected constraint. (See Figure 14.58.) For example, if you select an angle constraint, units are displayed in degrees. Or if you select a mate constraint, units are displayed in inches or mm depending on the design. The following identifies each of the components available in the **Drive Constraint** dialog box:

- **Start**   Use this drop-down list to enter, select, or measure the desired start point of the drive. By default the start position is set at the value you specify when you create the con-

FIGURE 14.57   A simple example of a driven constraint application.

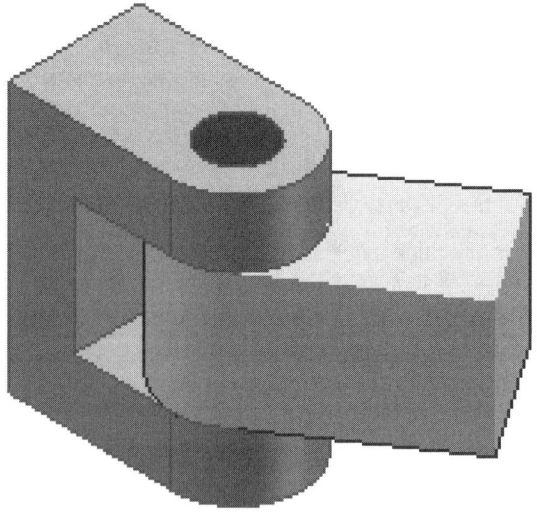

FIGURE 14.58 The
Drive Constraint
dialog box.

Start drop-down list

End drop-down list

Pause Delay edit box

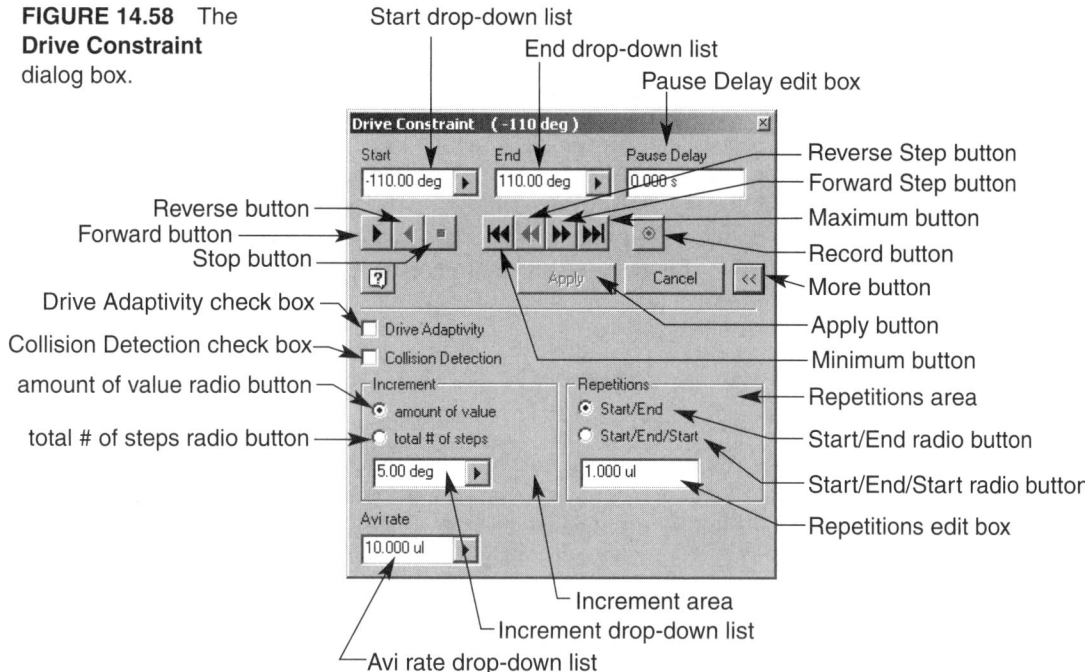

Reverse Step button
Forward Step button
Maximum button
Record button
More button
Apply button
Minimum button
Repetitions area
Start/End radio button
Start/End/Start radio button
Repetitions edit box

Reverse button
Forward button
Stop button
Drive Adaptivity check box
Collision Detection check box
amount of value radio button
total # of steps radio button

Increment area
Increment drop-down list
Avi rate drop-down list

straint, such as the offset or the angle. For example, when working with a mate constraint, if you specify an offset of 5mm, the start value will be 5mm. Figure 14.59 shows a possible start value for an angle constraint.

■ **End**   This drop-down list is used to enter, select, or measure the desired end position of the drive. By default the end point is set at the offset or angle value you specify when you create the constraint, plus 10. For example, when working with a mate constraint, if you specify an offset of 5mm, the start value will be 15mm. Figure 14.59 shows a possible end value for an angle constraint.

■ **Pause Delay**   Driven constrains move a certain number of steps. For example, you can set an angle constraint to move 5° per step. Use the **Pause Delay** edit box to specify the number of seconds between steps. The default value is 0, which results in no pause between steps.

■ **Forward**   Pick this button to drive the constraint forward. The **Forward** button is available only when both the **Start** and **End** edit boxes have values and the constraint is driven to an appropriate location.

■ **Reverse**   Pick this button to drive the constraint in the reverse direction. The **Reverse** button is available only when both the **Start** and **End** edit boxes have values and the constraint is driven to an appropriate location.

■ **Stop**   This button is available when a constraint is being driven and allows you to stop the component movement.

■ **Minimum**   Use this button to automatically return to the start position.

■ **Reverse Step**   Pick this button to drive the constraint in the reverse direction, a single step in the movement progression.

FIGURE 14.59 The
Video Compression
dialog box.

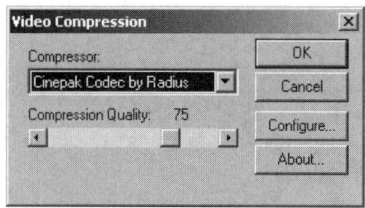

- **Forward Step**   Pick this button to drive the constraint in the forward direction, a single step in the movement progression.

- **Maximum**   Use this button to automatically return to the end position.

- **Record**   Pick this button to record the driven constraint movement as an .avi file. When you record a drive, a copy of the specified AVI increments is created and stored. When you select the **Record** button, the **Open** dialog box appears, allowing you to specify a location for the animation file, followed by the **Video Compression** dialog box, shown in Figure 14.59, which, as the name implies, allows you to compress, or not compress, the video file using a specified video compressor.

- **Drive Adaptivity**   Pick this check box to drive the adaptive properties of components within relation to the constraint. When you select this check box, the **Collision Detection** check box is not available because of the adaptive nature of the drive.

- **Collision Detection**   You can select this check box to automatically identify a collision between the driven constraint and moving components. As shown in Figure 14.60, the components involved in the collision are highlighted, and the collision detection message is displayed. Pick the **OK** button to accept the collision and modify the drive.

When the **Collision Detection** is selected, the **Drive Adaptivity** check box is not available.

Field Notes

The **Analyze Interference** tool discussed earlier in this chapter functions similarly to collision detection.

- **Increment**   Use this area to specify the incremental movement of the drive. Pick the **amount of value** radio button; if you want, define a certain increment value, such as 10° for a driven angle constraint or 5″ for a driven mate constraint. Then specify or select the value from the **Increment** drop-down list. A high amount of value results in fewer increments and quicker movement. Pick the **total # of steps** radio button if you want to define a certain number of steps, such as 5 or 10. This means the total amount of drive movement is divided into 5 or 10 steps. Use the **Increment** drop-down list to specify the number of steps. In contrast to the amount of value option, a high number of steps results in more increments and slower movement.

- **Repetitions**   This area allows you to define the repetition options for the drive. Pick the **Start/End** radio button to drive the constraint from the start position to the end position. To cause the drive to return back to the start after it meets the end position, pick the **Start/End/Start** radio button. Then identify the number of times you want the drive to move

FIGURE 14.60   An example of a drive collision detection application.

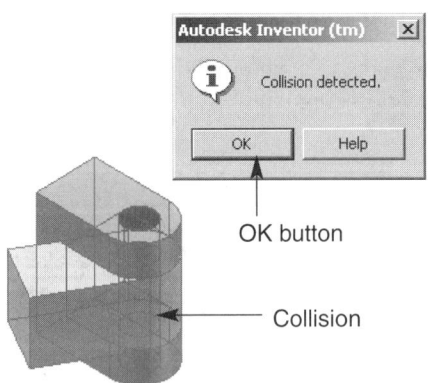

from start to end, or from start to end to start, by specifying a value in the **Repetitions** edit box.

■ **Avi rate**   Use this drop-down list to specify or select an AVI rate, which as previously mentioned, defines the number of increments copied when you record the constraint using the **Record** button.

■ **Apply**   When you have finished the driving a constraint, you can pick the **Close** or **Cancel** button to exit the command. Or you can pick the **Apply** button to save the drive settings and apply them to the constraint, and components.

Field Notes

> While constraints are being driven, you can use tools such as **Rotate, Pan, Zoom,** and **Look At** to better analyze the design movement.

Exercise 14.14

1. Continue from EX14.13, or launch Autodesk Inventor.
2. Open EX14.13.
3. Save a copy of EX14.13, as EX14.14.
4. Close EX14.13 without saving and open EX14.14.
5. Suppress the flush iMate constraint located in EX14.12 and EX14.10 (when you suppress one, both iMate occurrences are suppressed).
6. Establish the constraint shown using the selections, and specifications:

   Constraint: Assembly
   Type: Angle
   Angle: 0
   Solution: Unselected (see arrows)

7. Right-click on the angle constraint you just created, and select the **Drive Constraint** menu option.
8. Specify the following parameters in the **Drive Constraint** dialog box:

   Start: 0°
   End: 360°
   Pause Delay: 0
   Collision Detection check box: Selected
   Amount of value radio button: Selected
   Amount of value: 10 degrees
   Start/End radio button: Selected
   Repetitions: 3 ul
   Avi rate:10 ul

9. Pick the **Forward** button to observe the drive.
10. When the forward operation has ended, pick the **Reverse Step** button to rotate the component 10° in reverse.
11. Pick the **Apply** button.
12. Return the components to the original positions by unsuppressing the flush iMate constraint.
13. Resave EX14.14.
14. Exit Autodesk Inventor or continue working with the program.

Exercise 14.14 continued on next page

Exercise 14.14 continued

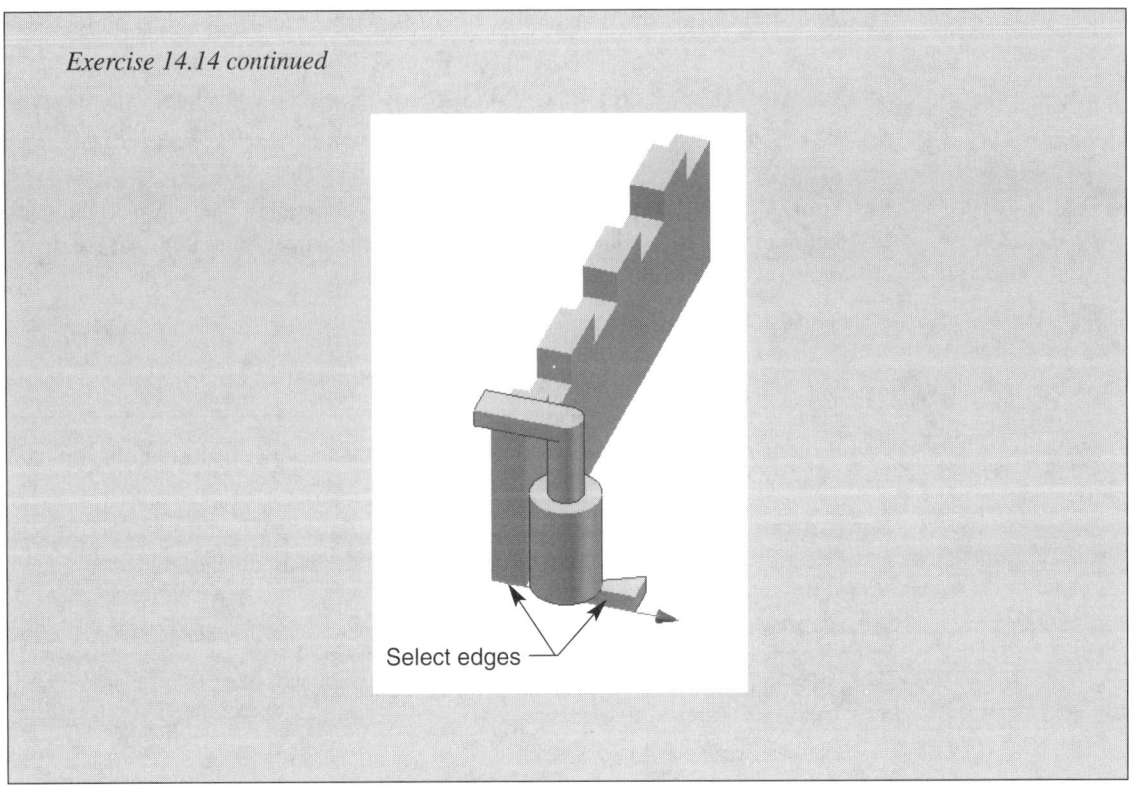

Select edges

Patterning Components

Often assemblies contain many of the same parts, or even subassemblies, that must be arranged in a certain pattern, such as in Figure 14.61.

You can use the tools discussed earlier in this chapter to place separate components and then individually constrain the components to other components in a pattern. However, Autodesk Inventor provides you with the **Pattern Component** tool, which allows you to quickly generate a circular or rectangular pattern from a single component or group of existing components. A reoccurrence of components, in a designated configuration, is known as a *pattern component.* Pattern components and the **Pattern Component** tool are very similar to pattern features and the feature patterning tools available in the part-modeling environment, discussed in Chapter 6. However, the **Pattern Component** tool allows you to associate the assembly component pattern with a feature component pattern. For example, the nuts and bolts patterned in the assembly shown in Figure 14.61 are associated with the holes in the cover plate and elbow. As a result, any changes

FIGURE 14.61   An example of an assembly with a pattern of components (circular pattern).

made to the individual component with the feature pattern are parametrically translated to the assembly component pattern.

Assuming you have at least one existing component in the assembly file that you want to pattern, such as a screw, and have the corresponding components necessary for generating an associative pattern, if desired, you can begin the component patterning process. First, access the **Pattern Component** tool, using one of the following techniques:

✓ Pick the **Pattern Component** button on the **Assembly** panel bar.

✓ Pick the **Pattern Component** button on the **Assembly** toolbar.

✓ Select the **Pattern Component...** option of the **Insert** pull-down menu.

The **Pattern Component** dialog box is displayed when you access the **Pattern Component** command. See Figure 14.62. Select one of the following tabs, and use the information provided to create the desired component pattern:

■ **Associative**   This tab, shown in Figure 14.62, is initially active when you access the **Pattern Component** dialog box and is probably the most versatile assembly component patterning option. Use the **Associative** tab to pattern an existing component, or group of components, in a rectangular or circular fashion, and in relation to an existing component feature pattern. For example, Figure 14.63 shows a component pattern of screws arranged in a rectangular pattern in reference to the slide bar hinge connector. To create an associative component pattern, pick the **Component** button if it is not already selected. Then choose the component, such as a bolt, or group of components, such as a bolt, washer, and nut, that you want to pattern. The component in Figure 14.63 is the inserted and constrained screw. Once you have selected the component or components to pattern, pick the **Associative Feature Pattern** button, and pick the part feature pattern, that will define the assembly component

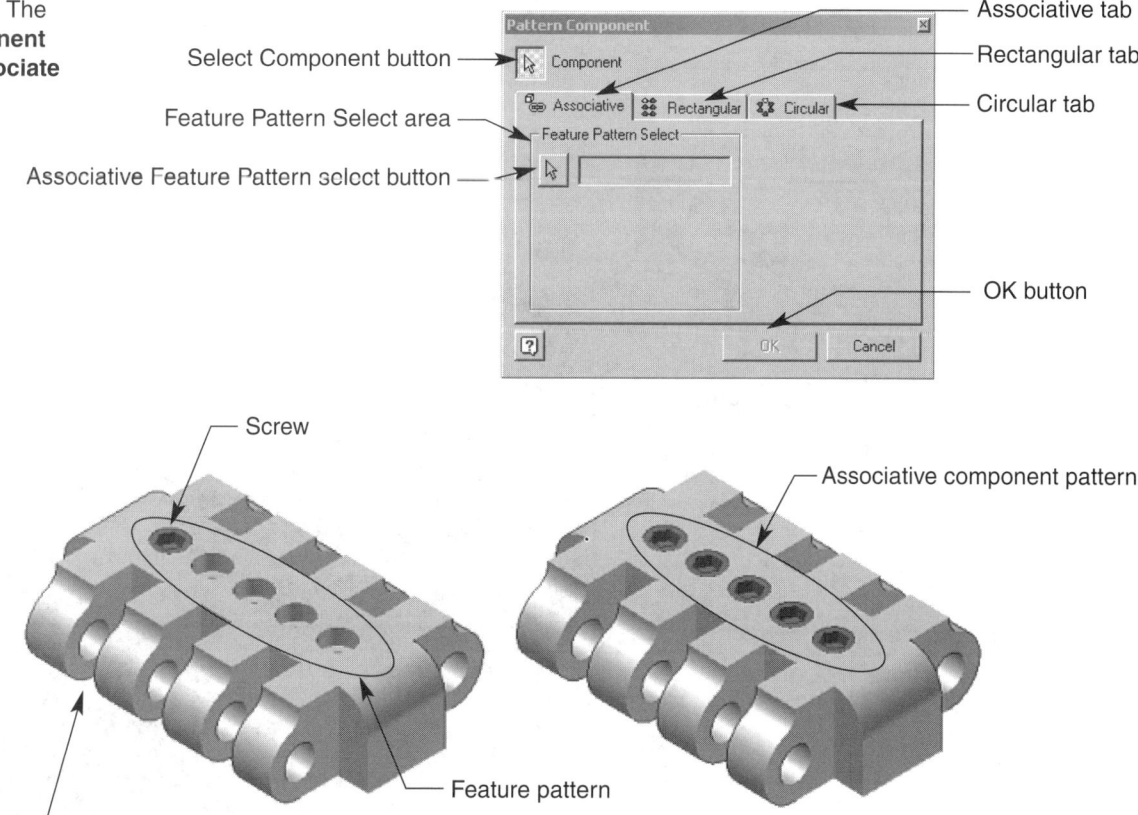

FIGURE 14.62   The **Pattern Component** dialog box (**Associate** tab selected).

Select Component button

Feature Pattern Select area

Associative Feature Pattern select button

Associative tab

Rectangular tab

Circular tab

OK button

Screw

Associative component pattern

Feature pattern

Slide bar hinge connector

FIGURE 14.63   An example of an associative component pattern.

pattern. The associative feature pattern in Figure 14.63 is the pattern of holes on the slide bar hinge connector. When you select the associative feature pattern, the name of the pattern is shown in the **Feature Pattern Select** display box, and a preview of the component pattern is displayed. If the correct feature pattern and preview is shown, pick the **OK** button to generate the component pattern.

■ **Rectangular**   This tab, shown in Figure 14.64, allows you to create a rectangular pattern of a component or series of components, without identifying an associative feature pattern. For example, Figure 14.65 shows a component pattern of three brackets arranged in a rectangular pattern, which will later be linked with other components created in-place. Generating a rectangular component pattern is very similar to creating a rectangular feature pattern, discussed in Chapter 6. However, this section reviews the process. To create a rectangular component pattern, pick the **Component** button if it is not already selected. Then choose the component, or group of components, that you want to pattern. The component in Figure 14.65 is the inserted and grounded bracket. Once you have selected the component or components to pattern, pick the **Column Direction** button and select the direction of the column of patterned components. You can select any feature edge, axis, or work axis available in the model, including the default reference axes located in the **Browser** bar **Origin** folder. When you choose a direction, an arrow shows you the specified pattern path, and a preview dis-

FIGURE 14.64   The **Pattern Component** dialog box (**Rectangular** tab selected).

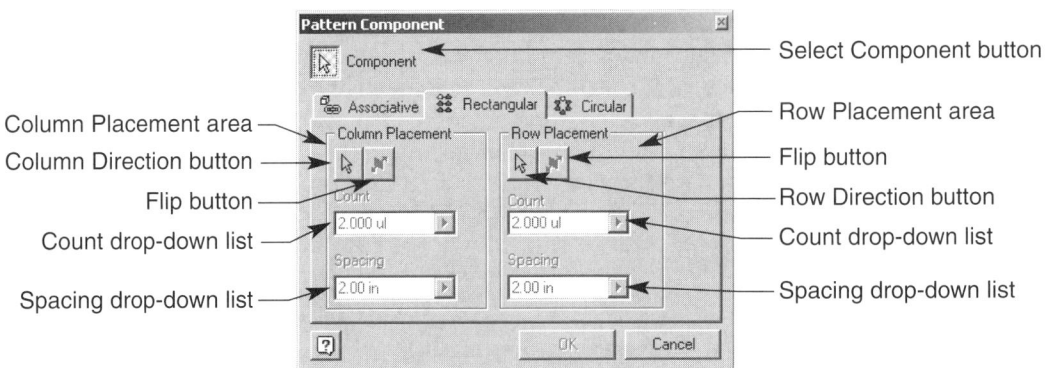

FIGURE 14.65   An example of a rectangular component pattern.

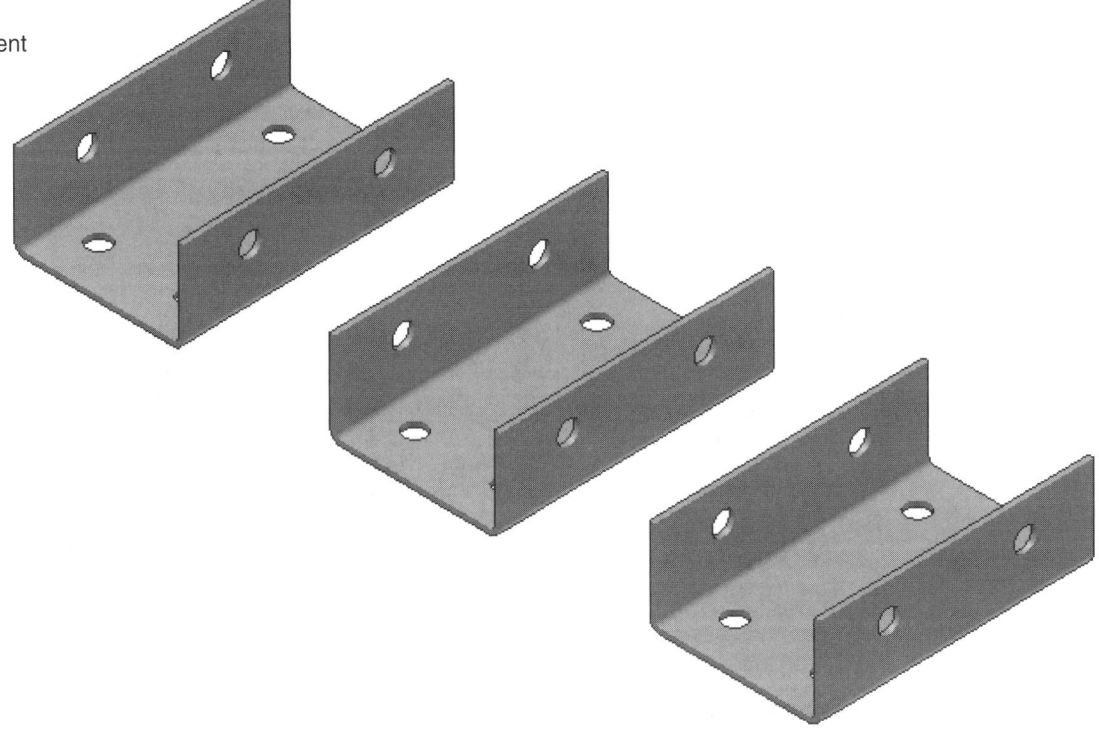

plays the pattern operation. If the direction is not correct, pick the **Flip** button. The next step is to specify how many copies of the component or series of components you want to create by entering or selecting a count value from the **Count** drop-down list. Then define the distance between the copies by entering or selecting a spacing value from the **Spacing** drop-down list. Spacing is measured by the width of the selected features and the distance between the copies. It is not just the space between features. For example, if you want to pattern a block that is 1″ wide, and you want a ½″ space between pattern copies, you must specify a 1½″ spacing.

Once you have fully defined the **Column Direction,** repeat the steps previously described for placing rows. Pick the **Row Direction** button, and select the desired direction. Again, you can select any component edge, axis, or work axis available in the model, including the default reference axes located in the **Browser** bar **Origin** folder. Theoretically you can choose the same edge or axis you chose for the column direction, or a parallel edge or axis. However, for most applications, you should select an edge or axis that is perpendicular, or at least nonparallel to the first direction. When you choose the row direction, an arrow shows you the specified pattern path, and a preview displays the pattern operation. After you define the row direction, specify how many copies of the components or series of components you want to create by entering or selecting a count value from the **Count** drop-down list. Then define the distance between the copies by entering or selecting a spacing value from the **Spacing** drop-down list. Again, spacing is measured by the width of the selected components and the distance between the copies. It is not just the space between components. Finally, pick the **OK** button to generate the rectangular pattern.

 Field Notes

To "turn off" and hide certain occurrences of patterned components, expand the component pattern in the Browser, right-click the desired occurrence, and pick the **Suppress** menu option.

■ **Circular**   This tab, shown in Figure 14.66, allows you to create a circular pattern of a component or series of components, without identifying an associative feature pattern. For example, Figure 14.67 shows a component pattern of six studs arranged in a circular pattern, which, when fabricated, will be welded to the plate. Generating a circular component pattern is very similar to creating a circular feature pattern, discussed in Chapter 6. However, this section reviews the process. To create a circular component pattern, pick the **Component** button, if it is not already selected. Then choose the component, or group of components, that you want to pattern. The component in Figure 14.67 is the inserted and constrained stud. Once you have selected the component or components to pattern, pick the **Axis Direction** button and select the axis about which the patterned components will be generated around. You can select any existing component or work axis, including the work axes available in the

FIGURE 14.66 The **Pattern Component** dialog box (**Rectangular** tab selected).

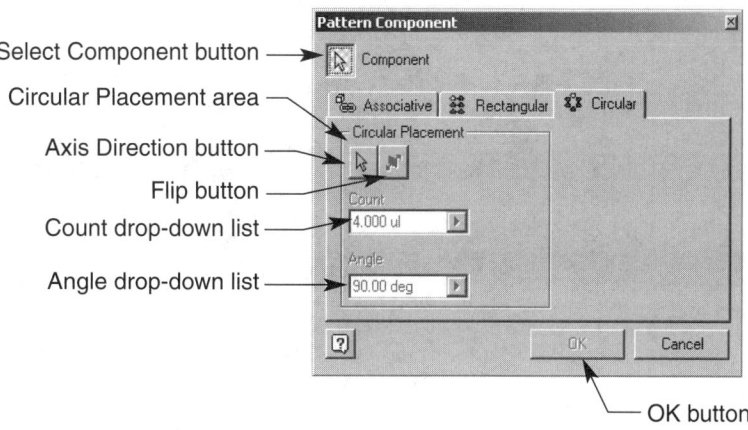

Select Component button
Circular Placement area
Axis Direction button
Flip button
Count drop-down list
Angle drop-down list

OK button

FIGURE 14.67 An example of a circular component pattern.

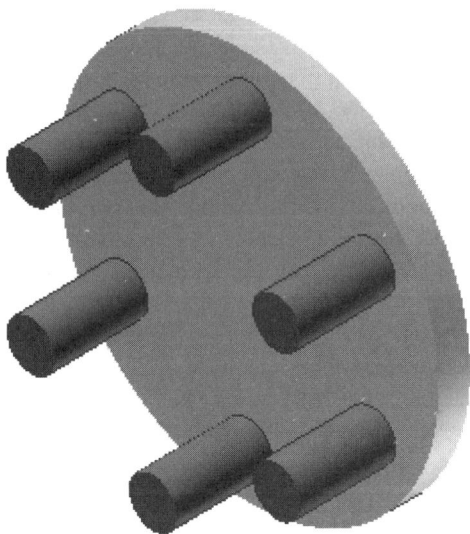

Browser bar **Origin** folder. When you choose a direction, an arrow and axis shows you the specified pattern path, and a preview displays the pattern operation. If the direction is not correct, pick the **Flip** button.

The next step is to specify how many copies of the component or series of components you want to create, by entering or selecting a count value from the **Count** drop-down list. Then define the spacing between the components by entering or selecting an angle value in the **Angle** drop-down list. For example, if you want the copies to be equally distributed in a 360° circle, divide 360 by the count. Finally, pick the **OK** button to generate the circular pattern.

Field Notes

To "turn off" and hide certain occurrences of patterned components, expand the component pattern in the Browser, right-click the desired occurrence, and pick the **Suppress** menu option.

Exercise 14.15

1. Continue from EX14.14, or launch Autodesk Inventor.
2. Open a new inch assembly file.
3. Access the **Place Component** tool, and insert EX6.3.
4. Access the **Create Component** tool.
5. Specify a new file name of EX14.15.ipt, a part file type, and a location and template of your choice.
6. Pick the **OK** button.
7. Specify the sketch location, by picking the face shown in A.
8. Project the hole edge as shown.
9. Extrude the projected edge .75″ in a negative direction.
10. Open a new sketch on the top of the cylinder you just created.
11. Project the hole edge as shown in B.
12. Extrude the projection .375″ as shown.
13. Finish the edit.

Exercise 14.15 continued on next page

Exercise 14.15 continued

14. Access the **Pattern Component** tool.
15. The **Component** button, and **Associative** tab should be active.
16. Select the new component, EX14.15.ipt, for patterning.
17. Pick the **Associated Feature Pattern** select button, and select the existing feature pattern of holes.
18. Pick the **OK** button to generate the pattern.
19. The final exercise should look like the assembly shown.
20. Save the assembly as EX14.15.
21. Exit Autodesk Inventor or continue working with the program.

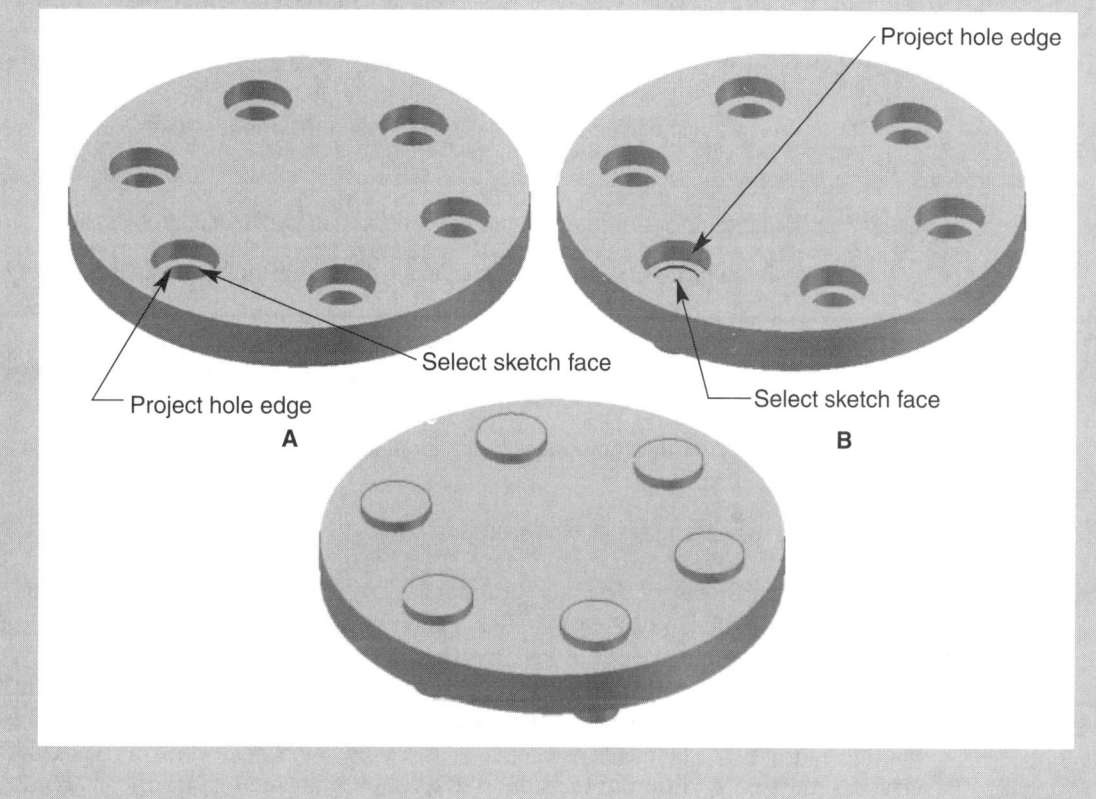

Manipulating Assembly Components

Once you have developed assemblies by inserting, creating, and constraining components, the assembly can be modified using a number of tools and techniques. As previously mentioned, you can edit the actual part or subassembly component file by opening the file, and modifying sketches, features, and other model specifications. You can open individual component files from inside the assembly by right-clicking on the component name in the Browser, or the actual component in the graphics window and selecting the **Open** menu option. Component geometry can also be modified inside the assembly file using one of the following methods to activate component specifications:

✓ Right-click on the component name in the Browser, and select the **Edit** menu option.

✓ Right-click on the component in the graphics window, and select the **Edit** menu option.

✓ Double-click the component name in the Browser.

✓ Double-click the component in the graphics window.

✓ Select the **Modeling Tasks** option of the **Browser Filters** button, discussed earlier in this chapter.

In addition to accessing individual model information and modifying specific component factors, such as an extrusion distance or hole width, you can also use other tools and techniques to further manipulate existing assembly components. These modification commands and methods allow you to replace, move, rotate, and change the color of individual components. In addition, sectioning tools allow you to view a variety of assembly sections.

Replacing Components

In many applications, a new or modified component design is developed for existing assemblies. Often the new component is similar to the old component and may even be constrained exactly the same as other assembly components. Although you can delete the old component, insert the new component, and constrain the new component to other assembly elements, an easier solution is to replace the old component. Autodesk Inventor provides you with the **Replace Component** and **Replace All** tools that allow you to replace a single component with a different component, or each occurrence of the same component, with a different component. For example, if an assembly contains eight of the same 1″ bolts, you can replace one of the 1″ bolts with a 2″ bolt, or you can replace all the 1″ bolts with 2″ bolts.

To replace a single component, access the **Replace Component** tool using one of the following techniques:

- ✓ Pick the **Replace Component** button on the **Assembly** panel bar (contained in a flyout button).

- ✓ Pick the **Replace Component** button on the **Assembly** toolbar (contained in a flyout button).

- ✓ Select the **Replace** option of the **Edit** pull-down menu.

- ✓ Right-click on the component name in the Browser, and select the **Replace Component** menu option.

- ✓ Right-click on the component in the graphics window, and select the **Replace Component** menu option.

- ✓ Press the **Ctrl** and **H** keys of your keyboard.

When you access the **Replace Component** tool, you must first select the component you want to replace. If you used one of the right-clicking access techniques, the component will already be selected. Once the desired component is chosen, the **Open** dialog box is displayed, allowing you to locate the new component you want to use to replace the old component. Find the desired component, and pick the **Open** button to replace the component. The replacement component changes to the specified color of the old component and is positioned in the same location as the old component. However, if the new component geometry is significantly different than the old, as shown in Figure 14.68, constraints may no longer be established between components.

To replace all the instances of a component, such as six of the same bolts, access the **Replace All** tool using one of the following techniques:

- ✓ Pick the **Replace All** button on the **Assembly** panel bar (contained in a flyout button).

- ✓ Pick the **Replace All** button on the **Assembly** toolbar (contained in a flyout button).

- ✓ Select the **Replace All** option of the **Edit** pull-down menu.

- ✓ Press the **Alt, Ctrl** and **H** keys of your keyboard.

When you access the **Replace All** tool, you must first select the component you want to replace. Once the desired component is chosen, the **Open** dialog box is displayed, allowing you to locate the new component you want to use to replace the old components. Find the desired component, and pick the **Open** button to replace the components. The replacement components change to the specified color of the old components and are positioned in the same location as the old components. However, if the new component geometry is significantly different than the old, constraints may no longer be established between components.

FIGURE 14.68 An example of a very different replaced component, with constraints automatically removed on placement.

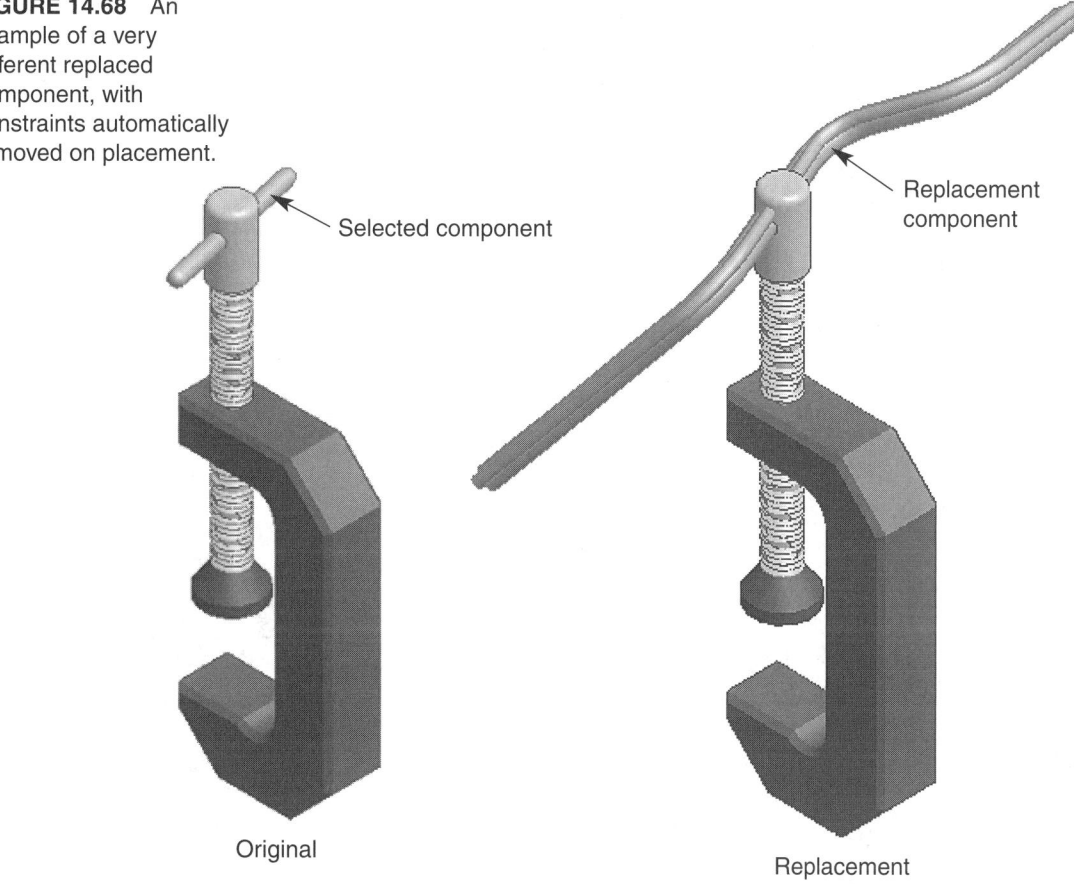

Original

Replacement

Exercise 14.16

1. Continue from EX14.15, or launch Autodesk Inventor.
2. Open the part file, P5.1.ipt.
3. Save a copy of P5.1.ipt as EX14.16.ipt.
4. Close P5.1.ipt without saving, and open EX14.16.ipt.
5. Edit the length of Extrusion1, from 52mm to 70mm.
6. Resave EX14.16, and close the file.
7. Open the assembly file, EX14.7.iam.
8. Save a copy of EX14.7.iam as EX14.16.iam.
9. Close EX14.7.iam without saving, and open EX14.16.iam.
10. Drag two more copies of P5.1.ipt from the Browser into the graphics window.
11. Access the **Replace All** tool.
12. Pick one of the existing short pins, P5.1.ipt.
13. Select EX14.16.ipt in the **Open** dialog box, and pick the **Open** button.
14. The longer pin, EX14.16.ipt, should replace both of the existing short pins, P5.1.ipt.
15. The final exercise should look like the assembly shown.
16. Save the assembly as EX14.16.
17. Exit Autodesk Inventor or continue working with the program.

Exercise 14.16 continued on next page

Exercise 14.16 continued

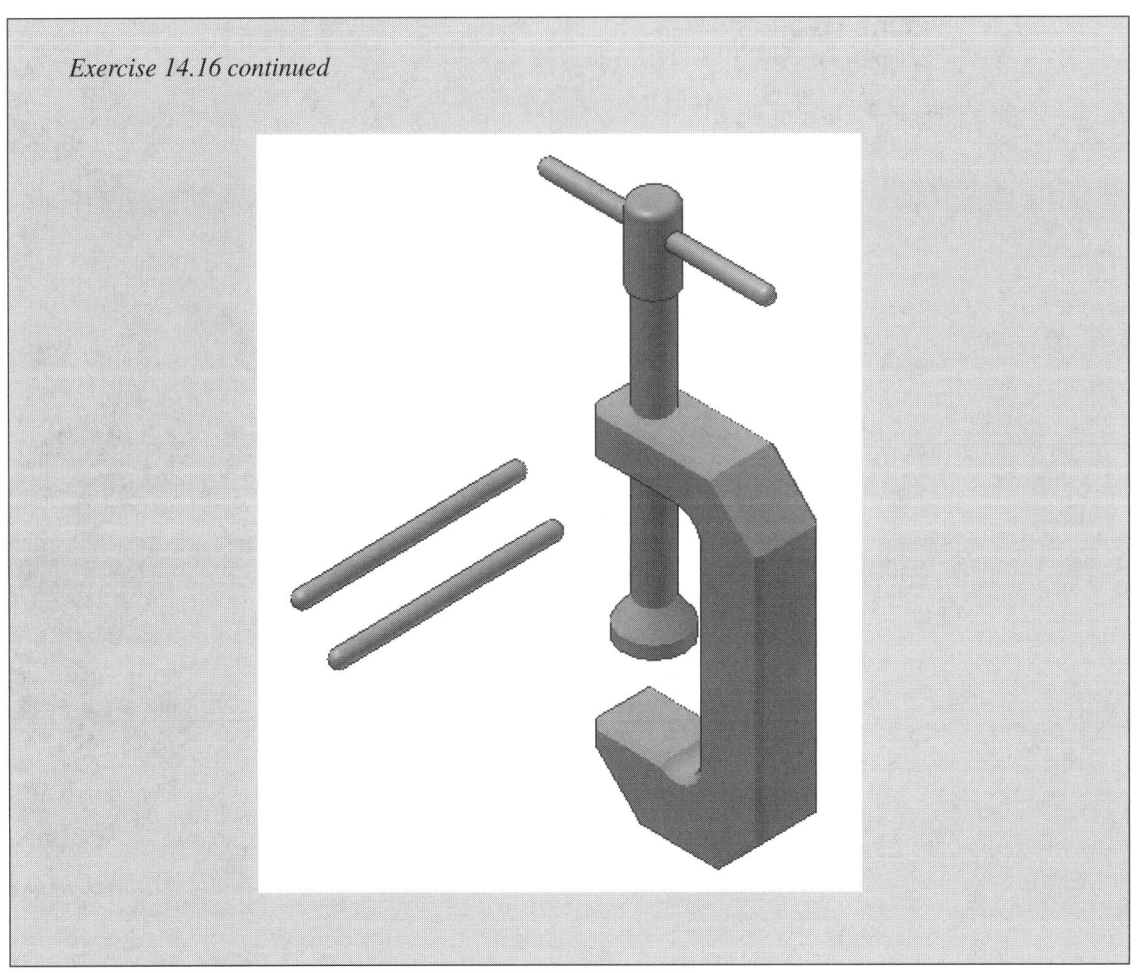

Moving Components

You are already familiar with moving components using dragging and driving options. You can select and drag a component to observe freedom of movement, assuming constraints are not set in place that that do not allow movement. You can also drive constrains to view component movement. When you move a component that is not constrained, the component remains in the moved location, unless you move it back to the old location or undo the move. Use the **Move Component** tool to move a fully constrained, partially constrained, or unconstrained component anywhere in space. The **Move Component** tool is typically used to move components for clarity and display purposes. Even grounded features can be moved using this tool.

Access the **Move Component** tool using one of the following techniques:

✓ Pick the **Move Component** button on the **Assembly** panel bar.

✓ Pick the **Move Component** button on the **Assembly** toolbar.

Once you access the **Move Component** tool, pick and hold down the left mouse button on the component you want to move, and drag it to the desired location. Release the left mouse button to drop the component in the new position. Continue moving components as required. Then right-click and select the **Done** menu option, pick anywhere in space, or access a different tool to exit the **Move Component** command. See Figure 14.69. When you update the assembly, the move is overridden if constraints are present, bringing components back to their original location in space and in reference to other components. For example, the moved screw in Figure 14.69 will return to its original position when the assembly is updated. However, unconstrained components remain in the moved position, and grounded components are grounded to the moved location.

FIGURE 14.69
Moving a component.

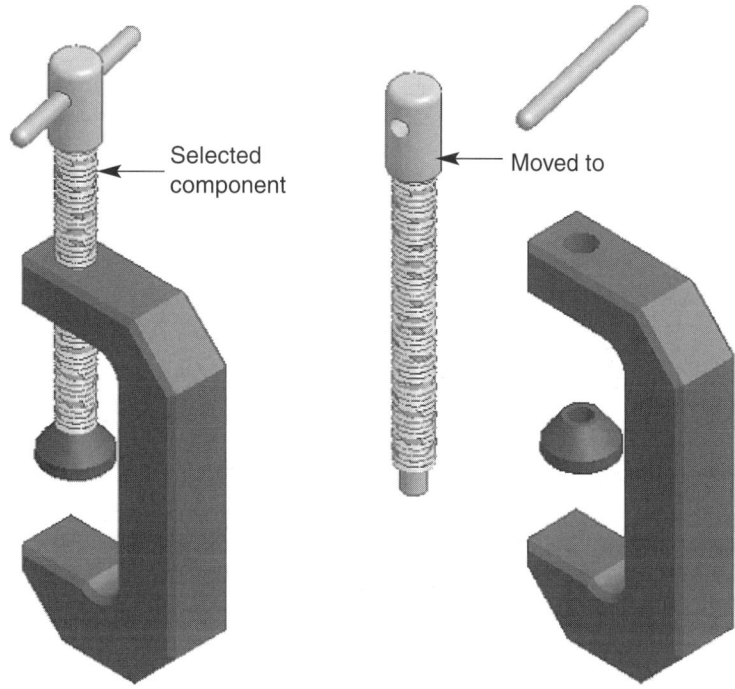

Selected component

Moved to

Rotating Components

When you rotate components using the **Rotate** command, discussed in previous chapters, you are rotating the entire model in space and not actually changing the model position. However, you can rotate a single component, changing its position in space and in relation to other components. When you rotate a component that is not constrained, the component remains in the rotated position, unless you rotate it back to the old location or undo the rotate. Use the **Rotate Component** tool to rotate a fully constrained, partially constrained, or unconstrained component, anywhere in space. Like the **Move Component** tool, the **Rotate Component** tool is typically used to rotate components for clarity and display purposes. Even grounded features can be rotated using this tool.

Access the **Rotate Component** tool using one of the following techniques:

✓ Pick the **Rotate Component** button on the **Assembly** panel bar.

✓ Pick the **Rotate Component** button on the **Assembly** toolbar.

Once you access the **Rotate Component** tool, pick the component you rotate. From this point on, rotating a single component is just like rotating an entire model using the **Rotate** command. To select another component to rotate while still in the **Rotate Component** command, position the rotate icon over the desired component, and pick. Continue rotating components as required. Then, right-click and select the **Done** menu option, pick anywhere in space, or access a different tool to exit the **Rotate Component** command. See Figure 14.70. When you update the assembly, the rotate is overridden if constraints are present, bringing components back to their original position in space and in reference to other components. For example, the rotated screw in Figure 14.70 will return to its original position when the assembly is updated. However, unconstrained components remain in the rotated position, and grounded components are grounded to the rotation.

FIGURE 14.70
Rotating a component.

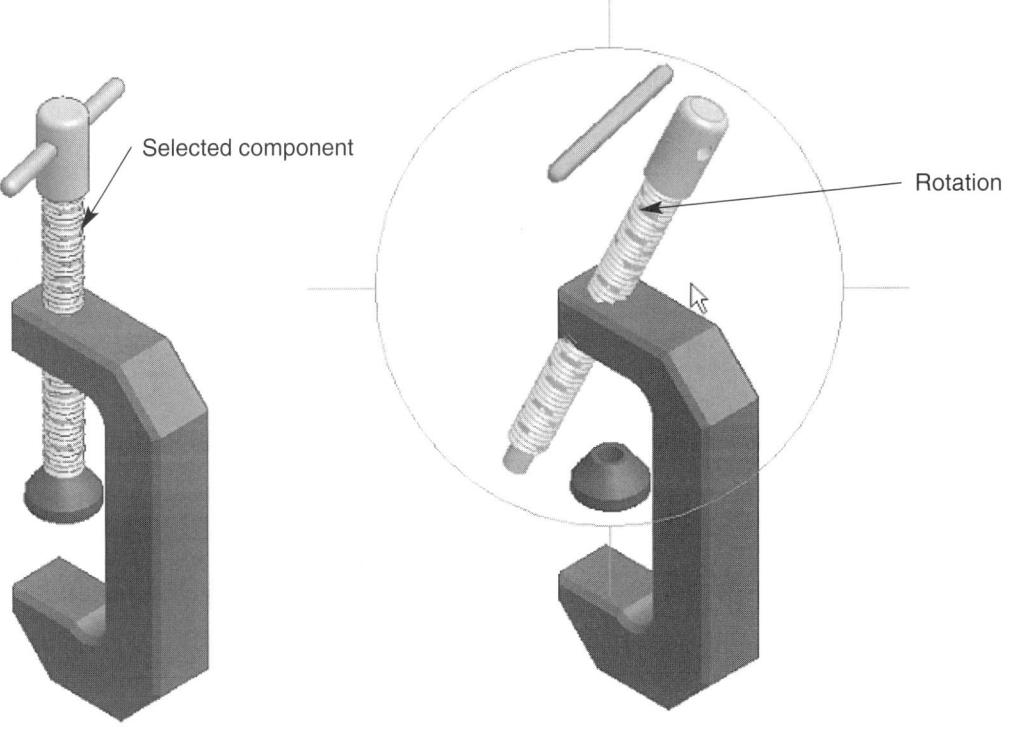

Selected component

Rotation

Exercise 14.17

1. Continue from EX14.16, or launch Autodesk Inventor.
2. Open the assembly file, EX14.16.iam.
3. Access the **Move Component** tool.
4. Move the c-clamp screw away from the other components.
5. Pick the **Update** button on the **Command** bar to return the screw to its constrained location.
6. Access the **Rotate Component** tool.
7. Rotate the c-clamp body so it does not contact any other components.
8. Pick the **Update** button on the **Command** bar to reposition the components constrained to the c-clamp body.
9. Close EX14.16 without saving.
10. Exit Autodesk Inventor or continue working with the program.

Changing Assembly Component Colors

Assembly components are inserted and created using either the default material color or another specified color defined by the material property. Assembly components can be made out of the same material, and consequently they are the same color. Or assembly components can be made from separate materials, but the material colors are similar. As shown in Figure 14.71, it is difficult, and sometimes impossible, to distinguish one component from another. Fortunately, you change the color of specific components without changing the material properties of the component. Changing component colors is almost necessary for large assemblies with multiple components, but can even be important for small assemblies as shown in Figure 14.71. When component colors are modified to contrast other components, designs become much clearer, and easier to understand.

To change the color of a component while in an assembly file, select the component in the graphics window or the component name in the Browser. Then select the desired color from the **Color** drop-down

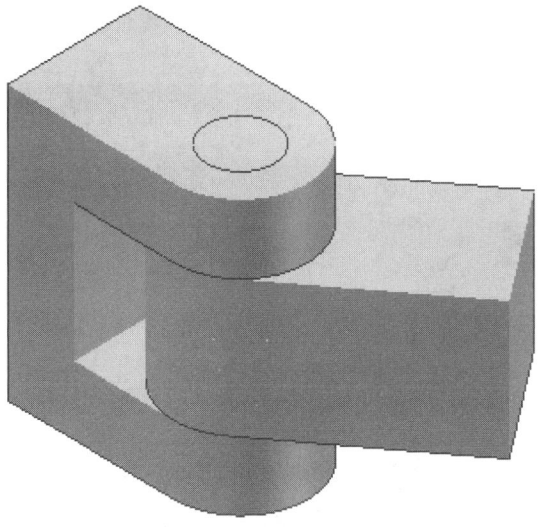

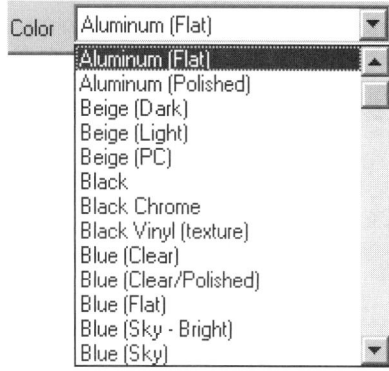

FIGURE 14.71 Distinguishing components of the same color.

FIGURE 14.72 The **Color** drop-down list.

list available in the **Command Bar.** See Figure 14.72. If the assembly components are made out of different materials, with colors that already contrast, pick the **As Material** color option, which displays the color of the component as the specified material color.

Field Notes

For an alternative color option, if assemblies contain parts with multiple, similar types of materials, you can slightly modify the material colors and leave the component colors as As Material. For more information regarding color modification refer to Chapter 4.

Exercise 14.18

1. Continue from EX14.17, or launch Autodesk Inventor.
2. Open the assembly file, EX14.4.iam.
3. Save a copy of EX14.4.iam as EX14.18.iam.
4. Close EX14.4.iam without saving, and open EX14.18.iam.
5. Access the **Place Constraint** tool, and link the axis of the lock washer, washer, and bolt, to the eye bolt axis, using a mate assembly constraint.
6. Using the information presented, change each of the assembly component colors to different colors of your choice.
7. The final exercise should look like the assembly shown.
8. Resave EX14.18.
9. Exit Autodesk Inventor or continue working with the program.

Exercise 14.18 continued on next page

Exercise 14.18 continued

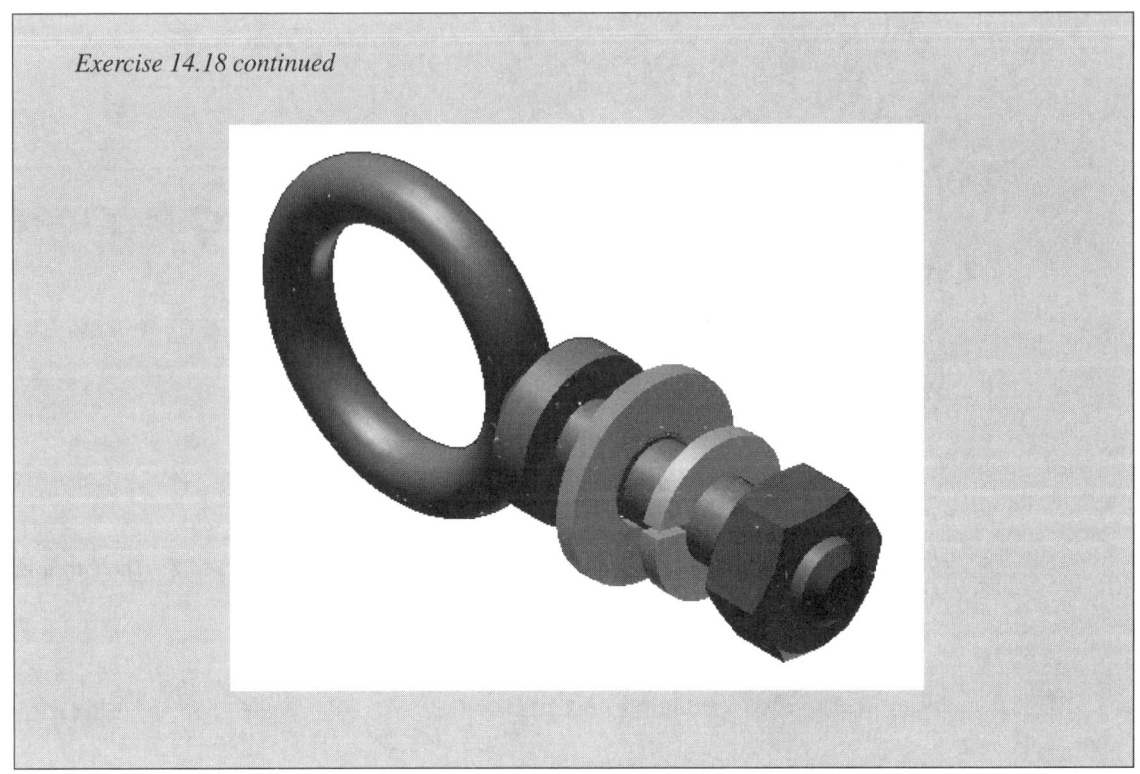

Sectioning Assemblies

In many assemblies it is difficult to clearly observe a design, visualize assembly component relationships, or access component elements when constraining existing components or creating components in place. As a result, a number of assembly sectioning tools are available. Just as in a two-dimensional drawing, sections in an assembly "cut away" material to display hidden component features. See Figure 14.73. Depending on the application and the amount of material you want to remove, you can create a quarter section view, a half section view, or a three-quarter section view. Assembly section views are created by selecting the appropriate existing face, faces, work plane, or work planes, depending on the type of section view. Each of the tools that allow you to generate the three types of section views are contained in a flyout button and can be accessed using one of the following techniques:

✓ Pick the desired section view tool button from the **Section View** flyout button on the **Assembly** panel bar.

FIGURE 14.73 An example of a sectioned assembly (half section).

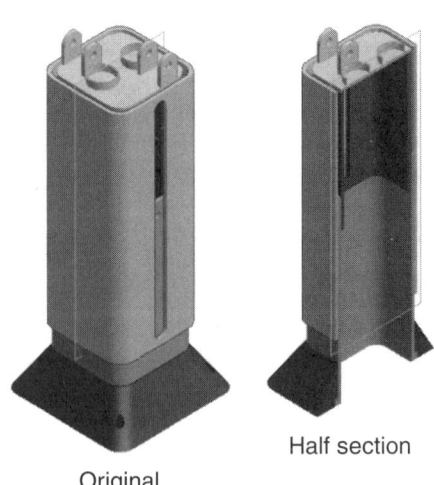

Half section

Original

✓ Pick the desired section view tool button from the **Section View** flyout button on the **Assembly** toolbar.

Once you access one of the available section view tools, use the following information to create the section:

- **Quarter**   A *quarter section* removes all but a quarter of material from two intersecting faces or planes. Once you access the **Quarter Section View** tool, create the section by picking two intersecting component faces or work planes. In addition, the side of the face or plane you select defines which quarter of material is left. See Figure 14.74. If the section looks acceptable, right-click and select the **Done** menu option, press the **Esc** key on your keyboard, or access a different tool to exit the **Quarter Section View** command. If the section is not correct, you may be able to solve the problem by right-clicking and selecting the **Flip** section menu option to change the section of removed material.

- **Half**   A *half section* removes one-half of material from a selected face or plane. Once you access the **Half Section View** tool, create the section by picking the desired component face or work plane. As in quarter section views, the side of the face or plane you select defines which half of material is removed. See Figure 14.75. If the section looks acceptable, right-click and select the **Done** menu option, press the **Esc** key on your keyboard, or access a different tool to exit the **Half Section View** command. If the section is not correct, you may be able to solve the problem by right-clicking and selecting the **Flip** section menu option to change the section of removed material.

- **Three Quarter**   A *three-quarter section* removes only a quarter of material from two intersecting faces or planes. Once you access the **Three Quarter Section View** tool, create the section by picking two intersecting component faces or work planes. Again, the side of the face or plane you select defines which quarter of material is removed. See Figure 14.76. If the section looks acceptable, right-click and select the **Done** menu option, press the **Esc** key on your keyboard, or access a different tool to exit the **Three Quarter Section View** command. If the section is not correct, you may be able to solve the problem by right-clicking and selecting the **Flip** section menu option to change the section of removed material.

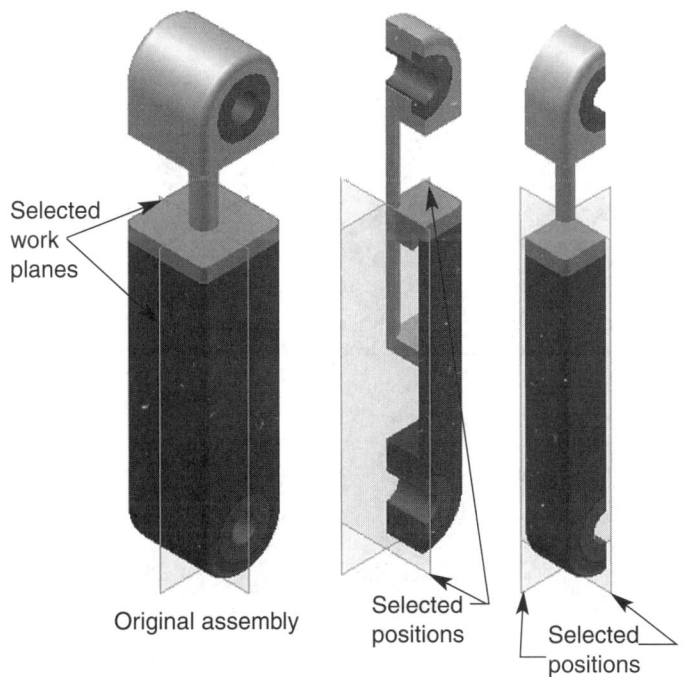

FIGURE 14.74   Creating a quarter section view.

FIGURE 14.75   Creating a half section view

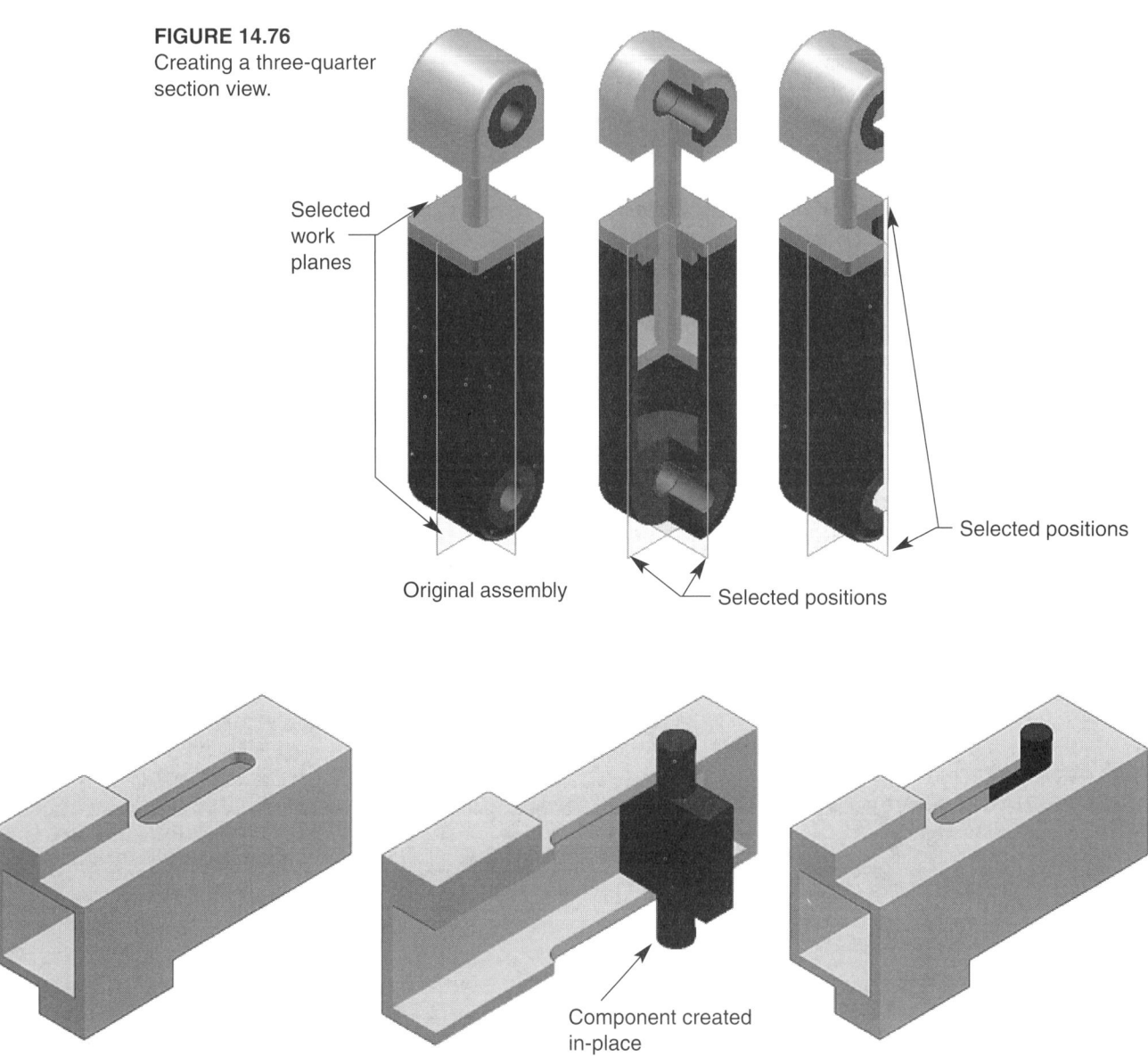

FIGURE 14.76
Creating a three-quarter
section view.

FIGURE 14.77   Using section views to help create components in-place.

Once you have defined a section view, you can now visualize the design and better observe assembly component interaction, as previously mentioned. In addition, you can sketch and create additional components in space that would have been difficult with a section. See Figure 14.77. When you have finished using the section view, pick the **End Section View** button from the **Section View** flyout button to return to the complete assembly.

Design Views

A *design view* is a type of assembly copy that establishes a certain assembly display. A design view takes a "picture" of an assembly complete with the specified view position, such as isometric, the defined set of component colors and styles, zoom, and component selection and visibility status. Clearly design views can potentially have multiple uses because of the variety of assembly content they store. For example, you can create design views with different-colored components, with a section view of the assembly, or with visibility of certain components turned off.

All design views are contained in a separate design view file, .idv, automatically named the same as the assembly file. Each individual design view does not require its own file, and they are not individual

FIGURE 14.78 The **Design Views** dialog box.

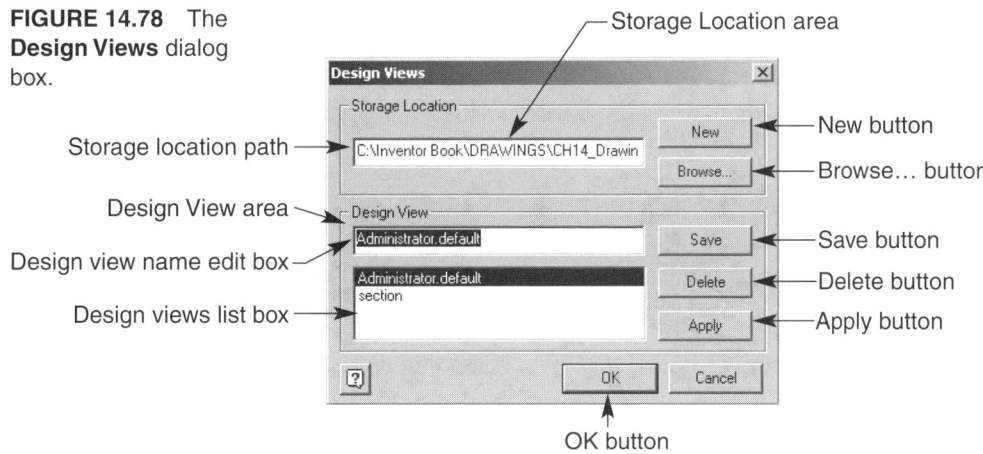

Storage Location area — New button — Browse... button — Storage location path — Save button — Design View area — Delete button — Design view name edit box — Apply button — Design views list box — OK button

assembly files. Design view files consist only of display information. You may have already noticed a default design view is generated when you save an assembly file. This design view is used to save and display assembly specifications when you reopen an assembly.

To create your own design view, first identify how you want the view to look by defining the colors, visible components, or the purpose the design view serves. Then use one of the following techniques to access the **Design Views** dialog box, shown in Figure 14.78:

✓ Pick the **Design Views** button on the Browser.

✓ Pick the **Design View** flyout button on the Browser, and select the **Other** menu option.

✓ Select the **Design Views...** option of the **View** pull-down menu.

The storage location for the design views is typically appropriate for the assembly file in which you are currently working. As a result, you can usually skip ahead to the **Design View** area. However, you can find another storage location by entering a path in the **Storage Location Path** edit box, or by picking the **Browse** button and locating another existing .idv file. To create a new design view file, .idv, specify a storage location path and file name in the **Storage Location Path** edit box, then pick the **New** button. Next, specify the design view name in the **Design View Name** edit box. Create the design view by picking the **Save** button, followed by **Apply.** If you are modifying an existing design view, select the design view name and pick the **Apply** button. You can also delete design views by selecting the design view you want to remove and selecting the **Delete** button. Once the design view is created, you can quickly access the saved display by selecting the desired design view from the **Design Views** flyout button.

Field Notes

> You may want to assign design view names such as half section or bracket-off, to relieve any design view confusion.

Exercise 14.19

1. Continue from EX14.18, or launch Autodesk Inventor.
2. Open the assembly file, EX14.7.iam.
3. Save a copy of EX14.7.iam as EX14.19.iam.
4. Close EX14.7.iam without saving, and open EX14.19.iam.
5. Change the color of each of the c-clamp components.
6. Use the **Work Plane** tool to place the work planes shown.
7. Use the sectioning tools discussed to create the section views shown.
8. Resave EX14.19.
9. Exit Autodesk Inventor or continue working with the program.

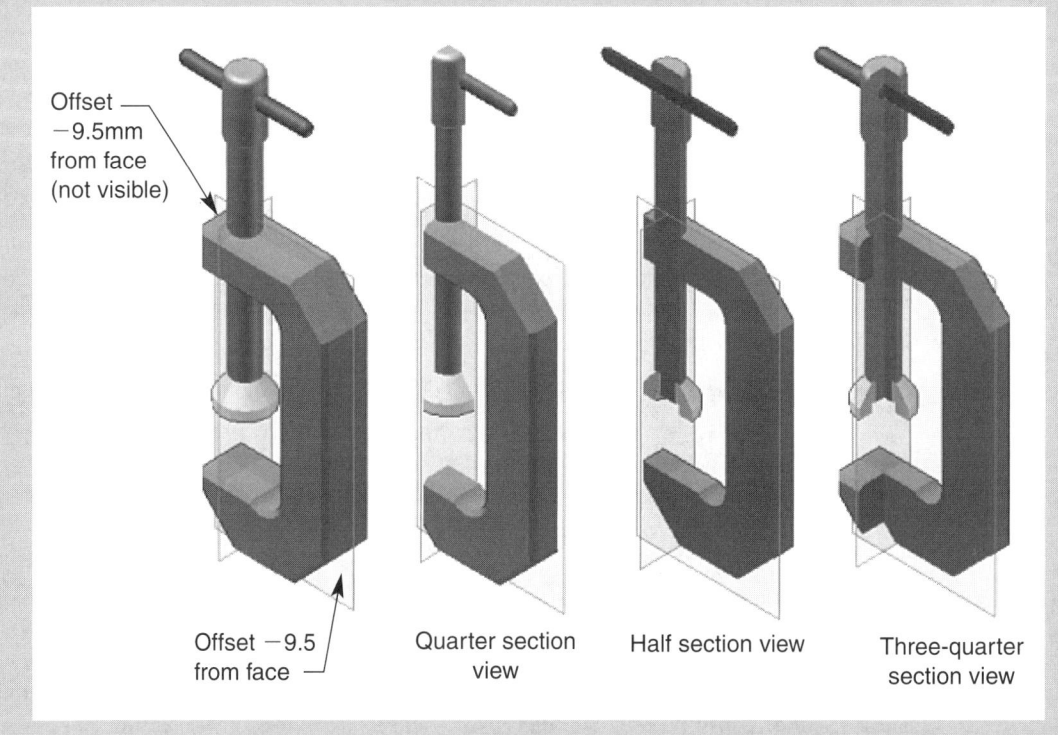

Offset —9.5mm from face (not visible)

Offset —9.5 from face

Quarter section view

Half section view

Three-quarter section view

|||||||||||| CHAPTER TEST

Answer the following questions on a separate sheet of paper.

1. What is an assembly?

2. What are subassemblies?

3. Give the assembly file extension and the purpose of assembly files.

4. What does the browser for a typical assembly display?

5. What is the purpose of the Analyze Interference tool?

6. By default, what does an assembly bill of materials contain?

7. Briefly give the general function of assembly tools.

8. What is a grounded part?

9. What happens to the first component that you place in an assembly drawing?

10. How do you automatically continue placing additional base components, and what do you do after placing all the desired components?

11. Identify at least three situations that may have occurred if you receive the Resolve Link dialog box when you try to insert an existing component.

12. How do you ensure that you do not receive the Resolve Link dialog box?

13. Explain how to insert components from one open window to another.

14. Discuss how to use Windows Explorer to open an assembly file.

15. What is shared content?

16. Why is shared content also referred to as I-drop?

17. Explain the option for using I-drop where you open the assembly file that you want to use to place the part into using your Web browser.

18. Discuss the option for using I-drop where you open the assembly file that you want to use to place the part into and then specify the shared content link.

19. Identify at least two advantages to creating components in-place.

20. Discuss the function of the Create Component tool.

21. Give the basic steps used to create a part in-place.

22. List the two ways that subassemblies can be created.

23. Dragging and dropping components into and out of assemblies and subassemblies can become confusing. Explain how to confirm the location of components.

24. Discuss the characteristics of assembly constraints.

25. Briefly explain the purpose of an assembly constraint.

26. Briefly discuss the purpose of a motion constraint.

27. Briefly explain the purpose of a transitional constraint.

28. Discuss the function of the Place Constraint tool.

29. What is a mate constraint?

30. Give the purpose of an angle constraint.

31. Briefly discuss the purpose of a tangent constraint.

32. Briefly explain the purpose of an insert constraint.

33. Identify the types of features or objects that can contain motion constraints.

34. Explain the function of the Rotation button.

35. What does a 3:1 rotation ratio mean?

36. Give the function of the Rotation-Translation button.

37. Give an example of a transitional constraint.

38. Identify another name for the dragging constraint method, and give its basic function.

39. How do you toggle between different types of constraints using shortcut keys?

40. Identify the shortcut keys used to change the current constraint to an angle constraint; and how do you change the angle direction if required?

41. Identify the shortcut keys used to change the current constraint to an insert constraint; and how do you change the insert direction if required?

42. Identify the shortcut keys used to change the current constraint to a mate constraint.

43. Identify the shortcut keys used to change the current constraint to a rotation motion constraint.

44. Identify the shortcut keys used to change the current constraint to a transition motion constraint.

45. Identify the shortcut keys used to change the current constraint to a tangent constraint.

46. Identify the shortcut keys used to change the current constraint to a transitional constraint.

47. Discuss the use of the Alt key and left mouse button when performing a dragging mate constraint operation.

48. What automatically happens when you sketch on an existing component face?

49. What automatically happens when you create a component in-place, using existing component geometry?

50. Explain what iMates are and briefly explain how they work.

51. Briefly discuss the most difficult aspect of working with iMates.

52. What do you do to create an iMate before accessing the Create iMate tool?

53. What does the iMate symbol specify, and when is the symbol is no longer visible.

54. When does the Place Component tool not work well when using iMates to create constraints between components?

55. When you use the Place Component tool to insert components that contain iMates, when do the inserted components become constrained to each other?

56. Give the steps used to apply iMates to the assembly; and what is the importance of the Use iMate check box?

57. Explain how to use the dragging method to place iMates.

58. When can you use the Place Constraint tool to insert iMates?

59. Discuss the process for using the Place Constraint tool to insert iMates.

60. What is the function of suppressing a constraint and how is a suppression performed?

61. What does driving a constraint allow you to do?

62. How do you access the Drive Constraint dialog box?

63. How do you get a driven constraint to move a certain number of steps?

64. What is the function of the Collision Detection check box and how does it work?

65. Briefly explain the function of the Pattern Component tool.

66. What is a reoccurrence of components in a designated configuration called?

67. Give the steps used to create an associative component pattern.

68. Briefly explain the function of the Replace Component and Replace All tools.

69. Give the steps for using the Replace All tool.

70. When using the Replace All tool, what can happen if the new component geometry is significantly different than the old?

71. Briefly explain the function of the Move Component tool.

72. Briefly discuss the function of the Rotate Component tool.

73. How do you change the color of a component while in an assembly file?

74. Explain what you get from a quarter section view, and how the Quarter Section View tool works.

75. Explain what you get from a half section view, and how the Half Section View tool works.

76. Explain what you get from a three-quarter section view, and how the Three Quarter Section View tool works.

77. How do you exit the Half Section View command if the section looks acceptable?

78. What is a design view and what is its function?

|||||||||||| PROJECTS

Instructions:

- Open a new assembly file, or the specified existing assembly file.

- Insert the specified parts or subassemblies.

- Create the specified parts or subassemblies.

- Apply the required constraints.

- Change the component colors as indicated, or to a color of your choice.

- When sketches, dimensions, and other component and assembly geometry are not provided, use specifications of your choice.

1. Name: Bottle Assembly

   Units: Inch

   Assembly File: New, inch

   Inserted Files: P5.8.ipt and P6.8.ipt.

   Shared Content Files: None

   New Components: None

   Constraints Used: Assembly Insert

   Bottle Color: Plastic (gray)

   Cap Color: Plastic (white)

   Save As: P14.1

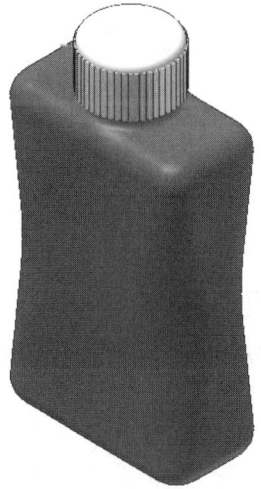

2. Name: Pipe Assembly

   Units: Metric

   Assembly File: New, metric

   Inserted Files: (2) P6.9.ipt and (2) P7.2.ipt.

   Shared Content Files: Fasteners/ Screws and Threaded Bolts/Hex Head Types/Heavy Hex Bolt (Regular – Metric) M16 X 2 X 40 and Fasteners/Nuts/Hex Nut Style 1 (Regular Thread – Metric) M16 X 2

   New Components: P14.2.ipt (300mm total length pipe, shown)

   Constraints Used: Assembly Insert, Flush, and Mate

   Pipe Color: Metal-Steel

   Fastener Color: Metal-Steel (polished)

   Save As: P14.2

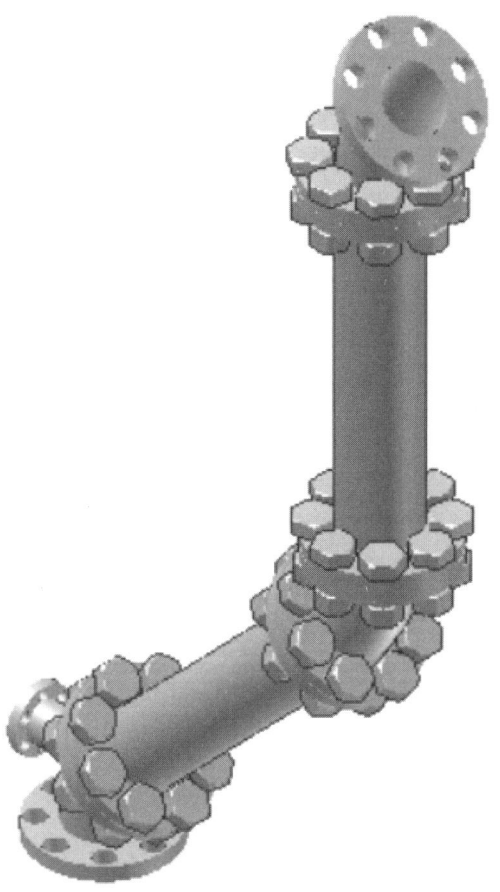

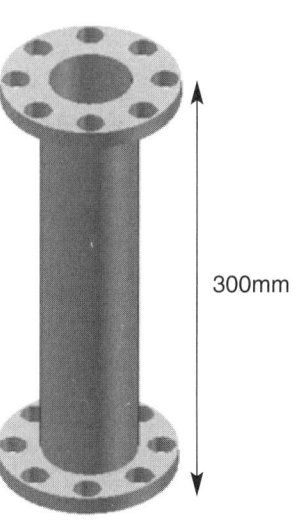

300mm

P14.2

3. Name: Rotating Hinge Assembly

   Units: Inch or Metric

   Assembly File: New, inch or metric

   Inserted Files: None

   Shared Content Files: None

   New Components: P14.3a.ipt, (2) P14.3b, and P14.3c

   Constraints Used: Assembly Insert

   Housing Color: Blue (Clear) for clarity

   Bushing Color: Yellow (Flat)

   Pin Color: Beige (Light)

   Save As: P14.3

   Special Instructions:

      Create each of the components in the assembly file, using dimensions of your choice. Begin with the housing (P14.3a.ipt), then create the bushing (P14.3b.ipt) using the housing for reference (adaptive). Finally, build the pin (P14.3c.ipt). Insert another copy of P14.3b, and constrain the components to each other.

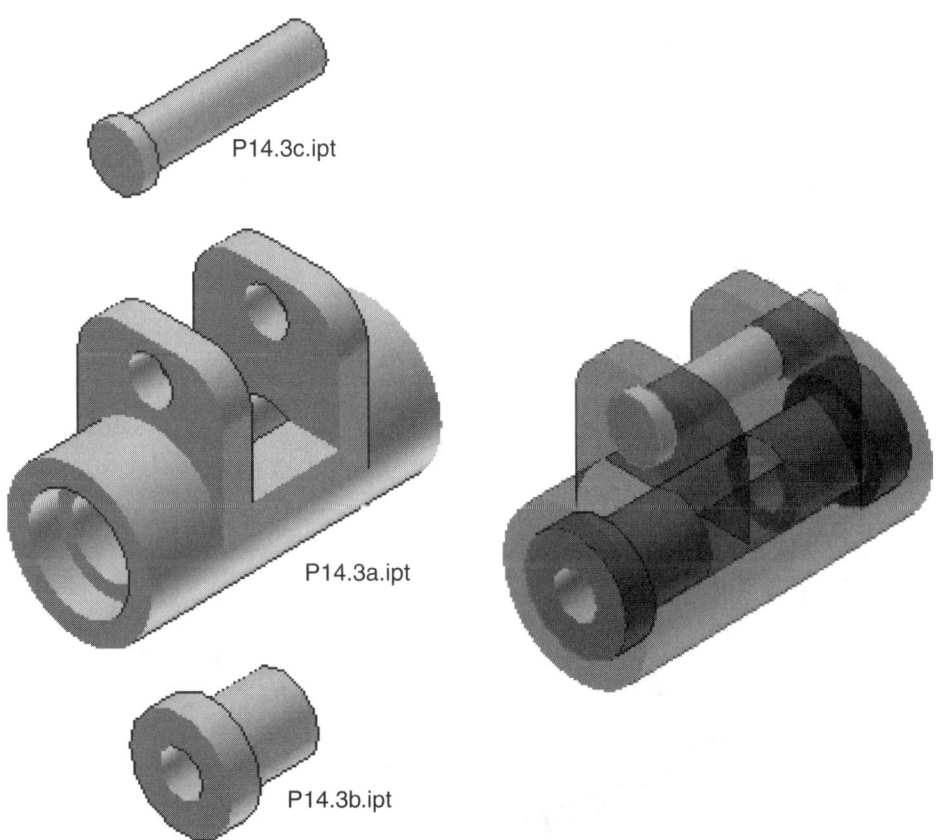

P14.3c.ipt

P14.3a.ipt

P14.3b.ipt

4. Name: Cam Assembly

   Units: Inch

   Assembly File: Save a copy of EX14.9 as P14.4

   Cam Color: Red (Flat)

   Bracket Color: Aluminum (flat)

   Save As: P14.4

   Special Instructions:

   Open the saved copy of EX14.9 (P14.4). Remove adaptivity of EX14.9b, and replace the mate constraint between the bracket axis and cam axis. Suppress the flush constraint between the two cam teeth. Drive the angle constraint to observe cam rotation.

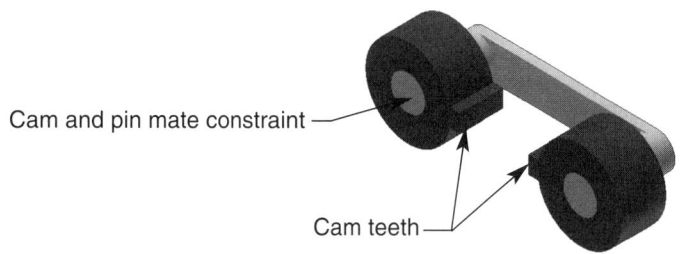

Cam and pin mate constraint

Cam teeth

5. Name: Fork Assembly

   Units: Metric

   Assembly File: New, metric

   Inserted Files: P8.5.ipt

   Shared Content Files: None

   New Components: P14.5.ipt (90mm total length fork handle, shown)

   Constraints Used: Assembly Flush

   Fork Color: As Material

   Handle Color: Brown (Flat)

   Save As: P14.5

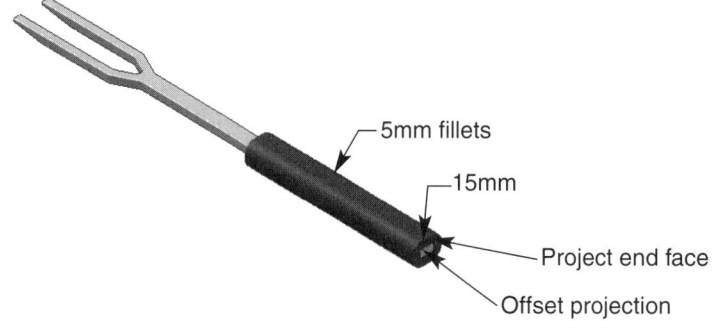

5mm fillets

15mm

Project end face

Offset projection

6. Name: Housing Assembly

Units: Inch

Assembly File: New, inch

Inserted Files: P10.2.ipt and EX14.9.iam

Shared Content Files: None

New Components: P14.6.ipt (housing cover, shown)

Constraints Used: Assembly Insert, Flush, and Mate

Housing Color: As Material

Housing Cover Color: Custom

Fasten colors: Custom

Save As: P14.6

Special Instructions:

First, insert P10.2.ipt, and then insert EX14.9.iam. Expand the children of EX14.9.iam in the Browser, and drag a copy of each of the assembly components (eye bolt, washer, lock washer, nut) into the graphics window. Delete EX14.9.iam in the Browser. Then add the housing cover as shown using projected edges for reference. Be sure you use a sheet metal template. Next, constrain a single fastener (eye bolt, washer, lock washer, nut) to one of the holes. Finally, apply a rectangular component pattern, using counts of 2 and spacing of 3″ and 7″.

|||||||||||||||||||||||||||||||||||||||||||||||||||||||||||||||||||||||||||||||||||||||||||||||||||||||||||||||||||||||||||||

CHAPTER 15

Adapting Parts and Assemblies

LEARNING GOALS

After completing this chapter, you will be able to:

◎ Work with adaptivity to define the size and location of geometry.

◎ Use adaptive work environment options.

◎ Adapt part sketches and features.

◎ Adapt parts and subassemblies.

Working with Adaptivity

Often when developing a product design, part and subassembly geometry changes to reflect new design intent. For example, you may need to specify a larger diameter hole for a certain part. Consequently, you must also specify a new diameter for the shaft that fits inside the hole. Using this example, one option for defining a larger hole and shaft size is to individually edit the hole and shaft feature sketches, which specify the feature diameters. However, Autodesk Inventor contains adaptive capabilities; in the previous example, this automatically changes the diameter of the shaft when the size of the hole is modified, or conversely, the diameter of the hole changes when the size of the shaft is redefined. As the name implies, *adaptive* design elements, including part feature sketches, part feature specifications, and subassemblies, change or adapt to other assembly components. Adapting geometry is another way to define the location and size of design elements.

Adapting sketches, features, and subassemblies further contributes to the parametric nature of Autodesk Inventor. When you modify adaptive geometry, elements that correspond to the modified features also change to ensure correct design intent and product development. In addition, if you produce a longer part, for example, in an assembly by adapting the part to another component, both the assembly and part files update to accept the new design. Another way to understand adapting geometry is to think about sketch, feature, and assembly constraints. Constraints remove the ability of geometry to change in reference to other design components. When a sketch is not fully constrained, or a feature is adapted, the elements are free to reconstrain or become fully constrained to other components.

Figure 15.1A shows a part model of a 12″ section of angle iron inserted into an assembly file and partially constrained to an 18″ long plate. However, the design requires the angle to be the same length as the plate. As previously mentioned, you could edit the angle extrusion from 12″ to 18″. However, this process does not ensure design intent or future parametric relationships. Figure 15.1B shows how if the angle is adaptive and a flush constraint is placed between the two components, the angle automatically changes to reflect the new design requirements. In addition, as shown in Figure 15.2, if you change the size of the plate, the length of the angle iron will also change.

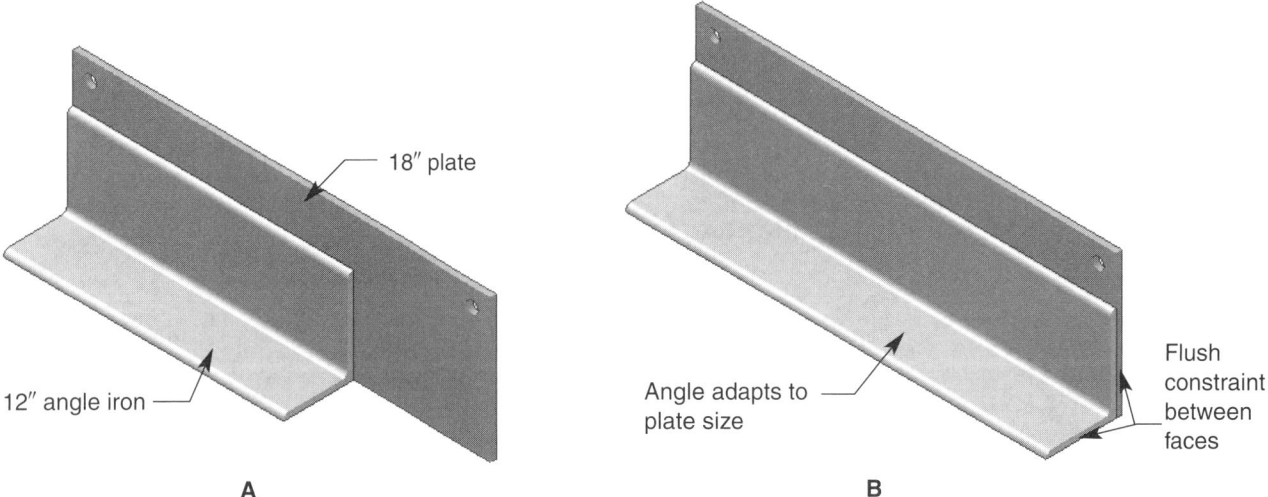

FIGURE 15.1   An example of using adaptivity.

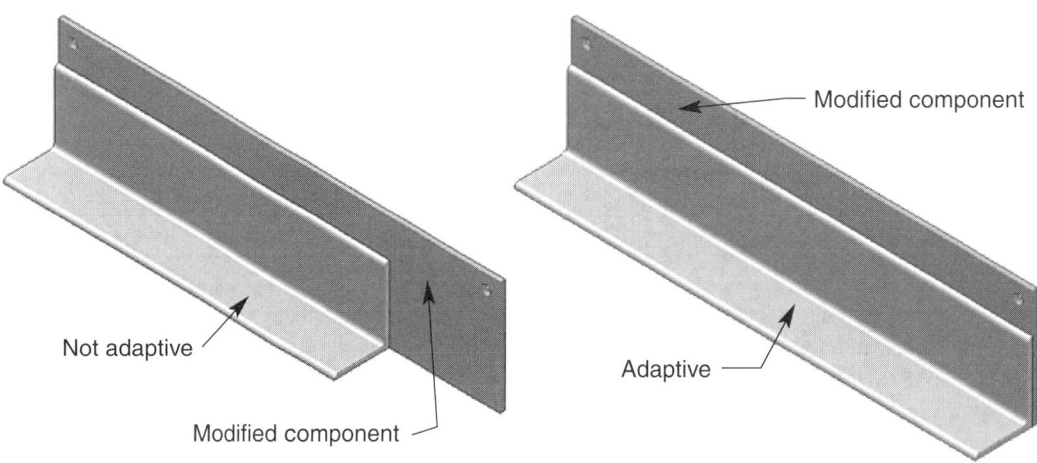

FIGURE 15.2   An example of using adaptivity, and an adaptive parametric relationship.

Adaptive Work Environment Options

As previously mentioned, part sketches, part features, and subassemblies can be adapted in reference to other assembly components. As a result, adaptive tools, options, and specifications are available in both the part model and assembly work environments. However, depending on the type of element, feature or sketch, and type of file you are working with, the options and process of adapting geometry are slightly different. The adaptive capabilities of sketches, features, parts, and subassemblies are discussed throughout this chapter. However, you may want to become familiar with some of the following interface options available in both part and assembly files:

■ **Browser**   The browser displays the adaptive characteristics of sketches, features, parts, and subassemblies, using the adaptive symbol. See Figure 15.3. In addition to identifying adaptive elements, as you will see, you can also specify item adaptivity using the sketch, feature, or subassembly name in the Browser.

FIGURE 15.3 The browser display for an assembly with an adaptive subassembly, adaptive parts, and adaptive sketches.

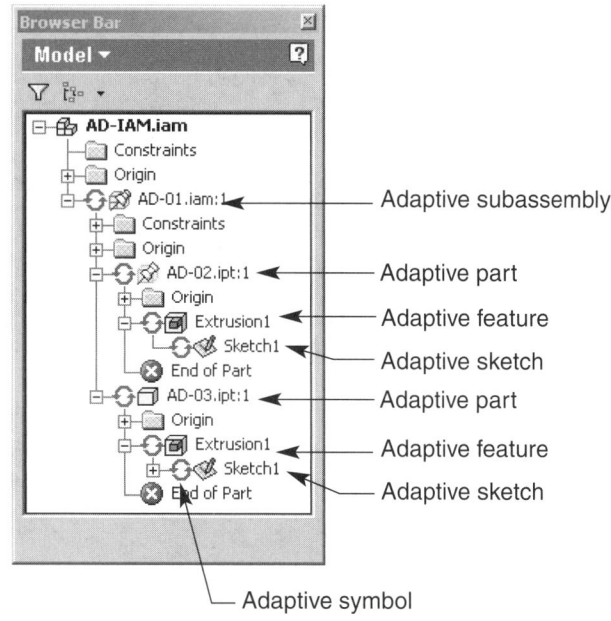

Adaptive subassembly

Adaptive part

Adaptive feature

Adaptive sketch

Adaptive part

Adaptive feature

Adaptive sketch

Adaptive symbol

Field Notes

Additional assembly browser options are discussed throughout this chapter.

■ **Adaptivity used in assembly** This check box is located in the **Modeling** tab of the **Document Settings** dialog box, available by picking the **Document Settings...** option of the **Tools** pull-down menu. See Figure 15.4. The **Adaptivity used in assembly** check box is typically always selected, but you can pick the check box if you do not want the geometry to be adaptive. For example, if you clear this check box while in a part file, then insert the part into an assembly, the part will not adapt, even if you have specified the part as adaptive.

FIGURE 15.4 The **Document Settings** dialog box (**Modeling** tab displayed).

Adaptivity used in assembly check box

Modeling tab

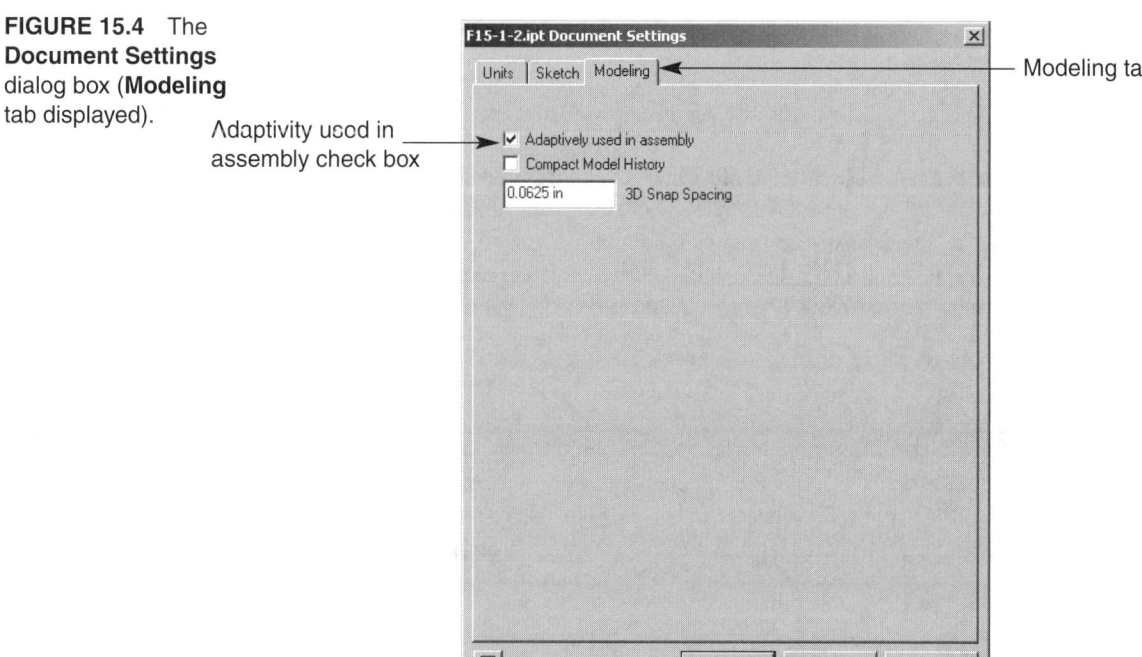

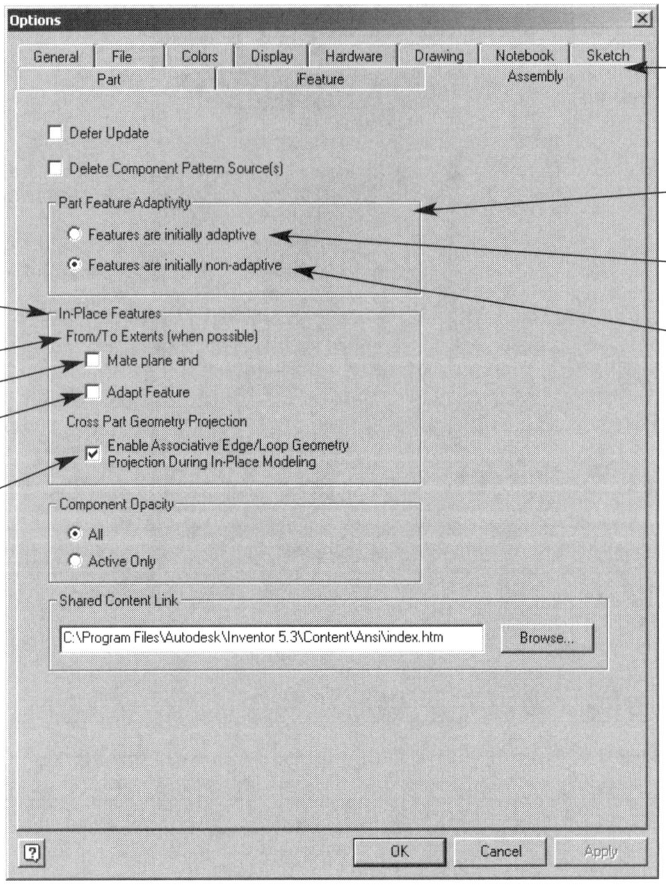

FIGURE 15.5   The
Options dialog box
(**Assembly** tab displayed).

Assembly tab

Part Feature
Adaptivity area

Features are initially
adaptive radio button

Features are initially
non-adaptive radio
button

In-Place Features area

From/To Extents (when possible)

Mate Plane and

Adapt Feature check boxes

Cross Part Geometry
Projection Enable Associative
Edge/Loop Geometry
Projection During In-Place
Modeling check box

- **Part Feature Adaptivity**   This area is located in the **Assembly** tab of the **Options** dialog box, available by picking the **Application Options...** option of the **Tools** pull-down menu. See Figure 15.5. Use the **Part Feature Adaptivity** area to define the adaptivity of assembly components you create. Pick the **Features are initially adaptive** radio button if you want to automatically make new component features adaptive, or select the **Features are initially non-adaptive** radio button if you want to manually control the adaptivity of component features.

- **In-Place Features**   This area is also located in the **Assembly** tab of the **Options** dialog box, available by picking the **Application Options...** option of the **Tools** pull-down menu. See Figure 15.5. Use this area to specify preferences for creating components inside an assembly file. Select the **Mate Plane** check box if you do not want features to be adaptive, but do want a mate constraint to occur between the feature and the plane. Pick the **Adapt Feature** check box to automatically adapt features to changing geometric places. Choose the **Enable Associative Edge/Loop Geometry Projection During In-Place Modeling** check box to generate a sketch from selected geometry when creating a new feature.

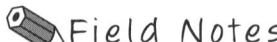

Field Notes

Additional assembly pull-down menu system options, and those not previously discussed, are identified throughout this chapter in the sections where they apply.

Exercise 15.1

1. Launch Autodesk Inventor.
2. Open a new inch or metric assembly file.
3. Explore the basic assembly file interface options and work environment, as previously discussed.
4. Exit Autodesk Inventor without saving, or continue working with the program.

Adapting Sketches

One option for creating adaptive assembly components is to use adaptive sketches that define sketched features. As discussed in Chapter 3, when you add geometric constraints and dimensions to a sketch, you are constraining the sketch. Typically, you should fully constrain your sketches to ensure dimensions and geometry are correct. However, you may not want to completely constrain the sketch if you will not know the appropriate specifications of the sketched feature until the part is placed inside an assembly, or you want the sketched feature to adapt to other components. In these situations, an adaptive sketch is used. Adaptive sketches do not become fully constrained until they are placed inside an assembly and components are constrained together.

The sketch shown in Figure 15.6A is fully constrained and cannot adapt to other geometry. When the sketch is extruded, as displayed in Figure 15.6B, only the extrusion can be adapted, because all the degrees of sketch movement have been removed. See Figure 15.6C. Adapting features, like the extrusion in Figure 15.6, is discussed later in this chapter. However, if one or more sketch dimensions or sketch constraints are

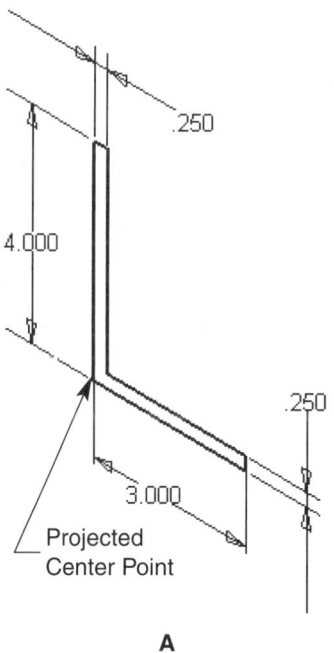

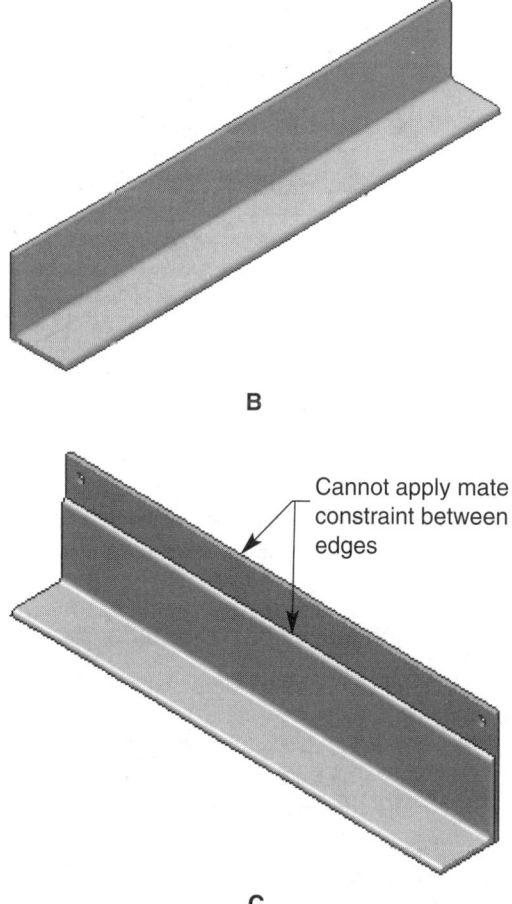

FIGURE 15.6 (A) An example of a fully constrained sketch, in which none of the sketch dimensions or constraints can adapt to other component geometry. (B) The extruded sketch. (C) Sketched feature cannot adapt.

FIGURE 15.7 Using adaptive sketches.

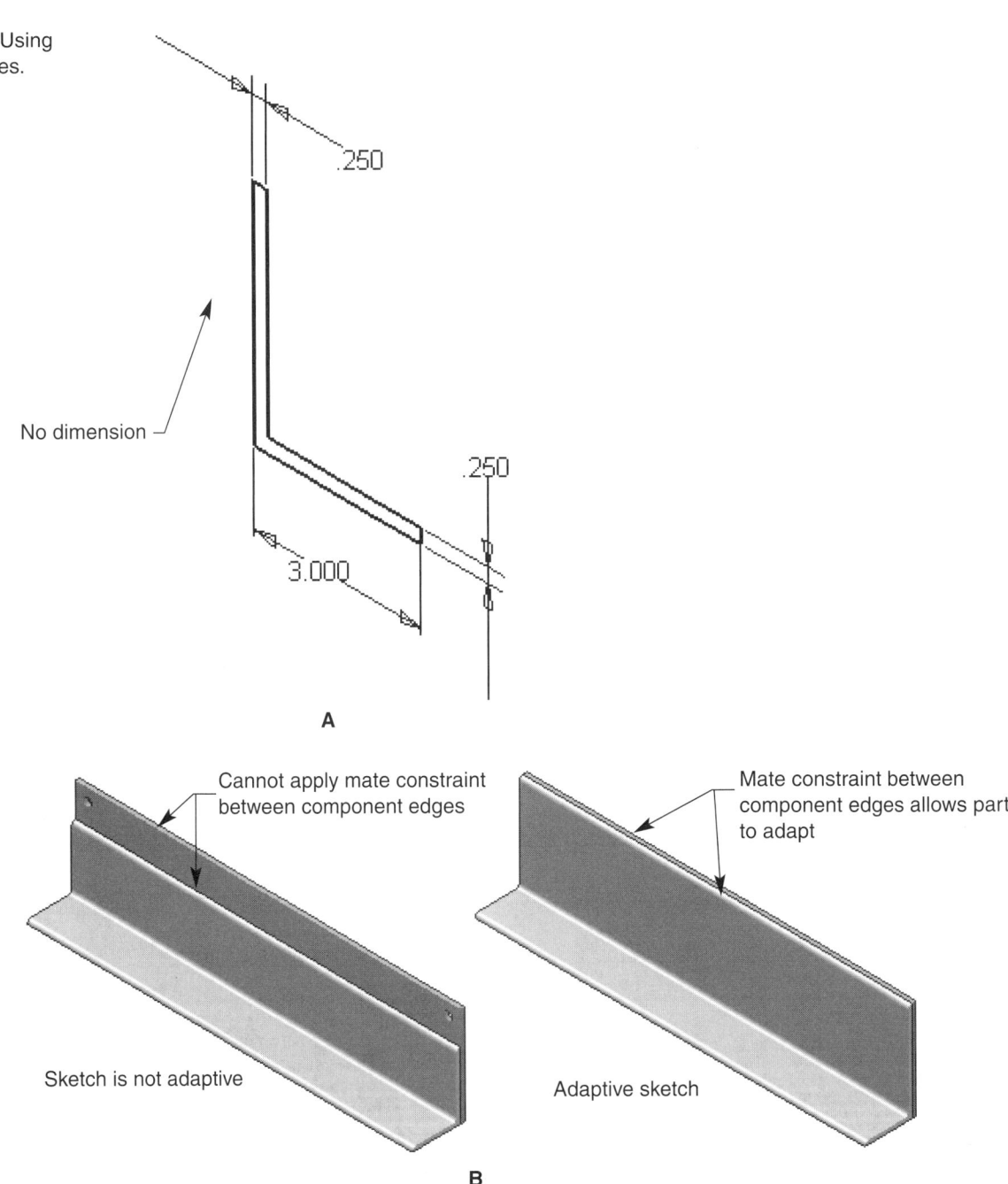

not initially added or are removed, as shown in Figure 15.7A, the sketched feature is free to adapt to other component geometry. See Figure 15.7B.

Creating Adaptive Sketches

The key to developing an adaptive sketch and producing a sketched feature that can adapt to other component geometry is to not fully constrain the sketch. An extreme example is shown in Figure 15.8A. As you can see, this sketch does not contain any dimensions or geometric constraints. When extruded and placed inside an assembly, the sketch is free to constrain to other component geometry and create a part that is fully adaptive. See Figure 15.8B. You do not need to leave a sketch completely unconstrained, as shown in Figure 15.8A. Typically, only one or two dimensions are left off a sketch to allow the part to adapt to specific component geometry. You must determine what sketch pieces you want to fully define, and which you want to adapt, depending on the design intent, and based on existing and future components.

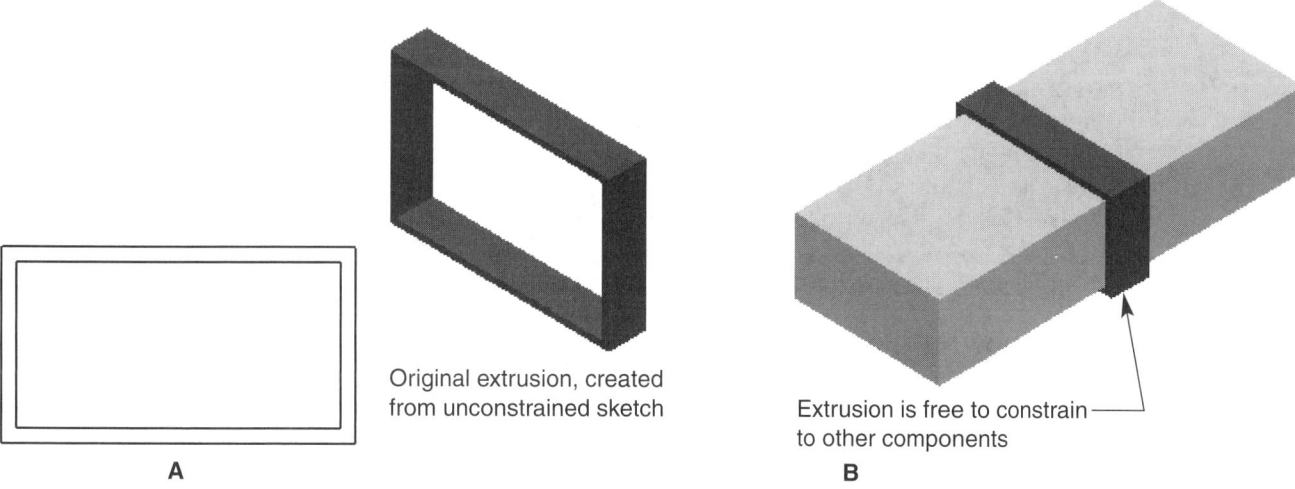

FIGURE 15.8 (A) Creating and (B) using an adaptive sketch.

Typically, a sketch is used to create a sketch feature, which is often also adaptive. As a result, when the feature becomes adaptive, the sketch also automatically becomes adaptive. However, if you want to make a sketch adaptive, right-click on the sketch name in the Browser, or the sketch geometry in the graphics window, and select the **Adaptive** menu option.

Exercise 15.2

1. Continue from EX15.1, or launch Autodesk Inventor.
2. Open a new inch part file.
3. Open a new sketch on the XY Plane, and sketch the geometry shown.
4. Delete the dimension shown.
5. Finish the sketch.
6. Right-click on the sketch name in the Browser, or the actual sketch in the graphics window.
7. Select the **Adaptive** menu option to make the sketch adaptive.
8. The adaptive symbol should be displayed next to the sketch name in the Browser.
9. Save the part file as EX15.2.
10. Exit Autodesk Inventor or continue working with the program.

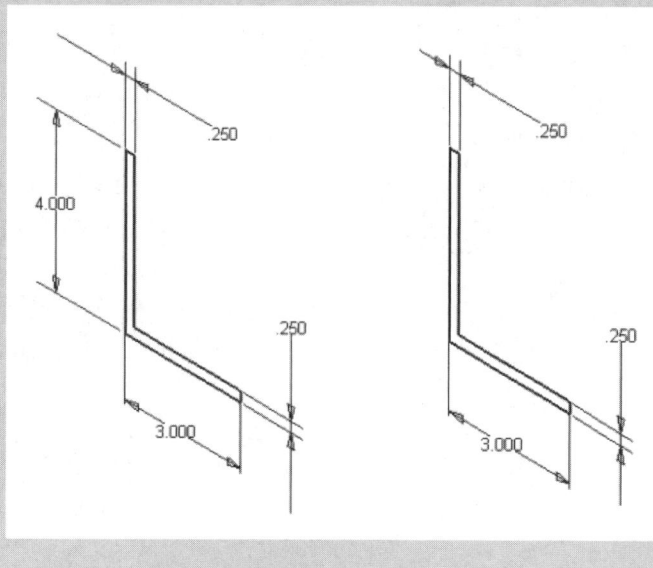

In-Place Adaptive Sketches

In Chapter 14, you were introduced to adaptive sketches when creating new components in-place. As discussed, using the **Create Component** tool, you can locate a new component sketch plane on an existing component face, and then reference existing component feature geometry to define the new component. For example, if you want to add a shaft that is the same size as a part hole, you can place the new component sketch plane on the existing part face, and then project the hole edge. This type of sketch is known as a *reference sketch*. See Figure 15.9A. When you use this technique for developing a new component, the sketch and the new part automatically become adaptive, and an adaptive reference of the projected feature element is defined. See Figure 15.9B.

Field Notes

> The **Adapt Feature** check box in the **Options** dialog box must be selected to automatically specify adaptivity when creating components in-place.

For many applications, a reference sketch is appropriate. Still, you cannot dimension reference geometry like the sketch shown in Figure 15.8A, because it is essentially part of the existing component. However, you can remove the relationship between the existing parent feature and the new sketch by right-clicking on the reference in the Browser and selecting the **Break Link** menu option. This action allows the sketch to become independent of the parent feature, even though the sketch is still referenced. In order for you to dimension and constrain the sketch, just like any other sketch, you must also select the sketch and pick the **Normal** option from the **Command** bar **Style** drop-down list.

When you finish a sketch created in-place using parent feature geometry, you will notice an assembly constraint is automatically placed between the existing component and the new sketch plane. See Figure 15.10. This constraint ensures the parametric, adaptive relationship between the existing component, the new component sketch plane, and the new component future face, edge, or point. The constraint is usually required to establish a link between two components, but can be removed if necessary.

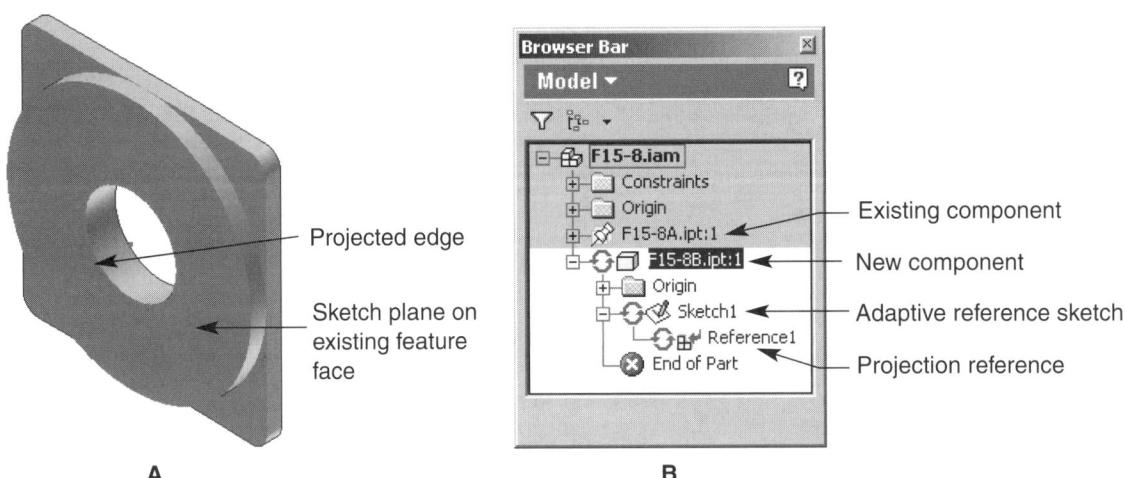

FIGURE 15.9 (A) Using existing feature geometry to define an adaptive sketch. (B) The Browser display of a component sketch created in-place.

FIGURE 15.10 A constraint identifies the relationship between the sketch plane and the selected component face.

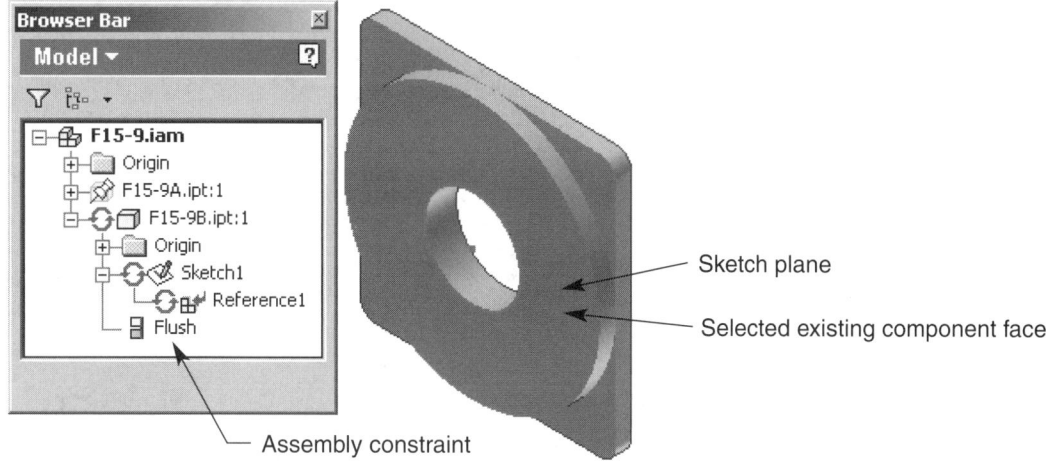

Exercise 15.3

1. Continue from EX15.2, or launch Autodesk Inventor.
2. Open a new metric assembly file.
3. Access the **Create Component** tool.
4. Specify a new file name of EX15.3A.ipt, a part file type, and a location and template of your choice.
5. Pick the **OK** button.
6. Open a new sketch on the XY Plane, and sketch a 100mm×100mm square.
7. Finish the sketch.
8. Extrude the sketched square 25mm in a positive direction.
9. Open a new sketch on the top face, and place a **Hole, Center Point** in the middle of the face as shown.
10. Add a ⌀25mm hole, using the **Through All** termination option.
11. Finish the edit.
12. Access the **Create Component** tool.
13. Specify a new file name of EX15.3B.ipt, a part file type, and a location and template of your choice.
14. Pick the **OK** button.
15. Select the top face of the extrusion to sketch the new component.
16. Project the hole edge.
17. Finish the sketch.
18. In the Browser, you should notice EX15.3B.ipt is adaptive, along with an adaptive reference sketch.
19. Finish the edit.
20. Save the assembly file as EX15.3.
21. Exit Autodesk Inventor or continue working with the program.

Exercise 15.3 continued on next page

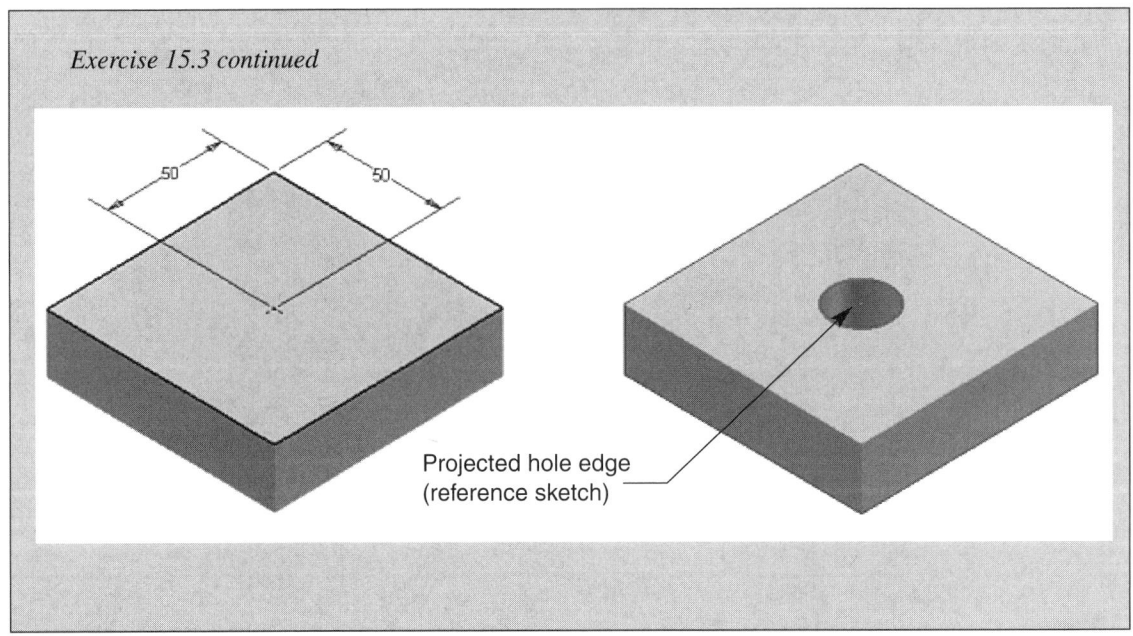

Exercise 15.3 continued

Projected hole edge
(reference sketch)

Adapting Features

Another way to adapt a part is by adapting the part feature. Often it is difficult to differentiate between an adaptive sketch and an adaptive feature, because a sketch is required to create a base feature. Typically, an adaptive sketch is not fully defined using dimensions or geometric constraints. For example, you may not add a width dimension to a sketched rectangle. This would allow the width of the rectangle and feature the rectangle creates to adapt. As this example shows, when you adapt the sketch of a sketched feature, the feature is also adaptive. In contrast, to an adaptive sketch, an adaptive feature may be fully defined. For example, if you extrude a $1'' \times 3''$ rectangle to a distance of $5''$, the extrusion is fully defined. However, if you adapt the extrusion and place assembly constraints, the distance is free to change according to the constraints.

You can adapt both a sketch and a feature together by not fully defining the sketch and by specifying the feature as adaptive. However, you do not have to create a sketch that is unconstrained and adaptive, to adapt a feature. For example, an assembly requires $\varnothing.5''$ rod, but the rod length will not be defined until it is placed in the assembly, or the rod length must change to reflect a new design. Even though the adaptive extruded rod feature is created by extruding a sketched circle, the sketch is fully defined using a $\varnothing.5''$ circle. See Figure 15.11.

FIGURE 15.11
Working with an
adaptive feature.

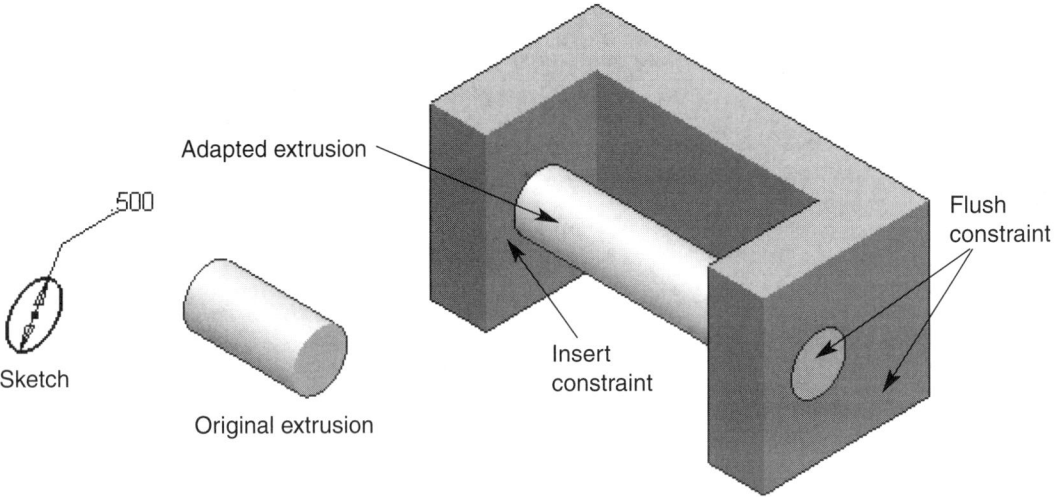

Adapted extrusion

Flush
constraint

.500

Insert
constraint

Sketch

Original extrusion

Extrusions, revolutions, holes, and work planes can all be made adaptive. The way these features adapt depends heavily on the design and the assembly constraints used. As previously mentioned, if you create a feature inside an assembly file that references existing component features, and the **Adapt Feature** check box in the **Options** dialog box is selected, the feature automatically becomes adaptive. In addition, if you set a sketch as adaptive, the corresponding sketched feature also automatically becomes adaptive. However, if a feature is not adaptive, you can use the techniques described to adapt the following features:

■ **Extrusions**   Adapt these features using one of the following options:

✓ Right-click on the extrusion in the graphics window or the extrusion name in the Browser and select the **Adaptive** menu option. When you use this option, the extrusion, the extrusion sketch, and the entire part become adaptive.

✓ Right-click on the extrusion in the graphics window or the extrusion name in the Browser, and select the **Properties** menu option. When you use this option, the **Feature Properties** dialog box is displayed. See Figure 15.12. This dialog box contains an **Adaptive** area, with the following check boxes that allow you to specify the adaptive properties of the feature:

■ **Sketch**   Pick this check box to specify the feature sketch as adaptive.

■ **Parameters**   Select this check box to specify the extrusion parameters as adaptive. For example, if you specify an extrusion distance, then select the **Parameters** check box, the distance becomes adaptive, allowing you adapt the extrusion distance to other components.

■ **From/To Planes**   Pick this button to adapt an extrusion that was created using the **From To** termination option.

Field Notes

Only certain check boxes will be available inside the **Feature Properties** dialog box depending on the type of extrusion. Fro example, if you create an extrusion using a distance termination option, the **From/To Plane** check box will not be available.

■ **Revolutions**   Adapt these features using one of the following options:

✓ Right-click on the revolution in the graphics window or the revolution name in the Browser and select the **Adaptive** menu option. When you use this option, the revolution, the revolution sketch, and the entire part become adaptive.

✓ Right-click on the revolution in the graphics window or the revolution name in the Browser, and select the **Properties** menu option. When you use this option, the **Feature**

FIGURE 15.12 The **Feature Properties** dialog box for an extrusion.

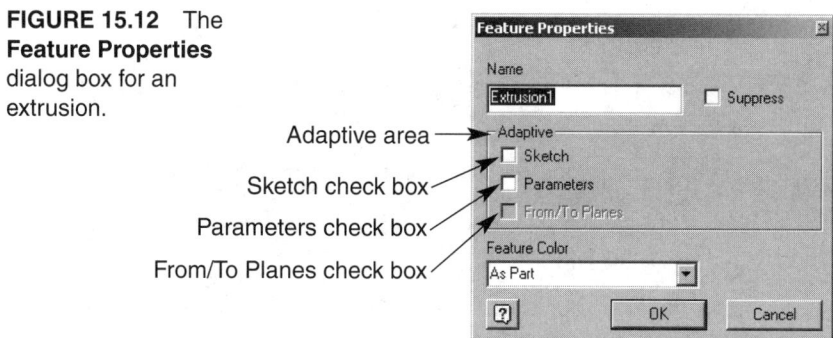

Adaptive area

Sketch check box

Parameters check box

From/To Planes check box

Properties dialog box is displayed. See Figure 15.12. As with an extrusion, this dialog box contains an **Adaptive** area, with the following check boxes, that allow you to specify the adaptive properties of the feature:

- **Sketch**   Pick this check box to specify the feature sketch as adaptive.

- **Parameters**   Select this check box to specify the revolution parameters as adaptive. For example, if you specify a revolution angle, then select the **Parameters** check box, the angle becomes adaptive, allowing you adapt the revolution angle to other components.

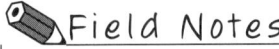 Field Notes

The **From/To Plane** check box is available for revolved feature properties, because **From To** is not a revolution termination option.

- **Holes**   Adapt these features using one of the following options:

  ✓ Right-click on the hole in the graphics window or the hole name in the Browser, and select the **Adaptive** menu option. When you use this option, the hole, the hole sketch, and all the hole parameters become adaptive.

  ✓ Right-click on the hole in the graphics window or the hole name in the Browser, and select the **Properties** menu option. When you use this option, the **Feature Properties** dialog box is displayed. See Figure 15.13. This dialog box contains an **Adaptive** area, with the following check boxes, that allow you to specify the adaptive properties of the feature:

    - **Sketch**   Pick this check box to specify the feature sketch as adaptive.

    - **Hole Depth**   Select this check box to adapt the depth of the hole.

    - **Nominal Diameter**   Choose this check box if you want to adapt the nominal diameter of the hole.

    - **Counterbore Diameter**   Pick this check box to specify the counterbore diameter as adaptive.

    - **Counterbore Depth**   Select this check box to adapt the counterbore depth.

FIGURE 15.13   The **Feature Properties** dialog box for a hole.

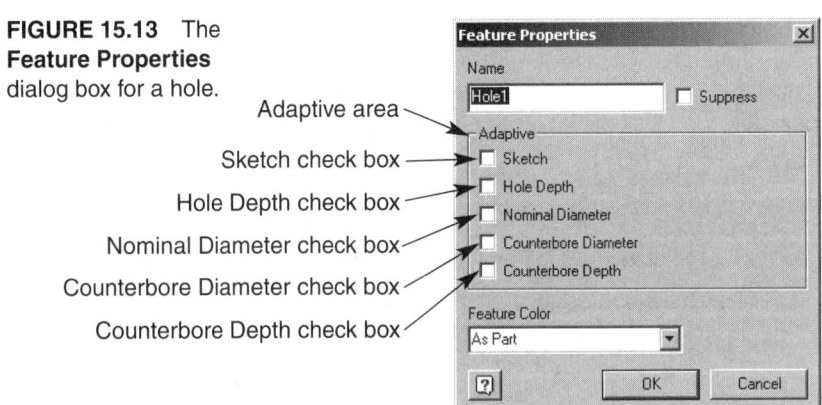

Field Notes

> Only certain check boxes will be available inside the **Feature Properties** dialog box depending on the type of hole you have created.

- **Work Planes** Adapt these features by right-clicking on the work plane in the graphics window or the work plane name in the Browser, and select the **Adaptive** menu option.

Once you set features as adaptive, feature parameters are able to change in reference to other components and specified constraints. If a feature is not adaptive, you cannot specify assembly constraints in an effort to modify, or adapt, the feature. If a feature is not set as adaptive, you must edit the actual feature to modify its parameters, which does not necessarily ensure parametric relationships between certain components.

Exercise 15.4

1. Continue from EX15.3, or launch Autodesk Inventor.
2. Open EX15.2, and save a copy as EX15.4.
3. Close EX15.2 without saving, and open EX15.4.
4. Extrude the sketch 10″ in a positive direction.
5. The sketch was made adaptive in EX15.2. As a result, the extrusion automatically becomes adaptive when created.
6. To observe the extrusion is adaptive, right-click on the feature name in the Browser, or the actual feature in the graphics window, and notice the **Adaptive** menu option is selected.
7. Then right-click on the feature name in the Browser, or the actual feature in the graphics window, and select the **Properties** menu option.
8. The **Sketch** and **Parameters** check boxes should be selected.
9. Resave EX15.4.
10. Exit Autodesk Inventor or continue working with the program.

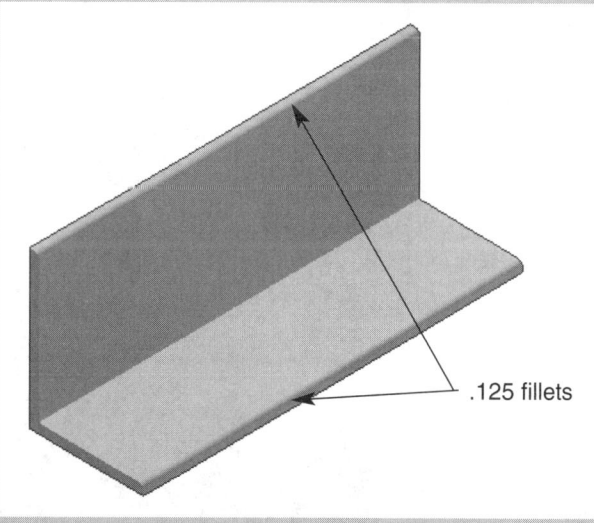

.125 fillets

Adapting Parts in an Assembly

Parts are adapted in assemblies using assembly constraints. For more information regarding assemblies and assembly constraints, refer to Chapter 14. The constraints you establish between adaptive parts and other assembly components depend on the application, design intent, and all other part and assembly model creation factors. Regardless of the application, in order for a part to adapt to other assembly components, the part must be adaptive. Even if you have identified a part sketch or feature as adaptive in the part environment, you must also define the part as adaptive in the assembly environment. If a feature is not adaptive, it is known as a *rigid body*. This means you cannot specify assembly constraints in an effort to modify, or adapt, the feature. If a feature is not set as adaptive, you must edit the actual sketch or feature to modify its parameters, which does not necessarily ensure parametric relationships between certain components.

Use one of the following options to make a part adaptive in an assembly:

✓ Right-click on the part in the graphics window or the part name in the Browser, and select the **Adaptive** menu option.

✓ Right-click on the part in the graphics window or the part name in the Browser, and select the **Properties** menu option. When you use this option, the **Properties** dialog box for the specific part is displayed. See Figure 15.14. Pick the **Occurrence** tab, and then select the **Adaptive** check box in the Properties area to make the part adaptive.

Assuming the part sketches and/or features are adaptive, once you specify part adaptivity, the part is free to be constrained to other components.

Figures 15.15, 15.16, and 15.17 represent a few different examples of using adaptive parts.

If you have difficulty constraining adaptive parts, particularly if you continually receive the "constraint is inconsistent with another" error message, one of the following problems may be the cause:

■ You are trying to adapt a sketch that contains too many dimensions or geometric constraints. Solve the problem by removing the required dimension or geometric constraint.

■ A sketch or feature of the part you are trying to constrain has not been made adaptive. Solve this problem by specifying the sketch or feature as adaptive.

■ The wrong constraint is being used. You may want to try an alternative constraining option.

■ The part has not been made adaptive. Adapt the part to solve the problem.

FIGURE 15.14 The **Properties** dialog box for a part (**Occurrence** tab shown).

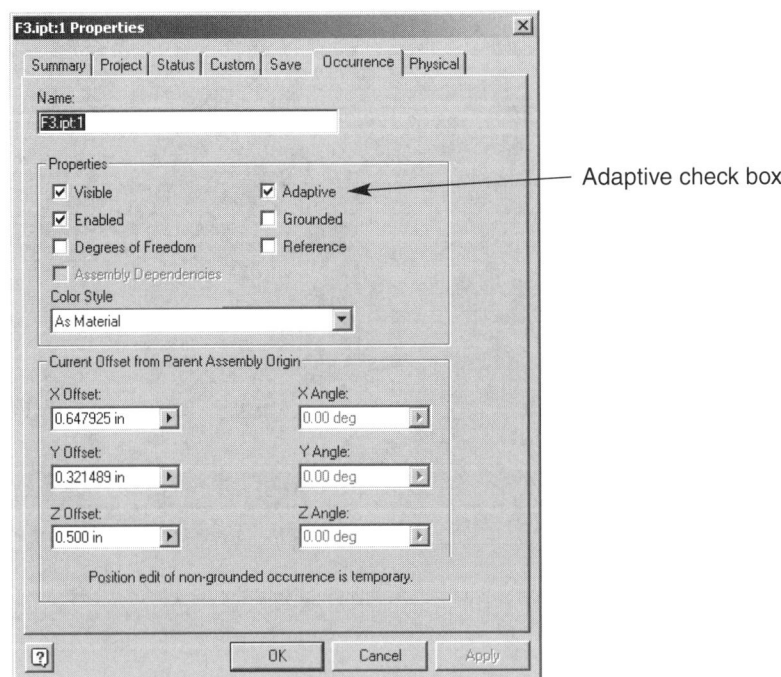

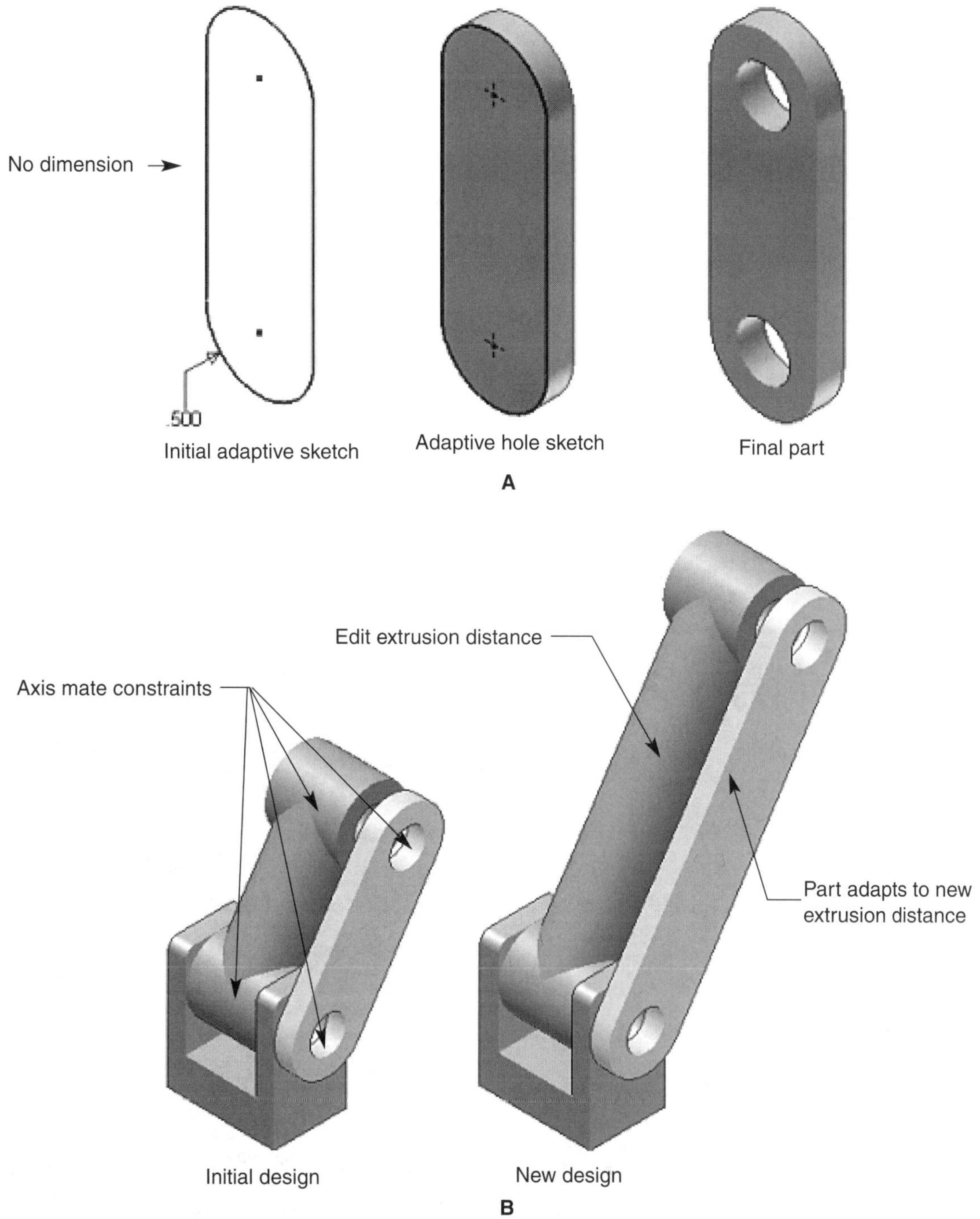

No dimension →

.500

Initial adaptive sketch

Adaptive hole sketch

Final part

A

Edit extrusion distance

Axis mate constraints

Part adapts to new extrusion distance

Initial design

New design

B

FIGURE 15.15 (A) Creating an adaptive part with adaptive sketches. (B) Adapting the part to a new design using constraints.

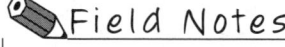 Field Notes

If an assembly does not change to reflect new constraints or relationships between adaptive parts, try an update. In some situations, more than one update may be necessarily.

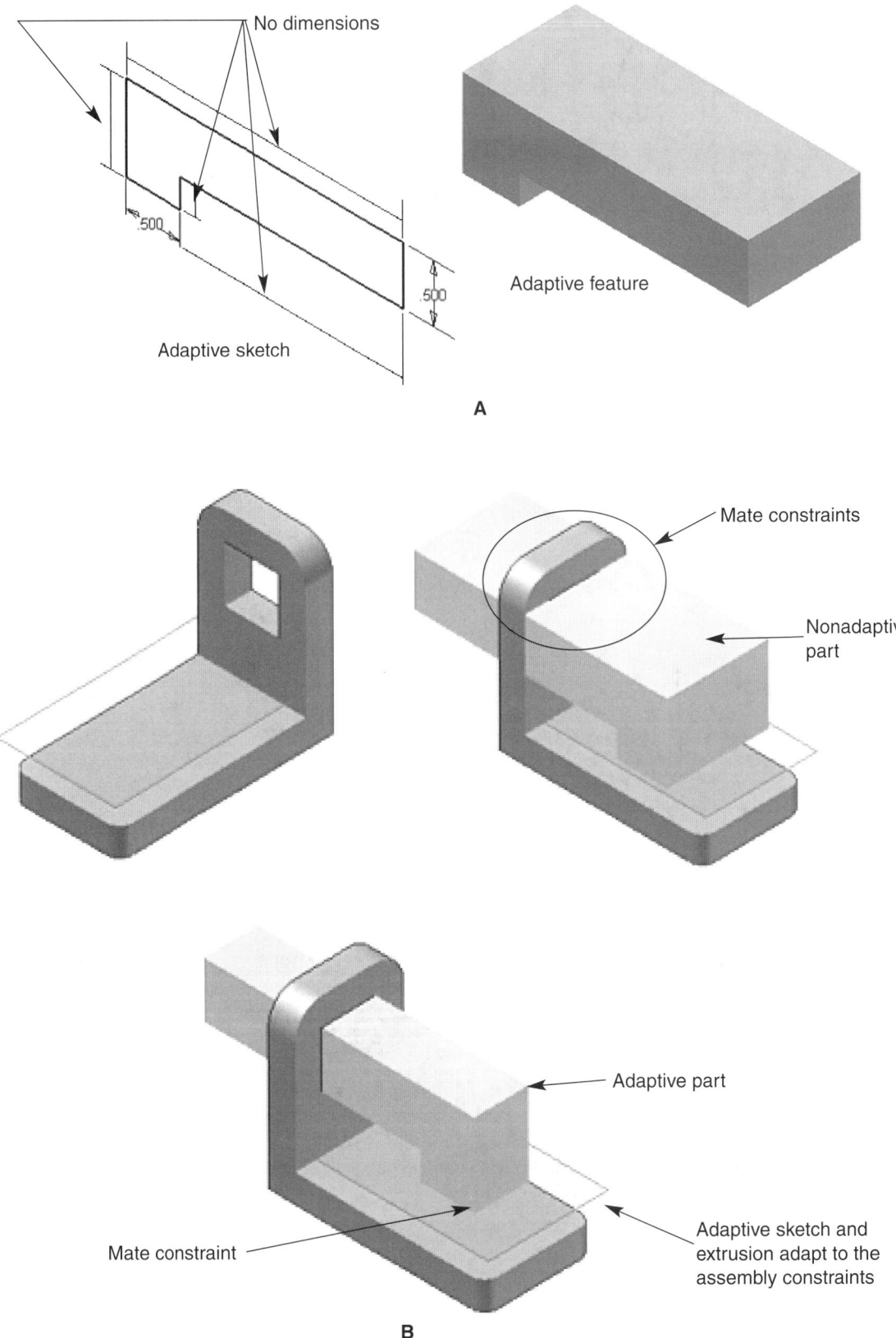

FIGURE 15.16 (A) Creating an adaptive part with an adaptive sketch and an adaptive feature. (B) Adapting the part to another component using constraints.

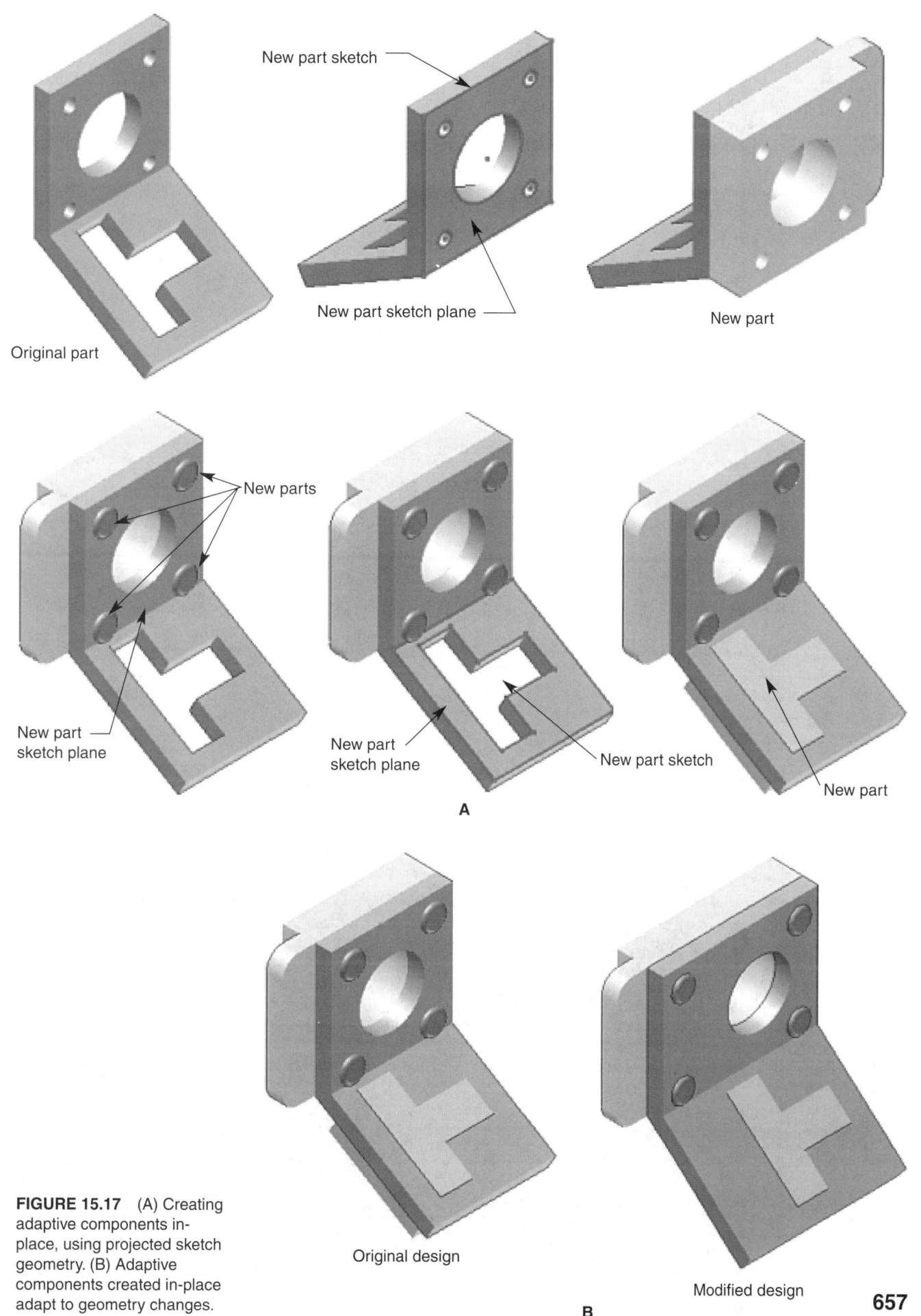

New part sketch

New part sketch plane

New part

Original part

New parts

New part
sketch plane

New part
sketch plane

New part sketch

New part

A

Original design

Modified design

FIGURE 15.17 (A) Creating adaptive components in-place, using projected sketch geometry. (B) Adaptive components created in-place adapt to geometry changes.

B

657

If an assembly contains several of the same parts, only one occurrence can be defined as adaptive and adjusted to other component geometry. For example, if you insert two 120mm × 35mm × 50mm blocks and adapt one of the blocks to change length from 120mm to 70mm, both occurrences of the block will change length to 70mm. In addition, because the parts are parametrically controlled, the separate part file will also reflect the change in length. To overcome this problem, save a copy of the original part—in this example, the 120mm × 35mm × 50mm block. You may also want to save a separate copy of the part that is not adaptive.

Exercise 15.5

1. Continue from EX15.4, or launch Autodesk Inventor.
2. Open a new inch assembly file.
3. Access the **Create Component** tool.
4. Specify a new file name of EX15.5.ipt, a part file type, and a location and template of your choice.
5. Pick the **OK** button.
6. Open a sketch on the XY Plane, and sketch a 6″×24″ rectangle.
7. Finish the sketch.
8. Extrude the rectangle .25″ in a positive direction.
9. Add holes similar to the holes shown.
10. Finish the edit.
11. Access the **Place Component** tool.
12. Insert EX15.4.
13. Create the constraints shown using the selections, and specifications:

    (1) Constraint: Assembly
    Type: Mate
    Offset: 0
    Solution: Mate

    (2) Constraint: Assembly
    Type: Mate
    Offset: 0mm
    Solution: Flush

    (3) Constraint: Assembly
    Type: Mate
    Offset: 0mm
    Solution: Flush

14. Make the part EX15.4.ipt adaptive using one of the techniques discussed.
15. Create the constraints shown using the selections and specifications:

    (1) Constraint: Assembly
    Type: Mate
    Offset: 0
    Solution: Mate

    (2) Constraint: Assembly
    Type: Mate
    Offset: 0mm
    Solution: Flush

16. The part should adapt to the component, because of the specified sketch and feature adaptivity.
17. The final exercise should look like the assembly shown.
18. Save the assembly file as EX15.5.
19. Exit Autodesk Inventor or continue working with the program.

Exercise 15.5 continued on next page

Exercise 15.5 continued

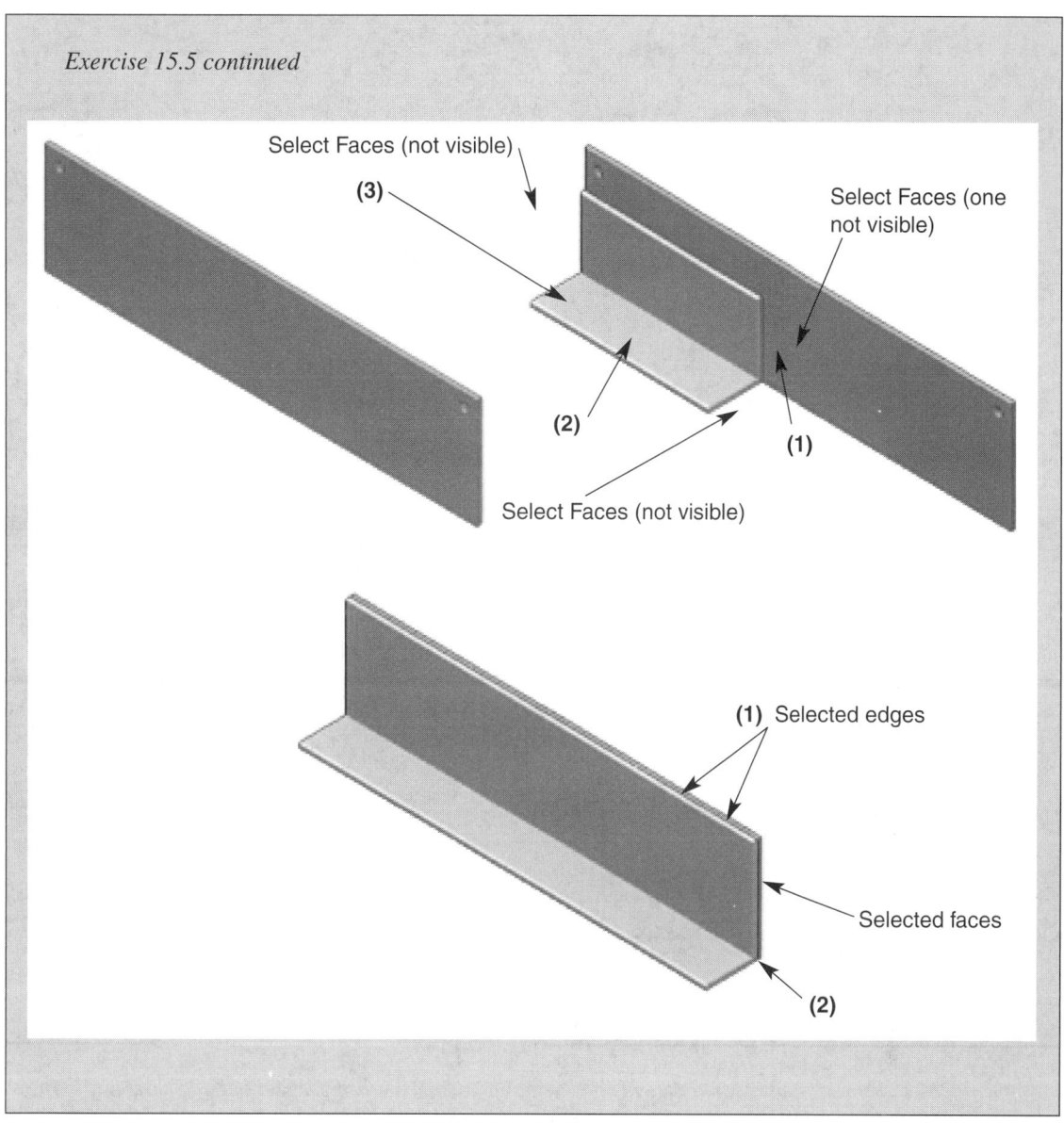

Select Faces (not visible)

(3)

Select Faces (one
not visible)

(2)

(1)

Select Faces (not visible)

(1) Selected edges

Selected faces

(2)

Exercise 15.6

1. Continue from EX15.5, or launch Autodesk Inventor.
2. Open EX15.3, and save a copy as EX15.6.
3. Close EX15.3 without saving, and open EX15.6.
4. Edit EX15.3B.
5. Extrude the sketch 50mm in both directions.
6. Finish the edit.
7. To observe the adaptive nature of the components, edit EX15.3A.
8. Change the hole diameter from 25mm to 50mm.
9. Finish the edit.
10. The final exercise should look like the assembly shown.
11. Resave EX15.6.
12. Exit Autodesk Inventor or continue working with the program.

Adapting Subassemblies in an Assembly

Subassemblies are adapted in assemblies using the same techniques as adapting parts. When you adapt a part in an assembly, a sketch and/or a feature must be adapted. In the same respect, when you adapt a subassembly in an assembly, the subassembly parts must be adaptive, and the individual part sketches and/or features must be adaptive. In addition, subassemblies are also adapted using assembly constraints. For more information regarding assemblies and assembly constraints, refer to Chapter 14.

The constraints you establish between adaptive subassemblies and other assembly components depend on the application, design intent, and all other part, subassembly, and assembly model creation factors. Regardless of the application, in order for a subassembly to adapt to other assembly components, the subassembly must be adaptive. Even if you have identified individual subassembly parts, part sketches, and part features as adaptive in the part environment and subassembly file, you must also define the subassembly as adaptive in the assembly environment. If a subassembly is not adaptive, the components are not free to move, and you cannot specify assembly constraints in an effort to modify, or adapt, the subassembly. If a subassembly is not set as adaptive, you must edit the actual sketches, features, parts, and constraints that define the subassembly, to modify its parameters, which does not necessarily ensure parametric relationships between certain components.

Use one of the following options to make a subassembly adaptive in an assembly:

✓ Right-click on the subassembly in the graphics window or the subassembly name in the Browser, and select the **Adaptive** menu option.

✓ Right-click on the subassembly in the graphics window or the subassembly name in the Browser, and select the **Properties** menu option. When you use this option, the **Properties** dialog box for the specific subassembly is displayed. Pick the **Occurrence** tab and then select the **Adaptive** check box in the **Properties** area to make the subassembly adaptive.

Assuming the individual part sketches and/or features are adaptive, and the subassembly parts are adaptive, once you specify subassembly adaptivity, the subassembly is free to be constrained to other components. See Figure 15.18.

If you have difficulty constraining adaptive subassemblies, particularly if you continually receive the "constraint is inconsistent with another" error message, one of the following problems may be the cause:

- You are trying to adapt a sketch that contains too many dimensions or geometric constraints. Solve the problem by removing the required dimension or geometric constraint.

- A sketch or feature of one or more of the individual subassembly parts you are trying to constrain has not been made adaptive. Solve this problem by specifying the individual part sketches and/or features as adaptive.

- The wrong constraint is being used. You may want to try an alternative constraining option.

- A certain part or group of parts has not been made adaptive. Adapt the parts to solve the problem.

- The subassembly has not been made adaptive. To solve the problem adapt the subassembly.

Field Notes

If an assembly does not change to reflect new constraints or relationships between adaptive subassemblies, try an update. In some situations, more than one update may be necessary.

FIGURE 15.18 An example of adapting a subassembly.

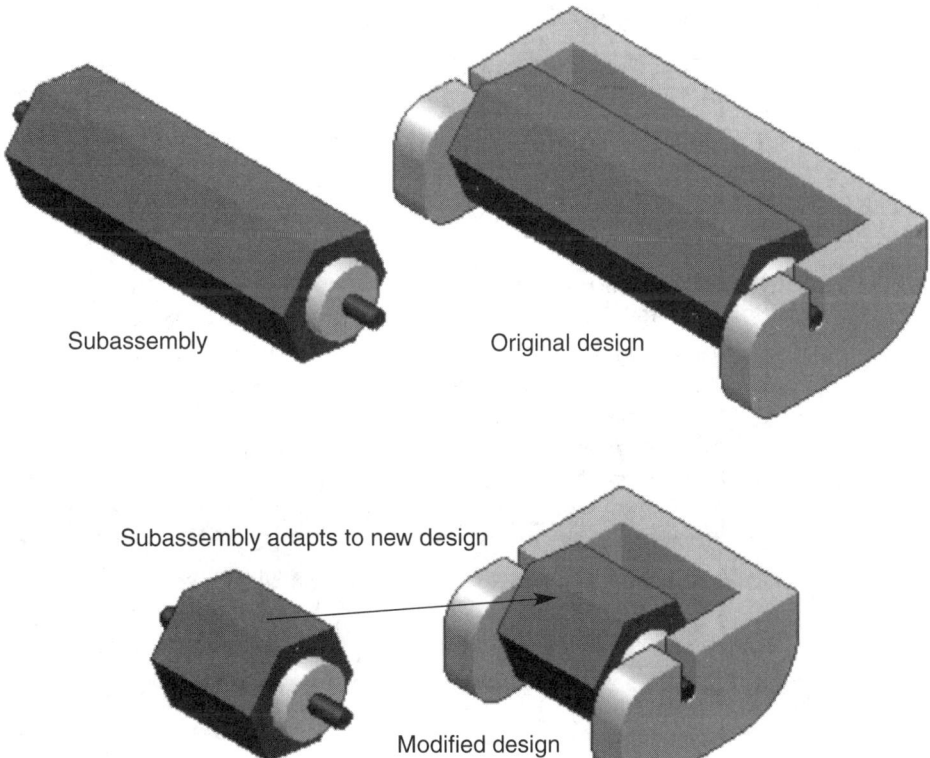

If an assembly contains several of the same subassemblies, only one of the occurrences adjusts to other component geometry. Consequently, because the subassemblies are parametrically controlled, the separate assembly and individual part files will also reflect the change in geometry. To overcome this problem, save a copy of the original subassembly and the individual parts using different names. You may also want to save a separate unadaptive copy of the subassembly and subassembly parts.

Exercise 15.7

1. Continue from EX15.6, or launch Autodesk Inventor.
2. Open a new inch assembly file.
3. Access the **Create Component** tool.
4. Specify a new file name of EX15.7A.ipt, a part file type, and a location and template of your choice.
5. Pick the **OK** button.
6. Open a sketch on the XY Plane, and sketch a rectangle. Do not add dimensions.
7. Finish the sketch.
8. Extrude the rectangle 1″ in a positive direction.
9. Add a Hole, Center Point near the center of the extrusion. Do not dimension the Hole, Center Point.
10. Cut extrude the circle through all.
11. Finish the part edit.
12. Access the **Create Component** tool.
13. Specify a new file name of EX15.7B.ipt, a part file type, and a location and template of your choice.
14. Pick the **OK** button.
15. Open a sketch somewhere in space, without referencing any existing features or planes.
16. Sketch a circle with no dimensions.
17. Use a mate/flush constraint between the two planar faces, a mate/mate constraint between the two axes, and an inside tangent constraint between the two tangent edges, as shown.
18. Finish the part edit, and close the assembly file after saving.
19. Open a new inch assembly file.
20. Access the **Create Component** tool.
21. Specify a new file name of EX15.7C.ipt, a part file type, and a location and template of your choice.
22. Pick the **OK** button.
23. Open a sketch on the XY Plane, and sketch a 4″×4″ rectangle.
24. Finish the sketch.
25. Extrude the rectangle .5″ in a positive direction.
26. Add a Hole, Center Point at the center of the extrusion.
27. Add a 3″ hole, through all.
28. Finish the part edit.
29. Access the **Place Component** tool.
30. Insert the subassembly EX15.7A.iam.
31. Make the subassembly, EX14.7A.iam, adaptive.
32. Add four mate/flush constraints to adapt the subassembly to the part as shown.
33. Add the mate/mate constraint to mate the subassembly and part axes together as shown.
34. Add a mate/mate constraint between the subassembly and part planar faces as shown.
35. Add the tangent constraint between the subassembly and part tangent faces, as shown.
36. The final exercise should look like the assembly shown.
37. Resave EX15.7.iam.
38. Exit Autodesk Inventor or continue working with the program.

Exercise 15.7 continued on next page

Exercise 15.7 continued

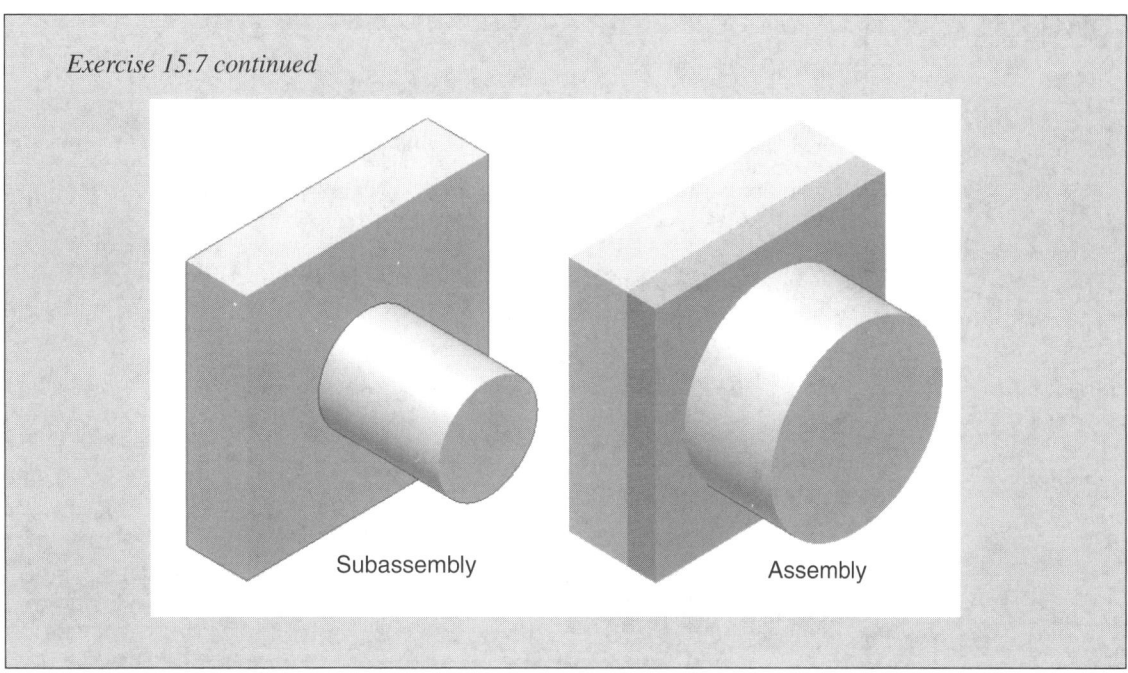

Subassembly   Assembly

|||||||||||| CHAPTER TEST

Answer the following questions on a separate sheet of paper.

1. What happens when you insert a part into an assembly when the Adaptivity used in assembly check box is unselected?

2. Name the check box that you would select if you do not want features to be adaptive, but do want a mate constraint to occur between the feature and the plane.

3. Name the check box that automatically adapts features to changing geometric places.

4. Under what conditions do you not want to completely constrain a sketch?

5. What is a reference sketch?

6. How do you remove the relationship between an existing parent feature and the new sketch?

7. When you finish a sketch created in-place using parent feature geometry, you will notice an assembly constraint is automatically placed between the existing component and the new sketch plane. What does this constraint ensure?

8. Give the function of the Sketch check box in the Feature Properties dialog box.

9. Explain the function of the Parameters check box in the Feature Properties dialog box.

10. How do you adapt work planes?

11. What is a rigid body?

12. List at least three possible problems if you have difficulty constraining adaptive parts, particularly if you continually receive the "constraint is inconsistent with another" error message.

13. If an assembly contains several of the same parts, how many occurrences can be defined as adaptive and adjusted to other component geometry?

14. What can you do if an assembly does not change to reflect new constraints or relationships between adaptive subassemblies?

15. If an assembly contains several of the same subassemblies, only one of the occurrences adjusts to other component geometry. Consequently, because the subassemblies are parametrically controlled, the separate assembly and individual part files will also reflect the change in geometry. How can you overcome this problem?

|||||||||||| PROJECTS

Instructions:

- Create the parts, subassemblies, and assemblies similar to those shown in A.

- Using the information presented in this chapter, adapt the sketches, features, parts, and sub-assemblies as needed.

- Modify the assembly component sketches and feature parameters to observe the adaptivity.

1. Name: Shock Absorber Subassembly

    Units: Inch or Metric

    Save As: P15.1

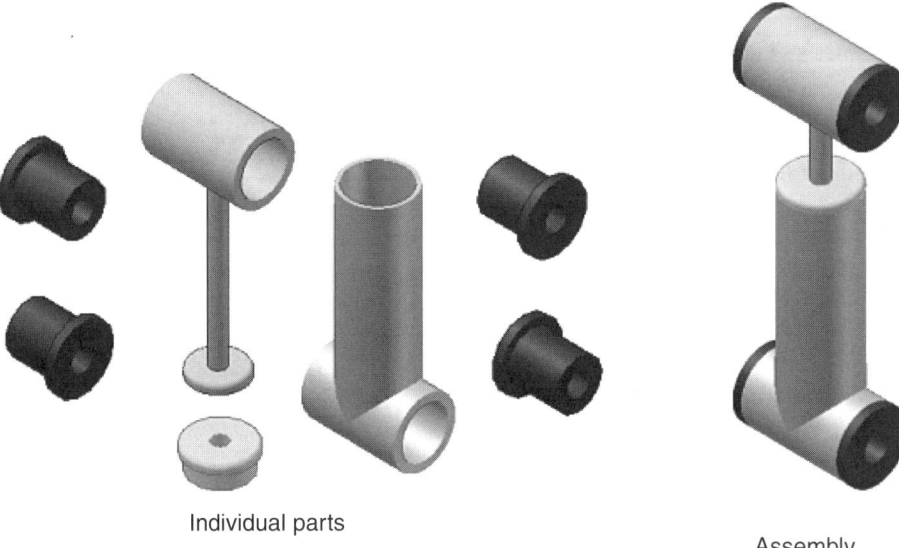

Individual parts

Assembly

2. Name: Shock Absorber Assembly

    Units: Inch or Metric

    Save As: P15.2

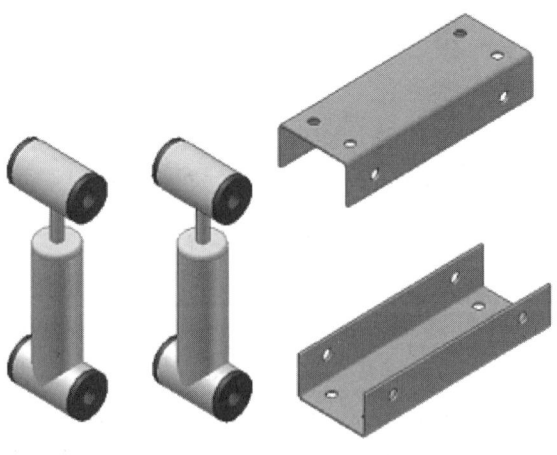

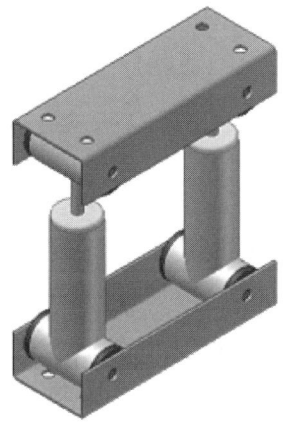

Subassemblies   Parts

Assembly

3. Name: Lever Assembly

   Units: Inch or Metric

   Save As: P15.3

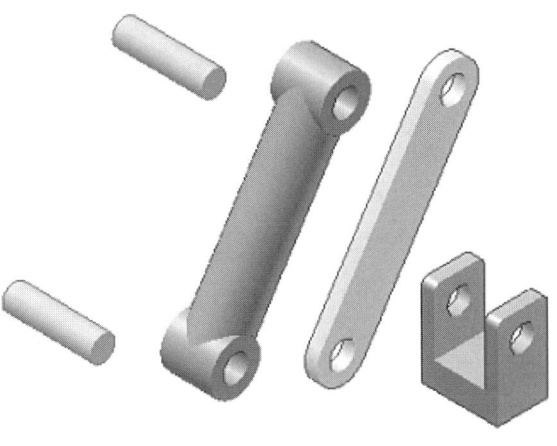
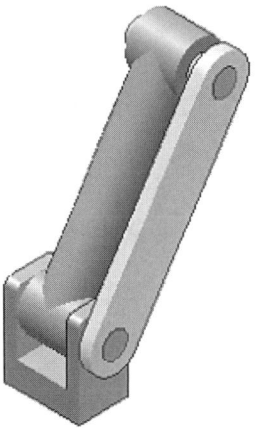

Individual parts   Assembly

4. Name: Dispenser Assembly

   Units: Inch or Metric

   Save As: P15.4

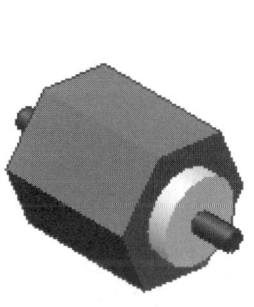
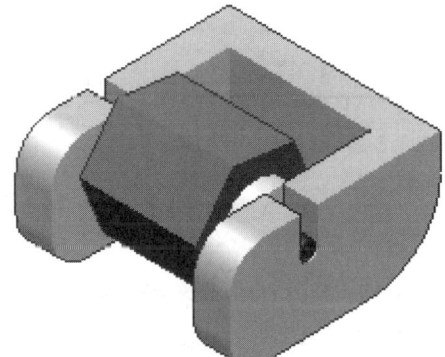

Subassembly

Assembly

CHAPTER 16

Working with Presentations

LEARNING GOALS

After completing this chapter, you will be able to:

◎ Create presentations files.

◎ Work in the presentation environment.

◎ Create exploded presentation views.

◎ Use the **Precise View Rotation** tool.

◎ Animate presentation views.

◎ Place additional presentation views.

Working with Presentations

A *presentation* is a display of a complete assembly that can be exploded, animated, and used for two-dimensional presentation drawings. See Figure 16.1. Only assemblies can be used for presentations, because parts consist of only one component, whereas assemblies may contain several components. Presentations are very similar to assemblies, but do not contain assembly tools. Therefore, you cannot modify component parameters and do not have access to assembly options, such as work features, in the presentation environment.

The purpose of a presentation, as previously mentioned, is to display assemblies, in ways they are not shown in the actual assembly file. Presentations are basically *exploded* assemblies. This means the individual assembly components are moved away from other components in an unassembled or ungrouped fashion. You can use presentation views to record design intent and help with assembly visualization. For example, if parts and subassemblies hide other components or certain features in the assembly file due to constraints and assembly design parameters, you can use a presentation to clearly show all the assembly components. A presentation can also be used to visually explain how assembly components fit together, interact in reference to other components, and display how the assembly is built. In addition, before you can generate a two-dimensional presentation drawing, as discussed in Chapter 17, you must create a presentation file.

As discussed in Chapter 1, Autodesk Inventor presentation files are separate files and carry the extension .ipn. Presentation template files are used to create presentations by inserting assemblies and assembly design views into a presentation file. You can create multiple individual presentation views in the single presentation template file. For example, you may want to show a complete assembly in one presentation view, but only a couple components from the entire assembly in another presentation view.

FIGURE 16.1 An example of a c-clamp presentation, created using a c-clamp assembly.

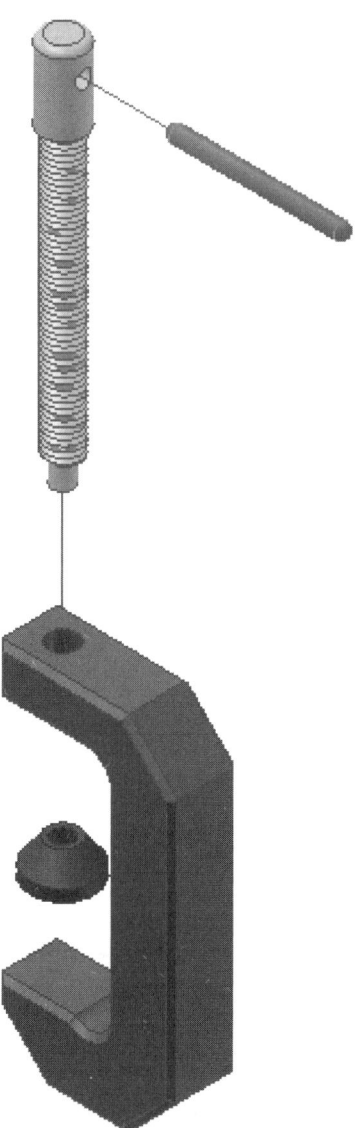

Working in the Presentation Environment

The work environment and interface is different for each stage of the design process. Some of the tools and interface components available in the presentation environment are also available in other design environments. Still, most presentation tools are specific to working with presentations. By default, when you open a new Autodesk Inventor presentation, the work environment should look like the display in Figure 16.2. When you open a presentation template, you are ready to insert an assembly file to create the presentation view. Throughout this chapter you will explore specific tools, commands, and options that allow you to develop Autodesk Inventor presentations. Still, you may initially want to become familiar with some specific work environment options for future reference.

Field Notes

Many of the interface options that relate to working with parts, drawings, and assemblies are available inside the presentation environment. For information on these pull-down menu system options, refer to Chapters 3 and 4. Specific pull-down menu options available only while in the presentation environment are explained in the section where they apply.

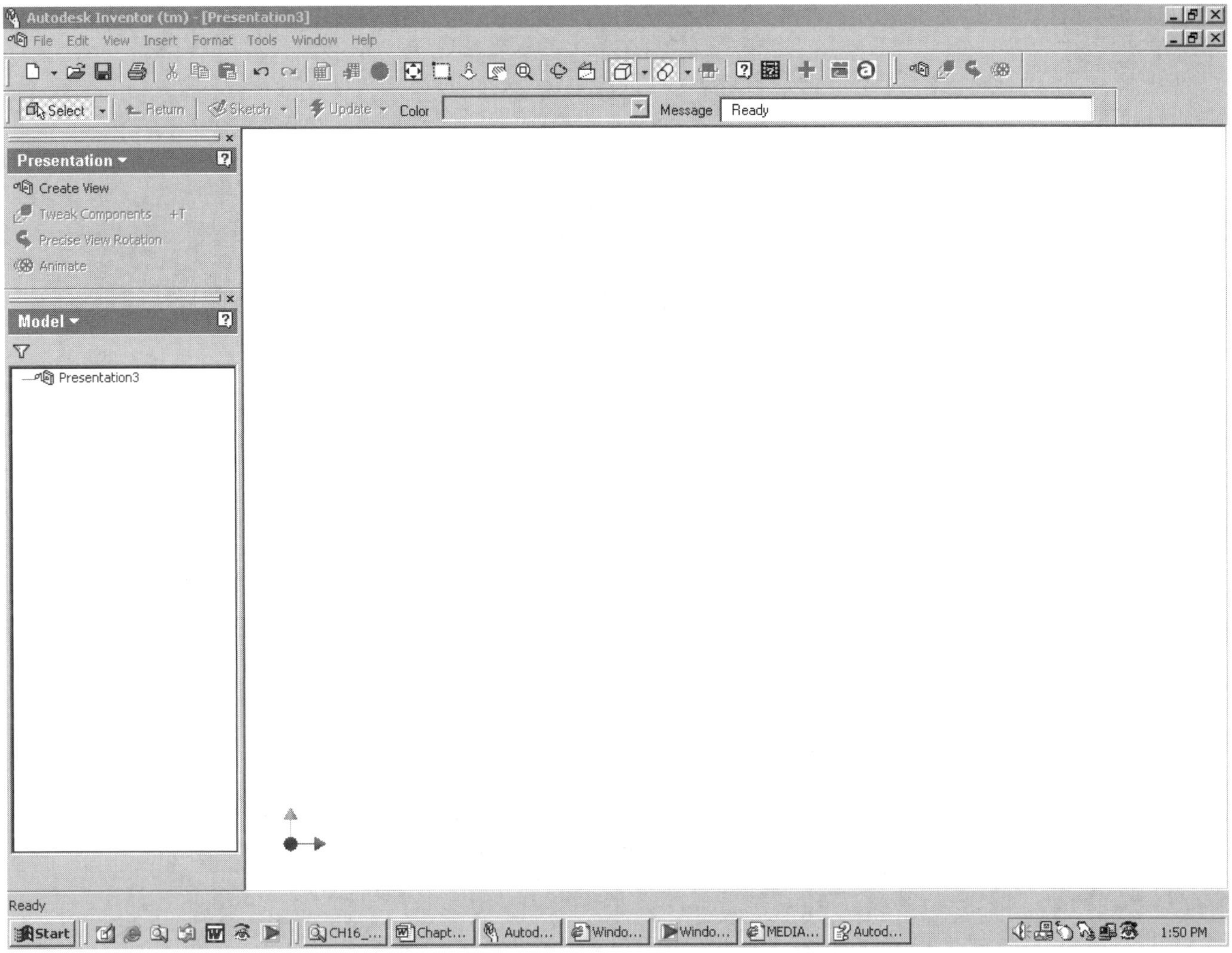

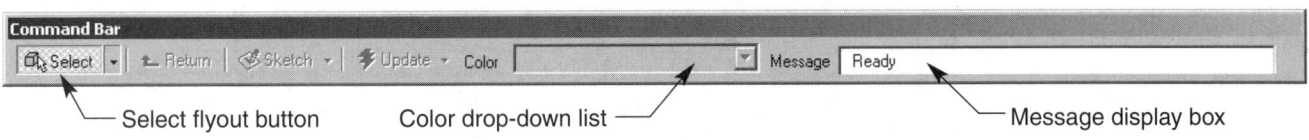

FIGURE 16.2 The default Autodesk Inventor presentation work environment.

Using the Command Bar for Assemblies

In previous chapters you have become familiar with the Autodesk Inventor command bar, shown in Figure 16.3. The command bar options are the same when working in presentations as they are when working with assemblies. However, because you cannot create or modify model specifications in a presentation, the **Return, Sketch,** and **Update** buttons do not apply. In addition, you can specify only **Component Priority** or **Leaf Part Priority,** from the **Select** flyout button. Still, the **Color** drop-down list and **Message** display box function the same in a presentation file as they do in an assembly file. You can use the **Color** drop-down list to change the color of specific assembly components.

Using the Browser in a Presentation

The browser for a typical presentation displays the presentation icon and presentation file name, followed by each of the presentation views. The individual presentation views contain the assembly file and the separate assembly components. Finally, each of the presentation parameters or tweaks is listed in each component. Figure 16.4 shows the browser for a c-clamp presentation file named F16.1.iam that contains two presenta-

FIGURE 16.3 The command bar.

FIGURE 16.4 The browser display for a typical presentation.

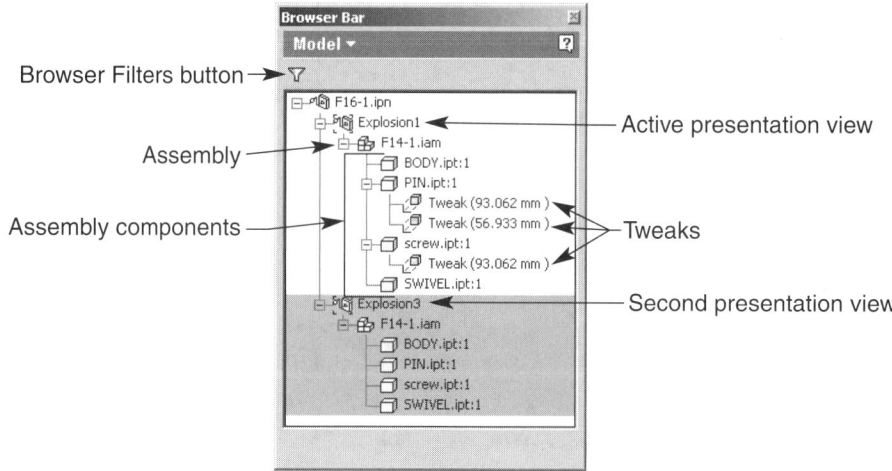

Browser Filters button

Assembly

Assembly components

Active presentation view

Tweaks

Second presentation view

tion views. As in Autodesk Inventor part and assembly files, the assembly browser contains a **Browser Filters** button. The following Browser filter options are specific to the presentation environment. For information regarding other Browser options available in multiple work environments, refer to Chapter 4.

Field Notes

Only one presentation view can be active: the view that is not active is shaded.

- **Assembly View**   This option is set as default and displays the hierarchy shown in Figure 16.4. The assembly hierarchy shows the presentation view, followed by the assembly file, and each of the assembly components found in the assembly. In addition, the tweaks related to each assembly component are identified under the specific components.

- **Tweak View**   Select this option if you want to display a tweak hierarchy to clearly identify presentation tweaks, which specify the explosion characteristics of the presentation. The tweak specifications, used to explode the assembly in each presentation view, are displayed first, and then the assembly and assembly components are shown. See Figure 16.5.

- **Sequence View**   Select this option if you want to display a presentation task hierarchy to clearly identify which components and tweaks are used to create the exploded presentation. Often this option is specified when working with presentation animations, or if you want to change the presentation explosion arrangement. The task specifications, used to identify how an assembly is assembled and disassembled in an animation in each presentation view, are displayed first. Next, the task sequences are shown, which identify the order of tweak groups used to explode the assembly in the presentation. Finally, the assembly and assembly components are shown. See Figure 16.6.

FIGURE 16.5 The presentation browser display with a tweak view hierarchy.

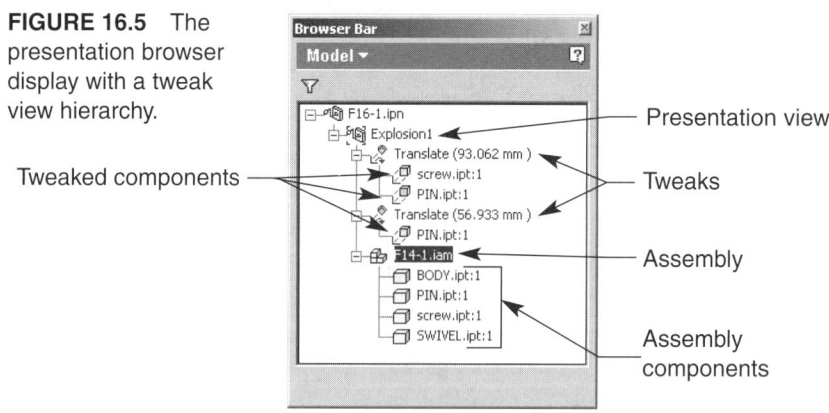

Tweaked components

Presentation view

Tweaks

Assembly

Assembly components

FIGURE 16.6 The presentation browser display with a sequence view hierarchy.

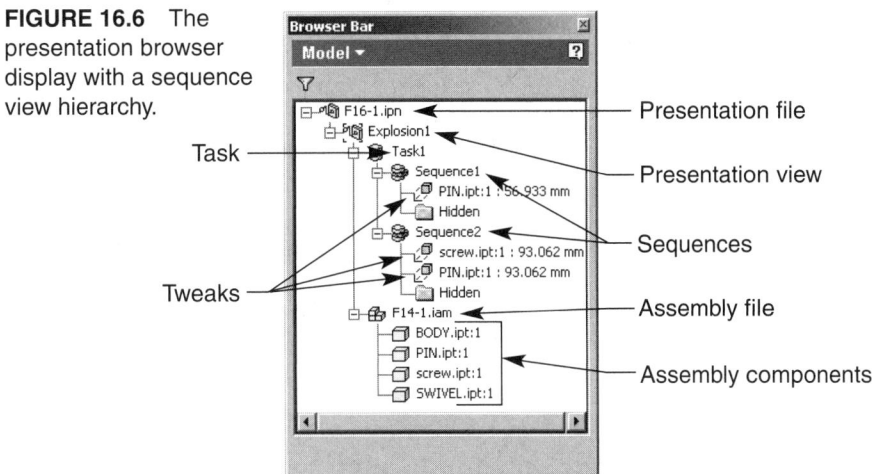

In addition to the **Browser Filters** button, you may want to become familiar with the following options, which are only available in a presentation file when you right-click on an item in the Browser or a component in the graphics window:

- **Activate**   This menu option is available when two or more presentation views are available and when you right-click on an inactivate view. Pick the **Activate** menu option to make active the specified view. Activating a presentation view opens and displays the view in the graphics window, allowing you to work with the presentation view. You can also activate a presentation view by double-clicking on the view name in the Browser.

- **Save Camera**   Pick this option to set the specified presentation view display as the default view. Saving a camera view allows you to identify the desired position and zoomed display of the presentation view. You can then return to the saved view if modifications are made, such as rotating the presentation view or zooming in or out.

- **Restore Camera**   Select this option to return the presentation view display to the previously saved camera view.

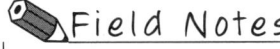 Field Notes

> The available presentation options depend on the specified browser filter selection. Additional presentation browser options are discussed throughout this chapter.

Using Presentation Tools

Presentation tools are used to develop presentations by inserting, exploding, rotating views, and animating presentations. These tools are accessed from the **Presentation** toolbar, shown in Figure 16.7, or from the **Presentation** panel bar. In addition, many of the tools can be accessed from the **View, Insert,** and **Tools** pull-down menus. When you are in the presentation work environment, the panel bar is in presentation mode. See Figure 16.8. Unlike other work environments, presentations contain only one panel bar mode, which is presentation mode. By default, the panel bar is displayed in learning mode, as seen in Figure 16.8. To change from learning to expert mode, as seen in Figure 16.9 and discussed in Chapter 2, right-click any of the panel bar tools or left-click the panel bar title, and select the **Expert** option.

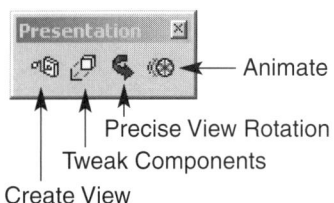

FIGURE 16.7   The **Presentation** toolbar.

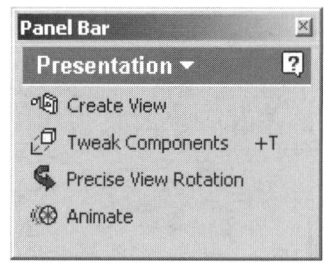

FIGURE 16.8   The panel bar in presentation (learning) mode.

FIGURE 16.9   The panel bar in presentation (expert) mode.

Exercise 16.1

1. Launch Autodesk Inventor.
2. Open a new inch or metric presentation file.
3. Explore the basic presentation file interface options, and work environment, as previously discussed.
4. Exit Autodesk Inventor without saving, or continue working with the program.

Creating Presentation Views

As previously mentioned, when you open a new presentation file (.ipn) you are ready to insert an assembly and generate a presentation view. The initial presentation view must be developed from an assembly file (.iam), because the assembly file identifies the model to be used for the presentation. As discussed later in this chapter, when you create additional presentation views in the same presentation file, the new view references the initially specified assembly model. This means that although you can place multiple presentation views in a single presentation file, the views must all come from the same assembly file. You cannot place different assembly files in a presentation.

To place the initial presentation view, access the **Create View** tool, which is the only tool available when you open a new presentation file. The **Create View** tool is available using one of the following techniques:

✓ Pick the **Create View** button on the **Presentation** panel bar.

✓ Pick the **Create View** button on the **Presentation** toolbar.

✓ Select the **Create View...** option of the **Insert** pull-down menu.

✓ Right-click inside the graphics window or the Browser, and select the **Create View...** menu option.

When you access the **Create View** tool, the **Select Assembly** dialog box, shown in Figure 16.10, is displayed. The **Select Assembly** dialog box is similar to the **Open** dialog box, discussed in previous chapters, and allows you to locate the assembly you want to insert in to the presentation file. In addition, the **Select Assembly** dialog box allows you to define the method of presentation explosion. Use the following options to begin the presentation creation process:

■ **Assembly**   This area allows you to locate the assembly file or design view you want to use to create the presentation view. If you are placing the first presentation view in the presentation file you must specify an assembly file. For additional presentation views, you can specify an existing design view. To insert the initial presentation view, enter or select an assembly file from the **File** drop-down list, or use the **Browse** button located to the right of the **File** drop-down list to locate the desired assembly file.

FIGURE 16.10 The **Select Assembly** dialog box.

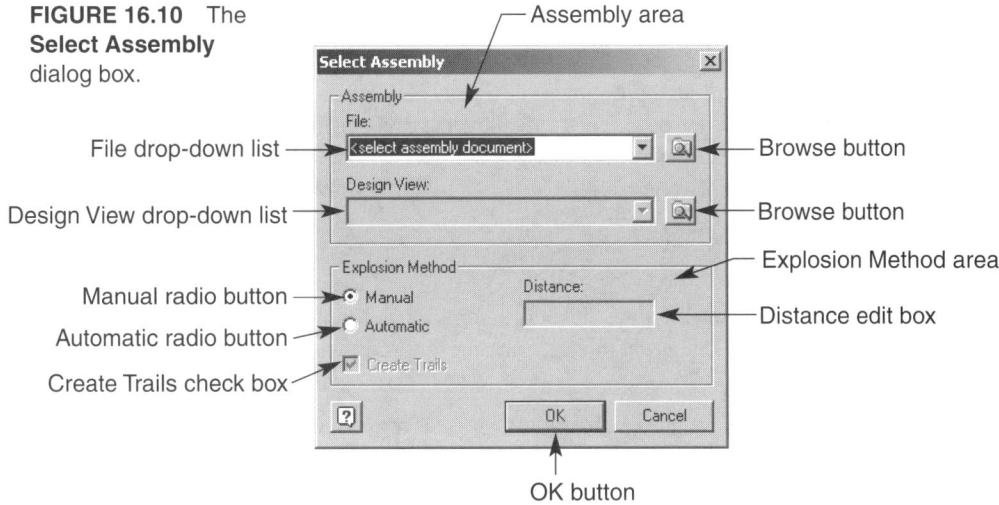

Assembly area

File drop-down list

Browse button

Design View drop-down list

Browse button

Explosion Method area

Manual radio button

Distance edit box

Automatic radio button

Create Trails check box

OK button

As discussed later in this chapter, to insert additional presentation views, enter or select a design view file from the **Design View** drop-down list, or use the **Browse** button located to the right of the **Design View** drop-down list to locate the desired design view file.

- **Explosion Method** Once you have identified the assembly file you want to use to create the presentation view using the options in the **Assembly** area, the next step is to define the explosion characteristics. As the name implies, if you pick the **Manual** radio button, you must manually explode the assembly components in the presentation view using the **Tweak Components** tool. You may choose to automatically explode the presentation when the presentation view is created by picking the **Automatic** radio button. When you select the **Automatic** radio button, the **Distance** edit box and **Create Trails** check box become available. Use the **Distance** edit box to enter the desired explosion distance for each of the components. An explosion distance is the tweak offset from the assembled presentation to the exploded presentation, and depends on how far you want exploded components to be spaced in reference to other exploded components. For example, the cylinder shown in Figure 16.11 has a tweak distance of 1″. Select the **Create Trails** check box if you want trails to be displayed in the presentation. *Trails,* as shown in Figure 16.11, identify the bond between individual components and the assembly by displaying the distance and location of the explosion.

FIGURE 16.11 (A) The assembled presentation. (B) The exploded presentation.

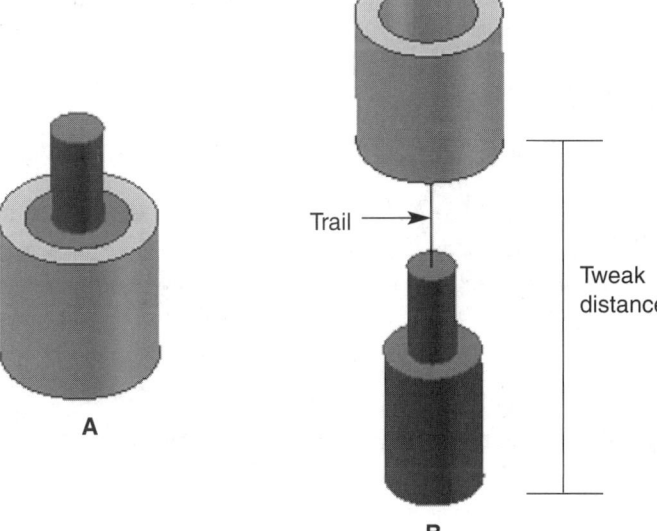

Trail

Tweak distance

A

B

Field Notes

Automatically exploded presentation view display is based on assembly constraints. An automatic explosion may not be correctly displayed, or occur at all, for large or complexly constrained assembly files. For these situations you will need to manually explode the presentation view, or modify the assembly constraints.

Once you have selected the desired assembly file and specified the explosion characteristics of the presentation, pick the **OK** button to create the presentation view.

Exercise 16.2

1. Continue from EX16.1, or launch Autodesk Inventor.
2. Open a new inch presentation file.
3. Access the **Create View** tool to open the **Select Assembly** dialog box.
4. Use the file **Browse** button to open EX15.2.iam. The file should be displayed in the **File** drop-down list.
5. Select the **Automatic** radio button and the **Create Trails** check box.
6. Enter a tweak distance of 6 in the **Distance** edit box.
7. Pick the **OK** button to generate the presentation view.
8. The final exercise should look like the assembly shown.
9. Save the presentation as EX16.2.
10. Exit Autodesk Inventor or continue working with the program.

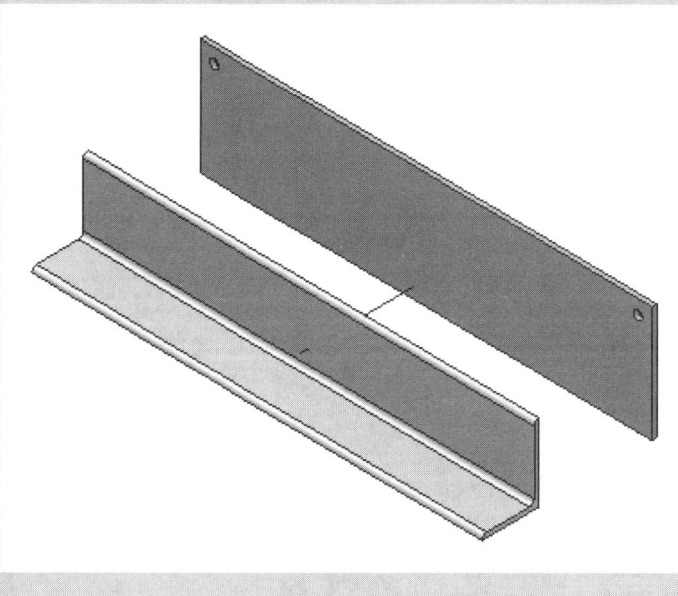

Exercise 16.3

1. Continue from EX16.2, or launch Autodesk Inventor.
2. Open a new metric presentation file.
3. Access the **Create View** tool to open the **Select Assembly** dialog box.
4. Use the file **Browse** button to open EX14.19.iam. The file should be displayed in the **File** drop-down list.
5. Select the **Manual** radio button.
6. Pick the **OK** button to generate the presentation view.
7. The final exercise should look like the presentation shown.
8. Save the presentation as EX16.3.
9. Exit Autodesk Inventor or continue working with the program.

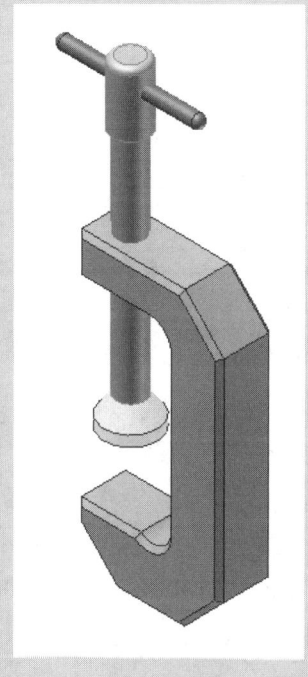

Exploding Presentations

The main function of an assembly file is to develop an assembly of parts and subassemblies by constraining components to each other and defining a complete product. In contrast, one of the major purposes of a presentation is to display assemblies in a partially unconstrained or exploded fashion. Exploding an assembly in the presentation environment allows you to see all the assembly components removed from and aligned with each other. Just as presentation views have many purposes, based on your application, exploding an assembly in a presentation view, also known as an illustrated parts breakdown, is useful for instructions manuals, identifying maintenance and assembly procedures, and multiple other purposes. Autodesk Inventor allows you to explode an assembly in the presentation view automatically or manually. An automatic explode separates and aligns all the assembly components at one time, using a specified distance between the components. *Tweaks,* which are modifications made to the originally constrained assembly, are automatically specified when you automatically explode. When you manually explode an assembly in the presentation file, you separate, or tweak, individual components from other components.

After creating a presentation view using the **Create View** tool, you have discovered that you can manually explode a presentation using the **Tweak Components** tool, discussed later in this chapter, or automatically explode a presentation. As discussed, one option for automatically exploding the assembly is to select the **Automatic** radio button in the **Explosion Method** area of the **Select Assembly** dialog box. Refer to the previous section for information regarding this technique. Another way to automatically explode an assembly in a

FIGURE 16.12 The
Auto Explode dialog
box.

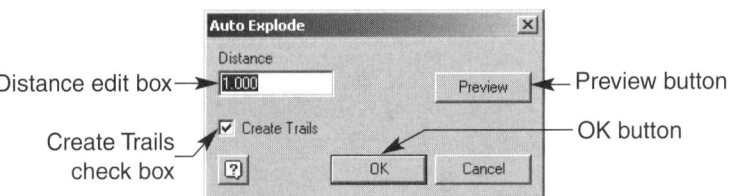

presentation view is to right-click on the assembly file name or assembly icon in the Browser, and then select
the **Auto Explode** menu option. When you pick the **Auto Explode** menu option, the **Auto Explode** dialog box
is displayed. See Figure 16.12. Exploding an assembly using the **Auto Explode** dialog box is the same as
exploding an assembly when you insert the presentation view using the **Select Assembly** dialog box and the
Create View tool. To use the **Auto Explode** dialog box to explode the view, enter the desired explosion distance for each of the components in the **Distance** edit box. Again, an explosion distance is the offset components are removed from other components. Next, select the **Create Trails** check box if you want trails to be
displayed in the presentation. As shown in previous figures, trails identify the relationship between component
distance and location. Then pick the **Preview** button to display the intended explosion. If the preview looks
acceptable, pick the **OK** button to apply the settings, and explode the assembly.

Exercise 16.4

1. Continue from EX16.3, or launch Autodesk Inventor.
2. Open EX16.2.ipn and save a copy as EX16.4.ipn.
3. Close EX16.2.ipn without saving and open EX16.4.ipn.
4. Expand all children in the Browser.
5. Delete the tweak created in EX16.2. The presentation should return to the assembled display.
6. Right-click on the assembly EX15.2.iam in the Browser.
7. Select the **Auto Explode** menu option.
8. Specify a distance of 10 in the **Distance** edit box.
9. Deselect the **Create Trails** check box.
10. Pick the **Preview** button to preview the tweak.
11. Pick the **OK** button to exit the **Auto Explode** command.
12. The final exercise should look like the presentation shown.
13. Resave EX16.4.ipn.
14. Exit Autodesk Inventor or continue working with the program.

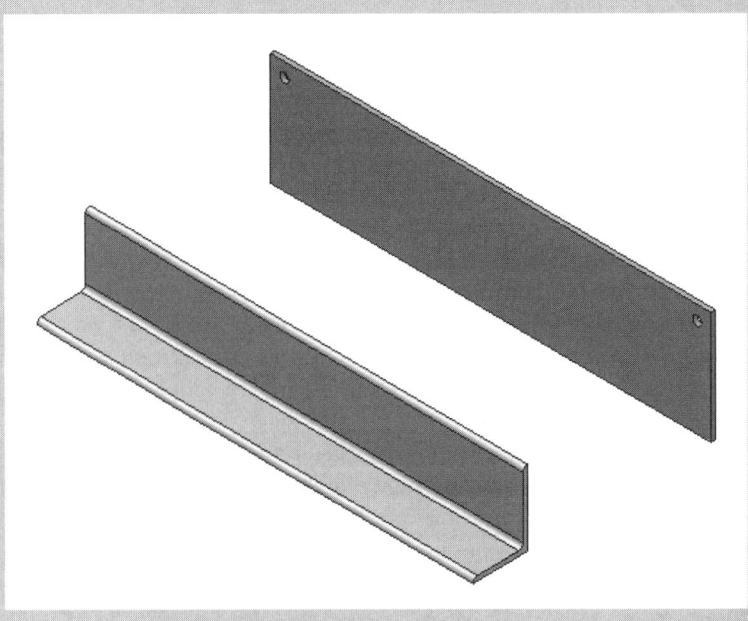

Using the Tweak Components Tool

Although one way to explode a presentation view is to use one of the automatic options previously discussed, often it is necessary to individually separate components, or manually explode an assembly. Manually exploding an assembly is done using the **Tweak Components** tool, which establishes presentation tweaks between the selected components. Access the **Tweak Components** tool using one of the following techniques:

✓ Pick the **Tweak Components** button on the **Presentation** panel bar.

✓ Pick the **Tweak Components** button on the **Presentation** toolbar.

✓ Press the **+** and **T** keys on your keyboard.

✓ Select the **Tweak Components...** option of the **Insert** pull-down menu.

✓ Right-click on an assembly or component in the Browser and select the **Tweak Components...** menu option.

✓ Right-click on a component, or any location in the graphics window, and select the **Tweak Components...** menu option.

When you activate the **Tweak Components** tool, the **Tweak Component** dialog box, shown in Figure 16.13, is displayed. The **Tweak Component** dialog box allows you to identify the direction of separation, the components you want to move apart, and trial and tweak specifications. The first step to tweaking a component is to identify the tweak direction. To specify the direction, pick the **Direction** button if it is not currently selected. Then, as the direction icon shows, you can select any edge, face, or feature to identify the desired direction of movement. See Figure 16.14A. When you select an edge, face, or feature, one of the axis indicators is highlighted, indicating the axis is selected, and the component will move along the highlighted axis. See Figure 16.14B.

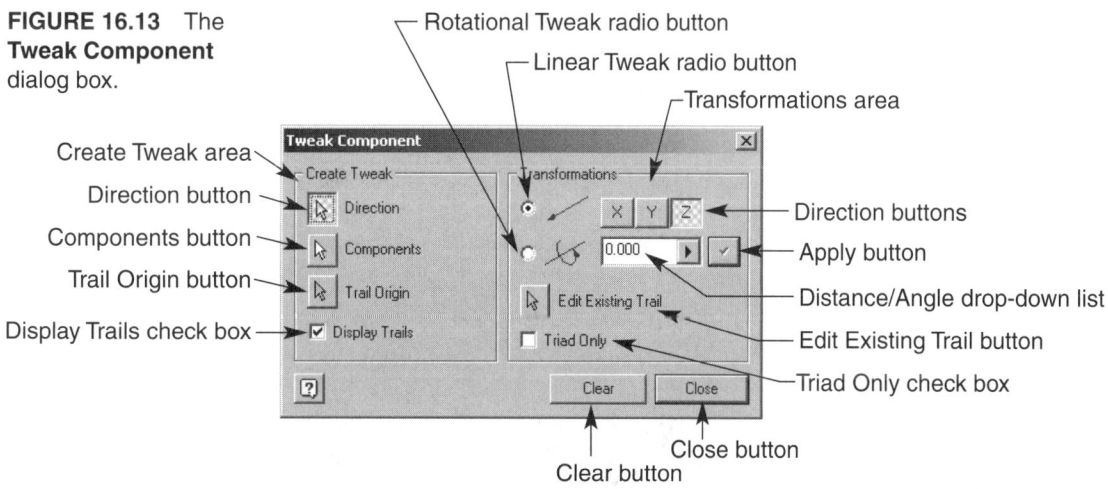

FIGURE 16.13 The **Tweak Component** dialog box.

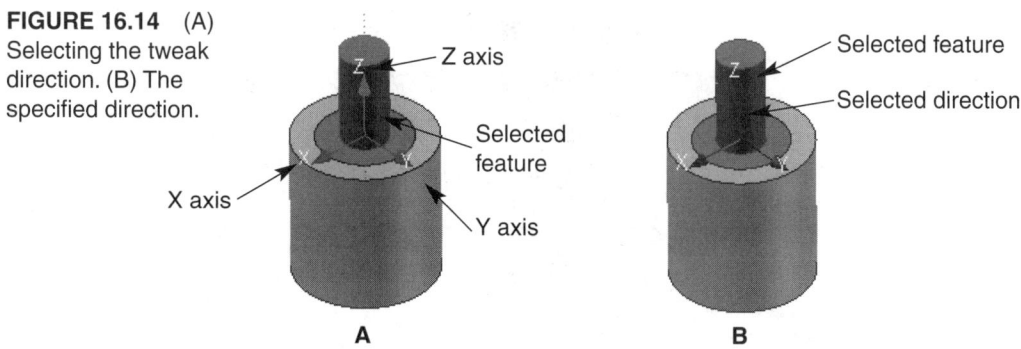

FIGURE 16.14 (A) Selecting the tweak direction. (B) The specified direction.

Once you have identified the tweak direction, the **Components** button becomes active, allowing you to select the component or components you want to move. If you accessed the **Tweak Components** tool by right-clicking on a specific component or a component name, one component will already be selected. You still need to select any additional components you want to tweak at the same time. In the same respect, you can deselect components by holding down the **Ctrl** key on your keyboard and picking the component you want to remove from the tweak. Figure 16.15 shows how a single or multiple components can be selected depending on the presentation application.

Field Notes

In addition to selecting components, when the **Components** button is active, you can select a different axis on the axis indicator to redefine the direction.

After you define the components you want to tweak, pick the **Trail Origin** button, then select a point on the assembly to identify the origin, or beginning, of the trail. If you do not specify a trail origin, the trail will begin at the center of the selected component mass. If you do not want to show a trail at all, deselect the **Display Trail** check box.

The next step to tweaking components is to manage the following options in the **Transformations** area of the **Tweak Component** dialog box:

FIGURE 16.15
Selecting components to tweak.

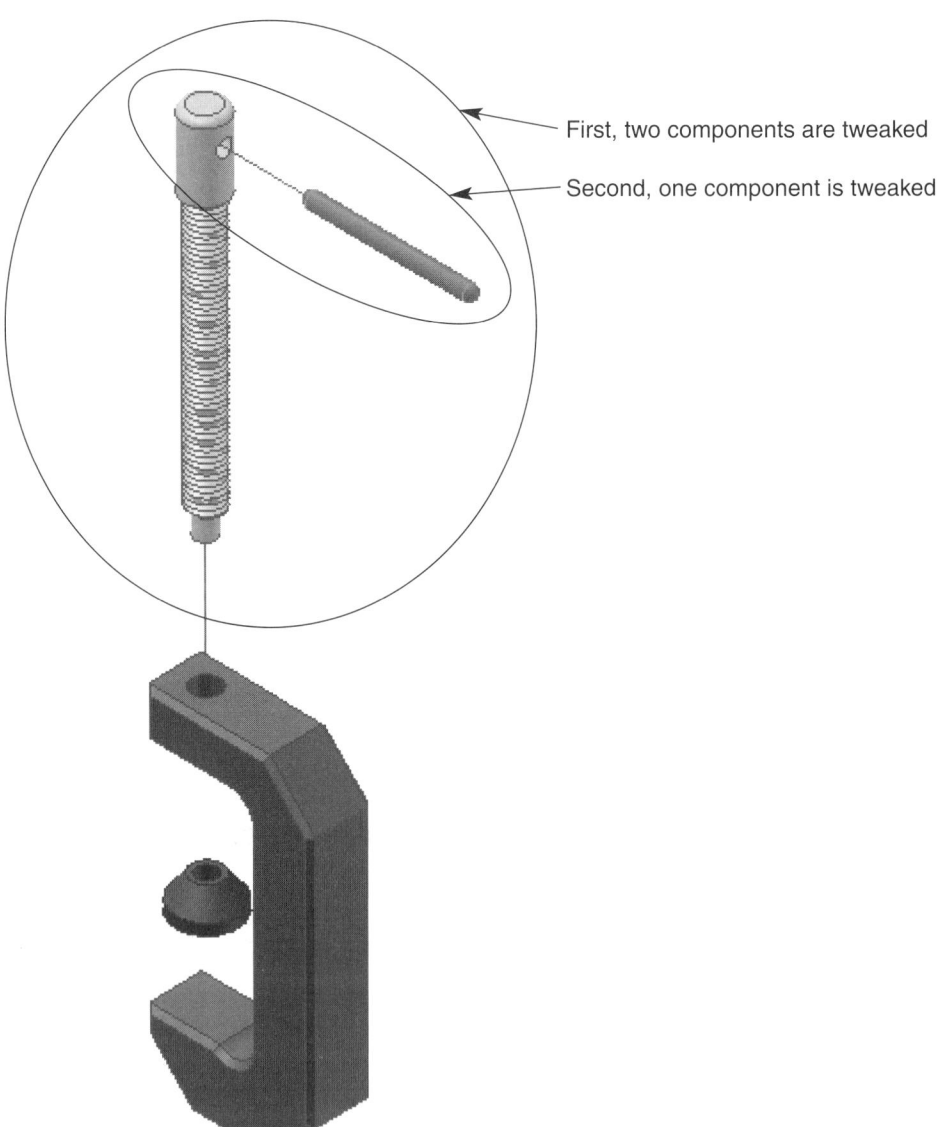

First, two components are tweaked

Second, one component is tweaked

- **Triad Only**   Pick this check box if you want to rotate the axis indicator triad, shown in Figure 16.14, without actually rotating a component. To manipulate the axis indicator triad, ensure the **Triad Only** check box is selected, then enter an angle in the **Distance/Angle** edit box, and finally pick the **Apply** button to rotate the triad. You can also rotate the triad by holding down on one of the axes and dragging/rotating the triad to the desired angle.

- **Linear Tweak**   Select this radio button if you are moving the component in a linear fashion along the specified axis. A linear tweak moves the component an identified distance. See Figure 16.16.

- **Rotational Tweak**   Pick this radio button to define a rotational tweak, which rotates the component a specified angle, instead of a distance, as in a linear type of tweak. A rotational type of tweak is effective anytime you want a component to revolve around the selected axis, such as when a nut threads onto a bolt. See Figure 16.17.

- **Direction**   Using these buttons is another way to adjust the specified tweak direction axis. The button of the axis that is currently selected will be activated, but you can choose the **X,** **Y,** or **Z** buttons, depending on the tweak axis you want to define.

- **Distance/Angle**   If you are creating a linear type of tweak, use this edit box to enter the distance you want the component to be offset from the assembly and other components. If you are creating a rotational type of tweak, use this edit box to enter the angle of rotation. For example, if you want a hub to fully rotate, enter a value of 360°.

Field Notes

Although entering a specific distance ensures a linear tweak is created using a specific offset, you can hold down a component and linearly drag it to the desired location. You will notice the **Distance** edit box will identify the distance you drag the component.

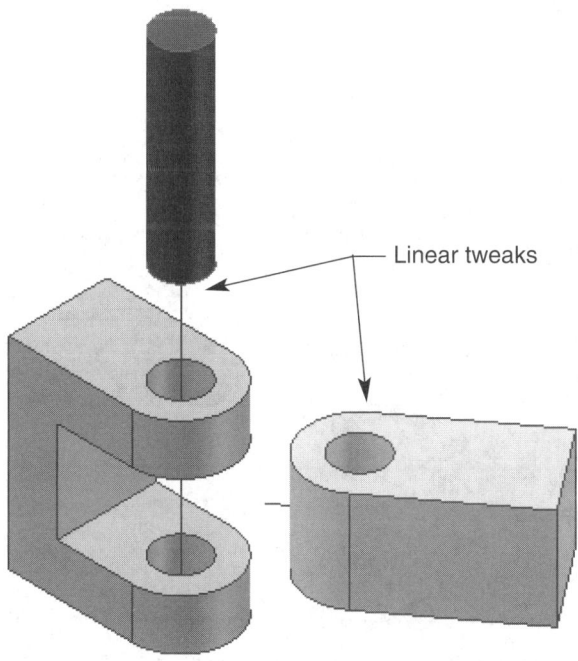

FIGURE 16.16   An example of using a linear type of tweak.

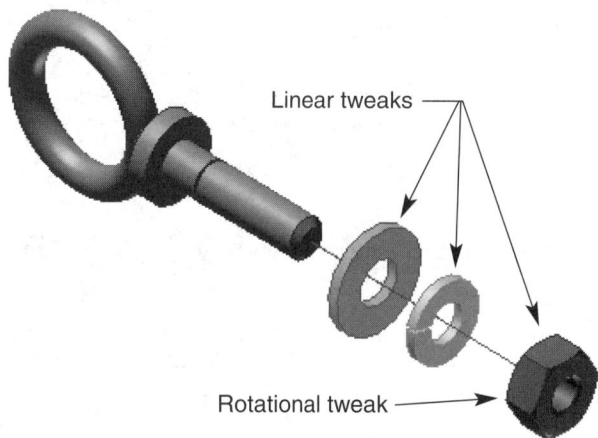

FIGURE 16.17   An example of using a rotational type of tweak (bolt rotates onto stud).

- **Apply**   Pick this button to create the specified tweak. When you pick the **Apply** button, the Tweak Component dialog box does not close, allowing you to continue applying tweaks.

- **Clear**   Select this button to reset the **Tweak Component** dialog box, so you can apply an additional tweak.

- **Close**   Once you have specified the desired number of tweaks, pick this button to close the dialog box.

- **Edit Existing Trail**   Select this button when you want to modify an existing tweak trail. Once you pick the **Edit Existing Trail** button, select the trail you want to modify, then edit the distance or angle by dragging the component trail, or entering a different value in the **Distance/Angle** edit box.

Exercise 16.5

1. Continue from EX16.4, or launch Autodesk Inventor.
2. Open EX16.3.ipn and save a copy as EX16.5.ipn.
3. Close EX16.3.ipn without saving and open EX16.5.ipn.
4. Access the **Tweak Component** dialog box.
5. Select the c-clamp body face, and pick the Y axis as shown in A.
6. After you specify the direction, the **Components** button should be come active.
7. Hold-down the **Ctrl** key on your keyboard, and select the c-clamp screw and pin.
8. Pick the **Trial Origin** button, and select a point on the bottom face of the c-clamp swivel as the trail origin.
9. Right-click and select the **Continue** menu option.
10. Ensure the **Linear** radio button is selected, and enter a distance of 85 in the **Distance** edit box.
11. Pick the **Apply** button to create the tweak shown in B.
12. Pick the **Clear** button.
13. Again, select the c-clamp body face as shown in A to define the direction. However, this time, pick the X axis.
14. After you specify the direction, the **Components** button should become active.
15. Pick the c-clamp pin.
16. Ensure the **Linear** radio button is selected, and enter a distance of 50 in the **Distance** edit box.
17. Pick the **Apply** button to create the tweak shown in C.
18. Select the Z axis on the triad, enter a distance of -50, and pick the **Apply** button to create the tweak shown in D.
19. Pick the **Clear** button.
20. Again, select the c-clamp body face as shown in A to define the direction. However, this time, pick the Z axis.
21. After you specify the direction, the **Components** button should be come active.
22. Pick the c-clamp swivel.
23. Ensure the **Linear** radio button is selected, and drag the swivel to a location similar to the location shown in E.
24. Pick the **Apply** button to create the tweak.
25. Select the X axis on the triad, and then drag the swivel to a location similar to the location shown in E.
26. Pick the **Apply** button to create the tweak shown in E.
27. Select the Z axis on the triad, enter a distance of -50, and pick the **Apply** button to create the tweak shown in D.
28. The final exercise should look like the presentation shown in E.

Exercise 16.5 continued on next page

Exercise 16.5 continued

29. Resave EX16.5.ipn.
30. Exit Autodesk Inventor or continue working with the program.

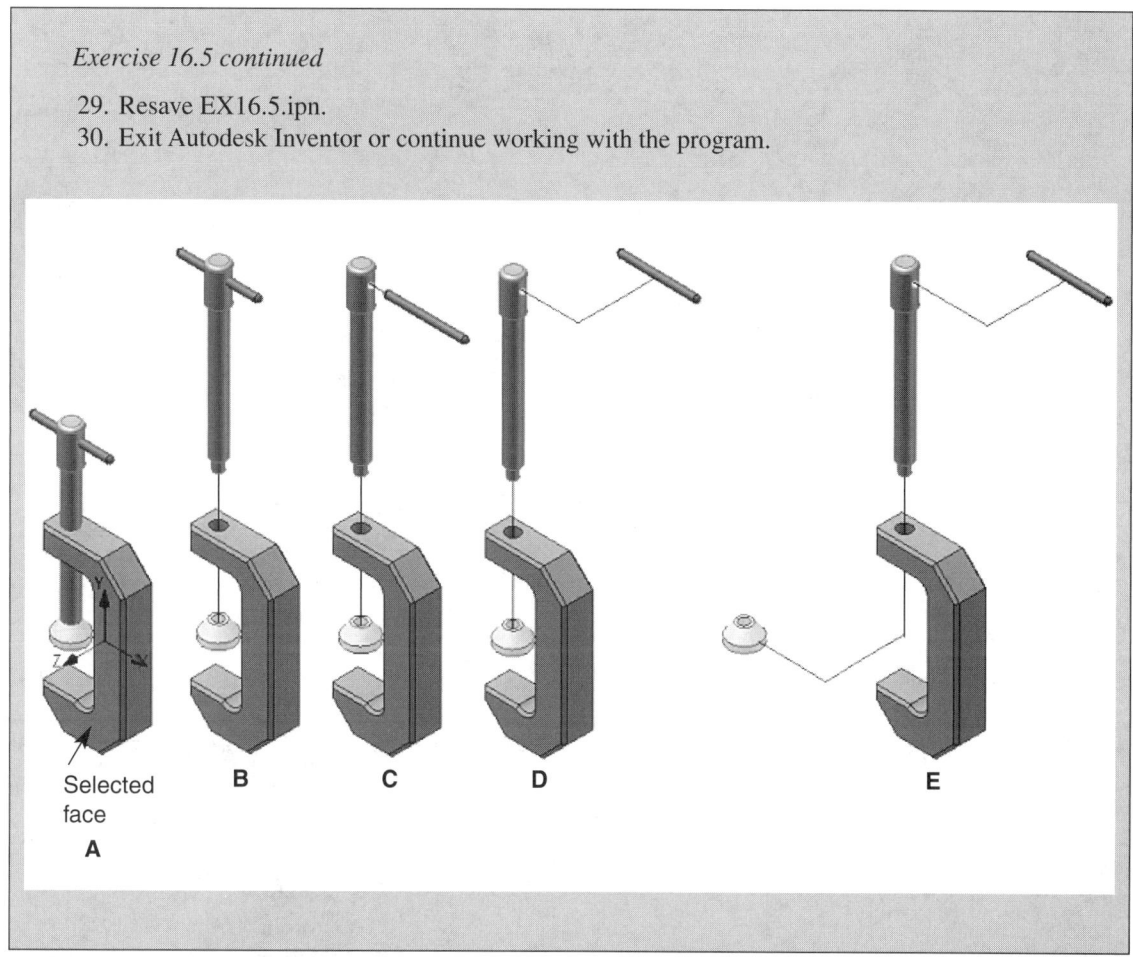

Selected face

A   B   C   D   E

Editing Tweaks and Trails

Like any other Autodesk Inventor design application, you must often edit presentation parameters once they have been defined. Typically the only presentation factors that may require modification are tweaks and trails. The location of the available tweak and trail editing options depends on the specified Browser filter option. In addition, if you define a **Sequence View** filter, you will be able to modify tweak sequence parameters that are not available when the **Assembly View** filter is set. However, as previously mentioned, sequence parameters are typically used for animation purposes and will be discussed later in this chapter. This section identifies specific tweak and trail editing options available in the Browser, but also other, more general, tweak and trail modification tools.

When you right-click on a component in the graphics window, or in the Browser, that has been tweaked, the following edit options are available:

- **Delete Tweaks** This option displays a cascading submenu that allows you to delete tweaks. If a component has more than one tweak, as shown in Figure 16.18A, and you want to delete the most recently created tweak, select the **Last** cascading submenu option. See Figure 16.18B. If you want to remove all the tweaks associated with a component, pick the **All** cascading submenu option. See Figure 16.18C.

- **Add Trail** Use this option to display an additional tweak trail. Once you select the **Add Trail** option, pick one or more points on the component. Each point you select will create a new trail. The only point you can select is on the specified component, because the trail is automatically referenced to other assembly components. To redefine the point or points you select, right-click and choose the **Reselect** menu option, then pick the new point. When finished selecting points for new trails, right-click and select the **Done** menu option to generate the new trail or trails. See Figure 16.19.

- **Hide Trails** Select this option to "turn off," or hide, the tweak trails so they are not visible.

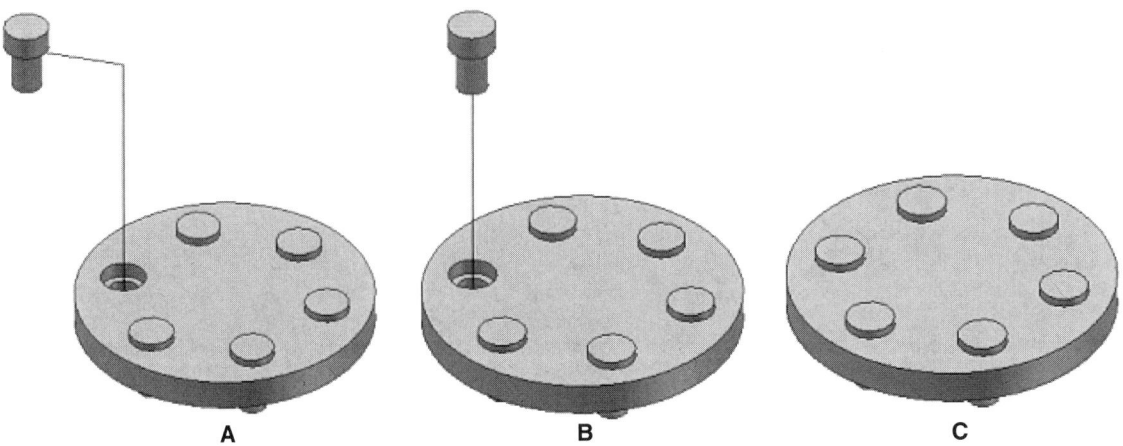

FIGURE 16.18   (A) An example of a component with two tweaks. (B) Deleting the last
tweak. (C) Deleting all the tweaks.

FIGURE 16.19
Placing an additional
trail.

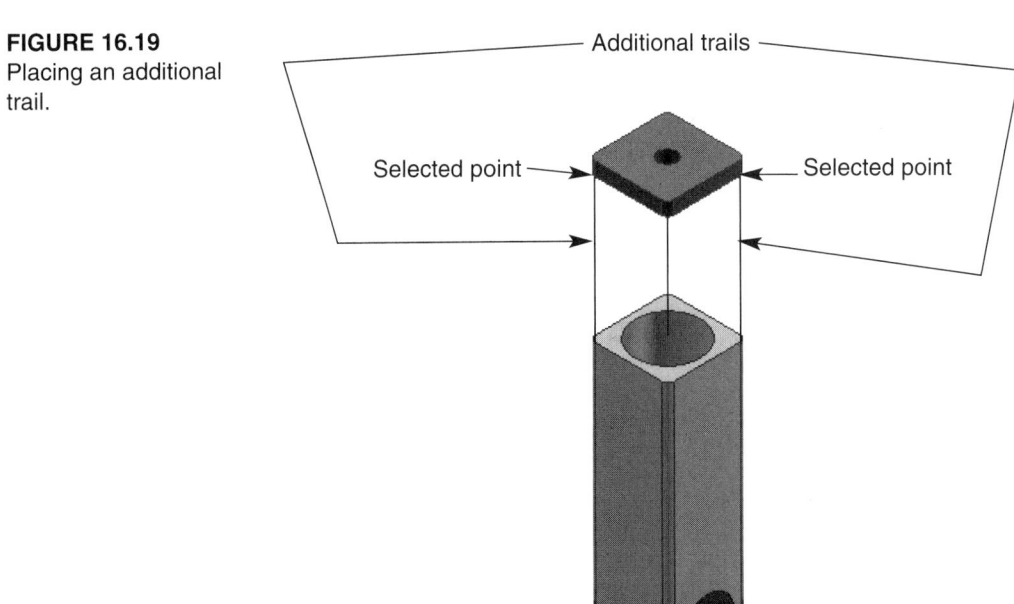

Field Notes

> The **Hide Trails** option is also available when you right-click directly on a tweak trail.

When you right-click on a tweak trail in the graphics window, the following edit options are available
including the **Hide Trails** option previously discussed:

- **Delete**   This option functions similarly to the **All** selection of the **Delete Tweaks** cascading
  submenu, and removes the entire tweak, not just the trail.

- **Visibility**   This option functions the same as the **Hide Trails** choice, and "turns off" tweak
  trails.

- **Edit...**   Select this option to open the **Tweak Component** dialog box for the specified
  tweak. Here you can redefine the tweak parameters.

Field Notes

You can also open the **Tweak Component** dialog box for a specified component tweak by double-clicking on a component, or component trail.

The **Delete** and **Visibility** options are also available when you right-click on a tweak in the Browser.

As mentioned, you can reaccess the **Tweak Component** dialog box for a selected component by right-clicking on the tweak trail, or double-clicking on the component or trail. Once you open the **Tweak Component** dialog box, you can modify the tweak parameters. However, if all you want to do is extend the trail length by increasing the tweak distance, or modify the rotational angle, you can use one of the following options:

- **Drag**   To lengthen the explosion offset, pick the trail of the tweak you want to modify. Then drag the green dot to the desired location. See Figure 16.20.

- **Select Tweak in the Browser**   When you pick a tweak in the Browser, the distance or angle of the tweak is shown in an edit box. See Figure 16.21. You can use this edit box to change the distance or angle of the tweak by highlighting the existing value and entering a new value.

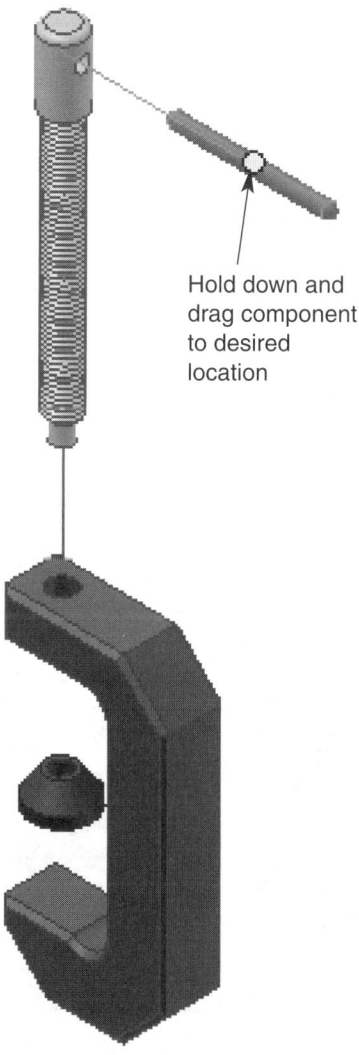

Hold down and drag component to desired location

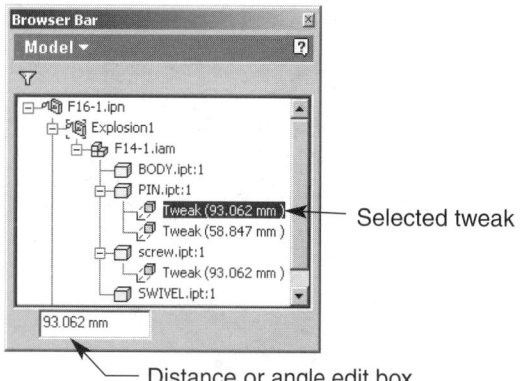

Selected tweak

Distance or angle edit box

FIGURE 16.20   Dragging a component to expand the tweak distance.

FIGURE 16.21   Modifying tweak distance or angle specifications.

Exercise 16.6

1. Continue from EX16.5, or launch Autodesk Inventor.
2. Open EX16.5.ipn and save a copy as EX16.6.ipn.
3. Close EX16.5.ipn without saving and open EX16.6.ipn.
4. Use the editing techniques previously discussed, and the figure shown, to:

   Delete tweaks
   Add a trail
   Turn off a trail visibility
   Hide trails
   Edit a tweak
   Lengthen a tweak by dragging
   Select a tweak in the Browser to edit the distance

5. The final exercise should look similar to the presentation shown.
6. Resave EX16.6.ipn.
7. Exit Autodesk Inventor or continue working with the program.

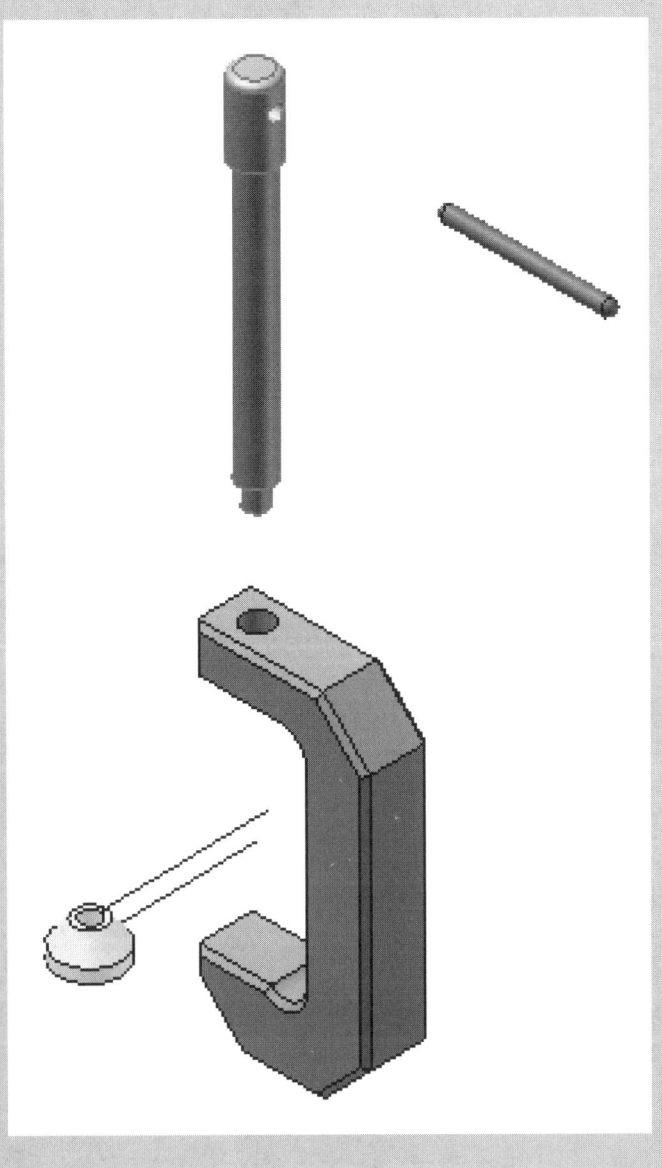

Using the Precise View Rotation Tool

The presentation environment contains all the view tools you have become familiar with, including zoom, pan, rotate and look at. Any of these tools can be used to set the desired *camera view,* which is the term used to define the position and zoom parameters of the presentation view. However, these view tools do not allow you to rotate the presentation view using a specified incremental value, such as 10° in a certain direction. To manage this type of view manipulation, the **Precise View Rotation** tool is used.

Access the **Precise View Rotation** command using one of the following techniques:

✓ Pick the **Precise View Rotation** button on the **Presentation** panel bar.

✓ Pick the **Precise View Rotation** button on the **Presentation** toolbar.

✓ Right-click in the graphics window, and select the **Precise View Rotation...** menu option.

✓ Select the **Precise View Rotation** option from the **View** pull-down menu.

When you access the **Precise View Rotation** command, the **Incremental View Rotate** dialog box is displayed. See Figure 16.22. This dialog box functions the same as the **Incremental View Rotate** dialog box, discussed in Chapter 12, and used for creating a custom drawing view.

The **Incremental View Rotate** dialog box allows you to precisely rotate the presentation view, using a specified angle of rotation in certain direction. Use this dialog box by first entering the amount, or angle, of rotation in the **Increment** edit box. Then pick the rotate buttons to rotate the view to specific position. Each time you select one of the rotate buttons, the view rotates in the specified direction at the increment you define. If you want to return to the original view, pick the **Reset** button. Finally, when you have rotated the view as desired, pick the **OK** button to close the **Incremental View Rotate** dialog box, and exit the **Precise View Rotation** command.

Field Notes

> Once you define a specific presentation camera view using the **Precise View Rotation** tool, you may want to use the **Save Camera** option, discussed earlier, to save the current presentation view display.

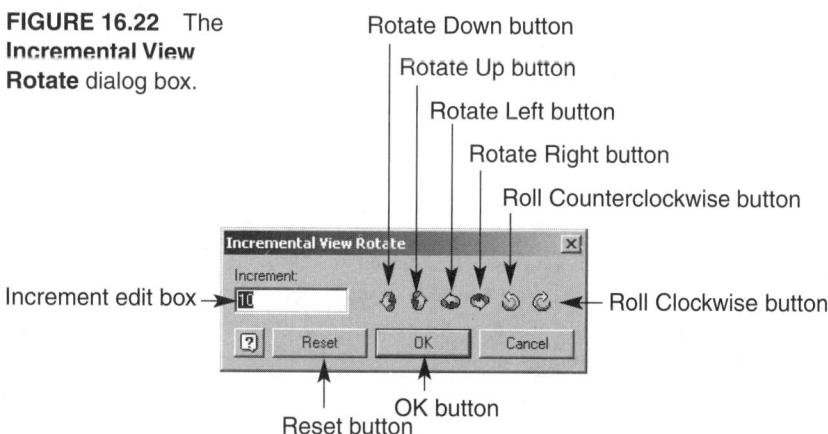

FIGURE 16.22 The **Incremental View Rotate** dialog box.

Rotate Down button
Rotate Up button
Rotate Left button
Rotate Right button
Roll Counterclockwise button

Increment edit box →

← Roll Clockwise button

Reset button
OK button

Exercise 16.7

1. Continue from EX16.6, or launch Autodesk Inventor.
2. Open EX16.2.ipn.
3. Access the **Precise View Rotation** command.
4. Specify an increment of 10 in the **Increment** edit box of the **Incremental View Rotate** dialog box.
5. Pick each of the rotate buttons and observe the effects.
6. Select the **OK** button to exit the command.
7. Close EX16.2.ipn without saving.
8. Exit Autodesk Inventor or continue working with the program.

Animating Presentation Views

Once you have created and exploded a presentation view, you can display the assembly characteristics and the tweaked component parameters by animating the presentation. An *animation* is useful for visualizing how assembly components fit together. In addition, a presentation animation allows you to further display product assembly and disassembly characteristics for multiple purposes, including maintenance and assembly requirements. The **Animate** tool is the principle command used for animating presentation views. This tool allows you to fully define, play, and record an exploded presentation view animation. Another animation option is to set the Browser filter as <u>**Sequence View**</u> and use the sequence view options to show the animation. However, using the sequence view technique does not allow you to edit the animation and is typically used in conjunction with the **Animate** tool.

Using Browser Sequence View Options

As previously discussed, the **Sequence View** browser filter option is typically used to help develop task and sequence specifications, especially for animating presentation views. When the **Sequence View** option is specified, presentation view tasks and sequences are displayed. A *task* is a group of sequences that define a certain assembly or disassembly relationship. A *sequence* is a group of tweaks that identify tweak relationships between components. Tasks describe how groups of components function in the animation, while sequences describe how each component functions in reference to other components during the animation. For example, Figure 16.23 shows a presentation view with two tasks. The first task contains three sequences. The first sequence aligns the washer with the bolt shaft, the second sequence aligns the lock washer with the bolt shaft, and the third sequence aligns the nut with the bolt shaft. The second task also contains three sequences, which assemble the washer, lock washer, and nut, with the eyebolt as specified with tweaks.

Field Notes

Although the presentation view in Figure 16.23 does contain two tasks, all the sequences from both tasks can be placed in one task. As shown in the Browser display of Figure 16.23, the sequences are not really separate, as evidenced by the sequence numbers 1–6.

Typically multiple tasks are used for larger assembly presentations, where distinct components or subassemblies are animated.

As you can tell, task and sequencing of presentation view animations can be complicated and difficult to understand, especially for large, complex presentation views. The key to animation is to effectively manage tweaks, tasks, and sequences. To view the effects of a task or a sequence, right-click on the task or sequence in the Browser, and select the **Edit...** menu option. When you edit a task or sequence, the **Edit Task & Sequences** dialog box is displayed, with the specified task or sequence active. See Figure 16.24.

FIGURE 16.23 An example of a presentation view with two tasks.

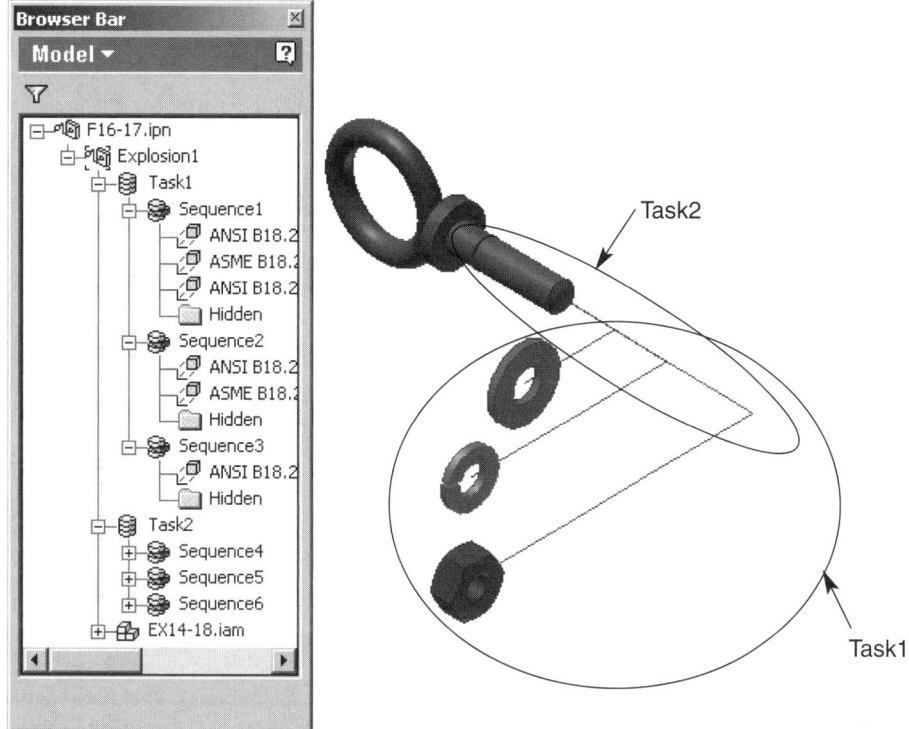

FIGURE 16.24 The **Edit Task & Sequences** dialog box.

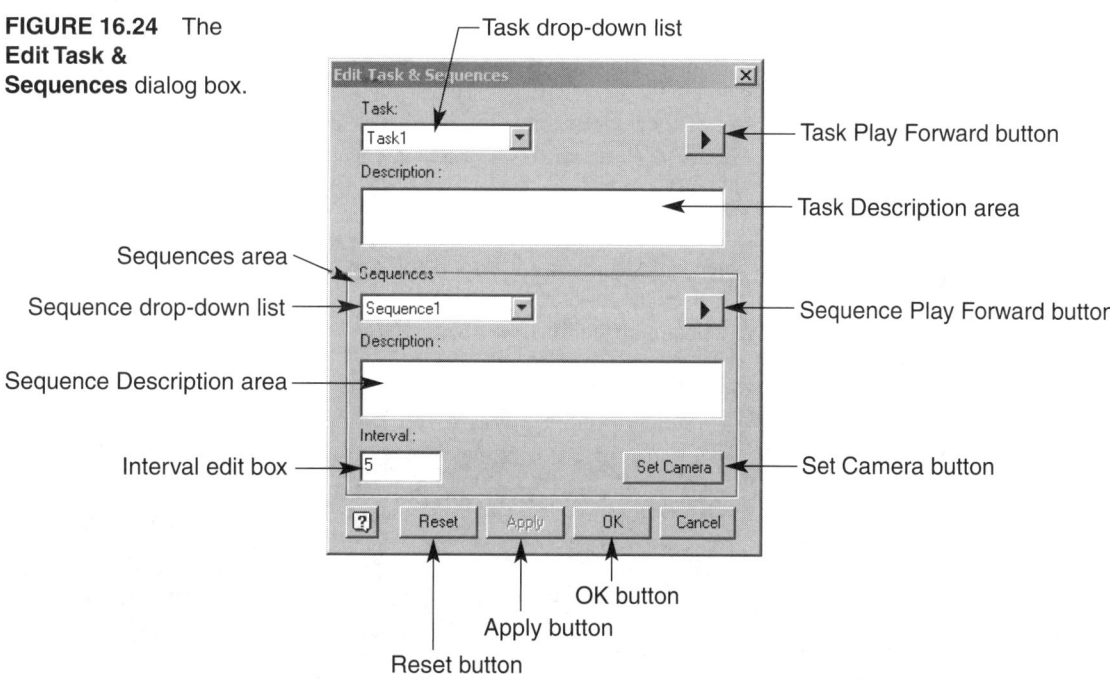

As mentioned, if you edit a specific task, the task will be active and displayed in the **Task** drop-down list. Similarly, if you edit a certain sequence, the sequence will be active and displayed in the **Sequence** drop-down list. Still, you can choose to edit a different task or sequence by picking a task or sequence from the **Task** drop-down list or the **Sequence** drop-down list. Then you can enter a description of the animation task or sequence in the corresponding **Description** area. Descriptions are typically used for clarification and future task and sequence editing reference. To observe the specified task or sequence operation, pick either the task or sequence **Play Forward** button. After you play the animation, pick the **Reset** button to return to the exploded view.

In addition to observing the animation operation of certain tasks or sequences and defining task and sequence descriptions, use the **Edit Task & Sequences** dialog box to specify the sequence interval and sequence camera view. The sequence interval is the playing speed of the animation. To create a faster playing speed, enter a higher interval value in the **Interval** edit box. Conversely, to create a slower playing speed, enter a lower interval value in the **Interval** edit box.

Although you may have set a camera view for the entire presentation view, you can also create a separate sequence camera view. When you define a sequence camera view, then animate the presentation, the animated display automatically zooms in or out and rotates according to the specified camera view. To identify a sequence camera view, first rotate and zoom in on the presentation view as desired, using viewing tools. Then pick the **Set Camera** button to create the sequence camera view. Once you have specified all the options in the **Edit Task & Sequences** dialog box, pick the **Apply** button to apply the descriptions and camera setting. Then continue working with the dialog box, or pick the **OK** button to exit the command.

In addition to accessing the **Edit Task & Sequences** dialog box from the sequence view presentation Browser, you can also accomplish the following processes:

- **Create a new task**   To create a new task, right-click on the presentation view in the Browser, such as "Explosion1," and select the **Create Task** menu option. Once the new task is present, you can add sequences by creating additional tweaks or drag existing sequences into the new task.

- **Create a new sequence**   As you have seen, sequences contain tweaks. If a sequence contains more than one tweak, you can create a new sequence by right-clicking on a tweak and selecting the **New Sequence** menu option. The new sequence will become the first sequence of the first task and contains the specified tweak. As a result, the tweak is removed from the original sequence.

- **Modify sequence order**   Modifying the sequence order is one of the most useful purposes of the sequence view option of the presentation Browser and allows you to redefine the animation operation. Changing sequence order is accomplished much the same way as dragging a sequence into a different task. To edit sequencing, drag the sequence above or below another sequence, depending on the requirement. For example, if a washer is set as Sequence1 and a lock washer is set as Sequence2, but you want the lock washer to assemble first, drag Sequence2 above Sequence1. In this example, Sequence2 will automatically become Sequence1.

- **Group sequences**   If you want to add the tweaks of two or more sequences together, you can group the sequences. This operation is done by holding down the **Ctrl** key on your keyboard, and selecting each of the sequences you want to group. Then right-click and select the **Group Sequence** menu option.

Exercise 16.8

1. Continue from EX16.7, or launch Autodesk Inventor.
2. Open EX16.5.ipn.
3. Set the Browser filter option as sequence view.
4. Expand all children.
5. Explore the sequence view options and tools previously discussed.
6. Close EX16.5.ipn without saving.
7. Exit Autodesk Inventor or continue working with the program.

Using the Animate Tool

Once you have fully created the presentation view, you can use the **Animate** tool to define animation characteristics, play the animation, and record an exploded presentation view animation, using specified task and sequencing information. Access the **Animate** command using one of the following techniques:

✓ Pick the **Animate** button on the **Presentation** panel bar.

✓ Pick the **Animate** button on the **Presentation** toolbar.

✓ Right-click in the graphics window, and select the **Animate...** menu option.

✓ Select the **Animate...** option from the **Tools** pull-down menu.

When you access the **Animate** command, the **Animation** dialog box is displayed. See Figure 16.25. Use the following areas in the **Animation** dialog box to set the animation for the presentation view:

■ **Parameters**   This area is used to set the interval and repetition characteristics of the animation. The ***animation interval*** functions similar to the sequence interval discussed earlier and specifies the playing speed of the animation. The interval specifies how many steps are required to move a component the specified distance. For example, if a tweak distance is 5″, and you define an interval of 5, each step moves the component 1″. To create a faster playing speed, enter or select a higher interval value in the **Interval** edit box. Conversely, to create a slower playing speed, enter or select a lower interval value in the **Interval** edit box. The ***animation repetition*** is the number of times the animation goes through the animation process. The default repletion of 1 allows you to view the assembly or disassembly animation one time. Enter or select the desired number of repetitions from the **Repetitions** edit box.

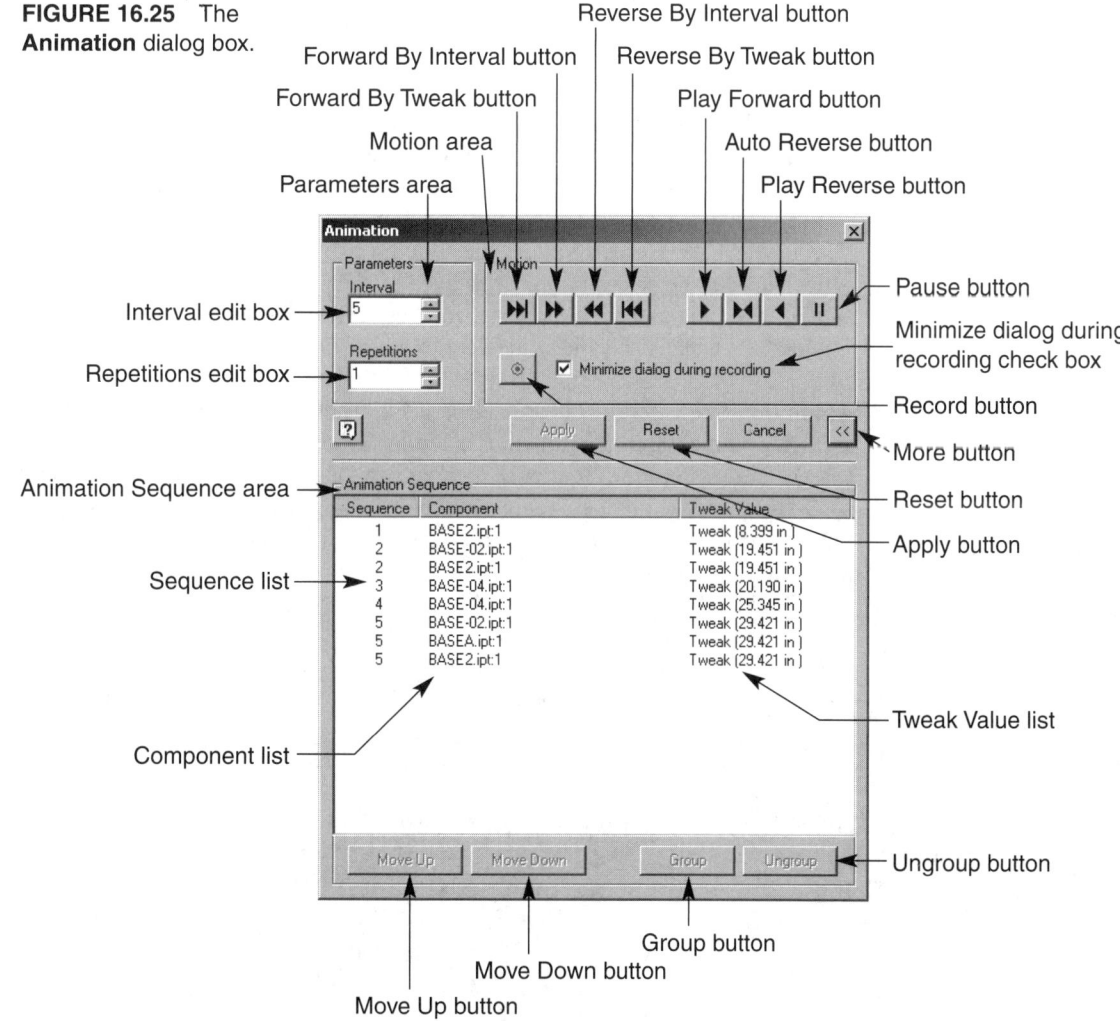

FIGURE 16.25 The **Animation** dialog box.

Field Notes

After you modify the interval and repetition specification, you must pick the **Apply** button before you can set the animation in motion.

■ **Motion**   This area allows you to play and record the animation. The following options are available in the **Motion** area:

■ **Forward By Tweak**   Pick this button to move the animation forward by one tweak.

■ **Forward By Interval**   Select this button to advance the animation forward by one interval step. Using the previous example of a 5″ tweak distance and an interval of 5, the component will move forward 1″.

■ **Reverse By Interval**   Use this button to move the animation back by one interval step. You can only move an animation in reverse if it has been advanced forward.

■ **Reverse By Tweak**   Select this button to reverse the animation by one tweak. Again, you can only move an animation in reverse if it has been advanced forward.

■ **Play Forward**   Pick this button to play the entire animation forward.

■ **Auto Reverse**   Select this button to play the animation forward, then once it has finished the forward cycle, the animation will automatically play in reverse.

■ **Play Reverse**   Pick this button to play the entire animation in reverse. If you have not advanced the animation forward, when you select the **Play Reverse** button, the presentation will begin as the assembled product and then disassemble.

■ **Pause**   Select this button to pause the animation.

■ **Record**   Pick this button to record the presentation view animation as an .avi file. Before you select the **Record** button, you may want to determine whether or not you want the **Animation** dialog box to be displayed in the recording. If you do not want the **Animation** dialog box show, deselect the **Minimize dialog during recording** check box. When you select the **Record** button, the **Open** dialog box appears, allowing you to specify a location for the animation file, followed by the **Video Compression** dialog box, shown in Figure 16.26, which, as the name implies, allows you to compress, or not compress, the video file using a specified video compressor.

■ **Animation Sequence**   This area is available when you select the **More** button and is used to set the animation sequence, much like the sequence view Browser options previously discussed. The **Animation Sequence** area contains **Sequence, Component,** and **Tweak Value** lists. The sequence numbers represent the order in which the components and corresponding tweaks are animated. Components that have the same sequence number are grouped as tasks and move together during the animation sequence. You can remove a sequence from a task by selecting the sequence and picking the **Ungroup** button. Conversely, you can add a sequence or a number of sequences together to form a task, or add sequences to an existing task. To apply this technique, hold down the **Ctrl** key on your keyboard and select each of the sequences you want to group together. Then pick the **Group** button to generate the task.

FIGURE 16.26 The **Video Compression** dialog box.

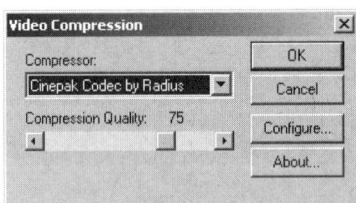

Field Notes

When you pick a sequence in the Animation dialog box, the tweak trail is highlighted in the graphics window.

After you modify the animation sequence specifications, you must pick the **Apply** button before you can set the animation in motion.

Anytime while working with the **Animation** dialog box, pick the **Reset** button to reset the animation and return the presentation view to its original display. Once you have finished the animation session, pick the **Cancel** button to exit the **Animate** command.

Exercise 16.9

1. Continue from EX16.8, or launch Autodesk Inventor.
2. Open EX16.5.ipn.
3. Access the **Animation** dialog box.
4. Animate the presentation view using the techniques discussed.
5. Close EX16.2 without saving.
6. Exit Autodesk Inventor or continue working with the program.

Placing Additional Presentation Views

As mentioned initially in this chapter, a presentation file can contain as many presentation views as needed. For some presentations only one presentation view is required. However, for other presentations, multiple presentation views are necessary. Although there are numerous reasons for having more than one presentation view, typically additional presentation views allow you to display different views of the assembly and create different types of animations. For example, you may want to show an assembly with different component colors, or with some components suppressed. You may want to display an animation of an alternative product assembly technique.

Creating additional presentation views is very similar to placing the initial presentation view. To place another presentation view, access the **Create View** tool, previously discussed. Again, the **Create View** tool is available using one of the following techniques:

✓ Pick the **Create View** button on the **Presentation** panel bar.

✓ Pick the **Create View** button on the **Presentation** toolbar.

✓ Select the **Create View...** option of the **Insert** pull-down menu.

✓ Right-click inside the graphics window or the Browser, and select the **Create View...** menu option.

When you access the **Create View** tool, the **Select Assembly** dialog box is displayed. However, the **File** drop-down list, and the file **Browse** button are not available, because the assembly file used for the presentation was already defined when the initial presentation view was created. You cannot define presentation views from different assembly files in one presentation file. To create a presentation view of a different assembly, you must open a new presentation file. Although the **File** drop-down list is not available, the **Design View** drop-down list is displayed. To insert additional presentation views, enter or select a design view file from the **Design View** drop-down list, or use the **Browse** button, located to the right of the **Design View** drop-down list, to locate the desired design view file.

A new presentation view is identified as "Explosion2" in the browser, assuming only one other presentation view already exists. For example, if you have three existing presentation views, the new view will be named "Explosion4." As with other Autodesk Inventor applications, you may find it helpful rename the presentation views by picking and highlighting the name and entering a new name. Once created, additional presentation views can be manipulated as required, using the tools discussed in this chapter and previous chapters. As previously discussed, the active presentation view is displayed in the graphics window, while the inactive presentation views are shaded in the Browser. Again, to active a different presentation view, right-click on an inactivate view and select the **Activate** menu option. Or activate a presentation view by double-clicking on the view name in the Browser.

Field Notes

As you can see, assembly design views actually allow you to define the display of a presentation view and can be very useful for creating additional presentation views.

Exercise 16.10

1. Continue from EX16.9, or launch Autodesk Inventor.
2. Open EX16.5.ipn and save a copy as EX16.10.ipn.
3. Close EX16.5.ipn without saving and open EX16.10.ipn.
4. Access the **Create View** tool to open the **Select Assembly** dialog box.
5. There may only be one available design view for this exercise.
6. Select the **Manual** radio button, and pick the **OK** button to generate the presentation view.
7. Resave EX16.10.ipn.
8. Exit Autodesk Inventor or continue working with the program.

|||||||||||| CHAPTER TEST

Answer the following questions on a separate sheet of paper.

1. What is a presentation?

2. How are presentations different from assemblies?

3. What does it mean to say that presentations are basically exploded assemblies?

4. What does the browser for a typical presentation display?

5. What does the assembly hierarchy show?

6. Give the general function of presentation tools.

7. Discuss the general function of the Select Assembly dialog box.

8. What are trails?

9. Autodesk Inventor allows you to explode an assembly in the presentation view automatically or manually. Briefly explain the difference between automatically or manually exploding an assembly.

10. Discuss the general function of the Tweak Component dialog box.

11. Give the first step to tweaking a component.

12. How do you deselect components that you want to remove from the tweak?.

13. After you define the components you want to tweak, pick the Trail Origin button, then select a point on the assembly to identify the origin, or beginning, of the trail. Where does the trail begin if you do not specify a trail origin?

14. What do you do if you do not want to show a trail?

15. How do you reaccess the Tweak Component dialog box for a selected component?

16. What is a camera view?

17. Give the basic function of the Incremental View Rotate dialog box.

18. Discuss how an animation is useful.

19. What is the basic function of the Animate tool?

20. What is a task?

21. What is a sequence?

22. What is the animation interval?

23. How do you create a faster playing speed?

24. What is the animation repetition?

25. What does the default repletion allow?

|||||||||||| PROJECTS

Instructions:

- Open a new presentation file.

- Create a presentation view using the specified file.

- Use the information discussed in this chapter to develop presentation files.

- Explode and animate the initial presentation view.

- Create at least one more presentation view.

- Explode, manipulate, and animate the additional presentation views so they differ from the initial view.

- Save the file with the specified name.

1. Name: Bottle Assembly

   Units: Inch

   Assembly File: P14.1.iam

   Save As: P16.1

2. Name: Pipe Assembly

   Units: Metric

   Assembly File: P14.2.iam

   Save As: P16.2

3. Name: Rotating Hinge Assembly

   Units: Inch or Metric

   Assembly File: P14.3.iam

   Save As: P16.3

4. Name: Cam Assembly

   Units: Inch

   Assembly File: P14.4.iam

   Save As: P16.4

5. Name: Fork Assembly

   Units: Metric

   Assembly File: P14.5.iam

   Save As: P16.5

6. Name: Housing Assembly

   Units: Inch

   Assembly File: P14.5.iam

   Save As: P16.6

7. Name: Eye Bolt Assembly

   Units: Inch

   Assembly File: EX14.18.iam

   Save As: P16.7

|||||||||||||||||||||||||||||||||||||||||||||||||||||||||||||||||||||||||||||||||||||||||||

CHAPTER 17

Assembly and Multiple Sheet Drawings

LEARNING GOALS

After completing this chapter, you will be able to:

◎ Create assembly drawings and work with assembly drawing views.

◎ Dimension assembly drawings.

◎ Add and manipulate balloons.

◎ Insert and edit parts lists.

◎ Create and work with multiple sheet drawings.

◎ Print drawings.

Working with Assembly and Multiple Sheet Drawings

In Chapters 12 and 13 you learned how Autodesk Inventor combines three-dimensional solid modeling with powerful, parametric two-dimensional (2D) drawing capabilities. Chapter 12 explored how to create a single 2D part drawing, while Chapter 13 discussed how to add part drawing dimensions and annotations. Just as part drawings are created quickly and easily from existing part models, assembly drawings can be generated from existing assembly models. Assembly drawings are produced in the exact same drawing work environment as a part drawing. As a result, you can use all the drawing tools discussed in Chapters 12 and 13 to place assembly views and add projected, isometric, auxiliary, section detail, and broken views. In addition, you can fully define the drawing with a variety of dimensions, notes, and text, including balloons and parts lists. See Figure 17.1.

This chapter explores the entire process of creating a complete set of *working drawings,* which includes a fully defined assembly drawing with a parts list and a detail drawing for each part (monodetail), or multidetail drawings. A single drawing file, .idw, can be used to generate working drawings, because you can add multiple drawing sheets to the file. As when creating a part drawing, there are a number of methods for developing a set of working drawings using Autodesk Inventor. For example, you may want to initially create all the detail part drawings and then add an assembly drawing. Conversely, you may want to begin with the assembly drawing and then develop the detail drawings. This chapter approaches the creation of a set of working drawings in the following order:

1. Create an assembly drawing.

2. Add balloons and a parts list to the assembly drawing.

3. Using a single drawing file, add additional detail drawings.

FIGURE 17.1 An example of a two-dimensional assembly drawing created using Autodesk Inventor.

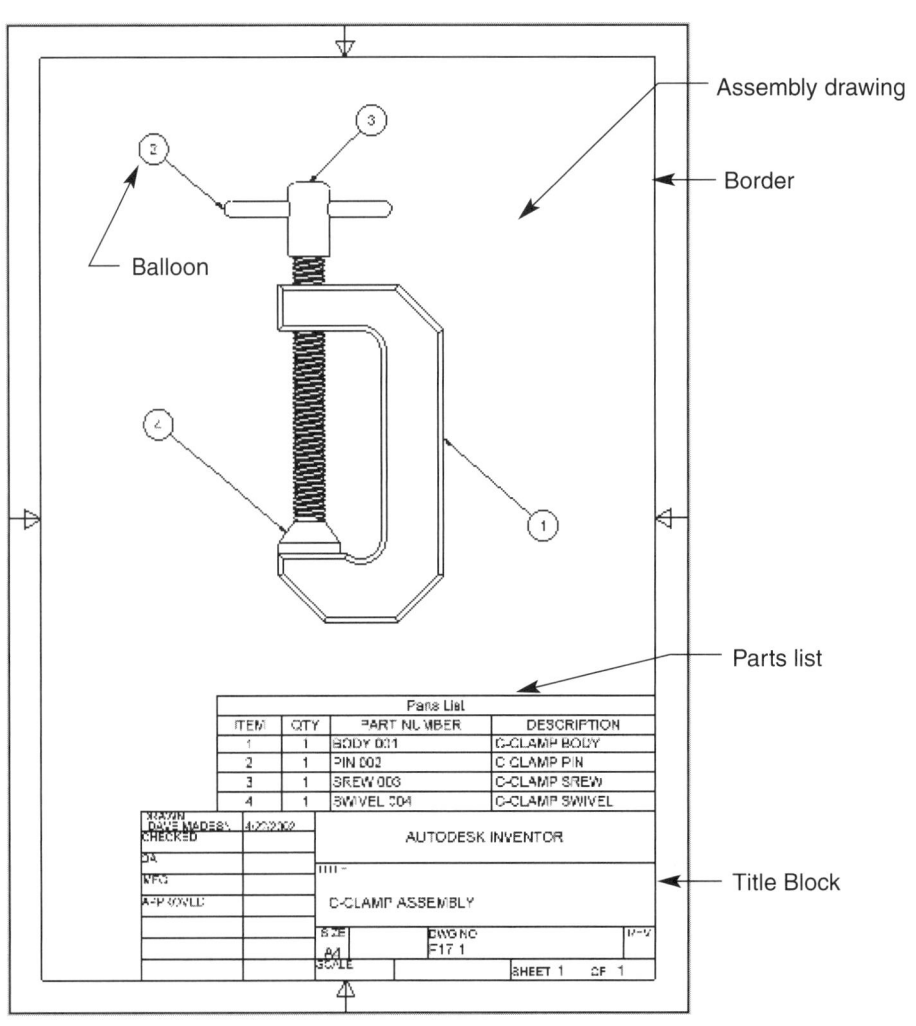

Assembly drawing

Border

Balloon

Parts list

Title Block

Parts List			
ITEM	QTY	PART NUMBER	DESCRIPTION
1	1	BODY 001	C-CLAMP BODY
2	1	PIN 002	C-CLAMP PIN
3	1	SREW 003	C-CLAMP SREW
4	1	SWIVEL 004	C-CLAMP SWIVEL

AUTODESK INVENTOR

C-CLAMP ASSEMBLY

Field Notes

For more information on the basic process of developing a single part 2D part drawing, refer to Chapters 12 and 13.

Working in the 2D Drawing Environment

As previously mentioned, the same Autodesk Inventor drawing file, .idw, used to create part drawings is used to produce working drawings. Throughout this chapter you will explore specific tools, commands, and options that allow you to develop assembly drawings, and multiple drawing sheets using one drawing file. As discussed in Chapter 12, a sheet is the base of a drawing and represents the physical limits of the drawing area, or the paper size. Think of the sheet as the piece of paper on which you are drawing. Drawing sheets contain all the drawing information and components, including a border, a title bock, and one or more drawing views. In addition, special symbols, dimensions, and other annotations displayed on the drawing are located on the sheet.

When you create a single part drawing, typically only one drawing sheet is used. However, when producing a set of working drawings, multiple sheets are created. This does not mean that you open several drawings files, because each of the working drawing sheets is contained in one drawing file. For example, a single drawing file may contain a sheet with an assembly drawing and several additional sheets that hold the individual part drawings. As you will see later in this chapter, multiple drawing sheets can be created anytime during the drawing process and are listed in the drawing Browser, just like the initial drawing sheet.

Field Notes

Refer to Chapters 12 and 13 for information regarding drawing work environment and interface options not discussed in this chapter.

Additional part drawing browser options are discussed throughout this chapter.

Assembly Drawings

An *assembly drawing* is a two-dimensional representation of an assembly. As you know, an Autodesk Inventor assembly file is used to show how multiple part and subassembly models fit together. In the same respect, an assembly or presentation file is used to produce a 2D assembly drawing, which also shows how components fit together. You can create many different types of assembly drawings depending on the application. Assembly drawings are created in a way that displays how each component relates to other components and the entire assembly. As a result, some assembly drawings may require only a single front view of the assembly, while others may require multiviews, sections, and any other views that are necessary to clearly define the component and assembly characteristics. The following represents a number of assembly drawings that can be created using Autodesk Inventor:

- A *general assembly drawing* is, as the name implies, the most common type of assembly drawing. This form of assembly drawing shows only the assembly, in a fully assembled form. A general assembly drawing may contain any views required to display each of the assembly components within the assembly, including multi, section, and detail views. An example of a general assembly drawing is shown in Figure 17.1. General assembly drawings are created from existing assembly models and use balloons and parts lists to define components.

- A *working-drawing, or detail, assembly drawing* is the same as a general assembly drawing in that it shows the assembly in a fully assembled form and may contain any views required to display each of the assembly components within the assembly. However, a working-drawing, or detail, assembly drawing also displays drawings, or details, of each of the assembly components on the same page as the assembly drawing. Working-drawing, or detail, assembly drawings are created from existing assembly models and part models. Then dimensions, annotations, balloons, and parts lists are used to define components and details.

- An *erection assembly drawing* is the same as a general assembly drawing but also includes dimensions and fabrication specifications. Erection assembly drawings are created from existing assembly models and part model. Then dimensions, annotations, balloons, and parts lists are used to define component and assembly specifications.

- A *subassembly drawing* is typically a general assembly drawing, but could be a detail, or erection assembly drawing. As you know from Chapter 14, subassemblies are groups of parts that are typically standard product components. Subassembly drawings do not show the entire assembled product, only the individual subassembly. In a set of working drawings, a subassembly drawing is followed by each of its own detail drawings. Subassembly drawings are created from existing assembly models and use balloons and parts lists to define components.

- A *pictorial assembly drawing* shows a pictorial view of an assembly, such as a single isometric view, instead of multiviews, as in a general assembly drawing. Usually a pictorial assembly drawing is used to clearly show an assembled product for sales, maintenance, catalog, or other similar purposes. Pictorial assembly drawings are created from existing assembly models or presentations. They may or may not use balloons, parts lists, and other annotations to define components.

- An *exploded assembly drawing* is the same as a pictorial assembly drawing, but it allows you to display assemblies in ways they are not shown when fully assembled. As in a presentation, discussed in Chapter 16, assembly components are moved away from other components in an unassembled or ungrouped fashion. Exploded assembly drawings, which clearly show all the assembly components, are typically used to help with assembly visualization, and for maintenance and assembly purposes. Exploded assembly drawings are created from existing presentations and may or may not use balloons, parts lists, and other annotations to define components.

Creating Assembly Drawings

As with part drawings, an assembly drawing sheet has a border, title block, and drawing views. The process for setting up an assembly drawing sheet is the same as the process of developing a part drawing sheet. You may want to approach the creation of an assembly drawing sheet, using the following methods discussed in Chapter 12:

1. Define drafting standards if they have not already been specified.

2. Specify drawing sheet parameters.

3. Use a default drawing border, or create your own border.

4. Place a default title block, or develop your own title block.

5. Place and create the required assembly drawing views.

6. Add balloons and a parts list.

7. Add any other required annotations or information.

In addition, you may want to use sheet formats; you can work with all the drawing management tools and options available in the drawing work environment.

 Field Notes

> You can add a parts list before balloons, or balloons before a parts list. Both techniques are effective.
>
> For more information regarding drawing sheet specifications and creating a drawing, refer to Chapter 12.

As you can see, the procedure for developing an assembly drawing is very similar to creating a part drawing, as discussed in Chapter 12. The only main difference is that you are using an existing assembly or presentation file to generate the assembly drawing, and balloons and parts lists are added. Balloons and parts lists are discussed later in this chapter.

Creating Assembly Drawing Views

As with part drawings, there are three ways to generate assembly drawing views. You can create assembly views by actually sketching an assembly using sketching tools, or by inserting views automatically into a sheet, using the sheet formats located in the **Sheet Formats** folder of the **Drawing Resources** folder in the **Browser** bar. These approaches may be appropriate for some designs, but the typical method for creating assembly views is to use drawing management tools.

As discussed in Chapter 12, drawing management tools allow you to insert a view from an existing file, project views, and create auxiliary, section, broken, and detail views. The techniques for establishing each of these drawing views is exactly the same when creating assembly drawings, except for inserting the initial assembly drawing view using the **Create View** tool. When you access the **Create View** tool to place the initial assembly view, instead of selecting a part file, as when developing a part drawing, you select either an assembly or presentation file, depending on the application.

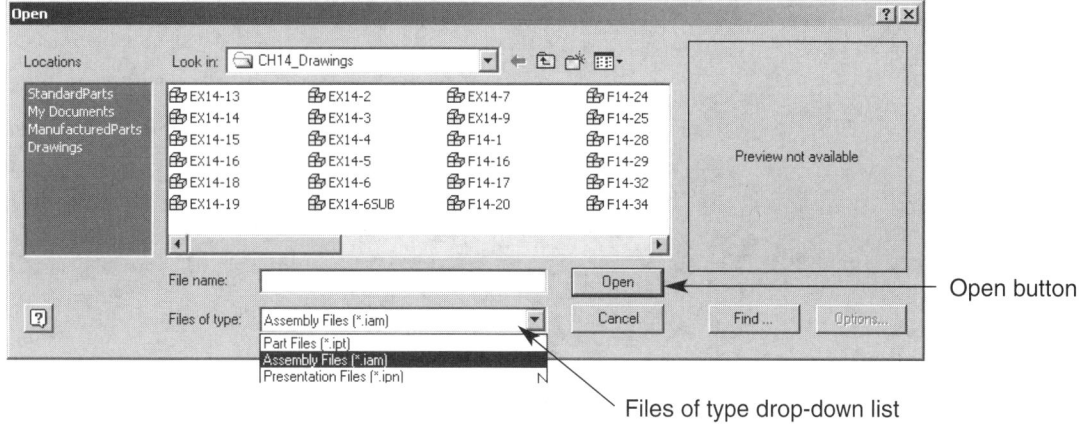

FIGURE 17.2   Selecting an assembly or presentation file to create an assembly drawing view.

Once you access the **Create View** command, the first step in creating an assembly drawing view is to select the existing assembly or presentation file from which you want to reference the drawing view. Pick the **Explore** button to access the **Open** dialog box. Then select either "Assembly Files (*.iam)" or "Presentation Files (*.ipn)" from the **Files of type** drop-down list. See Figure 17.2. Again, an assembly file is used to create most assembly drawings, such as a general assembly, while a presentation file would be used to create a pictorial or exploded assembly drawing. Once you locate the file you want to use, pick the **Open** button. The file and its directory should be displayed in the **File** drop-down list.

You will notice the **Design View** drop-down list is available when you select an assembly file, and the **Presentation View** drop-down list is available when you select a presentation file. The **Design View** drop-down list allows you to select a specified design view to use for the assembly drawing, while the **Presentation View** drop-down list allows you to select a specified presentation view to use for the assembly drawing. Once you locate and select the desired assembly or presentation file, and design view or presentation view, the **View, Style, Reference Data Display,** and **Options** areas function the same as when creating a part drawing. Again, refer to Chapter 12 for more information. Once you have selected a component to insert into the drawing sheet, specified the desired view, and defined any other view options, pick the location on the sheet where you want to place the view.

Field Notes

> Once the initial assembly view has been placed using the **Create View** tool, you can use the **Project View, Auxiliary View, Section View, Detail View,** and **Broken View** tools, discussed in Chapter 12, to create additional views.

Exercise 17.1

1. Launch Autodesk Inventor.
2. Open a new metric drawing file.
3. Right-click on Sheet:1 and select the **Edit Sheet...** menu option.
4. Use the following specifications in the **Edit Sheet** dialog box.

   Name: C-Clamp Assembly
   Size: A4
   Orientation: Bottom right corner radio button selected and **Portrait** radio button selected.
   Options: None selected.

5. Pick the OK button.

Exercise 17.1 continued on next page

Exercise 17.1 continued

6. Delete the default title block, and add a smaller default, or custom title block.

7. Ensure the part number and title property fields in the title block are "model properties." As described in Chapter 12, model properties relate to the specific model used to generate the drawing view, while "design properties" relate the drawing file in which you are currently working.

8. Access the **Create View** tool and specify the following options:

   File: EX14.7.iam
   View: Front
   Scale: 1:1
   Style: Hidden Line Removed

9. Place the assembly drawing in the middle of the drawing space.

10. Add properties such as a company name, title, scale, and other properties not directly related to the assembly or the individual assembly parts.

11. The part number should be displayed as EX17.7, if the part number property field is set to model properties.

12. The final exercise should look similar to the drawing shown.

13. Save the drawing file as EX17.1.

14. Exit Autodesk Inventor or continue working with the program.

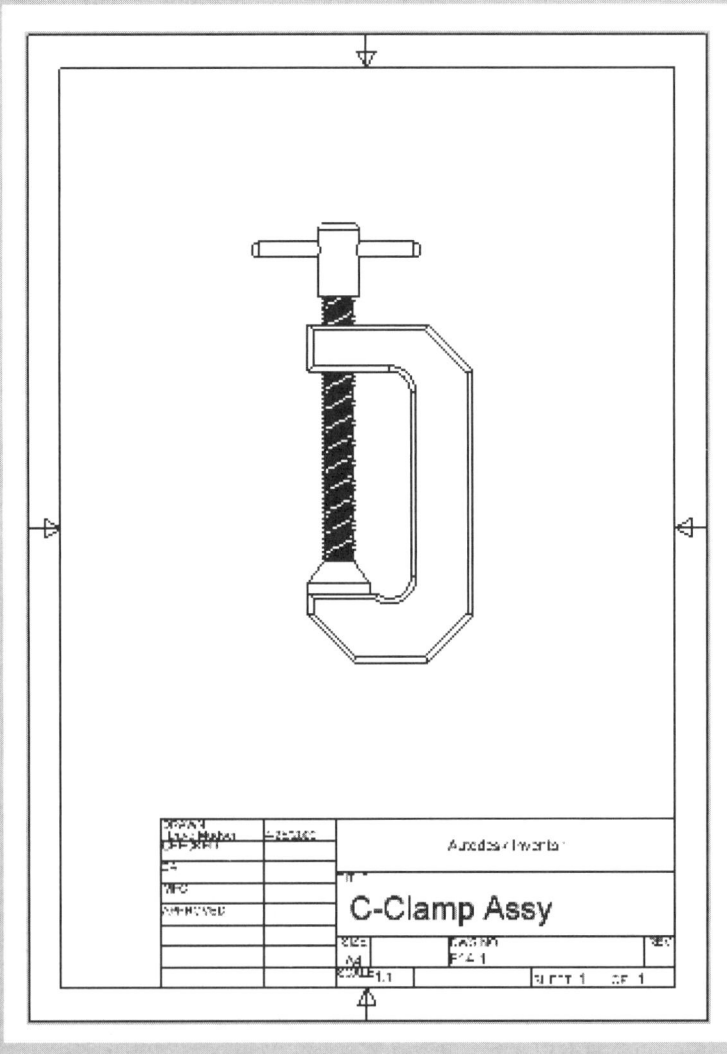

Working with Assembly Drawing Views

Once you have placed the required assembly drawing views, some specific options that relate only to assembly drawings are available, in addition to the tools and techniques discussed in Chapter 12. These options allow you to change design views and work with assembly components. In addition, you may want to become familiar with the preference that allows you to exclude components from an assembly section view. The following information describes each of these functions:

■ If you create an assembly drawing using a specified design view, you have the option of redefining the drawing with a different design view. To change the design views used for the drawing, right-click on an assembly drawing view in the graphics window, or an assembly drawing view in the Browser. Then select the **Apply Design View...** menu option to open the **Apply Design View** dialog box. See Figure 17.3. The **Apply Design View** dialog box displays the assembly file and design view file currently being used to generate the assembly drawing. To use a different design view, select a design view from the **Design View** drop-down list or pick the **Browse** button to locate a view. You will notice that each of the assembly drawing views are listed in the **Assembly Views** list box, along with the design view used to create the drawing view. To apply the new design view selection to the current drawing view the apply option must be set to "Yes." If you do not want to redefine the design view for a certain drawing view, pick the "Yes" option to change its value to "No." Once you have fully defined the specifications in the **Apply Design View** dialog box, pick the **OK** button to change the design views used for the drawing.

■ Typically, it is difficult to manipulate and access assembly drawing view preferences. For example, you cannot show assembly drawing model dimensions unless the assembly components are available. To access the components and features used to define the assembly and the assembly drawing, right-click on an assembly drawing view in the graphics window, or an assembly drawing view name in the Browser. Then select the **Show Contents** menu option. When you show contents, all the assembly components used in the assembly drawing view are displayed in the Browser. See Figure 17.4. As you can see from Figure 17.4, the contents of an assembly drawing can be extensive depending on the number of assembly components and drawing views. Once the content is shown, you can work with and manipulate specific components for assembly drawing purposes. For more information on component options refer to Chapter 12.

■ Assembly drawing section views are created the same way as part drawing section views, described in Chapter 12. If you do not want to display the geometry of a certain component in a section display, such as the pin shown in Figure 17.5, you must first specify the **Show Contents** option previously discussed. Then right-click on the component you do not want to show in the section view, and select the **Section** menu option.

FIGURE 17.3 The
Apply Design View
dialog box.

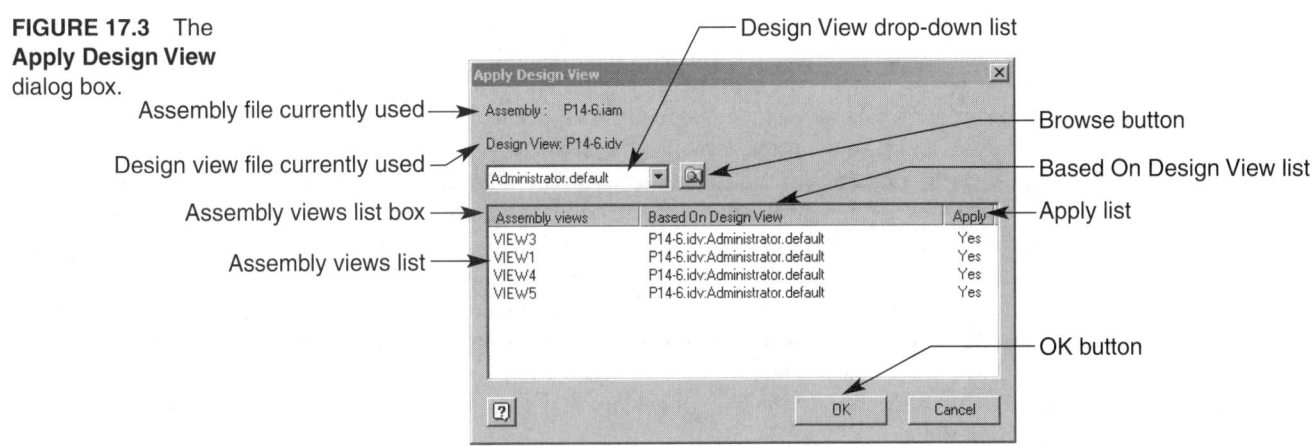

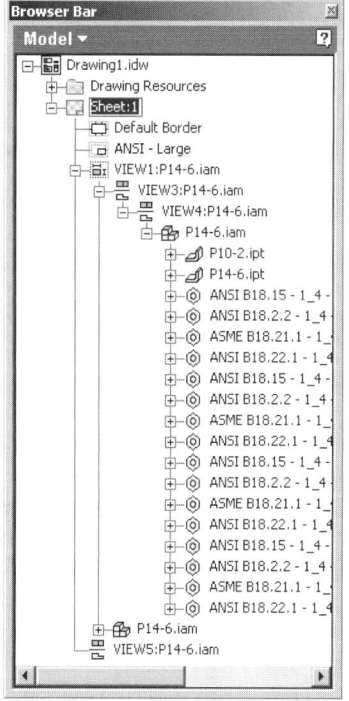

FIGURE 17.4   Showing assembly drawing content.

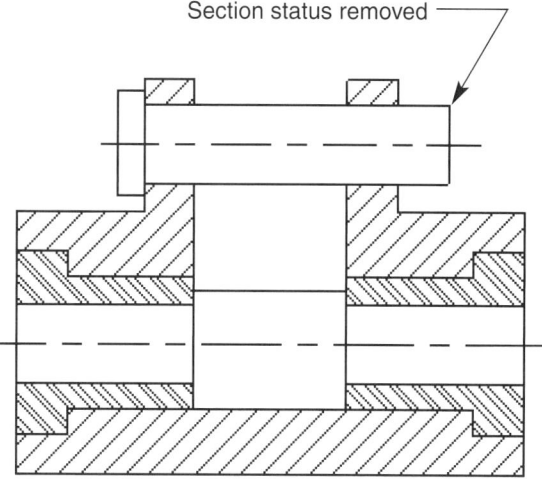

Section status removed

FIGURE 17.5   Removing component section view priority.

Dimensioning Assembly Drawings

The amount and type of dimensions and annotations added to an assembly drawing depends on the application. For example, an erection assembly drawing contains dimensions and fabrication specifications, while a pictorial assembly drawing may have no annotations. For information regarding specific dimensioning tools, refer to Chapter 13. Still, most assembly drawings do have balloons and a parts list (discussed next in this chapter).

In general, a *parts list* records and displays the parts and subassemblies used to create the assembly. Then balloons link the parts list to the drawing, by identifying the assembly components in relation to the parts list specifications. See Figure 17.6. You can add a parts list before you place balloons, or create balloons before you add a parts list. Both techniques are effective; this chapter introduces balloons first, followed by parts lists.

Adding Balloons

The term *balloon* refers to a circle that is connected to an assembly component with a leader. Inside the balloon is an identification number, typically known as the item number. As the name "item number" implies, usually numbers are used to define assembly components. However, some companies may choose to use letters. Balloons allow you to reference assembly components to the parts list. As a result, when you change item number order and value in the parts list, the corresponding balloon identification number changes to reflect the parts list modification. In addition, because balloons are connected to the parts list, you can either place balloons before you add the parts list, or place balloons after you insert the parts list.

Before you create balloons, you may want to adjust some of the balloon properties in the **Drafting Standards** dialog box. Access the **Drafting Standards** dialog box by picking the **Standards...** menu option from the **Format** pull-down menu. As discussed in Chapter 13, the **Terminator** tab of the **Drafting Standards** dialog box allows you to adjust the terminator and arrowhead options for balloon leaders, in addition to hole/thread notes, surface texture symbols, weld symbols, feature control frames, feature identifier symbols, datum identifier symbols, text leaders, and datum target leaders. For more information regarding the **Terminator** tab, refer to Chapter 13.

FIGURE 17.6 An example of an assembly drawing with a parts list and balloons.

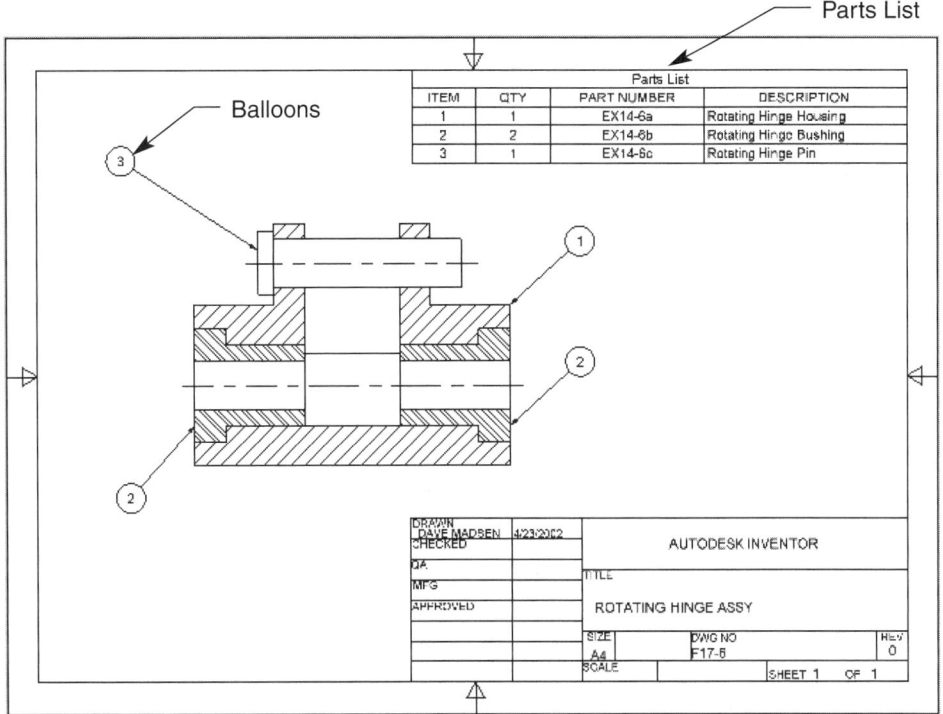

Field Notes

Individual balloon leaders can be modified once they are placed, but all balloons and balloon leaders in the drawing should be the same.

In addition to the leader options located in the **Terminator** tab, the **Balloon** tab contains specific balloon properties. See Figure 17.7. The following properties are available in the **Balloon** tab and allow you to define default balloon options:

■ **Circular with 1 entry**   Pick this button to create a balloon with a single identification number inside a circle. A single entry circular balloon is probably the most common type of balloon and identifies the corresponding item number in the parts list. Single-entry circular balloons are shown in Figure 17.6.

■ **Circular with 2 entries**   Select this button to create a balloon with two entries inside a circle. As shown in Figure 17.8, the top number in the balloon represents the item number, while the bottom number specifies the quantity of items.

■ **Hexagon**   Select this button to create a balloon with a single identification number inside a hexagon. See Figure 17.9.

■ **Property Field**   Select this button if you want to use a certain property or even a group of properties in place of the balloon. Figure 17.10 shows an example of the part number and quantity being used in place of a typical balloon. When you pick the **Property Field** button, the **Bill Of Material Column Chooser** dialog box is displayed. See Figure 17.11. The **Bill Of Material Column Chooser** dialog box allows you to reference properties used in the parts list for component identification purposes. The first step in using a property instead of a balloon is to select the property fields you want to have access to from the **Select available fields from** drop-down list. The types of properties you select are then shown in the **Available Properties** list box. If you pick the **All Properties** option, all types of properties are

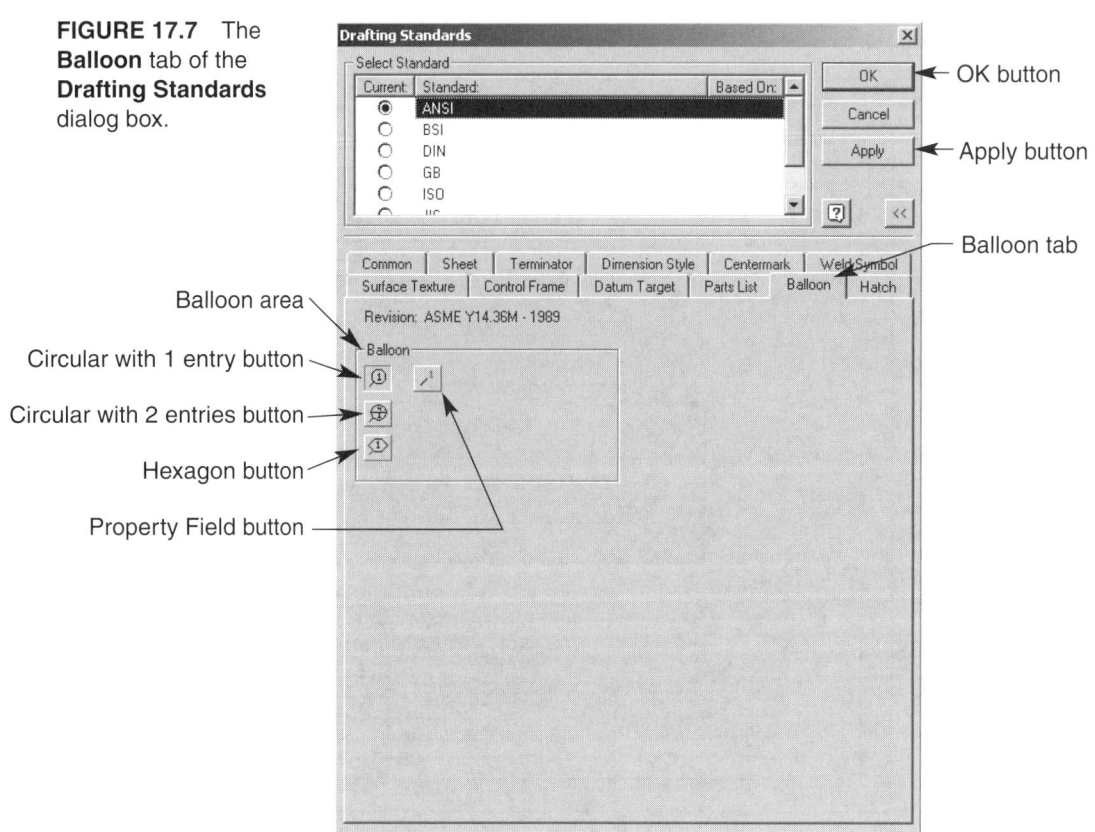

FIGURE 17.7 The **Balloon** tab of the **Drafting Standards** dialog box.

OK button

Apply button

Balloon tab

Balloon area

Circular with 1 entry button

Circular with 2 entries button

Hexagon button

Property Field button

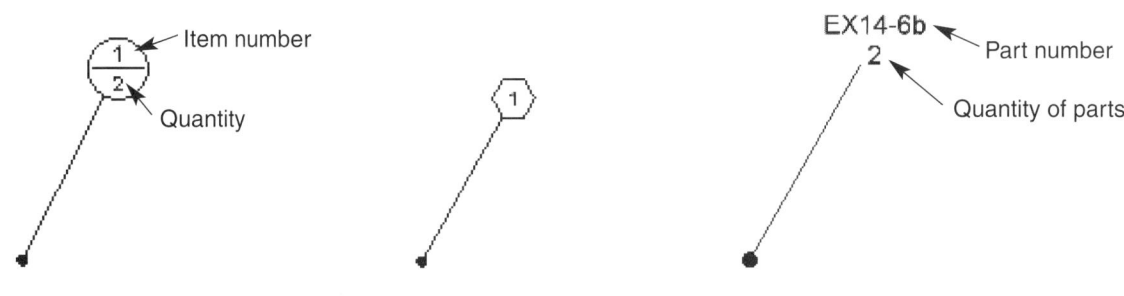

Item number

Quantity

FIGURE 17.8 An example of a double entry circular balloon.

FIGURE 17.9 An example of a single entry hexagon balloon.

EX14-6b

Part number

2

Quantity of parts

FIGURE 17.10 An example of using property fields instead of a typical balloon.

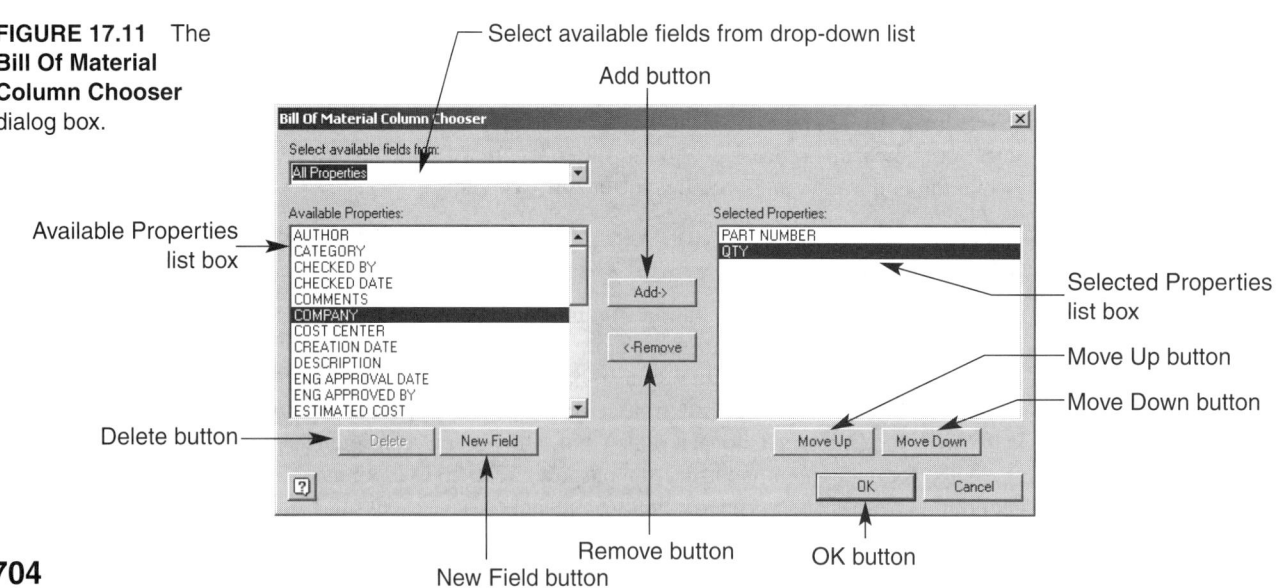

FIGURE 17.11 The **Bill Of Material Column Chooser** dialog box.

Select available fields from drop-down list

Add button

Available Properties list box

Selected Properties list box

Move Up button

Move Down button

Delete button

Remove button

OK button

New Field button

FIGURE 17.12 The **Define New Field** dialog box.

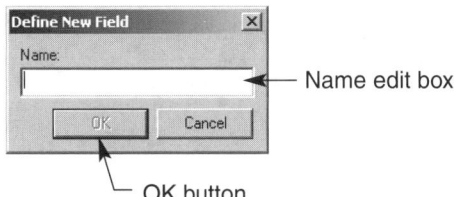

Name edit box

OK button

shown in the **Available Properties** list box. To create a custom field, pick the **New Field** button, which opens the **Define New Field** dialog box. See Figure 17.12. Enter the new property name in the **Name** edit box, and pick the **OK** button to create the property. To remove the newly made property, highlight the property and pick the **Delete** button.

The next step in defining a property field is to specify which property or properties you want to use in place of the balloon. To accomplish this task, pick the property and then select the **Add** button to add the property to the **Selected Properties** list box. When two or more properties are specified, the properties are displayed in the graphics window in the same order they are shown in the **Selected Properties** list box. To move a property up in the list, highlight the property and pick the Move Up button. Conversely, to move a property down in the list, highlight the property and select the **Move Down** button. Once you have finished specifying property information in the **Bill Of Material Column Chooser** dialog box, pick the **OK** button to define the property field balloon style.

Field Notes

In order for the properties you specify in the **Selected Properties** list box of the **Bill Of Material Column Chooser** dialog box to be displayed in the drawing, the properties must also be listed in the parts list. Parts lists are discussed later in this chapter.

When you have fully defined the balloon standards in the **Balloon** tab of the **Drafting Standards** dialog box, continue specifying standards, or pick the Apply button to set the standards, and the **OK** button to exit the **Drafting Standards** dialog box.

Field Notes

The balloon properties defined in the **Balloon** tab of the **Drafting Standards** dialog box cannot be individually overridden once balloons are placed in the drawing.

There are two different options for adding balloons to an assembly drawing. You can place balloons individually using the **Balloon** tool, or automatically add balloons to the entire drawing using the **Balloon All** tool. Both the **Balloon** and **Balloon All** tools are contained in a flyout button.

To add assembly drawing balloons individually, access the **Balloon** tool using one of the following techniques:

✓ Pick the **Balloon** tool button from the **Drawing Annotation** panel bar.

✓ Pick the **Balloon** tool button from the **Drawing Annotation** toolbar.

✓ Press the **+** and **B** keys on your keyboard.

Once you access the **Balloon** tool, use the cursor to select the point where you want the balloon leader to begin. If a parts list has not already been inserted, the **Parts List – Item Numbering** dialog box pops up. See Figure 17.13. This dialog box has options that allow you to begin to define the parts list. The

FIGURE 17.13 The **Parts List – Item Numbering** dialog box.

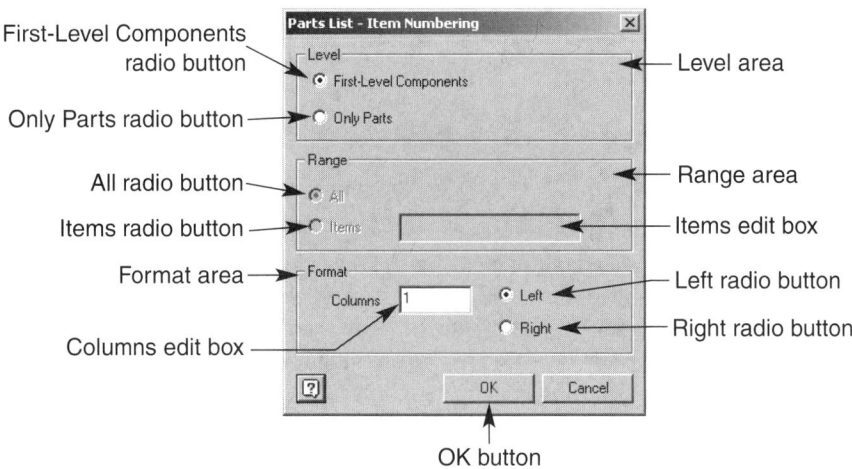

dialog box opens when you try to create a balloon in a drawing that does not have a parts list, because balloons and parts lists reference each other. The following areas are available inside the dialog box:

- **Level**   This area specifies the types of components displayed in the parts list, and as a result the components that can be ballooned. Pick the **First-Level Components** radio button if you want to display only first-level assembly components in the parts list. If you select this radio button, only subassemblies and parts not related to subassemblies are shown. The parts used to create a subassembly are not shown. If you want all the parts used in the assembly to be displayed, including those used in a subassembly, pick the **Only Parts** radio button.

- **Range**   This area is available when the **Only Parts** radio button of the **Level** area is selected and allows you to identify a specified arrangement of parts to display in the parts list. Pick the **All** radio button to show all the parts, or select the **Items** radio button to display only a certain range of parts. When you choose the **Items** radio button, you must specify the parts you want to show in the parts list, by entering the item numbers, separated by commas, in **Items** edit, or by selecting actual parts from the graphics window.

- **Format**   Use this area to identify the parts list layout. For assembly drawings with large parts lists, you can enter the number of parts list columns you want to display in the **Columns** edit box. The assembly components are divided amongst the specified number of columns. So far, the parts lists shown in this chapter contain only one column. Figure 17.14 shows an example of a parts list with two columns. Then, specify where you want the cursor to be connected to the parts list when inserted, by picking the **Left** or **Right** radio button. When you select the **Left** radio button, the cursor is attached to the upper right corner of the parts list. When you choose the **Right** radio button, the cursor is attached to the upper left corner of the parts list. Again, parts lists are fully discussed later in this chapter, but you must begin the process of creating a parts list before you can place balloons.

Field Notes

You cannot create columns that do not have any component information. For example, if you specify ten columns, but there are only seven parts in the assembly, only two columns will be displayed.

ITEM	QTY	PART NUMBER	DESCRIPTION	ITEM	QTY	PART NUMBER	DESCRIPTION
				3	1	EX14-6c	
1	1	EX14-6a		2	2	EX14-6b	
ITEM	QTY	PART NUMBER	DESCRIPTION	ITEM	QTY	PART NUMBER	DESCRIPTION
Parts List				Parts List			

FIGURE 17.14 An example of a parts list with two columns.

Once you have defined the options in the **Parts List – Item Numbering** dialog box, pick the **OK** button to continue the process of adding balloons to the assembly drawing.

The point on the assembly you selected to connect the leader and balloon defines both the component and the leader start point. A piece of the selected component, such as an edge or line, is highlighted, identifying the component. The next step in defining the balloon is to drag the balloon to the desired location and pick the position to specify the leader length. Depending on the leader style, you can continue selecting points as required, such as adding a leader shoulder. See Figure 17.15. After you select a point on appropriate drawing geometry and locate the balloon, press the enter key on your keyboard, or right-click and select the **Continue** menu option to create the balloon. The **Balloon** tool is still active, allowing you to continue placing balloons as required. To exit the **Balloon** command, right-click and select the **Done [Esc]** menu option, press the **Esc** key on your keyboard, or access a different tool.

If you would rather add all the required assembly drawing balloons at one time, using a single selection, access the **Balloon All** tool using one of the following techniques:

✓ Pick the **Balloon All** tool button from the **Drawing Annotation** panel bar.

✓ Pick the **Balloon All** tool button from the **Drawing Annotation** toolbar.

Once you access the **Balloon All** tool, use the cursor to select the drawing view to which you want to add balloons. As when using the **Balloon** tool previously described, if a parts list has not already been inserted, the **Parts List – Item Numbering** dialog box will pop up. Refer to the previous discussion for information regarding this dialog box.

Once you have selected a drawing view, and after you exit the **Parts List – Item Numbering** dialog box, or if a parts list has already been inserted, all the required balloons will be displayed in the selected drawing view. See Figure 17.16. The **Balloon All** tool adds balloons to only a single drawing view. To place additional balloons you must reselect the **Balloon All** command.

As you can see from Figure 17.16, typically when you use the **Balloon All** tool, balloons may be randomly placed in the drawing. In addition, even if you use the **Balloon** tool to individually place balloons, you may have to reorganize the balloons so they are arranged in an easy-to-read pattern. To redefine balloon location, select the balloon you want to modify. Then drag the green dots to the desired location. Figure 17.17 shows how to reposition balloons so that they are correctly aligned and easy to read.

In addition to repositioning balloons and balloon leaders, use the following information to add or further manipulate balloon specifications:

■ Remove a balloon   Like most Autodesk Inventor applications, if you want to remove an existing balloon from the drawing, right-click on the balloon you want to remove and select the **Delete** menu option. Or pick the balloon and press the **Delete** key on your keyboard.

■ Editing individual balloon arrowheads   Typically, all the balloon leaders in a drawing should be the same. However, if you want to change the leader style of a specific balloon

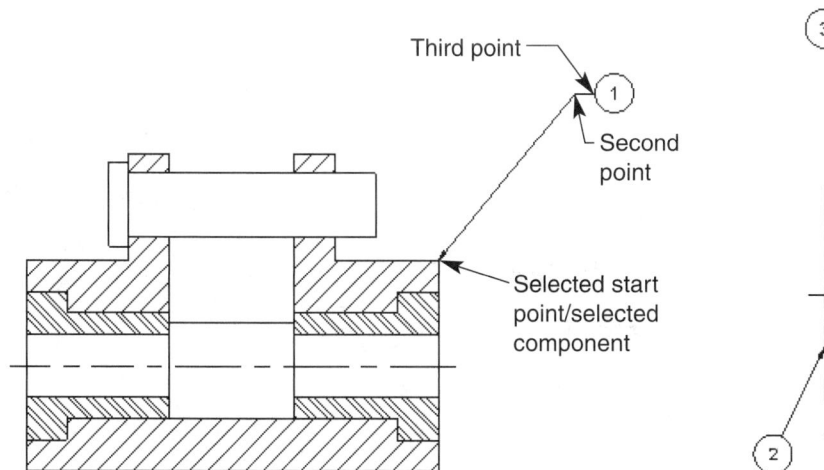

FIGURE 17.15 Placing a balloon using the **Balloon** tool.

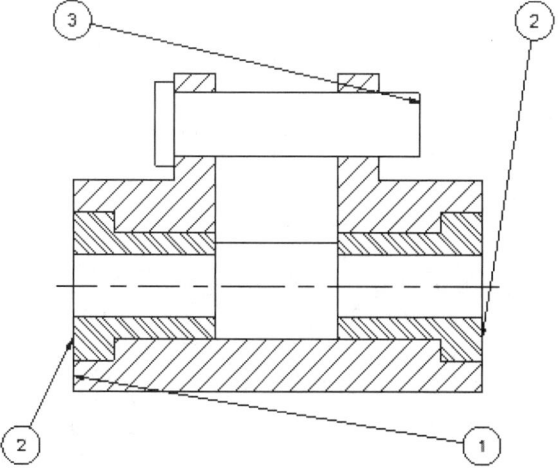

FIGURE 17.16 Adding balloons to the entire drawing view using the **Balloon All** tool.

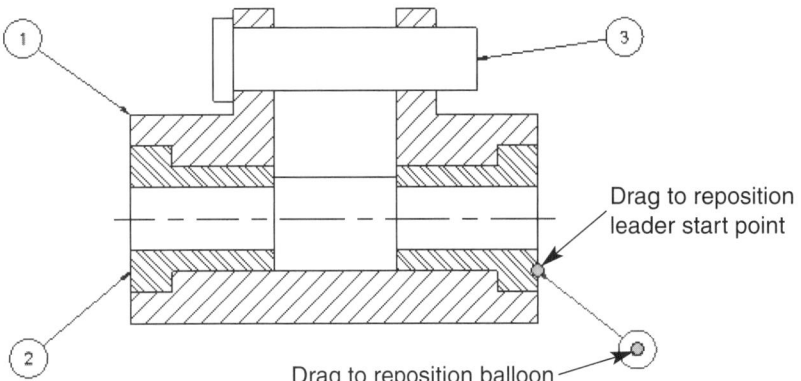

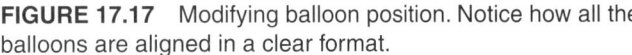

Drag to reposition
leader start point

Drag to reposition balloon

FIGURE 17.17   Modifying balloon position. Notice how all the
balloons are aligned in a clear format.

Apply
button

FIGURE 17.18   The **Change Arrowhead**
dialog box.

leader, right-click on the leader you want to modify, and select the **Edit Arrowhead...** menu
option to display the **Change Arrowhead** dialog box. See Figure 17.18. To specify a different arrowhead style, select the arrowhead from the drop-down list, then pick the **Apply** button to create the change.

■ Grouping balloons   Balloons may be grouped together for closely related clusters of assembly components, such as a bolt, a washer, and a nut. As shown in Figure 17.19, *grouped balloons* share the same leader, which is typically connected to the most obviously displayed
component, such as the bolt mentioned in the previous example. To group two or more balloons, right-click on an existing balloon to which you want to add balloons, and pick the
Attach Balloon menu option. Then select the component that the balloon will identify. The
new balloon is attached to the existing balloon, but the final step is to select the side you
want to place the new balloon. If you want to remove an attached balloon, right-click on the
balloon you want to delete, and select the **Remove Balloon** menu option, which is only
available if two or more balloons are attached to each other.

■ Place an additional segment or vertex   Although usually only one leader is attached to a
balloon, or group of balloons, you can add multiple leaders if required. To place an additional leader, right-click on the existing balloon or balloon leader and select the **Add Segment
or Vertex** menu option. Then pick the start point of the new leader, followed by selecting the
existing balloon to which you want to connect the new leader. See Figure 17.20.

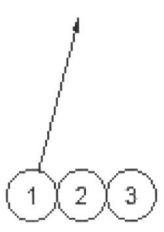

FIGURE 17.19   An
example of grouping
balloons.

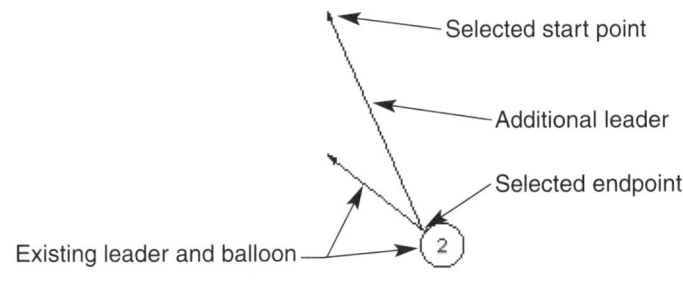

Selected start point

Additional leader

Selected endpoint

Existing leader and balloon

FIGURE 17.20   Adding a leader to a balloon.

Exercise 17.2

1. Continue from Exercise 17.1 or launch Autodesk Inventor.
2. Open EX17.1.idw and save a copy as EX17.2.idw.
3. Close EX17.1.idw without saving, and open EX17.2.idw.
4. Open the **Drafting Standards** dialog box, and explore the **Balloon** tab. Do not change the default setting.
5. Access the **Balloon** command.
6. Add balloons as shown. When the **Parts List – Item Numbering** dialog box pops up, pick the **Right** radio button, and then select the **OK** button.
7. The final exercise should look similar to the drawing shown.
8. Save the drawing file as EX17.2.
9. Exit Autodesk Inventor or continue working with the program.

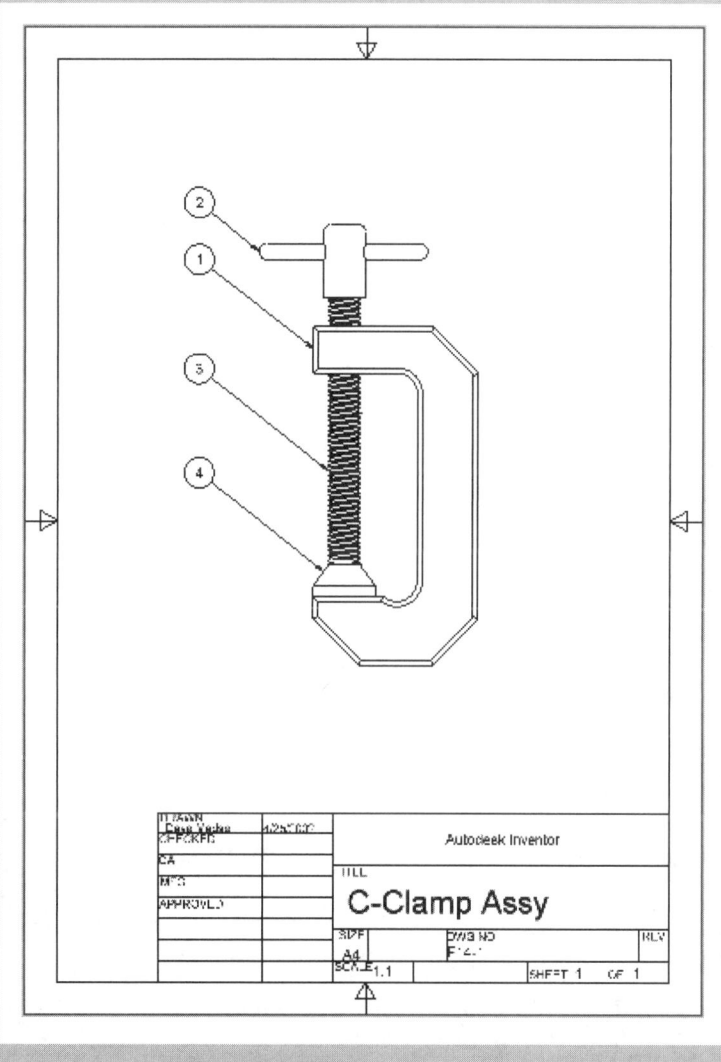

Exercise 17.3

1. Continue from Exercise 17.2 or launch Autodesk Inventor.
2. Open EX17.1.idw and save a copy as EX17.3.idw.
3. Close EX17.1.idw without saving, and open EX17.3.idw.
4. Access the **Balloon All** command.
5. Select the drawing view. When the **Parts List – Item Numbering** dialog box pops up, select the **OK** button.
6. Reorganize the balloon similar to the organization shown.
7. The final exercise should look similar to the drawing shown.
8. Save the drawing file as EX17.3.
9. Exit Autodesk Inventor or continue working with the program.

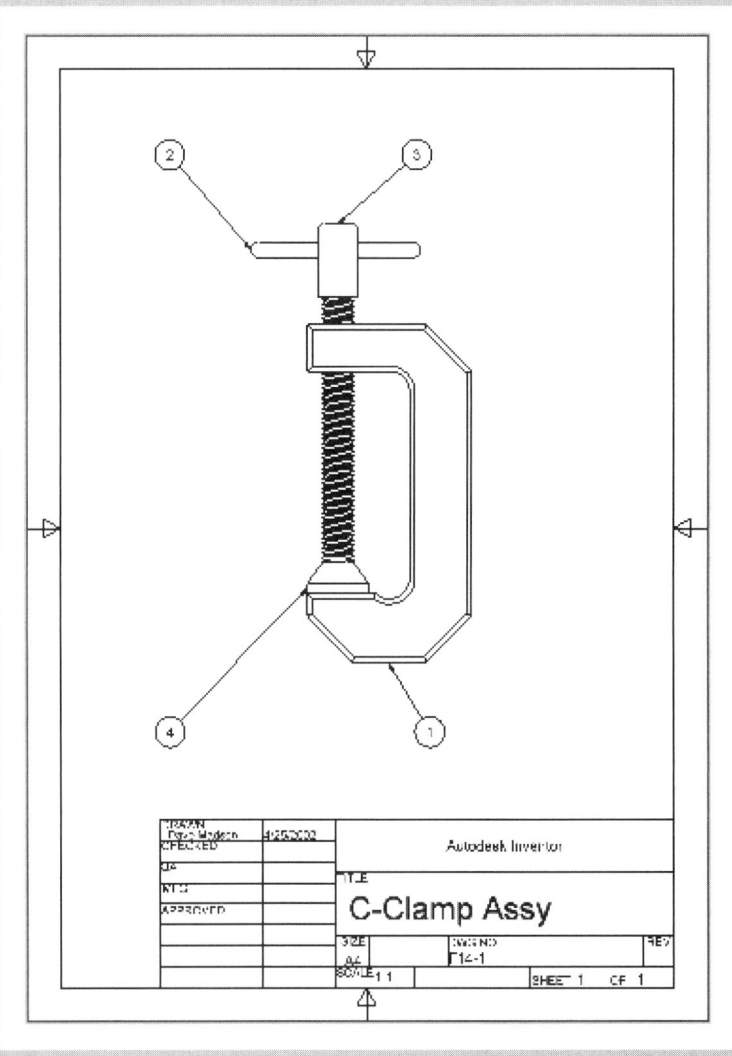

Inserting Parts Lists

A *parts list* records and displays the parts and subassemblies used to create the assembly, and keys, or links balloons to component information. See Figure 17.21. Parts lists and parts list information differs greatly depending on the application and specific company standards. Parts lists are tables that contain vertical columns and horizontal rows. A typical parts list may contain some or all the following components in a table (column and row) format:

1. Item Number   This column contains the item numbers, linked to the balloons, for each of the assembly components.

2. Quantity   This column identifies the number of each component, or items, used in the assembly.

3. Part Number   Use this column to display the part number assigned to the component. The part number is referenced to the individual detail drawing, and may be used in addition to or substituted by the sheet or drawing number.

4. Drawing Number   This column displays the drawing number on which the detail is found. The drawing number may be used in addition to or substituted with the part number previously mentioned.

5. Description   Use this column to provide descriptions of each item.

6. Material   This column identifies the material used to create the item.

7. Additional columns   A parts list may also contain any additional company specific columns, such as vendor and part purchasing information.

Parts lists are similar to title blocks discussed in Chapter 12, because much of the title block information is automatically entered in the columns based on the assembly characteristics. For example, if three of the same items are displayed in the assembly, a quality of three is automatically established in the quality area for the item. Or, the part numbers are referenced from the component name. Other part list entries are controlled by specific component properties, set in the **Properties** dialog box, discussed in Chapter 2. For example, if you enter "Housing" in the description edit box of the **Properties** dialog box, "Housing" is displayed in the parts list description area for the item.

Before you insert a parts list, you may want to adjust some of the parts list properties in the **Drafting Standards** dialog box. Access the **Drafting Standards** dialog box by picking the **Standards...** menu option from the **Format** pull-down menu. Parts list standards are located in the **Parts List** tab. See Figure 17.22. The following properties are available and allow you to define default parts list standards:

- **Output Direction**   These buttons control the order of items in the parts list. Pick the **Add new parts to bottom** button to create a parts list that provides data from the bottom of the sheet to the top. See Figure 17.23. Select the **Add new parts to top** button to create a parts list that provides data from the top of the sheet to the bottom. See Figure 17.21.

FIGURE 17.21   An example of a parts list.

Parts List			
ITEM	QTY	PART NUMBER	DESCRIPTION
1	1	P10-2	Housing
2	1	P14-6	Housing Cover
3	4	ANSI B18.15 - 1/4 - 20. Shoulder Pattern Type 2 - Style A	Forged Eyebolt
4	4	ANSI B18.2.2 - 1/4 - 20	Hex Nut
5	4	ASME B18.21.1 - 1/4 Regular. Carbon Steel	Helical Spring Lock Washer
6	4	ANSI B18.22.1 - 1/4 - narrow - Type A	Washer A

FIGURE 17.22 The **Parts List** tab of the **Drafting Standards** dialog box.

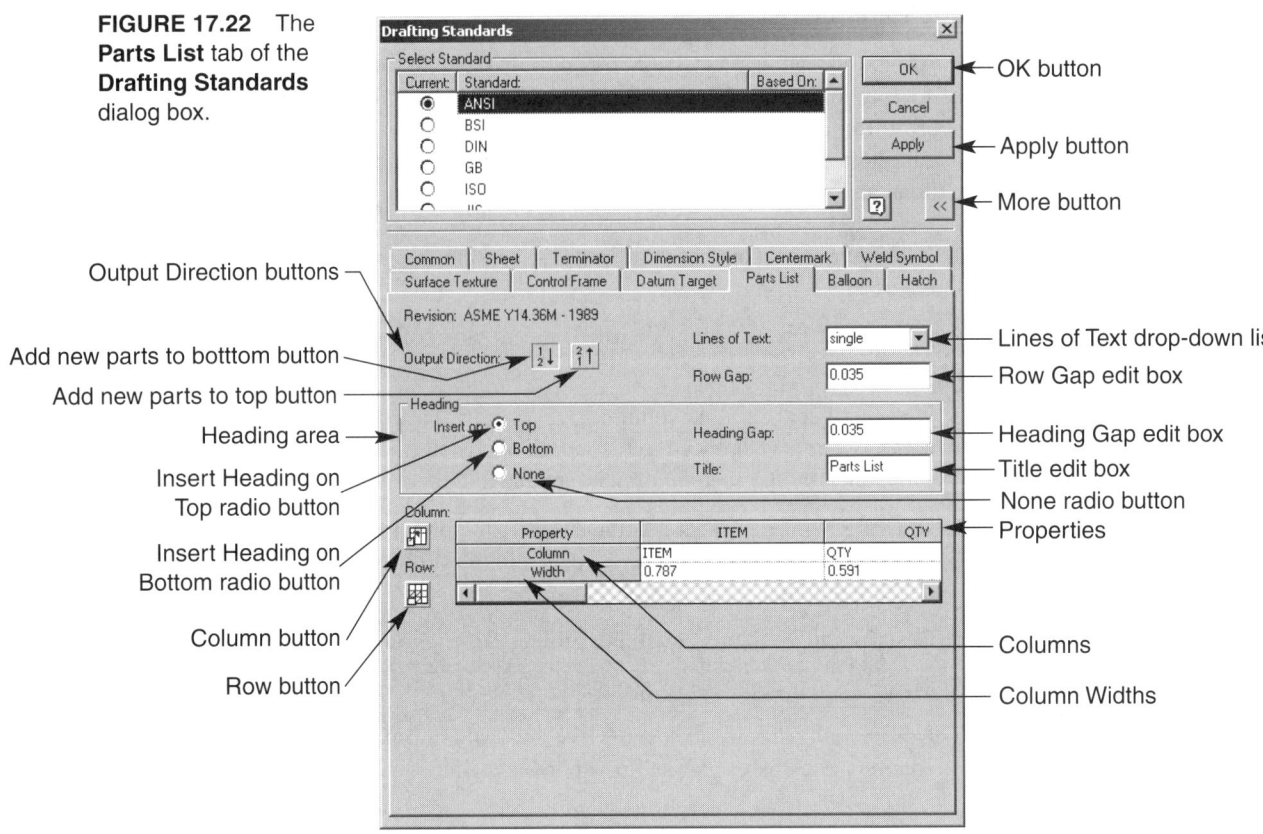

FIGURE 17.23 An example of a parts list where new part data is added above existing data.

6	4	ANSI B18.22.1 - 1/4 - narrow - Type A	Washer A
5	4	ASME B18.21.1 - 1/4 Regular. Carbon Steel	Helical Spring Lock Washer
4	4	ANSI B18.2.2 - 1/4 - 20	Hex Nut
3	4	ANSI B18.15 - 1/4 - 20. Shoulder Pattern Type 2 - Style A	Forged Eyebolt
2	1	P14-6	
1	1	P10-2	Housing
ITEM	QTY	PART NUMBER	DESCRIPTION
Parts List			

- **Lines of Text** This drop-down list allows you to select and specify the spacing between multiple lines of text in the parts list.

- **Row Gap** Use this edit box to enter the value of the space between the parts list text and the cell edge. A *cell* is the individual box, or rectangle, that displays parts list information and defines the extents of the rows and columns.

- **Heading** This area contains options that allow you to set the parts list heading information. The default heading is "Parts List"; it is inserted on the top of the parts list, because the **Top** radio button is selected. To insert the heading at the bottom of the parts list, pick the **Bottom** radio button, and if you do not want a heading at all, select the **None** radio button. Then enter the value of the space between the parts list heading text and the heading cell in the **Heading Gap** edit box. The parts list tile can also be changed by entering a different name in the **Title** edit box.

- **Column** Pick this button to display the button to display the **Parts List Column Chooser** dialog box, shown in Figure 17.24. The **Parts List Column Chooser** dialog box has a simi-

FIGURE 17.24 The **Parts List Column Chooser** dialog box.

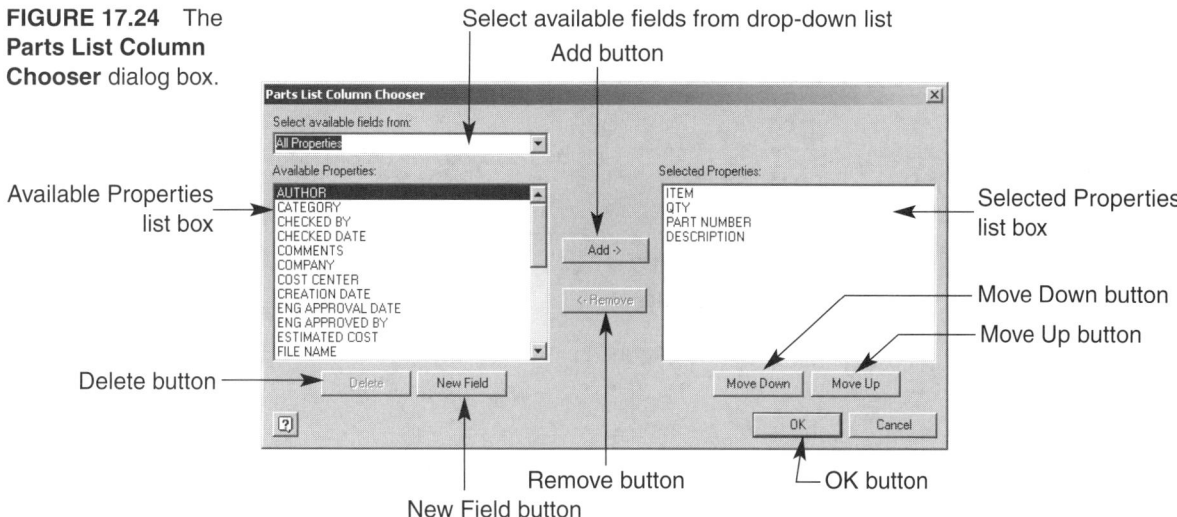

Select available fields from drop-down list

Add button

Available Properties list box

Selected Properties list box

Move Down button

Move Up button

Delete button

Remove button

OK button

New Field button

lar purpose as the **Bill Of Material Column Chooser** dialog box discussed earlier and allows you to add columns to the parts list by referencing properties. The first step in adding a property column is to select the property fields you want to have access to from the **Select available fields from** drop-down list. The types of properties you select are then shown in the **Available Properties** list box. If you pick the **All Properties** option, all types of properties are shown in the **Available Properties** list box. To create a custom field, pick the **New Field** button, which opens the **Define New Field** dialog box. Enter the new property name in the **Name** edit box and pick the **OK** button to create the property. To remove the newly made property, highlight the property and pick the **Delete** button.

The next step in defining a new column is to specify which property or properties you want to add to the parts list. To accomplish this task, pick the property and then select the **Add** button to add the property to the **Selected Properties** list box. When two or more properties are specified, the properties are displayed in the graphics window in the same order they are shown in the **Selected Properties** list box. To move a property up in the list, highlight the property and pick the **Move Up** button. Conversely, to move a property down in the list, highlight the property and select the **Move Down** button. Once you have finished specifying property information in the **Parts List Column Chooser** dialog box, pick the **OK** button create the new column.

Field Notes

If you use properties in the place of a typical balloon, as described in the balloon section, you must ensure properties are placed in a parts list column. This means properties listed in the **Bill Of Material Column Chooser** dialog box must also be listed in the **Parts List Column Chooser** dialog box.

■ **Row** Use this button to access the **Row Keys** dialog box. When you have several parts with the same part number, but some of the parts have different properties, you may want to be able to display parts separately. For example, if two brackets used in an assembly have the same part number, but they come from different vendors or their materials are different, you can display the brackets in the parts list as separate items. This process is accomplished using the **Row Keys** dialog box shown in Figure 17.25. When two or more items have the same content, use the **Row Keys** dialog box by selecting the first property that will differentiate the items from the **First Key** drop-down list. Then, if needed, select a second and even third property to differentiate the items from the **Second Key** and **Third Key** drop-down lists.

FIGURE 17.25 The
Row Keys dialog box.

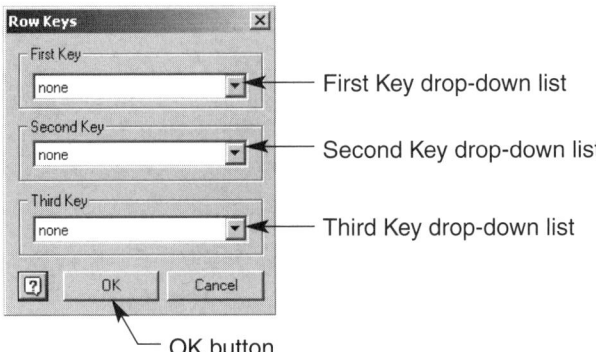

First Key drop-down list

Second Key drop-down list

Third Key drop-down list

OK button

- ■ **Property**   This row shows all the properties used in the parts list.

- ■ **Columns**   This row identifies the name used in the parts list for each of the properties. To change the name, highlight the name you want to modify, enter a new name, and then select outside of the cell.

- ■ **Column Widths**   Use this row to define the width of each of the parts list property columns. To change the width, highlight the value you want to modify, enter a new width, and then select outside of the cell.

When you have fully defined the parts list standards in the **Parts List** tab of the **Drafting Standards** dialog box, continue specifying standards, or pick the **Apply** button to set the standards and the **OK** button to exit the **Drafting Standards** dialog box.

Field Notes

Many of the parts list properties defined in the **Parts List** tab of the **Drafting Standards** dialog box can be individually overridden once a parts list is placed in the drawing.

To insert an assembly drawing parts list, access the **Parts List** tool using on of the following techniques:

✓ Pick the **Parts List** tool button from the **Drawing Annotation** panel bar.

✓ Pick the **Parts List** tool button from the **Drawing Annotation** toolbar.

If you have not already inserted balloons, the **Parts List – Item Numbering** dialog box, shown in Figure 17.13 and discussed in the balloon section, pops up. Again, this dialog box has options that allow you to begin to define the parts list. Refer to the previous discussion on balloons for information regarding the **Parts List – Item Numbering** dialog box. Once you access the **Parts List** tool and specify the **Parts List – Item Numbering** dialog box if no balloons exist, the next step is to select the drawing view from which to generate the parts list. A rectangle representing the extents of the parts list is displayed. See Figure 17.26. Then use the cursor to select the location point where you want the parts list to be placed and to generate the parts list.

Editing Parts Lists

Once you have created a parts list, you can edit much of its information and its position. A parts list can be moved much like other Autodesk Inventor applications. When you select the parts list, green buttons are displayed at each corner. Use these buttons to move the parts list to a desired location either in space, or by linking a corner of the parts list to an existing drawing point, such as the corner of a border or title block.

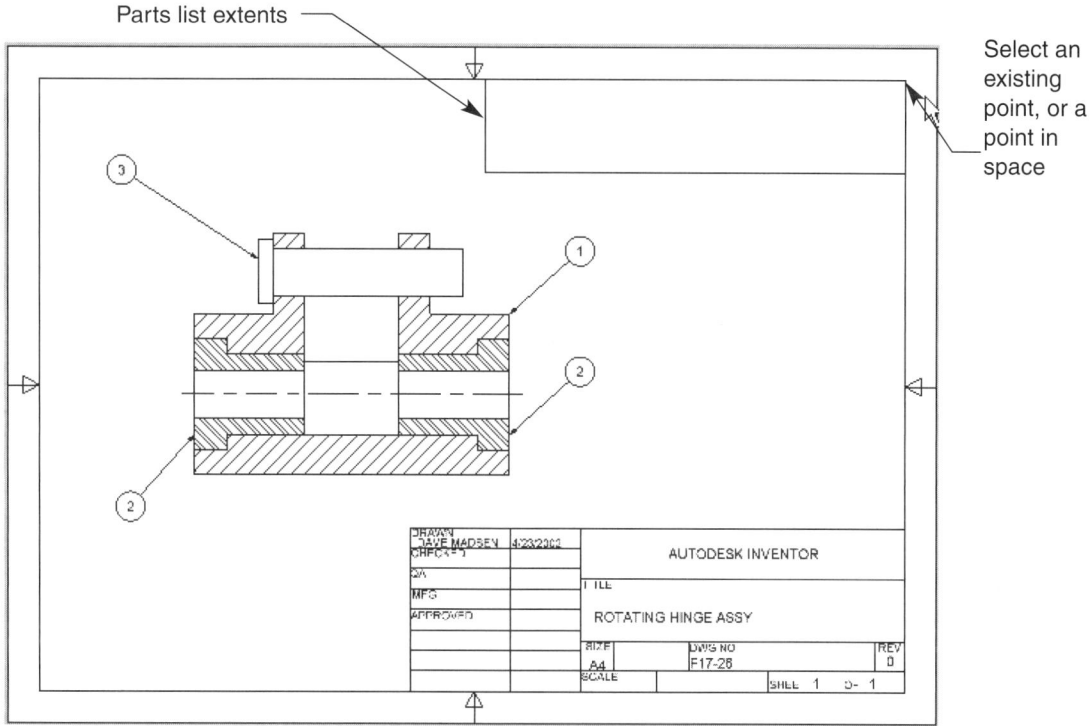

Parts list extents —

Select an existing point, or a point in space

FIGURE 17.26 Placing a parts list.

Field Notes

Although you can drag a parts list to a new location, you cannot stretch or otherwise manipulate the parts list using the green indicator buttons.

Most parts list editing options are controlled in the **Edit Parts List** dialog box, shown in Figure 17.27; they are available by right-clicking on the parts list and selecting the **Edit Parts List...** menu option. The following options are available in the **Edit Parts List** dialog box and allow you to manipulate the characteristics of an existing parts list:

- **Compare** Use this button to compare and identify if any parts list properties have been added or modified. If the value of a current property is different from the value specified in the parts list, the corresponding spread sheet cell is highlighted. For example, if you specify a description property of a component as "Bushing" in the component file, the parts list description will display "Bushing." Then, if the component description is modified to "Poly Bushing," and you select the **Compare** button, "Bushing" will be highlighted.

- **Column Chooser** Select this button to open the **Parts List Column Chooser** dialog box, shown in Figure 17.24. Again, use the **Parts List Column Chooser** dialog box to add columns to the parts list by referencing properties.

- **Sort** When you select this button, the **Sort Parts List** dialog box is displayed. See Figure 17.28. The **Sort Parts List** dialog box allows you to modify the current parts list order. Select the property you want to sort first by selecting a property from the **Sort by** drop-down list. Then specify whether you want the property displayed in ascending order or descending order by selecting either the **Ascending** or **Descending** radio button. Next, if required, identify which property you want to sort second, when the first property is not available, by selecting a property from the **Then by** drop-down list. Then specify whether you want the property displayed in ascending order or descending order by selecting either the **Ascending**

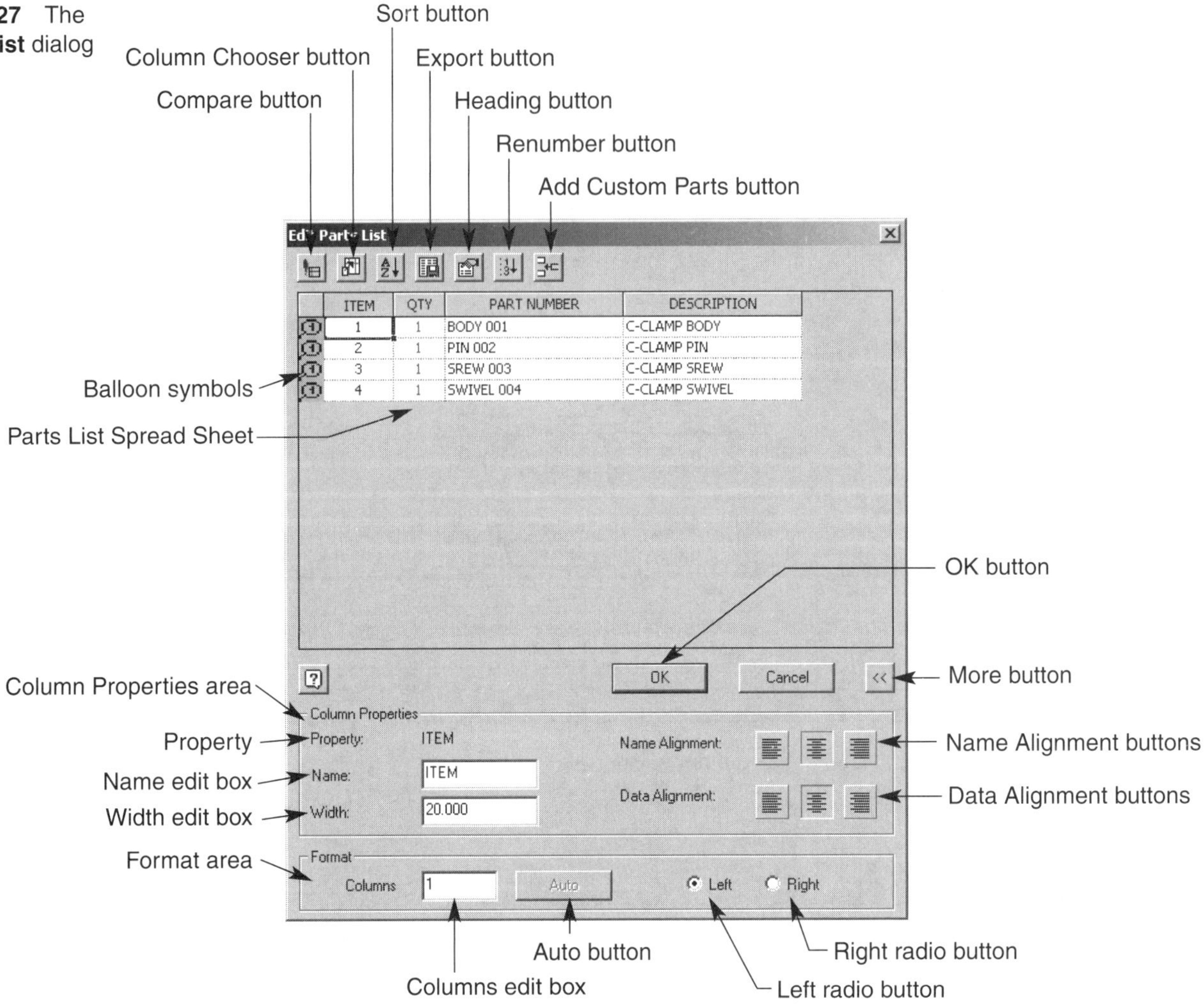

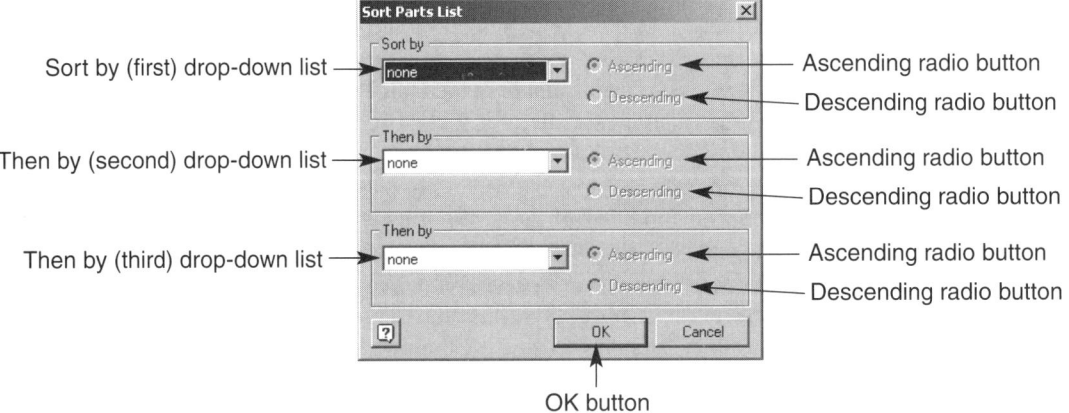

FIGURE 17.27 The **Edit Parts List** dialog box.

FIGURE 17.28 The **Sort Parts List** dialog box.

or **Descending** radio button. Finally, if needed, identify which property you want to sort third, when the first and second properties are not available, by selecting a property from the second **Then by** drop-down list. Then, specify whether you want the property displayed in ascending order or descending order by selecting either the **Ascending** or **Descending** radio button.

■ **Export**   Pick this button if you want to export, or save, the parts list to an external, or non-Inventor type of file. When you select the **Export** button, the **Export Parts List** dialog box

FIGURE 17.29 The
Export Parts List
dialog box.

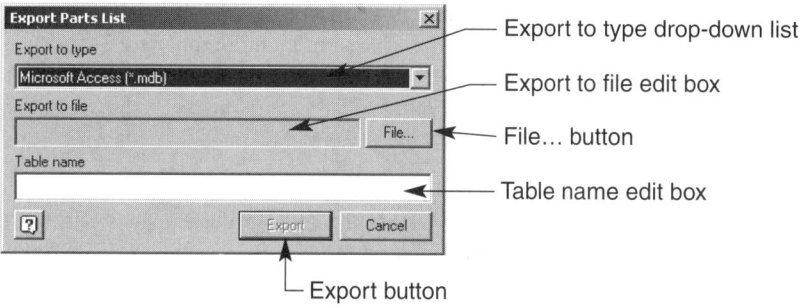

is displayed. See Figure 17.29. To use the **Export Parts List** dialog box, first select the type of file to which you want to export the parts list from the **Export to type** drop-down list. Then define the name of the file by entering a name in the **Export to file** edit box, or by locating a folder and defining a file name in the **Save As** dialog box, which opens when you select the **File...** button. Finally, enter the name of the table in the **Table name** edit box, and pick the **Export** button to finish the process and exit the dialog box.

■ **Heading**   Choose this button to open the **Parts List Heading** dialog box, which allows you to edit the parts list heading. See Figure 17.30. Select the **Top** radio button to insert the heading on the top of the parts list. To insert the heading at the bottom of the parts list, pick the **Bottom** radio button, and if you do not want a heading at all, select the **None** radio button. Then, if required, enter a new parts list tile in the **Title** edit box.

■ **Renumber**   This button renumbers the parts list columns consecutively, as defined in the parts list standards. Pick this button to override the previously discussed sort options, value edits, or other numerical modifications, in order to renumber the parts list.

■ **Add Custom Parts**   When you pick this button, a new custom row is placed in the parts list and is shown in the **Edit Parts List** dialog box spread sheet. You can use the spread sheet to define the custom part.

Field Notes

For most applications custom part list parts should not be used.

■ **Parts List Spread Sheet**   The **Parts List Spread Sheet** displays the parts list and allows you to modify parts list values. Many of the options associated with the parts list spread sheet have already been described. However, the spread sheet also contains many other parts list manipulation options. The balloon symbols identify which parts list items have corresponding balloons in the assembly drawing. Also, you can edit any of the information in the cells by selecting inside the cell, or by highlighting and redefining the existing value. The column widths can also be edited, by selecting the line in between two columns and drag-

FIGURE 17.30 The
Parts List Heading
dialog box.

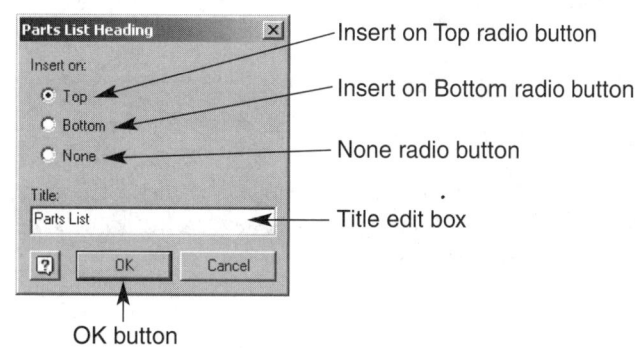

ging the column to the desired width. In addition, if you right-click on any of the spread sheet cells, the following menu options are listed:

- **Update Value**   If the part property shown in the specified cell has been modified after the parts list was created, pick this option to update and display the modified property value.

- **Update All**   If the part properties shown in multiple highlighted cells have been modified after the parts list was created, pick this option to update and display the modified property values in each of the cells.

- **Keep Value**   If the part property shown in the specified cell has been modified after the parts list was created, but you want to keep the existing parts list cell value, pick this option.

- **Keep All**   If the part properties shown in multiple highlighted cells have been modified after the parts list was created, but you want to keep the existing parts list values, pick this option.

- **Visible**   When this option is selected, the row is visible. Conversely, deselect the **Visible** option to "turn off" the row display.

- **Remove**   Select this option to remove a custom part row.

- **Column split**   Select this button to split or divide the column at the specified row. This option functions the same as the **Columns** edit box in the **Parts List – Item Numbering** dialog box. For assembly drawings with large parts lists, you may want to split columns so that the assembly components are divided among the specified number of columns.

- **Column Properties**   This area controls specific column display options. To use this area, select the column or a cell in the column you want to modify. The column property should be displayed next to "Property." Then enter a different column name in the **Name** edit box, and the desired column width in the **Width** edit box. You can also use the **Column Properties** area to specify the alignment characteristics of the column name and column data. Specify the alignment of the column name, such as "QTY" or "ITEM," by selecting the **Align Left, Center,** or **Align Right** button. Define the alignment of the column data, or information, such as 1, 2, or 3, by selecting the corresponding **Align Left, Center,** or **Align Right** button.

- **Format**   Use this area to modify the parts list layout. For assembly drawings with large parts lists, you can enter the number of parts list columns you want to display in the **Columns** edit box. The assembly components are divided among the specified number of columns. This process is similar to splitting a column in the spread sheet. If you do split a column, you can return to an equal column division by selecting the **Auto** button. Then specify how you want the parts list extents to expand when you increase or decrease the number of rows, or column width, by picking the **Left** or **Right** radio button. When you select the **Left** radio button, the parts list expands to the left. When you choose the **Right** radio button, the parts list expands to the right.

When you have made all the necessary parts list modifications, pick the **OK** button to generate the changes and exit the **Edit Parts List** dialog box.

Field Notes

Although the properties of a single parts list can be modified using the **Edit Parts List** dialog box, to set the default parts list standards, use the **Parts List** tab of the **Drafting Standards** dialog box, previously discussed.

Exercise 17.4

1. Continue from Exercise 17.3 or launch Autodesk Inventor.
2. Open EX17.2.idw and save a copy as EX17.4.idw.
3. Close EX17.2.idw without saving, and open EX17.4.idw.
4. Open the **Drafting Standards** dialog box and explore the **Parts List** tab. Do not change the default setting.
5. Access the **Parts List** command.
6. Place the parts list in the upper right corner as shown.
7. You may need to open the individual part files used to create the assembly and add a description in the **Description** edit box of the **Properties** dialog box in order for a description to be displayed.
8. Right-click and select the **Edit Parts List...** menu option.
9. Explore the options in the **Edit Parts List** dialog box.
10. Modify some of the options as discussed.
11. Select the **Cancel** button.
12. The final exercise should look similar to the drawing shown.
13. Save the drawing file as EX17.4.
14. Exit Autodesk Inventor or continue working with the program.

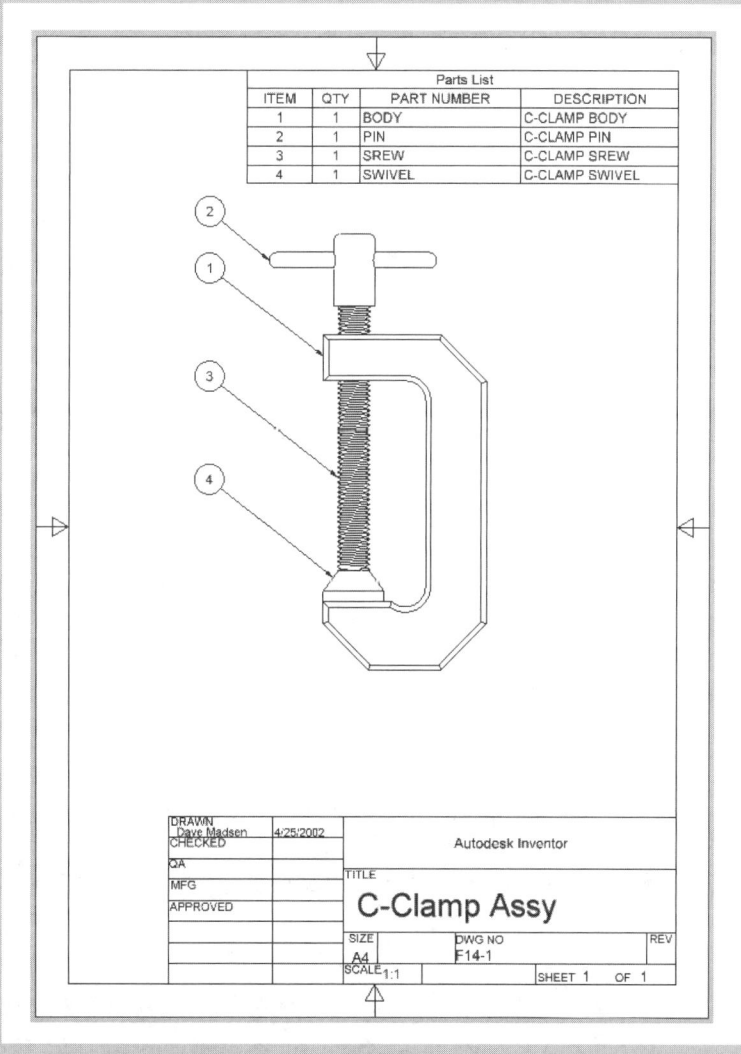

Multiple Sheet Drawings

A set of working drawings typically contains a multiple number of sheets. In general, a fully defined assembly drawing, with a parts list, is printed on one sheet. Then each detail drawing is printed on a separate sheet. For example, depending on company standards, five separate sheets of paper may be used to create a set of working drawings for an assembly that has four parts. Individual Autodesk Inventor drawing sheets are used in the same way that individual pieces of paper are used to document an assembly and each of the assembly components. This does not mean that you have to open several drawings files, .idw, because each of the working drawing sheets is contained in one drawing file. You should be able to document an entire design with one drawing file and multiple drawing sheets.

In Chapters 12 and 13 you learned how to create a single 2D part drawing on one sheet. In this chapter you create a single 2D assembly drawing on one sheet. For more information regarding specific sheet settings and specifications, refer to Chapters 12 and 13. A set of working drawings is developed by adding the assembly drawing and the detail drawings together in one part drawing file. The process of generating multiple sheet drawings can be done in a number of ways. Probably the most effective way of creating multiple sheet drawings is to open a new drawing file, create the assembly drawing, then create the detail drawings. However, you may choose any approach you find effective. You can even insert existing sheets from a drawing file into the multiple drawing sheet file.

Placing Additional Drawing Sheets

There are a number ways to create additional drawing sheets, based on the application. You can insert a new blank sheet, a new sheet with the same layout as the existing sheet, a new sheet with a different predefined layout, or a copy a drawing sheet from one drawing file to another, as follows:

- ■ To create a new blank sheet, select **Sheet...** from the **Insert** pull-down menu. When you select this menu option, the **New Sheet** dialog box is displayed. See Figure 17.31. A new blank sheet does not have any sheet specifications, a border, a title block, or any other information. The **New Sheet** dialog box, similar to the **Edit Sheet** dialog box discussed in Chapter 12, allows you to define the new sheet parameters. The following options are available inside the **New Sheet** dialog box:

  - ■ **Size**   This drop-down list allows you to specify the sheet size. There are a number of standard sheet sizes available such as A, B, C, D, E, F, or you can define your own sheet size by selecting the **Custom Size** option for the desired units (**Custom Size [inches]** or **Custom Size [mm]).** Once you choose a custom size option, the **Height** and **Width** edit boxes become active, allowing you to enter the desired height and width of the custom sheet size.

  - ■ **Title Block Orientation**   Title block orientation is located in the **Orientation** area and consists of four radio buttons. The radio buttons correspond to the corner where the title block is placed. To change the orientation of the title block, pick the desired radio button. A preview of the action is displayed.

FIGURE 17.31   The **New Sheet** dialog box.

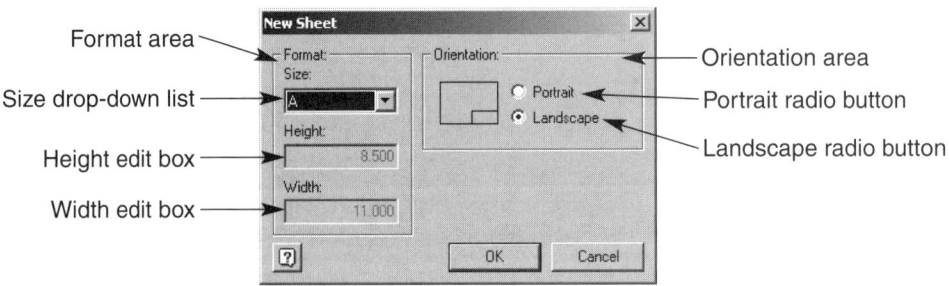

The title block orientation radio buttons are available only when there is an existing title block on the sheet.

- **Portrait**   This radio button allows you to apply a portrait or vertical sheet orientation.

- **Landscape**   This radio button allows you to apply a landscape or horizontal sheet orientation.

    Once you have defined the desired sheet options, pick the **OK** button to generate the sheet.

■ To create a new sheet with the same layout as the existing sheet, use one of the following techniques:

✓ Right-click in the graphics window background and select the **New Sheet** menu option.

✓ Right-click in the Browser and select the **New Sheet** menu option.

✓ Pick the **New Sheet** button on the **Drawing Annotation** panel bar.

✓ Pick the **New Sheet** button from the **Drawing Annotation** toolbar.

    When you use this sheet creation option, a copy of the existing sheet is added and includes the same sheet parameters, border, and title block as the existing sheet. If you do not want to use the same title block, delete the title block or select the **Undo** button until the title block is removed. Similarly, if you do not want to use the same border, delete the border or select the **Undo** button until the border is removed.

■ If you want to create a new sheet with a predefined layout that is different from the existing sheet in the drawing file, you can copy a different sheet from another drawing file. This process the discussed later in this chapter.
    Another option for using a predefined layout is a sheet format. As described in Chapter 12, a *sheet format* is a template stored in the drawing file. When you use a sheet format, a new sheet is created, and views from a specified model are inserted onto the sheet along with a border and title block if they are part of the sheet format. To use an existing sheet format, expand the **Drawing Resources** folder and the **Sheet Formats** folder in the **Browser** bar. Then double-click the sheet format you want to use, or right-click the sheet format and select the **New Sheet** menu option. Once the sheet format is selected, the **Select Component** dialog box appears. See Figure 17.32. If you have previously inserted a component, the model is available in the **Document Name** drop-down list. Otherwise, you need to pick the **Browse** button to display the **Open** dialog box, and locate the model you want to insert. Once you have located the file pick the **OK** button to create the sheet complete with a border, title block, and the specified views.

For more information regarding sheet formats refer to Chapter 12.

FIGURE 17.32   The **Select Component** dialog box.

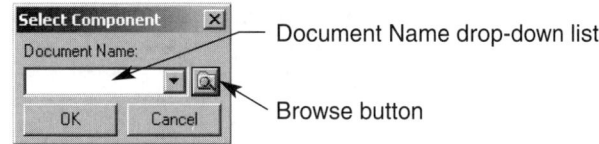

Document Name drop-down list

Browse button

■ If you want to copy an existing drawing sheet from a drawing file into another file, first open the drawing file where the drawing sheet is located. Then right-click on the drawing sheet in the graphics window or the sheet name in the Browser, and select the **Copy** menu option. Or, select the sheet and press the **Ctrl** and **C** keys on your keyboard. Next, open the drawing file where you want to place the copy, and right-click on the graphics window background or the drawing file name in the Browser, and select the **Paste** menu option. Or, select the sheet and press the **Ctrl** and **V** keys on your keyboard.

Working with Multiple Sheets

Once a new sheet is added to a drawing file, you can use any of the tools and techniques discussed in this chapter and Chapters 12 and 13 to create a complete set of working drawings. In addition to the tools and drawing processes discussed in these chapters you may want to become familiar with some particular concepts related to multiple sheet drawings. You can number each sheet, change the order of sheets, and move drawing views between sheets.

Figure 17.33 shows the Browser display of a drawing file with five drawing sheets. The contents of the active sheet are displayed in the graphics window. Only one sheet can be active at a time. The active sheet in Figure 17.33 is Sheet:1, and all other sheets are shaded. To activate a different sheet double-click the sheet name or right-click on the sheet name and select the **Activate** menu option. The sheet number or page number is identified in the Browser by the order of drawing sheets. As in other Autodesk Inventor applications, you can drag a sheet above or below another sheet to change the sheet order.

The order of the sheets in the Browser also defines the drawing title block page information, when drawing properties are used. For example, if there are five sheets in the drawing file, and each sheet has a title block with the "Number of Sheets" drawing property field, the first sheet in the Browser will have a sheet number of "1 of 5," while the last sheet in the Browser will have a sheet number of "5 of 5." In addition, as discussed in Chapter 12, title block "Model Properties" correspond to the model in the drawing view while "Design Properties" correspond to the properties defined in the **Properties** dialog box for the drawing file in which you are currently working.

Although you cannot copy a sheet within the same drawing file, in some situations you may want to copy or move a view from one sheet to another. Copying a view is similar to copying an entire sheet. First, right-click on the drawing view in the Browser, or open the sheet with the view and right-click on the drawing view and select the **Copy** menu option. Or, select the view and press the **Ctrl** and **C** keys on your keyboard. Next, right-click on the drawing sheet in the Browser, where you want to place the view, and select the **Paste** menu option. Or, select the sheet and press the **Ctrl** and **V** keys on your keyboard. To move a view from one sheet to another, drag the view name from its current location in the Browser to the new sheet location in the Browser. When you move a base or dependent view from one sheet to another, a shortcut is created. The shortcut and the shortcut icon in the Browser identify that there is a relationship between two views in different sheets. To view the relationship, right-click on the shortcut icon, and select the **Go To...** menu option to open the corresponding sheet.

FIGURE 17.33   An example of the browser display for a multiple-sheet drawing.

Drawing sheets

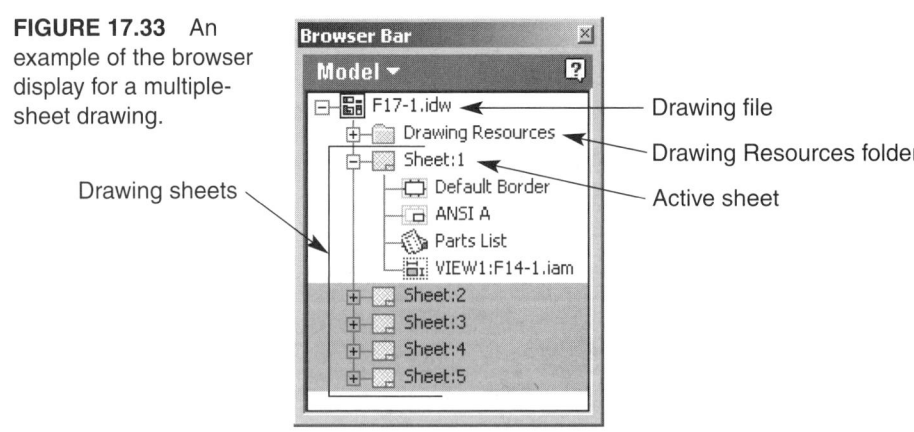

— Drawing file

— Drawing Resources folder

— Active sheet

Exercise 17.5

1. Continue from Exercise 17.4 or launch Autodesk Inventor.
2. Open EX17.4.idw and save a copy as EX17.5.idw.
3. Close EX17.4.idw without saving, and open EX17.5.idw.
4. Using the information provided, create a second drawing sheet using the same border and title block as the first sheet.
5. If the title block sheet number property field is set as a "sheet property" the second sheet should read 2 of 2.
6. If you specify the part number and title in the title block as "model properties" the fields will be blank until you create a drawing view.
7. Access the **Create View** tool and specify the following options:

   File: P5.6.ipt
   View: Front
   Scale: 1:1
   Style: Hidden Line

8. Place the part drawing view in the middle of the drawing space.
9. Add dimensions, notes, and annotations as required. If the title and part number fields in the title block have been set as "model properties," the drawing title and part number of the c-clamp body should be shown.
10. Rename Sheet:2 as C-Clamp Body.
11. Create a third drawing sheet, using the same border and title block as the first sheet.
12. If the title block sheet number property field is set as a "sheet property" the third sheet should read 3 of 3.
13. If you specify the part number and title in the title block as "model properties" the fields will be blank until you create a drawing view.
14. Access the **Create View** tool and specify the following options:

    File: P5.3.ipt
    View: Bottom
    Scale: 1:1
    Style: Hidden Line

15. Place the part drawing view in the middle of the drawing space.
16. Add dimensions, notes, and annotations as required. If the title and part number fields in the title block have been set as "model properties," the drawing title and part number of the c-clamp screw should be shown.
17. Rename Sheet:3 to C-Clamp Screw
18. Create a fourth drawing sheet, using the same border and title block as the first sheet.
19. If the title block sheet number property field is set as a "sheet property" the third sheet should read 4 of 4.
20. If you specify the part number and title in the title block as "model properties" the fields will be blank until you create a drawing view.
21. Access the **Create View** tool and specify the following options:

    File: P5.1.ipt
    View: Left
    Scale: 1:1
    Style: Hidden Line

22. Place the part drawing view in the middle of the drawing space.
23. Add dimensions, notes, and annotations as required. If the title and part number fields in the title block have been set as "model properties," the drawing title and part number of the c-clamp pin should be shown.

Exercise 17.5 continued on next page

Exercise 17.5 continued

24. Rename Sheet:4 to C-Clamp Pin
25. Open the drawing P12.4.idw.
26. Right-click on Sheet:1 in the Browser, and select the **Copy** menu option.
27. Close P12.4.idw without saving. EX17.5.idw should be displayed.
28. Right-click on EX17.5,idw in the Browser and select the **Paste** menu option.
29. Edit Sheet:5 using the following specifications:

   Name: C-Clamp Swivel
   Orientation: Lower right corner radio, and **Portrait** radio buttons selected.

30. Replace the existing title block with the title block used on the other drawing sheets.
31. Add dimensions, notes, and annotations as required. If the title and part number fields in the title block have been set as "model properties," the drawing title and part number of the c-clamp swivel should be shown.
32. Drag the sheet "C-Clamp Swivel" above the sheet "C-Clamp Pin," to reorder the drawing sheets. "C-Clamp Swivel" should now be sheet 4 of 5, and "C-Clamp Pin," should be sheet 5 of 5.
33. Resave the drawing EX17.5.idw.
34. Exit Autodesk Inventor or continue working with the program.

Printing Drawings

Once you have created your design drawings, you will probably want to print or plot the drawings onto a sheet of paper. Plotting a drawing is very similar to printing a model as discussed in Chapter 2. However, there are some differences between printing a model and a drawing, because it is crucial that a drawing is printed correctly, and to scale. In order to print a drawing, you may first want to define the print specifications, by selecting **Pr̲int Setup...** from the **F̲ile** pull-down menu. Selecting the **Print Setup** option opens the **Print Setup** dialog box shown in Figure 17.34.

The following options are available inside the **Print Setup** dialog box:

■ **Printer** area   This section of the **Print Setup** dialog box contains several options as follows:

   ■ **Name** drop-down list   The **Name** drop-down list allows you to select the printer you would like to use to plot your document.

   ■ **Status**   This displays the status of the print job—for example, printing, canceled, complete, etc.

FIGURE 17.34   The **Print Setup** dialog box.

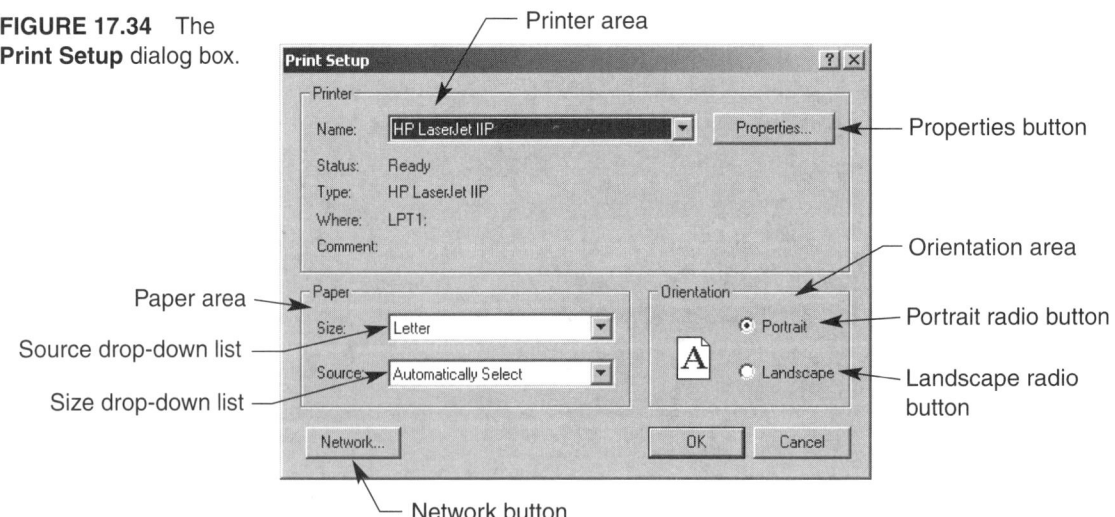

- **Type**   Displays the type of printer currently being used.

- **Where**   Displays the printer port currently being used by the specified printer.

- **Properties** button   Selecting the **Properties** button opens the **Document Properties** dialog box for the specified printer. See Figure 17.35. Here you can adjust the page layout, paper source, print quality, and other options that may apply to the specified print device.

- **Paper** area   The following options are available in the **Paper** area of the **Print Setup** dialog box:

  - **Size** drop-down list   This drop-down list lets you select the size of paper to use for the document.

  - **Source** drop-down list   This drop-down list lets you select where the paper comes from.

- **Orientation** area   Inside this area of the **Print Setup** dialog box, you can select the **Portrait** radio button or the **Landscape** radio button. The **Portrait** radio button allows you to rotate and print the document in a portrait orientation. The **Landscape** radio button allows you to rotate and print the document in a landscape orientation.

- **Network** button   Selecting the **Network** button opens the **Connect to Printer** dialog box, shown in Figure 17.36. If you are working in a networked setting and are not connected to a local printer, use this dialog box to connect to a shared network printer and plot to that device.

To preview the intended print job with the current print specifications set, access the **Print Preview** dialog box, shown in Figure 17.37, by selecting **Print Preview** from the **File** pull-down menu.

From the print preview area, you can initiate the print by selecting the **Print...** button, review other pages that will print with the current print job by selecting the **Next Page, Prev Page,** or **Two Page,** and zoom in or zoom out to further review the print preview, by choosing the **Zoom In** or **Zoom Out** buttons. In addition, you can select the **Close** button to terminate the print preview.

Although the **Print Setup** and **Print Preview** dialog boxes can be useful and the **Print Preview** dialog box allows you to print a drawing, by selecting the **Print...** button, usually you must define additional drawing print specifications such as scale in order to print the drawing correctly. To define printing parameters and print a drawing, access the **Print** dialog box, shown in Figure 17.38, by selecting **Print...** from the

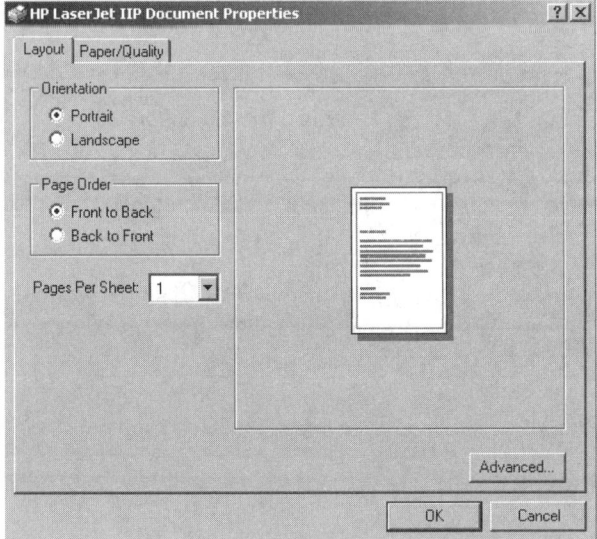

FIGURE 17.35   The **Document Properties** dialog box.

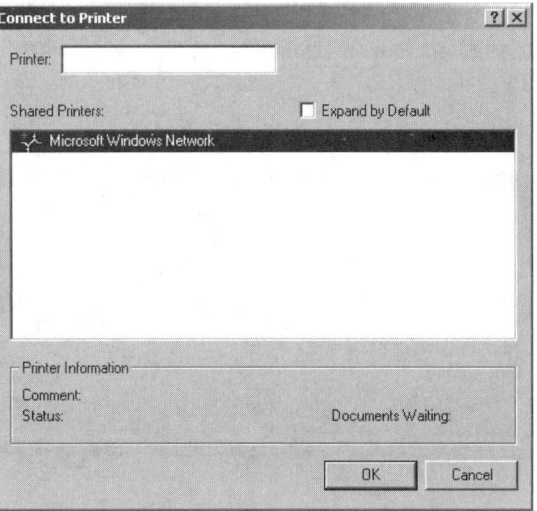

FIGURE 17.36   The **Connect to Printer** dialog box.

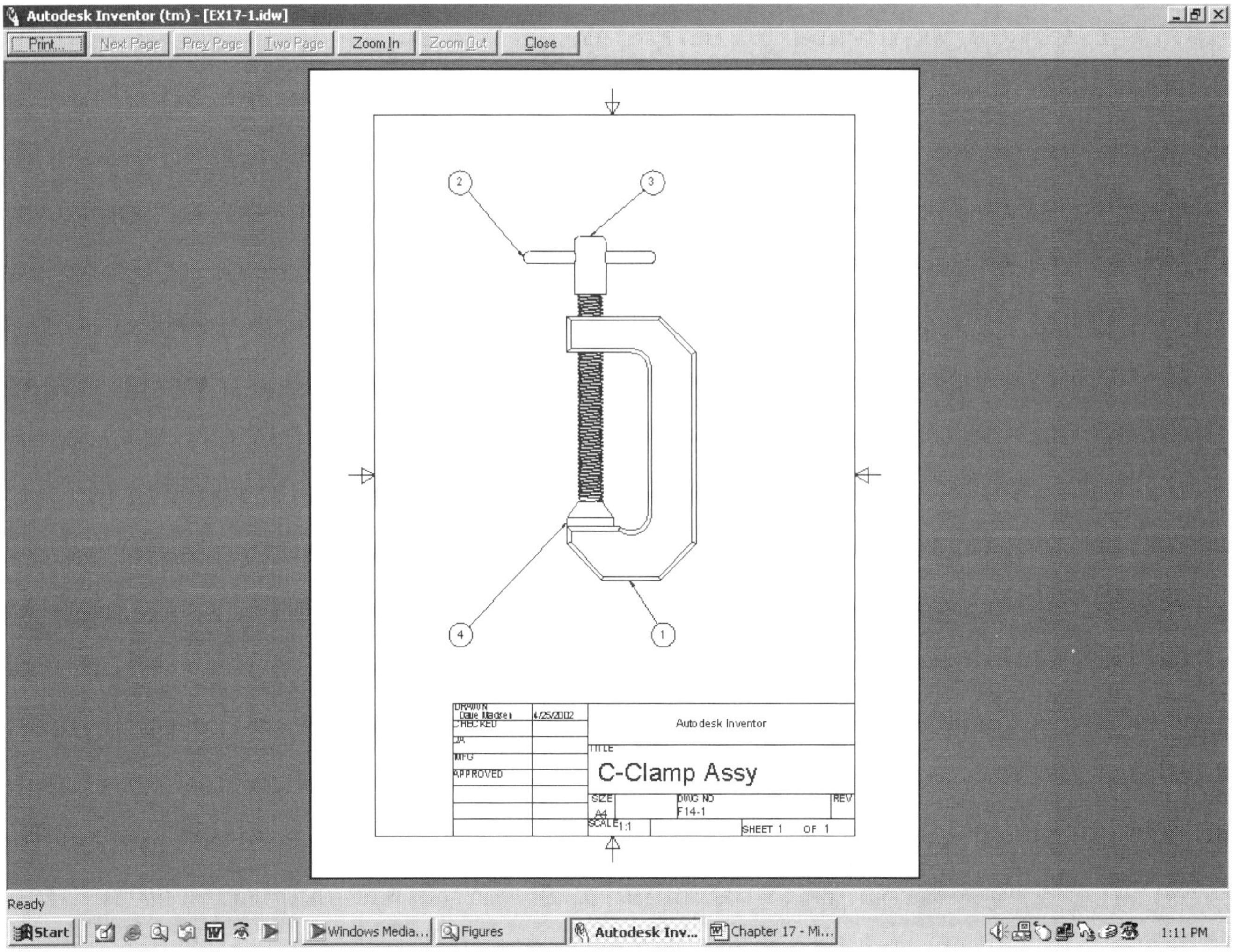

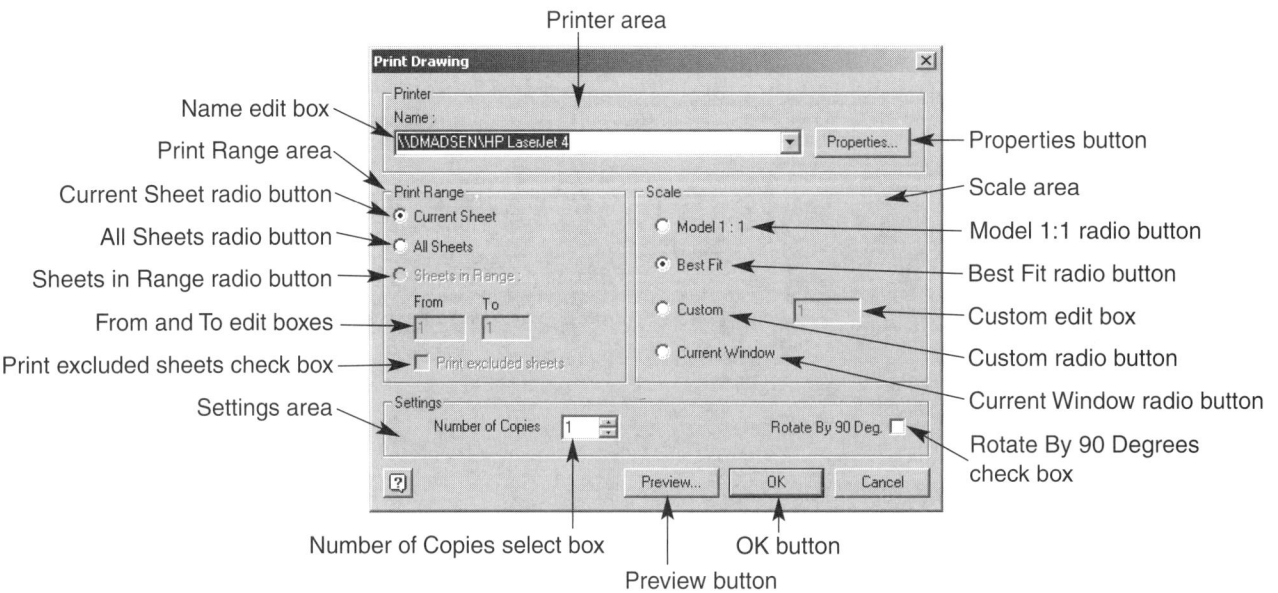

FIGURE 17.37 A preview of your print job.

FIGURE 17.38 The **Print** dialog box for printing a drawing.

File pull-down menu. As you can see, the **Print** dialog box is the most powerful printing tool and allows you to set up and preview the print. The following options are available inside the **Print** dialog box:

- **Printer**   This area allows you to define the printer to use for the print job. Use the **Name** drop-down list to select a printer. Then pick the **Properties** button to access the **Document Properties** dialog box previously discussed.

- **Print Range**   This area is use to define which sheets in the drawing file you want to print. Select the **Current Sheet** radio button if you want to print only the active drawing sheet, or select the **All Sheets** radio button to print all the sheets in the drawing file. You can also choose to print a certain range, or group, of sheets. To print a range of sheets, pick the **Sheets in Range** radio button and then enter the sheet you want to print first in the **From** edit box, and the sheet you want to print last in the **To** edit box. If you want to override the **Exclude From Print** option specified in the **Edit Sheet** dialog box and print the sheets specified as "excluded" select the **Print excluded sheets** check box.

- **Scale**   Use this area to specify the drawing print scale. Select the **Model 1:1** radio button to print the drawing sheet full-size. If the paper is too small for the specified sheet size, part of the sheet will not be displayed, but the views will be printed at full scale. If you do not want to apply a certain scale factor, select the **Best Fit** radio button, which scales the sheet according to the specified paper size. To apply a certain scale factor for the drawing, pick the **Custom** radio button, and then specify the scale in the **Custom** edit box. For example, if you enter a value of 2, the drawing will be printed at a 2:1 scale factor. Or, if you enter a value of .5, the drawing will be printed as a 1:2 scale factor. You can also choose the **Current Window** radio button to print the display of the current window without a specified scale factor.

- **Settings**   Use this area to set the number of print copies and rotation factor. Define the number of copies by entering or selecting a value from the **Number of Copies** select box. Then, if you need to rotate the drawing, select the **Rotate By 90 Degrees** check box.

Once you have fully defined the print options available in the **Print** dialog box, select the **Preview...** button to display the **Print Preview** dialog box previously discussed and preview the intended print job. Finally, pick the **OK** button to print the drawing.

Exercise 17.6

1. Continue from Exercise 17.5 or launch Autodesk Inventor.
2. Open EX17.5.idw.
3. Using the information provided, print each of the drawing sheets at full scale.
4. Exit Autodesk Inventor or continue working with the program.

|||||||||||| CHAPTER TEST

Answer the following questions on a separate sheet of paper.

1. What is found in a set of working drawings?

2. What is an assembly drawing?

3. What is a general assembly drawing?

4. Briefly discuss the difference between developing an assembly drawing and a part drawing.

5. Give the function of the Apply Design View dialog box.

6. What are balloons, what is contained in a balloon, and what is their function?

7. Identify the button used to create a balloon with a single identification number inside a circle.

8. Identify the button used to create a balloon with two entries inside a circle.

9. Give the basic function of the Bill Of Material Column Chooser dialog box.

10. Identify the two different options for adding balloons to an assembly drawing.

11. Explain the function of the Level area in the Parts List – Item Numbering dialog box.

12. Discuss the function of the First-Level Components radio button in the Parts List – Item Numbering dialog box.

13. Identify the button to pick if you want all the parts used in the assembly to be displayed, including those used in a subassembly.

14. What does the point on the assembly that you select to connect the leader and balloon define?

15. How many drawing views receive balloons when using the Balloon All tool?

16. How do you redefine balloon location?

17. What are grouped balloons?

18. How do you group two or more balloons?

19. What is the function of a parts list and what items are found in a parts list?

20. Give the function of the Output Direction buttons.

21. Briefly give the function of the Parts List Column Chooser dialog box.

22. How do you move a parts list?

23. Briefly describe the function of the Sort Parts List dialog box.

24. Give the function of the Parts List Spread Sheet.

25. Briefly discuss the function of the New Sheet dialog box.

|||||||||||| PROJECTS

Instructions:

- Open a new drawing file for each of the projects.

- Create a complete set of working drawings using each of the specified files.

- The first sheet should be an assembly with balloons and a parts list.

- Additional sheets are created for each of the details.

- Use a border and title block for each drawing sheet.

- Fully dimension, note, and annotate each drawing as required.

- Use the information provided to create the drawings shown.

- Define the drawing properties in the drawing and individual model **Properties** dialog box, so the properties are displayed in the title block.

1. Name: Plate and Angle Drawings

   Units: Inch

   Sheet Size: D

   Drawing View Scale: 1:1

   Border: Default or custom

Title Block: Default or custom

Assembly File: (Presentation) EX16.2.ipn

Detail Files: EX15.4.ipt and EX15.5.ipt

Special Instructions: The assembly drawing is an exploded assembly drawing and is created with a presentation file. Insert the presentation drawing view used an isometric right display.

Save As: P17.1

2. Name: Hub Drawings

   Units: Inch

   Sheet Size: B

   Drawing View Scale: 1:1

   Border: Default or custom

   Title Block: Default or custom

   Assembly File: EX14.15.iam

   Detail Files: EX6.3.ipt and EX14.15.ipt

   Save As: P17.2

3. Name: Bottle Drawings

   Units: Inch

   Sheet Size: B

   Drawing View Scale: 1:1

   Border: Default or custom

   Title Block: Default or custom

   Assembly File: P14.1.iam

   Detail Files: P5.8.ipt and P6.8.ipt

   Save As: P17.3

4. Name: Rotating Hinge Drawings

   Units: Metric

   Sheet Size: A3

   Drawing View Scale: 1:1

   Border: Default or custom

   Title Block: Default or custom

   Assembly File: P14.3.iam

   Detail Files: P14.3a.ipt, P14.3b.ipt, and P14.3c.ipt.

   Save As: P17.4

5. Name: Housing Drawings

   Units: Inch

   Sheet Size: C

   Drawing View Scale: 1:1

   Border: Default or custom

Title Block: Default or custom

Assembly File: P14.6.iam

Detail Files: P14.6.ipt, P10.2.ipt, and eye bolt, washer, lock washer, and nut shared content files.

Save As: P17.5

||||||||||||||||||||||||||||||||||||||||||||||||||||||||||||||||||||||||||||||||||||

CHAPTER 18

Additional Modeling Tools and Techniques

LEARNING GOALS

After completing this chapter, you will be able to:

◎ Create and work with derived parts.

◎ Create and work with derived assemblies.

◎ Work with derived components.

◎ Develop an iPart template.

◎ Create and work with iPart factories.

◎ Place and use iParts in assemblies.

Additional Modeling Tools and Techniques

Throughout this text you have explored tools, commands, techniques, and options for developing models and fully documenting the design process using Autodesk Inventor. In addition to the tools and options you have used, you may want to become familiar with derived components and iParts for additional model development purposes. For many applications, you may want to use derived components to generate different models from existing part or assembly geometry. Or, you may want to use iParts to create models that are similar, but have different parameters and specifications. Although derived component and iPart tools are different, this chapter discusses working with these two useful model creation options.

Derived Components

As described in Chapter 2, derived components are essentially elaborate catalog features. As you may recall, catalog features are existing features you create and then store as a single design element in a catalog. Then, when inserted into other models, catalog features, or iFeatures, can retain their original size and position values, or can be modified to fit a different design application. Derived components are similar to catalog features in that they are created, saved, and available to use in other designs. As the name implies, *derived components* derive, or obtain, geometry from the original part or assembly for use in a new model. However, in contrast to design elements, derived components are more complicated than design elements and may contain a complete model consisting of several features and even an assembly containing several parts. Still, a derived component, no matter how geometrically complicated, is a single part feature and does not represent each of the features used to create the original part. See Figure 18.1.

Derived components are used to develop the base, or initial, model component by inserting sketches, features, parts, and groups of parts. The geometry of a derived component itself cannot be modified because a derived component is a single body. However, you can modify the parent model used to create the derived component, and then update the derived component to observe the changes. In addition, you can manipulate a derived component by adding new features to the solid body. For example, you can cut a hole through a derived component, or even add a fillet to an edge. When features are added to a derived component, the

FIGURE 18.1 An example of a part feature derived from a part file. The individual features are grouped together to form one feature.

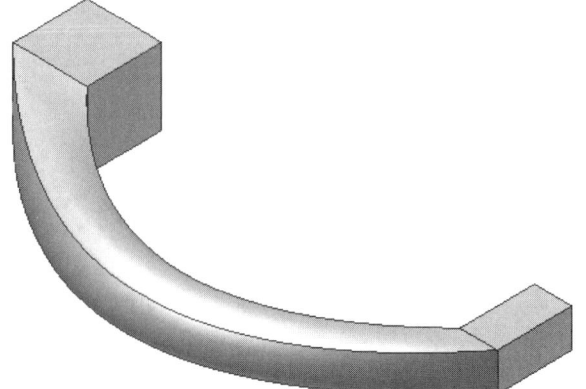

Not represented as individual features

parent component is not modified, because the derived component is essentially a copy of the original component. Using the previous example, if you cut a hole through a derived component, a hole is not added to the original parent component.

Like other Autodesk Inventor options, the uses for derived components vary greatly depending on the application. The following represents a number of options for using derived components:

- Derive a solid casting or machined part from an assembly containing several distinct components.

- Display various machining options for the same part blank, such as adding holes, fillets, chamfers, and other features.

- Make a scaled copy of an existing part.

- Define a weldment complete with post assembly holes and other features.

- Create two part halves from a single component.

- Mirror components using the existing work planes.

Similar to catalog features, derived components originate from an existing part or assembly file. Consequently, the first step in using a derived component is to create, and save, the part or assembly model you want to later specify as a derived component. Once you have a model saved, you can create a derived component. You should begin to think about derived components as part features that can originate from complete parts or assemblies. See Figure 18.2. However, a derived component is used as the base feature of a part and can only be created in the part work environment of a part or assembly file. Theoretically, an assembly cannot be created from a derived component, although individual assembly part components can be created from derived components.

Usually, derived components are placed in blank part or assembly files. If you add a derived part to a file with existing components, the derived component is inserted as a derived work body. See Figure 18.3.

A ***derived work body,*** also know as a derived surface, is a complex work feature and can be used just like any other work features to help you generate and position features. As discussed later in this chapter,

FIGURE 18.2 An example of (A) a part feature derived from a part and (B) a part feature derived from an assembly. In both, the derived assembly geometry becomes one part feature.

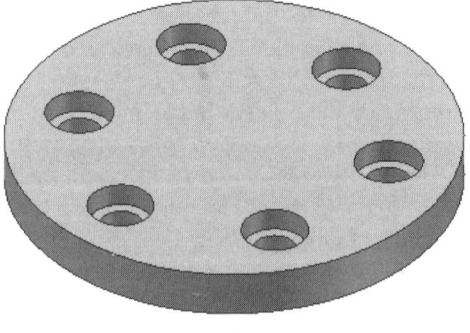
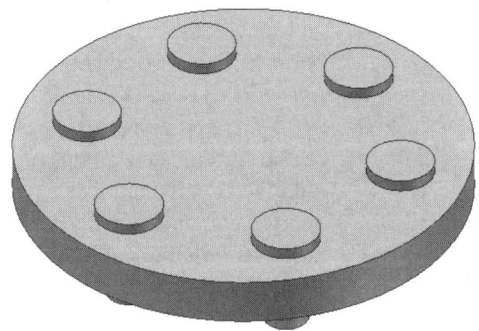

A   B

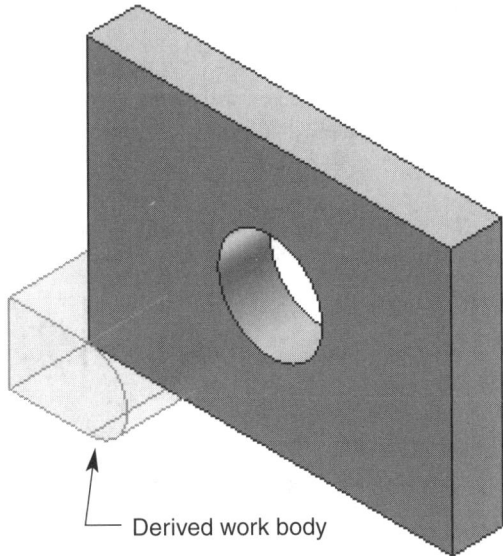

FIGURE 18.3 An example of a derived work body.

Derived work body

you can create a derived work surface for any derived feature, even if the file is not blank, by selecting **Body as Work Surface** in the **Derived Part** dialog box when you create the derived part feature.

Developing Derived Parts and Assemblies

As previously mentioned, derived parts are created from existing part or assembly models within the part work environment. To begin the process of creating a derived component feature, open a new part file or open a new assembly file and use the **Create Component** tool to enter the part work environment. Next, if a sketch is active, finish the sketch, and delete the sketch if it is not needed. Once you exit sketch mode, you are ready to create a derived component feature.

Access the **Derived Component** command using one of the following techniques:

✓ Pick the **Derived Component** button on the **Feature** panel bar.

✓ Pick the **Derived Component** button on the **Feature** toolbar.

The **Open** dialog box is displayed when you access the **Derived Component** command. See Figure 18.4. For creating derived parts, the **Open** dialog box functions much as it does for other Autodesk Inventor applications. Use the **Open** dialog box to select the part or assembly file from which you want to derive the part feature. Once you locate the file you want to use, pick the **Open** button to insert the derived component feature in the graphics window and display the **Derived Part** dialog box. See Figure 18.5. Creating a derived part and the options available in the **Derived Part** dialog box are different when deriving a part than when deriving an assembly. The following information describes each of these processes.

FIGURE 18.4 The **Open** dialog box.

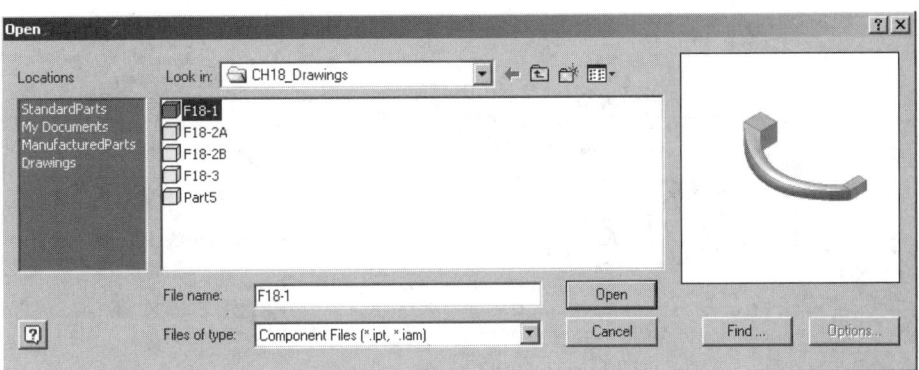

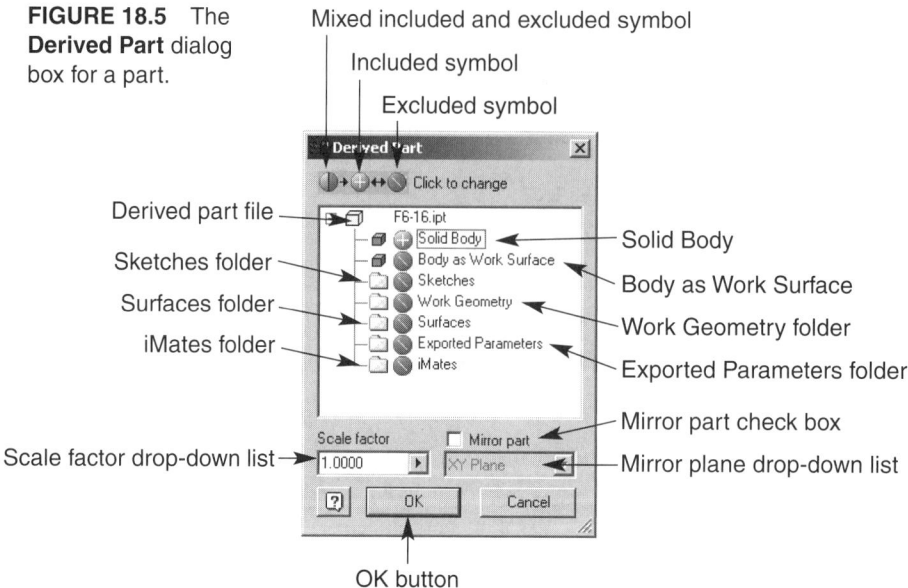

FIGURE 18.5 The **Derived Part** dialog box for a part.

Creating a Derived Part Feature

To derive a feature from a part file, select the desired part file from the **Open** dialog box. When you pick the **Open** button, the **Derived Part** dialog box for the part is displayed. See Figure 18.5. As you can see from Figure 18.5, part sketches, work features, surfaces, parameters, and iMates can be included in or excluded from the derived part feature, depending on the design requirements.

By default, when you access the **Derived Part** dialog box, any unconsumed sketches, work features, surfaces, parameters, and iMates are listed in the corresponding folder and are included in the selected derived part geometry, as indicated by the included symbol. If you want to exclude certain information from the derived part feature, select the included symbol to change to an excluded symbol. You can select the inclusion or exclusion symbol for any of the listed geometry. However, only geometry that is available will be displayed or removed from the derived part feature. For example, if a derived part does not have any unconsumed sketches, including sketches by selecting the **Included** symbol has no effect. Geometry that is available will be listed in the corresponding folder. See Figure 18.6. If a folder—for example, the sketches folder—contains more than one unconsumed sketch, you can choose to include or exclude certain sketches. As shown in Figure 18.7, a folder that contains both included and excluded geometry is identified by the **Mixed Included and Excluded** symbol. For additional clarity, the following information describes each of the geometry options and symbols available in the **Derived Part** dialog box.

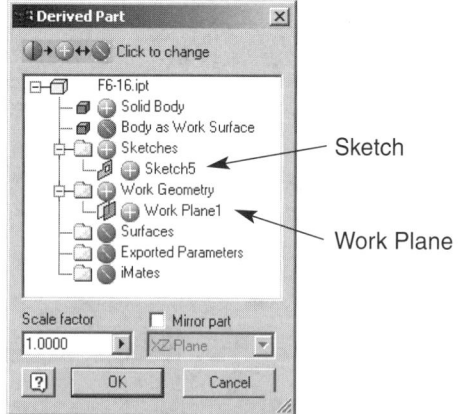

FIGURE 18.6 An example of deriving a part that contains additional geometry, such as an unconsumed sketch and a work plane.

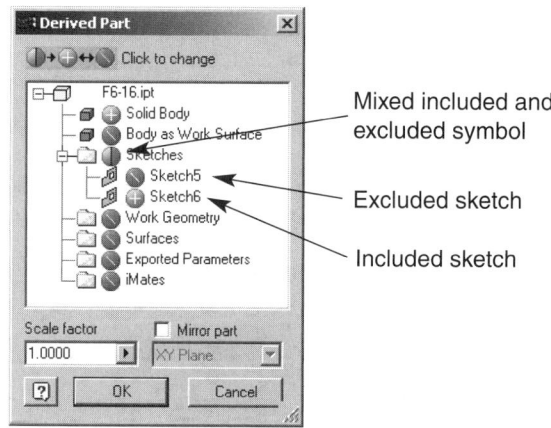

FIGURE 18.7 Mixed included and excluded geometry.

Field Notes

> Geometry marked with the excluded symbol is not displayed as part of the derived part feature and is not updated along with included geometry when changes are made.

- **Derived Part File**   Here you can see the original, or derived, part file used to create the derived part feature.

- **Solid Body**   Select the include button to use the derived part feature as a solid body. If **Solid Body** is excluded, **Body as Work Surface** must be included, or nothing happens.

- **Body as Work Surface**   Select the include button to use the derived part feature as a work surface, as previously discussed. Again, if **Body as Work Surface** is excluded, **Solid Body** must be included, or nothing happens.

- **Sketches**   Only visible sketches that are not shared or consumed in the original part file can be displayed in the derived part feature. If sketches are available, include them for use in the derived part feature, or exclude them if you do not want to use them in the derived part feature.

- **Work Geometry**   This folder contains any visible work features created in the original part. If included, the work feature will be available in the derived component feature.

- **Surfaces**   This folder lists any visible surfaces that are found in the original file and can be included or excluded from the derived part feature.

- **Exported Parameters**   If included, the exported parameters found in the original part file are available in the derived part feature.

- **iMates**   If included, any iMates found in the original part file are available in the derived part feature.

Field Notes

> While in the original part file, you can specify a parameter to be exported by selecting the check box next to the parameter value.

Once you have fully identified which pieces of geometry you want to include and exclude from the derived part feature, you can change the scale of the derived component and/or mirror the component, depending on the application. To scale the derived part, enter or select a scale factor from the **Scale factor** drop-down list. A scale factor of 1 does not change the size of the derived part in reference to the original component. To increase the size of the derived part, enter a value greater than 1, such as 2 or 3.5. Conversely, to decrease the size of the derived part, enter a value less than 1, such as .5 or .25.

If you want to flip, or mirror, the derived part feature, select the **Mirror** part check box. Then select the mirror plane, or line of symmetry, from the **Mirror Plane** drop-down list. The final step in creating a derived part is to select the **OK** button.

Exercise 18.1

1. Launch Autodesk Inventor.
2. Open a new inch file.
3. Save the part file as EX18.1.ipt. Saving the file should finish the sketch if a sketch is active.
4. Delete the default sketch if present in the file.
5. Access the derived component tool.
6. Locate and open the part file: EX5.5.ipt.
7. Explore the options in the **Derived Part** dialog box, as discussed.
8. Enter a scale factor of 2 in the **Scale** drop-down list edit box.
9. Select the **Mirror** part check box, and select XY Plane from the **Mirror Plane** drop-down list.
10. Pick the **OK** button to generate the derived part feature.
11. What you created is a solid body copy of EX5.5 that is twice the size. The copy or derived part feature can be used as a part blank that will be machined, or simply as a scaled copy of the original part.
12. Resave the part file.
13. The final exercise should look like the part shown.
14. Exit Autodesk Inventor or continue working with the program.

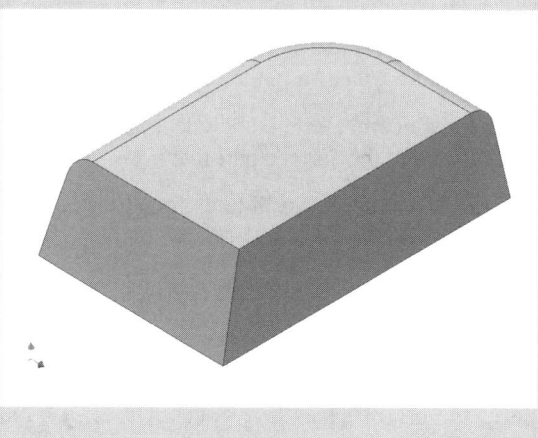

Creating a Derived Assembly Feature

To derive a feature from an assembly file, select the desired assembly file from the **Open** dialog box. When you pick the **Open** button, the **Derived Assembly** dialog box for the assembly is displayed. See Figure 18.8. As you can see from Figure 18.8, all the assembly components found in the derived assembly can be included, excluded, or subtracted from the derived assembly feature, depending on the design requirements.

By default, when you access the **Derived Assembly** dialog box, all assembly components are listed and are included in the selected derived assembly geometry, as indicated by the included symbol. If you want to exclude certain information from the derived part feature, select the included symbol to change to an excluded symbol. When you exclude a component, the component is not displayed in the derived assembly feature display and is not updated along with included geometry when changes are made.

You can also choose to remove a component from the derived assembly feature by changing to a subtract symbol. In contrast to the exclude option, the subtract option removes the component from the derived assembly feature, and if the subtracted component intersects another component a cavity is created. See Figure 18.9. Typically the subtract symbol is used when you create two halves of an assembly. For this process you create an assembly with two parts, and then create two derived assembly features by subtracting the top part on one derived part, and the bottom part on the other derived part.

You can select the inclusion, exclusion, or subtraction symbol for any of the listed components. If a derived assembly has components that are included and/or excluded and/or subtracted, the derived assembly file is identified by the **Mixed Included, Excluded, and Subtracted** symbol.

FIGURE 18.8 The **Derived Assembly** dialog box for an assembly.

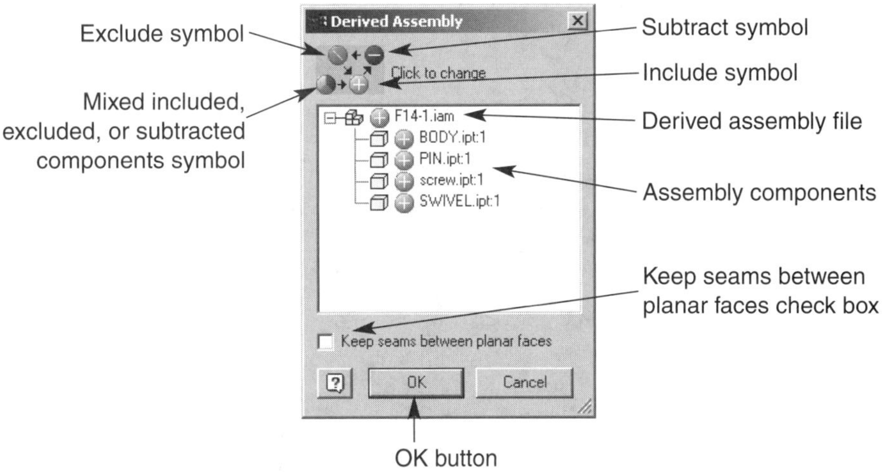

Exclude symbol

Mixed included, excluded, or subtracted components symbol

Subtract symbol

Include symbol

Derived assembly file

Assembly components

Keep seams between planar faces check box

OK button

FIGURE 18.9 Using the subtract and exclude options.

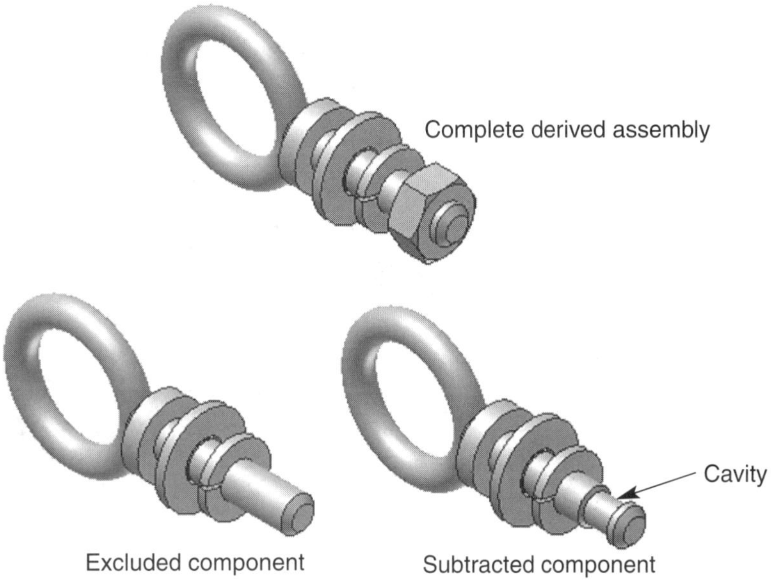

Complete derived assembly

Cavity

Excluded component

Subtracted component

Once you have fully identified which components you want to include and exclude from the derived assembly feature, you determine whether or not you want to display the seams, or edges, between coincident planar faces of two different components. *Seams* give the allusion that the derived assembly consists of more than one feature. If you do want to show the seams, as seen in Figure 18.10A, pick the **Keep seams between planar faces** check box. If you do not want to display the seams, as shown in Figure 18.10B, do not select this check box. The final step in creating a derived assembly is to select the **OK** button.

FIGURE 18.10 An example of (A) a displayed seam and (B) a removed seam.

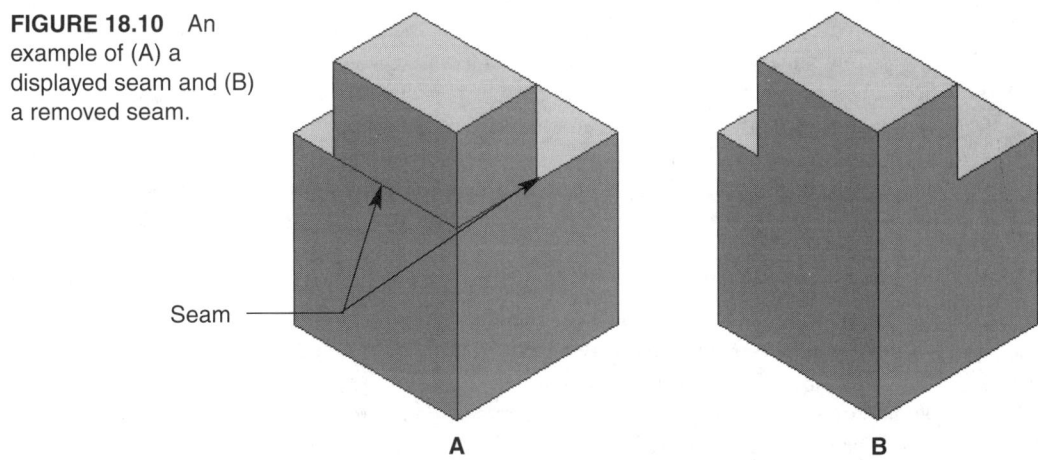

Seam

A

B

Exercise 18.2

1. Continue from Exercise 18.1 or launch Autodesk Inventor.
2. Open a new inch file.
3. Save the part file as EX18.2.ipt. Saving the file should finish the sketch if a sketch is active.
4. Delete the default sketch if present in the file.
5. Access the derived component tool.
6. Locate and open the assembly file: EX14.15.iam.
7. Explore the options in the **Derived Assembly** dialog box, as discussed.
8. Enter a scale factor of 1 in the **Scale** drop-down list edit box.
9. Pick the **OK** button to generate the derived assembly feature.
10. What you created is a solid body copy of EX14.15. The copy, or derived assembly feature, in this example is used as a weldment, which is a product assembled using welds. This derived component can be used as a base feature that can be machined after the welding process.
11. Resave the part file.
12. The final exercise should look like the part shown.
13. Exit Autodesk Inventor or continue working with the program.

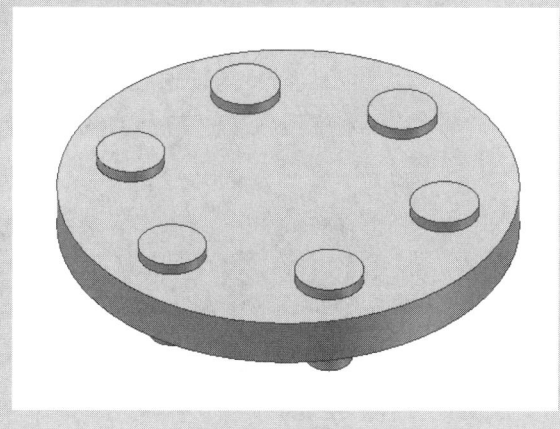

Working with Derived Components

Once you create a derived part or assembly, there are numerous ways of working with these features, depending on your specific applications. Several uses of derived components have already been discussed. No matter what you intend to do with a derived component, remember that a derived component is a solid body. The parent part may have been made of several features, or the parent assembly may have contained a number of parts, but the derived component is a single feature. As a result, anything you can do with a feature such as adding another feature to an extrusion or revolution can be done with a derived component.

Figure 18.11 shows an example of the Browser display for a derived component. Notice the derived component name and icon are listed first, followed by the solid body or surface, depending on the specification. Then, any additionally included geometry, such as a sketch and a work plane, is shown. When you add new features to a file with a derived component, the features are listed under the derived component, just as if you are working on a model without derived components. Adding additional features to a derived component is one of the most effective uses. For example, Figure 18.12A shows an assembly file weldment that has been inserted into a new part file as a derived assembly feature. A *weldment* is any product assembled using welds. In this example, the studs are welded to the plates. Figure 18.12B shows how holes are added to and then patterned around the weldment to indicate the postassembly operation of drilling through multiple components at the same time.

Although you can work with a derived component much the same as any other component, there are also some specific commands and options that allow you to edit derived parts by making changes to the parent component, edit derived component specifications, and unlink derived parts from the parent component. The following information discusses each of these processes:

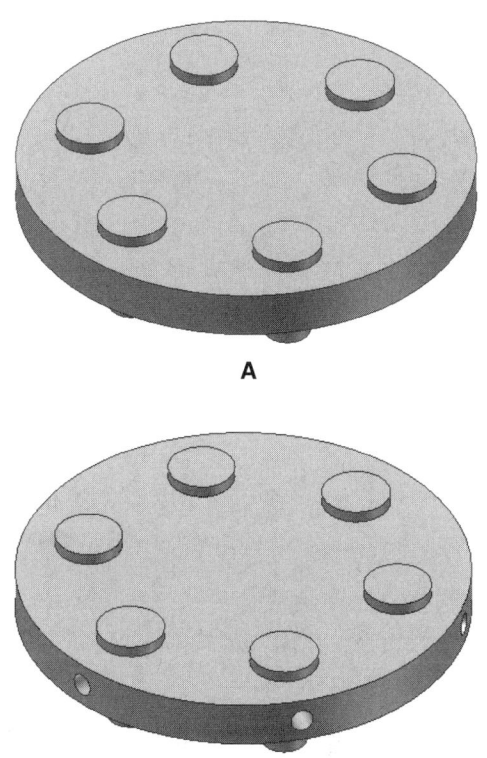

A

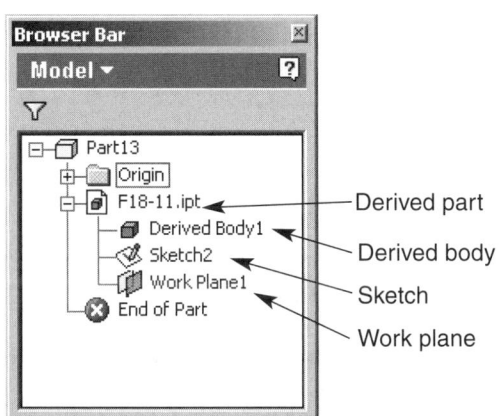

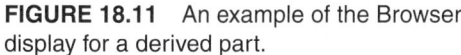

— Derived part

— Derived body

— Sketch

— Work plane

B

FIGURE 18.11 An example of the Browser display for a derived part.

FIGURE 18.12 An example of adding features (holes) to a derived component feature.

- Editing derived components   When you add features to a derived component the geometry changes, but the actual features used to create solid cannot be modified. As previously mentioned, there is no way to physically modify derived component geometry, unless you modify the parent component.

  You can open the original file for editing purposes, just as you would open a file for any other purpose, using the **Open** dialog box, or Windows Explorer. Or you can open the original component from inside the file that contains the derived component. To use this option, right-click on the derived component in the Browser and select the **Open Base Component** menu option, or double-click on the derived component in the Browser. When you use this option the base, or original, component used to generate the derived component is opened in a separate window. You can make the needed modifications to the component, then save and close the file just as any other time you would edit a model.

  When you modify the original model and then open the file containing the derived component, a red lighting bolt symbol is shown next to the derived component in the Browser. See Figure 18.13. The symbol indicates that the base component has been modified and that you must update the derived component to observe the changes. To update the derived part or assembly, pick the **Update** tool button on the **Command** bar.

- Editing derived component specifications   If you want to change the specifications used to create a derived component, right-click on the derived part or assembly in the Browser and select the **Edit Derived Part** or **Edit Derived Assembly** menu option. When you select one of these options, the **Derived Part** or **Derived Assembly** dialog box, previously discussed, will open, allowing you to edit the derived component characteristics.

- Unlinking derived components   As you have seen, when you edit the original component geometry used to define the derived component, the derived component can be updated to reflect the new design. However, if you do not want to continue this parametric relationship, you can break the link between the derived component and the original part. To break the connection between the two files, right-click on the derived part in the Browser and select

FIGURE 18.13 The **Update** symbol.

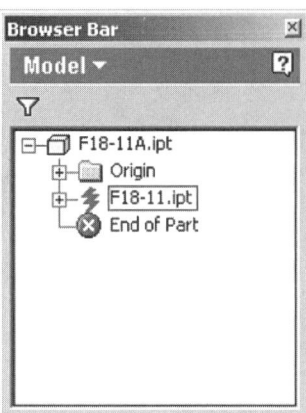

the **Break Link with Base Part** menu option for a part, or the **Break Link with Base Assembly** menu option for an assembly.

Exercise 18.3

1. Continue from Exercise 18.2 or launch Autodesk Inventor.
2. Open EX18.1.ipt, and save a copy as EX18.3.ipt.
3. Close EX18.1.ipt without saving and open EX18.3.ipt.
4. Use the **Hole, Fillet,** and **Chamfer** tools to machine the part blank similar to the part shown.
5. Resave the part file.
6. The final exercise should look like the part shown.
7. Exit Autodesk Inventor or continue working with the program.

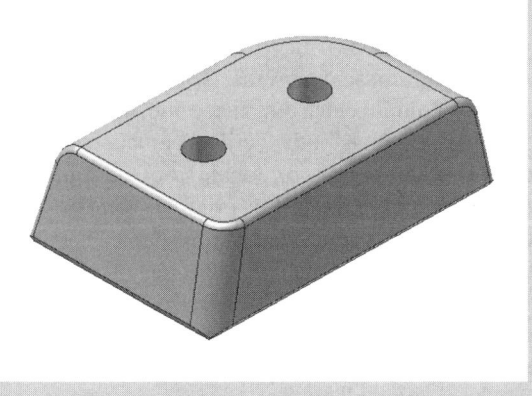

Exercise 18.4

1. Continue from Exercise 18.3 or launch Autodesk Inventor.
2. Open EX18.2.ipt and save a copy as EX18.4.ipt.
3. Close EX18.2.ipt without saving and open EX18.4.ipt.
4. Open a sketch on the **XY Plane**.
5. Project the **Center Point**.
6. Sketch a ∅.25″ hole fixed to the projected **Center Point**.
7. Cut-extrude the sketched circle through all, in both directions.
8. Access the **Circular Pattern** tool.
9. Select the cut extrusion as the feature, and the **Y Axis** as the rotation axis.
10. Specify a count of 6, and an angle of 360°.
11. Resave the part file.
12. The final exercise should look like the part shown.
13. Exit Autodesk Inventor or continue working with the program.

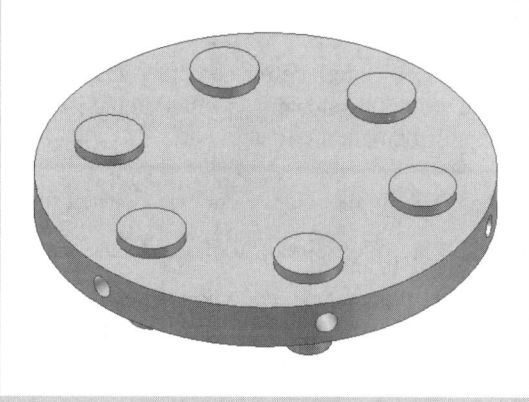

iParts

In many applications, products are designed that are very similar but have slightly different component requirements. For example, one bracket design may require a short bolt, while another similar bracket design may require a longer bolt. As you have learned throughout this text, you can create parts and assemblies quickly and easily with Autodesk Inventor's parametric modeling capabilities. Using the previous example, you can create two different brackets and make a copy of the bolt. Then one of the bolt designs can be edited to reflect the different bracket style. Another option is to adapt the bolt length to the bracket style to ensure design intent. Both of these options are effective and work for many applications. However, if you want to create a more extensive part library, you are constantly creating new, slightly different assemblies, or you want to enable multiple designs to be created and shared between several users, the previous options are not always the best solutions.

As a result, *iParts* and *iPart factories* are used to generate a set of similar parts that have one or more different parameters and properties. iParts are similar to shared content parts discussed in Chapter 14, because they can potentially be accessed by multiple users and allow you to insert a part with different parameters and properties. iParts are created in a template file called an iPart factory. iPart factories allow you to specify multiple-part parameters for a single part, using a spreadsheet. For example, you may identify several possible lengths of a ∅1″ steel rod that can be used in assemblies. Once an iPart factory is created, you can insert the iPart factory into an assembly and then select from the parameters you specified when creating the iPart factory to generate an iPart, which is a single instance of the part template. Using the previous example, when you insert the iPart factory of the rod, you can select which rod length you want to use for the assembly, and the new design, or iPart, is created.

As you can see, using iParts and iPart factories is essentially a two-step process—part authoring and part publishing. The first process, known as *part authoring,* involves the following steps:

1. Create a part that will be used as a template to define each instance of the part when inserted into an assembly.

2. Use the **Parameters** tool to define part geometry, and rename parameters used in the iPart factory.

3. Access the **iPart Author** tool, and define in a table which feature, or parameter, of the part you want change for each different instance. For example, you may define five different possible standard lengths and two different standard materials of a $\varnothing 1''$ steel rod. Although only one part file is used to create an iPart, theoretically multiple parts are generated in a spreadsheet based on the different size and physical properties you specify. The group of parts, or part variations, is known as a *family-of-parts*.

4. The part, or at this point the part template, is transformed into an iPart factory.

5. Save the part file in a library that is included in the active project, because typically iParts are used in multiple assemblies.

The second process, often referred to as *publishing* an iPart, involves inserting the iPart into an assembly using the desired parameters specified in the iPart factory. This stage is also known as *part placement* and requires the following four steps:

1. Open the assembly file where you want to insert the iPart.

2. Access the **Place Component** tool, and locate the iPart factory you want to insert.

3. Use the **iPart Placement** dialog box to select which specified size and property characteristics of the iPart you want to use in the assembly. For example, you can select one of the five $\varnothing 1''$ steel rod lengths, and one of the two materials.

4. Pick the location for the part in the graphics window.

Like other Autodesk Inventor tools, the uses for iParts vary greatly depending on the application. The following represent a number of options for using iParts:

- Develop parts libraries.

- Create a single part file with multiple versions.

- Create an external spreadsheet identifying the iPart parameters.

- Control designs by managing part versions and allowing only certain parts, with specified parameters, to be used in an assembly.

Developing iPart Templates

As mentioned, the first step to creating an iPart is to develop an iPart factory. The initial step in creating an iPart factory is to build a part just as you would develop any other part. In fact, an existing part can be transformed into an iPart factory at any time. The part used to generate an iPart factory is basically a template. This means that the part acts as a guide for developing multiple instances depending on the variable parameters that you chose to specify. Many of the dimensions, parameters, properties, and positioning information used to create the part will remain constant, while some parameters are modified when the iPart is placed in an assembly. For example, Figure 18.14 shows a section of L3×3×1/4×12 angle iron, which contains many parameters that define its length, width (leg), height (leg), thickness, and edge radius.

The part shown in Figure 18.14 will be used to create an iPart factory for L3×3×1/4 angle iron. In this example, the lengths will vary depending on the assembly requirement, because when the iPart factory is created, several possible lengths are specified, including 24″. As a result, all part parameters, except the length, will remain constant when the iPart is placed in the assembly. This means that you can define the

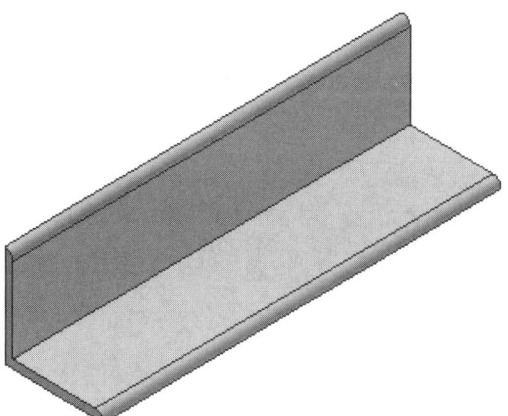

FIGURE 18.14 An example of part used to create an iPart factory.

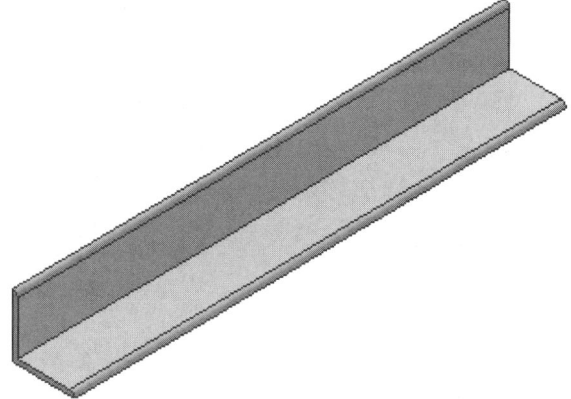

FIGURE 18.15 When you insert the part into an assembly, you can choose from a range of specified lengths, such as 24″ as shown. The length of 12″ used to create the part is disregarded.

length of the part using one of the available lengths, because when inserted into an assembly, the length of the part will change depending on the length you choose to use. See Figure 18.15.

If you do not define part parameters while you are in the process of building a part, you may want to specify parameters once the part is created, because parameter names can be very important for working with iParts and iPart factories. As discussed in previous chapters, parameters are size and structural limits placed on sketches and features, such as sketch dimensions or feature size. To access the **Parameters** dialog box/spreadsheet, select the **Parameters** button from the **Standard** toolbar, or choose the **Parameters** option from the **Tools** pull-down menu. A model parameter is automatically created every time you dimension a sketched object or build a feature, such as an extrusion.

Although you do not have to rename parameters for them to be used in an iPart factory, renamed parameters are easier to understand. In addition, as you will see, named parameters are automatically added to the iPart factory list of parameters when you create an iPart factory. Figure 18.16A shows the default parameters for the section of angle iron shown in Figure 18.14. These parameters are difficult to recognize, as compared to the parameters shown in Figure 18.16B, which have been named.

 Field Notes

> You will especially want to rename the parameter or parameters that will change when you insert the iPart factory into an assembly.

In addition to renaming model parameters, any properties you specify in the **Properties** dialog box, and any iMates you create, can be used in the iPart factory. Although you can define properties and add iMates after you create an iPart factory by editing the iPart factory specifications, you may want to establish these options before you create the iPart factory.

Working with properties and iMates for use with iParts is the same as when working with any other part model application. However, when you define properties and place iMates, these specifications can be included when you generate an iPart factory. Then properties can be assigned for a part when an iPart is inserted into an assembly, and iMates can constrain parts together, when another component, with corresponding iMates, is inserted.

FIGURE 18.16 The
Parameters dialog box
with (A) default
parameter names and
(B) specified parameter
names.

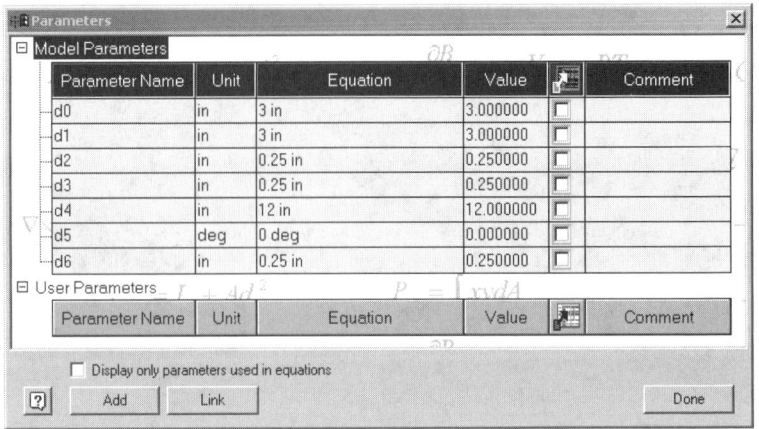

A

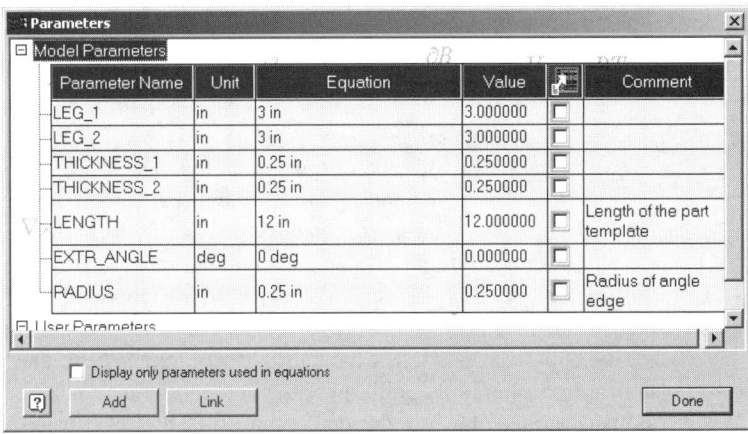

B

Exercise 18.5

1. Continue from Exercise 18.4 or launch Autodesk Inventor.
2. Open a new inch part file.
3. Open a sketch on the **XY Plane.**
4. Project the **Center Point.**
5. Sketch a 1″×2″ rectangle.
6. Use the **Offset** tool to offset the rectangle .125″ as shown.
7. Finish the sketch.
8. Extrude the sketch 12″ in a positive direction.
9. Open a sketch on the top face of the extrusion, and sketch the geometry shown.
10. Cut-extrude the sketch through all as shown.
11. Access the **Parameters** dialog box, and rename the parameters as shown.
12. Open the Properties dialog box and specify a part number of EX18.5 and a material of Aluminum-6061.
13. Save the part file as EX18.5.ipt.
14. Exit Autodesk Inventor or continue working with the program.

Exercise 18.5 continued on next page

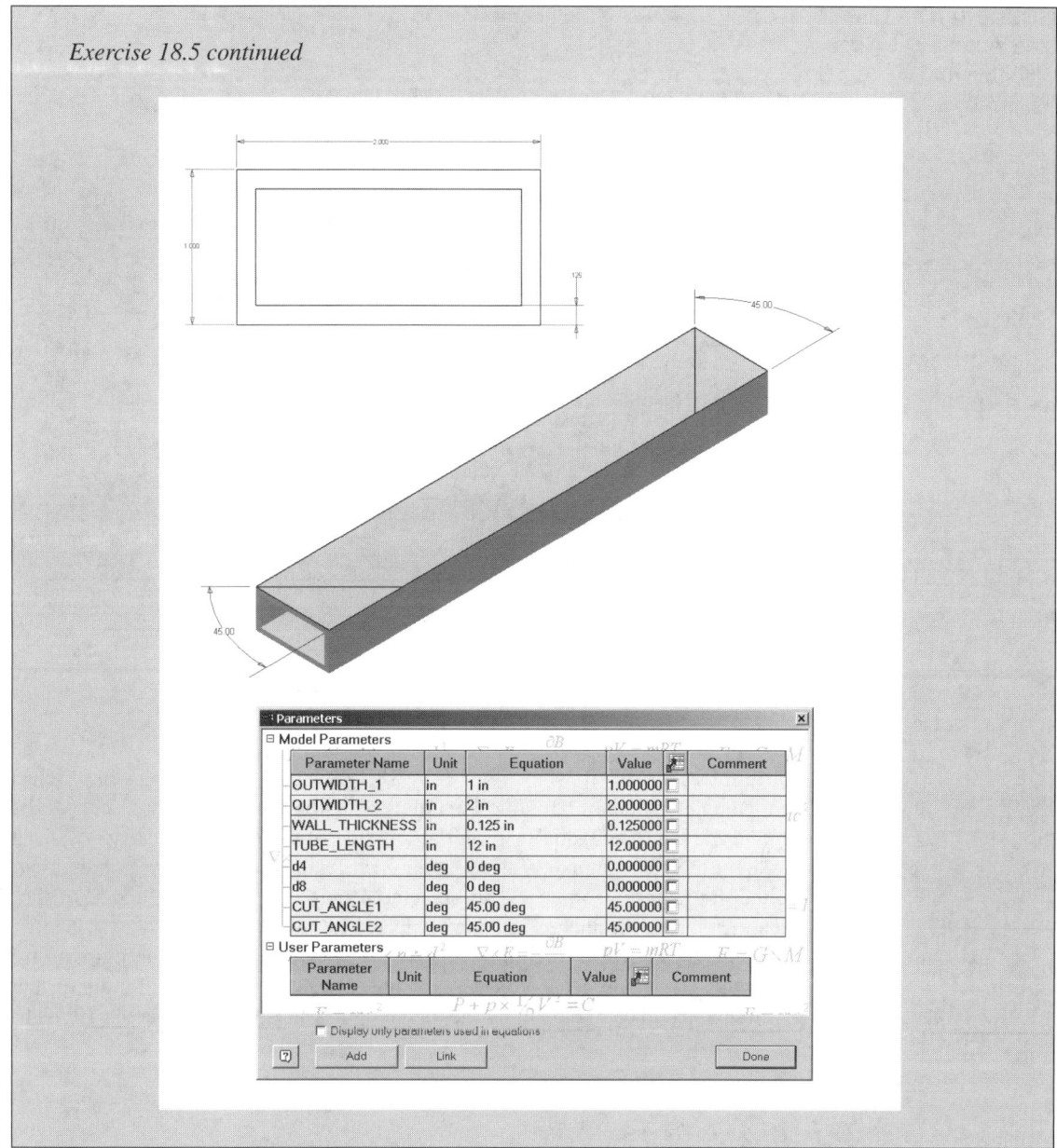

Exercise 18.5 continued

Creating iPart Factories

Once you have created a part that you will use as a template and transform into an iPart factory, you can begin the process of generating the iPart factory. Assuming the part file is opened, create the iPart factory by accessing the **iPart Author** command through the **iPart Author** tool button on the **Standard** toolbar. When you access the **iPart Author** tool, the **iPart Author** dialog box is displayed. See Figure 18.17.

The **iPart Author** dialog box has tabs and options that allow you to fully define an iPart factory. In general, this dialog box allows you to identify which model parameters, properties, features, iMates, and custom values you want to use in the iPart factory. Each of these preferences is specified using the corresponding tab. As you define iPart information using the tabs, columns that relate to the tab items are created in the iPart Table, in addition to a single row, which contains column values or cells. Then you can add rows to the iPart Table to define additional part values for defining different instances of the iPart when inserted into an assembly.

The following information describes the function of each of the tabs located in the **iPart Author** dialog box:

FIGURE 18.17 The **iPart Author** dialog box (**Parameters** tab displayed).

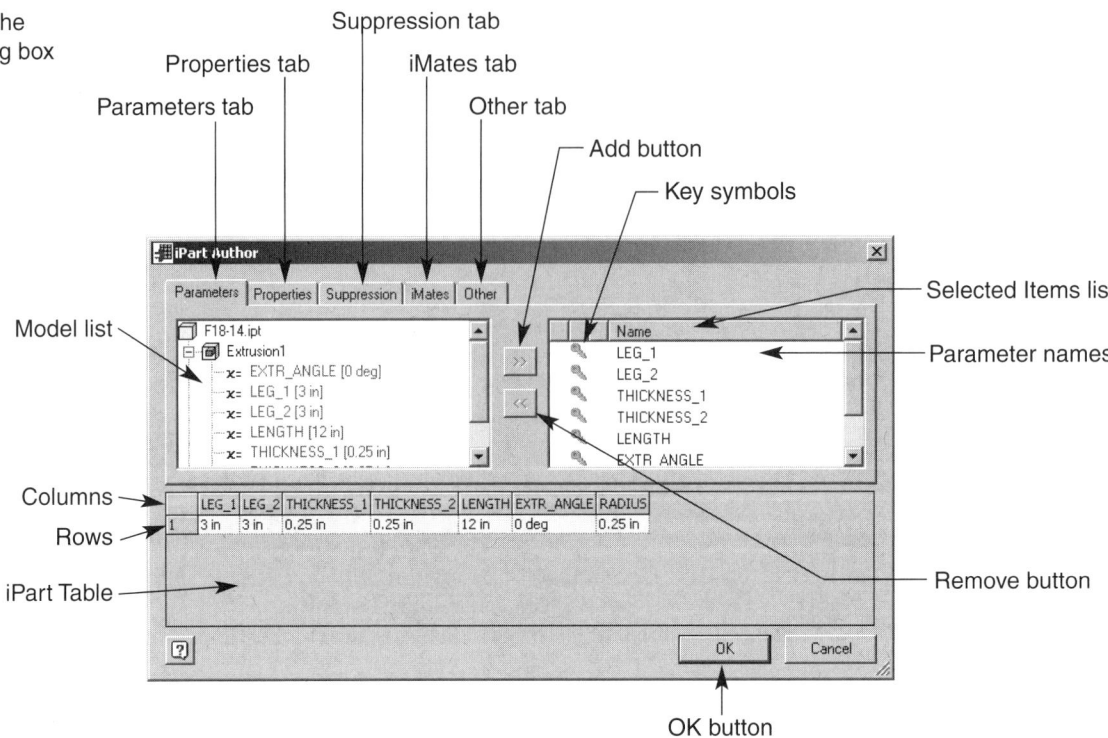

- **Parameters**   This tab, shown in Figure 18.17, displays and allows you to select model feature parameters identified in the **Parameters** dialog box. As you can see, two lists are available in the tab. The **Model** list, on the left side of the dialog box, displays the features that make up the part template. Listed under each feature are the corresponding parameters that define the feature geometry. The **Selected Items** list, on the right side of the dialog box, identifies which parameters you have selected to use for the iPart factory. Again, the selected items are typically those features you want to manipulate when the iPart is inserted into an assembly. For example, if you want to have various lengths of aluminum tube, you must ensure the parameter controlling the tube length is listed in the **Selected Items** list.

Field Notes

Even if a parameter value does not vary between instances, such as the diameter of an aluminum tube, you must include the parameter in the Selected Items list if you want the parameter to be available in assembly bills of materials and drawings.

As previously discussed, if you change the default names of parameters in the **Parameters** dialog box, the named parameters are automatically added to the iPart factory **Selected Items** list when you access the **iPart Author** dialog box. To remove a parameter in the **Selected Items** list, select the parameter and then pick the **Remove** button or right-click and select the **Delete Column** menu option. Similarly, to add a parameter from the **Model** list to the **Selected Items** list, select the parameter you want to add in the **Model** list and pick the **Add** button, or double-click on the item. The name of the parameter in the **Selected Items** list is defined in the **Parameters** dialog box and corresponds to the column header in the iPart Table, as you can see in Figure 18.17.

You will notice that there are key symbols displayed next to each of the parameters. Keys identify which parameters, properties, features, iMates, and custom items you want to use to define the part instance when you place an iPart. For example, when working with parameters, if you want to create a ⌀1″ diameter steel pipe with multiple lengths to select from, you can key the length parameter. You can specify nine keys in the entire iPart factory. This means that the nine keys can be spread throughout the four different tabs. Each tab does not have nine separate keys. Keys are listed in order when you insert the iPart factory into an assembly. Key 1 is known as the primary key, while keys 2 through 8 are secondary keys. To identify the primary key in the parameter tab, pick the key symbol or right-click on the parameter in the **Selected Items** list, and select the key 1 from the **Key** cascading submenu. If you select another parameter key symbol, key 2 is defined. Additional keys can also be specified by right-clicking on the parameter in the **Selected Items** list and selecting the desired key from the **Key** cascading submenu. Selected keys are blue and have a key number next to them. Remove a key by selecting the key symbol a second time, or by right-clicking on the parameter in the **Selected Items** list, and select the **Not A Key** option from the **Key** cascading submenu.

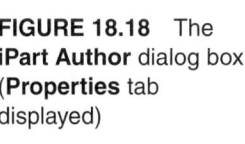

Field Notes

If you do not specify any keys, all parameters will be listed when you publish an iPart factory.

- **Properties** This tab, shown in Figure 18.18, displays and allows you to select model feature properties identified in the **Properties** dialog box. As with the **Parameters** tab, two lists are available in the **Properties** tab. The **Model** list, on the left side of the dialog box, displays the part model and all the model properties available and defined in the **Properties** dialog box. Listed under each property type is each related property. The **Selected Items** list, on the right side of the dialog box, identifies which properties you have selected to include in the iPart factory. Again, the selected items are typically those properties you want to manipulate when the iPart is inserted into an assembly. For example, if you want to have brackets made with various materials, you must ensure the physical property controlling the bracket material is listed in the **Selected Items** list.

Field Notes

Even if a parameter value does not vary between instances, such as the part number, you must include the property in the **Selected Items** list if you want the property to be available in assembly bills of materials and drawings.

FIGURE 18.18 The **iPart Author** dialog box (**Properties** tab displayed)

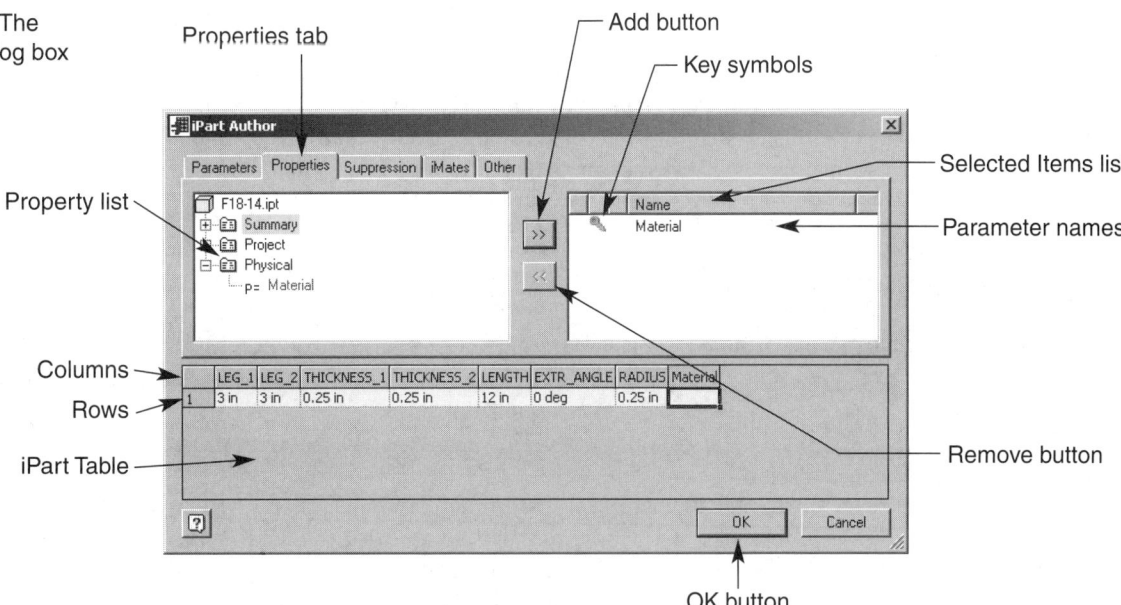

To add a property from the **Model** list to the **Selected Items** list, select the property you want to add in the **Model** list and pick the **Add** button, or double-click on the item. The name of the property in the **Selected Items** list is defined by in the **Properties** dialog box and corresponds to the column header in the iPart Table, as you can see by the **Material** column in Figure 18.18. If the property has a value such as aluminum for a material property, the specified value is listed in the table cell or cells. To remove a property in the **Selected Items** list, select the property, and then pick the **Remove** button or right-click on the property and select the **Delete Column** menu option.

As with parameters, described earlier, there are key symbols displayed next to each of the properties. These keys identify which parameters, properties, features, iMates, and other items you want to use to define the part instance when you place an iPart. For example, when working with properties, if you want to create a plate with multiple materials to select from, you can key the material property. Again, you can specify a total of nine keys in the entire iPart factory. So, to identify the primary key in the **Properties** tab, pick the key symbol or right-click on the parameter in the **Selected Items** list, and select key 1 from the **Key** cascading submenu. If you select another parameter key symbol, key 2 is defined, assuming no other keys have been established. Additional keys can also be specified by right-clicking on the parameter in the **Selected Items** list and selecting the desired key from the **Key** cascading submenu. Remove a key by selecting the key symbol a second time, or by right-clicking on the parameter in the **Selected Items** list, and select the **Not A Key** option from the **Key** cascading submenu.

 Field Notes

> If you do not specify any keys, all properties will be listed when you publish an iPart factory.

- **Suppression**   This tab, shown in Figure 18.19 displays and allows you to select specific features that you want to suppress or compute for part instances. A suppressed feature is "turned off," and not displayed, while a computed feature is "turned on," and shown when the part is inserted into an assembly.

  As with the **Parameters** and **Properties** tab, two lists are available in the **Suppression** tab. The **Model** list, on the left side of the dialog box, displays the part model and model features. "Compute" or "Suppress" is listed next to the features in the **Model** list depending on whether the feature is suppressed in the model. When a feature is computed, it means that it is calculated in the creation of the part and is actively displayed.

  The **Selected Items** list, on the right side of the dialog box, identifies which part features you have selected to include in the iPart factory. Again, the selected items are typically those features you want to manipulate when the iPart is inserted into an assembly. For example, if you want to suppress a certain feature when you insert the iPart, you must ensure the fillet feature is listed in the **Selected Items** list.

 Field Notes

> Even if a feature's suppression value does not vary between instances, you must include the feature in the **Selected Items** list if you want the feature suppression information to be available in assembly bills of materials and drawings.

To add a feature from the **Model** list to the **Selected Items** list, select the feature you want to add in the **Model** list and pick the **Add** button, or double-click on the item. The name of the feature in the **Selected Items** list is defined when you create the feature and corresponds to the column header in the iPart Table, as you can see by the **Fillet** column in Fig-

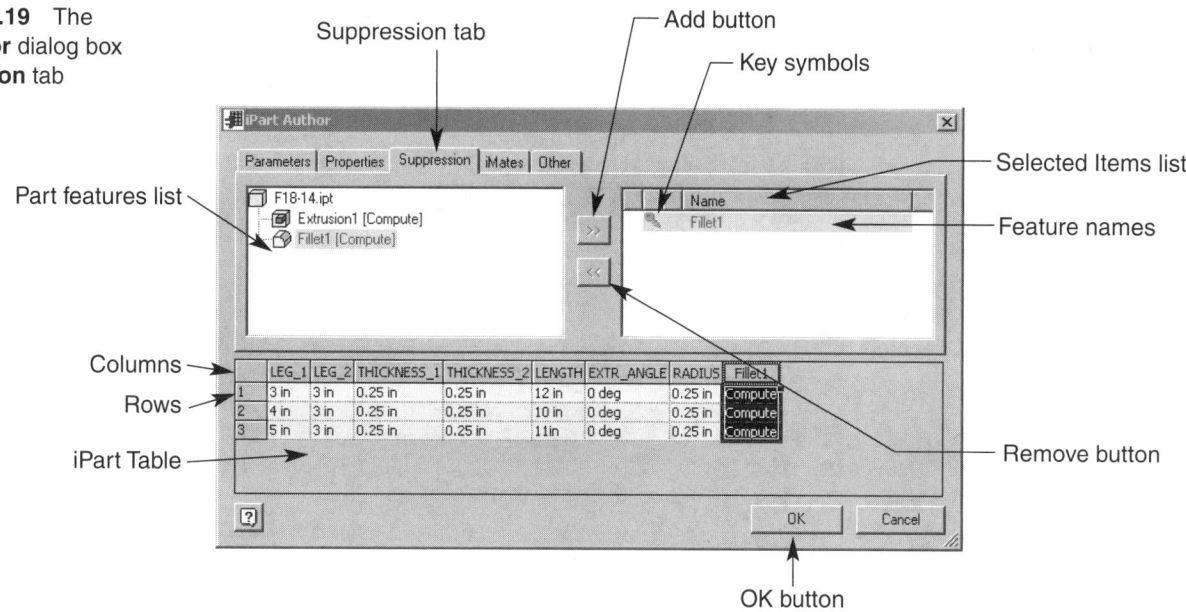

FIGURE 18.19 The **iPart Author** dialog box (**Suppression** tab displayed).

ure 18.19. If the feature is suppressed or calculated, the specified value is listed in the table cell or cells. To remove a feature in the **Selected Items** list, select the feature, and then pick **Remove** or right-click on the feature and select the **Delete Column** menu option.

As with parameters and properties, described earlier, there are key symbols displayed next to each of the selected features. These keys identify which parameters, properties, feature, iMates, and custom items you want to use to define the part instance you place an iPart. For example, when working with feature suppressions, if you want to create a plate with the option of suppressing a chamfer, you can key the chamfer feature. Again, you can specify a total of nine keys in the entire iPart factory. So, to identify the primary key in the **Suppression** tab, pick the key symbol or right-click on the feature in the **Selected Items** list, and select key 1 from the **Key** cascading submenu. If you select another suppression key symbol, key 2 is defined, assuming no other keys have been established. Additional keys can also be specified by right-clicking on the feature in the **Selected Items** list and selecting the desired key from the **Key** cascading submenu. Remove a key by selecting the key symbol a second time or by right-clicking on the feature in the **Selected Items** list, and select the **Not A Key** option from the **Key** cascading submenu.

 Field Notes

If you do not specify any keys, all features will be listed when you publish an iPart factory.

■ **iMates**   This tab, shown in Figure 18.20, displays and allows you to select specific iMates added to the part that you want to use to create part instances when placing iParts. When you create an iMate, as discussed in Chapter 14, you specify an offset. In addition, the iMate can be suppressed, computed, or renamed, and it has a sequence number. Each of these components can be manipulated and used when the iPart is inserted into an assembly.

As with the **Parameters, Properties,** and **Suppression** tabs, two lists are available in the **iMates** tab. The **Model** list, on the left side of the dialog box, displays the part model and model iMates. Listed under each iMate is the offset, suppression status, name, and sequence number. The iMate offset is the offset value you enter when creating the iMate and the suppression is either set as "Compute" or "Suppress." Also, the iMate name is set as default in a sequence when you include the iMate in the **Selected Items** list, with the

FIGURE 18.20   The
iPart Author dialog box
(**iMates** tab displayed).

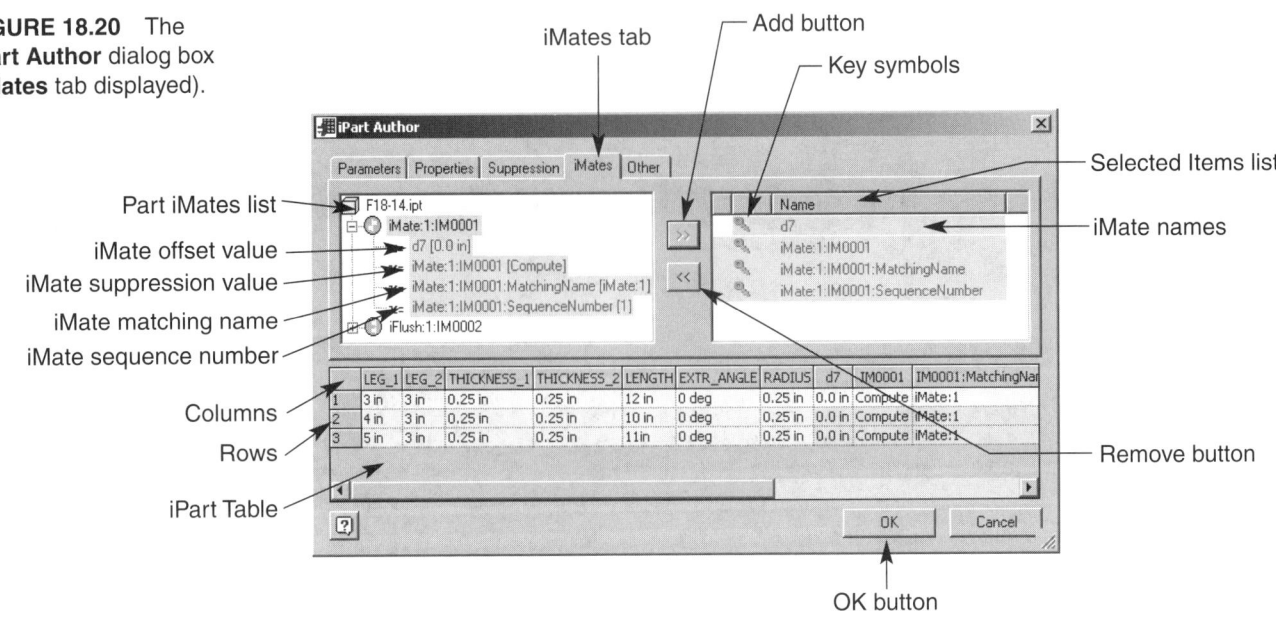

sequence number indicating the order in which the iMates will constrain. Although you cannot edit the default name, you can add information to the name such as for clarity.

Field Notes

iMate suppression can also be defined in the **Suppression** tab.

The **Selected Items** list, on the right side of the dialog box, identifies which part iMates, or iMate characteristics, you have selected to include in the iPart factory. Again, the selected items are typically those iMates or iMate properties you want to manipulate when the iPart is inserted into an assembly. For example, if you want to use a certain iMate to constrain the iPart to another feature when you insert the iPart, you must ensure the iMate is listed in the **Selected Items** list.

Field Notes

Even if an iMate does not vary between instances, you must include the iMate in the **Selected Items** list if you want the iMate information to be available in assembly bills of materials and drawings.

To add an iMate or iMate characteristic from the **Model** list to the **Selected Items** list, select the iMate or iMate attribute you want to add in the **Model** list and pick the **Add** button, or double-click on the item. The name of the iMate in the **Selected Items** list is defined when you create and edit the iMate and corresponds to the column header in the iPart Table, as you can see in Figure 18.20. The iMate model values are listed in the table cell or cells. To remove an iMate or iMate attribute from the **Selected Items** list, select the item, and then pick the **Remove** button, or right-click on the iMate and select the **Delete Column** menu option.

As with parameters, properties, and suppressions described earlier, there are key symbols displayed next to each of the iMate items. These keys identify which parameters, properties, feature, iMates, and custom items you want to use to define the part instance when you place an iPart factory. For example, when working with iMates, if you want to insert a bolt that constrains to a hole, you can key the bolt iMate. Again, you can specify a total of nine keys in the entire iPart factory. So, to identify the primary key in the **iMate** tab, pick the key symbol or right-click on the iMate in the **Selected Items** list, and select key 1 from the **Key** cascading submenu. If you select another iMate key symbol, key 2 is defined, assuming no other keys have been established. Additional keys can also be specified by right-clicking on the iMate in the **Selected Items** list, and selecting the desired key from the **Key** cascading submenu. Remove a key by selecting the key symbol a second time, or by right-clicking on the feature in the **Selected Items** list, and select the **Not A Key** option from the **Key** cascading submenu.

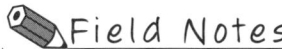Field Notes

If you do not specify any keys, all iMates will be listed when you publish an iPart factory.

■ **Other** This tab, shown in Figure 18.21, displays and allows you to create any other options to use to create part instances when placing iParts. Most part attributes are controlled by the specifications in the **Parameters, Properties, Suppression,** and **iMates** tabs of the **iPart Author** dialog box. Any other custom attributes that cannot be defined in the other four tabs can be created in the **Other** tab. For example, maybe you want to define various clients, departments, assembly information for the part, or any other possible required options.

To create a custom item, pick the **Click here to add value** button. As shown in Figure 18.22, a default "NewItem" name is added to the **Item** list, and a corresponding column with the same header name is created. To rename the item, ensure the **Edit** symbol is displayed by selecting the current item name. Enter the name of the custom attribute, then select the **Edit** symbol to apply the changes. You can remove a custom item by right-clicking on the item and selecting the **Delete Column** menu option.

Field Notes

Even if a custom attribute does not vary between instances, you must include the iMate in the **Selected Items** list if you want the item to be available in assembly bills of materials and drawings.

FIGURE 18.21 The **iPart Author** dialog box (**Other** tab displayed).

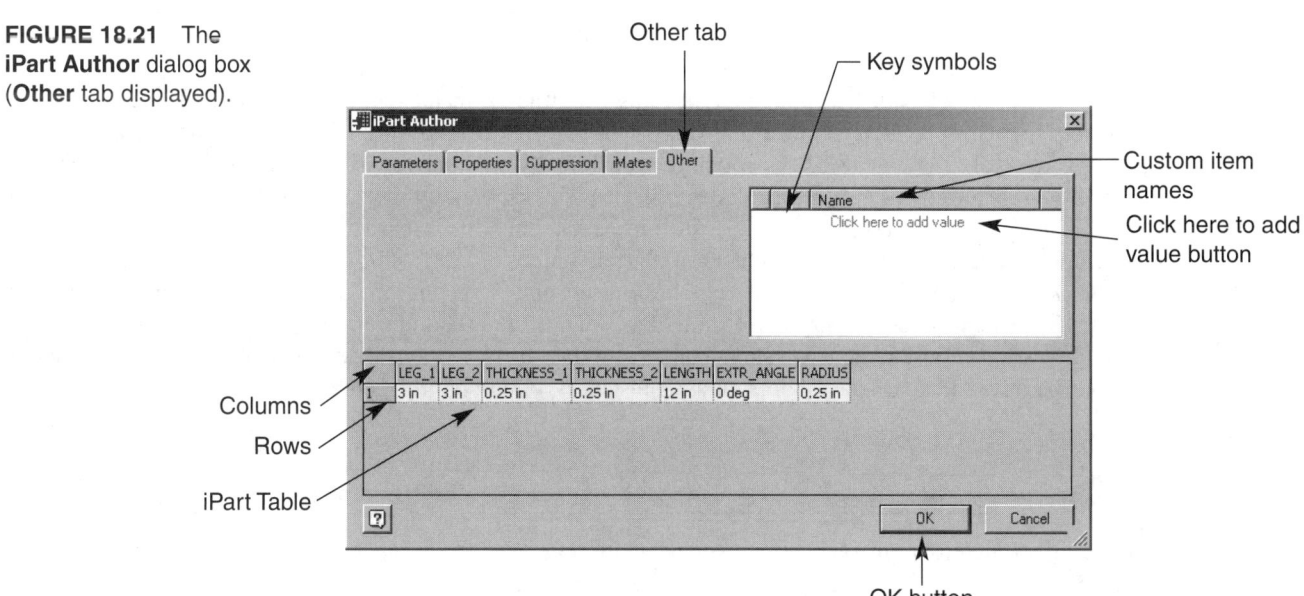

FIGURE 18.22
Creating a custom item.

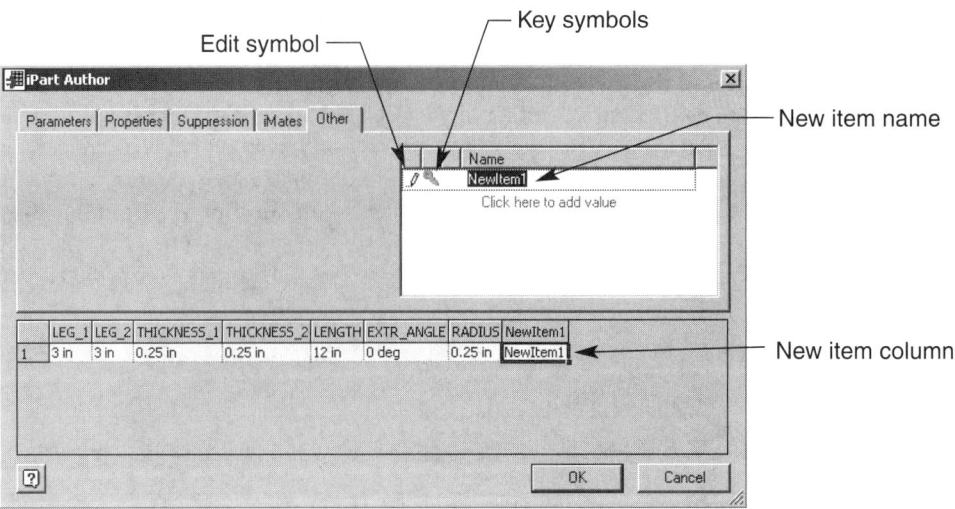

As with parameters, properties, suppressions, and iMates described earlier, there are key symbols displayed next to each of the custom items. These keys identify which parameters, properties, feature, iMates, and custom items you want to use to define the part instance when you place an iPart factory. For example, when working with a custom item named "Client," if you want to have multiple client names available, you can key the "Client" custom item. Again, you can specify a total of nine keys in the entire iPart factory. To identify the primary key in the **Other** tab, pick the key symbol or right-click on the item in the **Items** list, and select key 1 from the **Key** cascading submenu. If you select another item key symbol, key 2 is defined, assuming no other keys have been established. Additional keys can also be specified by right-clicking on the item in the **Items** list and selecting the desired key from the **Key** cascading submenu. Remove a key by selecting the key symbol a second time, or by right-clicking on the item in the **Items** list, and select the **Not A Key** option from the **Key** cascading submenu.

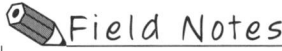Field Notes

If you do not specify any keys, all custom items will be listed when you publish an iPart factory.

Once you have fully identified all the items you want to manipulate when inserting an iPart, using the options in each of the **iPart Author** dialog box tabs you are ready to work with the iPart factory table, shown in Figures 18.17 through 18.22. As you have seen, initially only one row is created based on the options you specify in each of the **iPart Author** dialog box tabs. You can modify the values in each of the cells to reflect the desired part attributes. Still, a single row means that only one instance of the part can be inserted into an assembly file. For example, if you create an iPart of a piece of L3×3×1/4×12 angle iron, and only one row with no value range is set, you will be able to insert only a section of L3×3×1/4×12 angle iron into an assembly, which is not the purpose of using iParts. iParts are used to insert multiple types, or instances, of a similar part. As a result, the next step in defining an iPart factory is to add additional rows to the iPart factory table, because additional rows define multiple instances, or part family members.

Field Notes

If you do not specify a unit of measurement in a table cell, the default document units are applied.

To create another iPart factory table row, right-click on the existing row and select the **Insert Row** menu option. A copy of the row you selected with all the existing cell values will be added to the iPart factory table. See Figure 18.23. Just as with the initial table row, you can modify the values in each of the cells to reflect the desired part instance attributes. For example, if you want one part length instance option to be 12″ and another part length instance option to be 18″, enter 12″ in one of the length cells of the first row and 18″ in the other length cell of the second row. You can enter any value you want in the table cells. However, suppression values require certain expressions. To specify a feature as suppressed, you can enter "Suppress", S, s, OFF, Off, off, or O. To identify a feature as computed, you can enter Compute, U, u, C, c, ON, On, on, or 1. Continue adding rows and defining cell values as required, to generate the desired number of iPart instances. In addition, you can delete a cell row by right-clicking on the row you want to remove, and selecting the **Delete Row** menu option.

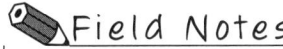 Field Notes

To define a row as the default instance of the iPart for inserting into a file, right-click and select the **Set as Default Row** menu option.

When you specify a certain value in a cell such as 3 or 5, these values are considered standard and cannot be modified when an iPart is published in an assembly. However, you can also identify custom values, which can be defined as any value, or a specific range of values, when the iPart is inserted into an assembly. You can establish a custom column in which all cell values can be modified, or a custom cell, in which only the specified cell value can be modified.

To create a custom column, right-click on the column header in the iPart factory table, or the column name in the **Selected Items** list. Then pick the **Custom Parameter Column** menu option. Now, when you insert the iPart into an assembly, you can specify any value for the parameter and create any size part based on the column information. Another option is to specify a certain range of values that can be used to build the part. To identify a range, right-click on the column header in the iPart factory table, or the column name in the **Selected Items** list that has been established as a custom parameter column. Then pick the **Specify Range For Column...** menu option to display the **Specify Range** dialog box. See Figure 18.24. The **Speci-**

FIGURE 18.23
Creating an additional iPart table row.

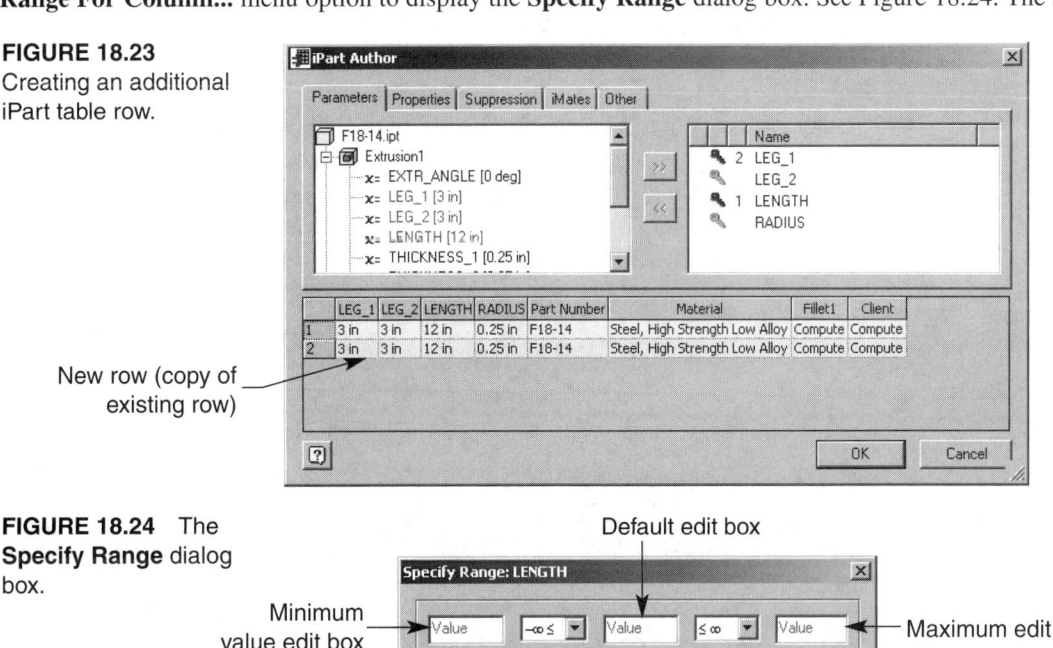

New row (copy of existing row)

FIGURE 18.24 The **Specify Range** dialog box.

Minimum value edit box · Default edit box · Maximum edit box · OK button

fy Range dialog box allows you to define a minimum limit, less than a specified value, less than or equal to a specified value, or infinitely less than the default value. You can also define a maximum limit, greater than a specified value, greater than or equal to a specified value, or infinitely greater than the default value. Finally, you can modify the default value in the **Default** edit box. When you have established a range of values, pick the **OK** button to exit the **Specify Range** dialog. Only the range of values you specify will be available when an iPart is inserted.

The process for creating a custom cell is similar to creating a custom column. To define a custom cell, right-click on the cell in the iPart factory table and pick the **Custom Parameter Cell...** menu option. Now when you insert the iPart into an assembly, you can specify any value for the cell parameter and create any size part based on the cell information. Another option is to specify a certain range of values that can be used to build the part. To identify a range, right-click on the cell in the iPart factory table that has been established as a custom parameter cell. Then pick the **Specify Range For Column...** menu option to display the **Specify Range** dialog box (see Figure 18.24).

Field Notes

You cannot specify a column or cell range for certain items including some properties, suppressions, iMates, and some custom items.

Custom columns and cells have blue backgrounds.

In addition to defining a custom column or cell, the following options are available when you right-click on a property, iMate, or custom column, and further allow you to control the instance of iParts inserted into an assembly:

- **File Name Column**   This option is primarily used for custom columns and allows you to assign the file name to the iPart when it is published in the assembly.

- **Display Style Column**   This option allows you to assign the color specified in the iPart factory to the iPart when it is published in the assembly.

- **Material Column**   Use this option to assign the default material, specified in the iPart factory, when you place the iPart in an assembly.

Columns can be removed using the options previously discussed, or by right-clicking on the column you want to remove and selecting the **Delete Column** menu option.

Once you have fully defined all the preferences in the **iPart Author** dialog box, pick the **OK** button to generate the iPart factory.

Exercise 18.6

1. Continue from Exercise 18.5 or launch Autodesk Inventor.
2. Open EX18.5ipt, and save a copy as EX18.6ipt.
3. Close EX18.5ipt without saving and open EX18.6ipt.
4. Access the **iPart Author** dialog box.
5. All the named parameters should be in the **Parameters Selected Items** list.
6. Pick the **Properties** tab, and expand "Physical."
7. Add "Material [Aluminum-6061]" to the **Selected Items** list.
8. Key material in the **Selected Items** list.
9. Right-click on the row in the iPart factory table and select the **Insert Row** menu option.
10. Continue inserting rows until there is a total of four rows.
11. Change the material in rows 2 and 4 to Steel, Mild.
12. Change the wall thickness in rows 3 and 4 to .25.
13. Right-click on the Tube_Length column and select the **Custom Parameter Column** menu option.
14. Right-click on the Tube_Length column again, and select the **Specify Range for Column** menu option.
15. Select the less than or equal to option, and enter a maximum value of 96.
16. Pick the OK button to generate the iPart factory.
17. Resave the part file.
18. Exit Autodesk Inventor or continue working with the program.

Working with iPart Factories

There are a number of Browser options related to working with iPart factories once they are created. Figure 18.25 shows an example of the Browser display for an iPart factory. Notice the iPart factory name and icon is listed first, followed by the iMates folder, if you placed iMates. Then the iPart factory table is shown. Listed under the table are all the iPart instances. As you can see in Figure 18.25 this iPart factory has three instances, which means there are three rows in the table. If keys are specified in the iPart factory, only the keyed attributes will be listed under the table. For example, in the iPart factory shown in Figure 18.25, "Length" was set as the primary key and was the only key identified. As a result, only "Length" is shown in each of the rows. If additional keys are established, the keyed attributes will also be available in the Browser. If no keys are established, all the attributes are available.

The check boxes next to row specifications indicate the iPart currently displayed in the graphics window. To observe a different part instance, double-click on the last attribute. For example, because "Length" is the only keyed trait shown in Figure 18.25, you can double-click on "Length – 6 in" to observe the 6″ part.

FIGURE 18.25 An example of the Browser display for an iPart factory.

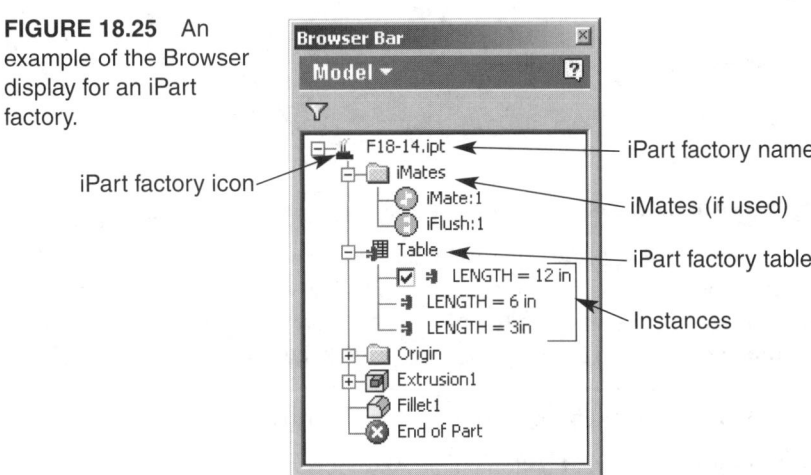

FIGURE 18.26 An example of an embedded Microsoft Excel spreadsheet created when you generate the iPart factory.

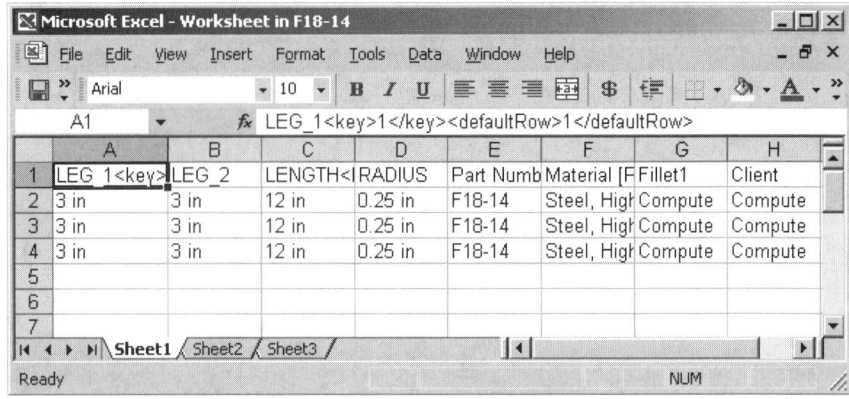

There are two options for editing iPart factories. One method is to right-click on the iPart factory table in the Browser and select the **Edit Table...** menu option. Using this technique reopens the iPart Author dialog box, allowing you to manipulate the initial iPart factory specifications. When you generate an iPart factory, a corresponding spreadsheet is also created. As a result, another editing method is to right-click on the iPart factory table in the Browser and select the **Edit via Spread Sheet...** menu option to access the embedded Microsoft Excel spreadsheet, as shown in Figure 18.26.

The final step in creating an iPart factory is to save the part file. Often iParts are used in multiple assemblies. Consequently, you should save the iPart factory in the active project.

Field Notes

> For more information regarding spreadsheets, you may want to access the Autodesk Inventor or Microsoft Excel help files.

Exercise 18.7

1. Continue from Exercise 18.6 or launch Autodesk Inventor.
2. Open EX18.6.ipt.
3. Right-click on the iPart factory table in the Browser, and select the **Edit Table...** menu option.
4. Pick the **OK** button.
5. Right-click on the iPart factory table in the Browser, and select the **Edit via Spread Sheet...** menu option.
6. Explore, and then close the spreadsheet.
7. Close EX18.6.ipt without saving.
8. Exit Autodesk Inventor or continue working with the program.

Placing iParts in Assemblies

Once you have created an iPart factory, you are ready to begin placing iParts in assemblies. This first step is to open a new assembly file or an existing assembly file that you want to use to receive the iPart. Next, use the **Place Component** tool, discussed in Chapter 14, to locate and insert the iPart factory. To review, access the **Place Component** tool using one of the following techniques:

✓ Pick the **Place Component** button on the **Assembly** panel bar.

✓ Pick the **Place Component** button on the **Assembly** toolbar.

✓ Type the **+** and the **P** keys on your keyboard.

✓ Select the **Existing Component...** option of the **Insert** pull-down menu.

When you access the **Place Component** tool, the **Open** dialog box, shown in Figure 18.27, is displayed. Use the **Open** dialog box to locate the desired iPart factory and select the **Open** button. As shown in Figure 18.28 if you select a standard iPart to insert, the **Place Standard iPart** dialog box is displayed, and the default part instance is shown and attached to the cursor. Again, a standard iPart has all possible part values fully specified and does not have any custom columns. In contrast, if you select a custom iPart to insert, the **Place Custom iPart** dialog box is displayed, as shown in Figure 18.29, and the default part instance is shown and attached to the cursor. As described, a custom iPart has part values that are not defined, or has a specified range of possible part values.

To use the **Place Standard iPart** dialog box and publish a standard iPart, select a value of the desired iPart by picking the default value in the **Keys** tab. When you select a value in the **Keys** tab, the **Select Value** list box is displayed. See Figure 18.30. To show all the specified values pick the **All Values** check box. When you select a value, the **Select Value** list box closes. To exit the **Select Value** list box without choosing a value, press the **Esc** key on your keyboard. Then, continue defining the part instance by selecting val-

FIGURE 18.27 The **Open** dialog box.

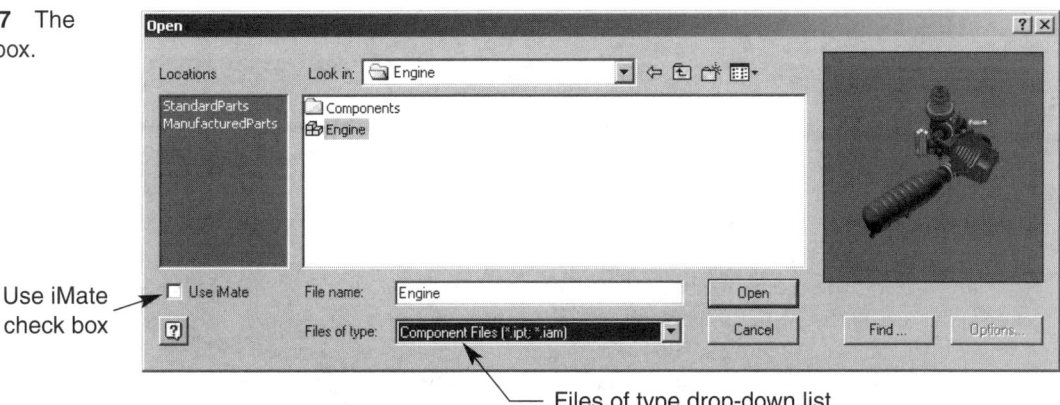

FIGURE 18.28 The **Place Standard iPart** dialog box (**Keys** tab shown).

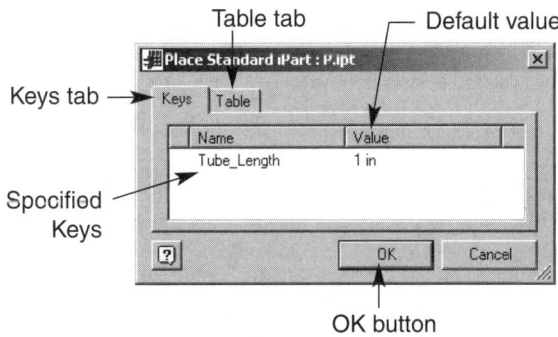

FIGURE 18.29 The **Place Custom iPart** dialog box (**Keys** tab shown).

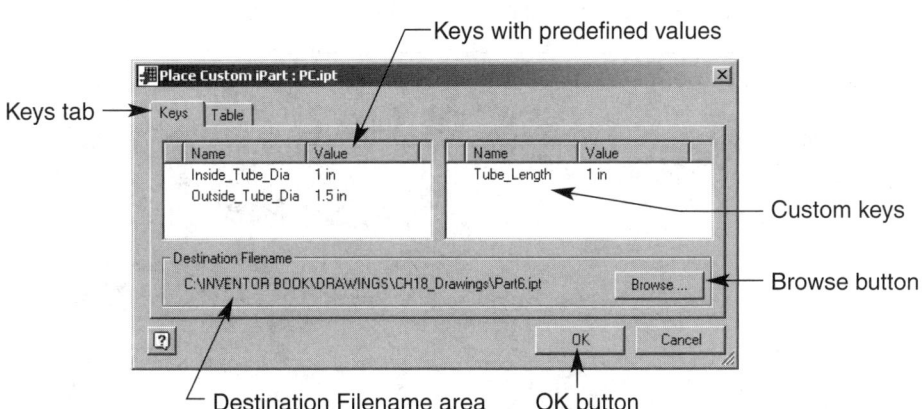

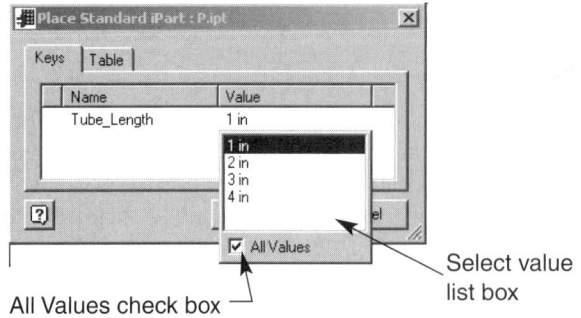

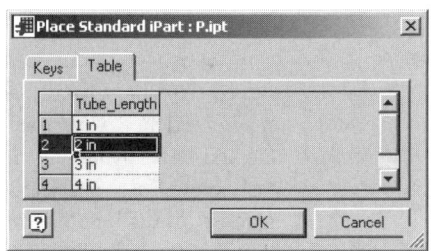

All Values check box

Select value
list box

FIGURE 18.30 The **Select Value** list box.

FIGURE 18.31 The **Place Standard iPart** dialog box (**Table** tab shown).

ues in the **Keys** tab until you have fully established the iPart parameters. Another option for assigning iPart values is to select the **Table** tab, shown in Figure 18.31. This tab displays the same table shown in the **iPart Author** dialog box, allowing you to select a row to define the iPart instance. To reorder the rows, right-click and select the **Sort Ascending** or **Sort Descending** menu option.

The next step in creating a standard iPart is to pick a location for the part in the graphics window. Then, you can continue placing as many iParts as needed, including different instances by selecting different values in the **Keys** tab, or rows in the **Table** tab. When you have finished placing iParts using the current iPart factory, pick the **Dismiss** button to exit the command.

The process for publishing a custom iPart and using the **Place Custom iPart** dialog box differs slightly from a standard iPart. As you can see in Figure 18.29, the **Place Custom iPart** dialog box has a list with pre-defined keyed values, and a list with custom values. To use the **Place Custom iPart** dialog box, first select the values of from the predefined value list, located in the left column of the **Keys** tab. When you choose a value in the predefined list of the **Keys** tab, the **Select Value** list box is displayed, as previously discussed and shown in Figure 18.30. Again, to show all of the specified values pick the **All Values** check box. When you select a value, the **Select Value** list box closes. To exit the **Select Value** list box without choosing a value, press the **Esc** key on your keyboard. Then, continue defining the predefined part instance values by selecting values in the left list of the **Keys** tab, until you have fully established the predefined iPart parameters.

The next step in establishing a custom iPart is to select the values from the custom value list, located in the right column of the **Keys** tab. Custom values are defined by highlighting the default value, and enter-ing a new iPart instance value, or selecting an item from a drop-down list if available. If a range has been set in the iPart factory, you will not be able to enter certain values, as indicated by red expressions. Continue defining custom part instance values until you have fully established the custom iPart parameters.

Another option for assigning predefined and custom iPart values is to select the **Table** tab, shown in Figure 18.32. This tab displays the same table shown in the **iPart Author** dialog box, allowing you to select a row to define the iPart instance. To reorder the rows, right-click and select the **Sort Ascending** or **Sort Descending** menu option.

Field Notes

Custom table columns and cells have blue backgrounds.

FIGURE 18.32 The **Place Custom iPart** dialog box (**Table** tab shown).

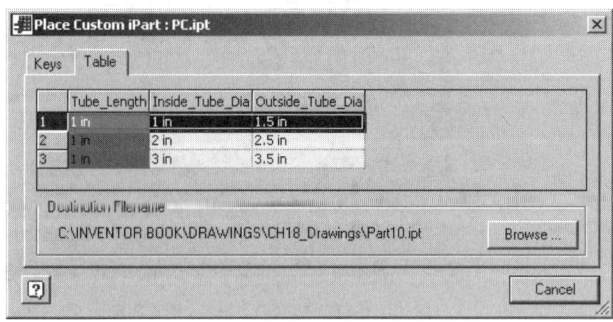

The next step in creating a custom iPart is to redefine the default file destination and file name if needed. The destination filename is the name of the new iPart part file, which is created by selecting new iPart values. You can keep the default filename and location, or select the **Browse** button in the **Destination Filename** area to locate a different file location, and enter a different filename. Next, pick a location for the part in the graphics window. Then you can place copies of the iPart as needed. You will notice the options in the **Place Custom iPart** dialog box are not available once you have defined a single iPart. To create a new iPart with different specifications and selected values, pick the **Browse** button in the **Destination Filename** area to locate a different file location, and enter a new filename. You can repeat the process of selecting values in the **Place Custom iPart** dialog box and placing iParts. When you have finished placing iParts using the current iPart factory, pick the **Dismiss** button to exit the command.

Exercise 18.8

1. Continue from Exercise 18.7 or launch Autodesk Inventor.
2. Open a new inch assembly file.
3. Save the assembly file as EX18.8.iam.
4. Access the **Place Component** tool.
5. Locate and open EX18.6.ipt using the **Open** dialog box.
6. Select the default material value "Auminum-6061" in the **Keyed Items** list.
7. Pick "Steel, Mild" from the **Value** list box.
8. Select the **Browse…** button in the **Destination File** area.
9. Save the iPart as EX18.8A.
10. Pick the **OK** button on the **Place Custom iPart** dialog box to place and create iPart EX18.8A.ipt.
11. Pick a different location to place another copy of EX18.8A.ipt.
12. Select the **Browse…** button again to define a new iPart.
13. Save the new iPart as EX18.8B.
14. Replace the default tube length value of 12″ with 24″ in the **Custom Items** list.
15. Pick a location in the graphics window to place the iPart.
16. Pick a location in the graphics window to place another copy of the 24″ iPart.
17. Pick the **Dismiss** button on the **Place Custom iPart** dialog box to exit the command.
18. Constrain the four iParts as shown.
19. Resave EX18.8.iam.
20. The final exercise should look like the exercise shown.
21. Exit Autodesk Inventor or continue working with the program.

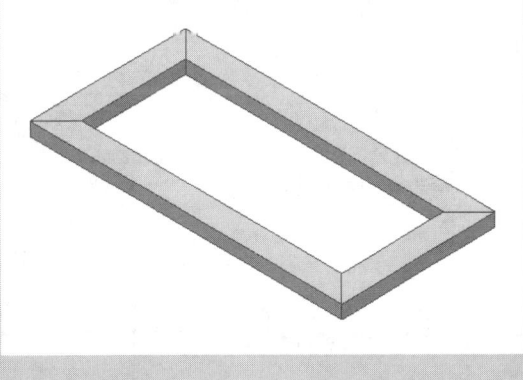

Working with iParts in an Assembly

Once you have published an iPart in an assembly file, there are number of Browser preferences related to working with iParts. In addition, there are some other iPart options you may want to become familiar with, such as storing iParts and updating iPart after changes have been made to an iPart factory.

Figure 18.33 shows an example of the Browser display for a standard iPart published in an assembly. The assembly name and icon are listed first, followed by the assembly origin folder. Then the iPart name and icon are shown. Inside the iPart is the embedded Microsoft Excel file, which is a 3rd party file. You can access and modify the iPart information using the embedded spread sheet by right-clicking on the spread sheet and selecting the **Edit** menu option. In addition, the default spread sheet name can be edited by highlighting the existing name and entering a new name. Listed next under the iPart are the iPart table and each of the standard iPart values, or instances. The selected value is identified by a check box.

The physical parameters of a standard iPart cannot be manipulated, or added to, unless you choose to replace the component with a different iPart instance. This means that you cannot add features to a standard iPart that is inserted into an assembly. To modify an existing standard iPart by identifying a different standard iPart instance, or value, right-click on the iPart table in the Browser, or an iPart version under the iPart table in the Browser. Then select the **Change Component** menu option. Using this technique reopens the **Place Standard iPart** dialog box, as previously discussed. Use the dialog box to choose a different iPart key value, pick the **OK** button, and the new iPart is generated.

Figure 18.34 shows an example of the Browser display for a custom iPart published in an assembly. The assembly name and icon are listed first, followed by the assembly origin folder. Then the iPart name and icon are shown. The iPart is a new part file that is placed in the same location as the iPart factory. The new iPart has a default part name, such as "Part2.ipt," and can be renamed. Inside the iPart is the embedded Microsoft Excel file, which is a 3rd party file. You can access and modify the iPart information using the embedded spread sheet by right-clicking on the spread sheet and selecting the **Edit** menu option. In addition, the default spreadsheet name can be edited by highlighting the existing name and entering a new name. Listed next under the iPart are the iPart table and each of the standard iPart values, or instances. The selected value is identified by a check box. However, the custom parameters are not shown.

Unlike a standard iPart some of the physical parameters of a custom iPart can be manipulated and added to, because a derived part is generated when you define custom iPart values. To add features to a custom iPart, right-click on the iPart name in the Browser, or the iPart in the graphics window. Select the **Edit** menu option to display the derived component and enter the features work environment. Refer to the previous discussion in this chapter for more information regarding derived components.

Field Notes

You can also edit a custom iPart by opening the iPart file externally or from inside the assembly.

In addition to editing a custom iPart, you can identify a different standard or custom iPart instance, or value, by right-clicking on the iPart table in the Browser, or an iPart version under the iPart table in the

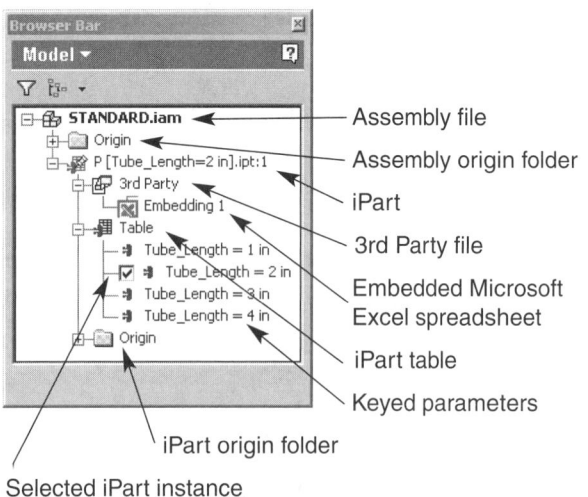

FIGURE 18.33   An example of the Browser display for a standard iPart in an assembly.

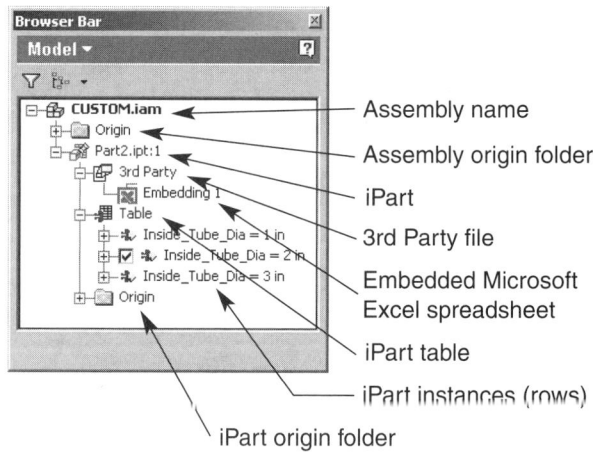

FIGURE 18.34   An example of the Browser display for a custom iPart in an assembly.

Browser. Select the **Change Component** menu option. Using this technique reopens the **Place Custom iPart** dialog box, as previously discussed. Use the dialog box to choose a different iPart key value or specify a different custom value, pick the **OK** button, and the new iPart is generated.

When you edit information in an iPart factory and then open an assembly with a published iPart, the new information may not be updated, especially if a table row used to define the iPart no longer exists. As a result, you may have to update the iPart to reflect the changes made to the iPart factory. You can update the file by picking the **Full Update** tool button on the **Command** bar. If an iPart needs to be redefined, an update symbol is shown next to the iPart. To update the iPart, use the **Change Component** tool previously discussed to redefine the iPart specifications.

iParts generated using an iPart factory are stored in the same location as the iPart factory, based on the specified search paths and active project. When you publish a standard iPart in an assembly, a folder is automatically created in the same location of the iPart factory, and with the same name as the iPart factory. New, separate iParts, which are contained in part files, are added to the folder. The new standard iParts are based on the standard specifications and values you identify when the iPart factory is created. When you publish a custom iPart in an assembly, a new part file is automatically created with a default name and is stored in the same location as the iPart factory.

Field Notes

For more information regarding storing iParts, search paths, and active projects, you may want to refer to Chapter 1, the Autodesk Inventor help files.

Exercise 18.9

1. Continue from Exercise 18.8 or launch Autodesk Inventor.
2. Open the iPart factory, EX18.6.ipt, and save a copy as EX18.9.ipt.
3. Close EX18.6.ipt without saving, and open EX18.9.ipt.
4. Right-click on the iPart factory table in the Browser and select the **Edit table...** menu option.
5. Pick the **Suppression** tab, and add "Extrusion2" to the **Selected Items** list.
6. Make the column "Extrusion2" custom.
7. Pick the **OK** button to redefine the iPart factory.
8. Save and close EX18.9.ipt.
9. Open a new inch assembly file.
10. Save the assembly file as EX18.9.iam.
11. Access the **Place Component** tool.
12. Locate and open EX18.9.ipt using the **Open** dialog box.
13. Replace the default tube length value of 12″ with 18″ in the **Custom Items** list.
14. Select the **Browse...** button in the **Destination File** area.
15. Save the iPart as EX18.9A.
16. Pick the **OK** button on the **Place Custom iPart** dialog box to place and create iPart EX18.9A.ipt.
17. Pick a different location to place another copy of EX18.9A.ipt.
18. Select the **Browse...** button again to define a new iPart.
19. Save the new iPart as EX18.9B.
20. Replace the default tube length value of 12″ with 24″ in the **Custom Items** list.
21. Pick a location in the graphics window to place the iPart.
22. Pick a location in the graphics window to place another copy of the 24″ iPart.
23. Select the **Browse...** button again to define a new iPart.
24. Save the new iPart as EX18.9C.

Exercise 18.9 continued on next page

Exercise 18.9 continued

25. Replace the default tube length value of 12″ with 14″ in the **Custom Items** list.
26. Pick a location in the graphics window to place the iPart.
27. Pick the Cancel button on the **Place Custom iPart** dialog box.
28. Expand EX18.9C.ipt:1 in the Browser.
29. Right-click on the EX18.9C.ipt:1 table in the Browser, and select the **Change Component** menu option.
30. Select the **Browse...** button again to define a new iPart.
31. Save the new iPart as EX18.9D.
32. Pick "Compute" in the **Custom Items** list, and select the **Suppress** from the drop-down list.
33. Pick the **OK** button to replace the iPart.
34. Constrain the five iParts as shown.
35. Resave EX18.9.iam.
36. The final exercise should look like the exercise shown.
37. Exit Autodesk Inventor or continue working with the program.

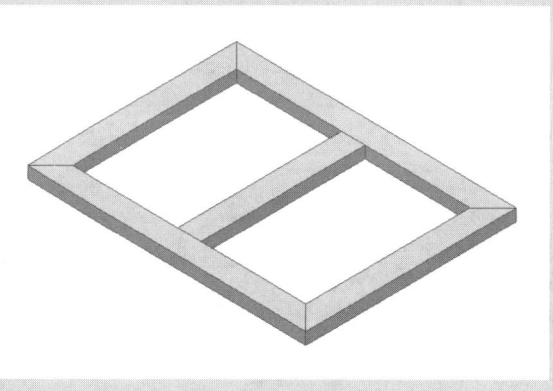

|||||||||||| CHAPTER TEST

Answer the following questions on a separate sheet of paper.

1. What are derived components and how are they different from design elements?

2. Give another name for a derived work body.

3. How do you use the Open dialog box to select a part or assembly file and open the Derived Part dialog box?

4. How do you exclude certain information from the derived part feature?

5. How is a folder that contains both included and excluded geometry identified?

6. How do you flip or mirror a derived part feature?

7. What are seams, and how do you show or not show them?

8. What is a weldment?

9. When is a red lighting bolt symbol displayed and what does it indicate?

10. What is an iPart and an iPart factory and what is their use?

11. Outline the five steps of part authoring.

12. Outline the four steps of part publishing or part placement.

13. What does it mean when it is said that the part used to generate an iPart factory is basically a template?

14. Give the basic function of the iPart Author dialog box.

15. How do you add a property from the Model list to the Selected Items list?

16. Explain the difference between a suppressed feature and a computed feature.

17. As with parameters, properties, suppressions, and iMates described earlier, there are key symbols displayed next to each of the custom items. What do these key symbols identify?

18. How do you create another iPart factory table row?

19. Explain how to modify the values in each of the cells to reflect the desired part instance attributes.

20. Discuss the certain expressions required by suppression values and give examples.

21. How do you delete a cell row?

22. How do you create a custom column?

23. How do you specify a certain range of values that can be used to build the part?

24. Discuss the basic function of the Specify Range dialog box.

25. Provide are two options for editing iPart factories.

26. Discuss how to use the Place Standard iPart dialog box, and publish a standard iPart.

27. Explain the process for publishing a custom iPart, and using the Place Custom iPart dialog box.

28. Discuss the use of the embedded Microsoft Excel file Inside the iPart.

29. When can the physical parameters of a standard iPart not be manipulated, or added to?

30. How do you modify an existing standard iPart by identifying a different standard iPart instance, or value?

|||||||||||||||||||||||||||||||||||||||||||||||||||||||||||||||||||||||||||||||||||||||||

CHAPTER 19

Additional Data and File Management Tools and Techniques

LEARNING GOALS

After completing this chapter, you will be able to:

◎ Use the Engineer's Notebook to add notes to models and model items.

◎ Work with Design Assistant to search for files, create file reports, work with the links between Autodesk Inventor files, and manipulate design properties.

◎ Compact files to remove unnecessary file information.

◎ Use Pack and Go to package files.

◎ Work with Windows NetMeeting to further collaborate on design issues.

The Engineer's Notebook

The ***Engineer's Notebook*** is an Autodesk Inventor tool that allows you to further document the design process. It allows you to add notes to part and assembly models in the part or assembly work environment. These notes, also known as ***design notes,*** are placed in the Engineer's Notebook and stored inside the model as a model element, similar to a part feature, for example. You can add as many notes as needed for the model, and attach them to most model geometry including sketches, features, edges, entire parts, entire assemblies, and assembly constraints.

Typically notes and the Engineer's Notebook are used as a collaborative tool, in which multiple users can share a wide variety of information, such as identifying design intentions. However, there are multiple other purposes for adding notes, such as documenting the process of creating a part from the sketch to the final features. See Figure 19.1. The following represent a few different options for using notes and the Engineer's Notebook:

- Communicating and documenting product and material testing information.

- Identifying Autodesk inventor tools and commands used to generate a model.

- Defining manufacturing and assembly requirements and processes.

FIGURE 19.1 An example of a note.

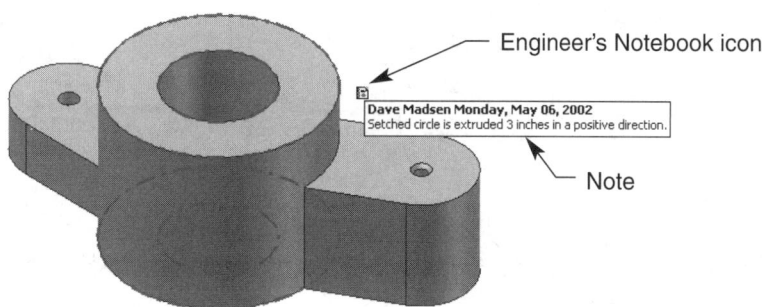

Engineer's Notebook icon

Dave Madsen Monday, May 06, 2002
Setched circle is extruded 3 inches in a positive direction.

Note

- Relating design history and model development, including model changes.

- Highlighting significant model information.

Before you add notes using the Engineer's Notebook, you may want to become familiar with the Engineer's Notebook preferences found in the **Options** dialog box. As discussed in previous chapters, the Options dialog box is available by selecting **Application Options...** from the **Tools** pull-down menu. Pick the **Notebook** tab of the **Options** dialog box to access the Engineer's Notebook options. See Figure 19.2. The following options are located in the **Notebook** tab:

- **Display in Model**   This area contains two check boxes that allow you to control the display of note information in the model. Select the **Note Icons** check box to show the note icon in the graphics window, as in Figure 19.1. Choose the **Note Text** check box to display the note as shown in Figure 19.1. When this check box is selected, the note pops up when you move your cursor over the note icon in the graphics window, or the note in the Browser.

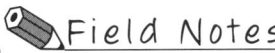Field Notes

Only one note icon is displayed for a single item. This means that if you add more than one note to a model item, only one note icon will be shown.

- **Color**   Use this area to set color preferences for Engineer's Notebook items, as shown in Figure 19.3. Select the **Text Background** color button to open the **Color** dialog box as dis-

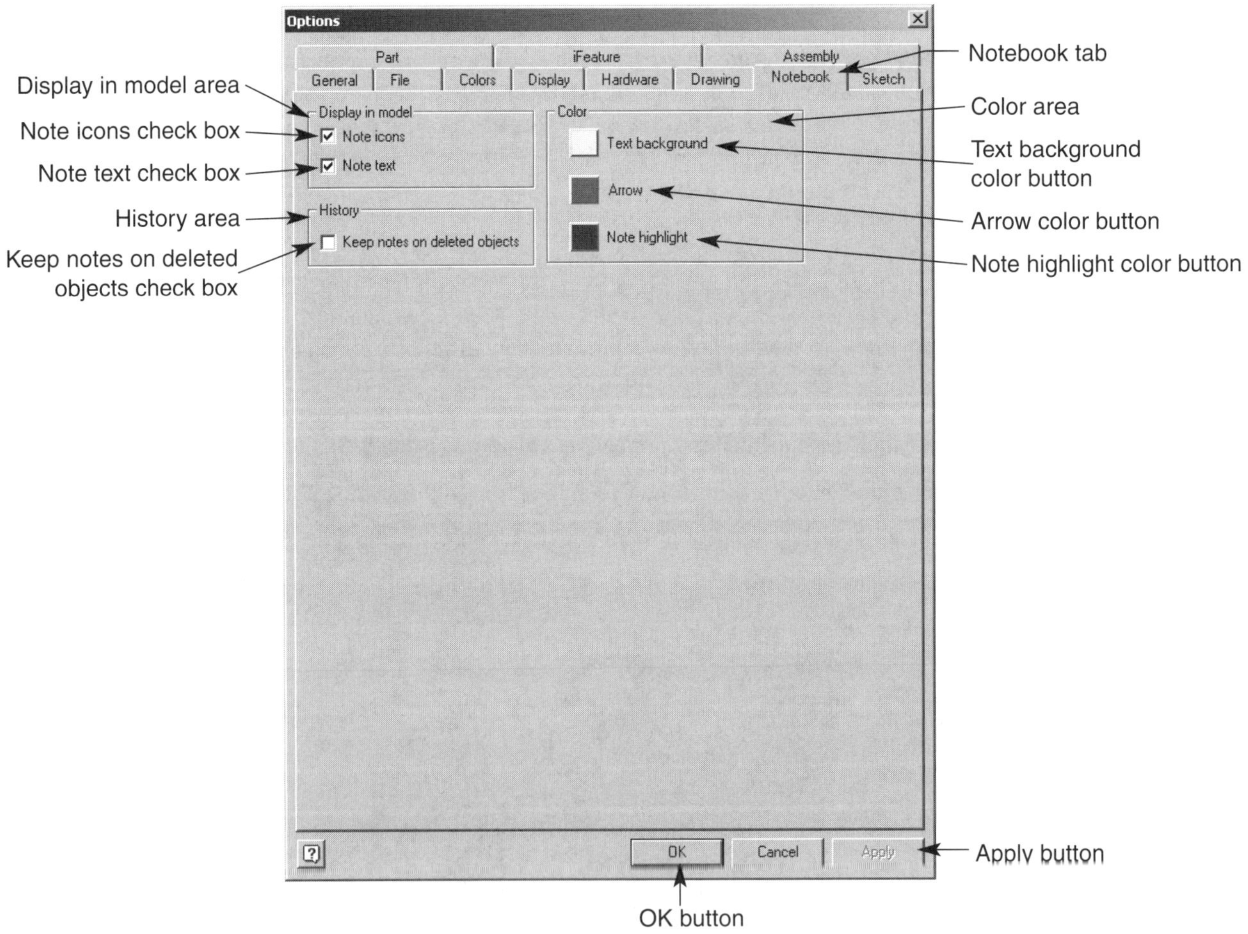

Display in model area

Note icons check box

Note text check box

History area

Keep notes on deleted objects check box

Notebook tab

Color area

Text background color button

Arrow color button

Note highlight color button

Apply button

OK button

FIGURE 19.2   The **Notebook** tab of the **Options** dialog box.

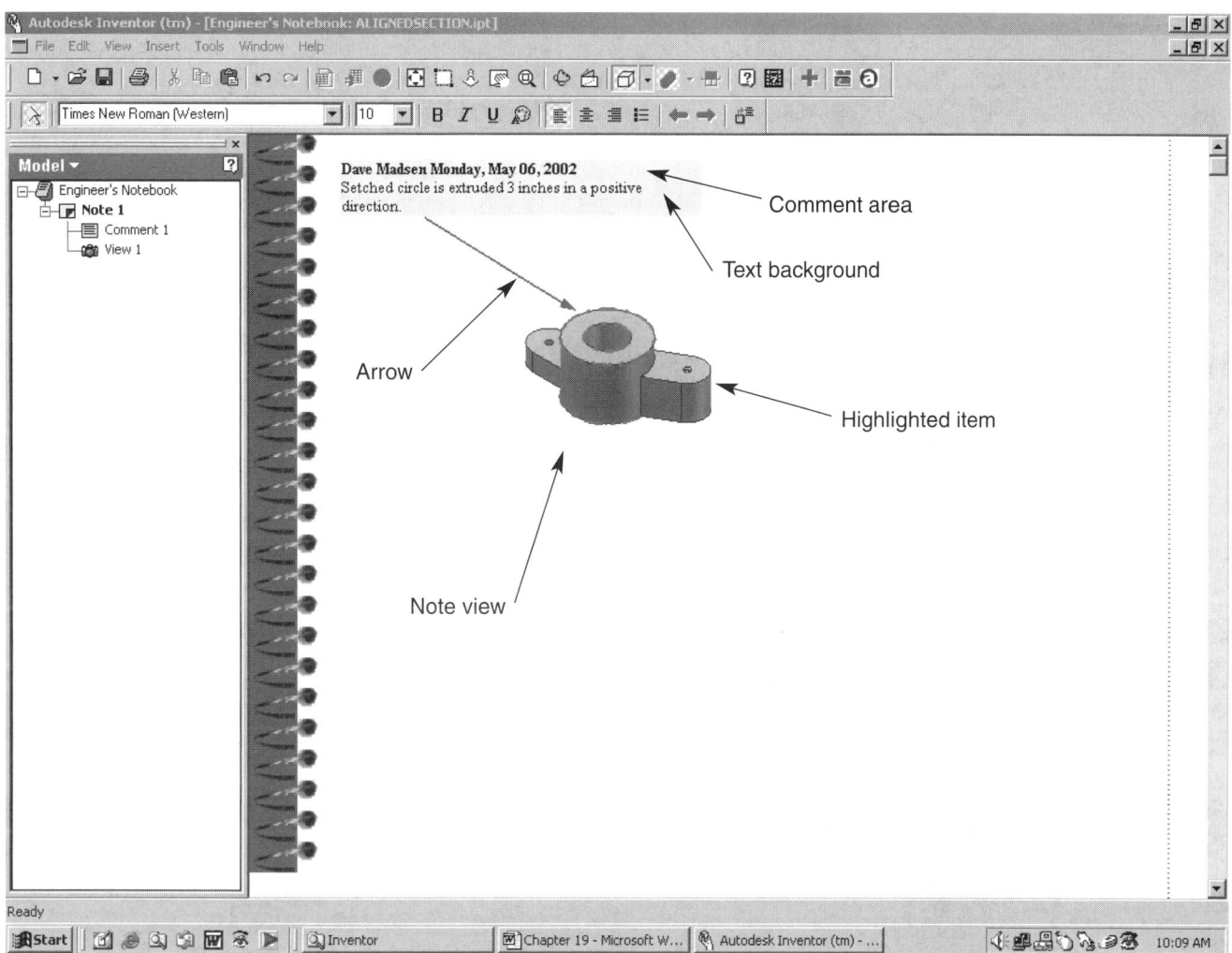

FIGURE 19.3   An example of the Engineer's Notebook work environment.

cussed in previous chapters. Choose the desired text background color, as identified in Figure 19.3. Choose the **Arrow** color button to open the **Color** dialog box and select an arrow color, as identified in Figure 19.3. Pick the **Note Highlight** color button to open the **Color** dialog box, and define a color for the highlighted note view item. Again, see Figure 19.3.

■ **History**   Often notes and the Engineer's Notebook are used to describe model development, or the model history. If you choose the **Keep notes on deleted objects** check box, located in the **History** area, notes are not deleted when geometry is deleted. This allows you to retain notes that were attached to deleted model items, for reference, and in order to keep an exact design history record.

Creating Design Notes

The first step in creating a note is to identify the model item to which you want to add a note. As discussed, you can add notes to a wide variety of part and assembly geometry, depending on the required note. Once you identify the geometry, use one of the following techniques to establish the design note:

✓ Right-click on specific geometry in the Browser and select the **Create Note** menu option.

✓ Right-click on the actual model geometry in the graphics window and select the **Create Note** menu option.

When you access the **Create Note** tool, a new window opens. See Figure 19.3. The new window is not a separate file. As indicated by the name [Engineer's Note Book: "file name".ipt or .iam], the window is part of the existing part or assembly file and is used to create notes in the Engineer's Notebook.

The Engineer's Notebook work environment is similar to other Autodesk Inventor work environments. As you can see in Figure 19.3, the Engineer's Notebook window contains a graphics window, which is displayed as a notebook, a Browser, toolbars, and pull-down menus. Many of the interface options available in a modeling window are also available in the Engineer's Notebook window. However, as shown in Figure 19.4, there are a number of tools and options that are specific to working with notes. Throughout this section you will discover specific options related to the Engineer's Notebook interface and work environment.

As you can see in Figure 19.3 and 19.4, an Engineer's Notebook note contains a comment box, which is the actual note, and a view box, which displays the model and the highlighted item. The actual design note is specified in the comment box by entering text. The first line of text is default and identifies the username as specified in the **General** tab of the **Options** dialog box, and the date. You can add as many additional lines of text as required to note the information. Move the comment box by moving your cursor near the edge of the box until you see the move icon, then drag the box to the desired location. Similarly, you can modify the size and shape of the comment box by dragging and expanding a corner or edge of the box. When you change the size and shape of the comment box, the note text adjusts to fit the new box. However, you cannot create a comment box that is too small to fit the note text.

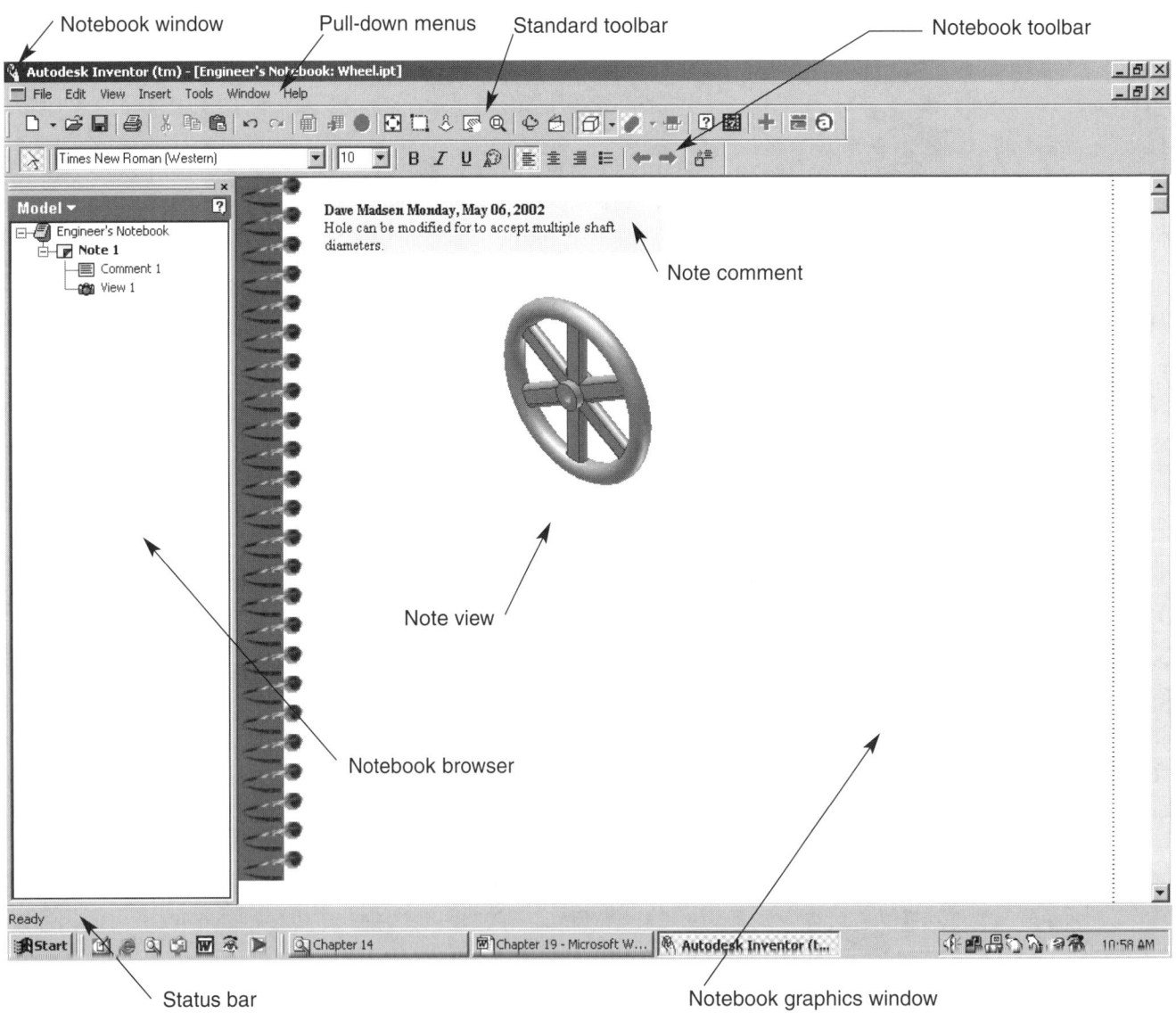

FIGURE 19.4 Engineer's Notebook interface options.

In addition to modifying the location, size, and shape of the comment box, you can redefine the text with the following format preferences of the comment box contents:

- **Font**   Modify the text font style by highlighting the note and selecting the desired font style from the **Font** drop-down list in the **Notebook** toolbar.

- **Font Size**   Change the text font height by highlighting the note and selecting the desired font size from the **Font Size** drop-down list in the **Notebook** toolbar.

- **Format**   You can redefine the text format by highlighting the note and selecting the **Bold, Italic,** or **Underline** button on the **Notebook** toolbar. This same operation can also be accomplished by highlighting the note, right-clicking, and selecting the **Bold, Italic,** or **Underline** option from the **Character** cascading submenu.

- **Color**   Change the text color by highlighting the note and selecting the desired color from the **Text Color** flyout button on the **Notebook** toolbar.

- **Justification**   To reposition, or modify, the justification of the note inside the comment box, highlight the note, and select the **Align Left, Center,** or **Align Right** button on the **Notebook** toolbar. This same operation can also be accomplished by highlighting the note, right-clicking, and selecting the **Align Left, Center,** or **Align Right** option from the **Paragraph** cascading submenu.

- **Bullets**   You can bullet the note text by highlighting the text you want to bullet and selecting the **Bullets** button on the **Notebook** toolbar, or by right-clicking and selecting the **Bullets** menu option.

By default one comment box is available for each note. However, in some situations, you may want to add another comment box to the same note. Use one of the following options to place an additional comment box for the same note:

✓ Right-click in the graphics window, outside of a comment box or view, and select the **Insert Comment** menu option.

✓ Select **Comment** from the **Insert** pull-down menu.

Once you access the **Comment** tool, use your cursor to select a start point and endpoint for the comment box, as shown in Figure 19.5. Once you pick the endpoint, or comment box corner, the comment box is created.

FIGURE 19.5
Creating an additional
notebook comment box.

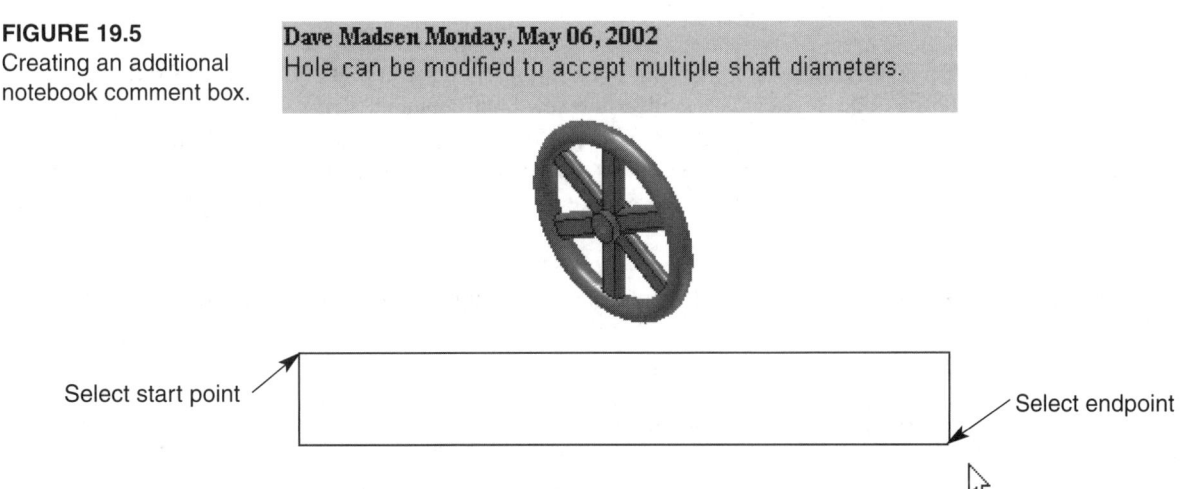

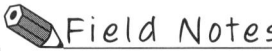**Field Notes**

Right-click inside a comment box and select the **Insert New Object...** menu option, or pick the **Object...** option from the **Insert** pull-down menu to link an external file, such as a document or spreadsheet, to the note.

The view box, shown in Figures 19.3 and 19.4, displays the model with the highlighted, selected note item. You can move and modify the size and shape of the view box using the same techniques discussed for the comment box. In addition, you can use viewing tools to redefine the display of the model in the view box, and use the following options available in the **View** pull-down menu.

- **Freeze**   Select this option if you do not want the note view to update when changes are made to the model. When you create a note, the view displayed in the notebook is linked to the state of the model when the note is created, but is also free to change when the model is modified. As a result, when you edit model geometry, the view in the notebook automatically updates to reflect the new design, unless you select the **Freeze** option, which retains the same view shown when the note was created.

- **Restore Camera**   Pick this option to override any other viewing operations, and return the note view display to the original camera scene.

Field Notes

The **Freeze** and **Restore Camera** options are also available by right-clicking on the note view display in the notebook graphics window, and selecting the **Freeze** or **Restore Camera** menu option.

As with the comment box, by default one view box is available for each note. However, in some situations, you may want to add another view box to the same note to additional displaying or noting purposes. Use one of the following options to place an additional view box for the same note:

- ✓ Right-click in the graphics window, outside of a comment box or view, and select the **Insert View** menu option.
- ✓ Select **View** from the **Insert** pull-down menu.

Once you access the **View** tool, use your cursor to select a start point and endpoint for the view box. Once you pick the endpoint, or view box corner, the view box is created.

Often it may be helpful to add a leader, or arrow, in the graphics window, to more effectively identify geometry in the view box and explain note content. As a result, the Engineer's Notebook contains an Arrow tool, which allows you to place a leader. See Figure 19.6. Use one of the following options to access the **Arrow** tool:

- ✓ Right-click in the graphics window, outside of a comment box or view, and select the **Insert Arrow** menu option.
- ✓ Select **Arrow** from the **Insert** pull-down menu.
- ✓ Select the Arrow button on the **Notebook** toolbar.

After you access the **Arrow** tool, use your cursor to select the start point of the arrow. Then pick an additional point including an endpoint. Once you have defined the arrow, right-click and select the **Done** menu option to create the arrow.

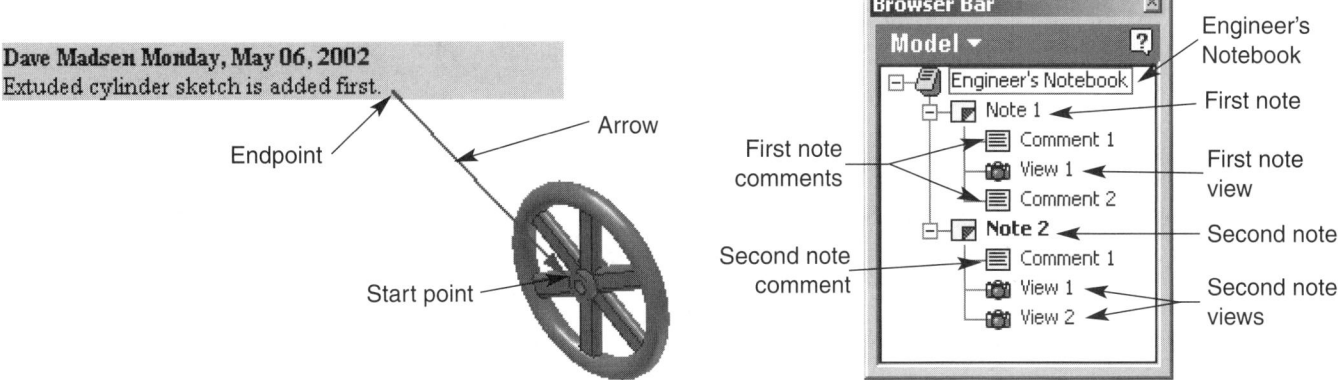

FIGURE 19.6   An example of a notebook arrow.

FIGURE 19.7   An example of Engineer's Notebook Browser for a part file.

An example of the Engineer's Notebook Browser is shown in Figure 19.7. As you can see, the Browser displays all the model notes created and contained in the Engineer's Notebook. For part models, the notes are listed in the order they were created; for assembly models, the notes are listed under components, and then in the order they were created. Nested under each note is the comment box or boxes and the view box or boxes. You may find it helpful to rename the notes, comments, and views in the Browser for clarity. Like items in the Browser of a model, you can drag notes, comments, and views, to different locations in the Browser as needed.

Field Notes

When notes are added to assemblies, folders are automatically generated for each assembly component.

In addition to moving and renaming the Engineer's Notebook Browser items, when you right-click on "Engineer's Notebook," or a folder in the Browser, the following options are available:

- **<u>A</u>rrange Notes**   This cascading submenu allows you to specify how to sort the notes inside the Browser; you can choose to arrange the notes by name, author, date, or text.

- **Insert <u>F</u>older**   Use this option to insert a new folder into the Engineer's Notebook. The note folder is added to the end of the current list of notes in the Browser. You can add new or existing notes to a folder.

- **Place <u>N</u>ew Notes Here**   Select this option to define a folder as the destination when you add new notes to the notebook. "Engineer's Notebook" is the main folder. You can choose to add new notes to it or another folder created using the **Insert <u>F</u>older** option.

Working with the Engineer's Notebook

The actual process for creating a note is the same in a part as it is in an assembly. However, there are some specific design note considerations for working in these two different environments. When you place notes in a part, the notes can be attached to a sketch, feature, edge, and the entire part. Notes added to part sketches or features are listed under the sketch or feature in the Browser. However, if you add a note to the entire part or to a feature edge, the note is listed after the Origin folder in the Browser.

When you place notes in an assembly, the notes can be attached to components, the entire assembly, or component sketches, features, and edges. If you insert a component that already has notes into an assembly file, the notes are listed inside the components in the Browser. As a result, the notes are available in both the part and assembly files. However, if you add a note to a component or component item while working in the assembly environment, the note is available only in the assembly file and is not displayed when you open the separate part file. When you develop a component in-place and add a note while creating the part, the note is available in both the assembly and part files.

Exercise 19.1

1. Launch Autodesk Inventor.
2. Open EX14.7.iam and save a copy as EX19.1.iam.
3. Close EX14.7.iam without saving, and open EX19.1.iam.
4. Explore the preferences in the **Notebook** tab of the **Options** dialog box, as discussed.
5. Right-click on P5.2.ipt (the c-clamp swivel) in the graphics window, or in the browser, and select the **Create Note** menu option.
6. In the comment box, enter: PEEN END AFTER ASSEMBLY WITH P7.6.
7. Reformat the note in the comment box using a RomanS font, and a font size of 11.
8. Edit the size and location of the comment box as shown.
9. Edit the size and location of the view box as shown.
10. Add another view as shown, and rotate the assembly to provide a display of the swivel and screw fit.
11. Add the arrows shown.
12. Right-click on P7.6.ipt (the c-clamp screw) in the graphics window, or in the browser, and select the **Create Note** menu option.
13. In the comment box, enter: LUBRICATE THREADS AFTER ASSEMBLY.
14. Reformat the note in the comment box using a RomanS font, and a font size of 11.
15. Edit the size and location of the comment box as shown.
16. Edit the size and location of the view box as shown.
17. Add the arrows shown.
18. Resave the assembly file.
19. Exit Autodesk Inventor or continue working with the program.

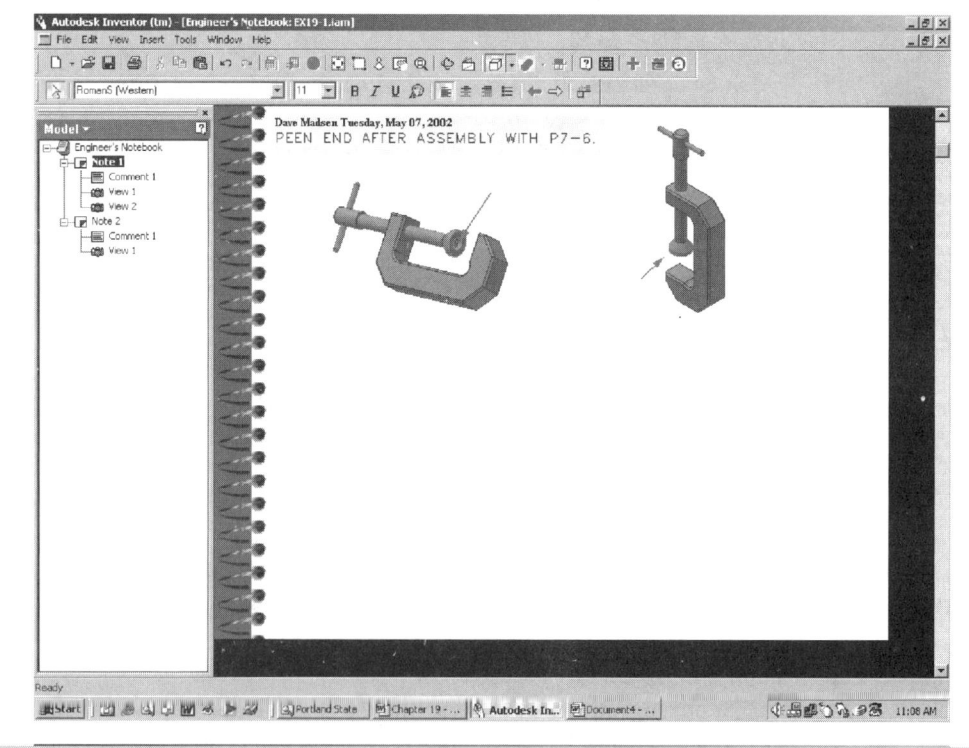

Exercise 19.1 continued on next page

Exercise 19.1 continued

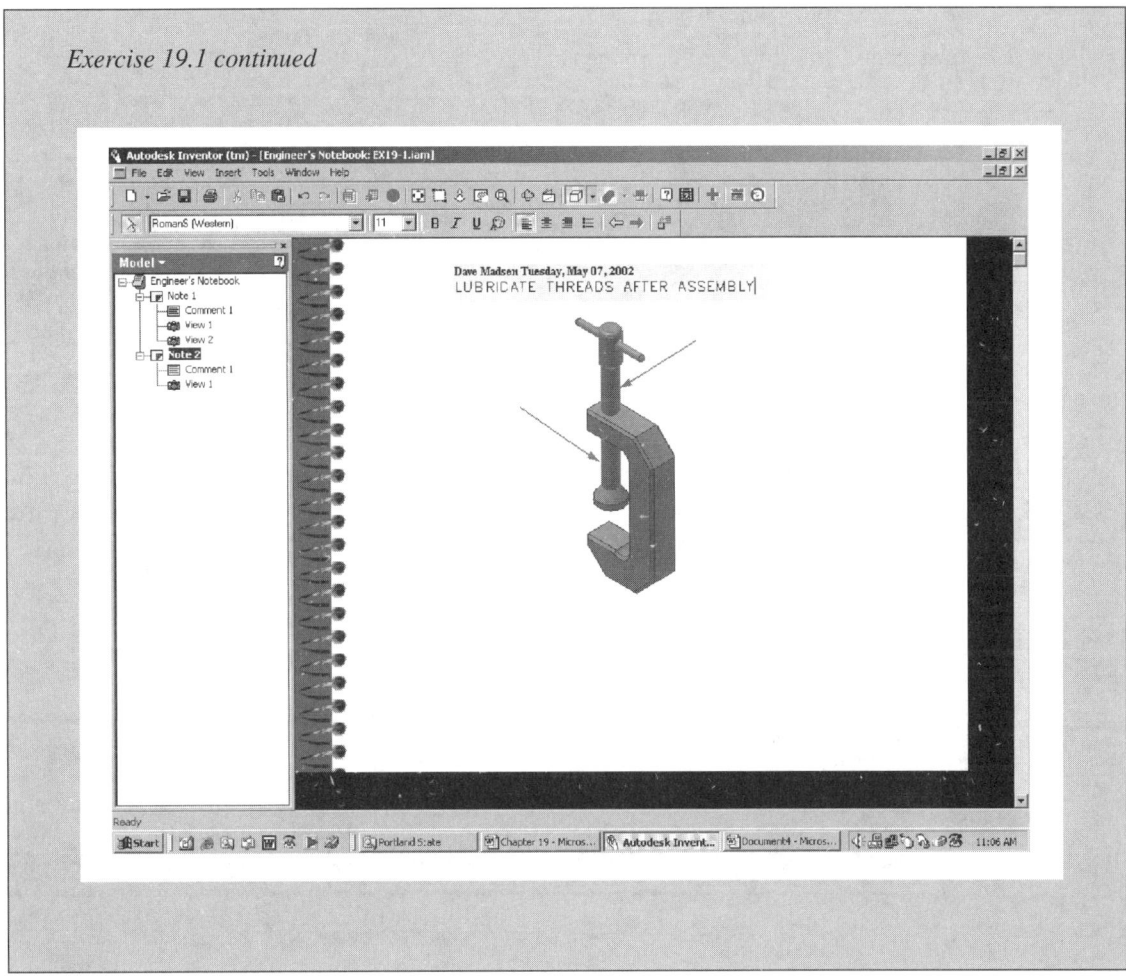

Design Assistant

Autodesk Inventor **Design Assistant,** shown in Figure 19.8, is a tool that allows you to locate, manage, and help modify Autodesk Inventor files. Using **Design Assistant,** you can search for files, create file reports, work with the links between Autodesk Inventor files, and manipulate design properties. Any additional files from other programs that may relate to your Autodesk Inventor files are also available in **Design Assistant**. You can access the Autodesk Inventor **Design Assistant** using one of the following techniques:

✓ Select **Design Assistant** from the **File** pull-down menu.

✓ Select the **Windows Start** menu button, followed by the **Programs** option, then the **Inventor 4** option, and finally **Design Assistant.**

✓ Right-click on an Autodesk Inventor file from inside your Windows Explorer, and select **Design Assistant**.

✓ Right-click on any folder inside your Windows Explorer, and select **Design Assistant.**

When you open **Design Assistant** from outside Autodesk Inventor using one of these methods, Autodesk Inventor does not have to be open and will not be launched. The following options are available when **Design Assistant** is opened outside of Autodesk Inventor:

■ **File** pull-down menu   Inside this menu, shown in Figure 19.9, you can open a specific file using the **Open File...** option. This command is also accessed by picking the **Open** button. In addition, you may want to open an entire folder using the **Open Folder...** option. Pick the

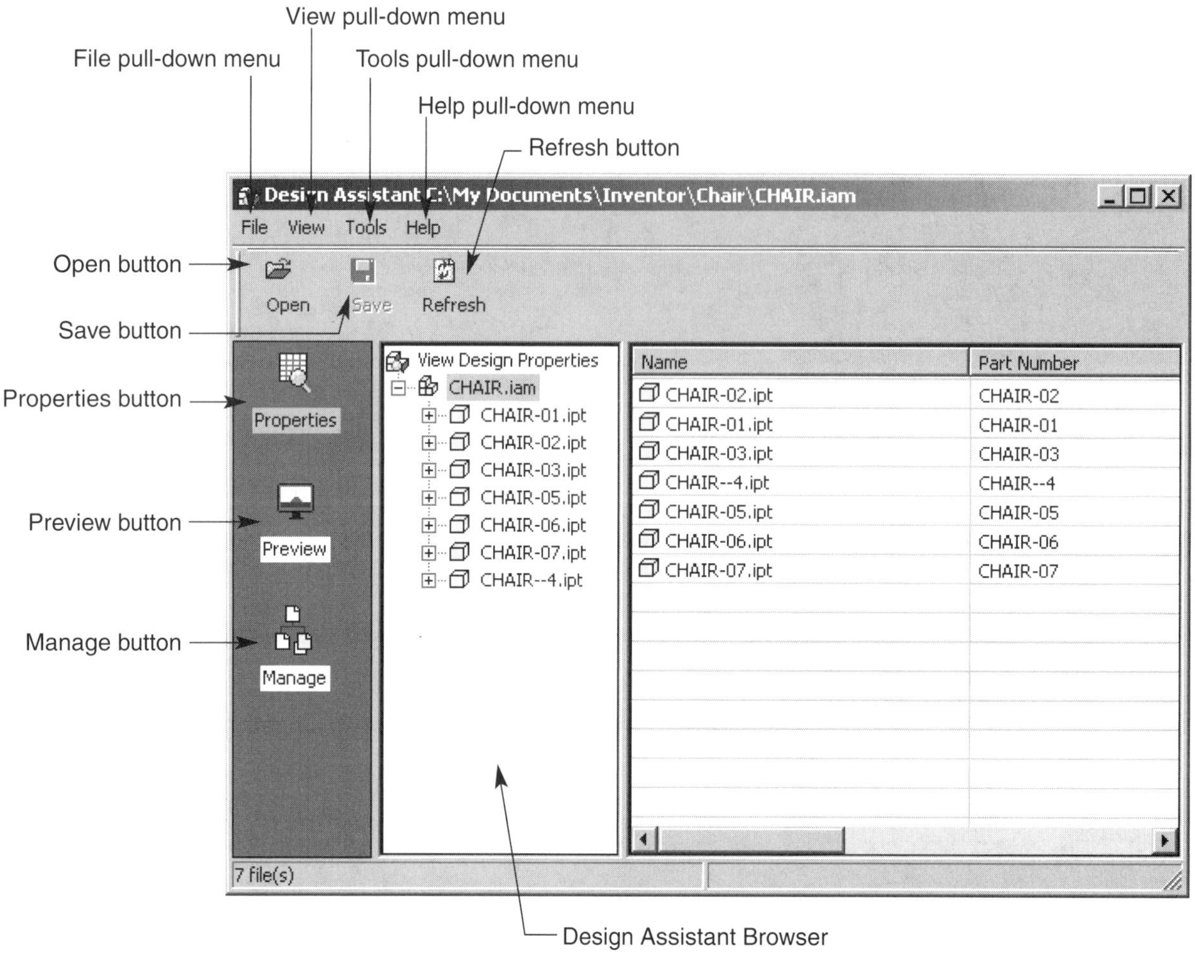

View pull-down menu

File pull-down menu

Tools pull-down menu

Help pull-down menu

Refresh button

Open button

Save button

Properties button

Preview button

Manage button

Design Assistant Browser

FIGURE 19.8   Autodesk Inventor **Design Assistant** (properties mode).

FIGURE 19.9   The Autodesk Inventor **Design Assistant, File** pull-down menu.

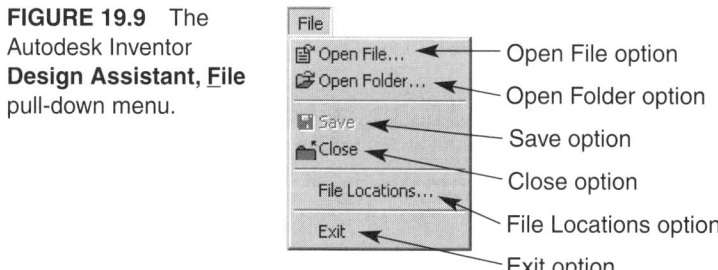

Open File option

Open Folder option

Save option

Close option

File Locations option

Exit option

Save selection to save any changes made while in manage mode. This option is also available by selecting the **Save** button. To close the current session without exiting **Design Assistant**, pick the **Close** option. Choosing the **File Locations...** option opens the Autodesk Inventor **Project Editor** discussed in Chapter 1. To completely exit **Design Assistant**, pick the **Exit** option.

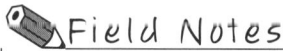 Field Notes

Changes are saved automatically while in properties mode.

FIGURE 19.10 The Autodesk Inventor **Design Assistant, View** pull-down menu.

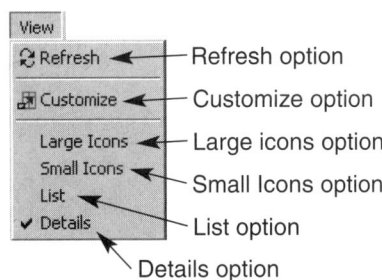

Refresh option — Refresh
Customize option — Customize
Large icons option — Large Icons
Small Icons option — Small Icons
List option — List
Details option — Details

- **View** pull-down menu   This pull-down menu, shown in Figure 19.10, contains a number of options that allow you to manipulate the **Design Assistant** display and viewing information. The following selections are available inside the **Design Assistant, View** pull-down menu:

   - **Refresh**   Pick the **Refresh** option to refresh the current **Design Assistant** session by showing changes you have made and collapsing or expanding files and folders. You can also access the refresh option by selecting the **Refresh** button.

   - **Customize**   Picking the **Customize** option opens the **Select Properties to View** dialog box. See Figure 19.11. This dialog box allows you to specify which of the properties found in the **Properties** dialog box, previously discussed, you would like to display.

      Choose the **Select properties group:** drop-down list to designate one of the groups of properties. These correspond to the **Properties** dialog box tabs. The properties for the specified properties group are displayed in the **Available Properties:** list box. To add one of these properties to those already listed, select the property and pick the **Add->** button. You will now see the additional property displayed in the **Selected Properties**: list box.

      If you would like to remove one of the selected properties, highlight the property and pick the **<-Remove** button. You also have the option of moving the selected properties up or down in the list by highlighting the property and choosing the **Move Up** or **Move Down** buttons. Moving a property changes its left-to-right position in **Design Assistant.**

      If you select the **Use Default Settings** button, all your customization and the **Select Properties to View** dialog box options return to the default settings.

      You can also create a new property by selecting the **New** button. Picking the new button opens the **Add New Property Column** dialog box, shown in Figure 19.12. To add a new property column, type the property name in the **Name:** edit box, and pick the **OK** button. If you would like to create an additional property, and keep the **Add New Proper-**

FIGURE 19.11 The Autodesk Inventor **Design Assistant, Select Properties to View** dialog box.

Select property group drop-down list
Add button

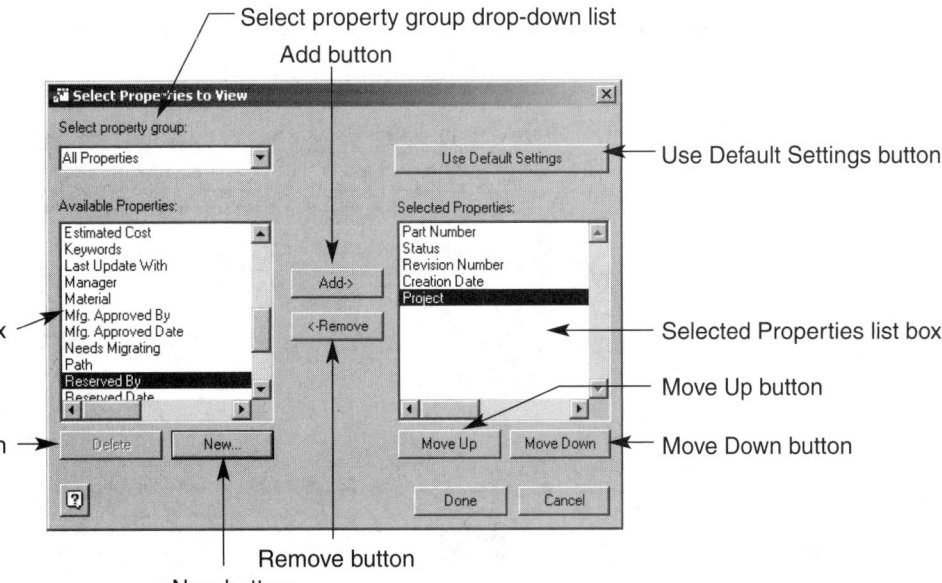

Use Default Settings button
Available Properties list box
Selected Properties list box
Move Up button
Delete button
Move Down button
Remove button
New button

FIGURE 19.12   The
**Add New Property
Column** dialog box.

Enter property name area

Name edit box

Select properties
from file area

File edit box

Name display box

Load button

Browse button

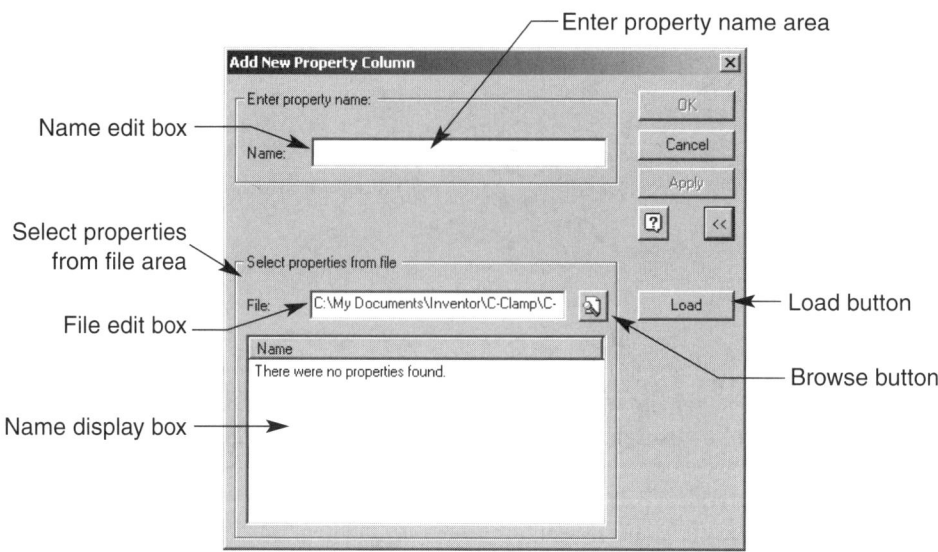

ty Column dialog box open, pick the **Apply** button, and type a new name in the **Name:**
edit box. Continue this process until you have created all of the properties, and pick the
OK button.

In addition, you can locate a new property column from another file by entering a
path in the **File:** edit box of the **Select properties from file** area. To help locate a file, you
may want to use the **Browse** button to open the **Add new property: Select File** dialog
box, shown in Figure 19.13. When you have located the file, pick the **Open** button, followed by the **Load** button.

Field Notes

To use customized user-defined properties inside an Autodesk Inventor file, provide the same
name in the **Custom tab** of the **Properties** dialog box, discussed earlier, as in the **Add New
Property Column** dialog box.

■ Display options   The display options found in the **View** pull-down menu allow you to
view the **Design Assistant** properties mode information four different ways. You can dis-

FIGURE 19.13   The
**Add new property:
Select File** dialog box.

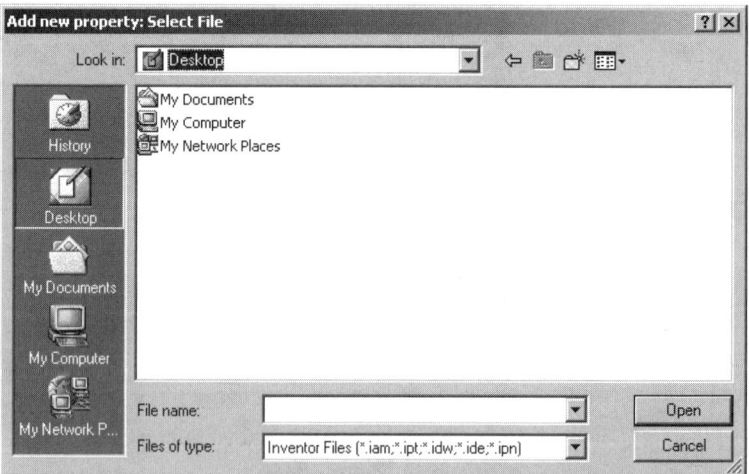

play information as **Large Icons, Small Icons, List,** or **Details.** A checkmark next to the option identifies the current display setting.

- **Tools** pull-down menu   The **Tools** pull-down menu has a few powerful options that allow you to create reports, copy design properties, and locate information. See Figure 19.14. The tools available are as follows:

  - **Reports**   The **Reports** cascading submenu allows you to create a **Hierarchy Report** or a **Design Property Report**. Hierarchy reports are used to assist you in file management by showing you folder and file information as it is developed throughout the design process. See Figure 19.15. They are developed for a folder, which displays the folder and all the subfolders, or an assembly file, which displays component and additional files referenced by the assembly.

    To create a hierarchy report, select **Hierarchy** from the **Reports** cascading submenu. See Figure 19.16. This will open the **Hierarchy Report** dialog box shown in Figure 19.17. Choose the level of hierarchy you would like to display from the **Expand Hierarchy to level:** drop-down list, and pick the **Next>>** button. Finally, specify a location to save the report in the **Report Location** dialog box, shown in Figure 19.18, and pick the **Save** button. An ASCII text file is created and saved in the specified location. Locate, and open the file to view your report.

    Design property reports are used to display folder and file design properties. See Figure 19.19. The information found in the **Summary, Project, Status,** and **Custom** tabs of the **Properties** dialog box, previously discussed, is available to create a design properties report. The process for creating a design property report is exactly the same as the process for creating a hierarchy report, discussed earlier.

  - **Copy Design Properties...**   Choosing the **Copy Design Properties...** option opens the **Copy Design Properties** dialog box, shown in Figure 19.20. The **Copy Design Properties** dialog box allows you to copy design properties from one file to another without opening each individual file and manually editing the properties. This is very helpful and time-saving tool when you are working on a project or assembly that contains many components that have similar design properties.

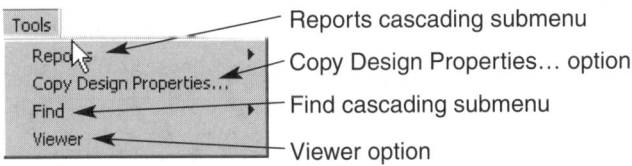

Reports cascading submenu

Copy Design Properties... option

Find cascading submenu

Viewer option

FIGURE 19.14   The Autodesk Inventor **Design Assistant, Tools** pull-down menu.

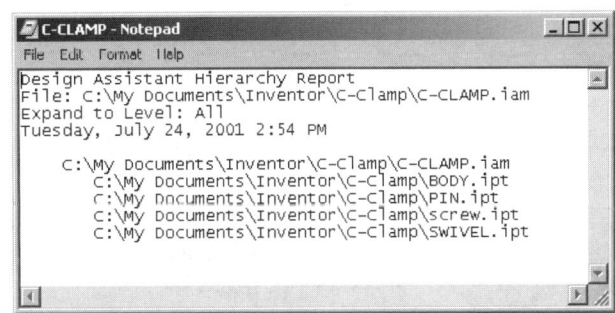

FIGURE 19.15   An example of a simple **Hierarchy Report**.

FIGURE 19.16   The **Hierarchy** option from the **Reports** cascading submenu.

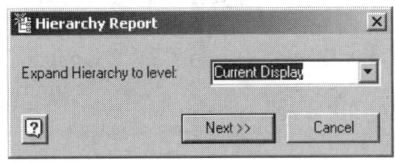

FIGURE 19.17   The **Hierarchy Report** dialog box.

FIGURE 19.18   The **Report Location** dialog box.

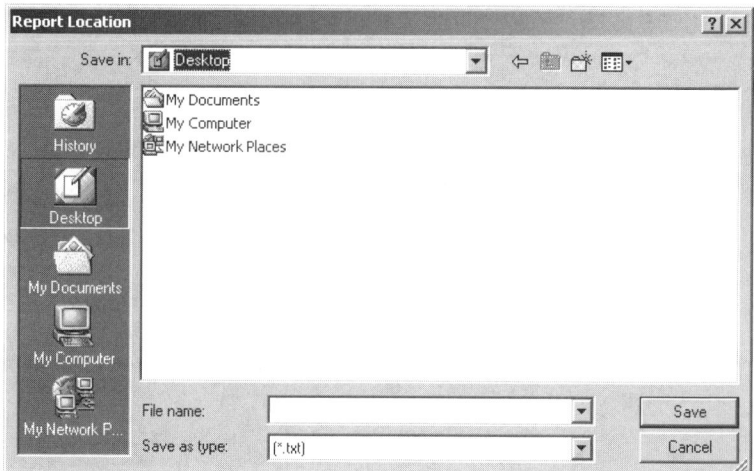

FIGURE 19.19   An example of a simple **Design Properties Report.**

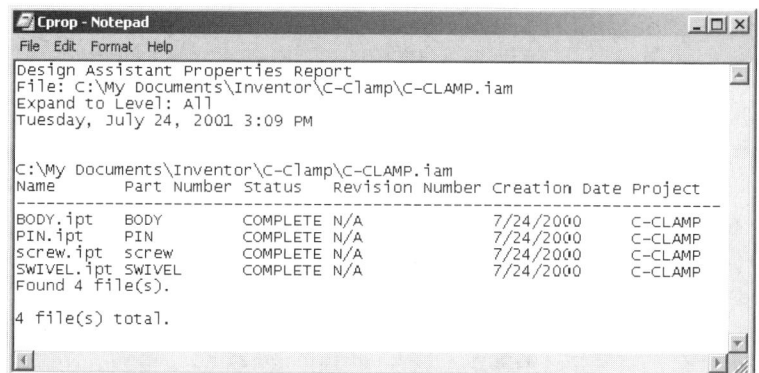

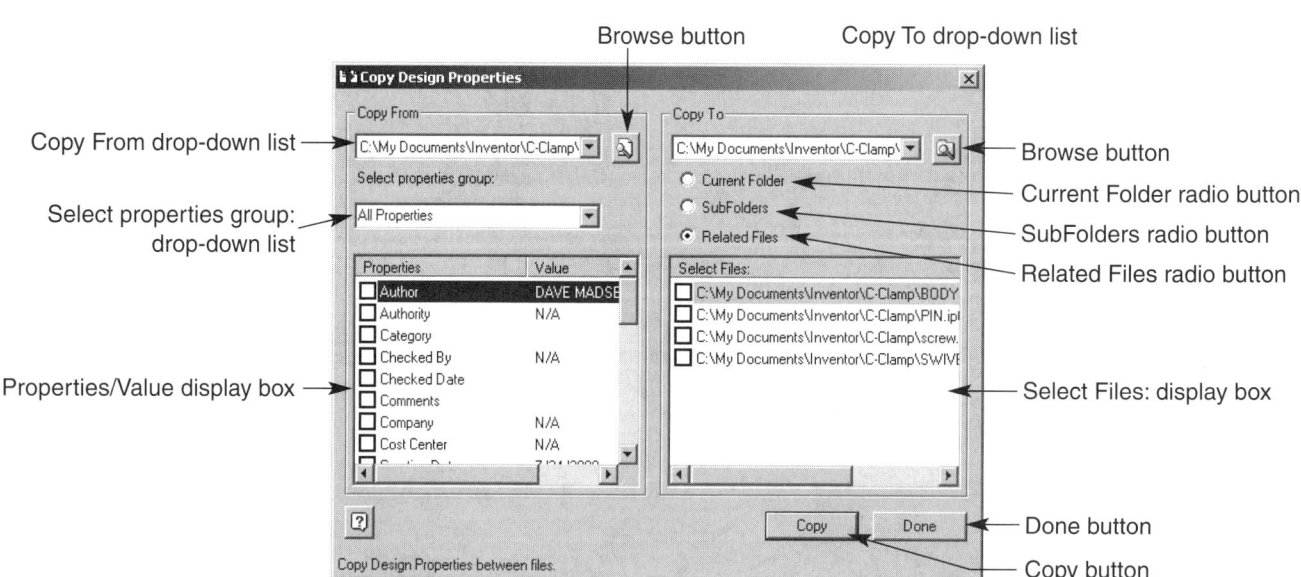

FIGURE 19.20   The **Copy Design Properties** dialog box.

To use the **Copy Design Properties** dialog box, first select the file where the properties are located from the **Copy** from drop-down list, or use the **Browse** button to locate a file. Then select the properties group from the **Properties Group** drop-down list. These correspond to the **Properties** dialog box tabs previously discussed. Next, mark the check boxes for the properties you would like to copy from the **Properties/Value** display box. Finally, specify the options in the **Copy To** area, by locating where you would like to copy the properties to. Use the **Current Folder, SubFolders,** or **Related Files** radio buttons to identify which folders you would like to display. Then mark the check boxes in the **Select Files:** display box to specify which files you want to copy design properties to, and pick the **Copy** button. When you are completely finished coping design properties, pick the **Done** button.

Field Notes

Copying design properties is a valuable tool. However, you may find it helpful to set all the properties for a project in your template files. You can use the template files to create new files that already contain the associated properties.

- **Find** The **Find** cascading pull-down menu option contains three selections that allow you to locate Autodesk Inventor files, strings, and where a file is used. The options are as follows:

  - **Autodesk Inventor Files** Picking this option opens the Find: Autodesk Inventor Files dialog box discussed in Chapter 1.

  - **String** Choosing this option opens the **Find String** dialog box, shown in Figure 19.21. This tool allows you to locate a string by typing the name in the **Search String** edit box.

  - **Where Used** Selecting this option opens the **Where Used** dialog box. See Figure 19.22. This tool displays each place a particular file is used. For example, if a part file is used in several different assemblies, you can locate each of the assemblies where the specified part is used. To use this tool, type the name of the file, or pick the **Browse** button to locate the file. Then select the Search Now button to begin the search. You can also include search paths to locate related documents by clicking the **<Click To Add...>** button in the **Path** display list and selecting the folder from the **Browse for**

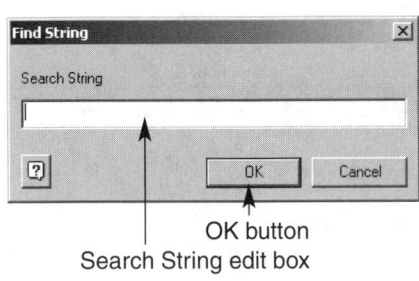

FIGURE 19.21 The **Find String** dialog box.

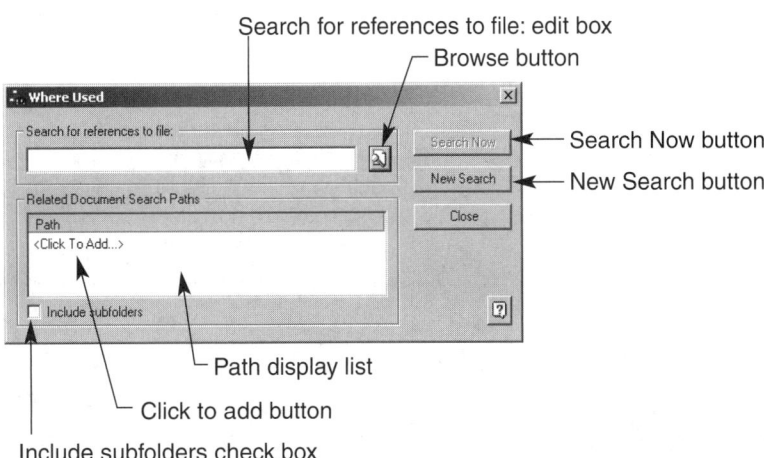

FIGURE 19.22 The **Where Used** dialog box.

Folder dialog box. Pick the **Include subfolders** check box if you want to include sub-folders in the path. If you would like to begin a new search, pick the **New Search** button to clear all existing information.

■ **Viewer**   Select this option to open **Volo View Express,** which allows you to view, mark up, and measure AutoCAD drawings and Autodesk Inventor images. For more information regarding **Volo View Express** refer to the Autodesk Inventor and Volo View help files.

■ <u>**Help**</u> pull-down menu   This pull-down menu, shown in Figure 19.23, allows you to access help, and receive further information regarding the Autodesk Inventor **Design Assistant.**

■ **Properties**   Selecting this button displays **Design Assistant** in properties mode. See Figure 19.8. Here you can open, locate, create reports, and work with Autodesk Inventor files and related files from other applications. The files you open, or the file you are currently working with, are displayed in the **Design Properties** browser. While in **Properties** mode, you can utilize many of the options previously discussed. Options that are not available are shaded and are not accessible. In addition, right-clicking on any of the files found in the **Design Properties** browser reveals the following additional options.

■ **Compact**   This option is similar to a purge command. When it is selected and the **Start** button is pressed, the specified file is "cleaned up" or compacted to remove unnecessary file information.

■ **Compact All**   This option is available for assemblies and functions the same as the **Compact** option. The only difference is that the assembly file and each of the component files are compacted.

■ **Pack and Go**   The **Pack and Go** option allows you to copy a file from one location to another without removing the source file. When you package a file, all the referenced files are included in the package and are copied to the new location. The copied files are placed in a parent file folder. As a result, project information for the copied files is lost.

Selecting the **Pack and Go** option opens the **Pack and Go** dialog box. See Figure 19.24. The **Source File:** display box shows the file where the information comes from and is specified by whatever file you right-clicked on. To use the **Pack and Go** dialog box, select a destination file by typing a path in the **Destination Folder:** edit box or picking the **Browse** button to locate a Folder. The total number of files that are going to be

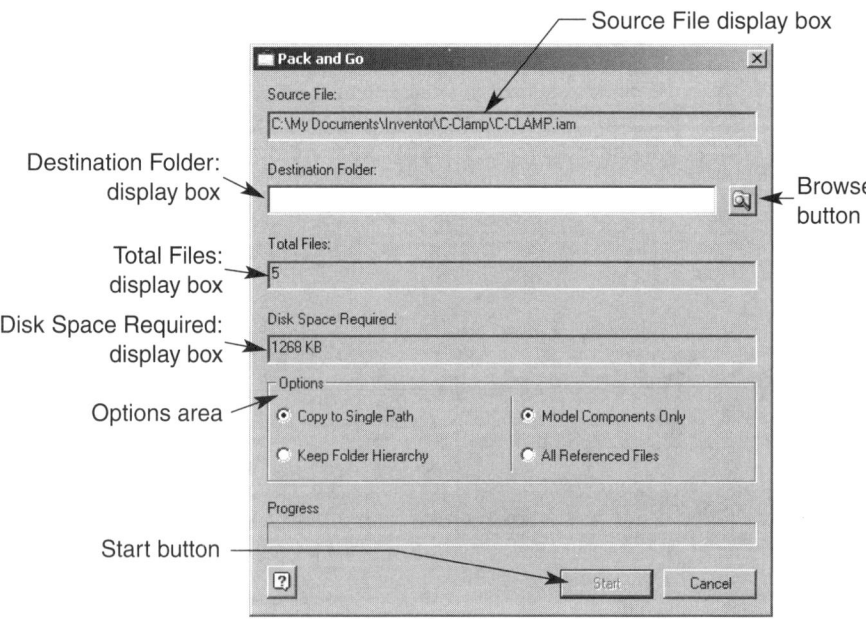

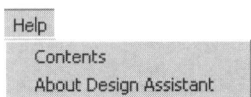

FIGURE 19.23   The
<u>**Help**</u> pull-down menu.

FIGURE 19.24   The **Pack and Go** display box.

copied is shown in the **Total Files:** display box, and the number of bytes they occupy is shown in the **Disk Space Required:** display box. Then choose the appropriate radio buttons from the **Options** area. Here you can choose to **Copy to Single Path** or **Keep Folder Hierarchy,** and package **Model Components Only** or **All Referenced Files.** When you have specified all your settings, pick the **Start** button to begin the packaging and copying of files.

- **Copy Design Properties**   Selecting this option opens the **Copy Design Properties** dialog box previously discussed.

- **Properties**   Selecting the **Properties** option opens the **Properties** dialog box for the specified file. The **Properties** dialog box was discussed earlier in this chapter.

■ **Preview**   Choose this button to display part models, assembly models, and assembly components. See Figure 19.25. After you select the **Preview** button, pick the part, assembly, or part component that you want to preview in the **Design Assistant** Browser. The model will be displayed in the preview area.

■ **Manage** mode button   Selecting the **Manage** mode button displays **Design Assistant** in manage mode. See Figure 19.26. *Manage mode* allows you to manage property information, file interaction, and file linkage for Autodesk Inventor files only. Manage mode is a great tool to use when you copy files and do not want to compromise the integrity of the source file. You can change, rename, and adjust any necessary information. The most effective way to manage your files in file mode is to double-click the information to access a shortcut menu or highlight the information for renaming, or to right-click the information to access the shortcut menu options.

Field Notes

Changes made to files while in manage mode are automatically adjusted for every linked file.

FIGURE 19.25
Autodesk Inventor
Design Assistant
(preview mode).

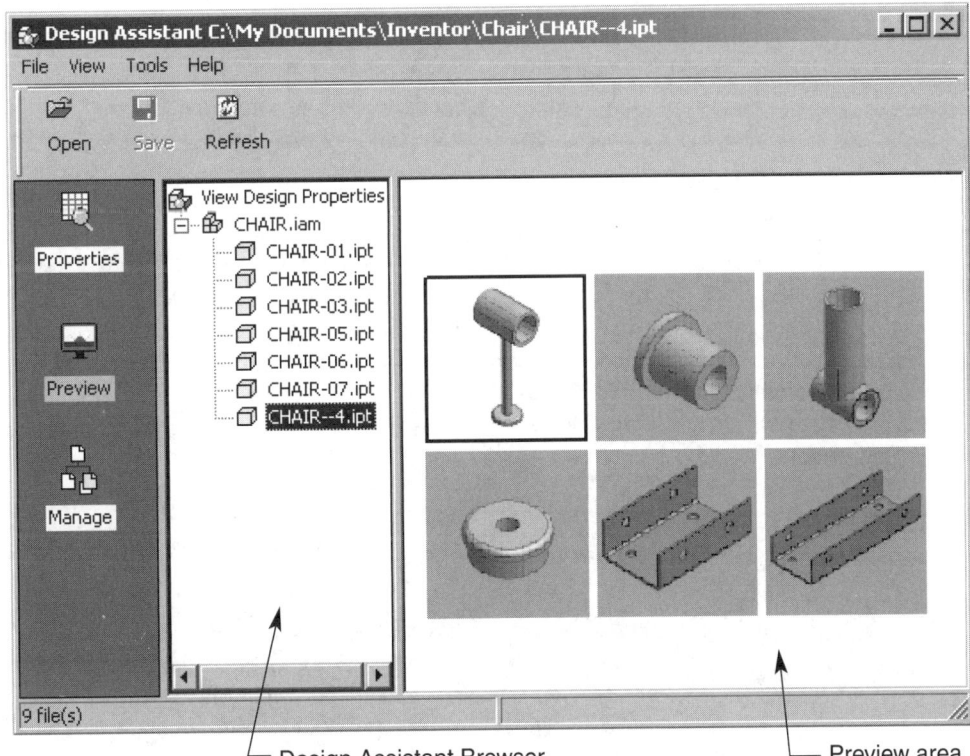

Design Assistant Browser   Preview area

FIGURE 19.26 An example of **Design Assistant** in manage mode.

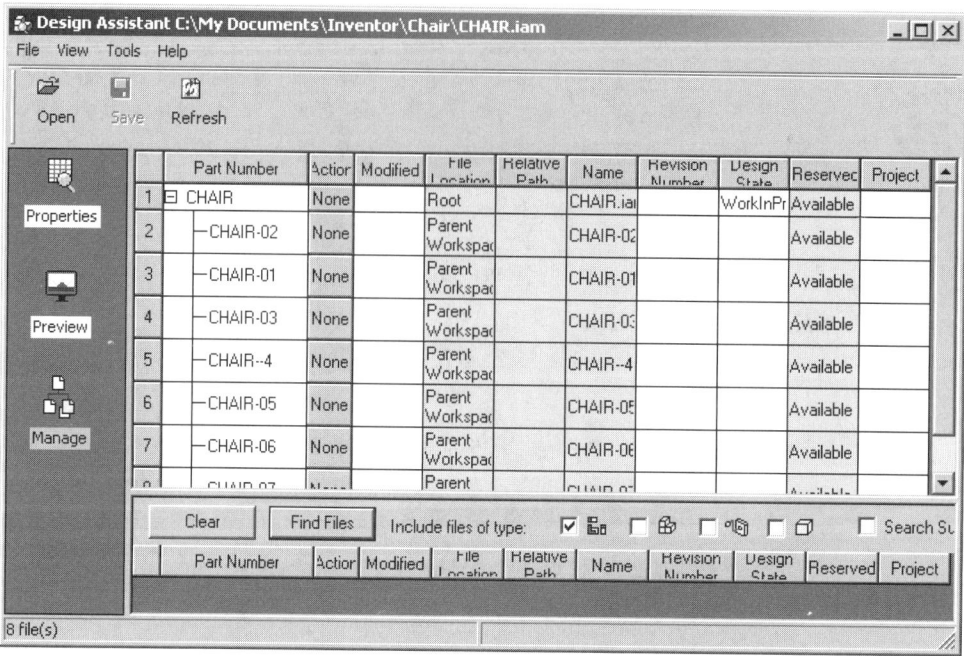

When you open **Design Assistant** from inside Autodesk Inventor, using the **File** pull-down menu, the current file information is displayed. Only the **File, View, Tools,** and **Help** pull-down menus are available, and they contain fewer available options. In addition, the **Open, Save,** and **Manage** buttons do not exist. As a result, if you access **Design Assistant** from inside Autodesk Inventor, you cannot open or save files or view file locations.

Exercise 19.2

1. Continue from Exercise 19.1, or launch Autodesk Inventor.
2. Open EX19.1.iam.
3. Explore each of the **Design Assistant** options previously discussed.
4. Close EX19.1.iam without saving.
5. Exit Autodesk Inventor or continue working with the program.

Compacting Files

As discussed in the previous section, you can compact a single file or multiple files to remove unnecessary file information. Compacting files deletes all previously saved file versions, and "cleans up" the file. Using the **Design Assistant** is an effective way to compact Autodesk Inventor files, but the process can also be done in Microsoft Windows Explorer. To compact a single part or assembly file, select the file, right-click, and choose the **Compact** menu option. To compact several files, an assembly file, and the assembly component files, or an entire folder, select the files, or folder, right-click, and choose the **Compact All** menu option.

Exercise 19.3

1. Continue from Exercise 19.2, or launch Autodesk Inventor.
2. Use **Design Assistant** or Windows Explorer to compact a single exercise file, such as EX19.1.iam, or compact all your exercises and projects.
3. Exit Autodesk Inventor or continue working with the program.

Pack and Go

Previously, you learned how **Design Assistant** allows you to package files using the **Pack and Go** command. Again, **Pack and Go** is an option that allows you to copy a file from one location to another without removing the source file. When you package a file, all the referenced files are included in the package and are copied to the new location. The copied files are placed in a parent file folder. As a result, project information for the copied files is lost. Typically, files are packaged so they can be sent to other users for reference, or in order to keep a record of a design. Still, **Pack and Go** can be used any time you want to clear the active project and bring files together, including all the files that reference the packaged file, without removing the source. This means that you can edit designs in packaged files without altering the source file, and separate files for any reason.

While using the **Design Assistant** is an effective way to package Autodesk Inventor files, the process can also be done in Microsoft Windows Explorer. To package a file, right-click and choose the **Pack and Go** menu option. Just as when using **Design Assistant,** the **Pack and Go** dialog box opens, allowing you to package files.

Exercise 19.4

1. Continue from Exercise 19.3, or launch Autodesk Inventor.
2. Use **Design Assistant** or Windows Explorer and the information presented in this chapter to package EX14.7.iam from its current location to a completely separate folder of your choice.
3. Exit Autodesk Inventor or continue working with the program.

Additional Collaboration Options

Throughout this text you have discovered multiple ways to work collaboratively with Autodesk Inventor. The collaborative tools, options, and commands allow you to work with other users and a design team in an effort to develop product designs using Autodesk Inventor. The following are a few of the collaboration tools that have already been explored:

- Projects
- File management and reservation
- Assembly shared content
- Catalog features and derived components
- iPart factories
- Engineer's Notebook

These powerful tools allow you to communicate ideas to others. In addition, you can use *Windows NetMeeting* to "meet" with other Autodesk Inventor users across the Internet or a network and discuss design issues.

A Windows NetMeeting consists of a host and at least one other person who is a participant in the meeting. The host identifies when a meeting is needed to discuss design issues and determines when the meeting begins. However, at any time, the host can turn over the control to a meeting participant. The host must have Autodesk Inventor files opened, must define a NetMeeting session, and is responsible for initiating the meeting. In contrast, a meeting participant must have Windows NetMeeting running but does not have to have Autodesk Inventor installed on his or her computer unless this person wants to operate Autodesk Inventor.

Typically, the first step in using the NetMeeting tool is for the host to begin a Windows NetMeeting session on his or her computer and define the host security options. Then, at the identified meeting time, the participants begin a Windows NetMeeting session on their computers. The host initiates the meeting using

Autodesk Inventor, causing a message to be displayed on the participant's computer indicating the start of the meeting. A window is displayed on the meeting participant's computer, which shows the host's Autodesk Inventor display.

If you want to host a Windows NetMeeting, use the following steps:

1. Tell people you want to have a meeting; they will launch Windows NetMeeting at a designated time.

2. Open Windows NetMeeting and go through the setup procedures if you have not already done so.

3. Open Autodesk Inventor and the files you want to discuss with others. As shown in Figure 19.27, the **Collaboration** toolbar contains many of the same options available in Windows NetMeeting.

4. From the Autodesk Inventor **Tools** pull-down menu, select **Meet Now...** from the **Online Collaboration** cascading submenu.

5. Use the **Place A Call** dialog box shown in Figure 19.28 to enter the computer address of the individual that you want to meet with (assuming Windows NetMeeting is also running on this person's computer).

6. Select the type of connection from the **Connection Type** drop-down list. When you choose the **Automatic** option, the connection method is automatically selected. Pick the **Network** option to identify a person on the same network to which you are connected. Or choose the **Directory** option to define the e-mail address of person on the directory server on which you are currently logged.

7. Select the **Call** button to connect to the specified meeting participant.

8. Once the first participant responds to the call, continue using the **Place A Call** dialog box to connect to other meeting members.

If you want to participate in a Windows NetMeeting, use the following steps:

1. Open Windows NetMeeting, and go through the setup procedures if you have not already done so.

2. When you see the message identifying that the host is calling your computer, pick the button to accept the call.

3. Once the window opens, and Autodesk inventor is shown, you can signal to the host that you want to operate Autodesk Inventor by double-clicking in the window. You can also pick the Whiteboard button on the **Collaboration** toolbar to use the **Whiteboard** tool, and pick the **Chat** button on the **Collaboration** toolbar to use the **Chat** tool.

Once you are finished with the meeting, select the **Hang Up** button on the **Collaboration** toolbar to exit the Autodesk Inventor collaborative environment. Then close Windows NetMeeting.

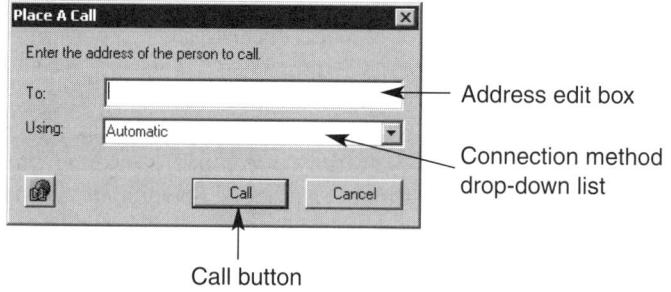

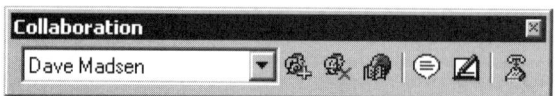

FIGURE 19.27   The **Collaboration** toolbar.

FIGURE 19.28   The **Place A Call** dialog box.

Field Notes

For more information regarding working collaboratively with Autodesk Inventor and Windows NetMeeting, refer the help files for both programs.

|||||||||| CHAPTER TEST

Answer the following questions on a separate sheet of paper.

1. What is the function of the Engineer's Notebook?

2. Identify at least four different options for using notes and the Engineer's Notebook.

3. Often notes and the Engineer's Notebook are used to describe model development or the model history.

4. Once you identify the geometry, identify two techniques to establish the design note.

5. By default one comment box is available for each note. However, in some situations, you may want to add another comment box to the same note. Give two options to place an additional comment box for the same note, and tell how the Comment tool is used to create a comment box.

6. Give the basic function of the view box.

7. Identify the basic function of the Arrow tool.

8. Explain the function of the Design Assistant.

9. Discuss the basic function of the Copy Design Properties dialog box.

10. What does manage mode allow you to do?

11. Explain the most effective way to manage your files in file mode.

12. Why do you want to compact files?

13. What is the function of Pack and Go and when do you want to use it?

14. What is Windows NetMeeting and how does it work?

15. Discuss the steps used if you want to participate in a Windows NetMeeting, and tell what to do when you are finished with the meeting.

Index